PHYSICS OF SOUND
IN THE SEA

PHYSICS OF SOUND
IN THE SEA

PENINSULA
PUBLISHING

WESTPORT, CONNECTICUT USA

PHYSICS OF SOUND

IN THE SEA

Reprint Edition Published by:
Peninsula Publishing
Westport, Connecticut, USA

E-mail: sales@PeninsulaPublishing.com
Telephone: 203-292-5621
Website: http://www.PeninsulaPublishing.com

Library of Congress Catalog Number: 88-062268

ISBN-10: 0-932146-24-4

ISBN-13: 978-0932146-24-3

Printed in the United States of America

NATIONAL DEFENSE RESEARCH COMMITTEE

James B. Conant, *Chairman*

Richard C. Tolman, *Vice Chairman*

Roger Adams   Army Representative[1]

Frank B. Jewett   Navy Representative[2]

Karl T. Compton   Commissioner of Patents[3]

Irvin Stewart, *Executive Secretary*

[1] *Army representatives in order of service:*

Maj. Gen. G. V. Strong	Col. L. A. Denson
Maj. Gen. R. C. Moore	Col. P. R. Faymonville
Maj. Gen. C. C. Williams	Brig. Gen. E. A. Regnier
Brig. Gen. W. A. Wood, Jr.	Col. M. M. Irvine
Col. E. A. Routheau	

[2] *Navy representatives in order of service:*

Rear Adm. H. G. Bowen	Rear Adm. J. A. Furer
Capt. Lybrand P. Smith	Rear Adm. A. H. Van Keuren
Commodore H. A. Schade	

[3] *Commissioners of Patents in order of service:*

Conway P. Coe	Casper W. Ooms

NOTES ON THE ORGANIZATION OF NDRC

The duties of the National Defense Research Committee were (1) to recommend to the Director of OSRD suitable projects and research programs on the instrumentalities of warfare, together with contract facilities for carrying out these projects and programs, and (2) to administer the technical and scientific work of the contracts. More specifically, NDRC functioned by initiating research projects on requests from the Army or the Navy, or on requests from an allied government transmitted through the Liaison Office of OSRD, or on its own considered initiative as a result or the experience of its members. Proposals prepared by the Division, Panel, or Committee for research contracts for performance of the work involved in such projects were first reviewed by NDRC, and if approved, recommended to the Director of OSRD. Upon approval of a proposal by the Director, a contract permitting maximum flexibility of scientific effort was arranged. The business aspects of the contract, including such matters as materials, clearances, vouchers, patents, priorities, legal matters, and administration of patent matters were handled by the Executive Secretary of OSRD.

Originally NDRC administered its work through five divisions, each headed by one of the NDRC members. These were:

Division A — Armor and Ordnance
Division B — Bombs, Fuels, Gases, & Chemical Problems
Division C — Communication and Transportation
Division D — Detection, Controls, and Instruments
Division E — Patents and Inventions

In a reorganization in the fall of 1942, twenty-three administrative divisions, panels, or committees were created, each with a chief selected on the basis of his outstanding work in the particular field. The NDRC members then became a reviewing and advisory group to the Director of OSRD. The final organization was as follows:

Division 1 — Ballistic Research
Division 2 — Effects of Impact and Explosion
Division 3 — Rocket Ordnance
Division 4 — Ordnance Accessories
Division 5 — New Missiles
Division 6 — Sub-Surface Warfare
Division 7 — Fire Control
Division 8 — Explosives
Division 9 — Chemistry
Division 10 — Absorbents and Aerosols
Division 11 — Chemical Engineering
Division 12 — Transportation
Division 13 — Electrical Communication
Division 14 — Radar
Division 15 — Radio Coordination
Division 16 — Optics and Camouflage
Division 17 — Physics
Division 18 — War Metallurgy
Division 19 — Miscellaneous
Applied Mathematics Panel
Applied Psychology Panel
Committee on Propagation
Tropical Deterioration Administrative Committee

NDRC FOREWORD

As events of the years preceding 1940 revealed more and more clearly the seriousness of the world situation, many scientists in this country came to realize the need of organizing scientific research for service in a national emergency. Recommendations which they made to the White House were given careful and sympathetic attention, and as a result the National Defense Research Committee [NDRC] was formed by Executive Order of the President in the summer of 1940. The members of NDRC, appointed by the President, were instructed to supplement the work of the Army and the Navy in the development of the instrumentalities of war. A year later, upon the establishment of the Office of Scientific Research and Development [OSRD], NDRC became one of its units.

The Summary Technical Report of NDRC is a conscientious effort on the part of NDRC to summarize and evaluate its work and to present it in a useful and permanent form. It comprises some seventy volumes broken into groups corresponding to the NDRC Divisions, Panels, and Committees.

The Summary Technical Report of each Division, Panel, or Committee is an integral survey of the work of that group. The first volume of each group's report contains a summary of the report, stating the problems presented and the philosophy of attacking them and summarizing the results of the research, development, and training activities undertaken. Some volumes may be "state of the art" treatises covering subjects to which various research groups have contributed information. Others may contain descriptions of devices developed in the laboratories. A master index of all these divisional, panel, and committee reports which together constitute the Summary Technical Report of NDRC is contained in a separate volume, which also includes the index of a microfilm record of pertinent technical laboratory reports and reference material.

Some of the NDRC-sponsored researches which had been declassified by the end of 1945 were of sufficient popular interest that it was found desirable to report them in the form of monographs, such as the series on radar by Division 14 and the monograph on sampling inspection by the Applied Mathematics Panel. Since the material treated in them is not duplicated in the Summary Technical Report of NDRC, the monographs are an important part of the story of these aspects of NDRC research.

In contrast to the information on radar, which is of widespread interest and much of which is released to the public, the research on subsurface warfare is largely classified and is of general interest to a more restricted group. As a consequence, the report of Division 6 is found almost entirely in its Summary Technical Report, which runs to over twenty volumes. The extent of the work of a Division cannot therefore be judged solely by the number of volumes devoted to it in the Summary Technical Report of NDRC: account must be taken of the monographs and available reports published elsewhere.

Any great cooperative endeavor must stand or fall with the will and integrity of the men engaged in it. This fact held true for NDRC from its inception, and for Division 6 under the leadership of Dr. John T. Tate. To Dr. Tate and the men who worked with him — some as members of Division 6, some as representatives of the Division's contractors — belongs the sincere gratitude of the Nation for a difficult and often dangerous job well done. Their efforts contributed significantly to the outcome of our naval operations during the war and richly deserved the warm response they received from the Navy. In addition, their contributions to the knowledge of the ocean and to the art of oceanographic research will assuredly speed peacetime investigations in this field and bring rich benefits to all mankind.

The Summary Technical Report of Division 6, prepared under the direction of the Division Chief and authorized by him for publication, not only presents the methods and results of widely varied research and development programs but is essentially a record of the unstinted loyal cooperation of able men linked in a common effort to contribute to the defense of their Nation. To them all we extend our deep appreciation.

Vannevar Bush, Director
Office of Scientific Research and Development

J. B. Conant, Chairman
National Defense Research Committee

FOREWORD

THIS VOLUME, together with Volumes 6, 7, and 9, summarizes four years of research on underwater sound phenomena. The purpose of this research was to provide a firmer foundation for the most effective design and use of sonar gear. It is generally true that wide basic knowledge is an important element in engineering practice. In the development of sonar gear, knowledge of how sound is generated, transmitted, reflected, received, and detected is clearly useful both in the design of new equipment and in the most efficient utilization of existing gear. As a result of the time delay between the design of new equipment and its use in service, the most important application of this basic information during World War II has been in suggesting how existing equipment could best be operated and tactically used.

The importance of basic information on underwater sound had been evident to both our own Navy and the British for some time. Practical experience had shown that the maximum distance at which a target could be detected with underwater sound was highly variable, even when the equipment was in good operating condition. Since it was realized that such variability might well be related to a variability in oceanographic conditions, the Navy brought this problem to the attention of the Woods Hole Oceanographic Institution and the Scripps Institute of Oceanography. To support an investigation, NDRC contracted in 1940 with the former institution to carry out studies and experimental investigations of the structure of the superficial layer of the ocean and its effect on the transmission of sonic and supersonic vibrations.

The work carried out under this contract, together with supporting information obtained elsewhere, emphasized the relation of such basic factors to the variable performance of sonar gear. Thus when some months later it was proposed to establish a section in NDRC to undertake research and development relating to the detection of submerged submarines, plans were made to increase substantially this research effort. To this end, the plans which were formulated by NDRC and approved by the Navy included research on underwater sound phenomena at the proposed laboratory at San Diego, to be operated under a contract with the University of California Division of War Research. This step not only in-creased the number of personnel engaged in this research and facilitated study of oceanic conditions peculiar to the Pacific area, but also most fortunately made it possible for the San Diego Laboratory to recruit certain of its staff from the Scripps Institution of Oceanography and to draw upon the director and staff of the Scripps Institution for very pertinent background information in oceanography. While the major source of the experimental data continued to be the Woods Hole and San Diego Laboratories, very pertinent data were from time to time obtained from other laboratories, notably New London, Harvard, the Massachusetts Institute of Technology, and the Underwater Sound Reference Laboratories.

Quite promptly, an analytical section, later known as the Sonar Analysis Group, was organized under a contract with Columbia University Division of War Research. The function of this group was to assist in the analyses of data being accumulated by Woods Hole, San Diego, and other laboratories, and, as it became possible to draw conclusions, to present these to other groups interested in operations or design. In this connection it should be emphasized that the seeming importance of this research to the Navy led to the assignment of naval personnel to follow the work actively. In particular, officers of the Sonar Design Section of the Bureau of Ships followed very closely the research of this analytical group, participating directly in much of the work.

The results obtained in this research and summarized in this and companion volumes found many important applications during World War II. The rules used for operating sonar gear were based in part on these results. Many tactical rules embodied in submarine and antisubmarine doctrine were directly based on information obtained in these basic studies of transmission, reflection, detection, and the like. As an example, the spacing between antisubmarine vessels in different tactical and oceanographic conditions was varied according to the measured temperature gradients in the upper layers of the ocean. In addition, the choice of operating frequency, pulse length, size, and power for new equipment, especially for submarines, was considerably influenced by such basic knowledge. It can be stated with considerable confidence that a detailed basic knowledge of underwater sound phenomena will be of increasing help in the design and operation of Navy sonar equipment.

Only a few of the scientists and others contributing to this war effort can be named. Mr. C. O'D. Iselin, Director of the Woods Hole Oceanographic Institution, and his staff brought to this research, to which they ably contributed, a sound background knowledge of oceanography. Dr. V. O. Knudsen, Dean of the Graduate School of the University of California at Los Angeles and for some time the Director of the Division's San Diego Laboratory, was one of this country's foremost scientists in the field of acoustics. Dr. Knudsen played a prominent part in organizing the research program, and after leaving the San Diego Laboratory he contributed actively and effectively to research work closely related to the subject of this volume. Dr. G. P. Harnwell, Chairman of the Department of Physics at the University of Pennsylvania, who succeeded Dr. Knudsen as Director at San Diego after having served some time as a technical aide to the Division, gave wise general direction to this research at San Diego. In operations at San Diego, Dr. Knudsen and Dr. Harnwell were ably supported by Dr. Carl Eckart, Professor of Theoretical Physics at the University of Chicago, who became Associate Director at San Diego, responsible for the planning and execution of the basic research there.

Dr. H. Sverdrup, Director of the Scripps Institution of Oceanography, and his staff also contributed significantly to this work. The U. S. Navy Electronics Laboratory at San Diego collaborated most helpfully in much of this basic research. The task of organizing the very important analytical work was assumed by Dr. W. V. Houston, Professor of Physics at the California Institute of Technology and Director of the Special Studies Group; he delegated the very large part of the responsibility to Dr. Lyman Spitzer, Jr., an outstanding member of the Departments of Physics and Astronomy at Yale University, who became Director of the Sonar Analysis Group.

As the reader will note, Dr. Spitzer undertook the responsibility for preparing this volume, and in this he had had the assistance not only of members of his own staff but also of naval personnel and of members of the Woods Hole and San Diego staffs. The Division appreciates the efforts of all those who have participated.

This research project secured most effective support from the Navy. The broad program of research and study which was proposed by Dr. Jewett and Dr. Bush, and which included this basic research on underwater sound, was supported by Rear Admiral S. M. Robinson, Chief of the Bureau of Ships, who took steps to provide facilities for this work at San Diego. Later, when Rear Admiral Van Keuren became Chief of the Bureau of Ships, he likewise strongly backed the program, which was still in its initial stages. Support of the program continued with Vice Admiral Cochrane as Chief of the Bureau of Ships, and most helpful liaison was provided by Captain Rawson Bennett, Jr., Commander J. C. Myers, Commander Roger Revelle, and others in the Bureau. The Coordinator of Research and Development and his staff continually gave support to this research. The results of much of this work were of special interest to the Tenth Fleet and very close contact was accordingly maintained with its staff, particularly with the Operations Research Group.

In presenting this volume the hope is expressed that research in this area will be energetically continued. It is also hoped that general interest in this field may be maintained by the distribution to the widest possible audience of this volume and other volumes which have been written from the standpoint of basic science.

JOHN T. TATE
Chief, Division 6

PREFACE TO THE REPRINT EDITION

PHYSICS OF SOUND IN THE SEA is a classic of the underwater acoustics literature. It was first produced in 1945 by the Subsurface Warfare Division of the National Defense Research Committee. In 1969 the book was reprinted by the Headquarters Naval Material Command of the U.S. Navy and went out of print about 1981. This volume is too valuable not to be made available to today's underwater acousticians. Accordingly, we have undertaken its current reprinting.

PREFACE

IN THE COURSE of prosubmarine and antisubmarine research carried out during World War II, a large amount of information was obtained on the propagation of underwater sound. Much of this was gathered in fairly random ways, such as while testing underwater sound equipment. Most of the useful information, however, was obtained by groups devoted primarily to the problem of underwater sound propagation. While valuable results had been found before World War II by the Naval Research Laboratory, the British, and other groups, most of the information on underwater sound transmission obtained during the war resulted from a program of studies organized by Division 6 of the National Defense Research Committee and carried out in collaboration with Navy laboratories at San Diego and elsewhere.

It should be kept in mind that these so-called fundamental programs were not fundamental in the usual scientific sense. They were not aimed at isolating and understanding the different factors at work, but were designed rather for the accumulation of information which would be useful in antisubmarine and prosubmarine operations. Thus effort was concentrated on the study of the transmission loss of sound generated with standard sonar gear under varying oceanographic conditions, rather than on a detailed study of each of the individual factors affecting underwater sound transmission. Similarly, the reflection of sound from actual submarines was studied rather than the individual mechanisms responsible for the origin of echoes from underwater targets.

During the war this approach was abundantly justified by its results. The information obtained on underwater sound propagation under different oceanographic and tactical situations was immediately applied to the more effective use of existing underwater sound equipment in different situations. The results of transmission, reverberation and other studies were usually used operationally much more rapidly than the results of equipment development.

Over a longer period, however, information on underwater sound can be most useful if the phenomena are not merely observed but also explained. An understanding of each of the basic factors affecting underwater sound propagation would make it possible to predict the transmission and reflection to be expected under conditions widely different from those prevailing when the original measurements were taken. While the primarily experimental research carried out during the war could be immediately applied to the gear then in existence, the development of new equipment for new and unforeseen tactical situations requires an understanding of the factors which influence underwater sound. The ultimate aim of basic underwater sound research, especially during peacetime, should be to develop such an understanding.

The present volume presents the essential results obtained in the studies of underwater sound up to the middle of 1945. This volume was written primarily from the fundamental viewpoint of scientific research; in other words, the data are presented against a framework of an attempted understanding of the factors involved rather than as an unadorned summary of the experimental results. Since the measurements were not carried out primarily to increase this understanding, this presentation of the subject leads to many obvious gaps. However, it is hoped that the overall scientific picture presented will be stimulating to any future research workers in this field. To aid those interested in application, practical summaries of the results are given at the end of each of the four parts comprising this volume.

Since our understanding of the details of underwater sound has not been sufficient in most cases to allow an elaborate comparison between theory and experiment, it has been possible in most of this volume to write the text on the level of a senior engineering student. A deliberate effort has been made to keep to this level wherever possible in order to make the results available to the widest possible group of readers. However, more elaborate theoretical developments have been included where it was believed that they were essential to an understanding of the full significance of current information.

The first two parts of this volume deal with the propagation of sound in the absence of targets. Part I discusses the transmission loss of sound sent out from a projector, while Part II deals with sound which has been scattered back to the vicinity of the original sound source. Part III deals with the echoes returned from submarines and surface vessels. Part IV discusses the transmission of sound through wakes and echoes received from wakes.

It should be emphasized that this work is essentially a report of the work carried out by the University of California Division of War Research in collaboration with the U. S. Navy Electronics Laboratory, formerly the U. S. Navy Radio and Sound Laboratory; and by the Woods Hole Oceanographic Institution. Both the Underwater Sound Reference Laboratories and the New London Laboratory of Columbia University Division of War Research, as well as the underwater sound laboratory of the Massachusetts Institute of Technology, have also made important contributions in special fields. All these groups, under contract with Division 6, have been very helpful in the preparation of this volume. They have at times supplied unpublished data and have made many helpful comments and suggestions for improving the presentation.

The direct preparation of this volume has been largely a cooperative enterprise of the Sonar Analysis Group, operating under different auspices at different times. This work was initiated under the Special Studies Group of Columbia University Division of War Research, under Contract OEMsr-1131. Most of the writing was done by the Sonar Analysis Group

under Contract OEMsr-1483; during this time, the Group operated under the auspices of the Office of Field Service but under the cognizance of Section 940 of the Bureau of Ships. Final preparation of the manuscript was completed while the Group formed part of the Woods Hole Oceanographic Institution under Contract Nobs-2083 with the Bureau of Ships.

The scientific staff of the Sonar Analysis Group engaged in this work were: P. G. Bergmann, E. Gerjuoy, P. G. Frank, A. N. Guthrie, Lieut. (jg) J. K. Major, USNR (Project Officer of the Group), J. J. Markham, L. Spitzer, Jr., R. Wildt, and A. Yaspan. The work was under the general supervision of the director of the Group, aided by A. N. Guthrie, Administrative Director. The editors were:

Part I   P. G. Bergmann and A. Yaspan
Part II   E. Gerjuoy and A. Yaspan
Part III   Lieut. (jg) J. K. Major, USNR
Part IV   R. Wildt

Because Part I was largely the result of cooperative effort by many members of the Group, as well as by C. Herring of the Special Studies Group, names of individual chapter authors of Part I are listed in the Table of Contents. Final assembly of the material was under the supervision of Mrs. E. E. Wagner, and of H. Birnbaum and M. Klapper.

LYMAN SPITZER, JR.
Director, Sonar Analysis Group

CONTENTS

CHAPTER   PAGE

PART I
TRANSMISSION

1 Introduction by *P. G. Bergmann* and *L. Spitzer, Jr.* . . . 3

2 Wave Acoustics by *P. G. Frank* and *A. Yaspan* 8

3 Ray Acoustics by *P. G. Frank*, *P. G. Bergmann*, and *A. Yaspan* 41

4 Experimental Procedures by *P. G. Bergmann* 69

5 Deep-Water Transmission by *L. Spitzer, Jr.* 86

6 Shallow-Water Transmission by *P. G. Bergmann* . . . 137

7 Intensity Fluctuations by *P. G. Bergmann* 158

8 Explosions as Sources of Sound by *C. Herring* 173

9 Transmission of Explosive Sound in the Sea by *C. Herring* 192

10 Summary by *L. Spitzer, Jr.* and *P. G. Bergmann* . . . 236

PART II
REVERBERATION

11 Introduction 247

12 Theory of Reverberation Intensity 250

13 Experimental Procedures 272

14 Deep-Water Reverberation 281

15 Shallow-Water Reverberation 308

16 Variability and Frequency Characteristics 324

17 Summary 334

CHAPTER

PAGE

PART III
*REFLECTION OF SOUND FROM SUBMARINES
AND SURFACE VESSELS*

18 Introduction 343
19 Principles 345
20 Theory 352
21 Direct Measurement Techniques 363
22 Indirect Measurement Techniques 379
23 Submarine Target Strengths 388
24 Surface Vessel Target Strengths 422
25 Summary 434

PART IV
ACOUSTIC PROPERTIES OF WAKES

26 Introduction 441
27 Formation and Dissolution of Air Bubbles 449
28 Acoustic Theory of Bubbles 460
29 Velocity and Temperature Structure 478
30 Technique of Wake Measurements 484
31 Wake Geometry 494
32 Observed Transmission Through Wakes 503
33 Observations of Wake Echoes 512
34 Role of Bubbles in Acoustic Wakes 533
35 Summary 541
    Bibliography 547
    Contract Numbers 557
    Service Project Numbers 558
    Index 559

PART I

TRANSMISSION

Chapter 1

INTRODUCTION

1.1 IMPORTANCE OF TRANSMISSION STUDIES

SINCE SOUND WAVES are transmitted through water very much more readily than radio and light waves, the use of underwater sound has become a basic part of subsurface warfare. There are always many different ways in which equipment can be designed and used. An intelligent choice between the different alternatives depends on accurate knowledge of the different factors affecting final performance. One of these factors is the extent to which sound is weakened in passing from one point to another; this weakening is called *transmission loss*. The present volume summarizes the information available in 1945 on transmission loss of underwater sound.[a] Much of the detailed discussion refers to a sound frequency of 24 kc since this is the frequency most commonly used in practical echo ranging, and most of the available data are at that frequency.

This information, although incomplete, is useful in a variety of ways. In particular, it is helpful both in the design of gear and in the development of operational doctrine.

It is evident that the intelligent design of new equipment requires reliable information on underwater sound transmission as well as on a variety of other factors. For example, the choice of frequency in any device usually involves a compromise between high frequency for the sake of directivity and low frequency for the sake of good transmission. It is possible to arrive at a suitable compromise by trial-and-error methods. However, the choice is made more quickly if routine methods can be used to predict the transmission loss at each frequency, the directivity,

and other factors, such as the noise level, which affect performance. These different predictions can then be combined to find which frequency gives the best results. The optimum frequency will, of course, depend on the purpose for which the equipment is designed, and on the limitations of size, available power, and other characteristics. Thus, in some types of echo-ranging equipment, low-frequency gear with a wide beam pattern and a long maximum range is used in searching for submarines, but tilting high-frequency gear is provided for tracking a submarine at close range during an attack.

The development of operational doctrine for the gear already in use also depends on the results of transmission studies largely because of the wide variability of underwater sound transmission. If a pulse of sound is sent into the water and received near the surface 3,000 yd away, the signal energy received will sometimes be only a millionth of the signal energy received at other times. This enormous variation is due mainly to changes in the vertical temperature gradients present in the water. These changes have a direct effect on the maximum range at which submarines can be detected by echo-ranging gear. When the maximum range is known to be short, the gear can be operated most effectively with a short keying interval, since more rapid keying increases the chances of finding a submarine which happens to be within the maximum range. If a long keying interval were used under these conditions, time would be wasted in listening for echoes during periods when no echoes would be possible.

Information on the change of sound transmission conditions with changing temperature conditions is useful in the choice of antisubmarine tactics as well as in the selection of rules for operating the sonar gear. When the transmission loss of sound is high and the maximum range of sonar gear is short, the spacing between surface vessels conducting an antisubmarine hunt must be reduced. Sharp temperature gradients at considerable depths may weaken sound passing

---

[a] This volume includes primarily those data applicable in the frequency range above 200 cycles. Sound of lower frequencies has not been used in sonar equipment and its transmission has not been investigated by Division 6 of the NDRC, except occasionally in connection with the transmission of explosive pulses.

through them, and reduce the maximum range on a deep submarine to much less than the maximum range on the same submarine at periscope depth. The maximum range at which two surface ships can obtain echoes from each other gives, by itself, no information on the maximum range that can be expected on a deep submarine. Thus, use of the bathythermograph is required to estimate the approximate maximum range obtainable on a submarine at evasive depths. Such an estimate is useful not only in the choice of spacing between antisubmarine vessels but also in evaluating the desirability of detaching escort vessels from a convoy to hunt a submarine reported sighted some distance away. When sound conditions are good, detaching antisubmarine vessels is less likely to endanger the convoy and more likely to sink the submarine than when sound conditions are bad.

Information on sound transmission conditions is also useful in the choice of submarine tactics. A submariner is free to choose his depth of operation, and one of the factors influencing this choice is the maximum range at which he is likely to be detected at each depth. In any case, the behavior of the submarine may be influenced by knowledge of the maximum range at which detection may be expected. At times, transmission conditions are so severe that the submarine cannot be detected even at 500 yd; such conditions, if they can be readily and reliably identified, provide opportunity for unusually aggressive action.

1.2   NATURE OF SOUND

Historically, the various types of physical phenomena were first defined in terms of the human senses. Physics was divided into the fields of (1) mechanics (dealing with touch and displacements effected by human muscle power), (2) light (dealing with the perception of objects by the eye), (3) sound (pertaining to hearing), (4) heat (dealing with the sensations of heat and cold), and other similar fields. Gradually, as the causes of the nerve stimuli became understood, the subject matter of physics was regrouped; classification in accordance with physiological perception was gradually replaced by classification according to the physical nature of the phenomena studied. Thus, optics became more and more a subdivision of the theory of electricity and magnetism, while heat and sound came to be treated as subdivisions of mechanics. The theory of heat is concerned with random motions of many particles. In contrast, sound is concerned with the formation and propagation of vibrations, primarily in a fluid,[b] at frequencies both within and above the range of audibility. This definition is purely arbitrary, dictated by practical considerations, and may be ambiguous under certain circumstances. Nevertheless, it is generally accepted.

The physics of sound is usually called acoustics. Although a major part of the work in acoustics deals with sound perceptible by the human ear (the acoustics of rooms, the physiology of sound, and similar subjects), inaudible sound, consisting of mechanical vibrations above the range of frequencies perceived by the ear, has come to play an important role in subsurface warfare. In this volume on the properties of sound in the ocean, more than half of the discussion will be devoted to the propagation of *supersonic* sound, that is, sound at frequencies well above those which can be heard.

1.2.1   Sound as Mechanical Energy

It must be understood that sound energy is a form of mechanical energy. The particles of a fluid in which sound is traveling are set in motion and temporary stresses are produced which increase and decrease during each vibration. The motion of the individual particles gives the fluid kinetic energy while the stresses induce potential energy. In acoustics, the sum of these two kinds of energy is called sound energy or acoustic energy. It is not always easy to separate the acoustic energy from other forms of mechanical energy possessed by the fluid.

A fluid obtains acoustic energy by some kind of energy transformation. As an illustration, consider a tuning fork in air. When this tuning fork is struck with a rubber hammer, its two prongs are set in rhythmic vibratory motion. The vibrating prongs of the tuning fork produce compressions and rarefactions in the surrounding air by pushing the adjacent air mass away and then permitting it to rush back. These alternating compressions and rarefactions are propagated through the air and may be detected as sound by a suitable instrument, such as the human ear or a microphone. The original source of energy was the rubber hammer, which had kinetic energy of translation. This energy was transformed, by means of a collision, to vibratory energy in the tuning fork,

---

[b] The term *fluid*, as used in physics and chemistry, means any liquid or gaseous substance. Thus air and water are fluids, but steel is not.

which was communicated to the air as acoustic energy.

The foregoing example illustrates the general process by which sound is generated and detected. A source of sound converts mechanical or electrical energy into energy of vibration and communicates this energy to the surrounding medium as acoustic energy. This acoustic energy travels through the medium to the receiving instrument where it is detected.

1.2.2 Production and Reception of Sound

Most types of sonar gear produce sound by converting electrical energy into acoustic energy and detect sound by converting acoustic energy into electrical energy. They do this by making use of one of two effects, *magnetostriction* or the *piezoelectric effect*.

When certain metals, such as nickel, are placed in a magnetic field, they contract (or expand) in the direction of the field; conversely, when they are subjected to a contracting (or expanding) force they become partially magnetized. Thus, if a nickel rod is made the core of a solenoid and if it is given a permanent magnetization by means of a direct current, then an alternating current passed through the winding will cause the magnetization to increase and decrease with the frequency of the current. As a result, the rod will contract and expand or, in other words, vibrate with the frequency of the impressed current. In this arrangement, electrical energy is converted into acoustic energy which is passed into the surrounding medium. Conversely, if a sound wave hits this instrument and causes the nickel rod to alternately expand and contract, the rod will be magnetized and demagnetized rhythmically, thus inducing an electromotive force in the surrounding solenoid. The resulting alternating current may be amplified and ultimately recorded in one form or another. Such a magnetostriction transducer may thus be used both as a source of sound and as a receiver of sound.

Certain crystals, such as quartz, Rochelle salt, and ammonium dihydrogen phosphate, exhibit the piezoelectric effect. If a slice is cut from such a crystal and if an electric potential difference is applied across such a slice, the crystal will either contract or expand, depending on which of the two faces is electrically positive. Conversely, if such a slice is compressed or expanded mechanically, the two opposite faces will develop a potential difference. Thus, a piezoelectric crystal, or an array of such crystals, may be used as a transducer. If an alternating voltage is applied to the opposite sides of the crystal slice, it will vibrate with the frequency of the applied voltage; and if it is placed in a fluid where the pressure is fluctuating, it will develop a fluctuating emf across its faces.

Other important sources of waterborne sound are underwater explosions, ships, submarines, waves, underwater ordnance, and biological sources.

1.2.3 Propagation of Sound

Chapters 1 through 10 are concerned with the propagation of sound in the ocean. The complexity of this problem is due to the great variability of the mechanical properties of the medium in which the propagation takes place, but the basic underlying physical concepts are fairly simple. These principles are discussed in the following sections.

DIRECTION OF PROPAGATION

Sound energy is propagated away from the source into a medium. If a single pulse of sound is considered, such as that produced by a sudden explosion, the course of the sound energy in the medium can be followed by placing a large number of recording microphones in the general vicinity and by noting the times at which they show the first response. Each will respond at a slightly different time. Some, placed behind obstructions, may not respond at all.

By using a sufficiently large number of such microphones, we can record all those points in space which are reached by the spreading sound pulse at the same time. We shall call the surface on which these points are located a *sound front* (a better expression will be introduced later). The progression of the pulse in space may then be described by a succession of sound fronts along with the statement of the time at which each front is activated. If the medium of propagation is homogeneous, the perpendicular distance between two sound fronts is proportional to the time it takes the sound pulse to travel from one to the other. In other words, in a homogeneous medium sound travels at a constant speed in a direction perpendicular to the sound front. This direction is called the direction of propagation.

These simple rules apply only if the sound beam meets no obstructions. If an obstruction is placed between source and microphone, the microphone usually registers some sound, but with a delay indicating that the sound pulse had to travel "around the corner" to reach the microphone. In that case, sound energy is obviously deflected around the obstruction; and it can be shown that this energy does not travel everywhere in a direction normal to the sound front. A rigorous treatment of these more involved cases shows, nevertheless, that a direction of propagation always can be defined in a natural and unique manner.

INTENSITY OF THE SOUND FIELD

Sound is weakened as it travels and at very great distances from the sound source cannot be detected. We specify the strength of the sound by its *intensity*. Sound intensity is defined as the rate at which sound energy passes through an area 1 centimeter square placed squarely in the path of the traveling sound.

In theoretical studies sound intensity is usually expressed in units of ergs per square centimeter. In engineering work, on the other hand, it is usually more practical to express intensities on a logarithmic scale both because of the very wide range of sound intensities in practice and because sound intensities are frequently the product of several factors. Use of the logarithmic scale narrows down the numerical range between very faint and very loud sounds and also simplifies the computation of many sound intensities by replacing multiplication by addition.

The logarithmic scale in general use is the *decibel* scale. This scale may be explained as follows. Suppose we want to compare two sound intensities I_1 and I_2. To find the decibel difference between I_1 and I_2, the common logarithm (base 10) of the ratio I_1/I_2 is multiplied by 10. As an example, suppose the intensity I_1 is 1,000,000 times the intensity I_2. The logarithm of 1,000,000, multiplied by 10, is 60. Thus the intensity I_1 is 60 db above the intensity I_2. In many studies it is the decibel difference between two different sounds rather than the absolute strength of any one sound, which is of most interest and can be most readily determined.

The decibel scale is also suitable for expressing absolute sound intensities. For this purpose, a standard intensity is first selected, called the *reference intensity* or *reference level*, and then all other sound field intensities are expressed in terms of decibels above (or below) the standard. Unfortunately, different standards have been used by different groups in underwater sound research. Sometimes, 10^{-16} watt per sq cm has been used as the standard since this is the usually accepted standard in air. More frequently, the

reference level has been expressed in terms of the sound pressure.

Since sound represents vibrations and since vibrations of a fluid (such as air or sea water) are associated with periodic changes in the local pressure, the deviation of instantaneous pressure from the hydrostatic or atmospheric pressure may be used as a measure of sound intensity. This excess pressure oscillates during each cycle; therefore, the intensity must be expressed in terms of some averaged quantity. Since the excess pressure is positive during one half of the cycle and negative during the other, its arithmetic mean vanishes. It is possible to obtain a nonvanishing average quantity by considering the rms excess pressure. In the case of a sinusoidal vibration, the rms excess pressure is equal to $1/\sqrt{2}$, or 0.7 times the maximum value of the excess pressure. It will be shown in Chapter 2 that in a given medium the sound intensity is proportional to the mean square excess pressure. Two standards based on pressure have been used in underwater sound studies. One is a sound intensity corresponding to an rms excess pressure of 0.0002 dyne per sq cm. This standard has been recently replaced by that of an intensity corresponding to an rms excess pressure of 1 dyne per sq cm. When sound field intensities are expressed on a decibel scale relative to some standard intensity, they are usually referred to as sound levels.

1.3   **PROPAGATION OF SOUND IN
THE SEA**

When the propagation of sound in the sea first became a matter of prime military importance, it was hoped and expected that sound would travel along straight lines from the source and that the sound field intensity would decrease in accordance with the simple inverse square law. However, this hope was not realized. Because of the peculiar characteristics of the ocean as a sound-transmitting medium, marked deviations occur from both straight-line propagation and inverse square intensity decay.

Straight-line propagation of sound is to be expected only if the velocity of propagation is constant throughout the medium. In the ocean this condition is usually violated primarily because of the variation of temperature with depth. There is almost always a layer in which the water temperature drops appreciably with increasing depth. This layer may begin right at the sea surface, or it may lie beneath a top layer of constant temperature. In such a region of temperature change, the sound paths are bent in the direction of lower velocity of propagation, in other words, in the direction of lower temperature. Even though the changes in sound velocity are small (about 1 per cent for a temperature drop of 10 F), the resultant bending of the sound path becomes appreciable over a distance of a few hundred yards. If, for instance, the drop in temperature begins directly at the surface of the water, and totals a degree or more in 30 ft of depth, most of the sound energy will travel along paths bent downward and will miss a shallow target at a range of 1,000 yd.

Because of this bending of sound by temperature gradients, some departure of sound intensity from the inverse square law is to be expected. The amount of this departure can be calculated if the temperature distribution in the ocean is known. However, even this more complicated process for computing the intensity is too simple. The effects of the boundaries of the medium (ocean bottom and surface), and of the absorption and scattering of sound in the body of the ocean must, also, be considered.

Both the sea surface and the sea bottom affect the sound field intensity. Some of the sound energy strikes these boundaries and is then partly reflected back into the ocean, partly permitted to pass into the adjoining medium (air or sea bottom). The portion of the energy which is reflected will return into the interior in a variety of directions. Also, little under-stood processes in the body of the ocean affect sound intensity. In some way, a certain amount of the passing sound energy is converted into heat (absorption of sound); and chance impurities such as fish, seaweed, plankton, and gas bubbles, tend to scatter a small amount of the passing sound energy in all directions out of its principal path.

For all these reasons, the propagation of underwater sound presents, at first, a rather confusing picture. Considerable progress has been made, however, in understanding the behavior of underwater sound and in utilizing this partial understanding in the design and tactical use of sound gear. The results which have been achieved are due to a combination of theoretical and experimental investigations. Chapters 2 through 10 discuss the background and progress of these investigations. Chapters 2 and 3 lay the theoretical groundwork for the physics of underwater sound. Chapter 4 leads toward the experimental results by reporting on the equipment and procedures employed in the experiments. Chapters 5 and 6 report experimental results on the propagation of sound, primarily sound generated by transducers. Chapter 7 is concerned with the observed short-term fluctuation of underwater sound intensity. Chapters 8 and 9 deal with the formation and transmission of explosive sound. Finally, Chapter 10 summarizes the results obtained to date and discusses possibilities for future research.

Chapter 2

WAVE ACOUSTICS

SOUND ENERGY takes the form of disturbances of the pressure and density of some medium. Therefore, the basic relationships between impressed forces and resulting changes in pressure and density are useful in an understanding of sound transmission. In this chapter we shall derive several such relationships, and shall combine them into one differential equation relating the time derivatives and space derivatives of the pressure changes to several physical constants of the medium itself. This differential equation is the foundation for the mathematical treatment of sound transmission to which the rest of the chapter is devoted.

We shall see that this mathematical approach cannot in itself furnish complete information on sound transmission in the ocean. The physical picture must necessarily be simplified to make mathematical description possible — and even this simplified scheme does not yield explicit results for the sound intensity in all cases. However, it is valuable to know the mathematical theories even if they are partially unsuccessful in predicting the qualities of sound transmission. Tendencies predicted by a simplified theory are often verified qualitatively in practice. Also, there is always the hope that by changes and amplifications an incomplete theory can be made much more useful.

2.1 BASIC EQUATIONS

In this section we shall derive the basic equations which will be put together to derive the fundamental differential equation of wave propagation, the *wave equation*. These equations are (1) the equation of continuity, which is the mathematical expression of the law of conservation of mass; (2) the equations of motion, which are merely Newton's second law applied to the small particles of a disturbed fluid; (3) force equations, which relate the fluid pressure inside a small volume of the fluid to the external forces acting on the periphery of the volume; (4) the equation of state, which relates the pressure changes inside a fluid to the density changes.

2.1.1 Equation of Continuity

The equation of continuity is simply a mathematical statement of the law that no disturbance of a fluid can cause mass to be either created or destroyed. In particular, any difference between the amounts of fluid entering and leaving a region must be accompanied by a corresponding change in the fluid density in the region.

To express this law in mathematical terms we must first derive an expression for the mass of fluid which passes through a certain small area of a surface in one second. Let the small surface element have the area A, as in Figure 1, and let the fluid move in a direction

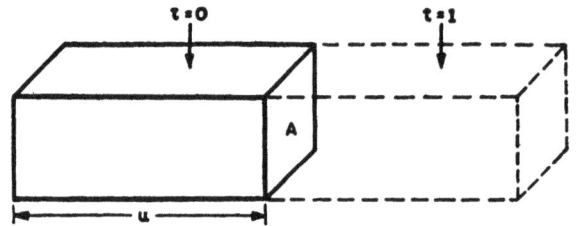

FIGURE 1. Passage of fluid through area element A.

perpendicular to A with the velocity u. In one second, a rectangular fluid element of base A and height u has passed through this element of area; that is, a volume of Au cubic units of fluid has traversed the area. The mass of fluid passing through the area per second will thus be ρAu, where ρ is the density at the point and time in question. If the fluid is moving not perpendicular to the element A, but in some other direction, the mass passing A per second will still be given by ρAu, if u is taken to be the velocity component in the direction perpendicular to A.

Now consider a small hypothetical box-shaped volume inside the fluid, and examine the amounts of fluid entering and leaving this box (pictured in Figure 2). For simplicity, we can assume that the edges of the box are parallel to the coordinate axes. Let the dimensions of the box be l_x, l_y, l_z, as shown in the diagram, and let the coordinates of the point H be (x,y,z). Let the components of the fluid velocity at the point H be u_x, u_y, u_z.

The mass of fluid entering the face $AHED$ in unit time is clearly the rate at which mass is moving in the x direction times the area of $AHED$, or $\rho u_x l_y l_z$. The mass of fluid leaving the box through $BCGF$ is a

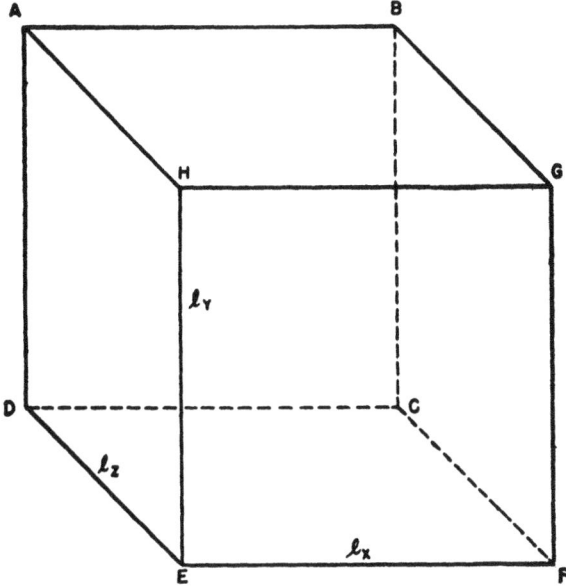

FIGURE 2. Infinitesimal cube of fluid.

similar expression, but with ρ and u_x measured at $(x + l_x, y, z)$. The value of ρu_x at $(x + l_x, y, z)$ is just its value at (x,y,z) increased by $l_x \partial(\rho u_x)/\partial x$ since l_x is very small. That is, the mass leaving in one second through face $BCGF$ is

$$\left[\rho u_x + \frac{\partial}{\partial x}(\rho u_x)l_x\right]l_y l_z.$$

Then the net increase per second in the mass inside the box caused by the flow through the two faces perpendicular to the x axis is

$$-\frac{\partial}{\partial x}(\rho u_x)l_x l_y l_z.$$

Similarly, the net increase per second caused by the flow through the two faces perpendicular to the y axis is

$$-\frac{\partial}{\partial y}(\rho u_y)l_x l_y l_z,$$

and through the two faces perpendicular to the z axis,

$$-\frac{\partial}{\partial z}(\rho u_z)l_x l_y l_z.$$

The total time rate of increase in the mass contained in the box is simply the sum of these three quantities, or

$$-\left[\frac{\partial(\rho u_x)}{\partial x} + \frac{\partial(\rho u_y)}{\partial y} + \frac{\partial(\rho u_z)}{\partial z}\right]l_x l_y l_z. \qquad (1)$$

Since no mass can be created or destroyed inside the box, this rate of deposit of mass must result in a corresponding change in the average density ρ inside the box. That is,

$$\frac{\partial}{\partial t}[\rho l_x l_y l_z] = -\left[\frac{\partial(\rho u_x)}{\partial x} + \frac{\partial(\rho u_y)}{\partial y} + \frac{\partial(\rho u_z)}{\partial z}\right]l_x l_y l_z.$$

Canceling out the constant factor $l_x l_y l_z$, we obtain the *general* equation of continuity

$$\frac{\partial \rho}{\partial t} = -\left[\frac{\partial(\rho u_x)}{\partial x} + \frac{\partial(\rho u_y)}{\partial y} + \frac{\partial(\rho u_z)}{\partial z}\right]. \qquad (2)$$

This equation can be simplified if it is assumed that all displacements and changes of density are so small that second-order and higher products of them can be neglected. The actual density ρ, then, will not be very different from the constant equilibrium density ρ_0. If σ is defined by

$$\sigma = \frac{\rho - \rho_0}{\rho_0}, \qquad (3)$$

then σ, the fractional change in density caused by the displacement of the fluid from equilibrium, will be a very small number. Henceforth σ will be called the fractional density change or *condensation*.

With this understanding, it is clear that

$$\frac{\partial(\rho u_x)}{\partial x} = \frac{\partial}{\partial x}[(\rho_0 + \rho_0\sigma)u_x]$$

$$= \frac{\partial}{\partial x}(\rho_0 u_x) = \rho_0\frac{\partial u_x}{\partial x},$$

since the second-order product σu_x can be neglected. By substituting this value of $\partial(\rho u_x)/\partial x$ and similar expressions for $\partial(\rho u_y)/\partial y$ and $\partial(\rho u_z)/\partial z$ into equation (2), the following *simplified* equation of continuity results:

$$\frac{\partial \sigma}{\partial t} = -\left(\frac{\partial u_x}{\partial x} + \frac{\partial u_y}{\partial y} + \frac{\partial u_z}{\partial z}\right). \qquad (4)$$

Equation (4) is the form of the equation of continuity which will be used in the derivation of the wave equation (27).

For later reference, we shall note what happens to the volume occupied by an infinitesimal mass of the fluid when the fluid is given a small displacement from equilibrium. If v_0 is the volume occupied by the small mass at equilibrium, and v is the volume at time t, then a fractional volume change ω can be defined by

$$\omega = \frac{v - v_0}{v_0}. \qquad (5)$$

From equations (3) and (5),

$$\rho = \rho_0(1 + \sigma), \tag{6}$$

$$v = v_0(1 + \omega). \tag{7}$$

Since the masses at equilibrium and at time t are equal,

$$\rho v = \rho_0 v_0. \tag{8}$$

By combining equations (6), (7), and (8)

$$(1 + \omega)(1 + \sigma) = 1.$$

The product $\omega\sigma$, a second-order term, can be neglected, giving

$$\omega = -\sigma. \tag{9}$$

That is, under the assumption of small displacements and small density changes, the fractional volume change ω is the negative of the fractional density change σ.

2.1.2 Equations of Motion

In this section, we shall apply Newton's second law of motion to the mass of fluid within the volume element v_0. This law states that the product of the mass of a particle by its acceleration in any direction is equal to the force acting on the particle in that direction.

Given the velocity distribution within the fluid as a function of the position coordinates and time,

$$u_x = u_x(x,y,z,t), \text{ etc.}; \tag{10}$$

then the distribution of acceleration within the fluid is to be calculated,

$$a_x = a_x(x,y,z,t), \text{ etc.} \tag{11}$$

We cannot immediately say that $a_x = \partial u_x/\partial t$. For, in order to calculate the acceleration at a particular point and a particular time, we must focus attention on one particular particle. At the end of a time increment dt, the particle has moved to a point $(x + dx, y + dy, z + dz)$, where it has the velocity component $u_x(x + dx, y + dy, z + dz)$. The difference between its new velocity and its original velocity, divided by the time interval dt, gives the desired acceleration component du_x/dt. This value is not exactly the same as the simple partial derivative $\partial u_x/\partial t$, because the latter does not focus attention on the change of velocity of a single particle, but instead compares the velocity of a particle at the point (x,y,z) and time t with the velocity of the particle which occupies the position (x,y,z) at the end of the time interval dt. However, du_x/dt and $\partial u_x/\partial t$ will be almost equal

under the assumptions that second-order products of displacements, particle velocities, and pressure changes are negligible. To show this, we note that

$$\frac{du_x}{dt} = \frac{\partial u_x}{\partial t} + \frac{\partial u_x}{\partial x}\frac{dx}{dt} + \frac{\partial u_x}{\partial y}\frac{dy}{dt} + \frac{\partial u_x}{\partial z}\frac{dz}{dt}, \tag{12}$$

which is the usual equation found in calculus texts relating the partial and total time derivatives of a function. The last three terms are second-order products, since dx/dt, dy/dt, dz/dt are merely u_x, u_y, u_z. Thus, the component of acceleration in the x direction may be approximated by $\partial u_x/\partial t$, and similarly for a_y and a_z. That is,

$$a_x = \frac{\partial u_x}{\partial t}; \quad a_y = \frac{\partial u_y}{\partial t}; \quad a_z = \frac{\partial u_z}{\partial t}. \tag{13}$$

The mass of fluid within the volume element v_0 is ρv_0. If F_x, F_y, F_z are the components of the forces acting on the element, then the equations governing the motion of this element are, in view of equation (13),

$$F_x = \rho v_0 \frac{\partial u_x}{\partial t}; \quad F_y = \rho v_0 \frac{\partial u_y}{\partial t}; \quad F_z = \rho v_0 \frac{\partial u_z}{\partial t}. \tag{14}$$

It is desirable to make the equations governing the motion of the small element independent of the particular value of the small volume v_0. For this reason, we rewrite equations (14) as

$$f_x = \rho \frac{\partial u_x}{\partial t}; \quad f_y = \rho \frac{\partial u_y}{\partial t}; \quad f_z = \rho \frac{\partial u_z}{\partial t}, \tag{15}$$

where

$$f_x = \frac{F_x}{v_0}; \quad f_y = \frac{F_y}{v_0}; \quad f_z = \frac{F_z}{v_0}.$$

The normalized force components may be regarded as the *force components per unit volume* acting on the small volume element.

The next section is concerned with calculating f_x, f_y, f_z in terms of the pressure or density changes occurring within the fluid.

2.1.3 Law of Forces in a Perfect Fluid

A fluid is called *perfect* if the forces in its interior are solely forces of compression and expansion, in other words, if the fluid is incapable of shear stress. If a fluid is perfect, the force on any portion of its surface is perpendicular to the surface. Fluids which can exhibit shear stress in response to shear deformation, in addition to responding to compressive and expansive forces, are called *viscous*.

Before the equations of motion (15) can be used, expressions must be derived for the force components f_x, f_y, f_z acting on the small box of Figure 2. Accordingly, we shall calculate these forces under the assumption that the fluid is perfect. According to this assumption, the box will move in the x direction if and only if the pressure on face $ADEH$ is different from the pressure on face $BCFG$. Similarly, it will move in the z direction only if the pressures on faces $ABCD$ and $EFGH$ are unequal. Motion in the y (vertical) direction is not quite so simple because of the hydrostatic, or gravity-produced pressure differences, which do not of themselves cause motion. The box will move in the y direction if and only if the pressure on face $DCFE$ is not exactly equal to the pressure on face $ABGH$ plus the total weight of the box. If the corrected pressure $p(x,y,z,t)$ is defined as the total pressure $P(x,y,z,t)$ minus the hydrostatic pressure at the point (x,y,z) when the fluid is at equilibrium, then the criterion for motion in the y direction may be restated as follows. Motion will occur in the y direction if and only if the corrected pressure at face $ABGH$ differs from the corrected pressure at face $DCFE$. We shall have little occasion to use the total pressure P since the hydrostatic pressures are seldom important in sound propagation.

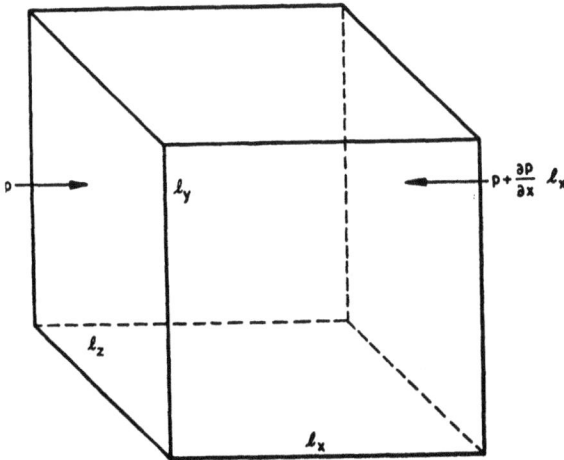

FIGURE 3. Pressure on opposite faces of infinitesimal fluid element.

Figure 3 is a duplication of the box of Figure 2 showing the forces acting in the x direction. If the pressure at the left-hand surface is p, the total force on that surface is $p l_y l_z$. The pressure at the right-hand surface is clearly $p + (\partial p/\partial x)l_x$; and the total force on that surface is therefore $[p + (\partial p/\partial x)]l_x l_y l_z$. Since the fluid is assumed to be perfect, these forces are

parallel and their resultant can be obtained by simple subtraction. Thus, the total force on the volume v_0 in the x direction is given by

$$F_x \equiv f_x v_0 = -\frac{\partial p}{\partial x} l_x l_y l_z. \quad (16)$$

Thus, the force per unit volume in the x direction f_x is given by

$$f_x = -\frac{\partial p}{\partial x}. \quad (16a)$$

Similarly,

$$f_y = -\frac{\partial p}{\partial y} \quad (16b)$$

$$f_z = -\frac{\partial p}{\partial z}. \quad (16c)$$

From equations (15) and (16a, b, c) we obtain

$$\frac{\partial p}{\partial x} = -\rho \frac{\partial u_x}{\partial t}; \quad \frac{\partial p}{\partial y} = -\rho \frac{\partial u_y}{\partial t}; \quad \frac{\partial p}{\partial z} = -\rho \frac{\partial u_z}{\partial t}. \quad (17)$$

2.1.4 Equation of State

Our aim is to derive a differential equation which will relate certain properties of the disturbed fluid (pressure changes, density changes) to the independent variables x,y,z,t. For effective use, this differential equation should contain only one dependent variable. The basic equations derived up to this point — (4), (15), and (16) — contain the dependent variables σ, p, ρ, u_x, u_y, u_z. σ and ρ are one variable since they are related by equation (3). It will be seen in Section 2.2.1 that the velocity components can be easily eliminated by use of the equation of continuity (4). However, we must consider the relationship between density and pressure before we obtain a differential equation for the propagation of sound. Such a relationship between density and pressure is obtained from the equation of state of the fluid.

The equation of state of any fluid[a] is that equation which describes the pressure of the fluid as a function of its density and its temperature,

$$P = P(\rho,T).$$

This function $P(\rho,T)$ must be determined experimentally for each fluid separately. In the case of sea water, it depends on the percentage of dissolved salts.

A relation between pressure and density is obtained from the equation of state by making two assumptions. First, it is assumed that a passing sound wave

---

[a] We shall follow the common usage of physicists and use the term *fluids* to denote both liquids and gases.

causes the fluid to deviate so slightly from its state of equilibrium that the change in pressure is proportional to the fractional change in density. Second, it is assumed that the changes caused by the passing of the sound wave take place so rapidly that there is practically no conduction of heat. We shall denote the fractional change in density as heretofore by σ; the change in pressure will be called *excess pressure* and denoted by p.

Thus we assume that the fractional change in density and the excess pressure caused by the passing sound wave are both small and that they are proportional to each other:

$$p = \kappa\sigma. \tag{18}$$

The constant of proportionality κ is called the *bulk modulus*. It depends not only on the chemical nature of the fluid (such as the concentration of salts in the sea), but also on the equilibrium temperature T, the equilibrium pressure P_0, and the equilibrium density ρ.

The temperature in the ocean varies from point to point, usually decreasing with increasing depth. The equilibrium pressure increases rapidly with the depth, and the density increases very slightly with depth. As a result, the bulk modulus κ is itself a function of all three coordinates x, y, and z, although its greatest changes take place in a vertical direction. To the extent that the temperature distribution of the ocean is subject to diurnal and seasonal changes, κ is also a function of the time t. However, in the following sections we shall usually simplify matters by disregarding these variations in space and time and by treating κ as a constant.

2.2 WAVE EQUATION IN A PERFECT FLUID

2.2.1 Derivation

If in a certain region of a fluid in equilibrium the pressure is changed from its equilibrium value, the fluid immediately produces forces which aim toward restoring the equilibrium state. Vibrations result, which are propagated as *waves* through the fluid. These waves are sound waves, and the fundamental differential equation governing their propagation will now be developed by using the basic equations derived in the preceding sections. The particular equations used are the equation of continuity (4), the

equations of motion (15), the law of forces (16), and the equation of state (18).

From equation (18) we have

$$\frac{\partial p}{\partial x} = \kappa\frac{\partial \sigma}{\partial x}; \quad \frac{\partial p}{\partial y} = \kappa\frac{\partial \sigma}{\partial y}; \quad \frac{\partial p}{\partial z} = \kappa\frac{\partial \sigma}{\partial z}.$$

By putting these values for $\partial\sigma/\partial x$, $\partial\sigma/\partial y$, and $\partial\sigma/\partial z$ in the law of forces (16) we obtain

$$f_x = -\kappa\frac{\partial \sigma}{\partial x}; \quad f_y = -\kappa\frac{\partial \sigma}{\partial y}; \quad f_z = -\kappa\frac{\partial \sigma}{\partial z}. \tag{19}$$

After these values for the components of the force on a small box are substituted into the equations of motion (15), we obtain the following relations.

$$\rho\frac{\partial u_x}{\partial t} = -\kappa\frac{\partial \sigma}{\partial x}$$
$$\rho\frac{\partial u_y}{\partial t} = -\kappa\frac{\partial \sigma}{\partial y} \tag{20}$$
$$\rho\frac{\partial u_z}{\partial t} = -\kappa\frac{\partial \sigma}{\partial z}.$$

Since we assume that density changes and velocity changes are all comparatively small, the expressions $\rho\partial u_x/\partial t$ and $\rho_0\partial u_x/\partial t$ will differ by a second-order term, and hence can be regarded as equal. Then equations (20) become

$$\rho_0\frac{\partial u_x}{\partial t} + \kappa\frac{\partial \sigma}{\partial x} = 0$$
$$\rho_0\frac{\partial u_y}{\partial t} + \kappa\frac{\partial \sigma}{\partial y} = 0 \tag{21}$$
$$\rho_0\frac{\partial u_z}{\partial t} + \kappa\frac{\partial \sigma}{\partial z} = 0.$$

In order to apply the equation of continuity (4), we differentiate the first equation of (21) with respect to x, the second with respect to y, and the third with respect to z. The equations are added, leading to

$$\rho_0\frac{\partial}{\partial t}\left(\frac{\partial u_x}{\partial x} + \frac{\partial u_y}{\partial y} + \frac{\partial u_z}{\partial z}\right) + \kappa\left(\frac{\partial^2 \sigma}{\partial x^2} + \frac{\partial^2 \sigma}{\partial y^2} + \frac{\partial^2 \sigma}{\partial z^2}\right) = 0. \tag{22}$$

From the equation of continuity, the first parenthesis is $-\partial\sigma/\partial t$; and equation (22) reduces to

$$\frac{\partial^2 \sigma}{\partial t^2} = \frac{\kappa}{\rho_0}\left(\frac{\partial^2 \sigma}{\partial x^2} + \frac{\partial^2 \sigma}{\partial y^2} + \frac{\partial^2 \sigma}{\partial z^2}\right). \tag{23}$$

Since $\sigma = p/\kappa$, from equation (18), where κ is constant, equation (23) implies

$$\frac{\partial^2 p}{\partial t^2} = \frac{\kappa}{\rho_0}\left(\frac{\partial^2 p}{\partial x^2} + \frac{\partial^2 p}{\partial y^2} + \frac{\partial^2 p}{\partial z^2}\right). \tag{24}$$

It can be shown that the velocity components u_x, u_y, u_z also satisfy a differential equation of the form of (23), provided the motion of the disturbed fluid is irrotational. That is, if sound is propagated in a perfect fluid in such a manner that no eddies are produced,

$$\frac{\partial^2 u_x}{\partial t^2} = \frac{\kappa}{\rho_0}\left(\frac{\partial^2 u_x}{\partial x^2} + \frac{\partial^2 u_x}{\partial y^2} + \frac{\partial^2 u_x}{\partial z^2}\right) \qquad (25)$$

with similar equations for u_y and u_z.

Equations (23), (24), and (25) are equivalent; that is, the fundamental laws of sound propagation can be deduced from any one of them by using the known relationship between σ, u_x, and p. In the following sections, the most frequent reference is to equation (24). Variation in the excess pressure is the most familiar and probably the most intuitive change in the disturbed fluid; also, the majority of hydrophones used in the reception of underwater sound respond directly to variations in excess pressure rather than to variations in particle velocity or condensation.

It is convenient, in equations (23) to (25), to set

$$c^2 = \frac{\kappa}{\rho_0}, \qquad (26)$$

so that the *wave equation* becomes

$$\frac{\partial^2 p}{\partial t^2} = c^2\left(\frac{\partial^2 p}{\partial x^2} + \frac{\partial^2 p}{\partial y^2} + \frac{\partial^2 p}{\partial z^2}\right). \qquad (27)$$

It will be pointed out in Section 2.3.1 that c, defined by equation (26), has the general significance of sound velocity.

The wave equation (27) gives the relationship between the time derivatives and the space derivatives of the pressure in the fluid through which the sound is passing. Relationships of this sort have been used by generations of mathematicians as a starting point for the development of physical theory. In the field of sound, these mathematicians have explored the methods by which the future course of pressure in a fluid can be calculated if only the initial distribution of pressure is given. Mathematically, this amounts to solving the wave equation (27) with given initial and boundary conditions. Once the distribution of pressure is known, the sound intensity at any point and time can be calculated by the methods of acoustics.

2.2.2 Initial and Boundary Conditions

The differential equation of a physical process gives a dynamical description of the process relating the various temporal and spatial rates of change, but does not of itself tell all we want to know. In the case of the wave equation, we desire knowledge of how the excess pressure varies in space and time. This information is obtainable, not from the wave equation itself, but from its mathematical solution. The general solution of a partial differential equation like the wave equation always contains arbitrary constants and even arbitrary functions. These arbitrary constants and functions are, in any individual problem, adjusted to make the solution fit the special conditions of the problem.

These special conditions are of two kinds: boundary conditions and initial conditions. In the problem of sound propagation, the two types of conditions can be defined as follows. *Boundary conditions* are fixed by the geometry of the medium itself. If the medium is finite, boundary conditions must always be considered. The excess pressure must fulfill certain conditions at a boundary such as the sea surface, sea bottom, or internal obstacle. The pressure may have to be zero at one boundary, or a maximum at some other boundary, or may satisfy some other condition.[b]

Initial conditions are concerned not with the fixed geometry of the fluid and its surroundings, but with the special disturbances which cause sound to be propagated. One type of initial conditions specifies the pressure distribution at a certain instant of time, $t = t_0$, over the whole fluid. That is, we are given a function $\bar{p}(x,y,z)$, and are told that

$$p(x,y,z,t_0) = \bar{p}(x,y,z). \qquad (28)$$

Another type of initial conditions specifies the pressure as a function of time at a fixed point (x_0,y_0,z_0) of the fluid. That is, we are given a function $\bar{p}(t)$ and are told that

$$p(x_0,y_0,z_0,t) = \bar{p}(t). \qquad (29)$$

Every actual case of sound propagation involves both initial conditions and boundary conditions. However, in our mathematical approximations to reality it is best to start with the simplest case, that is, where sound is propagated into a medium that is infinite. Of course, in theory every problem can be regarded as a problem in an infinite medium. We can consider the sea and air together as one medium, whose physical properties at equilibrium (elasticity,

---

[b] It will be seen in Section 2.6.1 that the excess pressure must be nearly zero at the boundary separating water and air and a maximum at the solid bottom of the sea.

density, and other properties) suffer a sharp change at the separating surface. However, the mathematical treatment of a medium with strongly variable physical properties is still more difficult than the treatment of boundary conditions. We are free to schematize the physical situation in the most convenient way. Accordingly, we shall start with the consideration of an infinite medium, where the elasticity and density at equilibrium are not strongly variable, and shall later specialize our treatment by the consideration of boundary conditions (Sections 2.6 and 2.7).

Initial conditions must always be considered since without them no sound could possibly originate. Unless the initial conditions are themselves very simple, the solution of problems even in an infinite medium is quite involved mathematically. The most practical procedure is to first find solutions under very simple initial conditions and use these solutions as building blocks for constructing solutions of problems with more complicated initial conditions.

2.3 SIMPLE TYPES OF SOUND WAVES

2.3.1 Plane Waves

It is convenient to start our study of the solution of the wave equation with the assumption that the disturbance is propagated in layers. We assume that at any time t the excess pressure p is a function of x only; that is, p is independent of y and z. With this understanding, the wave equation (27) reduces to

$$\frac{\partial^2 p}{\partial t^2} = c^2 \frac{\partial^2 p}{\partial x^2}. \qquad (30)$$

The eighteenth century mathematicians knew that if p is an arbitrary function of either $(t - x/c)$ or $(t + x/c)$, or a sum of two such functions, then p satisfies equation (30). The proof is easy. Assume that $p = f(t - x/c)$ where f is any function. Then,[c]

$$\frac{\partial^2 p}{\partial t^2} = f''\left(t - \frac{x}{c}\right); \quad \frac{\partial^2 p}{\partial x^2} = \frac{1}{c^2}f''\left(t - \frac{x}{c}\right).$$

The proof is identical for a function of the argument $(t + x/c)$. The sum of two solutions will itself be a solution because of the general theorem that the sum of two solutions of a homogeneous linear differential equation will also satisfy the equation.

[c] In accordance with usual calculus notations, $f''(t - x/c)$ represents the second derivative of $f(z)$ evaluated for $z = t - x/c$.

Also, it can be shown that any function of x and t which is not of the form

$$f_1\left(t - \frac{x}{c}\right) + f_2\left(t + \frac{x}{c}\right) \qquad (31)$$

cannot possibly satisfy equation (30). The proof is carried out as follows. Represent the unknown solution of equation (30) in the form

$$p = f(\xi,\eta), \ \xi = x - ct, \ \eta = x + ct.$$

If the differential equation (30) is written in terms of the new variables ξ and η, it reduces to

$$\frac{\partial^2 f(\xi,\eta)}{\partial \xi \partial \eta} = 0.$$

This equation implies that the first derivative $\partial f/\partial \xi$ must be a function of ξ only and independent of η, for otherwise the second derivative $\partial^2 f/\partial \xi \partial \eta$ could not vanish. Thus f itself must have the form

$$f(\xi,\eta) = f_1(\xi) + f_2(\eta).$$

Let us focus attention on all the solutions of equation (30) which have the form

$$p = f\left(t - \frac{x}{c}\right).$$

There are an infinite number of such solutions corresponding to the infinite number of possible choices of f. However, no more than one of these solutions can fit the special conditions of a particular physical situation since the actual pressure at a specified point and specified time can have only one value. Suppose that there is one member of the family of functions, denoted by $f_1(t - x/c)$, which satisfies the given initial and boundary conditions. Then a fixed value of $(t - x/c)$, say 4.13, will always be associated with some fixed value of the excess pressure, given by $f_1(4.13)$. If $f_1(4.13)$ is equal to 0.02, then the excess pressure will be 0.02 at those combinations of time and place where $(t - x/c) = 4.13$, that is, where

$$x = ct - 4.13c.$$

In other words, as the time increases, any fixed value of the excess pressure travels in the positive x direction with the speed c. This result is clearly true no matter what the form of f_1 or the particular value of the excess pressure. Such a process, in which a given pressure change travels outward through a medium, is referred to as *propagation of progressive waves*.

Similarly, if a function $f_2(t + x/c)$ is the sole member of the family of functions (31) which satisfies the imposed initial and boundary conditions, progressive waves will be propagated with the speed c in the negative x direction. If, however, an expression of the form (31) is the function describing the given physical situation, the situation is more complicated. The resulting distribution of pressure will be the mathematical resultant of the pressure distributions calculated for f_1 and f_2; and a given value of the excess pressure will no longer be propagated in a single direction with the speed c. Discussion of this more complicated type of wave propagation will be deferred until Section 2.7.

Now we consider two specific examples of the propagation of plane waves in an infinite homogeneous medium (no boundary conditions). In the first example, we assume as an initial condition that the exact pressure distribution is specified at the time instant $t = 0$ between the planes $x = 0$ and $x = x_0$, by $p(x,0) = \bar{p}(x)$; and also that the excess pressure is zero at $t = 0$ for all values of x less than 0 and greater than x_0. We assume that this initial disturbance gives rise to progressive waves traveling in the positive x direction, that is, that the solution is of the form $p = f(t - x/c)$. Then the solution of the wave equation with these conditions must be

$$p(x,t) = \bar{p}(x - ct) \qquad (32)$$

since first, it satisfies the initial conditions $p(x,0) = \bar{p}(x)$; second, it is a function of $(t - x/c)$ and therefore satisfies the wave equation (30); and third, there can be only one solution to this physical problem.

By the results of preceding paragraphs, we know that a given value of the excess pressure will be propagated in the x direction with the speed c. Thus, at the time t the initial disturbance will be duplicated between the planes $x = ct$ and $x = x_0 + ct$; and the excess pressure will be zero for $x < ct$ and $x > x_0 + ct$. The disturbance of the fluid remains of width x_0, remains unchanged in "shape," and is propagated with the speed c.

As another example, we suppose as an initial condition that the values of the excess pressure are specified only for the plane $x = 0$, but for the total time interval between $t = 0$ and $t = t_0$, by the equation $p(0,t) = \bar{p}(t)$ and that the excess pressure at the plane $x = 0$ is zero for $t < 0$ and $t > t_0$. Here, also, we assume that this disturbance causes progressive waves to be propagated in the positive x direction. Arguing as in the preceding example, the solution of the wave equation (30) with these imposed conditions is

$$p(x,t) = \bar{p}\left(t - \frac{x}{c}\right). \qquad (33)$$

The expression (33) differs somewhat in form from equation (32) because the initial conditions are extended in time instead of in space.

In this example, it is known that the excess pressure will be the same at all combinations of space and time where $x - ct = $ constant. Since $x - ct$ equals zero when $x = 0$, $t = 0$, the value of the excess pressure corresponding to $x = 0$, $t = 0$, must be assumed by the plane $x = ct$ at the time t. Further, since $x - ct$ equals $-ct_0$ at $t = t_0$, $x = 0$, the value of the excess pressure corresponding to $t = t_0$, $x = 0$, will be assumed by the plane $x = ct - ct_0$ at the time instant t. Thus, at time t the disturbance is confined between the planes $x = ct - ct_0$ and $x = ct$; that is, the region of disturbance is always of width ct_0 and is propagated along the positive x axis with the speed c.

SOUND VELOCITY

We have seen that in some simple situations the quantity c may be regarded as the velocity with which the disturbance is propagated in the medium, or more simply, the velocity of sound in the medium. It will be recalled that c was defined in equation (26) by

$$c = \sqrt{\frac{\kappa}{\rho_0}},$$

where κ is the bulk modulus and ρ_0 is the density of the fluid at equilibrium.

If the medium is a perfect gas, the relation of the sound velocity to the temperature and pressure can be expressed in a simple formula. The **pressure** changes produced by sound in a fluid are usually so rapid that they are accomplished without appreciable heat transfer, that is, they are practically adiabatic. For a perfect gas suffering adiabatic pressure changes the bulk modulus is γP, where P is the total pressure and γ is the ratio between the specific heat at constant pressure and the specific heat at constant volume. For a perfect gas suffering any kind of pressure change $P = \rho R T$. Thus, the simple result follows that

$$c = \sqrt{\gamma R T}.$$

Hence, in air at normal pressure, which is not far from a perfect gas, the sound velocity increases with the square root of the absolute temperature.

No such relationship can be derived for the velocity of sound in sea water since the pressure, density, and temperature of sea water are not related by any

FIGURE 5. Percentage variation of sound velocity with water temperature, salinity, and depth. A. Effect of temperature. B. Effect of salinity. C. Effect of depth.

simple formula. However, tables have been constructed which show the velocity of sound as a function of three variables which can be measured directly: the water temperature, pressure, and salinity. Although the relationship is not simple, these three variables determine precisely both the bulk modulus and the density, from which the sound velocity can be calculated from equation (26).

At 32 F, atmospheric pressure, and normal salinity (34 parts per thousand by weight), the velocity of sound in sea water is about 4,740 ft per sec. Increase of either temperature, pressure, or salinity causes the sound velocity to increase. The increase of sound velocity with temperature is about 8.5 ft per sec per degree F at 32 F, and about 4.0 ft per sec per degree F at 90 F. The increase of sound velocity with water depth, caused by the increase in pressure, is 1.82 ft per sec per 100 ft of depth. In the open ocean for the depths of interest in sonar operations the water temperature is the controlling factor in determining the velocity; since sonar gear is usually operated at

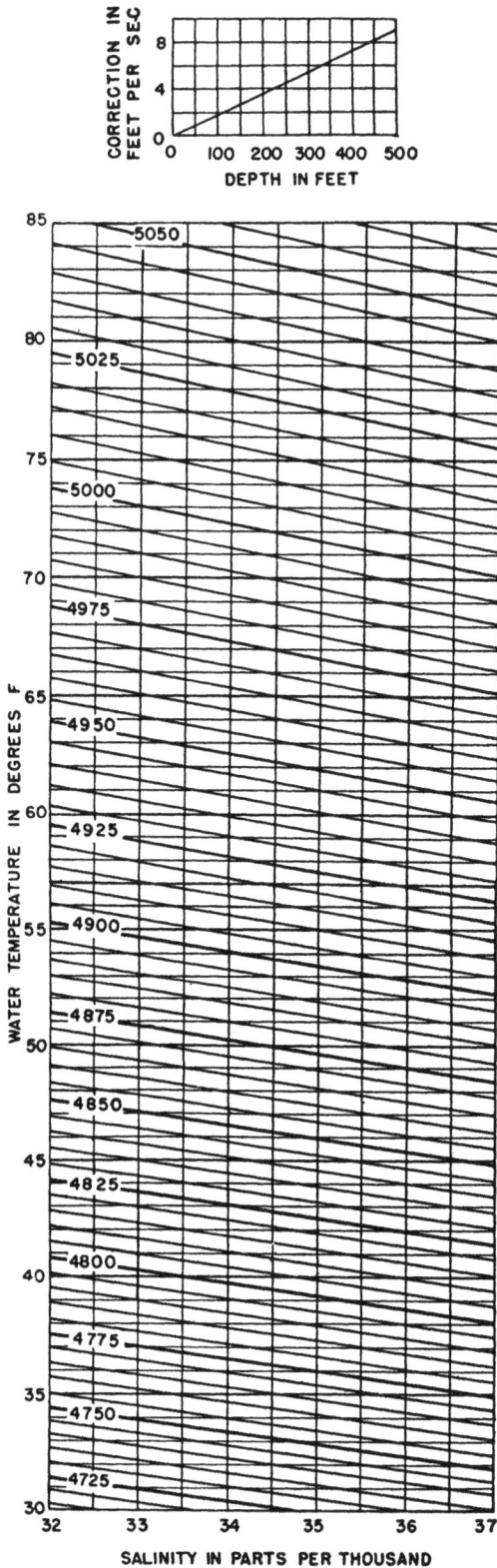

FIGURE 4. Speed of sound in sea water.

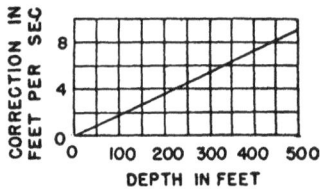

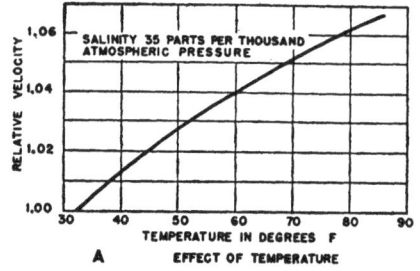

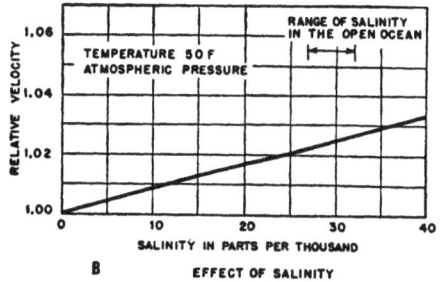

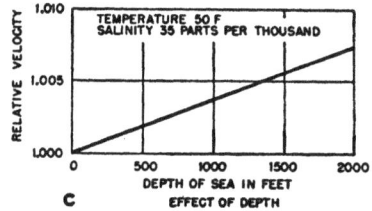

shallow depths, the pressure changes are relatively unimportant, and salinity changes in the open ocean are usually too small to matter much. Near the mouths of large rivers, however, where fresh water is continuously mixing with ocean water, the variations in sound velocity may be largely controlled by variations in salinity.

The quantitative dependence of sound velocity on temperature, pressure, and salinity is summarized in Figures 4 and 5. In Figure 4, obtained from a report by Woods Hole Oceanographic Institution [WHOI] [1] the value of the sound velocity at zero depth can be read from the main charts for any given combination of temperature and salinity. This velocity can then be corrected to the velocity at the actual depth by use of the curve in the small box. Figure 5 gives the percentage changes in sound velocity caused by specified absolute changes in the three determining variables. It will be shown in Chapter 3 that it is the relative changes in sound velocity which determine whether sound transmission is expected to be good or bad rather than the absolute changes.

The direct measurement of sound velocity in the ocean is very difficult. The intuitive method of dividing distance traveled by time is difficult since sufficiently accurate measurement of distances at sea is usually not feasible. The U. S. Navy Electronics Laboratory at San Diego, formerly the U. S. Navy Radio and Sound Laboratory [USNRSL], developed an acoustic interferometer for the determination of the wavelength of sound at a point in the ocean; [2] multiplication of this local wavelength by the known frequency gives the local sound velocity. This instrument was developed mainly for the purpose of checking whether the temperature changes indicated by the bathythermograph were correlated with the actual changes of sound velocity in the ocean. Good general agreement was observed between the velocity-depth plots obtained with the interferometer or velocity meter and those computed from bathythermograph observations. However, since the bathythermograph cannot follow rapid changes in water temperature with the detail possible with the velocity meter, a velocity microstructure was frequently recorded with the meter which deviated as much as 0.1 per cent from the velocity calculated from the simultaneous bathythermograph reading. That these deviations were due to the slow response of the bathythermograph rather than to physical factors was verified by correlating the velocity microstructure with the temperature microstructure obtained by a thermocouple recorder.

2.3.2   # Harmonic Waves

Up to now we have allowed the initial disturbances $\bar{p}(x,y,z)$ or $\bar{p}(t)$ to be arbitrary functions. However, most initial disturbances which occur in practice are of a very special type that originate in the elastic vibration of some medium. Such disturbances are produced by small displacements of some parts of the medium from their positions of equilibrium; these displacements in turn produce restoring forces which tend to restore the state of equilibrium. Such restoring forces are, in first approximation, proportional to the displacements.

It is well known that under such conditions (restoring forces proportional to displacements) the initial disturbance must be of the form of a harmonic vibration; that is, it must be representable by trigonometric functions of the time. In acoustics, such a vibration produces a pure tone of a definite frequency. Since echo-ranging pulses are very nearly pure tones, the importance of a study of harmonic vibrations is obvious. Also, harmonic vibrations are of crucial importance because they are the most convenient building stones of the more general solutions of the wave equation (see Section 2.7).

Suppose, in the second example under plane waves, that the initial disturbance of the plane $x = 0$ is a harmonic vibration. That is,

$$\bar{p}(t) = a \cos 2\pi f(t - \epsilon)$$

for values of t between 0 and t_0. One solution of the plane wave equation (30) under initial conditions $p(0,t) = \bar{p}(t)$ is always

$$p(x,t) = \bar{p}\left(t \pm \frac{x}{c}\right)$$

since $\bar{p}(t \pm x/c)$ satisfies the wave equation and also the imposed initial conditions. Thus, if we restrict our attention to progressive waves traveling in a single direction, the solution of the wave equation with the given initial conditions is

$$p = a \cos 2\pi f\left(t + \frac{x}{c} - \epsilon\right)$$

or

$$p = a \cos 2\pi f\left(t - \frac{x}{c} - \epsilon\right) \tag{34}$$

Clearly, the pressure changes represented by equation (34) are at most a; for that reason, a is called the *amplitude* of the disturbance. Also, it is clear that at a fixed point of space, p goes through f periods in one second; and so f is called the *frequency* of the disturbance. The quantity ϵ is called the *phase constant* because it fixes the position of the disturbance in time. Two vibrations of the same frequency and the same ϵ have their zeros simultaneously, also, their maxima. If they have different ϵ's, one has its zeros a fixed time interval ahead of the other, and we say that there is a *phase difference* between the vibrations.

2.3.3   Spherical Waves

The sound at large distances from an actual source resembles the sound from a point source more closely than it does the sound from an infinite plane. Hence, for some purposes it is more realistic to abandon the assumption of plane waves, and assume instead a point source at the origin which causes the pressure in the surrounding medium to be a function only of the distance r from the origin and of the time t. That is, the pressure is given by some function

$$p = p(r,t) \tag{35}$$

and is thus independent of the direction of the line joining the source to the point in question.

We shall now show that the wave equation (27) reduces, for the assumed case of spherical symmetry, to the simple form (37).

From simple analytic geometry,

$$r^2 = x^2 + y^2 + z^2; \quad \frac{\partial r}{\partial x} = \frac{x}{r}, \quad \frac{\partial r}{\partial y} = \frac{y}{r}, \quad \frac{\partial r}{\partial z} = \frac{z}{r}. \tag{36}$$

In order to transform the wave equation, the variables x,y,z must be eliminated, and the variable r inserted. In order to do this, we must use the relations (36) to calculate $\partial^2 p/\partial x^2$, $\partial^2 p/\partial y^2$, and $\partial^2 p/\partial z^2$ in terms of r and the derivatives of p with respect to r. This is done as follows.

$$\frac{\partial p}{\partial x} = \frac{\partial p}{\partial r} \frac{\partial r}{\partial x} = \frac{\partial p}{\partial r} \frac{x}{r}$$

because spatially p depends only on r. By differentiating again,

$$\frac{\partial^2 p}{\partial x^2} = \frac{\partial}{\partial x}\left[\frac{\partial p}{\partial r} \frac{x}{r} \right] = \frac{\partial p}{\partial r}\left[\frac{r^2 - x^2}{r^3} \right] + \frac{x}{r} \frac{\partial}{\partial x}\left(\frac{\partial p}{\partial r} \right)$$

$$= \frac{\partial p}{\partial r}\left[\frac{r^2 - x^2}{r^3} \right] + \frac{x^2}{r^2} \frac{\partial^2 p}{\partial r^2}.$$

Similarly, we can show that

$$\frac{\partial^2 p}{\partial y^2} = \frac{\partial p}{\partial r}\left[\frac{r^2 - y^2}{r^3} \right] + \frac{y^2}{r^2} \frac{\partial^2 p}{\partial r^2}$$

$$\frac{\partial^2 p}{\partial z^2} = \frac{\partial p}{\partial r}\left[\frac{r^2 - z^2}{r^3} \right] + \frac{z^2}{r^2} \frac{\partial^2 p}{\partial r^2}.$$

Addition of these expressions for $\partial^2 p/\partial x^2$, $\partial^2 p/\partial y^2$, and $\partial^2 p/\partial z^2$, in order to obtain the right-hand side of equation (27), gives

$$\frac{\partial^2 p}{\partial x^2} + \frac{\partial^2 p}{\partial y^2} + \frac{\partial^2 p}{\partial z^2} = \frac{\partial^2 p}{\partial r^2} + \frac{2}{r} \frac{\partial p}{\partial r}.$$

The latter expression is easily verified to be $(1/r)(\partial^2/\partial r^2)(rp)$, so we finally obtain

$$\frac{\partial^2 p}{\partial x^2} + \frac{\partial^2 p}{\partial y^2} + \frac{\partial^2 p}{\partial z^2} = \frac{1}{r}\frac{\partial^2}{\partial r^2}(rp),$$

and the general wave equation (27) reduces to

$$\frac{\partial^2 (rp)}{\partial t^2} = c^2 \frac{\partial^2}{\partial r^2}(rp). \tag{37}$$

This equation has a form similar to that of equation (30) for plane waves with p replaced by rp, and x by r. By using an argument similar to that in Section 2.3.1, it can be shown that equation (37) is satisfied by

$$rp = f\left(t \pm \frac{r}{c} \right),$$

where f is an arbitrary function of one variable. By dividing out the r, we get the following expression for $p(r,t)$ as the general solution of equation (37):

$$p(r,t) = \frac{f_1\left(t - \dfrac{r}{c} \right) + f_2\left(t + \dfrac{r}{c} \right)}{r}. \tag{38}$$

Assume that the following initial conditions are given. The initial disturbance is confined to a spherical shell of infinitesimal thickness at a distance $r = r_0$ from the origin. We suppose that the excess pressure in this spherical shell source is given by

$$p(r_0,t) = \frac{\bar{p}(t)}{r_0} \tag{39}$$

between the times $t = 0$ and $t = t_0$. We also suppose that the excess pressure at points outside this shell is zero at the time $t = 0$. The general solution of equation (37) with these initial conditions is

$$p(r,t) = \frac{1}{r}\left[c_1 \bar{p}\left(t - \frac{r}{c} + \frac{r_0}{c} \right) \right.$$
$$\left. + (1 - c_1)\bar{p}\left(t + \frac{r}{c} - \frac{r_0}{c} \right) \right], \tag{40}$$

because first, the right-hand expression is in the form of equation (38), and therefore satisfies equation (37); second, the right-hand expression satisfies the initial conditions imposed. In particular, if the spherical-shell source has a very small radius so that it approximates a point source at the origin, the following solution is obtained.

$$p(r,t) = c_1 \frac{\bar{p}\left(t - \dfrac{r}{c}\right)}{r} + (1 - c_1)\frac{\bar{p}\left(t + \dfrac{r}{c}\right)}{r}. \quad (41)$$

Physically, we can eliminate the solution $(1/r)\,\bar{p}(t + r/c)$. This solution corresponds to wave propagation in the negative r direction, with the speed c; in other words, to a wave which starts out at some negative time with a great radius and contracts into the point $x = y = z = 0$ at the time $t = 0$. The first solution is physically valid since it resembles actual propagation from a point source. It implies that the spherical wave spreads out from the point source into ever-increasing spheres with the speed c. Therefore, an initial ping of duration τ seconds will cause the resulting sound energy to be contained within an expanding spherical shell of thickness $c\tau$.

If the source is harmonic (emits a pure tone of the frequency f), the initial conditions are of the form

$$p = \frac{a \cos 2\pi f(t - \epsilon)}{r_0};$$

and if r_0 is nearly zero, the pressure at the distance r from the source and time t is given by

$$p = \frac{a \cos 2\pi f\left(t - \dfrac{r}{c} - \epsilon\right)}{r}. \quad (42)$$

The constants f and ϵ have the same physical significance as for the plane wave case; f is the frequency of the vibration and ϵ is the phase constant which orients the vibration in time. There is a difference, however, in the interpretation of a. In the plane wave case, a represents the maximum pressure change in the wave at all distances from the source; since a is a constant, all these pressure changes are equal. For spherical waves described by equation (42), however, it is clear that the maximum pressure change at the distance r is given by a/r, decreasing as r increases. The constant a is no longer the amplitude at all ranges, but merely the amplitude at the particular range $r = 1$.

2.4 PROPERTIES OF SOUND WAVES

2.4.1 Pressure versus Fluid Velocity

PLANE WAVES

For a plane wave we have, from equation (4),

$$\frac{\partial \sigma}{\partial t} = -\frac{\partial u}{\partial x},$$

where u is the particle velocity in the positive x direction. Because of equation (18), this equation can be transformed into

$$\frac{\partial p}{\partial t} = -\kappa \frac{\partial u}{\partial x}. \quad (43)$$

From equations (21) and (18), there results the following expression for $\partial p/\partial x$:

$$\frac{\partial p}{\partial x} = -\rho_0 \frac{\partial u}{\partial t}. \quad (44)$$

Equations (43) and (44) will be used to derive the general relationship between the excess pressure and the particle velocity in a plane wave. Assume that as initial conditions the plane $x = 0$ has its excess pressure given by $p(0,t) = \bar{p}(t)$ between $t = 0$ and $t = t_0$, and $p(0,t) = 0$ for all other values of t. Then the general solution of the plane wave equation, if we assume that the wave moves in the positive x direction, is given by

$$p(x,t) = \bar{p}\left(t - \frac{x}{c}\right).$$

By differentiating both sides of this equation with respect to t and also with respect to x, we obtain

$$\frac{\partial p}{\partial t} = \bar{p}'\left(t - \frac{x}{c}\right); \quad \frac{\partial p}{\partial x} = -\frac{1}{c}\bar{p}'\left(t - \frac{x}{c}\right).$$

This means that

$$\frac{\partial p}{\partial x} = -\frac{1}{c}\frac{\partial p}{\partial t}. \quad (45)$$

Combining equations (43) and (45) gives

$$\frac{\partial}{\partial x}\left(u - \frac{cp}{\kappa}\right) = 0, \quad (46a)$$

and by combining equations (44) and (45)

$$\frac{\partial}{\partial t}(p - \rho_0 c u) = 0. \quad (46b)$$

Equation (46a) means that $(u - cp/\kappa)$ is a function of t alone; and equation (46b) implies that $(p - \rho_0 c u)$ is a function only of x. But these two parentheses are proportional, differing by the factor $(-\rho_0 c)$ since $\rho_0 c^2 = \kappa$. Therefore, each parenthesis must be identically equal to some constant. This constant turns out to be zero for both since at any given point both p and u vanish before and after the disturbance has passed. The following relation between particle velocity and pressure results:

$$u = \frac{c}{\kappa}p = \frac{1}{\rho_0 c}p. \quad (47a)$$

Equation (47a) can also be shown to hold good for the wave moving in the positive direction if the initial conditions are for the space interval $0 < x < x_0$ at the time $t = 0$. If the wave is moving in the negative x direction, it can easily be shown by an argument similar to the above that the pressure and particle velocity are related by

$$u = -\frac{c}{\kappa}p = -\frac{p}{\rho_0 c}. \qquad (47b)$$

For the particular case of a plane harmonic wave, we have from equations (47a) and (47b)

$$u = \pm \frac{a}{\rho_0 c} \cos 2\pi f(t - \epsilon).$$

The following interesting result stems directly from equations (47). If the particles in a plane wave are moving in the direction of wave propagation, they are in a region of positive excess pressure; if they are moving in a direction opposite to the route of the wave, they are in a region of negative excess pressure; and if the particles are not moving, they are in a region of zero excess pressure. Also, we can argue from equations (47) that if the initial conditions fulfill neither

$$u(x,0) = \frac{1}{\rho_0 c}p(x,0)$$

nor

$$u(x,0) = -\frac{1}{\rho_0 c}p(x,0),$$

then waves are propagated in both a positive and a negative direction from the initial source of pressure disturbance.

SPHERICAL WAVES

At great distances from the source a small section of a spherical wave approximates a plane wave. For this reason, many of the foregoing results can be rewritten in a form valid for spherical disturbances far from the source. Since the mathematical proofs, though straightforward, are rather cumbersome they will not be reproduced here.

For a general spherical wave far from the source, the following relation exists between the excess pressure and particle velocity:

$$u = \pm\frac{c}{\kappa}p = \pm\frac{1}{\rho_0 c}p,$$

in analogy with equations (47). For a spherical harmonic wave it will be remembered that the maximum pressure change at the distance r from the source is

given by a/r. Thus, for the case of a spherical harmonic wave,

$$u_{\max} = \frac{1}{\rho_0 c}\frac{a}{r}. \qquad (48)$$

The relations (47) are not necessarily true for the general solution of the wave equation (27).

2.4.2 Acoustic Energy and Sound Intensity

The vibration of the particles of a fluid disturbed by wave propagation is a process which involves both kinetic and potential energy. The energy of vibration of the sound source is propagated through the fluid along with the sound wave. In a perfect fluid where frictional heat losses are zero, the energy content of the wave is unchanged as the wave travels. The energy passes from one region to another, "activating" the region through which the wave is passing. Thus, there are two quantities of interest. One is the energy found at any location as a function of time; the other is the rate at which energy is transported from one region to another as a function of time. In the following sections, both of these quantities are expressed in terms of the wave parameters we have introduced.

The kinetic energy possessed by a volume element v, whose volume was v_0 at equilibrium, and whose speed is u, is given by

$$\text{Kinetic energy of } v = \tfrac{1}{2}\rho_0 v_0 u^2. \qquad (49)$$

The potential energy possessed by the volume element v is the work which was done on it to change its volume from v_0 to v. This work can be calculated as follows: By equation (9), the relative change in volume produced by an infinitesimal alteration of condensation from σ to $\sigma + d\sigma$ is just $-d\sigma$. The total volume change caused by this infinitesimal alteration of σ is, to a first approximation, $-v_0 d\sigma$. The work done during this infinitesimal alteration is merely the pressure times the small volume change, that is, $pv_0 d\sigma$, which, because of equation (18), equals $\kappa\sigma v_0 d\sigma$. The total amount of work done on the volume element as its volume changes from v_0 to v can be obtained by integrating this infinitesimal amount of work between a condensation of zero and condensation of σ.

$$\text{Potential energy of } v = v_0 \kappa \int_0^\sigma \sigma d\sigma = \tfrac{1}{2}v_0 \kappa \sigma^2$$

$$= \frac{v_0 p^2}{2\kappa} \qquad (50)$$

because of equation (18). By adding equations (49) and (50), we obtain

$$\text{Total energy of } v = \frac{\rho_0 v_0 u^2}{2} + \frac{v_0}{2\kappa} p^2. \qquad (51)$$

By dividing equation (51) by the volume v_0:

Energy density at (x,y,z)

$$= \frac{\rho_0 u^2}{2} + \frac{p^2}{2\kappa}$$

$$= \frac{\rho_0}{2}(u_x^2 + u_y^2 + u_z^2) + \frac{p^2}{2\kappa}. \qquad (52)$$

We are now in a position to give a general expression for the intensity of a progressive sound wave, the characteristic which determines its loudness. Intensity is defined for a general progressive wave as the amount of energy which crosses a unit area normal to the direction of propagation in unit time. Since the energy travels at the same rate as the sound pulse, the instantaneous rate of energy flow will be equal to the energy density at the point in question times the sound velocity at this point. The intensity will be the time average of the instantaneous rate of energy flow, or

Intensity at (x,y,z)

$$= \text{Time average of } c\left(\frac{\rho_0 u^2}{2} + \frac{p^2}{2\kappa}\right)$$

$$= \frac{c\rho_0 \overline{u^2}}{2} + \frac{c\overline{p^2}}{2\kappa} \qquad (53)$$

where the bar over a quantity denotes the time average of that quantity.

We shall now attempt to calculate the intensities explicitly for various types of sound waves. We shall first consider plane waves and spherical waves, and then more general waves.

Plane Waves

A plane progressive wave satisfies equation (47); and therefore equation (50) reduces, for that case, to

$$\text{Potential energy of } v = \tfrac{1}{2}\rho_0 v_0 u^2 \qquad (54)$$

which is exactly equal to the expression (49) for the kinetic energy of v. We therefore get the result that for a plane progressive wave the kinetic and potential energies possessed by any small volume element at any time are equal. The kinetic and potential energies attain their greatest values at the spots where the particle velocity and excess pressure have their maxima or minima; and they vanish at the spots where

the particle velocity and the excess pressure are both zero.

Because of this equality of kinetic and potential energies for a progressive plane wave, equation (53) simplifies to

$$\text{Intensity} = c\rho_0\overline{u^2} = \frac{\overline{p^2}}{\rho_0 c} \qquad (55)$$

by equations (47) and (26).

In a plane progressive wave that is also harmonic the pressure is a sinusoidal function of the time with maximum value a. Since the average value of $\sin^2\theta$ over a complete period is $\tfrac{1}{2}$, $\overline{p^2} = a^2/2$.

$$\text{Intensity} = \frac{a^2}{2\rho_0 c}. \qquad (56)$$

Also, from equation (55) and the fact that u is also a sinusoidal function of the time,

$$\text{Intensity} = \tfrac{1}{2}\rho_0 c u_{\max}^2. \qquad (57)$$

Spherical Waves

For a spherical wave far from the source the formulas derived for plane waves are approximately true. We must be careful in applying them, however, to remember that the amplitude of the pressure vibration is no longer constant, but diminishes inversely with distance.

Using equations (57) and (48), we obtain

$$\text{Intensity} = \frac{1}{2\rho_0 c}\frac{a^2}{r^2} \qquad (58)$$

for harmonic spherical waves where a is the maximum pressure change at a distance one unit from the source.

Equation (58) is the familiar inverse square law of intensity loss for a spherical wave spreading out from a point source into an infinite homogeneous medium.

Let F represent the amount of energy radiated by the source into a unit solid angle[d] in one second. Then

$$\text{Total rate of emission} = 4\pi F. \qquad (59)$$

[d]Solid angle is the three-dimensional analogue to the ordinary, two-dimensional, plane angle. It measures the angular spread of such three-dimensional objects as a cone, a light beam, or the beam of a radio transmitter. Its measure is defined as follows. Construct a sphere of arbitrary size with the apex of the solid angle as its center. The solid angle will then cut out a certain area of the surface of the sphere. This area, divided by the square of the radius of the sphere, is the measure of the solid angle. It is dimensionless and does not depend on the sphere radius chosen. The unit solid angle is frequently called the steradian. The full solid angle, comprising all directions pointing from the apex, has the value 4π.

We can calculate F by means of equation (58). Since a sphere of radius r has the area $4\pi r^2$, the total energy crossing such a spherical surface per unit of time is merely the intensity times this area:

$$\text{Rate at which energy crosses sphere} = \frac{2\pi a^2}{\rho_0 c}. \quad (60)$$

Because of the assumption of conservation of sound energy, equation (60) must be equal to the amount of energy radiated by the source per unit of time. Dividing equation (60) by 4π gives

$$F = \frac{a^2}{2\rho_0 c}. \quad (61)$$

GENERAL SOUND WAVES

We now examine the transport of acoustic energy for the case of a general solution of the wave equation (27).

In the general case, it is useful to start with an equation of continuity for energy flow analogous to the exact equation of continuity (2) for mass flow. It will be recalled that equation (2) followed directly from the law of conservation of mass. The law of continuity for energy flow will follow from the law of conservation of energy in exactly the same fashion. For the mass density ρ, the energy density which may be denoted by Z is substituted. Also, for the instantaneous flow of matter with components u_x, u_y, u_z, the instantaneous flow of energy is substituted. The components of the instantaneous energy flow past normal unit area may be denoted by E_x, E_y, E_z. The equation of the continuity for energy flow becomes, in analogy with equation (2),

$$\frac{\partial Z}{\partial t} = -\left[\frac{\partial E_x}{\partial x} + \frac{\partial E_y}{\partial y} + \frac{\partial E_z}{\partial z}\right]. \quad (62)$$

Equation (62) is the mathematical expression of the assertion that the energy flow through a closed surface is equal to the decrease of energy inside this surface. A rather complicated argument must be used to calculate the components of energy flow E_x, E_y, E_z. Equations (21) are rewritten by using $p = \kappa\sigma$, as

$$\rho_0\frac{\partial u_x}{\partial t} = -\frac{\partial p}{\partial x}$$

$$\rho_0\frac{\partial u_y}{\partial t} = -\frac{\partial p}{\partial y} \quad (63)$$

$$\rho_0\frac{\partial u_z}{\partial t} = -\frac{\partial p}{\partial z}.$$

Also, from equations (4) and (18), we have the relation

$$\frac{1}{\kappa}\frac{\partial p}{\partial t} = -\left[\frac{\partial u_x}{\partial x} + \frac{\partial u_y}{\partial y} + \frac{\partial u_z}{\partial z}\right]. \quad (64)$$

Multiplying the first equation of (63) by u_x, the second by u_y, the third by u_z, and equation (64) by p, and adding them all up, we obtain

$$\frac{\partial}{\partial t}\left[\frac{\rho_0}{2}(u_x^2 + u_y^2 + u_z^2) + \frac{p^2}{2\kappa}\right]$$
$$= -\left[\frac{\partial(pu_x)}{\partial x} + \frac{\partial(pu_y)}{\partial y} + \frac{\partial(pu_z)}{\partial z}\right]. \quad (65)$$

Because of equation (52), we see that the left-hand sides of equations (65) and (62) are equal. Hence the right-hand sides are also equal, and we must have

$$E_x = pu_x; \quad E_y = pu_y; \quad E_z = pu_z. \quad (66)$$

The instantaneous energy flow E is the resultant of its three components E_x, E_y, E_z and is numerically equal to $\sqrt{E_x^2 + E_y^2 + E_z^2}$. Thus, we have the general result that

$$E = pu. \quad (67)$$

According to equation (66), this energy flow is always along the direction of the particle velocity.

The intensity I, which was defined as the time average of E, is therefore always given by the following formula:

$$I = \overline{pu}. \quad (68)$$

2.4.3   Complex Representation of Harmonic Vibrations

The complex number e^{iw} is defined by the equation

$$e^{iw} = \cos w + i \sin w$$

where $i^2 = -1$. The one-dimensional harmonic vibration

$$d = a \cos 2\pi ft$$

can therefore be regarded as the real part of $ae^{i2\pi ft}$. Similarly, the vibration

$$d = a \cos 2\pi f(t - \epsilon) \quad (69)$$

can be rewritten as

$$d = \text{real part of } ae^{i2\pi f(t-\epsilon)}.$$

The latter relation can be expressed in the following less cumbersome form

$$D = ae^{i2\pi f(t-\epsilon)} \quad (70)$$

if the conventions are adopted that the actual physical displacement is the real part of the complex displacement D and that the numerical value of this actual displacement is the real part of the right-hand side of equation (70). With this understanding, equations (70) and (69) represent one and the same physical process.

The complex form for a vibration simplifies some types of calculations and will be frequently used in the remainder of this chapter. We notice that equation (70) can be rewritten in the form

$$D = Ae^{i2\pi ft} \tag{71}$$

where A is the complex number $ae^{-i2\pi f\epsilon}$. A is called the *complex amplitude* of the vibration described by equation (69). It is apparent that

$$A = a\left[\cos 2\pi f\epsilon - i \sin 2\pi f\epsilon\right].$$

Thus, the complex amplitude has $a \cos 2\pi f\epsilon$ as its real component and $-a \sin 2\pi f\epsilon$ as its imaginary component.

As an example of the convenience afforded by the complex representation of a vibration, we shall use it to find the harmonic solution of the plane wave equation (30). We assume tentatively that

$$p(x,t) = Ae^{2\pi i(ft + mx)} \tag{72}$$

and see if we can find a value of m which will make equation (72) a solution of equation (30). Substituting equation (72) into equation (30), we have

$$(2\pi i f)^2 p = c^2(2\pi i m)^2 p.$$

In other words, a value of m equal to f/c or $-f/c$ makes the expression (72) a solution of equation (30). These two solutions are, explicitly,

$$p = Ae^{2\pi i f[t \pm (x/c)]}; \quad A = ae^{-2\pi i f\epsilon}. \tag{72a}$$

These two solutions, interpreted according to the convention of this section, are obviously identical with the "real" solutions (34).

Similarly, a point harmonic source in an infinite homogeneous medium gives rise to spherical harmonic waves according to the equation

$$p(r,t) = \frac{A}{r}e^{2\pi i f[t - (r/c)]}; \quad A = ae^{-2\pi i f\epsilon}. \tag{73}$$

2.4.4 Sound Sources

The wave equation (27) governs the manner in which disturbances will be propagated in the interior of a fluid, but does not say anything about the initial disturbances themselves. In this section we shall consider the various types of initial disturbances which can be produced by sound sources. First we shall discuss the quality of the sound put out by various sources, where *quality* refers to the frequency characteristics of the emitted sound. Next, since some sources radiate equally in all directions while others do not, we shall consider in a general way the directivity properties of sources.

FREQUENCY CHARACTERISTICS

Strictly speaking, the concept of frequency can be applied only to simple harmonic disturbances. A simple harmonic disturbance of the pressure in a fluid is described by an equation of the form

$$p = a \cos 2\pi f(t - \epsilon)$$

and gives rise to what is called a *pure tone*. Most of the echo-ranging transducers used at present produce sounds which are very nearly pure tones, but cannot be heard by the ear because the frequencies are too high.

If two or more pure tones are put into the water at the same time, the resultant is known as a *compound tone*. Some transducers, used mainly in research work, can produce compound tones. Any sound of this nature can be expressed as the sum of a finite number of harmonic vibrations.

Many sources, however, produce in their immediate vicinity an irregular change in pressure which cannot be represented as the sum of a finite number of sinusoidal vibrations. Such sound outputs are called *noises*. Ship sounds and torpedo sounds are examples of noises, and the reader can doubtless supply other examples. According to a mathematical theorem called the Fourier theorem, it is often possible to represent such an irregular sound output as an infinite sum of simple tones, whose intensities, frequencies, and phase relationships are such that they add up to the given noise. If most of the component frequencies lie in a narrow frequency range, the sound is called a narrow-band noise; otherwise it is called a wide-band noise.

Some types of echo-ranging gear put out a frequency-modulated signal. In this type of output, the pressure is at every instant a sinusoidal function of time, but the frequency changes during the signal in some designed way. In one type of frequency-modulated signal, called a "chirp" signal, the frequency increases linearly with time for the duration of the pulse:

$$p = a \cos 2\pi[(f_0 + at)t].$$

In a typical chirp, 100 msec long, the frequency may increase from 23.5 kc at the beginning of the pulse to 24.5 kc at the end of the pulse.

DIRECTIVITY CHARACTERISTICS

So far we have been mainly concerned with the simple point source which gives rise to a spherically symmetric disturbance in the immediate vicinity of the source. It is called a point source because the resulting sound field is discontinuous at only one point of space, at the source itself. If the discontinuity is of a more complicated nature, as in the case of a line source, the sound field will not, in general, be spherically symmetric in the neighborhood of the source; that is, the amounts of sound energy radiated into different directions will be different. In this subsection, sources giving rise to sound fields that are not spherically symmetric are discussed.

Double Sources. Suppose there are two point sources, S_0 and S_0', one at (x_0,y_0,z_0) and the other at (x_0',y_0',z_0'), as indicated in Figure 6. The resulting

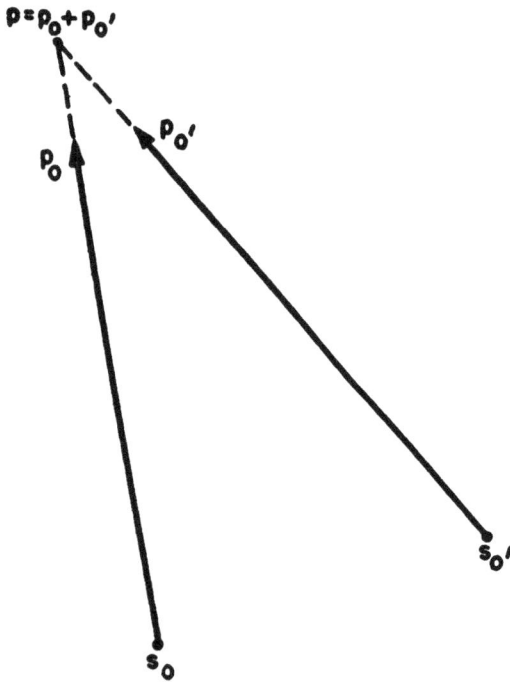

FIGURE 6. Resultant pressure produced by two separate sources.

pressure at any one point P and time t will be the algebraic sum of the pressures that would be produced by each source separately. That is, if f and f' are the two frequencies emitted, ϵ and ϵ' are the two phase constants, and A and A' are the two complex amplitudes, the resulting $p(r,t)$ is given by

$$p = \frac{A}{r}e^{2\pi i f[t-(r/c)]} + \frac{A'}{r'}e^{2\pi i f'[t-(r'/c)]}. \quad (74)$$

Also,

$$A = ae^{-2\pi i f/\epsilon}; \quad A' = a'e^{-2\pi i f'/\epsilon'}, \quad (75)$$

according to equation (71).

We shall restrict our attention to the case of two sound sources situated on the x axis, one at the origin and the other a small distance s away. We assume that these two sources produce initial pressure disturbances of equal real amplitude and equal frequency and that the initial disturbances are opposite in phase.

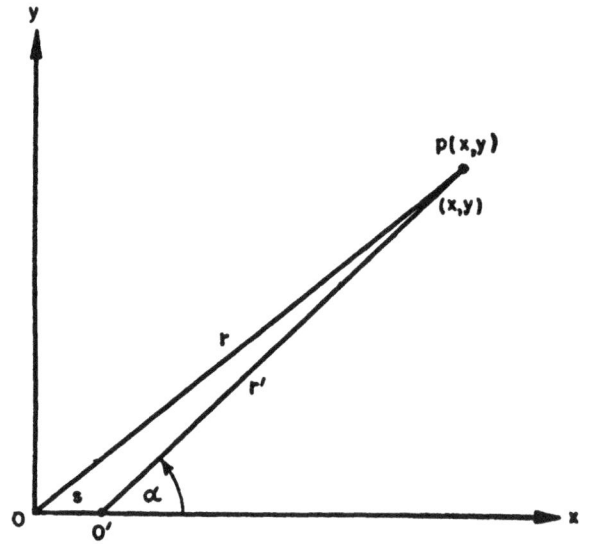

FIGURE 7. Resultant pressure produced by double source 00′.

This case, pictured in Figure 7, is a fairly good approximation to many sources occurring in practice, such as a vibrating diaphragm. Because of these assumptions, the following relationships exist among the quantities in equations (74) and (75):

$$a = a'; \quad f = f'; \quad \epsilon = 0; \quad \epsilon' = \frac{1}{2f}. \quad (76)$$

Also, $A = a$ and $A' = ae^{-\pi i} = -a$ because of equation (75).

Of particular interest is the extreme case where the distance between the two sources is very nearly zero, but where the real amplitude a of the individual disturbances is so large that the product as is an appreciable quantity. Such a combination of two single sources with very small separation and very large individual amplitudes is called a *double source*. A double source may be described by two quantities: the product as, and its axis, the direction of the line joining the two sources.

With these assumptions, equation (74) becomes

$$p = \frac{a}{r}e^{2\pi i f[t-(r/c)]} - \frac{a}{r'}e^{2\pi i f[t-(r'/c)]} \qquad (77\text{a})$$

where

$$r^2 = x^2 + y^2 + z^2; \quad r'^2 = (x-s)^2 + y^2 + z^2. \qquad (77\text{b})$$

If $F(r)$ is an arbitrary function of r, and if r and r' are very nearly equal, we have from simple calculus

$$F(r) - F(r') \approx (r - r')\frac{dF}{dr}.$$

The quantity $r - r'$ may be calculated as a function of x,s,r as follows:

$$r - r' = \frac{r^2 - r'^2}{r + r'} = \frac{r^2 - r'^2}{2r} \text{ because } r \approx r',$$

which equals sx/r from equation (77b). The quantity dF/dr may also be calculated:

$$\frac{dF}{dr} = \frac{\partial F}{\partial x}\frac{\partial x}{\partial r} + \frac{\partial F}{\partial y}\frac{\partial y}{\partial r} + \frac{\partial F}{\partial z}\frac{\partial z}{\partial r}.$$

As the origin changes from O to O' on the x axis, thereby changing r to r', the coordinates y and z of all points in space are unchanged. Thus, for all changes in r defined in this manner,

$$\frac{\partial y}{\partial r} = \frac{\partial z}{\partial r} = 0$$

so that

$$\frac{dF}{dr} = \frac{\partial F}{\partial x}\frac{\partial x}{\partial r} = \frac{\partial F}{\partial x}\frac{r}{x}$$

because of equation (36). Using these values of $r - r'$ and dF/dr,

$$F(r) - F(r') = s\frac{\partial F}{\partial x}$$

and equation (77) may be rewritten as

$$p = ase^{2\pi i ft}\frac{\partial}{\partial x}\left[\frac{1}{r}e^{-2\pi i f(r/c)}\right].$$

By calculating out the derivative of the bracket with respect to x, and by remembering that $\partial r/\partial x = x/r$, this equation becomes

$$p = ase^{2\pi i ft}e^{-2\pi i f(r/c)}\frac{x}{r^2}\left(\frac{2\pi f i}{c} + \frac{1}{r}\right).$$

If α is the angle between the x axis and the radius vector **OP**, $x/r = \cos\alpha$, and the preceding equation becomes

$$p = ase^{2\pi i ft}e^{-2\pi i f(r/c)}\frac{\cos\alpha}{r}\left(\frac{2\pi f i}{c} + \frac{1}{r}\right).$$

If r is very large compared with c/f, the second term in the brackets may be neglected, and as a result

$$p = \frac{2\pi f asi}{cr}\cos\alpha e^{2\pi f i[t-(r/c)]}.$$

Replacing the factor as by b, we obtain the final result

$$p = \frac{2\pi f bi}{cr}\cos\alpha e^{2\pi f i[t-(r/c)]}. \qquad (78)$$

By comparing equation (78) with equation (73), we see that the pressure changes produced at great distances by the double source are identical with the pressure changes produced by a single source, which is situated at the same place, vibrates with the same frequency, and has the following complex amplitude.

$$A = \frac{2\pi f bi}{c}\cos\alpha. \qquad (79)$$

It is clear from equation (75) that the real amplitude of the vibration is equal to the absolute value of the complex amplitude. From algebra, we know that the absolute value of a complex number A is just $\sqrt{A\overline{A}}$, where $\overline{A}$ is the conjugate complex of A. Let a be the real amplitude corresponding to equation (79). Then a is given by

$$a = \frac{2\pi f b}{c}\cos\alpha. \qquad (80)$$

With this definition of a, the actual pressure distribution defined by equation (78) is

$$p(r,\alpha,t) = a\cos 2\pi f\left(t - \frac{x}{r}\right). \qquad (81)$$

Since p is harmonic, its square averaged over a complete period is one-half the square of its amplitude (80); from equation (58) we have for the intensity at the distance r and angle α:

$$I(r,\alpha) = \frac{1}{r^2}\left(\frac{2\pi f b}{c}\cos\alpha\right)^2\frac{1}{2\rho_0 c}$$

$$= \frac{2\pi^2 f^2 b^2}{\rho_0 c^3 r^2}\cos^2\alpha. \qquad (82)$$

Thus, the sound intensity caused by a double sound source is directly proportional to the square of b and to the square of the cosine of the angle α of emission and is inversely proportional to the square of the distance from the sound source.

Let $F(\alpha)$ denote the average rate at which energy is emitted in the direction α. It is clear from Figure 8 that this average emission per unit solid angle is given by

$$F(\alpha) = \frac{I(r,\alpha)\,r^2 d\omega}{d\omega} = \frac{2\pi^2 f^2 b^2}{\rho_0 c^3}\cos^2\alpha \qquad (83)$$

where $d\omega$ is an infinitesimal solid angle in the direction α. The maximum value of $F(\alpha)$ occurs in the direction of the x axis, for which

$$F(0) = \frac{2\pi^2 f^2 b^2}{\rho_0 c^3}. \qquad (84)$$

Thus equation (83) can be rewritten as

$$F(\alpha) = F(0)\cos^2\alpha \qquad (85)$$

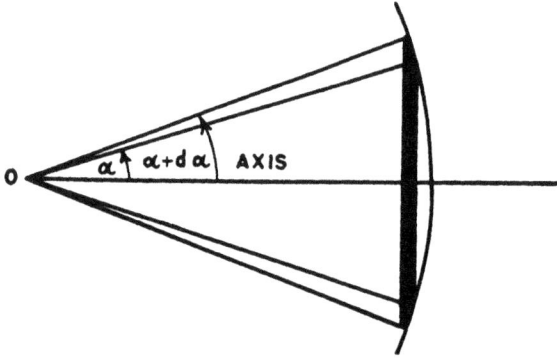

FIGURE 8.   Rotation of wedge $\alpha\alpha$ about axis.

and equation (82) as

$$I(r,\alpha) = \frac{F(0)\cos^2\alpha}{r^2}. \qquad (86)$$

In order to find the total energy emitted by the double source in one second, we calculate the total energy traversing the surface of a sphere of radius r in one second. This is clearly equal to the rate at which the source is putting out power. To get this total energy, it is necessary to integrate the average energy flow (86) over the whole sphere. Such an integral is in general multiple, but in this particular case it can be expressed as a single integral because the energy flow depends only on α. First consider the average rate at which energy is flowing through the small area element intercepted on the sphere by the two cones defined by the angles α and $\alpha + d\alpha$, as in Figure 8. This small element of the sphere has the area $2\pi r^2 \sin\alpha\,d\alpha$; and, therefore, it intercepts a solid angle of $2\pi(\sin\alpha)d\alpha$ units. By equation (86), the average rate of energy flow through this element is

$$F(0)\cos^2\alpha \cdot 2\pi \sin\alpha \cdot d\alpha.$$

The total emission in one second is this average rate

of energy flow integrated between the angles 0 and π; that is,

$$\text{Rate of emission} = 2\int_0^{\pi/2} F(0)\cos^2\alpha \cdot 2\pi \sin\alpha \cdot d\alpha$$

$$= \frac{4\pi F(0)}{3}. \qquad (87)$$

It will be remembered that $F(0)$ is the maximum rate of emission per unit solid angle, by the double source.

All sound projectors have pattern functions which describe the distribution of sound energy emitted in different directions. A general direction in space can be defined by the two coordinates (θ,ϕ), where θ is the angle of elevation of the direction OP relative to the horizontal xy plane, and ϕ is the polar angle in the xy plane between the x axis and the projection OP', as in Figure 9. Let $F(\theta,\phi)$ be the

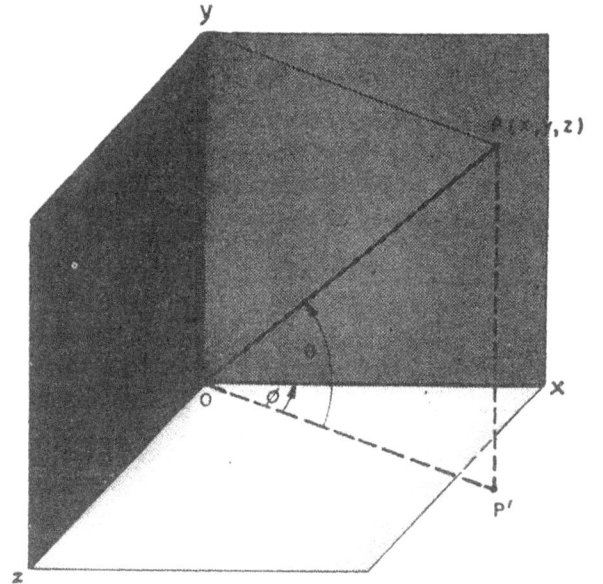

FIGURE 9.   Coordinates specifying direction OP.

emission per unit solid angle in the direction OP, and let $F_{\max}$ be the emission per unit solid angle in the direction of maximum emission, called the *acoustical axis* of the projector. Then the pattern function $b(\theta,\phi)$ is defined by

$$F(\theta,\phi) = F_{\max}\, b(\theta,\phi).$$

If we take the acoustical axis in the direction $(0,0)$, this becomes

$$F(\theta,\phi) = F(0,0)b(\theta,\phi). \qquad (88)$$

The pattern function b clearly depends only on the nature of the projector.

In analogy with the result (86) for the simple case of axial symmetry,

$$I(r,\theta,\phi) = \frac{F(0,0)\,b(\theta,\phi)}{r^2} \qquad (89)$$

at a distance r from the source much greater than a wavelength.

The rate at which the projector emits energy in all directions must be exhibited as a double integral in this general case. The area element on the sphere of radius r, intercepted between (θ,ϕ) and $(\theta + d\theta, \phi + d\phi)$, is $r^2 \cos\theta d\theta d\phi$; and the solid angle intercepted by this area element is $\cos\theta d\theta d\phi$. Thus, in view of equation (88),

Emission through area element
$$= F(0,0)\,b(\theta,\phi)\cos\theta d\theta d\phi \qquad (90)$$
and therefore

Total rate of emission
$$= F(0,0)\int_{-\pi}^{\pi} d\phi \int_{-\pi/2}^{\pi/2} b(\theta,\phi)\cos\theta d\theta. \qquad (91)$$

Equation (91) can be put in the following form, which is directly comparable to the law of emission (59) of a point source:

$$\text{Rate of emission} = 4\pi F(0,0)\delta \qquad (92)$$
where
$$\delta = \frac{1}{4\pi}\int_{-\pi}^{\pi} d\phi \int_{-\pi/2}^{\pi/2} b(\theta,\phi)\cos\theta d\theta. \qquad (93)$$

The factor δ is a constant depending on the nature of the source and may be called the *directivity factor* of the source. For the point source of equation (59), this directivity factor is 1; while for the double source of equation (87) it is $\frac{1}{3}$.

2.5 SOUND WAVES IN A VISCOUS FLUID

In a homogeneous perfect fluid, the decrease of sound intensity with increasing distance from the source is due only to spreading according to the inverse square law (58). However, sound intensity measurements show clearly that the intensity loss in the ocean tends to be much greater than the value predicted by equation (58). These extra losses above the theoretical loss, (58), due to the fact that the ocean is not a homogeneous perfect fluid, are called *transmission anomalies* or, more loosely, *attenuations*. In this section, we shall derive some results for sound intensity in a viscous fluid and see how much of the observed attenuation can be ascribed to fluid viscosity.

In the derivation of the wave equation (27) for a perfect fluid, we used the equation of continuity (4), the equations of motion (15), the equation of state (18), and the law of forces (16). The equation of continuity, the equations of motion, and the equation of state, it will be recalled, apply to any fluid, whether it is perfect or viscous. The law of forces (16), however, is valid only for a perfect fluid. The exact law of forces operating in a viscous fluid is quite difficult to derive since it depends on the theory of viscous fluid flow. It is sufficient to say that this complicated law of forces, combined with equations (4), (18), and (15), can be used to derive a general wave equation for viscous fluids, analogous to equation (27). Under the assumption that the resulting pressure distribution in the viscous medium depends only on the coordinate x, this general equation reduces to[3]

$$\frac{\partial^2 p}{\partial t^2} = \frac{\kappa}{\rho_0}\frac{\partial^2 p}{\partial x^2} + \frac{4\mu}{3\rho_0}\frac{\partial^2 p}{\partial x \partial t} \qquad (94)$$

where μ is the coefficient of shear viscosity. Equation (94) is the plane wave equation for a viscous fluid. In the absence of viscosity ($\mu = 0$), equation (94) reduces to equation (30).

Let us see whether we can find a solution to equation (94) of the form

$$p = Ae^{2\pi i(ft + mx)}.$$

By substituting this expression for p into equation (94), we obtain

$$m^2 = \frac{f^2}{\dfrac{\kappa}{\rho_0} + \dfrac{4\mu}{3\rho_0}2\pi i f}. \qquad (95)$$

If c and α are defined by

$$c^2 = \frac{\kappa}{\rho_0}, \qquad (96)$$

$$\alpha = \frac{8}{3}\pi^2 \frac{\mu f^2}{\rho_0 c^3}, \qquad (97)$$

the relation (95) becomes

$$m = \pm\frac{f}{c}\frac{1}{\sqrt{1 + \dfrac{i}{\pi}\dfrac{c}{f}\alpha}} = \pm\left(\frac{f}{c} + \frac{\alpha}{2\pi i}\right), \qquad (98)$$

according to the binomial theorem, if we assume that α is so small that α^2 and higher terms can be neglected.

In order to get the case of waves propagated in the positive x direction, the negative sign of equation (98) must be chosen and $2\pi imx$ becomes

$$2\pi imx = -2\pi i\frac{f}{c}x - \alpha x.$$

The corresponding solution of equation (94) is therefore

$$p = Ae^{2\pi i f t}\, e^{-2\pi i f(x/c)}\, e^{-\alpha x}.$$

We can write this expression for p in the form

$$p = Ae^{-\alpha x}e^{\,2\pi i f[t-(x/c)]}. \tag{99}$$

For $\mu = 0$, α vanishes, and equation (99) reduces to

$$p = Ae^{2\pi i f[t-(x/c)]} \tag{100}$$

which is just the solution for plane waves propagated harmonically into a perfect fluid. By comparing equations (99) and (100), we see that the effect of viscosity is to cause the amplitude of the pressure vibration to decay exponentially with distance, by the factor $e^{-\alpha x}$, where α is the positive real number defined by equation (97). A vibration of the type of (99) is referred to as a *damped* vibration, and $e^{-\alpha x}$ is called the *damping factor*.

To see whether this energy loss due to shear viscosity is the cause of the attenuation observed in the sea, one can first calculate α for sea water, by using the known values of ρ_0, c, μ for sea water, and the known frequency f of the sound source. The intensity loss is measured between two points so far from the sound source that the wave propagation between those two points approximates plane wave propagation. Then this observed intensity loss is compared with the theoretical intensity loss calculated from equations (97) and (99). It is found that only for very great frequencies (much higher than 100 kc) can an appreciable fraction of the observed attenuation be ascribed to shear viscosity; at lower frequencies, the theoretical loss from viscosity makes up only a very small part of the observed attenuation. Thus other causes must be sought for the extinction of sound energy in the sea. The sound transmission studies of Section 6.1 of NDRC have had as one of their primary objectives the discovery of the factors governing the intensity loss of sound in the sea. Although some progress has been made, the problems of attenuation in the sea have by no means been completely solved (see Chapters 5 to 10).

Since the observed attenuation of sound in the sea is much greater than the value indicated by equation (99), it appears possible that there is another type of viscosity, in addition to the classical shear viscosity, which may be responsible for part or all of the remaining attenuation. The classical theory of the flow of viscous fluids is based on Stokes' hypothesis that frictional forces within a fluid arise only from a change in the shape of a volume element; in other words, that a change in the size of a volume element, if its shape remained unaltered, would meet no resistance. The concept of a *compression viscosity* has been suggested to represent the resistance of the fluid to pure volume dilatation. Such a compression viscosity would not be discovered in a stationary flow of the type employed to measure shear viscosity because in these experiments the fluid acts essentially as an incompressible fluid. But in the transmission of sound this conjectural compression viscosity would contribute a term to the expression for α which would also be proportional to the square of the frequency f.

Actual determinations of the constant α at many different frequencies show that between 0 and 100 kc the attenuation increases less rapidly than the square of the frequency. There are no theoretical grounds for assuming any power law for the dependence of attenuation on frequency. If a power law is assumed, the empirical curve is best fitted by a 1.4th power dependence, but even this best fit is poor. It thus appears that factors other than viscosity must account for much of the attenuation of sound observed in the sea.

2.6   EFFECT OF A BOUNDARY

2.6.1 Conditions of Transition and Boundary Conditions

We shall now return to the assumption of a perfect fluid and turn our attention to the effects of boundaries. Consequently, we now drop the assumption that waves are propagated in a single homogeneous infinite medium. Instead, we shall consider the case that all space is filled up by two different homogeneous media separated by a plane, which we choose as the plane $y = 0$. For the one medium (the sea), at $y < 0$, we denote the density, excess pressure, bulk modulus, and sound velocity by ρ, p, κ, and c respectively; for the other medium, air, for example, at $y > 0$, we call these quantities ρ_1, p_1, κ_1, and c_1.

It is necessary, from a physical point of view, to assume that the pressure in both media is the same at the boundary. Otherwise, the force per unit mass at the interface would become infinite. We have, thus,

$$p = p_1 \text{ at } y = 0. \tag{101}$$

Also, if the two media are to remain in contact with each other at all times, the displacements normal to the boundary must have the same value in both media at the boundary. In symbols, if (S_x, S_y, S_z) are the components of particle displacement in the first medium, and (S_{1x}, S_{1y}, S_{1z}) are the components of particle displacement in the second medium,

$$S_y = S_{1y} \text{ at } y = 0. \tag{102}$$

No restrictions of the form of (102) can be placed on the displacements S_x and S_z because displacements parallel to the boundary will not cause loss of contact.

Since equation (102) holds for all time, the time derivatives of S_y and S_{1y} must also be equal at the boundary; in other words,

$$u_y = u_{1y}; \quad \frac{\partial u_y}{\partial t} = \frac{\partial u_{1y}}{\partial t} \quad y = 0. \tag{103}$$

Because of equation (17), equation (103) implies

$$\frac{1}{\rho}\frac{\partial p}{\partial y} = \frac{1}{\rho_1}\frac{\partial p_1}{\partial y} \text{ at } y = 0$$

or

$$\frac{\partial p}{\partial y} = \frac{\rho}{\rho_1}\frac{\partial p_1}{\partial y} \text{ at } y = 0. \tag{104}$$

We shall call equations (101) and (104) *conditions of transition*. In the general case, the propagation in one medium depends on the exact nature of the propagation in the other medium, because of the conditions of transition. In the case of the sea, however, conditions are often such that we can ignore the exact propagation in the surrounding medium; the transition conditions of the type (104) may then be reduced to boundary conditions for the sea itself. In the next section, the conclusion is reached that at the yielding boundary between sea and air the following condition holds:

$$p = 0 \tag{105}$$

and that at the solid boundary between sea and rock bottom we always have, approximately,

$$\frac{\partial p}{\partial y} = 0. \tag{106}$$

Relations of the type of (105) and (106) will, in many cases, suffice for calculating the sound field in the medium of interest. By use of such boundary conditions explicit consideration of the sound field beyond the boundaries may be made unnecessary.

2.6.2 Reflection and Refraction of Plane Waves

Consider now what happens to a plane wave when it hits the plane boundary $y = 0$ between two dis-

similar media, in one of which the sound velocity is c, and in the other of which it is c_1. For generality, we assume that the direction of propagation of the incident wave is oblique to the boundary, making an angle θ_i with the normal to the plane boundary. We can also assume, without losing generality, that the direction of propagation is parallel to the xy plane; y represents the vertical direction positive upward, x a horizontal direction, and z a front-back direction, as in Figure 10.

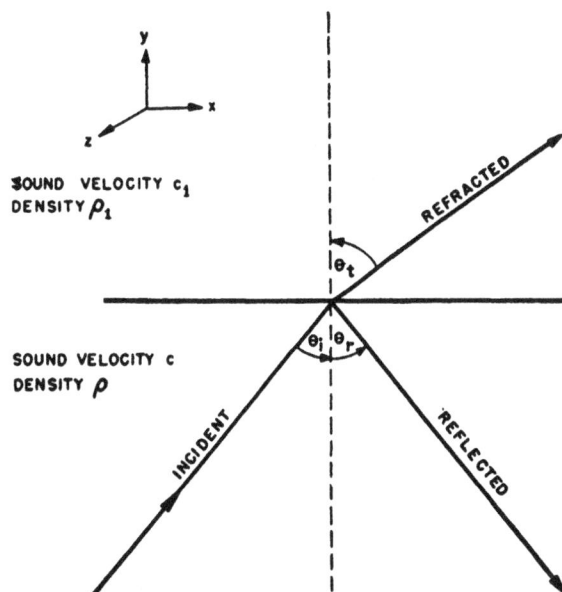

FIGURE 10. Splitting of plane wave at boundary between two media.

Since the incident wave is plane, it may be described by the equation (72a) with x replaced by

$$x \sin\theta_i + y \cos\theta_i,$$

in view of the oblique direction of propagation. That is, for the incident wave,

$$p_i = A_i e^{2\pi i f\left(t - \frac{x\sin\theta_i + y\cos\theta_i}{c}\right)}, \tag{107}$$

where p_i represents the sound pressure of the incident wave, and A_i its complex amplitude.

We can consider that the incident wave terminates its existence when it hits the boundary and expends its energy in producing a disturbance of the interface. Thus the boundary will act as a sound source, which vibrates with the frequency f of the incident wave. The vibration of the interface will send out sound waves of the frequency f into both media. We shall assume that these two waves are plane waves; this result is intuitively apparent, but can be proved only by a long tedious argument.

For brevity, the wave propagated by the boundary into the second medium will be called the *transmitted wave*, and the wave propagated back into the first medium is called the *reflected wave*. We shall now calculate the amplitudes and directions of propagation of the transmitted and reflected waves.

The pressure and complex amplitude of the transmitted wave are denoted by p_t and A_t; the same quantities for the reflected wave are denoted by p_r and A_r. Let the transmitted wave have the direction θ_t, relative to the normal, and the reflected wave have the direction θ_r, as indicated in Figure 10. The angles θ_t and θ_r are usually called the *angle of refraction* and *angle of reflection*, respectively, and the angle θ_i is called the *angle of incidence*.

Because the reflected and transmitted waves are plane,

$$p_r = A_r e^{2\pi i f \left(t - \frac{-x \sin \theta_r - y \cos \theta_r}{c} \right)}, \tag{108}$$

$$p_t = A_t e^{2\pi i f \left(t - \frac{x \sin \theta_t + y \cos \theta_t}{c_1} \right)} \tag{109}$$

The sign of y is different in equations (108) and (109) because in equation (108) y decreases with the time on the wave front; in equation (109) it increases.

Equation (109) gives the resultant total pressure in the second medium. The resultant pressure in the first medium is the sum of the pressures of the incident and reflected waves, which is obtained by adding equations (107) and (108). Denoting the resultant pressure in the first medium by p, we obtain

$$p = p_i + p_r = A_i e^{2\pi i f \left(t - \frac{x \sin \theta_i + y \cos \theta_i}{c} \right)}$$
$$+ A_r e^{2\pi i f \left(t - \frac{x \sin \theta_r - y \cos \theta_r}{c} \right)}. \tag{110}$$

The pressure must be the same on both sides of the boundary. Therefore, $p_i + p_r = p_t$ at $y = 0$; that is,

$$A_i e^{2\pi i f \left(t - \frac{x \sin \theta_i}{c} \right)} + A_r e^{2\pi i f \left(t - \frac{x \sin \theta_r}{c} \right)} = A_t e^{2\pi i f \left(t - \frac{x \sin \theta_t}{c_1} \right)}$$

or

$$e^{2\pi i f t} \left[A_i e^{-2\pi i f \frac{\sin \theta_i}{c} x} + A_r e^{-2\pi i f \frac{\sin \theta_r}{c} x} - A_t e^{-2\pi i f \frac{\sin \theta_t}{c_1} x} \right]$$
$$= 0 \tag{111}$$

for all values of t and x. Therefore, the bracket itself must be zero. Furthermore, the sum of three har-

monic functions of x can vanish for all values of x only if their periods are the same. It follows that

$$\frac{\sin \theta_t}{c_1} = \frac{\sin \theta_i}{c} = \frac{\sin \theta_r}{c}. \tag{112}$$

The second equation of (112) implies that $\theta_i = \theta_r$; that is, the angle of incidence is equal to the angle of reflection. The first equation may be rewritten as

$$\frac{\sin \theta_i}{\sin \theta_t} = \frac{c}{c_1}, \tag{113}$$

a relation which is well known in optics as Snell's law.

Because of equation (112), the exponential factor is the same f for all three terms in the bracket of equation (111) and can be divided out, giving

$$A_t = A_i + A_r. \tag{114}$$

The individual amplitudes A_t and A_r are calculated by making use of the transition conditions (104). By calculating $\partial p / \partial y$ from equation (110), and $\partial p_t / \partial y$ from equation (109) and by substituting these values into equation (104), we obtain

$$\frac{A_i \cos \theta_i}{c} e^{2\pi i f \left(t - \frac{x \sin \theta_i}{c} \right)} - \frac{A_r \cos \theta_r}{c} e^{2\pi i f \left(t - \frac{x \sin \theta_r}{c} \right)}$$
$$= \frac{\rho}{\rho_1} \frac{A_t \cos \theta_t}{c_1} e^{2\pi i f \left(t - \frac{x \sin \theta_t}{c_1} \right)}. \tag{115}$$

In view of equation (112), the exponential factor is the same for all three terms and may be divided out. Also, $\theta_i = \theta_r$. Thus, equation (115) becomes

$$\frac{\cos \theta_i}{c} (A_i - A_r) = \frac{\rho}{\rho_1} A_t \frac{\cos \theta_t}{c_1}. \tag{116}$$

Equations (114) and (116) are two linear equations in A_t and A_r. By solving them in terms of A_i, the amplitude of the incident wave, and by replacing c_1/c in the result with its equivalent from equation (113),

$$A_r = A_i \frac{\rho_1 c_1 \cos \theta_i - \rho c \cos \theta_t}{\rho_1 c_1 \cos \theta_i + \rho c \cos \theta_t} \tag{117}$$

$$A_t = A_i \frac{2 \rho_1 c_1 \cos \theta_i}{\rho_1 c_1 \cos \theta_i + \rho c \cos \theta_t}. \tag{118}$$

To eliminate the angle θ_t from equation (117), equation (113) is used which can be transformed by trigonometric identities into

$$\frac{\cos \theta_t}{\cos \theta_i} = \sqrt{1 + \tan^2 \theta_i \left(1 - \frac{c_1^2}{c^2} \right)} \equiv B.$$

Thus, equation (117) becomes

$$\frac{A_r}{A_i} = \frac{\rho_1 c_1 - \rho c B}{\rho_1 c_1 + \rho c B}. \tag{119}$$

EFFECT OF A BOUNDARY 31

Equation (119) gives interesting results when it is applied to the case of a sound wave in water hitting the surface separating water from air. The numerical values are (subscripts for air; no subscripts for water):

$$\frac{c}{c_1} = 4.3; \quad \frac{\rho}{\rho_1} = 770.$$

By substituting these values into equation (119),

$$\frac{A_r}{A_i} = \frac{1 - 3{,}311\sqrt{1 + 0.95 \tan^2 \theta_i}}{1 + 3{,}311\sqrt{1 + 0.95 \tan^2 \theta_i}}.$$

For perpendicular incidence θ_i vanishes, and A_r/A_i differs from -1 by less than one part in a thousand; for greater values of θ_i the approximation to -1 is even better. A wave in water reflected by air thus preserves its real amplitude almost exactly, that is, almost all the energy in the incident wave remains in the water. But it reverses its phase; this means that $p_r = -p_i$ at the boundary. This conclusion, that the resulting total pressure at this type of interface should be very nearly zero, was called a boundary condition in Section 2.6.1. The derivation of this section furnishes the justification for assuming this boundary condition, which was stated without rigorous proof in Section 2.6.1. Equation (119) provides an estimate of the error caused by replacing transition conditions at a boundary with the more simple boundary conditions. In the case of the interface separating water and air this error is clearly very slight.

Another interesting case is the incidence of underwater sound on a hard bottom like solid rock or tightly packed coarse sand. The treatment of sound waves in solids is rather more involved than the treatment of sound waves in fluids because a solid has two different kinds of elastic forces: those which resist changes in volume; and those which resist changes of shape (bulk modulus and shear modulus). Consequently, two different kinds of propagation of sound are possible in a solid. The two types are usually referred to as longitudinal waves and transverse waves. In the oblique incidence of underwater sound in a water-solid interface both types of waves are generated in the solid and the transition conditions are, therefore, more involved than those discussed previously. If, however, the solid is quite rigid — that is, if both bulk modulus and shear modulus are appreciably greater than the bulk modulus of water — then it may be assumed, in good approximation, that the interface will not permit

displacements perpendicular to itself. In other words, u_y will vanish approximately. If u_y at the interface is zero, then its time derivative vanishes as well; and by reason of the equations of motion (17),

$$\frac{\partial p}{\partial y} = 0 \text{ at } y = 0.$$

This is the boundary condition which is often assumed in the treatment of reflection from a hard bottom. If this boundary condition is realized, it can be shown that the incident and reflected waves will have equal amplitude and the same phase. Thus, when sound is reflected from a rock bottom, almost all the energy of the incident wave will be found in the reflected wave. For a soft bottom like mud, this boundary condition will no longer be satisfied, even in approximation, and considerable sound energy may be lost by transmission through the interface.

2.6.3 Homogeneous Medium with Single Boundary

POINT SOURCE NEAR SEA SURFACE

We shall now solve the problem of finding the solution of the wave equation which satisfies the boundary condition $p = 0$ at the interface $y = 0$, and corresponds to a sound wave radiated by a point source at the depth h. This situation is illustrated in Figure 11. The depth of the ocean is assumed to be

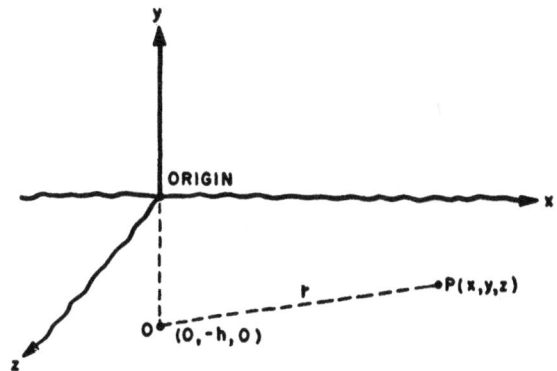

FIGURE 11. Pressure produced at location P by sound source O.

infinite. The initial conditions are specified by the assumption that in the immediate vicinity of the source, that is, for points whose distance from the source r,

$$r = \sqrt{x^2 + z^2 + (y + h)^2}$$

is very small, the pressure satisfies the relationship

$$rp(r,t) = F(t). \tag{120}$$

If it were not for the boundary condition $p = 0$ at $y = 0$, the problem would be solved by means of the expression

$$p(r,t) = \frac{1}{r}F\left(t - \frac{r}{c}\right) \qquad (121)$$

We shall have to modify this solution in order to satisfy the boundary condition as well. To this end, we resort to a trick. We solve a fictitious problem, one in which a source exactly like the first one is located at a distance h on the other side of the interface with the water extending through space and with the initial conditions

$$r'p'(r',t) = -F(t) \qquad (122)$$

at points very close to the new source. This problem has the solution

$$p'(r',t) = \frac{-1}{r'}F\left(t - \frac{r'}{c}\right) \qquad (123)$$

where

$$r' = \sqrt{x^2 + z^2 + (y - h)^2}.$$

Clearly, since $r = r'$ at $y = 0$, we have $p + p' = 0$ at $x = 0$. Thus, the wave given by the sum of the two disturbances described by equations (121) and (123), or

$$p = \frac{F[t - (r/c)]}{r} - \frac{F[t - (r'/c)]}{r'} \qquad (124)$$

satisfies the imposed boundary conditions. Also, equation (124) satisfies the initial conditions (120) because the expression (124) can be rewritten as

$$rp = F\left(t - \frac{r}{c}\right) - \frac{r}{r'}F\left(t - \frac{r'}{c}\right)$$

which reduces to equation (120) in the vicinity of the actual source, where $r \approx 0$. Finally, equation (124) satisfies the wave equation itself since the difference of two solutions of that equation is itself a solution.

If the source S executes a harmonic vibration, the solution (124) becomes

$$p = a\left\{\frac{\cos 2\pi f[t - (r/c)]}{r} - \frac{\cos 2\pi f[t - (r'/c)]}{r'}\right\}. \qquad (125)$$

Formula (125) fully describes the effect of surface reflection on harmonic waves emitted by a single source under the assumptions that air has negligible density and elasticity and that the sea surface is a perfect plane.

From the method of construction of the solution (124), it is possible to deduce that there should be a zone of low intensity near the surface. The reason is that, at points near the surface, r and r' will be nearly identical; and the two resulting fictional pressures will almost balance each other. This type of destructive interference near the surface is called the *Lloyd mirror* effect or *image interference* effect. The next few paragraphs will discuss the width of this low intensity zone and the intensity within this zone.

Consider the intensity measured by a receiver at the depth h_1, located at a horizontal range R from the source, as in Figure 12. We also assume that R

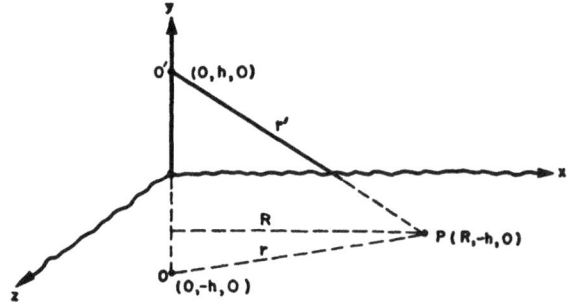

FIGURE 12. Fictitious scheme for solving wave equation and surface boundary conditions.

is so large compared with h and h_1 (the depths of source and receiver) that second-order products of h/R and h_1/R may be neglected.

By applying the Pythagorean theorem to Figure 12, then

$$r = R\sqrt{1 + \frac{(h_1 - h)^2}{R^2}} \quad r' = R\sqrt{1 + \frac{(h_1 + h)^2}{R^2}}.$$

Since h/R and h_1/R are small, these equations may be rewritten as

$$r = R\left\{1 + \frac{1}{2}\left[\frac{(h_1 - h)^2}{R^2}\right]\right\}$$
$$r' = R\left\{1 + \frac{1}{2}\left[\frac{(h_1 + h)^2}{R^2}\right]\right\} \qquad (126)$$

and as a result

$$\frac{1}{r} = \frac{1 - \frac{1}{2}\left(\frac{h_1 - h}{R}\right)^2}{R},$$

$$\frac{1}{r'} = \frac{1 - \frac{1}{2}\left(\frac{h_1 + h}{R}\right)^2}{R} \qquad (127)$$

because $1/(1 - \epsilon) = 1 + \epsilon$ if ϵ is small. Putting these in equation (125), we obtain

$$p = \frac{a}{R}\left[\cos 2\pi f\left(t - \frac{r}{c}\right) - \cos 2\pi f\left(t - \frac{r'}{c}\right)\right]$$

plus negligible terms, which may be rewritten as

$$p = \frac{a}{R}\left\{ -2 \sin\left[2\pi f\left(t - \frac{r+r'}{2c}\right)\right] \cdot \right.$$
$$\left. \sin\left[2\pi f\left(\frac{r'-r}{2c}\right)\right]\right\}$$

by the trigonometric identity for the difference of two cosines. This equation reduces approximately, because of equation (126), to

$$p = -\frac{2a}{R} \sin\left[2\pi\left(\frac{f}{c}\right)\frac{h_1 h}{R}\right] \sin\left[2\pi f\left(t - \frac{R}{c}\right)\right] \cdot (128)$$

At the point P, equation (128) tells us that the amplitude of the pressure variation with time is '

$$\text{Amplitude} = \frac{2a}{R} \sin 2\pi \frac{h_1 h}{R\lambda} \qquad (129)$$

where $\lambda = c/f$ is the wavelength of the sound. The resulting sound intensity at P, which is proportional to the square of the maximum acoustic pressure, will be very small if the argument of the sine in equation (129) is small. That is, the intensity will be low if $h_1 h/\lambda$ is very small compared with R, or, in other words, if

$$h_1 << \frac{\lambda R}{h} \cdot \qquad (130)$$

Assuming that equation (130) holds, the sine in equation (129) will be approximately equal to its argument, and we have

$$\text{Amplitude} = \frac{4\pi a h h_1}{R^2 \lambda} \cdot$$

In terms of the intensity, this means

$$\text{Intensity} \propto \frac{16\pi^2 a^2 h^2 h_1^2}{\lambda^2}\left(\frac{1}{R^4}\right). \qquad (131)$$

That is, in the layer of poor sound reception the sound intensity falls off inversely as the fourth power of the horizontal range at great ranges.

For smaller values of horizontal range R, we find that the amplitude vanishes wherever the argument of the sine in equation (129) is an integral multiple of π, or, in other words, where

$$\frac{2hh_1}{R\lambda} = j, j = 0, 1, 2, 3, \cdots$$

while the amplitude will show greatest values in the neighborhood of those points where the argument is $\pi/2, 3\pi/2 \cdots$; in other words, where

$$\frac{4hh_1}{R\lambda} = \kappa, \kappa = 1, 3, 5, \cdots$$

This sequence of interference minima and maxima is called the image interference pattern.

The image interference effect described here is only occasionally observed in the sea for reasons which are discussed in Section 5.2.1.

POINT SOURCE FAR FROM SEA SURFACE

We assume now that the source is located so far from the surface that the sound waves near the surface can be regarded as plane waves. Only incident waves propagated purely in the y direction are considered, that is, normal to the surface. Then equation (125) has to be replaced by

$$p = a\left[\cos 2\pi f\left(t - \frac{y}{c}\right) - \cos 2\pi f\left(t + \frac{y}{c}\right)\right]\cdot (132)$$

By applying to equation (132) the trigonometric formula for the difference of two cosines, we obtain

$$p = -2a \sin 2\pi\frac{y}{\lambda} \sin 2\pi ft. \qquad (133)$$

We notice a very curious thing about the disturbance described by equation (133). The acoustic pressure is zero over the entire fluid when ft is any integral multiple of $\frac{1}{2}$. Further, the acoustic pressure is zero for all time at points where y/λ is an integral multiple of $\frac{1}{2}$. Thus we see that the interference between two plane waves of equal amplitude and of the same frequency traveling in opposite directions produces, at least in this case, a disturbance of the medium for which at any instant all points have identical, or opposite phase. We no longer have progressive waves, but a phenomenon which we call *stationary* or *standing* waves. The points where the amplitude is zero for all time are called *nodes*; the points where the amplitude term of equation (133) is a maximum are called *loops* or *antinodes*.

This state of affairs is permanent as long as the source keeps vibrating. The nodes are permanent regions of silence; and the loops are permanent regions of maximum pressure amplitude. Such a state, in which all points of the medium perform vibrations of the form sin $2\pi ft$ with an amplitude dependent on position is called a *stationary state* of the medium.

REFLECTION FROM SEA BOTTOM

If water is separated by the plane $y = 0$ from a medium with a density much greater than its own, the boundary condition which must be fulfilled at this plane is

$$\frac{\partial p}{\partial y} = 0 \cdot$$

It turns out that a solution synthesized as was equation (124), but with a plus sign in equation (122) instead of a minus sign, will satisfy this boundary condition. This solution is

$$p = \frac{F\left(t - \frac{r}{c}\right)}{r} + \frac{F\left(t - \frac{r'}{c}\right)}{r'}. \quad (134)$$

We verify that equation (134) satisfies $\partial p/\partial y = 0$ by differentiating equation (134) with respect to y, and noting that $\partial r/\partial y = -\partial r'/\partial y$ at $y = 0$.

We now examine the possibility of stationary states for the case where the boundary is a hard sea bottom. Again, we assume a harmonic source so far from the bottom that waves reaching the bottom are plane and we assume perpendicular incidence. Then equation (134) must be replaced by

$$p = a\left[\cos 2\pi f\left(t - \frac{y}{c}\right) + \cos 2\pi f\left(t + \frac{y}{c}\right)\right] \quad (135)$$

which, by trigonometry, reduces to

$$p = 2a \cos 2\pi\frac{y}{\lambda} \sin 2\pi ft. \quad (136)$$

We easily see that equation (136) also represents a stationary state of our fluid. The nodes of utter silence are situated where $\cos 2\pi(y/\lambda)$ disappears, that is, at $y = \lambda/4,\ 3\lambda/4,\ 5\lambda/4, \cdots$; the loops of maximum sound intensity are located where $\cos 2\pi(y/\lambda)$ equals ± 1, that is, at $y = 0,\ \lambda/2,\ \lambda, \cdots$.

2.7 NORMAL MODE THEORY

2.7.1 Plane Waves in a Medium with Parallel Plane Boundaries

The problem of sound propagation in a medium bounded on two sides is extremely complicated and cannot be solved in general. The difficulty lies in the fact that the solution must satisfy not only the wave equation, but also the initial conditions and the boundary conditions at each boundary.

In Section 2.6 it was shown that certain definite and instructive results could be obtained for the case of a single boundary by considering the case of plane waves and assuming (1) perpendicular incidence and (2) an infinite change in density at the boundary. The result was a standing wave pattern whose geometrical properties depended on the wavelength and whose maximum amplitude depended on the energy in the incident wave.

We shall keep these two assumptions in this section and shall first find out under what conditions a stationary wave pattern of any sort can be set up in our bounded medium. The general expression for a standing wave pattern is

$$p = \psi(y) \cos 2\pi f(t - \epsilon) \quad (137)$$

where ψ is any function of y. In other words, equation (137) means that all points of the fluid perform vibrations with the same frequency f and phase constant ϵ, but with amplitude $\psi(y)$ depending on the position coordinate y. The immediate problem is to find out what sort of functions $\psi(y)$ are necessary to make equation (137) a solution of the plane wave equation

$$\frac{\partial^2 p}{\partial t^2} = c^2\frac{\partial^2 p}{\partial y^2} \quad (138)$$

and also a solution of the boundary conditions on the boundaries $y = 0$ and $y = L$. These boundary conditions are either equation (139a), (139b), or (139c).

$$p = 0 \text{ at both } y = 0 \text{ and } y = L \quad (139a)$$

$$p = 0 \text{ at } y = 0; \quad \frac{\partial p}{\partial y} = 0 \text{ at } y = L \quad (139b)$$

$$\frac{\partial p}{\partial y} = 0 \text{ at both } y = 0 \text{ and } y = L. \quad (139c)$$

It is immediately apparent that the condition for equation (137) to satisfy the plane wave equation (138) is

$$\frac{d^2\psi}{dy^2} + \frac{4\pi^2}{\lambda^2}\psi = 0. \quad (140)$$

First consider the case of the boundary conditions (139a). The boundary conditions (139a) can be restated as

$$\psi(0) = 0, \quad \psi(L) = 0. \quad (141)$$

The problem is thus reduced to the case of finding the solution of an ordinary differential equation with boundary conditions on both ends of an interval $0 \leqq y \leqq L$.

Equation (140) is a simple differential equation whose general solution is well known to be

$$\psi(y) = A \sin\frac{2\pi}{\lambda}y + B \cos\frac{2\pi}{\lambda}y \quad (142)$$

where A and B are arbitrary constants.

The condition $\psi(0) = 0$ implies that $B = 0$ and

$$\psi(y) = A \sin\frac{2\pi}{\lambda}y. \quad (143)$$

Since equation (143) must satisfy $\psi(L) = 0$,

$$\sin 2\pi\frac{L}{\lambda} = 0; \quad \frac{L}{\lambda} = \frac{1}{2}, \frac{2}{2}, \frac{3}{2}, \cdots \quad (144)$$

This means that under the given boundary conditions there cannot be a stationary state of the type of (137) unless the wavelength λ has one of a number of definite ratios to the depth L. The ratio λ/L must be either 2/1, 2/2, 2/3, 2/4, or in general 2/j. A set of wavelengths λ_j can be defined by

$$\lambda_j = \frac{2}{j}L, j = 1, 2, 3, \cdots \quad (145)$$

Then, if the actual wavelength in the problem is equal to one of these λ_j, the expression (143) will satisfy equation (141); and the stationary state equation (137), with this value of ψ, will satisfy both the wave equation (138) and the boundary conditions (139a). If the actual wavelength is not equal to one of the λ_j, then there can be no stationary state in the given medium in which the wave planes are parallel to the interfaces.

Mathematically, all this means that the total problem defined by equations (138) and (139a) can be solved only if the coefficient of ψ in equation (140) has certain definite values α_j defined by

$$\alpha_j = \frac{4\pi^2}{\lambda_j^2}. \quad (146)$$

These values, α_j, are called *characteristic values*. The solutions (143) corresponding to them, namely

$$\psi_j(y) = A \sin \frac{2\pi}{\lambda_j}y = A \sin \sqrt{\alpha_j}\, y \quad (147)$$

are called *characteristic functions* of the problem. Clearly, every characteristic value α_j corresponds to a possible frequency f_j and wavelength λ_j; by possible is meant that it gives rise to a stationary state, or *normal mode* of vibration. The characteristic function ψ_j determines the distribution of acoustic pressure within this normal mode of vibration.

If the boundary conditions are changed, the characteristic values and characteristic functions change also, although the differential equation which ψ must satisfy remains equation (140). If the boundary conditions (139b), which correspond most closely to actual conditions in the sea, are assumed, it is found that, by methods similar to those described previously, a normal mode can arise only if

$$\frac{L}{\lambda} = \frac{1}{4}, \frac{3}{4}, \frac{5}{4}, \cdots \quad (148)$$

That is,

$$\lambda_j = \frac{4L}{j} \text{ where } j = 1, 3, 5, \cdots \quad (149)$$

We notice that the characteristic wavelengths λ_j are different from the first case. The characteristic value and functions can be calculated by using equation (146) and (147) and by remembering that the λ_j must now be taken from equation (149).

Clearly a sum of two normal modes also satisfies the differential equation (138) and the imposed boundary conditions.

Suppose we have the general case where the boundary conditions determine an infinite number of normal modes

$$p = c\,\psi_j(y) \cos 2\pi f_j(t - \epsilon_j), j = 1, 2, 3, \cdots. \quad (150)$$

Suppose we have initial conditions on y at $t = 0$, of the nature

$$p(y,0) = D\psi_\kappa(y), \quad (151)$$

where D is some constant. For equation (150) to satisfy the initial conditions (151), we must choose $j = \kappa$, $\epsilon_\kappa = (1/2\pi f_\kappa)$ arc cos D/c. Since this can always be done, we have the result that if the initial pressure disturbance is a multiple of one of the characteristic functions, then one of the solutions of the boundary problem will also satisfy the initial conditions.

This result can be generalized. If the initial distribution of the pressure is a linear combination of several of the characteristic functions $\psi_\kappa(y)$, then we shall show that these initial conditions can be satisfied by a corresponding sum of normal modes. Suppose

$$p(y,0) = \sum_j D_j \psi_j(y) \quad (152)$$

where the D_j are any constants. Then the distribution of pressure given by

$$p(y,t) = \sum_j A_j \cos 2\pi f_j(t - \epsilon_j)\psi_j(y) \quad (153)$$

will satisfy the initial conditions (152) if only the A_j and ϵ_j are chosen so that

$$A_j \cos 2\pi f_j\epsilon_j = D_j. \quad (154)$$

The expression (153) also satisfies the wave equation and the boundary conditions since it is a sum of normal modes; hence, it is the solution to the problem when the initial pressure disturbance can be expressed as a finite sum of characteristic functions.

Now suppose the initial disturbance cannot be expressed in the form of (152), but is a general function

$f(y)$. It is a remarkable fact of mathematics that if we allow the sum to be an infinite one, then any function of y can be expressed in that form. That is, it is possible to express $f(y)$ as

$$f(y) = \sum_{j=1}^{\infty} c_j \psi_j(y)$$
$$= \sum_{j=1}^{\infty} c_j \sin \frac{2\pi}{\lambda_j} y \qquad (155)$$

where the c_j depend on the law of formation of the λ_j [whether (145) or (149), etc.], and usually become small very rapidly as j increases.

If the boundary conditions are (139a), then the λ_j are given by equation (145), and we have

$$f(y) = \sum_{j=1}^{\infty} c_j \sin \frac{\pi j}{L} y. \qquad (156)$$

In equation (156) the coefficients c_j are given, according to the usual laws of Fourier analysis, by

$$c_\kappa = \frac{2}{L} \int_0^L f(y) \sin \frac{\pi \kappa y}{L} dy. \qquad (157)$$

The c_κ are called *Fourier coefficients* of $f(y)$. Once we know these Fourier coefficients, we can solve our problem as in the finite case. We have

$$p(y,0) = \sum_{j=1}^{\infty} c_j \psi_j(y) \qquad (158)$$

as our initial condition; and

$$p(y,t) = \sum_{j=1}^{\infty} A_j \cos 2\pi f_j(t - \epsilon_j) \psi_j(y) \qquad (159)$$
$$A_j \cos 2\pi f_j \epsilon_j = c_j$$

as the set of solutions to the total problem.

While each term in the sum (159) represents a stationary state of vibration, the infinite sum is not stationary, in view of the fact that the terms have different frequencies f_j.

2.7.2 General Waves in a Medium with Parallel Plane Boundaries

Section 2.7.1 showed how the assumption of a stationary state led to a possible solution which satisfied the wave equation, the boundary conditions, and the initial conditions. However, the treatment in Section 2.7.1 was restricted to the case of plane waves moving perpendicular to two enclosing plane boundaries. This section explains how this method may be generalized for the case of general waves in a medium with two parallel plane boundaries.

We assume again a stationary state in the medium, of the form

$$p(x,y,z,t) = \cos 2\pi f t \cdot \psi(x,y,z). \qquad (160)$$

If this solution is substituted into the wave equation (27), we find that ψ satisfies a partial differential equation of the form

$$\frac{\partial^2 \psi}{\partial x^2} + \frac{\partial^2 \psi}{\partial y^2} + \frac{\partial^2 \psi}{\partial z^2} + \frac{4\pi^2}{\lambda^2} \psi = 0 \qquad (161)$$

which is time independent. In addition, ψ must satisfy the boundary conditions imposed at the bounding planes $y = 0$ and $y = L$. The boundary conditions may be of the form (139a), (139b), or (139c).

We shall treat only the case characterized by the conditions (139a). The treatment of the other cases is completely analogous. We attempt to find a solution of equation (161) of the form

$$\psi(x,y,z) = \sin \frac{2\pi y}{\lambda_y} G(x,z) \qquad (162)$$

in which the constant λ_y must have one of the values

$$\lambda_y = \frac{2L}{j}; \ j = 1, 2, 3, \cdots \qquad (163)$$

to satisfy the boundary conditions. For the $G(x,z)$ we have the equation

$$\frac{\partial^2 G}{\partial x^2} + \frac{\partial^2 G}{\partial z^2} + 4\pi^2 \left(\frac{1}{\lambda^2} - \frac{1}{\lambda_y^2} \right) G = 0. \qquad (164)$$

Any solution of this equation when multiplied by $\cos 2\pi f t \sin 2\pi y/\lambda_y$ is a solution of the wave equation (27) and also satisfies the boundary conditions (139a). Equation (164) will be satisfied by any plane wave solution in the xz plane with the wavelength λ^* given by

$$\frac{1}{\lambda^*} = \sqrt{\frac{1}{\lambda^2} - \frac{1}{\lambda_y^2}}.$$

Such a solution will be of the form

$$G = a_1 \cos \frac{2\pi}{\lambda^*} s + a^2 \sin \frac{2\pi}{\lambda^*} s; \ s = x \cos \theta + z \sin \theta \cdot \qquad (164a)$$

It is easily verified that this function G satisfies (164). If $\lambda_y < \lambda$, λ^* is imaginary instead of being real. In that case, the solution should be written in the form

$$G = b_1 e^{(2\pi/\lambda')s} + b_2 e^{-(2\pi/\lambda')s}$$

in which λ' is the real constant given by

$$\frac{1}{\lambda'} = \sqrt{\frac{1}{\lambda_y^2} - \frac{1}{\lambda^2}}.$$

G is thus the sum of two terms, one increasing exponentially with the distance s from the source, the other decreasing exponentially. The only solutions of this character which have physical significance are those for which the first (increasing) term is zero. If b_1 were not zero, the sound intensity would increase rapidly with the distance from the source and that is physically impossible.

Since the greatest possible value of λ_ν is $2L$, or twice the depth, it follows that sound of wavelength greater than $2L$ will have an exponential rather than a trigonometric solution; because b_1 must be zero, such sound will suffer an exponential pressure decay with increasing range. It is clear that the longer the wavelength, the more rapid will be the decay. A more detailed discussion of this type of transmission is given in a report by the Naval Research Laboratory [NRL], where the detailed properties of the bottom are taken into account.[4]

The angle θ in the solution (164a) may be chosen arbitrarily. Thus to any value of j in the equation (163) belongs an infinite set of characteristic functions. These characteristic functions can be combined to satisfy particular initial conditions; however, the rules for their combination are too involved to be presented here.

The derivation of the wave equation (27) was based partly on the assumption that the velocity of propagation was everywhere the same, in other words, that the medium was homogeneous. Let c be an arbitrary function of (x,y,z). In the ocean, the variation in sound velocity is due mainly to the variation in water temperature with depth.

To assume that the velocity is variable, amounts for most practical purposes to assuming that c in equation (27) is now a function of position. The method of normal modes can be applied to find a solution in that case just as in the case of constant sound velocity. As before, we assume a stationary state of form (160). Substituting equation (160) into the wave equation, we get as the time-independent differential equation the following:

$$\frac{\partial^2 \psi}{\partial x^2} + \frac{\partial^2 \psi}{\partial y^2} + \frac{\partial^2 \psi}{\partial z^2} + \frac{4\pi^2 f^2}{c^2(xyz)}\psi = 0 \qquad (165)$$

which differs from equation (161) only in that c is now variable. The solution of equation (165) satisfying the imposed initial and boundary conditions can be found as before by the superposition of an infinite number of normal modes; in this case of variable c, however, the computation of the characteristic

values and functions is more troublesome. An application of this type of analysis is discussed in Section 3.7.

2.7.3 Intensity as a Function of Phase Distribution

Whenever the sound source is harmonic, the pressure distribution resulting from given initial and boundary conditions can be written in the form

$$p = a(x,y,z)e^{2\pi i f[t-\epsilon(x,y,z)]} \qquad (166)$$

a and ϵ being real functions of x, y, and z. For some purposes it is convenient to set

$$\epsilon(x,y,z) = \frac{V(x,y,z)}{c_0} \qquad (167)$$

so that equation (166) becomes

$$p = a(x,y,z)e^{2\pi i f[t-(V/c_0)]} \qquad (168)$$

or, explicitly for the real pressure,

$$p = a(x,y,z) \cos 2\pi f\left(t - \frac{V}{c_0}\right). \qquad (169)$$

Since we know from Section 2.4.2 that $I = \overline{pu}$, we must derive an expression for pu. This is done by making use of equation (44), relating the derivatives of p and u_x:

$$\frac{\partial u_x}{\partial t} = -\frac{1}{\rho}\frac{\partial p}{\partial x}. \qquad (170)$$

From equation (169), then

$$\frac{\partial p}{\partial x} = a(\sin H)\frac{2\pi f}{c_0}\frac{\partial V}{\partial x} + \frac{\partial a}{\partial x}\cos H$$

where

$$H = 2\pi f\left(t - \frac{V}{c_0}\right). \qquad (171)$$

Therefore, from equation (170),

$$\frac{\partial u_x}{\partial t} = -\frac{a}{\rho}\sin H \frac{2\pi f}{c_0}\frac{\partial V}{\partial x} - \frac{1}{\rho}\cos H\frac{\partial a}{\partial x}.$$

Integrating this, we get

$$u_x = \frac{a}{\rho c_0}\cos H\frac{\partial V}{\partial x} - \frac{1}{2\pi f\rho}\sin H\frac{\partial a}{\partial x}. \qquad (172)$$

From equations (172) and (169), we obtain

$$pu_x = \frac{a^2}{\rho c_0}\cos^2 H \cdot \frac{\partial V}{\partial x} - \frac{a}{2\pi f\rho}\sin H \cos H\frac{\partial a}{\partial x}. \qquad (173)$$

The average energy flow in the x direction I_x, at the point x,y,z, is just the time average of equation (173) over a complete period. The time average of the square of the cosine is $\frac{1}{2}$; the time average of the product of the sine and cosine is zero. Thus,

$$I_x = \left(\frac{a^2}{2\rho c_0}\right)\frac{\partial V}{\partial x}\cdot$$

Similarly,
$$I_y = \left(\frac{a^2}{2\rho c_0}\right)\frac{\partial V}{\partial y}$$

$$I_z = \left(\frac{a^2}{2\rho c_0}\right)\frac{\partial V}{\partial z}\cdot \qquad (174)$$

Since $I = (I_x^2 + I_y^2 + I_z^2)^{\frac{1}{2}}$, we have the following expression for the intensity

$$I = \frac{a^2}{2\rho c_0}\left[\left(\frac{\partial V}{\partial x}\right)^2 + \left(\frac{\partial V}{\partial y}\right)^2 + \left(\frac{\partial V}{\partial z}\right)^2\right]^{\frac{1}{2}}\cdot \qquad (175)$$

The relations (174) and (175) will be found useful in Chapter 3 when the equivalence of wave acoustics and ray acoustics is investigated.

2.8 PRINCIPLE OF RECIPROCITY

The principle of reciprocity makes a statement concerning the interchangeability of source and receiver. Very crudely, the import of the statement is that if in a given situation the locations and orientations of source and receiver are interchanged, the sound pressure measured at the receiver will be the same as before. To be true, under the most general conditions, this statement has to be qualified in detail. The following is an attempt to formulate the General Reciprocity Principle.

Assume that a source of a given directivity pattern b and a receiver of a different directivity pattern b' are placed in a medium with a particular distribution of sound velocity $c(x,y,z)$, enclosed by boundaries of any given shape with any particular boundary conditions; let the output of the source on its axis be given by an amplitude A at one yard. The receiver will then record some pressure amplitude, corresponding to the amplitude B on its axis. Now let the source be replaced by a receiver having the same orientation of its axis and having the directivity pattern b'; assume also that the receiver is replaced by a source which has the same orientation of its axis, the directivity pattern b, and the output A on its axis. Then the new receiver will again register a pressure equivalent to that of a sound wave incident on its axis with an amplitude B.

The proof of this theorem is difficult in the general case, and will not be reproduced here. Instead, we shall give the exact proof for the simple case of a plane source emitting plane waves into a medium that satisfies boundary conditions of the type (139). We shall then indicate, roughly, how the proof can be generalized.

Suppose the pressure is a function of y and t only. The medium may be inhomogeneous, but both the density and bulk modulus are assumed to be a function of y only. Then, the sound velocity will be a function of y. The wave equation for this case therefore reduces to

$$\frac{\partial^2 p}{\partial t^2} = c^2(y)\frac{\partial^2 p}{\partial y^2} \qquad (176)$$

and its solution, by equation (160), will be

$$p(y,t) = \psi(y) \cos 2\pi ft \qquad (177)$$

where $\psi(y)$ is obtained from

$$\frac{d^2\psi}{dy^2} + k^2(y)\psi = 0, \quad k^2 = \frac{4\pi^2 f^2}{c^2(y)} = \frac{4\pi^2}{\lambda^2(y)}\cdot \quad (178)$$

We assume the medium is bounded by the planes $y = 0$ and $y = L$ and satisfies boundary conditions of the type (139a), (139b), or (139c).

We assume that the plane $y = a$ is a source of sound. If by $\psi_a(y)$ is meant the function defining the pressure amplitudes at every point of the medium, including near $y = a$, then ψ_a satisfies equation (178) everywhere except near $y = a$. That is, it satisfies

$$\frac{d^2\psi_a}{dy^2} + k^2(y)\psi_a = A(y) \qquad (179)$$

where $A(y)$ is very large in the immediate neighborhood of $y = a$, and zero everywhere else. Then the magnitude of the plane source S_a, located at $y = a$, will be defined by

$$S_a = \int_0^L A(y)dy = \int_{a-\delta}^{a+\delta} A(y)dy \qquad (180)$$

where it is understood that S_a is the limit of the integral as δ approaches zero.

In the same way, we define a plane source at $y = b$ by assuming an amplitude function $\psi_b(y)$ which satisfies equation (178) everywhere except near $y = b$, that is, it satisfies

$$\frac{d^2\psi_b}{dy^2} + k^2(y)\psi_b = B(y) \qquad (181)$$

where $B(y)$ is very large in the immediate neighborhood of $y = b$, and zero everywhere else. Then the magnitude of the source at $y = b$ will be

$$S_b = \int_{b-\delta}^{b+\delta} B(y)dy. \qquad (182)$$

By multiplying equation (181) by ψ_a, and equation (179) by ψ_b, and by subtracting the latter result from the former, we get

$$\psi_a \frac{d^2\psi_b}{dy^2} - \psi_b \frac{d^2\psi_a}{dy^2} = \psi_a B - \psi_b A$$

which may be rewritten as

$$\frac{d}{dy}\left(\psi_a \frac{d\psi_b}{dy} - \psi_b \frac{d\psi_a}{dy}\right) = \psi_a B - \psi_b A. \qquad (183)$$

Equation (183) holds if ψ_a and ψ_b are arbitrary functions of y, and A and B are defined by equations (179), (181). We can integrate (183) over the entire extension of the fluid between $y = 0$ and $y = L$, and get

$$\left(\psi_a \frac{d\psi_b}{dy} - \psi_b \frac{d\psi_a}{dy}\right)_0^L = \int_0^L (\psi_a B - \psi_b A)dy. \qquad (184)$$

Since ψ_a and ψ_b each satisfy some combination of $\psi = 0$ or $d\psi/dy = 0$ at $y = 0$, and $y = L$, the left-hand side vanishes identically, and

$$\int_0^L (\psi_a B - \psi_b A)dy = 0. \qquad (185)$$

Equation (185) is valid for all functions of ψ_a and ψ_b and satisfies equations (179), (181), and the boundary conditions (139).

Since $B = 0$ except near $y = b$, by equation (181),

$$\int_0^L \psi_a B dy = \psi_a(b) \int_{b-\delta}^{b+\delta} B dy = \psi_a(b)S_b$$

because of equation (182). Similarly,

$$\int_0^L \psi_b A dy = \psi_b(a) \int_{a-\delta}^{a+\delta} A dy = \psi_b(a)S_a.$$

Therefore, equation (185) becomes

$$S_b\psi_a(b) - S_a\psi_b(a) = 0. \qquad (186)$$

If both sources are of equal magnitude, then $S_a = S_b$, and

$$\psi_a(b) = \psi_b(a). \qquad (187)$$

That is, if two plane sources of equal strength are emitting plane waves into a "stratified" medium, where the sound velocity obeys an arbitrary law, and where boundary conditions are of the form of (139), then the first source (at $y = a$) produces at $y = b$ the same acoustic pressure which the source at $y = b$ would produce at $y = a$.

It is interesting to note that we have proved equation (187) without solving explicitly for the pressure amplitudes. We remember that even in this one-dimensional problem the equation (178) with boundary conditions usually cannot be solved for ψ in terms of elementary functions if the sound velocity is an arbitrary function of y. However, we found we could prove equation (187) merely by assuming that ψ_a and ψ_b were solutions of the wave equation with initial and boundary conditions, and by following up the consequences of that assumption.

In the general case, in which we no longer assume perpendicular incidence on plane parallel boundaries, the proof is more complex. Instead of equation (178), we have the more complicated form of (165). Equations (179) and (181) must be replaced by equations with the same left-hand sides as equation (165), but with right-hand sides which are different from zero only in the immediate vicinity of a particular point, (x_a,y_a,z_a) and (x_b,y_b,z_b), respectively. The distribution of the functions A and B about these two points determines the directivity of the source considered.

The integration (184) must be replaced by a volume integral, or rather, by an infinite series of volume integrals to account fully for the two directivity patterns, the left-hand sides of which can be shown to vanish. From there on, the proof runs similarly to the plane case.

The foregoing remarks have applied only to the case of propagation in a perfect fluid. It can be shown that the reciprocity principle holds, with additional qualifications, for propagation in a viscous fluid also.

2.9 INADEQUACY OF WAVE ACOUSTICS

In this chapter, we have set up a schematic picture of the transmission of sound in the ocean, and proceeded to derive a rigorous mathematical description of our schematic picture. Unfortunately, the results obtained cannot be used directly as a basis for the prediction of the performance of sonar gear. The schematic picture is not nearly complete enough from a purely physical point of view; furthermore, even the simplified schematic picture can be solved rigorously only for simple cases; and in the cases where solutions are possible, the calculations are very difficult.

The physical picture is inadequate on several counts. For one thing, boundary conditions like $p = 0$ of $\partial p/\partial y = 0$ at the boundaries are only a vague description of what actually happens at the boundaries. The surface is not a perfect plane, but is usually disturbed and uneven, with the result that even plane waves are not reflected according to the law of reflection; they are partly reflected in a direction depending on the direction of the surface and also partly scattered in all directions. Neither does the bottom obey the postulated conditions; it is never infinitely dense; at best it is rocky; at worst it is so muddy that it can hardly be called a boundary. The medium itself, the sea water, is not completely described by its density and its bulk modulus. There are many inhomogeneities in the sea volume, such as bubbles, floating plant and animal life, fish, and others. For all we know, such inhomogeneities may produce a very important part of the observed transmission loss, perhaps as important a part as the variations in sound velocity.

The mathematical difficulties should be apparent to anyone who has even glanced at the remainder of the chapter. Even when the boundary conditions can be formulated exactly, and initial conditions are simple, the exact solution of the problem usually cannot be presented. In the general case, it can be proved that a solution exists and is unique, but the solution cannot be written in a formula which would provide a practical basis for intensity calculations. The primary benefit of the rigorous approach is that one can derive certain very useful properties of the sound field, such as the principle of reciprocity and the dependence of intensity on the phase distribution, without going into the exact solution itself.

How, then, are we going to predict the sound field intensity? We certainly cannot go out and measure the intensity in all cases; such measurements are time-consuming, and provide information only about that particular part of the ocean at that particular time. We should have some method for estimating the intensity field, at least qualitatively, so that the observed intensity data can at least be interpreted according to a frame of reference; mere data without some reference to a theoretical scheme are useless.

In the theory of light, this problem was solved by using the methods of ray optics. The fundamental problems about optical instruments, like those for telescopes, can be solved by ray tracing methods without resorting to the exact solution of the wave equation. This ray theory is based on the assumption that light energy is transmitted along curved paths, called rays, which are straight lines in all parts of the medium where the velocity of light is constant, and which curve according to certain rules in parts where the velocity of light is changing. This light-ray theory is valid in all cases where obstacles and openings in the path of the radiation are much greater in size than the wavelength of light.

In the next chapter, we shall describe the application of ray methods to underwater sound transmission and shall also examine the validity of this type of approximation.

Chapter 3

RAY ACOUSTICS

Chapter 2 was devoted to the rigorous computation of the acoustic pressure p as a function of position in the fluid and of time. In situations where the acoustic pressure could be determined the sound intensity at an arbitrary spot and at an arbitrary time could be calculated. However, it was noted that, in most situations involving initial and boundary conditions similar to those met with in actual sound transmission in the ocean, this computation was at best very laborious and at worst completely impossible to carry out. Ray acoustics provides a more convenient though less rigorous approach.

In the study of sound the ray concept has not played so great a role as in optics. The reason for this is that the wavelengths of most audible sounds are not small compared to the obstacles in the path of the sound. Consequently, sounds audible to the ear do not travel straight-line or nearly straight-line paths; they bend around corners and fill almost all of any space into which they are directed. However, for the short wavelengths used in supersonic sound, the ray methods have an important, if limited, application. This chapter elaborates the theory of sound rays, describes the computation of sound intensity from the ray pattern, and finally, examines the conditions under which ray intensities may be expected to approximate the intensities calculated according to the rigorous methods of the second chapter.

3.1 WAVE FRONTS AND RAYS

3.1.1 Spherical Waves

The wave equation was solved explicitly for p in one very important case: an impulse sent out by a point source into a homogeneous medium under the assumption of spherical symmetry. The pressure as a function of time and space was found to be

$$p = \frac{1}{r} f\left(t - \frac{r}{c}\right). \tag{1}$$

In this expression, r is the distance from the source,

and the function $f(t - r/c)$ is determined by the output of the source. Specifically, the source can be characterized by the statement that, while the pressure itself at the source is infinite, the product rp in the immediate vicinity of the source has a finite value at every instant, namely $f(t)$.

Obviously, this function $f(t - r/c)$ determines when a pulse emitted by the source at a particular instant will arrive at a given point of space. If the pulse should, for instance, have started in time at some instant $t = \epsilon$ so that for all values of the argument less than ϵ the function $f(t)$ vanishes, then the onset of the disturbance at a distance r from the source will be observed at the time

$$t = \epsilon + \frac{r}{c}.$$

Likewise, we find that the front of the pulse will have reached at the time t a distance r, given by

$$r = c(t - \epsilon). \tag{2}$$

What has just been stated about the front of the pulse might just as well have been said about any other specified part of the pulse; only, ϵ would in that event characterize the time at which the specified part of the pulse was radiated by the source. What has been called, vaguely, part of the pulse, is often more concisely called *phase*, particularly in connection with harmonic pulses. The term ϵ then characterizes the phase of the pulse considered and is ordinarily referred to as a *phase constant*.

The surface defined by equation (2) is, of course, a sphere of radius $c(t - \epsilon)$ at the time t. As the time increases, the radius of the sphere increases at the rate of c units per second. The surface of this sphere of constant phase is called the *wave front*. The energy represented by the disturbance of equilibrium conditions clearly spreads out radially from the source. We may focus attention on the direction of energy flow by mentally drawing an infinite number of radii from the source to the wave front. These radii may be regarded as representing the paths of energy flow and may be called *sound rays*, in analogy with light rays. Sound energy may be regarded as traveling out along these rays with the speed c. The wave fronts assume in this description the secondary role of surfaces everywhere normal to the rays.

An individual sound ray cannot exist as a physical phenomenon. An isolated sound ray would mean a state of the fluid where the condensation was confined to the immediate neighborhood of a particular straight line. Beams of narrow cross section can be produced by directing a wave front onto a very narrow slit; but if the size of the slit becomes comparable to the wavelength of the sound, the sound leaving the slit is not a narrow beam, but a cone. This phenomenon is called *diffraction*, and will be discussed in Section 3.7. It is mentioned here only to indicate that the concept of a sound ray refers not to the propagation of a narrow beam with sharp edges, but merely to the direction of propagation of actual wave fronts.

Spherical wave fronts represent only one particular case of sound propagation, even in an infinite fluid. First, the wave front can be spherical only if the initial disturbance has no preferential direction (a vibrating bubble satisfies this condition). Second, the expanding surface remains a sphere concentric with the origin only if the sound velocity c is either constant throughout the fluid, or has spherical symmetry about the sound source.

The wave factor $f(t - r/c)$ in equation (1) is responsible for the conclusion that the disturbance is propagated with the velocity c. The remaining factor, $1/r$, is called the amplitude factor since it is responsible for the decrease in sound intensity as the distance from the source increases. The rate at which sound intensity is weakened with distance can be easily computed by using the concept of rays as carriers of sound energy, provided we assume that energy is generated only at the source and then flows through space without gain or loss. For reasons of symmetry, the energy flow from the source must take place along the radial sound rays. There will be a definite number of rays inside a unit solid angle. These rays will intercept an area of 1 sq ft on a sphere of radius 1 ft whose center is at the source, an area of 4 sq ft on a sphere of radius 2 ft, and generally r^2 sq ft on a sphere of radius r. Since the total energy flow is the same for all these spherical surfaces, the energy flow per unit area, or sound intensity, must be inversely proportional to the square of the distance of the unit area from the source.

3.1.2   General Waves

The frontal attack on the wave equation was the solution of the boundary problem by the method of normal modes. This method was found to be too complicated. It was shown in Section 3.1.1 that the method of sound rays gave a simple and plausible account of sound propagation for the case of spherical symmetry. A natural approach to the general problem would be to generalize the definition of sound rays, and see if light could thereby be thrown on the general case of variable sound velocity and arbitrary initial distributions of p.

First we must generalize the definition of wave fronts. In what follows we shall restrict ourselves to harmonic sound waves, that is, sound waves which have been produced by a sound source which undergoes single-frequency harmonic vibrations. In accordance with Section 2.4.3, the pressure at any point inside the fluid can be represented as the real part of an expression having the form

$$p = A(x,y,z)e^{i\theta(x,y,z,t)} \qquad (3)$$

in which the angle θ at each point in space increases linearly with time,

$$\theta = 2\pi f[t - \epsilon(x,y,z)]. \qquad (4)$$

We shall now call a wave front all those points at which the phase angle θ has a specified value, say θ_0. At any time t, this surface is defined by the equation

$$\epsilon(x,y,z) = t - \frac{\theta_0}{2\pi f}. \qquad (5)$$

For later convenience, we shall replace $\epsilon(x,y,z)$ by an expression $W(x,y,z)/c_0$, in which c_0 is the velocity of sound under certain designated standard conditions. Equation (3) then takes the form

$$p = A(x,y,z)\, e^{2\pi i f\left[t - \frac{W(x,y,z)}{c_0}\right]}, \qquad (6)$$

both A and W being real functions of the space coordinates. The defining equation (5) of an individual wave front assumes the form

$$W(x,y,z) = c_0(t - t_0) \qquad (7)$$

where
$$t_0 = \frac{\theta_0}{2\pi f}.$$

The term t_0 has different values for different wave fronts, but is constant both in space and in time for a given wave front. The function W clearly has the dimension of a length.

In order to make use of the concept of sound rays to describe the energy propagated by such generalized wave fronts, we must also generalize the definition of a ray. We can no longer assume that the rays are straight lines since we concede the possibility of refraction and reflection. We shall, however, retain the property that the rays are everywhere perpendicular to the wave fronts. It is, of course, by no means obvious that the results of this new approach will agree with results from a direct solution of the wave equation plus initial and boundary conditions. A comparison between the results from the ray pattern approach and the results from a rigorous treatment of the wave equation will be carried out in Section 3.6 once the ray method has been fully described. It will be found that in many practical situations these two approaches lead to similar results.

Geometrically, the rays and successive wave fronts can be constructed as in Figure 1. The wave front at

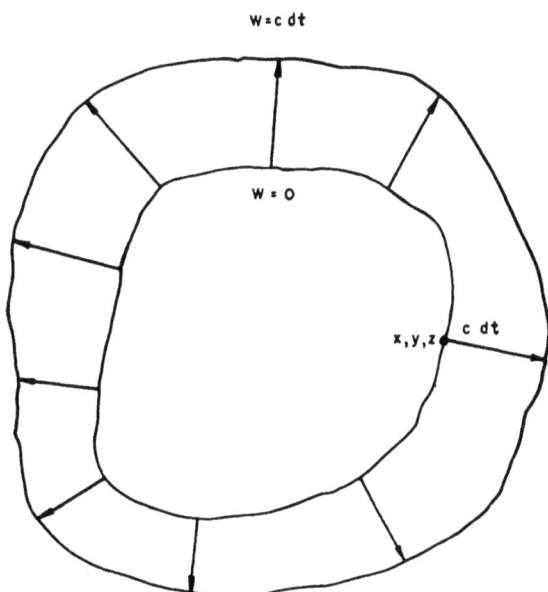

FIGURE 1. Huyghens' method for constructing successive wave fronts.

time $t = 0$ (whose equation is given by $W = -c_0 t_0$) is first drawn. In order to determine the wave front at the time dt, the small ray elements are drawn as straight-line segments perpendicular to the initial wave front, as at (x_1, y_1, z_1). In the time dt, the end point of the ray starting at (x_1, y_1, z_1) will have progressed to a point a distance $c\, dt$ from the initial wave front, where c is the velocity at the point (x_1, y_1, z_1). If this process is carried through for all the points on the initial wave surface, the end points of all the small ray elements will determine a second surface, which may be regarded as the wave front at

the time dt. By performing this process many times, the wave front can be obtained at any time t. This method of determining wave fronts by gradually widening an initial wave front was first suggested by the Dutch physicist, Huyghens, in the seventeenth century, for the solution of problems in optics.

3.2 FUNDAMENTAL EQUATIONS

3.2.1 Differential Equation of the Wave Fronts

Since the construction of wave fronts described in the preceding section is purely geometrical, it must be reformulated in mathematical terms for use in an algebraic analysis of the sort we are carrying out.

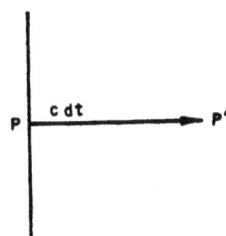

FIGURE 2. Differential ray path.

Let P in Figure 2 be any point on the wave front at time t. The equation of the wave front is given by equation (7). Let the coordinates of P be (x, y, z); let PP' be the ray element emanating from P at the end of a time interval dt; and let α, β, γ be the direction cosines of PP'. Then the coordinates of P' are $(x + \alpha cdt,\ y + \beta cdt,\ z + \gamma cdt)$. Further, the wave front at the time $t + dt$ is given by the equation

$$W(x + \alpha cdt, y + \beta cdt, z + \gamma cdt) = c_0(t - t_0 + dt). \quad (8)$$

If cdt is assumed to be very small, the left-hand side of equation (3) is very nearly equal to

$$W(x,y,z) + \left(\alpha \frac{\partial W}{\partial x} + \beta \frac{\partial W}{\partial y} + \gamma \frac{\partial W}{\partial z} \right) cdt.$$

If we substitute this expression into equation (8), and use equation (7), equation (8) reduces to

$$\alpha \frac{\partial W}{\partial x} + \beta \frac{\partial W}{\partial y} + \gamma \frac{\partial W}{\partial z} = \frac{c_0}{c}. \quad (9)$$

The direction cosines α, β, γ will next be eliminated from equation (9). It is a well-known theorem[1] of analytical geometry that the direction cosines of the normal to the surface $W = $ constant at the point (x, y, z) satisfy the proportion

$$\alpha : \beta : \gamma = \frac{\partial W}{\partial x} \cdot \frac{\partial W}{\partial y} \cdot \frac{\partial W}{\partial z}.$$

Because the sum of the squares of the direction cosines equals unity,

$$\alpha^2 + \beta^2 + \gamma^2 = 1,$$

the constant of proportionality in the multiple proportion above can be determined, and we obtain

$$\alpha = \left[\left(\frac{\partial W}{\partial x} \right)^2 + \left(\frac{\partial W}{\partial y} \right)^2 + \left(\frac{\partial W}{\partial z} \right)^2 \right]^{-\frac{1}{2}} \frac{\partial W}{\partial x} \quad (10)$$

$$\beta = \left[\left(\frac{\partial W}{\partial x} \right)^2 + \left(\frac{\partial W}{\partial y} \right)^2 + \left(\frac{\partial W}{\partial z} \right)^2 \right]^{-\frac{1}{2}} \frac{\partial W}{\partial y}, \text{ etc.}$$

By substituting these values of α, β, γ into equation (9) and squaring both sides,

$$\left(\frac{\partial W}{\partial x} \right)^2 + \left(\frac{\partial W}{\partial y} \right)^2 + \left(\frac{\partial W}{\partial z} \right)^2 = \frac{c_0^2}{c^2(x,y,z)}. \quad (11)$$

If we define n, the *index of refraction*, by

$$n(x,y,z) = \frac{c_0}{c(x,y,z)}, \quad (12)$$

equation (11) becomes

$$\left(\frac{\partial W}{\partial x} \right)^2 + \left(\frac{\partial W}{\partial y} \right)^2 + \left(\frac{\partial W}{\partial z} \right)^2 = n^2(x,y,z). \quad (13)$$

Equation (13), often called the *eikonal equation*, is the fundamental equation of ray acoustics. It is a partial differential equation satisfied by all functions W which can define wave fronts according to equation (7). Initial conditions for equation (13) are usually of the form that W has the value zero for all points (x,y,z) on a particular surface.

Once the solution W of equation (13) has been found, the ray pattern can easily be drawn. The direction cosines of the rays at every point of space can be computed from equation (10); more simply, if equations (10) and (13) are combined,

$$\alpha = \frac{1}{n} \frac{\partial W}{\partial x}; \quad \beta = \frac{1}{n} \frac{\partial W}{\partial y}; \quad \gamma = \frac{1}{n} \frac{\partial W}{\partial z}. \quad (14)$$

Later we shall eliminate the function W from equation (14), and derive a set of *ordinary* differential equations, which together determine the course of each individual ray. First, however, we shall give a simple example illustrating how the ray pattern may be calculated from the partial differential equation (13) for the wave fronts.

Let us consider the special case where the sound velocity c depends only on the vertical depth coordinate y. Thus, the sound velocity is assumed constant everywhere on a particular horizontal plane. We shall examine only the ray pattern in one vertical

plane, which we can take as the xy plane. Then equation (13) reduces to

$$\left(\frac{\partial W}{\partial x} \right)^2 + \left(\frac{\partial W}{\partial y} \right)^2 = n^2(y). \quad (15)$$

To find a simple solution of equation (15), we assume that $W(x,y)$ is the sum of a function of x and a function of y.

$$W(x,y) = W_1(x) + W_2(y).$$

Substituting this expression into equation (15), we obtain

$$\left(\frac{dW_1}{dx} \right)^2 + \left(\frac{dW_2}{dy} \right)^2 = n^2(y).$$

To obtain a family of solutions, we put $dW_1/dx = k$, where k is an arbitrary constant. Then, the differential equation will be satisfied if

$$\frac{dW_2}{dy} = \sqrt{n^2(y) - k^2}; \quad W_2 = \int_0^y \sqrt{n^2(y) - k^2} \, dy.$$

Therefore, in view of the assumed nature of W, the equation (15) will be satisfied by all functions W defined by

$$W(x,y) = kx + \int_0^y \sqrt{n^2(y) - k^2} \, dy \quad (16)$$

where k is any constant. A particular choice of k corresponds to a particular solution $W(x,y)$ and therefore to a particular set of wave fronts (7).

The direction cosines of the rays, corresponding to this choice of k, can be calculated from equations (14) and (16).

$$\alpha = \frac{k}{n}; \quad \beta = \sqrt{1 - \frac{k^2}{n^2}}.$$

We use these expressions to obtain the equations $y = y(x)$ of the sound rays. If we denote by dy/dx the slope of the direction of the ray at the point (x,y), then

$$\frac{dy}{dx} = \frac{\beta}{\alpha} = \sqrt{\frac{n^2(y)}{k^2} - 1}.$$

This equation integrates immediately to

$$x = \int_0^y \frac{dy}{\sqrt{\dfrac{n^2(y)}{k^2} - 1}} + x_0 \quad (17)$$

where x_0 is an arbitrary constant. Regard the k as fixed, and the x_0 as variable; then equation (17) gives an infinite set of curves which satisfy the definition of rays.

If n is a constant, that is, if the sound velocity is independent of depth, the rays (17) are clearly straight lines.

3.2.2 Differential Equations of Rays

It may be argued that the replacement of the wave equation by the ray treatment as represented by the differential equation (13) has little to recommend itself. It appears that one difficult partial differential equation has merely been replaced by another, which might resist attempts at solution as effectively as the first one.

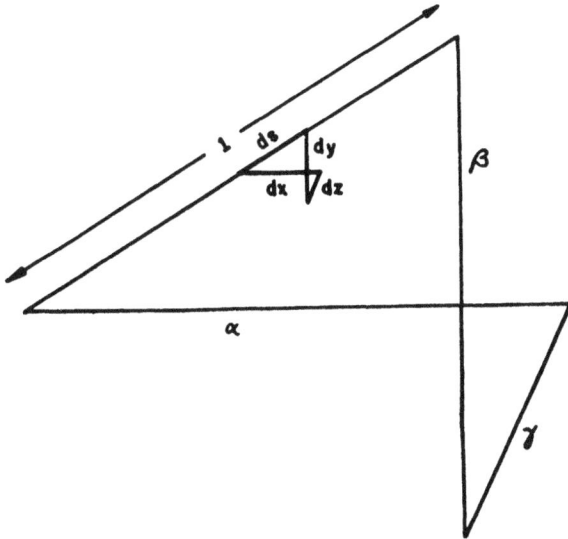

FIGURE 3.   Specification of direction of ray element ds by direction cosines.

Further examination shows, however, that the new equation (13) has two properties which tend to simplify its solution. First, equation (13) contains no time derivatives. This means that it describes the propagation of a disturbance in terms independent of the frequencies which make up this disturbance. Second, it is possible to set up *ordinary* differential equations that describe the path of individual rays; the latter equations will be derived in this section.

We start with the equations (14), from which we proceed to eliminate W. This can easily be done by use of the formulas $\partial^2 W/\partial x \partial y = \partial^2 W/\partial y \partial x$, etc. By differentiating the first equation of (14) with respect to y, the second with respect to x, and by equating the results, a relation between α, β, and n is obtained. Proceeding similarly with the other equations, we obtain the following relationships which must hold at any point of the ray pattern:

$$\frac{\partial(n\alpha)}{\partial y} = \frac{\partial(n\beta)}{\partial x} \; ; \; \frac{\partial(n\alpha)}{\partial z} = \frac{\partial(n\gamma)}{\partial x} \; ; \; \frac{\partial(n\beta)}{\partial z} = \frac{\partial(n\gamma)}{\partial y}.$$

$$(18)$$

These equations can be developed further to yield the changes of α, β, and γ along the path of an individual ray. If the arc length along the ray path from a given starting point is denoted by s, we have

$$\frac{d(n\alpha)}{ds} = \frac{\partial(n\alpha)}{\partial x}\frac{dx}{ds} + \frac{\partial(n\alpha)}{\partial y}\frac{dy}{ds} + \frac{\partial(n\alpha)}{\partial z}\frac{dz}{ds}. \quad (19)$$

We see from Figure 3 that

$$\frac{dx}{ds} = \alpha \; ; \quad \frac{dy}{ds} = \beta \; ; \quad \frac{dz}{ds} = \gamma. \quad (20)$$

Thus equation (19) turns into

$$\frac{d(n\alpha)}{ds} = \alpha\frac{\partial(n\alpha)}{\partial x} + \beta\frac{\partial(n\alpha)}{\partial y} + \gamma\frac{\partial(n\alpha)}{\partial z},$$

which, upon using the relations (18), becomes

$$\frac{d(n\alpha)}{ds} = \alpha\frac{\partial(n\alpha)}{\partial x} + \beta\frac{\partial(n\alpha)}{\partial x} + \gamma\frac{\partial(n\alpha)}{\partial x}$$

$$= (\alpha^2 + \beta^2 + \gamma^2)\frac{\partial n}{\partial x}$$

$$+ \left(\alpha\frac{\partial\alpha}{\partial x} + \beta\frac{\partial\beta}{\partial x} + \gamma\frac{\partial\gamma}{\partial x}\right)n. \quad (21)$$

The first parenthesis equals unity, because it is the sum of squares of direction cosines; while the second parenthesis, which is equal to one-half times the derivative of the first one, vanishes. Thus equation (21) simplifies to

$$\frac{d(n\alpha)}{ds} = \frac{\partial n}{\partial x}.$$

After similar calculations are carried out for $d(n\beta)/ds$ and $d(n\gamma)/ds$, we get the following set of three ordinary differential equations:

$$\frac{d(n\alpha)}{ds} = \frac{\partial n}{\partial x} \; ; \; \frac{d(n\beta)}{ds} = \frac{\partial n}{\partial y} \; ; \; \frac{d(n\gamma)}{ds} = \frac{\partial n}{\partial z}. \quad (22)$$

It is understood that n, the index of refraction, is a given function of x,y,z.

We now deduce an important result for the special case where the sound velocity is a function of the vertical depth coordinate y alone. We shall show that for this case the entire path of an individual ray lies in a plane determined by the vertical line through the projector and the initial direction of the ray.

Let the origin of coordinates be taken at the projector, and let the direction cosines of a ray leaving the projector be $\alpha_0, \beta_0, \gamma_0$, as in Figure 4. Since n depends only on y, equations (22) simplify to

$$\frac{d(n\alpha)}{ds} = 0 \; ; \; \frac{d(n\beta)}{ds} = \frac{dn}{dy} \; ; \; \frac{d(n\gamma)}{ds} = 0. \quad (23)$$

Thus, along any individual ray we have $n\alpha = $ constant, $n\gamma = $ constant, which in turn implies

$$\frac{\gamma}{\alpha} = \text{constant} = \kappa$$

along the ray. Then, the initial direction of the ray is $(\alpha_0, \beta_0, \kappa\alpha_0)$.

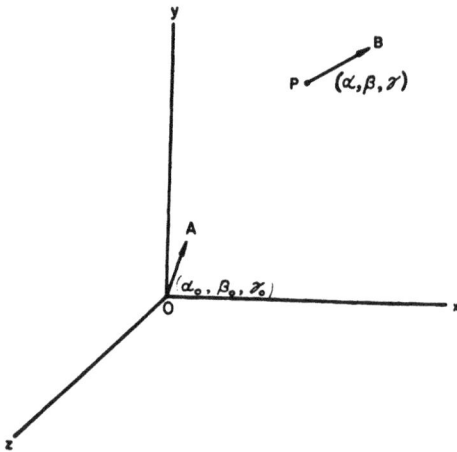

FIGURE 4. Change of ray direction between point $(0, 0, 0)$ and point (x, y, z).

The direction at a general point P along the ray will be characterized by the direction cosines $\alpha, \beta, \kappa\alpha$ because of the equations (23). It can easily be shown by the methods of analytical geometry that the normal to the plane determined by OY (direction cosines $0,1,0$) and OA (direction cosines $\alpha_0, \beta_0, \kappa\alpha_0$) has the direction cosines $\kappa/\sqrt{\kappa^2 + 1}, 0, -1/\sqrt{\kappa^2 + 1}$. The direction of the ray at P is characterized by the direction cosines $\alpha, \beta, \kappa\alpha$; thus the ray direction at P is perpendicular to the normal to the plane AOY; hence the segment PB lies in that plane. Since P was any point on the ray, the entire ray must lie in the plane AOY.

3.3 RAY PATHS FOR VERTICAL VELOCITY GRADIENTS

3.3.1 Derivation of the Equations of Ray Paths

We now solve the equations (22) for the special case where the sound velocity depends only on the vertical depth coordinate y and discuss this solution

in detail. It is intuitively obvious that if we carry through the solution for the xy plane, then the ray pattern in any other plane through the vertical (y) axis will be identical in size and shape.

Since the water depth increases in the downward direction, we shall take the y axis positive downward. We shall denote the angle which a direction in the xy plane makes with the positive x direction by θ, as in Figure 5. To avoid ambiguity, we must specify care-

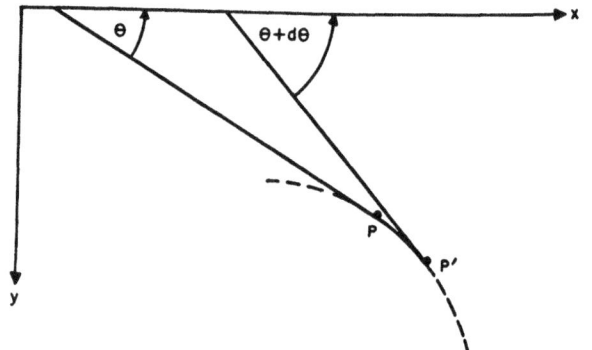

FIGURE 5. Change in ray direction over ray element PP'.

fully the sign of the angle θ. We shall be concerned only with rays moving in the direction of increasing x, in other words, to the right in the figure. If the ray is gaining depth with increasing range, we give the angle θ a positive sign; while if the ray is losing depth with increasing range, we give θ a negative sign. These conventions, illustrated in Figure 6, enable us to use

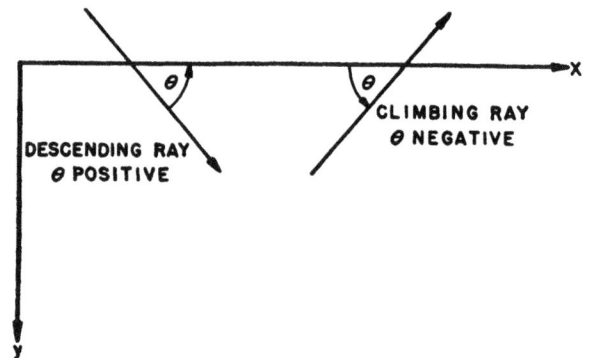

FIGURE 6. Conventions fixing sign of θ.

the following relations both for climbing and descending rays:

$$\alpha = \cos \theta \; ; \; \beta = \sin \theta \; ; \; \gamma = 0. \quad (24)$$

Since the sound velocity is assumed to depend only on y, we have

$$\frac{\partial n}{\partial x} = \frac{\partial n}{\partial z} = 0,$$

and by reason of relations (24) the equations (22) reduce to

$$\frac{d(n \cos \theta)}{ds} = 0 \; ; \quad \frac{d(n \sin \theta)}{ds} = \frac{dn}{dy}. \quad (25)$$

From the first equations it follows that $n \cos \theta$ has a constant value along a particular single ray. That is, if P and P' are two points on the ray, then

$$\frac{c_0}{c} \cos \theta = \frac{c_0}{c'} \cos \theta'.$$

If, in particular, P is located at the depth where $c(y) = c_0$, and if θ_0 is the direction of the ray at this point, this equation becomes

$$\frac{\cos \theta}{\cos \theta_0} = \frac{c}{c_0} \equiv \frac{1}{n}. \quad (26)$$

Equation (26) is identical in form with Snell's law in optics.

The second equation in (24) is used to compute the *curvature* of the ray at any point. The *curvature* of a curve at a point on it is defined as $d\theta/ds$, the angle through which the tangent turns as one travels along the curve for unit distance. Because of our conventions for the sign of the direction angle θ, upward bending is always associated with negative curvature, and downward bending with positive curvature.

From the relations (25), we have

$$\begin{aligned} \frac{dn}{dy} &= n \frac{d(\sin \theta)}{ds} + \sin \theta \frac{dn}{ds} \\ &= n \frac{d(\sin \theta)}{d\theta} \frac{d\theta}{ds} + \sin \theta \frac{dn}{dy} \frac{dy}{ds} \quad (27) \\ &= n \cos \theta \frac{d\theta}{ds} + \sin^2 \theta \frac{dn}{dy} \end{aligned}$$

since $dy/ds = \sin \theta$, from Figure 5. The solution of equation (27) for $d\theta/ds$ yields

$$\frac{d\theta}{ds} = \frac{1}{n} \frac{dn}{dy} \cos \theta = \frac{d(\log n)}{dy} \cos \theta. \quad (28)$$

Since $\log n = \log c_0 - \log c$, equation (28) can be rewritten as

$$\frac{d\theta}{ds} = -\frac{d(\log c)}{dy} \cos \theta. \quad (29)$$

We can use equation (29) to describe, qualitatively, what happens when a ray travels to a layer just above it $(dy < 0)$ of different sound velocity. If the new layer has higher sound velocity, the curvature $d\theta/ds$ has a positive sign, and the ray is bent downward. If the layer just above has lower sound velocity, the curvature $d\theta/ds$ is negative, and the ray is bent upward. We get the opposite result if the ray is traveling

to a layer just below it $(dy > 0)$ of different sound velocity. Thus we can say, in general, that a ray entering a layer of higher sound velocity is bent away from the layer, and a ray entering a layer of lower sound velocity is bent into the layer.

In the open ocean the vertical velocity gradient usually falls into one of two types, depending on the temperature-depth variation. If the temperature does not depend on the depth, the velocity is determined by the pressure, which increases with depth; therefore, in such isothermal water the sound velocity increases gradually with depth, and sound rays should possess slight upward bending. Another common case has the temperature decreasing with depth. Since velocity is much more sensitive to changes in temperature than to changes in pressure, the velocity will also decrease with depth, and the sound rays will bend strongly downward. The water temperature rarely increases with depth; when it does, the sound rays are bent strongly upward.

We shall now examine, quantitatively, the change of curvature along an individual ray, and derive certain relationships between the range and depth reached at time t by a ray leaving the projector at a certain angle. Assume that the projector is situated at the depth where $c = c_0$; thus the ray may be characterized by its initial angle θ_0 at the projector. Because of equation (26), equation (29) becomes

$$\frac{d\theta}{ds} = -\frac{dc}{dy} \frac{\cos \theta_0}{c_0}. \quad (30)$$

The advantage of the representation (30) is that it gives the curvature along a single ray as a function of dc/dy only, since θ_0 is constant for that particular ray.

We consider, in particular, the case where the velocity gradient has the constant value a; that is,

$$c = c_0 + ay, \quad (31)$$

if the origin of coordinates is taken at the projector. At all points on the ray, in view of equation (30),

$$\frac{d\theta}{ds} = -\frac{a \cos \theta_0}{c_0}. \quad (32)$$

We see from equation (32) that the curvature is constant along the ray; this means that the ray must be an arc of a circle. As the radius of curvature is the reciprocal of the curvature $d\theta/ds$, the radius r of this circle must be given by

$$r = \left| \frac{c_0}{a \cos \theta_0} \right|. \quad (33)$$

If a is positive, the curvature (32) is negative, and

the circular arc bends upward; but if a is negative, the circular arc bends downward.

We can determine the center of the circle defining the ray by a simple geometrical construction. Figure 7 shows the path of a ray leaving the projector at the

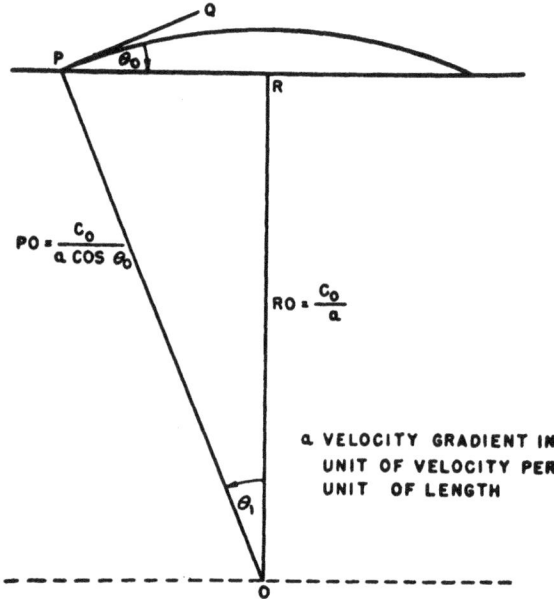

FIGURE 7.   Geometrical construction of ray path.

angle θ_0 into a medium of constant negative velocity gradient. The center of the circle is obtained by following the perpendicular to PQ down through the medium a distance $c_0/a \cos \theta_0$. It is a simple consequence of the geometry of the situation that this center will lie on the horizontal line a distance c_0/a below the projector. For, from the illustration, $RO = (c_0/a)(\cos \theta_1/\cos \theta_0)$, and θ_1 clearly equals θ_0. Similarly, if the constant velocity gradient is positive so that the rays are bent upward, the centers of the defining circles lie on a horizontal line a distance c_0/a *above* the projector. It is easily shown that the dashed horizontal "line of centers" in Figure 7 is at the depth where the velocity c would equal zero if the assumed linear gradient extended indefinitely.

An approximate solution in the general case where c is an arbitrary function of y can be obtained by repeated use of the solution for constant gradient. Even a complicated velocity-depth curve can be closely approximated, as in Figure 8, by dividing the depth interval into a relatively small number of segments in each of which the velocity is assumed to change linearly with depth. Within each layer the ray path is an arc of a circle; and the total ray path is a consecutive series of such arcs.

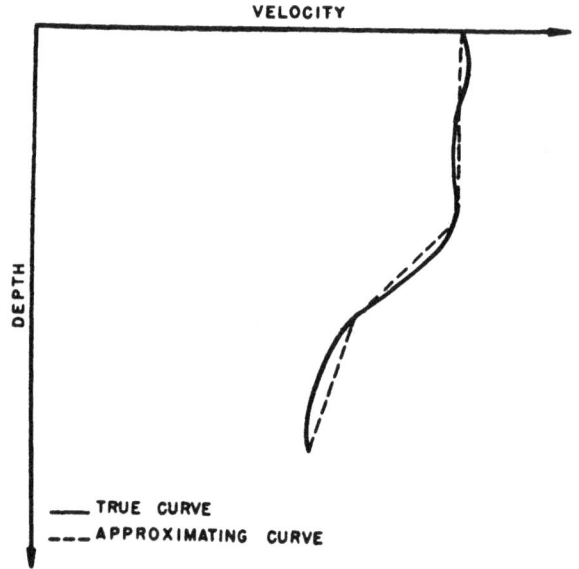

FIGURE 8.   Approximating velocity-depth curve by a succession of linear gradients.

In practice, the path of the ray cannot be conveniently plotted as a sum of circular arcs because the horizontal ranges are much greater than the depths of interest, and therefore their scale must be contracted. Instead, the ray is usually traced by calculating the angles θ_1 and θ_2 at which it enters and leaves a given layer, and the horizontal distance it travels in the layer. This calculation is illustrated in Figure 9,

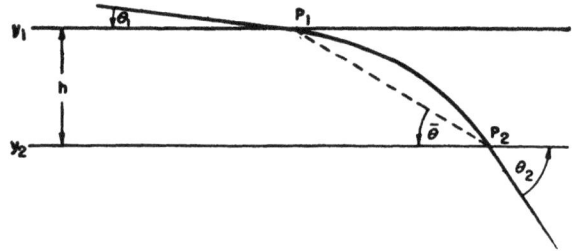

FIGURE 9.   Ray path in layer of linear gradient.

where the top of the layer is at depth y_1, the bottom is at depth y_2, and the thickness of the layer is h.

The ray leaves the projector at an angle θ_0, enters the layer at the angle θ_1, and leaves the layer at the angle θ_2. Then, by equation (26),

$$\theta_1 = \text{arc } \cos\left[\frac{c(y_1) \cos \theta_0}{c_0}\right]$$

$$\theta_2 = \text{arc } \cos\left[\frac{c(y_2) \cos \theta_0}{c_0}\right] \qquad (34)$$

where $c(y_1)$ and $c(y_2)$ are calculated from equation (31).

Consider now the chord P_1P_2 converting the end points of the circular arc. The direction $\bar{\theta}$ of this chord is by simple plane geometry $\frac{1}{2}(\theta_1 + \theta_2)$; and its length is therefore given by

$$P_1P_2 = \frac{h}{\sin \frac{1}{2}(\theta_1 + \theta_2)}. \qquad (35)$$

The increase in horizontal range due to the passage of the ray through the layer is $P_1P_2 \cos \bar{\theta}$, or

$$\longrightarrow \text{Range in layer} = h \cot \frac{1}{2}(\theta_1 + \theta_2). \qquad (36)$$

This result may be applied to the following problem. Suppose we have a sum of layers of the sort shown in Figure 10; and we wish to find the horizontal

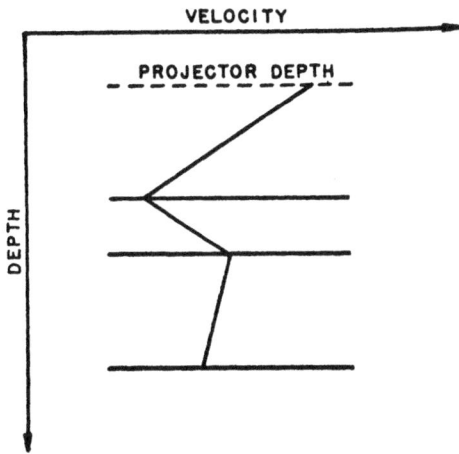

FIGURE 10. Succession of linear gradients.

range attained by the time the ray reaches the depth H below the projector. We let the bottom layer extend just to the depth H; suppose this is the third layer below the projector. We know θ_0 and we calculate $\theta_1, \theta_2, \theta_3$ by the relations (34). Then the horizontal range to the depth H will be the sum of terms of the form (36):

Horizontal range to $H = h_1 \cot \frac{1}{2}(\theta_0 + \theta_1) +$
$$h_2 \cot \frac{1}{2}(\theta_1 + \theta_2) + h_3 \cot \frac{1}{2}(\theta_2 + \theta_3). \quad (37)$$

The inverse problem is a little more complicated. Suppose we wish to find the depth reached by a ray of initial direction θ_0 by the time it has traveled a horizontal distance R in a stratified medium that consists of layers of thickness h_1, h_2, h_3, etc. We calculate the range R_1 in the first layer, R_2 in the second layer, and so on, until the sum of these partial ranges is greater than R:

$$R_1 + R_2 + R_3 < R$$
$$R_1 + R_2 + R_3 + R_4 > R.$$

Then the depth the ray reaches at range R will be

greater than $h_1 + h_2 + h_3$ and less than $h_1 + h_2 + h_3 + h_4$. Its value may be obtained with sufficient accuracy by interpolation.

The ray-tracing methods described in this section are too cumbersome to use in practice. A number of devices have been developed to facilitate the plotting of rays bent by known velocity gradients; these devices will be discussed in Section 3.5.1.

3.3.2 Application to Depth Correction

The ray-tracing methods described in Section 3.3.1 have had a valuable application in correcting the depths determined by the use of tilting beam sonar gear. These instruments are used on surface vessels to determine the true depths of submerged submarines. They employ a transducer with good vertical directivity and tiltable in the vertical plane, which sends out echo-ranging pings at various angles of depression. When velocity gradients are absent, the sound rays are straight lines, and the true depth of the target is just the slant range times the sine of the angle of depression at the orientation for which the target returns the loudest echo. The depth finder computes this latter product automatically. When velocity gradients are present, however, this simple method often leads to serious underestimation of the target depth. In this section, we shall describe a method for estimating the error produced.

For simplicity, we assume that the projector is at the surface, as in Figure 11. Let the apparent target

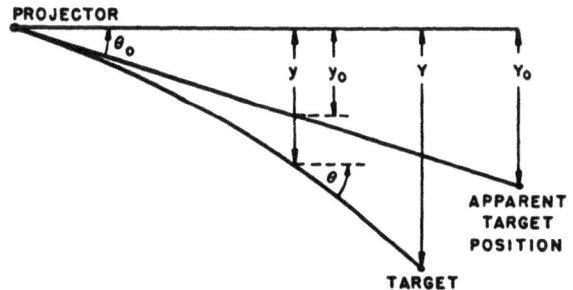

FIGURE 11. Error in target position due to refraction.

angle be θ_0, and the apparent target depth indicated by the depth finder be Y_0; the true depth of the target is designated by Y. Our aim is to derive an expression for $Y - Y_0$ in terms of the way the sound velocity c varies with depth.

Let y represent the actual depth attained by the sound ray at time t, and y_0 represent the apparent depth reached by the ray. Then

$$y_0 = c_0 \sin \theta_0 t \qquad (38)$$

where c_0 is the velocity of sound at the projector. Since the actual ray path is curved, all we can say is that y is some function of the surface velocity, the apparent angle θ_0, and the velocity-depth pattern. We can, however, give an exact expression for the increase in y during the time interval dt. If c is the sound velocity at the depth y, and θ is the inclination of the ray at the depth y, we have

$$dy = c \sin \theta dt. \qquad (39)$$

We now take differentials of both sides of equation (38), obtaining

$$dy_0 = c_0 \sin \theta_0 dt. \qquad (40)$$

By dividing equation (40) by equation (39) to eliminate the time,

$$\frac{dy_0}{dy} = \frac{c_0 \sin \theta_0}{c \sin \theta}. \qquad (41)$$

We eliminate θ from equation (41) by using Snell's law (26):

$$\sin \theta = \sqrt{1 - \cos^2 \theta} = \sqrt{1 - [(c/c_0) \cos \theta_0]^2}$$

so that equation (41) becomes

$$\frac{dy_0}{dy} = \frac{\sin \theta_0}{\frac{c}{c_0}\sqrt{1 - \left(\frac{c}{c_0} \cos \theta_0\right)^2}}. \qquad (42)$$

The quantity c/c_0 represents the variation of velocity with depth. If ϵ is defined by the relationship

$$\frac{c}{c_0} = 1 + \epsilon,$$

then ϵ represents the relative change in velocity as a function of depth. Rewriting equation (42) in terms of ϵ, we obtain

$$\begin{aligned}
\frac{dy_0}{dy} &= \frac{\sin \theta_0}{(1 + \epsilon)[1 - (1 + \epsilon)^2 \cos^2 \theta_0]^{\frac{1}{2}}} \\
&= \left\{(1 + \epsilon)^2 \csc^2 \theta_0 [1 - (1 + \epsilon)^2 \cos^2 \theta_0]\right\}^{-\frac{1}{2}} \quad (43) \\
&= [1 + 2(1 - \cot^2 \theta_0)\, \epsilon + (1 - 5 \cot^2 \theta_0)\, \epsilon^2 \\
&\qquad - 4 \cot^2 \theta_0\, \epsilon^3 - \cot^2 \theta_0\, \epsilon^4]^{-\frac{1}{2}}
\end{aligned}$$

upon multiplying out and collecting terms.

Since percentage changes of sound velocity are always small in the sea, the quantity ϵ is a very small fraction, almost always less than 0.02. Consequently, the terms with ϵ^2 or higher powers of ϵ in equation (43) may safely be neglected, giving approximately

$$\frac{dy_0}{dy} = [1 + 2\epsilon(1 - \cot^2 \theta_0)]^{-\frac{1}{2}}. \qquad (44)$$

If we define w by

$$w = 2(1 - \cot^2 \theta_0),$$

equation (44) may conveniently be rewritten as

$$dy_0 = \frac{dy}{(1 + w\epsilon)^{\frac{1}{2}}}. \qquad (45)$$

It may be noted that although ϵ is always much less than one, $w\epsilon$ is not necessarily so. We now integrate both sides of equation (45) between 0 and the true depth Y, obtaining

$$Y_0 = \int_0^Y \frac{dy}{(1 + w\epsilon)^{\frac{1}{2}}}. \qquad (46)$$

The expression (46) provides a functional relationship between the true depth Y, the apparent depth Y_0, and the velocity-depth variation $\epsilon(y)$. In any practical situation it is possible to determine Y_0 and θ_0, and $\epsilon(y)$ can be deduced from the temperature-depth variation indicated by the bathythermograph slide. Thus, all quantities in equation (46) are known except the true depth Y, which occurs only as the upper limit of integration. The value of Y may be estimated by using trial values for the upper limit of integration and by seeing which trial value yields a value for the definite integral closest to the known left-hand side Y_0. If the velocity-depth variation is not simple, the integrals must be evaluated by numerical integrations; but if $\epsilon(y)$ is a linear function of y, or a succession of linear functions of y, the integrals can easily be evaluated exactly.

Tables have been developed by the use of such methods for the depth errors expected in the presence of various types of velocity gradients. In preparing these tables it was assumed that the sound velocity versus depth curve could be approximated by judiciously chosen straight-line segments without introducing too much error in the calculated depth error.

Though equation (46) is the relation used in the construction of depth correction tables, it is interesting to carry the approximation two steps further. If we expand the integrand in powers of $w\epsilon$ and neglect all but the first two terms, we get

$$Y_0 = \int_0^Y (1 - \tfrac{1}{2}w\epsilon + \cdots)dy$$

which becomes

$$Y - Y_0 = \tfrac{1}{2}w \int_0^Y \epsilon\, dy + \text{terms in } (w\epsilon)^2.$$

To the same order of approximation, Y_0 may be substituted in place of Y as the upper limit of integration, which gives

$$Y - Y_0 = \tfrac{1}{2}w \int_0^{Y_0} \epsilon\, dy + \text{terms in } (w\epsilon)^2. \qquad (47)$$

When $w\epsilon$ is small the terms in $(w\epsilon)^2$ may be neglected; under these same conditions $Y - Y_0$ is small compared with Y_0. Equation (47) thus provides a useful approximation when the depth error is relatively small. By translating from ϵ back to c this equation becomes

$$Y - Y_0 \approx \frac{w}{2c_0} \int_0^{Y_0} (c - c_0)dy. \qquad (48)$$

The expression (48) has a simple interpretation in terms of the velocity-depth diagram. The integrand $(c_0 - c)dy$ is just the black area in Figure 12; thus

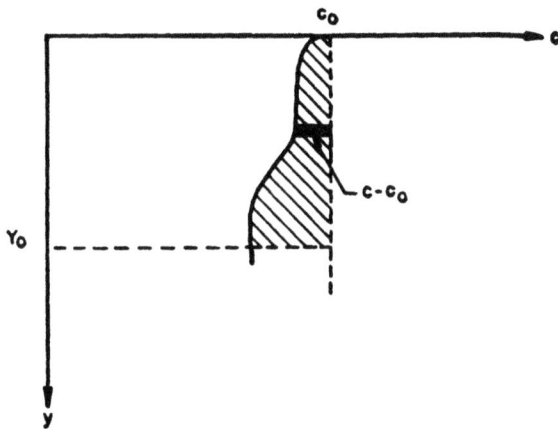

FIGURE 12. Depth correction as area under bathythermograph trace.

the integral from 0 to Y_0 is the shaded area between the velocity-depth curve, the vertical line $c = c_0$, and the horizontal line $y = Y_0$. Qualitatively, we may conclude that the depth correction will be large for steep gradients and larger if these steep gradients are located at shallow depth.

3.4 CALCULATION OF SOUND INTENSITY FROM RAY PATTERN

The foregoing sections were devoted exclusively to tracing the paths of individual rays and stated nothing about sound intensity in the ray pattern except for the special case of spherical waves. For that situation an assumption that the energy flows out radially along the sound rays led to the same inverse square law of intensity decay which was derived rigorously under "Spherical Waves" in Section 2.4.2. It is a plausible generalization to assume that energy always travels out along the rays even when the sound velocity is not constant and the rays are curves.

The assumption for the case of constant sound velocity and its generalization for the case of variable velocity are illustrated in Figure 13. In the left-hand

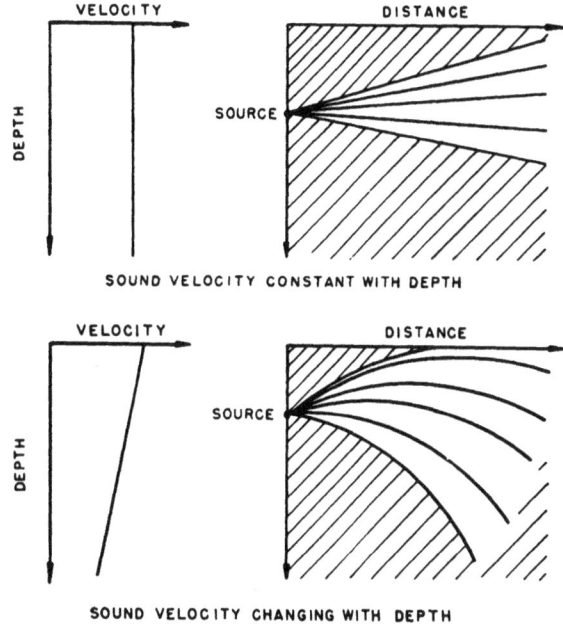

FIGURE 13. Effect of vertical velocity changes on ray paths.

drawing, the rays are straight lines; and the energy radiated by the source into a small solid angle is confined inside the indicated cone. Because of this assumption, we get the exact inverse square law of intensity loss. In the right-hand drawing, the rays are curves; and the energy radiated by the source into the same small solid angle is confined inside the horn-shaped surface displayed. In this general situation the energy flow through normal unit area depends not only on the distance r from the source, but also on the total cross-sectional area of the horn which, in turn, depends on the way the rays are bent. Thus it is clear that the inverse square law will not, in general, be predicted even by this simplified ray treatment.

3.4.1 General Formulas for Change of Intensity along a Ray

The prediction of shadow zones as described in the preceding section is only one part of the description of the expected intensity distribution. There remains the problem of calculating the intensity in regions traversed by the rays. We already know that this intensity loss will not exactly obey the inverse square law except in very special cases.

We restrict ourselves to the case where the sound velocity is a function only of the depth coordinate y. The ray pattern in the xy plane can be computed according to the methods of Section 3.3. We can get the entire ray pattern in space by rotating the ray pattern of the xy plane about the vertical (y) axis; because the velocity depends only on y, the ray patterns in every plane through the y axis will be identical in size and shape.

We assume that the projector is a point source located on the y axis at the depth y_0, which radiates energy at the rate of F energy units per unit solid angle per second. Then, energy will be projected into

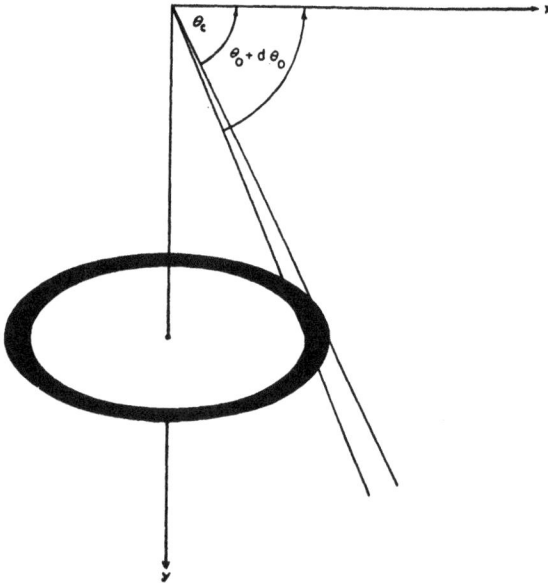

FIGURE 14.   Specification of solid angle.

a very small solid angle $d\Omega$ at the rate of $Fd\Omega$ energy units per second. The rays bounding this solid angle will curve in some fashion depending on $n(y)$ and the angle of emission; suppose at the point P somewhere out along the ray bundle, the cross-sectional area of the bundle is dS. Then the intensity at P will be the energy crossing this area dS per second, which equals $Fd\Omega$, divided by the cross-sectional area dS.

$$\text{Intensity at } P = F\frac{d\Omega}{dS}. \qquad (49)$$

Because of the cylindrical symmetry of the rays with respect to the y axis, we shall find it convenient to define our small solid angle as indicated in Figure 14. It is the solid angle swept out in space by rotating the portion of the xy plane between the angles θ_0 and $\theta_0 + d\theta$ about the y axis. On a unit sphere the

solid angle so defined intercepts a spherical zone of radius $\cos\theta_0$ and width $d\theta_0$ which is therefore of area $2\pi\cos\theta_0 d\theta_0$. Thus our solid angle $d\Omega$ is given by

$$d\Omega = 2\pi\cos\theta_0 d\theta_0. \qquad (50)$$

We wish to calculate the intensity for the ray of initial direction θ_0 in the xy plane, at the horizontal range R. By equations (49) and (50), this intensity is given by

$$I(R,\theta_0) = \frac{F\cdot(2\pi\cos\theta_0 d\theta_0)}{dS} \qquad (51)$$

where dS, the cross-sectional area, is clearly the area swept out when the segment PP' in Figure 15 is rotated about the y axis. We proceed to calculate dS.

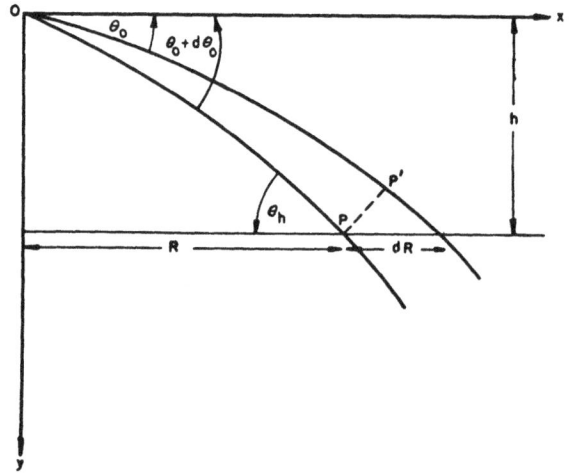

FIGURE 15.   Diagram used in deriving intensity formulas.

The horizontal range R is clearly a function of the depth h and θ_0:

$$R = R(h,\theta_0). \qquad (52)$$

Therefore, the horizontal separation dR at a fixed depth is given by

$$dR = -\frac{\partial R}{\partial\theta_0}d\theta_0. \qquad (53)$$

The minus sign is inserted because R at fixed depth decreases as θ_0 increases.

Let θ_h be the direction of the ray at the point (R,h) as in Figure 15. Then PP', the shortest (normal) distance between these two rays near P is $dR\sin\theta_h = -(\partial R/\partial\theta_0)d\theta_0\sin\theta_h$. By rotating PP' about the y axis we get dS, the desired cross section of our bundle. This area is clearly that of a spherical zone with radius R and thickness $-(\partial R/\partial\theta_0)\sin\theta_h d\theta_0$; its area dS is therefore

$$dS = -2\pi R\frac{\partial R}{\partial\theta_0}\sin\theta_h d\theta_0. \qquad (54)$$

Substituting this expression into equation (51), we obtain for the intensity I at the range R the expression

$$\frac{I(R,\theta_0)}{F} = - \frac{\cos\theta_0}{R\frac{\partial R}{\partial\theta_0}\sin\theta_h}. \qquad (55)$$

It is now necessary to obtain expressions for R and for the partial derivative of R with respect to θ_0. We begin by calculating the range R. We have

$$R = \int_{y_0}^{h}\frac{dx}{dy}dy = \int_{y_0}^{h}\cot\theta\,dy, \qquad (56)$$

which, upon substituting $\cos\theta = (c/c_0)\cos\theta_0$ from equation (26), becomes

$$R = \cos\theta_0\int_{y_0}^{h}\frac{c\,dy}{\sqrt{c_0^2 - c^2\cos^2\theta_0}}. \qquad (57)$$

We differentiate this expression for R as a product of functions of θ_0, assuming that the usual formulas for differentiating under the integral sign are valid. When the two resulting integrals are put over one denominator, the whole expression for the derivative simplifies to

$$\frac{\partial R}{\partial\theta_0} = -c_0^2\sin\theta_0\int_{y_0}^{h}\frac{c\,dy}{(c_0^2 - c^2\cos^2\theta_0)^{\frac{3}{2}}}. \qquad (58)$$

Substituting the expressions for R and for $\partial R/\partial\theta_0$, equations (57) and (58), into equation (55), we find for the intensity I the expression

$$\frac{I}{F} = \frac{1}{\sin\theta_0\sin\theta_h\int_{y_0}^{h}\frac{dy}{\sin\theta}\int_{y_0}^{h}\frac{dy}{\sin^3\theta}}, \qquad (59)$$

with

$$\sin\theta \equiv \sqrt{1 - \left(\frac{c}{c_0}\right)^2\cos^2\theta_0} \qquad (60)$$

and

$$\sin\theta_h \equiv \sqrt{1 - \left(\frac{c_h}{c_0}\right)^2\cos^2\theta_0}. \qquad (60a)$$

For application, this formula suffers from two defects. First, it is not sufficiently simple; second, it is not sufficiently general. The second point will be taken up later, where it will be seen that equation (60) does not cover the important class of conditions where the sound ray becomes horizontal anywhere en route. As for the first point, we shall simplify equation (59) for application to these cases where it is valid.

Under ordinary circumstances, c does not vary between the sea surface and operational depths by more than 5 per cent of the surface velocity. As a result, those rays which leave the projector at a moderate angle will not become so steep that the sine of the angle cannot be replaced in good approximation by the angle itself. Thus we may replace the expression $c_0^2 - c^2\cos^2\theta_0$, which appears three times in equation (59), by the approximation

$$c_0^2 - c^2\cos^2\theta_0 \approx c_0^2\left[1 - \left(\frac{c_0 + \Delta c}{c_0}\right)^2(1 - \theta_0^2)\right]$$
$$\approx c_0^2(\theta_0^2 - 2\epsilon), \qquad (61)$$

in which ϵ, as in Section 3.3.2, stands for the expression $(c - c_0)/c_0$. As a result, we can replace equation (60) by the approximate relation

$$\frac{I}{F} \approx \frac{1}{\theta_0\sqrt{\theta_0^2 - 2\epsilon}\int_{y_0}^{h}\frac{dy}{\sqrt{\theta_0^2 - 2\epsilon}}\int_{y_0}^{h}\frac{dy}{(\theta_0^2 - 2\epsilon)^{\frac{3}{2}}}}; \qquad (62)$$

while the range is given approximately by the expression

$$R = \int_{y_0}^{h}\frac{dy}{\sqrt{\theta_0^2 - 2\epsilon}}. \qquad (63)$$

In most transmission work, the sound field intensity is reported either as transmission *loss* or as transmission *anomaly*. The transmission loss H is defined as the ratio of the source strength F and the sound field intensity in decibels,

$$H \equiv 10\log\frac{F}{I}. \qquad (64)$$

The transmission anomaly A is defined as the ratio of the intensity predicted by the inverse square law and the sound field intensity I, also in decibels,

$$A = 10\log\frac{F/R^2}{I} = 10\log\frac{F}{R^2 I}. \qquad (65)$$

On the basis of this definition and equation (62) the transmission anomaly will be given by the approximate relationship

$$A \approx -10\log\int_{y_0}^{h}\frac{dy}{\theta(y)} + 10\log\int_{y_0}^{h}\frac{dy}{\theta^3(y)}$$
$$+ 10\log\theta_0 + 10\log\theta_h, \qquad (66)$$

where $\theta(y) = \sqrt{\theta_0^2 - 2\epsilon}$.

Where it applies, this formula is simple enough to lead readily to results of practical significance, as under "Layer Effect" in Section 3.4.2.

From its mode of derivation the expression (59) for

the intensity at a point P on a ray is not valid if at any place between the projector and P the ray has become horizontal. For, at a point where the ray is horizontal, $\theta = 0$, which implies that $c_0 - c \cos \theta_0 = 0$, by equation (26). This means that the integral in equation (57) becomes infinite at points where the ray is horizontal and cannot be differentiated under the integral sign. We therefore conclude that the expression (59) is valid only for rays that are always climbing or always dropping, but cannot be used, for example, to examine the intensity near places where the ray diagram predicts shadow zones.

We now derive an expression for $I(\theta_0, R)$ which will be valid even at points on a ray beyond where the ray has become horizontal. This will be done by deriving an expression similar to equation (58) in which the variable of integration is θ instead of y.

In all cases, we have, because of equation (56),

$$R = \int_{y_0}^{h} \cot \theta \, dy$$
$$= \int_{y_0}^{h} \cot \theta \frac{dy}{dc} \frac{dc}{d\theta} d\theta.$$

Since $\dfrac{dc}{d\theta} = \dfrac{d}{d\theta}\left[c_0 \dfrac{\cos \theta}{\cos \theta_0} \right] = -\dfrac{c_0}{\cos \theta_0} \sin \theta,$

it follows that $R = -\dfrac{c_0}{\cos \theta_0} \int_{\theta_0}^{\theta_h} \cos \theta \dfrac{dy}{dc} d\theta.$ (67)

This expression for R has the advantage over equation (56) in that the integrand does not become infinite for $\theta = 0$. This expression can, therefore, be differentiated with respect to θ_0 even when the ray passes through points at which it is horizontal.

The variable θ_0 occurs explicitly in the factor in front of the integral and as the lower limit of integration and implicitly in the terms θ_h and dy/dc. Taking this into consideration, it follows that

$$\frac{\partial R}{\partial \theta_0} = -\frac{c_0 \sin \theta_0}{\cos^2 \theta_0} \int_{\theta_0}^{\theta_h} \frac{dy}{dc} \cos \theta d\theta$$
$$- \frac{c_0^2 \sin \theta_0}{\cos^3 \theta_0} \int_{\theta_0}^{\theta_h} \frac{d^2y}{dc^2} \cos^2 \theta d\theta$$
$$- \frac{c_0 \sin \theta_0}{\cos^2 \theta_0}\left[\frac{\cos^2 \theta_h}{\sin \theta_h}\left(\frac{dy}{dc}\right)_h - \frac{\cos^2 \theta_0}{\sin \theta_0}\left(\frac{dy}{dc}\right)_0 \right]. \quad (68)$$

Though the expression (68) is much lengthier than the expression (58), it has the advantage of being valid at points on the ray where $\theta = 0$. The resulting intensity, calculated by using equation (55), will also be valid at all points on the ray. The quantities R and $\partial R/\partial \theta_0$ must be substituted from equations (67) and

(68). These expressions can be simplified by means of the assumption that all angles are small. With this assumption all cosines of angles can be replaced by one; the sines may be replaced by the angles themselves; and among a number of terms those multiplied by higher powers of the angles may be disregarded. The simplified expressions for R and for $\partial R/\partial \theta_0$ then take the form

$$R \approx -c_0 \int_{\theta_0}^{\theta_h} \frac{dy}{dc} d\theta \quad (69)$$

and

$$\frac{\partial R}{\partial \theta_0} \approx c_0\left[\left(\frac{dy}{dc}\right)_0 - \frac{\theta_0}{\theta_h}\left(\frac{dy}{dc}\right)_h \right]. \quad (70)$$

From equations (55) and (65) we have, as a general expression for the transmission anomaly A,

$$A = 10 \log \left(\frac{\frac{\partial R}{\partial \theta_0} \sin \theta_h}{R \cos \theta_0} \right). \quad (71)$$

If we assume that θ_0 and θ_h are both so small that $\cos \theta_0$ can be replaced by one, and $\sin \theta_h$ by θ_h, formula (71) becomes

$$A \approx 10 \log \left(\frac{\frac{\partial R}{\partial \theta_0} \theta_h}{R} \right). \quad (71a)$$

By putting in the approximate values of R and $\partial R/\partial \theta_0$ from equations (69) and (70), an explicit expression can be obtained for the transmission anomaly.

In the application of equations (67) to (70) one precaution must be taken. While the integrands of the integrals that occur remain finite when the angle of inclination becomes zero, these expressions approach infinity as the gradient approaches zero. They are, therefore, not suitable for the treatment of propagation through isovelocity layers.

Another method of computing the transmission anomaly that may be used whether or not a ray has become horizontal and is in a more convenient form for numerical computation is given under "Combination of Linear Gradients" in Section 3.4.2.

3.4.2 Applications

Section 3.4.1 was devoted to deriving formulas for the intensity out along a ray as a function of the horizontal range and the velocity-depth variation. These formulas involve line integrals and are too complicated to use for practical intensity computations.

The formulas are simplified in this section by using various simplifying assumptions concerning the velocity-depth variation.

DIRECT BEAM IN LINEAR GRADIENT

Let us assume that the sound velocity increases or decreases linearly with depth, with the gradient a. Then,

$$a = \frac{dc}{dy}. \tag{72}$$

Since the velocity is never constant with increasing depth in this case, the approximate equations (69) and (70) are applicable. Using these equations, we find that the range R is given by the expression

$$R = -\frac{c_0}{a}(\theta_h - \theta_0) \tag{73}$$

and that the derivative of the range with respect to θ_0 is given by

$$\frac{\partial R}{\partial \theta_0} = \frac{1}{a}\left(1 - \frac{\theta_0}{\theta_h}\right). \tag{74}$$

Substituting these expressions into equation (71a), we obtain for the transmission anomaly the expression

$$A = 10 \log 1 = 0. \tag{75}$$

The transmission anomaly vanishes, at least in this approximation. If we had used the rigorous expressions (67) and (68) for R and $\partial R/\partial \theta_0$, and the exact form (71) for the transmission anomaly, the following formula would have been obtained, which is rigorously correct,

$$A = 20 \log \cos \theta_0. \tag{76}$$

It may seem surprising that the transmission anomaly (76) does not depend on the sharpness of the velocity gradient. This seeming discrepancy results from the use of the horizontal range R in the definition (65) of the transmission anomaly instead of the slant range r.

The results (75) and (76) for this case of uniform downward refraction apply only to the sound field at points actually reached by the direct rays. If the water is very deep, there are portions of the ocean where no sound ray penetrates, as illustrated in Figure 24; in such regions the ray theory predicts a vanishing sound intensity and thus an infinite transmission anomaly.

REFLECTED BEAM IN LINEAR GRADIENT

We now calculate the intensity along a ray which has suffered one or more bottom reflections, for the same linear velocity gradient assumed under "Direct Beam in Linear Gradient" in Section 3.4.2. First we shall assume one bottom reflection. This situation is pictured in Figure 16A where the ray hits the bottom

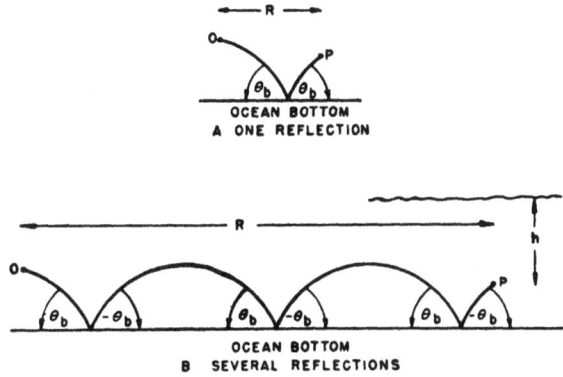

FIGURE 16. Reflection of sound ray from sea bottom. A. One reflection. B. Several reflections.

at an angle θ_b ($\theta_b > 0$) and is reflected at the angle $-\theta_b$. The rays will be refracted downward, as indicated; and the incident and reflected rays will be circular arcs with equal radii.

We can compute the horizontal range R by equation (69), which is valid for all cases where the ray path is made up of several arcs, provided care is taken in breaking up the interval of integration correctly.

$$\begin{aligned} R &= -c_0 \int_{\theta_0}^{\theta_b} \frac{dy}{dc}d\theta - c_0 \int_{\theta_b}^{\theta_h} \frac{dy}{dc}d\theta \\ &= -\frac{c_0}{a}(\theta_b - \theta_0) - \frac{c_0}{a}(\theta_h + \theta_b) \\ &= -\frac{c_0}{a}(2\theta_b + \theta_h - \theta_0). \end{aligned}$$

To use equation (71a), we must also calculate $\partial R/\partial \theta_0$. Using equation (74), we obtain

$$\begin{aligned} \frac{\partial R}{\partial \theta_0} &= \frac{c_0}{a}\left(1 - \frac{\theta_0}{\theta_b}\right) + \frac{c_0}{a}\left(1 - \frac{-\theta_b}{\theta_h}\right) \\ &= \frac{c_0}{a}\left(2 - \frac{\theta_0}{\theta_b} + \frac{\theta_b}{\theta_h}\right). \end{aligned}$$

Substitution of these expressions for R and $\partial R/\partial \theta_0$ into equation (71a) gives

$$A = 10 \log \left| \frac{\theta_h\left(2 - \dfrac{\theta_0}{\theta_b} + \dfrac{\theta_b}{\theta_h}\right)}{2\theta_b + \theta_h - \theta_0} \right|. \tag{77}$$

For the case of a ray suffering $n + 1$ reflections, pictured in Figure 16B, the procedure is similar except that n complete journeys from the bottom back to the bottom must be added to the interval of integration. The calculated transmission anomaly for a ray which leaves the source at the angle θ_0, suffers $n + 1$ bottom reflections, and strikes a receiving hydrophone at the inclination θ_h, turns out to be

$$A = 10 \log \left| \frac{\theta_h \left[2(n+1) - \frac{\theta_0}{\theta_b} + \frac{\theta_b}{\theta_h} \right]}{2(n+1)\theta_b + \theta_h - \theta_0} \right|. \quad (78)$$

LAYER EFFECT

When sound originates in an isovelocity layer or in a layer with a weak velocity gradient and then passes into a layer with a sharp negative gradient, the sharp refraction results in an extra spreading of the sound rays and a consequent drop in intensity. This phenomenon is called *layer effect*, and is of operational importance. We shall consider only rays which leave the projector in a downward direction, so that the formula (66) for the transmission anomaly will apply.

Two separate cases will be treated. First, we shall consider the velocity-depth pattern shown in Figure 17: an isovelocity layer, followed by a layer of

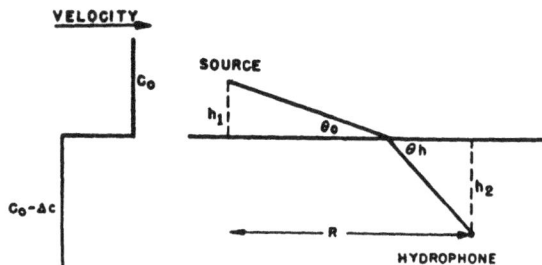

FIGURE 17. Bending of ray by temperature discontinuity.

negligible thickness with a very sharp gradient and a total drop of sound velocity of amount Δc, followed in turn by a second isovelocity layer with the velocity $c_0 - \Delta c$. If the ray direction in the first isovelocity layer is θ_0, and in the second isovelocity layer θ_h, we have by Snell's law (26)

$$\frac{\cos \theta_h}{\cos \theta_0} = \frac{c_0 - \Delta c}{c_0} = 1 - \frac{\Delta c}{c_0}.$$

If the angle θ is small, we may replace its cosine by its approximate equivalent $1 - \theta^2/2$. Using this approximation, and dropping the negligible term $(\Delta c/c_0)\theta_0^2$, the preceding equation becomes

$$\theta_h = \sqrt{\theta_0^2 + 2\frac{\Delta c}{c_0}}. \quad (79)$$

If h_1 is the height of the sound source above the abrupt velocity change, and h_2 is the depth of the hydrophone below the velocity change, we easily find from formula (66) that

$$A = -10 \log \left(\frac{h_1}{\theta_0} + \frac{h_2}{\theta_h} \right) + 10 \log \left(\frac{h_1}{\theta_0^3} + \frac{h_2}{\theta_h^3} \right)$$
$$+ 10 \log \theta_0 + 10 \log \theta_h$$
$$= 10 \log \left[\frac{\theta_h h_1}{R \theta_0^2} \left(1 + \frac{h_2}{h_1} \frac{\theta_0^3}{\theta_h^3} \right) \right], \quad (80)$$

since

$$R \approx \frac{h_1}{\theta_0} + \frac{h_2}{\theta_h}.$$

Next we shall consider the velocity-depth pattern shown in Figure 18: an isovelocity layer extending to

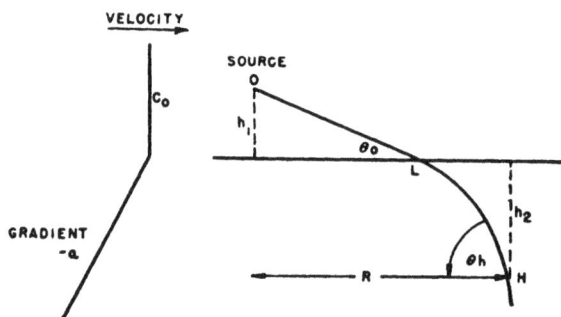

FIGURE 18. Bending of ray by deep thermocline.

a depth h_1 below the sound source, followed by a layer of indefinite extent with the constant velocity gradient $-a$. At a depth y' below the top of the gradient layer the sound velocity will be $c_0 + ay'$. We therefore obtain the following expression analogous to expression (79) for the ray direction $\theta(y')$ at the depth y'.

$$\theta(y') = \sqrt{\theta_0^2 - 2(a/c_0)y'}. \quad (81)$$

We shall use equation (66) to calculate the intensity of the sound received by a hydrophone at a depth h_2 below the top of the gradient layer. Since the ray direction in the isovelocity layer is constant at θ_0, the separate integrals in equation (66) have the values

$$\int_0^h \frac{dy}{\theta} = \frac{h_1}{\theta_0} + \int_0^{h_2} \frac{dy'}{\sqrt{\theta_0^2 - 2(a/c_0)y'}}.$$

The last term may be integrated directly, and with use of equation (81)

$$\int_0^h \frac{dy}{\theta} = \frac{h_1}{\theta_0} + \frac{(\theta_0 - \theta_h)}{a/c_0}$$
$$= \frac{h_1}{\theta_0} + \frac{h_2}{\frac{1}{2}(\theta_0 + \theta_h)},$$

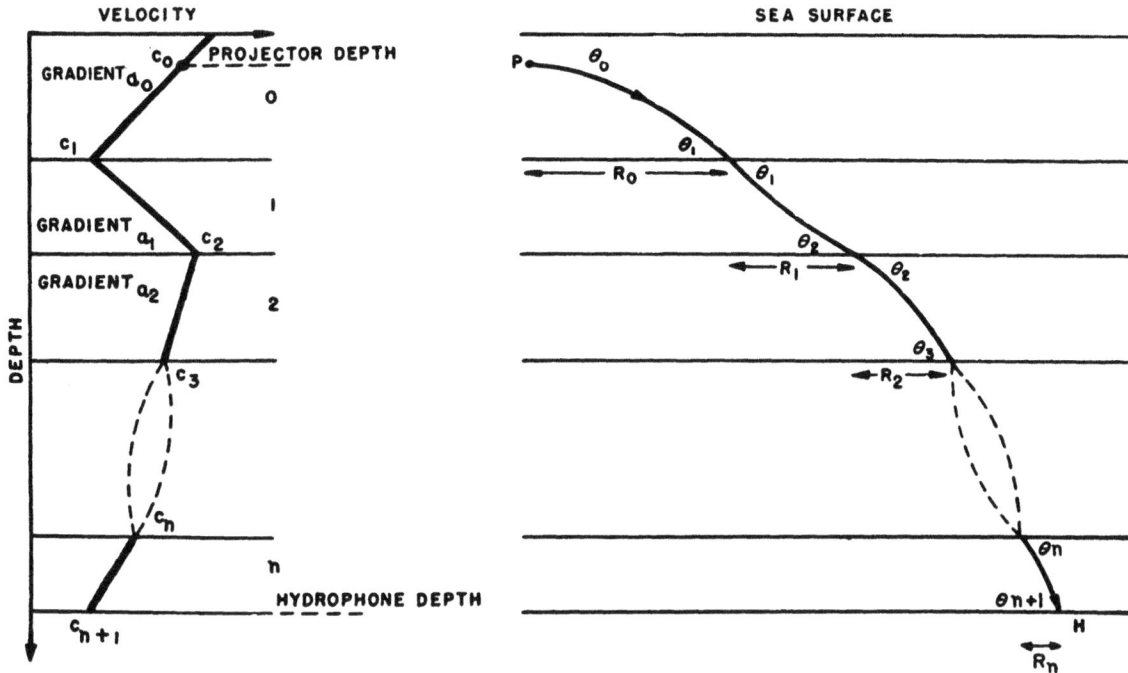

FIGURE 19. Ray path in succession of linear gradients.

since $\theta_0^2 - \theta_h^2$ has the value $2(a/c_0)h_2$ by equation (81). Similarly, we have

$$\int_0^h \frac{dy}{\theta^3} = \frac{h_1}{\theta_0^3} + \int_0^{h_2} \frac{dy'}{(\theta_0^2 - 2(a/c_0)y')^{\frac{3}{2}}} = \frac{h_1}{\theta_0^3} + \frac{(\theta_0 - \theta_h)}{(a/c_0)\theta_0\theta_h}$$

$$= \frac{h_1}{\theta_0^3} + \frac{1}{\theta_0\theta_h} \frac{h_2}{\frac{1}{2}(\theta_0 + \theta_h)}.$$

Substituting these expressions into equation (66), we find for the transmission anomaly

$$
\begin{aligned}
A &= 10 \log\left[\frac{\theta_0\theta_h}{R}\left(\frac{h_1}{\theta_0^3} + \frac{1}{\theta_0\theta_h} \frac{h_2}{\frac{1}{2}(\theta_0 + \theta_h)}\right)\right] \\
&= 10 \log\left[\frac{\theta_h h_1}{R\theta_0^2}\left(1 + \frac{h_2}{h_1} \frac{\theta_0^2}{\frac{1}{2}\theta_h(\theta_0 + \theta_h)}\right)\right],
\end{aligned}
\tag{82}
$$

since from equation (37),

$$R \approx \frac{h_1}{\theta_0} + \frac{h_2}{\frac{1}{2}(\theta_0 + \theta_h)}.$$

If the gradient is sharp, and if the range is considerable, the angle θ_0 will generally be small compared with the angle at the hydrophone, θ_h. Also, if the hydrophone is not too far down, we may assume that the fraction h_2/h_1 is not too large. In that case, the second term in the parentheses in both equations (80) and (82) is small compared with unity and may be omitted as negligible. In either case we have, then, as a rough estimate of layer effect, the simple relationship

$$A = 10 \log\left(\frac{h_1\theta_h}{R\theta_0^2}\right). \tag{83}$$

COMBINATION OF LINEAR GRADIENTS

In this subsection we shall derive a formula for the intensity along a ray which has passed through a succession of layers in each of which the sound velocity changes linearly with depth in the layer. This condition is of considerable practical importance since most velocity-depth curves can be replaced in good approximation by a number of linear gradients.

The assumed velocity-depth pattern is shown in Figure 19. There are $n + 1$ layers, labeled 0, 1, 2, 3, $\cdots n$, in which the velocity gradients are $a_0, a_1, \cdots a_n$ respectively; the term a_i represents the velocity change in the ith layer in velocity units per foot of depth increase in the layer labeled i, where i takes any integral value from 1 to n. The velocity at projector depth is c_0; at the top of layer 1, c_1; at the top of layer 2, c_2; and so on, as indicated in Figure 19. The ray direction is θ_0 at the projector, θ_1 at both the bottom of layer 0 and the top of layer 1, and so on; and, finally, θ_n at the bottom of the $(n + 1)$ layer, which is assumed to be the depth of the receiving hydrophone. The total horizontal range covered by the ray is R; the component of horizontal range covered in each layer is designated by $R_0, R_1, \cdots R_n$, as indicated.

We shall compute the intensity at the range R by means of the formula (71), which is generally applicable. To use this formula, we must first derive an explicit expression for $\partial R/\partial\theta_0$ in terms of parameters which may be calculated from the given velocity-depth distribution.

The ray path in the ith layer will be an arc of a circle whose radius is $c_i/(a_i \cos \theta_0)$ according to equation (33). Let the small ray element ds be inclined at the angle θ, as in Figure 20. In traversing this small distance, the

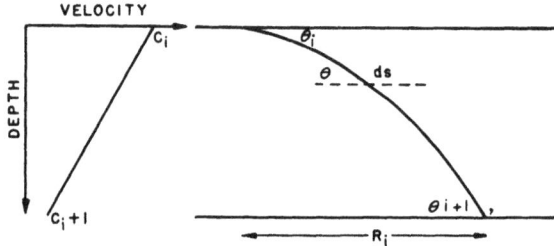

FIGURE 20. Ray path in ith layer.

ray travels horizontally a distance $ds \cos \theta$. By equation (32), we see that ds is given by the expression $-c_i d\theta/(a_i \cos \theta_i)$; and the element of horizontal range covered by the ray in a distance ds is

$$-\frac{c_i}{a_i \cos \theta_i} \cos \theta \, d\theta.$$

To get the horizontal range covered by the ray in its entire journey through the ith layer, we must integrate this result between θ_i and θ_{i+1}.

$$R_i = \int_{\theta_i}^{\theta_{i+1}} \frac{-c_i}{a_i \cos \theta_i} \cos \theta \, d\theta$$

$$= \frac{c_i}{a_i \cos \theta_i} (\sin \theta_i - \sin \theta_{i+1}) \qquad (84)$$

$$= \frac{c_0}{\cos \theta_0} \frac{\sin \theta_i - \sin \theta_{i+1}}{a_i}$$

because of Snell's law (26). The results of this paragraph apply without changes of sign both to layers where the velocity increases with depth and to layers where the velocity decreases with depth.

The total horizontal range R from the projector to the receiver is the sum of the range components in the $n + 1$ layers.

$$R = \sum_{i=0}^{n} R_i = \frac{c_0}{\cos \theta_0} \sum_{i=0}^{n} \frac{\sin \theta_i - \sin \theta_{i+1}}{a_i} \qquad (85)$$

where the symbol Σ indicates summation. By differentiating both sides of equation (85) with respect to θ_0

$$\frac{\partial R}{\partial \theta_0} = \frac{c_0 \sin \theta_0}{\cos^2 \theta_0} \sum_{i=0}^{n} \frac{\sin \theta_i - \sin \theta_{i+1}}{a_i}$$

$$+ \frac{c_0}{\cos \theta_0} \sum_{i=0}^{n} \frac{1}{a_i} \left(\cos \theta_i \frac{\partial \theta_i}{\partial \theta_0} - \cos \theta_{i+1} \frac{\partial \theta_{i+1}}{\partial \theta_0} \right)$$

which may be written

$$\frac{\partial R}{\partial \theta_0} = \frac{c_0 \sin \theta_0}{\cos^2 \theta_0} \sum_{i=0}^{n} \frac{1}{a_i} \bigg(\sin \theta_i - \sin \theta_{i+1}$$

$$+ \frac{\cos \theta_i \cos \theta_0}{\sin \theta_0} \frac{\partial \theta_i}{\partial \theta_0} - \frac{\cos \theta_{i+1} \cos \theta_0}{\sin \theta_0} \frac{\partial \theta_{i+1}}{\partial \theta_0} \bigg). \qquad (86)$$

For equation (86) to be usable, we must calculate $\partial \theta_i/\partial \theta_0$. By Snell's law,

$$\cos \theta_i = \frac{c_i}{c_0} \cos \theta_0. \qquad (87)$$

By differentiating both sides of this with respect to θ_0

$$\frac{\partial \theta_i}{\partial \theta_0} = \frac{c_i \sin \theta_0}{c_0 \sin \theta_i} \qquad (88)$$

$$= \frac{\sin \theta_0 \cos \theta_i}{\cos \theta_0 \sin \theta_i}.$$

By using $\partial \theta_i/\partial \theta_0$ from equation (88), and a corresponding expression for $\partial \theta_{i+1}/\partial \theta_0$, the expression in parentheses in equation (86) becomes

$$\sin \theta_i - \sin \theta_{i+1} + \frac{\cos^2 \theta_i}{\sin \theta_i} - \frac{\cos^2 \theta_{i+1}}{\sin \theta_{i+1}}$$

$$= \frac{1}{\sin \theta_i} - \frac{1}{\sin \theta_{i+1}}.$$

Thus equation (86) becomes

$$\frac{\partial R}{\partial \theta_0} = \frac{c_0 \sin \theta_0}{\cos^2 \theta_0} \sum_{i=0}^{n} \frac{1}{a_i} \left(\frac{1}{\sin \theta_i} - \frac{1}{\sin \theta_{i+1}} \right)$$

$$= -\frac{\sin \theta_0}{\cos \theta_0} \sum_{i=0}^{n} \frac{R_i}{\sin \theta_i \sin \theta_{i+1}}.$$

Putting this value of $\partial R/\partial \theta_0$ into formula (71), and noting that θ_h is simply θ_{n+1}, we obtain the final result

$$A = 10 \log \left[\frac{\sin \theta_0 \sin \theta_{n+1}}{R \cos^2 \theta_0} \sum_{i=0}^{n} \frac{R_i}{\sin \theta_i \sin \theta_{i+1}} \right]. \qquad (89)$$

The expression (89) is in a form well-suited for practical intensity calculations. The various angles θ_i can be computed from the known velocity-depth pattern by equation (26), and the R_i can be obtained either from equation (84) or equation (36).

FORMULAS FOR TRANSMISSION ANOMALIES

In this section, the formulas obtained for the transmission anomaly resulting from refraction will be summarized.

In the absence of a velocity gradient, or if the sound velocity increases or decreases linearly with depth below the projector, the transmission anomaly in the direct beam is negligible.

If the sound velocity decreases linearly with depth

below the projector, and if the ray has been reflected once by the ocean bottom, the transmission anomaly is given in good approximation by the equation

$$A = 10 \log \left| \frac{\theta_h \left(2 - \frac{\theta_0}{\theta_b} + \frac{\theta_b}{\theta_h} \right)}{2\theta_b + \theta_h - \theta_0} \right|$$

where θ_b is the angle of inclination at the bottom. In the case of multiple bottom reflections the transmission anomaly is given by

$$A = 10 \log \left| \frac{\theta_h \left(2(n+1) - \frac{\theta_0}{\theta_b} + \frac{\theta_b}{\theta_h} \right)}{2(n + 1)\theta_b + \theta_h - \theta_0} \right|$$

where the number of bottom reflections is $n + 1$.

The transmission anomaly for sound propagated through a thermocline (layer effect) is approximately given by

$$A \approx 10 \log \left(\frac{h_1 \theta_h}{R \theta_0^2} \right)$$

where h_1 is the height of the projector above the top of the thermocline, θ_0 is the inclination of the ray in the overlying isovelocity layer, and θ_h is the inclination of the ray at the receiver.

The transmission anomaly for sound propagated through a succession of layers, each of which possesses a constant gradient of velocity, is given by

$$A = 10 \log \left(\frac{\sin \theta_0 \sin \theta_{n+1}}{R \cos^2 \theta_0} \sum_{i=0}^{n} \frac{R_i}{\sin \theta_i \sin \theta_{i+1}} \right)$$

where the various terms are defined under "Combination of Linear Gradients" in Section 3.4.2.

3.5 RAY DIAGRAMS AND INTENSITY CONTOURS

3.5.1 Methods

The differential equations (25) which govern the path of a ray in a medium where the sound velocity depends only on one coordinate cannot be easily integrated if the velocity depends on depth in a very complicated manner. We have seen, however, that the integration can be accomplished, and the path of a ray with a specified initial direction calculated if the depth interval is divided into layers in each of which the velocity gradient is constant. For this reason, rays are traced in practice by replacing the actual velocity-depth curve with a series of straight-line segments, as in Figure 8.

The bending of sound rays in the ocean is too slight to be evident in a drawing that uses the same scale for range and depth. The deviation from straight-line propagation is never more than a few hundred feet in a mile; although such deviations are extremely important to a surface vessel seeking a submarine with echo-ranging gear, they do not show up well on paper unless the depth scale is expanded. For this reason, ray diagrams cannot be constructed geometrically in practice as the sum of circular arcs through the various layers. Instead, the change in range as the depth increases must be computed algebraically as described under "Combination of Linear Gradients" in Section 3.4.2, and the results plotted on a graph with suitably chosen scales.

A special circular slide rule has been invented to simplify this calculation.[2] This instrument, developed by WHOI early in the war, gives the horizontal range covered by a ray in its passage through a layer with a constant gradient. It does this by using several scales arranged for convenience as concentric circles. The thickness of the layer, the temperature at the beginning and the end of the layer, and the direction of the ray at the projector are given to start with. By use of the slide rule one calculates directly the direction of the ray when it enters and leaves the layer; from the average of the two directions the horizontal range covered in the layer can then be computed by use of equation (36). The instrument exactly duplicates the calculations described in Section 3.3 and avoids the necessity of consulting trigonometric tables. Since the direction of the ray as it enters the following layer is given as an intermediate step in calculating the range in the layer, the process may be duplicated until the ray has been traced through all the layers. This slide rule may also be used to compute intensities by integration along each ray since the scales provided may be used for evaluating equation (90), which appears later. Similar mathematical aids were developed by UCDWR, and by other research groups doing a large amount of ray tracing.[3]

Another instrument developed by NDRC for the facilitation of ray tracing is the sonic ray plotter,[4] pictured in Figure 21. The ray plotter is a device which integrates the differential equations (25) mechanically and exhibits the solution not as an algebraic function but as a curve denoting the ray path, drawn, of course, with a much expanded depth scale. The ray plotter has one advantage over the slide rule — it can plot the ray paths for any type of velocity-depth variation, no matter how complicated.

The accurate plotting of many rays is facilitated by use of a method called the method of proportions (see reference 2). When this method is used, a few rays are drawn with the aid of a slide rule; the positions of intervening rays can be estimated rapidly by an interpolating process. A full description of the method of proportions is given in reference 2.

If the ray plotter or the method of proportions is used, it is not difficult to obtain a ray diagram with many rays drawn for closely adjacent values of θ_0, the ray inclination at the projector. From such a diagram

FIGURE 21.   The sonic ray plotter.

the intensity at any point can be determined graphically by measuring the vertical separation between rays at that point. In most situations this is the simplest method for computing approximately the theoretical intensities in the sound field.

The basic equation used in this graphical procedure for determining the sound intensity may readily be derived from the analysis in Section 3.4.1. By combining equations (49) and (50) the following results for the intensity I

$$I = 2\pi F \cos \theta_0 \frac{d\theta_0}{dS}.$$

The area dS is given by

$$dS = 2\pi R \cos \theta dh,$$

where dh is the vertical distance between the two rays at the point where the intensity is measured, and R is the horizontal range. By combining these formulas, and by substituting into equation (65) for the transmission anomaly the following equation results

$$A = 10 \log \left(\frac{\cos \theta}{\cos \theta_0} \frac{dh}{d\theta} \right) - 10 \log R.$$

For most cases of practical importance, $\cos \theta$ and $\cos \theta_0$ may be replaced by one; dh and $d\theta$ may be replaced by finite increments Δh and $\Delta \theta$. Thus, we have, finally, the simple result

$$A = 10 \log \left(\frac{\Delta h}{\Delta \theta} \right) - 10 \log R. \qquad (90)$$

In equation (90), Δh and R must be expressed in yards; while $\Delta \theta$ must be given in degrees. Although this equation usually gives sufficiently accurate results, it is difficult to apply practically in regions where the intensity is changing rapidly, such as near the shadow boundary below an isothermal layer.

The practical application of equation (90) is given in reference 2, which includes a graph giving the theoretical intensity I in terms of the measured ray separation in feet at the range R and the initial angular separation of the rays in degrees.

3.5.2   Ray Diagrams for Various Temperature-Depth Patterns

In practice, the ray paths are usually computed not from the velocity-depth curve, but from the temperature-depth curve obtained with a bathythermograph. This is done because the sound velocity is very sensitive to changes in temperature of the magnitude usually encountered in the ocean and relatively insensitive to changes in pressure and salinity. The effect of pressure, although small, is usually allowed for in the drawing of rays because it is constant, causing an increase of 0.0182 ft per sec in sound velocity per foot increase of depth. The effect of salinity on the ray paths is usually ignored, except near regions where fresh water is continually mixing with ocean water; in such cases, the velocity-depth pattern must be calculated explicitly by use of both the bathythermograph record and the salinity-depth variation.

The following paragraphs describe ray diagrams for various commonly observed temperature-depth patterns. A more detailed explanation of ray diagrams along with explicit diagrams for some 380 temperature-depth patterns of the sort found in the ocean is given in a report by WHOI.[5]

VERY DEEP ISOTHERMAL WATER

In deep isothermal water all the rays show slight upward bending because of the constant effect of pressure. This bending, for a ray leaving the projector in a horizontal direction, amounts to about

50 ft in a distance of one mile. Figure 22 is the ray diagram for this case.

Isothermal Layer above Thermocline

The most common temperature-depth distribution observed in the ocean possesses a surface layer of reasonably constant temperature, which overlies a

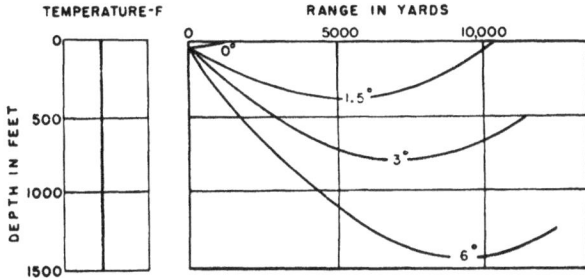

FIGURE 22. Ray diagram for deep isothermal water.

layer where the temperature decreases rapidly with increasing depth, called a thermocline. A ray diagram for such a temperature-depth pattern, with the surface mixed layer extending down to a depth of 100 ft, and an underlying thermocline is shown in Figure 23. It will be noted that all the rays which issue from the projector at higher angles than 1.44 degrees remain entirely within the top layer; the rays become horizontal at some depth less than 100 ft and bend back to the surface. All the rays leaving the projector at

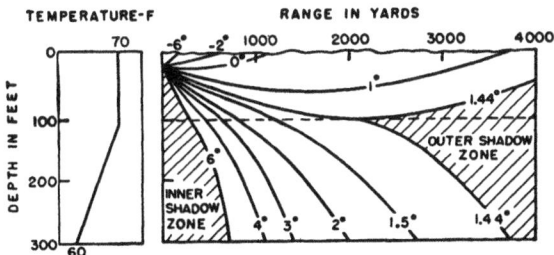

FIGURE 23. Ray diagram for isothermal layer above thermocline.

angles lower than 1.44 degrees reach the thermocline while still inclined downward; the thermocline progressively increases this downward bending. For theoretical reasons, then, the beam should split into two parts; one heads back toward the surface and the other heads down into the thermocline. Between these two beams the sound intensity should be very low according to the ray theory.

All velocity-depth patterns for which the ray theory predicts such a splitting of the beam have been called split-beam patterns. The existence of the predicted low-intensity zone which lies between two zones of higher intensity has been verified in experiments with explosive sound. The experiments are described in Chapters 8 and 9. However, experiments with single-frequency sound, which are designed to test whether or not the beam splits when predicted, have frequently failed to indicate any splitting of the beam at all. The reasons for this discrepancy are not completely understood. Diffraction of sound into the low-intensity zone, although predicted to a limited extent by wave acoustics, is not sufficient to explain why the beam does not split. Possibly the sound in the predicted low-intensity zone may be largely due to scattering of sound either by inhomogeneities in the predicted path of the rays, or by roughness of the sea surface, or by irregularities of the temperature distribution in the ocean.

Strong Negative Gradient

Negative temperature gradients are a frequent occurrence near the sea surface, especially when the surface is receiving more heat than it is losing. Under such temperature conditions the rays are bent strongly downward. Figure 24 gives a ray diagram

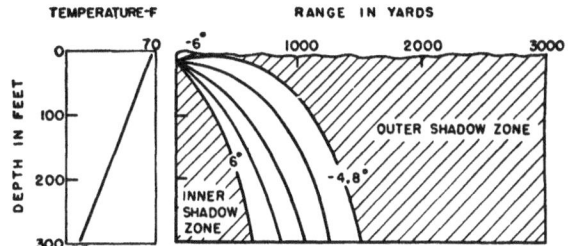

FIGURE 24. Ray diagram for strong negative temperature gradient.

drawn for a case where the negative temperature gradient amounted to about 8 F per 100 ft of depth. It is clear from the figure that the ray which left the projector horizontally has been bent down 400 ft by the time it has covered a horizontal range of 1,000 yd. The most important quality of the ray diagram for this case is the indicated shadow cast by the surface. It is clear from the figure that no ray leaving the projector can possibly penetrate into the zone marked shadow zone if the water is deep. All rays lower than the ray leaving the projector at a climbing angle of 4.8 degrees stay within that ray and are bent downward. The 4.8-degree ray itself becomes tangent to the surface and then bends downward. All rays higher than this "limiting ray" are reflected by the surface

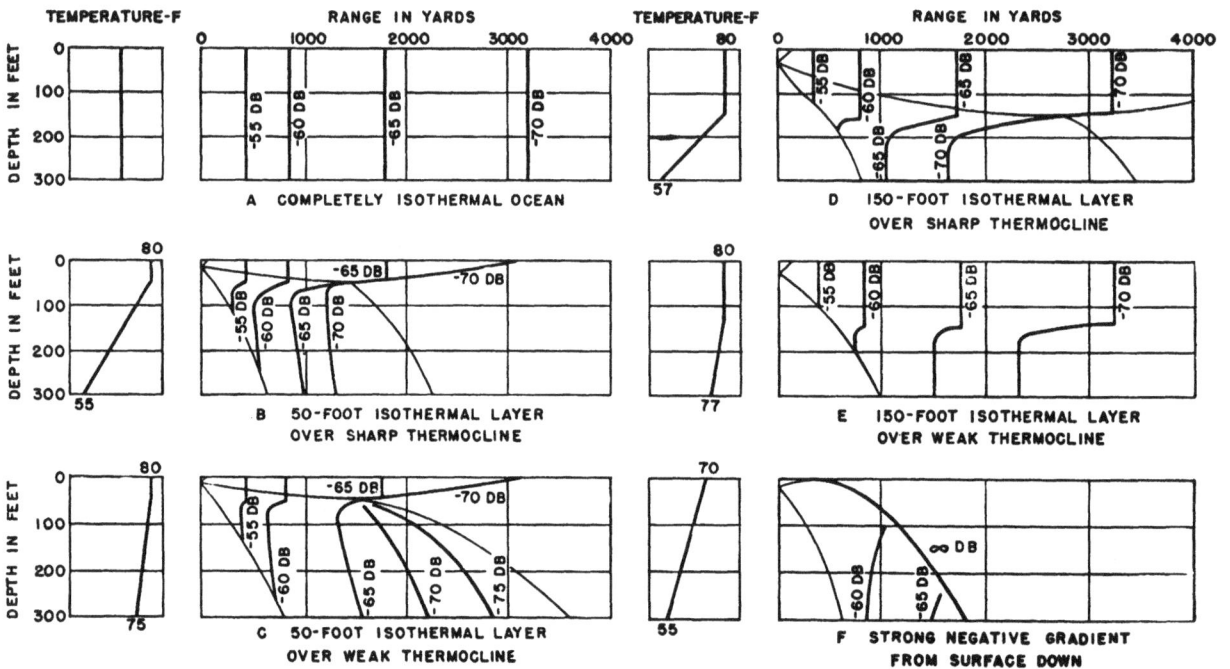

FIGURE 25. Sample intensity contour diagrams.

inside the place where the limiting ray hits it; and so all the surface-reflected rays remain inside the 4.8-degree ray, also.

If the actual sound intensity obeyed the predictions of the ray diagram, no sound at all would penetrate out to horizontal ranges of more than a couple of thousand yards in the top 500 ft of the ocean. Thus, a submarine further from the projector than 2,000 yd could be almost certain of escaping detection by sonar gear. In practice, as with split-beam patterns, the shadow zone in the strict mathematical sense of a region of zero intensity does not exist. However, unlike split-beam patterns, the transmission anomaly invariably increases sharply at or near the indicated separation of sound from shadow when the downward refraction is strong. As discussed in Section 5.4, such zones of weak sound are observed whenever the temperature gradient is strong enough to cause the predicted shadow zone to start nearer the projector than about 1,000 yd; the sound in them is about 30–40 db weaker than would be predicted by the inverse square law. If the negative gradient is weak, however, so that the predicted shadow zone does not begin until a range of about 1,500 yd or more, sound usually decreases gradually and at a more uniform rate; the shadow zone in such a case can scarcely be said to have any real existence. Possible mechanisms for the penetration of sound into predicted shadow zones are discussed in Section 3.7.

3.5.3   **Intensity Contours**

Not all the characteristics of the sound field become apparent from a glance at the ray diagram. Although the distribution of intensity is governed by the spreading of the rays, the degree of spreading cannot be accurately estimated visually, and it is even difficult to judge qualitatively. If a ray diagram is available, the intensity at a point may be quickly estimated by measuring the vertical separation of the rays nearest that point, in accordance with equation (90). Since it is assumed that the ray bending is in a vertical direction only, the predicted sound intensity will be directly proportional to the measured separation of the rays.

Intensity contours provide a very graphic method for displaying the results of such computations. In practice, the intensity loss is usually reported in decibels below the sound level on the axis at a distance of 1 yd from the projector. The exact value of the spreading loss at maximum echo range depends on many factors, such as the strength and directionality of the projector, the efficiency and operating condition of the gear, the intensity of background noise, and the amount of intensity loss due to absorption and scattering. In many cases, it is useful to know at what range the intensity loss due to spreading will be 55 db, at what range it will be 60 db, etc. It is clear that the range at which the spreading loss has a specified value will depend on the depth of the point to which the range is measured, and on refraction conditions, or more specifically on the temperature-depth variation indicated by the bathythermograph.

The intensity contour diagram is a set of lines drawn on a ray diagram indicating the intensity loss. On each contour the intensity loss has a constant value, in a fashion similar to the curves of constant barometric pressure on a weather map. The contours are obtained from a ray diagram by using one of the methods discussed in Sections 3.4 and 3.5. On each ray, or for each pair of adjacent rays, the intensity, or transmission anomaly, is computed at suitably chosen intervals. Then one finds, by interpolation, the points where the intensity loss is 55 db, 60 db, 65 db, and so on. After this process is carried through for all the rays, intensity contours can be drawn by joining the points of equal transmission loss on all the rays.

Sample intensity contour diagrams for the oceanographic situations treated in Section 3.5.2 are given in Figure 25. The contour diagram for isothermal water is shown for comparison since it indicates optimum sound-ranging conditions, that is, the intensity losses which would be observed if the water had no temperature gradients, and if there were no attenuation losses; for this situation, the intensity loss out to the range R is given by the inverse square law and amounts to $20 \log R$. The contour diagrams for the split-beam cases are identical with that for the isothermal case at depths near the sea surface and at short to moderate ranges; at depths below the thermocline, however, the predicted spreading loss is much increased; the amount of increase depends on the depth to the thermocline and the sharpness of the thermocline gradient. In the case of downward refraction, the intensity contours which denote large values of the intensity loss are piled together in the vicinity of the predicted shadow boundary.

A more detailed discussion of intensity contours with a derivation of some of the basic equations derived at the beginning of this chapter is given in a report by UCDWR.[6] Sample theoretical intensity contours for different temperature patterns are also discussed in this reference. A comparison of these predicted intensities with sound intensities found from explosive pulses is given in Chapter 9. The encyclopedia of ray diagrams in reference 5 includes intensity contours on most of the diagrams and thus may be used to find the type of predicted sound field for many different varieties of temperature-depth patterns.

It will be seen in Chapter 5 that the intensity predictions of the contour diagram are not, in general, sufficiently accurate to be trusted for the prediction of maximum echo ranges. However, they are useful for various special purposes, such as indicating how sound intensities should vary with depth at a fixed range.

3.6 VALIDITY OF RAY ACOUSTICS

In Sections 3.1 to 3.5 of this chapter the method of ray acoustics has been presented as an independent theory without much connection with the rigorous treatment of wave propagation presented in Chapter 2. We first noted in Section 3.1.1 that the important features of the propagation of spherical waves could be derived equally well by using the concept of wave fronts connecting points which have equal phase of condensation, and the concept of energy transported by rays perpendicular to these wave fronts. Then we generalized the definition of wave fronts and rays, derived differential equations for the ray paths from these definitions, and solved these differential equations for the ray paths and the resulting sound intensity.

It is important to remember, however, that the method of wave fronts for the general case placed no requirement on the wave front, except for stipulating that it be of the form (7) for some function $W(x,y,z)$. To make the idea of wave fronts intuitively significant, it was implied that the wave front should always join points of constant phase of condensation; but this implication was never used. The ray paths depended only on the form of the function W and the variation of c; the intensity calculations depended, in addition, on the assumption that energy is transported out along the rays. In this section, where we try to find a connection between ray acoustics and wave acoustics, we must assume a physical significance for the wave fronts. Accordingly, we shall make the explicit assumption that the wave fronts join points of equal phase of condensation since we already know that the assumption brings ray acoustics and wave acoustics into agreement for the case of spherical waves.

In this section, we shall examine whether wave acoustics and ray acoustics with this definition of wave fronts are equivalent in general or only under some special conditions. Since sound field calculations are much simpler by the ray method than by a rigorous solution of the wave equation, it will be extremely valuable to know when the ray theory can be applied without much error and when it will lead to definitely wrong results.

3.6.1 Eikonal Wave Fronts versus General Wave Fronts

It will be remembered that the entire method of rays was based on the eikonal equation (13), which in turn was based on the assumption that the wave fronts (7) "grow" perpendicularly to themselves. That is, the eikonal equation was derived by assuming that the wave front at time $t + dt$ is found from the wave front at time t by moving each point on the latter a distance $c\,dt$ along the outward normal. We shall now show that wave fronts ordinarily do not obey this law of propagation rigorously, but that the assumption often provides a good approximation.

It is intuitively apparent that wave fronts, defined purely as surfaces of constant phase without reference to the way they grow, exist in the exact case, at least when the dependence on time is harmonic. We shall define these wave fronts in the rigorous case by

$$V(x,y,z) = c_0(t - t_0) \qquad (91)$$

reserving the expressions W for those cases where the wave fronts grow perpendicularly to themselves, and where W therefore satisfies the eikonal equation. We shall call surfaces (91) *general wave fronts*, and surfaces defined by similar equations, with V replaced by W, *eikonal wave fronts*.

We know that in instances where the sound source vibrates harmonically with a single frequency f the solution of the wave equation can be expressed as the real part of the complex expression

$$p = A(x,y,z)e^{2\pi i f\{t - [V(x,y,z)]/c_0\}} \qquad (92)$$

This expression is identical with equation (6), except that we assume that the expression (92) with the function $V(x,y,z)$ is a rigorous solution of the wave equation, while the expression (6) with the function $W(x,y,z)$ was obtained by means of a Huyghens construction so that $W(x,y,z)$ would satisfy the eikonal equation.

We now shall see under what conditions the expression (92) can satisfy the wave equation and, simultaneously, $V(x,y,z)$ satisfy the eikonal equation. Suppose p satisfies equation (27) of Chapter 2, and simultaneously V satisfies the eikonal equation (13). The latter condition is

$$\left(\frac{\partial V}{\partial x}\right)^2 + \left(\frac{\partial V}{\partial y}\right)^2 + \left(\frac{\partial V}{\partial z}\right)^2 - n^2 = 0. \qquad (93)$$

The former condition may be simply calculated by noting that equation (92) may be written as

$$p = e^{\log A - 2\pi i f(V/c_0)}e^{2\pi i f t} \qquad (94)$$

Substitution of the expression (94) into the wave equation, performance of the indicated differentiations, and collection of terms is a straightforward calculation which will not be reproduced here. The real and imaginary parts must vanish separately; these parts are

$$\left(\frac{\partial V}{\partial x}\right)^2 + \left(\frac{\partial V}{\partial y}\right)^2 + \left(\frac{\partial V}{\partial z}\right)^2 - n^2 - \frac{\lambda_0^2}{4\pi}\left\{\frac{\partial^2(\log A)}{\partial x^2}\right.$$
$$+ \frac{\partial^2(\log A)}{\partial y^2} + \frac{\partial^2(\log A)}{\partial z^2} + \left[\frac{\partial(\log A)}{\partial x}\right]^2$$
$$\left. + \left[\frac{\partial(\log A)}{\partial y}\right]^2 + \left[\frac{\partial(\log A)}{\partial z}\right]^2\right\} = 0. \qquad (95)$$

and

$$\frac{\partial^2 V}{\partial x^2} + \frac{\partial^2 V}{\partial y^2} + \frac{\partial^2 V}{\partial z^2} + 2\left[\frac{\partial V}{\partial x}\frac{\partial(\log A)}{\partial x}\right.$$
$$\left. + \frac{\partial V}{\partial y}\frac{\partial(\log A)}{\partial y} + \frac{\partial V}{\partial z}\frac{\partial(\log A)}{\partial z}\right] = 0. \qquad (96)$$

Clearly, V will satisfy condition (93) only if

$$\lambda_0^2\left\{\frac{\partial^2(\log A)}{\partial x^2} + \frac{\partial^2(\log A)}{\partial y^2} + \frac{\partial^2(\log A)}{\partial z^2}\right.$$
$$\left. + \left[\frac{\partial(\log A)}{\partial x}\right]^2 + \left[\frac{\partial(\log A)}{\partial y}\right]^2 + \left[\frac{\partial(\log A)}{\partial z}\right]^2\right\} = 0. \qquad (97)$$

This can happen if λ_0 is zero, or if

$$B \equiv \frac{1}{A}\left(\frac{\partial^2 A}{\partial x^2} + \frac{\partial^2 A}{\partial y^2} + \frac{\partial^2 A}{\partial z^2}\right) = 0, \qquad (98)$$

since the expression in braces in (97) easily reduces to the above. This condition (98) is usually not satisfied. While it happens to be satisfied by the pressure wave of a point source in a homogeneous medium, it does not hold, for instance, for the radiation of a double source. In general, equations (93) and (95) will be rigorously equivalent only if the wavelength λ_0 vanishes.

3.6.2 Conditions for Nearly Eikonal Wave Fronts

We derived in Section 3.6.1 the conditions under which wave fronts, defined as expanding surfaces of constant phase of condensation, expand perpendicularly to themselves. It is more useful to know how large the frequency must be, relative to the other parameters of the problem, before the function $V(x,y,z)$ of equation (92) very nearly satisfies the eikonal equation; we will then know under what conditions the wave fronts are very nearly perpendicularly expanding.

Clearly the expression B of equation (98), the remainder term will be negligible compared with the other terms if

$$\lambda_0(\log A)' \ll V' \qquad (99)$$

$$\lambda_0^2(\log A)'' \ll (V')^2, \qquad (100)$$

where the prime denotes *any* spatial derivative, and $\ll$ means "is negligible compared with." If V even approximately satisfies the eikonal equation (13), then

$$V' \sim n, \qquad (101)$$

where the symbol $\sim$ signifies "is of the same order of magnitude as."

Another useful relation is obtained from equation (96). The functions A and V must satisfy equation (96) as long as the surface (91) has the significance of a general wave front. But equation (96) implies that

$$V'' \sim V'(\log A)', \qquad (102)$$

which in turn implies that

$$\lambda_0 V'' \sim V'\lambda_0(\log A)' \ll V'V' \qquad (103)$$

because of equation (99). Combining equations (103) and (101),

$$\lambda_0 V'' \ll n^2. \qquad (104)$$

In the ocean the index of refraction n is of the order of magnitude of unity. Then, the relation (104) may be stated in the following words. The first spatial derivative of V must not change much over a spatial distance of one wavelength. The first spatial derivatives of V give the direction of the rays; while the second derivatives, yielding the rate of change of ray direction, give the curvature of the rays. Therefore, the condition (104) becomes the following. The direction of the ray must not change much over a distance of one wavelength. In regions where the ray curves very strongly, ray acoustics cannot be applied safely.

Differentiating the eikonal equation (13), we get $V'V'' \sim nn'$ or $V'' \sim n'$ because of equation (101). In view of equation (104), this means that

$$\lambda_0 n' \ll n^2 \sim 1. \qquad (105)$$

In other words, the index of refraction must not change much over a distance of one wavelength.

We derive one more restriction — this time on the amplitude function A. From equations (102) and (104), we also have

$$\lambda_0(\log A)' < 1. \qquad (106)$$

The relation (106) means that $\log A$ must not change much over a distance of one wavelength. Since this change is very nearly $\lambda_0 A'/A$, this means that the

percentage change in A over one wavelength must be very small.

We can summarize our conclusions as follows. The eikonal equation usually will not lead to a good approximation (1) if the radius of curvature of the rays is anywhere of the order of, or smaller than, one wavelength, or (2) if the velocity of sound changes appreciably over the distance of one wavelength, or (3) if the percentage change in the amplitude function A is not small over the distance of one wavelength.

3.6.3 Comparison of Ray Intensities and Rigorous Intensities

It follows from the results of Section 2.7.3 that if the general wave fronts are defined by equation (91), and the instantaneous acoustic pressure by equation (92), then the rigorous intensity is given by

$$I = \frac{a^2}{2\rho c_0}\sqrt{\left(\frac{\partial V}{\partial x}\right)^2 + \left(\frac{\partial V}{\partial y}\right)^2 + \left(\frac{\partial V}{\partial z}\right)^2} \qquad (107)$$

and, further, that the direction of energy flow is characterized by the direction numbers $\partial V/\partial x : \partial V/\partial y : \partial V/\partial z$. The latter direction is perpendicular to the general wave front; thus, if the wave fronts are eikonal wave fronts, the energy flows along the rays in the rigorous case. If the wave fronts are approximately eikonal wave fronts, then the directions perpendicular to these wave fronts represent very nearly the true direction of energy flow.

Thus, if the conditions for eikonal wave fronts derived in Section 3.6.2 are satisfied, the energy emanating from the source into all solid angles will remain within the tubular confines assumed in deriving the ray intensity. We can therefore say, intuitively, that if the wave fronts are very nearly eikonal wave fronts, the ray intensity will be very close to the rigorous intensity. Further, we can say that in both cases the intensity will be given by

$$I = \frac{a^2}{2\rho c}, \qquad (108)$$

since $\left[\left(\dfrac{\partial V}{\partial x}\right)^2 + \left(\dfrac{\partial V}{\partial y}\right)^2 + \left(\dfrac{\partial V}{\partial z}\right)^2\right]^{\frac{1}{2}} \approx n = \dfrac{c_0}{c}.$

3.7 SHADOW ZONE AND DIFFRACTION

When the velocity decreases from the surface downwards, the ray theory predicts a sharp shadow boundary across which no sound ray penetrates; a typical ray diagram for such an instance is shown in Figure 24. At the shadow boundary the ray theory

predicts a discontinuous drop of intensity from a
finite value on one side to a zero value on the other.
It was shown in Section 3.6 that the ray theory cannot be trusted whenever it predicts such a rapid
change of intensity in a distance of only a few wavelengths. Thus, it is necessary to use the wave equation directly to compute the intensity of sound which
penetrates the so-called shadow zone.

The simplest case of a shadow zone is that produced by a screen in front of a light source. As shown
in Figure 26, the ray theory predicts that no light

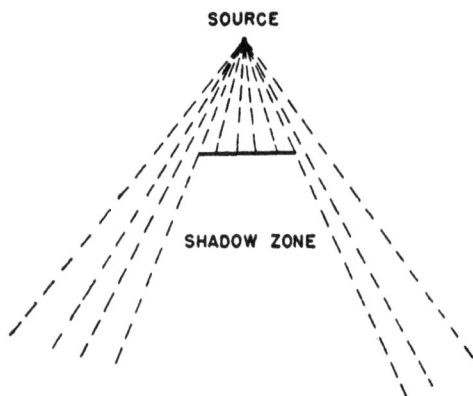

FIGURE 26.   Optical shadow zone produced by screen.

can reach the shadow zone behind the screen. When
the rays carrying the energy are curved, as in Figure
24, it is the surface of the ocean that intercepts the
curved rays and "casts a shadow." In either case,
however, some energy actually appears inside the
predicted shadow zone, and the wave is said to be
"diffracted."

The computation of diffracted sound in the shadow
zone is a rather complicated problem in the general
case. To indicate the type of analysis required, and to
show the general nature of the results, a simplified
problem will be considered here. As shown in Figure
27, a sound projector is assumed to be placed against
a vertical wall, which extends down to great depths.
The introduction of the wall simplifies the problem
without changing the final results essentially. The
water is assumed to be so deep that bottom-reflected
sound may be neglected. The projector face is assumed to be so wide that the horizontal spreading of
the sound beam may be neglected; thus, only the
two-dimensional problems need be considered. The
sound velocity c is assumed to vary according to the
law

$$c^2 = \frac{c_0^2}{1 + By} \tag{109}$$

where B is a constant, and y represents depth below
the surface. Since B is in practice very small, this
gradient is indistinguishable from a linear gradient
at depths of interest. The exact velocity gradient at
the depth y is given by

$$\frac{dc}{dy} = -\frac{c_0^2}{2c}\frac{B}{(1 + By)^2}. \tag{110}$$

Thus, at the surface, where $y = 0$, the velocity
gradient $-b$ is given by

$$-b = \frac{Bc_0}{2} = -\frac{dc}{dy}\bigg|_{y=0}. \tag{111}$$

The gradient (109) is chosen instead of a simple
linear gradient not for physical reasons, but because
it simplifies the following computations.

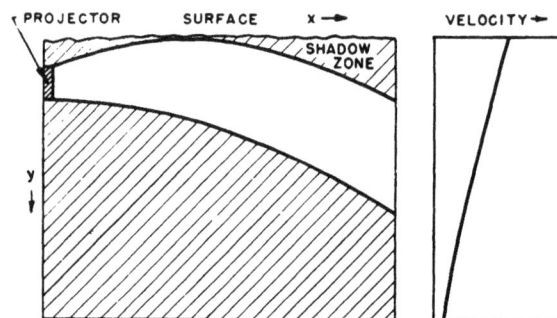

FIGURE 27.   Sound shadow cast by sea surface.

To solve the wave equation under these conditions,
it is necessary to use the method of normal modes
developed in Chapter 2. In particular, we must find a
solution to the wave equation (27) in Chapter 2 which
satisfies the boundary conditions we shall impose. As
in Section 2.7.2, we look for a solution which is the
product of three functions, one dependent only on
the time t, another dependent on the depth y, and the
third, a function of the horizontal distance x. The
coordinate z need not be considered in the two-
dimensional case under discussion.

Following the analysis of Section 2.7.2, we therefore write

$$p(x,y,z,t) = e^{2\pi i f t} F(y)G(x). \tag{112}$$

By substitution of equation (112) into the wave equation (27) of Chapter 2, and by dividing through by
C^2, it is found that F and G satisfy an equation of the
form

$$G\frac{d^2F}{dy^2} + F\frac{d^2G}{dx^2} + \frac{4\pi^2 f^2}{c^2}FG = 0. \tag{113}$$

If equation (113) is divided through by FG, and equation (109) used for c,

$$\left(\frac{1}{G}\frac{d^2G}{dx^2}\right) + \left[\frac{1}{F}\frac{d^2F}{dy^2} + \frac{4\pi^2f^2}{c_0^2}(1 + By)\right] = 0. \quad (114)$$

Since the first bracket depends only on x and the second only on y, equation (115) can be satisfied only if each bracket is constant. If we denote the first bracket by $-\mu^2$, the second bracket must be $+\mu^2$, and we have

$$\frac{d^2F}{dy^2} + \left[\frac{4\pi^2f^2}{c_0^2}(1 + By) - \mu^2\right]F = 0 \quad (115)$$

$$\frac{d^2G}{dx^2} + \mu^2G = 0. \quad (116)$$

The basic problem is to find solutions of equations (115) and (116) which satisfy the boundary conditions. First, we have the boundary conditions for equation (115). In the analysis in Section 2.7.2, these boundary conditions were that the pressure vanished both at the surface and at the bottom. Here, also, the pressure must vanish at the surface. However, the water is so deep that the condition at the bottom disappears. Instead, there is simply the condition that at some distance below the projector no sound is coming upwards; that is, any sound present at these depths is coming down from shallower depths. Although this boundary condition is somewhat complicated to formulate exactly, the general result is the same as that found in the solution of equation (161) of Chapter 2. In this earlier instance it was found that $\sin 2\pi y/\lambda_y$, corresponding to $F(y)$ in equation (112), when B is zero, satisfied the two boundary conditions only if λ_y had one of a number of fixed values. Similarly, the function $F(y)$ can satisfy the two boundary conditions only if μ has one of a certain number of values. These values, which are called characteristic values of μ, may be denoted by μ_1, μ_2, μ_3, and so on, or more generally by μ_j, where j can be any integral number. For each of the characteristic values μ_j, equation (115) has a particular solution $F_j(y)$ which satisfies the boundary conditions.

Once a value of μ_j has been chosen, the solution of equation (116) is very simple. For each value of μ_j,

$$G = A_j e^{-i\mu_j x} \quad (117)$$

where A_j is an arbitrary constant.[a] Thus the wave

equation is satisfied by any product of the type

$$p_j = A_j e^{2\pi i f t} F_j(y) e^{-i\mu_j x}. \quad (118)$$

Equation (118) satisfies the boundary conditions at the surface and at great depth since $F_j(y)$ satisfies these conditions. However, the boundary conditions at the vertical plane $x = 0$, the assumed vertical wall, must also be satisfied. These conditions are that the particle velocity at the sound projector must be $v_0 \cos 2\pi f t$, and that the particle velocity at all other points in the plane $x = 0$ must be zero.

To satisfy this boundary condition at the plane $x = 0$ requires a combination of an infinite number of possible solutions of the form (118). Each A_j must be chosen in such a way that the sum has the required properties. Methods for doing this have been developed, but are beyond the scope of this discussion. However, the final result is that the pressure p is the sum of many terms of the type (118) with $e^{2\pi i f t}$ the only common factor.

Within the direct sound field a large number of these terms are important, and an exact computation is necessary to find p. In the shadow zone, on the other hand, one term dominates, and the other terms may be neglected. This is because all the μ_j are partly real, partly imaginary, with the result that the absolute value of $\exp(i\mu_j x)$ decreases exponentially for sufficiently great values of x. It can be shown that the range at which only one term dominates is approximately the range to the shadow boundary computed from the ray theory. This dominant term is the one for which μ_j has the smallest imaginary part. Thus, the theory predicts that in the shadow zone the sound intensity falls off exponentially with increasing range, or, in other words, that the predicted transmission anomaly in the shadow zone increases linearly with increasing range.

Although the exact determination of the different characteristic values μ_j is somewhat involved, it is relatively simple to show how these values depend on the frequency f, the velocity gradient, and the sound velocity c_0 at the surface. This is useful since it indicates how the attenuation into the shadow zone may be expected to vary under different conditions. In order to investigate this dependence of μ_j on the other variables, we rewrite equation (115) in a simplified dimensionless form. Let

$$n = \frac{4\pi^2f^2}{c_0^2} - \mu^2 \quad (119)$$

and

$$y = \frac{u}{D} \quad (120)$$

---

[a] The negative sign must be taken in the exponent so that p_j in equation (118) will correspond to a wave moving away from the projector; that is, p_j must be a function of $2\pi f t - \mu_j x$, where μ_j is positive.

where D is an arbitrary constant to be determined later. Then equation (115) becomes, on dividing through by D^2,

$$\frac{d^2F}{du^2} + \left(\frac{n}{D^2} + \frac{4\pi^2f^2Bu}{c_0^2 D^3}\right)F = 0. \qquad (121)$$

If D is chosen so that

$$D^3 = \frac{4\pi^2f^2B}{c_0^2},$$

then equation (115) becomes

$$\frac{d^2F}{du^2} + (K + u)F = 0 \qquad (122)$$

where

$$K = \frac{nc_0^2}{(\pi^2f^2Bc_0)^{\frac{2}{3}}}. \qquad (123)$$

Equation (122) has solutions of the type desired only for certain characteristic values of K, denoted by the symbol K_i. The different values of K_i are determined only by the nature of the differential equation (122) and by the two boundary conditions, namely that the sound pressure is zero at the surface and that no sound is coming up from below the projector. Thus the values of K_i are independent of the frequency, sound velocity, and velocity gradient.

Once these characteristic values of K have been found, the corresponding values of μ to be used in equations (115) and (116) can be found directly. By substitution in equations (123) and (119), we find

$$\mu_i^2 = \frac{4\pi^2f^2}{c_0^2} - \frac{(\pi^2f^2Bc_0)^{\frac{2}{3}}}{c_0^2}K_i. \qquad (124)$$

The second term in equation (124) is always very much less than the first in cases of practical importance. Even for a temperature gradient as large as 1 F per ft of depth increase, and for a frequency of only 100 c, the second term is less than 1 per cent of the first for K_i less than 10, the region of practical interest. Thus we may take the square root of equation (124), expand in a series, and retain only the first two terms. This process gives

$$\mu_i = \frac{2\pi}{\lambda}\left[1 - \frac{K_i}{2}\left(\frac{Bc_0}{8\pi f}\right)^{\frac{2}{3}}\right]$$

$$= \frac{2\pi}{\lambda} - \frac{K_i}{4}\left(\frac{\pi fB^2}{c_0}\right)^{\frac{1}{3}}. \qquad (125)$$

Let K_1 be the characteristic value of K with the smallest imaginary part, and let this imaginary part be denoted by iK_1'. Let the theoretical sound pressure associated with the characteristic value K_1 be p_1. In the shadow zone the intensity is proportional to the square of p_1 since the sound pressures associated with the other characteristic values K_j may be neglected.

The intensity level found from equation (119) is

$$L = 20 \log p_1 = C - 20(\log_{10} e)\frac{K_1'}{4}\left(\frac{\pi fB^2}{c_0}\right)^{\frac{1}{3}}x \qquad (126)$$

where C includes A_1 and the other variables taken over from equation (118). While C changes gradually with position, it is nearly constant along the shadow boundary. Multiplying out terms in equation (126), and using equation (111) for B, we get, finally,

$$L = C - \frac{5.05K_1'f^{\frac{1}{3}}(-dc/dy)^{\frac{2}{3}}x}{c_0}. \qquad (127)$$

It should be emphasized that equations (126) and (127) apply only in the shadow zone. In the main beam other terms corresponding to other values of K_i must be considered.

The analysis in a report by Columbia University Division of War Research[7] considers the radiation in three dimensions sent out by a point source and is thus more general than the simple analysis presented here. However, the final result for the sound in the shadow zone is nearly identical with equation (127); the only difference is that the term $5.05K_1'$ becomes 25.7 in the exact computation of reference 7. With this substitution, we have the following formula for a, the attenuation coefficient beyond the shadow boundary in decibels per unit distance.

$$a = \frac{25.7f^{\frac{1}{3}}(-dc/dy)}{c_0}. \qquad (128)$$

In this equation f is the sound frequency in cycles per second, and dc/dy is the velocity gradient in feet per second per foot. If c_0 is in feet per second, formula (128) gives the attenuation in decibels per foot; if c_0 is in yards per second, the result is the attenuation in decibels per yard.

Since inverse-square spreading is quite negligible compared to the intensity drop at the shadow boundary, equation (128) gives the slope of the transmission anomaly at points beyond the shadow boundary. However, this equation cannot be used at shorter ranges and must therefore be regarded as an expression for the local attenuation coefficient in the shadow zone.

Equation (128) is compared with observational data under *Attenuation Coefficient at Shadow Boundary* in Section 5.4.1, where it is shown that the observed local attenuation coefficients beyond the shadow boundary are not more than about half the predicted values. In other words, in practice much more sound appears in the shadow zone than is predicted by equation (128).

Chapter 4

EXPERIMENTAL PROCEDURES

THE PRECEDING chapter was concerned chiefly with the development of the ray-tracing technique, the earliest theoretical approach which led to practical results in the prediction of maximum ranges. This method was, however, only partially successful. Its chief accomplishment was the prediction of the shadow zone boundary in the presence of pronounced negative gradients at the surface.

Predicted maximum echo ranges computed by ray-tracing methods agreed with the available observed range data to a fair degree of accuracy, but it was clear that these prediction methods were too simple. The evidence relating maximum observed ranges to temperature conditions was too incomplete to be analyzed with a view to improving range-prediction methods. Navy vessels could not often be made available for range determinations under carefully controlled conditions, and the scattered observations made in the course of routine operations were inconclusive. It was decided, therefore, to initiate a program in which the sound field produced with standard Navy echo-ranging gear would be measured in much greater detail than before. It was contemplated that this study would place the prediction of sound ranges on a firmer basis and in general would lead to a better understanding of the basic factors important in transmission of sound through the ocean. Subsequently, this program was broadened to include sound of frequencies between 100 and 60,000 c, and to cover situations somewhat different from those encountered in routine operation of standard gear. Only such a broad experimental investigation of the propagation of sound under various conditions can possibly furnish an adequate insight into the mechanisms determining the sound field in the sea.

This chapter deals with the experimental methods which have been developed in connection with the sound field program. The results obtained will be discussed in Chapters 5 and 6.

4.1 QUANTITIES CHARACTERIZING TRANSMISSION

Before launching into a detailed discussion of these experimental methods, it will be necessary to review briefly the principal quantities which characterize the transmission of sound energy in the sea. In general, sound power is transmitted at a particular frequency or in a specified frequency band; all statements in this section concerning power, intensity, and sound level refer to the frequency or frequency band once specified.

Let F denote the power output per unit solid angle on the axis of symmetry of the sound source; at a moderate distance r from the source, the sound intensity on the axis therefore equals F/r^2. The power output per unit solid angle in any other direction will be given by bF, where b, the *pattern function* defined in Section 2.4.4, is a function of the direction; by definition, b equals unity for the direction of the projector axis.

Since decibels are commonly used in sound field measurements, we shall transform F into a more convenient quantity, the *source level S*. At a point on the axis at a distance of 1 yd from a point source, the sound intensity I_0 will be proportional to F. The source level S is defined as this sound intensity at 1 yd in decibels above a suitably chosen reference intensity I_r:

$$S = 10 \log\left(\frac{I_0}{I_r}\right). \tag{1}$$

The reference intensity I_r is usually chosen as that corresponding to an rms pressure of 1 dyne per sq cm.

Actual sound sources, such as a battleship generating propeller and machinery noise, frequently have large spatial extensions, and the sound level 1 yd from the source is not well defined. However, at distances large compared with the linear dimensions of

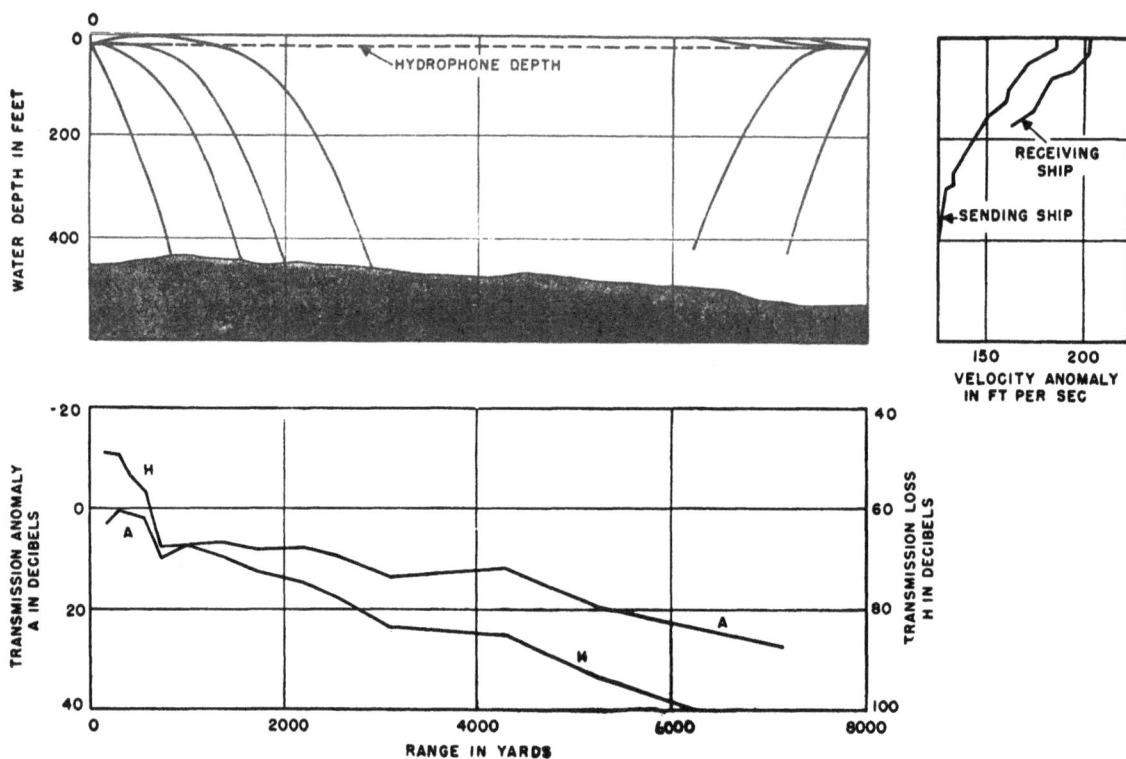

FIGURE 1.   Transmission loss (H) and transmission anomaly (A).

such a source, the sound field intensity drops off like that of a point source producing similar sounds; in other words, for intensity calculations at long ranges, the actual source may be replaced by an equivalent point source. The reported source level of an extended source is nothing but the sound level of the equivalent point source at a range of 1 yd. For echo-ranging transducers, the sound level is often measured at a distance of a few yards and then extrapolated to a distance of 1 yd by means of the inverse square law.

Consider now the sound field intensity I at some specified location, presumably at a fair distance from the sound source. This intensity is commonly expressed as a sound pressure level L in decibels above an rms pressure of 1 dyne per sq cm (decibels of a pressure level are defined as twenty times the logarithm of the ratio between the rms acoustic pressure and 1 dyne per sq cm). If the sound source is highly directional, like an echo-ranging projector, it is usually understood that the projector is trained, that is, rotated about its vertical axis, toward the point at which the transmission loss is to be determined. But in the absence of a tilting device, the axis ray

leaves the projector in a horizontal direction and may then be refracted to a depth different from that of the recording hydrophone. The difference $S - L$ will therefore depend, in general, on the directivity pattern of the transmitter. As long as the distribution of acoustic pressure does not approach the conditions of explosive sound, $S - L$ will be independent of the absolute power output of the transducer. The difference $S - L$ in decibels is called the *transmission loss* and is denoted by the symbol H.

Frequently, the transmission loss is represented in terms of its deviation from the law of inverse square spreading. If this law were valid, the transmission loss should amount to 20 log R, where R is the horizontal range from the transmitter to the chosen point. The expression $H - 20$ log R is called the *transmission anomaly* and is denoted by A. Figure 1 shows the experimentally determined values of H and A in a particular run with plots of the sound velocity against depth and of the computed ray diagrams.

H and A are quantities depending on the transmission characteristics of the path under consideration and on the directivity pattern of the transmitter. They are independent of the power output of the

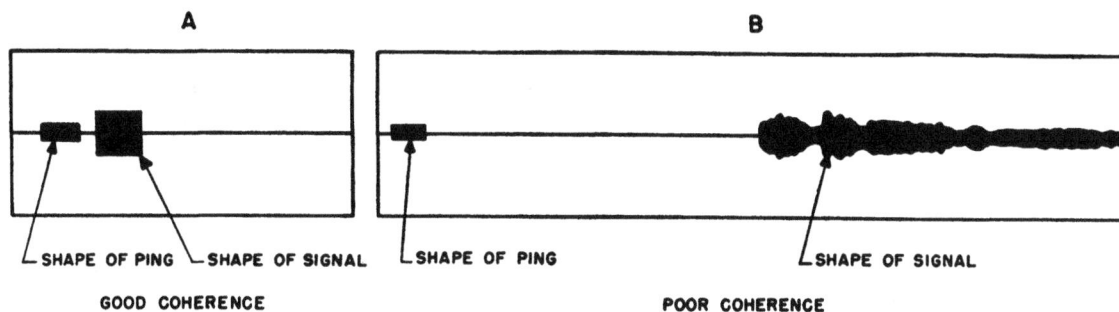

FIGURE 2. Examples of good and poor coherence.

transmitter; moreover, A is an absolute quantity which is independent of the system of units chosen.[a]

Transmission loss and transmission anomaly are the principal quantities which characterize the propagation of sound from the source to any point of interest. For some purposes, it is also desired to obtain information on the steadiness of the transmitted signal and on its "coherence." Slow changes in signal strength that occur in the course of several minutes are called *variation*. Changes that take place in the course of seconds are called *fluctuation*. The *coherence* of a signal may be loosely defined as the degree of fidelity with which the envelope of the transmitted signal is duplicated by the envelope of the received signal. If transmission conditions in the ocean did not change rapidly, one would be a perfect copy of the other, except for a negligible transient. Actually, conditions sometimes change so rapidly that the shapes of the transmitted and received signal resemble each other only slightly. Figure 2 shows (A) a case of good coherence, and (B) a case of poor coherence. Both of these figures show oscillograms of received supersonic signals recorded on the same equipment.

A detailed discussion of variation, fluctuation, and coherence will be given in Chapter 7.

4.2 DETERMINATION OF TRANSMISSION LOSS

Information on transmission loss has been obtained by three distinct methods: first, transmission runs; second, echo-ranging runs; and third, the statistical analysis of observed echo and listening ranges. Sound transmission runs include all investigations in which the source of sound is separate from the receiving instrument and in which the sound travels from source to receiver without suffering reflection from a target; slanting reflection from the surface or the bottom of the sea is, however, not excluded. While various setups have been used for transmission runs, the most common one involves the use of two ships. One ship carries the sound source, whereas the other ship is equipped with hydrophones whose outputs are recorded. In echo-ranging runs, the same transducer is used as both source and receiver. The sound is propagated to a target and then reflected back toward the point of origin. The target may be a vessel, but is more frequently an artificial target, that is, a device used exclusively for research and training purposes. The observed range information has been furnished to the research groups in the form of log books and patrol reports by naval craft on active duty.

Of the three methods of investigation mentioned, transmission runs have proved by far the most powerful and reliable tool. The other two methods, analysis of observed ranges and echo runs, are now merely subsidiary.

4.3 TRANSMISSION RUNS

The characteristic feature of the transmission run is the employment of separate devices for transmitting and receiving the sound. It is, therefore, possible to measure the transmission over any type of path by varying the depth of the projector, the depth of the hydrophone, and the horizontal distance between them. Depending on the instrumentation, it is further possible to vary other important acoustic parameters, such as signal frequency and signal length, or to employ signals composed of several frequencies or a continuous range of frequencies. The temperature

---

[a] A is ten times the logarithm of the ratio between the power flow per unit solid angle close to the source and the power flow per unit solid angle at the specified location; both of these quantities should be expressed in the same units. The apex of the solid angle is in both cases formed by the sound source.

distribution in the ocean during transmission, the depth and nature of the sea bottom, and in many instances factors such as condition of the sea surface, wind velocity, the presence of ocean currents, and the presence of salinity gradients, will affect the transmission characteristics of the ocean; these must be recorded along with the geometry of the transmission path itself. In recent experiments, not only the level but also the coherence and the degree of fluctuation of the received signal have been studied. All in all, the number of variables determining a signal is almost overwhelming; also, the characteristics of the resulting signal are quite complex. In any given investigation, both field procedure and the analysis of data are necessarily concerned with only part of the complete picture.

Ordinarily, the sound source in transmission runs is a transmitter driven by a harmonic oscillator through suitable amplifying stages, so that single-frequency sound is put into the water. Frequencies used range from 200 c up to 100 kc and more, but more runs have been carried out at 24 kc than at any other frequency. A second ship carries the receiving gear, hydrophone, amplifiers, and recorders. The hydrophones are usually cable-mounted hydrophones, which can be lowered to any desired depth from a few feet to several hundred feet below the surface.

In the most common form of run, the depth of the hydrophone or hydrophones is kept constant during one run. The range, however, is varied during the run from 100 or 200 yd to several thousand yd, by having the sending ship either approach or recede from the receiving vessel. The run is usually completed in less than half an hour. It is hoped no major changes in temperature distribution or other oceanographic variables will have taken place during that time. A more detailed description of the field procedure will be given in Section 4.3.2. First, however, a brief description of sound-transmitting and sound-receiving equipment will be given.

4.3.1 Sound Sources and Receivers

A sound source suitable for transmission runs should satisfy several requirements. It should be easily controlled. It should be capable of being mounted on a ship or towed by a ship. Its output should be stable. Its frequency characteristics should be simple, that is, it should produce either single-frequency sound or wide-band noise with a smooth

spectrum; and the acoustic power output should be high so that even at long ranges the received signal will usually be above the background. In practice, most results have been achieved with the use of single-frequency sources, such as echo-ranging projectors. Some work has also been done with noise makers of the type used for acoustic mine sweeping.

Single-frequency sources have been of three kinds, electromagnetic or dynamic speakers for sonic frequencies, and magnetostrictive and piezoelectric projectors for supersonic frequencies. Work has been reported by UCDWR at 200, 600, and 1,800 c, and 14, 16, 20, 24, 40, 45, and 60 kc, and by WHOI at 12 and 24 kc. More transmission runs have been carried out at 24 kc than at all the other frequencies combined because echo-ranging gear used by the Navy was designed for use at approximately that frequency. Occasionally, transmission runs have been made with "chirp" signals; these are frequency-modulated signals in which the frequency rises linearly from 23.5 to 24.5 kc or some similar frequency range during a pulse.

Other important parameters of the sound source are its directivity and its power output. The directivity may be reported in the form of pattern plots in the horizontal plane and in the vertical plane. Ten times the logarithm of b, the pattern function of the projector, is plotted on a circular graph against the angle from the axis. These plots are incomplete since no information is given concerning the value of b off the two planes plotted. Most echo-ranging projectors, however, approach rotational symmetry with respect to the axis so that a single plot including the axis gives adequate information on the pattern in all directions. Figure 3 shows the directivity pattern of the JK crystal projector which has been used by UCDWR for many transmission runs at 24 kc.

Frequently, the directivity of a projector is reported by means of a single quantity, the directivity index D. The directivity index is defined (see Section 2.4.4) by means of the equation

$$D = 10 \log \left(\frac{1}{4\pi} \int_{\Omega} b d\Omega \right) \tag{2}$$

in which Ω denotes the full solid angle. The units are decibels. The directivity index so defined has the value of zero decibel for a spherically symmetric sound source. Since the axis for echo ranging is invariably the direction of greatest power output, b nowhere exceeds unity, and D is a negative quantity. For the standard Navy sound heads QC (magneto-

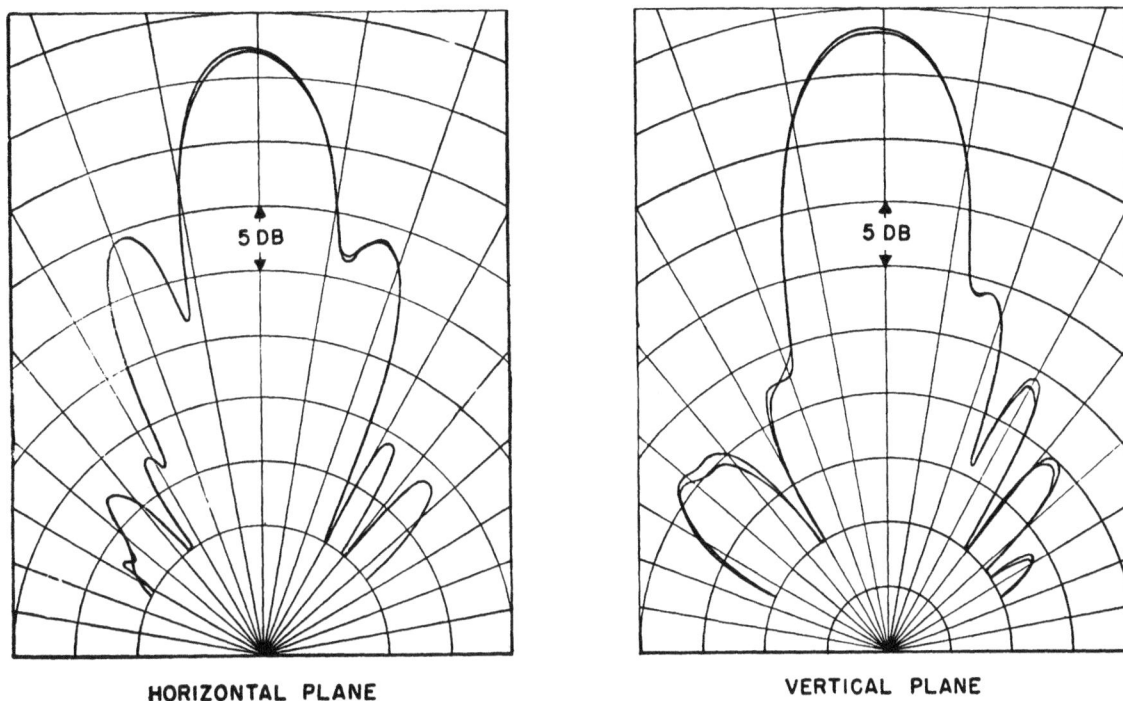

HORIZONTAL PLANE   VERTICAL PLANE

FIGURE 3. Directivity patterns of the JK SK4926 at 24 kc.

strictive) and JK (X-cut Rochelle salt), the directivity index is approximately -23 db at 24 kc.

The total power output of a projector is usually of less interest than the power output per unit solid angle on the axis. This quantity is customarily reported in terms of the source level S, which has already been defined in Section 4.1. The source level of the gear used in the UCDWR transmission studies is about 107 db above 1 dyne per sq cm 1 yd from the projector face.

The receiving instruments in transmission runs are usually cable hydrophones. The sound head consists of the electroacoustical element itself and sometimes contains a preamplifier which boosts the output voltage before it passes through the cable to the main sound stack. The electroacoustical element itself may be either a crystal element (as in the CN8 series used for a long time at UCDWR), or it may be a magnetostrictive device (similar to the Harvard Underwater Sound Laboratory [HUSL] B19–H).

The receiving response of a hydrophone is defined as the ratio between the rms voltage across the output terminals of the hydrophone or the preamplifier and the sound pressure of a plane wave incident on the axis of the hydrophone. It is ordinarily reported in decibels above 1 volt per unit sound pressure and then denoted by s. Waves incident in directions not

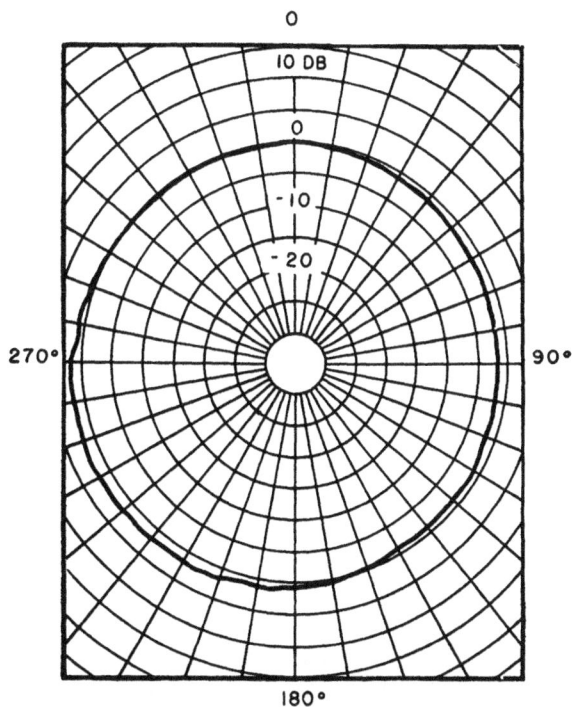

FIGURE 4. Response pattern of the CN-8-2 No. 597 hydrophone at 24 kc in horizontal plane.

parallel to the axis will produce lower voltages than sound waves of the same amplitude incident on the axis of the hydrophone; in other words, most hydrophones discriminate against off-axis sound inputs.

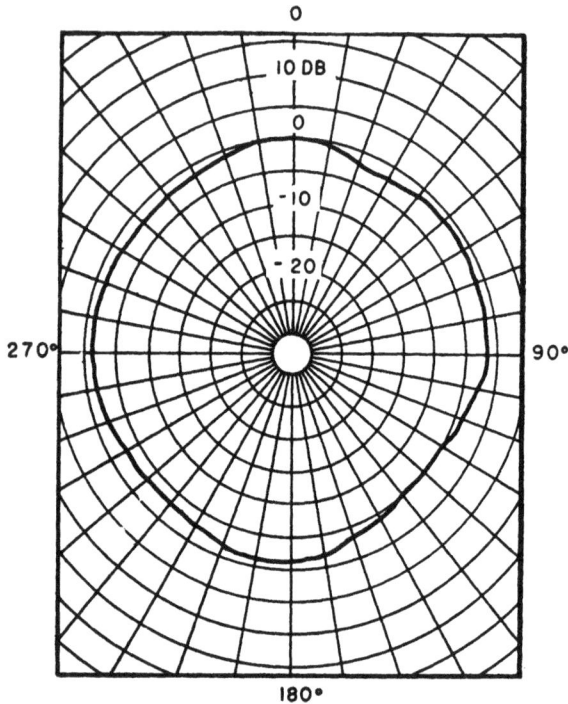

FIGURE 5. Response pattern of the CN–8–2 No. 597 hydrophone at 60 kc in horizontal plane.

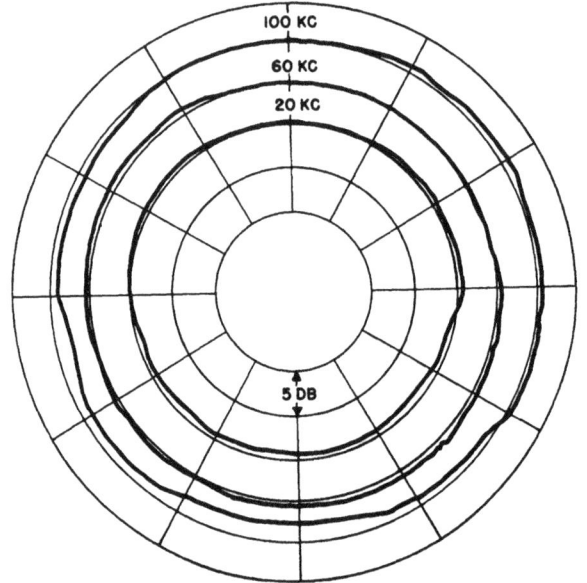

FIGURE 6. Horizontal directivity pattern of the Harvard B19–H hydrophone at 20, 60, and 100 kc.

The degree of discrimination is reported in a manner analogous to the statement of the directivity pattern of a projector. The ratio of sensitivity in a given direction to the sensitivity on the axis is denoted by b', which is often plotted in decibels relative to axis sensitivity. The degree of discrimination of the hydrophone may also be reported as a single quantity, its directivity index, defined by the equation

$$D' = 10 \log \left(\frac{1}{4\pi} \int_{\Omega} b' d\Omega \right), \qquad (3)$$

which is completely analogous to equation (2).

Cable hydrophones cannot be trained since they are freely suspended from their cables. It is, therefore, extremely important for a cable hydrophone to be nondirectional in the horizontal plane; otherwise appreciable unknown errors in the received intensity result. Several models, which have been used in research, come fairly close to nondirectionality. Figure 4 shows the horizontal directivity pattern of the CN–8 crystal hydrophone, used extensively for transmission runs at both UCDWR and WHOI. This pattern was determined at a frequency of 24 kc. Figure 5 shows the horizontal directivity pattern of the same hydrophone at 60 kc. It will be noted that at this frequency the CN–8 is quite noticeably directional in the horizontal plane. More recently the HUSL

B19–H magnetostrictive hydrophone has found favor because of its great stability and high degree of nondirectionality in the horizontal plane at a wide range of frequencies. Figure 6 shows the horizontal directivity of the B19–H at 20 kc, at 60 kc, and at 100 kc.

The response of a hydrophone is a measure of the strength of the electrical signal which will be passed into the cable with a given intensity of incident sound. Since the cable and subsequent amplifying stages will produce a certain amount of instrumental background noise, the response alone may, under exceedingly favorable external conditions, determine the level of the minimum detectable signal. The thermal noise in a 1-c band is determined by the receiving response and by the effective resistance G of the hydrophone according to the following equation.[1]

$$N = 10 \log G - s - 195. \qquad (4)$$

G is measured in ohms while N represents decibels above 1 dyne per sq cm.

The output of the hydrophone, or preamplifier, is transmitted through the hydrophone cable into the receiving sound stack. There the signal is filtered, amplified, possibly rectified or heterodyned, and then fed into the recorder. Most commonly used for recording are cathode-ray or galvanometer oscillographs with a very nearly linear[b] response, which re-

---

[b] The circuit is said to respond linearly if the amplitude of the recorded signal is proportional to the amplitude of the incident sound field.

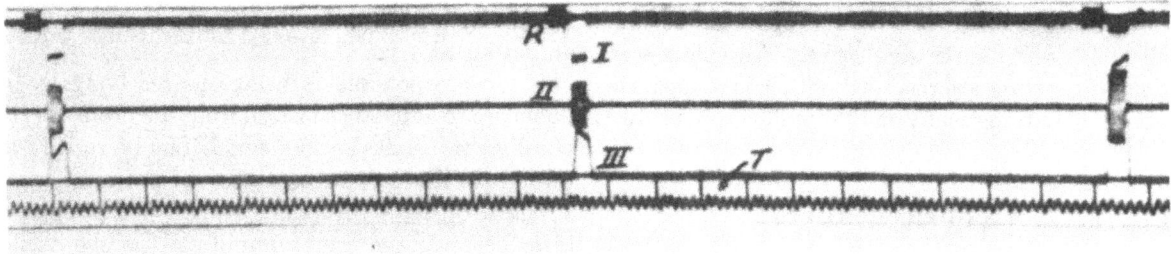

FIGURE 7. Oscillograph record of received 50-msec single-frequency signals; R is the radio signal, T is the 60-c timing trace, I and III are rectified traces, and II is the heterodyned trace. The range is approximately 80 yd.

cord the received signal on film or sensitized paper moving past the oscillograph at a constant speed. A timing trace, usually provided, permits accurate measurements of time intervals on the film or paper strip.

The signal is transmitted not only as an underwater sound signal, but also as an airborne radio signal over a radio link, usually FM, between the two ships. At the ranges involved, the radio signal arrives practically without time delay and without distortion. In addition to providing a convenient monitoring device, the radio signal serves as a means of accurately determining the range between the two ships. Since it is reproduced as a separate trace on the oscillograph record, it is an easy matter, with the help of the timing trace, to measure the time interval between the arrivals of the radio signal and the sound signal, and thus determine the distance traveled by the sound for each separate transmitted pulse. The resulting record will be similar to the strip in Figure 7. The top trace is the radio trace, the bottom trace the timing trace, and the three traces labeled I, II, and III, are the outputs of three different hydrophones, recorded simultaneously. The outputs of I and III were rectified before being recorded, and the output of II was heterodyned down to 800 c before recording, but not rectified. The range can be read off the record with an error of less than 15 yd. In the example shown in Figure 7 the range is approximately 80 yd.

In the past, transmission runs have also been recorded by means of power level recorders. These recorders are electromechanical recording instruments with a logarithmic response. A stylus records on a moving paper strip the received signal level in decibels above the reference level. Although these records are much easier to read than oscillograph records, they suffer from a certain unreliability of the instrument. Frequently, the stylus "sticks"; that is, it follows a change in signal level only when this

change exceeds an appreciable threshold value. Furthermore, the stylus travels only a certain number of decibels per sec (50 to 500 db per sec, depending on the model); the instrument, therefore, cannot record correctly the level of very short signals. For this reason, power level recorders have not been used in recent transmission work.

WHOI has developed an electronic device designed to combine the advantages of both oscillograph and power level recorder. It consists essentially of a rectifier, an amplifier with logarithmic response over a range of approximately 80 db, and a galvanometer oscillograph. This device has a time constant of roughly 0.5 msec.[2] Up to the present, it has been used only for reverberation studies; whether it will prove useful in transmission work remains to be seen. The amplifier used in this device has been improved since reference 2 was published.

The output of the hydrophone is passed through filters at some stage before it reaches the recording instrument. The purpose of the filter is to improve the signal-to-background ratio. All the unwanted background (see Division 6, Volume 7) contains energy in a very broad frequency band. A band-pass filter centered at the signal frequency will discriminate against the broad-band background in favor of sound at the signal frequency. Most of the filters used are approximately $\frac{1}{2}$ kc wide. Such a width leaves an adequate margin for possible drift of the driving oscillator in the sending stack and for doppler.

Both the amplifiers and the recording instruments will be linear and otherwise satisfactory only in a limited range of signal amplitude. On the other hand, actual signal levels are likely to change by as much as 80 db between short and long ranges of transmission. For this reason, step attenuators are provided. These attenuators are usually operated by hand; however, in the most recent installation at UCDWR the attenuator is automatically actuated when the received signal level rises above or drops below certain limits for several successive signals. All changes in attenuator setting are recorded, either in a separate log book, or automatically on the oscillograph record.[3]

4.3.2   Field Procedures [4]

In this section the field procedures used in transmission runs will be described. First, a number of oceanographic facts are ascertained and recorded, either immediately preceding or immediately following each transmission run. These include the depth of the ocean, the type of bottom (in shallow water), the state of the sea, the swell, the wind strength, and, most important, the vertical temperature distribution in the ocean. The bathythermograph, an instrument which measures vertical temperature gradients, is in general use in the Navy wherever echo ranging is involved. It is a recording device which can be lowered into the water down to considerable depth (as much as 450 ft for the "deep" model) and which, upon being returned to shipboard, indicates the temperature versus depth distribution as a trace marked on a smoked slide. Ordinarily, a bathythermograph is lowered on each of the two vessels participating in a transmission run; the source vessel makes its lowering at the point of greatest distance from the receiving vessel and frequently one or two lowerings at intermediate points. Figure 8 shows a blank which contains the oceanographic information belonging to a simple transmission run. This blank has been used at UCDWR.

Another subsidiary step is the calibration of equipment. In this chapter the term calibration will be used with a definite meaning. Calibration is a procedure which translates sound field data taken off the oscillograph trace into the transmission loss. The transmission loss was defined in Section 4.1 as the difference in decibels between the source level S of the projector and the sound level L at the hydrophone. Since the source level of the projector is defined in turn as the sound pressure level at a range of 1 yd, the transmission loss is then the difference in decibels between the sound levels at a range of 1 yd, and the range r of the hydrophone. In theory, then, one would obtain the transmission loss according to the following formula.

$$H = 10 \log \left(\frac{a}{a_1}\right)^2,$$

where a is the sound pressure amplitude in the water at the hydrophone, and a_1 is the sound pressure amplitude at 1 yd.

If it were possible to bring the receiving hydrophone up to a distance of 1 yd from the projector, the absolute transmission loss could thus be readily determined without knowing either the projector source strength or the hydrophone response. The squared ratio between the signal amplitude (on the oscillogram or on the tube screen) at 1 yd and the signal amplitude at R yards would give the transmission loss, provided the design of the receiving stack guarantees proportionality between received pressure amplitude and recorded trace amplitude. Actually, it is next to impossible to bring the projector of the sending vessel and the hydrophone of the receiving vessel closer together than about 30 to 50 yd without inviting a maritime catastrophe. Correction of the observed signal level at 30 or 50 yd back to the presumed level at 1 yd has at times been done by straightforward application of the inverse square law. However, this method is probably too simple. There is some evidence that, even at ranges of 50 yd, the transmission loss cannot always be expected to follow the inverse square law of spreading.[4]

Several more complicated methods have been employed, which in theory should enable determination of the absolute transmission loss. Although none of these suggested calibration procedures have proved completely satisfactory, some may be preferable to the simple correction by means of the inverse square law. The following paragraphs are devoted to a description of some of these more refined calibration procedures.

During a substantial part of its supersonic transmission program, UCDWR carried out runs called calibration runs at very short range, approximately 100 yd. During these runs, both the sending vessel and the receiving vessel were permitted to drift. The signal level at 100 yd, obtained from this run, was arbitrarily assigned a transmission anomaly value of zero; and all other transmission data obtained on the same day were referred to the 100-yd level obtained in the calibration run. Somewhat later, these special runs were discontinued. Instead, an average was taken of all the short-range data accumulated during the day, and a value of zero for the transmission anomaly was assigned to this average. In these two methods no attempt is made to calibrate in terms of a distance of the order of 1 yd; that is, no test is made which would relate

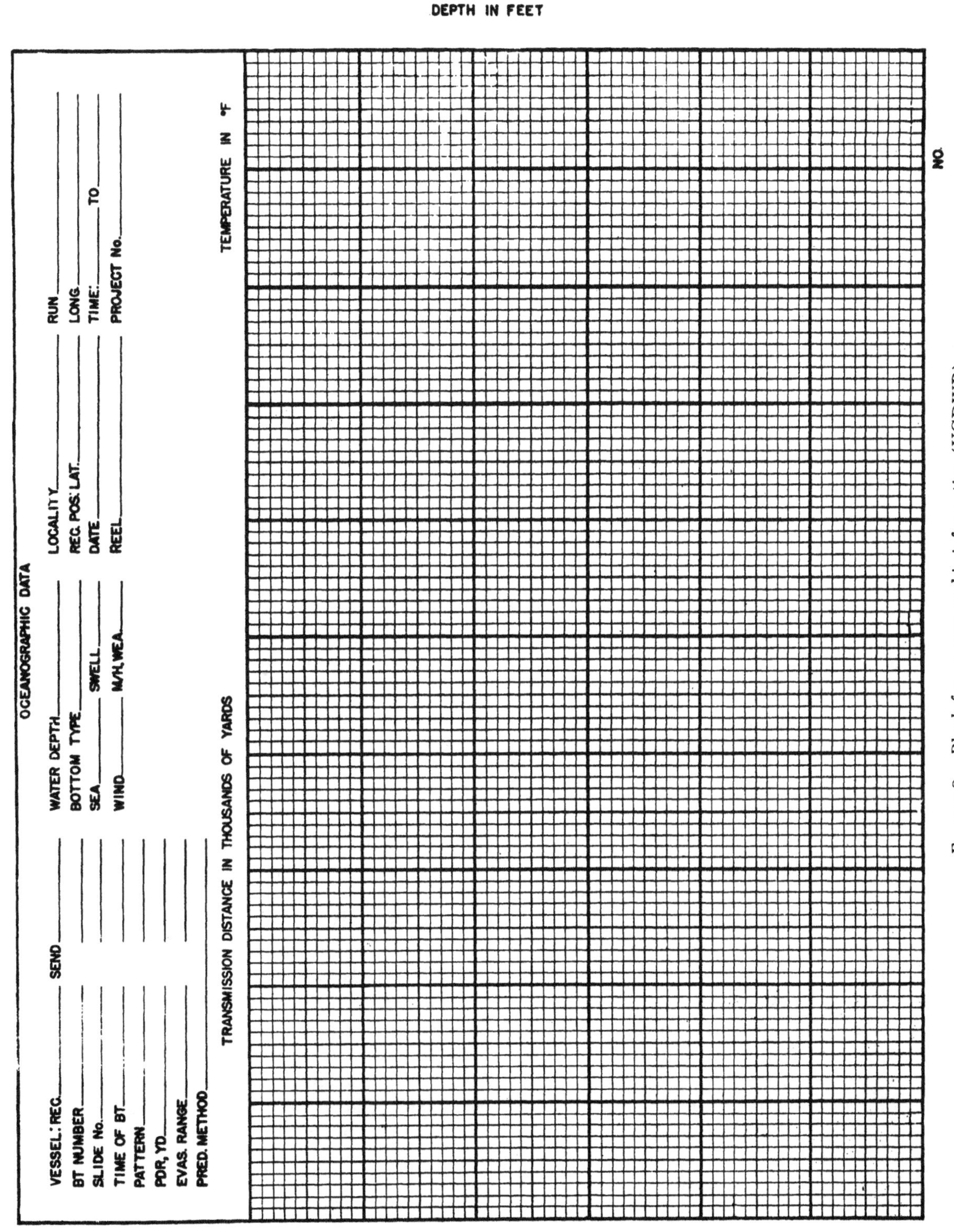

DEPTH IN FEET

FIGURE 8. Blank for oceanographic information (UCDWR).

the sound level at 100 yd to the source level as defined in Section 4.1.

WHOI used a related method of calibration until the summer of 1945. For each individual run, the observed sound field levels were each increased by $20 \log R$ and plotted against range. The resulting points between a range of 100 yd and the range where the observed intensity was 40 db less than the intensity at 100 yd were then fitted by inspection with a straight line. This line was extrapolated to a range of 1 yd to give the presumable sound level (in decibels above 1 μv) at that range.

More recently, both institutions have put into use new methods, which are designed to obtain a calibration directly in terms of the short-range (1 to 10 yd) sound level. These methods are of two kinds, which may be described as unaided calibrations and as calibrations with the help of standards. As an example of unaided calibration, WHOI floats one of the receiving hydrophones out to a distance where it can be picked up safely by the crew of the sending ship. The hydrophone remains connected by cable with the receiving ship, and its output is recorded by the same equipment used in the transmission runs. The hydrophone is then secured at a measured distance of a few yards from the face of the projector. The projector is trained on the hydrophone, and a signal is put into the water and received by the hydrophone. At this short range, the effect of the surface is minimized by the directivity of the projector, and tests have shown that the sound field obeys the inverse square law within the limits of observational accuracy. This method of calibration would be expected to yield the most accurate and most trustworthy results. A substantially identical method is occasionally employed by UCDWR for checking the results of other calibration methods. The unaided methods are not always practical in the field since in a heavy sea the transfer of a hydrophone from one ship to the other may not be possible. At best, the hydrophone transfer is a cumbersome and time-consuming maneuver. Nevertheless, unaided calibration is standard procedure at WHOI.

Most of the field calibrations at UCDWR are now made with the help of calibration standards, sound units which are designed primarily for this purpose and which are more stable than other units. Typical is a UCDWR procedure which involves the use of two OAX transducers; these transducers are HUSL designs. One of these two transducers is kept aboard the receiving vessel, the other aboard the sending ship. Aboard the sending ship, where the projector source strength S is to be determined, the OAX is used as a hydrophone. It is attached to a boom which can be swung over the side and which insures that the OAX is always at the same distance (13 ft) from the face of the keel-mounted JK projector. The projector is trained on the OAX, and the voltage generated across the terminals of the OAX unit by the sound field of the JK projector is measured. Aboard the receiving ship, where the hydrophone response S is sought, the second OAX unit is hung over the side amidships to a depth where it clears the keel. The receiving hydrophone is also hung over the side, on the other side of the ship, and is lowered to the same depth as the OAX. The distance between the two units is, therefore, with fair accuracy, the beam width of the receiving ship. The OAX is then energized as a projector with a standard power input, and the generated voltage across the terminals of the receiving hydrophone measured. If these two tests lead to the same results or very nearly the same results day after day, it is assumed that all four units are constant. Large jumps (several decibels) are presumably indications that either one of the four sound heads or the electrical follow-up (amplifiers and associated equipment) has changed its characteristics. If the change cannot be assigned to the electrical follow-up, it is assumed that the standard OAX units have remained unchanged and that either the JK power output or the receiving hydrophone response has changed. In a word, the characteristics of the projector and hydrophone used in the transmission runs are measured in terms of auxiliary standards, which are presumed to be stable. The standards themselves are thoroughly tested every few months at a special calibration station.

For a time, WHOI also used a similar method of calibration that depended upon the use of HUSL monitor standards. This method of calibration was later abandoned by that group in favor of unaided calibration.

Clearly, not one of the calibration methods which have been described is both convenient and wholly satisfactory as a method for translating observed hydrophone voltages into accurate estimates of the absolute transmission loss. Yet, until electroacoustical equipment is developed which can be relied on to remain stable, field calibration remains a necessity. It is to be hoped that rapid and adequate procedures of calibration will be developed in the future.

Once the equipment is calibrated, the transmission

loss can be determined by measuring the received signal level at the range and depth of the hydrophone. There are several types of runs. It is possible, for instance, to make a *vertical* transmission run, in which the range between the two ships is kept very nearly constant and in which the hydrophone is slowly raised or lowered, so that the transmission loss is determined as a function of depth at a fixed range. In *horizontal* runs, the depth of the hydrophone is kept fixed, while the range is changed. Horizontal

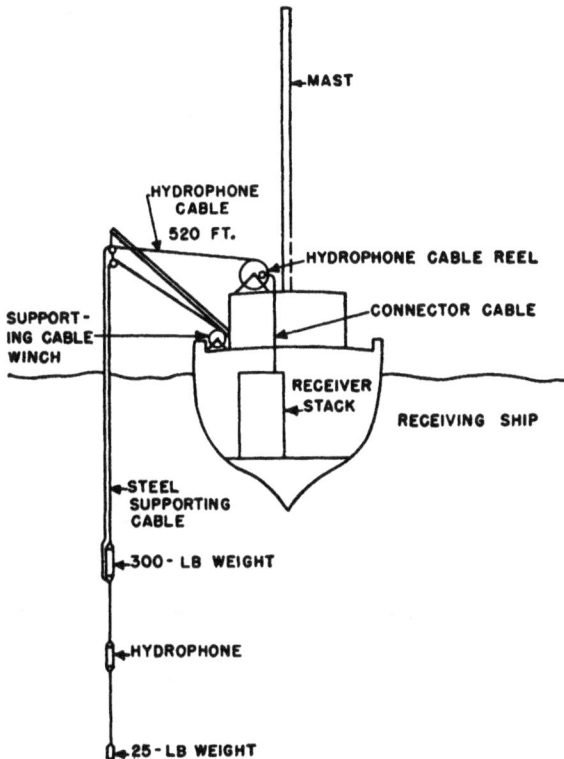

FIGURE 9. Method of suspension of deep hydrophones.

runs are much easier to carry out than vertical runs. In a vertical transmission run, the depth of the hydrophone can be changed only slightly each time. One or several pings are transmitted while the hydrophone is kept at a constant depth; and then the hydrophone depth is changed again. Also, whenever the hydrophone is moving through water, the flow of the water past the hydrophone gives rise to noise, which may effectively mask the signal. In a horizontal run, the receiving ship is permitted to drift, or, in shallow water, anchored. The noise due to water current thus is minimized and the hydrophone cable tends to hang straight down. Proper training and control of depth is thus facilitated. The sending vessel then runs either toward or away from the receiv-

ing ship. In this manner the range can be varied continuously by an amount of several thousand yards without ever interrupting the transmission of signals. A supersonic transmission run of the horizontal type takes, on the average, about 20 or 30 minutes. If it is desired to obtain the transmission loss at several depths, two or three hydrophones can be suspended at various depths from the receiving vessel. The overwhelming majority of transmission runs made up to the present have been horizontal runs.

In all transmission runs, elaborate precautions have always been taken to keep hydrophones at their nominal depth. Because of the wind drift of the receiving vessel, and because of ocean currents going in different directions at different depths, deep hydrophones, which are lowered occasionally as far as 450 ft below the surface, will rise to a much shallower depth unless special care is taken to make them hang straight. To this end, a 300-lb weight is suspended from a strong steel cable; the hydrophone hangs down from this weight and is held down by an additional 25-lb weight as shown in Figure 9. The hydrophone cable carries relatively little weight in this arrangement.

Horizontal transmission runs can be either approaching runs or receding runs, that is, the sending ship can either close or open the range. In the receding run, the wake of the sending vessel is located between the two ships. Since it has been found that wakes are capable of absorbing sound, the sending ship usually changes its course from time to time in a

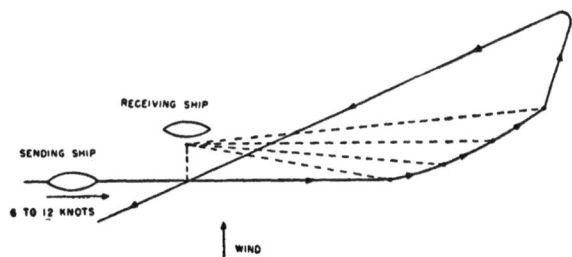

FIGURE 10. Course followed during a receding run.

manner illustrated in Figure 10 in order that the direct sound path between the two ships may never pass through the wake laid by the sending ship. However, over shallow bottoms where change of course would result in a changing bottom type, this procedure is sometimes not followed. During an approach run, the sending ship remains between its wake and the receiving ship, and it may, therefore, run along a straight course and pass the receiving vessel at a range of approximately 100 yd. During a run, the sending vessel keeps its projector trained at all times on the receiving vessel. This aiming is done by means of a pelorus, mounted either on the flying bridge or vertically above the sound projector to eliminate parallax. In a recently developed installation, selsyn repeaters cause the sound projector to follow automatically the changes in bearing of the pelorus. In former installations, an operator in the ward room of the sending ship had to match the two bearings by hand. At WHOI, projector training is now completely automatic, with the help of a radio compass.

In transmission work at supersonic frequencies, ping lengths are usually either about 50 msec or about 10 msec. It is believed that as the signal length decreases below 50 msec aural perception of the resulting echoes becomes more and more unsatisfactory (see Volume 9 of Division 6). Very short pings have an important use, however, in transmission studies. If the water is fairly shallow and the ping length is sufficiently short, the directly transmitted and the bottom-reflected signals can be examined separately. This is possible when the time resolution of the follow-up circuit is sufficient to resolve the time difference of arrival between direct signal and bottom-reflected signal. The minimum requirements of resolution in a specific case will depend on the geometry of the paths, depth, range, and refraction pattern of the ocean.

For very long ranges, the signals often arrive with badly distorted envelopes and with tails known as *forward reverberation*. When such tails are present, no resolution of the electrical circuit will result in satisfactory separation of the two sound paths. In the absence of such tails, resolution is frequently possible even with fairly long, square-topped signals; in other words, it is possible to distinguish three portions of the received signal trace, the direct signal alone, the composite signal, and the bottom-reflected signal alone.

Signals are emitted at a rate of about one signal per second. Once a minute pinging is interrupted, and a long signal of 10 seconds' duration is sent out. These long signals serve two purposes. First, a received long signal often provides a very instructive, graphic illustration of the degree of coherence of the transmission. Furthermore, this long signal makes it possible to correlate the received short sound signals with the radio signal, and thus to determine the range at which the signal was received. In a transmission run carried out at 5,000 yd, by the time a sound signal arrives at the receiving ship, three additional signals have already been put into the water. The once-a-minute breaks facilitate the identification of particular signals.

The overall accuracy of the determination of the transmission loss of an individual signal has been improved steadily in the course of the transmission program. A distinction must be made between the determination of the *absolute* transmission loss, and the determination of the *difference* in transmission loss between two signals received at the same range, or one signal received at different ranges. The determination of relative loss is not affected by errors of calibration, while the determination of the absolute loss is affected by calibration errors. The uncertainty of calibration in the earlier data taken both by UCDWR and WHOI is very large and probably exceeds 10 db in many instances. Improvement in procedure has cut this uncertainty down to approximately 1 db. Both absolute and relative determinations are affected by training errors of the projector and by the horizontal directivity of the hydrophone. Training errors are small at long range where the bearing is changing slowly. At ranges of the order of 100 yd, where the bearing changes rapidly, training errors can be significant even when great care is used in following the target. The uncertainty of training has been almost eliminated by improved instrumentation and is probably well within 1 db at the present time. Even in earlier work, training errors probably never caused an error in received sound level much in excess of 1 db. The most recent hydrophone models in use are practically nondirectional at 24 kc, but the directivity of the CN–8 model used in earlier studies introduced errors of about 2 db. Thus, the experimental error of most of the transmission loss determinations at UCDWR is probably about 2 db, while for the most recent data the experimental error is probably considerably smaller.

4.3.3   **Analysis of Data**

This section will be concerned with the analysis of data in which single-frequency supersonic sound is received by one of the recording systems with a linear response. The procedure used in the analysis of runs carried out with sonic sound will also be sketched.

Figure 11 shows that the received sound intensity is subject to rapid changes in intensity, which obviously cannot be related to observed changes in range or temperature distribution. Figure 12 shows the received amplitude of a continuous, 10-sec, 24-kc signal.

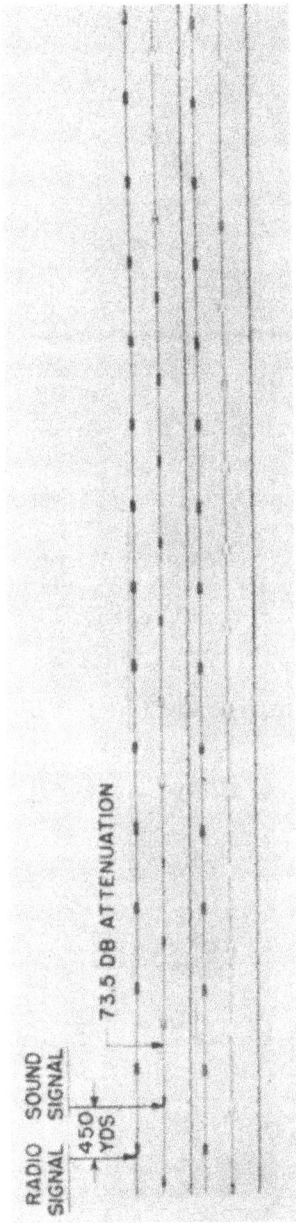

FIGURE 11. Records of received signals showing fluctuation.

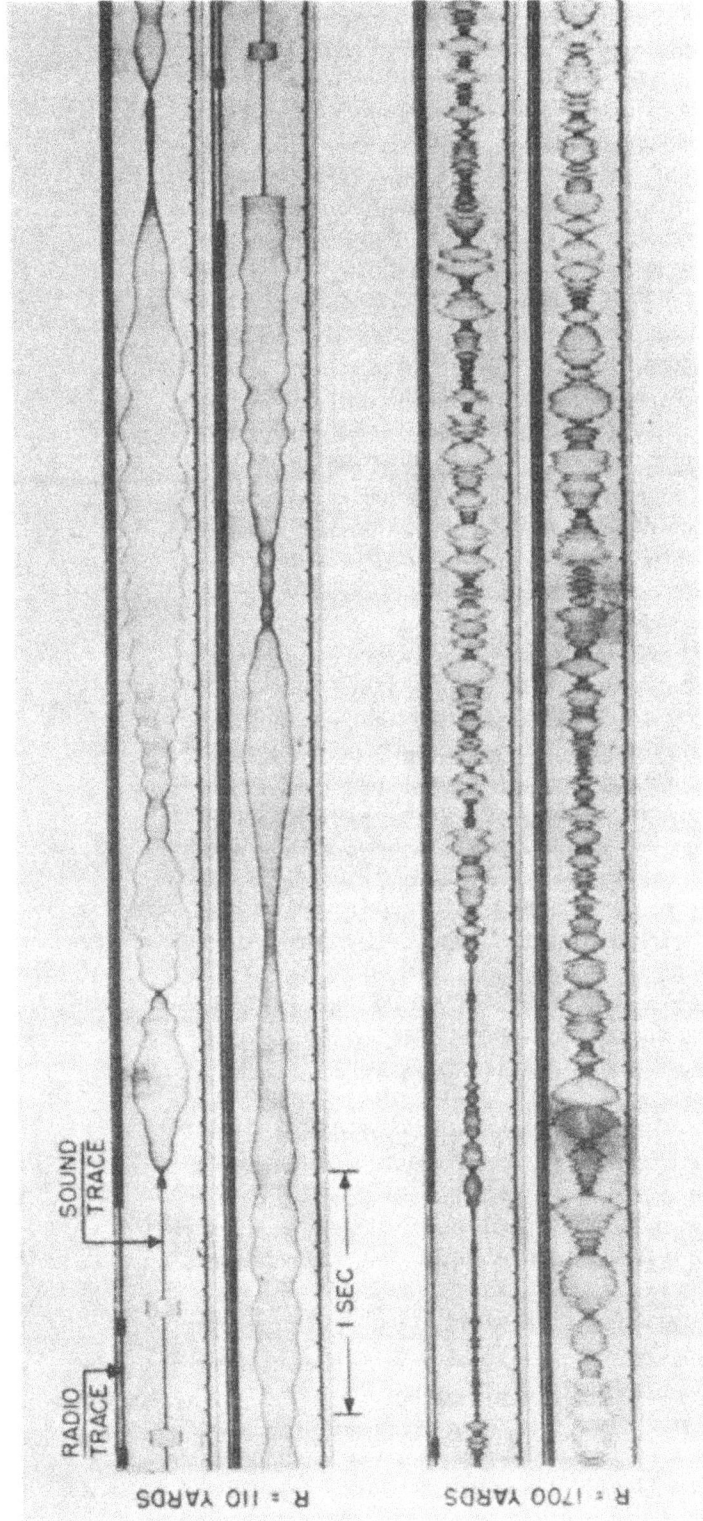

FIGURE 12. Two 10-sec signal oscillograms.

The upper strip was obtained at a range of 110 yd in the direct sound field, while the lower strip is a typical record of sound received at a long range, 1,700 yd, in the shadow zone. This fluctuation of received sound field intensity has become the subject of special investigations, which are summarized in Chapter 7. The principal purpose of transmission runs, however, is to obtain the average transmission properties of the ocean with a given set of oceanographic conditions.

To obtain a representative average, it is necessary to select a sample of signals, assign to each signal an individual sound field amplitude, and then to strike an average. The final result of these steps, the average sound field intensity or sound field level, will depend not only on the record obtained, but also on the details of the sampling and averaging procedure employed. UCDWR has standardized these procedures to insure intercomparability of results obtained at different times and by different research groups. The procedure is described in a report by UCDWR[5] and will be briefly recapitulated in the following paragraphs.

In the selection of a sample several requirements must be satisfied. The sample of individual signals must be large enough so that the standard deviation of the average is not much larger than of the order of 1 db. Moreover, the benefit of averaging will be obtained only if the sample covers a period of time in which the transmission passes through a number of maxima and minima, for otherwise the average would be an average of individual signals most of which may be relatively high or relatively low. On the other hand, the period of time covered by the sample must be short enough so that it corresponds to a negligible change of range between the two ships and a negligible change in the large-scale temperature structure.

The standard procedure for supersonic work, designed to strike a compromise between these requirements, has been to select five signals, equally spaced during a period of 20 sec. Since the standard deviation of an individual signal from average intensity is between 2 and 4 db in most samples, the standard deviation of the arithmetical average of five signals from the average of a very large number of signals is between 1 and 2 db, $(1/\sqrt{n-2}$ times the standard deviation of the individual signals).

At WHOI, the rule has been to use as a sample ten consecutive signals. Since signals are transmitted about 1.2 sec apart, a sample extends over a period of 12 sec. This method, although slightly different from that employed at UCDWR, leads to averaged ampli-

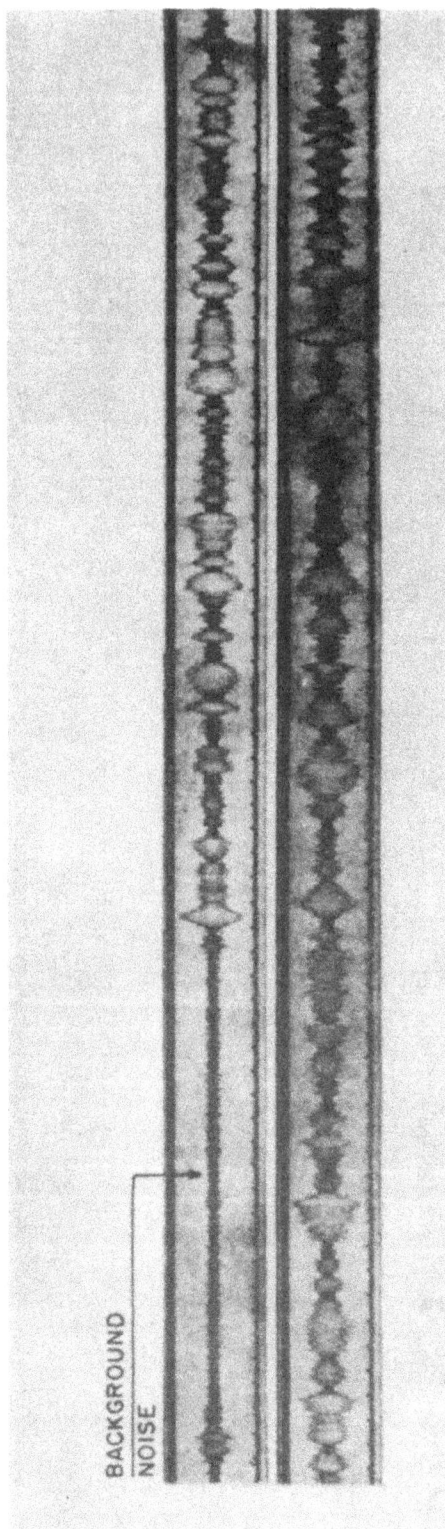

FIGURE 13.   Signal with high noise background.

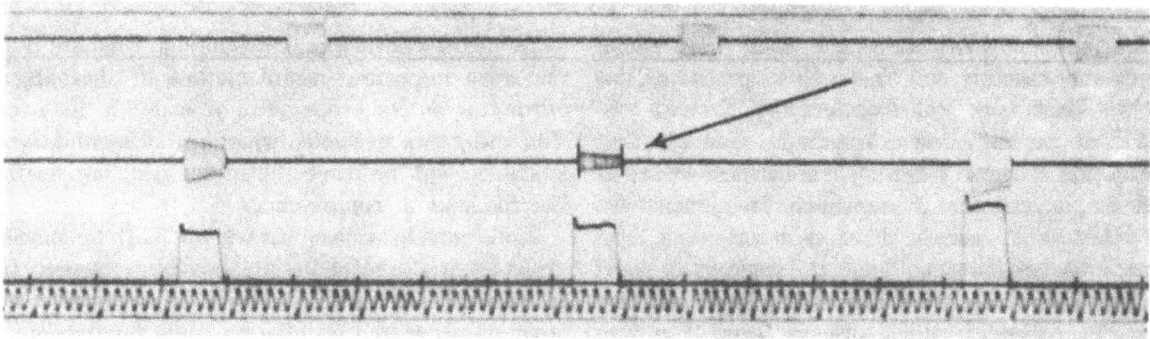

FIGURE 14.   Signal with end spikes.

tudes which differ from those obtained by the other method, but probably by no more than the internal spread of either method.

The next step consists of the assignment of an amplitude to each member of the selected sample. If the received signal were square-topped, like the emitted signal, this step would raise no questions. However, received signals are often far from square-topped. It was decided at UCDWR to assign to each signal the peak value of the amplitude registered anywhere during one signal, with two qualifications. One concerns noise received simultaneously with the signal. At low signal levels, the noise which is received continuously shows up as a very striking "spiny" record (shown in Figure 13). Such spines superimposed on the signal are disregarded. This rule presupposes that noise spikes can be distinguished from rapid signal fluctuations. It has been found that all persons competent to evaluate record film are able, with but little practice, to make that distinction. The other qualification concerns "end spikes." Frequently, there is interference between sound traveling via two different routes, for example, direct and surface-reflected sound. As the two paths do not have exactly the same length, the intensity at the beginning and end of the signal may be markedly different from the intensity during the signal. In the case of destructive interference, the signal then assumes the shape shown in Figure 14. The end spikes appearing in such signals are also disregarded.

The rules just outlined have certain advantages and certain drawbacks. The principal advantage is that the peak amplitude of a signal can be read much more rapidly than such quantities as mid-signal amplitude; also, it is unambiguous. The drawbacks appear when the signal envelope is not smooth. In

that case the sound emitted during a short signal interval arrives at the receiver during a much longer period of time, with the result that the energy received during an interval equal to the signal length is substantially less than all the energy received. This effect will be quite conspicuous for very short signals, but negligible for continuous transmission (10-sec signals). As a result, the average amplitude is very definitely a function of ping length, when peak amplitudes are used; it very likely would not be a function of ping length if the amplitudes of individual signals were defined in a different manner. One possible solution has been suggested by the group which is carrying out transmission experiments at WHOI. They have constructed an integrating circuit. If the received signal is squared and fed directly into this integrating circuit, the recording instrument shows the total energy received. This would be strictly proportional to the signal length and would thus provide a measure typical for the ocean and its overall transmission properties. Any deviation from strict proportionality would be indicative of nonlinear transmission and would, therefore, be of the greatest importance. At the time of this writing, no such experiments had been carried out.

Once individual amplitudes have been assigned to the five signals that comprise one sample, the average amplitude is found by taking the arithmetical mean of the five individual amplitudes. This procedure has the advantage of simplicity. Alternatively, one could compute the mean level or the mean intensity (squared amplitude). A very rough estimate shows that in a typical record the averaging of amplitudes and of intensities will lead to results which are different by about 1 db. While this difference depends on the assumed distribution function of amplitudes, it is not likely to be a dominant cause of error in a determination of the transmission loss.

Transmission work at single sonic frequencies began only recently, and the analysis procedure has not yet been very well standardized. Records obtained up to the present appear to indicate that fluctuation of signal intensity is much less severe at sonic frequencies than at supersonic frequencies. On the other hand, because of image interference, systematic changes in signal level are observed at short ranges which vary so rapidly with range that any averaging procedure would obscure them. For this reason, in transmission work at frequencies from 200 to 1,800 c individual signal levels rather than sample averages are reported.

In the records obtained at sonic frequencies the envelope of the signal trace, as a rule, is badly serrated. The fuzziness of the envelope is probably caused by the unfavorable signal-to-noise ratio, which is about 1 db for 1.8 kc and lower, somewhat higher for 22.5 kc, and by the relative narrowness of the filters used in the recording channels. If random noise is received through a wide filter, the oscillograph trace has a typical "spiked" appearance, that is, the noise is characterized by sudden high peaks of short duration. If the filter is narrow, as it must be in sonic transmission work, the individual peaks are lowered and broadened, and their separation from the single-frequency signal is more difficult. For this reason, the person reading the film record does not attempt to measure the "peak" level, which would be fictitious, but estimates and reports the average amplitude. It has been found that the uncertainty introduced by this estimate is less than 1 db, on the average.

The final step in the processing of a transmission run consists of the recording of the computed signal intensity. Since in most transmission runs the range is altered by a factor of 10 to 100, the signal levels change in the course of a run by a large number of decibels. It has, therefore, been found useful not to plot signal level in decibels below transducer output directly, but to take out the bulk of the variability by plotting the transmission anomaly. Usually, the transmission anomaly is plotted as the ordinate downward, with range as the abscissa. Theoretically, this curve should pass through zero for zero range. In view of the great experimental difficulties involved in the determination of the signal level at very short ranges, the curves usually stop at a range of 100 yd or more.

4.4   ECHO RUNS

As mentioned before, transmission runs are by far the most important useful method of obtaining information on the propagation of sound in the ocean. The other two methods, which are of secondary importance, will be discussed in the next two sections for the sake of completeness.

Echo runs have been carried out both on specially designed standard bodies and on chance targets, such as wrecks, in order to study the dependence of echo level on range and in order to study fluctuation and coherence. The principal purpose of echo runs has usually been to study not the propagation of sound between echo-ranging transducer and target, but rather the effect of certain targets on the received echo. (See Chapters 18 to 26 of this volume.)

The equipment used in echo runs differs from that used in transmission runs in that sending and receiving stacks are aboard the same ship and the same sound head is used both for sending and receiving. A change-over relay connects the sound head first with the sending stack and then, immediately following the emission of the signal, with the receiving stack.

Artificial targets have been developed for research and training purposes. Natural targets usually reflect very differently at different aspects; most artificial targets are designed to minimize this kind of directionality without sacrificing too much overall reflecting power. There is one geometrical shape which remains the same regardless of any twisting of the cable from which the target is suspended. That is the sphere. From the point of view of constant reflecting power, spheres constitute ideal artificial targets.

Unfortunately, the reflecting power of a sphere, while constant, is fairly small. To obtain useful echoes from spheres at distances similar to the ranges commonly encountered in practical echo ranging, one would have to use spheres with a diameter greater than 30 ft. It was found, however, that a 10-ft sphere was almost unmanageable at sea. The only spheres which could be handled with ease were spheres with a diameter of 2 or 3 ft.

In the search for an artificial target with a large target strength, the best solution found so far has been the triplane[6,7] shown in Figure 15, which combines ease of handling with a reflecting power comparable to that of a submarine. It is a well-known fact, sometimes used in optical signaling, that a ray which has been reflected from three planes which are mutually perpendicular leaves in a direction exactly opposite to the incident direction. The action of a triplane can, therefore, be compared with that of a single plane perpendicular to the incident rays. That is why a triplane reflects a larger percentage of the incident energy back into the transducer than any other body of equal size.

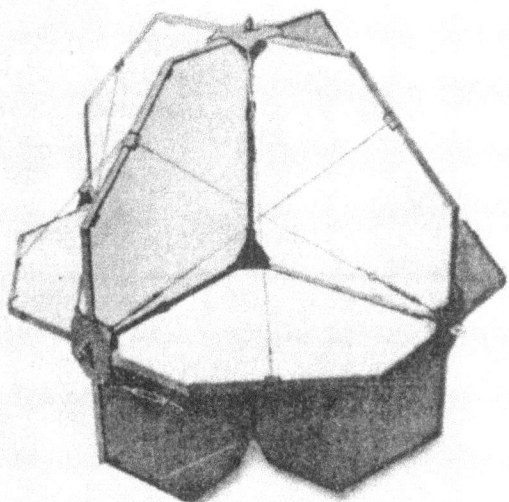

FIGURE 15. Triplane.

Another type of artificial target is the so-called echo repeater. This is a device which acts essentially as a relay. It consists of a transducer with power output proportional to the incident sound energy received by a hydrophone. Echo repeaters have been used only for training purposes. A full description of the echo repeater can be found in two UCDWR reports.[8,9]

4.5 INFORMATION OBTAINABLE FROM REPORTED RANGES

The research methods described in Sections 4.3 and 4.4 of this chapter are of comparatively recent origin. During the first year of the war, the only information available consisted of observed maximum echo and listening ranges obtained by surface ships on escort or patrol duty and by research vessels on ocean cruises.

In testing the performance of echo-ranging gear, several workers recognized the strong variability of achieved maximum echo ranges.[10,11,12] Vessels at-

tached to the West Coast Sound School at San Diego found in practice maneuvers that ranges in the afternoon compared unfavorably with ranges in the morning of the same day. These observed maximum ranges gave the first clue to the existence of shadow zones in the presence of sharp downward refraction. Maximum ranges obtained by echo ranging on submarines above and below the depth of the layer revealed the existence of a layer effect (see Section 5.3.3). Also, the tabulation of observed ranges over various types of ocean bottom in shallow water gave clues as to the effect of the bottom on sound transmission.

Because of the unexplained variability of observed maximum ranges, it was decided to set up the investigation of the underwater sound field as a research program, and the quasi-laboratory methods of transmission runs and echo runs were developed. Since the inception of the transmission program, observed echo ranges have rarely served as scientific evidence; they have continued to serve as a stimulus for the investigation of new problems and as signposts on the road to solutions. The SS *Nourmahal*, a converted yacht with a deep projector and with unusually quiet machinery, has reported extreme echo ranges in the presence of very deep isothermal layers; as a result, the sound field in deep mixed layers was investigated. To give another example, earlier experience had shown that the sound field in shallow water over MUD bottoms is very nearly the same as the sound field in deep water, because MUD reflects sound rather poorly. Unexpectedly long echo ranges in certain areas in which the bottom was classified as MUD led to a new program aimed at a differentiation of the various bottom sediments now called MUD. Also, data obtained at WHOI seem to show that attenuation increases at very high wind forces.

These observed ranges have been obtained both by naval vessels in regular operations and by research vessels. A number of naval vessels have sent to WHOI records of observed maximum echo ranges along with bathythermograph slides. Additional observed ranges were obtained by research vessels on extended cruises in various parts of the world. The range data thus obtained do not permit any detailed conclusions concerning the transmission loss as a function of range but serve to indicate whether sound transmission was good, fair, or poor. A summary of the theory of maximum echo ranges is presented in Volume 7 of Division 6.

Chapter 5

DEEP-WATER TRANSMISSION

TRANSMISSION IN DEEP WATER, where bottom-reflected sound is unimportant, is somewhat simpler to study than transmission in shallow water. Even when the effects of the bottom have been eliminated, however, sound transmission in the ocean remains an exceedingly complex phenomenon. The theoretical results based on the elementary ray theory and on an idealized ocean stratified in uniform horizontal layers are seldom realized exactly in the sea. Sometimes this simple picture leads to erroneous results, even qualitatively. Moreover, the only constant element in underwater sound transmission is change. No one ping resembles the preceding. In this chapter, the rapid fluctuation of transmitted sound from one second to the next is ignored and reference is made throughout to averages based on many consecutive pings. However, even these averages sometimes vary considerably.

Although the theory developed in the previous sections is admittedly imperfect and may be incorrect in principle, this theory is nevertheless retained as the framework on which to hang the discussions of the observational material. The theory is believed valuable, partly in indicating which results may be expected to have general validity beyond the particular conditions under which the results were obtained. Even more important, a discussion of the interrelation between facts and theories is essential for an intelligent formulation of research programs. In the long run, progress in any scientific problem can be achieved most efficiently by formulating hypotheses and then testing them in critical experiments. To lay the groundwork for such future research is, in large part, the objective of the present chapter.

5.1 FACTORS AFFECTING DEEP-WATER TRANSMISSION

In principle, the propagation of sound can be completely determined if the nature of the medium through which the sound passes is known. In the present section a description is given of the known properties of the sea which are believed to influence underwater sound transmission.

5.1.1 Meaning of "Deep Water"

For the purposes of this chapter, water is deep when the bottom has a negligible effect on underwater sound propagation. From a theoretical standpoint this has the very simple meaning that the bottom is ignored; the ocean is thought of as extending to infinite depths. From the observational standpoint, this means that only those observations will be considered here on which the bottom is believed to have no effect.

In general, the bottom can have several effects on underwater sound. Sound energy reaching the bottom may be partly reflected back at various angles into the body of the sea and partly transmitted into or absorbed by the bottom. The relative amounts reflected and absorbed depend on the depth and the nature of the bottom, prevailing refraction conditions, and sea state. This dependence and, generally, the effect of the bottom on sound transmission will be discussed in detail in Chapter 6. Furthermore, the presence of the bottom affects the background. Some of the sound reflected backward by the bottom reaches the receiver and gives rise to a ringing sound known as reverberation.

For most of the observations discussed in this chapter, short pulses of sound are used. With this technique, sound that has traveled to the bottom and has been reflected toward the hydrophone can readily be distinguished from sound that has traveled directly from projector to hydrophone. Thus, for most observations the bottom-reflected sound can readily be distinguished. If the bottom is rough, an appreciable amount of sound may reach the hydrophone after having been scattered from various portions of the sea bottom so that the signal is followed by a long single-frequency train of forward reverberation. Usually, the direct signal is so far above this background of scattered sound that forward reverberation is negligible in the evaluation of transmission observations.

It is of importance to know when the ocean is effectively deep in practical applications of underwater sound. With present echo-ranging gear, an echo from

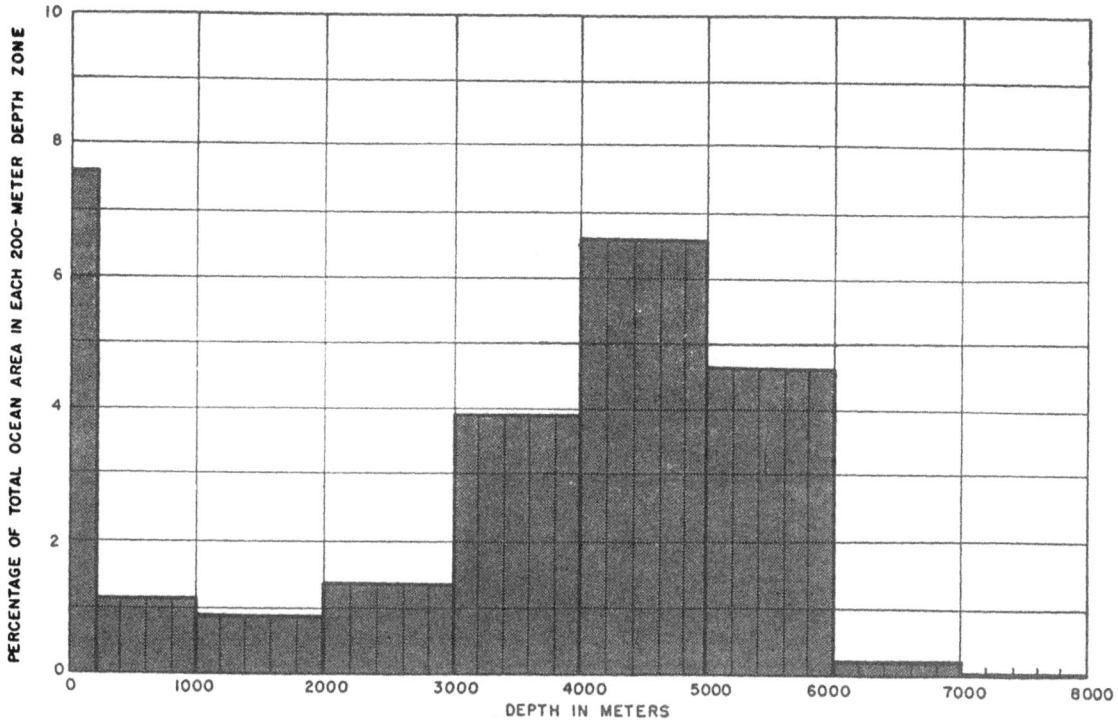

FIGURE 1.   Distribution of depths in the sea.

a submarine or typical surface vessel cannot ordinarily be detected more than 3,000 yd from the projector in water of any depth, except under unusually favorable conditions. All supersonic projectors are highly directional with not much energy radiated at angles more than 6 degrees from the axis. If the bottom is more than 150 fathoms below the projector, and the water is isothermal, very little of the energy in an echo-ranging pulse will reach the bottom at ranges less than 3,000 yd or return to the surface at ranges less than 6,000 yd. Thus, the echo from targets near maximum range will contain very little bottom-reflected sound; however, the background for such echoes may contain some bottom reverberation. If sharp temperature gradients are present in the upper layer of the ocean, the sound beam will be bent down more sharply, and a considerable amount of bottom-reflected sound could reach a target 3,000 yd away in water 150 fathoms deep. The bottom reverberation in such conditions may be quite intense even at 1,500 yd. To insure that bottom-reflected sound cannot return an echo in practical echo ranging, a depth of more than 200 fathoms is required, while twice this depth is required to eliminate bottom reverberation. For various types of tilting beam equipment, sound scattered from the bottom can be important even in

somewhat deeper water. For most echo-ranging situations, however, 100 or 150 fathoms is a more representative dividing line between deep and shallow water.

It makes very little difference whether the point of division is taken as 150 fathoms, or 100 fathoms, as has been done in the manuals of echo-ranging prediction issued by the Navy,[1,2] or 200 fathoms, as has been suggested. Water depths between 100 and 1,500 fathoms are quite uncommon. Figure 1 shows the distribution of depths in the sea.[3] It is evident from Figure 1 that almost all of the ocean bottom is either less than 100 fathoms, about 200 meters, below the surface, or more than 1,500 fathoms, about 3,000 meters, below the surface.

Water which is deep for echo ranging may be shallow for sonic listening, since average listening ranges are so much longer than average echo ranges, and since sonic listening gear is nondirectional. Listening ranges are often greater than 10,000 yd. Except in the deepest parts of the ocean, sound arriving from such long ranges will contain bottom-reflected sound. Sonic gear is usually nondirectional in a vertical plane, at least at low frequencies, and bottom-reflected sound in 2,000 fathoms may contribute appreciably to the received signal. Thus, for sonic lis-

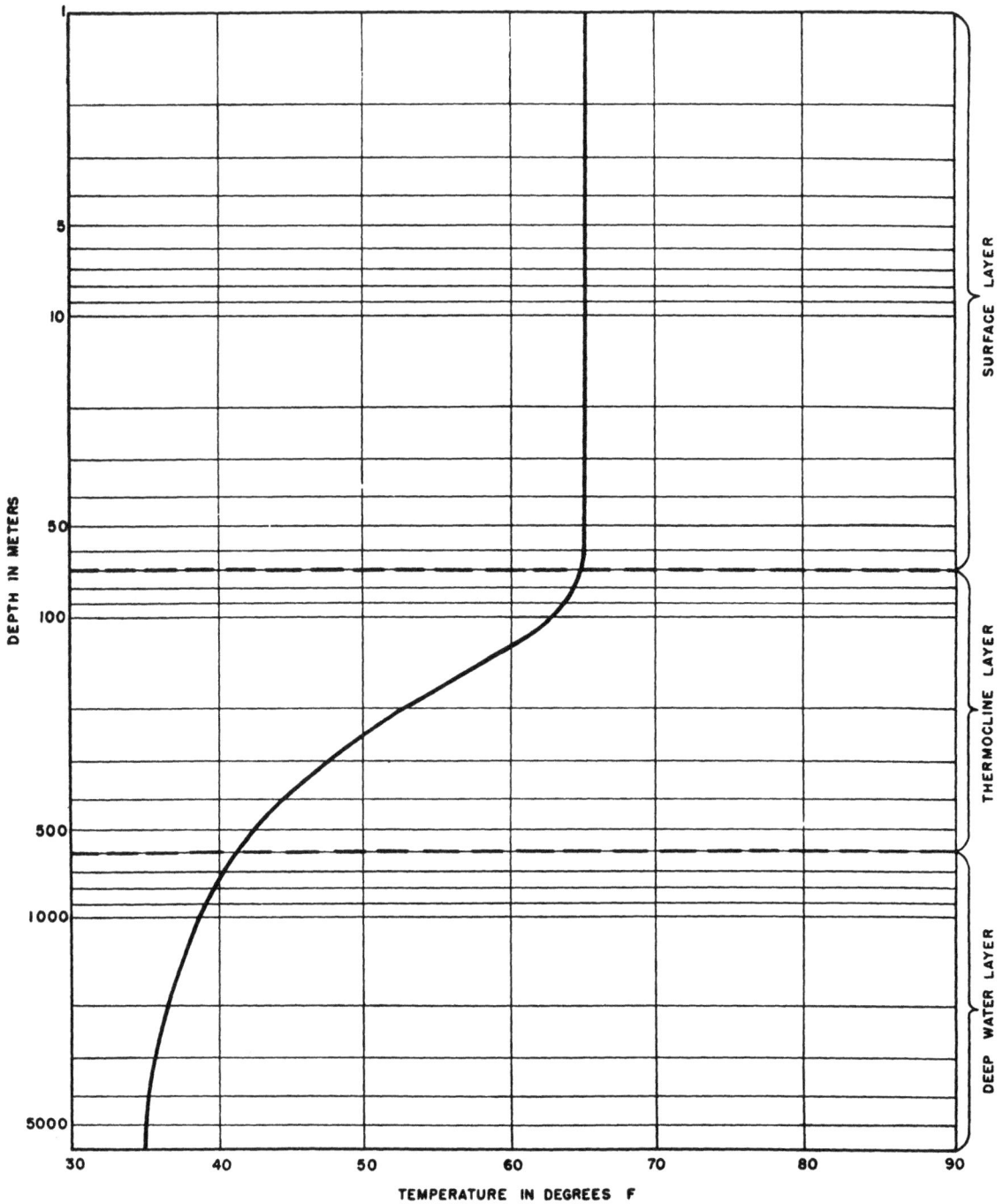

FIGURE 2.   Typical temperature-depth curve.

tening at long ranges the ocean is rarely if ever deep.

Supersonic listening gear, on the other hand, is usually sharply directional in the vertical plane and will discriminate against bottom-reflected sound from ships in the same way that it will discriminate against bottom-reflected echoes. Thus water which is deep

for supersonic echo ranging will also be deep for supersonic listening, even though supersonic listening ranges may exceed sonic ranges.

Evidently, at supersonic frequencies most of the ocean is effectively deep for practical purposes. Even at low sonic frequencies, transmission in water deeper than 1,500 fathoms is practically deep-water transmission under many conditions. Thus the study of sound transmission in deep water is of considerable practical importance.

5.1.2 Vertical Temperature Structure and Computed Ray Diagrams

The temperature distribution in the ocean largely determines the sound velocity distribution, which we have seen is an important factor in sound intensity. For this reason, measurement of ocean temperatures at various depths has been an integral part of the research on sound transmission and is also important in the tactical use of sonar equipment.

The temperature in the ocean is affected by the absorption of radiation from the sun and sky, by the cooling of the surface layer by evaporation, by displacements due to currents and upwelling, and by the addition of fresh water near shore. Usually a water column in the deep sea can be divided into three principal layers, shown by the sample temperature-depth plots in Figure 2: (1) a relatively warm surface layer, which is subject to daily and seasonal changes in thickness and vertical temperature gradients, (2) a layer of transition at mid-depths called the *thermocline*, in which the temperature decreases rapidly with depth, and (3) the cold deep-water layer, in which the temperature decreases only gradually with depth. A detailed discussion of the temperature distribution in the ocean is given in Volume 6 of Division 6. Here, only a few of the basic temperature-depth patterns are described and their expected influence on underwater sound transmission briefly discussed.

It will be pointed out in later sections that the transmission loss is least and sound ranges are longest when the surface layer is reasonably isothermal and deeper than about 100 ft. Such deep isothermal layers tend to occur when the water at the surface is losing more heat than it is gaining, as in midwinter in the high latitudes. The colder surface water will be heavier than the water just beneath and will mix with it. As a result, a surface layer of more or less constant temperature will be formed. In midwinter the isothermal surface layer is usually several hundred feet deep, except in tropical waters, where this depth varies from 50 to 500 ft depending on ocean currents and other factors. In very high latitudes the isothermal layer may extend down to the ocean bottom in February or March.

The ray diagram for sound transmission in the case of an isothermal surface layer above a thermocline has approximately the characteristic shape shown in Figure 3. According to the simple ray theory, the sound beam should split at the bottom of the isothermal layer, with the upper portion bending gradually

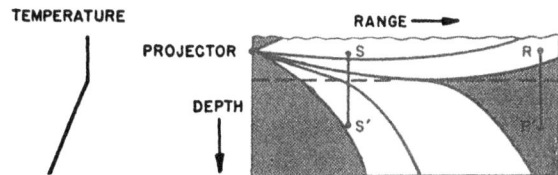

FIGURE 3. Ray diagram for isothermal water above thermocline.

back to the surface because of the effect of pressure and the lower portion bending sharply down into the thermocline. Temperature-depth patterns resulting in such a predicted ray diagram are called "split-beam" patterns. If the intensity of the sound field were measured along the vertical line SS' in Figure 3, the intensity immediately below the isothermal layer should decrease sharply with increasing depth. After having reached a minimum, the sound field intensity should begin to increase slowly with increasing depth until finally the edge of the main sound beam is reached.

At longer ranges with split-beam patterns a measurement of intensity, as along the vertical line RR' in the diagram, should indicate a substantial amount of sound in the isothermal layer, but in or below the thermocline very little sound should appear. As shown in Section 5.3.2, these predictions of theory for long range are not confirmed by the observations, which show no clear trace of the predicted shadow boundary below the layer.

It may be pointed out that the shaded area in Figure 3 is usually not a true shadow zone, even in theory. The temperature-depth graph usually curves continuously from the isothermal layer into the thermocline, rather than breaking at the sharp angle shown in Figure 3. As a result of this curvature, some direct sound always theoretically penetrates the "shadow zone" in this case, but this theoretical sound is much weaker than the sound actually observed.

Heating of the surface layer of the ocean sometimes produces a temperature gradient which extends all the way to the sea surface. Such conditions are most common at high latitudes during the summer months, when the surface water is gaining more heat than it is losing. Surface heating and marked temperature gradients in the top 30 ft of the ocean are particularly marked on summer afternoons with calm seas and cloudless skies. At night, or during periods of high winds, gradients near the surface tend to disappear.

The ray diagram computed for a temperature gradient extending up to the surface is shown in Figure 4. The decrease of sound velocity with in-

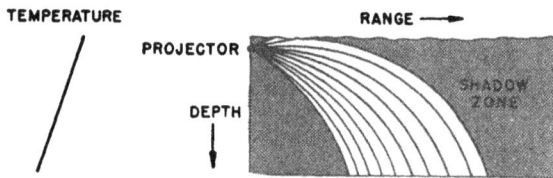

FIGURE 4.   Ray diagram for negative gradient extending to surface.

creasing depth bends the entire sound beam downward, as discussed in Chapter 3. Beyond a certain limiting range, which increases with increasing depth, no sound ray can penetrate, and a shadow zone of complete silence should result. Observations of underwater sound transmission confirm the presence of this shadow zone when the temperature gradient is sufficiently strong, about 1 degree or more in the top 30 ft. The only sound reaching such a shadow zone is the scattered sound, which also produces reverberation back at the echo-ranging projector. When the surface gradients are weak, however, other effects become important, and shadow zones do not appear.

In addition to these two basic but simplified temperature-depth patterns, innumerable varieties of intermediate situations occur. Complicated temperature structure is especially likely in the surface layer; and the accurate computation of a ray diagram from a temperature-depth record can be very laborious. Since the observations usually do not confirm the detailed predictions of the simple theory, which is based on small details of the temperature structure, the computing of ray diagrams for these intermediate cases is of limited usefulness.

Sharp positive temperature gradients are extremely rare in deep water. Such gradients are stable only when accompanied by positive salinity gradients. Salinity gradients may also affect sound velocity, but their effect is usually negligible compared to that of temperature gradients. Salinity gradients may be appreciable in some near-shore areas, where large rivers drain into the sea, and at the coastwise margins

of the permanent ocean currents, such as the Gulf Stream. In such regions, sharp positive temperature gradients may occur. In the open ocean, however, they are usually less than a few tenths of a degree in 30 ft. Because of the rarity of sharp positive gradients, there is a complete absence of data on sound transmission in deep water through regions of strong upward refraction.

5.1.3   Variability of Vertical Temperature Gradients

The way in which ocean temperature changes with depth is variable from time to time and from place to place. Gradual changes from day to day and from one geographical region to another have an important effect on the performance of sonar gear. These changes are discussed in detail in Volume 6 of Division 6, and form the basis for the Sound-Ranging Charts[4] and the Submarine Supplements.[5]

In some areas these changes are so rapid that they greatly complicate the study of underwater sound transmission. In the coastal waters off San Diego, a bathythermograph lowered at one end of a transmission run frequently showed marked differences from the bathythermogram obtained at the other end, with wholly different ray diagrams resulting. Two samples of such records are presented in Figure 5. Some of this variation represents a change with time, while much of it arises from changes with location. In early comparisons between transmission data and the computed range to the shadow boundary, an average was taken of the ranges computed from several bathythermograph records. More recently, a single bathythermograph record taken on the receiving vessel has been used at UCDWR in studying the relation between the transmitted sound intensity and the temperature-depth record.

TEMPERATURE MICROSTRUCTURE AND EFFECTS

In addition to these large temperature changes over several thousand yards, smaller changes take place over much smaller distances. These changes may affect the way in which the sound beam travels through the water. In Chapter 3 the predictions of the ray theory were discussed for a sound beam passing through an ocean in which the sound velocity depends only on depth, but decreases gradually with depth. In such an ideal ocean an exact temperature-depth record would be similar to that shown in Figure 6. A plot of temperature against range at any depth would give the horizontal lines shown in the figure. Under such temperature conditions a shadow zone is predicted at a certain limiting range.

In practice, the ocean is never stratified in plane parallel layers, each of uniform temperature. Instead, an exact temperature-depth record might be

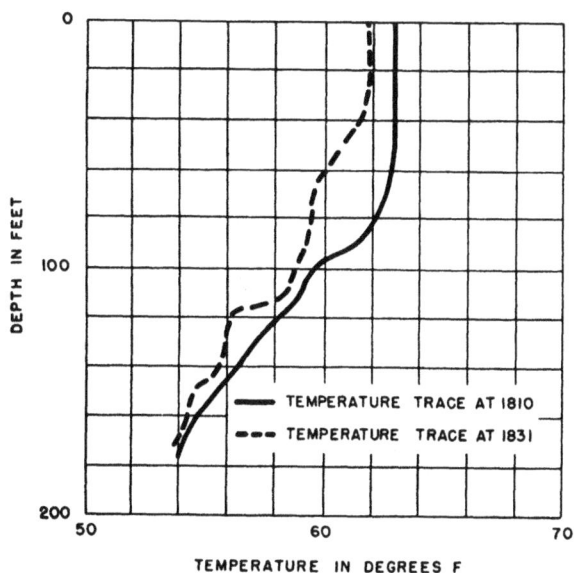

FIGURE 5. Temperature conditions at beginning and end of transmission run.

similar to that shown in Figure 7. Plots of temperature against range at different depths would be curves similar to the wavy curves also shown. This "temperature microstructure" must be taken into account in any explanation of observed underwater sound transmission.

The evidence available on thermal microstructure is very limited. Measurements on a surface ship are difficult to interpret because of the rise and fall of the measuring instrument through the water. Some of this vertical motion arises from the roll and pitch of the measuring ship, and some from the distortion of the temperature-depth pattern by the surface waves. For this reason, a very small-scale microstructure is difficult to measure from a surface ship, although changes over a hundred yards or so can usually be disentangled from the more rapid changes resulting from roll and pitch. Measurements from a submarine show conclusively the presence of complicated thermal microstructure.[6] In Figure 11 of reference 6, fluctuations of the vertical gradient are shown which amount to about 0.020 degree per ft over patches about 100 yd long. This result was obtained with

large temperature gradients present near the surface. When the bathythermograph shows mixed water to more than 100 ft, the microstructure is much less marked.

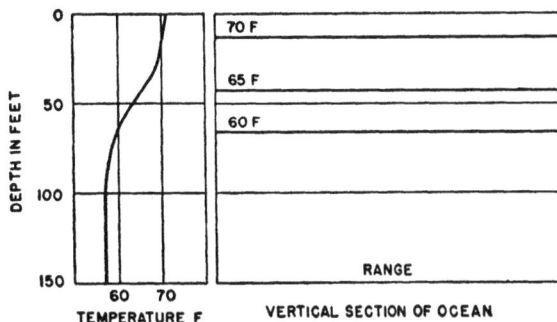

FIGURE 6. Temperature distribution in ideal ocean.

A general theory of underwater sound transmission which takes microstructure into account has not yet been formulated. However, certain general results seem apparent. These temperature fluctuations are usually fairly small compared to the smoothed gradient, and on the whole the actual temperature-depth

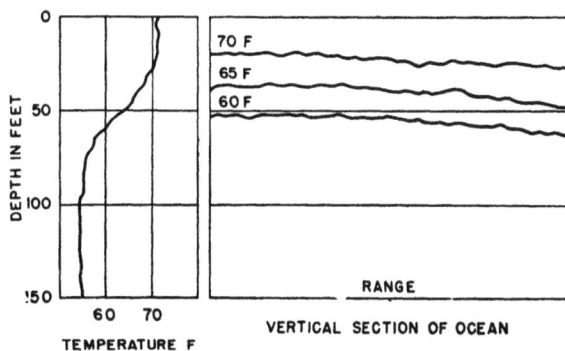

FIGURE 7. Temperature distribution in actual ocean.

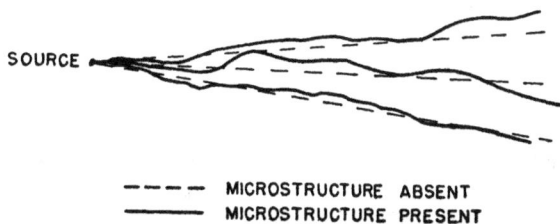

FIGURE 8. Distortion of sound beam by microstructure.

pattern portrayed in Figure 7 does correspond to the ideal pattern shown in Figure 6. Thus some correspondence may be expected between observed sound transmission data and predictions based on the smoothed temperature-depth pattern.

The chief effect of temperature microstructure is to introduce irregularities into the path of the individual sound rays. They will be slightly bent away from the average ray path in random fashion, as indicated in Figure 8. Considering a sound beam as a whole, we may expect that microstructure will very slightly broaden the beam pattern, although such broadening effects have never been determined with assurance. Within the sound field, local intensities may show deviations from the average values which would be observed in the absence of microstructure; these local deviations will be discussed in Chapter 7. Another effect of these irregularities is that sound may penetrate with a small but observable intensity into regions which are shadow zones according to the large-scale ray pattern.

It is possible to estimate the effect which microstructure will have on the ray trajectories of individual sound rays. Theoretical analysis shows that with certain simplifying assumptions the rms lateral displacement Δy of a sound ray because of microstructure is given by the formula[7]

$$\Delta y = \tfrac{1}{3}\sqrt{b}GR^{\frac{3}{2}}.$$ (1)

In this equation G is the rms value of the fractional velocity gradient caused by microstructure, R is the range, and b is a quantity having the dimension of a length, which may be called the *patch size* of the microstructure. Roughly speaking, b is the average distance over which the vertical velocity gradient caused by microstructure retains the same sign. To derive this formula, an expression was first obtained for the lateral displacement of a ray passing through a given microstructure. This expression was then squared and averaged. The square root of the final result gave expression (1).

It has already been noted that fluctuations of the vertical temperature gradient, amounting to 0.02 F per ft over patches about 100 yd in length, have been reported. If these values for G and for b are substituted into equation (1), it is found that at a range of 1,000 yd the rms lateral spreading of the sound beam amounts to about 20 ft; while at 2,500 yd it amounts to 70 ft and at 4,000 yd at 150 ft. These figures indicate that at these ranges random spreading of the transmitted sound beam, because of microstructure, will obscure bending of the sound rays due to large-scale vertical temperature structure if the vertical gradient is of the order of 0.1 F in 30 ft. Actual observation shows that even negative gradients of four times this magnitude often fail to produce clearly

recognizable shadow zones, although the sound does weaken gradually with increasing range. It is not known at present whether microstructure will frequently have a magnitude appreciably in excess of that assumed for the estimate of lateral beam spread. If not, some other cause must be invoked for an explanation of why weak negative gradients do not produce shadow zones.

Since no complete theory exists at the present time capable of explaining in detail the results obtained in transmission runs, much of the discussion of underwater sound transmission must be empirical in character. It is possible, for example, that some of the empirical relationships found between the smoothed temperature-depth curves and the measured transmission anomalies result primarily from an oceanographic correlation between the temperature microstructure and the smoothed distribution of temperature with depth. Such observed empirical relationships are valuable, but until their basic physical cause is explained they should be used with caution since they may be valid only for the particular time and place in which the observations were made.

5.1.4 Classification of Bathythermograms

For practical use of temperature-depth information some simple method of classifying bathythermograph records is essential. Even if the predictions of ray theory were exactly fulfilled, practical requirements would probably rule out the time and effort required to construct ray diagrams and to compute theoretical intensities. Thus, a set of rules has been devised to classify temperature-depth records by the properties which are acoustically significant.

Such classifications have also proved useful in transmission research. Since the simple ray theory was clearly inadequate, some other basis was required for comparing measured anomaly curves with the corresponding bathythermograms. In view of the complexity of possible temperature-depth curves, no classification can be entirely satisfactory. All such classifications must be regarded as preliminary until sufficient acoustic information is available to indicate exactly what features of the temperature-depth pattern are significant in any situation.

Present systems of classification are primarily designed to correspond to different types of transmission loss for a shallow projector, about 15 ft. When the temperature increases with depth sufficiently for the temperature at some depth below the projector to be greater than the projector temperature, rays leaving the projector at slight downward inclinations will in theory be bent back up to the surface again; the transmission anomaly for a shallow hydrophone should therefore be low, although experimental data on this point are lacking. When such positive gradients are present, the temperature pattern is called positive, sometimes denoted by PETER.

Such patterns may be more completely characterized by the depth of the layer of maximum temperature and the difference between maximum temperature and the temperature at projector depth. The sharpness of the underlying thermocline may also be acoustically significant.

Other temperature-depth records are classified by the temperature difference in the top 30 ft. If this difference is 0.3 F or less, the water is said to be *isothermal*, and the temperature pattern is called mixed, sometimes denoted by the word MIKE. When this difference is greater than 1/100 of the surface temperature the computed range to the shadow boundary is less than 1,000 yd, for projector at 15 ft, hydrophone at 30 ft. For this temperature condition, the predicted shadow zone is commonly observed, and transmission to a shallow hydrophone becomes poor for ranges greater than 1,000 yd. Such a temperature distribution is called a sharp negative pattern, sometimes denoted by NAN. Temperature differences intermediate between MIKE and NAN tend to be somewhat variable and are classified as weak negative or changing patterns, denoted by CHARLIE.

One exception is included in this relatively simple scheme. When the temperature difference from 0 to 30 ft is large enough to give a NAN pattern, but the temperature difference from 15 to 50 ft is 0.2 F or less, the pattern is classified as CHARLIE. With such an extremely shallow and negative gradient and with the projector in isothermal or nearly isothermal water, good but variable sound conditions may be expected. This type of pattern is the most favorable for the formation of a sound channel.

With MIKE and CHARLIE patterns the depth and sharpness of the thermocline would be expected to affect the transmission of sound to a deep hydrophone. Appropriate methods for characterizing these quantities are discussed in Section 5.3, where the acoustic measurements made with a hydrophone in or below the thermocline are summarized.

Another more detailed system of classification, which supplements the classification of negative gradients into MIKE, CHARLIE, and NAN patterns, has been devised at UCDWR. This system utilizes the depths at which the temperature is 0.1, 0.3, 1.0, 5.0, and 10 F below the surface temperature. These depths are called, respectively, D_1, D_2, D_3, D_4, and D_5.

For statistical analysis, these depths are given code numbers between 0 and 9 by the following numerical scale.

Code digit	Depth D in feet
0	$0 \leqq D < 5$
1	$5 \leqq D < 10$
2	$10 \leqq D < 20$
3	$20 \leqq D < 40$
4	$40 \leqq D < 80$
5	$80 \leqq D < 160$
6	$160 \leqq D < 320$
7	$320 \leqq D$
9	D greater than greatest depth reached by bathythermograph

Any bathythermogram may then be coded by giving the code digits corresponding to D_1, D_2, D_3, D_4, D_5. The surface temperature T is also coded by giving $T/10$ to the nearest whole number and by placing this digit after the other five and separating it by a decimal point. The code numbers for D_1 through D_5 and also $T/10$ are denoted by d_1, d_2, d_3, d_4, d_5, and d_6, respectively. The series of numbers is then written as $d_1 d_2 d_3 d_4 d_5.d_6$, as for example 23 457.6. The accurate determination of d_1 is very difficult because of the wide trace made by the bathythermograph near the surface; since the variability of this small temperature difference will usually be high, there is some question whether this quantity is usually significant.

Examples of bathythermograms classified by the two methods are given in Figure 9. These two systems of classification supplement each other and should probably be used together. The code system is probably most useful for surface gradients, where considerable detail is provided. For example, it is shown in subsequent sections that the transmission of sound to a shallow hydrophone depends markedly on d_2 for different NAN patterns. On the other hand, for a fixed d_2, transmission to a shallow hydrophone differs markedly between NAN and MIKE patterns. For deep gradients the code system is somewhat less useful, owing to the very expanded depth scale. For example it is frequently not clear from the present code whether a deep hydrophone is above or below the thermocline. It seems likely that when more com-

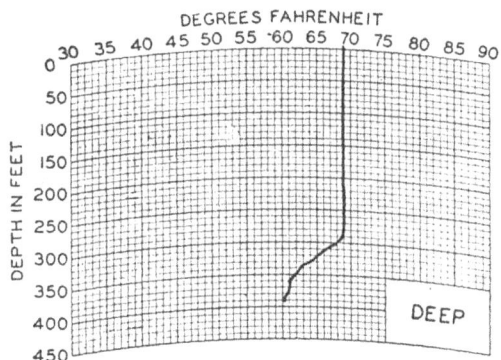

MIKE
CODE SYMBOL 66 679.7

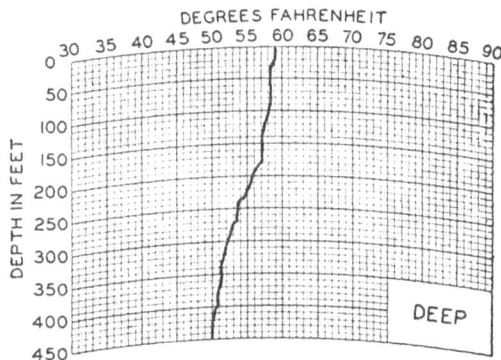

CHARLIE
CODE SYMBOL 22 569.6

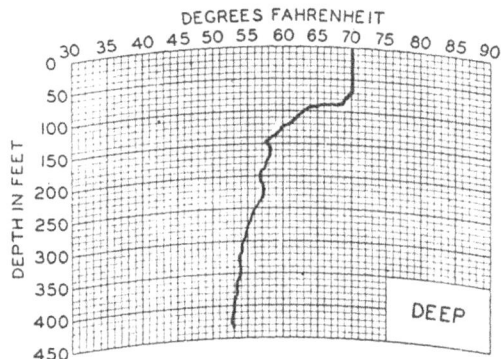

MIKE
CODE SYMBOL 44 555.7

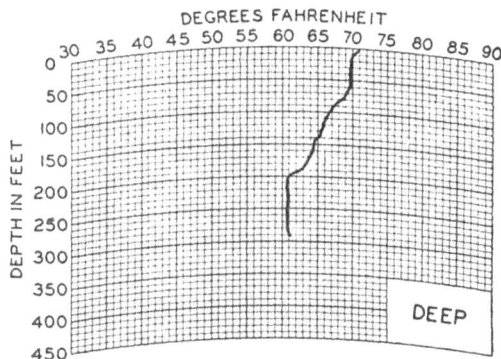

NAN
CODE SYMBOL OO 156.7

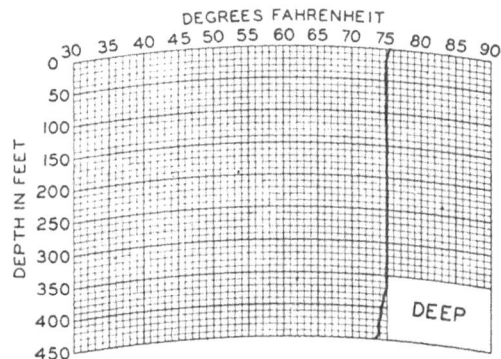

CHARLIE
CODE SYMBOL 02 799.8

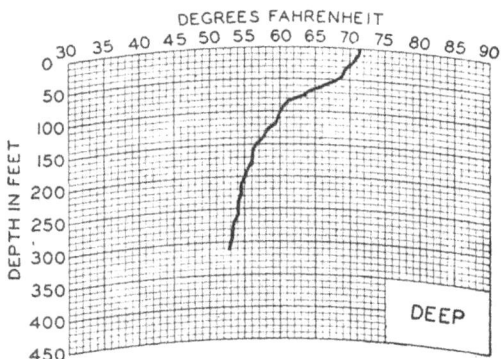

NAN
CODE SYMBOL 22 345.7

FIGURE 9.   Classification of bathythermograph records.

plete acoustic information is available, a modification of the code system, more closely related to the typical structure of the thermocline, will prove desirable.

5.2 TRANSMISSION IN ISOTHERMAL WATER

When temperature and salinity gradients are absent, the transmission of sound may be expected to be relatively simple. If pressure did not affect sound velocity, sound would travel outward in straight lines, and the inverse square law of intensity decay would be directly applicable. Since the effect of pressure on sound velocity is in fact very small, one may expect straight-line propagation to provide a reasonably good first approximation to the actual situation. An examination of the ray diagram given in Figure 23 of Chapter 3 shows that the ray leaving horizontally from a projector 15 ft deep in isothermal water is not bent up to the surface until it reaches a range of about 1,000 yd. On the other hand, in an isothermal layer 100 ft thick the range to the shadow boundary is always greater than 2,200 yd. Thus, one may expect that, certainly for ranges less than 1,000 yd and probably also for ranges up to 2,000 yd, the assumption that sound travels in straight lines in isothermal water is legitimate. This expectation is justified by the observations. It will be shown later that the assumption of straight-line propagation agrees with some of the data out to very much greater ranges than might be expected. The reason for this is not known. In the following discussion, sound transmission measurements in an isothermal surface layer of the ocean will, therefore, be discussed as though the sound rays did, in fact, travel in straight lines in such a layer.

Even with all sound rays traveling in straight lines, two influences act to disturb the ideal inverse square law discussed in Chapter 2. In the first place, sound is reflected from the sea surface; in the second place, various impurities, and possibly also the water itself, absorb energy passing through the sea, converting this energy to heat. The effects of surface reflection and absorption on the transmission of sound are discussed later.

5.2.1 Image Effect

It was shown in Section 2.6.3 that sound reflected from the surface can reduce the sound intensity close to the surface to a very small value. This effect arises from the phase reversal suffered by a wave when it is reflected at a free surface. If p_1 is the pressure amplitude at 1 yd from the source, and if h_1 and h_2 are the depths of source and receiver respectively, we see from equation (129) of Chapter 2 that the pressure amplitude at the range R is given by

$$\text{Amplitude} = \frac{2p_1}{R} \sin 2\pi \frac{h_1 h_2}{R\lambda}.$$

If the simple inverse square law for intensity were satisfied, the pressure amplitude at the range R would be proportional to $1/R$, that is,

$$\text{Amplitude} = \frac{p_1}{R}.$$

The transmission anomaly, which is the transmission loss in decibels above that predicted by the inverse square law for intensity, is therefore given by

$$A = -20 \log \left[2 \sin 2\pi \frac{h_1 h_2}{R\lambda} \right]. \tag{2}$$

If the surface is assumed to reflect only a fraction γ_a^2 of the sound energy incident on it, the analysis in Section 2.6.3 must be modified. With a little mathematical manipulation, the transmission anomaly for this case becomes

$$A = -10 \log \left[1 - 2\gamma_a \cos 4\pi \frac{h_1 h_2}{R\lambda} + \gamma_a^2 \right]. \tag{3}$$

The transmission anomalies resulting from this formula for different values of γ_a are plotted in Figure 10.

This analysis is of doubtful validity for short wavelengths since neglect of the surface water waves in heavy seas is probably not legitimate for sound waves only a few inches long. With calm seas, however, the interference patterns predicted by the above analysis have occasionally been observed at 24 kc at close ranges. Equation (3) must be used with caution for ranges much greater than 1,000 yd. The upward refraction caused by the pressure effect, as well as the variation in travel time caused by thermal microstructure, distort and obscure the interference pattern predicted by the elementary theory. However, the exact limits of validity of equations (2) and (3) must be determined empirically.

The available data show that for sufficiently low frequencies, equation (2), or equation (3) with an amplitude reflection coefficient γ_a nearly equal to unity, provides an approximate description of the observed transmission. In particular, beyond a range R', corresponding to a path difference of a half wavelength, the transmission anomaly increases steadily

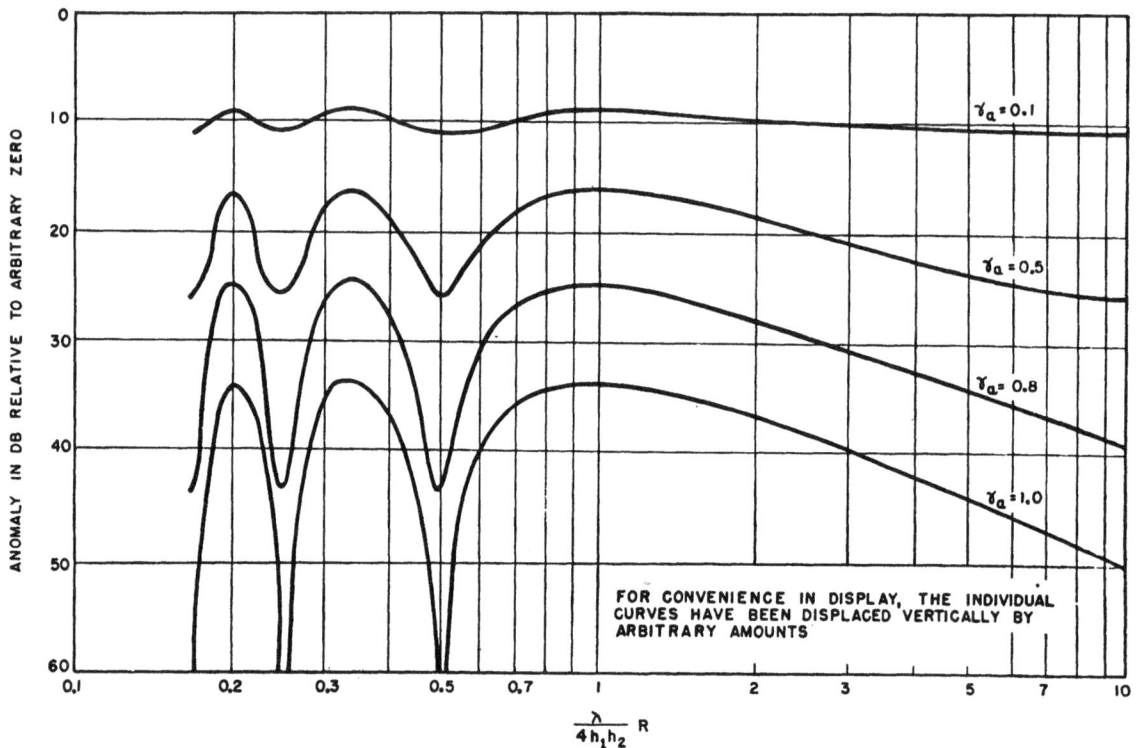

FIGURE 10.   Theoretical transmission anomalies for different values of the reflection coefficient of the surface.

from its minimum value and, beyond a range of about $2R'$, increases equally with $20 \log R$. This range R' is given by the equation

$$R' = \frac{4h_1h_2}{\lambda}. \qquad (4)$$

More specifically, equation (2) is applicable for frequencies of less than 200 c. For this low frequency, this equation may be used out to ranges of 1,000 to 2,000 yd, beyond which bottom-reflected sound, even in 2,000 fathoms, is usually stronger than the direct sound. At 600 c the correspondence between theory and observation is not so good, although the general tendency predicted by equation (3) is definitely present; possibly the best value of γ_a for 600-c sound is around ½ to ¾. At frequencies greater than several thousand cycles, no definite trace of image effect has been consistently observed in the open sea.

One of the earliest sources of observational information on this subject consisted of a set of transmission runs made jointly in 1943 by the Columbia University Division of War Research at the U. S. Navy Underwater Sound Laboratory, New London [CUDWR–NLL], University of California Division of War Research [UCDWR], and Massachusetts Institute of Technology Underwater Sound Labora-

tory [MIT–USL]. [8–10] An acoustic minesweeper was used as the source, and the signal was received with band-pass filters centered at different frequencies. The water depth was 600 fathoms. Unfortunately, the temperature near the surface was not isothermal to 100 ft; in fact, in some of the runs sharp negative gradients extended to the surface, and for some runs the hydrophone was below a sharp thermocline. Since the image effect was shown consistently at 250 and 700 c at ranges between 100 and, 300 yd, where bending by temperature gradients rarely affects measured sound intensities, the data may be taken as an indication that the same effects would also appear in isothermal water.

Some transmission runs at 2 kc and at 8 kc also showed some leveling off of the transmission anomaly at higher ranges.[8] However, the steeper slope was never so marked at ranges less than 500 yd that it could be attributed to image effect rather than to downward refraction.

A detailed comparison between theory and observation for the sound of lower frequencies is made in reference 10. The theory takes bottom-reflected sound into consideration and achieves rather good agreement with the observational data. However, the bottom-reflected sound comes in at such close range that the results do not cast much light on the problems of transmission of low-frequency sound in deep water at ranges greater than a few hundred yards.

A much more complete set of measurements on low-frequency transmission in deep water is given in a progress report from UCDWR.[11] That report summarizes the first results obtained in a long-range program designed to investigate sound transmission at frequencies of 200, 600, 1,800, 7,500, and 22,500 c. Since short pulses of sound were used in this work, the direct and bottom-reflected sound can be distinguished by the difference in travel time, provided the range is not too great. Also, specially designed transducers were employed with the result that the power output was high and usually remained constant during each transmission run.

Sample results for individual transmission runs at 200, 600, and 1,800 c are shown in Figure 11. As is evident from the bathythermograph code given with each plot, these data were obtained with isothermal water at the surface. Each point in these plots represents the transmission anomaly found for a single sound pulse. The curved lines represent the values found from equation (3) with the amplitude reflection coefficient γ_a chosen to give the best fit to the observations. For each frequency, the measured transmission anomalies at different ranges are moderately accurate relative to each other, while the absolute values are less reliable; hence each set of observed anomalies in Figure 11 has been shifted vertically to give the best agreement with the theoretical curves.

The agreement between the plotted points and the theoretical curves is typical of the results generally obtained in deep water with an isothermal layer. At 200 c, image effect is usually marked, and at 1,000 to 2,000 yd it reduces the sound level about 15 db on the average below the inverse square value; this corresponds to an effective reflection coefficient γ_a of 0.8, somewhat lower than the value of 0.9 for the 200-c curve in Figure 11. At 600 c, the effect is less marked, corresponding to an effective reflection coefficient γ_a in the neighborhood of 0.7, and an average transmission anomaly of only about 10 db at 1,000 yd. However, the dip in intensity shown in Figure 11 at about 60 yd, in agreement with theoretical prediction, suggests that image effect is in fact present. At 1,800 c, the predicted minimum between 150 and 200 yd is apparently present, but the reduction in intensity at ranges between 1,000 and 2,000 yd is quite small, corresponding to a value of about 0.5 for γ_a. Thus at

frequencies above about 1,000 c, it appears that image effect is relatively unimportant. This is in general agreement with theoretical expectations.[12] At ranges greater than 2,000 yd, bottom-reflected sound is usually dominant, even in water several thousand fathoms deep.

The reflection coefficients to which the different curves in Figure 11 correspond should not be taken as a measure of the amount of sound reflected by the surface. It is, of course, virtually certain that most of the sound reaching the ocean surface is reflected back into the ocean in some direction, except possibly when strong winds produce absorbing bubbles close to the surface. However, some of this sound may be reflected in directions quite different from the sound reflected at an ideally flat, horizontal surface. Also, the relative phases of the direct and surface-reflected sound may be altered by the irregularities in the ocean surface. As a result of these two factors, the image effect to be expected for a flat, perfectly reflecting surface may be modified. Equation (3) is then useful as a semi-theoretical, semi-empirical formula for fitting the observed data.

A somewhat different manner of presentation, which includes all the data available at the time reference 11 was written, is shown in Figure 12. Here, all available anomalies at certain fixed ranges are plotted on a linear range scale. In such a plot, the short ranges are too compressed to show the interference patterns characteristic of the image effect. However, such plots are very suitable if emphasis on the data at longer ranges is desired.

These data provide supporting evidence for the general statements made in the discussion of Figure 11. The rise of the median curves for 600 and 1,800 c is probably not real, but simply a result of observational selection; at the long ranges, the received signals are difficult to distinguish from noise, and only those few signals can be measured which, because of fluctuation, rise far above the noise level. Thus the median curves in Figure 12 are probably considerably higher at long range than they would be if all the runs yielding data at short range could have been continued successfully to long range.

5.2.2   **Absorption**

At frequencies above several thousand cycles, image effect is usually unimportant, and in the absence of temperature and salinity gradients, absorption becomes the chief effect modifying the inverse square

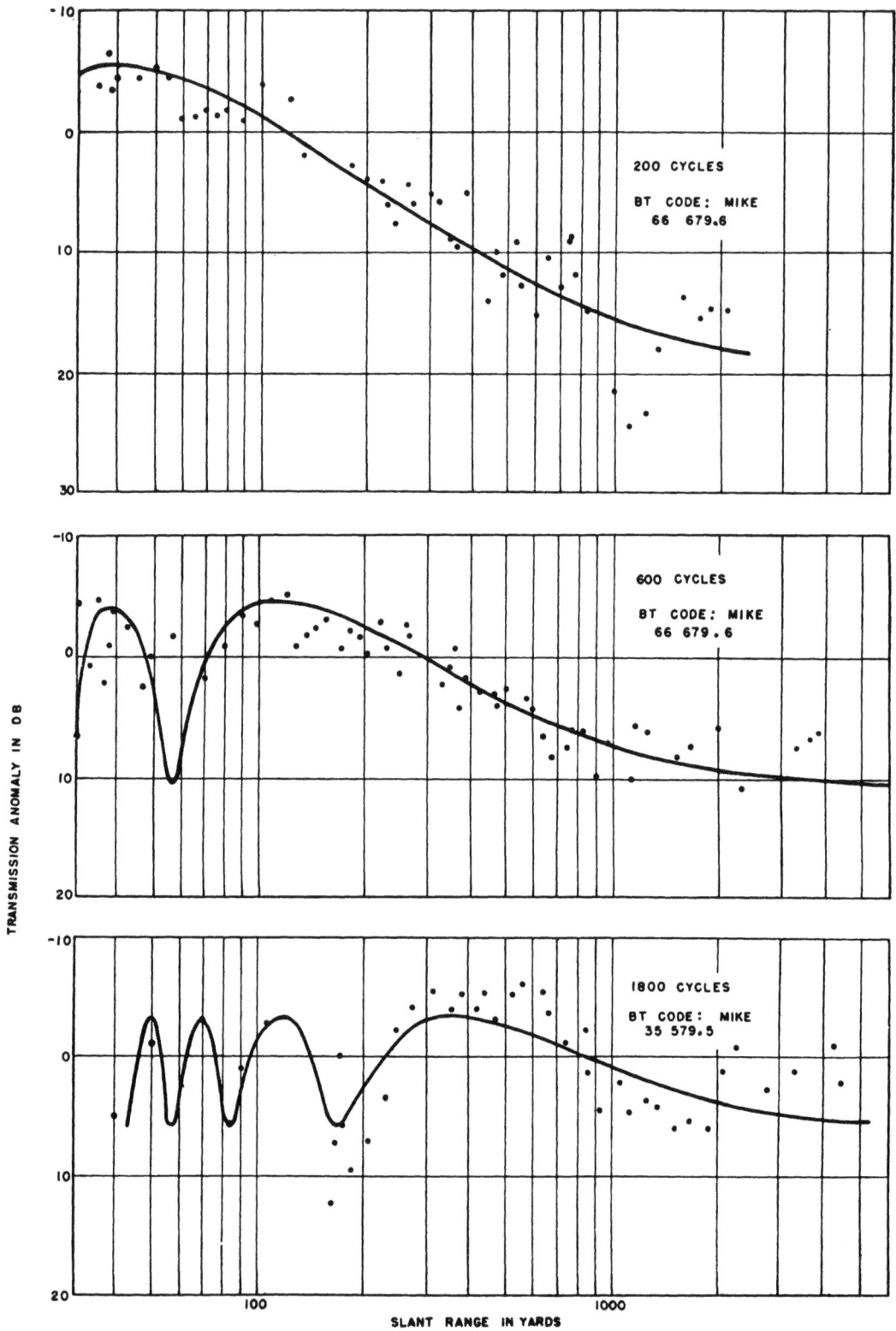

FIGURE 11. Typical transmission anomalies at sonic frequencies, source at 14 feet, hydrophone at 50 feet.

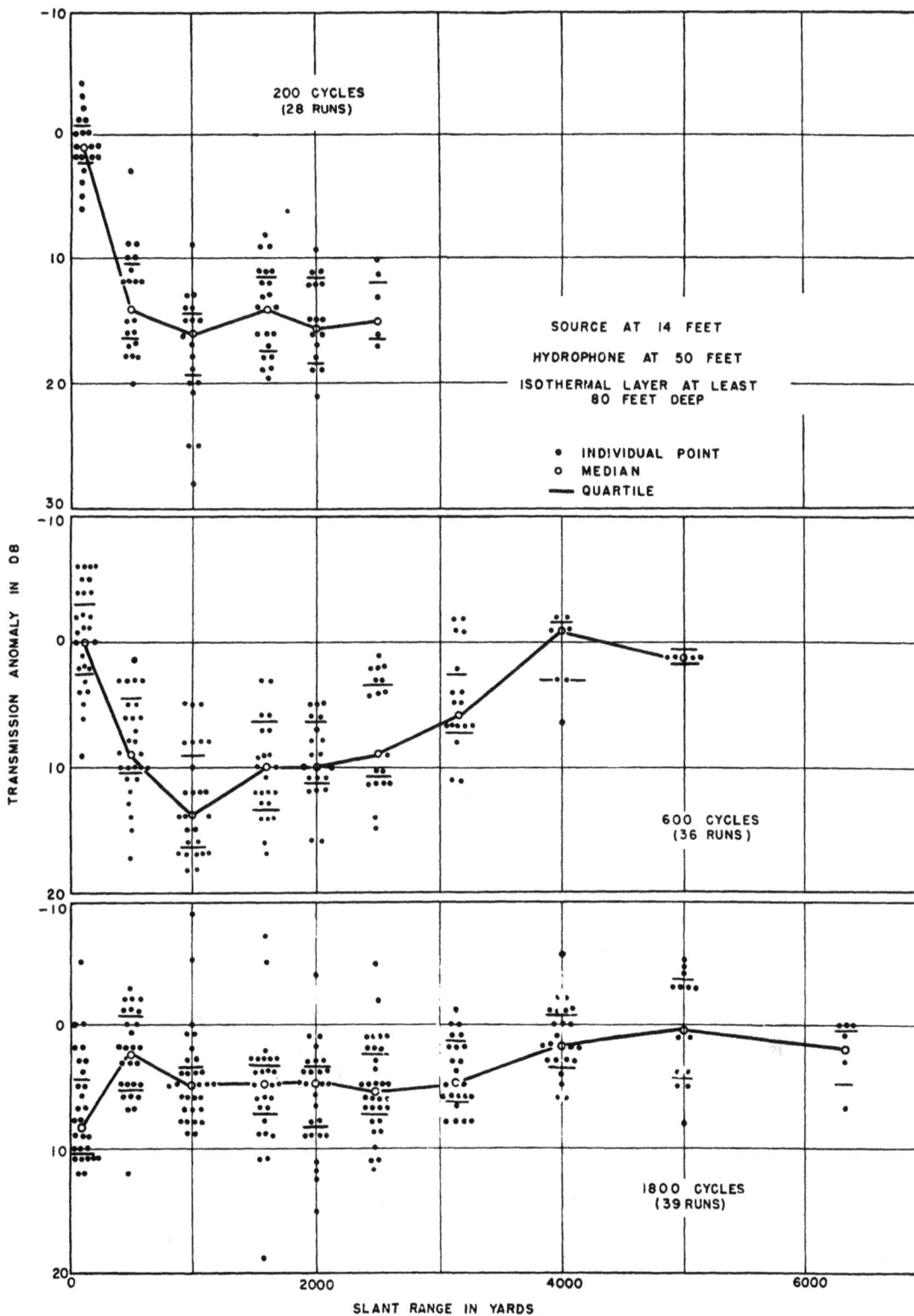

FIGURE 12. Average transmission anomalies at sonic frequencies.

law of simple geometrical spreading. A detailed analysis was given in Section 2.5 for the absorption resulting from viscosity. However, it was pointed out in that section that the observed absorption of sound in the ocean is far greater than can be accounted for by viscosity.

The effect of absorption on underwater sound transmission is shown most simply by considering the propagation of sound in an unbounded homogeneous ocean. Let J be the total energy proceeding outward from a sound source in each second. If the ocean were not at all absorbing, this same amount of energy would spread to all ranges. If the source is nondirectional, this energy would be spread over an area of $4\pi R^2$ at a range R, and the sound intensity I would be given by the equation

$$I = \frac{J}{4\pi R^2} = \frac{F}{R^2} \tag{5}$$

where F, defined in Chapter 2, is given by

$$F = \frac{1}{4\pi}J.$$

In the presence of absorption, a constant fraction n of the sound energy is absorbed in each yard of sound travel. Thus in a distance dr, an amount of energy $4\pi nF\, dr$ will be absorbed per second, and converted into heat energy. The constant n is called the *absorption coefficient* of the water. The decrease $4\pi\, dF$ of the sound energy over this distance will equal the energy absorbed, or $4\pi nF\, dr$. Thus we have the equation

$$\frac{dF}{dR} = -nF, \tag{6}$$

which has the familiar exponential solution

$$F = F_0 e^{-nR}. \tag{7}$$

The sound intensity is then given by the equation

$$I = \frac{F_0 e^{-nR}}{R^2}. \tag{8}$$

In terms of decibels, this equation may be written

$$10 \log I = 10 \log F_0 - 20 \log R - \frac{aR}{1000} \tag{9}$$

where $a/1000 = 10n \log_{10} e = 4.34n$; a expresses the absorption in decibels per kiloyard and is called the coefficient of absorption. The transmission loss H, as defined in Chapter 4, is $10 \log F_0 - 10 \log I$. The transmission anomaly A is simply $H - 20 \log R$.

Thus we have the simple equation

$$A = \frac{aR}{1000}. \tag{10}$$

Hence, in an unbounded medium, absorption produces a transmission anomaly which increases linearly with the distance covered by the sound beam.

Scattering of sound is more complicated than absorption. When scattering rather than absorption is present, equation (10) may still be used to describe the decay of the unscattered sound. However, the sound that has been scattered must also be considered in computing the expected sound intensity. It is shown in Section 5.4.1 that scattering is probably not very important in isothermal water. However, since the exact role of scattering in isothermal water is not certain, and since some forms of scattering may be very important when temperature gradients are present near the surface, it is customary to refer to the combined effects of absorption, scattering, and similar phenomena as attenuation. The quantity a, determined by direct measurement of A and use of equation (10), is then called the *coefficient of attenuation*. Attenuation, as so defined, includes all effects which may produce a transmission anomaly.

Extensive observations at a number of laboratories indicate that in isothermal water the transmission anomaly does, in fact, increase linearly with increasing range, in accordance with equation (10). Thus, the attenuation coefficient a for each frequency is a constant for any one run. The data at 24 kc, in a report on attenuation issued by UCDWR, provide a check of this point.[13] Of the many runs available in deep water off the coast of southern California and Lower California, at the time reference 13 was written, 65 were made when the temperature difference from the surface to a depth of 30 ft was 0.1 F or 0.0 F. For all these runs the graphs of transmission anomaly against range could "reasonably be approximated by straight lines beyond a range of 1,000 yards." Two sample plots of transmission anomaly, with the corresponding temperature-depth records, are shown in Figure 13. Each point represents the average amplitude of five different pings.

The linearity of the observed points is evident in Figure 13. On the average, about half of the plotted points lay within 2 db of the straight-line curve drawn for each run. Thus it is reasonable to conclude that in water which is isothermal from the surface to 30 ft, the transmission anomaly increases linearly with range from 1,000 to more than 6,000 yd. Since

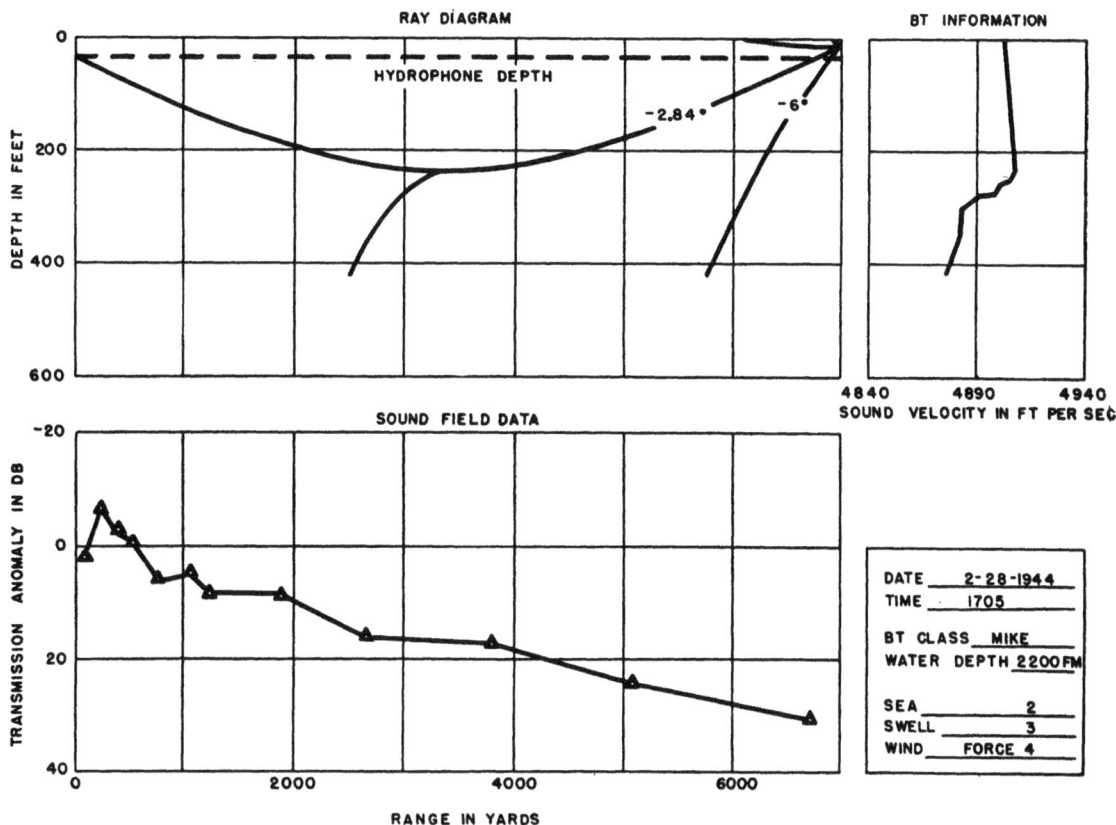

FIGURE 13A. Sample transmission anomaly in isothermal water.

only about half the runs were made with shallow hydrophones, between 16 and 30 ft, and the other half with deeper hydrophones, usually below the thermocline, this result apparently applies for sound transmitted below the thermocline as well as for sound in the isothermal layer. This linearity of the transmission anomaly for deep hydrophones is discussed again in Section 5.3.2.

It is perhaps surprising that the transmission anomalies should be straight lines out to long ranges when the isothermal layer is at most a few hundred feet thick. In a completely isothermal layer the upward bending of the sound, caused by the increase of pressure with depth, should give rise to a shadow zone near the surface at 3,000 yd for a thermocline starting at 150 ft below the projector and at 6,000 yd for one starting at a depth of 600 ft below. While sound reflected from the surface would penetrate this shadow zone computed for the direct sound, some drop in transmission might nevertheless be expected at the shadow boundary.

It is possible that slight negative gradients of about 0.1 F in 30 ft were present in most of these measurements since slight gradients are common off San Diego and are very difficult to measure exactly with the bathythermograph. Such a slight gradient would offset the effect of pressure on sound velocity and give nearly straight-line propagation out to considerable range.

The observed results could also be qualitatively explained on the assumption that the temperature in the isothermal layer is not completely constant, but varies irregularly from point to point. It was shown in Section 5.1.3 that the microstructure observed in regions of sharp temperature gradient can broaden the sound beam in the vertical direction by a hundred feet in several thousand yards. If microstructure of similar effectiveness were present in the isothermal layer, this alternate up-and-down bending from microstructure would wash out the pressure effect entirely and would enable some direct sound to travel to an indefinite range in the isothermal layer. Since a fraction of this sound would be bent down into the thermocline at all ranges, the linearity of the transmission anomaly curve at depths below the thermocline might also be explained on this basis.

RAY DIAGRAM
HYDROPHONE DEPTH Δ

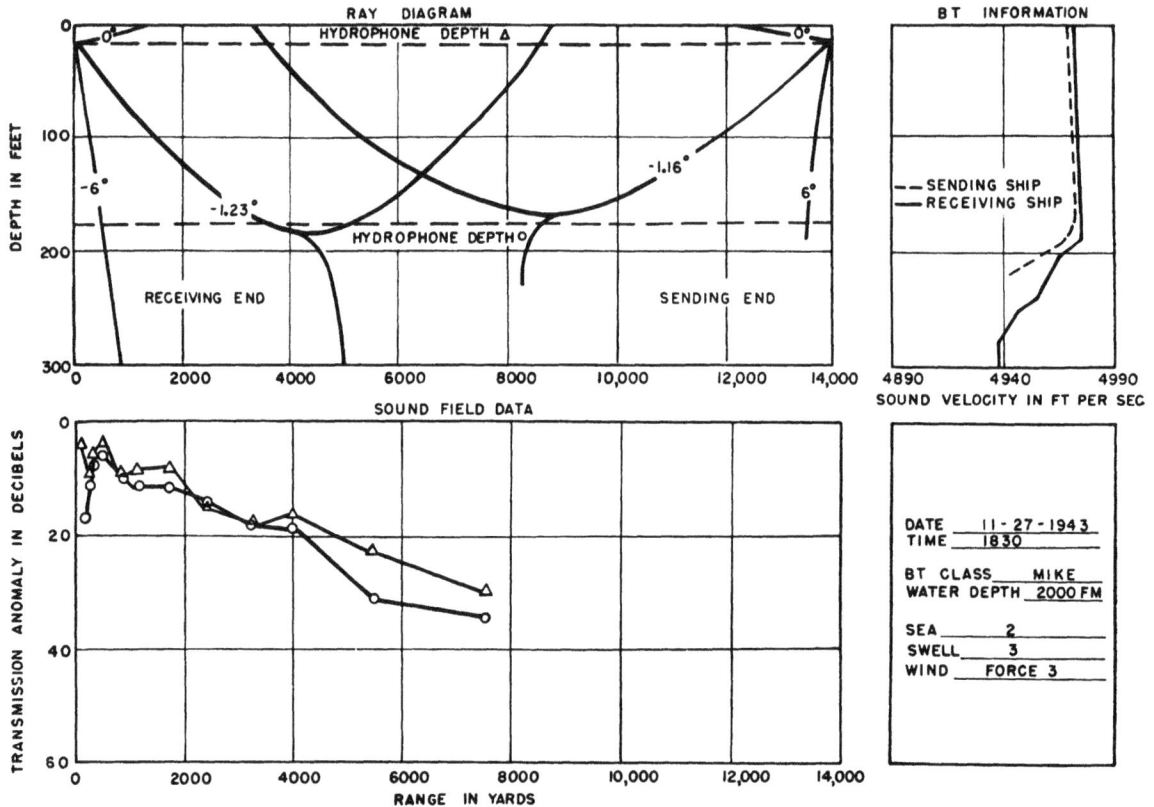

FIGURE 13B. Sample transmission anomaly in isothermal water.

However, since there are no extensive measurements on small-scale thermal structure in the isothermal layer, no conclusions about these problems can be reached at the present time.

The straight line of best fit drawn on a plot of measured transmission anomaly against range gives, with moderate precision, the attenuation coefficient for the time and place of the measurements. Because the transmission anomaly usually approximates a straight line for measurements at fixed depth in isothermal water, the probable error of the attenuation coefficient found in this way is only about ½ db per kyd, for runs extending out to about 6,000 yd.

The detailed variation of the attenuation coefficient in isothermal water from place to place and from time to time has not been thoroughly explored; and so it is most useful to deal with an average attenuation coefficient. The evidence on variation of transmission loss in isothermal water will be given in Section 5.2.3.

The most complete investigation of the average attenuation coefficient at 24 kc in isothermal water is apparently that presented in a report by UCDWR.[14]

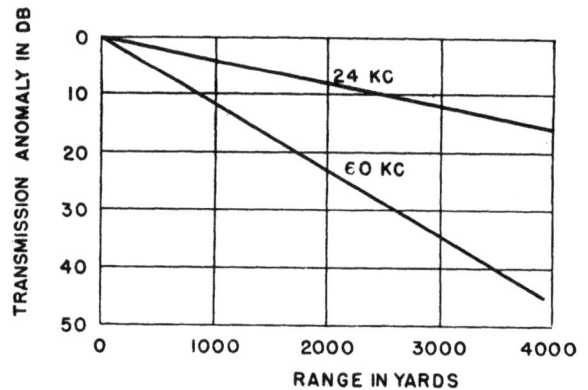

FIGURE 14. Average transmission anomalies in isothermal water.

All deep-water runs in isothermal water are analyzed together. This analysis shows that if the water is essentially isothermal to about 100 ft the measured transmission anomalies do not depend on the temperature at greater depths or on the state of the sea surface, at least out to ranges of 3,000 yd.

Transmission runs made under these conditions have been combined to yield the average curves of

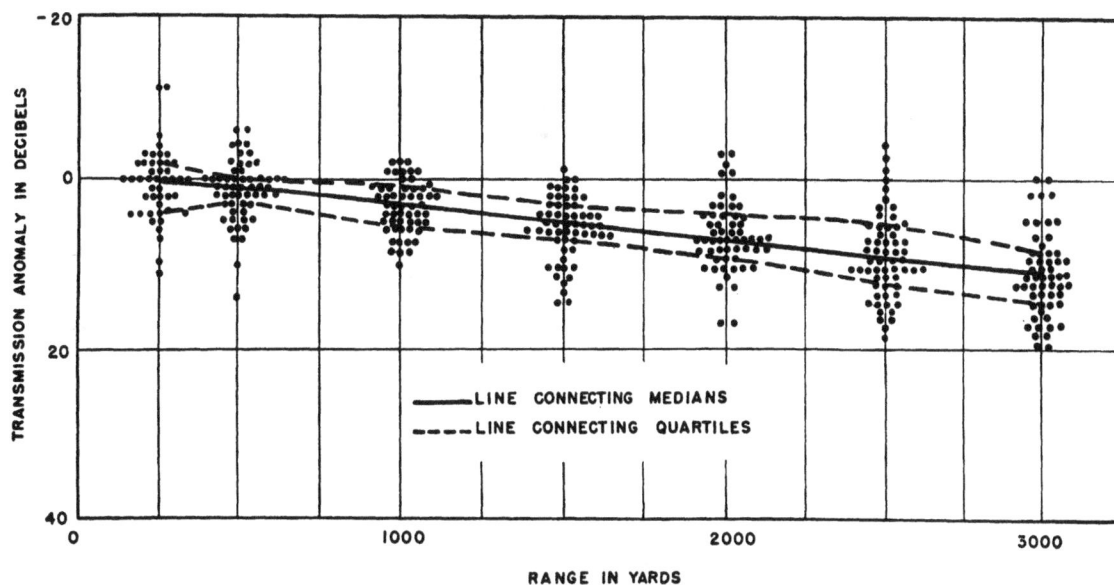

FIGURE 15.   Individual transmission anomalies in isothermal water.

transmission anomaly shown in Figure 14. Only runs with a hydrophone at 16 ft have been included. In the curve for 24 kc, taken from reference 14, runs made with a total temperature change of less than 0.3 degree between the surface and 80 ft were included. Since a general analysis shows that temperature differences as small as 0.2 degree in the top 30 ft do not affect transmission anomalies appreciably, probably much the same curve would be obtained if only those runs in water isothermal to less than 0.1 F were considered. In the curve for 60 kc (taken from Volume 7 of Division 6), all runs made with an approximately isothermal surface layer were included. The slopes of the two average curves in Figure 14 give attenuation coefficients of 4.0 db per kyd at 24 kc and 13.5 db per kyd at 60 kc.

In Figure 15 are plotted all the individual points used at 24 kc, with solid lines connecting the median points and dashed lines connecting the upper and lower quartiles. The average quartile deviation shown in Figure 15 is 2.6 db. This is about the same as the deviation of any single transmission anomaly curve in isothermal water from a straight line, and is also about the same as the experimental error, discussed in Chapter 4, in the determination of the transmission loss from the average of five observations.

Other sets of measurements give similar values of the attenuation coefficient a. Many transmission runs were made some time before the war by NRL. [15-17] In these runs, the range was commonly opened from less than 1,000 to more than 10,000 yd. These data were originally plotted in terms of intensities rather than transmission anomalies. It was found that in a considerable number of runs, plots of sound intensity (in decibels) against range gave essentially straight lines beyond 1,000 yd. The measurements were reanalyzed and the results reinterpreted in terms of transmission anomalies. [13]

For the data reported in reference 16, all the straight-line graphs were obtained when the water was isothermal (to ± 0.1 C) from the surface to more than 100 ft. The average attenuation coefficients found at 17.6, 23.6, and 30 kc were 1.8, 4.4, and 6.5 db per kyd, respectively. Half the observed values lay within ± 1 db per kyd of these average values.

The NRL and UCDWR measurements described above are the only ones in deep water which have been analyzed in terms of the temperature structure present at the time the measurements were made. A considerable body of other measurements have been made to determine the attenuation coefficient, but these are less reliable.

Measurements of bottom-reflected sound in shallow water have been used to determine the attenuation coefficient resulting from absorption and scattering in the volume of the ocean. In particular, sound has been sent out vertically, and the strength of the echo received in different depths of water used as a measure of attenuation in the water. Since these measurements depend entirely on the reflection coefficient of the bottom, they can give results on attenuation only if it is assumed that the reflection coefficient of

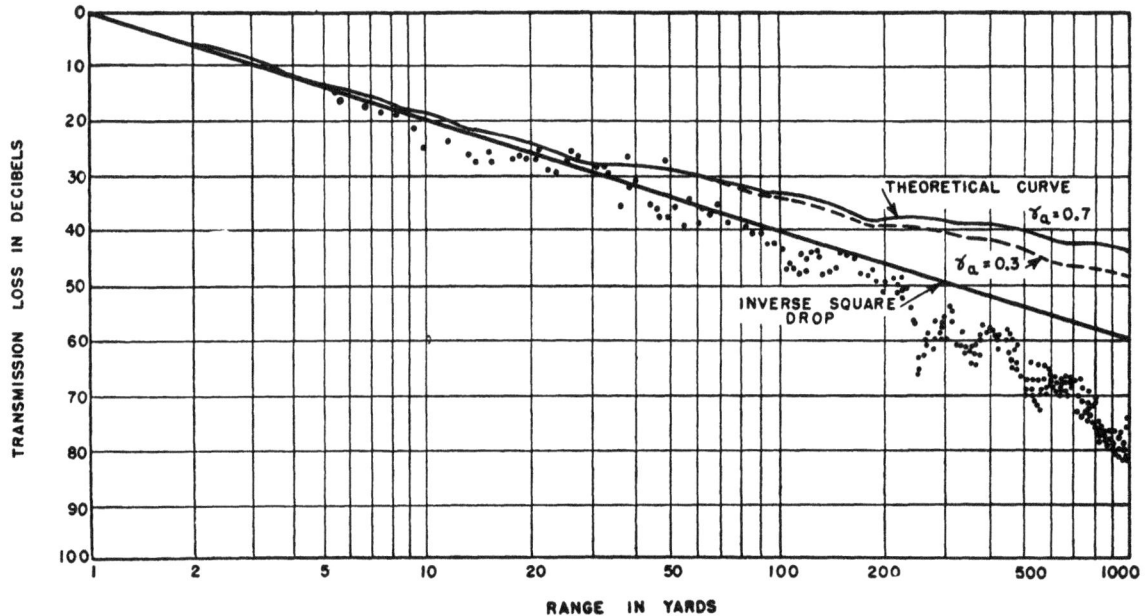

FIGURE 16.   Measured and computed intensities in shallow water.

the bottom is independent of depth and location. There is no evidence for this assumption, especially since the nature of the bottom in the deep ocean is believed to be somewhat different from that in shallow water. Moreover, the attenuation coefficient measured at considerable depths has no necessary connection with the coefficient in the surface layers of the sea. Thus, these data are of little use, except for predicting the levels of sound reflected vertically from the bottom in different depths of water.[13]

Values of the attenuation coefficient have also been found from measurements of horizontal transmission in shallow water. Since these depend on the numerical value of the bottom-reflection coefficient, they do not give an accurate indication of the attenuation in deep water of constant temperature. However, at high frequencies, the error introduced into the results by an incorrect value of the bottom-reflection coefficient becomes small, and these values are relatively trustworthy.

An investigation along these lines is described in a report issued by UCDWR.[18] In this work, sound was transmitted from a dock in San Diego harbor through water 30 to 50 ft deep. As the range was changed, from 10 to 1,000 yd, the recorded sound intensities fluctuated rapidly, since the direct, surface-reflected, and bottom-reflected rays interfered, first constructively, then destructively. The exact computation of these interference effects would be very complicated. Instead, only the highest peaks reached by the

fluctuating sound were considered; for these peaks it was assumed that the different rays involved were all in phase, all interfering constructively. These measured values were then compared with theoretical curves, computed without regard for absorption. These curves were found by adding the calculated direct sound to the calculated sound from all the images resulting from successive surface and bottom reflection; the directivity of the sound projector was taken into account, and all the different rays were assumed to arrive in phase. The difference between the observed and computed curves was then used as a measure of the absorption. A sample plot showing the observed peaks of the sound intensity and the theoretical curves for different values of γ_a, the amplitude reflection coefficient of the bottom, is given in Figure 16 for a sound frequency of 100 kc. For comparison, the inverse square curve expected in a deep, ideal ocean is also shown in the figure. Unfortunately, no temperature measurements were made during this work. Measurements made in the same location one year later showed that the temperature gradients were usually small because of strong tidal currents. The temperature difference between 0 and 20 ft was found to be less than 0.2 F, 94 per cent of the time, and less than 0.1 F, 75 per cent of the time. Thus, it is probably legitimate to take the attenuation coefficients of this study as representative of isothermal or nearly isothermal water.

At 60, 80, and 100 kc, this work gave attenuation

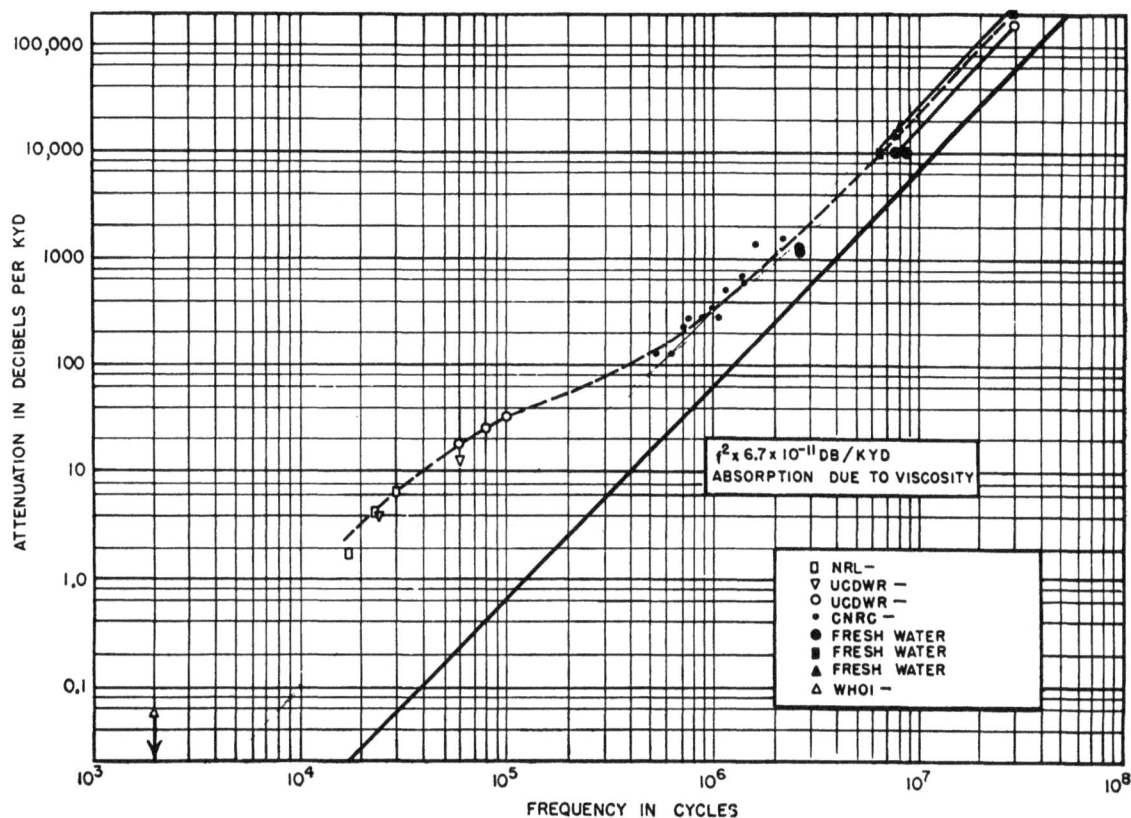

FIGURE 17.   Dependence of attenuation coefficient on frequency. The points on the curve are from the following references: □ NRL, 13 and 16; ▽ UCDWR, 14; ○ UCDWR, 18; • CNRC, 19; ● Fresh Water, 20; ■ Fresh Water, 22; ▲ Fresh Water, 23; △ WHOI, Chapter 9 of this book.

coefficients of 18, 26, and 32 db per kyd, respectively, for an assumed amplitude reflection coefficient of the bottom of 0.5 (energy loss of 6 db per reflection), corresponding to the SAND–AND–MUD bottoms over which the measurements were made. A change of the reflection coefficient by 0.2 in either direction changes the attenuation coefficient by about 2.5 db per kyd in the same direction; this variation may be taken as a rough estimate of the probable error of the results. Owing to this high probable error, the values of 2.0 and 7.0 db per kyd, found at 24 and 40 kc respectively, are of relatively low weight and may be disregarded.

At frequencies between 500 and 2,500 kc, extensive measurements of attenuation have been made by the Canadian National Research Council.[19] A projector was mounted on a dock in Vancouver Harbor in 13 to 25 ft of water. The receiver was also mounted on the same dock at distances varying up to 100 ft. As a result of the high directivity of the projector, surface and bottom-reflected sound were largely eliminated. The slope of the transmission anomaly was measured

to give an attenuation coefficient at each frequency. Relative probable errors of these coefficients, estimated from the reproducibility of the results, averaged about 7 per cent. No temperature measurements were made. Over such short ranges any gradients would have had a negligible effect.

No measurements at frequencies above 3 mc are available for sound in the ocean. However, such measurements have been made in the laboratory.[20-23] Those of reference 20 extend down to 2.8 mc, where the values found are of the same order of magnitude as those determined in the ocean. Other determinations of absorption in fresh water in the frequency range between 200 and 4,000 kc are about ten times as high as those found in the sea.[24-26] These fresh-water measurements are not in good agreement with each other and may be affected by systematic errors. Since the sea-water values taken from reference 16 were made over a much greater sound path, these should be much more reliable, and in any case, constitute better evidence for the attenuation of sound in the sea; the fresh-water measurements in references

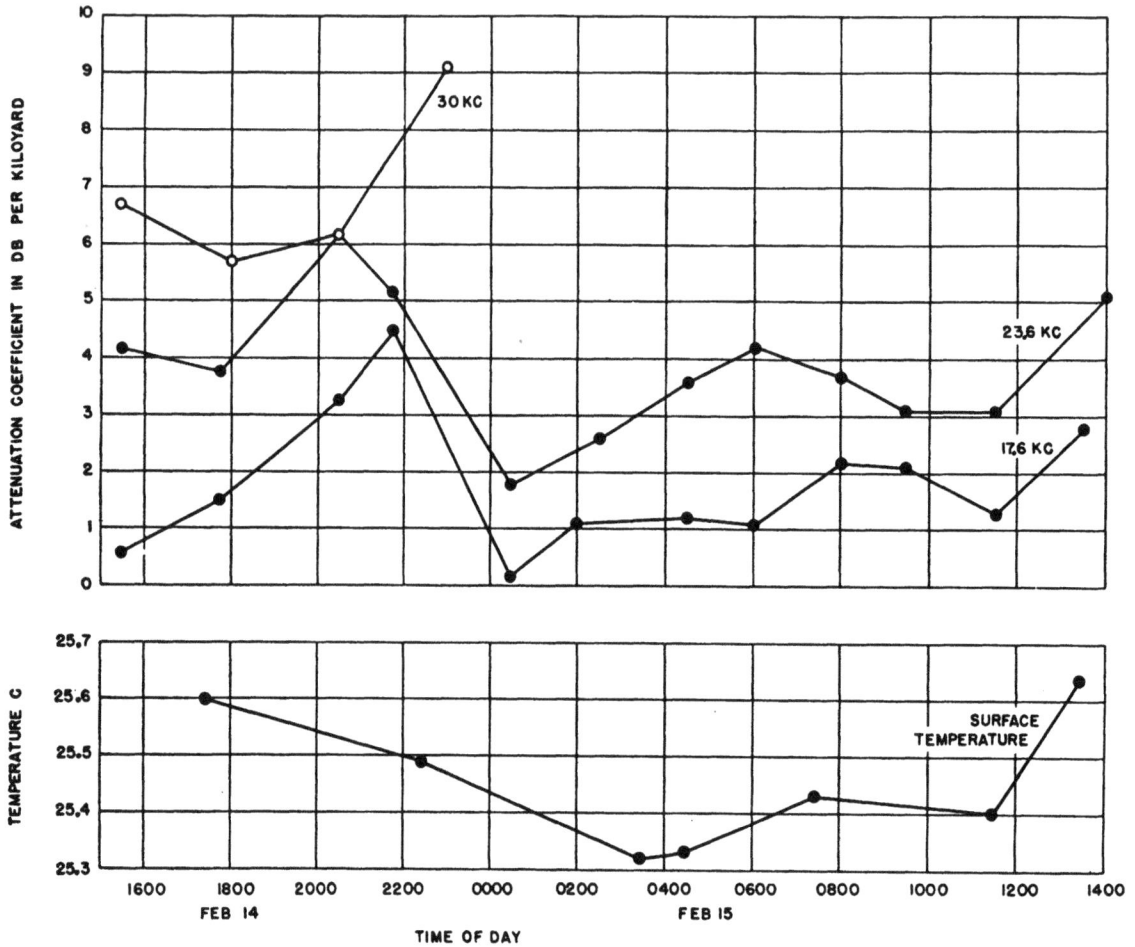

FIGURE 18.   Variability of attenuation coefficient during one day.

24, 25, and 26 may therefore be ignored in the present discussion.

At frequencies below 14 kc no data on deep-water attenuation in the surface layers of the sea are available. The long-range measurements of explosive sound propagation, discussed in Chapter 9, give an upper limit on the attenuation coefficient deep in the ocean. Although the data are uncertain, the results quoted in Chapter 9 indicate that the attenuation at 2,000 c is probably less than 5×10^{-2} db per kyd.

These different determinations of the attenuation coefficient are combined in the plot of a against frequency shown in Figure 17. The dashed line gives a curve of best fit drawn through the plotted points. The solid line in this figure gives the value of a to be expected from viscous damping, taken from Section 2.5. Evidently, at frequencies above 1 mc the attenuation coefficient is three to four times the classical value. In a UCDWR memorandum,[27] this discrepancy is attributed to an additional viscous force

proportional to the rate of compression of the water. Such a force has usually been neglected in hydrodynamics, since ordinarily water flows like an incompressible fluid. Although no tests of such an hypothesis have been suggested, it is entirely possible that this "compressional viscosity" may be responsible for the observed values of a at frequencies above 1 mc.

No explanation has yet been advanced for the observed values of the attenuation at somewhat lower frequencies. Since all the measured values between 10 and 100 kc were obtained in temperate latitudes in water well above the freezing temperature, it is possible that these values are not applicable for all oceanographic conditions. It is still not wholly certain that the attenuation observed for supersonic frequencies in isothermal water is entirely the result of absorption rather than scattering; however, the weakness of scattered sound observed for backward scattering (reverberation) and for forward scattering (incoherent sound measured in shadow zones) makes this highly probable.

At frequencies below 15 kc the attenuation of sound is largely conjectural. The shallow water measurements discussed in Chapter 6 show that a is not greater than about 1 db per kyd at frequencies below 2 kc. On the other hand, if the attenuation in this frequency range were as low as 0.1 db per kyd, high-speed warships could be heard consistently many hundreds of miles away. Since such listening ranges are apparently not obtainable, it may be inferred that the attenuation coefficient in the surface layers of the ocean is greater than 0.1 db per kyd at all sonic frequencies. Such a high value is not necessarily inconsistent with the much lower value observed in the deep sound channel since the attenuation at low frequencies near the ocean surface may result primarily from scattering of sound out of the isothermal layer into the thermocline, where it is bent sharply downward, and is lost.

5.2.3 Variation of Transmission Loss

The previous section has discussed average values of the attenuation coefficient at each frequency, but has ignored changes in this coefficient. It has already been noted that from a single run out to 6,000 yd the attenuation coefficient can be determined with a probable discrepancy of only about ½ db per kyd from its true value at that particular time and place. Since the scatter of the observed values exceeds this, it may be inferred that the attenuation coefficient in sea water is probably not constant.

In the measurements reported in reference 16, for example, half the attenuation coefficients at 24 kc differed by more than 1 db from the average value of 4.4 db per kyd. Corresponding variations also appeared at the other two frequencies (17.6 and 30 kc). However, those runs in which more than one frequency was used show a good correlation (correlation coefficient r between 80 and 85 per cent) in the variation of the coefficients for the three frequencies.

A good illustration of this correlation is provided by the runs made during one 24-hr period (February 14–15, 1944), analyzed in reference 13. Seven measurements of the vertical temperature structure were made during this period. In each case, no variation of temperature of more than 0.2 F was noted down to depths of 120 ft, but this was also the limit of accuracy of the thermometer on these days. The surface temperature changed appreciably during the period,

however, probably as a result of the changing position of the vessels. The variation of the attenuation coefficients for sound at 17.6, 23.6, and 30 kc is plotted in Figure 18 with the measured surface temperature. Evidently the attenuation coefficients at the three frequencies changed very substantially; however, the difference in the attenuation coefficients between the different frequencies was more nearly constant.

Under some conditions, however, the attenuation coefficient during a 48-hr period is less variable. A series of transmission measurements was made by UCDWR in the deep water off Point Conception, California, where a persistent well-mixed layer was to be expected. Measurements were carried out during and after a storm, with winds of force 3 to 6 (Beaufort scale). The surface layers of the sea were probably better mixed during these transmission runs than for any other reported transmission experiments. A typical temperature-depth record taken during these measurements is shown in Figure 19.

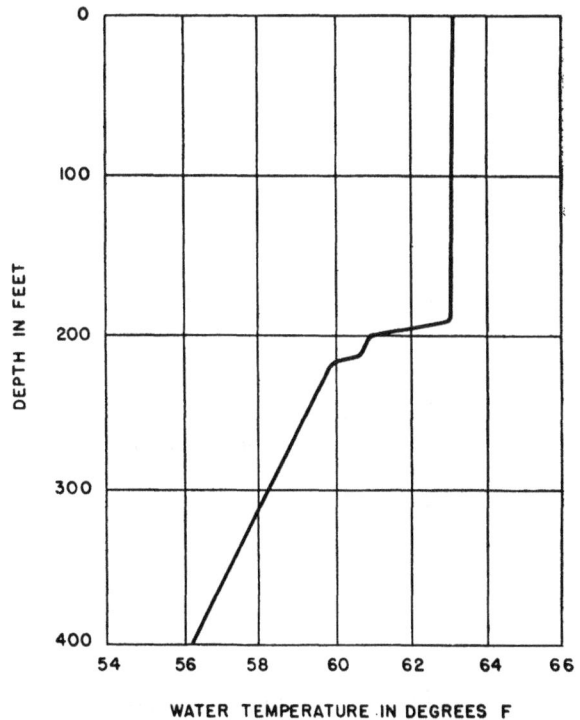

FIGURE 19. Depth record for Point Conception runs.

The cumulative distribution of attenuation coefficients for the data taken with the shallow and deep hydrophone is shown in Figure 20. These are plotted on probability paper, so that a normal, or Gaussian, distribution of plotted points would lie on a straight line. For the data shown in this figure, half of the observed values of a fall within about 0.6 db per kyd of the average values. This is so close to the probable error of 0.5 db per kyd for a single determination of a that the result may be entirely due to observational errors. Certainly the reduced scatter can be attributed to the relatively uniform conditions prevailing during these tests. The observed attenuation coefficients for the deep hydrophone will be discussed in Section 5.3. In addition, the relatively small scatter shown in Figure 15, about the same as the observational error in the measurement of transmission loss,

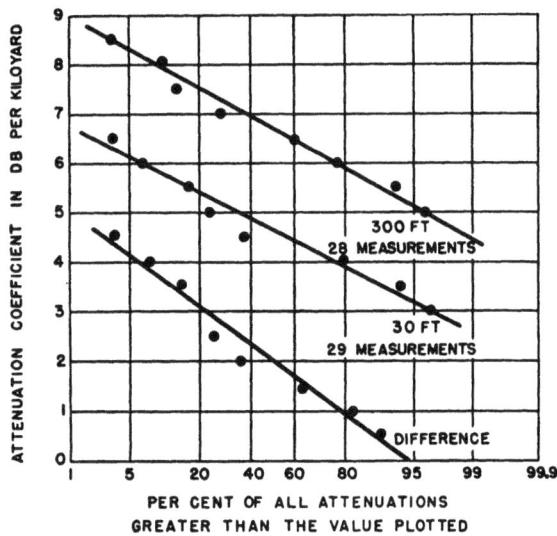

FIGURE 20. Distribution of attenuation coefficients for Point Conception runs.

suggests that in isothermal water the attenuation may be relatively constant. Further information is required, however, before any very definitive conclusions can be drawn about the variability of the attenuation coefficient when the temperature of the water in the upper 100 ft of the sea is approximately constant. Most of the UCDWR data have not been analyzed with this purpose in mind. In reference 13, where a high variability of a is found, the runs in isothermal water are not treated separately. Also, some of the runs included do not extend out very far; when marked peaks in the anomaly curve are present, the attenuation coefficient for such runs may be as low as 1 db per kyd. An examination of a sample run of this type, shown in Figure 21, indicates that such values are not necessarily indicative of the attenuation over longer ranges. Thus, these data do not cast much light on the variability of the attenuation coefficient in isothermal water.

It is tempting to assume that the values shown in Figure 17 represent pure absorption, and that any variations from these values found in approximately isothermal water represent distortion of wave front by temperature gradients too small to be detected reliably on the bathythermograph. More accurate data on transmission in mixed water and more accurate thermal measurements would be required to test such an hypothesis.

5.2.4 Short-Range Transmission

The goal of transmission studies is to relate the sound intensity at any range to the sound output of the source. Most transmission measurements at sea, however, do not measure the sound level closer than about 100 yd from the source. In principle it should be possible to measure the absolute sound level in the water, and also to measure the absolute level 1 yd from the projector; in practice, the measuring equipment has apparently not been sufficiently stable to make these absolute measurements possible. Thus transmission measurements give only relative sound levels and may be used to give the sound level at long range relative to the level at several hundred yards. The methods used in computing transmission anomalies are discussed in some detail in Chapter 4. For most of the data discussed here, the transmission anomaly at short range, usually about 100 or 200 yd, has been taken equal to zero. Thus, to find true transmission anomalies at long range requires information on the true value of the transmission anomaly at ranges around 100 yd.

Since refraction can be ignored at such close ranges, the transmission anomaly for the sound passing directly from projector to hydrophone must be very close to zero. If the surface were perfectly flat, surface-reflected sound would, on the average, double the sound intensity at ranges of several hundred yards, provided that the intensity is averaged over the interference pattern discussed in Section 2.6.3; the transmission anomaly A at these ranges would then be -3 db. Sound intensity measurements between 1 yd and several hundred yards would then show a gradually decreasing transmission anomaly as the range increased and as surface-reflected sound approached the same average strength as the direct sound.

However, the sea surface is never perfectly flat, and this fact may be expected to alter the simple relationships to be expected for a flat surface. Al-

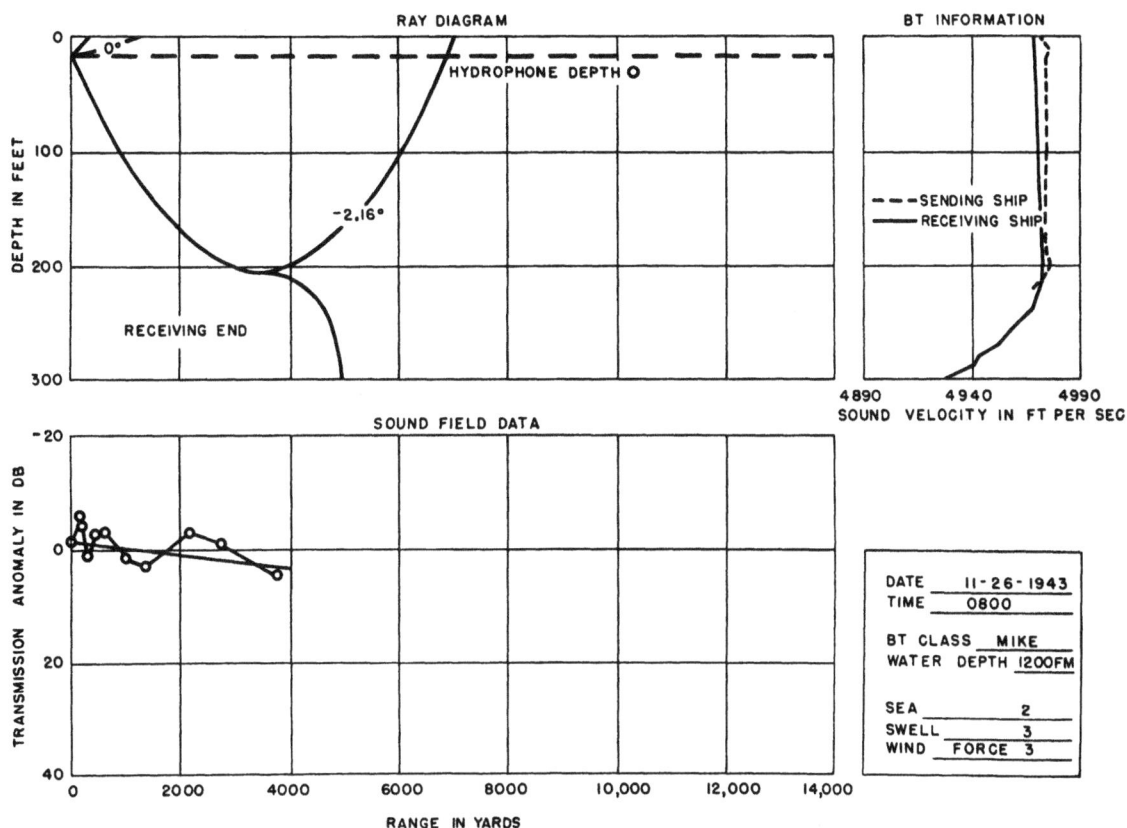

FIGURE 21.   Sample transmission anomaly out to short range.

though practically all the sound which strikes the surface will be reflected back into the water, its direction will usually be affected by the water waves on the surface. A glance at sunlight reflected from the ocean surface shows how a sound beam may be reflected in a variety of directions at a rough surface. It is possible, for example, that the surface reflects sound predominantly downward, with little surface-reflected sound reaching a shallow hydrophone at ranges of several hundred yards or more. While some observational and theoretical studies of this problem have been attempted, the transmission anomaly at several hundred yards is still uncertain. One would expect theoretically that the anomaly might lie anywhere from 0 to −3 db. This corresponds to an uncertainty of 6 db in computed echo levels from targets of known target strength. This important gap in transmission information will presumably be filled in when more transmission measurements have been made with the help of one of the several calibration methods discussed in Section 4.3. In most of the runs made up to 1944, however, neither the instrumentation nor the calibration procedure was completely

reliable. For example, the observed values of the anomaly at 500 yd vary from −6 db to +15 db when the water is isothermal to a depth of at least 40 ft. Thus, one may readily believe that the absolute values of the transmission anomaly for the majority of the runs available may be systematically in error by as much as 3 db.

5.3   **TRANSMISSION FROM AN ISOTHERMAL LAYER THROUGH A THERMOCLINE**

As pointed out in Section 5.1, the ocean is rarely isothermal to great depths. In the more typical case, an isothermal layer overlies a sharp negative gradient, or thermocline, whose top may be at a depth anywhere from less than a hundred to many hundreds of feet. When the isothermal layer is a hundred feet thick or more, sound transmission above the thermocline is apparently independent of the depth or sharpness of the thermocline, at least out to ranges of several thousand yards. Sound transmitted to points in or below the thermocline may be appreciably weakened, however. This section discusses sound received by a hydrophone in or below the thermocline; emphasis is placed primarily on thermoclines a hundred feet thick or more. Almost all the data available for these conditions are at a frequency of 24 kc. Although some observations have been made at higher frequencies, especially 60 kc, practically no relevant information is available at sonic frequencies.

5.3.1   Echo-Ranging Trials

The importance of the thermocline in weakening sound which passes through it was first shown in practical echo-ranging runs. The earliest and most extensive data of this type were collected by the British.[28] In each run, a surface vessel echo-ranged on a submarine at continually increasing or decreasing range. The maximum range at which echoes could be obtained was noted, together with the temperatures and salinities at fixed depth intervals. Among various effects produced by refraction, the most striking was the reduction in maximum echo range resulting when a submarine submerged below the thermocline. This reduction in range is called *layer effect*.

The quantitative importance of layer effect is evidenced by the fact that in 40 out of 68 trials reported in reference 28 the maximum range decreased as the submarine dove from periscope depth down to about 100 ft. Of the 28 trials in which layer effect did not appear, all but 5 were made in water with very weak temperature gradient, and 3 of these 5 exceptions occurred in shallow water. In 12 of the 68 trials there was a difference of more than 9 F between the temperature at projector depth and the temperature at the top of the deep submarine. In these 12 trials the maximum range on the deep submarine varied between 20 and 90 per cent of the range found at periscope depth, the average being 65 per cent.

Similar results were obtained in echo-ranging trials made by the USS *Semmes* (AG24, ex-DD189) on four fleet-type American submarines.[29] Below the isothermal layer, which was 150 ft thick, was a sharp thermocline, as shown in Figure 22. When the submarine submerged to 250-ft keel depth or deeper, the maximum echo range was consistently about half the maximum echo range observed when the submarine was at periscope depth.

5.3.2   Sample Transmission Runs

Although the detailed interpretation of these echo-ranging results involves many complicated factors,

such as the change of reverberation level with range, the general explanation is that the sound intensity below the layer is less than above. This result is borne out by detailed transmission measurements. A sample plot of measured transmission anomalies for deep

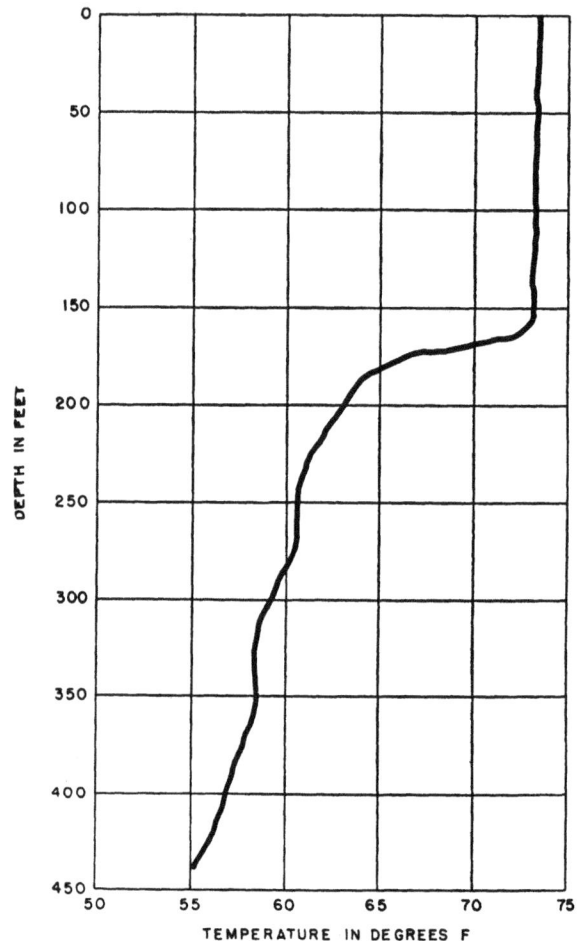

FIGURE 22.   Temperature-depth record for Semmes tests (deep layer).

and shallow hydrophones is shown in Figure 23, representing a run made by UCDWR. The computed ray diagram is also shown.

The signals measured to give Figure 23 were coherent[a] at all ranges, reproducing moderately well the outgoing pulse. Some of the weaker signals received below the layer were characterized by "reverberation tails," representing incoherent sound arriv-

---

[a] A received signal is called *coherent* if its envelope reproduces faithfully the outgoing pulse. An *incoherent* received signal will in general have a ragged envelope and a length in excess of the length of the original outgoing pulse. Since no received signal portrays the envelope of the outgoing signal completely without distortion, coherence is a question of degree.

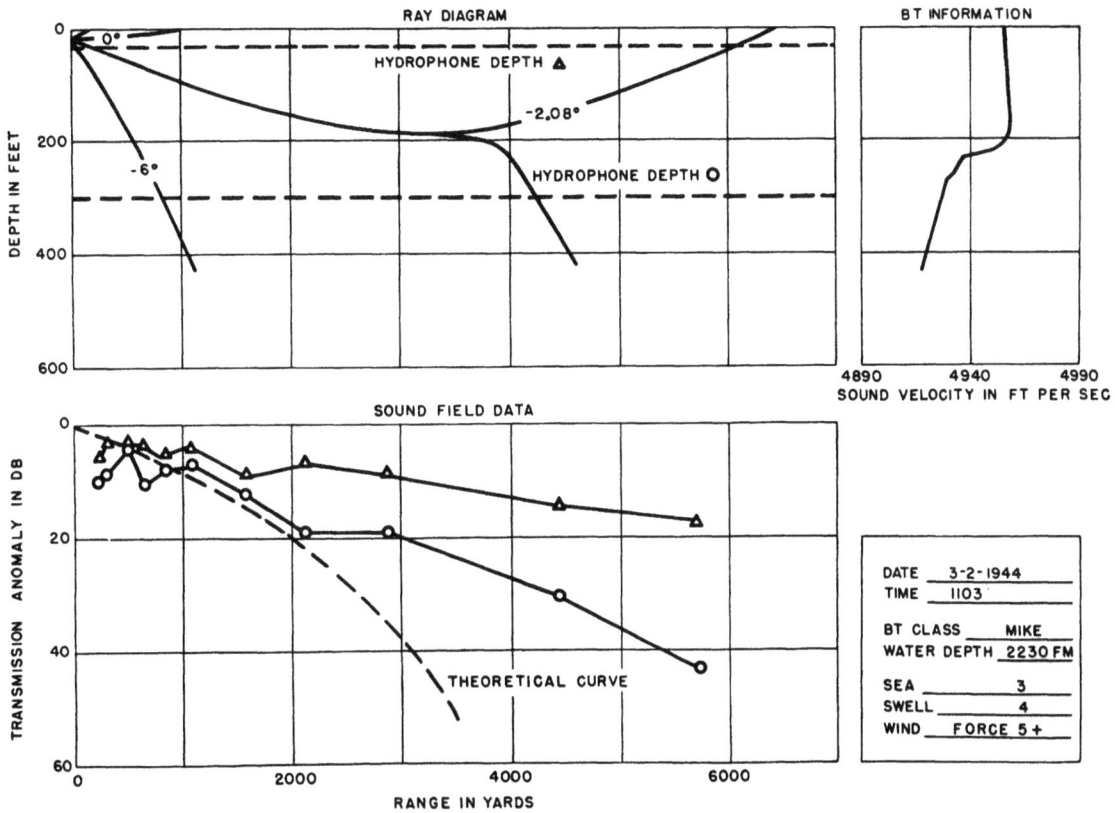

FIGURE 23. Sample transmission anomalies above and below thermocline.

ing after the direct signal. This scattered sound usually appears whenever the amplification is made sufficiently great. Since this incoherent sound is relatively more prominent when sharp downward refraction is present, it is discussed in Section 5.4.1.

It is evident from Figure 23 that the plot of transmission anomaly against range below the layer was approximately a straight line. This result is quite general, as noted in reference 13, where it was found that for all runs made with isothermal water in the top 30 ft of the ocean the transmission anomaly plots were approximately straight lines. This analysis included both deep and shallow hydrophones, and the conclusions are therefore valid for hydrophones either above or below the layer.

On the basis of the simple ray diagrams shown in Section 5.1.2, this result is rather surprising since the transmission anomaly should rise to a very high value at the shadow boundary, as shown in Figure 23 by the dashed line, taken from a UCDWR internal report.[30] This dashed line represents the anomaly computed by ray tracing methods, as explained in

Section 3.4.2. An absorption of 4 db per kyd has also been included. The consideration of sound reflected from a flat ocean surface will not change the computed anomaly appreciably. In particular, the shadow boundary will not be much affected; for the isothermal layer shown in Figure 23, rays leaving the projector at an upward angle less than 2.08 degrees will, after reflection, become horizontal before reaching the bottom of the isothermal layer, while the steeper rays will penetrate the thermocline at ranges of not more than about 3,000 yd.

Thus, to explain the straight-line anomaly curves found below the layer, some mechanism must be involved which is not included in the simple ray theory. Either an irregular sea surface or thermal microstructure explains the results qualitatively since either mechanism will take sound from the isothermal layer at all ranges and deflect some of it sufficiently sharply so that it will pass out of the surface layer into the thermocline below. At present, it is not possible to state whether either mechanism can explain the facts quantitatively.

5.3.3   Average Layer Effect

Differences in transmission anomaly between shallow and deep hydrophones are characteristic of isothermal water overlying a thermocline. Some evidence for a correlation of this difference with oceanographic conditions has been found. These results are

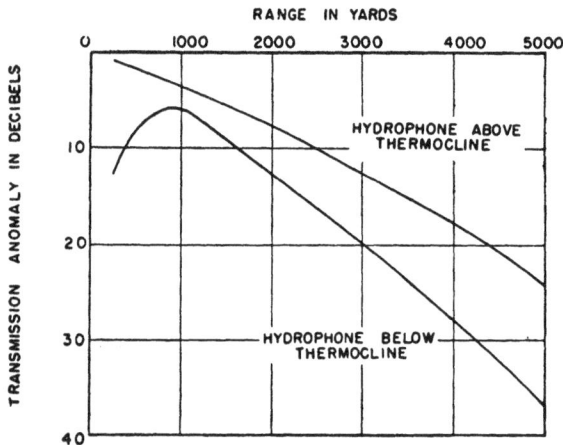

FIGURE 24. Average transmission anomalies, above and below thermocline.

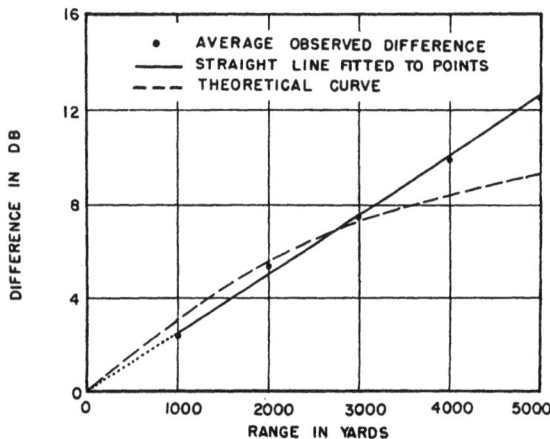

FIGURE 25. Difference in transmission anomaly above and below thermocline.

given later in this section. Since these results are subject to some uncertainty, it is useful to obtain an average value for this difference in transmission between a shallow hydrophone in isothermal water and a deep hydrophone below the layer. Such an average has been obtained by averaging together all UCDWR runs made off San Diego under these conditions.

The average curves found at 24 kc are shown in Figure 24. These curves include all runs in which the water was isothermal to more than 40 ft. The probable error of each curve, determined from the

quartile deviation of the individual points, divided by the square root of the number of runs, is about 1 db. The increased anomaly at short ranges for the deep hydrophone results from the vertical directivity of the sound projector. The difference between these two curves is plotted as a function of range in Figure 25. It is evident that this difference, in decibels, increases linearly with range. The dotted line at less than 1,000 yd indicates the difference in anomaly that would presumably be found for a nondirectional projector. The dashed line represents the semi-empirical formula (13) discussed below.

5.3.4   Studies of Layer Effect at 24 kc

The average effect is large and significant. The way in which thermoclines of different depth and sharpness weaken the sound intensity below them is a detailed problem of both scientific and practical interest. First, the theoretical expressions for sound intensity below a thermocline are discussed and applied to the average observational results. Secondly, the effect which thermocline depth and sharpness has on the measured sound intensity at depth is discussed in detail. Some early UCDWR studies are reported which fail to show the expected effects; a more detailed study is then given which indicates that general theoretical expectations are, in fact, fulfilled.

THEORY

One might expect that at least in some cases the theory developed in Chapter 3 could be used to predict the sound intensity below a layer of sharp temperature gradient. This expectation is supported by the discussion in Section 9.2.2, which shows that the intensities of explosive pulses agree rather well with the intensity calculations based on the ray theory. It is evident from Figure 23, on comparison of the solid line drawn through the circles with the dashed theoretical curve, that at ranges less than 2,000 yd, layer effect can in fact be explained on the basis of the simple ray theory. The predicted decrease of intensity results from the increased divergence of rays, which are bent sharply downward on passing through a temperature gradient. This increase of divergence is shown in the idealized diagram in Figure 26. The rays are close together in the ideal isovelocity layer, and the intensity is therefore high; but below the layer of sharp gradient they are much further apart, resulting in much reduced intensity.

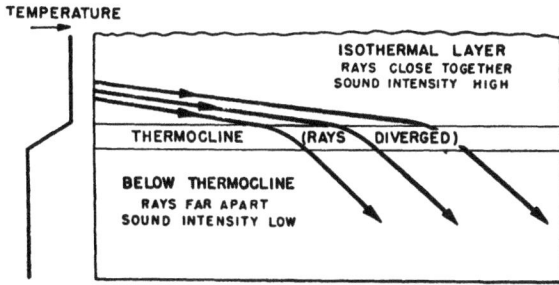

FIGURE 26. Cause of layer effect.

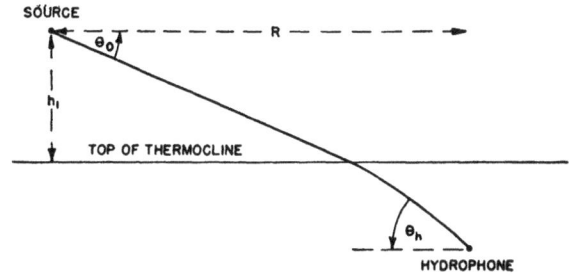

FIGURE 27. Diagram clarifying layer effect formula.

Layer effect was examined theoretically, from the standpoint of ray acoustics, under "Combination of Linear Gradients" in Section 3.4.2. The approximate formula obtained in Chapter 3 was

$$A = 10 \log \left(\frac{h_1 \theta_h}{R \theta_0^2} \right). \tag{11}$$

In this formula, A is the transmission anomaly, h_1 is the vertical distance between sound source and top of the thermocline, θ_h is the angle of inclination of the sound path at the hydrophone, R the range, and θ_0 is the angle of inclination of the sound path at the sound source, as in Figure 27. To facilitate comparison with experiment, this expression can be further simplified. If the upward bending caused by the increasing pressure is unimportant, which will be the case at ranges which are not too long, then to an adequate approximation, θ_0 may be replaced by h_1/R. If θ_h is calculated by means of equation (81) of Chapter 3, equation (11) then becomes

$$A = 10 \log \sqrt{1 + \frac{2\Delta c R^2}{c_0 h_1^2}}$$
$$= 5 \log \left(1 + \frac{2\Delta c R^2}{c_0 h_1^2} \right). \tag{12}$$

In most situations, the thermocline is not confined to a thin layer but is a hundred feet or more in thickness. Extensive intensity computations for simple types of bathythermograph slides are included in a report by WHOI.[31] Some of these have been reproduced in Figure 25 of Chapter 3. It is evident from a study of these figures that most of the drop in intensity takes place in the top part of the gradient. With increasing hydrophone depth in the thermocline, the increase of total temperature change begins to be offset by the larger inclination with which the sound ray enters the thermocline on its way to the hydrophone. More detailed theoretical calculations show that the temperature gradient in approximately the top $D/3$ ft of the thermocline should be important, where D is

the depth to the top of the thermocline, sometimes called layer depth. Since the exact choice of depth interval should not be very critical, it has been customary to use the temperature change ΔT in the top 30 ft of the temperature gradient in computations of the intensity change to be expected theoretically.

When there are several temperature gradients present, or when the sharpness of the gradient increases with depth, the theoretical intensity depends in a complicated way on the temperature pattern. However, an empirical study of the numerical intensity computations summarized in reference 31 shows that the following procedure usually gives a moderately good approximation to the theoretical intensity found by ray tracing methods. Consider separately each 30-ft interval of the thermocline. Take the largest value of A found from equation (12); this then gives the theoretical intensity for a given initial ray inclination θ when the ray reaches a depth about 4/3 of the depth to the top of this 30-ft interval. This computed intensity also applies in theory to somewhat greater depths, since especially for deep thermoclines the intensity increases only relatively slowly with increasing depth below the depth of minimum intensity.

Only that part of attenuation which is due to the sharp refraction at the top of the thermocline is taken into account in the expression (12) for the transmission anomaly to points below the layer. If we assume that the attenuation due to absorption is 4 db per kyd, which is a reasonable estimate for 24-kc sound in an isothermal layer, the formula (12) is replaced by the following more realistic formula

$$A = 0.004R + 5 \log \left(1 + \frac{2\Delta c R^2}{c_0 h_1^2} \right). \tag{13}$$

In formula (13), R is the horizontal range to the point where A is measured, Δc is the temperature decrease in the top 30 ft of the thermocline, and h_1 is the height of the sound projector above the top of the thermocline. The first term on the right in formula (13) represents the attenuation caused by absorption, and the second term represents the attenuation due to refraction.

For the data plotted in Figures 24 and 25, the average depth to the top of the thermocline is about 150 ft, which gives a value of 135 ft for h_1. The average value of Δc is about 15 ft per sec, which for a surface temperature of 70 F corresponds to a temperature decrease of 3 F in the top 30 ft of the thermocline. If these numerical values are substituted into equation (13), and this term plotted against the range R, the dashed curve of Figure 25 results. The agreement between theory and observation is fairly close at ranges between 1,000 and 4,000 yd.

This agreement is rather surprising since equation (13) is theoretically not valid at ranges so large that the upward bending in the isothermal layer becomes important and θ_h in equation (11) is no longer equal to h_1/R. A more detailed theory, which takes into account this upward refraction and assumes reflection of sound from a flat surface, would predict a shadow boundary at about 3,000 yd, with a very large anomaly at greater ranges. Possibly, sound reflected from the irregular ocean surface, temperature microstructure, or small systematic negative gradients near the surface, discussed in Section 5.4.2, might explain why equation (13) agrees so well with the facts beyond its expected range of validity. Regardless of the explanation, however, equation (13) may be regarded tentatively as a semi-empirical formula which may be used to predict the sound intensity below a thermocline of given depth and sharpness. A detailed comparison between this equation and the observational data is given in Section 5.3.4.

EARLY STUDIES

One would expect from equation (13) that the difference in intensity above and below the thermocline would depend both on the depth of the layer and on the magnitude of the temperature change in the thermocline. Two early studies along this line were made at UCDWR. Although these studies have not been conclusive and are largely superseded by the more recent results in the following section, they are given here for completeness.

A preliminary plot of the intensity difference above and below the thermocline was made in a UCDWR internal report,[32] using data obtained in 10 vertical runs, during which the hydrophone depth was slowly changed. A least squares solution, with A

as the dependent variable and log $(1 + 2\Delta cR^2/c_0h^2)$ as the independent variable, gave the relation

$$A = -0.3 + 2.85 \log\left(1 + \frac{2\Delta cR^2}{c_0h^2}\right) \quad (14)$$

with h set equal to the thermocline depth, and c to the total velocity change from the isothermal layer to the measuring hydrophone. The measurements extended over a spread of 10 to 5,000 for $2\Delta cR^2/c_0h^2$. Use of the velocity change in the top 30 ft of the thermocline would probably not have changed the results appreciably because of this large spread in $2\Delta cR^2/c_0h^2$. Thus these data indicated a difference of transmission anomaly only about half of that predicted by equation (13); also, the scatter from the mean curve was very great. However, the temperature gradients near the surface were not specified, and an analysis of runs with isothermal water at the surface might be expected to give better agreement. Of possible importance also is the fact that during the appreciable time required for vertical runs the temperature pattern could change appreciably.

More recent analyses have dealt not with transmission anomalies but with values of R_{40}, the range at which the sound intensity is 40 db below the intensity measured with the hydrophone at 16-ft depth at a range of 100 yd. Further, a single parameter is frequently useful to characterize each transmission run, especially when a preliminary analysis of many runs is being attempted. For these reasons R_{40} has been widely used in analyses of transmission data.

Studies have been made of ΔR_{40}, the difference in the R_{40} values determined above and below the layer. Since the values of ΔR_{40} are based on differences between intensities measured simultaneously in the isothermal layer and below the thermocline, it was hoped that these values would be less influenced by variability than individual values of R_{40}. However, the study of ΔR_{40}, given in a UCDWR internal report,[33] has yielded relatively few results, apart from providing general confirmation of the presence of layer effect. For isothermal layers deeper than 40 ft, the average value of ΔR_{40} was 800 yd at 24 kc. However, no correlation of ΔR_{40} at 24 kc could be found with the depth or sharpness of the thermocline underlying the isothermal layer, or with any other feature of the temperature distribution.

This result may be attributed in part to the fact that R_{40} above the thermocline apparently shows some correlation with both the depth and the sharpness of the thermocline. The data in reference 32 sug-

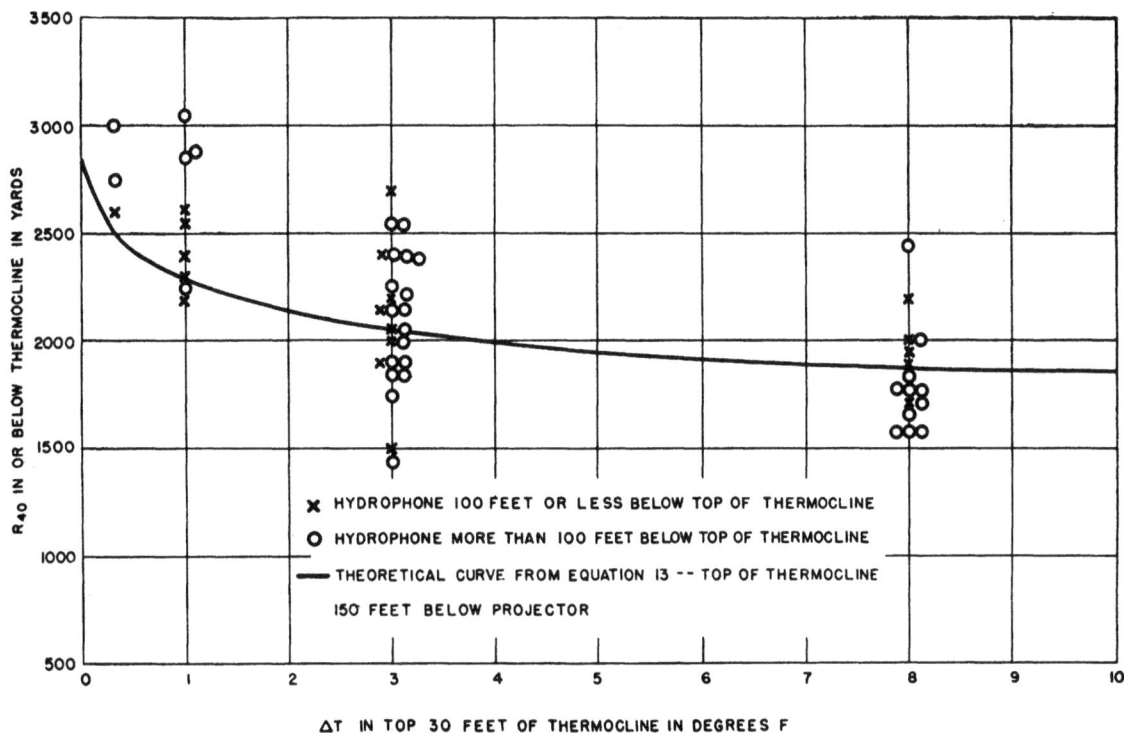

FIGURE 28. Correlation between R_{40} and sharpness of thermocline.

gest that this variation is sufficiently similar to the variation of R_{40} below the thermocline that ΔR_{40} would not be expected to show much predictable change with hydrographic conditions. However, the data analyzed in reference 32 are not sufficiently complete to allow definite conclusions.

The large scatter of the observed data may also contribute to the failure to find any significant correlations in the study of ΔR_{40}. Only half the observed values of ΔR_{40} at 24 kc lay between 450 and 1,150 yd, corresponding to a quartile deviation of 350 yd. This is about what would be expected from an observational error of 2 db in the measured transmission losses. In the isothermal layer, the value of R_{40} is changed about 300 yd by a 2-db change in transmission loss. Below the thermocline, the same change in transmission loss produces a change of only 175 yd in R_{40}; because of the steeper slope of the transmission-loss curve below the thermocline a smaller change of range is required to offset a change of transmission loss than in the isothermal layer. The square root of the sum of the squares of these two quantities is about 350 yd, in agreement with the observed quartile deviation. This close agreement is somewhat surprising since some of the observed scatter is presumably due to variations in hydrographic conditions. In

any case, since the values of R_{40} in the isothermal layer introduce so much scattering in the values of ΔR_{40}, it is reasonable to expect that the analysis of ΔR_{40} is not an appropriate method for investigating the way in which R_{40} below the thermocline depends on detailed oceanographic conditions.

CORRELATION WITH DEPTH AND SHARPNESS OF THERMOCLINE

Examination of the values of R_{40} below the thermocline shows that these are in fact correlated with both the sharpness and the depth of the thermocline. First, the data will be presented on the change of R_{40} with thermocline sharpness. All the values of R_{40} obtained by UCDWR when the depth to the top of the thermocline was between 100 and 200 ft and the hydrophone was below the thermocline are plotted in Figure 28 for different intervals of ΔT, the temperature change in the top 30 ft of the layer. The values of ΔT shown are only approximate, as a result of the grouping of the recorded data into four different groups, as follows: ΔT less than 0.7 F; ΔT between 0.7 and 1.5 F; ΔT between 1.6 and 4.0 F; and ΔT between 4.1 and 12.5 F.

The crosses represent runs in which the hydrophone was 100 ft or less below the top of the thermocline while for the circles the hydrophone depth was more than 100 ft below the top. From intensity contour diagrams like those shown in Figure 25 of Chapter 3, it is evident that for thermoclines within less than 100 ft from the surface, the computed intensity increases appreciably as the hydrophone goes from just below the layer to considerably greater depths. No simple formula has been derived for the increase of intensity in this case. The points plotted in Figure 28 indicate that for these deeper thermoclines also, the value of R_{40} tends to increase somewhat with increasing depth of hydrophone below the top of the thermocline, provided the value of ΔT is less than about 2 degrees. This result is also in general accordance with the predictions of intensity-contour diagrams.

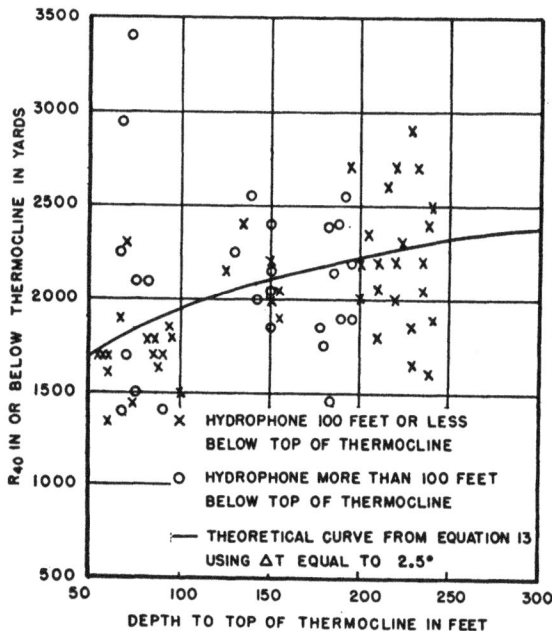

FIGURE 29.   Correlation between R_{40} and depth to thermocline.   Temperature difference in top 30 feet of thermocline, 1.6 degrees to 4.0 degrees.

The curve in Figure 28 is computed directly from equation (13), with a thermocline depth of 150 ft and a surface temperature of 70 F. It is evident from Figure 28 that the change of R_{40} with changing temperature difference is, if anything, somewhat greater than can be explained on the basis of equation (13). This result is the reverse of that found in the empirical equation (14). The many different points plotted in Figure 28 are not all completely independent since many were taken on the same day. Thus, the sampling error may be larger than might be expected from

the number of points plotted. However, the data shown in Figure 28 are more extensive than those used in reference 32, and the result should therefore be more reliable.

Similar data may be used to show the dependence of R_{40} on thermocline depth. Values of R_{40} obtained with hydrophones below the thermocline, and with temperature differences of 1.6 to 4.0 F in the top 30 ft of the thermocline are shown in Figure 29. The circles and crosses have the same meaning as before. It is apparent that, for the shallower layers, the increase in intensity at depths well below the thermocline can become quite marked; this is in accordance with the theoretical expectations schematically presented in the intensity-contour diagrams of reference 31.

The curve in Figure 29 shows theoretical values, computed from equation (13), with a velocity difference Δc of 12 ft per sec, corresponding to a temperature change of 2.5 F at a surface temperature of 70 F. The change in the median R_{40} is approximately that predicted by equation (13). The quartile deviation, however, is of the same order of magnitude as the increase in median R_{40} when the depth of the isothermal layer is increased from 70 ft to 200 ft.

The general trend in Figures 28 and 29 seems to indicate that equation (13) gives a rough approximation to the median observed transmission. Though the spread is large, the data are not in disagreement with the predictions of that equation about the effect of changes in layer depth and thermocline sharpness. Thus equation (13) gives a moderately good fit to UCDWR transmission data.

Additional data are required, of course, for more conclusive results. In particular, the number of variables that might enter the problem is so great that other factors may be responsible for the apparent agreement between observations and the simple theory. Nevertheless Figures 28 and 29 indicate that equation (13) provides a moderately satisfactory empirical fit for the present data.

Some of these same results have been obtained in greater detail in an analysis of average transmission anomaly curves. This analysis[14] classifies the data according to the temperature code discussed in Section 5.1.3. The average anomaly curves for hydrophones in or below the thermocline with d_2 equal to 4 and to either 5 or 6 are given in Figures 30 and 31, respectively. These correspond to isothermal layers between 40 and 80 ft thick, and between 80 and 320 ft thick, respectively. In Figure 30, curve III, with the hydrophone between 20 and 160 ft below the top of

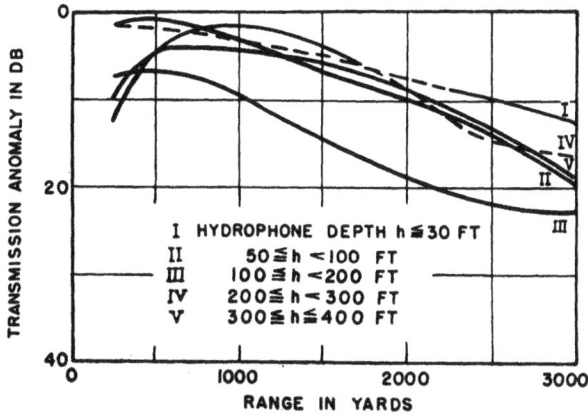

FIGURE 30. Average transmission anomalies for isothermal layer 40 to 80 feet thick.

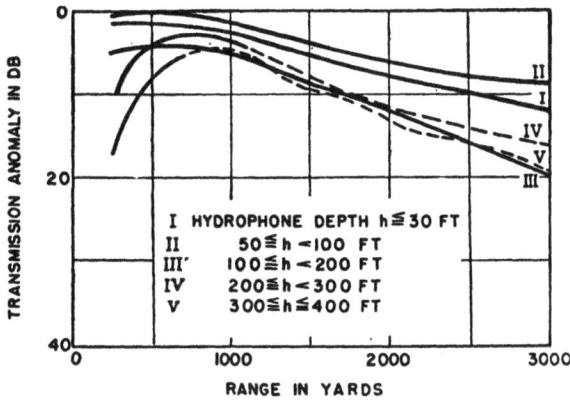

FIGURE 31. Average transmission anomalies for isothermal layer more than 80 feet thick.

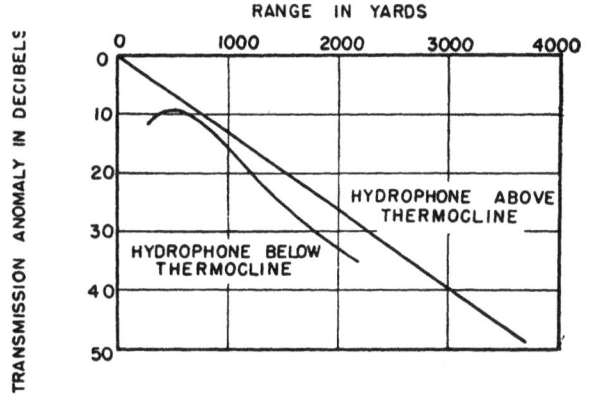

FIGURE 32. Average transmission anomalies at 60 kc, above and below thermocline.

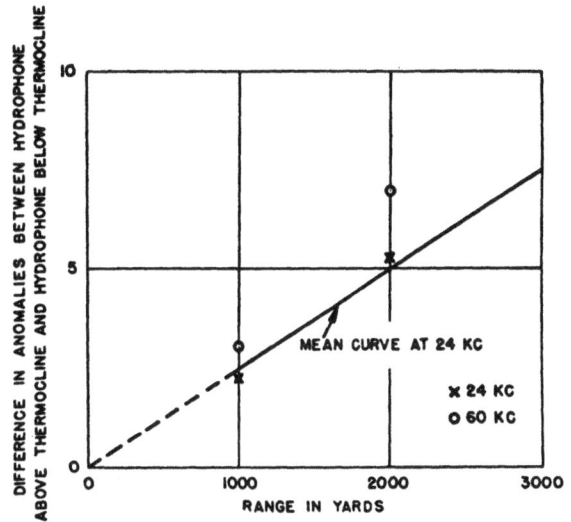

FIGURE 33. Difference in transmission anomalies above and below thermocline.

the thermocline, is significantly lower than curves IV and V, with the hydrophone between 120 and 360 ft below the top. Curve II apparently combines some runs with the hydrophone above the layer with others below the layer and is thus intermediate between curve I (hydrophone in the isothermal layer) and curve III (hydrophone below top of thermocline). In Figure 31, curves III, IV, and V agree with each other, and apparently represent an average anomaly curve for a hydrophone below the thermocline. In agreement with equation (13), the anomalies are not so great as those found for a hydrophone just below a shallow layer. An analysis relating the average transmission anomaly below a thermocline to changes in the depth and sharpness of the thermocline might give more useful information than can be obtained with the temperature code used in Figures 30 and 31.

5.3.5 Transmission at 60 kc

The only other frequency for which data are available on transmission below the thermocline is 60 kc.

The average transmission anomalies both in the isothermal water above the thermocline and in the thermocline are shown in Figure 32. The difference between these curves at 1,000 and 2,000 yd is plotted in Figure 33; for comparison, the corresponding differences at 24 kc are also shown, which were taken from Figure 24. On the ray theory, these two curves should be identical if θ_0 is the same for the two frequencies, since the increased divergence resulting from downward bending should be independent of frequency. The agreement between the 24-kc crosses and the 60-kc circles in Figure 33 is not too close; however, such a comparison of data cannot be reliable unless the measurements were made under similar thermal conditions.

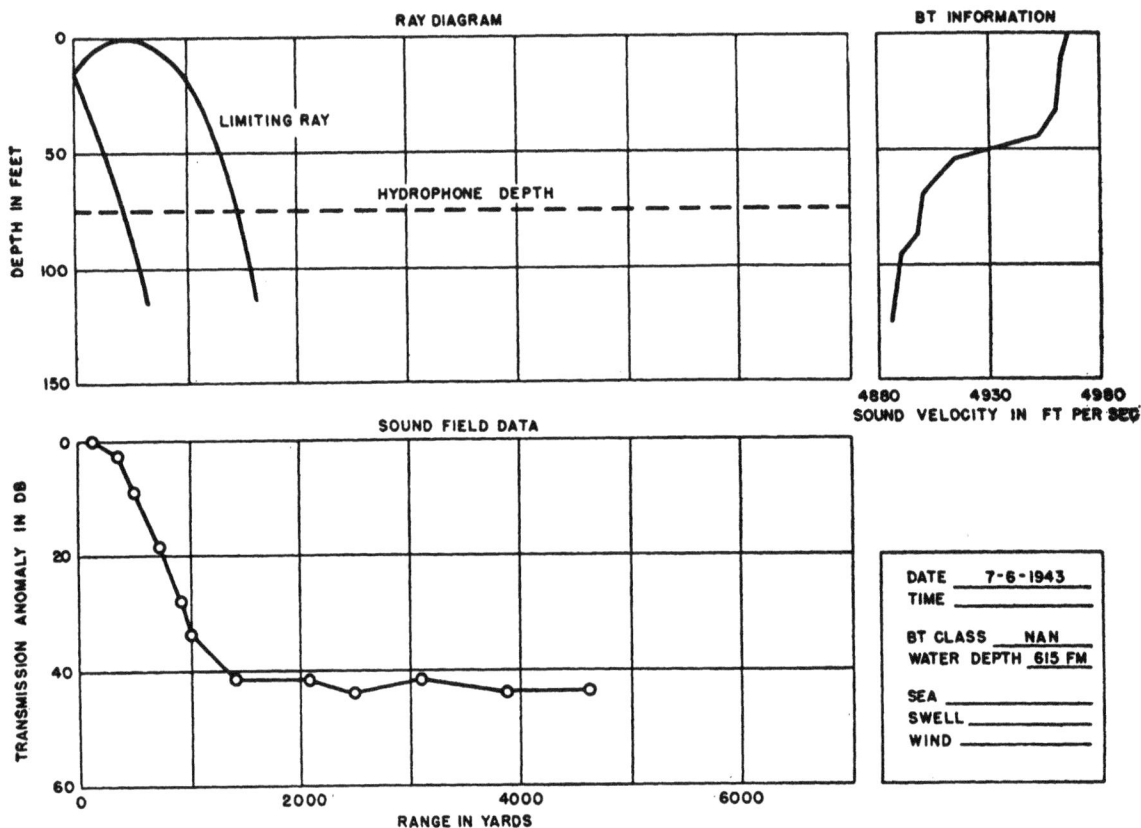

FIGURE 34.   Sample transmission anomaly plot.

5.4   TRANSMISSION WITH NEGATIVE GRADIENTS NEAR SURFACE

In some regions at certain times, negative temperature gradients in the upper 50 to 100 ft of the ocean are very common. Especially in coastal waters and during the summer months the temperature difference between the surface and 30 ft frequently exceeds 0.3 F. Temperature patterns of this type are usually highly variable.

Many of the transmission measurements at UCDWR were made with negative gradients at the surface. The present section discusses these data. While emphasis is placed on temperature patterns for which the temperature difference from the surface to 30 ft is 0.4 F or more (NAN and CHARLIE patterns), some discussion is included of those cases where the top 30 ft is isothermal but the top of the thermocline is less than 100 ft below the surface. A more rigid division of the acoustic data by the different types of temperature-depth patterns is not possible, since the different analyses that have been made have used somewhat different classifications.

In the following pages, a discussion will first be given of the general types of transmission anomalies that are found with temperature gradients near the surface. Subsequently, more detailed discussions will be given, first for those situations where the temperature gradient in the top 30 ft is large, about 0.7 F or more, and second, for smaller gradients in the top 30 ft.

Because of an almost complete lack of analyzed data at other frequencies, this discussion is largely confined to results obtained at 24 kc. A few results obtained at 60 kc are described at the end of this section. The bulk of the results come from two reports by UCDWR, one issued in 1943[31] and the other in 1945.[14] Use is also made of individual transmission anomaly curves obtained from UCDWR which have not been published.

An examination of the transmission anomalies measured with temperature gradients near the surface shows that most of the observational curves may be divided into two types. In the first of these, the anomaly does not change rapidly with range at very close ranges or at very long ranges, but drops very rapidly at a range somewhere between 500 and 2,000 yd. In the second type, the transmission anomaly in-

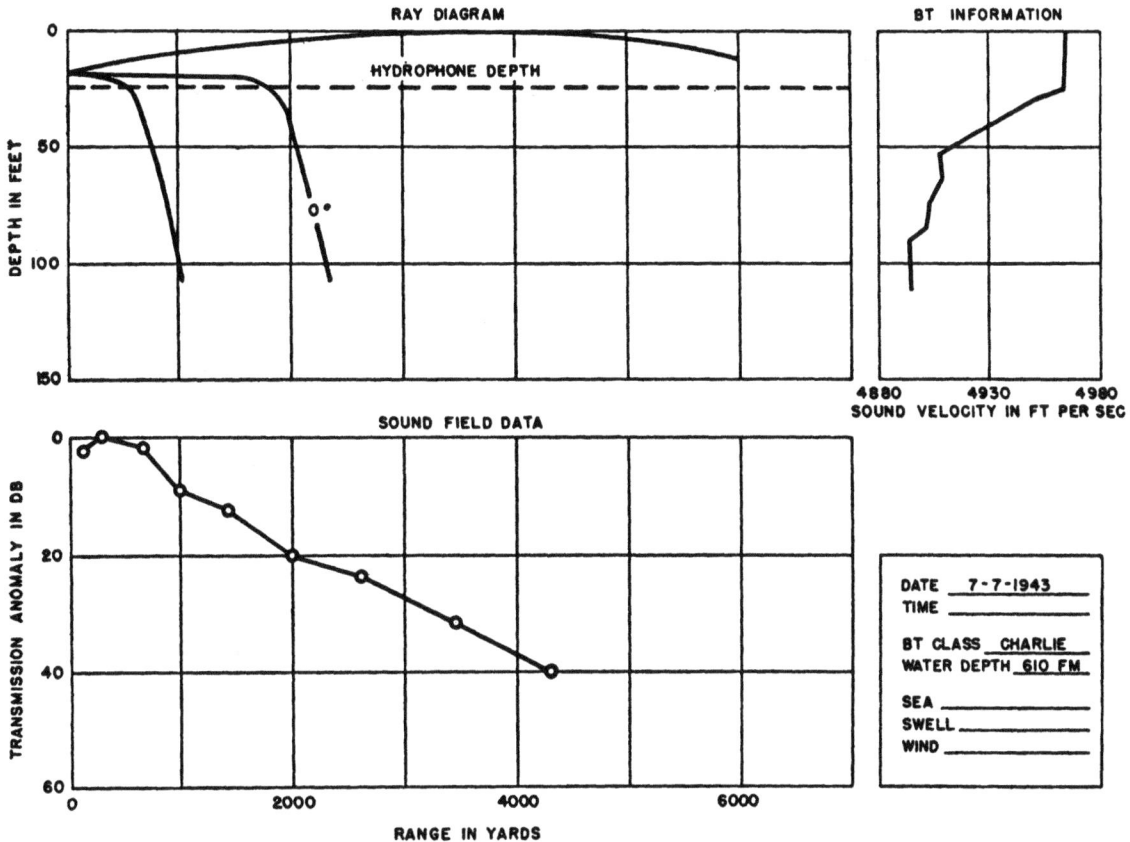

FIGURE 35.   Sample transmission anomaly plot.

creases linearly with range out to long ranges. Sample types of these two curves are shown in Figures 34 and 35, respectively. With each figure is included a smoothed temperature-depth plot and a corresponding ray diagram.

The relative number of straight-line anomaly curves for different hydrographic conditions has been investigated.[13] The number of straight-line graphs was found for different values of the temperature difference from 0 to 30 ft and for different values of the computed limiting range at the hydrophone depth used. The results of this study are given in Table 1. The accuracy with which observed curves could be fitted by straight lines has already been discussed in Section 5.2 where the high probability of straight-line graphs in isothermal water was also pointed out.

The classification of bathythermograms into MIKE, CHARLIE, and NAN patterns on the basis of the temperature difference from 0 to 30 ft has already been discussed in Section 5.1.4. These patterns are indicated in Table 1; the dividing line between NAN and CHARLIE patterns, usually defined as 1/100 of the surface temperature, is about 0.7 F for

TABLE 1.   Relative number of straight-line anomaly graphs.

Temperature difference 0 to 30 ft in °F	Number of straight graphs	Number of graphs not straight	Temperature pattern
0.0	40	0	MIKE
0.1	25	0	MIKE
0.2	18	2	MIKE
0.3	5	1	MIKE
0.4	8	7	CHARLIE
0.5	4	1	CHARLIE
0.6	3	7	CHARLIE
0.7 or more	8	42	NAN
Total	111	60	
Range to shadow boundary at hydrophone depth in yards			
0–500	0	8	
500–1,000	10	22	
1,000–1,500	5	16	
1,500–2,000	12	6	
2,000–3,000	15	5	
3,000–4,000	31	1	
4,000 or more	38	2	
Total	111	60	

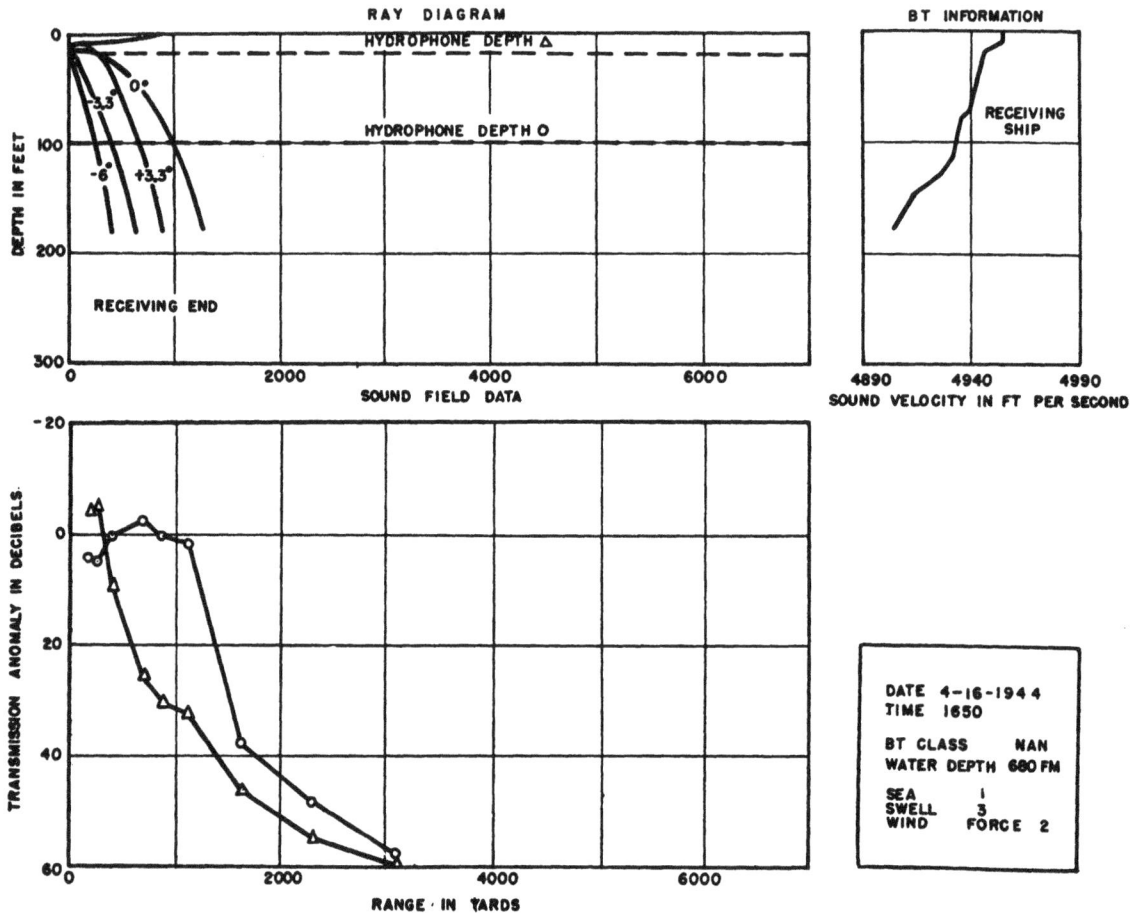

FIGURE 36.   Sample transmission anomaly plot.

most of the runs made off San Diego. While these patterns are by no means fundamental, Table 1 suggests that they provide a natural classification for the data.

It is evident from these tables that when sharp temperature gradients are present at the surface, straight-line graphs are unlikely. In the following discussion, transmission runs made with sharp temperature gradients present in the top 30 ft are therefore treated separately from those made in the presence of weaker gradients.

5.4.1 Sharp Temperature Gradients in Top 30 ft

When the temperature gradient in the top 30 ft is sharp and extends all the way to the surface, theory predicts a sharp downward bending of the sound beam, and a shadow zone of low sound intensity at fairly close ranges. This expectation is generally ful-

filled, provided that the difference of temperature between the surface and 30 ft is more than 0.7 F, and the hydrophone is above 100 ft. Samples of such curves are shown for the shallow hydrophones in Figures 36 and 37 as well as in Figure 34. Since the surface temperature for most of the UCDWR transmission runs off San Diego was in the neighborhood of 70 F, this critical value is 1/100 of the surface temperature, which is the dividing line between NAN and CHARLIE patterns.

While the ray theory can explain the qualitative features of the sound field observed with NAN patterns, it cannot predict exactly the transmission anomalies to be expected under different conditions. The detailed dependence of sound transmission under these conditions on the temperature structure and on the hydrophone depth is an important subject on which considerable data are available. This information is summarized in the following pages. First, the correlation between transmission data and the com-

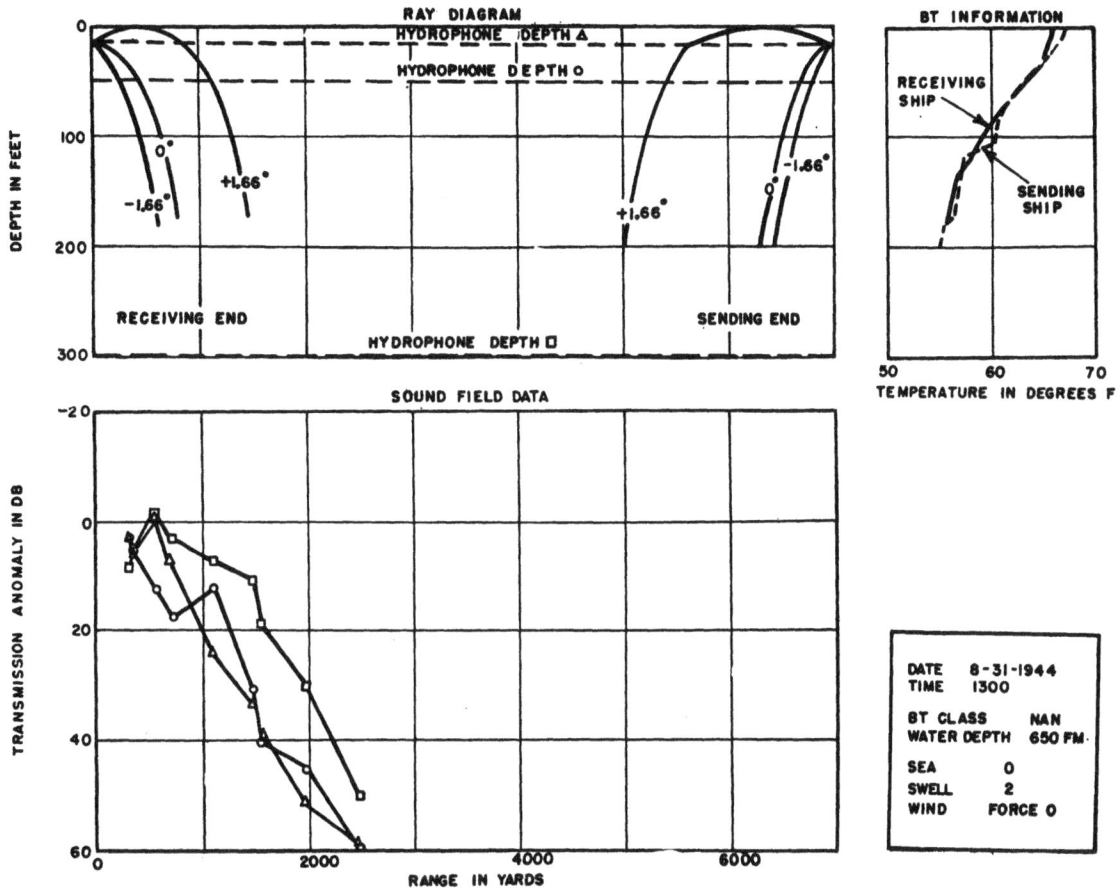

FIGURE 37. Sample transmission anomaly plot.

puted range to the shadow boundary is discussed. This is followed by a more detailed statistical investigation of average transmission anomaly curves for different types of temperature patterns and for different hydrophone depths. Finally, separate discussions are given of the observed slope of the transmission anomaly near the shadow boundary and of the sound observed in computed shadow zones.

COMPUTED RANGE TO THE SHADOW BOUNDARY

When the temperature gradient extends right up to the surface, and a shadow zone is predicted from the simple theory, the range to the break in the observed transmission anomaly curve might be expected to equal the computed range to the shadow boundary. More frequently, however, the range to the break is less than this value, as shown in Figures 34, 36, and 37. A detailed comparison between the ray diagram and the transmission anomaly curves is complicated by the variability of temperature conditions, already discussed in Section 5.1.3. However, a

brief analysis of some of the data has been carried out, which averages the range to the shadow boundary computed from bathythermograms taken at the beginning and end of the transmission run.

The resulting plot of these data is shown in Figure 38, taken from reference 34. This figure includes data taken during two days of measurements. The dashed line has been fitted visually to the observed points. The correlation between the observed and the theoretical values is moderately good, considering the variability of the temperature structure. However, the observed ranges to the shadow boundary seem to be systematically less than the predicted values, an effect which is difficult to attribute to changes in ocean temperature or errors in reading the bathythermograph slide. The data presented in Section 9.2.3 on the change in shape of the pulse in the shadow zone seem to indicate that the range to the break in the transmission anomaly curve is correctly identified as the boundary of the shadow zone. Thus, Figure 38 may present a real discrepancy, although the amount

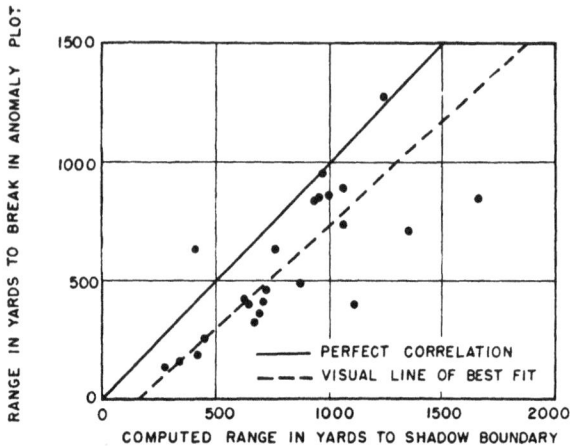

FIGURE 38.   Correlation between break in transmission anomaly plot and computed range to shadow boundary.

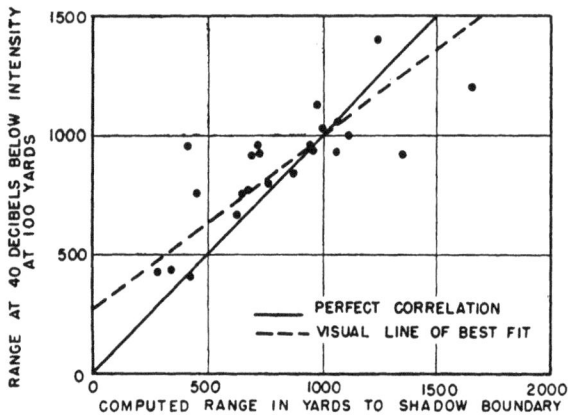

FIGURE 39.   Correlation between R_{40} and computed range to shadow boundary.

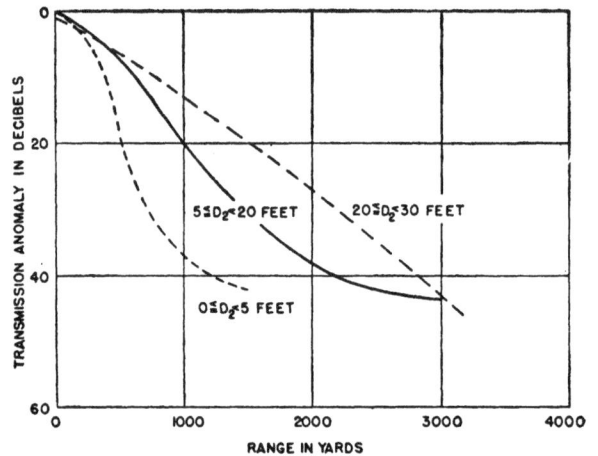

FIGURE 40.   Average transmission anomalies for NAN patterns, shallow hydrophone.

of data is too limited to permit very definite conclusions. Possibly the presence of water waves on the ocean surface lowers the effective level of the surface and brings the shadow boundary in to closer range than would be expected for a flat surface.

A somewhat more practical quantity found from the transmission curves is the range R_{40} at which the sound intensity is 40 db below the intensity at 100 yd. The significance of this quantity has already been discussed in Section 5.3.4. A plot of R_{40} against computed limiting range is shown in Figure 39. The correlation is again only fair. The dashed curve of best fit was chosen visually, as in Figure 38.

Figures 38 and 39 cannot be used to give reliable average results because of the paucity of data included. They indicate in a general way the correlation that may be expected between detailed computations of limiting rays and observed transmitted sound intensities.

AVERAGE TRANSMISSION ANOMALY

A statistical average of the measured transmission anomalies for different temperature conditions is given in reference 14, based on the temperature-depth code described in Section 5.1.4. The data for shallow hydrophones (16 to 30 ft) are discussed first. Corresponding data for deeper hydrophones are given in the next section. Three curves for shallow hydrophones in water with sharp temperature gradients in the top 30 ft (NAN patterns) are given in Figure 40. Each curve represents an average for a different value of D_2, the depth at which the temperature is 0.3 F less than the surface temperature. An examination of the bathythermograph data shows that for D_2 less than 5 ft, the main thermocline in every case extended all the way to the surface, resulting in very sharp downward bending of the beam. For D_2 between 5 and 20 ft, a layer of small gradient overlay the thermocline, which usually extended up to within 20 ft of the surface, while for D_2 between 20 and 30 ft the main thermocline was always deeper than 20 ft and usually between 20 and 30 ft from the surface. To indicate the scatter of the individual points making up these average curves, all anomalies for D_2 between 5 and 20 ft are plotted in Figure 41. Half of the points lie within about 4 db of the mean curve. The scatter shows some tendency to increase with range and is significantly greater than that found with a hydrophone in deep isothermal water (see Figure 15). There is no significant change either in the mean curve or in the scatter if values of D_2 between 5 and 10 ft and between 10 and 20 ft are considered separately. The curve for D_2 between 20 and 30 ft is based on rela-

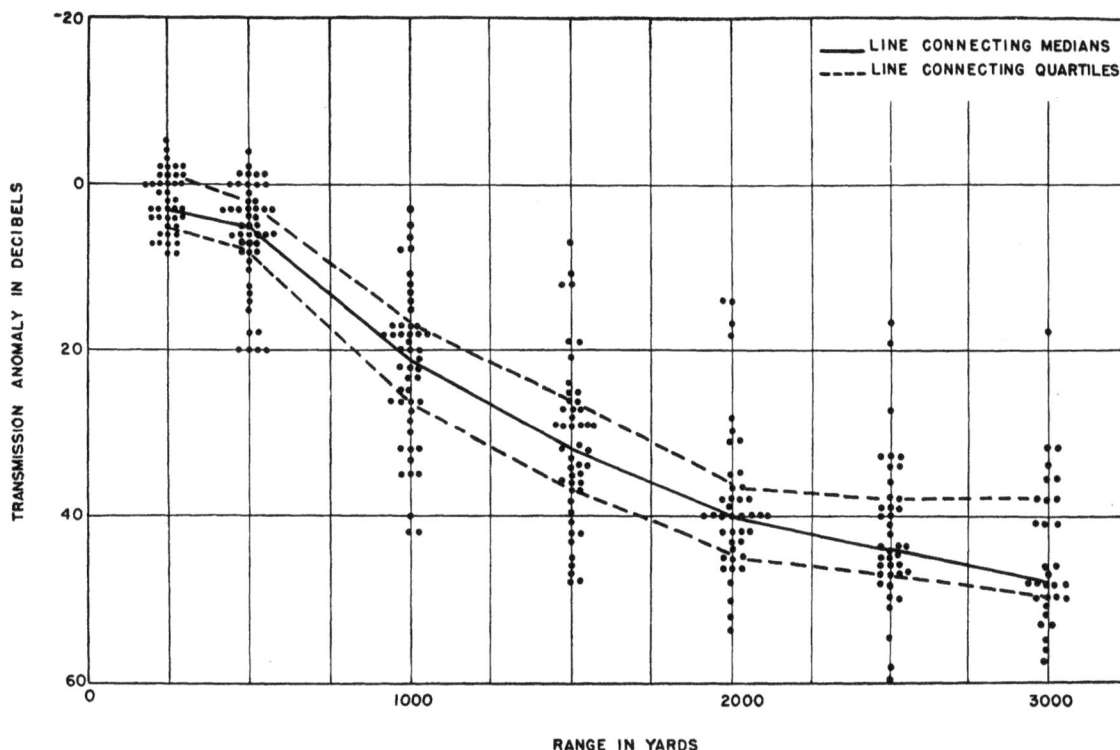

FIGURE 41. Plot of individual anomalies for D_2 between 5 and 20 feet.

tively few points, and the individual points show a large scatter. Thus, this curve is not very reliable.

Average anomaly curves have also been computed for different values of D_1, the depth at which the temperature is 0.1 F less than the surface temperature. For a fixed value of D_2, these average curves show no systematic variation with changes in D_1. With NAN patterns, gradients below 40 ft also have a negligible effect on the average transmission anomalies measured with a shallow hydrophone.

The change of sound transmission with changing D_2 is to be expected on theoretical grounds. It is evident from Figure 35 that the range to the computed shadow boundary becomes much extended when a shallow layer of weak gradient overlies the major temperature decrease. When thermal microstructure or surface reflections are taken into account, sound in a thin surface layer of weak gradient can evidently be propagated out to substantial ranges, with some sound continually being bent down into the layer of sharp gradient. Thus no shadow zone is to be expected in this situation, and straight-line graphs of the type shown in Figure 35 are likely. The high attenuation observed in this thin layer is also generally found in shallow isothermal layers, and is discussed again in Section 5.4.2.

DEPENDENCE ON HYDROPHONE DEPTH

In accordance with expectations, the range to the observed shadow zone — either R_{40} or the range to the break in the transmission anomaly curve — increases with increasing hydrophone depth. This effect, when present, results in a predicted increase of maximum echo range with increasing submarine depth; this increase was observed in early practical echo-ranging trials.[35,36] The same effect is shown clearly in Figures 36 and 37; the drop in intensity for the deep hydrophones is found at ranges substantially larger than for the shallow hydrophone. When a deep isothermal layer is present below the sharp surface gradient, this effect would be expected to be very marked, on the basis of ray theory. A sample run in one of the rare observed situations of this type is shown in Figure 42, with the accompanying temperature-depth plot.

The statistical analysis of the data reported in reference 14 provides an indication of the average change of intensity with depth. The average curves found for hydrophones at different depths below 50 ft are reproduced in Figure 43. The upper and lower plots correspond, respectively, to intervals of 0–10 and 10–20 ft for D_2, the depth at which the temperature is 0.3 degree less than the surface temperature

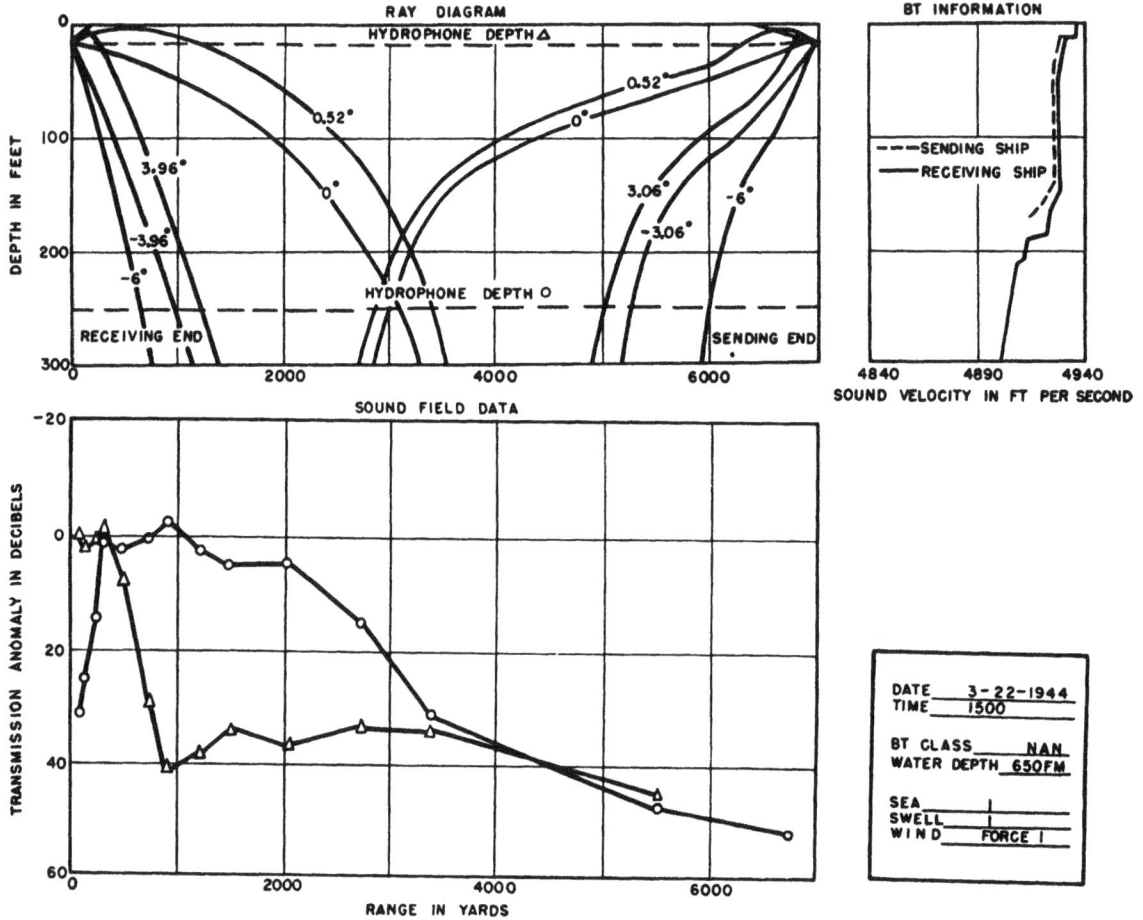

FIGURE 42.　Sample transmission anomaly plot; NAN pattern with deep isothermal layer below.

Insufficient data are available for NAN patterns with D_2 between 20 and 30 ft to yield much information on the change of transmission anomaly with hydrophone depth for that case. The scatter of the individual points averaged to yield Figure 43 is moderately large. Since only about ten runs were averaged to give each curve, these average curves have a probable error of between 2 and 3 db.

ATTENUATION COEFFICIENT AT SHADOW BOUNDARY

The mechanism by which sound penetrates the shadow zone is of theoretical and practical interest. Information about this mechanism may be obtained by comparing the rate of increase of transmission anomaly near the shadow boundary with that computed from the diffraction of sound. The slope of the transmission anomaly beyond the break is very high, usually between 20 and 70 db per kyd when the sharp temperature gradient extends all the way to the surface (D_2 less than 10 ft). We shall call this slope a' and

may regard it as a sort of "local attenuation coefficient," that is,

$$a' = \frac{dA}{dR}. \tag{15}$$

The local attenuation coefficient defined by equation (15) is not to be confused with the actual attenuation coefficient characterizing the transmission from the sound source to the range R, defined as $1,000A/R$.

The observed slope beyond the break is to be compared with the local attenuation coefficient at the shadow boundary which would result from diffraction. From Section 3.7, we have the following formula for a' in the case of a linear velocity gradient.

$$a' = \frac{25.7}{c}\left(-\frac{dc}{dy}\right)^{\frac{2}{3}} f^{\frac{1}{3}}. \tag{16}$$

In formula (16) c is the sound velocity in yards per second; dc/dy is the velocity gradient in feet per second per foot; and a' is in units of decibels per yard.

If the temperature difference between 0 and 30 ft is about 1 F, and the surface temperature is about 70 F, then for sound of 24,000 cycles equation (16) gives an attenuation of about 0.1 db per yd, or about 100 db per kyd. This is at least twice as great as the values usually obtained. The discrepancy seems somewhat too great to be explained as observational error, although the fact that observed and predicted attenuations are of the same order of magnitude is suggestive. It is possible that the presence of thermal microstructure may explain this difference. No attempt has been made to correlate the observed attenuation across the shadow boundary with either the frequency f or the velocity gradient dc/dy at the surface.

SCATTERED SOUND IN THE SHADOW ZONE

Most transmission anomaly plots for NAN patterns are characterized by a nearly constant transmission anomaly well beyond the computed boundary of the shadow zone, amounting to between 40 and 60 db and extending out to the limit of measurement at about 5,000 yd.

An examination of the oscillograph record shows that the signal received at ranges between 1,500 and about 3,000 yd bears very little relation to the shape or length of the original pulse. Figure 44 shows the signals received at different ranges when a marked negative gradient was present at the surface. At moderately short ranges, less than 1,000 yd, the received signal reproduces the outgoing pulse rather faithfully. At moderate ranges into the shadow zone, however, the signal has an appearance similar to that of reverberation, and is much prolonged, as shown in trace C made at 1,340 yd. At these intermediate ranges, the use of peak amplitudes in reading each signal gives a value about 7 db higher than the use of average intensities. For coherent 100-msec signals, the difference between peak amplitude and rms amplitude is negligible. At slightly longer ranges even the few traces of the direct pulse, visible for some of the signals in trace C, completely disappear. At the longest ranges the signal begins again to resemble the emitted pulse, as shown in trace D.

The intensity of the sound received in the shadow zone increases with increasing pulse length. The observed difference in intensity for 100-msec and 10-sec pulses is shown in Figure 45. In the shadow zone at intermediate range, where the received signal is prolonged and incoherent, the effect is large, amounting to between 4 and 8 db. At very long

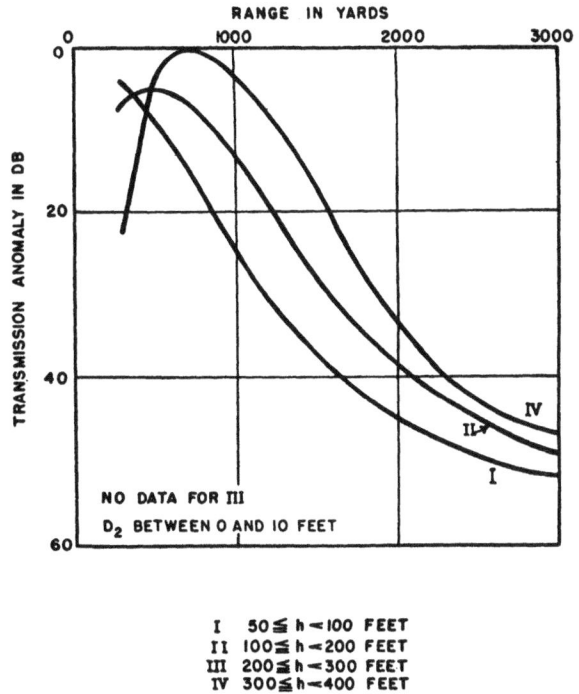

I $50 \leq h < 100$ FEET
II $100 \leq h < 200$ FEET
III $200 \leq h < 300$ FEET
IV $300 \leq h < 400$ FEET

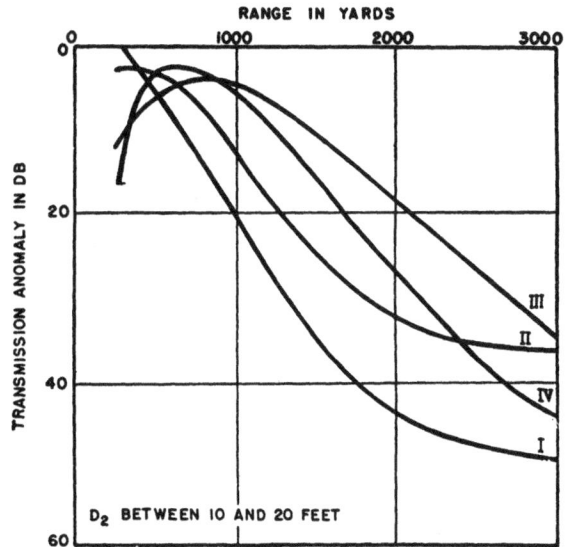

FIGURE 43. Average transmission anomalies for NAN patterns with hydrophone deeper than 50 feet.

ranges and within the direct sound field at shorter ranges the signal is coherent and the difference is small. This small difference results from the fact that the peak amplitude of a long signal, which fluctuates from minimum to maximum values many times, tends to be greater than the peak amplitude of a short signal, which may be altered by fluctuation to a low value.

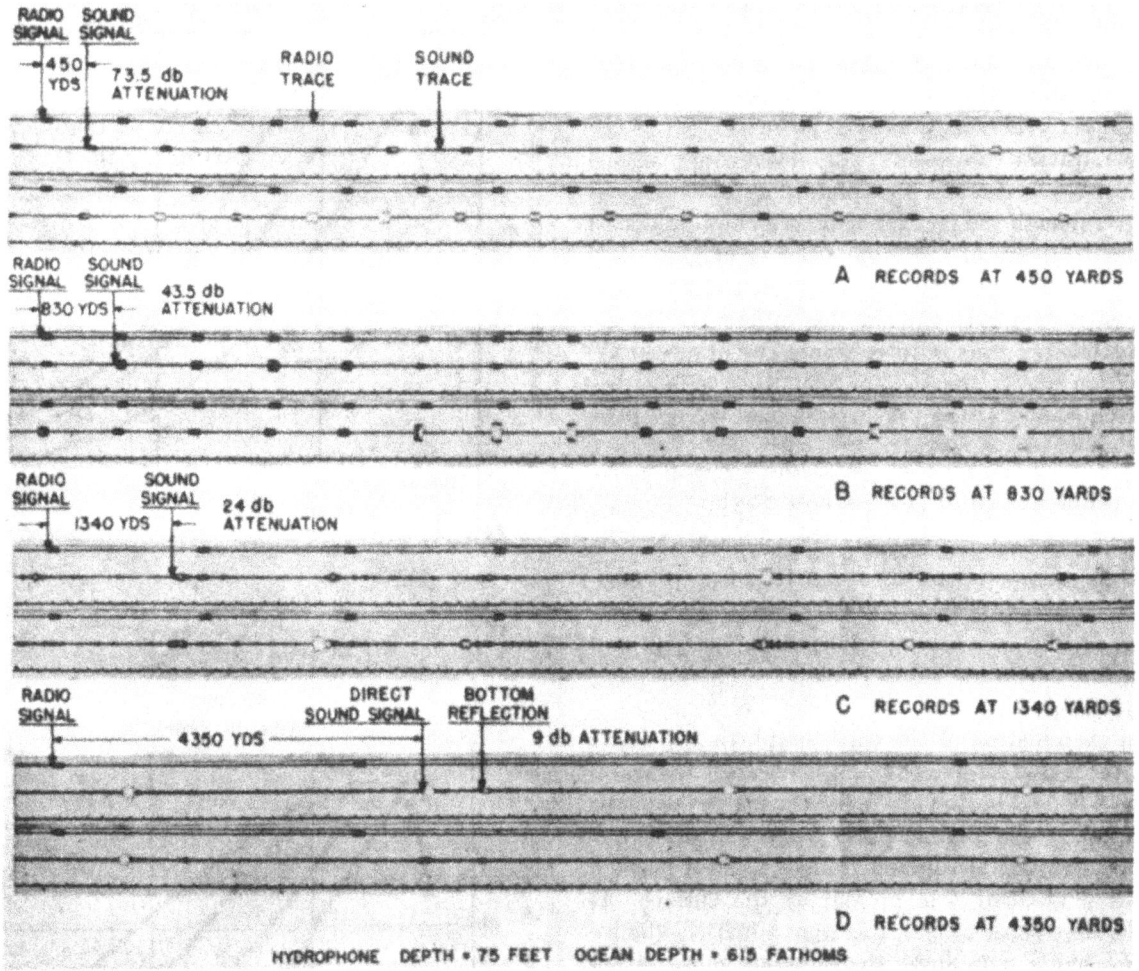

FIGURE 44.   Records of received signals at various ranges under downward refraction.

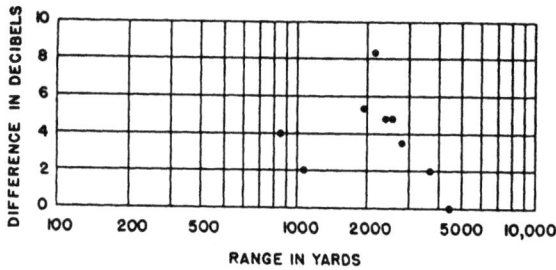

FIGURE 45.   Increase of average peak amplitudes for long pulses.

The sound received in the shadow zone, at least out until about 2,500 or 3,000 yd, is probably sound scattered from the main beam at considerable depth up to the hydrophone near the surface. The scattering coefficient required to explain the observations can be readily estimated from the transmission anomalies of the long pulses. The transmission anomaly for 100-msec pulses in the shadow zone is usually between 40 and 60db. Figure 45 shows that about 4 to 8 db must be subtracted from these values to find the transmission anomaly of the 10-sec pulse, which may be regarded as essentially a continuous tone in these observations. On the other hand, 7 db must be added to find the anomaly in terms of average intensity instead of average peak amplitude. Since these two corrections about cancel out, 40 to 60 db is the transmission anomaly for the intensity of long pulses in the shadow zone.

A theoretical value for this transmission anomaly may be computed on a somewhat simplified picture. Although the scattered sound is itself refracted by the prevailing vertical velocity gradient, most of the scattered rays are inclined so steeply that they may be regarded as straight lines. With this assumption,

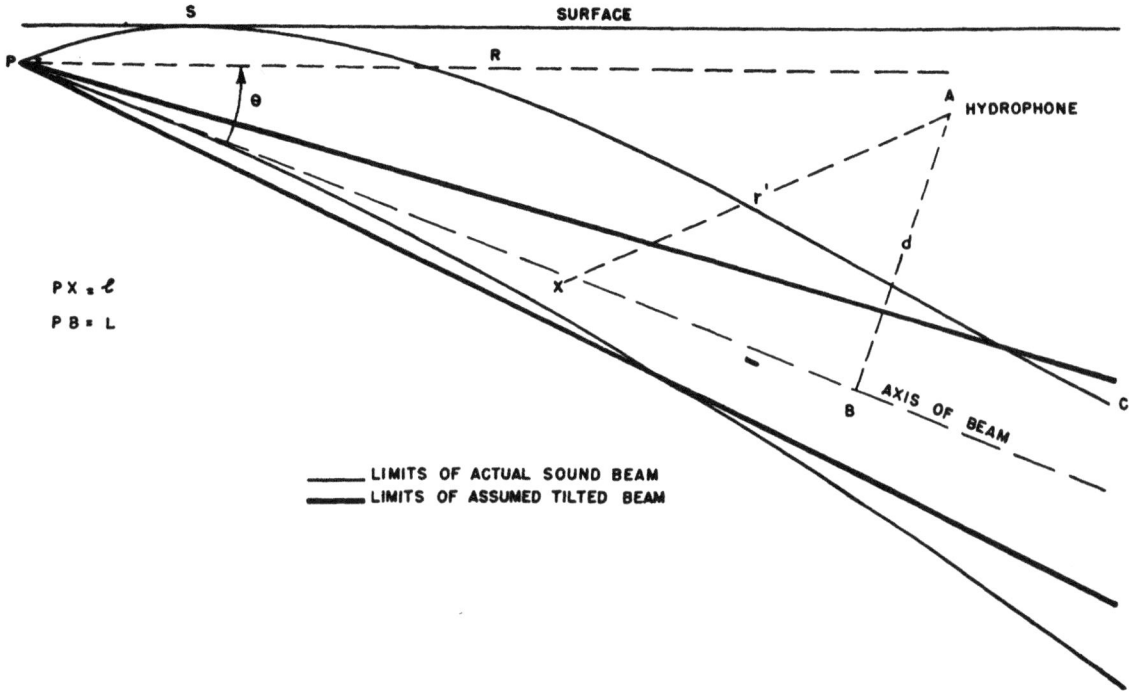

FIGURE 46. Diagram used in calculating scattered sound intensity.

the scattered sound is transmitted to the entire ocean as if refraction were not operating. Also, for the purpose of calculating the sound scattered into the shadow zone, the actual direct beam may be replaced by a tilted beam traveling in a straight line, as in Figure 46.

To calculate the scattered sound received when a long (10-sec) pulse is sent out, it may be assumed that sound scattered from the entire length of the beam is received at the hydrophone. Let the total initial power in the beam be denoted by J. If attenuation is neglected, this energy will remain constant as the sound travels outward. If the scattering coefficient is m per yard, a fraction m of the sound energy will be scattered per yard of travel of the beam (see Chapter 2 of Part II). This energy will be scattered in all directions; and the intensity of the sound scattered from this cross section of the beam 1 yd thick and reaching the hydrophone at a distance r' yd away will be $mJ/4\pi r'^2$. While in actual fact the sound scattered from the lower side of the beam will be more weakened than that scattered from the upper side, the distance r' from the hydrophone to a point on the axis of the beam should be a reasonable approximation for each separate cross section of the beam. Let l represent the distance from the projector to the point where scattering is taking place. The sound scattered

between l and $l + dl$ is thus $(mJ/4\pi r'^2)dl$; and the total scattered sound received at the hydrophone is

$$I_s = \int_0^\infty \frac{mJ}{4\pi r'^2}dl = \frac{mJ}{4\pi}\int_0^\infty \frac{dl}{(L-l)^2 + d^2} \quad (17)$$

where the quantities L and d have the meanings shown in Figure 46. The integration yields approximately

$$I_s = \frac{mJ}{4\pi}\left(\frac{\pi}{d}\right) = \frac{mJ}{4d}. \quad (18)$$

While in the general case m will be a complicated function both of position in the ocean and of the direction in which the scattered sound is measured, here m is assumed to be constant. Equation (18) thus refers in practice to an average value of m.

The expression (17) does not take into account the transmission anomaly resulting from absorption or refraction. When a sound beam is refracted sharply downward, the intensity in the direct sound field is not reduced much below the inverse square value. The scattered sound, which reaches the hydrophone at steep angles, is also relatively unaffected by refraction. The absorption must be considered, however. For points in the sound beam to the left of the point B in Figure 46, the sum of the absorption loss for direct and scattered sound will not depend much on whether the sound was scattered close to the projector or some distance out. Sound scattered from points to the right of the point B will suffer a two-way absorption loss, and may be neglected. Therefore, to calculate the sound intensity at the hydrophone, taking absorption into account, we must first integrate equation (17) with the infinite upper limit replaced by L, obtaining approximately $mJ/8d$; then we must multiply this value by some factor to take the absorption into account. Because we are neglecting the sound scattered to the right of B, we may consider that the sound reaching the hydrophone has traveled a total path length, on the average, equal to R, the range from the projector to the hydrophone. If a is the attenuation coefficient in decibels per yard, the intensity I_s is therefore given by

$$I_s = \frac{mJ}{8d} 10^{-aR/10} \qquad (19)$$

Equation (19) was derived without considering the possibility that sound could be scattered up to the surface by points to the left of B and reflected back to the hydrophone. We can allow for this extra intensity due to surface reflection by multiplying the expression (19) by 2. Our final result is

$$I_s = \frac{mJ}{4d} 10^{-aR/10}. \qquad (20)$$

It is convenient to restate equation (20) in decibels:

$$10 \log I_s = 10 \log m + 10 \log J - 10 \log (4d) - aR. \qquad (21)$$

The total emitted power J is related to F, the power output per unit solid angle in the direction of the projector axis, by the formula

$$10 \log J = 10 \log (4\pi F) + D, \qquad (22)$$

where D is the directivity index of the projector. Combining equations (22) and (21) gives

$$10 \log I_s = 10 \log m + 10 \log F + D - 10 \log d$$
$$- aR + 10 \log \pi. \quad (23)$$

The transmission anomaly A of the scattered radiation is given by

$$A = 10 \log F - 20 \log R - 10 \log I_s.$$

If $R \sin \theta$ is substituted for d, from Figure 46, and I_s is taken from equation (23), the transmission anomaly A becomes

$$A = - 10 \log R - 10 \log m - D + aR - 10 \log \pi$$
$$+ 10 \log (\sin \theta). \quad (24)$$

For a total temperature decrease of about 20 degrees in the thermocline, the limiting ray PSC in Figure 46 bends downward below the thermocline at an angle of 12 degrees. Thus, a typical value of θ is 12 degrees. For a directional transducer of the type normally used in echo ranging, the directivity index D is -23 db. If an absorption coefficient a of 4 db per kyd is used in equation (24), and D and θ are set equal to -23 db and 12 degrees, respectively, then A is nearly constant from 1,000 to 3,000 yd, and equals $-10 \log m - 14$. Since A is observed to lie between 40 and 60 for transmission to points well inside the shadow zone, $10 \log m$ is between -54 and -74. Because the receiver is directional in a vertical plane, these values for $10 \log m$ must be increased somewhat to take account of the directivity pattern of the receiver. An examination of the receiver patterns in reference 34 indicates that this correction should be about 6 db. Thus, we finally have for $10 \log m$ a value between -48 and -68 db.

This result is in general agreement with the value of -60 ± 10 db for the scattering coefficient of volume reverberation given in Chapter 4 of Part II. A value greater than -40 db seems definitely ruled out by the observations. Thus one may conclude that the scattering coefficient for sound at angles between roughly 10 and 120 degrees is not more than about 10 db greater than for the backward scattering which gives rise to reverberation. It is possible that the scattering of sound by the volume of the sea is the same in all directions. More exact conclusions would require simultaneous determinations of reverberation and sound scattered into the shadow zone. In addition, the change of scattering coefficient with depth, frequently observed in the deep scattering layers discussed in Chapter 14, would demand consideration. The present very rough analysis is adequate, however, to indicate that the attenuation observed in deep isothermal water is not the result of scattering, unless one makes the improbable hypothesis that scattering in the isothermal layer is very much greater than the scattering in the thermocline. If an attenuation coefficient of 4 db per kyd or 4×10^{-3}db per yd is attributed entirely to scattering, the scattering coefficient m would be $10 \log_{10} e$ times a or 1.7×10^{-2}, giving more than -20 db for $10 \log m$. This is 20 db greater than the maximum possible value of m consistent with the low intensity of sound observed in the shadow zone. If not all the sound in the shadow zone is due to scattering, the disparity becomes even greater.

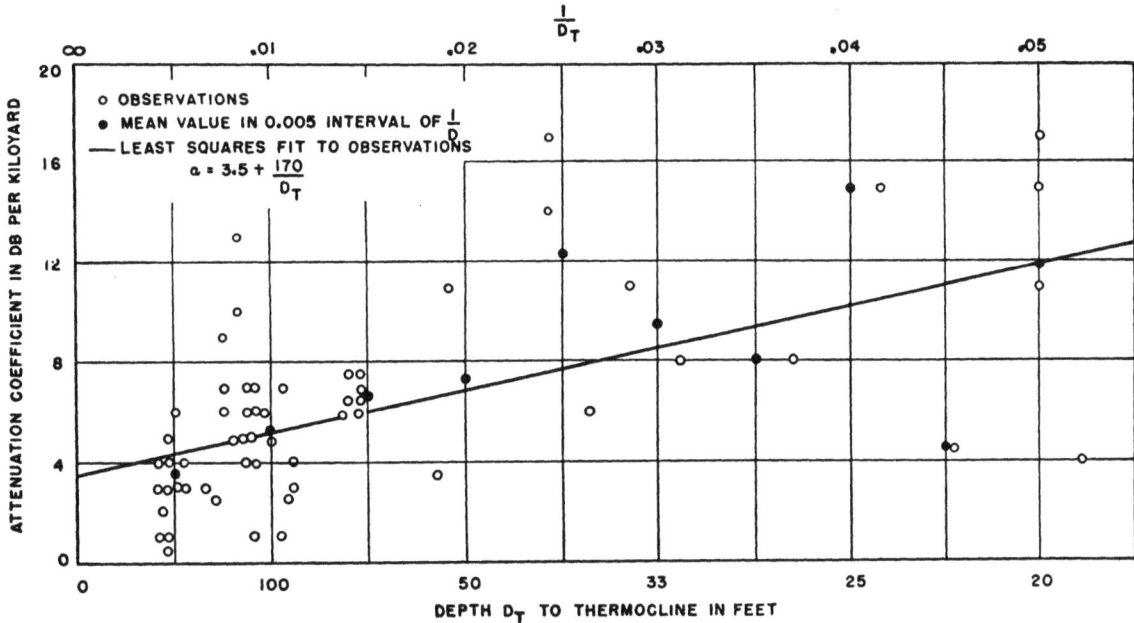

FIGURE 47. Attenuation coefficient above the thermocline.

Although the scattered sound observed at ranges between 1,000 and 2,500 yd is readily explained, the coherent signals received in the shadow zone at 4,000 yd are less simply explained. These coherent signals could be produced by a suitable variation of the scattering coefficient m with depth similar to those found in deep scattering layers. A preliminary UCDWR analysis of reverberation measurements made with a vertical projector at the same time as transmission measurements indicates that this hypothesis is correct; the scattering coefficients found by these two methods agree to within a few decibels.

5.4.2 Weak Temperature Gradients in Top 30 ft

When the temperature gradients in the top 30 ft are intermediate—CHARLIE patterns—it has been clearly demonstrated in Table 1, that at least half of the transmission anomaly curves are approximately straight lines while the others have more complicated shapes. Thus the type of transmission likely to be encountered is highly unpredictable. This observational result may be in part caused by the rapid variability of temperature conditions for this type of pattern; small changes of temperature, of the sort very common near the surface, can change the theoretical ray diagram completely in a matter of minutes. Various methods have been developed for analyzing the trans-

mission conditions to be expected with these patterns. Since the average transmission loss observed with shallow MIKE patterns is very similar to that observed for CHARLIE patterns, these two temperature types are combined in the present discussion. Most of this section refers to average results obtained at 24 kc. Some special temperature distributions are discussed at the end of this section.

ATTENUATION COEFFICIENTS

In reference 13, an attenuation coefficient was found from the slope of each straight-line transmission anomaly graph. Attempts to correlate these coefficients with various temperature differences either in the surface layers or in the thermocline were not very successful. However, a significant correlation was found with the depth D_T to the thermocline. The plots showing this correlation are reproduced in Figures 47 and 48. A least-squares solution gave the following equations of best fit:

$$\text{Above the thermocline } a = 3.5 + \frac{170}{D_T} \cdot \quad (25)$$

$$\text{Below the thermocline } a = 4.5 + \frac{260}{D_T} \cdot \quad (26)$$

Although these mean curves are unquestionably significant, only half of the individual points lie within 2 db of the values predicted from equations (25) and (26).

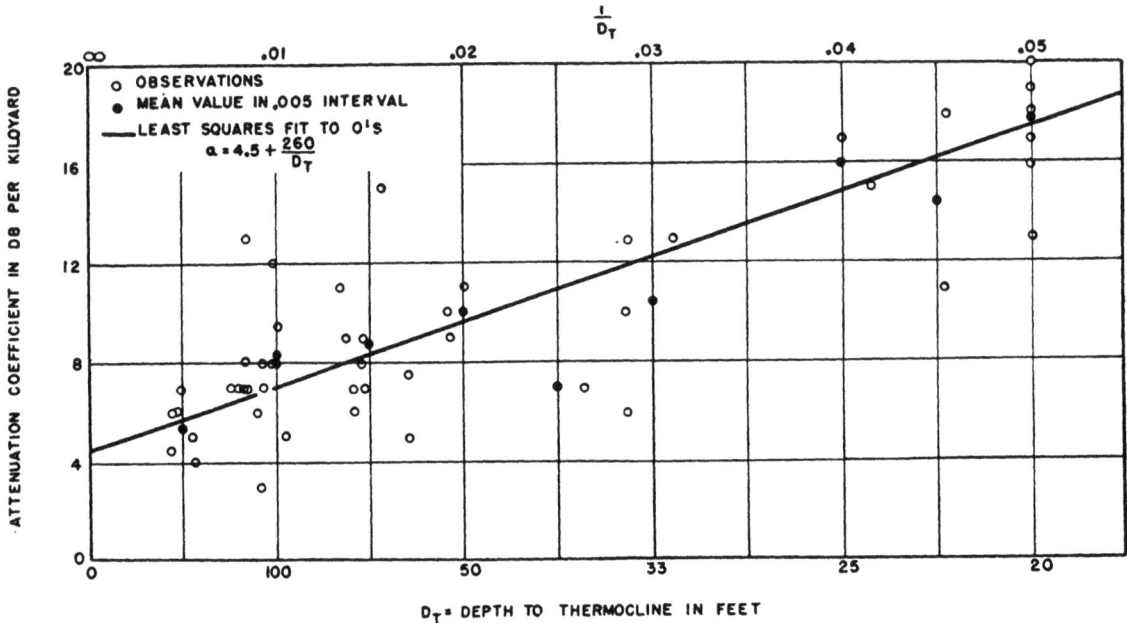

FIGURE 48.   Attenuation coefficient below the thermocline.

Figure 48 shows an increase of attenuation below the thermocline with decreasing thermocline depth. This effect has already been noted in Section 5.3.4 for thermocline depths greater than 40 ft; the change of R_{40} with layer depth (Figure 29) was shown to be consistent with the theoretical curve based on equation (13). It is apparent from Figure 47 that this same effect persists to much shallower layers. A comparison of Figures 47 and 48 indicates that layer effect increases steadily with decreasing depth of the thermocline. This result is also consistent with expectations based on equation (13).

The increase of the attenuation coefficient in the isothermal layer as the depth of the layer decreases is quite marked in Figure 47. It may be noted that the values shown for D_T between 20 and 30 ft are not inconsistent with the attenuation coefficient of about 13 db per kyd found from the upper curve in Figure 40, drawn for D_2 between 20 and 30 ft. The origin of this high attenuation is hard to explain. As sound travels along through the layer of nearly isothermal water, with the sound rays continually reflected from the surface and distorted by temperature microstructure, it may be expected that a certain fraction of the sound would be bent out of the isothermal layer in each yard of sound travel. Any such sound reaching the thermocline will be bent down so sharply that it is unlikely to return to the isothermal layer. It is possible that a quantitative theory along these lines, based

on more accurate information on the properties of the isothermal layer, may explain the observed decrease of attenuation with increasing thickness of the layer. In any case there is little question as to the reality of the effect noted in Figures 47 and 48.

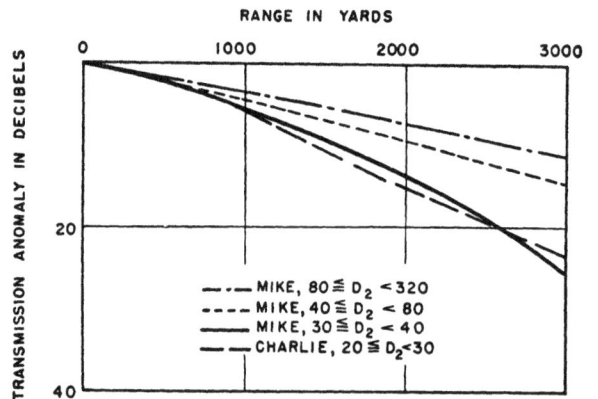

FIGURE 49.   Average transmission anomalies for MIKE and CHARLIE patterns (hydrophone shallow).

Whether the scatter evident in these figures is greater than can be explained by the observational scatter of all observed transmission anomalies is not evident from an examination of reference 13. It has already been noted, in Chapter 4, that for most of the UCDWR data transmission anomalies determined by averaging 5 successive received pings have a probable error of about 2 db as a result of hydrophone directivity, training errors, and sampling errors; the errors

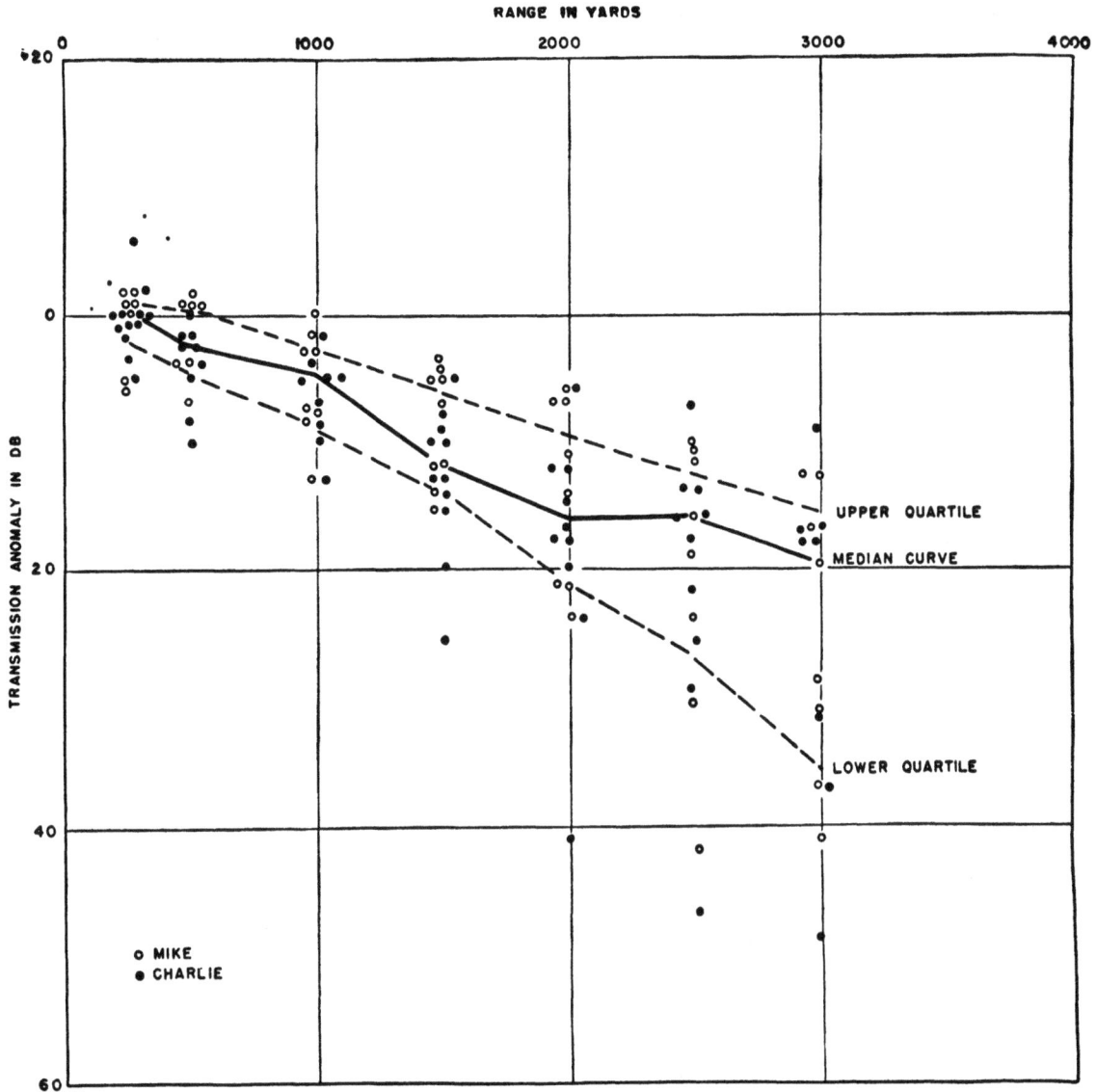

FIGURE 50. Individual anomalies for D_2 between 20 and 40 feet.

due to calibration do not affect the accuracy of the attenuation coefficient determined by fitting a straight line to the observed anomalies. With such a scatter, the uncertainties in the attenuation coefficient a can be very substantial for those runs in which the length of run was short. However, it does not seem likely that observational error can account for all of the scatter shown in Figures 47 and 48.

The attenuation coefficients shown in Figures 47 and 48 should probably not be used for estimating transmitted sound intensities when temperature gradients are present in the top 30 ft. The results are valid on the average when the transmission anomaly graph is a straight line, but unfortunately there is no way of predicting whether or not the measured sound intensities will yield a straight line, except when the top 30 ft are isothermal. Exclusion of those situations where the transmission anomaly curve is not a straight line may be expected to give results systematically different from those obtained when runs are classified only by the temperature distribution. Therefore the average anomaly curves given below are preferable as a tool for estimating the sound intensities to be expected in any situation.

AVERAGE TRANSMISSION ANOMALIES

The average transmission anomalies obtained with MIKE and CHARLIE patterns have been combined in reference 14 to give average curves. The curves for a shallow hydrophone (16 to 30 ft) are reproduced in Figure 49. The curves are again plotted for different values of D_2.

To illustrate the scatter of the individual observations, all individual anomalies averaged to give the curve for D_2 between 20 and 40 ft are shown in Figure 50. The open circles represent the anomalies for MIKE patterns, the solid circles those for CHARLIE patterns; no systematic difference is apparent between these two sets of points. The upper and lower quartiles of the distribution are shown by dashed lines. The increase of spread with increasing range is very marked and is an evidence of the unpredictability of transmission conditions for such shallow isothermal layers. The quartile spread at short range is much smaller and represents the more normal scatter, apparent also in Figures 15 and 41.

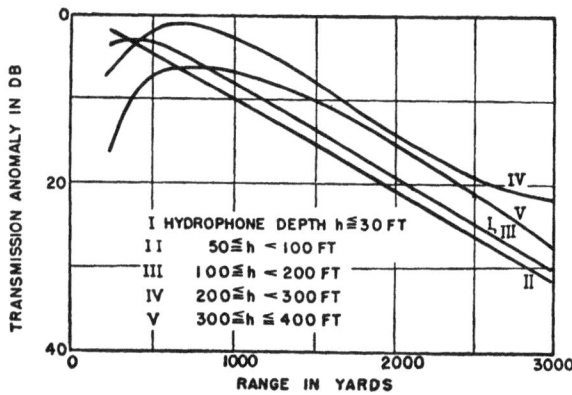

FIGURE 51. Average transmission anomalies for D_2 between 20 and 40 feet.

The change of transmission anomaly with changing hydrophone depth has not been analyzed separately for MIKE, NAN, and CHARLIE patterns. The average results for D_2 between 20 and 40 ft, combining results for all three patterns, is shown in Figure 51. The change with depth down to 200 ft is negligible, but at greater depths an appreciable increase in sound intensity is noted. This change is greater than the probable error of each curve resulting from the internal scatter of the points and is therefore probably significant, even though different temperature patterns were present when different hydrophone depths were used.

The corresponding plot for D_2 between 40 and 80 ft has already been given in Figure 30. For such temperature structure, the intensity first decreases with increasing depth as the hydrophone goes below the

thermocline, and then increases. As pointed out in Section 5.3.4, the transmission anomaly below a sharp thermocline is likely to show better correlation with the depth and sharpness of the thermocline than with the temperature code used in Figures 49 and 51. In fact the limited results available are consistent with the belief that for MIKE and CHARLIE patterns in general the average transmission anomaly below a thermocline is approximately given by equation (13) in Section 5.3.4.

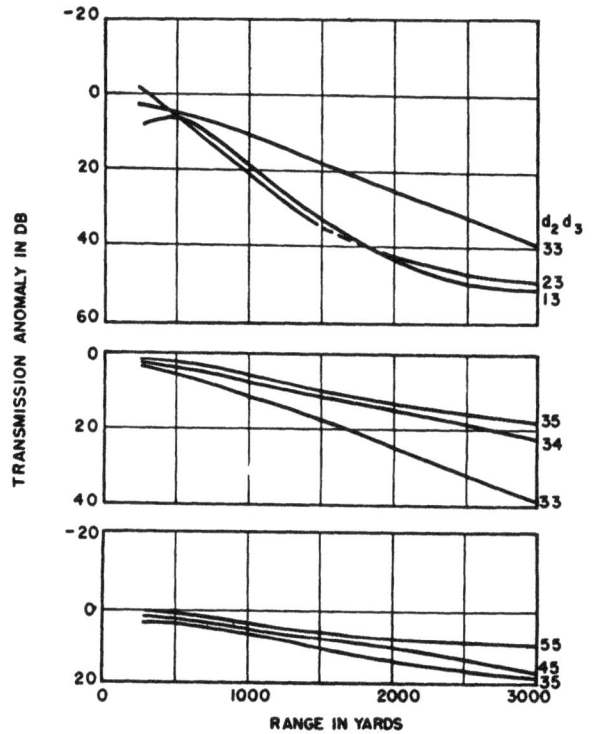

FIGURE 52. Average transmission anomalies for hydrophone 50 to 100 feet deep.

An example of the way in which transmission anomalies change with changing temperature structure is shown by the set of average curves for hydrophones between 50 and 100 ft deep shown in Figure 52, again taken from reference 14. Most of the data used in these curves were actually obtained with hydrophones at 50 ft. The curves are in terms of the digits in the temperature-depth code explained in Section 5.1.4. The successive curves may be interpreted as follows:

d_2d_3

13   Temperature decreases to 0.3 F below surface temperature between 5 and 10 ft; between 20 and 40 ft it has decreased to 1 F below surface temperature. This is a moderately sharp NAN pattern with the sharp gradient extending practically up to the surface, and

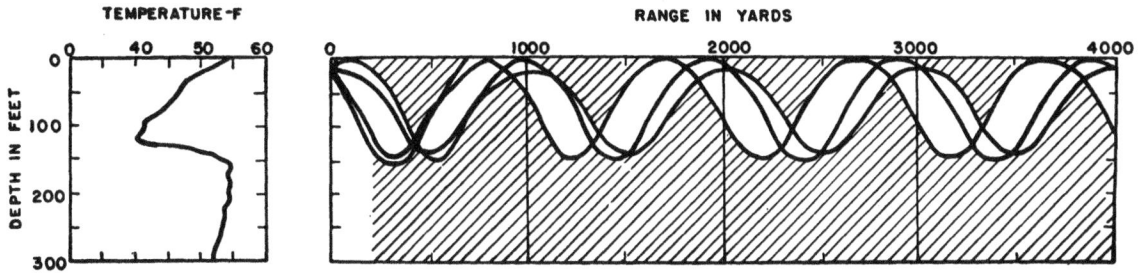

FIGURE 53. Sound channel ray diagram, extreme case.

shows the corresponding rapid rise in the transmission anomaly.

23 This is a much the same as above except that the gradient in the upper 10 ft is somewhat weaker. This curve and the preceding one are closely similar.

33 This is a NAN, MIKE, or CHARLIE pattern, but in any case the thermocline is shallow and the attenuation high.

34 The main thermocline is deeper here since the temperature is within 1 F of the surface temperature down to at least 40 ft. The hydrophone may be below or above the thermocline.

35 The top of the main thermocline is not much above 80 ft, and the hydrophone is either close to the top or above it. However, there are gentle gradients above the hydrophone, and these act to reduce the sound intensity. The reduction in sound intensity produced by the weak gradient between the projector and the hydrophone may be regarded as an example of layer effect.

45 The gradients above the hydrophone are weaker and transmission is improved.

55 The water is virtually isothermal down to 80 ft, and the results discussed in Section 5.2 are applicable. The deviation of this curve from a straight line is probably not significant.

SOUND CHANNELS

When the sound velocity at the projector is less than the velocities above and below, rays leaving the projector at sufficiently small angles will, in theory, curve back and forth within two fixed depths of equal sound velocity, giving rise to the curious ray diagram shown for an extreme case in Figure 53. This situation is called a *sound channel*, and should in theory give rise to high sound intensity at long ranges. When sharp negative temperature gradients are present over sharp positive gradients, such sound channels should be persistent and very marked. However, very few measurements have been made with positive temperature gradients present in the water.

In the absence of positive temperature gradients, the effect of pressure on sound velocity can produce a positive velocity gradient below the projector. However, this gradient is very small, and a temperature

decrease of only 0.3 F (at 60 F) between the projector and the isothermal layer will bend the sound rays down so sharply that an isothermal layer 100 ft thick is required to bend the rays back up again. Moreover, as a result of this small gradient, rays bent down into the isothermal layer would return to the surface only at ranges of many thousands of yards. This bending is so gradual that the presence of thermal microstructure might be expected to mask completely any sound channel effects resulting from upward bending in nearly isothermal water. However, since some striking acoustic effects are observed with shallow gradients overlying isothermal layers, and since thermal microstructure has never been measured under such conditions, it is instructive to examine what the sound field would be like in truly isothermal water underlying slight gradients at the surface.

If the projector were in such a hypothetical layer of completely isothermal water, the effects of the sound channel would not be particularly noticeable since the sound that has curved first up into the negative gradient and then down into the isothermal layer would be indistinguishable from the rays that have traveled through the isothermal layer for their entire path. In fact downward bending by a very shallow surface gradient above the projector is probably very similar to reflection by the surface.

To produce marked effects the negative temperature gradient at the surface must extend below the projector depth, so that the entire sound beam is bent downward, resulting in low sound intensities measured by a shallow hydrophone at short range. Then when the rays are curved back to the surface thousands of yards out, the sound intensity should show a marked increase. On the basis of the simple ray theory, which neglects thermal microstructure and diffraction, the theoretical intensity at the projector depth is infinite at the range where the axial ray from the projector becomes horizontal again; this singularity results from the crossing of many adjacent rays at this point. Although of course diffraction and ther-

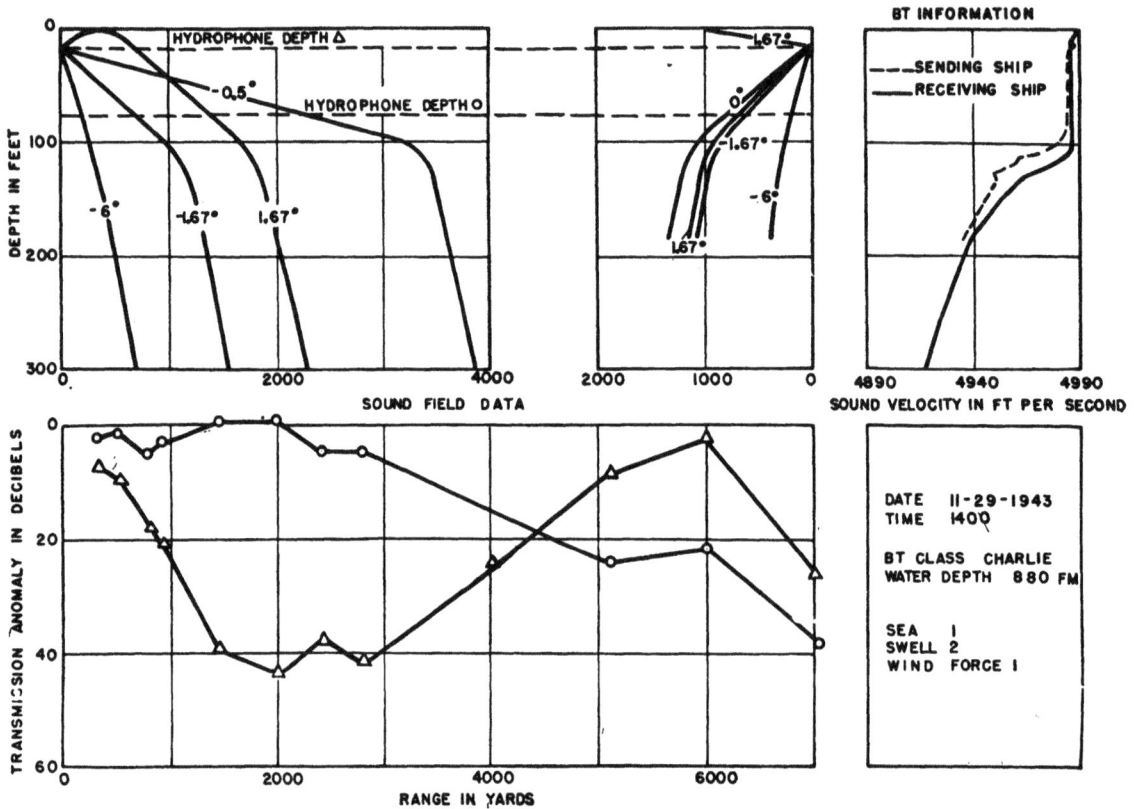

FIGURE 54. Peaked transmission anomaly possibly resulting from sound channel ray diagram.

mal irregularities will reduce this theoretical intensity, sound intensities considerably above normal would be expected at certain ranges if a sound channel were produced by a slight temperature gradient lying above rigorously isothermal water.

The conditions for a sound channel, when no positive temperature or salinity gradients are present, are thus rather critical. There must be a temperature gradient at the surface which extends somewhat below the projector depth. Below this must be a layer of completely isothermal water a hundred feet or more in depth. The temperature difference between the isothermal layer and projector depth must be not less than about 0.1 F but not greater than about 0.3 F, the exact limits depending on the surface temperature as well as on the depth of the layer. Thus, even if no thermal microstructure were ever present, sound channels of this type would normally be quite transitory, appearing during the development of surface heating in deep isothermal water; they would be expected to become prominent when the temperature difference between the projector and the isothermal layer increased to 0.1 F, and to disappear as the gradient extended downward and the temperature

at projector depth gradually increased by another one or two tenths of a degree.

Sound transmission measurements at the UCDWR laboratory in San Diego show transitory effects similar to those which may be expected to result from sound channels. Because of the difficulty in reading the bathythermograph slide accurately to 0.1 F, and because of the high variability of thermal conditions in space as well as in time, it is not possible to predict from the bathythermograms exactly when a sound channel may be present. However, marked peaks in the measured transmission anomalies are occasionally found when thermal conditions are appropriate.

A good example of the type of effect that can be observed is shown in Figure 54. As shown in the accompanying temperature-depth record, a sharp negative gradient extends from about 15 ft to the surface while below this a nearly isothermal layer extends down to 100 ft. A careful reading of the trace indicated a slight negative gradient in this layer (about 0.3 F in 100 ft) giving a constant sound velocity below the projector. Moreover, the sharp surface gradient did not extend below the projector depth. Thus, the ray diagram in Figure 54 does not show a sound

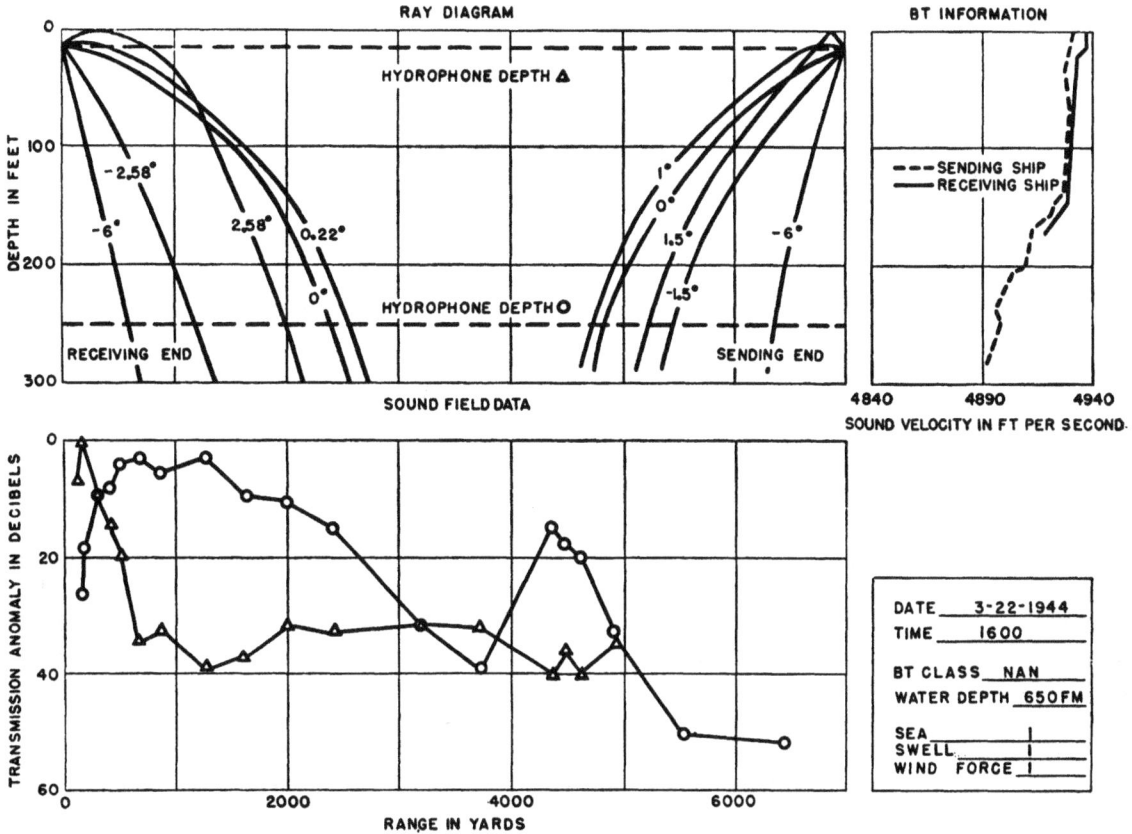

FIGURE 55. Peaked transmission anomaly with sound channel unlikely.

channel although barely perceptible changes in the temperature-depth record would make it so. Nevertheless, the anomalously high intensity at long range in the shallow hydrophone is very striking. If an absorption coefficient of 4 db per kyd is assumed, the increase of intensity at 6,000 yd shown in Figure 54 is about 25 db. Until a more accurate means for determining ocean temperatures is available, it cannot be decided whether peaks of this type are actually the result of the focusing action predicted in sound channel theory.

However, it is suggestive that an examination of all cases in which the anomaly curves show peaks of 10 db or more, shows that the temperature-depth curves for most of these are very similar to the curves which would, on the simple theory, give rise to sound channels. Out of the many hundreds of runs made off San Diego, only about 25 show these peaks. In almost all of these the thermocline is below 100 ft with nearly isothermal water above, and the receiving hydrophone is at shallow depth. Moreover, in most cases slight negative gradients are present close to the sur-

face. A few significant exceptions are present, as for example the run shown in Figure 55 where the deep hydrophone shows a peak of 20 db at 4,500 yd. As an example of the transitory nature of most such peaks, Figure 55 may be compared with Figure 42, which plots the run immediately preceding and shows no trace of any peaks. Also on one day (March 15, 1944) the shallow hydrophone showed a marked peak throughout the day while the temperature records only occasionally showed the deep layer of constant temperature required for a sound channel. Thus, while the evidence suggests that sound channels may in fact occur, there may quite possibly be other factors, still unexplored, that play a part in producing anomalously high intensities at certain ranges.

5.4.3 Transmission at 60 kc

The effects produced by temperature gradients on the transmission of underwater sound have been thoroughly explored only at 24 kc. A few measurements are available, however, at 60 kc.

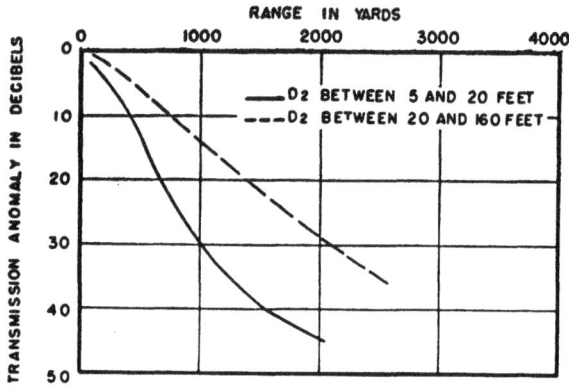

FIGURE 56.   Average transmission anomalies at 60 kc.

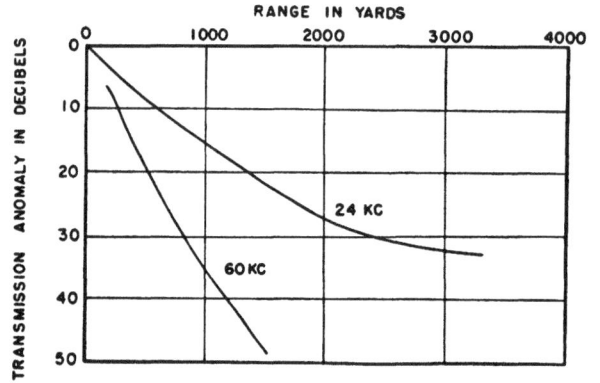

FIGURE 57.   Simultaneous transmission at 24 and 60 kc.

Average transmission anomalies for a shallow hydrophone at 60 kc are shown in Figure 56 for different values of D_2. All runs for D_2 between 20 and 160 ft are combined in the upper curve, since no systematic variation of transmission anomaly with changing D_2 was noted for these data. The agreement between the curves for D_2 between 20 and 40 ft and for greater D_2 is in marked contrast to the differences shown at 24 kc (see Figure 49).

Another, more conclusive indication of the complicated differences between the two frequencies is shown by simultaneous measurements at both frequencies on a single shallow hydrophone. Measurements were made with a CHARLIE pattern (temperature difference about 0.5 F in the top 30 ft) and a thermocline

at 50 ft. The resulting curves are shown in Figure 57. The difference of anomalies between the two frequencies increases by 15 to 20 db per kyd, as compared to a corresponding difference of not more than about 10 db per kyd in isothermal water; see Figure 17. If data from many more runs under somewhat similar conditions are used, however, the average difference of transmission anomaly between 60 and 24 kc has a slope of about 9 db per kyd. This is in close agreement with the difference of 8.5 db per kyd found in Section 5.2.2 for isothermal water. Further data on the difference of transmission loss between different frequencies are required to show how much and in what way this difference varies.

Chapter 6

SHALLOW-WATER TRANSMISSION

IN CHAPTER 5, it was shown how the propagation of sound in deep water is affected by temperature gradients in the sea and by the sound frequency. In shallow water, these factors continue to operate; added to them is the effect of the bottom. The bottom affects the sound field in two different ways. Some of the sound incident on the bottom will be reflected and may penetrate into shadow zones. Also, some of the sound incident on the bottom will be scattered backward and will form part of the reverberation background against which an echo must be recognized in echo ranging. This latter effect of the bottom will be considered in Chapters 11 to 17 of this volume. In this chapter, only the transmitted sound reaching a receiving hydrophone will be considered.

6.1 PRELIMINARY CONSIDERATIONS

In deep water, it was found that the most important single factor determining the transmission of sound of a given frequency is the vertical temperature structure of the ocean. The roughness of the surface of the sea plays a poor second, and nothing is known concerning the effects of other oceanographic variables on sound transmission. In shallow water, the number of factors which may conceivably affect sound transmission is greater; it would be impractical to make a large number of sound transmission runs and then obtain rules of sound propagation empirically merely by subjecting the data amassed to an unprejudiced statistical analysis. Rather, it was found necessary to assess beforehand the possible effects of bottom character, roughness of the sea surface, and refraction conditions, and then to analyze the transmission run data purposefully. This procedure proved successful in bringing order into a mass of data, and it will also be followed in this discussion.

6.1.1 Effects of Sea Bottom

If bottom-reflected sound is added to the sound field which reaches the receiving hydrophone (or the target in echo ranging), interference between the signals transmitted via the different possible paths

may be either constructive or destructive, depending on the geometry of the paths. However, if the sound field intensity is averaged over a volume of the ocean sufficiently large to include several maxima and minima of the interference pattern, the averaged sound field intensity will be the algebraic sum of the intensities of sound resulting from each path by itself. In this sense, averaged sound field intensities in shallow water are always higher than sound field intensities in deep water under otherwise identical conditions. The extent to which bottom-reflected sound will increase the "deep water" sound field intensity and to which it will eliminate shadow zones depends on a number of factors, which will be treated in this chapter. One of these factors is the reflectivity of the sea bottom.

A theoretical treatment of bounding surfaces indicates that the reflectivity of a surface is determined by two factors: the degree of roughness of the surface itself, and the density and elastic moduli of the two adjoining media, such as sea water and granite. For the special case of two fluid media, it was shown in Section 2.6.2 that the percentage γ_e of reflected energy depends on the ratio of the densities as well as the angles of incidence and refraction, according to the formula

$$\gamma_e = \left[\frac{\rho_1 c_1 - \rho c \sqrt{1 + \tan^2\theta \left(1 - c_1^2/c^2\right)}}{\rho_1 c_1 + \rho c \sqrt{1 + \tan^2\theta \left(1 - c_1^2/c^2\right)}} \right]^2 \quad (1)$$

in which ρ and ρ_1 are the densities of the two adjoining media, c and c_1 are the sound velocities, and θ is the angle of incidence. This quantity γ_e is called the coefficient of reflection of the separating surface. The coefficient of reflection equals unity when the angle of incidence exceeds the critical angle for total reflection.

Equation (1) is based on the assumption that a smooth plane interface separates two perfect fluids. This assumption is not entirely correct for either the surface or the bottom of the ocean. The surface of the ocean is not smooth. With high winds, it may contain a large number of air bubbles (whitecaps), which absorb and scatter sound. The bottom often consists of material capable of shear stress, like rock, and is frequently rough. It is, therefore, simpler to determine the coefficient of reflection experimentally than to attempt to obtain it from the mechanical parameters of the two substances separated by the interface.

A rough bottom will not only give rise to a transmitted sound wave, which disappears from the ocean, and a specularly[a] reflected wave, but will also scatter sound in a random fashion. The sound beam resulting from the reflection of a plane wave incident on a bottom of moderate roughness has a certain directivity pattern. If the roughness is not excessive, this directivity pattern will show an intensity maximum in the direction which corresponds to specular reflection from the bottom. The smoother the bottom, the more highly directive or collimated is the reflected sound beam.

Any nonspecularity of the reflection at the bottom has essentially the same effect as would a broadening of the directivity pattern of the projected beam. This increased divergence will not, however, always cause a decrease in the peak level of the received signal. If the bottom is smooth or only moderately rough, and if projector and receiver are not too highly directional, there should be little or no decrease in the peak signal level. This is because, on the average, as much energy will be deflected toward the hydrophone by nonspecular reflection as will be lost out of the main beam through the same mechanism. Excessive roughness of the bottom, however, should cause a decrease in the peak signal level unless the projector is nondirectional and a long pulse is used. The reason for specifying a long pulse is that some of the sound scattered toward the hydrophone by a very rough bottom will travel paths much longer than the path corresponding to specular reflection. Thus, when short pulses of sound are projected over a rough bottom, the received signal will last longer than the transmitted signal. Although the total energy received at the hydrophone may be the same as if the bottom were smooth, the peak signal level will be lower.

6.1.2   Velocity Gradients and Wind Force

The magnitude of the contribution of bottom-reflected sound to the total sound field will depend not only on the acoustic properties of the bottom, but

---

[a] Specular reflection is reflection for which angles of incidence and reflection are equal.

also on refraction conditions in the body of the sea and on the reflectivity of the sea surface.

The bottom should probably be of little importance if upward refraction in the sea volume prevented most of the sound energy from ever reaching the bottom. We may therefore tentatively predict that in the presence of positive gradients there will be no difference between deep-water transmission and shallow-water transmission. On the other hand, in the presence of downward refraction the bottom should usually play a role of some importance. If the bottom is a very poor reflector of sound, then the sound field should not differ significantly from the sound field in deep water under similar refraction conditions, since bottom-reflected sound will make only a slight contribution to the total sound field. But if the ocean bottom is a good reflector, then the contribution of bottom-reflected sound will be significant. This contribution will increase in importance as the downward refraction becomes sharper and removes more and more energy from the direct sound field at long range.

In addition, if the ocean bottom reflects sound fairly well, the sound field intensity at long range will probably be increased appreciably by sound which has been reflected several times between the ocean surface and the ocean bottom. We should, therefore, look for a dependence of the shallow-water sound field on the roughness of the sea surface.

The quantitative prediction of sound field levels in shallow water, by combining the information on bottom and surface reflectivity with that on refraction conditions, would be very difficult. The bulk of the reliable information on shallow-water transmission has been obtained directly by means of transmission runs. The qualitative considerations of this section, however, have been valuable in planning these transmission runs and in interpreting the resulting sound field data.

6.1.3   Effects of Frequency on Spreading Factor

It was shown in Section 5.2.2 that the attenuation of sound in deep water depends strongly on the frequency. It has been tentatively suggested that the observed dependence of attenuation on frequency might be fitted by a 1.4th power law.[1] It might be expected that a formula of the form

$$H = aR + 20 \log R \qquad (2)$$

would not be applicable for transmission in shallow

water since this formula was derived in Section 5.2.2 without taking into account the contribution of sound reflected from bottom and surface. For the higher supersonic frequencies, this fear is frequently unjustified. As a matter of fact, in the presence of a well-reflecting bottom, equation (2) provides a better fit to the observations, considering all types of refraction conditions, than in deep water. This apparent paradox can be explained easily. For 24-kc sound, for instance, it is known that the attenuation in the direct beam amounts to 4 or 5 db per kyd. Beyond a range of 2,000 yd, the second term in the expression (2) increases less rapidly than the first term. As a result, modifications in the second term of equation (2) due to changes in the geometry of spreading are insignificant compared with the first, or absorption, term — at least at ranges where the effect of the bottom might be expected to become noticeable. On the other hand, deviations from equation (2) in deep water are common in the presence of downward refraction in the shadow zone. These deviations are mitigated by the appearance of bottom-reflected sound in the shallow-water sound field at long range. It is, therefore, convenient to plot and to analyze transmission anomaly in supersonic shallow-water transmission, since at short range the sharp inverse square drop is taken out of the transmission loss, while at long range the variations of transmission loss and transmission anomaly are not very different (see Figure 1 of Chapter 4).

At sonic frequencies the situation is different, since the attenuation as determined in deep-water transmission experiments is very small, certainly less than 1 db per kyd. As a result, the first term of expression (2) does not overpower the second term even at ranges of the order of 10,000 yd. Moreover, sonic sources and receivers tend to be nondirectional, and bottom-reflected sonic sound tends to become important at shorter ranges than does bottom-reflected supersonic sound. It may, therefore, be expected that the contribution of bottom-reflected sound will significantly affect sonic transmission at all ranges of operational importance for all refraction conditions except sharp upward refraction. At sonic frequencies, a modification of the inverse square law to take bottom-reflected sound into account thus is more necessary than at frequencies above 10 kc. If both the surface and the bottom were perfectly reflecting, sound energy would spread only in two dimensions, and as a result, the sound field decay at long range should be approximated by an inverse first power law. Actual interfaces permit sound to "leak" across, and the power law of sound field decay must be obtained by fitting a curve to the observations. Even under these circumstances, however, the consideration of transmission anomalies based on the inverse square law should reveal the essential features of sound transmission and may be preferred on the grounds of uniformity of approach. In addition, a plot of transmission anomaly has the practical advantage that a more open decibel scale is possible than for transmission loss.

6.2 SUPERSONIC TRANSMISSION

To study the acoustic properties of various sea bottoms, both UCDWR and WHOI have carried out transmission and reverberation runs in shallow water. The purpose of these experiments has been both to measure specific parameters characterizing the sea bottom and to obtain information on the overall properties of the sound field encountered in shallow water. Reverberation experiments are discussed in detail in Chapters 11 to 17 of this volume; but it is necessary to refer to them in this chapter, because they have incidentally furnished tentative values for the reflection coefficients of sea bottoms for slant incidence.[2]

6.2.1 Acoustic Properties of Sea Bottoms

TYPES OF SEA BOTTOMS

Analysis of observed echo and listening ranges, which began in 1941, indicated that ocean bottoms could be roughly subdivided into a few geological types with fairly consistent reflection characteristics for each type. The classification of bottoms for sound ranging purposes has been standardized and includes the following: SAND, SAND–AND–MUD, MUD, ROCK, STONY, and CORAL. These bottom types are described as follows.[3]

SAND	Firm, relatively smooth bottom.
SAND–AND–MUD	Relatively firm, smooth bottom.
MUD	Soft, smooth bottom.
ROCK	Rough, broken bottom. Includes bedrock, outcrops, and areas covered by boulders.
STONY	Hard bottom, commonly rough. Predominantly cobbles, gravel, and shells. Varying amounts of sand and mud commonly present.
CORAL	Hard bottom, with sandy patches, irregular to smooth. Includes various marine forms which secrete masses of lime covering the bottom.

In the actual classification of sea bottoms, the criterion established for estimating the relative firmness or softness of the bottom was grain size, as determined by mechanical analysis. The size limits were set as follows (Division 6, Volume 6).

MUD	90% by weight smaller than 0.062 mm.
SAND–AND–MUD	Between 10% and 90% smaller than 0.062 mm.
SAND	Less than 10% smaller than 0.062 mm and 90% smaller than 2.0 mm.
STONY	Rounded or angular pieces of rock more than 2.0 mm and less than 10 cm, which appear to represent glacial drift or other transported material.
ROCK	Rocks of a size greater than 10 cm or pieces broken from rock ledges or where bottom photographs show projecting rocks or rock ledges.
CORAL	Samples containing calcareous masses of coral, algae, or other lime secreting organisms, or bottom photographs showing their existence.

This classification has been reasonably satisfactory from the acoustical standpoint except in the case of mud, for which it is now likely that texture alone is not an adequate criterion.

Although the bulk of experimental work on sound transmission in shallow water has consisted of transmission runs, some special experiments have been made to determine numerical reflection coefficients of sea bottoms.

REFLECTION COEFFICIENTS

It may be gathered from the discussion in Section 6.1.1 that different values of the reflection coefficient are to be expected for sound incident vertically on the bottom and for slant rays. Although the determination of reflection coefficients for vertical incidence has some interest at sonic frequencies, the most important situations at all frequencies, from an operational point of view, involve slant rays. To obtain the reflectivity of the sea bottom for slant incidence, three different experimental methods have been considered.

The most direct method uses transmission runs with very short signals (10 msec), which often permit the separate reception of the direct signal and the bottom-reflected signal. (The surface-reflected signal cannot be resolved for the usual projector depth of 16 ft, but would be resolvable if the projector could be lowered to several hundred feet.) No coefficients of reflection resulting from these experiments have been reported. It is possible to make a very crude estimate of the reflection coefficient from published standard transmission runs in those cases where the curve showing the transmission anomaly plotted against range indicates at least two well-marked peaks corresponding to single and double bottom reflections. Reading the level difference between consecutive peaks and correcting for transmission loss due to absorption between reflection (say 4 or 5 db per kyd) leads to the following estimates: (1) for SAND, the loss through reflection amounts to between 0 and 6 db per reflection, corresponding to intensity reflection coefficients between 1 and 0.25; (2) for MUD, the loss is between 10 and 30 db per reflection, corresponding to coefficients of intensity reflection between 10^{-3} and 10^{-1}. The wide spread in each of these estimates indicates both the uncertainty of the estimate in a given case and the wide variability among ocean bottoms falling within one classification.

The second method involves the measurement of bottom reverberation. If an echo-ranging transducer is tilted downward about 30 degrees, a peak of the reverberation is received at the range at which the sound beam strikes the bottom. Sometimes, over well-reflecting bottoms, a secondary peak is observed, at a range at which the sound beam, specularly reflected first by the bottom and then by the surface, strikes the bottom a second time. This secondary reverberation peak has been observed over a coarse sand bottom, and the average amplitudes of principal and secondary peaks have been determined by UCDWR.[4] If reasonable assumptions are made concerning the transmission loss between the primary and the secondary reverberation peak, and if the reflection at the sea surface is assumed to be perfect (a calm sea and a low wind), then the reflection coefficient of the sea bottom can be estimated. It was found be somewhere between 0.25 and 1.0, with the most probable value 0.5. These values were obtained for coarse sand, probably the bottom with the highest reflectivity.

Reflection coefficients have been estimated as 0.031 for foraminiferal SAND, between 0.005 and 0.025 for SAND-AND-MUD, and 0.0017 for MUD.[5] These determinations are not very reliable, because they involve unrealistic assumptions concerning the transmission loss between consecutive reflections. In these computations, it was assumed that the transmission anomaly amounted to 1.6 db per kyd of vertical path

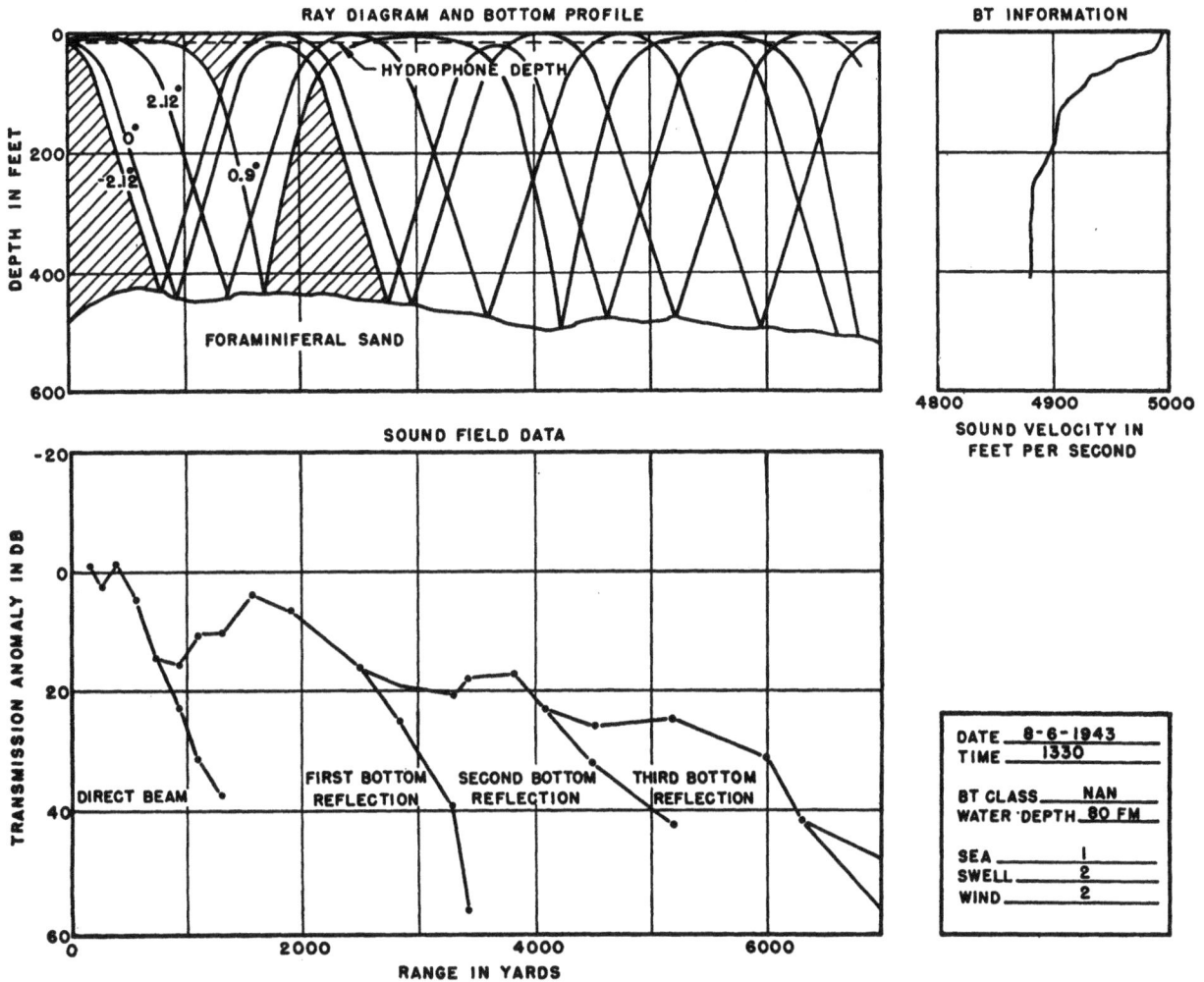

FIGURE 1. Typical transmission run over SAND bottom.

traveled. This value is probably too low, and, therefore, the reflection coefficients are too low, if the attenuation of vertical 24-kc pulses is anything like the attenuation of horizontal 24-kc pulses.

WHOI has recently developed a new method for measuring bottom-reflection coefficients by means of a determination of the interference pattern found at short ranges when both transmitter and receiver are very close to the bottom.[6] This experiment is carried out with a cable-suspended transmitter which must be nondirectional in the horizontal plane, for the same reason that cable-suspended receiving hydrophones must be nondirectional in the horizontal plane (see Chapter 4). If the hydrophone is moved up and down while a constant distance is maintained between transmitter and hydrophone, then the output of the hydrophone goes through a series of maxima and minima. If the bottom were specularly reflecting,

then the ratio between the pressure at the maxima and minima should equal $(1 + \gamma_a)/(1 - \gamma_a)$, where γ_a is the amplitude-reflection coefficient of the bottom; the amplitude-reflection coefficient is the square root of the intensity-reflection coefficient γ_e, which is usually employed. The method yields values for the reflection coefficient which may be too low if the reflection from the bottom is not completely specular. Experiments over a SAND–AND–MUD bottom have led to a value of the amplitude-reflection coefficient of 0.3 ± 0.1, corresponding to an intensity-reflection coefficient of about 0.1.

6.2.2 Analysis of 24-kc Transmission Runs

Transmission runs in shallow water and at supersonic frequencies have been made by UCDWR since 1943, and by WHOI since 1944. Runs carried out in

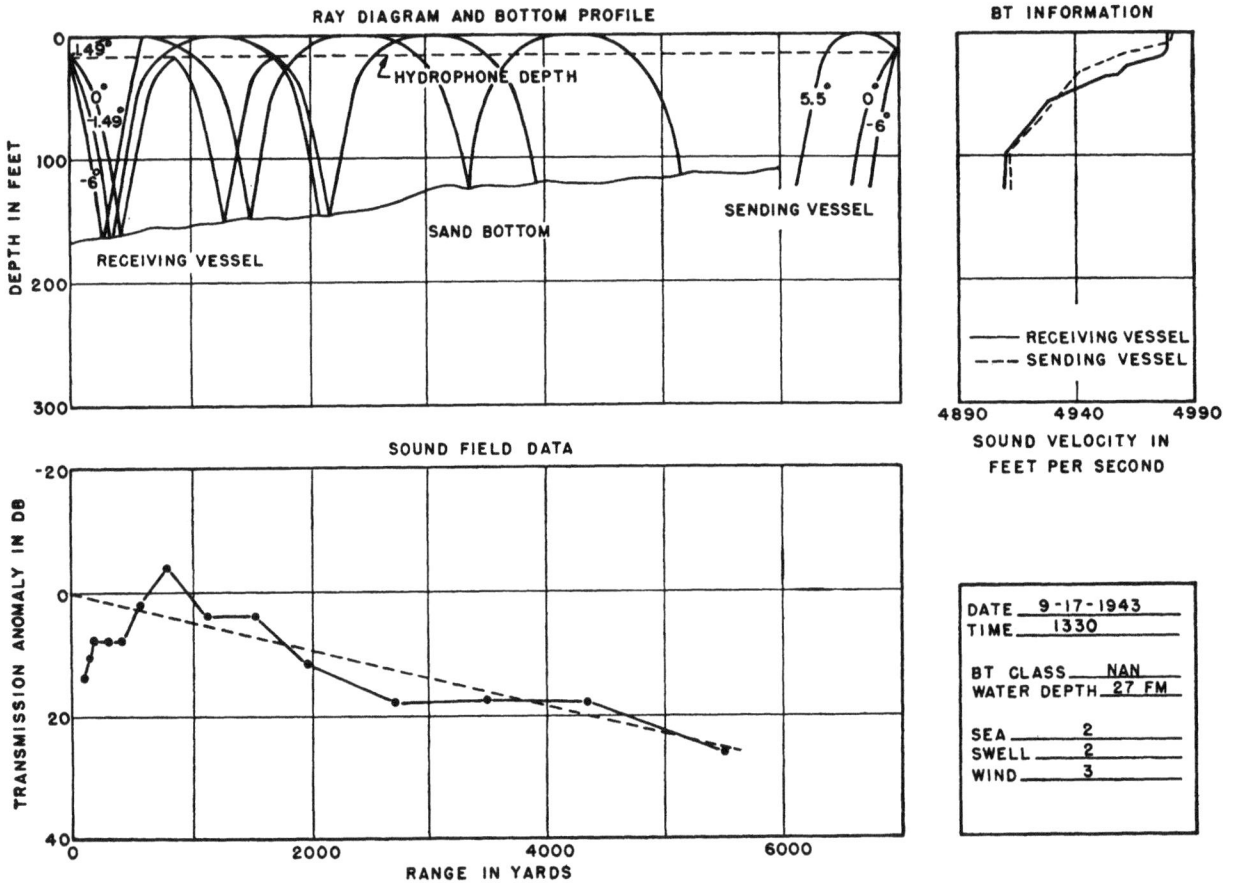

FIGURE 2.   Transmission run over sand showing linear transmission anomaly.

the presence of negative gradients have furnished the most valuable information on the reflectivity of sea bottoms, because of the large contribution of bottom-reflected sound under these conditions, even at relatively short ranges. Fortunately, the Pacific Ocean off southern California has sharp thermoclines most of the year, and the bulk of shallow-water transmission runs by UCDWR off San Diego were made in the presence of downward refraction. Figure 1 is a data sheet from a typical transmission run in shallow water over a SAND bottom. On the sheet, the sound data are plotted as transmission anomaly against range.[b] The transmission anomaly vs range diagram would be a horizontal straight line if the sound field intensity obeyed the inverse square law. If the transmission obeyed a law of the form of equation (2), the transmission anomaly would be represented by

a slanting straight line, whose slope would be a, the coefficient of attenuation. It has already been mentioned that in many cases the transmission anomaly can be approximated reasonably well by a straight line. Such a straight line, fitted by inspection, is drawn as the dashed line of Figure 2. The slope of this line is approximately 4.5 db per kyd.

Experience has shown that reasonably linear transmission anomalies are typical of well- and fairly well-reflecting bottoms. The slope of the transmission anomaly curve depends markedly on the degree of reflectivity of the sea bottom, at least for supersonic sound, but much less on the exact shape of the temperature distribution, as long as the downward refraction is strong enough to force the direct sound field out of the depth of the receiving hydrophone.

EFFECT OF VELOCITY GRADIENTS

This section deals with an analysis of several hundred shallow-water transmission runs, which were obtained by the UCDWR and by the WHOI

---

[b] It will be recalled that transmission anomaly is defined as the excess of the transmission loss in decibels over the value computed in accordance with the inverse square law of spreading.

laboratory groups. Plots of these runs were analyzed with respect to three factors: refraction pattern, depth of the receiving hydrophone and bottom character. Two other parameters which are also significant, the water depth and wind force, were not taken into account in order not to split the sample into too many small divisions.

No analysis is at present available concerning the effect of water depth on transmission. It is known that in very shallow water (5 fathoms) transmission is poorer in the presence of downward refraction than it is over the same type of bottom in deeper water.[7] For the purposes of the analysis reported below, runs in water of less than 10 fathoms and runs in water of more than 200 fathoms have been omitted. As for wind force, a separate analysis has been made at UCDWR, which will be reported at the end of this section.

The following types of bottoms have been treated separately: SAND, SAND–AND–MUD (including SAND–MUD and MUD–SAND), MUD (including only the soft muds), CLAY (the plastic muds), ROCK (including ROCK and CORAL), and STONY (including gravel, cobbles, and similar notations on the original sheets). Sand-and-shells was treated as SAND.

The depths of the receiving hydrophone were subdivided into three classes: shallow (0 to 16 ft), intermediate (17 to 100 ft), and deep (more than 100 ft). These classes were chosen for convenience and uniformity. A division of hydrophone depths into depths above and below the thermocline might have been preferable from a theoretical point of view; but on many bathythermograph traces the location of the thermocline is not uniquely determined. Therefore, a more mechanical division on the basis of hydrophone depth in feet was decided on.

The bathythermograph patterns were divided into the usual classes, described in Section 5.1.4 as NAN, CHARLIE, MIKE, and PETER. All patterns were classified as in deep water, that is, the classification BAKER (used for most conditions in shallow water) was never used. The MIKE patterns were subdivided into two classes, DEEP MIKE, consisting of all patterns in which the water was isothermal to at least 100 ft below the surface, and SHALLOW MIKE, including all other MIKE cases.[c] In the case of certain

well-reflecting bottoms, NAN and CHARLIE were combined into one class.

As a preliminary step, median and quartile R_{40} ranges[d] were determined for all combinations of the three parameters considered in this analysis; quartiles were omitted wherever the number of runs was 7 or less. Table 1 lists the results obtained for the UCDWR runs in shallow water. Table 2 lists the results obtained for the WHOI runs available at the time of the analysis.

For each class of runs, two figures are supplied in the upper right-hand corner of the box for median values of R_{40} in order to indicate the size and extent of the sample. The first of these two figures is the total number of runs making up the sample. The second number, which shall be called the "adjusted number of days" and is separated from the total number of runs by a slant line, indicates how widely distributed the sample is in time. The latter figure is supplied because it has been found that the acoustic data obtained on a particular day and at a particular location resemble each other more closely than data which have been obtained on different days, even though the oceanographic conditions are closely similar. Instead of simply noting the number of different days on which the various runs making up the sample were obtained, it was decided to give an "adjusted" number, computed as follows. If the number of runs made on k different days are denoted by n_1, $n_2, \cdots, n_k$, then the adjusted number of days K is defined as the expression

$$ K = \frac{\left(\sum_{i=1}^{k} n_i \right)^2}{\sum_{i=1}^{k} (n_i^2)}. \qquad (3) $$

K equals the number of days k if all the n_i are very nearly equal; in other words, if the sample is evenly distributed over the various days on which runs in this classification were obtained. But if some of the days furnish only one or two runs with other days contributing large numbers of runs, the days with very few runs will not be counted fully. To give a

---

[c] This division of the MIKE patterns was made for this analysis only. The designations DEEP MIKE and SHALLOW MIKE have no official standing in Navy doctrine.

[d] These ranges represent that range at which the transmitted sound level is 40 db below the level at 100 yd. In some cases, the level at 100 yd was ascertained by extrapolating in from several hundred yards. In the case of WHOI data, R_{40} is determined with reference to the sound level at 100 yd at the depth of the hydrophone in question, and R_{40} is thus nothing but a measure of the slope of the transmission anomaly vs range; while for UCDWR data, reference is made to the sound level at 100 yd at a depth of 16 ft below the sea surface.

TABLE 1. Median and quartile values of R_{40} for UCDWR transmission runs at 24 kc in shallow water.

BT Pattern	Hydrophone depth	SAND			ROCK			STONY			SAND-AND-MUD			MUD		
		Lower quartile	Median	Upper quartile	Lower quartile	Median	Upper quartile	Lower quartile	Median	Upper quartile	Lower quartile	Median	Upper quartile	Lower quartile	Median	Upper quartile
NAN	Shallow		See below								1,850	$2,020^{16/4.3}$	2,235	650	$1,020^{29/4.4}$	1,775
	Intermediate										1,575	$1,840^{9/2.5}$	2,325	760	$900^{15/2.4}$	1,700
	Deep										1,333	$1,680^{14/3.4}$	2,363	1,470	$1,700^{13/1.6}$	1,910
CHARLIE	Shallow											$2,100^{7/3.8}$		800	$1,400^{11/3.5}$	2,500
	Intermediate											$1,800^{7/3.8}$			$1,250^{9/1}$	
	Deep											$1,660^{6/2.8}$			$1,285^{4/1.6}$	
NAN AND CHARLIE	Shallow	2,100	$2,700^{75/8.3}$	3,100	1,910	$2,220^{37/3.0}$	2,515	2,105	$2,965^{12/2}$	3,755						
	Intermediate	2,300	$2,820^{61/7.5}$	3,205	2,000	$2,310^{23/3.1}$	2,700		$3,500^{6/1.9}$							
	Deep	2,100	$2,475^{32/3.7}$	2,780	1,780	$2,150^{25/3.3}$	2,625									
SHALLOW MIKE	Shallow	2,050	$2,700^{23/2.6}$	3,450	2,145	$2,410^{21/2.8}$	2,655				1,840	$2,400^{2/5}$	3,245	2,500	$3,200^{27/3}$	3,590
	Intermediate	1,735	$2,350^{23/3.4}$	3,290	2,225	$2,300^{9/3.0}$	2,575					$2,000^{5/2.8}$		1,740	$2,290^{10/2.8}$	3,130
	Deep		$2,650^{2/2.0}$		2,050	$2,190^{9/3.0}$	2,480					$1,950^{3/1.8}$		1,800	$2,000^{9/2.5}$	2,185
DEEP MIKE	Shallow													2,655	$3,300^{8/3.2}$	3,810
	Intermediate														$3,500^{1/1}$	
	Deep															

TABLE 2. Median and quartile values of R_{40} for WHOI transmission runs at 24 kc in shallow water.

BT Pattern	Hydrophone depth	SAND Lower quartile	SAND Median	SAND Upper quartile	ROCK Lower quartile	ROCK Median	ROCK Upper quartile	STONY Lower quartile	STONY Median	STONY Upper quartile	SAND-AND-MUD Lower quartile	SAND-AND-MUD Median	SAND-AND-MUD Upper quartile	MUD Lower quartile	MUD Median	MUD Upper quartile
NAN	Shallow										925	$1,100^{17/3.1}$	1,550	600	$825^{10/2.3}$	1,113
	Intermediate											$1,950^{3/1}$				
	Deep															
CHARLIE	Shallow										1,150	$1,300^{26/2.2}$	1,450	837	$925^{10/3}$	1,838
	Intermediate														$1,900^{4/1.6}$	
	Deep															
NAN AND CHARLIE	Shallow	1,275	$2,275^{32/6.4}$	2,500		$2,625^{4/2.7}$			$2,975^{2/1}$							
	Intermediate	1,900	$2,375^{4/4.0}$	2,575		$1,900^{4/1.8}$										
	Deep		$2,700^{4/1.5}$													
SHALLOW MIKE	Shallow		$2,775^{2/2.0}$						$3,700^{/2}$			$2,300^{6/1.5}$			$1,587^{4/2.0}$	
	Intermediate		$2,650^{1/1.0}$						$3,500^{2/1}$			$1,800^{3/1}$				
	Deep		$2,300^{1/1.0}$						$3,950^{2/1}$							
DEEP MIKE	Shallow		$2,650^{2/1}$						$3,400^{2/1}$			$2,775^{2/1}$			$3,400^{1/1.0}$	
	Intermediate								$3,225^{2/1}$			$2,750^{2/1}$				
	Deep															

numerical example, assume that a total number of 26 runs were obtained on 4 days. If the numbers of runs on these four days were 7, 7, 6, and 6, then the value of K for this case would be 4.0. If, on the other hand, two of the four days contributed 12 runs each, and the other two days only one run apiece, K would be found to equal 2.3.

The results in Tables 1 and 2 indicate that SAND, ROCK, and STONY are well-reflecting bottom types; they lead to values of R_{40} in excess of 2,000 yd in the great majority of cases, regardless of refraction conditions. CLAY also appears to be a well-reflecting bottom, although it should be emphasized that all the CLAY runs by UCDWR were made at a single location off San Francisco, and that a generalization of the results obtained should be based on a more adequate sample. In this method of analysis, no systematic dependence on hydrophone depth can be discovered, although the samples for deep hydrophones are mostly too small for the results to be considered conclusive. Transmission over STONY bottoms appears somewhat better than over any other type of bottom.

For the classification MUD, it is apparent that the dependence of R_{40} on the conditions of refraction is similar to the situation in deep water. The WHOI SAND–AND–MUD data resemble the MUD data in this respect. The R_{40} ranges are long when the water is isothermal, and short when the sound beam is bent downward by negative temperature gradients. In the classification SHALLOW MIKE, there is some evidence of layer effect over the poorly reflecting bottoms. Layer effect is much weaker, if present at all, for SAND, ROCK, and STONY bottoms. There is also some evidence that in the case of MUD bottoms the transmission for the deep hydrophone is better than for the intermediate and shallow hydrophones. The transmission results obtained by UCDWR and by WHOI appear to be in fair agreement with each other except for the SAND–AND–MUD bottoms. In this classification, the transmission observed by WHOI is significantly poorer than the transmission observed by UCDWR. This disparity is not too surprising in view of the fact that the SAND–AND–MUD classification covers a wide variety of bottoms, namely, all those bottoms in which very fine particles are mixed with sand grains and in which the percentage of sand grains lies between 10 and 90 per cent. It appears reasonable to assume that the SAND–AND–MUD bottoms investigated by UCDWR contained a larger percentage of sand grains and were

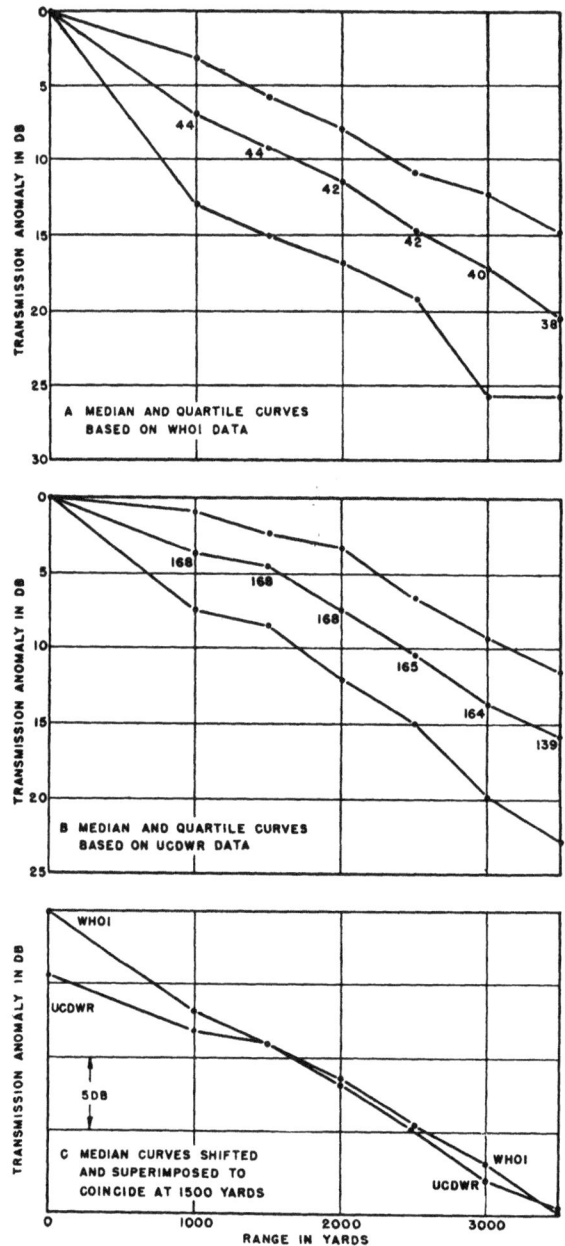

FIGURE 3. Comparison of WHOI and UCDWR transmission data over sand with downward refraction.

harder, on the average, than the SAND–AND–MUD bottoms investigated by WHOI.

The value of R_{40} is a useful parameter for the description of transmission; but in view of the frequent nonlinearity of the transmission anomaly range curve, no single parameter can be relied on to adequately characterize the transmission from very short to very long ranges. A more adequate method for describing a sample of transmission curves is the computation of a curve of median transmission anomaly. To obtain such a curve, values of the transmission anomaly are read off at a number of predetermined ranges, such as every 500 yd. At each range, the median transmission anomaly is noted. If these median values are plotted against the range and the resulting points connected, the resulting curve will have the property that at each of the ranges at which values were read it separates the actual curves into two equally numerous portions. In a similar fashion, upper and lower quartile curves of transmission anomaly may be obtained.

In general, the median anomaly curve will look smoother than the individual curves making up the sample, and "bumps" in the individual curves will not show up if they do not all appear at the same range. However, the median curve will be representative of average transmission conditions, and the quartile curves will provide a graphical measure of the spread of the sample.

Obviously, median curves of transmission anomaly are valuable only if they are based on a fair-sized sample. For this reason, separate median curves were not constructed for all the classifications which were established in the analysis reported here.

Not all individual transmission curves extend out to the same range. When transmission conditions are relatively poor, the reading of the traces must be stopped at a rather short range. When an appreciable number of curves in the sample cannot be read at the longer ranges, the median curve for the remainder apparently turns upward; since this upward turn has no physical significance, the median curve is stopped short in such cases.

In the computations summarized in this chapter, median and quartile transmission anomalies were determined at 1,000 yd and every 500 yd from there on; the number of curves in the sample was noted at each of these ranges.

As a first problem, the degree of consistency between the UCDWR and the WHOI runs was investigated. For this purpose, all runs over SAND with downward refraction (NAN and CHARLIE), regardless of hydrophone depth, were collected for each institution separately, and median and quartile curves of transmission anomaly plotted. These curves are shown in parts (A) and (B) of Figure 3. There is some evidence that the discrepancy of about 5 db between the curves is due, in part, to a different method of calibration. While UCDWR has usually referred transmission anomalies to the transmission level re-

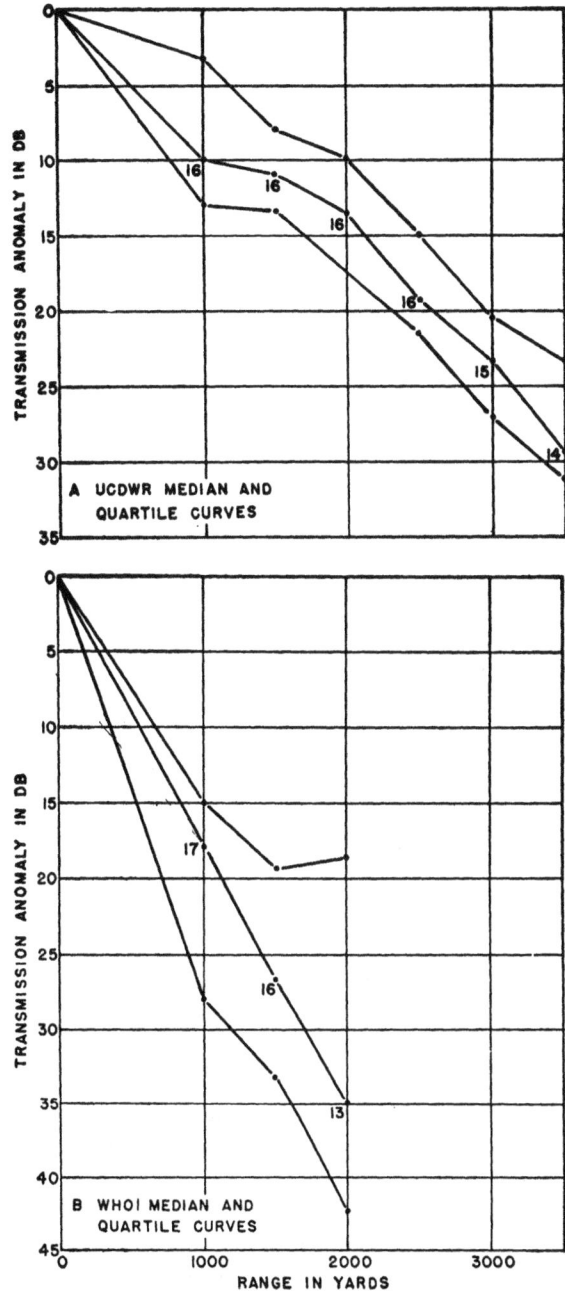

FIGURE 4. Comparison of UCDWR and WHOI transmission data over sand-and-mud bottoms with strong downward refraction.

corded at about 100 yd, WHOI has frequently obtained the reference level by extrapolating the measured relative transmission anomalies backward. To illustrate the relatively good fit which results from vertical shifting of the curve, part (C) of Figure 3 shows the two median curves shifted so that they coincide at a range of 1,500 yd.

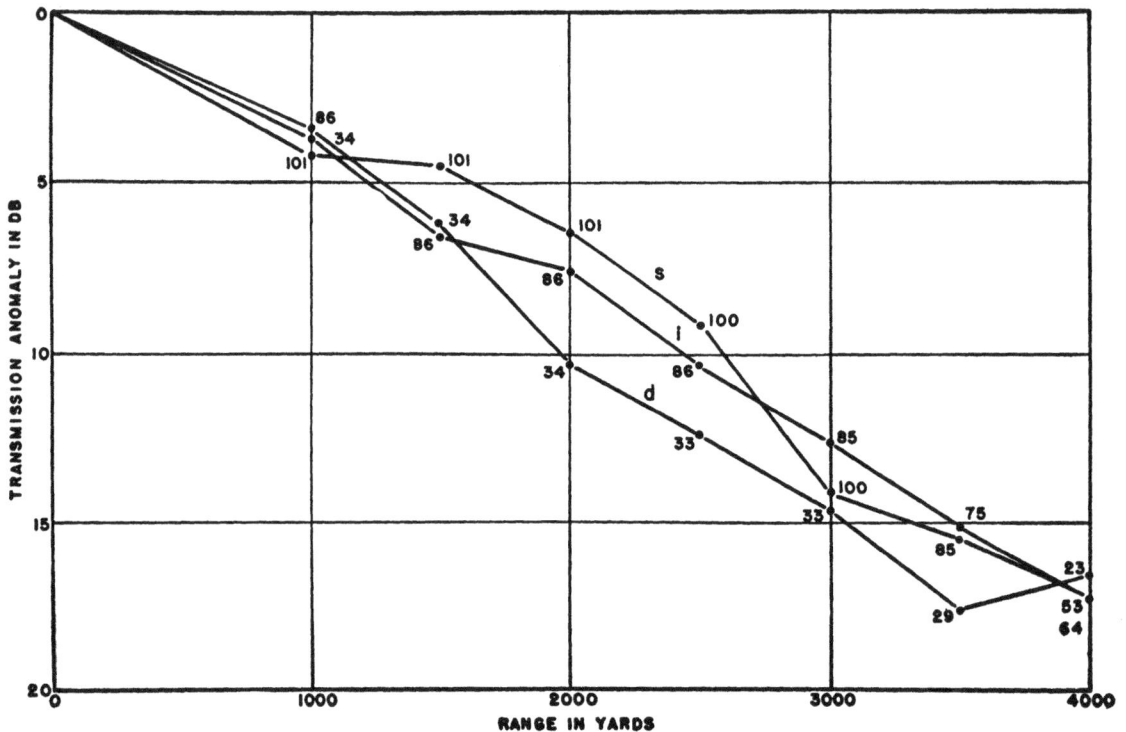

FIGURE 5.   Transmission over SAND for different hydrophone depths.

We have already noted, from consideration of the R_{40} values, that for SAND–AND–MUD bottoms the agreement between WHOI and UCDWR is very poor. This discrepancy is confirmed by the median and quartile transmission curves of those runs over SAND–AND–MUD which were carried out with a shallow hydrophone in the presence of NAN pattern (strong downward refraction), shown in Figure 4. In this case, the discrepancy is undoubtedly real and not caused by different calibration methods; for not only are the transmission anomalies at a given range different, but the WHOI median curve has a much steeper slope. The slope of the median UCDWR is roughly between 8 and 10 db per 1,000 yd, while the slope of the WHOI median curve is about 18 db per 1,000 yd.

No other comparisons were made between WHOI and UCDWR transmission data because most of the WHOI samples were too small for such comparisons. All the median curves to be discussed later are based exclusively on UCDWR runs.

To examine the effect of hydrophone depth over a well-reflecting bottom, median curves over SAND were determined for the three classes of hydrophone depth without regard to refraction pattern. In Figure 5, the three resulting curves are superimposed on each other, identified as s (shallow), i (intermediate), and d (deep). Table 1 shows that the bulk of these runs were carried out in the presence of downward refraction, with about one-fourth of the BT patterns showing a shallow mixed layer above the thermocline. In Figure 5 there are no significant differences between the three curves.

The quartile curves have not been reproduced, but they are all fairly similar, deviating from the median curve by about 5 db at 3,000 yd. The transmission anomaly over SAND can be represented fairly well by a straight line passing through zero at zero range and having a slope of 5 ± 2 db per kyd. This numerical estimate is also good for the median and quartile curves shown in Figure 3, which do not include the SHALLOW MIKE cases forming part of the sample used in constructing Figure 5.

Figure 6 shows median curves, for shallow and for deep hydrophones, of all runs obtained over ROCK bottoms. It will be noted that the transmission over ROCK is not quite so good as over SAND, the average slope for ROCK being 6 db per kyd. The quartiles deviate from the median, in this case also, by roughly 2 db per kyd.

Figure 7 shows the median and quartile curves for all runs obtained by UCDWR over STONY bottoms.

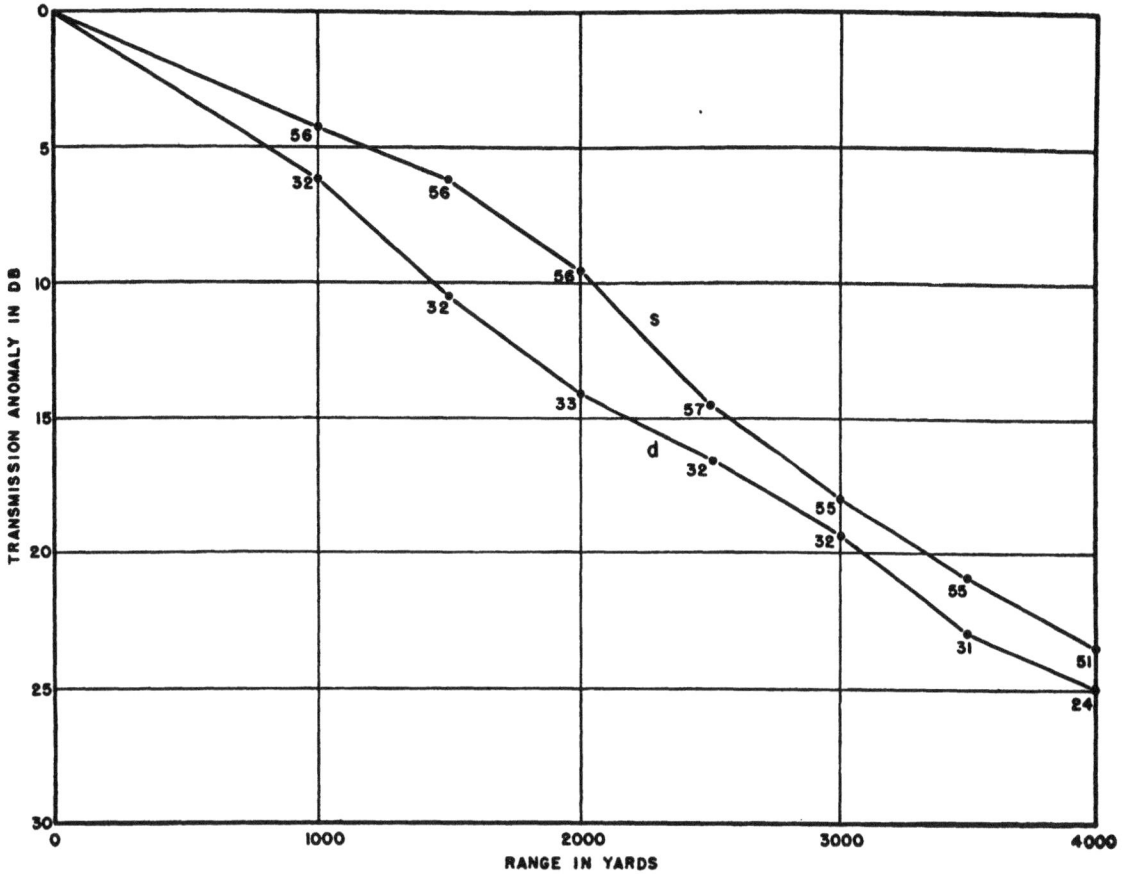

FIGURE 6.   Transmission over ROCK for shallow and deep hydrophones.

Transmission appears to be better out to 2,000 yd for STONY bottoms than for any other type of bottom, but deteriorates rapidly from 2,000 yd on out. The wide spread between the median and quartile curves is an indication that these bottoms are acoustically less uniform than SAND or ROCK bottoms.

Figures 8 and 9 show two typical transmission anomaly plots, which were obtained over a ROCK and over a STONY bottom respectively. These runs were carried out in the presence of pronounced negative gradients from the surface of the sea down to well below the depth of the projector. The ray diagrams, which are shown in the upper parts of the figures, indicate that in deep-water transmission conditions would be confidently predicted to be poor. Because of the well-reflecting bottom, however, the observed transmission is comparable to that in a deep mixed layer in deep water.

MUD bottoms were originally defined as bottoms in which the average particle was too small in size to be classified as SAND. However, evidence accumu-

lated indicating that from an acoustic point of view there are two different types of bottoms which are composed of very small particles. These two types of bottom can be characterized by their consistency as soft and as plastic, and they have been designated in this analysis as MUD and as CLAY. Some evidence concerning the difference in acoustic properties of these two types of bottoms has been collected and published by WHOI.[8] This evidence for separating MUD bottoms into MUD and CLAY was apparently borne out by the analysis of the transmission data obtained by UCDWR, but doubts as to the correct classification of the CLAY samples involved detract from the value of this evidence.

Figure 10 shows median and quartile transmission curves over MUD in the presence of negative gradients, separated according to hydrophone depth. It appears that the quartile spread is appreciably reduced for the shallow and deep hydrophone depths by this separation. Regardless of hydrophone depth, the transmission anomaly at 3,000 yd is approxi-

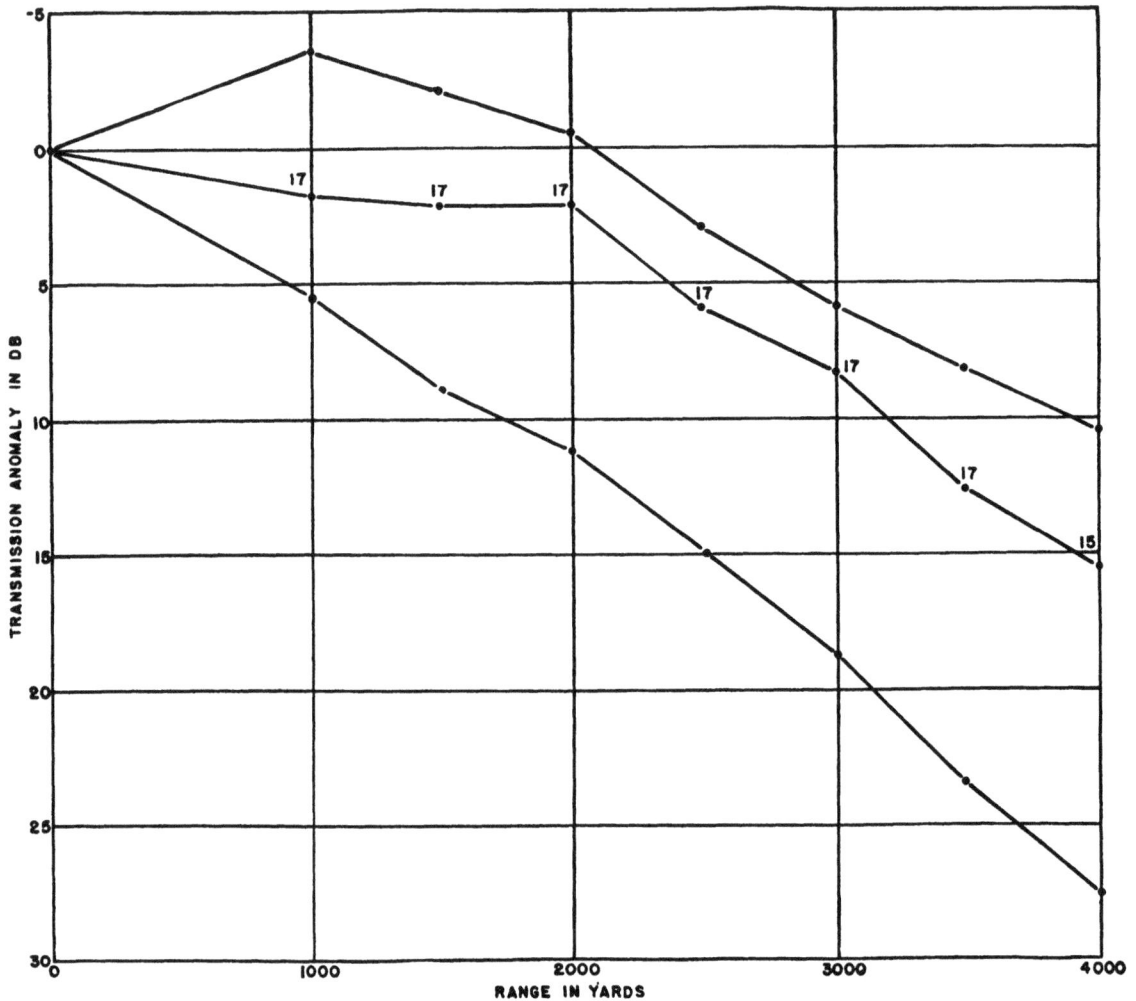

FIGURE 7. Transmission over STONY bottoms.

mately 30 db. For shorter ranges, there is a significant difference between the anomalies at different hydrophone depths. With the hydrophone deeper than 100 ft, the transmission anomaly is almost linear and increases at the rate of about 10 db per kyd. For the more shallow hydrophone depths, there is a much more precipitate drop at short range. With the hydrophone at 16 ft, the median transmission anomaly at 1,000 yd is 26 db. From 1,000 to 3,000 yd, it drops only another 8 db, resembling in this respect the transmission of sound in the shadow zone in deep water.

Figure 11 shows a typical run over a soft MUD bottom in the presence of a pronounced negative temperature gradient. Just as in deep water, the sound level at shallow depth begins to drop rapidly at a shorter range than does the level at considerable depth. At all hydrophone depths, the transmission

anomaly increases sharply at the approximate range of the predicted shadow zone boundary. A slight recovery of the sound level recorded by the two shallow hydrophones is noted at almost exactly the range at which the axis of the reflected beam rises to the depth of the hydrophone. This recovery is, however, not very pronounced. While it increases the sound level at 2,400 yd to approximately 10 db above the level which would have been recorded in deep water under similar circumstances, the transmission anomaly still amounts to about 25 db.

EFFECT OF WIND FORCE

UCDWR has carried out an analysis of the effect which the roughness of the surface has on sound transmission in shallow water over well-reflecting bottoms such as ROCK and SAND.

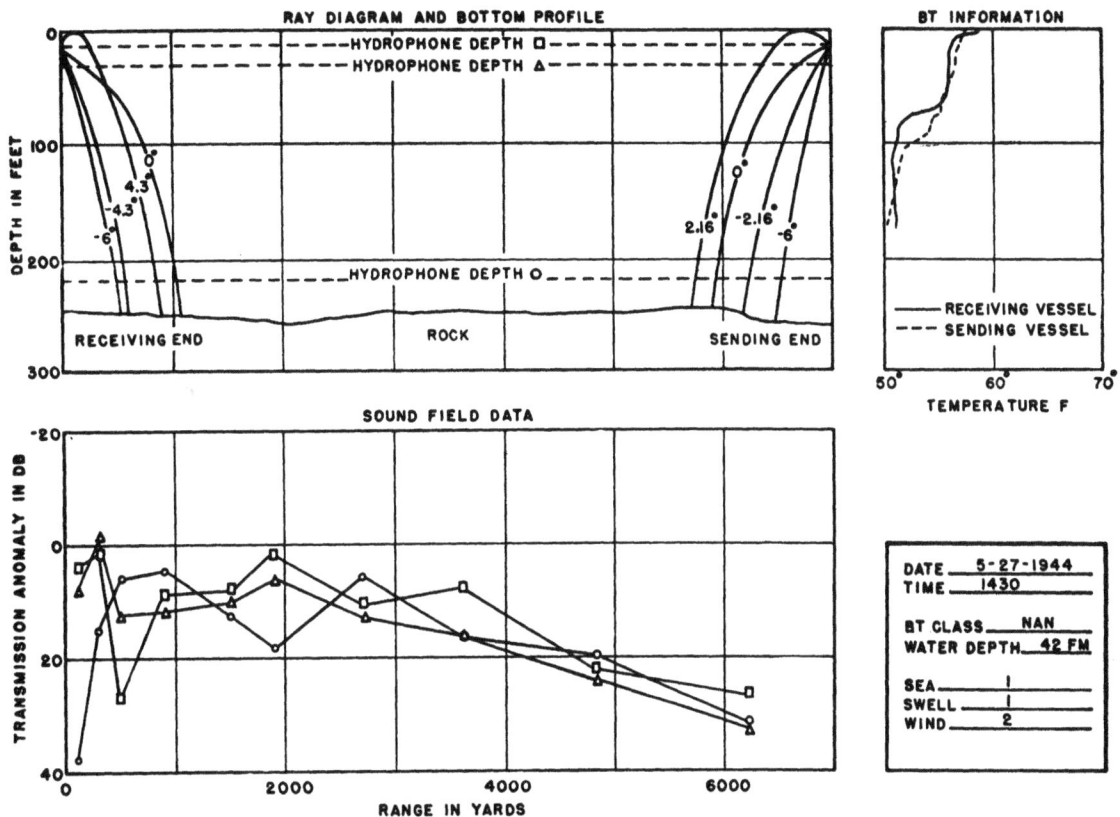

FIGURE 8.   Transmission run over ROCK.

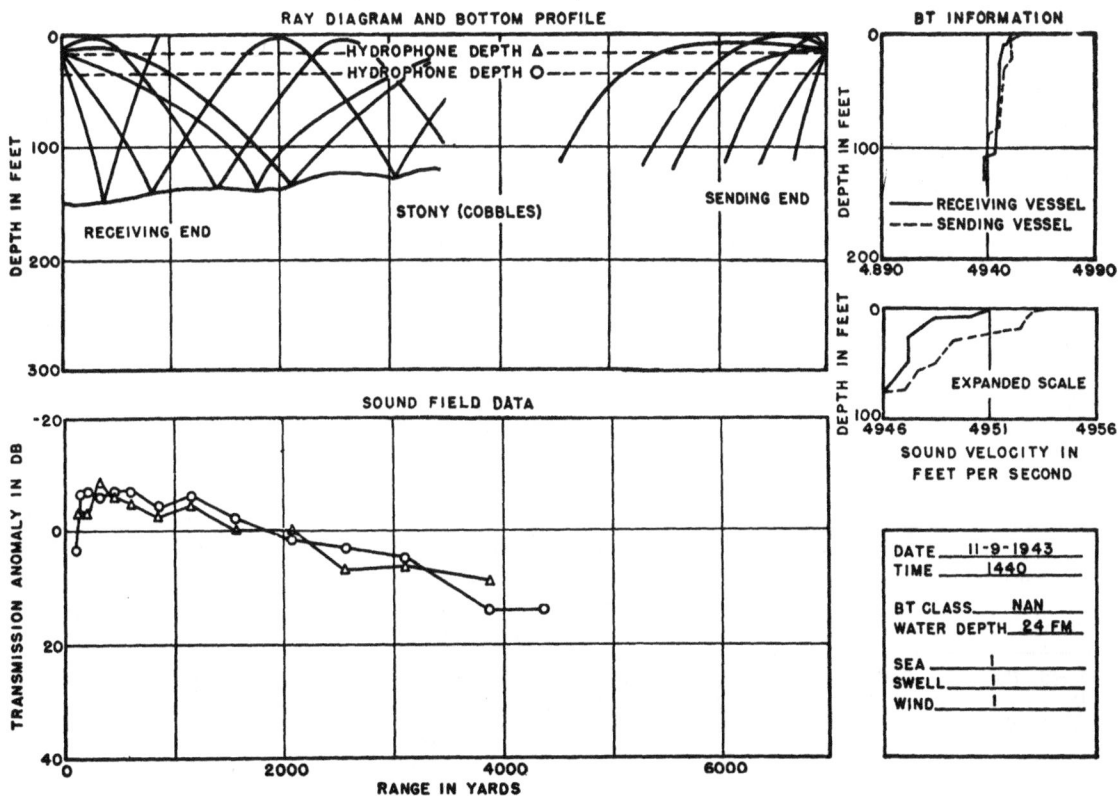

FIGURE 9.   Transmission run over a STONY bottom.

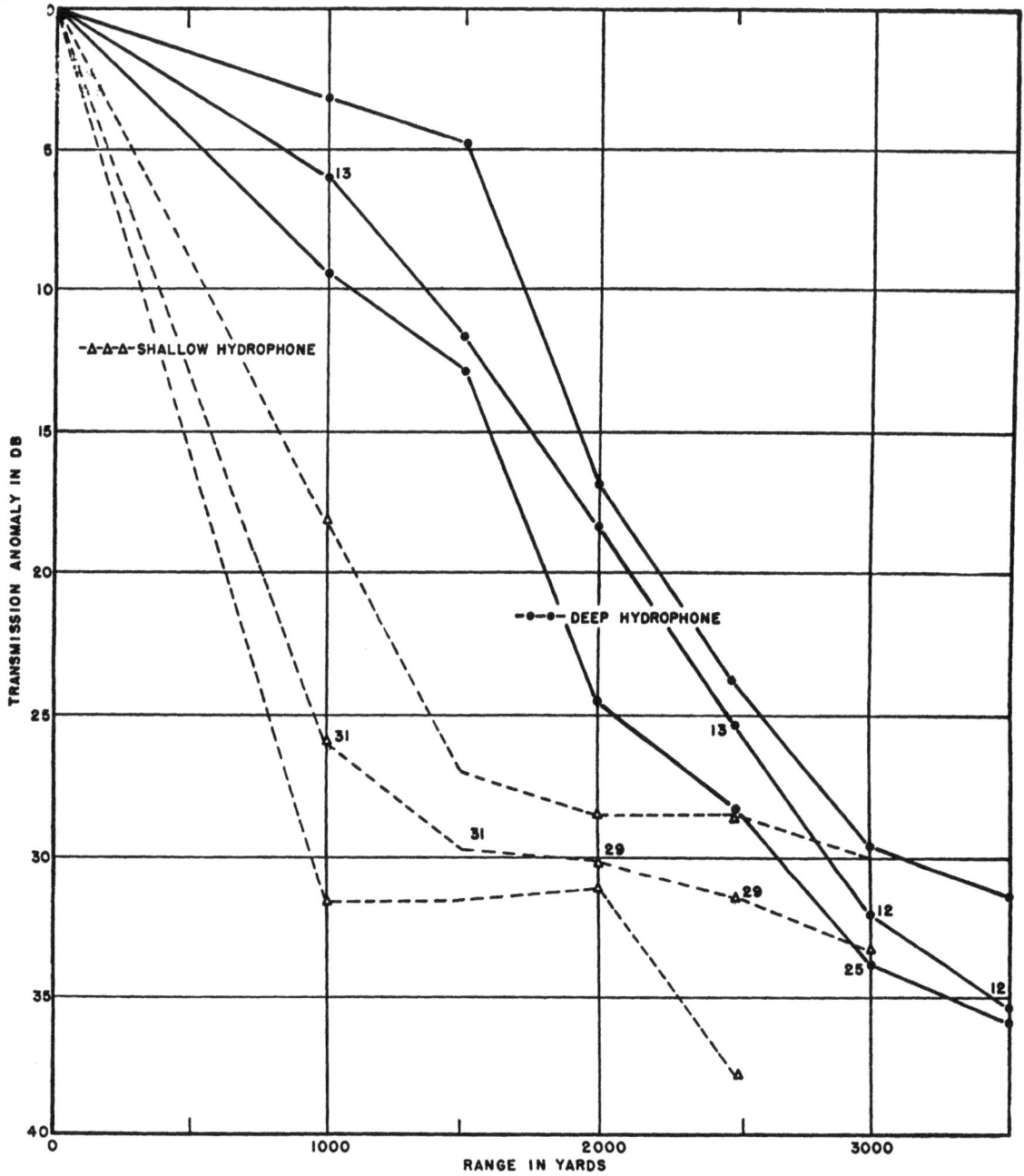

FIGURE 10.   Transmission over MUD with NAN pattern.

TABLE 3. R_{40} versus wind force over ROCK.

Wind force (Beaufort)	1	2	3	4
Number of runs	29	54	31	21
Lower quartile R_{40}	2,050	2,150	1,550	1,500
Median R_{40}	2,950	2,400	2,100	1,750
Upper quartile R_{40}	3,150	2,650	2,300	2,100

One hundred thirty-five runs were carried out over ROCK bottom at fairly constant depth. In Table 3 are listed the number of runs made at each wind force and the median and quartile values of R_{40}. Figure 12 shows the complete distribution.

One hundred seventy-four runs were carried out over SAND bottoms in water of more than 6 fathoms. Table 4 shows the same data for these runs as Table 3 does for runs over ROCK.

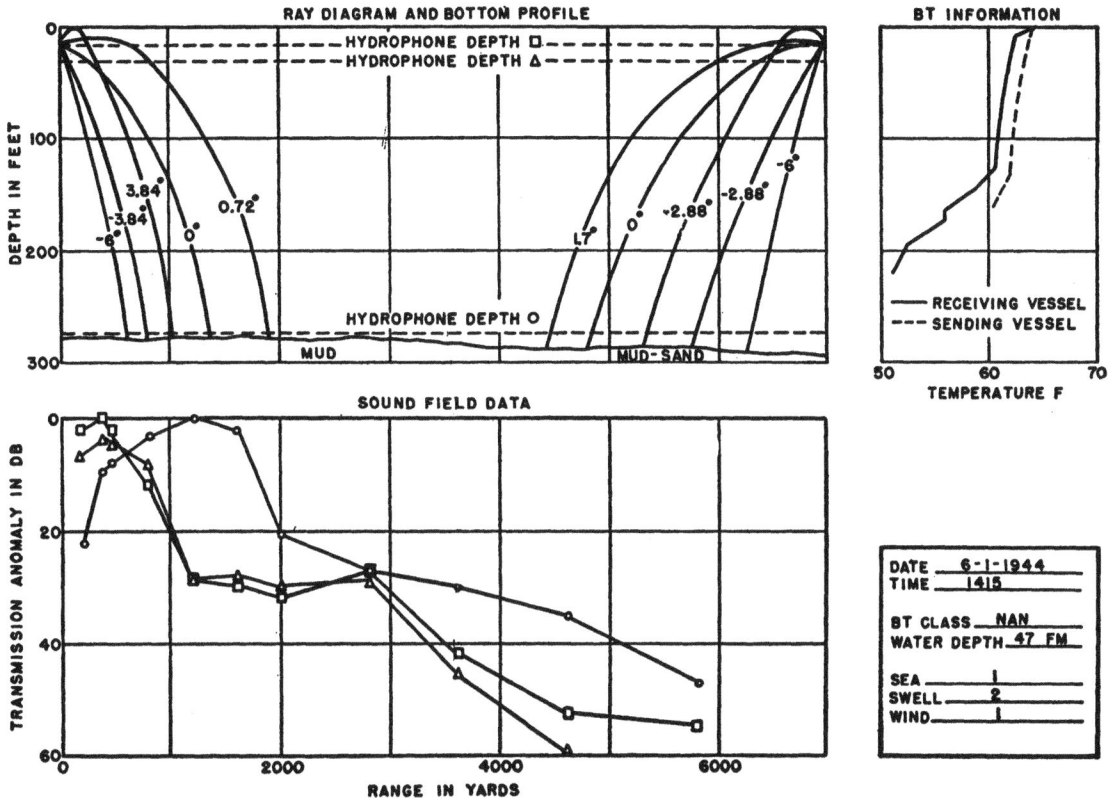

FIGURE 11.   Transmission run over MUD with NAN pattern.

Forty-six runs were made in water of 6 fathoms or less and over SAND bottoms. These runs were analyzed as a separate group. Table 5 summarizes the results. Figure 13 shows the complete distribution of SAND runs.

The majority of the runs used in this analysis were carried out in the presence of downward refraction, but no attempt was made to separate the runs with downward refraction from those with the projector located in a mixed layer. This method of analysis may account for the wide quartile spread and more particularly for the great upper quartile spread for wind force 4. The reduction of R_{40} between wind force 0 or 1 and 3 amounts to an increase in the slope of the transmission anomaly curve of roughly 1 db per kyd.

6.2.3   Summary

Transmission experiments at 24 kc indicate that the sea bottoms can be roughly divided into well-reflecting bottoms comprising ROCK, CORAL, STONY, SAND, and CLAY bottoms, and poorly reflecting bottoms, mostly MUD and some of the SAND–AND–MUD. Most of the SAND–AND–MUD bottoms are intermediate between well and

TABLE 4. R_{40} versus wind force over SAND in water depth greater than 6 fathoms.

Wind force (Beaufort)	0	1	2	3	4
Number of runs	18	33	51	60	12
Lower quartile R_{40}	2,850	3,100	2,900	2,100	1,400
Median R_{40}	3,450	3,250	3,500	2,450	1,800
Upper quartile R_{40}	4,000	3,500	3,700	2,950	2,500

TABLE 5. R_{40} versus wind force over SAND in water depth 6 fathoms or less.

Wind force (Beaufort)	2	3	4
Number of runs	16	13	17
Lower quartile R_{40}	1,400	1,250	950
Median R_{40}	1,550	1,400	1,100
Upper quartile R_{40}	1,700	1,750	1,650

poorly reflecting bottoms. Present evidence indicates that in shallow water at least 10 fathoms deep and in the presence of downward refraction, transmission anomalies over SAND and STONY bottoms increase with the range by 5 ± 2 db per kyd, and over ROCK bottoms by 6 ± 2 db per kyd. The transmission is not significantly affected by hydrophone

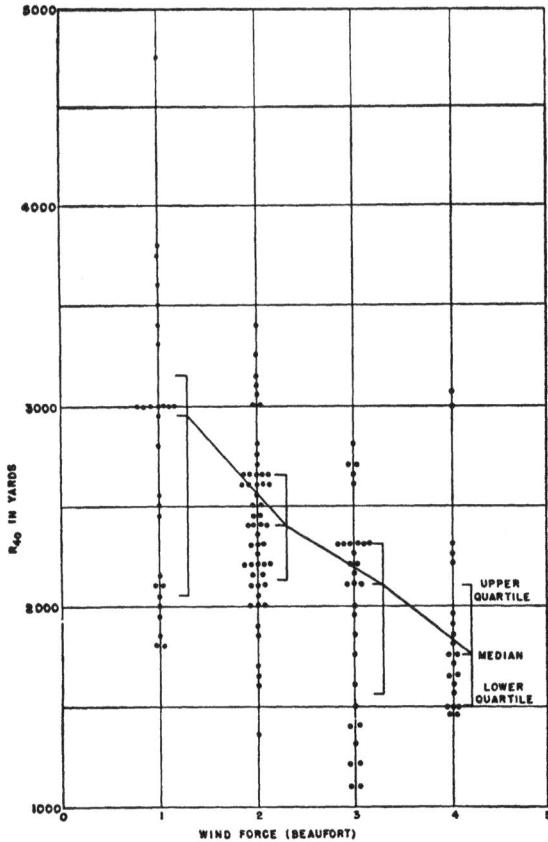

FIGURE 12. R_{40} versus wind force over ROCK.

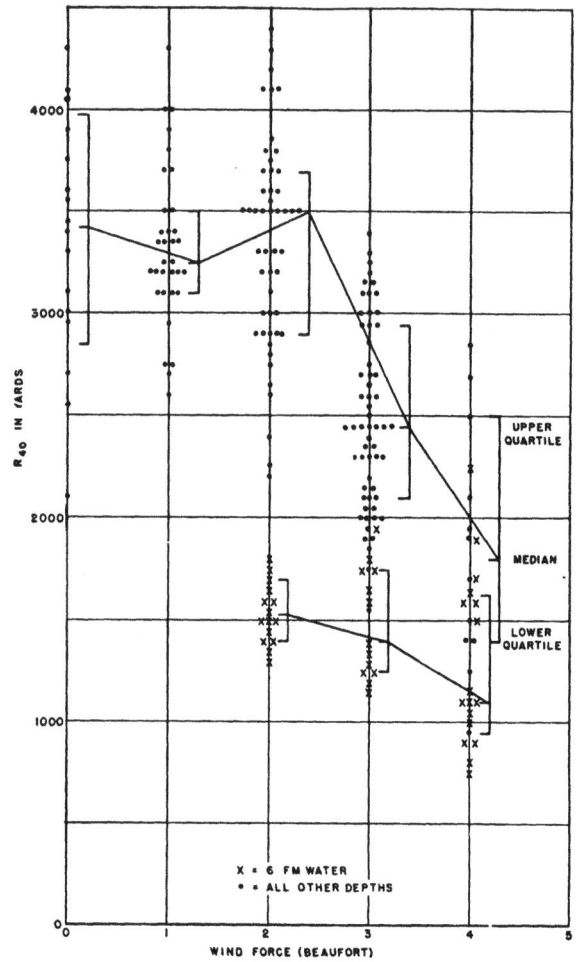

FIGURE 13. R_{40} versus wind force over SAND.

depth or bottom depth. (In very shallow water, less than 10 fathoms deep, transmission is inferior to that found in moderately shallow water.) Transmission over MUD differs but little from transmission in deep water; secondary peaks due to bottom-reflected sound are not likely to raise the level more than 10 db above the level that would be observed in a deep-water shadow zone. In isothermal water or with upward refraction, transmission over all bottoms is about as good and sometimes slightly better than deep water.

Transmission anomalies with negative gradients over the well-reflecting bottom types are affected adversely by heavy seas. For sea state 3, transmission anomalies are likely to be at least 1 db per kyd higher than in calmer seas.

6.3   SONIC TRANSMISSION

Sonic transmission differs from supersonic transmission primarily in that dissipative processes within the water are much less important. The probable value of the absorption or attenuation coefficient at

sonic frequencies has been discussed in Chapter 5. It has been estimated[9] that at the lower sonic frequencies (2,000 c and less) the attenuation of sound in sea water at a depth of several hundred fathoms is less than 1 db in 20,000 yd. While there is reason to believe that close to the surface, absorption at these low frequencies is appreciably higher,[e] it is probably no more than about 0.5 db per 1,000 yd. As a result, sonic sound in shallow water may show evidence of a spread less than that predicted by the inverse square law. This section summarizes the results which were obtained by UCDWR,[10–12] and by CUDWR–NLL.[13,14] In these experiments, CUDWR–NLL used a single-frequency source, with higher

---

[e] If dissipative processes near the surface at low sonic frequencies were as low as they were estimated at great depths in reference 10, then listening ranges on noisy surface targets should be of the order of 100 miles in the presence of deep mixed layers; actual listening ranges rarely exceed 20 miles even with the best sonic listening gear available.

harmonics present because of overloading, while UCDWR used a noise source of the type employed for acoustic minesweeping. The receiving equipment consisted of hydrophones, whose output was amplified and sometimes put through band filters to be recorded by means of power level recorders. Transmission was always continuous throughout the run.

6.3.1 Long Island Area Survey

Long Island Sound is mostly shallow, less than 15 fathoms deep, and the bottom is predominantly sandy, although some runs were made over MUD, SAND–AND–MUD, and STONY bottoms. All runs were made with a single-frequency source. Frequencies used were 0.6, 2, 8, and 20 kc. Geographically, the survey was divided into three areas: the Fisher's Island area, the New York Harbor approaches, and Block Island Sound. In all three areas, hard bottoms were predominant. Depths varied from about 50 ft to 200 ft. During the New York Harbor and Fisher's Island area surveys, refraction was mostly upward, owing in part to salinity gradients. Off Block Island, some negative gradients were found. Sea states were low with the exception of the New York Harbor runs where sea states up to 4 were encountered. Table 6 summarizes the results obtained. To obtain this table, the investigators attempted to fit each run by a formula of the form

$$H = n \cdot 10 \log r + ar \qquad (4)$$

in which n is the power of spreading and a represents the attenuation in decibels per kiloyard. Since it is difficult to determine both n and a simultaneously by a best fit calculation, n was chosen arbitrarily to assume the values 1, 1.5, and 2. The best value of a was then determined by inspection for each of the three assumed values of n.

The three fits for $n = 1$, 1.5, and 2 were classified in order of decreasing preference as I, II, III. In addition, the individual fits were graded on an absolute standard as "good" (g), "fair" (f), "poor" (p). Table 7 is a summary of the goodness of the fit obtained for these runs. Despite the equal standard deviation values of Table 6, the value 1 for n seems to be most frequently the best fit to the data, although 1.5 is probably the best average, especially at the higher frequencies.

This survey resulted in the following general conclusions. Higher frequencies were attenuated more than the lower frequencies. At high frequencies the transmission loss increased with increased disturbance

TABLE 6. Statistical analysis of empirically determined values of a and n, poor fits omitted.

Survey Area	Frequency	Total number of runs	$n = 1$ Number of cases	$n = 1$ Median a^*	$n = 1$ Average a	$n = 1$ Standard deviation σ	$n = 1.5$ Number of cases	$n = 1.5$ Median a	$n = 1.5$ Average a	$n = 1.5$ Standard deviation σ	$n = 2$ Number of cases	$n = 2$ Median a	$n = 2$ Average a	$n = 2$ Standard deviation σ
FI	600 c	9	10	1.2	1.8	1.0	8	0.2	0.8	1.0	3	Insufficient Data	Insufficient Data	
FI	2 kc	11	11	1.2	1.4	1.0	8	0.2	0.6	1.0	9	Insufficient Data	Insufficient Data	
FI	8 kc	13	10	2.8	2.9	1.0	11	1.2	1.7	1.0	11	Insufficient Data	Insufficient Data	
FI	20 kc	12	13	5.2	5.3	1.0	13	4.2	4.1	1.0	11	2.9	2.8	1.9
BI	600 c	22	19	3.5	3.5	1.5	18	2.0	2.3	1.5	12	0.8	1.8	1.7
BI	8 kc	18	11	4.3	5.2	2.7	15	3.5	3.6	2.5	14	3.0	2.6	2.2
BI	20 kc	22	17	7.7	9.0	4.3	19	6.5	7.1	3.9	17	5.1	5.5	3.1
NY	600 c	17	18	1.5	2.5	1.5	19	1.0	0.8	0.5	8	0.2	0.2	0.3
NY	2 kc	10	8	2.5	3.5	2.0	9	2.7	1.4	1.0	2	Insufficient Data	Insufficient Data	
NY	8 kc	16	12	5.5	5.2	2.5	15	4.2	3.1	2.7	11	4.0	2.1	2.4
NY	20 kc	19	18	8.1	8.8	3.8	17	7.6	7.9	3.1	15	5.0	5.3	5.0

* Values of a and σ are given in decibels per kiloyard.

of the sea for upward refraction, but there was no such effect at the lower frequencies. With downward refraction the state of the sea did not influence the transmission. Except at the lowest frequencies and over the softest bottoms, the type of bottom did not appreciably affect the transmission loss. Bottom types were MUD, SAND, and GRAVEL. Shoal areas and areas over sea valleys showed high transmission losses. The attenuation was virtually independent of depth for flat bottoms. Some correlation was found between the empirical value of n and refraction conditions; the power of spreading tended to assume large values in isothermal water.

TABLE 7. Summary of shallow-water results (BI,FI) number of fits in indicated classification.

n	Classified	0.6 kc	2.0 kc	8.0 kc	20 kc
1.0	I	18	7	6	18
	II	5	3	9	4
	III	8	1	16	13
	g	12	2	11	20
	f	17	9	9	10
	p	2	0	11	5
1.5	I	9	4	15	9
	II	22	7	16	26
	III	0	0	0	0
	g	10	1	15	18
	f	16	7	11	14
	p	5	3	5	3
2.0	I	4	0	10	8
	II	3	1	6	5
	III	24	10	15	22
	g	4	0	7	6
	f	11	2	18	22
	p	16	9	6	7

6.3.2 Pacific Ocean Measurements

Twenty transmission runs were made in the coastal waters off volcanic islands in the Pacific (see reference 11). Measurements were made of overall transmission in the 1- to 3-kc band over SAND and CORAL bottoms. These measurements permitted the following conclusions. In the 1- to 3-kc band, the transmission loss from 100 to 3,000 yd could best be fitted by n equal to 1.5 and a equal to 2.5 db per kyd, under most hydrographic conditions. These conditions included slight upward refraction in water of depth about 200 ft, slight upward refraction over sloping bottoms, and downward refraction in water of depth about 100 ft. For downward refraction over a sloping bottom, however, the transmission loss at ranges above 1,000 yd was much greater. For this case, the best fit above 1,000 yd was estimated to be

10 db per kyd for the attenuation with the spreading factor n uncertain. This result is in agreement with ray theory,[15] which predicts that sound multiply reflected from the bottom under these conditions should run downhill, following the bottom slope and leaving a shadow zone near the surface.

Also, some runs were made in the Thirteenth Naval District.[11] These runs were made in water less than 300 ft deep over coarse gravel or rocky bottoms. Velocity gradients were slight and the sea calm. No correlation with computed limiting ranges was observed. The majority of the runs were best approximated by zero attenuation and n equal to the values given in Table 8. One run through a tide rip was best approximated with n equal to 1 and the values of a given on the right-hand side of Table 8.

TABLE 8. Summary of shallow-water results (Thirteenth Naval District).

Ordinary Runs			Special run through tide rip	
Frequency	Average n	a	n	a
0.1	1.4	0	1	3.0
0.6	1.3	0	1	1.5
2.0	1.5	0	1	3.5
8.0	2.5	0	1	8.0
20.0	3.2	0	1	8.0

6.3.3 Summary

Recent experiments carried out by UCDWR with pulses of sonic single-frequency sound have not yet been reported; they are, therefore, not included in this summary. This summary lists conclusions which were reached in the spring of 1945 on the basis of data available then.[16,17] It should be pointed out, however, that none of the conclusions reached at that time have been invalidated by later information.

In shallow water, a distinction must be made between transmission over MUD bottoms (which resembles deep-water transmission) and transmission over all other bottom types. No significant differences were discovered in sonic experiments between any of the other bottom types including MUD–AND–SAND. Over sloping bottoms, a significant dependence on refraction pattern has been observed: with downward refraction transmission tends to be poor, while in isothermal water it is as good as in deep water.

Over level bottoms, with isothermal water or in the presence of downward refraction, the transmission loss can be most adequately represented by an equation having the form,

$$H = 15 \log r + a, \qquad (5)$$

where a, the coefficient of attenuation in decibels per kiloyard, depends on f, the frequency in kilocycles, according to

$$a = 0.25(f - 2) \qquad (6)$$

above 2 kc. Below 2 kc the attenuation is very small. Equation (6) is believed to be adequate up to about 20 kc.

Equation (6) represents merely the average dependence of the attenuation coefficient on frequency.

TABLE 9.   Variation of attenuation with frequency.

Source	a at 2.0 kc	a at 8.0 kc	a at 20.0 kc	a at 8.0 less a at 2.0	a at 20.0 less a at 2.0
San Diego	0.0	2.5	4.2	2.5	4.2
Fisher's Island	0.6	1.7	4.1	1.1	3.5
Block Island	2.3	3.6	7.1	1.3	4.8
New York Harbor	1.4	3.1	7.9	1.7	6.5
Average	1.1	2.7	5.8	1.6	4.7

In the portion of the sea fairly near to the surface, which is the only region of interest in sonic listening, the absorption coefficient probably depends on highly variable factors, such as bubble content; thus large deviations from equation (6) may be expected to occur quite frequently.

There appears to be little correlation at sonic frequencies between transmission loss and refraction conditions, depth of the water, and surface roughness. With strong upward refraction, an increase of attenuation with increasing sea state has been observed, undoubtedly caused by the poor reflectivity of a rough and aerated surface.

At short ranges, out to approximately the range equal to the depth of the water, image interference maxima and minima have frequently been measured. However, except possibly at very low frequencies, the inverse fourth power decay has not been observed because of the disruptive effect of bottom-reflected sound at the ranges where the fourth power decay might be expected.

In general, reliable information on sonic transmission is scanty and is less consistent than the information on the transmission of 24-kc sound. In the future an increasing amount of stress is likely to be laid on the investigation of sonic transmission. However, sonic transmission will probably remain a more difficult field for investigation than supersonic transmission, because of the low directivity of most sources of sonic sound.

Chapter 7

INTENSITY FLUCTUATIONS

IT IS CLEAR from the preceding chapters that the sonar officer or the research worker cannot predict with precision the sound field intensity in the vicinity of a calibrated sound source, no matter how complete his information on oceanographic conditions. Chapter 5, in particular, mentions the wide range of sound field levels which are recorded under identical or nearly identical oceanographic conditions.

This chapter will be concerned with the variability of the sound field which is found when a succession of single-frequency signals are transmitted over the same path and received and recorded through the same receiving sound head and stack. This variability within a single sequence of sound signals has been subdivided into *fluctuation*, changes in intensity observed to occur during seconds or fractions of a second; and *variation*, a slow drift of the average intensity, which becomes noticeable in the course of minutes. This division between short-term and long-term variability can be justified on practical grounds. Variation may well be correlated with those large-scale changes in the thermal structure of the ocean which would be revealed by a continuously recording bathythermograph. Fluctuation is caused by mechanisms which cannot be observed by means of any oceanographic instrument in current use. This chapter will be concerned, exclusively, with the short-term variability of the sound field. The longer-term variability has already been discussed in Chapters 5 and 6.

The first section of this chapter will set forth the mathematical concepts commonly used in the description of fluctuation and will report the results of fluctuation experiments. In the second section, the significance of these experimental results will be assessed, and the contribution of various mechanisms to the observed fluctuation will be estimated tentatively.

7.1   OBSERVED FLUCTUATION

7.1.1   Magnitude of Fluctuation

In describing fluctuation quantitatively, we need expressions which characterize both the magnitude of fluctuation — roughly the amount by which an individual signal deviates from the mean for the run — and the time rate at which the sound field intensity changes. This subsection will be concerned with the magnitude of fluctuation.

Three different quantities are commonly used to express the magnitude of a received signal: the pressure amplitude (in dynes per square centimeter), the intensity (in watts per square centimeter), and the level (in decibels above some standard). When we consider a sequence of N signals received under apparently identical conditions, we can characterize this sequence by three sets of figures: amplitudes, intensities, and levels of all the individual members of the sample. Each of these three sets of figures describes the sample. Depending on our particular viewpoint, we may prefer one or another.

These three sets of figures can be converted one into another by means of the two equations

$$I = \frac{a^2}{2\rho c}, \tag{1}$$

$$L = 10 \log \frac{I}{I_0} = 20 \log \frac{a}{a_0}, \tag{2}$$

in which a stands for the pressure amplitude, I for the intensity, and L for the level in decibels. To each of these three sets we may assign as an average quantity the arithmetical mean, such as

$$a = \frac{1}{N}(a_1 + a_2 + \cdots + a_N), \tag{3}$$

and refer to these quantities as the mean amplitude, the mean intensity, and the mean level of the sample. These average quantities are no longer related by the equations (1) and (2).

Individual amplitudes will, of course, deviate from the mean amplitude. But some of these deviations will be positive, others negative, and it can be shown very easily that their sum vanishes. To express the spread of the amplitudes of the sample about the mean amplitude, a common procedure is to square the deviation of each individual amplitude from the mean amplitude and to average these squared deviations. The square root of the mean of the squared

158

deviations has the same dimension as an amplitude. It is called the root-mean-square (rms) deviation of the amplitude or, more briefly, the standard deviation of the amplitude. If it is divided by the mean amplitude, the resulting dimensionless quantity is called the relative standard deviation of the amplitude; this quantity is often expressed in per cent.

The analogous quantities formed with intensities and levels bear analogous names. These names are, in fact, common in all fields of statistics. If the relative standard deviation of the amplitude is very small compared with unity, the relative standard deviation of the intensity is about twice the relative standard deviation of the amplitude, while the absolute standard deviation of the level is approximately 4.34 times the relative standard deviation of the intensity.[a] The fluctuation of underwater sound is usually so large that these relationships between the standard deviations do not hold.

Relative standard deviations of the amplitude have been determined for transmitted signals of underwater sound under various conditions.[1-3] Most of the available data were taken at 24 kc. From the data at that frequency, an analysis was made of the dependence of the relative standard deviation on refraction conditions.[2] It was found that in the presence of strong downward refraction the median of the relative standard deviation, for the 29 samples collected, was 38 per cent.[b] For eleven samples, in which the receiving hydrophone as well as the projector were located within a mixed layer above a thermocline, the median of the relative standard deviation was 47 per cent. Seventeen samples, in which the hydrophone was in the thermocline beneath a mixed layer, showed a median relative standard deviation of 41 per cent, not much higher than the fluctuation in the presence of strong gradients from the surface down. Although these differences are probably significant, they should not be overestimated, in view of the wide spread within each of the

groups of samples discussed previously. The lower and upper quartiles in the group of strong downward refractions are 46 per cent and 36 per cent, respectively, while the quartiles for the isothermal group are 61 per cent and 44 per cent. It is probably justifiable to say that, on the average, the amplitude fluctuation in isothermal water is significantly higher than the amplitude fluctuation in the presence of strong downward refraction. The width of the quartile spread shows that even under similar conditions the magnitude of the fluctuation itself fluctuates from sample to sample. In view of the large number of signals making up a sample, usually between 50 and 200, this variability is not to be explained as sampling error but represents an actual change in the transmission conditions as they affect signal fluctuation. The high degree of variability of fluctuation is an indication of the complexity of the underlying mechanism or mechanisms as well.

Some information is available concerning the dependence of the relative standard deviation of the amplitude on frequency. One set of experiments, carried out at UCDWR, involved the simultaneous transmission of signals at two supersonic frequencies.[3] The frequency pairs used were 14 and 24 kc, 16 and 24 kc, 24 and 56 kc, and 24 and 60 kc. In 17 runs one frequency was either 14 or 16 kc, while the other was 24 kc. It was found that the mean of the relative standard deviations at the lower frequency (14 or 16 kc) was 38.8 per cent and at 24 kc 37.7 per cent. The difference is well within the root-mean-square spread, and is thus not significant. For the individual samples themselves, the difference between the fluctuations at the two frequencies is considerable for some runs, amounting to 19.2 per cent in one case. The root mean square difference is 8.5 per cent. As a result, it may be concluded that the average fluctuation is the same at 15 and at 24 kc, but that for any individual run the fluctuation may be considerably different at these two frequencies. In the majority of cases, however, high fluctuation at one frequency is associated with high fluctuation at the other, and unusually small fluctuation at one frequency tends to be associated with small fluctuation at the other. An analysis of the runs carried out with the frequency pairs 24 and 56 kc, and 24 and 60 kc, leads to similar conclusions for these frequencies.[c]

[a] 4.3429 is 10 log e where the log is to the base 10 and e is the base of the natural logarithms.

[b] In the discussion of the spread of a given set of data, it is very convenient to use the terms "median" and "quartile." Their meaning is as follows. If all the determinations of a certain quantity are arranged in the order of increasing magnitude, the value corresponding to the midpoint of the array is called the median value of the spread. The point separating the lowest quarter of determinations from the rest is called the lower quartile, and the point which separates the highest quarter of all determinations from the rest is called the upper quartile. These terms will be used occasionally in the remainder of the chapter.

[c] The correlation coefficient between the magnitude of the fluctuation for the frequency pair 14, 16 to 24 kc was found to be 0.65 and for the frequency pair 24 and 56 or 60 kc, 0.68. For a definition of the coefficient of correlation, see Section 7.2.3.

At frequencies below 10 kc, it was found both at San Diego and at the New London laboratory of CUDWR that the magnitude of the fluctuation decreases with frequency. No quantitative data are available. A particularly interesting result was obtained in a single transmission run at 5 kc at San Diego. It was found that at moderately short range the relative standard deviation of the amplitude was 47 per cent, while beyond the computed last maximum of the image interference pattern (see Chapter 5) it dropped to 10 per cent.

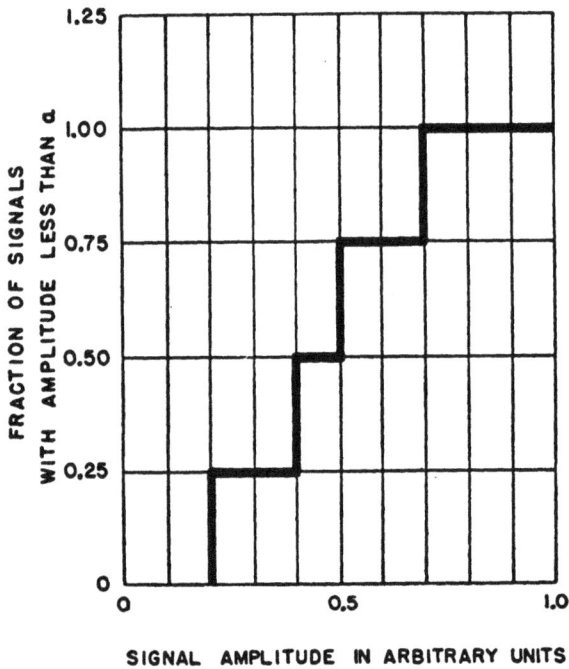

FIGURE 1. Cumulative distribution function of four signals.

Complete evidence is not available concerning the dependence of the magnitude of fluctuation on range. It is known that at distances of a few feet fluctuation of transmitted sound is negligible. From 100 yd out to very long ranges the average magnitude of the fluctuation appears to be the same at all ranges. No analyses have been made comparing the magnitude of fluctuation at different ranges under identical thermal conditions.

There have been recent experiments at UCDWR designed to determine the possible dependence of fluctuation magnitudes on the depths of the projector and receiver. In these experiments, a cable transducer was used as a projector which could be lowered to various depths up to 300 ft. When the projector depth was kept constant at 16 ft, the magnitude of the

fluctuation was found to be independent of hydrophone depth (except for MIKE patterns).[2] When both the projector and receiver are deep, it is possible to distinguish between the direct and the surface-reflected signal. Two runs were carried out with the projector at a depth of 140 ft and the hydrophone at a depth of 300 ft, and the direct and surface-reflected signals were analyzed separately. For the direct signal, the relative standard deviation of the amplitude was 9.8 per cent for the first run and 6.8 per cent for the second run; while for the surface-reflected signal the two fluctuations were 57 per cent and 51 per cent respectively. With both projector and hydrophone at a depth of 300 ft, the fluctuation of the direct signal amplitude was 6.0 per cent and of the surface-reflected signal 50.5 per cent. These results indicate that much if not most of the observed fluctuation is caused by mechanisms operating at or near the sea surface. The remaining fluctuation is probably caused at least in part by the slight directivities of the cable-mounted projector and receiver used in these experiments.

7.1.2   Probability Distributions

The probability distribution of a set is that function which tells how many members of the set lie between two specified values. Suppose, for instance, that we consider a sample of signals transmitted consecutively over the same transmission path. After these samples have been rearranged in order of increasing amplitude, it is then easy, by mere counting, to say how many of those signals have amplitudes less than a_1, how many have amplitudes less than a_2, and so on. If we divide these numbers by the total number of members of the sample, we obtain the fraction of signals with amplitudes less than a as a function of a, say $P(a)$. $P(a)$ vanishes for $a = 0$, equals unity for $a = \infty$, and increases steadily between these limits. This function is called the cumulative or integrated distribution function. As a very simple case, Figure 1 shows the integrated distribution of four signals with amplitudes of 0.2, 0.4, 0.5, and 0.7. In the theory of statistics, it is usually assumed that if the number of members of the set is increased without limit, the shape of the function $P(a)$ approaches a limiting shape in better and better approximation. It is this limiting shape to which a fundamental physical significance is ascribed. If we assume that the limiting function can be differentiated, then

$$p(a) = P'(a) = \frac{dP(a)}{da} \qquad (4)$$

is the fractional density of members of the set at a. In other words, $p(a)\,da$ is the fraction of signals with amplitudes between a and $a + da$. The function $p(a)$ is often called the differential distribution function. Analogous concepts can be formed for intensity and level distributions which have here been sketched for amplitude distributions. If we call the intensity distributions $Q(I)$ and $q(I)$ (the capital denoting again

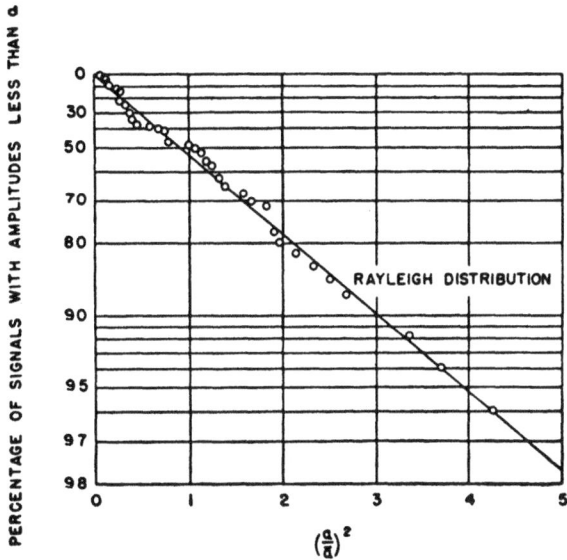

FIGURE 2. Cumulative distribution of the amplitudes of 50 signals.

the integrated and the lower-case symbol the differential distribution), and likewise the level distributions $W(L)$ and $w(L)$, then the integrated distribution functions are, of course, related to each other very simply by the equation

$$W(L) = W\left(20 \log \frac{a}{a_0}\right) = P(a) = Q(I) = Q\frac{a^2}{2\rho c}, \qquad (5)$$

since the fraction of signals with an intensity less than I is identical with the fraction of signals having an amplitude less than a if a and I are related to each other by means of equation (1). For the differential distributions it follows that

$$w(L) = \frac{dW(L)}{dL} = \frac{dP(a)}{da}\frac{da}{dL} = p(a)\frac{da}{dL},$$

since the amplitude and level are related by equation (2); thus,

$$w(L) = p(a)\frac{a}{8.69}. \qquad (6)$$

Similarly

$$q(I) = \frac{dQ(I)}{dI} = \frac{dP(a)}{da}\frac{da}{dI} = p(a)\frac{\rho c}{a} \qquad (7)$$

because the amplitude and intensity are related by equation (1).

Distribution functions can be determined experimentally, and the limiting distribution function will be approximated by the experimentally found distribution more and more closely as the size of the sample

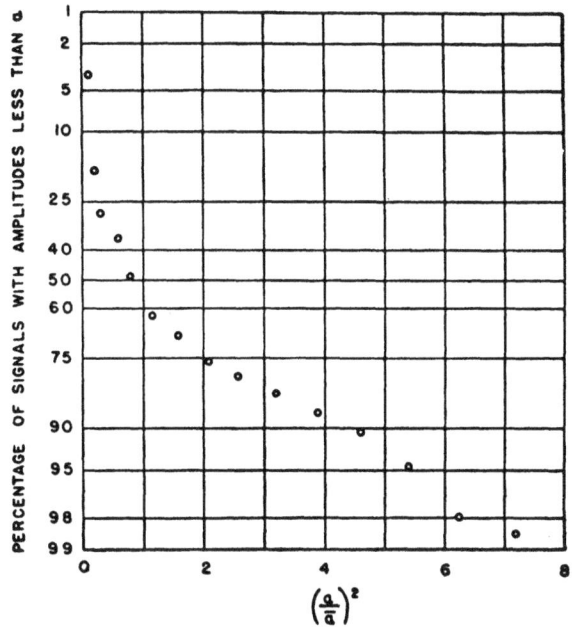

FIGURE 3. Cumulative distribution of the amplitudes of 287 signals.

is increased. On the other hand, distribution functions can also be predicted theoretically by assuming that fluctuation is caused by certain assumed mechanisms. Figures 2 and 3 show two integrated distribution functions which were obtained from actual samples. One of these samples is plotted on probability paper, on which any Gaussian distribution becomes a straight line.[d]

Two theoretically predicted distribution functions will be discussed here. The first of these is the so-called *Rayleigh distribution*. Let us consider, as an

---

[d] A Gaussian distribution is one in which the density $p(a)$ is given by the function

$$p(a) = \frac{1}{\sqrt{2\pi}\delta}e^{-(a-a_0)^2/2\delta^2}$$

A Gaussian distribution will usually result if a large number of random processes affect the value of the argument a. The two parameters a_0 and δ are the average value and the standard deviation of a respectively.

example of Rayleigh distribution, the intensity which will result if a very large number of randomly located scatterers return a single-frequency signal to an echo-ranging transducer. This situation is significant, because it is probably the most realistic model of volume reverberation. Each of these scatterers will return a weak echo, with definite amplitude but random phase. The resultant of all these individual echoes interfering with each other in a random manner will be the reverberation recorded. We shall not give a rigorous derivation of the resulting distribution function but shall sketch the argument leading to it.

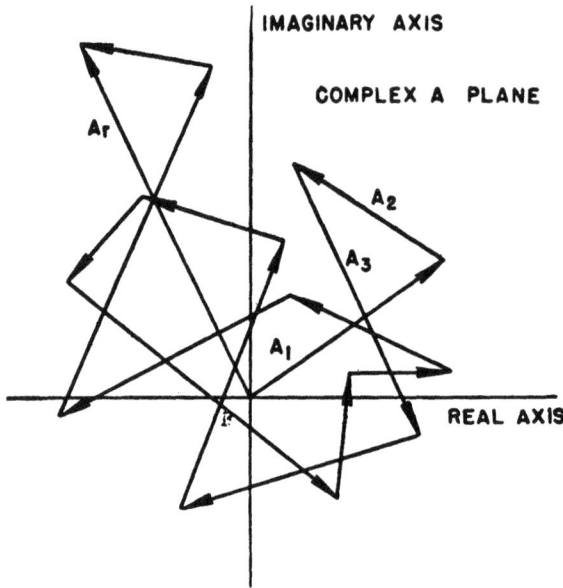

FIGURE 4. Reverberation amplitude produced by many individual echoes.

In Chapter 2, it was explained how the amplitude plus phase may be combined into a single quantity, the "complex amplitude," which is designated by A to distinguish it from the real amplitude a. Obviously, a is the absolute value of A. In an interference problem the complex amplitude of the resultant is the sum of the complex amplitudes of the interfering components. If we have a large number of interfering components with random phases, we may plot the individual complex amplitudes $A_1, A_2, \cdots, A_n$ and the resultant complex amplitude A_r in the complex A plane as illustrated in Figure 4. The direction of each individual component is completely random, while its magnitude is fixed. Obviously, the direction (phase) of the resultant A_r will be random. As for its magnitude, it is well to consider at first only its component in one direction, say the x axis. This

component of A_r will be the algebraic sum of the x components of the individual complex amplitudes $A_1, A_2, \cdots$. In the mathematical theory of probability it is shown that the distribution function for the sum of a large number of random terms is usually a Gaussian distribution, centered in this case about the zero point. In other words, the probability of the x component of A_r having a value between x and $x + dx$ is $(1/\sqrt{2\pi}\delta^2)e^{-x^2/2\delta}dx$, and the combined probability of having the x component and the y component of A_r in specified brackets of infinitesimal width is

$$p(x,y)dxdy = \frac{1}{2\pi\delta^2}e^{(-1/2\delta^2)(x^2+y^2)}dxdy. \quad (8)$$

It is convenient to introduce the polar coordinates a and θ in the complex A plane;

$$a^2 = x^2 + y^2 ,$$
$$\tan \theta = \frac{y}{x}. \quad (9)$$

Equation (8) then assumes the form

$$p(a,\theta)d\theta da = \frac{1}{2\pi\delta^2}e^{-(a^2/2\delta^2)}ad\theta da. \quad (10)$$

If we wish to disregard the dependence on the phase angle θ, we may integrate over θ from zero to 2π, with the result

$$p(a)da = \frac{1}{\delta^2}e^{-(a^2/2\delta^2)}ada = \frac{\rho c}{\delta^2}e^{-(\rho cI/\delta^2)}dI. \quad (11)$$

This last expression can be simplified by the introduction of the average intensity $\bar{I}$. By definition, this average intensity is given by the formula

$$\bar{I} = \int_{I=0}^{\infty} Iq(I)dI, \quad (12)$$

in which $q(I)$ is, according to equation (11),

$$q(I) = \frac{\rho c}{\delta^2}e^{-(\rho cI/\delta^2)}. \quad (13)$$

Carrying out the integration, we find for $\bar{I}$

$$\bar{I} = \frac{\delta^2}{\rho c}, \quad (14)$$

which means that we have for $q(I)$ and for $Q(I)$

$$q(I) = \frac{1}{\bar{I}}e^{-(I/\bar{I})} \quad (15)$$

and

$$Q(I) = 1 - e^{-(I/\bar{I})}, \quad (16)$$

respectively. Whenever a signal is the resultant of a large number of components with random phase relations, then the distribution function can be predicted except for one single parameter, and this parameter is the average intensity. The Rayleigh distribution differs in this respect from a Gaussian distribution, which contains two adjustable parameters, the average value and the standard deviation.

The other distribution function to be discussed here is the *image interference distribution*. It is calculated on the assumption that all the fluctuation of transmitted signal intensity is caused by the random interference of the sound transmitted directly and the sound reflected from the surface of the sea. The extent to which this assumption is justified will be discussed in Section 7.2.2. Let us assume that the amplitude of the direct signal alone is a_1, while the amplitude of the surface-reflected signal by itself is a_2. If the phase angle between these two components is denoted by θ, the resultant amplitude a will be given by the expression

$$a^2 = a_1^2 + a_2^2 + 2a_1a_2 \cos \theta. \qquad (17)$$

With the large-scale geometry given, the values of a_1 and a_2 will not vary significantly; but the phase difference between the two paths, θ, will change randomly because of the action of waves and because of the minute changes in position of both vessels. In other words, while a_1 and a_2 will be treated as fixed parameters (the values of which can, however, be specified only if the depths of source and receiver and their distance are known), θ will be assumed to take all values between $-\pi$ and π with equal probability. Since the value of a does not depend on the sign of θ, we shall restrict ourselves to values of θ between 0 and $+\pi$. The probability that θ exceeds a certain value θ^* equals $1 - \theta^*/\pi$, that is, the cumulative distribution function for A satisfies the equation

$$P(a) = 1 - \frac{\theta}{\pi} \qquad (18)$$

where a and θ are related to each other by means of equation (17). In other words, the fraction of signals for which the phase angle exceeds the value θ is identical with the fraction of signals with an amplitude less than a. By differentiating both sides of equation (18) with respect to a, we obtain an equation for the differential distribution, $p(a)$,

$$p(a) = -\frac{1}{\pi}\frac{d\theta}{da} \qquad (19)$$

with

$$\frac{d\theta}{da} = \frac{-2a}{\sqrt{-(a_1^2 - a_2^2)^2 + 2(a_1^2 + a_2^2)a^2 - a^4}} \qquad (20)$$

from equation (17). We find, then, for $p(a)$ the expression

$$p(a) = \frac{2}{\pi}\frac{a}{\sqrt{-(a_1^2 - a_2^2)^2 + 2(a_1^2 + a_2^2)a^2 - a^4}}$$
$$|a_1 - a_2| \leq a \leq a_1 + a_2. \qquad (21)$$

Outside the limits indicated, $p(a)$ vanishes since the amplitude cannot be greater than $a_1 + a_2$ nor less than $|a_1 - a_2|$. For $P(a)$ we find, by means of the relationship

$$P(a) = \int_{a=|a_1-a_2|}^{a} p(a)da, \qquad (22)$$

the expression

$$P(a) = \frac{1}{2} + \frac{1}{\pi} \arcsin \frac{a^2 - (a_1^2 + a_2^2)}{2a_1a_2}$$
$$= \frac{2}{\pi} \arcsin \sqrt{\frac{a^2 - (a_1 - a_2)^2}{4a_1a_2}}$$
$$|a_1 - a_2| \leq a \leq a_1 + a_2 \qquad (23)$$

by trigonometric transformations. Both expressions (21) and (23) become much simpler if it is assumed that the reflection from the sea surface is perfect, that is, if $a_1 = a_2$. We have, then,

$$p(a) = \frac{2}{\pi}\frac{1}{\sqrt{4a_1^2 - a^2}} \quad 0 \leq a \leq 2a_1, \qquad (24)$$

and

$$P(a) = \frac{2}{\pi} \arcsin \frac{a}{2a_1} \quad 0 \leq a \leq 2a_1. \qquad (25)$$

At UCDWR, some experimentally obtained cumulative distribution functions of transmitted signals have very nearly the form of equation (25), while others are approximated by a Rayleigh distribution. All the distribution functions published at UCDWR are plotted as integrated distributions. This has been done because with a limited size of the sample the differential distributions would be very difficult to compute with any degree of reliability. Integrated distributions are reasonably accurate in the central part of the curve, but the "tails" at both ends are necessarily based on very few experimental data. This is unfortunate, because the gross features of integrated distributions, and particularly the central portions, are not very sensitive to changes in the character of the distribution. By definition, all integrated distributions are functions which increase steadily from zero at $-\infty$ to 1 at $+\infty$. The central portions of two different distribution functions will be determined in their gross appearance by the location of the median point and by the slope with which the curve passes through the median point. The central portion of an integrated distribution function gives, de facto, no more information than is contained in the statement of two parameters of the distribution, such as the mean and the standard deviation.. The additional information which is represented by the shapes of the two "tails," must frequently be discounted because of the small number of signals which determine these shapes.

It is true that the mere existence of a tail at the high amplitude end permits certain conclusions although these conclusions are mostly negative. If fluctuation were brought about exclusively by the interference of two signals, each having a fixed amplitude, then there should be a cutoff at an amplitude equal to the algebraic sum of the amplitudes of the components, corresponding to constructive interference. According to equation (25), for instance, $P(a)$ should reach its maximum of 1 at an amplitude twice a_1. The fact that there is a percentage of amplitudes, however small, exceeding that value proves that interference between two signals of equal amplitude cannot be the only cause of fluctuation.

The variability of fluctuation magnitudes, which was touched on in Section 7.1.1, is reflected in the variability of the observed amplitude distributions. Even if large samples were processed consisting of thousands of signals for each sample, there is every reason to believe that their distribution functions would differ appreciably. At the present stage of the theory, the details of observed distribution functions do not lend themselves readily to theoretical interpretation.

Additional plots of observed distributions can be found in references 1 and 2, while additional theoretical distributions are discussed in a memorandum from HUSL.[4]

7.1.3 Rapidity of Fluctuation

So far, we have discussed only the typical deviations which individual signals show from the average. In this section, we shall be concerned with the time pattern of the fluctuation. Two sequences of signals could have the same relative standard deviation of amplitude, but could differ utterly in the nature of their fluctuations. For example, in one sequence the signal amplitudes might be distributed throughout the sequence in random fashion, so that a small amplitude signal is as likely to be followed by another small

amplitude signal as by a large amplitude signal; while in the other sequence, each signal amplitude might be only slightly different from the amplitude of the preceding or following signal. In the second sequence, the total spread of amplitudes can be just as large as in the first one, if a rising or falling tendency is maintained through a number of consecutive signals.

The self-correlation coefficient is the mathematical tool by means of which the time pattern of fluctuation can be expressed in quantitative form.

THE COEFFICIENT OF SELF-CORRELATION

Let us consider a sequence of signals which are received under apparently identical conditions. It is, of course, conceivable that each signal is completely unaffected by the strength of the preceding signal; this would mean that the distribution function of all those signals which follow immediately after signals of intensity I_1 are identical with the distribution function of all signals (without restriction). On the other hand, it may be found that the signals immediately following signals with the intensity I_1 have a distribution function which depends on the choice of I_1. Both of these situations seem to occur in practice. If the signals directly following those with intensity I_1 tend to have intensities not too much different from I_1, it is said that, in the sequence considered, consecutive signals have a positive correlation.

In order to obtain some numerical measure for the degree of correlation in a given sequence, we shall compare the difference between two consecutive signals with the difference between two signals picked at random. Focusing our attention on intensities, for instance (we might as well consider amplitudes or levels without changing the mathematics), we shall compare the mean squared intensity difference between two signals chosen at random with the mean squared intensity difference between a signal and its immediate predecessor. We are then concerned with the expression

$$S_1 = \overline{(I_n - I_m)^2} - \overline{(I_n - I_{n-1})^2} \equiv 2\overline{(I_n I_{n-1} - \bar{I}^2)},$$
(26)

in which n and m are to be varied independently of each other. The expression on the right-hand side can be obtained as follows. We have

$$\overline{(I_n - I_m)^2} - \overline{(I_n - I_{n-1})^2}$$
$$\equiv \overline{I_n^2} - 2\overline{I_n I_m} + \overline{I_m^2} - \overline{I_n^2} + 2\overline{I_n I_{n-1}} - \overline{I_{n-1}^2}.$$

In this expression, all the squared term averages, $\overline{I_n^2}$, $\overline{I_m^2}$, and $\overline{I_{n-1}^2}$, are equal and cancel each other. The expression $I_n I_m$ equals by definition the double sum

$$\overline{I_n I_m} \equiv \frac{1}{N^2} \sum_{n,m=1}^{N} I_n I_m,$$

which in turn can be written as the product of two single sums,

$$\frac{1}{N^2} \sum_{n,m=1}^{N} I_n I_m \equiv \frac{1}{N^2} \sum_{n=1}^{N} I_n \sum_{n=1}^{N} I_m = \bar{I}_n \bar{I}_m = \bar{I}^2.$$

If there is no correlation between consecutive signals, then the two terms $\overline{(I_n - I_m)^2}$ and $\overline{(I_n - I_{n-1})^2}$ in equation (26) are equal, and S_1 vanishes. If the correlation between consecutive signals is positive, then the rms difference between two signals picked at random will be greater than the rms difference between two consecutive signals, and S_1 will be positive. If S_1 should turn out to be negative, that would mean that the average difference square between consecutive signals exceeds the random value; the correlation between consecutive signals would then be said to be negative.

The quantity S_1 has the dimension of an intensity squared. If it is desired to obtain a measure of correlation which is dimensionless, it appears reasonable to divide S_1 through by the mean squared random difference,

$$\overline{(I_n - I_m)^2} = 2(\bar{I}^2 - \bar{I}^2). \tag{27}$$

For if the correlation were perfect (that is, if each signal pulse had the same intensity as its predecessor, a situation which can obviously not be realized exactly), this ratio would equal unity, while for negative self-correlation, the ratio can be shown never to drop below the value -1. Hence, it is customary to measure the self-correlation of consecutive signals by means of the quantity

$$\rho_1 \equiv \frac{\overline{I_n I_{n-1}} - \bar{I}^2}{\bar{I}^2 - \bar{I}^2}, \tag{28}$$

which is called the coefficient of self-correlation for unit step interval. In close analogy to this quantity, we may define the self-correlation coefficient for an interval of s steps, ρ_s, by means of the expression

$$\rho_s = \frac{\overline{I_n I_{n-s}} - \bar{I}^2}{\bar{I}^2 - \bar{I}^2}. \tag{29}$$

The averaging in the first term of the numerator is to be carried out by averaging over all values of the index n while keeping the step interval s fixed. For $s = 0$, the self-correlation coefficient equals unity, by definition. It can be shown that for all values of s, ρ_s lies between -1 (complete anticorrelation) and 1

(complete correlation). Furthermore, ρ_s is an even function of its argument s, that is:

$$\rho_s \equiv \rho_{-s}. \tag{30}$$

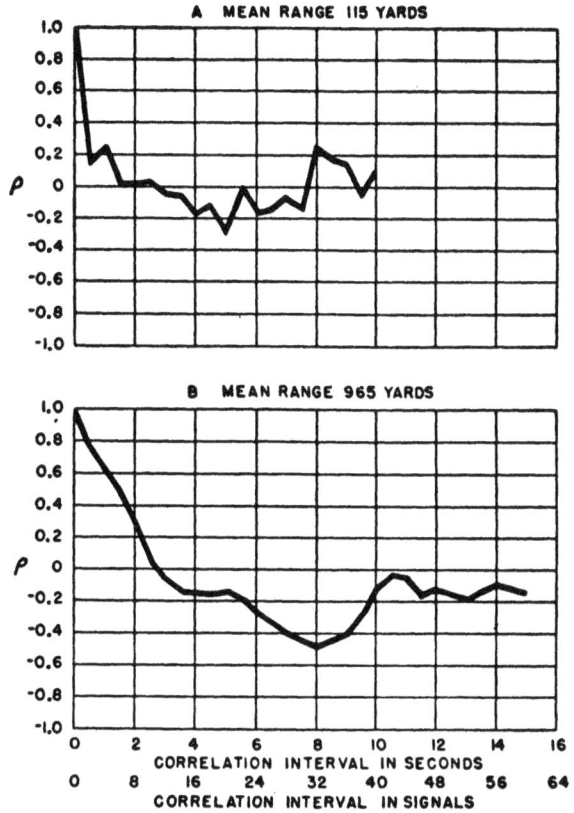

FIGURE 5. Self-correlation coefficients of two sequences of supersonic signals.

Figure 5 shows two self-correlation coefficients which were obtained at UCDWR and which were computed from samples at different ranges. In both cases, the receiving hydrophone was in the direct sound field. The abscissa represents the step interval, marked both in terms of the number of pulses s and in terms of the time in seconds. These two plots, which are typical of the others obtained, show that there is a marked positive self-correlation for step intervals of a few seconds. It appears that the longer the range, the longer is the step interval of positive correlation (that is, the slower is the fluctuation), although the evidence on that point is too scanty to be considered conclusive.

For many of these plots, the self-correlation becomes negative for some step interval before it drops down to zero. This anticorrelation has not yet been explained, although it is observed more often than not.

When the sound intensity is measured well inside a so-called shadow zone, there is usually no self-correlation for step intervals even as short as one second. As illustrated in Chapter 4, Figure 2B, the amplitude, or intensity, varies so rapidly that no correlation can be expected between consecutive signals with the usual keying intervals. However, the same figure illustrates another possibility for treating coherence in a quantitative manner. If we consider, instead of a sequence of signal pulses, the amplitude fluctuation in a continuous signal, we may define the self-correlation coefficient of the amplitude as a function of the continuously variable interval τ as follows:

$$\rho(\tau) = \frac{\frac{1}{T}\int_0^T A(t)A(t+\tau)dt - \bar{A}^2}{\overline{A^2} - \bar{A}^2} \quad (31)$$

where the interval of integration T must be large compared with the step interval τ. Figure 6 shows the self-correlation coefficient which was found during

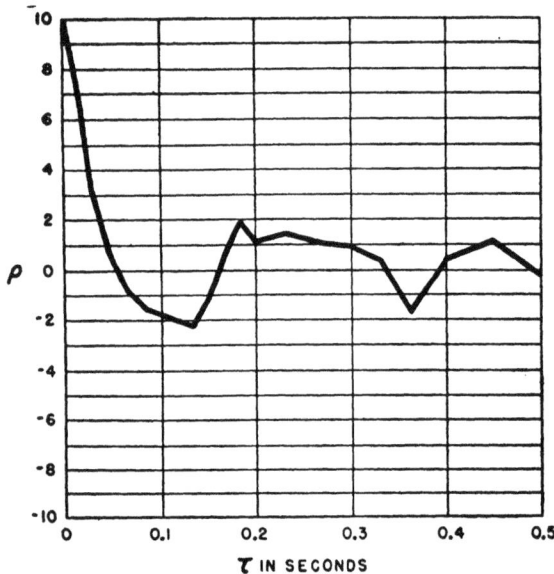

FIGURE 6. Self-correlation coefficient of a 10-sec signal received in the shadow zone.

one run for sound transmitted into the shadow zone. The appearance of this function is similar to those in Figure 5, except for the enormous change in the time scale. Figure 7 shows the self-correlation coefficient which has been predicted theoretically for the intensity of reverberation produced by a square-topped single-frequency signal of length t_0. The expression obtained by Eckart[5] for this coefficient is as follows:

$$\rho(\tau) = \begin{cases} \left(1 - \frac{|\tau|}{t_0}\right)^2 & \text{for } |\tau| \le t_0 \\ 0 & \text{for } |\tau| \ge t_0. \end{cases} \quad (32)$$

HIDDEN PERIODICITIES

In the preceding section, the coefficient of self-correlation was introduced primarily as a mathematical measure of the coherence of the transmitted signal or, in other words, as a measure of the rapidity of fluctuation. In addition, the self-correlation coef-

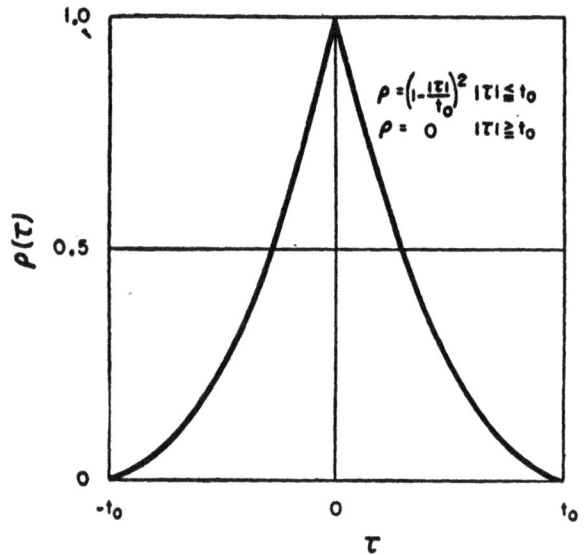

FIGURE 7. Theoretical self-correlation coefficient of the intensity of reverberation from a square-topped signal.

ficient provides a powerful tool for discovering "hidden periodicities." A hidden periodicity is essentially nothing but a tendency of the fluctuation pattern to repeat itself with a fixed period, a tendency which is modified by nonperiodic disturbances. Consider, for instance, an ordinary pendulum which is subject to random forces. This pendulum will be moved to carry out periodic swings, but the periodicity will not be strict since both amplitude and phase of its vibration are subject to random changes. But if we were to plot the motion of the pendulum for a long time (large compared with its period), we should find that the self-correlation coefficient will have a maximum (although not quite $+1$) for an interval equal to the period of the pendulum and a minimum (although not quite -1) for an interval equal to one-half the period of the pendulum. Extending the self-correlation analysis to longer intervals, we should find another minimum at 3/2 the period, again a maximum at twice the period, and so on, these consecutive minima and maxima becoming gradually more shallow until the self-correlation coefficient effectively approaches zero. In other words, hidden periodicities are revealed by the location of maxima and minima of the self-correlation vs interval curve.[e]

It was believed, at one time, that part of the observed signal fluctuation could be explained as training errors due to the roll and pitch of the transmitting vessel. Self-correlation coefficients were studied primarily with a view toward finding the fluctuation periodicities which would coincide with the known periodicity of the vessel. Although the results of these studies were at first disappointing, it is possible that future work will lead to more positive results.

7.1.4 Space Pattern of Fluctuation

In order to discover to what extent the observed fluctuation varies in space, an analysis was made of the fluctuations observed in the simultaneous outputs of two hydrophones.[2] The two hydrophones were kept either at the same or at two different recorded depths. No determination of their horizontal separation was made, but they are believed to have had a horizontal separation of between 5 and 25 ft. Thus, both hydrophones were at about the same distance from the projector, which emitted 24-kc signals. The number of samples analyzed is too small to establish any quantitative law, but indications are that the correlation between the simultaneous outputs tends to become weaker as the distance of the two hydrophones from each other or as their joint distance from the sound source is increased. However, in the majority of samples analyzed, the correlation remains significant, even at the maximum vertical separation of the two hydrophones, which was 300 ft.[f]

---

[e] This property of the self-correlation coefficient is incorporated in a mathematical theorem frequently quoted as Khintchine's theorem, which states that the coefficient of self-correlation is the Fourier transform of the (normalized) squared frequency spectrum of the time sequence considered. If, as in the case of the pendulum, the time sequence has a tendency to repeat its functional pattern, its spectrum will have a maximum at that frequency. This peak in the spectrum of the time sequence may remain undiscovered if the time sequence is inspected directly because of the changing phase relations. The squared spectrum, however, contains the absolute values of the squared frequency amplitudes, without regard for phase relationships. Consequently, its Fourier transform, the self-correlation coefficient, reveals the "hidden periodicities" more clearly than the original time sequence.

[f] With one hydrophone at 16 and the other at 300-ft depth, the correlation coefficient was 0.34 at a range of 950 yd and

7.2 CAUSES OF FLUCTUATION

It has not yet been possible to develop a theory of fluctuation which would permit the prediction of its magnitude and time rate as functions of oceanographic or other parameters. Nevertheless, it is of interest to consider the various mechanisms which have been considered responsible for fluctuation. These mechanisms may be described under three headings: roll and pitch of the vessel, interference mechanisms, and thermal microstructure of the ocean.

7.2.1 Roll and Pitch of Transmitting Vessel

Except in very calm weather, the transmitting vessel is subject to considerable roll and pitch, with the result that the bearing of the transmitter relative to the target and relative to the surface of the sea is not constant. Because of the directivity of the transmitted sound beam, a change in bearing will bring about a change in received signal intensity if the change exceeds a few degrees. This change may come about merely because the principal beam may miss the target during one phase of the roll and hit it during another phase. A more involved hypothesis considers the interference between direct and surface-reflected sound. In the presence of a slight upward refraction, the direct and surface-reflected rays to the target leave the projector at appreciably different angles. The sound received at the hydrophone is the result of interference between these two rays. If the training of the projector is changed slightly, the relative intensity of the two rays will also change, as the projector will discriminate first against one and then against the other. If the two rays are out of phase by nearly 180 degrees, the resultant change in the intensity distribution of the interference pattern may become very appreciable, even with comparatively minor changes in the relative intensity of the two component rays.

The chief argument in favor of roll and pitch as a cause of signal fluctuation was that in the earlier studies the self-correlation coefficient seemed to indicate a period of fluctuation similar to the known period of roll of the transmitting ship. Subsequent

---

0.38 at a range of 1,750 yd. The number of signals in each sample was 40. For a definition of correlation coefficient, see Section 7.2.3.

work has indicated, though, that the time rate of fluctuation is range-dependent, that is, the time rate decreases as the range increases, at least in the direct sound field. In addition, observations made when the

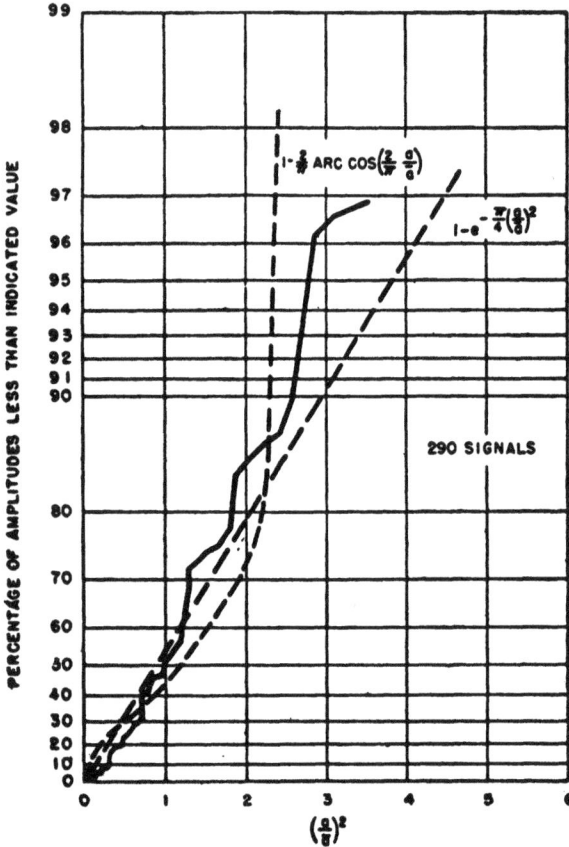

FIGURE 8. Observed cumulative distribution suggesting image interference fluctuation.

roll of the transmitting ship was less than 2 degrees show about the same fluctuation as other data. Thus, while no definitive conclusions can be drawn at the present time, it seems unlikely that roll and pitch are dominant causes of the observed fluctuation in underwater sound transmission.

7.2.2 Interference

If the sound signal received at the hydrophone were the resultant of several individual signals transmitted over two or more paths, any change in the relative phases and amplitudes would cause a change in received signal strength. If the properties of the transmission paths were subject to random variations, the resulting variability of the received signal would depend on the characteristics of these variations.

Several different models of underwater sound transmission have been studied which involve multiple paths of transmission.

Two Paths

We shall consider, first, interference between the direct and the surface-reflected signal. If the geometry of one or the other path could be changed randomly, the result should be a fluctuation in the relative amplitudes or, at least, in the relative phases of the two interfering signals.

Some evidence has been accumulated which indicates that interference between the direct signal and the surface-reflected signal is at least a major contributing cause for the observed fluctuation. Figure 8 shows a distribution of amplitudes similar to several which were observed at UCDWR. The theoretical curve corresponding to the expression (25) (with a_1 equal to $\pi/4$ times the mean amplitude $\bar{a}$) is superimposed on the observed points. The moderate agreement indicates that during the run from which this distribution of amplitudes was obtained, random interference between two equally strong signals could have been the principal cause of fluctuation. On the other hand, a large number of observed amplitude distributions fail to conform to the expression (25), suggesting that the assumed mechanism is not always the principal cause of fluctuation or, at least, that it is frequently modified by other mechanisms.

Another argument in support of the hypothesis that fluctuation is caused, in part, by interference between the direct and the surface-reflected signal is provided by the absence of regular image interference patterns for most transmission runs at supersonic frequencies. While traces of the pattern are regularly observed for transmission at low sonic frequencies (see Section 5.2.1), they are almost never found beyond 100 yd at frequencies exceeding 20 kc. This absence has usually been explained by the size of the irregularities of the sea surface. While at very low frequencies the wavelength of underwater sound is large compared with most of the water waves, this is not true for supersonic sound. Irregularities of the sea surface may well replace the theoretical image interference pattern by an image fluctuation; this conjecture is supported by the fact that fluctuation at sonic frequencies is often markedly lower in magnitude than it is at supersonic frequencies. A very striking plot of a transmission run at 20 kc showing both image effect and image fluctuation has been pub-

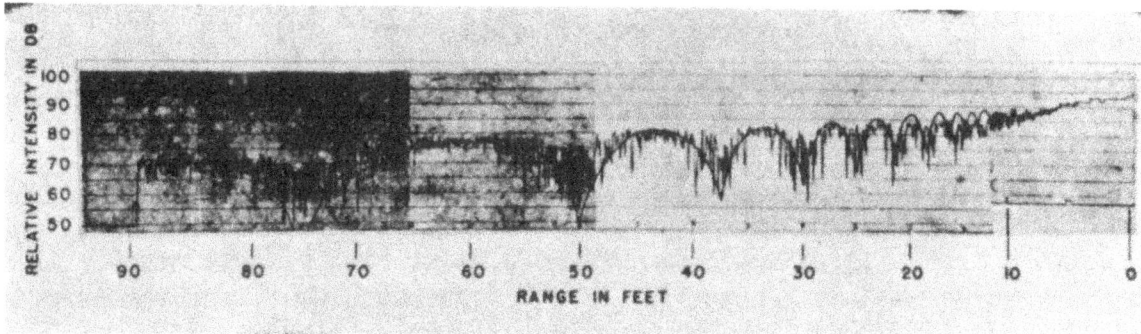

FIGURE 9. Transmission run showing image interference effect.

lished by UCDWR[6] and is reproduced in Figure 9. The signal level was recorded by a sound-level recorder. It will be noted that the ranges are very short, extending to not more than 90 ft. The recorder trace shows clearly that the amplitude of the signal fluctuation is greatest near the minima of the regular interference pattern (drawn in as a theoretical curve) where the magnitude of the resultant would be most sensitive to phase shifts of the components. This record was taken in shallow harbor water, and the surface was undoubtedly quite smooth. Otherwise, the interference pattern might not have been so noticeable.

Finally, attention is called to the experiments carried out with a deep transducer, which were mentioned in Section 7.1.1. These experiments indicate that the fluctuation is often reduced to a fraction of its usual magnitude when both the sound source and receiving hydrophone are so deep that the direct signal can be separated from the surface-reflected signal. It is true that some fluctuation remains, even when interference with the surface-reflected sound is eliminated; this small fluctuation may be the result of imperfect equipment. In all cases, however, the fluctuation of the direct signal is reduced drastically when it can be separated from the surface-reflected signal. The fluctuation of the surface-reflected signal by itself is somewhat higher than the fluctuation of the combined signal usually observed with a shallow projector.

SEVERAL PATHS

In shallow water, or even in fairly deep water with a transmitter of low directivity, sound will reach the receiving hydrophone not only over the direct path and through one surface reflection, but also through one bottom reflection, one bottom and one surface

reflection, etc. The number of possible paths is, strictly speaking, infinite, and small changes in geometry may bring about random phase shifts between the different arrivals. Nevertheless, the distribution cannot be expected to approach the Rayleigh case, because the intensity for the paths drops rapidly as the number of reflections is increased, both because there are recurring energy losses on reflection and because the high-order paths are steep-angle paths and therefore discriminated against by the transducer. Only very few of the theoretically possible paths of transmission will, therefore, be effective in contributing to the resultant signal. It has been found that in the presence of bottom-reflected sound the rapidity of fluctuation increases, as shown by the oscillograph trace reproduced in Figure 10. Unfortunately, no quantitative information is available concerning the decrease in the self-correlation coefficient due to the contribution of bottom-reflected sound.

MANY PATHS

Figure 2 shows a distribution function obtained at UCDWR, and superimposed on the experimental points is a curve representing the Rayleigh distribution. The fit is good. A model of sound transmission was set up in an attempt to explain this observed approximation to Rayleigh distribution. The model is based on the thermal microstructure which has been found to exist in the ocean[7] and which is described in Chapter 5. On the basis of ray acoustics, it was suggested that the irregular thermal structure of the ocean may give rise, simultaneously, to more than one ray path connecting the transmitter with the receiving hydrophone. It seems reasonable to assume that these paths will have different travel times and that the signals transmitted along them are, therefore, not in phase with each other. If the phase dif-

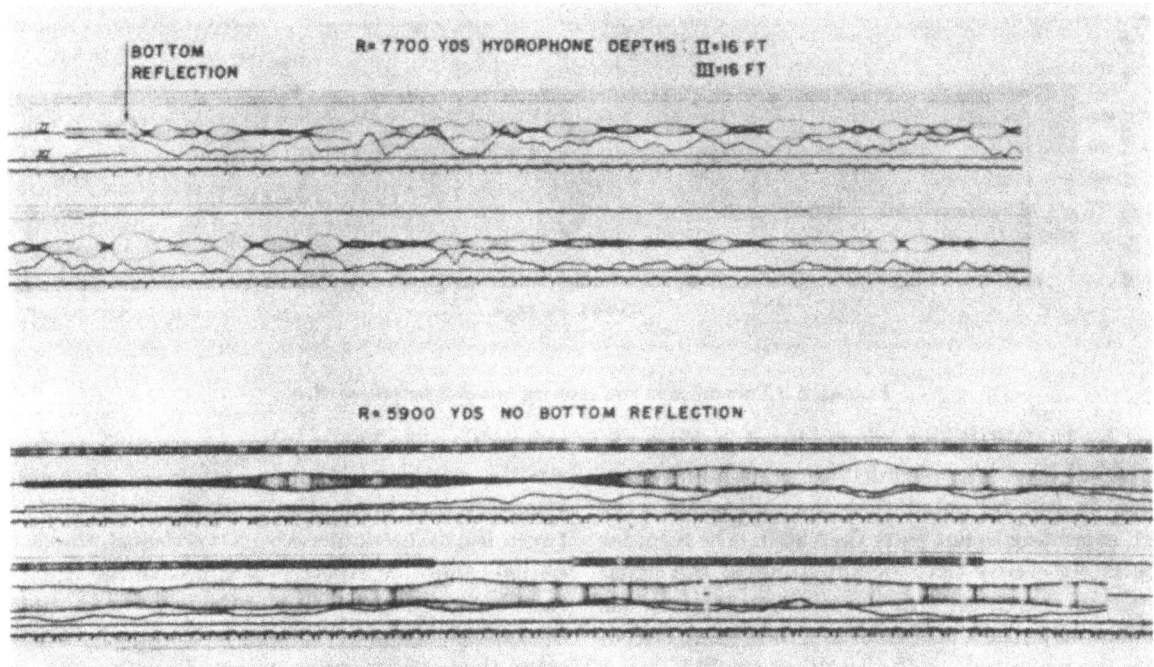

FIGURE 10.   Oscillograph trace showing the effect of bottom-reflected sound.

ferences were random and if the average number of paths were sufficiently great (at least five or six), the resulting distribution of intensities should approach the Rayleigh distribution very closely.

Against this proposed mechanism two principal objections have been raised: one, experimental, the other, theoretical. The experimental objection is simply that later research has revealed that the Rayleigh distribution is only occasionally a very good fit to observed transmitted intensities.

The theoretical objection concerns the phase shifts expected from the observed microstructure. It is possible to compute the root-mean-square difference in acoustical path length between two alternative paths through the interior of the ocean on the basis of the average parameters of the observed microstructure.[8] It turns out that the magnitude of this variation in path length is too small to produce the random phase shifts as required for Rayleigh distribution. This argument is not entirely conclusive, because the microstructure parameters reported in Section 5.1.3 were obtained on a single run and have not been confirmed by a repetition of the experiment.

No critical evaluation has as yet been made of the multiple path hypothesis on the basis of wave acoustics. The multiple path hypothesis is based, conceptually, on ray acoustics, and it may be that the ray concept has been stretched in this case be-

yond the limits of its validity. A similar analysis for a different problem has been made by CUDWR.[9]

7.2.3   Lens Action of Microstructure

If light passes through a medium with variable index of refraction, such as the turbulent heated air above a tarred road on a hot summer day, objects seen through this medium are often blurred. If sunlight falls on a white screen after having passed through such a medium, say the hot gases surrounding an open flame, the surface of the screen is mottled, with bright and dark patches alternating and changing rapidly as the thermal microstructure of the transmitting medium is varied. This random fluctuation in the brightness of the illuminated screen can be explained by means of the lens action of patches of above-average and below-average velocity of propagation of light. A similar explanation has been suggested to account for part of the fluctuation of transmitted sound intensity in the sea.[8] This role of the thermal microstructure in fluctuation is quite different from the hypothetical interference effect discussed in Section 7.2.2. While the interference effect is based on the coexistence of several distinct paths through the interior of the ocean, fluctuation because of refraction will be produced even over a single path. The theoretical treatment, not reproduced here,

leads to the result that if the refracting properties of
the microstructure were alone responsible for fluctuation, the magnitude of fluctuation should increase
with range. At moderate ranges the magnitude of the
fluctuation should be proportional to the 1.5th power
of the range. Since this hypothesis is based on ray
acoustics, the fluctuation should be independent of
the frequency, as long as the wavelength is short
enough for ray acoustics to be applicable.

The dependence of fluctuation on range predicted
by this hypothesis has not been confirmed by observations, although the variability of the magnitude
of the fluctuation is so great that a small effect might
not have been discovered. For that reason, some dependence of fluctuation on range cannot be definitely
ruled out. The theoretical formula connecting the
magnitude of the predicted fluctuation with the
parameters of the microstructure appears to lead to
a fluctuation of a magnitude much smaller than observed. There is, however, one feature which appears
to suggest that refraction by microstructure is at
least a contributing cause of the observed fluctuation.
It was pointed out that fluctuation caused by microstructure should be frequency-independent for a wide
range of frequencies. In this respect it differs from
hypotheses based on interference, since interference
leads to fluctuation which is critically dependent on
frequency. It has been possible to check the dependence of fluctuation on frequency by transmitting
signals simultaneously at two widely separated
frequencies and by noting the correlation between
their instantaneous amplitudes.[3] These trials indicated a partial but significant correlation between
the fluctuations at two widely separated frequencies.

To understand the significance of this result, it is
necessary to explain in a few words the mathematical
meaning of the term correlation coefficient. If there
are two time series, say $K_1, K_2, \cdots$, and $L_1, L_2, \cdots$,
then the correlation coefficient between them is defined (in close analogy to the self-correlation coefficient of one time series, introduced earlier in this
chapter) as the expression

$$\rho_{K,L} \equiv \frac{\overline{K_n L_n} - \overline{KL}}{\sigma_K \sigma_L}, \sigma_K^2 \equiv \overline{K^2} - \overline{K}^2. \quad (33)$$

This expression equals unity if there exists a relationship

$$L_n = \alpha K_n + \beta, \alpha > 0, n = 1, 2, 3, \cdots \quad (34)$$

or, in other words, if L is a linear function of K with

a positive slope. If $\alpha < 0$, $\rho_{K,L}$ will equal -1. If there
is some tendency of large values of L to be coupled
with large values of K, and small values of L to be
coupled with small values of K without the existence
of a rigorous linear relationship (34), then $\rho_{K,L}$ will
have a positive value less than 1; conversely, a
negative value of $\rho_{K,L}$ (greater than -1) will signify
a coupling of large values of L with small values of
K and vice versa. If $\rho_{K,L}$ vanishes, then the deviations
of individual K values from $\overline{K}$ are statistically independent of (uncorrelated with) the deviations of the
corresponding L values from $\overline{L}$.

It was found that the correlation coefficient between simultaneous signals at two different frequencies varied from 0 to 0.75 with an average of 0.3.
In other words, while there was some tendency for
strong 24-kc signals to be coupled with strong 56-kc
signals, the simultaneous signal amplitudes at these
two frequencies were far from proportional to each
other. The same statement holds for each of the three
other frequency pairs at which experiments were
performed. It must be concluded that the observed
fluctuation is caused by a combination of mechanisms, of which some operate independently of the
signal frequency (refraction and roll and pitch),
while others depend on the transmitted frequency
(interference).

7.2.4   Summary

The experiments carried out with a deep sound
source and a deep hydrophone indicate that most of
the observed fluctuation disappears if the whole
transmission path is more than 100 ft below the surface. They also show that the surface-reflected signal
by itself (without interference from another path)
fluctuates more strongly than the composite signal
observed in shallow transmission, at least when the
incidence at the surface is not glancing. Unfortunately, these findings are not helpful in a choice between the various mechanisms which have been
considered.

If roll and pitch contributed significantly to fluctuation, its effect on a cable-supported transducer
would be very much less noticeable than the effect
on a transducer rigidly connected with the hull of the
ship; but there are not yet enough data with a shal-
low-cable transducer to permit any conclusions.
Image interference fluctuation will cease to operate
when the surface-reflected sound can be separated
from the direct signal. Microstructure will probably

be very appreciably reduced below the region of strong thermal gradients.

Nevertheless, the composite evidence indicates that image interference is probably the most important single factor contributing to the observed fluctuation. There may also be interference between the beamlets into which the irregular surface of the sea breaks up the incident coherent beam. In the deep transducer experiments, the surface-reflected signal showed a very high degree of fluctuation, but at glancing incidence the path differences may not be large enough to bring about random interference.

If all interference effects could be eliminated, total fluctuation would probably be cut in half.[g]

The remaining fluctuation is essentially frequency-independent. It may be due in part to pitch and roll and in part to the lens effect of microstructure. Elimination of either of these effects is possible in principle, but would require very elaborate additions to present sound gear.

---

[g] Fluctuation by interference can be effectively eliminated by transmitting supersonic sound in a broad frequency band. The width of that band should probably exceed 5 kc in order to obtain maximum benefits.

Chapter 8

EXPLOSIONS AS SOURCES OF SOUND

8.1 INTRODUCTION

Explosive sound differs from sinusoidal sound both in the intensity which can be achieved with it and in the fact that it consists of one or more pulses of extremely short duration. These characteristics have prompted many suggestions for the employment of explosive sound in communication and echo ranging, few of which, however, have so far been utilized in practice. The survey given in this chapter and the next of what is known about explosive sound is partly designed to facilitate an understanding of the possibilities and limitations of explosive sound in such applications. The study of explosive sound can be useful in another way, however, in that it can supply valuable additions to our information about the nature of the sea and its bottom, and about the causes of many of the phenomena observed in sound transmission. The possibilities of explosive sound as a research tool have accordingly been kept in mind in the selection of material for these chapters.

To understand the complex phenomena which accompany the propagation of explosive sound in the sea one ought to begin by finding out as precisely as possible just what the explosive disturbance is like, originally, before it has been propagated to any appreciable distance. Fortunately, much has been learned about explosions and the pressure disturbances which they create in the water near them. A detailed survey of what is known about underwater explosions would require a volume in itself; however, an effort will be made in this chapter to summarize briefly those parts of our knowledge of underwater explosions which have a bearing on the use of explosions as sources of sound. In this chapter, therefore, we shall be concerned with the disturbance at comparatively short ranges from the explosion, where its characteristics are presumably little affected by the departures of sea water from the concept of a pure homogeneous fluid. Most of the information in this field has been obtained in the course of experiments directed toward the elucidation of the damag-

ing effects due to explosions. With this information as background, we shall be able, in Chapter 9, to discuss the propagation of explosive sound through sizable distances in the sea where departure of the medium from homogeneity, effects of the bottom, and other factors are important.

8.2 SEQUENCE OF EVENTS IN UNDERWATER EXPLOSIONS

An explosion is a process by which, in an extremely short space of time, a quantity of "explosive" material is converted into gas at very high temperature and pressure. This conversion is due to a chemical reaction which converts the explosive material from a thermodynamically unstable state to a more stable one with the evolution of a great amount of heat. This reaction, when initiated at one point of a mass of explosive, propagates itself rapidly until all the mass is involved. The propagation may take place in either of two ways, called respectively *burning* and *detonation*. In burning, the contact of the hot gaseous products of the reaction with the untransformed portion causes a reaction to take place at the surface of the latter, the rate of transformation being slow enough so that the boundary between transformed and untransformed material advances with a speed slower than the speed with which the pressure generated by the reaction is propagated through the mass. In detonation, on the other hand, the reaction takes place so rapidly that it can keep up with the pressure wave, which in this case is known as a detonation wave. These two processes, which will be discussed more fully later, permit explosive materials to be divided into two rather well-defined classes: explosives which detonate, commonly called *high explosives*, and explosives which merely burn, for which we shall use the term *propellants* since the most important explosives of this type are used to propel projectiles from guns, or as rocket fuels. A given quantity of high explosive will radiate considerably more acoustic energy when it is set off than will a like amount of a propellant; for this reason nearly all the material to be presented in these chapters concerns sound generated by high explosives.

Let us therefore consider what happens when a quantity of high explosive is set off under water. First, detonation is initiated at some point of the explosive; this may be done, for example, by using

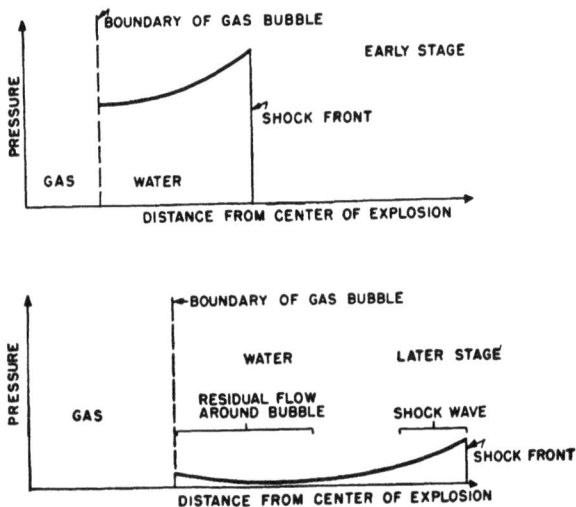

FIGURE 1. Pressure distribution in the water at two instants of time following detonation of a charge of high explosive.

a detonating cap containing a small quantity (about a gram) of an especially sensitive explosive traversed by a fine wire which can be suddenly heated to incandescence by a current of electricity. From the point of initiation a detonation wave spreads out in all directions through the explosive with a velocity of several thousand meters a second. In front of the detonation wave the material is in exactly the same state as before the explosion, while behind the wave front it is gas at a pressure of ten to a hundred thousand atmospheres and a temperature of several thousand degrees centigrade. When the detonation front reaches the boundary between the explosive and the water, this pressure is transmitted to the water, and a wave of intense compression starts outward through the water. If the pressure were not so enormous, this wave would be an ordinary sound wave. However, because of its great amplitude, the wave differs in a number of ways from ordinary sound waves and is called instead a *shock wave*; it bears somewhat the same relation to sound waves that a large breaker on the beach bears to an ordinary water wave. A shock wave is characterized by an almost discontinuous rise of pressure to a high value at the front of the ad-

vancing wave and travels with a speed greater than the normal velocity of sound. The reasons for these characteristics will be discussed in the following sections.

The pressure in the shock wave from an explosion dies off fairly rapidly behind the shock front, and by the time the shock wave has advanced to a distance of the order of ten times the radius of the original mass of explosive it has become a fairly well localized disturbance, advancing outward and practically independent of the motion of the water and gas in regions nearer to the center. Figure 1 shows schematically how the pressure may be expected to vary with distance from the original site of the explosion at two successive instants of time as this state of affairs is becoming established.

Although in these later stages it no longer affects the main part of the shock wave, the motion of the gas-filled cavity and the water immediately around it is by no means unimportant. At times such as those shown in Figure 1 the pressure in the gas cavity, hereafter called the *bubble*, is still quite high, and the water around it is rushing outward with a very high velocity. Because of the inertia of the water, this outward motion continues long after the force of gas pressure, which initiated it, has become negligible. As the gas bubble expands, the pressure in it drops, and eventually becomes far less than the normal hydrostatic pressure of the surrounding water. This excess of pressure on the outside finally brings the expansion to a halt, but not until the bubble has reached a radius which may be several dozen times the initial radius of the explosive. A contraction now sets in, and again, because of the inertia of the water, the bubble overshoots its equilibrium radius and the contraction does not stop until a very high pressure has been built up in the gas bubble. Several cycles of this expansion and contraction may take place before the oscillation dies out. The period of these bubble oscillations is of the order of a thousand times the duration of the pressure which the shock wave exerts as it passes a particular point in the water; it is usually of the order of 1/30 to 1 sec, depending on the size of the charge and its depth. At each contraction a new pressure wave is sent out into the water; these so-called "secondary pulses" are many times less intense than the shock wave, but as they have a duration many times longer, they may contain a greater amount of impulse, and a comparable though smaller amount of energy. A quantitative theory of this phenomenon will be sketched in Section 8.6.

Most of the commonly used high explosives are remarkably similar to one another in the amount of energy they release per unit mass, and in the relative amounts of energy which go into the shock wave and the oscillations of the bubble. Of the total work done by the gas on the water in its initial expansion, about 40 to 50 per cent remains as kinetic and potential energy in the oscillations of the gas bubble and surrounding water, part of this energy being ultimately converted into heat by dissipative actions in the neighborhood of the bubble and part being radiated as acoustic energy in the secondary pulses. The remaining 50 or 60 per cent of the original energy is at any stage divided between energy present in the shock wave and energy which has been converted into heat by dissipative processes occurring at the shock front. Dissipation of the latter kind is especially rapid in the early stages so that by the time the shock wave has advanced a distance of the order of ten or twenty times the original radius of the charge, about a quarter of the original energy has been dissipated into heat, and the other quarter continues to be radiated outward in the shock wave. From this time on, the dissipation is much slower, although not negligible.

In the preceding discussion, the phenomena have been described without reference to the size of the charge of explosive which is used. This is possible because explosions of all sizes are similar. If the range is not too great, the intensity and form of the shock wave, and many of the features of the bubble oscillation, can be predicted exactly for one quantity of explosive if they are known for another quantity of the same explosive substance. To give a precise statement of the rule by which this prediction can be made: suppose two experiments are carried out with the same explosive material, the shape of the charge and the position of the detonation being the same in both cases, but the linear dimensions of the second charge being β times as great as those of the first. Then the rule states that if the pressure is p and the velocity of the water is u at a distance r from the first charge, at time t after the detonation starts, the same pressure p and velocity u will obtain at a distance βr in the corresponding direction from the second charge, at a time βt after the detonation starts. This rule can be applied to the shock wave provided that the range from the explosion is sufficiently short so that the dissipative or dispersive effects responsible for slowing the time of rise to maximum pressure (see Section 9.2.1) have not had an appreciable effect on

the pressure-time curve. The applicability of the rule to the later oscillation of the bubble is more limited and will be discussed in Section 8.6. The physical basis of the rule will be taken up in Section 8.4.3.

The phenomena which occur when a propellant charge is set off under water are similar to those just described for high explosives, with the important exception that because of the comparatively slow burning of the explosive, the pressure transmitted to the water builds up gradually over a period of time, and does not usually create a steep-fronted shock wave. Thus instead of the sort of pressure-distance graph shown in Figure 1, a propellant would give a graph more like Figure 2. The division of the disturbance

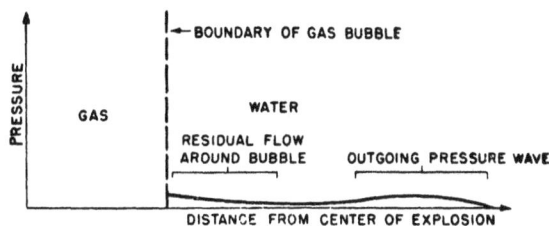

FIGURE 2. Pressure distribution in the water a short time after ignition of a propellant charge.

into an outgoing pressure pulse and a residual bubble oscillation can usually still be made, but the proportion of the total energy which appears in the pressure pulse from a propellant is much smaller than that which appears in the shock wave from a high explosive, and the maximum of the pressure is very much smaller.[1] The exact characteristics of the pressure pulse depend upon the rate of burning of the charge, which varies greatly depending on the type of propellant and the grain size.

The following sections discuss in greater detail the previously mentioned features of the disturbance due to a high explosive.

8.3 SHOCK FRONTS

The steep-fronted shock waves mentioned in Section 8.2 represent a form toward which all very intense pressure disturbances tend to develop. In this section and the following section we shall show why this is true and shall show that many of the characteristics of shock waves, such as the velocity of propagation of the shock front and the rate of dissipation of energy into heat, can be expressed as functions of the pressure jump, that is, the amount by which the pressure immediately behind the shock front exceeds the pressure in the undisturbed water in front of it. Characteristics of shock waves from explosions which depend on other factors beside the pressure jump will be treated in Section 8.5.

It is a familiar fact that the laws of acoustics are a limiting case of the laws of hydrodynamics for a compressible fluid and are valid only in the limit of very small amplitudes of pressure and velocity. According to these laws of acoustics, all pressure disturbances are transmitted as waves with velocity $c = (dp/d\rho)^{\frac{1}{2}}$, the derivative being understood, in the case of all ordinary fluids, to relate to the change of pressure p with density ρ in an adiabatic change. For the special case of a plane wave traveling in the positive x direction, the pressure in the acoustic approximation is simply a function of $(x - ct)$, where t is the time; any such wave is thus propagated forward with velocity c without change of shape. For a disturbance of large amplitude this is no longer true. The shape of the wave will, in general, change as it progresses. The way in which the changes of shape take place can be calculated by a method of reasoning due to Riemann, the mathematical form of which will be given later in Section 8.4.1. Here we shall be content to give a simple qualitative explanation of Riemann's ideas.

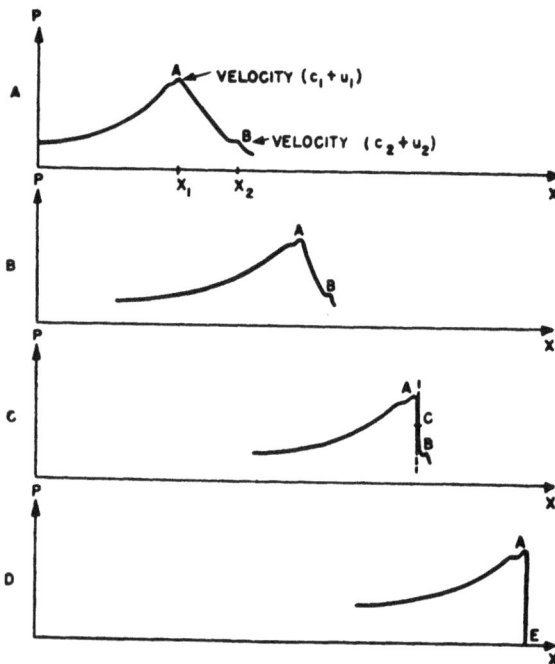

FIGURE 3. Development of a shock wave in one dimension.

Suppose we have a plane wave of the form shown in Figure 3A advancing in the positive x direction. Let the particle velocity at $x = x_1$ be u_1, and let that at $x = x_2$ be u_2. If we use acoustic theory as a first approximation, we have $u_1 \approx p_1/c$, $u_2 \approx p_2/c$, where p_1 and p_2 are the pressures at x_1 and x_2 respectively.[a] Now imagine an observer moving with the velocity u_1, so that to this observer the fluid at the point x_1 is instantaneously at rest. Any small additional disturbance, such as the bump A, will seem to this observer to be propagated forward with the velocity $c_1 = (dp/d\rho)^{\frac{1}{2}}_{p=P_1}$. Relative to the original system of reference, therefore, this bump will advance with velocity $(c_1 + u_1)$. Similarly the bump B will advance with velocity $(c_2 + u_2)$. Now the fact that $p_1 > p_2$ implies, as shown above for the acoustic approximation, that $u_1 > u_2$; moreover, the equation of state of all ordinary fluids is such that this fact also implies $c_1 > c_2$. Thus $(c_1 + u_1) > (c_2 + u_2)$ and bump A advances faster than B, so that after a short interval of time the pressure pulse will have somewhat the form shown in Figure 3B. This illustrates Riemann's result, that in an advancing wave the parts of higher amplitude move faster than those of lower amplitude. If continued long enough, this difference in velocity would cause the high pressure point A to overtake the low pressure B; however, before this occurs, the curve of p against x will acquire a vertical, or nearly vertical, tangent at some intermediate point C, as shown in Figure 3C. In the neighborhood of this point the pressure gradient and velocity gradient will be very large, and it will no longer be permissible to neglect the effects of viscosity and heat conduction, which have been omitted from Riemann's equations. It turns out that viscosity and heat conduction, by converting mechanical energy into heat, slow up the rate of advance of the high pressure regions, the amount of this slowing up becoming greater the larger the gradient of pressure and velocity. Thus a stage will eventually be reached, as in Figure 3D, where the steepness of the rise from A to E is just sufficient to keep A from overtaking E. This state of affairs constitutes a shock wave. Since in practice this limiting value of the time of rise is extremely short, the shock wave begins with an almost instantaneous rise of pressure.

The importance of this phenomenon of Riemann's in the understanding of explosive sound is not merely that it explains the origin of shock waves, which could simply be taken for granted, but that it also explains how the characteristics of the disturbance behind a shock front change with the time. This variation will be discussed in Section 8.5.

[a] Throughout this chapter the symbol $\approx$ will be used to denote "is approximately equal to."

From what has been said previously it would appear that any mathematical theory of the propagation of a shock front would have to be based on hydrodynamical equations of sufficiently complicated form to include the effects of viscosity and heat conduction. Fortunately, however, a very simple analysis based on the laws of conservation of mass, momentum, and energy suffices to determine the relation between pressure, particle velocity, temperature, and similar factors, just behind the shock front, and the velocity of propagation of the front. A detailed application of the laws of viscosity and heat conduction turns out to be necessary only if we are interested in phenomena in the shock front itself, that is, phenomena taking place in the very thin layer of water within which the abrupt rise of pressure takes place.

The simple analysis just mentioned, due to Rankine and Hugoniot, will be described at length in Section 8.4.2. It will be shown there, among other things, that in ordinary fluids a negative shock is impossible, in other words, that a discontinuity in pressure and density can only be propagated toward the region of lower pressure, and that the velocity V with which a shock front advances, relative to the undisturbed fluid in front of it, is greater than the velocity of sound [the value of $(dp/d\rho)^{\frac{1}{2}}$] in the undisturbed fluid ahead of the shock front.

The thickness of the region within which the pressure rises from p_0 to p_1 is of course determined by the dissipative phenomena, namely, viscosity, heat conduction, and any other sources of dissipation, such as bubbles, which may be present in sea water. A precise mathematical treatment of these factors would be difficult, but order-of-magnitude considerations to be given in Section 8.4.4 indicate that close to the explosion this thickness should be very small; at a distance where the pressure jump is 100 atmospheres, for example, as is the case at a range of about 1 ft from a Number 8 detonating cap, the shock front should be less than 0.001 cm thick, perhaps much less. In this region the thickness of the shock front is a function only of the magnitude of the pressure jump and decreases as the latter increases. It might be supposed that the time of rise would be connected with the time required for the detonation wave to travel across the explosive charge, but this is not the case unless the charge is extremely elongated, since the Riemann "overtaking effect" will succeed in making the shock front vertical before the shock wave has advanced an appreciable distance from the charge.

With increasing distance from the charge the pressure amplitude becomes small, and eventually the overtaking effect will become negligible in comparison with the dissipative processes which tend to smooth out the abrupt rise in pressure. In homogeneous sea water, however, the thickness of a shock front should remain quite small until it has traveled a considerable distance. Thus, for example, it can be shown that the thickness of the shock front at 50 yd from the explosion of a detonating cap should probably be only a fraction of a centimeter, corresponding to a few microseconds or less for the time of rise of the pressure at a given point.

Experimental information on the thickness of shock fronts, or equivalently the time of rise of the pressure at a given point, must be treated with caution. The measured value of time of rise can easily be completely falsified by inadequate frequency response characteristics of the hydrophone and recording equipment; in particular, the finite size of the hydrophone seems to have rather more effect on the time of rise than one would at first suppose. A very careful series of experiments has been conducted at NRL[2] on shock waves from detonating caps containing about half a gram of high explosive, at ranges from 1 to 30 ft. In these experiments the time taken for the pressure in the shock wave to rise to its maximum value was measured under the best conditions as about 0.3 microsecond (μsec) at all ranges, and since this was about the same as the estimated resolving time of the apparatus used, these experiments support the theoretical expectation of the preceding paragraph that the time of rise should be exceedingly minute. Other experiments supporting this expectation have been made at the Underwater Explosives Research Laboratory at Woods Hole,[3] using ½-lb charges and ranges of the order of 10 ft; however, the resolving time of the apparatus for these experiments was only about 4 μsec. Measurements made at ranges of the order of hundreds of feet, however, seem to show quite an appreciable time of rise;[4-6] this effect, which is probably instrumental but may possibly be related in some way to oceanographic conditions, will be discussed at length in Section 9.2.1.

8.4 THEORY OF NONLINEAR PRESSURE WAVES AND SHOCK FRONTS

The four following sections will be devoted to a mathematical discussion of some of the topics which have been treated briefly in the preceding sections. Although this material is essential to a complete understanding of explosive phenomena, the continuity of the chapter will not be impaired by omission of these sections provided the reader is willing to accept on faith those results from them which have already been quoted.

A more complete account of the theory of waves of finite amplitude and shock waves is given in a report issued by the Applied Mathematics Panel,[7] and in textbooks on hydrodynamics.[8]

8.4.1 Riemann's Theory of Waves of Finite Amplitude

In Sections 2.1.2 and 2.1.3 the equations of motion of a perfect fluid were derived on the assumption that the amplitude of the disturbance was small, so that the acceleration of a particle of the fluid could be approximated by the partial derivative of the velocity with respect to time. Since we wish in this section to treat disturbances for which this approximation will not be valid, we must start from a more exact form of the equations of motion. As before, let x,y,z be three rectangular coordinates in space, t the time, and u_x, u_y, u_z the three components of particle velocity in the fluid. As shown in Section 2.1.2, the x component of the acceleration of a particle of the fluid is

$$\frac{\partial u_x}{\partial t} = \frac{\partial u_x}{\partial t} + u_x\frac{\partial u_x}{\partial x} + u_y\frac{\partial u_x}{\partial y} + u_z\frac{\partial u_x}{\partial z}, \quad (1)$$

and this, when multiplied by the density ρ, must equal f_x, the x component of force per unit volume. According to Section 2.1.3, f_x is related to the distribution of pressure p by

$$f_x = -\frac{\partial p}{\partial x}.$$

Thus the exact equation of motion for the x component of velocity is

$$\frac{\partial u_x}{\partial t} + u_x\frac{\partial u_x}{\partial x} + u_y\frac{\partial u_x}{\partial y} + u_z\frac{\partial u_x}{\partial z} = -\frac{1}{\rho}\frac{\partial p}{\partial x}, \quad (2)$$

and applying the same reasoning to the y and z components, we get the remaining equations of motion

$$\frac{\partial u_y}{\partial t} + u_x\frac{\partial u_y}{\partial x} + u_y\frac{\partial u_y}{\partial y} + u_z\frac{\partial u_y}{\partial z} = -\frac{1}{\rho}\frac{\partial p}{\partial y}, \quad (3)$$

$$\frac{\partial u_z}{\partial t} + u_x\frac{\partial u_z}{\partial x} + u_y\frac{\partial u_z}{\partial y} + u_z\frac{\partial u_z}{\partial z} = -\frac{1}{\rho}\frac{\partial p}{\partial z}. \quad (4)$$

Let us now consider a disturbance in which the pressure and velocity are functions only of x and t, inde-

pendent of y and z. For such a disturbance the equation of continuity [equation (2) of Section 2.1.1] becomes

$$\frac{\partial \rho}{\partial t} + \frac{\partial(\rho u_x)}{\partial x} = 0, \quad (5)$$

and equation (2) becomes

$$\frac{\partial u_x}{\partial t} + u_x\frac{\partial u_x}{\partial x} = -\frac{1}{\rho}\frac{\partial p}{\partial x}. \quad (6)$$

Riemann discovered that the two equations (5) and (6) could be put into a symmetrical and useful form by introducing the variable

$$\psi = \int_{\rho_0}^{\rho}\left(\frac{dp}{d\rho}\right)^{\frac{1}{2}}\frac{d\rho}{\rho}, \quad (7)$$

where ρ_0 is a reference value of the density which is most conveniently chosen equal to the density of the undisturbed fluid. By using this equation and the abbreviation $c = (dp/d\rho)^{\frac{1}{2}}$, equation (5) becomes

$$0 = \frac{d\rho}{d\psi}\frac{\partial\psi}{\partial t} + u_x\frac{d\rho}{d\psi}\frac{\partial\psi}{\partial x} + \rho\frac{\partial u_x}{\partial x},$$

or

$$\frac{\partial\psi}{\partial t} + u_x\frac{\partial\psi}{\partial x} = -\rho\frac{d\psi}{d\rho}\frac{\partial u_x}{\partial x} = -c\frac{\partial u_x}{\partial x} \quad (8)$$

and equation (6) becomes

$$\frac{\partial u_x}{\partial t} + u_x\frac{\partial u_x}{\partial x} = -\frac{1}{\rho}\frac{dp}{d\rho}\frac{d\rho}{d\psi}\frac{\partial\psi}{\partial x}$$

$$= -c\frac{\partial\psi}{\partial x}. \quad (9)$$

Adding equations (8) and (9) gives

$$\frac{\partial(\psi + u_x)}{\partial t} + (u_x + c)\frac{\partial(\psi + u_x)}{\partial x} = 0, \quad (10)$$

and subtracting equation (9) from equation (8) gives similarly

$$\frac{\partial(\psi - u_x)}{\partial t} + (u_x - c)\frac{\partial(\psi - u_x)}{\partial x} = 0. \quad (11)$$

Equation (10) states that the quantity $(\psi + u_x)$ is propagated in the x direction with velocity $(u_x + c)$ while equation (11) states that the quantity $(\psi - u_x)$ is propagated with velocity $(u_x - c)$, that is, since ordinarily $c > u_x$, is propagated in the negative x direction with velocity $(c - u_x)$. These are Riemann's results.

The significance of these equations can be seen by considering a disturbance which is initially confined within the range $a < x < b$. For such a disturbance both $(\psi + u_x)$ and $(\psi - u_x)$ are initially zero below $x = a$ and above $x = b$. The region in which $(\psi + u_x)$

is different from zero will advance toward increasing x while the region in which $(\psi - u_x)$ differs from zero will recede in the opposite direction. Eventually these two regions will separate and leave between them an interval within which both $(\psi + u_x)$ and $(\psi - u_x)$ are zero so that the fluid is at rest at its normal density ρ_0. The original disturbance has thus been split up into two progressive waves traveling in opposite directions. In the wave which travels in the positive x direction $(\psi + u_x)$ is finite while $(\psi - u_x)$ is zero; in this wave, therefore, $\psi = u_x$ and both the density and the particle velocity are propagated forward with the velocity $(u_x + c) = (\psi + c)$. Since for all ordinary fluids both ψ and c increase with increasing pressure, this velocity of propagation will be greater the greater the pressure, and any disturbance will ultimately develop as shown in Figure 3. After the wave front becomes very steep, of course, the basic equation (2) or (6) is no longer valid and must be modified to take account of the effects of viscosity and heat conduction, which are negligible for waves of more gradual profile.

Most practical applications, such as pressure waves produced by explosions, involve spherical waves diverging from a source rather than plane waves of the type we have been discussing. It can be shown, however, that spherical waves have properties very similar to those just established for plane waves in that the high-pressure regions travel faster than the low-pressure regions and tend to overtake them. This overtaking effect becomes slower and slower as the wave advances farther from its source because of the decreasing amplitude of the disturbance due to spherical spreading. For this reason, a pressure pulse radiating from a small source has to be extremely intense if it is to develop a shock front by means of the overtaking effect; rough calculations[9] have shown that pressure pulses of the amplitudes ordinarily obtained from echo-ranging transducers will be only very slightly distorted by the overtaking effect and will not develop shock fronts.[3] However, in the case of transmissions at supersonic frequencies this slight distortion of the wave profile might be detectable by a receiver tuned to twice the original frequency.

8.4.2   The Rankine-Hugoniot Theory of Shock Fronts

We have seen in the preceding sections and Figure 3 how any pressure wave of sufficiently large amplitude ultimately develops an extremely steep shock front

within which the motion of the fluid will be strongly influenced by factors such as viscosity and heat conduction which do not appear in the equations of motion of perfect fluids. Certain characteristics of such a shock front can be predicted only by a theory which takes account of these additional factors explicitly; one such characteristic, which will be discussed in Section 8.4.4, is the thickness of the region within which the abrupt rise in pressure takes place. Fortunately, however, it was discovered by Rankine and Hugoniot in the last century that many valuable conclusions could be drawn merely by applying the laws of conservation of mass, momentum, and energy to the motion of the fluid, without bothering at all about the details of phenomena in the shock front.

To show how this can be done, let us consider the mass of fluid contained in a flat cylinder having unit cross-sectional area and having its end planes parallel to, and respectively ahead of and behind, the shock front. A side view of such a cylinder is shown at a particular time t by the full line $ABCD$ in Figure 4,

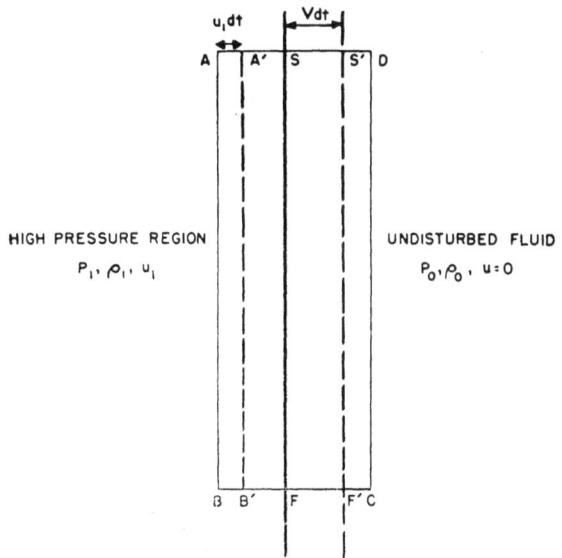

FIGURE 4.   Cylinder of fluid traversed by a shock front.

AB and CD being projections of the end faces. At a time dt later, the fluid which was originally in $ABCD$ will occupy the cylinder $A'B'CD$, shown with a dotted boundary. Now if the pressure variation in the shock wave is similar to that shown in Figure 3D, the pressure, density, and other factors will change very rapidly in a very thin region near the plane SF, but will change much more gradually everywhere else. If this is the case, we can assume the thickness

AD of the cylinder to be very small but still very much greater than the thickness of the shock front, that is, of the region within which the rapid rise of pressure takes place. It will then be legitimate to treat the pressure, density, and velocity as having constant values p_1, ρ_1, u_1, over that part *ABFS* of the cylinder which lies behind the shock front. Ahead of the shock front, of course, the fluid is at rest with undisturbed values p_0, ρ_0, of the pressure and density. Remembering that the cylinder has unit cross section, the mass of fluid in it may be written as

$$\rho_1 \overline{AS} + \rho_0 \overline{SD}.$$

If we let the boundary of the cylinder move with the water, this cannot change in the course of time. Now, after the interval dt, shown in the figure, the mass is

$$\rho_1 \overline{A'S'} + \rho_0 \overline{S'D}$$

and since

$$\overline{SD} - \overline{S'D} = V dt$$

and

$$\overline{A'S'} - \overline{AS} = (V - u_1) dt$$

where V is the speed with which the shock front advances, the two expressions given for the mass can be equal only if

$$\rho_1 (V - u_1) = \rho_0 V. \qquad (12)$$

Similar equations can be derived by applying, instead of the law of conservation of mass, the laws of conservation of momentum or energy. Thus, the change in the momentum of the cylinder of Figure 4 in the time dt is

$$\rho_1 u_1 (V - u_1) dt$$

and this must be equated to dt times the force $(p_1 - p_0)$ acting on the cylinder which gives

$$\rho_1 u_1 (V - u_1) = p_1 - p_0. \qquad (13)$$

For the energy equation, if we denote the internal energy per unit mass of the fluid in front of and behind the shock front respectively by ϵ_0 and ϵ_1 and remember that the moving part of the fluid has kinetic energy $\frac{1}{2}u_1^2$ per unit mass, we can write for the change in the total energy of the cylinder during the time dt

$$\rho_1 \left(\epsilon_1 + \frac{u_1^2}{2} \right) (V - u_1) dt - \rho_0 \epsilon_0 V dt.$$

This must be equal to the product of the pressure p_1 by the distance $u_1 dt$ through which the rear boundary of the cylinder has been pushed. Thus, we get the final equation

$$\rho_1 \left(\epsilon_1 + \frac{u_1^2}{2} \right) (V - u_1) - \rho_0 \epsilon_0 V = p_1 u_1. \qquad (14)$$

The three equations (12), (13), and (14), when augmented by known relations between the thermodynamic parameters of the fluid, can be shown to determine all the quantities $p_1, \rho_1, u_1, V, \epsilon_1$ in terms of any one of them, when ρ_0 and p_0 are given. The equations may be put in a more explicit form as follows. From equation (12),

$$u_1 = \frac{(\rho_1 - \rho_0)}{\rho_1} V. \qquad (15)$$

Inserting this and equation (12) into equation (13),

$$\rho_0 \frac{\rho_1 - \rho_0}{\rho_1} V^2 = p_1 - p_0$$

$$V = \sqrt{\frac{\rho_1}{\rho_0} \left(\frac{p_1 - p_0}{\rho_1 - \rho_0} \right)} \qquad (16)$$

whence, from equation (15),

$$u_1 = \sqrt{\frac{(p_1 - p_0)(\rho_1 - \rho_0)}{\rho_0 \rho_1}}. \qquad (17)$$

Finally, if we insert the expression (12) into the first term of (14), and the expression (15) into the right-hand side,

$$\rho_0 V \left(\epsilon_1 + \frac{u_1^2}{2} \right) - \rho_0 V \epsilon_0 = \rho_0 V \frac{p_1(\rho_1 - \rho_0)}{\rho_0 \rho_1}$$

whence, using equation (17),

$$\epsilon_1 - \epsilon_0 = p_1 \frac{\rho_1 - \rho_0}{\rho_0 \rho_1} - \frac{u_1^2}{2}$$

$$= p_1 \frac{\rho_1 - \rho_0}{\rho_0 \rho_1} - \frac{(p_1 - p_0)}{2} \left(\frac{\rho_1 - \rho_0}{\rho_0 \rho_1} \right),$$

or

$$\epsilon_1 - \epsilon_0 = \tfrac{1}{2}(p_1 + p_0) \left(\frac{1}{\rho_0} - \frac{1}{\rho_1} \right). \qquad (18)$$

In the discussion leading to these equations, we have spoken of the region behind the advancing shock front as the "high-pressure region," although all the equations which have been written would still be valid if p_1 were less than p_0 instead of greater. However, it is easy to show from the energy equation (18) that in ordinary fluids a "rarefactional shock wave," that is, one for which $p_1 < p_0$, cannot exist. To prove this, consider the pressure-volume diagrams shown in Figure 5. The state of the undisturbed fluid of the shock front is represented by the point S_0. If the fluid were gradually and adiabatically compressed to density ρ_1, it would reach the state S_1'. Now, according to the second law of thermodynamics, any sudden compression to this density involving irreversible processes must leave the fluid in a state which is hotter than S_1', in other words, since pressure normally increases with increasing temperature, the point S_1 corresponding to the state of the fluid behind the shock front must lie above S_1', as shown. That this is entirely consistent with equation (18) for a compressional shock wave can be seen from the upper half of Figure 5. The right side of equation (18) represents the area of the trapezoid under the straight line $S_1 S_0$, while the energy difference between S_1' and S_0 is represented by the area under the adiabatic curve between these two points. If the adiabatic curve is concave upward, as it is for all normal fluids, the area under the trapezoid will exceed the area under the curve, and this is consistent with the known fact that S_1 has a higher temperature, hence a higher energy, than S_1'. For a rarefactional shock wave, on the other hand, the energy of S_1' is lower than that of S_0 by an amount represented by the area under the adiabatic curve between these two points in the lower half of Figure 5, and since the area of the trapezoid is again greater than this, equation (18) could not be satisfied unless S_1 had less energy than S_1'. Thus, a rarefactional shock is impossible for a normal fluid,[b] as is indeed to be expected from the fact, proved in Section 8.4.1, that regions of high pressure advance faster than those of low pressure.

It is instructive to compare the equations (16) and (17) with the corresponding relations of acoustical theory to which they reduce in the limit. To verify the latter statement we may note that for a disturbance of infinitesimal amplitude, equation (18) reduces to the law of adiabatic compression, while equations (16) and (13) become respectively[c]

$$V \sim \left(\frac{dp}{d\rho}\right)^{\frac{1}{2}} = c$$

and

$$u_1 = \frac{p_1 - p_0}{\rho_1(V - u_1)} \sim \frac{p_1 - p_0}{\rho c}.$$

For disturbances of large amplitude, however, it is

[b] A British report[10] questions this conclusion on the basis of certain theoretical calculations and of some photographs of rarefactional waves. However, the computed example cited there of a "negative shock front" is merely a normal compressional shock when viewed in a system of reference in which the fluid ahead of it is at rest. Moreover, the rarefactional waves which have been photographed must be regarded as mere acoustic disturbances; the resolution of the photographs is insufficient to distinguish a discontinuous shock front from a gradual pressure front which has a fairly appreciable thickness.

[c] Throughout this chapter the symbol $\sim$ will be used to denote asymptotic equality; in other words, it implies that the quantity on the left equals the quantity on the right plus other terms whose ratio to the quantity written approaches zero in the limiting process being considered.

easily seen from the top half of Figure 5 that for all ordinary fluids the value of V given by equation (16) is greater than the value c_0 of $(dp/d\rho)^{\frac{1}{2}}$ in the undisturbed fluid. For the quantity under the radical in equation (16) is just $1/\rho_0^2$ times the slope of the straight line $S_1 S_0$ while c_0 is $1/\rho_0^2$ times the slope of the tangent to the adiabatic curve at S_0. By a similar argument it can be shown that for a fluid such as

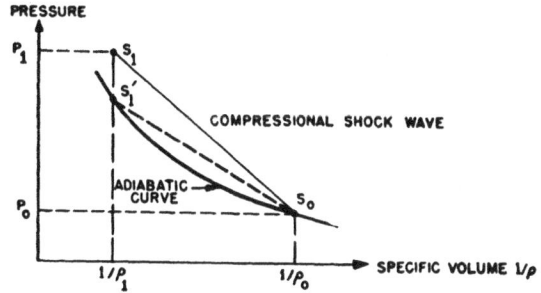

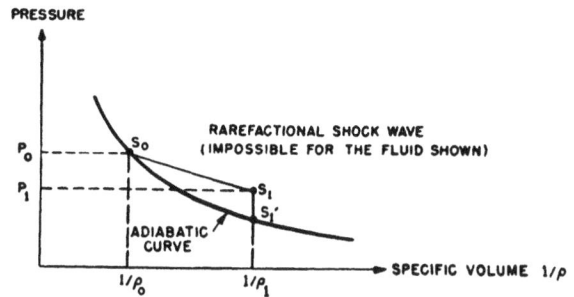

FIGURE 5. Pressure-volume diagram of the changes occurring in a shock front.

water, for which S_1' and S_1 are very close together $(V - u_1)$ is less than c_1, the value of $(dp/d\rho)^{\frac{1}{2}}$ immediately behind the shock front. These results mean that small disturbances created behind the shock front may overtake it, but that no small disturbance can be propagated from the shock front into the undisturbed fluid ahead.

In fluids such as water and air, the dissipative phenomena taking place in the shock front gradually convert the mechanical energy of a shock wave into heat, causing the amplitude of the wave to decrease as it advances by an amount additional to the familiar decrease due to spherical divergence. A sufficiently intense shock wave traversing a high explosive, however, can maintain itself indefinitely because of the energy supplied by the chemical conversion of the explosive into gaseous products; such a shock wave would constitute a detonation wave.

8.4.3   The Law of Similarity

According to the theory outlined in Sections 8.4.1 and 8.4.2 the disturbance created in the water by an explosion is uniquely determined by:

1. The forces which the explosive gases exert on the water near them.

2. The Hugoniot equations (16), (17), and (18), which hold across the advancing shock front.

3. The equation of continuity [equation (2) of Section 2.1.1].

4. The equations of motion (2), (3), and (4), which hold true to a very good approximation at all points of the water except points in the shock front.

5. The thermodynamic properties of the water, that is, the relations, such as the equation of state, between pressure, density, and energy. Moreover, from what has been said previously concerning the nature of detonation waves, it is likely that the course of events within the explosive material itself is determined by a similar set of equations, so that factor 1 can be derived from laws of the same type as 2, 3, 4, and 5. Now suppose a disturbance to be given which satisfies all these equations, and is described by

$$p = p(x,y,z,t)$$
$$\rho = \rho(x,y,z,t)$$
$$u_x = u_x(x,y,z,t)$$
$$u_y = u_y(x,y,z,t)$$
$$u_z = u_z(x,y,z,t)$$

Then it can easily be verified by substitution that all the equations mentioned in 2, 3, and 4 outlined previously and the laws 5 as well, will be satisfied at all points except those in a thin layer at the shock front, by a disturbance described by p',ρ',u_x',u_y',u_z' where

$$p'(x,y,z,t) = p(\beta x,\beta y,\beta z,\beta t)$$
$$\rho'(x,y,z,t) = \rho(\beta x,\beta y,\beta z,\beta t)$$
$$u_x'(x,y,z,t) = u_x(\beta x,\beta y,\beta z,\beta t)$$

etc., that is, by a disturbance identical with the first except that the distance and time scales have been changed by a factor β. A scale relationship of this sort may be expected to hold both in the explosive material and in the water, and if the linear dimension D' of the explosive in the second case is made equal to β times the corresponding dimension D in the first case, the disturbances in both the explosive and the water can be scaled together.

This law of similarity relating to disturbances pro-duced by different quantities of the same explosive has been fairly accurately verified experimentally at ranges for which the peak pressure in the shock wave is of the order of an atmosphere and above.[6] Provided the shape of the explosive charge and the position at which the detonation is initiated are the same on the two scales, an appreciable departure from the similarity law could be caused only by failure of the equations of motion (2), (3), and (4) to hold behind the shock front in the water, or by a failure of either the water or the explosive to have thermodynamic properties independent of the scale of times involved. A phenomenon of the latter type might conceivably occur, for example, in an aluminized explosive, if the reaction of the grains of aluminum with the hot gases were so slow that the reaction occupied an appreciable part of the volume behind the detonation front. However, the fact that significant departures from the scaling law have not been observed at the ranges mentioned indicates that such phenomena are not serious. As for the possibility of failure of the basic assumptions to be fulfilled in the water, such a failure could be caused only by bubbles or other extraordinary dissipative mechanisms; as far as is known, the effect of these only becomes appreciable at very long ranges (see Section 9.2.1). Ordinary viscosity and heat conduction can be shown to have a negligible effect on the scale used in experimental work. It should be remembered, of course, that the derivation we have given of the similarity rule does not apply in the very thin region of the shock front in which the abrupt rise of pressure takes place; as will be shown in the next section, the thickness of this region does not ordinarily scale proportionally to the factor β used previously.

8.4.4   Theoretical Thickness of a Shock Front

We have seen in the preceding sections that it is for many purposes unnecessary to consider the details of phenomena occurring in a shock front, and that many useful conclusions can be drawn by thinking of a shock front merely as a surface in the fluid across which the pressure and other quantities change discontinuously. However, these conclusions will be valid only if the thickness of the region in which the pressure rise takes place is very small; and to make our theoretical discussion complete we should verify that this is the case. Moreover, a study of the factors influencing the thickness of a shock front can be valuable in that it may help us to evaluate and interpret experiments which purport to measure the time of rise of the pressure at the front of a shock wave.

As has been explained previously, the Riemann overtaking effect tends to make any pressure pulse develop an infinitely steep front in the course of time, and this tendency can be counteracted only by factors neglected in the equations of motion of a perfect fluid, on which Riemann's analysis is based. These factors, of which viscosity and heat conduction are the most obvious, must include the dissipative phenomena responsible for the fact that the internal energy of the fluid behind the shock front, as given by equation (18), is greater than that which the fluid would have if it were compressed reversibly and adiabatically to the density ρ_1. This fact provides a clue which we can use to get a rough estimate of the thickness of the shock front. For, the amount of energy dissipated into heat per unit time by any given dissipative mechanism will be dependent on the thickness of the shock front, in other words, dependent on the rapidity with which the pressure changes from p_0 to p_1. This dissipated energy must be equal to the product of the mass of water crossing the shock front per unit time by the amount of energy dissipated per unit mass; this quantity being for all practical purposes simply the difference in energy between the states S_1 and S_1' in Figure 5. As will be shown later, the latter quantity can be calculated from equation (18) in terms of the pressure jump $(p_1 - p_0)$ and the known properties of water; for small amplitudes it is proportional to the cube of the pressure jump. Since all reasonable dissipative phenomena create heat more rapidly the more suddenly they are made to take place, the greater the pressure jump the thinner the shock front must be in order to dissipate the required amount of energy. Thus if we can show that the shock front should be quite thin for a fairly weak shock wave, it must be even thinner for a strong one.

We shall therefore begin by calculating the approximate value of the dissipated energy for a weak shock wave. Referring to the upper diagram in Figure 5, we wish to calculate the energy difference $(\epsilon_1 - \epsilon_1')$ between the states S_1 and S_1'.

Since by equation (18) the difference $(\epsilon_1 - \epsilon_0)$ equals the area of the trapezoid under the line S_0S_1 while the difference $(\epsilon_1' - \epsilon_0)$ equals the area under the adiabatic curve from S_1' to S_0, we must have

$(\epsilon_1 - \epsilon_1') =$ area of region between S_1S_0 and adiabatic curve

$=$ area of segment between $S_1'S_0$ and adiabatic curve $+$ area of triangle $S_1'S_0S_1$.

Now the area of the triangle $S_1'S_0S_1$ is equal to half the product of its altitude $(1/\rho_0 - 1/\rho_1)$ by its base $(p_1 - p_1')$. Since the latter quantity is in the limit of small amplitudes proportional to $(\epsilon_1 - \epsilon_1')$, we can make the area of the triangle as small as we like compared to $(\epsilon_1 - \epsilon_1')$ by taking the amplitude of the shock wave sufficiently small. Thus, for sufficiently weak shock waves

$(\epsilon_1 - \epsilon_1') \sim$ area of segment between $S_1'S_0$ and adiabatic curve

$$\sim k\left(\frac{1}{\rho_0} - \frac{1}{\rho_1}\right)^3 \qquad (19)$$

where the constant k is proportional to the curvature of the adiabatic curve and has the numerical value 1.5×10^{10} in cgs units for pure water.

Let us now consider the mechanism by which dissipation of energy occurs in the shock front. For any assumed mechanism the rate of this dissipation can be calculated, at least roughly, as a function of the thickness of the shock front and the magnitude of the pressure jump. Of the two most obvious mechanisms, viscosity and heat conduction, the former gives much the greater dissipation, and accordingly we shall carry through the calculation only for the case of dissipation by viscosity. According to the theory of viscous fluids, the mechanical energy converted into heat per unit time in a fluid having a coefficient of shear viscosity μ is given, for one-dimensional motion such as that in a plane wave traveling in the x direction, by

Dissipation per unit volume per unit time

$$= 2\mu\left(\frac{\partial u_x}{\partial x}\right)^2 = 2\mu\left(\frac{1}{\rho}\frac{d\rho}{dt}\right)^2 \qquad (20)$$

by the equation of continuity. If δ is the thickness of the shock front — the thickness of the region within which most of the change in density from the value ρ_0 to the value ρ_1 takes place — we have, roughly, for a weak shock wave,

$$\frac{1}{\rho}\frac{d\rho}{dt} \approx \frac{c(\rho_1 - \rho_0)}{\rho_0\delta}.$$

Multiplying equation (20) by the thickness δ gives a rough value for the dissipated energy per unit time per unit area of the shock front:

Dissipation per unit area per unit time

$$\approx \frac{2\mu c^2(\rho_1 - \rho_0)^2}{\delta\rho_0^2}. \qquad (21)$$

This must be equated to the product of the expression (19) by the mass of water which unit area of the shock front traverses in unit time, that is, since we are assuming the shock wave to be weak, by $\rho_1 c$:

$$\rho_1 c k \left(\frac{1}{\rho_0} - \frac{1}{\rho_1} \right)^3 \approx \frac{2\mu c^2 (\rho_1 - \rho_0)^2}{\delta \rho_0^2}.$$

By solving this for δ

$$\delta \approx \frac{2\mu c \rho_1^2 \rho_0}{k(\rho_1 - \rho_0)} \sim \frac{2\mu c \rho_0^3}{k(\rho_1 - \rho_0)}. \qquad (22)$$

With the value given above for k and the value $\mu = 0.01$ cgs unit characteristic of pure water at room temperature, equation (22) becomes

$$\begin{aligned} \delta \text{(in cm)} &\approx \frac{2 \times 10^{-7}}{(\rho_1 - \rho_0)\text{(in gm/cc)}} \\ &= \frac{4.5 \times 10^3}{(p_1 - p_0)\text{(in dynes/cm}^2)}. \end{aligned} \qquad (23)$$

More refined calculations of this type have been made[11] and indicate that, in fact, nearly all the calculated jump in pressure occurs within an interval of the thickness given by equation (23).[8]

If we are to regard the relations (22) and (23) as giving a valid estimate of the order of magnitude of the thickness of a shock front in sea water, we must assume three things:

1. That the Hugoniot equation (18) is sufficiently accurate.

2. That the curvature of the adiabatic for sea water is roughly the same as for pure water.

3. That the rate of dissipation of energy is of the same order as that due to shear viscosity.

The last of these assumptions is known to be true for sinusoidal disturbances in pure water at frequencies from 1 to 100 mc.[12] Since, according to Section 5.2.2, the absorption in sea water seems to approach that in pure water at frequencies near 1 mc, it is reasonable to expect this assumption to hold in the sea for values of δ between say 0.1 and 0.001 cm, and perhaps for much smaller values. The second assumption would fail if the water contained many bubbles, but calculations show that this would happen only for an unreasonably large concentration of bubbles. The Hugoniot equation depends for its validity on δ being very small. Although no reliable calculation of its range of validity has been made, it is not hard to show that the equation (19) derived from the Hugoniot equation should be correct as to order of magnitude for explosive waves from ordinary sources when the pressure amplitude $(p_1 - p_0)$ is

greater than about 100 atmospheres. Unfortunately, at 100 atmospheres amplitude the value of δ given by equation (23) is 4.5×10^{-5} cm, a value so small that it is conceivable that assumption (3) might fail. Thus, about all that can be concluded from the preceding analysis is that with a typical explosive source the thickness of the shock front at a distance where the pressure amplitude is 100 atmospheres is probably not greater than about 0.001 cm and may be much smaller.

At greater distances from the explosion, a probable upper bound to the thickness of the shock front can be set by neglecting the Riemann overtaking effect, which tends to make the shock front steeper, and imagining the pressure pulse to be propagated outward according to the laws of acoustics, subject to the same attenuating mechanisms as sinusoidal sound. By the method of Fourier analysis (see Section 9.2.4 and Figure 13 in Chapter 9) it can be estimated that to avoid inconsistency with the values given in Section 5.2.2 for the attenuation at high frequencies the thickness of the shock front should not exceed a fraction of a centimeter after it has traveled 50 yd through homogeneous sea water. A greater thickness could be produced only by inhomogeneity of the medium or by a highly nonlinear absorption, that is, by some mechanism which would be much more effective in absorbing energy from a disturbance of large amplitude than from a weak disturbance.

8.5   STRUCTURE AND DECAY OF SHOCK WAVES

When a shock wave from an explosion passes a given point in the water, the initial behavior of the pressure as a function of time consists ordinarily in a roughly exponential dropping off, as shown schematically in Figure 1. The time required for the pressure to fall to $1/e$ times its value just behind the shock front is of the order of 15 μsec for a Number 6 detonating cap and 600 μsec for a 300-lb depth charge. This decay time depends somewhat upon the type of explosive being used, and it may depend slightly upon the range; for any given explosive it varies as the cube root of the charge weight, in accordance with the similarity rule given in Sections 8.2 and 8.4.3. After the pressure has fallen to $\frac{1}{15}$ or $\frac{1}{10}$ its peak value, however, the decay of pressure is much more gradual than would correspond to an exponential law. Theories of the shock wave[13] predict a "tail" in which the pressure dies off with the time t approximately as $t^{-4/5}$. This law cannot of course hold indefinitely, since the momentum integral $\int p\,dt$ must be finite; the theory of bubble motion to be discussed in Section 8.6 predicts that the excess pressure should eventually go through zero and become weakly negative as the gas bubble expands. For many purposes this tail is unimportant, but its contribution to the momentum integral may exceed that of the earlier part of the shock wave. Detailed experimental information on the tail is almost entirely lacking since spurious signals due to the impact of the shock wave on the cables leading to the pressure gauge usually mask the latter part of the tail.

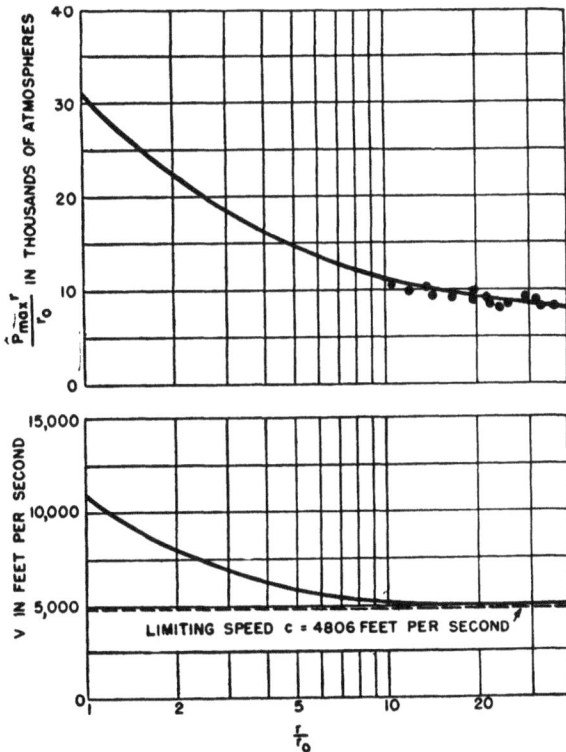

FIGURE 6. Peak pressure and speed of the shock front near a spherical charge of cast TNT.

Under some conditions small secondary peaks or fluctuations may appear in the measured pressure-time curve. These may be due to any of a variety of causes. Sometimes the effect is spurious, being due to instrumental factors — for example, diffraction of the pressure pulse around the hydrophone and its supports or shock excitation of vibrations in the hydrophone.[2] Irregularities genuinely present in the explosive wave itself have been observed, however. Sometimes these occur under exceptional circumstances, such as for shots fired on the bottom,[14] for

cylindrical charges detonated at one end rather than in the center,[15,16] for charges surrounded by an air pocket,[16,17] and for charges which fail to detonate completely because of inadequate boostering.[18] Moreover, even for spherical charges detonated at the center, the tail of the shock wave shows a small but reproducible hump or shoulder, whose magnitude depends upon the type of explosive.

In accordance with the theory of Section 8.4.2, it is to be expected that the speed of advance of a shock wave at great distances from the explosion will be the normal speed of sound, but that at close distances the speed of advance will be considerably greater. Moreover, we should not be surprised to find other departures from the usual laws of acoustics. The upper diagram of Figure 6 gives some experimental values of the peak pressure in the shock wave as a function of the distance r from an explosion and shows fitted to these points a theoretical curve which, though only approximate, should give a reasonably reliable extrapolation of the peak pressure for smaller values of r.[19] If the disturbance followed the ordinary laws of acoustics, the curve would be a horizontal line. The lower diagram of Figure 6 shows the velocity of the shock front as a function of distance from the charge, this curve being related to that of the upper diagram by equation (16) of Section 8.4.2. It will be seen that as r increases the pressure becomes more and more nearly inversely proportional to r, as it should be in the acoustic approximation. It has been shown theoretically, however, that even in a practically ideal fluid the peak pressure in a shock wave is not asymptotically proportional to $1/r$ at large distances, but that instead

$$p_{\max} \sim \frac{\text{Constant}}{r\left[\log\left(\dfrac{r}{r_1}\right)\right]^{\frac{1}{2}}} \qquad (24)$$

where r_1 is a quantity of the same order as the initial radius r_0 of the explosive charge.[13] This deviation from acoustical laws is due to the dissipation of energy in the shock front. The relation (24) has been confirmed experimentally[2] at NRL for No. 6 detonating caps at ranges of 1 and 31.3 ft. The ratio $(rp_{\max})1\text{ ft}/(rp_{\max})31.3\text{ ft}$ was found in these experiments to be 1.31 ± 0.04, while the ratio

$$\frac{\left[\log\dfrac{r}{r_0}\right]^{\frac{1}{2}}31.3\text{ ft}}{\left[\log\dfrac{r}{r_0}\right]^{\frac{1}{2}}1\text{ ft}}$$

is 1.32. It is not to be expected, however, that equation (24) will hold true indefinitely as the range is increased, for its theoretical derivation assumes the rise in pressure at the shock front to be instantaneous, and neglects any dissipation of energy behind the shock front. At large ranges neither of these assumptions is valid, and we should expect the decrease of pressure with distance to obey a law similar to the attenuation of sinusoidal sound (see Sections 2.5 and 9.2.1).

The theory just mentioned also predicts that the duration of the pressure in a shock wave should slowly but continually increase as the range increases. This effect is due to the fact that, according to Riemann's theory, the more intense front part of the wave should travel faster than the less intense tail. Specifically, the theory predicts that if at large values of the distance r from the charge the pressure in the wave is approximated by an exponential $p = p_{max}e^{-t/\theta}$, then the duration parameter θ should be given by

$$\theta \sim \text{constant}\left[\log \frac{r}{r_1}\right]^{\frac{1}{2}} \qquad (25)$$

where as before r_1 is of the order of the radius r_0 of the explosive charge. The experimental verification of this relation is less conclusive than for the preceding relation (24). Although a decided decrease in θ has been observed when r is decreased below a value corresponding to $p_{max} = 1,000$ atmospheres,[20] the experiments of reference 2, which covered a range from about 3 to 80 atmospheres, showed no measurable increase of duration; yet an increase of the amount given by the relation (25) should have been measurable.

Under most conditions, given complete detonation, the peak pressures and pressure-time curves obtained at a given range from a charge of a given size are quite accurately reproducible from shot to shot, the deviations of individual peak pressures from the mean being of the order of 2 per cent. However, for asymmetrical charges, such as long cylinders with detonation initiated at one end, both the peak pressure and the duration of the shock wave, when measured at a given distance, are different in different directions. Experiments on long cylindrical charges have shown that the peak pressure is greatest approximately at right angles to the axis of the cylinder and is least on the axis off the cap end. Differences as large as 40 per cent have been observed.[15]

The duration of the shock wave varies in the opposite sense from the peak pressure, and in fact the impulse $\int pdt$ contained in the early part of the shock wave is slightly greater off the cap end, where the peak pressure is least, than in any other direction. As one would expect, charges fired with an air cavity on one side also show an anisotropy.

When a charge is fired on the sea bottom, the peak pressure received at a point above the charge is somewhat greater than it would be in the absence of the bottom. For sand and gravel bottoms this increase in peak pressure has been found to be 10 to 15 per cent.[14] In directions near the horizontal the amplitude tends to become smaller, as one would expect from the shadowing effect of irregularities on the bottom. The pressure-time curves from shots fired on the bottom are not only likely to be rather irregular, as mentioned previously, but tend to be less consistent from shot to shot than is the case for explosions in free water.

8.6 SECONDARY PRESSURE WAVES

In Section 8.2 it was mentioned that because of the inertia of the water which has been pushed radially outward by an explosion, the gas bubble undergoes radial oscillations. Many features of this oscillatory motion can be explained by a very simple theory which treats the water as an incompressible fluid and the radial flow as spherically symmetrical.[21] Let $u(r,t)$ be the radial velocity of the water, measured positive outward at distance r from the center and at time t. The volume of water which passes outward in unit time across the surface of a sphere of radius r is $4\pi r^2 u$. At any given instant the flux across any two concentric spheres must be the same, since otherwise the amount of water in the shell between these two spheres would be increasing or decreasing. We must therefore have

$$4\pi r^2 u = \text{function of } t, \text{ independent of } r.$$

If $r_b(t)$ is the radius of the gas bubble at time t, we must have

$$u(r_b,t) = \frac{dr_b}{dt} = \dot{r}_b$$

and so $\qquad u(r,t) = \frac{r_b^2 \dot{r}_b}{r^2}. \qquad (26)$

We are now ready to apply the principle of conservation of energy. The energy of the water and gas bubble consists of three parts, kinetic energy, potential energy due to compression of the gas in the bubble, and potential energy representing work done

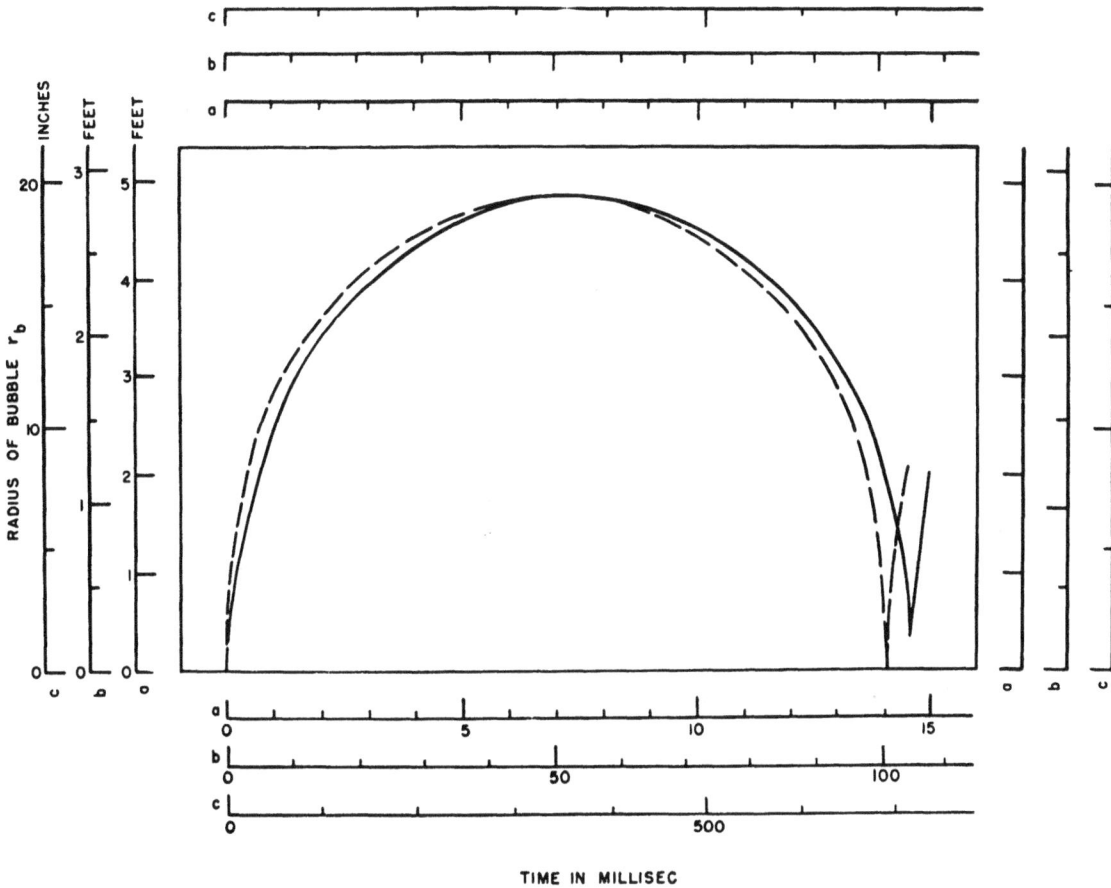

FIGURE 7. Ideal radius-time curve for the gas bubble from an underwater explosion. Full curve computed assuming pressure of gas in bubble to be given by

$$p = 0.064 \left(\frac{r_{\max}}{r_b}\right)^4 \text{ atmospheres.}$$

Dashed curve computed assuming pressure of gas in bubble to be zero. Scales: (a) No. 8 cap at 50-foot depth; (b) 1 lb TNT at 50-foot depth; (c) 300 lb TNT at 50-foot depth.

against the surrounding hydrostatic pressure, which we shall denote by p_∞, since it represents the pressure in the water at a great distance from the bubble at the same level. Consider the kinetic energy first; since the mass of the gas in the bubble is negligible, practically all the kinetic energy resides in the water, and the amount per unit volume is $\frac{1}{2}\rho u^2$. The total kinetic energy is thus

$$\int_{r_b}^{\infty} \frac{1}{2}\rho u^2 \cdot 4\pi r^2 dr = 2\pi\rho r_b^3 \dot{r}_b^2 \qquad (27)$$

by equation (26). The potential energy of the gas is a function of its volume, and can be represented by a function $G(r_b)$; it is a negligible fraction of the total energy except when the radius of the bubble is small. The work which has been done against the external pressure is represented by p_∞ times the volume of the bubble, or $(4/3)\pi p_\infty r_b^3$. The sum of these three terms must be constant in time in the approximation we are using

$$2\pi\rho r_b^3 \dot{r}_b^2 + \frac{4}{3}\pi p_\infty r_b^3 + G(r_b) = W. \qquad (28)$$

The behavior of r_b as a function of the time is thus determined by solving equation (28) for $\dot{r}_b$ and integrating. The result can be expressed in a form which is independent of the amount of explosive involved, manifesting a similarity rule of the same form as that given in Section 8.2 for shock wave pressures. For if, as before, we let r_0 be the radius (or equivalent linear dimension) of the original charge of explosive, G will be r_0^3 times a function of the ratio r_b/r_0, and W, which represents that part of the original energy which is not dissipated or carried away by the shock wave, will be proportional to r_0^3. Using these facts, it can be seen from equation (28) that the solution obtained for one quantity of explosive will be valid for any other quantity if the scales of r_b and t are changed in proportion to r_0. The full curve of Figure 7, taken from a report issued by the David Taylor Model Basin,[22] shows the form of the radius-time curve obtained from integration of equation (28), using a function $G(r_b)$ comparable with that which would obtain for a charge of high explosive; while the dotted curve is the one which would result from equation (28) assuming the same maximum value of r_b but setting $G = 0$. Several scales are given appropriate to several sizes of charge. The variation of r_b with t near the minimum of the contraction is too rapid to show on the scale of the figure. It would not be worth while to show this portion in greater detail, however, since, as will be explained presently, the motion of the bubble in this stage is strongly influenced by gravity and other asymmetrical factors, and also, although less strongly, by the finite compressibility of the water. These effects prevent the present simple theory from being even approximately correct near the minimum. When r_b is greater than three or four times the minimum radius shown in Figure 7, $G(r_b)$ becomes small and the motion is practically the same as it would be if there were no gas in the bubble at all, as can be seen from the agreement of the dotted curve with the full one. For this portion of the curve a change in p_∞ is equivalent to a change in W combined with a change in time scale; thus, over most of the period of the oscillation Figure 7 applies not only to charges of different sizes, but to charges at different depths, provided suitable time and radius scales are used; these scales can be deduced from equation (29) below.

The period of the motion, in the approximation neglecting gas pressure, is easily deduced from equation (28) and turns out to be

$$t_0 = 1.135\rho^{\frac{1}{2}}p_\infty^{-\frac{5}{6}}W^{\frac{1}{3}} = 1.829\rho^{\frac{1}{2}}p_\infty^{-\frac{1}{2}}r_{b\ max} \qquad (29)$$

where
$$r_{b\ max} = \left(\frac{3W}{4\pi p_\infty}\right)^{\frac{1}{3}}$$

and is the radius of the bubble at its maximum size. This expression, in spite of its neglect of gas pressure and the other effects to be discussed later, has been found to agree with measured values of the period to within a few per cent, provided the explosion takes place in open water well away from bounding surfaces, which exert a perturbing effect discussed later. With the same proviso, the variation of the period with depth and size of charge agrees with the exponents in equation (29) to within the accuracy of

measurement. The measured values of the bubble period are found to be reproducible from shot to shot to within a per cent or less.[23, 24]

The simple theory just outlined ignores the compressibility of the water and any influences which may make the motion asymmetrical. The compressibility of the water is important near the minimum of the contraction, when the pressure in the bubble is very high. During this stage of the motion the compression of the water near the bubble initiates a pressure wave which travels outward as an acoustic pulse. This is known as a "secondary pulse" or "bubble pulse," to distinguish it from the original shock wave. The acoustic energy carried away by this secondary pulse is usually small but may, under exceptionally symmetrical conditions, be appreciable compared with the total energy W of the oscillation. Loss of energy through this effect and through turbulence has the consequence that the next oscillation, although conforming generally to the theory of the preceding paragraph, is of lower amplitude than the first one, since it has an energy W_1 which is smaller than W. Energy is radiated in a similar manner at each succeeding contraction.

Several causes may act to prevent the motion from having true spherical symmetry. Most important of these, because it is always present, is gravity. The bubble, being buoyant, tends to rise; the rate of rise is limited by the inertia of the water around it. The rise is usually rather slight during the first expansion of the bubble, but as the bubble contracts again the rise is enormously accelerated and may result in a large portion of the total energy W being retained as kinetic energy in the water at the time the radius of the bubble is a minimum; since this energy is not available to compress the gas, the minimum radius of the bubble will not be as small as it would be if gravity were absent, and the secondary pressure pulse will be correspondingly weaker. Another effect of the rapid rise is to produce turbulence in the contracted stages; this turbulence dissipates energy and is probably the most important factor in the damping out of the oscillations.

This rapid rise in the contracted stages has been the object of many theoretical studies.[22,25-27] The explanation of the phenomenon rests on the fact that a spherical cavity moving through a fluid possesses an "effective inertia" equal to half the mass of the water it displaces. The buoyancy of the gas bubble causes it to acquire an ever-increasing amount of vertical momentum, and to conserve this momentum during the contraction, when the effective mass is greatly decreased, the velocity of rise must increase.

Besides gravity, other effects such as proximity to the free surface of the water or to the bottom can cause departures from spherical symmetry. The effects of such surfaces become appreciable when the distance from the bubble to the boundary surface is a few times the maximum radius of the bubble, and are of two sorts. In the first place, the period of the oscillation is shortened by proximity to the free surface and lengthened by proximity to an unyielding surface; secondly, a free surface repels the bubble and a rigid surface attracts it. This translational motion, which becomes very rapid in the contracted stage, weakens the secondary pressure pulse for the same reason that the rise due to gravity does. Much theoretical work has been done on the period and migration of the bubble,[27,28] and the results are in generally good agreement with experiment.[25] If gravity can be ignored, asymmetrical motion due to proximity to free or rigid surfaces obeys the scaling law of Section 8.2, the distance from the surface being changed in the same ratio as other linear dimensions. The gravity effect, however, does not scale in the same manner; gravity is relatively more important the larger the charge and the smaller the external hydrostatic pressure p_∞. Most features of the motion as affected by gravity can be approximately expressed in a form which is independent of the size of the charge, by using a unit of length $\Lambda = (W/g\rho)^{\frac{1}{4}}$, a unit of time $\sqrt{\Lambda/g}$, and a unit of pressure $\rho g\Lambda$.[25]

From the foregoing it can be seen that the form and strength of the secondary pressure pulses depend greatly on gravity and on proximity of surface, bottom, or objects to the explosion. The peak pressure in the first bubble pulse, for example, has been measured at values as large as 0.25, and as low as 0.06, times the peak pressure in the shock wave.[23,24,28,29] By contrast, the impulse $\int pdt$ contained in any one of the secondary pulses is not very sensitive to these factors. The amount of impulse contained within a few half-widths of the main pressure peak is of the same order as that in a corresponding portion of the shock wave; however, just as was the case with the shock wave, this impulse is considerably less than the amount contained in the "tails," which in the case of the secondary pulses extend to both directions in time. The total impulse in a secondary pulse is probably roughly equal to the amount which would be calculated from the simple theory which assumes

the water to be incompressible. It can be shown [21] that this impulse is

$$I = \frac{2^{1/6}r_{b\max}^2}{r}\sqrt{\rho p_\infty}. \qquad (30)$$

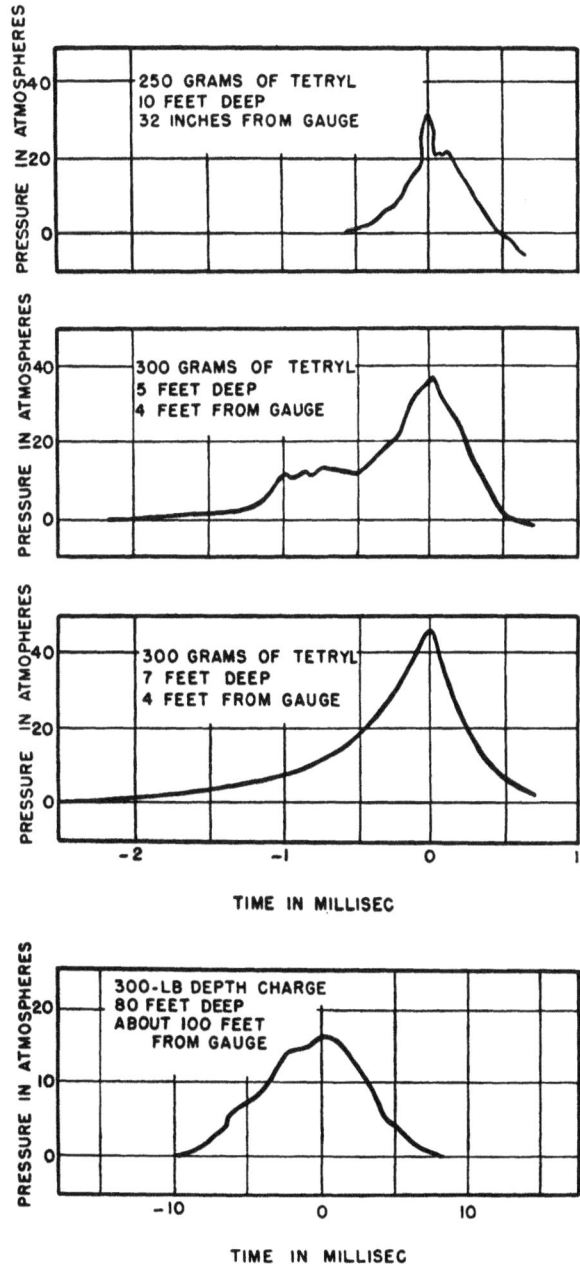

FIGURE 8. Typical pressure-time records for the first bubble pulse.

At all ordinary depths this is five or ten times as great as the impulse in the exponential part of the shock wave; however, one would expect from theory that the impulse in the tail of the shock wave would be about half of the quantity (30). In between the end of the tail of the shock wave and the beginning of the tail of the first bubble pulse there is a long period during which the pressure is below normal; this period occupies most of the time consumed by the first oscillation, and the negative impulse delivered during it is just equal to the expression (30). For most applications, however, this negative pressure and the tail parts of the shock and bubble pulses can be neglected.

In cases where migration of the bubble is slight, the bubble pulses show a fairly regular rise and fall of pressure. When migration is rapid, irregularities are more apt to occur, and sometimes two or more fairly

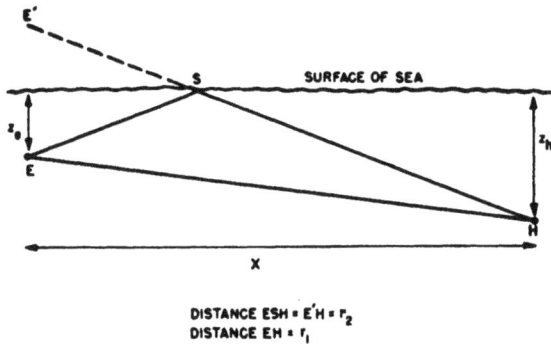

DISTANCE ESH = E'H = r_2
DISTANCE EH = r_1

FIGURE 9. Superposition of direct and surface-reflected pulses.

well-separated peaks are observed.[24] It has been suggested that these multiple impulses may be due to breaking of the bubble into several separate bubbles, which emit distinct pressure peaks in the contracted stage, but which coalesce when they expand again. A few typical oscillograms of bubble pulses, taken from references 23, 24, and 29, are shown in Figure 8. The first bubble pulse is usually by far the strongest; for small charges as many as eight or ten pulses have been counted, but for large charges usually only two or three bubble pulses are measurable. For some reason, a charge fired on or very close to the bottom usually gives a very weak bubble pulse, and the number of measurable pulses is less than for shots in open water.

A caution should be added concerning the interpretation of oscillograms of bubble pulses; because of the relatively long duration of these pulses the negative pulse reflected from the free surface of the water will often overlap the direct pulse, making the recorded pressure appreciably different from that due to the direct pulse alone. The statements given previously apply to the latter only.

8.7 SURFACE REFLECTION AND CAVITATION

When a hydrophone is placed in the water at some distance from an explosive charge, the shock wave and secondary pulses received are modified by reflection at the free surface of the water. This reflection is most conveniently described by the principle of images, which we have encountered in another application in Section 2.6.3. This principle is applicable whenever the pressure amplitudes are small enough for the laws of ordinary acoustics to apply. Referring to Figure 9, the pressure produced at any instant at

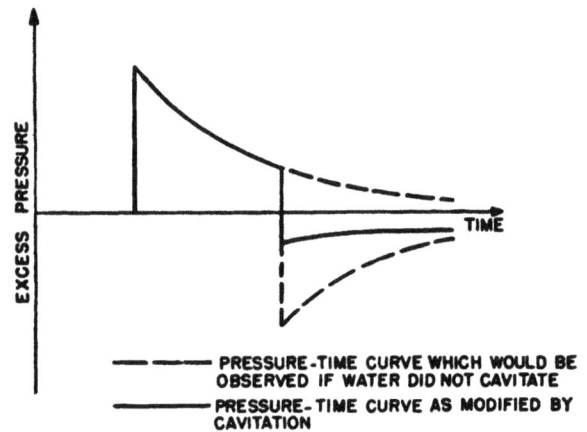

——— PRESSURE-TIME CURVE WHICH WOULD BE OBSERVED IF WATER DID NOT CAVITATE
———— PRESSURE-TIME CURVE AS MODIFIED BY CAVITATION

FIGURE 10. Modification of a surface-reflected pulse by cavitation.

a hydrophone H by an explosion or other source of sound at E is the sum of the pressure due at that instant to the direct wave EH and the pressure due at the same instant to the reflected wave ESH. According to the image principle, the latter pressure is exactly equal to the negative of the pressure which would be produced in the absence of a surface by a source E' which is the mirror image of E in the surface and which has the same time variation. If z_e is the depth of the explosion, z_h the depth of the hydrophone, and x the horizontal range, the path difference between these two waves is

$$r_2 - r_1 = \sqrt{(z_e + z_h)^2 + x^2} - \sqrt{(z_e - z_h)^2 + x^2}$$

$$= \frac{4z_e z_h}{r_1 + r_2}. \tag{31}$$

When the distance ES from the explosion to the portion of the surface at which the reflection takes place is so small that the incident pressure at S is appreciably in excess of one atmosphere, the simple image law just stated has to be modified; for, sea water is apparently incapable of sustaining a tension of any appreciable magnitude, and cavitation will therefore set in at any point where the pressure becomes negative. At short ranges, therefore, the pressure-time curve for a shock wave and its reflection usually looks like the full line in Figure 10, instead of following the dotted curve as it would if there were no cavitation.

Experiments on explosion waves in sea water[30] strongly suggest that the water begins to cavitate as soon as the pressure becomes negative, and that the cavitation can develop sufficiently rapidly to prevent negative pressures from persisting even as long as 10 μsec. This is to be expected if there are even a few tiny bubbles in the water. A theory of the propagation of cavitation fronts has been given by Kennard.[31,32]

Chapter 9

TRANSMISSION OF EXPLOSIVE SOUND IN THE SEA

INTRODUCTION

Having established in Chapter 8 the nature of explosions as sources of sound, we are ready to consider the results of experiments showing how pulses of explosive sound are affected by transmission through moderate and long distances in the ocean. The amount of experimental data available in this field is scanty by comparison with that which has been accumulated on sinusoidal sound, and since accurate recording of explosive pulses is possible only if very careful precautions are taken, one must be cautious in drawing conclusions from this work. Nevertheless, these experiments demonstrate strikingly the utility of explosive sound as an aid to fundamental research on the nature of the ocean as an acoustic medium. Whereas in experiments using long pulses of sinusoidal sound the signal received at the hydrophone is usually inextricably compounded out of directly transmitted sound, scattered sound, and sound reflected from the surface or bottom, the extremely short duration of explosive pulses makes it possible, in many cases, to distinguish between the contributions of these different mechanisms by virtue of the differences in time of arrival. Another characteristic difference between explosive and sinusoidal sound is that dispersion effects, which depend upon the phases of the various component frequencies in the arriving sound, can easily be studied with a transient disturbance, but are practically impossible to measure with single-frequency sound. The dispersion accompanying the transmission of sound through sea water alone, although doubtless present, is very minute, and has never been detected; in shallow-water transmission, on the other hand, dispersion phenomena are important and can be made to give useful information about the bottom. Other advantages of explosive sound which are significant in certain types of experiments include the high intensity attainable and the fact that explosive sources are relatively easy to manipulate and can be fired at great depths.

The experiments to be discussed in this chapter shed light on a variety of problems of sound propagation in the ocean. For example, the variation from shot to shot in the sound intensity received at a distance is found in Section 9.2.5 to be much smaller than that which is observed for sinusoidal sound, especially when the path of the sound lies entirely in isothermal water. This suggests that most of the variation observed with sinusoidal sound is due to some sort of interference phenomenon. Another example is provided by the estimates of attenuation of low-frequency sound, or at least of an upper bound to it, given in Section 9.3.2; these estimates are made on the basis of experiments which include ranges up to hundreds of miles. Other interesting results of experiments with explosive sound up to the present include the occurrence of a reflection coefficient near unity for the free surface of the ocean at directions of incidence surprisingly close to the horizontal (Section 9.2.1), the comparison of observed intensities with the predictions of ray theory (Section 9.2.2), the occurrence of diffraction (Section 9.2.3), the deduction of details of the bottom strata from shallow-water experiments (Section 9.4), etc.

Before discussing the experimental material in detail it will be worth while to say a few words regarding the technique of measuring and recording explosive sounds. A systematic discussion of experimental techniques would be beyond the scope of this volume; but to enable the reader to form a balanced opinion on past and future experiments with explosive sound, some of the pitfalls and stumbling blocks in this field should be pointed out. In the first place, if actual pressure-time curves are to be recorded, special attention has to be given to uniformity of the response of the hydrophone and recording circuit, both in amplitude and in phase, over a wide range of frequencies. Because of the very short time scale when small charges are used, trouble may be caused by the finite time of transit of the sound wave across the hydrophone, and by diffraction around the hydrophone and its supports. Changes in the orientation of the hydrophone sometimes have a surprisingly large effect on the form of the recorded pressure-time curve. Tiny quantities of gas occluded on the face of the hydrophone or included in water-proofing or insulating materials can slow up the response to a steep-fronted pulse, and make the behavior of the hydrophone nonlinear. Natural resonances in the hydrophone can be shock-excited by a steep-fronted pulse, causing spurious wiggles in the pressure-time curve, and in some cases making the pulse appear to last many times longer than it actually does. At short ranges, where relatively insensitive hydrophones may be used, emf's due to the impact of the pressure wave on the connecting cable may give spurious signals. With long cables, impedance matching and dielectric losses may have to be considered. These and many other points are discussed at length in other reports.[1-6]

In the following sections we shall first consider propagation of explosive pulses through the water alone, and later, in Section 9.4, shall take up pulses reflected from or transmitted through the bottom.

9.2 SHORT-RANGE PROPAGATION IN DEEP WATER

9.2.1 Attenuation and Change in Form of the Pulse

As the earlier chapters of this volume have shown, the most important single factor affecting the shape and strength of a sound pulse of given frequency traveling through the ocean is the variation of the velocity of sound from point to point, due chiefly to temperature gradients but produced also to some extent by pressure and salinity gradients. Since to a first approximation the velocity of sound is a function simply of the depth and to this approximation can be calculated from bathythermograph records, it will be convenient to separate, as far as possible, those features of explosive sound propagation which are due to this variation of velocity with depth from those features which are due to other properties of sea water and which would be encountered even when the bathythermograph record indicates no appreciable refraction. In this section we shall consider the latter features, recognizing, however, that undetected small-scale fluctuations in the velocity of sound may possibly be an important factor in accounting for them.

One of the most interesting features to be found in the measurements of explosive pulses at ranges from 30 to 2,000 yd is that with increasing range there is an increase in the time required for the pressure in the initial pulse to rise to its peak value. At shorter ranges, it will be remembered, this initial pulse is a shock wave and its time of rise is less than the resolving time of any measuring apparatus which has been used (see Section 8.3). Unfortunately, the measurements which have been made of the time of rise are not sufficiently detailed to establish the cause of this variation with range. Table 1 summarizes the experimental information to date; this information was taken from two NDRC reports.[7,8] In this table "time of rise" is defined as the interval between the first measurable increase of pressure and the maximum of the pressure-time curve. "Resolving time" is defined as the value of time of rise which the system would record for an instantaneous rise in pressure in the water. For the first set of observations this time was measured directly from records of shots at close range; for the other two sets it was merely estimated from acoustical and electrical characteristics of the hydrophone and circuit.

The data given in Table 1 have been chosen to exclude any cases where the hydrophone was in or near the shadow zone predicted from bathythermograph data. They therefore presumably represent an effect which occurs in the absence of large-scale refraction, although it is not impossible that through inaccuracy of the computed ray diagrams some of the shots at the longer ranges may have been close enough to the shadow zone to increase the time of rise by virtue of the shadow-zone effect discussed in Section 9.2.3 and shown in Figure 9. The deep-water data of reference 8, which are given in the table, are plotted in Figure 9 of that section, for comparison with similar time-of-rise data taken in the shadow zone. It is worth noting that in these experiments no marked dependence of time of rise on the depth of the explosion was found for those cases where the hydrophone was not in or near the shadow zone; a slight increase in time of rise with increasing depth was observed, but this was not significantly greater than the experimental error. Slightly more than half of the observations of reference 8 fell within ± 2 μsec of the means given in Table 1.

For the deep-water shots off San Diego, the variation of apparent time of rise with range is approximately what one would expect if the resolving time of the apparatus were actually about 10 μsec and if the

TABLE 1. Measured times of rise of the pressure in the initial pulse from an explosion, at various ranges.

Location, reference, and comments	Resolving time of recording system in microseconds	Water conditions	No. of shots	Charge	Range in yards	Depth of charge in feet	Depth of hydrophone in feet	Average measured time of rise in microseconds
Woods Hole Harbor Reference 7 results possibly influenced by nonlinear response of gauge	13 (Obs at short range)	Unknown (winter)	1	300 gm loose tetryl	33	20	10	50
			1	"	133	20	10	60–70
			1	9-lb TNT	367	20	10	50
				300 " "	167			50–60
Deep water off San Diego Reference 8	5 (Est)	Sea state usually 1 or 2 Negative temperature gradient near surface usually present	8	No. 8 Cap	120–133	20	12	11
			5	"	240	20–100	11	11
			2	"	320	40	12	12
			51	"	510–644	10–300	11–54	13
			25	"	700–900	20–400	11–12	15
			54	"	1,100–1,400	20–400	10–12	21
			11	"	1,800–1,860	20–400	11–12	27
San Diego Harbor Reference 8 data considered doubtful because of unexplained characteristics of p-t curves	5 (Est)	As above	7	"	110	10–35	12	5
			6	"	275	20–55	12	6
			2	"	385	10–55	12	6
			17	"	530	10–55	12	7

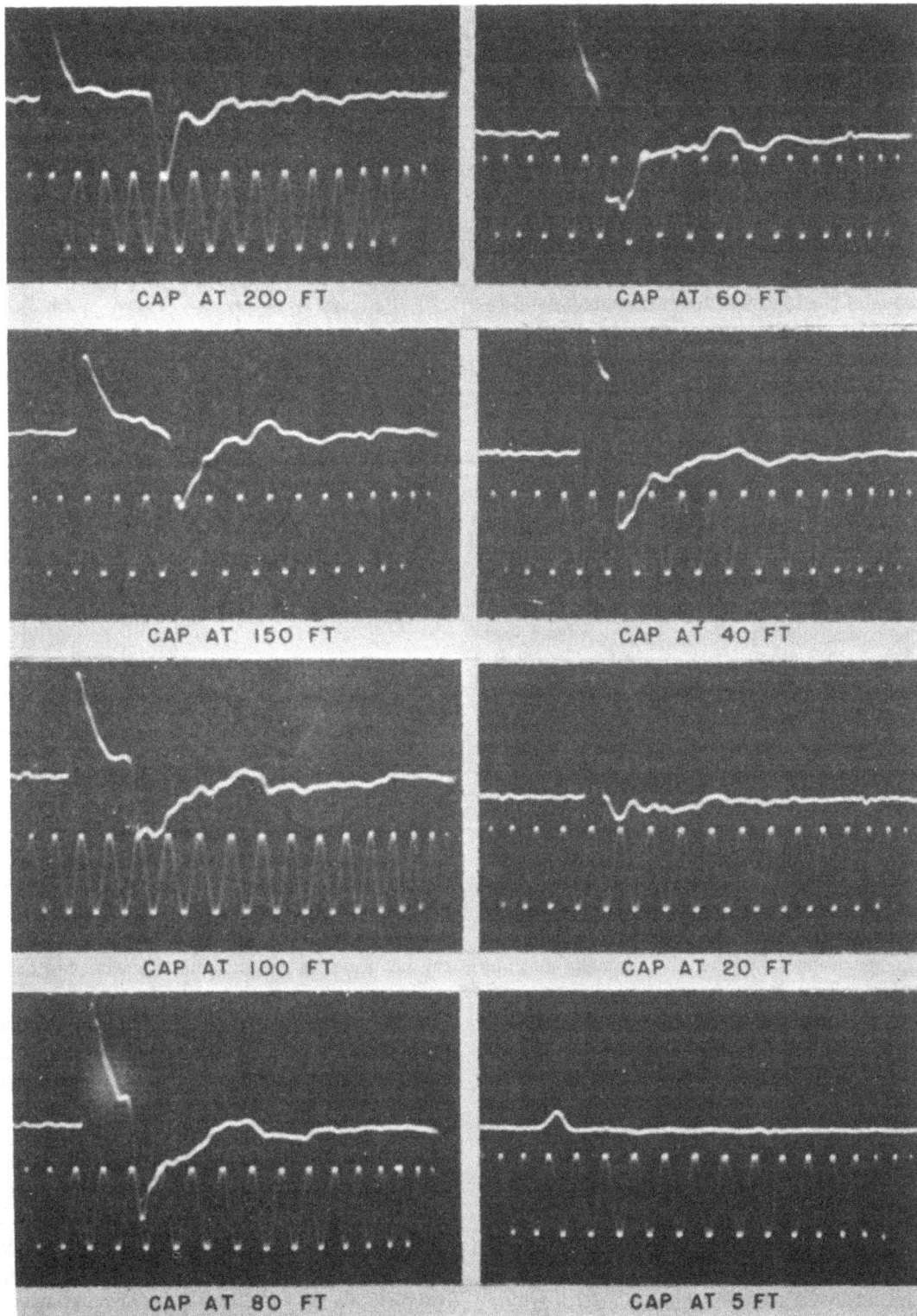

FIGURE 1. Shock wave pulses received via direct and surface-reflected)paths. Source: No. 8 blasting cap at depths indicated. Hydrophone at depth 11 feet and range 1,100 yards. Date: Feb..26, 1942.

explosive pulse is subject to the same frequency-dependent attenuation law as has been observed for supersonic sound (see Section 5.2.2). A more precise comparison of theory with observation would, however, require knowledge not only of the exact response of the hydrophone and recording system to a discontinuous change of pressure, but also of the slight dependence of the sound velocity on frequency. The WHOI data, on the other hand, are much less understandable. The increase in recorded time of rise from 13 μsec at short ranges to 50 μsec at 100 ft implies an attenuation of the high-frequency components of the explosive pulse, which is orders of magnitude greater than that encountered for supersonic sound. The comparative slowness of the increase in time of rise at greater ranges shows that this attenuation is nonlinear; and the most plausible suggestion seems to be that it has its origin in the coating of the hydrophone, rather than in the sea water (see Section 8.3).

So far we have considered only the direct pulse. The surface-reflected pulse may be expected to have a longer time of rise, or more correctly, time of fall, because of the diversity of possible paths from explosion to hydrophone involving reflection from various wave troughs. Whether the smearing out of the wave due to this effect exceeds that responsible for the time of rise of the direct wave will of course depend upon the geometry and especially upon the roughness of the sea. Some typical oscillograms of shots made in an average calm sea are shown in Figure 1, taken from a report by UCDWR[9] and from reference 8. These shots show that the roughness of the surface has surprisingly little effect on the first part of the reflected pulse, although the tail shows irregularities which are probably due in part to nonspecular reflection. The most remarkable thing about these records, however, is that as grazing incidence is approached the effective reflection coefficient of the surface remains close to unity far beyond the point at which the crests of the waves are in the geometric shadow created by the troughs. Figure 2 shows the variation of the reflected amplitude with angle of incidence. This quantity was estimated for a number of shots, including those shown in Figure 1, by taking the difference between the reflected peak and the estimated pressure in the direct pulse at the same time. For comparison, it may be noted that for a train of surface waves traveling in the direction from source to receiver, half of the surface of the sea in the region where reflection occurs would be in geometric shadow from source or receiver, or both,

if the ratio of crest-to-trough amplitude to wavelength is about one-third the angle which the incident ray makes with the horizontal. Since the sea was not unusually calm on the day the shots were made, Figure 2 strongly suggests that the sea surface acts as a flat reflecting plane for supersonic sound even when only the wave troughs are in the direct sound field. An effect of this sort has been predicted theoretically for the case of sinusoidal sound.[10]

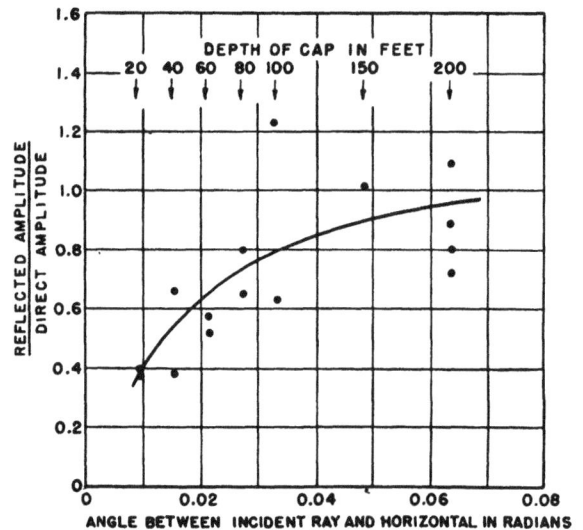

FIGURE 2. Variation of peak amplitude of surface-reflected pulse with angle of incidence. Horizontal range, 1,100 yds; depth of hydrophone, 11 ft.

Measurements of the time interval between the direct and surface-reflected pulses have been reported in a memorandum by UCDWR[11] and in reference 9, and agree with the values calculated from the geometrical formula (31) of Section 8.7 to within the accuracy of measurement of the depths of cap and receiver.

In the absence of refraction, one would expect the peak pressure of an explosive pulse having a time constant θ to be attenuated at long ranges at approximately the same rate as sinusoidal sound of a suitably chosen frequency, this "effective frequency" being probably a few times smaller than $1/\theta$. A detailed relationship between the attenuation of a weak explosive pulse and that of sinusoidal sound could be worked out by the methods of Fourier analysis; however, such a relationship would be considerably affected by dispersion, that is, by any variation of the velocity of sound with frequency. This is a phenomenon which must occur if there is a frequency-dependent attenuation. A brief discussion of attenuation data from the standpoint of Fourier analysis will be given in Section 9.2.4. In comparing the attenuation of explosive sound with that of sinusoidal sound, however, it must be kept in mind that the mechanism responsible for attenuation of the peak pressure in the initial pulse from an explosion is somewhat different at short and long ranges. At long ranges one may expect that linear absorption and dispersion will account for the decay and change of form of the pulse; while at short ranges, as explained in Section 8.3 to 8.5, the nonlinear Riemann overtaking effect plays the predominant role, causing the time constant of the pulse to increase with time and causing an attenuation of the peak pressure whose magnitude is independent of the specific mechanism responsible for the dissipation of energy. The range at which the transition from the latter type of attenuation to the former takes place is probably roughly the range at which the attenuation given by the short-range law (24) of Section 8.5 equals the attenuation computed by Fourier analysis from the linear laws of absorption and dispersion which hold for weak sinusoidal sound.

Observations on the variation of peak pressure with range do not suffice to determine the magnitude of the attenuation of this quantity, or even to establish that it is different from zero. For example, the data of Figures 3 and 4 of Section 9.2, taken from UCDWR experiments cited in reference 8, are in good agreement with intensity calculations which ignore attenuation. It would hardly be reasonable, however, to assume that there was practically no attenuation in these experiments, since the increase of time of rise with increasing range indicates that dissipative processes were appreciable. Measurements on a larger scale have been made at CUDWR–NLL.[12] These show an attenuation of about 2 db per kyd, that is, slightly more than that predicted by equation (24) of Section 8.5 for shock waves in an ideal fluid. Because of the difficulty of correcting accurately for the effect of refraction on the intensity, and because of the possibility of nonlinear behavior of the hydrophone in CUDWR–NLL experiments, none of these results can be given much weight.

Pulses have been propagated to very long ranges in the strata of the ocean where the velocity of sound is less than at shallower or deeper depths. These will be discussed in Section 9.3.2. Attenuation measurements have been made for these pulses with the use of recording equipment responsive to particular bands of frequencies. Because of the limited frequency response of the equipment, the results cannot be interpreted in terms of peak pressure; however, as will be seen in Section 9.3.2, they indicate that the low-frequency part of the pulse is transmitted with very low attenuation.

9.2.2 Effects of Refraction

This section and the next will be concerned with effects which can be correlated with the variation of the velocity of sound with depth, as determined from bathythermograph data, at ranges up to a few thousand yards. At these ranges few if any of the ray paths will cross one another. In Section 9.3, on the other hand, we shall consider propagation over long ranges in a layer of minimum sound velocity where many different ray paths can be found leading from the source to the receiver. Most of the results discussed in this section and the next will be taken from experiments conducted by UCDWR, and described in references 8, 9, and 11. Similar though less detailed results have been obtained in England at His Majesty's Anti-Submarine Experimental Establishment, Fairlie.[5]

In the UCDWR experiments considerable attention was devoted to the securing of bathythermograph data at as nearly as possible the same time as the firing of the shots. From these data ray paths were computed and graphs of predicted intensity as a function of depth were prepared for various values of the ranges, the intensities being computed from the geometrical divergence of the rays by the methods described in Chapter 3. Figure 3 shows a typical comparison between computed and observed peak pressures for a day when the upper layers of water were nearly isothermal. The pressure levels are all plotted in decibels, that is, the abscissas are $10 \log_{10} p_{max}^2$. It will be seen that in the direct zone the observations are in reasonable agreement with the ray theory but that they would not agree at all well with an inverse square law. It is a little surprising that the agreement with ray theory should be so good, since the theoretical intensities were computed without any allowance for attenuation. Particularly noteworthy is the decrease in intensity as the cap is raised into the shadow zone from below. The 3,600-yd points are all in the shadow zone. Figure 4 shows a similar comparison for another day. On this day conditions were rather variable. Of the three bathythermograph runs taken during the morning, one showed a very shallow split-beam pattern while the other two

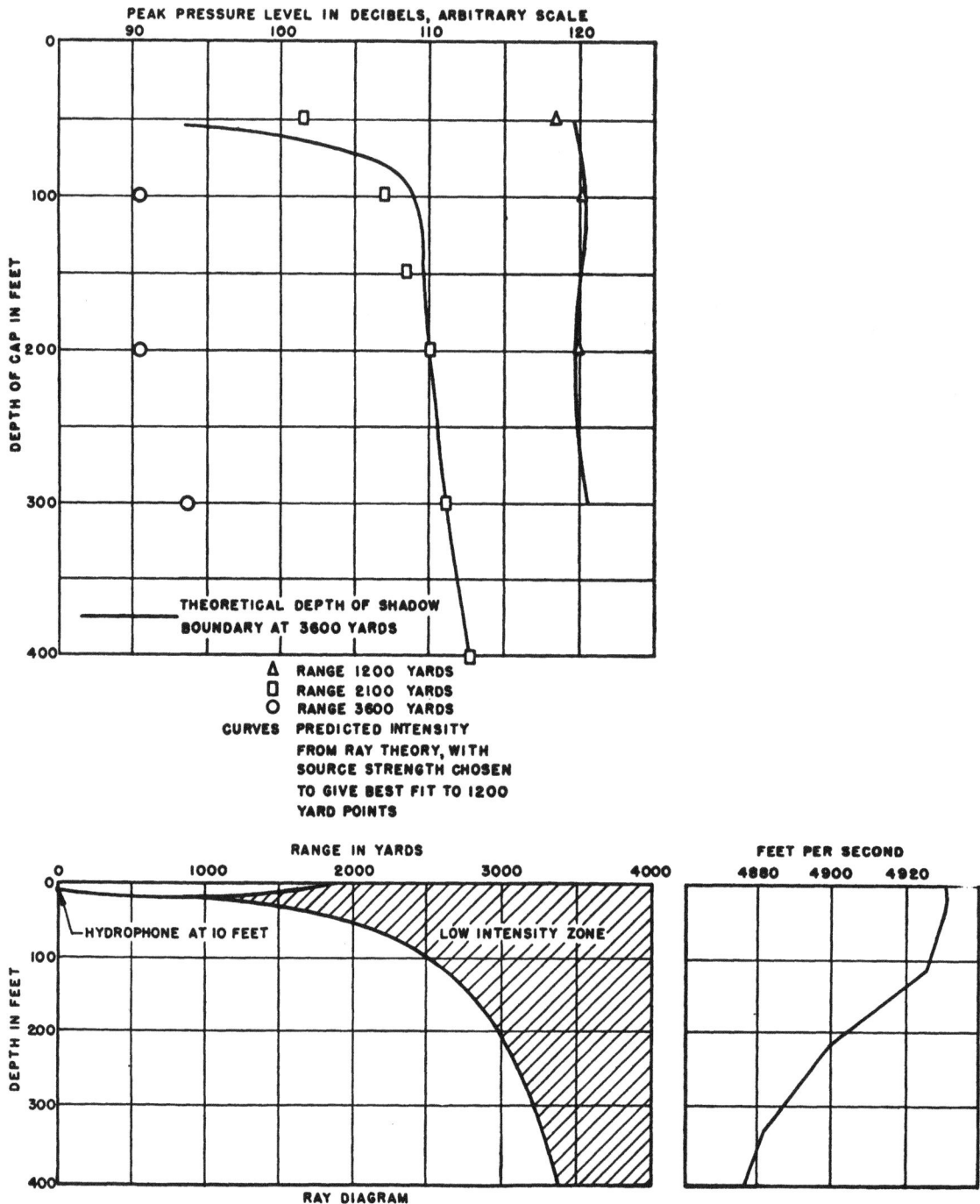

FIGURE 3.   Comparison of observed peak pressures with values calculated from ray theory for a split-beam pattern.

showed a weak negative gradient extending to the surface; in the afternoon there was a strong negative gradient at the surface. The ray diagram and theoretical intensities shown for the morning shots were constructed from an average of the three temperature-depth curves taken during the morning, and

thus are only a rough approximation to the truth at any one time; however, the error should not be serious except near the boundary of the shadow zone. It will be seen that the agreement of the theoretical and observed intensities is again fairly good. The reduction of intensity in the shadow zone for this

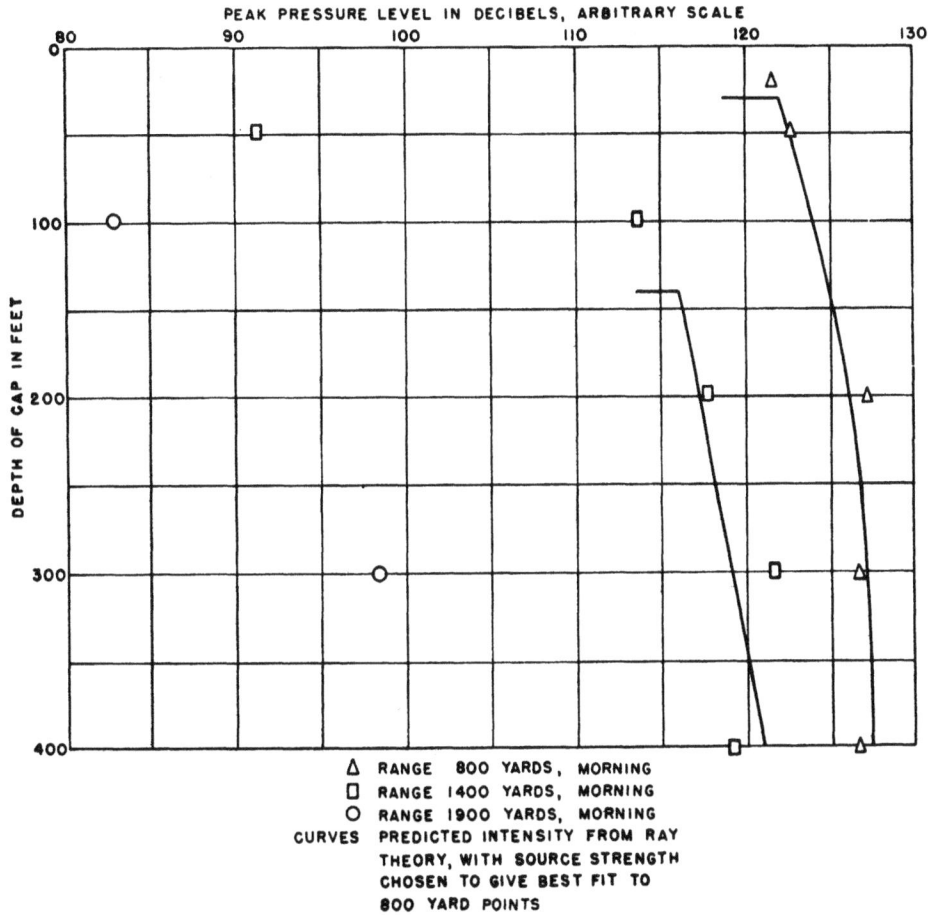

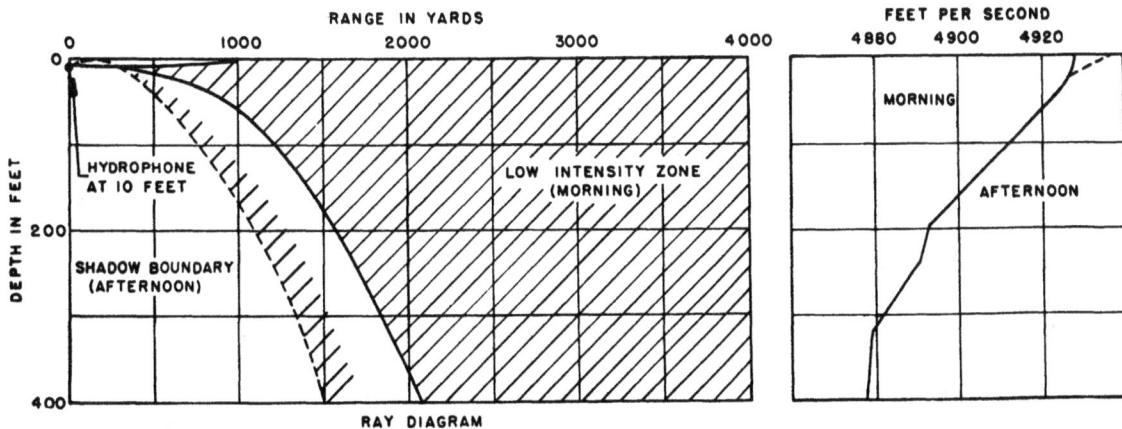

FIGURE 4. Comparison of observed peak pressures with values calculated from ray theory for a negative gradient extending to or almost to the surface.

case is, as one would expect, more pronounced than for Figure 3.

A particularly interesting variation of intensity with range is shown in Figure 5.[13] This series of shots was made at a single depth at various ranges, on a

day when there was a moderate negative temperature gradient at the surface. The velocity-depth curve and ray diagrams are shown in Figure 6. The hydrophone was placed at a depth of 54 ft, just below the knee of the velocity-depth curve, which comes at 48 ft. This causes a peculiar irregularity in the ray diagram, as shown in Figure 6; in this figure rays are drawn for initial inclinations in 0.1° steps from 1.0° to 2.5°. Rays whose vertices lie below 48 ft (1.0°, 1.1°, 1.2°) are bent downward strongly by the strong negative gradient below the knee. Rays rising above 48 ft (1.3°, 1.4°, etc.), however, are bent downward only weakly by the weak negative gradient above the knee, and diverge more rapidly; their divergence is

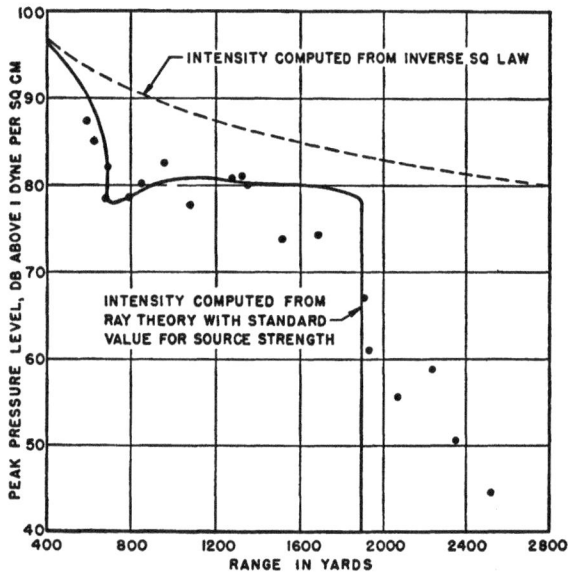

FIGURE 5. Observed and calculated variation of peak pressure with range for a negative temperature gradient. Hydrophone depth, 54 ft; cap depth, 100 ft; sound conditions as shown in Figure 6.

further increased when they curve back down through the knee. The result of this "double layer effect" is a "hole" in the sound field immediately beyond the 1.2-degree ray, i.e., as a sharp dip in the intensity-range curve, as shown in Figure 5. The theoretical curve for this figure was not fitted to the points, but was computed from the known absolute strength of the source. At its minimum this theoretical intensity is some 14 db lower than that which would be predicted by the inverse square law, and it is therefore quite significant that the observed intensities follow it so closely. As the shadow boundary is approached the observed intensities drop markedly before the computed shadow boundary is reached. This might be due to a departure from ray theory or to a slight error in the assumed temperature distribution near the surface which would cause the computed shadow boundary to be too far out. While data on the time of rise of the pressure to its peak value (see Table 2

on page 204) favor the latter interpretation, the systematic tendency of the observed shadow boundary to lie closer than predicted, discussed in Section 5.4.1, suggests that some other cause must be found for this apparent discrepancy.

9.2.3 Shadow Zones and Diffraction

As we have seen in Figures 3, 4, and 5 of Section 9.2.2, signals of appreciable intensity are received in places where no rays on the sound ray diagram penetrate. This phenomenon is familiar in experiments with sinusoidal sound and has been discussed in Section 5.4. This and other departures of observed intensities and pulse forms from those computed by applying ray theory to bathythermograph observations may be due to any of several causes. In the first place, the concept of propagation of sound along ray paths is only approximate; a more exact application of acoustical theory predicts that some sound should penetrate into the shadow zone by diffraction, and that in and near the shadow zone the shape of the pulse should be somewhat different from its shape close to the source. In the second place, it is known that the temperature in surface layers of the ocean is not simply a function of the depth, but varies appreciably from one position to another in the same horizontal plane. Thus a set of rays which were really accurately constructed would differ in many features from the rays which one computes from the assumption that temperature is a function of depth alone. Thirdly, the water is not homogeneous but contains bubbles, fish, etc., which can scatter sound and cause its velocity to vary with the frequency. Finally, the water is not at rest; portions of it may be set in motion relative to the rest by waves and swells, by tidal or other currents, and by the motion of ships, fish, etc. These irregularities in velocity, although small in comparison with the velocity of sound, may easily cause appreciable alterations in the shape and strength of an explosive pulse at ranges of the order of a thousand yards.

Unfortunately, the experimental data so far available are not sufficiently complete to enable many sure conclusions to be drawn about the mechanisms responsible for the various effects observed. However, a few tentative conclusions can be reached regarding the origin of the sound which is found in the shadow zone in experiments such as those of references 8 and 9 which have been discussed in the preceding section. Referring to Figure 5, it will be seen that in the

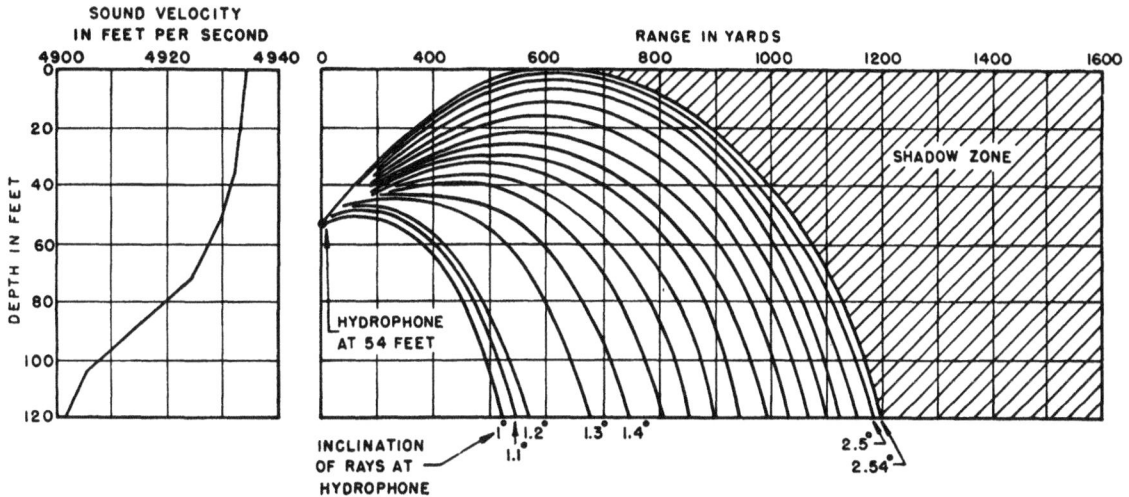

FIGURE 6. Sample velocity-depth curve and ray diagram for the day on which the data of Figures 5, 7, 8, 9, 12, 14, 15, and 16B were taken.

shadow zone the intensity, defined as the square of the pressure at the first positive peak, decreases at a rate of 35 or 40 db per kyd, down to a level which is about 30 db below the value which would be obtained by extrapolating the pressures obtained at short ranges according to the inverse square law. Less complete intensity data obtained on other days give comparable values for the decrease in intensity. This decrease is of the same order as that which would be expected for diffracted sound in an ideal medium in which the velocity of sound varies with depth in the manner shown in Figure 6.[14] However, as has been noted in Section 5.4, a very similar decrease is observed for the case of 24-kc sinusoidal sound;[15] for this case, however, the rapid decrease of intensity with increasing range ceases after the intensity has fallen to about 40 db below the inverse square extrapolation, and beyond this point the decrease in intensity seems once again to be described by an inverse square law. For this and other reasons the supersonic signal received at a considerable distance inside the shadow zone is believed to arrive there by some sort of scattering process, rather than by diffraction. It does not seem likely, however, that scattering contributes appreciably to the observed intensities of the explosive pulses plotted in Figure 5 or in Figures 3 and 4. For the disturbance produced at the hydrophone by scattered sound is a superposition of the disturbances produced by various scattering centers, and since these have different times of travel, the number of scattering centers which can contribute to the disturbance at the hydrophone at a

given instant increases with the length of the pulse. The explosive pulse is so short that one would expect the scattered intensity to be lower by 30 db, at the very least, than that from a 100-msec pulse of sinusoidal sound having the same initial amplitude and a frequency of the same order as that which predominates in the explosive pulse in the shadow zone. It is thus hard to see how the scattered intensity could be comparable with the shadow zone intensities observed in Figures 3, 4, and 5. Thus we are forced to consider diffraction as the mechanism by which an explosive pulse penetrates the shadow zone, at least in cases such as Figure 5 and the afternoon shots of Figure 4, where a true shadow zone is produced by downward refraction. For a split-beam pattern like Figure 3, the existence of a shadow zone in the ray diagram is due to the fact that the assumed velocity-depth curve has a discontinuity in slope; since the true variation of velocity with depth is undoubtedly represented by a smooth curve, it is better in this case to speak of a zone of low intensity, rather than of a shadow zone, and the argument just given for the occurrence of diffraction is less compelling.

The diffraction hypothesis receives support from a study of the shapes of pulses received in or near the shadow zone. Figure 7 shows the oscillograms for some of the shots plotted in Figure 5. It will be seen that as the shadow boundary is approached, the direct and surface-reflected pulses merge, and that within the shadow zone the pulse is oscillatory. The time of rise to the first maximum begins to increase suddenly at about the position of the shadow bound-

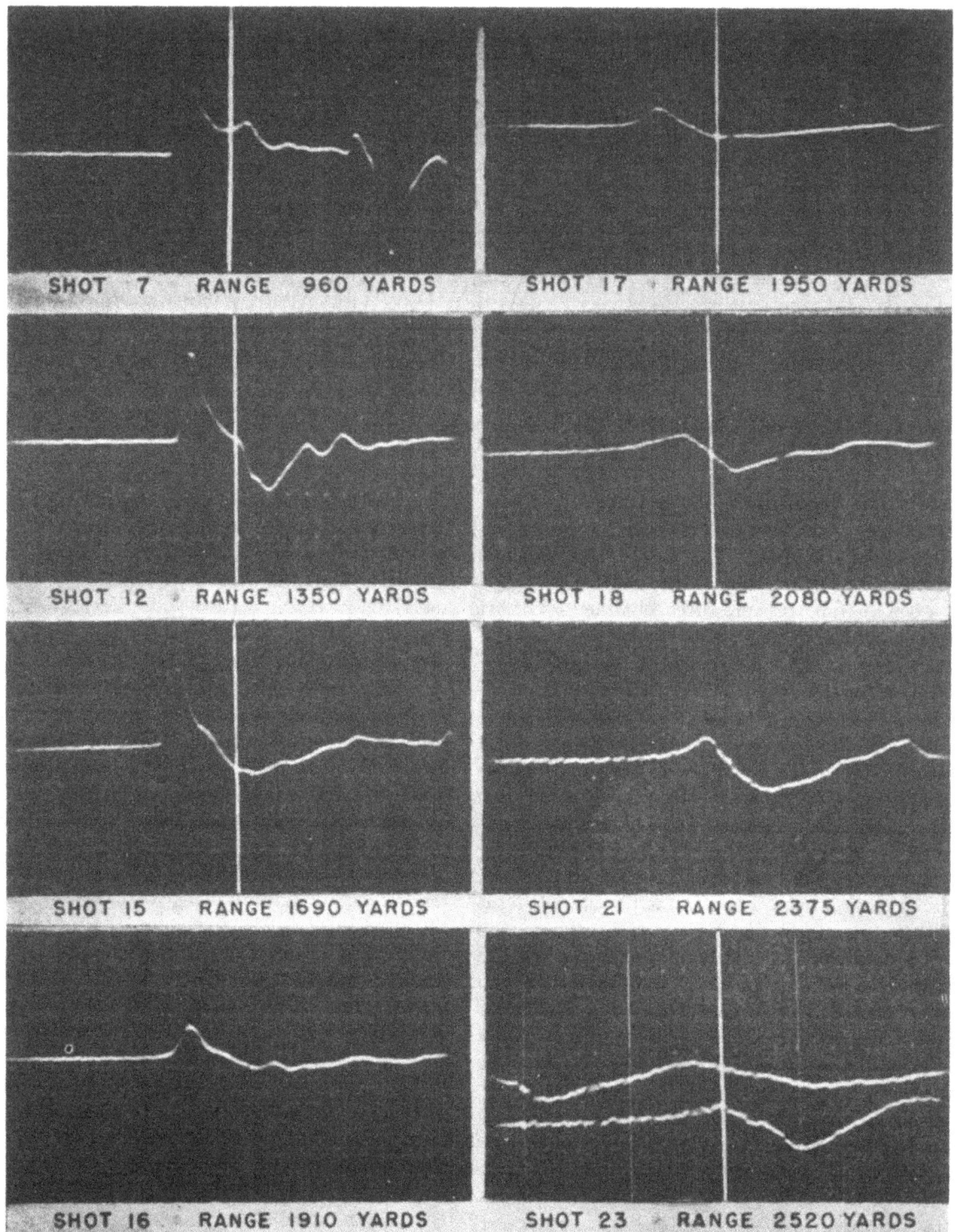

FIGURE 7.   Changes in the shape of an explosive pulse on passing from the direct zone into the shadow zone.   Source: No. 8 cap at depth 100 feet and at ranges indicated.   Hydrophone at depth 54 feet.   Date: Apr. 3, 1942; gain of recording system sometimes changed between shots.

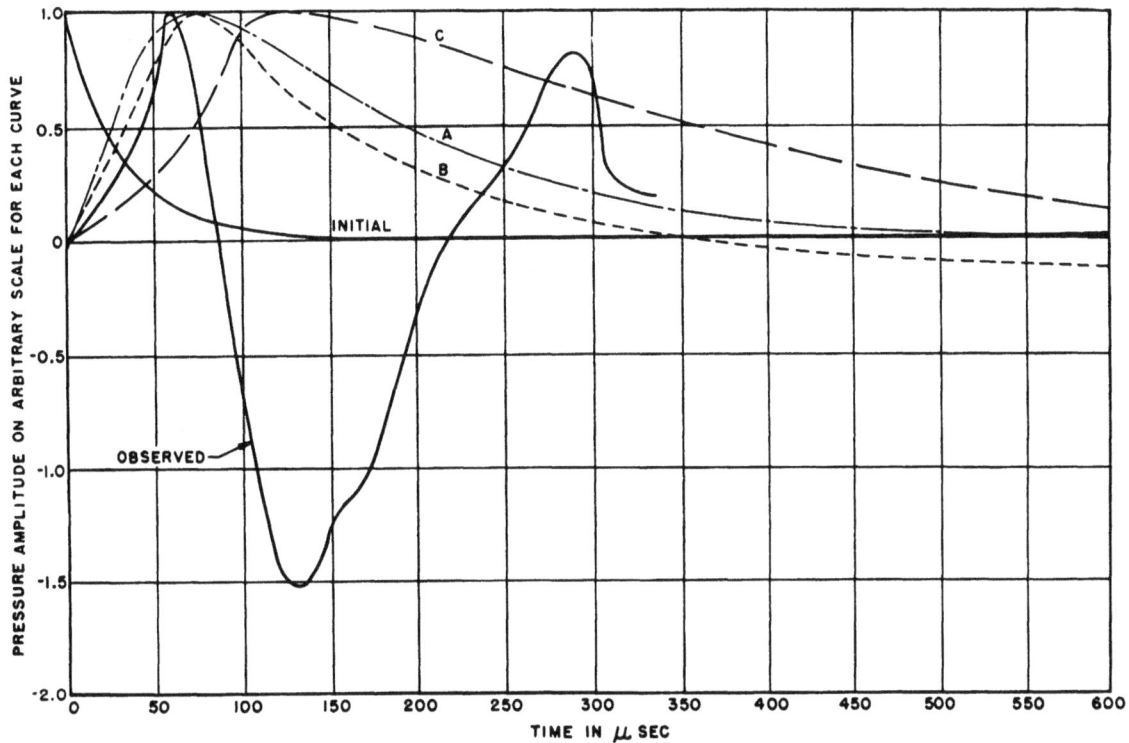

Curves *A* through *C* are computed from diffraction theory, assuming the explosive pulse near the source to have the form $p = p_0 \exp(-3 \times 10^4 t_{sec})$ shown in the curve labeled initial. The conditions assumed in the calculations are compared with those obtaining in the experiment in the following table:

Curve	Velocity gradient		Depth of source	Depth of receiver	Horizontal distance from shadow boundary
A	Constant {	0.1 sec^{-1}	50 ft	100 ft	500 yd
B		0.1	50 "	50 "	500 "
C		0.2	50 "	50 "	500 "
Observed	See Figure 6		50 "	100 "	460 "

FIGURE 8.   Observed and computed pressure-time curves in the shadow zone.

ary and continues to get larger and larger the farther one goes into the shadow zone. These features were observed on all occasions when shots were made in a shadow zone produced by downward refraction, and occasional repeat shots showed that the first cycle or so of the oscillatory pulse observed in the shadow zone was quite reproducible (see Figure 16C). It is interesting to compare these oscillograms with theoretical pressure-time curves for sound diffracted by an ideal medium which has a plane surface and in which the velocity of sound depends only on the depth. Such theoretical pressure-time curves for explosive sound in the shadow zone have been com-

puted by CUDWR;[14] because of mathematical complications in the theory, however, the theoretical calculations have not been made for exactly the same conditions as any of the shots of Figure 7. Figure 8 shows the comparison with shot 21 of Figure 7. The agreement is good as regards time of rise, but the amplitude of the negative part of the pulse is much greater for the observed than the theoretical case, and the observed wavelength is much shorter. The discrepancy would probably be reduced if the calculation could be carried through for a velocity distribution approximating the observed one more closely. However, it is quite possible that diffraction theories

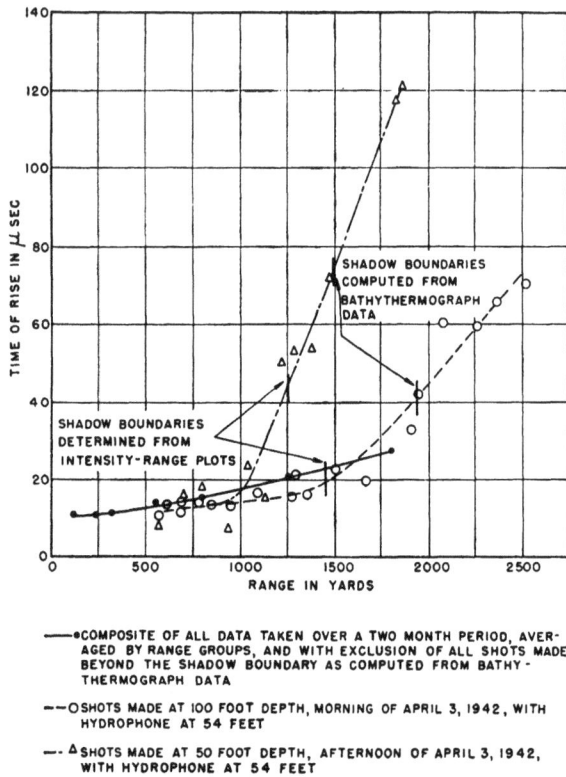

——●COMPOSITE OF ALL DATA TAKEN OVER A TWO MONTH PERIOD, AVER-
   AGED BY RANGE GROUPS, AND WITH EXCLUSION OF ALL SHOTS MADE
   BEYOND THE SHADOW BOUNDARY AS COMPUTED FROM BATHY-
   THERMOGRAPH DATA

—─○SHOTS MADE AT 100 FOOT DEPTH, MORNING OF APRIL 3, 1942, WITH
   HYDROPHONE AT 54 FEET

—·△SHOTS MADE AT 50 FOOT DEPTH, AFTERNOON OF APRIL 3, 1942,
   WITH HYDROPHONE AT 54 FEET

FIGURE 9.   Dependence of time of rise on range, showing influence of the shadow zone.

observed and computed shadow boundaries are marked on the figure. The abrupt increase in time of rise on crossing the observed shadow boundary is quite conspicuous, and was noticed in UCDWR experiments on all days when strong downward refraction was present. For comparison, Figure 9 also shows as a full curve the average values given in Table 1 of Section 9.2.1, which at each range represent the time of rise when refraction conditions are such that the hydrophone is in the direct zone. It will be seen that out to the shadow boundary all values agree to within the fluctuations of the data.

The sharpness of the increase in time of rise as the shadow boundary is crossed suggests using a plot like Figure 9 to determine the location of the shadow boundary. Table 2, taken from reference 9, gives a comparison of the range to the shadow boundary determined in this way with the range as deduced from the bathythermograph measurements, and also with the range as deduced from a plot of peak intensity against range, such as Figure 5. It will be seen that time of rise and peak pressure always give very nearly the same position for the shadow boundary, but that this position does not always agree well with that given by the bathythermograph. This is not surprising, since such things as surface waves and small changes of temperature very close to the surface can

TABLE 2.   Comparison of three methods of determining the range to the shadow boundary.*

Date, 1942 Depth of explosion in feet	April 2 100	April 3 50	April 3 100	April 8 50	April 9 50
Range from bathythermograph (yd)	1,800	1,490	1,940	1,080	1,010
Range from time of rise (yd)	1,170	1,100	1,430	1,150	1,100
Range from peak pressure (yd)	1,200	1,250	1,450	1,170	1,170

* Depth of hydrophone, 54 ft in all cases.

based on the concept of a horizontally stratified medium will prove inadequate to explain the experimental results, and that some more complicated process must be considered.

Figure 9 shows how the time of rise to the first pressure maximum is affected by crossing the shadow boundary. The lower dashed curve is for the same shots as Figures 5 and 7, while the upper dot-dash curve is for shots at shallower depth at a different time on the same day. Velocities and ray diagrams for this day have been given in Figure 6. Since the shadow boundaries computed from the temperature data do not agree very well with the boundaries determined empirically from the behavior of sound intensity as a function of range (see Table 2) both the

have quite an appreciable influence on the position of the shadow boundary, and the shadow boundary is determined by the distribution of temperature over a large area of the sea, while the bathythermograph measures temperatures on only one vertical line.

Many of the oscillographic pressure-time records obtained of explosive pulses are much less simple and comprehensible than the examples which have been selected for discussion in the preceding paragraphs. Some of the irregularities can apparently be explained in terms of multiple ray paths, while others are more puzzling. Figure 10 shows some typical oscillograms obtained on a day when the bathythermograph showed that there were alternate layers of large and small temperature gradients, which should have pro-

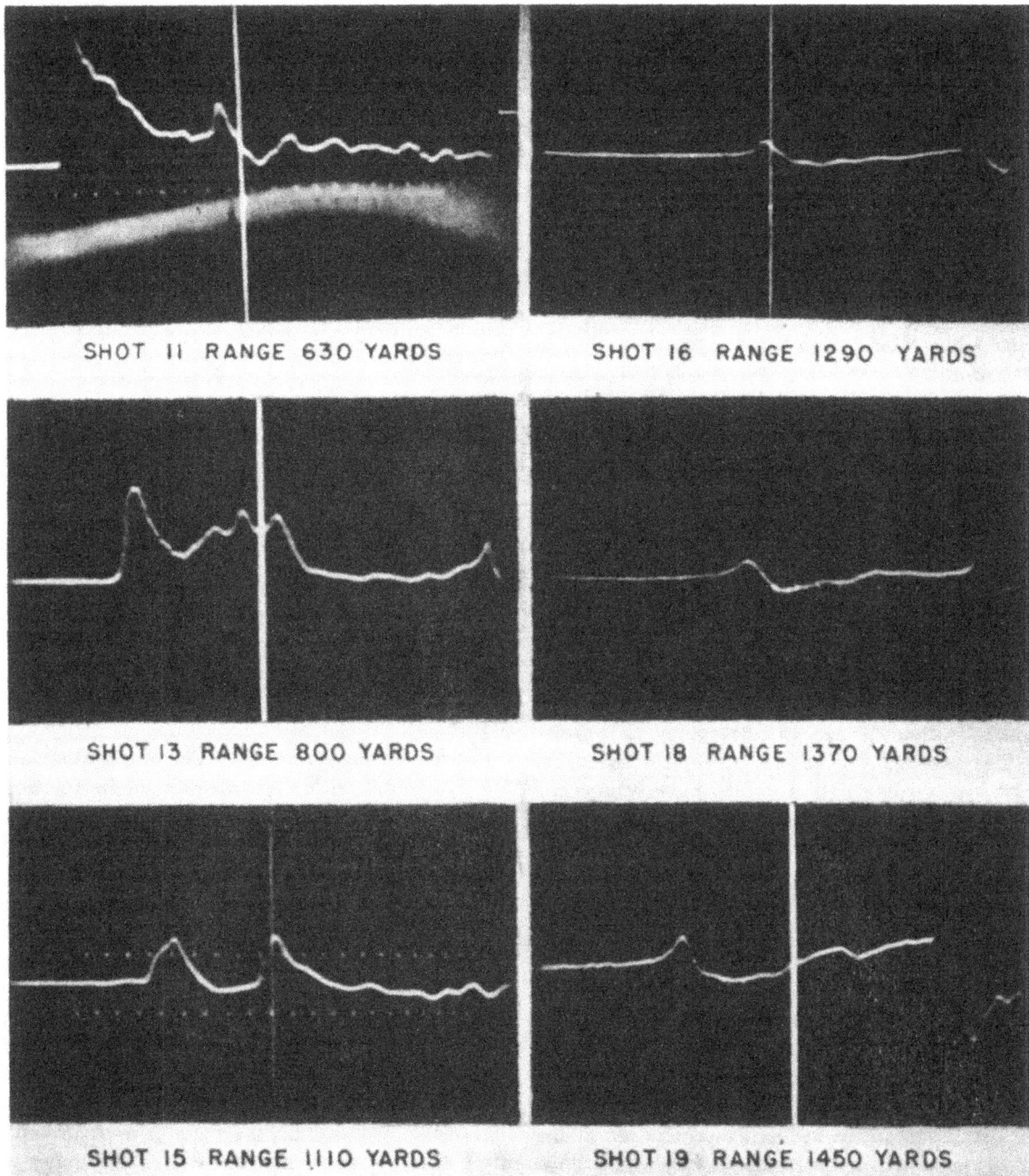

FIGURE 10. Records showing multiple path effects. Source: No. 8 cap at depth 100 feet and at ranges indicated. Hydrophone at depth 54 feet. Date: Apr. 2, 1942. Gain of the recording system was sometimes changed between shots.

duced considerable crossing of the ray paths.[9] The 630-yd oscillogram has very much the form one would expect if there were two ray paths leading from cap to hydrophone. That for 800 yd suggests a number of ray paths, but the arrivals are less sharp. It is diffi-

cult to correlate these features in detail with the calculated ray diagram, however, and there appear to be variations which make it hard to pick out systematic trends in the characters of the oscillograms as the range is gradually increased or decreased.

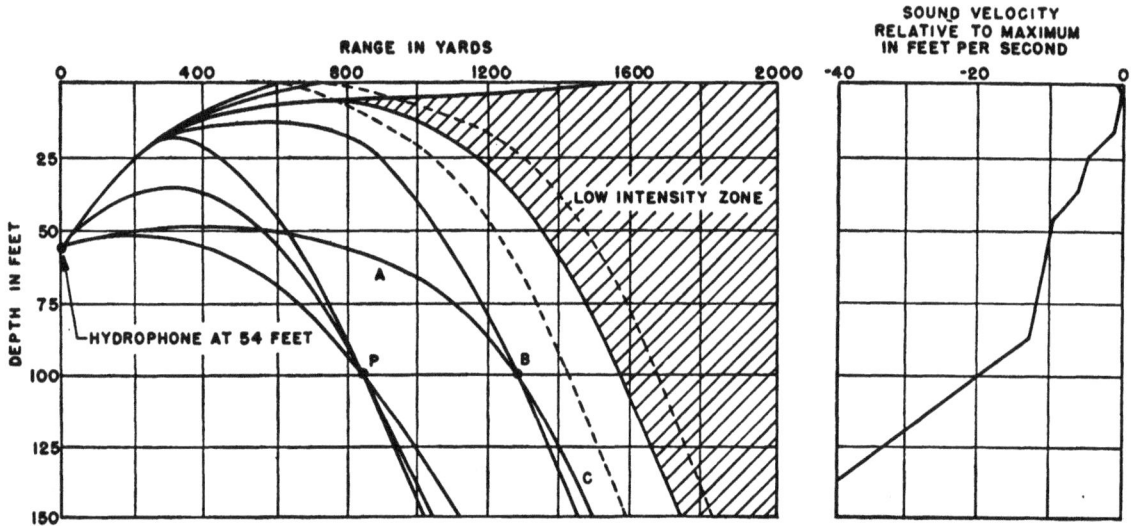

FIGURE 11.   Velocity-depth curve and sample ray paths for Figure 10.

The three oscillograms on the right of Figure 10 do seem to show a regular trend, however, in that with increasing range the first pulse becomes rapidly weaker in comparison with the second, and its time of rise increases. Comparison with the oscillograms of Figure 7, which show the same trends, suggests that the first pulse may have reached the hydrophone by diffraction, while the second corresponds to a direct ray. The possibility of a phenomenon of this sort is suggested by calculations which have been made on ray paths for this day, a few of which are shown in Figure 11. The ray ABC crosses the 100-ft depth line at a greater value of the horizontal range than do rays of slightly greater or slightly smaller inclinations so that this ray and its neighbors have an envelope, or caustic, passing through B. This caustic forms a shadow boundary as far as rays of inclinations near that of ABC are concerned,[16] although it happens in this case that rays having considerably greater inclinations fall outside the caustic. The complete ray diagram would of course be quite complicated, and attempts at a detailed correlation of the oscillograms with the bathythermograph data have not been very successful.

The oscillograms shown in Figure 12 are for shots made on the same day as those in Figure 7; bathythermograph data and ray diagrams for this day have been given in Figure 6. These oscillograms show that even when no crossing of rays is predicted, multiple peaks and similar irregularities can still occur, although these features are less pronounced than in Figure 10. As was cautioned in Section 9.1, instru-

mental sources for such irregularities must always be suspected; however, it seems likely that many of the unexplained irregularities found in UCDWR experiments are due in some way to oceanographic conditions.

9.2.4   Results of Fourier Analysis

So far we have discussed the propagation of explosive sound with little mention of its relation to sinusoidal sound waves. There is an important relation between the two, however. Any pulse of arbitrary shape can be approximated as accurately as desired by a linear superposition of a sufficiently large number of sine waves, of suitably chosen frequencies, amplitudes and phases. If the laws of propagation of sound are linear in the amplitude of the disturbance, as we believe them to be when the amplitude is sufficiently small, we can predict the changes in the size and shape of the pulse as it travels through the water from the changes in amplitude and phase which each of the sine waves would undergo if it were present by itself. Conversely, if the behavior of the pulse were accurately known, the attenuation and dispersion of all the component sine waves could be calculated.[a]

An analysis of this sort can be extremely useful in

---

[a] That this is indeed a practical possibility has been demonstrated in some experiments at NRL[17] in which the relative calibration curve of the two hydrophones was determined over the range from 5 to 100 kc by Fourier analysis of their responses to detonating caps. The resulting curve was found to be in excellent agreement with one determined by standard CW methods of calibration.

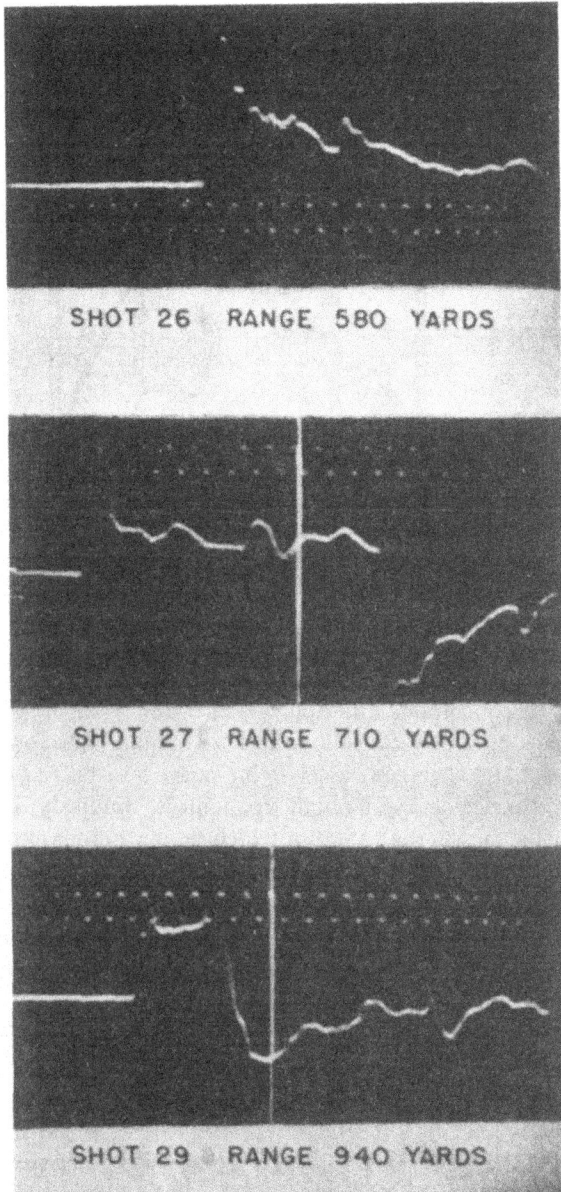

FIGURE 12. Samples of irregular pressure-time curves. Source: No. 8 detonating cap at depth 50 feet and ranges indicated. Hydrophone at depth 54 feet. Date: Apr. 3, 1942.

elucidating the physical mechanisms operative in the propagation of sound in the sea and in predicting the response of resonant or band-pass receiving systems to explosive sound.

The representation of an arbitrary disturbance as a superposition of sine waves is described mathematically by Fourier's theorem. The most useful form of this theorem for our present purpose is the integral

iorm, which states that if $p(t)$ is any function of an independent variable t such that $\int_{-\infty}^{+\infty} |p|^2 dt$ converges, then for all values of t except points at which p is discontinuous

$$p(t) = \int_{-\infty}^{\infty} \phi(f) e^{2\pi i ft} df \qquad (1)$$

where

$$\phi(f) = \int_{-\infty}^{\infty} p(t) e^{-2\pi i ft} dt. \qquad (2)$$

It can also be shown that

$$\int_{-\infty}^{\infty} |p|^2 dt = \int_{-\infty}^{\infty} |\phi|^2 df. \qquad (3)$$

When these mathematical theorems are applied to the pressure $p(t)$ in a pulse of sound, the physical interpretation of the results is simple. The integral on the left of equation (3), when divided by ρc, represents the total energy in the pulse per unit area. The quantity f represents frequency, measured for example in cycles per second, and so the integrand $|\phi|^2$ on the right of equation (3), when divided by ρc, represents the energy per unit area per unit frequency range. The spectrum level of the pulse at frequency f, as measured for example by the energy received from it by a receiving system sensitive only to a narrow band of frequencies in the neighborhood of f, is

$$U(f) = 10 \log_{10} |\phi(f)|^2 \qquad (4)$$

and U will be in decibels per cycle above 1 dyne per sq cm, if p was measured in dynes per sq cm and f in cycles per second.

In evaluating the expression (2) for an experimentally obtained pressure-time curve it is of course not possible to extend the upper limit of integration to infinity; in UCDWR work described in reference 9, for example, the oscillographic record obtained only lasted for a few hundred microseconds, and over the latter part of this range it was hard to estimate the position of the zero line accurately. The integrals which were actually evaluated were therefore

$$\int_0^{t_1} p(t) \sin 2\pi ft\, dt$$

and

$$\int_0^{t_1} p(t) \cos 2\pi ft\, dt,$$

where the origin of time is taken as the time of arrival of the first perceptible pressure and where t_1 is a few hundred microseconds, i.e., is of the order of the duration of the traces which were shown in Figures 7, 10, and 12. Such a curtailment of the upper limit

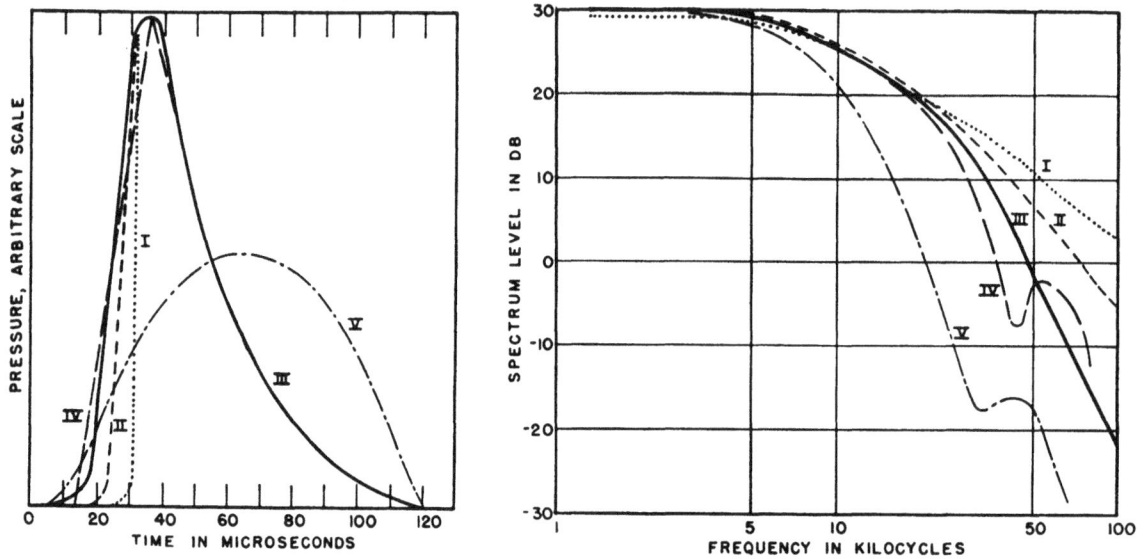

FIGURE 13.   Influence of time of rise on spectrum of a single pressure pulse.

of integration results, unfortunately, in omission of the surface-reflected pulse when its time of arrival exceeds t_1. As a result the computed ϕ will rise to a maximum value as the frequency approaches zero, whereas if the surface reflection were included ϕ would approach a small value or zero. Even if there were no surface reflection we should expect the computed value of ϕ to be considerably in error at very low frequencies through neglect of the tail of the shock wave, which has been discussed in Section 8.5. At frequencies large compared with $1/t_1$ the neglect of the surface reflection is unimportant, although it may result in the absence of some small-scale ripples from the curve of spectrum level against frequency, which would be present if the surface reflection were included. When the surface reflection arrives well within the time t_1, of course, the error due to cutting off the integration at this time is usually negligible, although of course if there are bottom reflections arriving after time t_1 the computed spectral distribution will be that which would apply in the absence of a bottom.

At very high frequencies the spectrum level depends primarily upon the time of rise, and the computed value may be in error if the response time of the hydrophone and recording system does not permit faithful reproduction of the rising portion of the pulse. The extent to which the time of rise affects the spectral distribution is shown by some sample calculations given in reference 9, for hypothetical pulses, which are presented in Figure 13. Note that pulse I, which has a vertical rise, has the highest spectrum

level at high frequencies. It has been shown by mathematicians that the Fourier transform of a finite but discontinuous function p is of order $1/f$ at large values of the frequency f, and we should therefore expect the spectrum level U for pulse I to decrease at 6 db per octave at high frequencies. Similarly, it can be shown that a pulse which is continuous but has discontinuities in slope, such as IV in the figure, should have a spectrum level which decreases at 12 db per octave at sufficiently high frequencies. The spectrum of a perfectly smooth pulse, such as III, should decrease still more rapidly.

Since actual oscillograms usually show irregularities in the tails of the pressure pulses, and since these irregularities are usually not very reproducible from shot to shot, it is pertinent to inquire how much influence they have on the spectral distribution. Sample calculations for hypothetical pulses have shown that the principal effect of a fairly smooth "satellite" peak is to introduce irregularities into the curve of spectrum level against frequency, without much alteration of its general trend.[9]

Figure 14 gives curves of spectrum level U against frequency f, computed from the oscillograms of Figure 7. In these curves the irregularities which appear in the curves directly computed from (4) and (2) have been arbitrarily smoothed out; from what has been said in the last paragraph these irregularities probably have little significance, and eliminating them makes it easier to follow the changes in the spectrum as the range of the shot is increased. All the

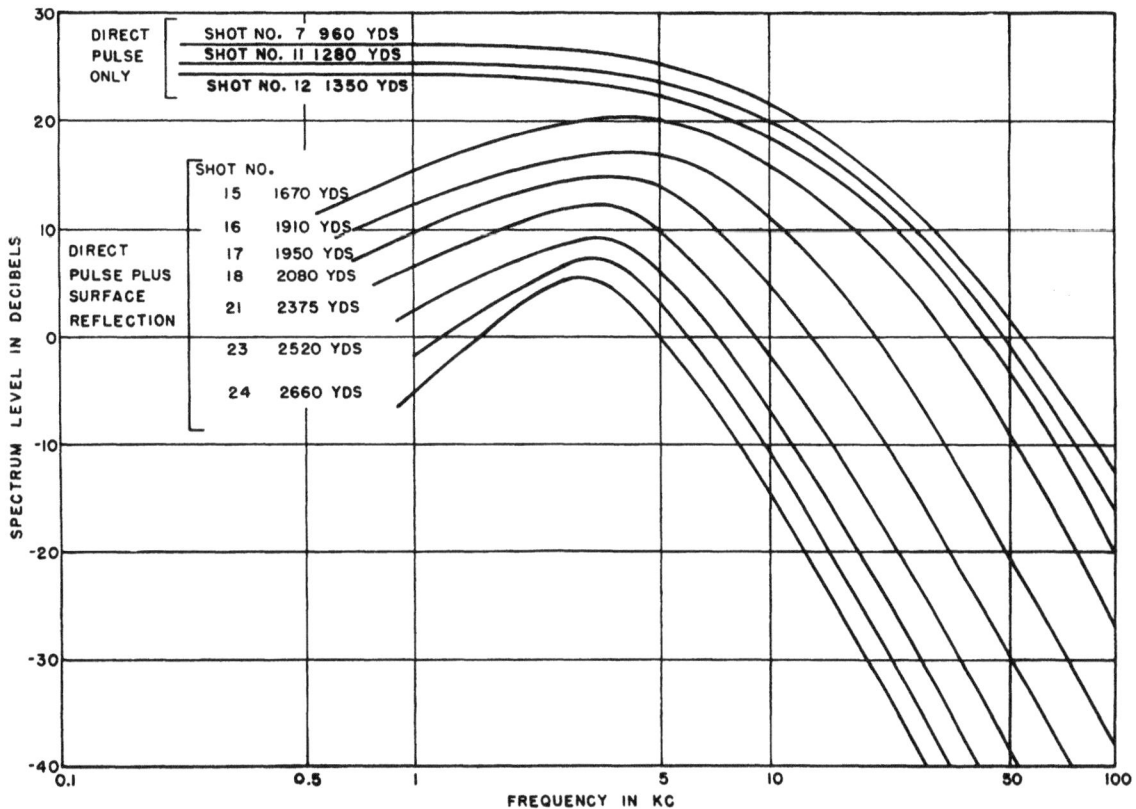

FIGURE 14. Spectral distribution of energy in explosive pulses received at various ranges. The curves shown were obtained by smoothing the values calculated by Fourier analysis from a number of shots made on Apr. 3, 1942, at a depth of 100 ft and received by a hydrophone at 54 ft.

curves except the first three show a maximum, since in the direct zone the surface reflection makes the spectrum level decrease at low frequencies, and in the shadow zone a similar effect is produced by the oscillatory nature of the pulse, although a separation into direct and reflected parts is no longer possible. For the first three pulses the integration was not extended over a sufficiently long time to include the surface reflection; if it had been, the curves would form a continuous family. The slopes of the curves for the pulses received in the direct zone are about 12 or 13 db per octave at 50 kc; in the shadow zone this slope increases to 17 or 18. This change is of course due to the increase in time of rise on entering the shadow zone.

As the range increases in the shadow zone there seems to be a slight decrease in the frequency at which U is a maximum. This shift, although hardly greater than the experimental error, is probably also due to the increase in time of rise. If the curves for the direct zone had been computed with inclusion of the surface reflection, however, they would have shown a trend

in the opposite direction, since the frequency below which the cancellation effect is felt is one for which a quarter of a cycle is of the same order as the time interval between the direct and reflected pulses, and this interval decreases with increasing range.

The frequency at which the spectrum level for shots in the shadow zone is a maximum shows variations from one day to another which are much greater than the variations from shot to shot on a given day. Table 3 gives values, taken from reference 9, of this frequency observed on a number of days for shots well inside the shadow zone.

TABLE 3. Frequency of maximum spectrum level in the shadow zone.

Date, 1942	Frequency of maximum, kc
March 12	7.1
March 19	8.0
April 3	3.3
April 8	9.0
April 9	15.0

An interesting application of this type of analysis is to compute the attenuation which sound of dif-

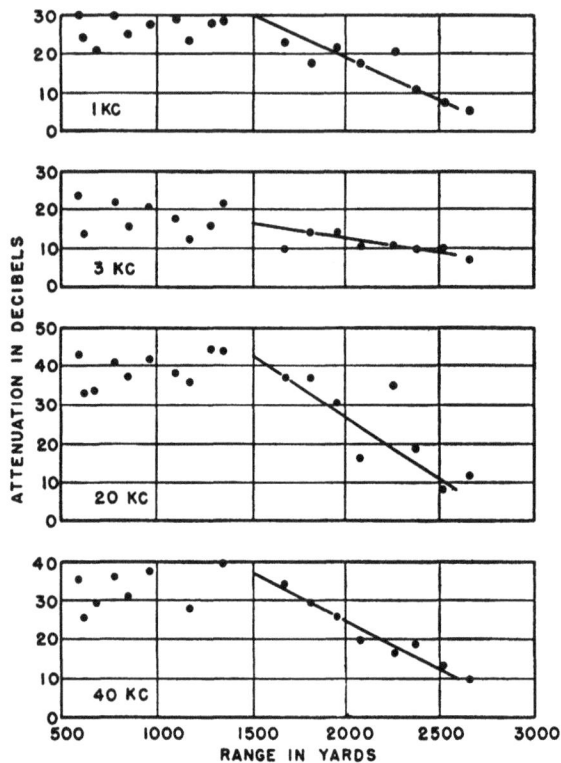

FIGURE 15. Shadow zone attenuation of various frequencies, as obtained by Fourier analysis of explosive pulses. Ordinate of each plot is $U(f) + 20 \log r + a(f)r/1,000$ referred to an arbitrary level, where $U(f)$ is the spectrum level, r is the range in yards, and $a(f)$ is the assumed absorption at the frequency f, as given in Section 5.22. Hydrophone depth, 54 ft; cap depth 100 ft; sound conditions as shown in Figure 6.

ferent frequencies suffers as the range is increased. The decrease of spectrum level with increasing range is due to geometrical divergence, which can be approximately but not accurately described by the inverse square law, to absorption and scattering, which are known to increase with frequency, and to the effect of the shadow boundary. To show the latter effect more clearly it is convenient to plot the quantity

$$U(f) + 20 \log r + \frac{a(f)r}{1,000}$$

against the range r, using values of the attenuation constant $a(f)$ appropriate to sinusoidal sound of frequency f. This is done for several frequencies in Figure 15 for the shots at 100-ft depth of the same series as has already been discussed in connection with Figures 5, 6, 7, 8, 9, and 14. The plots shown have been obtained by applying a correction to the points given in reference 9 to bring the assumed at-

tenuation $a(f)$ into line with the more up-to-date values given in Section 5.2.2.

Within the direct zone, the points for all frequencies lie roughly along horizontal lines, indicating that divergence and normal attenuation suffice at least approximately to explain the changes in size and shape which the pulse undergoes. Because of the deviations from the inverse square law discussed in Section 9.2.3 and shown in Figure 5, of course it is not to be expected that the points in the direct zone will follow a horizontal line very precisely. At the shadow boundary the spectrum levels start to decrease sharply; it is worth noting that the onset of the sharp decrease occurs at 1,400 to 1,600 yd for all cases and that, as one would expect, this range agrees much better with the value of distance to the shadow boundary derived from the time of rise and peak pressure in Table 2 than with the value computed from the bathythermograph data. Table 4 gives the

TABLE 4. Shadow zone attenuation at various frequencies on April 3, 1942.

Frequency in kc	Attenuation in db per kiloyard
1	21 ± 2
3	8 ± 1
20	29 ± 2
40	27 ± 2

magnitude of the additional attenuation beyond the shadow boundary, as obtained from the slopes of the lines in the figure; the probable errors quoted are based merely on the internal consistency of the data shown in Figure 15, and may not represent the overall accuracy of the calculation. The minimum of attenuation at 3 kc is striking, and corresponds of course to the maximum shown by the curves of Figure 14. As can be seen from Table 3 observations on other days, if treated in the same way, would have given quite a different dependence of attenuation on frequency. It is not known to what extent these differences can be correlated with thermal conditions, range, and the depths of source and receiver.

In making a comparison between the attenuation suffered by sinusoidal sound and the attenuation of the various frequencies making up an explosive pulse, we must keep in mind the limitations imposed by the short duration of the explosive pulse, or rather by the short period of time which is covered by the record of it. For example, the measured values of attenuation for a long pulse of sinusoidal sound can be greatly influenced by scattering, whereas sound scattered through any sizable angle would arrive too late to be recorded on oscillograms like those of Figure 7.

9.2.5 Variations

Ideally it should be possible to determine the magnitude and time scale of the fluctuations and variations in transmission by firing a number of caps in rapid succession from the same place. Unfortunately, no systematic experiments of this sort have been carried out. In the UCDWR work,[8,9] a few repeat shots were made at intervals of a few minutes; however, the number of such repeat shots was curtailed by the need for obtaining data at different ranges and depths in a time short enough so that oceanographic conditions could be assumed constant. Most of the material in the following paragraphs represents inferences obtained when some of the oscillograms for these experiments were restudied in the course of preparing material for the present report; because of the paucity of the data, these inferences must be regarded with caution.

In isothermal water, peak pressures from successive shots seem to vary but little out to ranges of over 1,000 yd. Most of the shots studied were consistent to within 1 or 2 per cent, though occasional shots deviated by 5 or 10 per cent. These variations are of the same order as those which are found at short ranges and attributed to nonuniformity of the caps themselves. The fact that they are so small is evidence of the uniformity of adjustment of the hydrophone and recording system.

When the cap is in the thermocline and the hydrophone in an isothermal layer above it, or when there is a negative temperature gradient at all depths, the fluctuations in peak pressure seem to be distinctly greater; for these cases successive shots at a few hundred yards range often differ by 20 per cent or more. In the experiments with single-frequency 24-kc sound, which were reported in Section 7.1.1, the fluctuation was found to decrease somewhat if the hydrophone was placed beneath the thermocline. Apart from the fact that neither the evidence on explosive sound nor that on single-frequency sound is based on an adequate number of samples, the apparent contradiction may be readily explained by the fact that in 24-kc single-frequency work the surface-reflected signal usually cannot be resolved from the direct signal, while in the experiments reported here, these two signals are generally received one after the other. Both in the present case and in the preceding the surface-reflected pulse seems to be a little more variable than the direct pulse, although the evidence for this is not very conclusive because of the irregularities, real and instrumental, which are present in the tail of the direct pulse on which the reflection is superposed.

Beyond a shadow boundary successive shots are often surprisingly consistent. The difference between the first pressure peak and the first trough, for example, has been observed in several cases of repeat shots to be reproducible to within 20 per cent or so although at least one case of a much larger fluctuation has been observed.

Figure 16 shows some typical oscillograms of shots made a few minutes apart, for three types of transmission conditions. One must be cautious in attributing physical reality to all the differences in detail which appear in successive oscillograms of this sort. For example, it has been demonstrated that slight changes in the orientation of a hydrophone from one shot to the next can sometimes produce considerable changes in the recorded pressure-time curve.[1] Other variable factors mentioned in reference 1 which can have an appreciable effect include scattering of sound by supports and other bodies near the hydrophone, and the possible presence of bubbles or other foreign matter on the hydrophone. Thus differences between successive records may or may not be real. On the other hand, any feature which is consistently reproduced in all records made under a given set of conditions, and for which an instrumental origin can be ruled out by virtue of its nonappearance under most other conditions, is probably a reproducible characteristic of the true pressure-time curves under the given conditions. A feature of this type is the shape of the first cycle or so of the oscillatory pressure-time curve in a shadow zone, as shown in Figure 16C.

9.3 LONG-RANGE SOUND CHANNEL PROPAGATION IN DEEP WATER

It has long been known that explosive sound from a charge of moderate size can be detected at ranges of the order of tens of miles, and this fact has received practical application in acoustic position finding.[18,19] As will be shown later, ranges of thousands of miles can be achieved by proper arrangement of source and receiver. It is to be expected that many of the characteristics of the signals received at long ranges will be determined by refraction and by re-

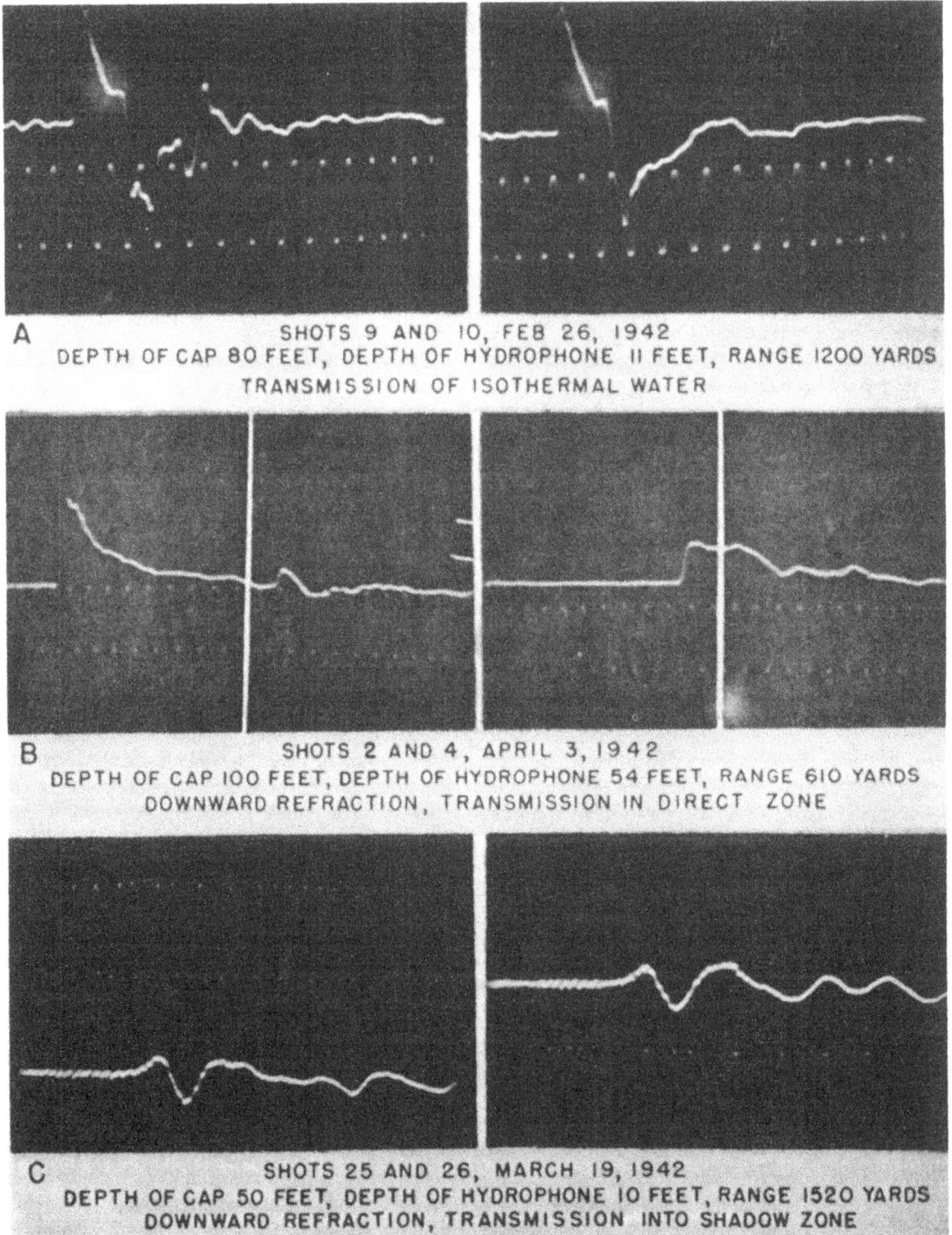

FIGURE 16. Typical examples of the reproducibility of records of explosive pulses for shots fired a few minutes apart.

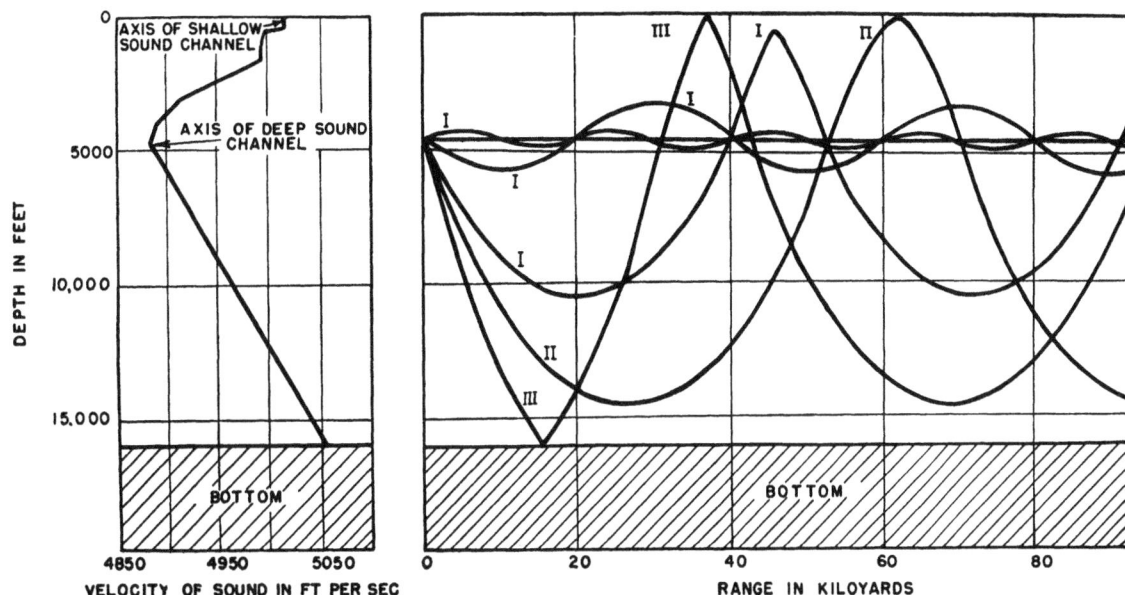

FIGURE 17.   Velocity distribution and schematic ray paths for a typical deep sound channel.

flection from the bottom. Leaving the detailed consideration of bottom reflections until Sections 9.4.1 and 9.4.2, we shall consider here a number of phenomena which are due to the presence of a "sound channel" at a certain depth in the ocean. A sound channel can be briefly described as a stratum of the ocean within which the velocity of sound first decreases with increasing depth, passes through a minimum, and then increases. The importance of such a region is due to the fact that if a source of sound is placed in it, all rays emitted within a certain range of initial directions will remain confined to the sound channel, executing periodic oscillations in depth.

All, or nearly all, of the experimental results which have been obtained so far on long-range transmission in deep water can be interpreted in terms of ray paths. Ray paths which lie in a sound channel are of especial importance, and since these rays have rather complicated characteristics, Section 9.3.1 will be devoted to a theoretical discussion of them. By using the facts established there as a foundation, the experimental results to be given in Section 9.3.2 can be concisely discussed and interpreted.

9.3.1   Deep Sound Channels

Sound channels may occasionally occur at shallow depth, when there is either a layer of isothermal water or a layer with a positive temperature gradient underneath a surface layer in which the temperature gradient is negative. The more common configuration of an isothermal layer immediately beneath the surface is, moreover, very similar to a sound channel, in that many rays undergo alternate upward refraction and surface reflection, and so remain confined to the isothermal layer out to indefinitely large ranges. Of more importance, however, for long-range transmission in deep water, is a deep sound channel which always occurs, except in polar regions, at a depth of the order of 4,000 ft or less. This deep sound channel is due to the fact that at all ordinary latitudes there is an extensive thermocline within which the temperature gradient is negative and below which the temperature gradient is so slight that the effect of increasing pressure suffices to make the velocity of sound increase with depth.

Figure 17 shows velocity minima of both shallow and deep types, and several ray paths emanating from a source located at the depth of the lower minimum. It will be convenient to refer to the ray which becomes horizontal at the depth of minimum velocity as the "axis" of a sound channel. If the thermal gradient is discontinuous, as in Figure 17, the number of different rays connecting any two points on the axis is infinite, since the range per cycle, that is, the horizontal distance traversed by a ray while traversing 1 c of its oscillation in depth, approaches zero as the ray approaches the horizontal. This is illustrated by the three rays labeled I. If the velocity-depth curve is smooth near the minimum, this will

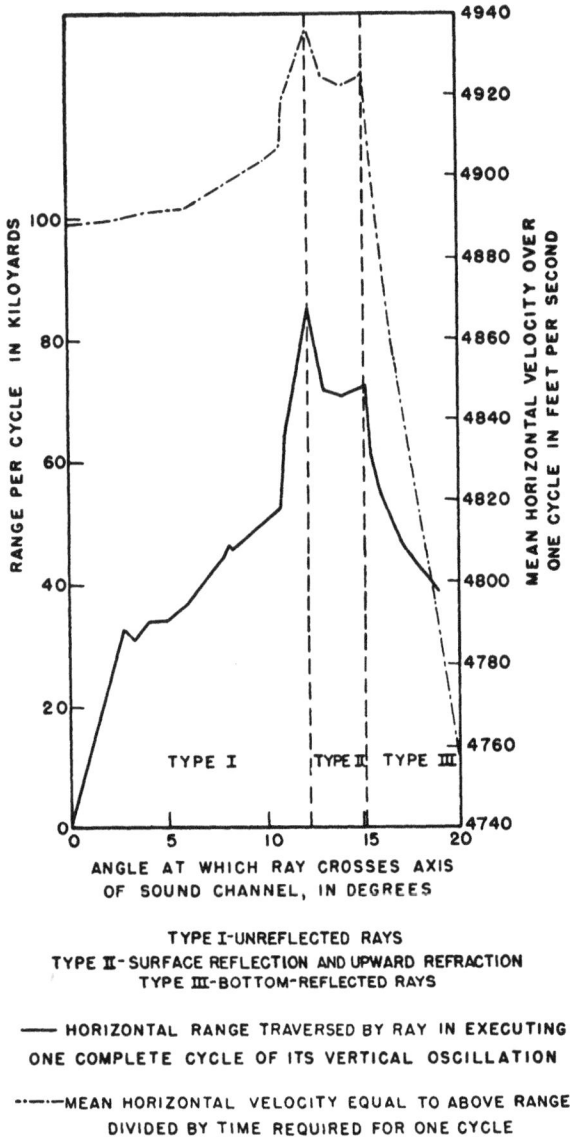

FIGURE 18. Mean horizontal velocity and range per cycle for rays oscillating about a sound channel.

not be true. In any case, however, whether source and receiver are on the axis of the sound channel or not, the number of different rays connecting them increases with increasing horizontal range. The number of such rays decreases, however, with increasing distance of source or receiver from the axis. If either source or receiver is too far from the axis of the sound channel, no ray can get from source to receiver without reflection from the surface or the bottom.

To study these phenomena quantitatively, and to compute times of arrival for the various rays, curves like those shown in Figure 18 are very helpful. Whatever its point of origin, any ray which traverses the

sound channel can be characterized by the angle at which it crosses the axis of the sound channel; any two rays which cross the axis at the same angle must be congruent, differing only by a horizontal displacement. Figure 18 shows how the horizontal range per cycle and the mean horizontal velocity, that is, the quotient of horizontal range per cycle by time per cycle, depend upon the angle of crossing the axis.

For angles less than a certain critical value, equal to 12.2 degrees in the example shown, the ray oscillates up and down without reaching either the surface or the bottom. For rays of this type (type I in Figure 17) it will be seen that the mean horizontal velocity is least for the axial ray and increases as the angle with the horizontal, and hence the range per cycle, increases. The consequences of this are especially interesting when both source and receiver are on the axis of the sound channel. For this case the first impulse to arrive will come along a ray for which the number of oscillations in depth has the smallest value consistent with avoidance of surface and bottom reflections. When the range is small, this ray will have only one half-cycle between source and receiver, but with increasing range more and more half-cycles are required, since the range per complete cycle can never be greater than a certain value, equal to 85 kyd in Figure 18. Rays with more and more oscillations will arrive later and later, and if for the moment we ignore reflected rays, the last one to arrive will be the straight axial ray. Thus, the early arrivals will be separated by considerable intervals of time, but later arrivals will be closer and closer together, finally merging into an unresolvable crescendo, followed, if we neglect reflected rays, by a sudden silence. Figure 19A shows the times of arrival of these sound channel rays, as computed from the data in Figure 18 for a particular value of the range.

The total time between the first and last of these arrivals can be computed from the spread in mean horizontal velocities for the sound channel rays; for the case plotted in Figure 18 the total time in seconds comes out to be 0.012 times the range in miles.

It will be noticed that the early arrivals in Figure 19A come in groups of three. The explanation of this is shown schematically in Figure 20 for the simplest case of the first arrivals at a very short range. Each oscillating ray travels much farther in a lower half-cycle than in an upper one; consequently the mean horizontal velocity of a ray between source and receiver will be principally a function of the amplitude of its lower half-cycles which in turn is principally

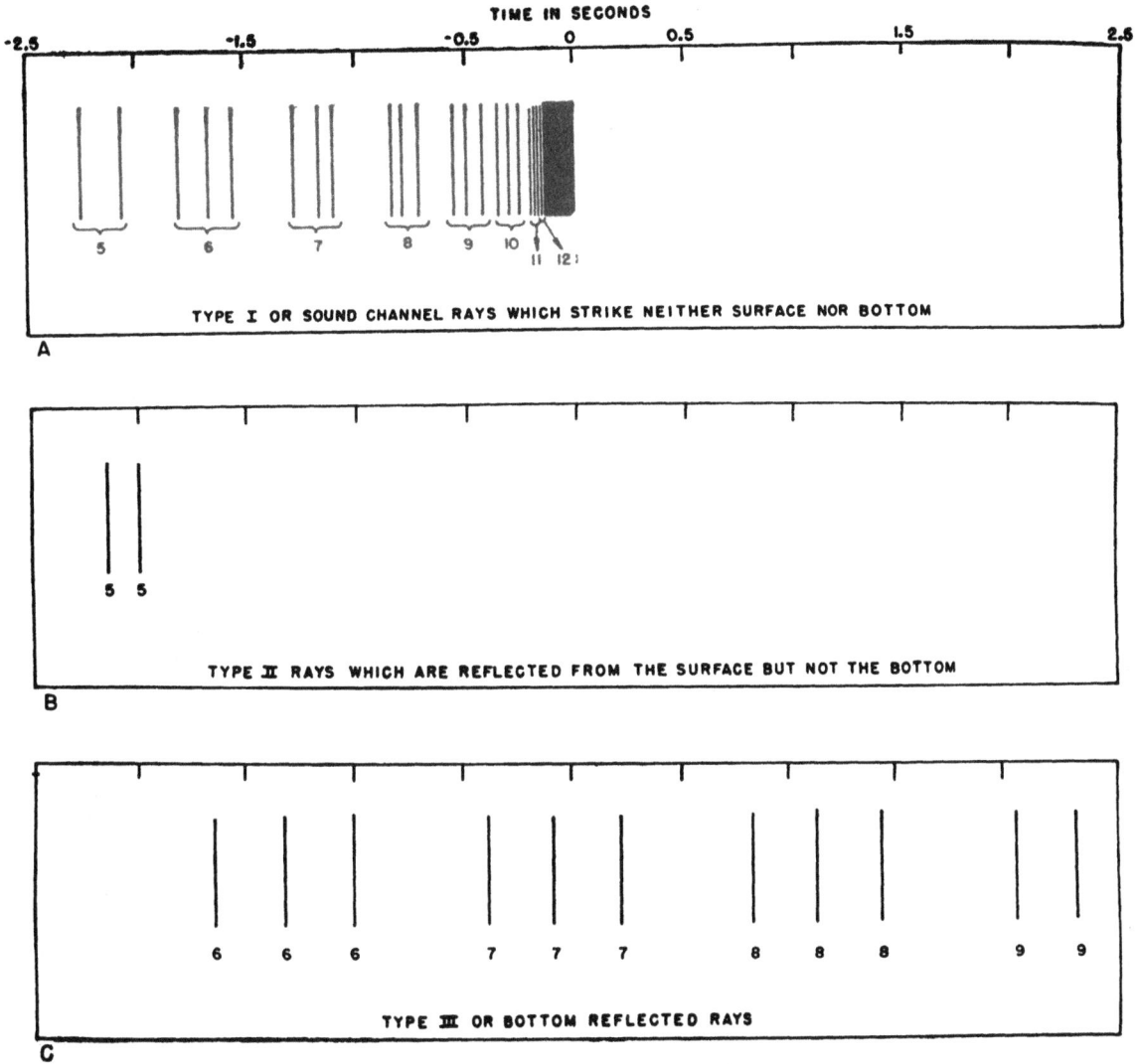

FIGURE 19. Times of arrival for the various rays connecting two points on the axis of a sound channel. Range, 400 kyd = 197 miles. Velocity-depth curve assumed same as for Figures 17 and 18. The numeral below each arrival gives the number of lower half-cycles in the corresponding ray path. The zero of time is taken as the time of arrival of the axial ray.

dependent on the number of lower half-cycles which occur during the passage from source to receiver. In the example of Figure 20, there are four rays which have two lower half-cycles between source and receiver; however, two of these four, namely, the ones with two upper half-cycles, arrive at the same time so there will be only three resolvable arrivals. When source and receiver are at different depths, the rays analogous to those in Figure 20 will all arrive at different times, and the hydrophone will receive pulses in groups of four. For the later arrivals, the upper half-cycles have more nearly the same travel time as the lower half-cycles, and the pulses no longer arrive in clearly separated groups of three.

When either the source or the receiver is at some distance from the axis of the sound channel, the piling-up effect shown in Figure 19A will not occur, since only a limited number of rays will be possible between source and receiver.

So far we have not considered sound which arrives at the receiver by paths involving surface or bottom reflection. The rays which undergo surface reflection and upward refraction without reaching the bottom (type II in Figure 17) have mean horizontal velocities which, according to Figure 18, are slower than the fastest unreflected, or type I rays, but faster than the axial ray. Thus the arrivals for rays of this sort are mixed in with those of type I, but when source and

RANGE

FIRST ARRIVAL OF GROUP

RANGE

SECOND AND THIRD ARRIVALS,
COINCIDENT IF, AS SHOWN HERE, SOURCE
AND RECEIVER ARE AT THE SAME DEPTH

RANGE

RANGE

● SOURCE   LAST ARRIVAL OF GROUP
○ RECEIVER

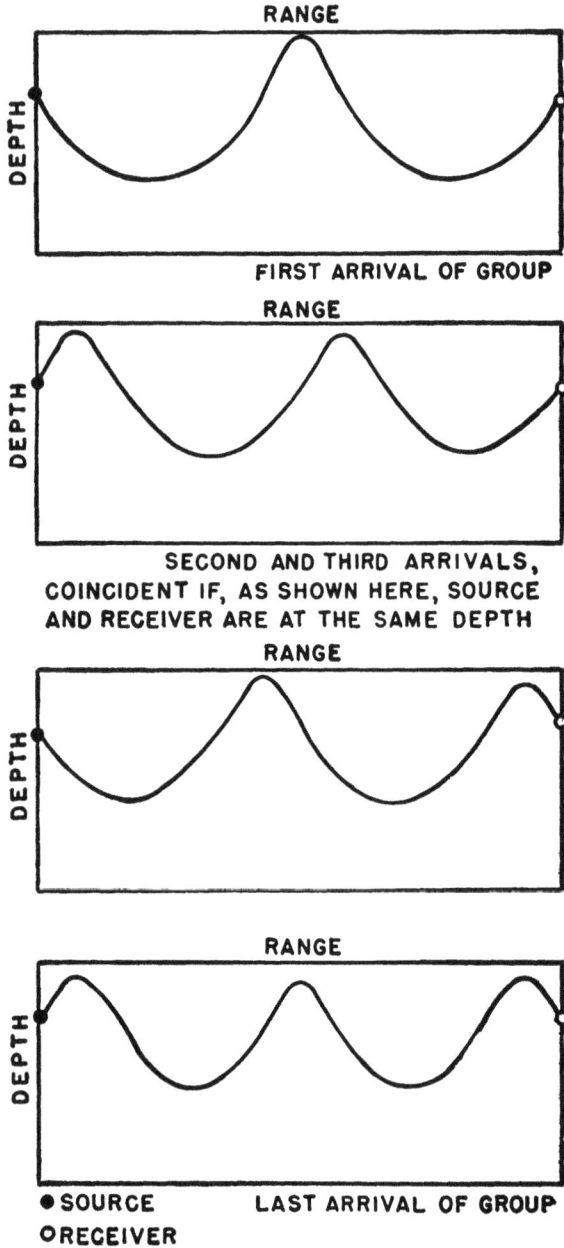

FIGURE 20.   Grouping of arrival times for sound channel rays. The four ray paths shown each have two
lower half-cycles, and the corresponding times of travel
are therefore close together, so that the four arrivals
form a group.

receiver are both on the axis they cease before the
"piling up" of the sound channel rays. This is shown
in Figure 19B for the particular set of conditions
chosen for that figure.

The bottom-reflected rays (type III in Figure 17)
have times of arrival which are also interspersed
among those of type I, but which continue after the

latter have ceased. Figure 19C shows these arrivals
for the example treated. The grouping for these rays
is again in threes or fours, according to whether
source and receiver are at the same or different
depths.

Let us now consider the energies and intensities of
the system of impulses arriving at the receiver. First
of all, it may be noted that all the energy emitted
by the source in directions giving rays of type I, i.e.,
for the cases of Figure 18 in directions within $\pm 12.2°$
of the horizontal, is propagated along the sound
channel and cannot disappear except by volume
absorption or scattering in the water. If the latter
processes are neglected, the total energy in the
system of impulses transmitted by these unreflected
rays, that is, the system exemplified by Figure 19A,
must be inversely proportional to the horizontal
range. This contrasts with propagation in an infinite
homogeneous medium, where the energy given to a
receiver varies inversely as the square of the distance.
The intensities of the individual arrivals can be calculated in the usual way from ray theory, which should
be applicable to the earlier arrivals, before there is
appreciable overlapping of consecutive pulses. It will
be apparent after a little pondering on Figure 17 that
in general these individual intensities must vary approximately as the inverse square of the range, the
slower rate of decay of the total energy being due
to the increase in the number of arrivals as the range
increases. For certain positions of source and receiver, however, some of the arrivals may have an
anomalously high intensity due to the fact that two
rays of infinitesimally different initial inclinations
are tangent at the receiver. This condition will be
more closely approached for the latest sound channel
arrivals to reach the receiver than for the earlier ones,
and accordingly these latest arrivals should be the
strongest.

The energy traveling along paths involving reflection from the surface or the bottom is channeled in a
similar manner. The bottom-reflected rays, however,
lose a considerable part of their energy at each reflection, and therefore die out more rapidly with increasing range than the others. (See Figure 21 of
Section 9.3.2 and Section 9.4.1.) For the same reason
successive arrivals of this type have progressively decreasing intensities.

9.3.2   Experimental Results

Two series of experiments on long-range transmission in deep water have been conducted by WHOI.[20, 21]

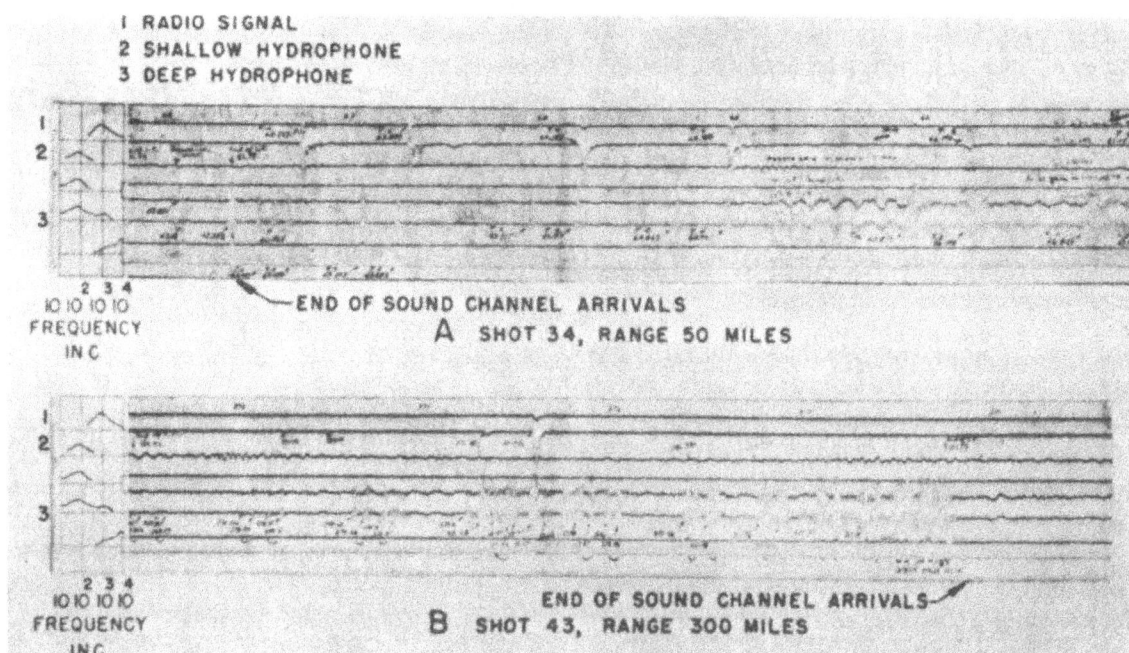

FIGURE 21. Typical records of explosive sound received at long ranges. Times marked along top of each oscillogram are in seconds. Curves at left give relative amplitude response of each channel to the various frequencies.

The first series was conducted just outside the Bahama Islands and consisted in the recording of impulses from shallow explosions on two hydrophones, one at shallow depth and the other at 1,600 ft, at various ranges all less than 30 miles. The second series was made some time later in regions extending northeastward and eastward from the same locality. In addition to the shallow shots, explosions at 4,000 ft, near the axis of the main sound channel, and a hydrophone near this depth as well as a shallow one were used. The ranges for this series extended out to 900 miles.[b] Data pertaining to the conditions of these experiments are given in Table 5. The velocity-depth curve for the first series is the one shown previously in Figure 17; that for the second series is very similar, and the curves of Figure 18 can be applied with little error to either series.

Figure 21 shows some typical records obtained in these experiments along with thumbnail sketches of the frequency response characteristics of the various recording channels used. The following paragraphs point out a number of features of these records which agree with the predictions of ray theory as outlined in Section 9.3.1.

TABLE 5. Experimental arrangements used in long-range transmission studies by WHOI.

	First series (from reference 21)	Second series (from reference 20)
Depth shallow hydrophone in feet	80	80
Depth deep hydrophone in feet	1,600	3,500
Charge weights and depths	½ lb, 50 ft	½ lb, 50 ft 4 lb, 4,000 ft 200 lb, 300 ft
Ranges, nautical miles	2.7–26	20–900
Depths of sound channels in feet	75, 4,500	4,100 (average)
Depths of water in feet	16,000	15,000–18,000 (usually)

Identification of the paths by which the various pulses arrive is usually difficult because of the large number of arrivals and because the predicted time for any arrival can be appreciably influenced by uncertainties in the depths of source and receiver and by small variations of the velocity-depth curve along the route. However, the general appearance of the records is very much as one would expect from the considerations given in Section 9.3.1. Thus, in Figure 21B the arrivals at the deep hydrophone come in groups of four while for the shallow hydrophone the four pulses show up as two, each of which is presumably double but unresolved because of the short-

---

[b] More recent experiments which have not yet been reported in full have yielded detectable signals at a range of 2,300 miles.

ness of the interval between any pulse and its reflection from the surface near the hydrophone. In Figure 21A a similar grouping occurs for the bottom-reflected pulses, which continue after the last sound channel arrival, but the range for this case is so short that the sound channel arrivals have not yet had time to form into well-defined groups.

In the records obtained when both source and receiver were shallow, each theoretical group of four is of course entirely unresolved and appears as a single pulse.

When source and hydrophone are both deep the total duration of the sound channel arrivals, that is, the time interval between the first arrival and the last arrival via the sound channel is found to agree nicely with the duration predicted by Figure 18. At the longer ranges the last sound channel arrival is easily spotted by the conspicuous "piling-up" effect which occurs just before it, as shown in Figure 21B. At shorter ranges, however, as in Figure 21A, the last sound channel arrival can only be identified by its high intensity and the fact that, like all the arrivals which do not involve reflection from the surface, it is absent from the record of the shallow hydrophone. For short ranges the number of sound channel arrivals is small because of the short range and the fact that neither source nor receiver is on the axis of the sound channel, and the bottom reflections, which come in both before and after the last sound channel arrival, may mask the abrupt termination of the sound channel arrivals even when a piling up occurs. At ranges beyond about 300 miles, on the other hand, the bottom reflections become so weak that they no longer appear on the records, and the piling up of the sound channel rays is followed by sudden silence. This disappearance of the reflected pulses is of course due to the fact that according to Figure 18 any ray traveling by bottom reflections cannot go more than about 70 kyd between successive reflections; and since an appreciable amount of energy is lost at each reflection, pulses traveling along such rays are much more rapidly attenuated than those which travel in the water alone.

According to Table 5 at the time of the first series of experiments there was a shallow sound channel with its axis at a depth of about 75 ft. That transmission to considerable distances near the surface was possible at this time is shown by a comparison of the velocity of propagation of the first arrival with the velocity for the bottom-reflected rays. The ratio of these velocities is found to agree nicely with the ratio of the velocity of sound at the surface to the velocity given by Figure 18 for the particular bottom reflection studied, showing that the first arrival does indeed come by a path lying entirely in the region near the surface.

The most significant results obtained in these experiments have to do with attenuation and with the reflection coefficients of bottom and surface. To study quantitatively the variation of intensity with distance and with number of bottom reflections, some sort of measurements must be carried out on the oscillograms. The most obvious thing to measure would be the peak pressures or momentums of individual arrivals. However, in many cases the individual arrivals of a group were not completely resolved, and in all cases the pressure-time curves may have been distorted by small-angle scattering or off-specular reflection. For these reasons it was concluded that the most suitable characteristic of the records from which to estimate attenuations and reflection coefficients is the energy of a poke, or of a group of two pokes, rather than the peak deflection, this energy being assumed to be a constant times the integral of the square of the deflection. One may hope that this quantity will represent a suitably weighted average of the spectrum level of the pressure pulse in the water in the region of frequencies covered by the recording channel being used. It is of course not strictly true that the "energy" measured in this way on an oscillogram represents this weighted average, since, for example, the phase of the transient disturbance produced by the first arrival at the time of the second arrival will determine whether the second arrival increases or decreases the amplitude. However, we may expect that the desired correspondence will be valid for an average over many pokes.

Because of the very large ranges covered by the second series of experiments, it was possible to measure the very small attenuations suffered by sound at the comparatively low frequencies to which the receiving channels responded, frequencies at which no other measurements of attenuation have been obtained. The results, based on the total energy of all the sound channel arrivals taken together, are summarized in Table 6. The data beyond about 200 miles are fairly consistent, as the sample plot given in Figure 22 shows. At shorter ranges the measured energies vary erratically, perhaps because the number of sound channel rays is too small to give a uniform spatial distribution of energy.

The interpretation which should be given to these

TABLE 6.   Attenuation coefficients for explosive sound received in various frequency bands.

Location of shots	Frequency limits of recording channel 6 db below peak sensitivity in cycles per second	Attenuation in db per kiloyard	Number of observations used	Spread of ranges used in kiloyards
Line from Latitude 26° N, Longitude 76° W to Latitude 39° N, Longitude 67° W	22–175	0.005	5	400–1,600
	2,300–10,000	0.013	4	200–1,000
Line running east from Latitude 25° N, Longitude 76° W	14–75	0.025	5	200–550
	56–350	0.043	4	300–550
	600–4,000	0.035	4	300–550
	56–350	0.050	4	300–550

attenuation figures is rather uncertain, since the receiving channels each cover a fairly broad band and since it is likely that the attenuation varies strongly with frequency. Moreover, it can hardly be decided yet whether the attenuation is due to absorption, to scattering, or to variations in the depth of the sound channel with geographic position. The latter factor would have an influence on the observed intensities similar to that of changing the depth of the explosion, the important variable being merely the distance of the explosion from the axis of the sound channel. In spite of all these uncertainties, however, the figures in Table 6 probably do give a significant upper limit to the order of magnitude of the absorption at sonic frequencies.

Measurements of a similar sort carried out on the bottom-reflected pulses of the first series of experiments give values for the reflection coefficient of the bottom which will be presented later in Table 7 (see Section 9.4.1). In these experiments no difference could be noticed between pulses of the same group whose ray paths differed by one in the number of surface reflections undergone. This shows that the reflection coefficient of the surface was unity, to within an accuracy of five or ten per cent, at the angles of incidence involved, which ranged from nearly normal incidence down to about 10 degrees from the horizontal.

It has been suggested that triangulation based on the times of the last sound channel arrivals at several stations might be of practical use in the accurate location of a boat or plane on the ocean. Extrapolation of the intensities so far measured for the crescendo formed by the last sound channel arrivals suggests that a few pounds of high explosive may be heard above background at ranges of ten or twenty thousand miles or more, if shoals or land masses do not intervene to cast a shadow.[20]

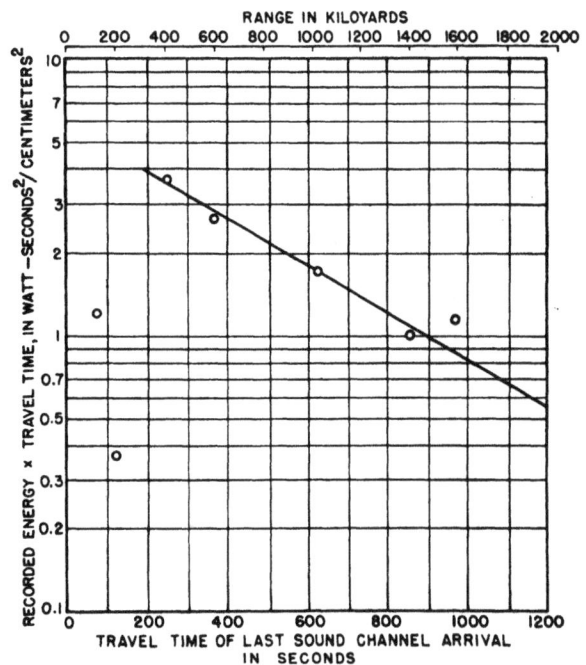

FIGURE 22.   Sample plot of the dependence on range of the total energy recorded for all sound channel arrivals. Source: 4-lb TNT bomb. Location of shots: line from latitude 26° N, longitude 76° W, to latitude 39° N, longitude 67° W. Recording channel within 6 db of peak sensitivity in range 22 to 175. Line shown corresponds to attenuation of 0.0050 db/kyd.

9.4   BOTTOM REFLECTION AND SHALLOW-WATER TRANSMISSION

The bottom of the ocean can influence the transmission of an explosive pulse in several closely related ways. When a pulse traveling through the water strikes the bottom, it is partly reflected and partly transmitted. If the bottom consists of two or more successive strata with different acoustic properties, the transmitted pulse may itself be partially reflected and partially transmitted at the boundaries between the strata, and a complicated sequence of multiple reflections may take place. Finally, the pulse may be transmitted horizontally through the bottom, the disturbance of the bottom at each point being accompanied by a corresponding disturbance in the water. In this phenomenon the impact of the explosive wave on the bottom below the charge sets the bottom into vibration, and this vibration is propagated radially outward like a surface wave on the water, or, to use a more accurate analogy, like a surface-bound earthquake wave, its frequency and velocity being influenced, however, by the water overlying the bottom.

The three following subsections deal in turn with the simpler and the more complex aspects of these phenomena. Section 9.4.1 treats ordinary reflections, using the concept of sound rays, and discusses arrival time data for certain parts of the "earthquake wave," since these data can also be interpreted in terms of rays. Sections 9.4.2 and 9.4.3 discuss the detailed form of the pulses transmitted by the "earthquake wave," which can be understood only by abandoning the ray concept and treating water and bottom as a single dynamical system.

In the theoretical portions of all these sections it will be assumed for simplicity that the bottom is smooth and horizontally stratified; and an effort will be made to interpret the experimental material in terms of this idealization. It must be remembered, however, that there may often be small-scale irregularities in the bottom which will scatter the explosive pulse, and large-scale departures from horizontal stratification, which will complicate the transmission phenomena.

9.4.1 Reflection Coefficients and Times of Arrival

When a pulse of sound strikes a plane boundary between two media of different acoustic properties, the reflected pulse has a lower amplitude than the incident pulse and in general a different phase. A theoretical derivation of the amplitude and phase relations to be expected at the boundary between two ideal fluid media has been given in Section 2.6.2. Actual ocean bottoms may differ in their properties from the ideal media considered there, however. To describe completely the reflecting properties of a given bottom, one should specify the amplitude reduction and phase shift for all frequencies and all

angles of incidence. An equivalent description, which could be related to this by the methods of Fourier analysis (see Section 9.2.4) would be provided by recording the exact form of the reflected pressure-time curve for an explosive pulse for all angles of incidence. So far, however, no pressure-time curves have been recorded for explosive pulses reflected from the bottom. The only quantitative data on bottom reflections which are available are those obtained at WHOI[17] in connection with the long-range propagation studies discussed in Section 9.3.2. These data will now be described.

As mentioned in Section 9.3.2, the series of experiments for which analyses of bottom reflections were carried out was made using a shallow hydrophone at 80 ft and a deep hydrophone at 1,600 ft, with shots of $\frac{1}{2}$-lb TNT fired at depths of the order of 50 ft at ranges up to 30 miles. Two recording channels with different frequency responses were used for the shallow hydrophone, and five for the deep hydrophone. The reflection coefficients of the bottom were determined for each of these channels by making plots similar to that in Figure 22 for the pulses undergoing respectively one, two, and three bottom reflections and then measuring the vertical displacements between the lines corresponding to different numbers of reflections. The values obtained are given in Table 7.

TABLE 7. Reflection coefficients for the bottom in the region near Latitude 26°46′ N, Longitude 76°25′ W.

Hydrophone	Recording channel	Frequency limits of recording channel 6 db below peak response	Average reflection coefficient
80 ft	2	1,900–6,200	0.04
	3	240–2,400	0.36
1,600 ft	4	42–230	0.72
	5	42–230	0.60
	6	160–2,400	0.33
	7	200–2,800	0.33

This method of analysis, while probably the best that can be applied to the data available, is rather crude in that the angle of incidence of the rays on the bottom changes with range and also with the order of the reflection; if, as is often the case, the reflection coefficient varies strongly with angle of incidence, only a vague average over a range of angles will be obtained.

The values given in Table 7 suggest a decrease of reflection coefficient with increasing frequency, an effect which would not occur at a plane boundary between two ideal acoustic media. Unfortunately these results cannot be compared with data for sinusoidal sound, since the character of the bottom in the locality of the experiments is at present unknown, and since measurements with sinusoidal sound at low frequencies are not very complete.

Information can also be obtained from explosive sound regarding the geological strata beneath the bottom. Figure 23 shows a typical ray diagram for

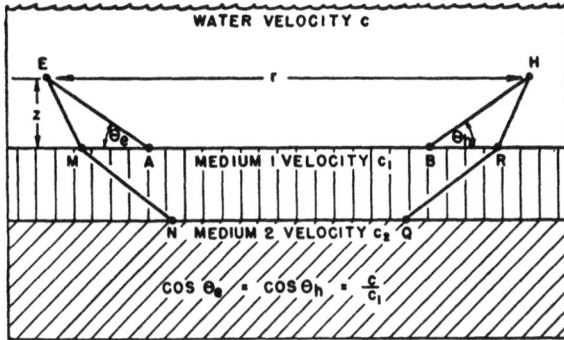

FIGURE 23.   Ray paths in a stratified bottom.

sound originating in the water over a stratified bottom, in which each successive layer has a higher sound velocity than the one above it. Without bothering about the detailed form of the pressure pulse received at a distant hydrophone, a subject which will be discussed fully in the two following subsections, we may study the way in which the time of arrival of the first measurable disturbance varies with the range from the explosive to the hydrophone. If this range, EH in Figure 23, is sufficiently short, the first disturbance will arrive by a path which lies entirely in the water. But if EH is greater than a certain value r_0, which depends upon the depths of source and hydrophone and upon the velocity of sound in the top layer of the bottom, sound traveling along the ray $EABH$ will arrive before the direct sound wave through the water; in such a case the fact that the sound velocity over the path AB is greater than that in the water more than compensates for the fact that $EABH$ is longer than the direct path EH. To find out when this occurs, let c be the velocity of sound in the water (assumed uniform for simplicity), c_1 the velocity in the top layer of the bottom, r the horizontal range, and z the height of explosive and hydrophone above the bottom, assuming for simplicity that both are at the same level. We shall first show that the positions of A and B, which minimize the time of travel, are those for which the angles EAB and ABH obey the refraction law of ray theory, and

shall then derive an expression for the value of r at which the time of travel via $EABH$ becomes shorter than via the direct route EH.

The time required for a pulse of sound to travel a path such as $EABH$ in Figure 23 is

$$t = \frac{z(\csc \theta_e + \csc \theta_h)}{c} + \frac{r - z(\cot \theta_e + \cot \theta_h)}{c_1}. \quad (5)$$

If this time has a minimum as the position of point A is varied, this minimum must occur when $dt/d\theta_e = 0$, that is, when

$$-\frac{z}{c} \csc \theta_e \cot \theta_e + \frac{z}{c_1} \csc^2 \theta_e = 0,$$

which is equivalent to

$$\cos \theta_e = \frac{c}{c_1}.$$

This is the well-known expression for the angle at which the transition from refraction to total reflection occurs. Similarly, the requirement that t be a minimum with respect to displacements of point B gives

$$\cos \theta_h = \cos \theta_e = \frac{c}{c_1}. \quad (6)$$

Eliminating the angles from equation (5) by use of this relation, we have for the time of arrival by the shortest path through the bottom

$$t = \frac{2z}{c\sqrt{1 - c^2/c_1^2}} + \frac{r}{c_1} - \frac{2z}{c_1} \frac{c/c_1}{\sqrt{1 - c^2/c_1^2}}$$

$$= \frac{r}{c_1} + \frac{2z\sqrt{c_1^2 - c^2}}{cc_1}. \quad (7)$$

This equals the arrival time r/c of the direct pulse through the water when

$$r = 2z\sqrt{\frac{c_1 + c}{c_1 - c}}. \quad (8)$$

Now if the time interval between the explosion and the first signal at the hydrophone is plotted against the range r, the graph will start out as a straight line passing through the origin and of slope $1/c$; and at the range given by equation (8) the slope will change abruptly to $1/c_1$. Thus all the quantities c, c_1, and h could be determined from this plot. If the plot is continued to larger values of r, another abrupt change of slope may occur when the travel time via a path $EMNQRH$ lying partly in a denser stratum (medium 2) becomes shorter than via $EABH$. If the bottom contains still deeper strata with higher sound velocities, further changes of slope will occur. By methods similar to those outlined above, the depths

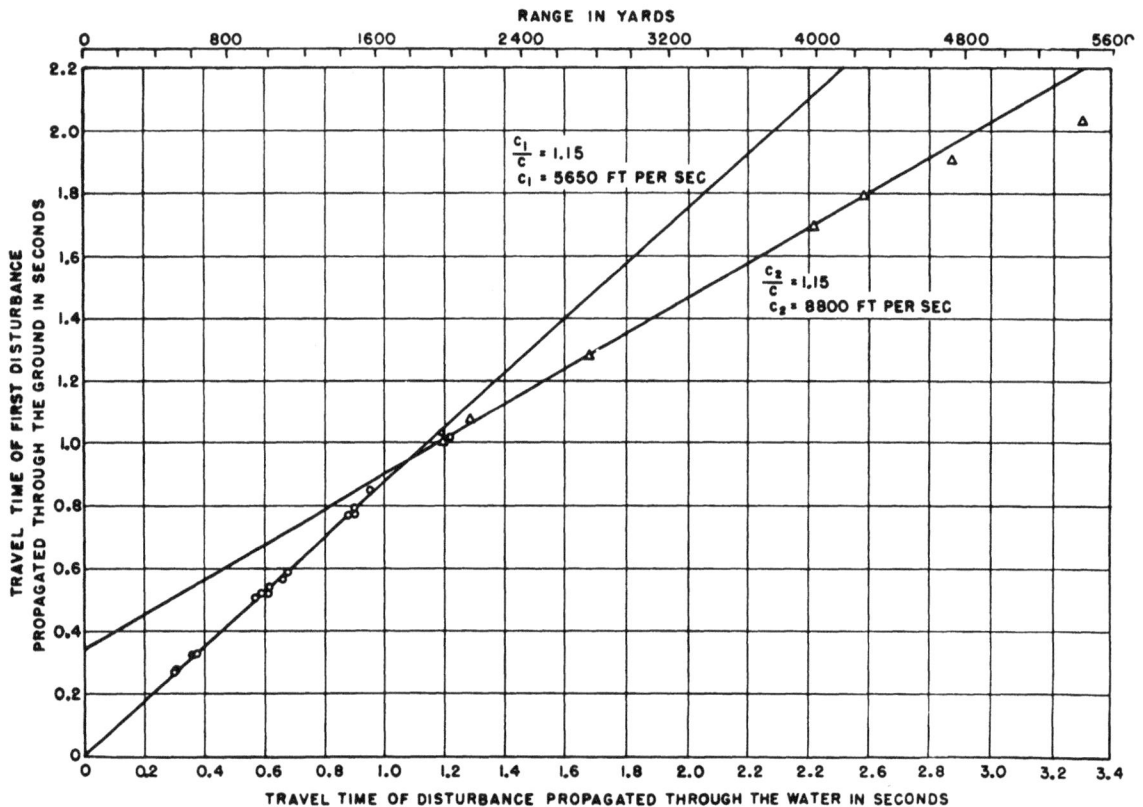

FIGURE 24. Typical plot of travel time against range showing layer structure of the bottom. Location near Solomons, Md., at mean depth of 52 ft, charge and hydrophone both resting on bottom. Lines cross at $r/c = 1.08$ seconds. Depth of upper layer in bottom $= r/c \sqrt{c_2 - c_1/c_2 + c_1} = 1,240$ feet, where r = range at which lines cross, c = velocity of sound in water, c_1 = velocity of sound in upper layer of bottom, and c_2 = velocity of sound in lower layer of bottom.

as well as the velocities of all these strata can be determined from this plot of arrival times. This type of analysis has long been familiar in geophysical prospecting.

Figure 24 shows a typical plot of arrival times constructed from some of the data obtained by WHOI,[22] with the layer depths and velocities deduced from it. The shots were made with both the charge and the hydrophone on the bottom, so the depth of the upper layer of the bottom can be calculated from equation (8) by replacing c by c_1 and c_1 by c_2. When more than two layers are involved, the plot of times of first arrival will still consist of straight-line segments, but the calculation of the depths of the second and deeper layers involves more complicated formulas in that case.

The representation of the plotted points by two straight lines is fairly easy for this case; however, data are often obtained for which the times of first arrival seem to form an almost smooth curve. This may sometimes be due to the absence of any well-defined layer structure in the bottom, as might be the case for example for a thick mud bottom whose compactness increases continuously with depth. It will be shown in Section 9.4.3, however, that there are many cases where there are recognizable layers in the bottom but where fluctuations of one sort or another prevent them from being accurately identified from mere arrival time data. For such cases the proper choice of straight lines to fit a plot such as Figure 24 may sometimes be facilitated by a study of the predominant frequencies in the first and subsequent arrivals (see Figure 32, Section 9.4.3).

9.4.2 Simplified Theory of Normal Modes

We have seen in the preceding Section 9.4.1 that if the range is sufficiently long compared with the distances of source and receiver above the bottom, the first sound to arrive must come by a path lying within the material of the bottom over most of the distance, as shown in Figure 23. One might at first suppose that refraction of this sort would be similar to refraction in the water alone, and that the received pulse would be a replica of the pressure wave emitted by the source, with an intensity which could be calculated by ray theory. It is easily shown, however, that this is not the case. We shall first show that when the bottom is acoustically uniform, so that rays in the bottom are straight lines, the intensity predicted by ray theory for a ray such as $EABH$ in Figure 23 is zero. Figure 25 shows a ray having inclination θ in the water, θ_1 in the bottom, together with a neighboring ray. By Snell's law of refraction we have

$$\cos \theta_1 = \frac{c}{c_1} \cos \theta \qquad (9)$$

where c and c_1 are the velocities of sound in water and bottom respectively. Now the energy which leaves the source E in an interval $d\theta$ of inclinations and in a fixed narrow interval of azimuth is partly reflected and partly transmitted, and the transmitted part is distributed over the region between the rays AP and $A'P'$ in Figure 25. If $R(\theta)$ is the reflection

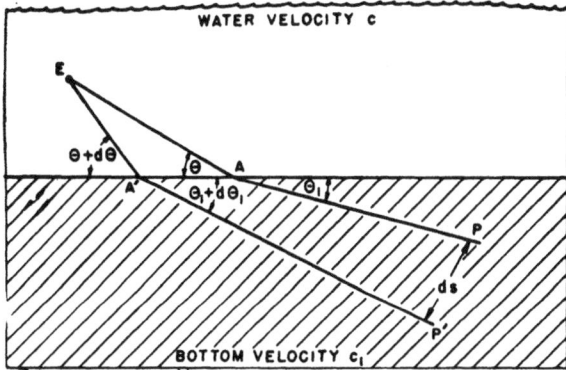

FIGURE 25. Spreading of adjacent sound rays on entering the bottom.

coefficient of the bottom at the angle θ, this transmitted energy is proportional to $[1 - R(\theta)]d\theta$. If the range $r = AP$ is large compared to EA, the distance between P and P' will be approximately

$$ds \approx rd\theta_1 = r\frac{c}{c_1}\frac{\sin \theta}{\sin \theta_1}d\theta \qquad (10)$$

by equation (9). By introducing another factor r to allow for azimuthal spreading, the energy received at P per unit area is then proportional to

$$\frac{[1 - R(\theta)]d\theta}{rds} = \frac{1}{r^2} \cdot [1 - R(\theta)] \cdot \frac{c_1 \sin \theta_1}{c \sin \theta}. \qquad (11)$$

As the ray AP approaches the horizontal, $\sin \theta_1$ approaches zero, and according to equation (11) the intensity at P must do likewise. This conclusion is made even stronger by the fact that, according to Section 2.6.2, $R(\theta)$ approaches unity as θ approaches the angle for total reflection. Thus, ray theory cannot account for the sound received via a path like $EABH$ in Figure 23, when the bottom is uniform.

The argument just given to show the inapplicability of ray theory to arrivals of the type shown in Figure 23 would of course not be strictly correct if there were a gradual increase of the velocity of sound with depth in the bottom, a situation which is quite common, especially for soft bottoms. It will be instructive to consider briefly the sound ray paths for this case, since the limitations of ray theory can be most clearly seen by studying this case where it is partially applicable.

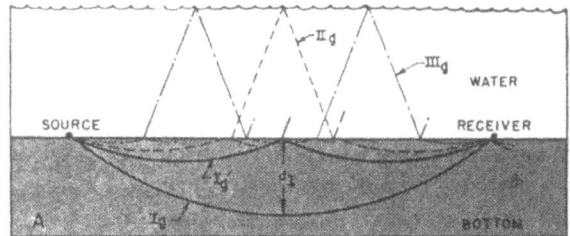

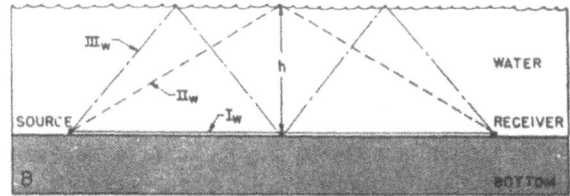

FIGURE 26. Ray paths in and over a bottom giving weak upward refraction.

Figure 26A shows a family of rays connecting a source and hydrophone, both of which are lying on a bottom characterized by weak upward refraction. According to ray acoustics the first signal to reach the receiver H will arrive via the path I_g. This will be followed almost immediately by arrivals along other paths, such as I'_g, which likewise lie in the bottom but which involve one or more reflections at the interface between bottom and water. Some time later another group of arrivals will be received, each of which comes along a path involving one reflection from the surface of the water. One path of this type is shown in Figure 26A and labeled II_g; many other such paths, not shown, are also possible; some of them involving additional reflections from the water-bottom interface as was the case for I_g'. This second group of arrivals will in turn be followed by another group, exemplified by III_g in Figure 26A, involving two reflections from the free surface, and so on. Mixed in with these arrivals along paths which enter and leave the bottom will be those along paths lying wholly in the water, shown in Figure 26B. These paths have for simplicity been drawn for the case where the velocity of sound in the water is uniform. In practice most of the experiments performed so far have encountered isothermal water with consequent upward refraction; this case, which will be discussed later, is in most respects little different from the uniform case considered here. The first arrival among these water rays will be along the direct path I_w, the second along the surface-reflected path II_w, the third along a path III_w involving one reflection from the bottom, and so on.

Thus if the predictions of ray acoustics were valid, we should expect the signal received at the hydrophone to consist of a number of evenly spaced groups of pulses of diminishing strength, corresponding to the "ground rays" shown in Figure 26A, plus a number of individual pulses starting at a later time and separated by gradually increasing intervals, which correspond to the "water rays" shown in Figure 26B. Of these various arrivals, some are positive pressure pulses, others negative, according to the number of phase-changing reflections each has suffered.

The extent to which the predictions of ray theory can be trusted in a case of this sort can be estimated by resolving the explosive pulse into a superposition of sine waves by use of Fourier's theorem, as described in Section 9.2.4, and then applying the criteria given at the end of Section 3.6.2 for applicability of the eikonal equation to sinusoidal waves. It is clear from these criteria that the condition for ray theory to be applicable to a sine wave along a path of the type I_g, II_g, etc., in Figure 26A, is that the maximum depth of the path, shown as d_I for ray I_g, should be large compared to the wavelength of the sound in the bottom. This condition will be fulfilled by the highest frequencies in the Fourier resolution of the explosive pulse, but not by the lowest frequencies; moreover, the frequency above which ray theory is applicable recedes to higher and higher values for the successive arrivals I_g, II_g, III_g, etc. As is to be ex-

pected, this critical frequency approaches infinity as the magnitude of the velocity gradient in the bottom decreases to zero, since the depths of penetration d_I, etc., of the rays approach zero.

A similar consideration of the disturbance propagated through the water suggests that ray theory should fail for frequencies of the order of c/h and smaller, where h is the depth of the water. This limit has little meaning, since this frequency can be expected to be lower than the frequency at which the ray picture fails for the ground rays, and we cannot make a clear separation between the ground disturbance and the water disturbance after we have abandoned the ray concept.

We may thus expect the pressure variation which would be recorded by a very high-fidelity receiver at H to consist of the succession of pulses which ray theory would predict plus a correction which is made up almost entirely of low frequencies. For the disturbance due to the shock wave from the explosion, the times of the various ray arrivals can, ideally at least, be identified on the oscillogram of the received pressure by the occurrence of sharp jumps in the pressure; these jumps, due to the sudden rise at the front of the shock wave, cannot easily be obliterated by the low-frequency correction (see Figure 13). Since, as explained above, the intensities of the arrivals predicted by ray theory are zero for a uniform or downward-refraction bottom and are small for a bottom with weak upward refraction, we should not be surprised to find the disturbance received by the hydrophone to be dominated by the low-frequency portion, with only a few detectable traces of the ray arrivals.

To determine the nature of the low-frequency correction just mentioned, it is necessary to study solutions of the wave equation similar to those considered in Section 2.7.2. In a report prepared by CUDWR,[23] it is shown how the normal modes of vibration of water and bottom can be computed and superposed to correspond to the disturbance produced by explosive source. The mathematical details are too complicated to be given here;[c] however an attempt will be made below to explain in a simplified manner the physical basis for some of the most important results of reference 23. In particular, it will be shown how many characteristics of the signal received at the

---

[c] The reader who wishes to study the mathematical theory of normal modes will find it profitable to study also the treatments devoted primarily to single-frequency sound in deep water[25] and electromagnetic waves in the atmosphere.[26]

hydrophone can be interpreted in terms of a simple dispersion law, i.e., a propagation of different frequencies with different velocities.

The physical reasons underlying the dispersion phenomena just mentioned can be seen by considering the simple case of a progressive wave of a single frequency f. Let us try to construct such a wave by assuming the pressure disturbance to be

$$p = e^{2\pi i(x/\lambda - ft)} M(z), \qquad (12)$$

where x is a horizontal coordinate, and z is the depth below the free surface of the ocean. This function p must satisfy the wave equation

$$c^2\left(\frac{\partial^2 p}{\partial x^2} + \frac{\partial^2 p}{\partial z^2}\right) = \frac{\partial^2 p}{\partial t^2} \qquad (13)$$

in the water; that is, when z is less than the depth h and must satisfy the analogous wave equation[d]

$$c_1^2\left(\frac{\partial^2 p}{\partial x^2} + \frac{\partial^2 p}{\partial z^2}\right) = \frac{\partial^2 p}{\partial t^2} \qquad (14)$$

in the bottom, that is, when z is greater than h. In addition, p must satisfy boundary conditions at the free surface and at the interface between water and bottom. These conditions are

$$p = 0 \text{ at } z = 0 \qquad (15)$$

$$p_{\text{water}} = p_{\text{bottom}} \text{ at } z = h. \qquad (16)$$

$$\left(\frac{1}{\rho}\frac{\partial p}{\partial z}\right)_{\text{water}} = \left(\frac{1}{\rho_1}\frac{\partial p}{\partial z}\right)_{\text{bottom}} \text{ at } z = h. \qquad (17)$$

Assuming for simplicity that water and bottom are uniform, so that c and c_1 are independent of z, we have, on inserting expression (12) into equation (13),

$$\frac{\partial^2 M}{\partial z^2} = 4\pi^2\left[\frac{1}{\lambda^2} - \frac{f^2}{c^2}\right]M \text{ for } z < h. \qquad (18)$$

To satisfy this equation and the condition (15), we must take

$$M = A \sin\left(2\pi\sqrt{\frac{f^2}{c^2} - \frac{1}{\lambda^2}}z\right) \text{ for } z < h. \qquad (19)$$

This equation is formally correct regardless of whether the quantity under the radical is positive or negative. However, it will be shown below that this quantity must be positive if equation (12) is to represent a physically possible disturbance. Similarly, to satisfy the wave equation in the bottom we must have

$$\frac{\partial^2 M}{\partial z^2} = 4\pi^2\left[\frac{1}{\lambda^2} - \frac{f^2}{c_1^2}\right]M \text{ for } z > h. \qquad (20)$$

Now, if $[(1/\lambda^2) - (f^2/c_1^2)]$ is negative, M will be a periodic function of z in the bottom, and according to equation (12) the pressure disturbance in the bottom will consist of progressive waves going diagonally up or down. The disturbance created by an explosion will consist in part of a superposition of progressive waves of this type which travel diagonally downward in the bottom; these waves are, however, a relatively unimportant part of the signal received in the water at a great distance, since their energy spreads out in a downward direction and thus decreases fairly rapidly with distance in the horizontal plane. The part of the signal which is most important in the present application consists, instead, of a superposition of waves of the form (12), for values of λ and f which make $[(1/\lambda^2) - (f^2/c_1^2)]$ in equation (20) positive. The two independent solutions of equation (20) for this case will be exponential functions of z, one increasing to infinity as z increases, the other decreasing to zero. The former of these is physically inadmissible; so we may conclude that if a pressure wave of the desired form exists at all, it must be of the form

$$M = B \exp\left(-2\pi\sqrt{\frac{1}{\lambda^2} - \frac{f^2}{c_1^2}}z\right) \text{ for } z > h, \qquad (21)$$

and of course of the form (19) for $z < h$. However, it is easily shown that no matter what values are given to the constants A and B, it is not possible to satisfy both of the boundary conditions (16) and (17) unless λ and f are related in a particular way. For, on inserting expressions (19) and (21) into these conditions, and using the abbreviations

$$\mu = 2\pi\sqrt{\frac{f^2}{c^2} - \frac{1}{\lambda^2}}, \quad \nu = 2\pi\sqrt{\frac{1}{\lambda^2} - \frac{f^2}{c_1^2}}$$

we obtain

$$A \sin \mu h = B e^{-\nu h} \qquad (22)$$

$$\frac{\mu A}{\rho} \cos \mu h = -\frac{\nu B}{\rho_1} e^{-\nu h}. \qquad (23)$$

Dividing the first of these equations by the second eliminates A and B, and gives the following relation which must be satisfied by f and λ.

$$\frac{\rho}{\mu} \tan \mu h = -\frac{\rho_1}{\nu}. \qquad (24)$$

If this relation is satisfied, a suitable choice of the ratio B/A will insure that both equations (22) and (23) are satisfied. It is easily verified that if $c_1 > c$

---

[d] The theory presented here ignores shearing stresses in the bottom and thus treats the bottom as a fluid rather than as a solid. This assumption, although reasonable for MUD bottoms, is of course not true for ROCK. However, many features of the disturbance predicted by the present theory would doubtless also be observed over a rock bottom.

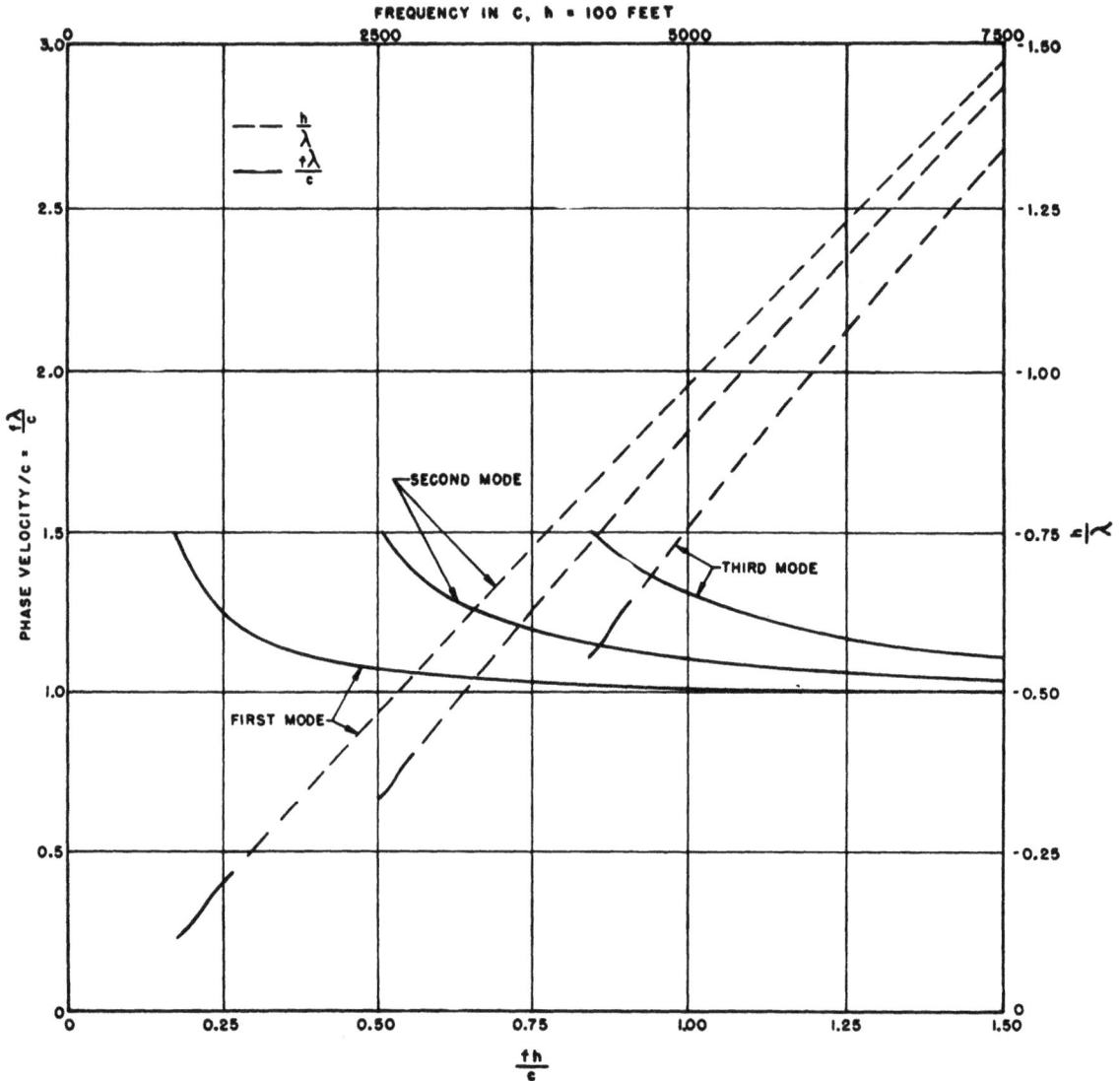

FIGURE 27.   Variation of wavelength and phase velocity with frequency for normal modes in shallow water. Velocity of sound in bottom assumed to be 1.5 × c. Density of bottom assumed 2 times density of water.

f = Frequency.
h = Depth of water.
c = Velocity of sound in water.
λ = Horizontal wavelength of disturbance.

and $\rho_1 > \rho$, equation (24) cannot be satisfied if μ is imaginary; this justifies the statement made in the second sentence following equation (19).

Graphs of the solutions of equation (24) are given in Figure 27 for a typical set of values of c_1, ρ_1, and h. Typical curves of the variation of pressure along a vertical line are given in Figure 28, corresponding to particular points on the graphs of Figure 27. As was explained in Section 2.7.1, it is customary, by analogy with the terminology used in the theory of vibrating systems of particles, to use the term "normal mode"

to describe a state of vibration of the water and bottom in which the pressure distribution is given by equation (12); for modes of the present type this is equivalent to a disturbance of the type shown in Figure 28 and having an amplitude represented by a horizontally moving sine wave. It is convenient to identify families of these normal modes by the number of horizontal planes in the water, including the free surface on which the pressure is always zero. This number is called the *order* of the normal mode, and is indicated by the labels "FIRST MODE,"

M(z) ARBITRARY SCALE

FIRST MODE $\frac{fh}{c} = 0.5$

M(z) ARBITRARY SCALE

MINIMUM FREQUENCY
FIRST MODE

M(z) ARBITRARY SCALE

SECOND MODE $\frac{fh}{c} = 1.5$

M(z) ARBITRARY SCALE

THIRD MODE $\frac{fh}{c} = 2.0$

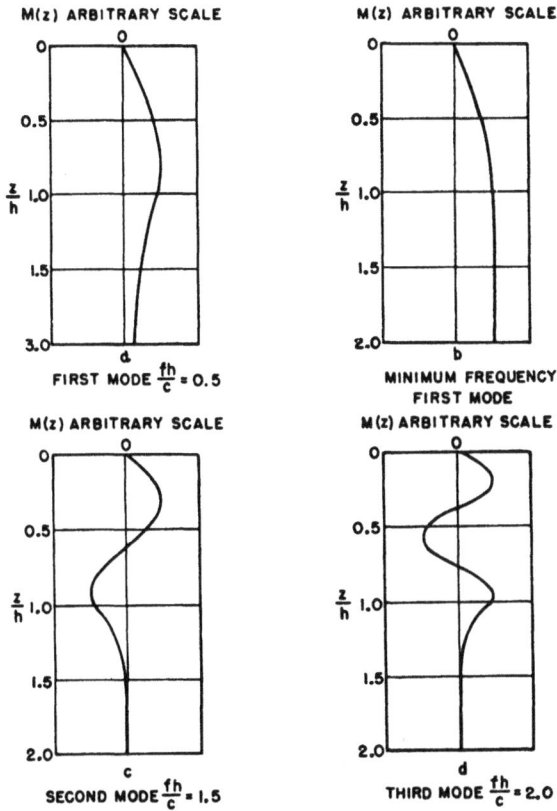

f = Frequency.
h = Depth of water.
c = Velocity of sound in water.
z = Distance below surface of water.
$M(z)$ = Pressure amplitude at depth z.

FIGURE 28. Variation of pressure with depth along a vertical line for various normal modes. Velocity of sound in bottom assumed to be $1.5 \times c$. Density of bottom assumed 2 times density of water.

"SECOND MODE," and so on, in Figures 27 and 28.

A noteworthy fact is that for modes of any given order there is a minimum frequency below which no value of λ can be found which will satisfy the boundary conditions. At this frequency the quantity ν, which is inversely proportional to the depth of penetration of the disturbance into the bottom, goes to zero; and the pressure distribution takes a form such as that shown in Figure 28B. As the mathematically inclined reader can verify for himself from equation (24), the minimum frequency for the first mode has a half-period equal to the interval which ray theory would predict between the arrivals of types I_g, II_g, III_g, etc., of Figure 26. This half-period is given by

$$\frac{T}{2} = 2h\sqrt{\frac{1}{c^2} - \frac{1}{c_1^2}}. \qquad (25)$$

Since these ray arrivals are alternately positive and negative, the period of the disturbance given by ray theory is the same as that for the minimum frequency. It is also noteworthy that the minimum frequency for the νth mode is $(2\nu - 1)$ times its value for the first mode. This has the very important consequence that any simple harmonic disturbance of low frequency which is propagated over a large horizontal range can be represented by a superposition of a finite number of normal modes of low order.

Let us now consider the velocity of propagation of a disturbance which consists of a superposition of normal modes of a given order but distributed over a narrow range of frequencies. It is easy to show that such a disturbance, considered as a function of horizontal distance x or of time t, will form a wave train. For each component normal mode has a phase factor proportional to $e^{2\pi i(x/\lambda - ft)}$. At any given time t there will be some value of x for which most of the component modes are approximately in phase; in the neighborhood of this value of x the pressure disturbance will therefore be large. If x differs very widely from this value, on the other hand, the phases of all the component normal modes will be rather randomly distributed since the different modes have slightly different wavelengths λ; for such values of x the pressure disturbance will be small. If we watch the motion of the wave train in the course of time, we shall find that the region of large amplitude moves with a certain velocity, commonly called the "group velocity." Now, if the center of the wave train is to be near x_1 at time t_1, and near x_2 at time t_2, the phase change $(x_2 - x_1/\lambda) - f(t_2 - t_1)$ must be very nearly the same for all the different normal modes contained in the wave train, in order that they may continue to reinforce one another. This implies, in the limit where only a very narrow range of frequencies is involved,

$$\frac{d}{df}\left[\frac{x_2 - x_1}{\lambda} - f(t_2 - t_1)\right] = 0,$$

or, if V is the group velocity,

$$V = \frac{(x_2 - x_1)}{(t_2 - t_1)} = \frac{df}{d\left(\frac{1}{\lambda}\right)}. \qquad (26)$$

Now the phase velocity of any single-frequency component, defined as the speed of advance of a point having a given constant value of the phase $2\pi(x/\lambda - ft)$, is equal to λf. If this quantity were a constant independent of frequency, as is the case for sound propagated in a single homogeneous medium, the expression (26) would simply equal the phase velocity. For the disturbances we are considering here, however, λf is not independent of frequency, as a glance at Figure 27 will show.

The importance of the result (26) in shallow-water transmission is that it enables us to understand the dispersion phenomena in the ground and water waves. The initial disturbance can be represented as a superposition of normal modes having a very wide range of frequencies. However, since the group velocity is different for different frequencies, the different frequencies in this superposition will get sepa-

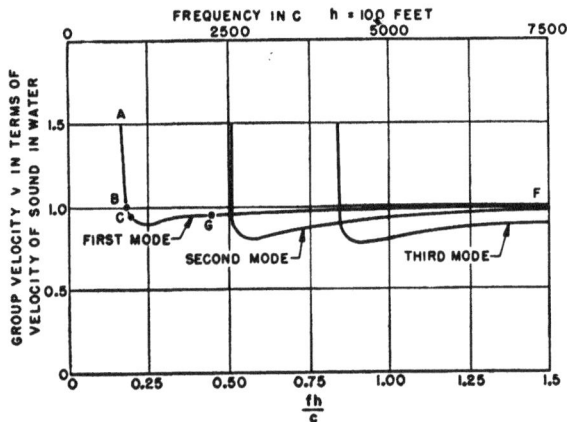

FIGURE 29. Dispersion in group velocities of normal modes. Frequency in f; depth of water, h; velocity of sound in water, c.

rated out somewhat at long ranges, and each band of normal modes of a given order and a given narrow range of frequencies will be propagated with its own group velocity. This effect is shown quantitatively for a typical set of conditions in Figure 29. The curves of this figure are derived from those of Figure 27 by differentiation. Note that the group velocity varies from the ground velocity c_1 at the low cutoff frequency to the water velocity c at very high frequencies, but has a minimum at an intermediate frequency. The existence of this minimum produces an interesting effect, which will be described later.

The main features of the disturbance received at a distance from an explosive source can be explained most simply by concentrating attention on one of these curves, say that for the first mode. This will not only be illustrative of the main characteristics shared by all the normal modes, but will in fact provide a rough prediction of what some of the actual records to be discussed in Section 9.4.3 should look like. For it has been pointed out that there exists a minimum

frequency for each normal mode and that the frequency for the first mode is the lowest. Thus if the disturbance produced by an explosion is received with equipment responsive only to sufficiently low frequencies, the resulting signal can be interpreted in terms of the first-order modes alone. Even when high-fidelity recording equipment is used, the first mode should dominate the initial or ground wave portion of the disturbance, since it can be shown theoretically that the amplitude of the first mode is greater than the amplitude of higher modes in this region.[24]

Let us therefore suppose that we have a source of sound which generates a transient disturbance consisting entirely of a superposition of first-mode vibrations of various frequencies. Since according to Figure 29, the highest group velocity occurs for the lowest frequencies above the cutoff, the first sound to arrive at a distant hydrophone will be a wave train whose frequency corresponds very nearly to point A of Figure 29. The disturbance arriving a little later will consist of frequencies having a slightly slower group velocity, that is, of slightly higher frequencies. Thus, the frequency of the received disturbance will gradually increase with time until the value corresponding to point B is reached. At this moment, the very highest frequencies present in the original disturbance start to come in, traveling in the limit with the velocity c. From this time onward, the received disturbance consists of a low-frequency part and a high-frequency part superposed, the former continuously increasing its frequency along the branch BC of the dispersion curve, and the latter continuously decreasing its frequency along the branch FG. Eventually these two coalesce, and the disturbance dies out at an intermediate frequency.

All these characteristics are apparent in the theoretical pressure-time curve of Figure 30, which shows the contribution of the first mode to the disturbance produced under a typical set of conditions by a source which emits a single positive-pressure pulse of short duration. The portions of the curve corresponding to the points A, B, C, F, G of Figure 29 are labeled with these letters. Similar curves showing the contributions of normal modes of higher order are given in reference 23. These have lower amplitudes than that for the first mode, especially during the "ground wave" phase, that is, the portion of the disturbance which has traveled with a velocity greater than c and thus lies to the left of B and F. According to the present theory, which idealizes the bottom as a homogeneous fluid, the variation of the pressure at

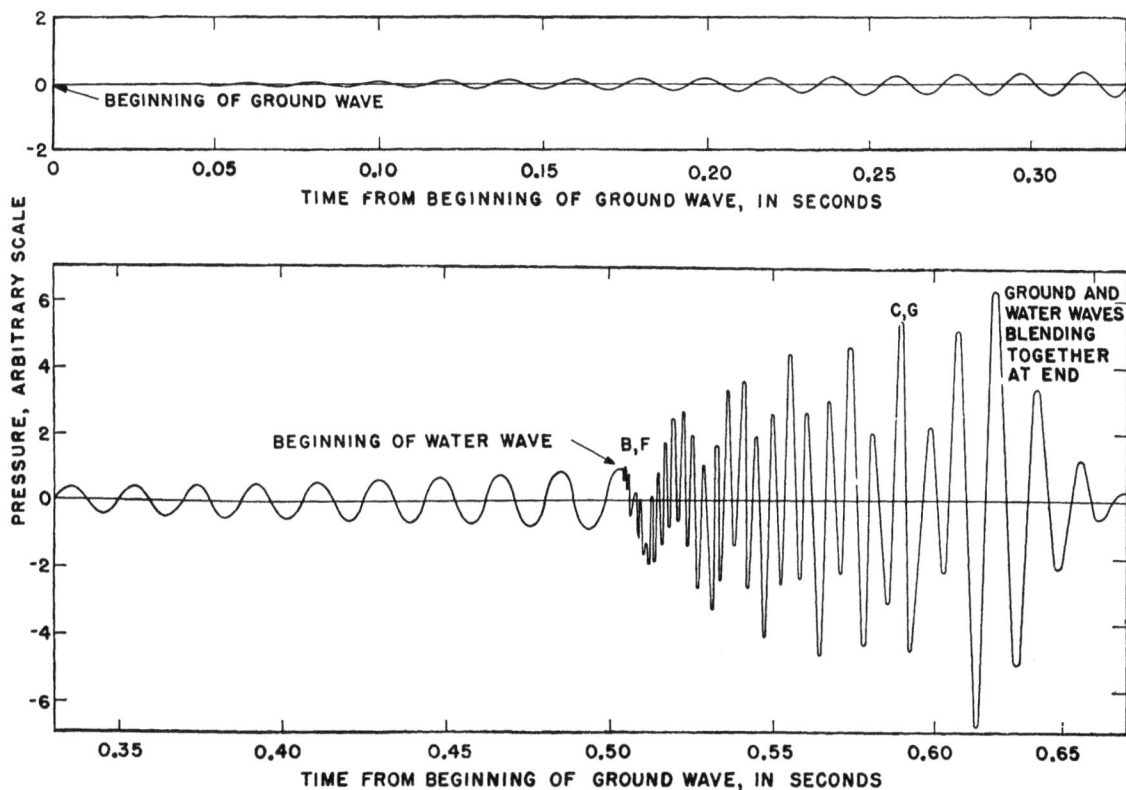

FIGURE 30. Theoretical contribution of the first mode to the disturbance at a distance from an explosion in shallow water. Source and receiver both assumed to be on the bottom. Range: 9,200 yards. Depth of water: 60 feet. Velocity of sound in bottom = 1.1 × velocity in water. Density of bottom = 2 × density of water. (*Note*, A should appear at point indicating beginning of ground wave.)

the hydrophone with time should be given by the sum of these contributions from all the normal modes, plus certain additional terms whose magnitude decreases rapidly with increasing range, so that they become negligible at very long ranges. This complete pressure-time curve would of course show sharp jumps at the positions corresponding to the arrivals predicted by the ray picture.

In the following section we shall compare these theoretical predictions with observations. In this comparison certain factors have to be taken into consideration which for simplicity have been neglected in this section, such as the modification of the received disturbance by the frequency response characteristics of the recording equipment, and the fact that instead of delivering a single impulse, an explosion gives out a shock wave followed by several bubble pulses (see Section 8.6.).

9.4.3 Analysis of Experimental Records

The Woods Hole Oceanographic Institution has obtained a large number of oscillographic records of

sound from explosions in shallow water at distances between 0.25 mile and 30 miles.[2b] Several series of experiments were conducted at widely separated places with bottoms of mud, sand, and coral. The depths of the water at the sending and receiving positions were usually similar and in the range 40 to 180 ft; some shots were made at greater depths. The hydrophones used were in all cases placed on the bottom, while the charges were usually on the bottom but sometimes at mid-depth. Charges of ½-lb TNT to 300-lb TNT were used. At all stations the water was very nearly isothermal, so that sound rays in the water were refracted slightly upward.

Figure 31 shows some typical oscillograms of the sound received in these experiments. Each record consists of eight traces simultaneously recorded. The first of these, labeled "time break," is used merely to record the instant at which the charge was set off; the others record the disturbance received, as modified by the frequency responses shown at the left for the various recording channels. Most of the interesting features show up best on the two channels labeled "Mark II low frequency," which record the same

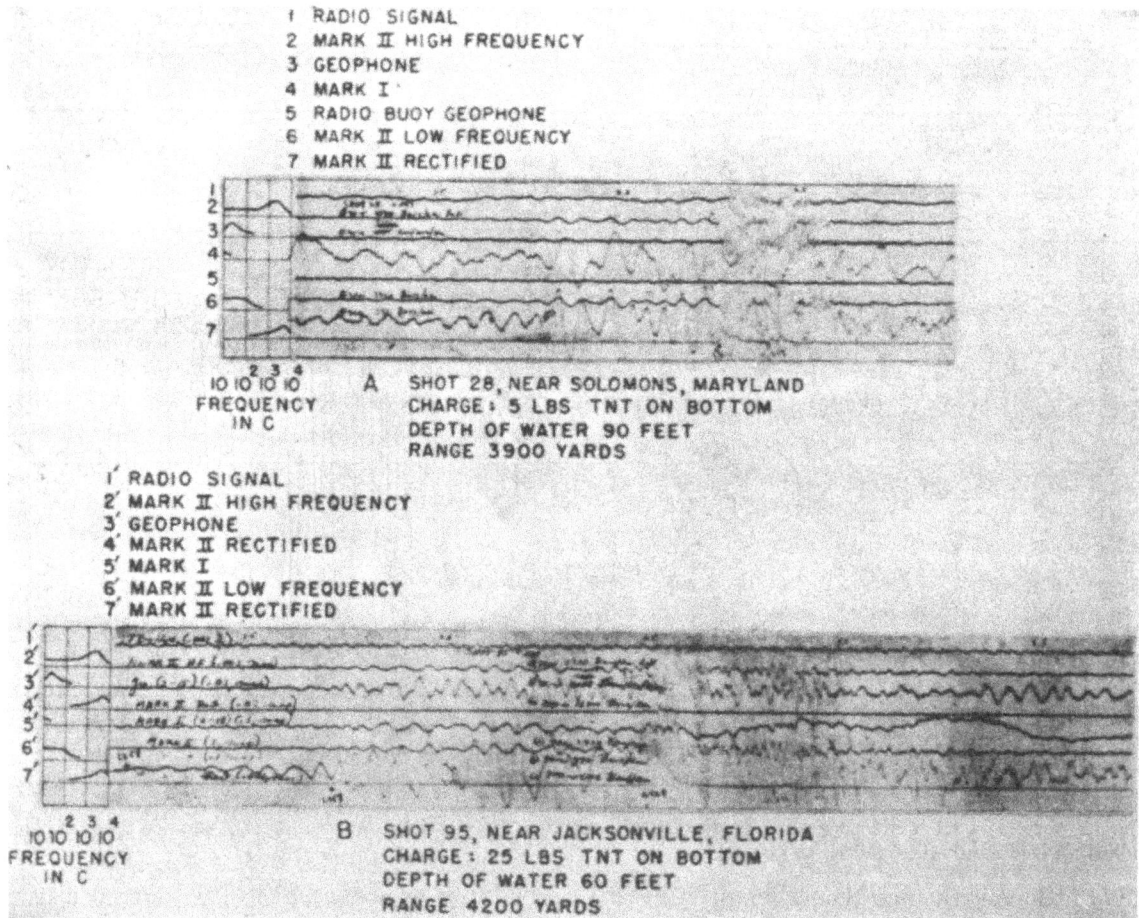

FIGURE 31. Typical records of explosive sound transmissions in shallow water. Times marked along the top of each oscillogram are in seconds. Curves at left give relative amplitude response of each channel to the various frequencies.

signal at two different amplitude levels. However, the rectified traces show most clearly the times of the various water wave arrivals, namely, shock wave and bubble pulses. The arrival time of the first of these is especially useful, since the range can be determined for any shot by multiplying by c the interval between the detonation and this arrival.

The interpretation of records like those of Figure 31 is often complicated by the fact that each observed trace represents a superposition of the disturbances produced by the shock wave and all the bubble pulses. According to the theory of normal modes, the amplitude of the disturbance produced by any one such pulse of very short duration should be proportional to the impulse $\int \rho dt$ of the pulse; since this quantity is of the same order of magnitude for the shock wave and the first one or two bubble waves, the resulting superposition can become very compli-

cated. However, as was mentioned in Section 8.6, the bubble pulses are often much weaker when, as was usually the case in the WHOI experiments, the charge is fired in contact with the bottom. Records (A) and (B) of Figure 31 are fairly typical examples of shots on the bottom; the former shows a strong bubble pulse and the latter a very weak one. Note that the separation of the first two high peaks in the ground wave of Figure 31A is just equal to the bubble period as read from the rectified trace. Since the periods of the oscillations are long compared with the duration of the impulse sent out by the explosions, the only noticeable effects of increasing the size of the charge are to increase the amplitude and to alter the time lag in the arrival of the bubble pulse effects. Changing the position of the charge from bottom to middepth also seems to have very little effect.

Let us begin the detailed discussion of the ob-

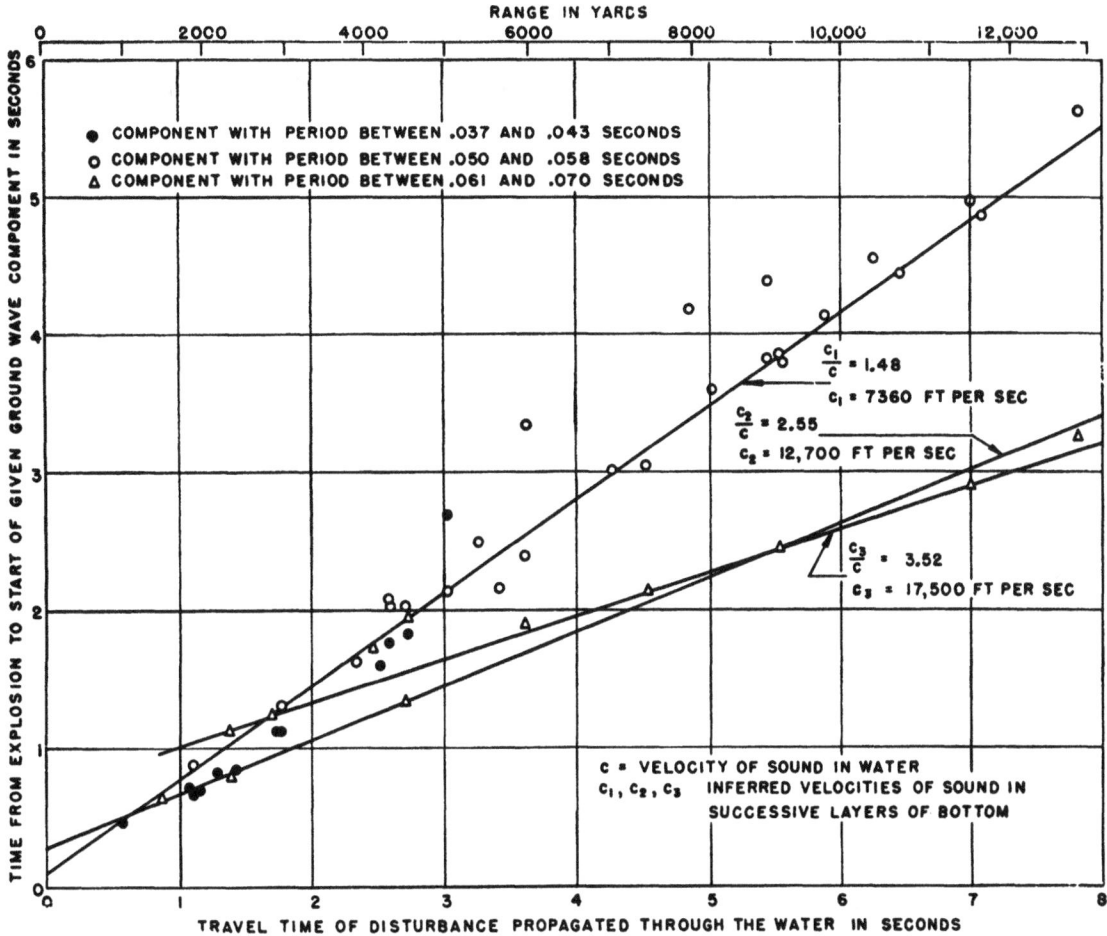

FIGURE 32. Typical plot of travel time against range for components of different frequencies in the ground wave. Location, near Jacksonville, Florida, at mean depth of 57 feet, charge and hydrophone both resting on bottom.

served records by considering the initial or ground wave phase of the disturbance. All the records agree in a general way with the predictions of the theory of reference 23 as outlined in Section 9.4.2 (see Figure 30) in showing a gradual increase of frequency between the beginning of the disturbance and the arrival of the water wave. According to equation (25) of Section 9.4.2, when the bottom is uniform the period of the disturbance at its beginning is a function of the velocity c_1 of sound in the bottom. Extension of the theory to cases where the bottom consists of a deep firm stratum overlaid by a slower one gives the result that at sufficiently long ranges the beginning of the ground wave should have a frequency dependent in a complicated way upon the velocities in both layers, but that, if the upper layer is sufficiently thick in comparison with the depth of the water, a strong new disturbance of distinctly higher frequency will arrive some time later, the arrival

time and frequency of this new disturbance being approximately the same as for the ground wave which would occur if the upper layer were infinitely thick. These theoretical predictions suggest that noting the frequencies of the first arrival and any subsequent arrivals in the ground wave may provide useful information about the different strata in the bottom. This is illustrated in Figure 32, which may be compared with Figure 24. Here many of the records show a recognizable new arrival of different frequency from the first which comes some time later. Complete interpretation of the data shown in Figure 32 is difficult, but there is definite evidence for a layer in which the velocity of sound is about 1.5 times the velocity in water, as well as of one or two layers of higher velocity which determine the times of the first arrivals. It is noteworthy that this dependence of the frequency of a ground wave arrival upon the velocity of sound in the layer chiefly responsible for the ar-

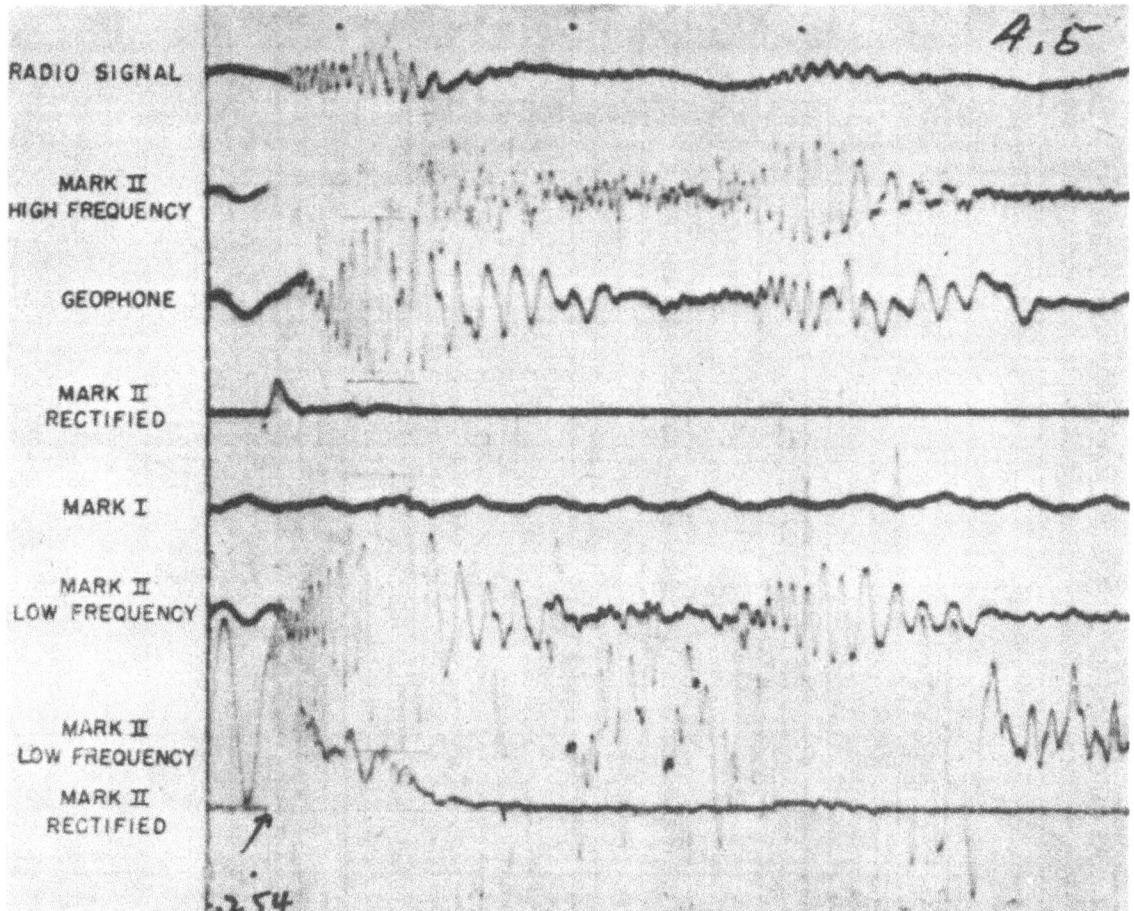

FIGURE 33.   Dispersion in the water wave produced by an explosion in shallow water. Shot 90, near Jacksonville, Fla.;
charge, 5 lb TNT on bottom; mean depth of water, 57 ft; range, 7,000 yd; frequency response of channels as shown in
Figure 31.

rival in question can be observed even in the first
arrivals at fairly short range. Thus in the data from
which Figure 24 was constructed, the period of the
first arrival was between 0.024 and 0.036 sec for the
points to the left of the intersection of the two
straight lines, and was between 0.050 and 10 sec for
the points to the right, except for two very close to
the intersection.

On some records the ground wave dies out quite
noticeably before the arrival of the water wave. The
theory of reference 23 indicates that if the bottom is
uniform to all depths, the amplitude of the ground
wave should increase steadily until the water wave
arrives, and that a decrease in the strength of the
ground wave in this region implies the presence of
layers of different materials. In the latter case, there
may be a secondary ground wave arrival of the sort
mentioned in the preceding paragraph, which dies

out considerably before the arrival of the water wave.

Let us now consider the disturbance after the arrival
of the water wave. Figure 33 shows, in more detail
than Figure 31, the dispersion phenomena occurring
in this stage. The third trace from the bottom shows
most clearly the ground wave just before the arrival
of the water wave, and the gradual development of
the water wave from a disturbance of very low amplitude and high frequency, superposed on the ground
wave, to the final, so-called Airey phase where
ground wave and water wave fuse at an intermediate
frequency and die out. The similarity of this record
to Figure 30 is quite striking. The resemblance is not
nearly so close for the second trace from the top,
since this trace was recorded with more fidelity at the
high frequencies, so that many normal modes higher
than the first contribute significantly to it.

About 0.2 sec after the main disturbance has died

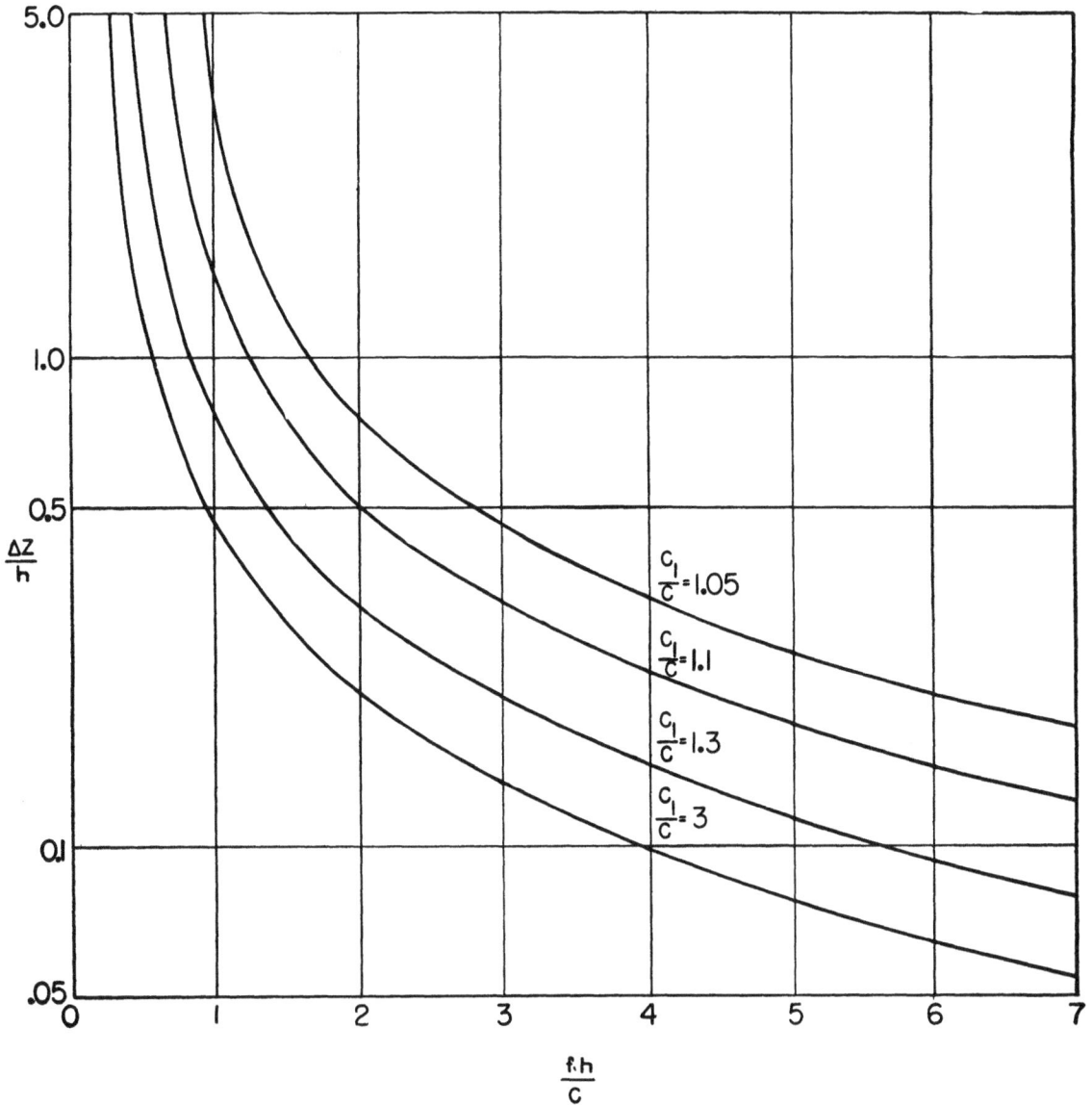

h = Depth of water.
f = Frequency of contribution of first normal mode.
c = Velocity of sound in water.
c_1 = Velocity of sound in bottom, assumed uniform.

Δz = Depth below the surface of the bottom above which 99% of the wave energy of the first mode in the bottom is included.

FIGURE 34. Typical curves of the frequency dependence of the depth of penetration of the first normal mode into the bottom. Density of bottom assumed 2 times density of water.

out, another disturbance is recorded, weaker than the first and due to the bubble pulse. On the low-frequency trace this second water wave looks similar to the first; but on the high-frequency trace it is very different. Frequencies above about 200 cycles are absent in the bubble pulse disturbance but strong in the primary disturbance. This is probably due to the fact that, as indicated in Figure 8 of Chapter 8,

the pressure delivered by the bubble in its contracted stage has a duration of several milliseconds and is thus lacking in high-frequency components.

A detailed analysis of the dispersion phenomena in the water wave can provide useful information on the characteristics of the upper layers of the bottom.[23] Unless shots are made at very short ranges, the arrival times and frequencies of ground waves furnish infor-

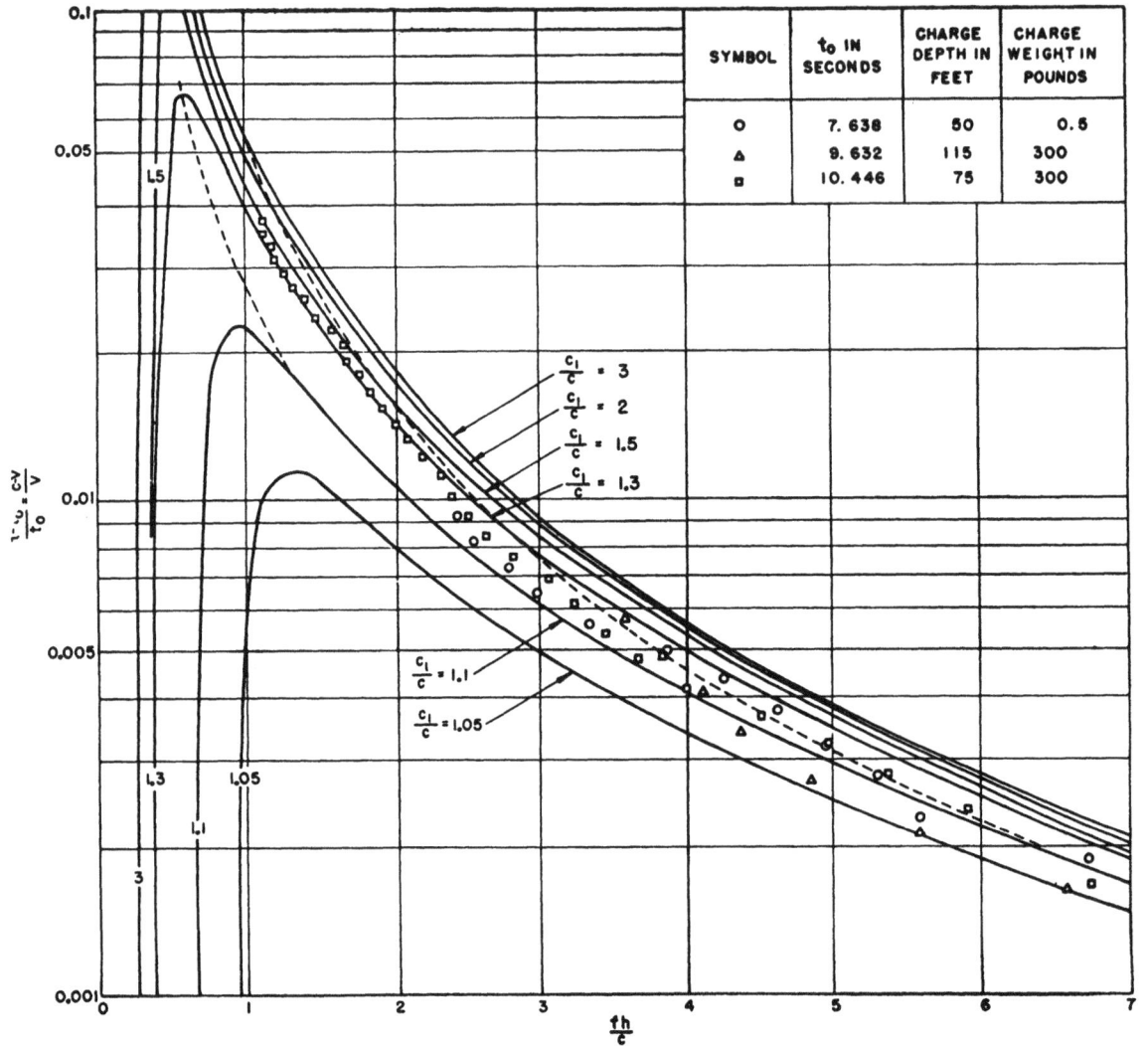

SYMBOL	t_0 IN SECONDS	CHARGE DEPTH IN FEET	CHARGE WEIGHT IN POUNDS
o	7.638	50	0.5
△	9.632	115	300
▢	10.446	75	300

h = Depth of water.
f = Frequency of contribution of first normal mode.
t = Time between explosion and arrival of frequency f.
t_0 = Time between explosion and first water wave arrival.
V = Group velocity for frequency f.
c = Velocity of sound in water.
c_1, c_2 = Velocities of sound in bottom layers.

——— Theoretical dispersion in the first normal mode for various uniform bottoms all of density 2.
— — — Same for layer of thickness $0.1h$ and $c_1/c = 1.1$ underlain by infinite layer with $c_2/c = 3$, both of density 2.
- - - - Same for layer of thickness h and $c_1/c = 1.1$ underlain by infinite layer with $c_2/c = 3$, both of density 2.

FIGURE 35. Theoretical and observed dispersion in the water wave. Shots were made off Jacksonville, Fla., where the depth of water was 115 to 120 feet. Hydrophone was on the bottom for all shots. Observed frequencies are taken from the Mark II low-frequency record whose response is shown in Figures 31 and 33.

mation only on those layers of the bottom which are reasonably thick, in comparison with the depth of the water; the water wave, on the other hand, can supply information on the uppermost layers even when they are much thinner. This is a consequence of the fact that for normal modes of any order the higher the frequency the more rapidly the disturbance dies out

with increasing depth. Since the frequencies in the water wave are much higher than those in the ground wave, the water wave will not penetrate so deeply into the bottom, and will thus be less affected by the characteristics of deep layers and more affected by the characteristics of the top layers. Figure 34 shows how the depth of penetration varies with frequency for the first mode for various types of bottoms. Figure 35 shows the theoretical dependence of the frequency of the first mode on the time, for these same types of bottoms, and shows also the observed frequencies as recorded in a particular region on the low-frequency Mark II channel, whose frequency response has been shown in Figures 31 and 33. Because of its suppression of high frequencies, the record of this channel should consist principally of the contribution of the first mode. It will be noticed that at the highest frequencies the slope of the theoretical frequency-time curve is nearly independent of the assumed structure of the bottom, and that the observed points show the same slope. Note that the frequency corresponding to a given group velocity over a uniform bottom varies inversely as the depth of the water, a relation which is easily verified from equation (24). This relation is fairly well confirmed experimentally.

The observed points in Figure 35 do not follow any of the curves for a uniform bottom but indicate rather that the velocity of sound increases with depth. A rough estimate of the scale and nature of this increase can be obtained by studying Figures 34 and 35 together. Thus at higher frequencies than that corresponding to fh/c equal to 4.5, the points of Figure 35 scatter evenly about the curve for c_1/c equal to 1.1. According to Figure 34, for fh/c equal to 4.5, 99 per cent of the wave energy in the bottom is confined within a layer of thickness 0.2 times the depth of the water. We may therefore say that the velocity of sound in the bottom is $1.1c$ down to a depth of the order of 20 ft. A similar reading of the two figures at fh/c equal to 1.5 gives the result that some sort of weighted average of the velocities over a depth in the bottom of the order of half the depth of the water has a value intermediate between $1.3c$ and $1.5c$. These rough conclusions are confirmed by the fact that the points of Figure 35 fall between the dotted and dashed curves. Information obtained from study of the ground waves regarding the deeper layers of the bottom should, of course, be borne in mind when studying the water waves in this way.

Measurements have been made on the maximum intensity of the water wave as recorded by the various channels. These indicate a decrease of the recorded maximum intensity with distance, which in different regions varies from an inverse fourth or fifth power to an inverse 2.4 power. Theoretically, if there is no absorption or scattering, the energy in any normal mode should vary inversely as the first power of the distance, because the normal modes spread out horizontally but not vertically. Since the duration of the signal increases proportionally to the range, the peak intensity of any normal mode should decrease about as the inverse square of the range; the more refined calculations of reference 24 show an inverse 5/3 power dependence. Thus, although the experimental data are scattered and hard to interpret, there appears to be a discrepancy between theory and experiment. One would, of course, not expect perfect agreement with a theory which neglects absorption and scattering in the bottom, especially since most of the experiments were conducted over soft bottoms.

At one of the stations where shots were made, near the Orinoco delta, it was found that no frequencies below about 300 c appeared in the water wave, although a normal dispersion record was observed in deeper water in the same locality. The ground wave on the anomalous records was fairly normal.

A few shots were made near the Virgin Islands with land between the shot and the hydrophone. These showed a ground wave similar to that which would have been observed in the absence of the land, but the water waves were entirely absent. A related observation is that blasting explosions on land gave weak ground waves at a hydrophone in the sea offshore, but no water waves.

Shots made near Solomons, Md., produced low-frequency disturbances of periods from 0.1 to 0.3 sec which were propagated with a low velocity, about 1,700 ft per sec. These disturbances have been tentatively ascribed to the so-called Rayleigh wave, which is a surface-bound wave in the bottom whose propagation involves shearing stresses. Such waves fall outside the province of the theory of the preceding Section 9.4.2, which idealizes the bottom as a fluid medium.

Chapter 10

SUMMARY

RESEARCH ON SOUND transmission during World War II was concerned almost exclusively with the investigation of sound fields which were operationally important. More than half of the experimental work was devoted to the sound field of standard echoranging transducers operating at frequencies around 24 kc. The purpose of the work was primarily to provide information which could be used to increase the effectiveness of Navy gear already in use on submarines and antisubmarine vessels. The instrumentation used for research differed as little as possible from standard operational gear; what modifications were made usually represented the minimum necessary for quantitative evaluation of the data obtained. Questions which did not seem important operationally, such as the physical cause of the observed attenuation of supersonic sound in the sea, received scant attention in these studies.

In the sections which follow, the essential results of these experiments on underwater sound transmission are summarized. Section 10.1 lists the definitions of the most important quantities used in describing underwater sound fields. Sections 10.2 and 10.3 summarize what is known concerning the average transmission of supersonic and sonic sound in the sea. In Section 10.4, data on the fluctuation and variation of seaborne sound are summarized. Finally, Section 10.5 provides a brief discussion of probable trends in future research on sound transmission.

10.1 BASIC DEFINITIONS

10.1.1 Sound Pressure and Sound Field Intensity

A sound wave in a fluid can be described conveniently in terms of the pressure disturbance which arises in the vicinity of a sound source, travels through the fluid, and is finally received by a hydrophone. The *instantaneous sound pressure* is the difference between the instantaneous value of the pressure at a chosen location and the mean or equilibrium pressure at the same point. The rms value of the instantaneous sound pressure is usually called the *rms sound pressure*. Usually, the average is carried out over a time interval which is long compared with the periods of the principal frequencies making up the sound signal. In the case of single-frequency sound, the average is extended over one period (or an integral number of periods). Unless specified otherwise, "sound pressure" as used in the technical literature is short for rms sound pressure. Except in the case of standing waves, the rms sound pressure is an excellent measure of the energy carried by the sound wave. At the present time, sound pressure values are uniformly reported in units of dynes per square centimeter.

The *sound field intensity* is defined as the averaged power carried by a sound wave per unit cross section of a wave front. The units in present use are watts per square centimeter. If the radii of curvature of the wave fronts are large compared with the wavelength, then the rms sound pressure and the sound field intensity are connected in excellent approximation by the formula

$$I = 10^{-7}\frac{p^2}{\rho c}, \tag{1}$$

in which p is the rms sound pressure, ρ is the density of the fluid in grams per cubic centimeter, c is the sound velocity in the fluid in centimeters per second, and I is the sound field intensity.

10.1.2 Sound Level

The sound field intensity is usually reported on a logarithmic scale. The most common scale for this purpose is the decibel scale. The quantity L,

$$L = 20 \log p \tag{2}$$

in which the rms sound pressure p is expressed in units of dynes per square centimeter, is called the *sound pressure level* or simply the *sound level*. As defined by equation (2), L is the sound level in decibels above a standard which corresponds to a sound pressure of 1 dyne per sq cm. In the past, sound levels were frequently reported in decibels above 0.0002 dyne per sq cm.

10.1.3 Source Level

The source level is a measure of the power output of a sound source on the decibel scale. Briefly, it is the sound level due to a point source at a distance of 1 yd, in decibels above 1 dyne per sq cm. If a point source is located in a homogeneous, nondissipative medium which is infinitely extended in all directions, the intensity of the sound field is inversely proportional to the square of the distance from the source,

$$I = \frac{F}{r^2}. \tag{3}$$

This law is called the inverse square law. In terms of the sound level, equation (3) becomes

$$L = S - 20 \log r. \tag{4}$$

In these equations, F and S are constants which depend on the power output of the source, and r denotes the distance (slant range) from the source. That S is the source level as defined above can be verified by setting r equal to 1 in equation (4).

For real sound sources in real media, equations (3) and (4) are not everywhere valid. Because of the finite extension of an actual sound source, the inverse square law fails at ranges of the order of the dimensions of the source. Because of absorption of the sound in the medium and because of scattering and reflection from bounding surfaces, it fails at very long ranges. However, there is frequently an intermediate range interval for which equation (4) holds. If there is such an interval, then the constant S is considered the source level, even though S may not be the actual sound level at a distance of 1 yd.

For a highly directional sound source, such as a standard echo-ranging transducer, the definition of the source level is further specified by the condition that the sound measurements are to be carried out on the axis, that is, the radial line of greatest sound field intensity.

10.1.4 Transmission Loss and Transmission Anomaly

The *transmission loss H* at the range r is defined by the formula

$$H(r) = S - L(r), \tag{5}$$

where S is the source level, and L is the sound level defined by equation (2). The transmission loss defined in this way measures the drop of the sound level with increasing distance from the source and has the virtue of being independent of the particular power output of the source. Other parameters of the source, such as operating frequency and directivity pattern, are known to affect the value of the function $H(r)$. The units of H are decibels.

The *transmission anomaly A* is the deviation of the transmission loss from that functional behavior demanded by the inverse square law of spreading. The defining equation for $A(r)$ is

$$A(r) = H(r) - 20 \log r = S - L(r) - 20 \log r. \tag{6}$$

The transmission anomaly vanishes if the inverse square law of spreading is satisfied, and it is positive if the sound level drops off more rapidly than $20 \log r$. Large positive transmission anomalies, therefore, correspond to poor sound conditions.

In sound transmission work, it has been customary to train the projector in a horizontal plane on the receiving hydrophone, but not to tilt the acoustic axis away from the horizontal. Hence, measured transmission anomalies will be large for a close deep hydrophone beneath the sound beam.

In supersonic transmission work, it has been found that when successive signals are transmitted a few seconds apart over the same transmission path, the received sound intensity is subject to irregular fluctuations. Reported transmission anomalies always represent values which have been obtained by averaging over a number of signals received during a brief period so that much of this fluctuation is smoothed out.

10.1.5 Variance of Amplitudes

The standard deviation of the individual pressure amplitudes in a sample of signals, divided by the average pressure amplitude for the sample, is called the *variance of amplitudes* for the sample. This variance is used as a measure of the fluctuation of received sound intensity. Observed values of the variance are summarized in Sections 10.4.1 and 10.4.2.

10.1.6 Deep and Shallow Water

Water is effectively deep when bottom-reflected sound is much weaker than the direct sound; otherwise, the water is effectively shallow. Over the continental shelf (depth less than 100 fathoms) the water is effectively shallow for most situations. Away from the continental shelf, the ocean is always deep when sharply directional sound is used (as in echo-ranging at supersonic frequencies), but may be shallow when listening at audible frequencies to a target at long range.

10.2   DEEP-WATER TRANSMISSION

The transmission loss in the open ocean depends on the way the velocity of sound changes with position in the sea, since velocity gradients distort the sound beam. These velocity gradients change with time and location, but in any localized region at any given time depend primarily on depth and relatively little on horizontal position within that region. Changes in sound velocity in deep water closely follow changes in water temperature; the effect of pressure changes is relatively slight and usually need not be considered except for transmission to great depths.

The following subsections tell of the transmission anomalies expected for various common temperature-depth distributions in the ocean.

10.2.1   Isothermal Water

When the top 50 ft of the ocean are isothermal, transmission anomalies are determined by two major effects, absorption and surface reflection.

SONIC FREQUENCIES

At low sonic frequencies, sound is reflected from the sea surface in somewhat the same way as from a flat, perfectly reflecting mirror. The partial cancellation of direct and surface-reflected sound reduces the sound intensity at long range near the surface. The transmission anomaly at any range may be computed from the equation

$$A = -10 \log \left(1 - 2\gamma_a \cos \frac{4rh_1h_2}{R\lambda} + \gamma_a^2\right), \quad (7)$$

where h_1 is the depth of the sound source, h_2 is the depth of the receiving hydrophone, R is the range from source to hydrophone, and λ the wavelength. The quantity γ_a, called the *effective reflection coefficient* of the surface, is a semi-empirical parameter; its average value for different frequencies is given in Table 1.

TABLE 1.   Effective reflection coefficient of the surface.

Frequency in cycles	200	600	1,800
γ_a	0.8	0.7	0.5

Absorption has little effect on sound transmission at frequencies below 2,000 c.

HIGH SONIC AND SUPERSONIC FREQUENCIES

At frequencies above 2,000 c, the value of γ_a to be used in equation (7) is seldom greater than 0.5 in the open sea and is frequently so small that image interference can scarcely be said to exist. Absorption plays an increasingly important role as the frequency increases. The transmission anomaly A may be computed from the relation.

$$A = \frac{ar}{1,000}, \quad (8)$$

where r is the range in yards and where a is the attenuation coefficient in decibels per kiloyard. Average values of a at a number of frequencies are given in Table 2. At frequencies above 1,000 kc, the attenua-

TABLE 2.   Attenuation coefficient in the sea.

Frequency in kc	20	24	30	40	50	60	80	100	500	1,000
a in db per kiloyard	3	4	6	10	13	18	26	35	150	300

tion coefficient is about three times the value predicted from the viscosity of the water. At frequencies of 24 kc and below, a is more nearly 100 times this theoretical value.

10.2.2   Thermocline below Isothermal Layer

When sound from an isothermal layer passes at grazing angle into a thermocline or temperature layer, where the temperature decreases sharply with increasing depth, the sound rays are bent downward and become more spread out. The increased distance between sound rays in and below the thermocline reduces the sound intensity; this phenomenon is known as *layer effect*. The transmission anomaly below the thermocline, at ranges out to 4,000 yd, may be computed from the equation

$$A = 5 \log \left(1 + \frac{2\Delta c - r^2}{c_0 h_1^2}\right) + \frac{ar}{1,000}, \quad (9)$$

where Δc is the change in sound velocity in the top 30 ft of the thermocline. (If several thermoclines lie above the hydrophone or if the gradient in the thermocline increases with depth, Δc is the velocity change in the 30-ft interval giving the maximum value of $\Delta c/h_1^2$.) c_0 is the sound velocity in the surface layer; h_1 is the height of the sound projector above the top of the thermocline, that is, above the top of the 30-ft interval in which Δc is measured; r is the range from projector to hydrophone. The last term on the right is taken over from equation (8); the values of the attenuation coefficient used are given in Table 2. Equation (9) has been checked in detail at 24 kc only, but presumably gives an approximate indication of the anomalies expected below the thermocline at all frequencies above a few hundred cycles. At increasing depths below the thermocline, the anomaly decreases somewhat, the decrease being most marked for the shallower thermoclines.

10.2.3 Temperature Gradients near Surface

When temperature gradients are present in the top 50 ft of the ocean, the transmission loss from a projector at 16 ft to a distant hydrophone is correlated with the following variables: the sharpness and depth of the gradients (for practical purposes, the decrease of temperature from the surface down to 30 ft); and D_2, the depth at which the temperature is 0.3 F less than the surface temperature. For a deep hydrophone, the temperature gradients at intermediate depths are also of importance.

SHARP SURFACE GRADIENTS

When the temperature change in the top 30 ft is more than 1/100 times the surface temperature, the sound beam is bent downward by the decrease of sound velocity with increasing depth. The plot of transmission anomaly against range usually shows three different regions as follows:

1. *The direct sound field* from the projector out to the shadow boundary. The anomaly within the direct sound field is primarily the result of absorption, and equation (8) is applicable.

2. *The near shadow zone.* Beyond the shadow boundary, the sound intensity decreases very rapidly for some distance. Representative values for this decrease are 50 db per kyd at 25 kc and about one-third this at 5 kc. These coefficients of attenuation in the shadow zone are apparently about half the values estimated from the theory of diffraction by a smooth velocity gradient. The range to the shadow boundary increases with depth in accordance with ray theory, but seems to be systematically somewhat less than predicted.

3. *The far shadow zone.* With standard echo-ranging gear and pulses 100 msec long, the transmission

anomaly of scattered sound at ranges of several thousand yards is about 50 db. Thus when the transmission anomaly of the direct or diffracted sound in the shadow zone exceeds about 50 db, the observed sound is scattered sound, with an anomaly which does not depend strongly on further increases in range. This scattered sound is incoherent. For short pulses the intensity of this scattered sound is proportional to the pulse length; it becomes negligibly small for explosive sound.

To predict the anomalies expected under given temperature conditions, it is simplest to use curves of average anomalies for such conditions. Since unexplained deviations are frequently found between individual anomalies and the predictions of ray theory, use of average curves gives results about as accurate as the more elaborate methods. An example of this approach is Figure 40 of Chapter 5, where average curves are given for different values of D_2, the depth at which the temperature is 0.3 F less than the surface temperature.

WEAK SURFACE GRADIENTS

When the temperature change in the top 30 ft is less than 1/100 of the surface temperature, but gradients are present in the top 50 ft, the division of the sound field into the three regions described previously is usually not observed. Since a small change in such temperature conditions may lead to a large change of transmission anomaly, the observed anomalies are highly variable and can neither be compared with theory nor predicted practically with much accuracy. Average anomalies for different values of D_2 are given in Figure 49 in Chapter 5 for a shallow hydrophone. For a deep hydrophone, below the thermocline, equation (9) may be used for approximate results.

10.2.4 Sound Channels

When the velocity of sound above and below the sound source is appreciably greater than the velocity at the source, the sound rays which leave the source with small inclinations will propagate out indefinitely without surface or bottom reflections, bending back and forth but always remaining within some fixed interval of depths.

SURFACE SOUND CHANNELS

When the sound projector lies below a sharp negative gradient and above a sharp positive gradient, sound channel effects should be marked, with regions of alternately high and low anomaly found out to considerable ranges. When a sharp gradient lies just above the projector and a layer of nearly isothermal water 100 ft or more in thickness lies below, ray theory predicts that the sound bent back up by the positive velocity gradient in the isothermal layer should be focused at shallow depths and long ranges, thus giving anomalously high intensities. Observations made under these conditions show that the transmission anomaly on a shallow hydrophone is sometimes as much as 40 db less than normal over a narrow range interval several thousand yards away. The details of these observed effects are not in good agreement, however, with the exact predictions of ray theory.

DEEP SOUND CHANNELS

At a depth of several thousand feet there is usually a deep sound channel. The effect of pressure on sound velocity increases the velocity at greater depths, and a thermocline usually present closer to the surface increases the velocity at shallower depths. Sound of frequency less than 200 cycles, for which the absorption is very low, has been observed to propagate out for several thousand miles in such a deep channel. With small explosive charges, the arrivals of the different pulses agree with the different rays predicted theoretically. The largest number of arrivals, with the highest observed intensity, occur just before the observed sound stops entirely; these last arrivals are the rays coming almost straight along the axis of the channel.

10.3 SHALLOW-WATER TRANSMISSION

In shallow water, the transmission of underwater sound is determined primarily by the character of the bottom, and by the frequency of the transmitted sound. The state of the sea is a much more important factor than in deep water. Temperature gradients are of secondary importance. There are two situations in which sound conditions do not differ appreciably from those found in deep water: (1) soft MUD bottom; (2) strong positive velocity gradients (PETER pattern) below a directional sound source. In both these cases, transmission is very nearly the same as in deep water with the same thermal conditions.

10.3.1 Sonic Frequencies

Most of the information on the transmission of sonic sound in shallow water was obtained in harbor surveys. The data obtained may be summarized as follows.

No systematic difference was found between different types of bottoms, with the exception of soft MUD, which turned out to be a poor reflector. All other bottoms apparently reflect equally well.

Transmission over sloping bottoms in the presence of downward refraction tends to be poor, in agreement with theoretical predictions.

Over flat bottoms, at ranges greater than the water depth and out to several thousand yards, average sound transmission can be best represented by an inverse 1.5th power law of spreading plus an attenuation which appears to increase roughly linearly with the frequency up to about 20 kc. The transmission loss is, thus, given roughly by a formula

$$H = 15 \log r + \frac{1}{4,000}(f - 2)r + C, \qquad (10)$$

where r is the range in yards, f is the frequency in kilocycles, and C is a constant independent of the range r.

10.3.2 Twenty-four Kilocycles

In moderately shallow water, and in the presence of any bottom but MUD and soft SAND-AND-MUD, the transmission anomaly can usually be represented in fairly good approximation by a straight line. For wind forces 0 to 2, transmission anomalies increase with the range at a rate of 5 db per thousand yards over STONY and SAND bottoms, and at a rate of 6 db per thousand yards over ROCK bottoms. About half of all the runs carried out yield values which differ from these average values by no more than 2 db per kyd.

The following special results are also worth noting. (1) For heavy seas, transmission is somewhat worse than for light seas. For wind force 3, about 1 db per kyd should be added to the attenuation coefficients given above. (2) Over sloping bottoms and in the presence of negative gradients, transmission is poor. The transmission anomaly may increase with the range at a rate exceeding 10 db per thousand yards. (3) In shallow isothermal water, transmission is at least as good as in deep isothermal water. (4) In very shallow water (5 fms deep), a series of experiments carried out over SAND gave very poor transmission; the anomaly increased at the rate of about 16 db per kyd.

10.3.3 High Supersonic Frequencies

As far as is known, transmission in shallow water at high supersonic frequencies is similar to that at 24 kc, except for greatly increased absorption losses; transmission anomalies in the presence of negative gradients are linear and their slopes are somewhat higher than those in deep isothermal water.

10.4 FLUCTUATION AND VARIATION

The transmission loss measured at any instant in the ocean will usually differ from the value found several seconds earlier. This rapid change of sound level is called fluctuation. Measured transmission losses and transmission anomalies are averaged to smooth out this fluctuation.

Fluctuation is invariably observed in the transmission of single-frequency supersonic signals transmitted over a path at least 100 yd long. Fluctuation is negligible over transmission paths of the order of 5 yd. Little is known concerning fluctuation over intermediate path lengths. For frequencies of 5 kc and less, fluctuation appears to be less pronounced than at frequencies of 10 kc and higher. The summary which follows is concerned only with the fluctuation of supersonic signals transmitted over paths at least 100 yd in length.

10.4.1 Variance with Shallow Projector

For a projector at a depth of 16 ft, the direct sound from an echo-ranging projector cannot be distinguished from the surface-reflected sound. The fluctuation is large and inexplicably variable. Observed values of variance average 40 per cent with an interquartile spread of about 20 per cent. The variance at 24 kc is significantly correlated with the variance at 16 kc or 60 kc, the coefficient of correlation being about 0.7.

10.4.2 Variance with Deep Projector

For a deep projector and a deep hydrophone, the direct signal can be resolved from the surface-reflected signal. The observed fluctuation of the direct signal is small; observed values of the variance at 24 kc lie between 5 and 10 per cent and may result from the variability of the measuring equipment. The surface-reflected pulse is highly variable with a variance between 50 and 70 per cent.

With explosive pulses, the direct sound can be resolved from the surface-reflected pulse even at shallow depths. The observed variance for the direct pulse is about 1 or 2 per cent if the transmission path lies wholly in an isothermal layer, but up to 20 per cent if part of the transmission path lies in the thermocline.

10.4.3 Rapidity of Fluctuation

The time during which the sound level is not likely to change appreciably is also variable, but seems to increase with increasing range. At a fixed range of less than a few hundred yards, the transmission loss for a shallow sound projector changes by about 20 per cent on the average during 0.5 sec. At a fixed range of several thousand yards in the direct sound field, the average time for a 20 per cent change might be 2 sec; while in the shadow zone, this average time is likely to be nearer 0.02 sec.

10.4.4 Variation

Slow changes in the (averaged) transmission of sound in the sea, which take place in several minutes and which cannot be explained in terms of observable changes in the vertical temperature pattern, are called variations. It has been found that at 24 kc the variation between two transmission runs about 20 minutes apart has an average value of about 4 db if only pairs of transmission runs are considered in which the bathythermograph pattern is significantly the same. This average value for the variation does not appear to depend significantly on range.

10.5 FUTURE RESEARCH

During World War II research on the transmission of underwater sound has been largely devoted to the empirical investigation of certain practical problems. A wealth of detailed information has been accumulated on the transmission loss of sound from a standard echo-ranging projector under conditions likely to be observed in practice. Although this information has been useful in subsurface warfare, it has not led to any complete understanding of the physical processes involved in underwater sound transmission. For example, the average attenuation in deep isothermal water near San Diego has been extensively measured, but the causes of this attenuation are completely unknown.

In the years to come, research in this field will probably change its character. The quest for empirical data on some particular situation has been carried about as far as usefulness requires, and future studies will most profitably be directed to a more fundamental investigation of the basic factors underlying the observed data of underwater sound transmission. Without such a reorientation of the basic research program, it will be impossible to predict the behavior of underwater sound under new and unexplored conditions. Suppose, for instance, that sound gear using a nondirectional supersonic projector were to be proposed. The transmission loss for the sound from such a system could not be predicted definitely from present data, which are all obtained with directional supersonic sources. To make such predictions would require some knowledge of the importance of the scattering of sound through small angles. Similarly, the attenuation of sound transmitted from a deep projector to a deep hydrophone cannot be predicted from the present empirical data taken with shallow projectors, but might be estimated if the basic causes of attenuation were known.

In principle, the answer to any practical question about underwater sound transmission could be obtained by a program of measurements planned wholly for the purpose of answering that question. When haste is required, this is frequently the quicker method. When time is available, however, such answers can most efficiently be provided by a broad program designed to yield a physical understanding of what is happening. Such a program makes it ultimately possible to answer not one but a large number of practical questions. Thus, in the long run, improved technology can best be based on a foundation of long-term fundamental research.

This final section gives a brief discussion of some of the basic physical factors that may be expected to be important in underwater sound transmission and also treats the type of observations that might be expected to give meaningful information on these different factors.

10.5.1 Basic Factors

The wave equation, equation (27) of Chapter 2, presumably governs in good approximation the propagation of sound waves in the interior of the ocean. It appears reasonable at first to investigate solutions of the simple wave equation, taking account of the presence of velocity gradients in the sea and of the reflections from sea surface and sea bottom. If the results are in flagrant disagreement with observations, then the effects of the approximations entering

into the derivation of the wave equation must be investigated in detail. Apart from the validity of the wave equation as such, it is known that the body of the ocean contains scatterers (their nature uncertain) which deflect a fraction of the sound energy from its original direction of propagation. Furthermore, the observed absorption at supersonic frequencies far exceeds the value predicted on the basis of viscosity alone, necessitating the assumption of additional dissipative processes.

The most important problems of underwater sound transmission may thus be summarized under the following four headings.

1. The effects of velocity gradients in the sea.
2. Absorption and scattering in the volume of the sea.
3. Surface reflection.
4. Bottom reflection.

Each of these topics is discussed in the following subsections.

SOUND VELOCITY

The velocity of sound is known as a function of temperature, pressure, and salinity and thus can be calculated at any point in the ocean where these physical quantities are known. The refraction effects produced by smooth vertical changes of temperature have been extensively investigated theoretically, and the results are in general qualitative agreement with the observations. Since the agreement is not complete, however, other effects must also play an important part. While the pressure is known as a function of depth, changes in temperature and salinity over distances of a few feet have not been extensively measured, and the acoustic effects to be expected from such changes have not been thoroughly explored. Microstructure of temperature and perhaps also of salinity may have an important effect on sound transmission, especially when the smoothed vertical gradient of sound velocity is small. Also, microstructure probably accounts for some part at least of the observed fluctuation of transmitted sound.

ABSORPTION AND SCATTERING

The attenuation observed in deep isothermal water is presumably the result of absorption, that is, some dissipative process which converts sound energy into heat. Since the attenuation observed at 24 kc exceeds by a factor of about 100 the value predicted on the basis of shear viscosity alone, the principal cause of the observed attenuation must be some other mechanism. Among the dissipative mechanisms considered are compression viscosity (which, however, certainly is not the principal factor at 24 kc), gas bubbles present in the water, fish bladders, plankton, and thermodynamically irreversible chemical reactions, such as the hydrolysis of dissolved salts.

Gas bubbles and other inhomogeneities would not only absorb but also scatter sound. That scatterers are present in the sea is known. Scattering may account for part of the attenuation of highly collimated beams and also is probably responsible for most of the sound observed in predicted shadow zones in the presence of negative velocity gradients.

All hypotheses concerned with the cause of the absorption of sound as well as with the role of volume scattering on sound transmission are at present largely speculative. Until further experimental and theoretical work has provided a scientific understanding of the mechanisms involved, it will not be possible to predict with confidence the attenuation under many different conditions.

SURFACE REFLECTION

The change in density at the sea surface is known and is so large that for most practical purposes the density of air may be set equal to zero; that is, the surface is almost a perfect reflector of sound. The complexity of surface-reflected sound arises from the complicated form of the ocean surface. In principle, it is simply a mathematical problem to compute the sound reflected from any surface of known properties. In practice, observations are unquestionably required. A thorough understanding of this topic would be important in studies both of fluctuation and of the average transmission anomaly in the surface layer.

BOTTOM REFLECTION

The ocean bottom may have a topography equally as complicated as the ocean surface. In addition, the relative change in the elastic parameters and in density across the interface is much less extreme than across the ocean surface, and the detailed values of these changes must be considered. Since the physical properties of the bottom may vary with position, both vertically and horizontally, the problem of bottom-reflected sound can be very complicated physically as well as mathematically. In certain regions, where the bottom is flat, and of uniform composition, the acoustic phenomena are perhaps capable of being understood. Bottom-reflected sound is obviously im-

portant in many situations, especially when the direct sound is weakened by temperature gradients.

10.5.2 Methods

To understand the physics of underwater sound transmission, each problem must be given separate consideration. The following methods may be applicable, however, to the investigation of a considerable variety of problems.

OCEANOGRAPHIC MEASUREMENTS

An important part of any basic research on sound in the sea must be the investigation of the physical properties of the medium in which the sound is transmitted. It is in terms of these properties that the acoustic data are presumably to be interpreted.

In the first place, detailed measurements of the factors influencing sound velocity seem desirable, especially temperature measurements showing the full detail actually present in the sea. In the second place, detailed measurements of the shape of the ocean surface are required before any attempt can be made to explain surface-reflected sound; in particular, statistical information on the spectrum of the surface water waves present during any interval seems desirable. In the third place, complete physical data on the ocean bottom (on topography, composition, porosity and compactness, etc.) are required to interpret physically the data on bottom-reflected sound. Finally, it may be necessary to make a variety of physical measurements on ocean water as part of the attempt to identify the cause of absorption.

CONTROLLED ACOUSTIC MEASUREMENTS

The experimental techniques of underwater acoustics research will probably be developed in a number of directions. Greater emphasis may be expected on detailed accuracy of the acoustic data; probable errors of several decibels for a transmission anomaly can presumably be considerably reduced. Measurements involving smaller samples of the ocean may perhaps be anticipated with relatively complete oceanographic data obtained for the small samples investigated. Some such experiment might be devised for measuring the sound absorption in a relatively small volume. Another possible development is along the lines of multiple measurement, in which many different items are measured almost simultaneously. For example, the inclination of the wave front might be measured at the same time as its intensity with simultaneous recordings at a number of different frequencies. Increasing complexity of the necessary equipment may probably be anticipated.

It is possible that explosive sound may be useful as a research tool. As pointed out in Chapter 9, short explosive pulses provide resolution of the direct and reflected pulses even at nearly grazing angles and also reveal clearly any multiple ray paths that may be present. By means of Fourier analysis, it is possible also to obtain with explosive sound many of the results which could be obtained by simultaneous transmission of many single frequencies over the entire spectrum. Finally, the high sound intensities possible with explosive pulses can provide data at longer ranges than are possible with standard sound projectors. Thus, explosive sound would appear to be a valuable tool of underwater sound research, deserving wider application than it has had in the past.

Regardless of what specific technique is used, the primary requirement for any basic experiment is that it be devised to give answers to certain physical questions rather than to operational problems. To satisfy this requirement, the theory underlying each experiment must be studied in detail before the experiment is actually performed to make sure that the results obtained will be significant. Considerable ingenuity may be required to find means for isolating the effects of the different factors involved in order to investigate them separately. It is only by such carefully designed experiments that our general understanding of sound in the sea can be continually increased.

PART II

REVERBERATION

Chapter 11

INTRODUCTION

11.1 DEFINITION OF REVERBERATION

IN ANY ECHO-RANGING or listening device, the wanted signal is always received in the midst of a certain amount of extraneous noise. This background noise is a mixture of noises from many sources, most of which are operative whether the transducer is being used for echo ranging or for listening. Examples of noises which appear in both echo-ranging and listening backgrounds are noise from breaking waves, sounds produced by such marine organisms as snapping shrimp, noise from the forming and collapsing of bubbles around the screws of the ship, noise due to the local motion of the water in the vicinity of the moving sound head, and the general din of the motors of the ship and auxiliaries. In both listening and echo ranging these noises, in varying degrees depending on conditions, are present to confuse the operator.

Echo-ranging backgrounds, however, have another component all their own, which is directly due to the pulse put into the water. This component is called "reverberation," and is the topic for discussion in Chapters 11 to 17.

Reverberation is evident to the operator of echo-ranging gear as a quavering ring, which sets in as soon as the gear is rigged for reception, that is, as soon as the period of sound emission is ended.[1] This rolling sound has about the same pitch as the outgoing pulse. Reverberation is very loud a fraction of a second after the ping is sent out, but diminishes rapidly thereafter if the receiver amplification is not changed. However, this rapid decrease does not necessarily mean that echoes from distant targets will always be audible above the reverberation coming in at the same time. Wanted echoes also become fainter as the time interval between ping emission and echo reception increases so that echoes which would be audible over the remainder of the noise background are often masked by reverberation.

Although reverberation in the sea shows some similarity to the well-known reverberations in a room, in many respects it is quite different. In a closed room, a sound is reflected with diminishing intensity back and forth between the walls and floor and ceiling until it is finally absorbed. Since the absorption of sound in air is relatively small, the sound energy disappears in appreciable amounts only at the boundaries of the room; and the decay of the reverberation is simply a measure of the decay of acoustic energy in the volume of air enclosed by the room. In the sea, there are these important differences. The sea has boundaries similar to a ceiling and floor (sea surface and bottom), but nothing like walls to interrupt the free passage of sound in a horizontal direction. Further, the sound-transmitting properties of sea water differ from those of air. The sea volume both absorbs and scatters sound energy in appreciable amounts. Thus, the behavior of reverberation in the sea is a special problem upon which little light can be cast by the known properties of reverberation in a room. When an echo-ranging pulse is sent into the water, some of its energy does return back to the transducer; but the amount of returning sound depends on many factors besides the rate at which sound energy is being removed from the volume of the ocean by the boundaries.

11.2 ELEMENTARY PROPERTIES OF REVERBERATION

There is every reason to believe that reverberation is almost always the resultant of a large number of very weak echoes arising from small bodies or irregularities in the path of the ping. These tiny targets may be called "scatterers." They may be air bubbles, suspended solid matter, minor irregularities of the ocean surface and bottom, local fluctuations of water temperature, or any other inhomogeneities in the sea.

Let us suppose, to begin with, that the scatterers producing reverberation are all identical, and are uniformly distributed throughout the volume of the ocean, and that a steady sound signal is sent out into this idealized medium. If we ignore, for the moment, the wave character of sound, and regard a traveling sound impulse as just a steady stream of energy, then each scatterer will return sound energy at a uniform rate, and in a receiver set up to measure the returning sound the reverberation will be heard as a steady ring of constant intensity. It is fairly obvious that in this situation the intensity of the received reverberation should be directly proportional to the intensity of the outgoing signal, since an individual scatterer returns a fixed fraction of the sound energy incident on it.

If, instead of a continuously projected signal, a very short pulse is put into our ideal medium, the resulting received reverberation will be quite different in character. Intuitively, we would guess that the reverberation should gradually taper off to a small value in a few seconds, since by that time the projected energy is already miles away from the receiver. It is not difficult to estimate the rate of decay of reverberation in this simple situation if we explicitly neglect the absorption of sound energy in the sea. As the sound pulse travels away from the projector, it spreads out over a larger and larger volume; but, if we neglect absorption, the amount of energy included in the pulse does not change. By assuming, as is approximately the case, that the rate at which energy is scattered is simply proportional to the energy in the pulse, it follows that as the pulse travels outward from the source its energy is scattered by the sea volume at a constant rate. However, this scattered sound must make the return trip back to the transducer before it is heard as reverberation; and on this return it is weakened by inverse square spreading. The reverberation from a short pulse, then, is inversely proportional to the square of the range of the scatterers producing it. Since the range is proportional to the time, this means to the man with the earphones that the reverberation decays inversely as the square of the time.

This result suggests that the intensity of reverberation is a mathematically smooth function of the time, of the sort indicated by the broken line in Figure 1. Such a curve implies that a short time after the cessation of the projected sound, the intensity of reverberation is great and that it fades away smoothly and gradually, becoming always fainter and fainter as time goes on. Everyone who has listened to echo ranging knows that this is not the case. The most obvious property of reverberation is its variability,

its alternation of bursts and silences, as schematically illustrated by the solid line of Figure 1.[2] This variability is associated with the phenomenon of interference.

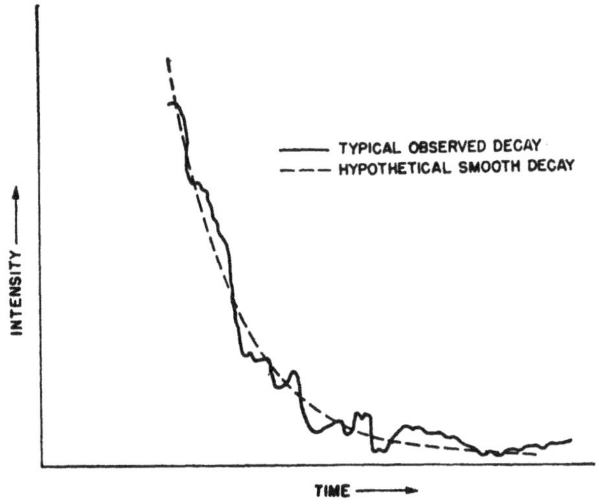

FIGURE 1.   Decay of reverberation intensity.

Interference arises because of the wave-like character of sound. Because of interference, the reverberation from n similar scatterers illuminated by the ping does not always have n times the intensity which would be observed if only one scatterer were present. At a particular instant, the sounds returning from the n scatterers may interfere destructively at the hydrophone so that the n sounds annul one another completely. Or, the n individual sounds may all combine constructively, so as to give n^2 times the intensity which would have been due to one of the scatterers alone. These are two extreme cases; but in general the n scatterers together may produce composite intensities ranging all the way between these two limits. The resultant intensity that occurs in a given situation depends in a critical way on the exact positions of the scatterers relative to one another. Since in the actual ocean the separations of the scatterers change from one portion of the ocean to another, it is plain that the reverberation from a ping should not change smoothly with time; rather, it should change irregularly, with bursts where the interference of the individual tiny echoes is primarily constructive, and relative silences where the interference is primarily destructive.

There is yet another complication. If the echo-ranging ship and the scatterers were all fixed in position, the pattern of bursts and silences, although complex, would not change from one ping to the next.

However, the echo-ranging vessel is usually rolling and pitching; and the scatterers in the ocean, that is, the air bubbles, the suspended solid matter, the turbulent regions, and the regions of temperature fluctuation are all free to move. Thus the interference pattern will vary widely from one ping to the next. Since the exact distribution of the scatterers in the ocean cannot be predicted, the irregularities of reverberation can be described adequately only by statistical methods. That is, if we are to be realistic, we can attempt to assign values only to the *average* reverberation intensity, and the *average* reverberation variability. For example, the inverse square law for the decay of reverberation from the volume of the sea, which was developed previously, could only be valid for the average reverberation from a series of pings; it would be nonsense to expect the reverberation from an individual ping to decay smoothly in exact agreement with this law.

In order to make clear the meaning of reverberation, the scatterers in the water, such as air bubbles, suspended solid particles, and the like, have been assumed to be very nearly uniformly distributed throughout the sea volume. We have really been describing what is known as *volume reverberation*, that is, reverberation due to scatterers in the body of the water. However, there is also reverberation due to scatterers at the ocean surface and ocean bottom. These three types of reverberation, volume, surface, and bottom reverberation, behave quite differently on the average. For example, they set in at different times. Volume reverberation is evident at the moment the ping is put into the water, and surface and bottom reverberation do not set in until the sound has had time to travel to these bounding surfaces and return to the transducer. There are other differences as well, which are discussed in the main body of the text.

11.3   PREVIEW

The next six chapters summarize the reverberation studies carried out under the auspices of NDRC through the spring of 1945. Chapter 12 derives theoretical formulas for the average reverberation intensity on the basis of assumptions which are explicitly stated and whose validity is critically examined. In that chapter, the expected intensities of reverberation from the volume, surface, and bottom are examined separately for their theoretical dependence on many other variables besides time; some of these other variables which play a major part in determining the reverberation intensity are the directivity characteristics of the transducer, the transmission loss between the projector and the scatterers, the intensity and duration of the projected signal, and the scattering power of the portion of the ocean under consideration. Chapter 13 describes the field and laboratory methods which have been developed for the measurement of reverberation intensity and the analysis of the resulting data. Chapters 14 and 15 summarize the observational information on volume, surface, and bottom reverberation which has been obtained by use of these experimental and analytical techniques and compare these results with the theoretical predictions of Chapter 12. In Chapter 16 the variability of reverberation is examined, both theoretically and in the light of the observed data, and the frequency characteristics of reverberation are described. Finally, in Chapter 17 the most important results presented in the body of Part II are summarized.

Chapter 12

THEORY OF REVERBERATION INTENSITY

THE AMOUNT and character of the sound heard or recorded as reverberation depends not only on the properties of this sound in the water, but also on the nature of the gear in which the reverberation is received. The intensity of the reverberation actually heard or recorded, after the sound in the water has been converted to electrical energy by the receiver, amplified, and passed to the ear or recording scheme, will be called the "reverberation intensity," and will be given the symbol G. As so defined, G equals the watts output across the terminals of the receiving gear. Although in practice the reverberation may be measured in terms of volts, or the height of a line on a motion picture film, or some other convenient quantity, these measurements can always be converted to watts output by the use of known parameters of the receiver system. In general, the reverberation intensity G is a function of time and is related to the sound intensity in the water by such parameters of the receiver system as receiver directivity and receiver gain.

Since G depends on the gear parameters, its absolute magnitude is usually not of great significance in research. For this reason it is customary to relate G to the reverberation intensity which would be registered under certain standard conditions. This standard reverberation intensity, in decibels, is called the "reverberation level" and will be defined precisely later. Reverberation levels are more useful than reverberation intensities for comparing measurements made with different systems.

This chapter is devoted to a theoretical analysis of expected reverberation intensities and reverberation levels. Formulas will be derived giving the dependence of these quantities on various gear parameters, oceanographic conditions, and elapsed time since emission of the signal. First, however, we must discuss the scattering of sound, since scattering is usually regarded as the fundamental source of reverberation.

12.1 SCATTERING OF SOUND

The analysis in this chapter is based on some very definite assumptions about the nature of reverberation. It is assumed that not all of the sound in the outgoing ping proceeds outward in accordance with the elementary theory in Chapters 2 and 3, but that some of the sound is "scattered" in other directions. The reverberation is thought to be that part of this scattered sound which returns to the transducer. Volume, surface, and bottom reverberation are assumed to result, respectively, from scattering in the volume of the ocean, at the surface of the ocean, and at the ocean bottom.

In an ideal unbounded fluid in which the sound velocity is everywhere the same, it is shown in Chapters 2 and 3 that sound always travels outward from its source along straight lines. In such a medium, then, scattering never occurs and no reverberation should be heard. There is no reason to doubt the validity of this theoretical result. Scattering arises because the ocean is not an ideal unbounded medium with constant sound velocity. It can be shown theoretically that whenever a sound wave travels through a portion of a fluid where the density or sound velocity varies with position, some energy is radiated in directions differing from the original direction of the wave. Similarly, whenever a sound wave in the ocean impinges on a new medium, for example a bubble, in which the density or sound velocity differ from their values in the surrounding water, energy is radiated in directions differing from the original wave direction.

Whether or not this deviated energy is called "scattered energy" is to some extent a matter of definition. If the inhomogeneity in density or sound velocity extends over a large region of space, a sound beam traveling through the medium may be changed in direction with little or no loss of energy from the beam; if this happens the sound wave is not regarded as scattered. Such changes in beam direction occur, for example, when the beam is refracted by a temperature gradient which is a function of water depth only. Similarly, reflection from an infinitely smooth plane surface sharply changes the direction of the beam; but since all the energy theoretically remains in the beam the process is not termed scattering.

However, there are inhomogeneities of small size, such as air bubbles, or small irregularities in the ocean surface, which cause energy to be "detached" from the main sound beam, that is, to travel in a different direction from that of the main beam. This detached energy, which differs in direction from the main beam and which results from local inhomogeneities in the ocean or bounding surfaces, is called scattered energy. It is apparent from this discussion that the distinction between scattered energy and nonscattered energy is not always too clear; for example, bottom reverberation received from underwater cliffs might more properly be called reflected energy rather than scattered energy. This possible confusion in nomenclature is of no immediate concern. The important point is that the existence of reverberation is predicted by theory from the known inhomogeneity of the ocean.

The magnitude of the reverberation reaching the water near the receiver is calculable, in principle, by solving some differential equation which, with appropriate boundary conditions, takes into account the inhomogeneity of the ocean. Since temperature gradients and density gradients are small in the body of the sea, the differential equation would be the wave equation

$$\frac{\partial^2 p}{\partial t^2} = c^2 \left(\frac{\partial^2 p}{\partial x^2} + \frac{\partial^2 p}{\partial y^2} + \frac{\partial^2 p}{\partial z^2} \right), \qquad (1)$$

where p is the sound pressure, and c is the velocity of sound at the time t at the point whose coordinates are (x,y,z); this equation was derived and its application discussed in Chapters 2 and 3. The presence of solid particles and air bubbles, and the nature of the roughnesses in the ocean surface and bottom, are described by the boundary conditions; these conditions give the positions at each instant of all the surfaces at which the density and sound velocity change discontinuously and the amounts of these changes.

Because of the complexity of the ocean, neither the function $c(x,y,z,t)$ nor the boundary conditions are precisely known. The bathythermograph gives the broad outlines of the temperature distribution. However, the locations and magnitudes of small local temperature gradients cannot be determined with the bathythermograph, although such "thermal microstructure" is known to exist.[1] Furthermore, the positions of bubbles, solid particles, waves, etc., change continually and unpredictably. Even if the sound velocity and the boundary conditions were specified exactly, it would be an insuperable mathematical problem to solve equation (1) rigorously for p. Thus, theoretical formulas for reverberation cannot be derived by solving equation (1) with boundary conditions.

Instead, we shall base our mathematical analysis of reverberation on several simplifying assumptions. Since a great deal is known about the general properties of solutions of equation (1), reasonable assumptions can be made about reverberation, even though a complete solution of equation (1) cannot be obtained. The principal assumptions which we shall use are:

1. Reverberation is scattered sound.

2. Scattering from an individual scatterer begins the instant sound energy begins to arrive at the scatterer and ceases at the instant sound energy ceases to arrive at the scatterer.

3. Multiple scattering (rescattering of scattered sound) has a negligible effect on the intensity of the received reverberation. In other words, all but a negligible portion of reverberation is made up of sound which has been scattered only once.

4. The intensity of the sound scattered backward from a small volume element dV is directly proportional to each of the three following quantities: the volume occupied by dV, the intensity of the incident sound, and a "backward scattering coefficient" designated by m, which depends only on the properties of the ocean in the neighborhood of dV.

5. The average reverberation intensity, which is a function of the time elapsed since the emission of the ping, is the sum of the average intensities received from the individual scatterers in the ocean. To express this assumption in mathematical form, let $g(t)dV$ represent the average intensity, t seconds after the emission of a ping, of the reverberation resulting from scattering in the volume element dV only. Then the average intensity of the reverberation received from the entire ocean, at the time instant t, is given by

$$\bar{G}(t) = \int g(t)dV \qquad (2)$$

where the integral is taken over the entire ocean. It

will be seen later that the function $g(t)$ is zero everywhere in the ocean except inside a thin, roughly spherical shell; this shell has the projector as its center, an average radius depending on the value of t, and a thickness depending on the length of the emitted signal.

With these assumptions, it is possible to derive theoretical formulas for the reverberation intensity as a function of gear parameters and oceanographic conditions. This chapter will be concerned only with the *average* reverberation intensity to be expected in a series of pings. No attempt will be made in this volume to predict the level of the reverberation from one specified ping. The observed levels of the reverberation from pings only a few seconds apart often differ by many decibels. Discussion of the average value of this fluctuation will be deferred until Chapter 16.

Although the above assumptions can be defended, they are by no means obvious and require elaboration. In particular, it is necessary to specify carefully the meaning of the terms "backward scattering coefficient" and "average reverberation intensity," which are introduced in assumptions 4 and 5. The average reverberation intensity is defined as the average from ping to ping. That is, if we measure the reverberation intensity on a succession of n pings with each measurement performed at a definite time t after midsignal, then the average reverberation intensity at time t is

$$\overline{G}(t) = \frac{1}{n}\sum_i G_i(t) \qquad (3)$$

where $G_i(t)$ is the reverberation intensity measured on the ith ping; the symbol Σ means summation over all the pings.

The number of pings averaged must be large enough to smooth out the effects of fluctuation, yet not so large that such external factors as wave height, water depth, and amount of suspended matter in the ocean can change materially during the series of pings. In practice, the number of pings averaged has usually been between 5 and 12, with not more than about 60 seconds between the first ping and the last. Some discussion of the validity of this averaging procedure is given in Chapter 16.

Also, we must specify more exactly the meaning of the backward scattering coefficient m. If we consider a volume V made up of many small volume elements dV, then, strictly speaking, dV can scatter sound only if sound energy reaches it and if it contains some scattering substance. Thus, if dV lies entirely within some rigid scatterer, such as a bit of metallic dust, practically no sound reaches dV because almost all the sound impinging on the scatterer is scattered at the surface of the scatterer. Another difficulty is that there is no way to predict the locations of the scatterers on any one ping. For these reasons it is impossible to predict how much scattering from a specified volume element dV will occur on any one ping. We can, however, speak of the average scattering power of the ocean in the neighborhood of dV. The backward scattering coefficient m for a volume V, in the neighborhood of and including dV, is defined as follows. Let V be insonified by a plane wave of unit intensity n times in succession. Let b_i be the energy scattered per second per unit solid angle in the backward direction, during the ith trial, by the volume V. Then m for V is defined by

$$\frac{m}{4\pi}V \equiv \frac{m}{4\pi}\int dV = \frac{1}{n}\sum_i b_i. \qquad (4)$$

The factor 4π is introduced so that in cases where the scattering is the same in all directions, the average amount of energy scattered per second in all directions will be just mV. With the definition of m given by equation (4) that the average energy scattered by dV per second per unit incident intensity per unit solid angle in the background direction is just $(m/4\pi)dV$, it also follows that $(m/4\pi)dV$ is just the intensity of the scattered sound from dV at unit distance from dV when the incident sound has unit intensity.

Evidently the volume V in equation (4) cannot be chosen arbitrarily if the definition of m is to have any significance. V must be chosen small enough that m can vary with position in the ocean and can thereby indicate the variation with position of the average number and strength of the scatterers. However, since it is desired that m not vary discontinuously from point to point, V must not be chosen too small. Because so little is known about the scatterers responsible for reverberation, it is difficult to formulate the conditions on V any more precisely than this. Some further discussion of the significance of assumption 4, as well as the other previous assumptions, is given in Section 12.5. That section is not a complete treatment of the problems involved, but may assist the reader to understand the physical ideas underlying the derivation of the theoretical formulas for reverberation.

12.2 VOLUME REVERBERATION

Volume reverberation is defined as sound scattered back to the transducer by scattering centers in the volume of the sea. Let a transducer be located at O in deep water far from both the sea surface and sea bottom. This transducer sends out a pulse of sound, or ping, of duration τ. Because the ping is of finite duration, a large part of the sound energy at a time $t/2$ seconds after midsignal will be contained between two closed surfaces S_1, S_2, portions of which are shown in Figure 1. If the sound velocity c in the

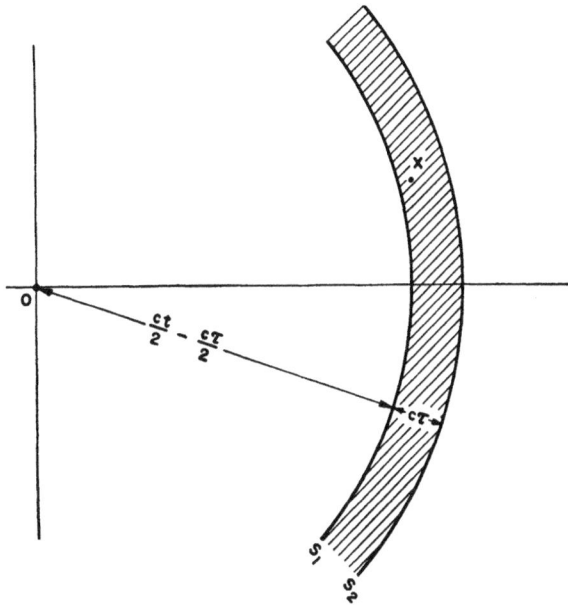

FIGURE 1. Portion of ocean scattering sound at time $t/2$.

ocean is everywhere the same, these surfaces are spheres centered at O, with radii $ct/2 - c\tau/2$ and $ct/2 + c\tau/2$. With refraction, however, the surfaces may be far from spherical. The reason for saying that a "large part" instead of "all" the energy in the ocean lies in the volume between S_1 and S_2 is that the very existence of reverberation shows that some sound, scattered earlier out of the ping, does not lie between these two surfaces at the time $t/2$. However, according to assumption 3 in Section 12.1, the amount of the previously scattered sound which is rescattered back to the receiver is negligible. Therefore, because of assumption 2, the only scattering taking place at time $t/2$, which is important in producing reverberation, occurs at those scatterers located within the volume S_1S_2 defined by the transmission laws of ray acoustics.

Now consider the sound scattered at time $t/2$ by the volume S_1S_2. Obviously, all this sound will not return to the receiver at the same instant since the sound scattered near S_1 travels a shorter distance than does sound scattered near S_2. It is shown in Section 12.5.5 that it can be assumed as a consequence of Fermat's principle,[2] that the average travel time of sound along the path from the transducer O to any point X in S_1S_2 equals the average travel time from X back to O. Using this result, we can readily delimit the region where the sound returning to the transducer at the time t is scattered backward. If τ is the signal duration, and if all times are measured from the middle of the signal, the signal emission starts at time $-\tau/2$ and ends at time $\tau/2$. The sound emitted first, at time $-\tau/2$, and returning as reverberation at time t, is scattered backward at some definite time which we shall call t'. The corresponding travel time T_1 out to the scatterers must be $t' + \tau/2$; the travel time back to the receiver must have an equal value because of our assumptions; and the sum of these two travel times and the time of emission $-\tau/2$ must equal t, the time at which the reverberation is received. Thus,

$$2\left(t' + \frac{\tau}{2}\right) - \frac{\tau}{2} = t; \quad t' = \frac{t}{2} - \frac{\tau}{4}$$

and therefore

$$T_1 = \frac{t}{2} + \frac{\tau}{4}. \tag{5}$$

Similarly, the sound emitted last, at time $\tau/2$ and returning as reverberation at the time t, must be scattered at a time t'' given by

$$t'' = \frac{t}{2} + \frac{\tau}{4};$$

and the corresponding travel time T_2 is $t'' - \frac{\tau}{2}$, or

$$T_2 = \frac{t}{2} - \frac{\tau}{4}. \tag{6}$$

Thus, all the scatterers effective in producing the reverberation at time t must lie between a pair of surfaces out to which the one-way travel times are, respectively, $t/2 - \tau/4$ and $t/2 + \tau/4$. These surfaces are indicated in Figure 2 by the labels S_1' and S_2'. If there is no refraction, these surfaces S_1' and S_2' are spheres with radii $c(t/2 - \tau/4)$ and $c(t/2 + \tau/4)$ respectively; the volume $S_1'S_2'$ is thus a spherical shell with average radius $ct/2$ and thickness $c\tau/2$. In general, even with refraction present, the volume $S_1'S_2'$ is about half the volume S_1S_2; that is, the volume in which the effective scatterers lie is about half the volume illuminated by the ping at time $t/2$.

We shall now determine the intensity of the reverberation which reaches O at time t from the volume element dV, located at X in Figure 2. We use the

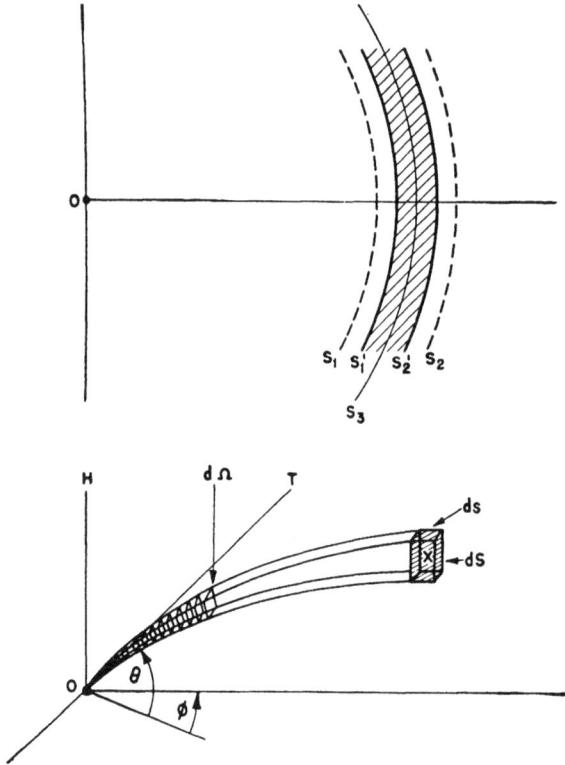

FIGURE 2. Diagrams used in developing volume reverberation formulas.

system of coordinate axes indicated in Figure 2; the origin is at O, and the ray from O to X leaves O with spherical coordinates (θ,ϕ) defined by the tangent to the ray at the origin. As drawn in Figure 2, θ is the angle made by the ray OT with the horizontal plane; thus θ is the complement of the polar angle made by OT with the vertical direction OH. The amount of energy which the projector radiates per second into the solid angle $d\Omega$ in the direction (θ,ϕ) is just $Fb(\theta,\phi)d\Omega$, where F is the emission per unit solid angle in the direction of maximum emission and $b(\theta,\phi)$ is the pattern function of the projector defined in Part I, Section 12.4.4. The sound intensity at unit distance (1 yd) from O, along this ray, is therefore just $Fb(\theta,\phi)$. If I_z is the intensity at the point X, the "intensity diminution" between the point one yard from the projector and the point X may be denoted by h and defined by

$$h = \frac{I_z}{Fb(\theta,\phi)}. \tag{7}$$

The quantity h is simply related to the (positive) decibel transmission loss H between the point 1 yd from the projector and the point X by the formula

$$H = -10 \log h. \tag{8}$$

The small tube of rays emitted by the projector into the solid angle $d\Omega$ will have, at the point X, a cross-sectional area, perpendicular to the sound rays, which may be denoted by dS. Let the volume element dV at X be defined as a cylindrical element whose base area is dS and whose height is ds, an infinitesimal extension along the direction of the wave propagation (see Figure 2). The volume included by dV is therefore given by

$$dV = aSds. \tag{9}$$

On the average, the sound which returns to O from X traverses the same ray traced out by the sound which was incident on dV; this assertion, a consequence of Fermat's principle, will be defended in Section 12.5.5. Therefore, the scattered sound giving rise to reverberation has been scattered directly "backward." By the definition of the backward scattering coefficient m, the intensity at a point, 1 yd from dV, of the sound which returns to O from dV is just $m/4\pi$ times the incident sound intensity at X, times the volume of dV, or

$$\frac{m}{4\pi}hFb(\theta,\phi)dSds.$$

If we now define h' as the intensity diminution between a point 1 yd from dV and the receiver at O, the intensity of the sound reaching O from dV is

$$\frac{m}{4\pi}hh'Fb(\theta,\phi)dSds. \tag{10}$$

The expression (10) gives the intensity of the sound scattered backward from dV in the water at the receiving hydrophone. Let F' be the output of the receiver in watts when a plane wave of unit intensity is incident on the receiver in the direction of its maximum response. A plane wave of unit intensity from some other direction (θ,ϕ) will stimulate the receiver to an output of $F'b'(\theta,\phi)$ where $b'(\theta,\phi)$ is called the *pattern function* of the receiver. Finally, a plane wave of intensity J incident on the hydrophone from the direction θ,ϕ will cause a watts output at the terminals of the receiver of

$$J \cdot F'b'(\theta,\phi). \tag{11}$$

With customary receivers of ordinary dimensions, the scattered sound returning from the ranges of interest (say greater than 50 yd) is for all practical purposes a plane wave at O. Thus, using the results (10) and (11), we have

Watts output from $dV =$

$$\frac{m}{4\pi}h^2 F \cdot F' b(\theta,\phi)b'(\theta,\phi)dSds. \quad (12)$$

In equation (12), h' has been set equal to h. This assumption that the transmission loss is the same on the outgoing and returning journeys will be justified on the basis of the laws of acoustics in Section 12.5.5. In addition, $b(\theta,\phi)$ and $b'(\theta,\phi)$ are very nearly equal for most transducers.

Using assumption 5 of Section 12.1, we next obtain an expression for the average reverberation intensity $\bar{G}(t)$ at time t. Integrating equation (12) over the volume between the surfaces S_1' and S_2' of Figure 2 gives

$$\bar{G}(t) = \frac{F \cdot F'}{4\pi}\iint mh^2 b(\theta,\phi)b'(\theta,\phi)dSds. \quad (13)$$

In equation (13), the dependence on range is contained principally in dS, h, and m; these quantities also depend on the direction θ,ϕ at which the ray which reaches a particular volume element leaves the projector. However, equation (13) can be simplified as follows. To a good approximation, the extension of any ray between the surfaces S_1' and S_2' can be considered equal to $c_0\tau/2$, where c_0, the average sound velocity along the ray, is always only slightly different from the sound velocity at O. Equation (13) can therefore be rewritten as

$$\bar{G}(t) = \frac{c_0\tau}{2}\frac{F \cdot F'}{4\pi}\int mh^2 b(\theta,\phi)b'(\theta,\phi)dS, \quad (14)$$

where the integral is to be evaluated on some average surface perpendicular to all the rays. It can usually be assumed that this representative surface is the surface reached at time $t/2$ by the sound emitted at midsignal; in Figure 2 this surface is labeled S_3.

This assumption for the surface of integration in equation (14) will not be valid if the average value of $mh^2 b(\theta,\phi)b'(\theta,\phi)dS$ along any ray does not occur near S_3. For example, this assumption fails when the ping length is not small compared to the range of the reverberation. By using the simplifying assumption that the sound intensity decays inversely as the square of the distance, it is easy to show directly from equation (13) that at close range (t not much

greater than $\tau/2$), the reverberation intensity may not be regarded as proportional to $c_0\tau/2$; rather it is proportional to the factor

$$\frac{c_0\tau}{2}\frac{1}{c_0^2 t^2 - \dfrac{c_0^2\tau^2}{4}}. \quad (15)$$

Another situation for which the average value of $mh^2 bb'dS$ may not occur near S_3 occurs when the rays are curving very sharply. For most oceanographic conditions, this error introduced by ray bending is not appreciable. However, when the layer effect discussed in Section 5.3 is present, the error might be significant. In that oceanographic situation, the ping travels out of an isothermal layer into an underlying region of sharp temperature gradient; and the sound scattered from parts of $S_1'S_2'$ below the isothermal layer has a higher transmission loss to the transducer than sound scattered from above the layer.

Although equation (14) cannot be used as it stands for the calculation of volume reverberation levels, it nevertheless is significant. It implies that irrespective of the directivity pattern of the transducer, and of the oceanographic conditions, the average intensity of the received volume reverberation should be proportional to the ping length. This important conclusion is based, of course, on the various assumptions made previously.

In equation (14), write $dS = (dS/d\Omega)d\Omega$, where $d\Omega$ is the element of solid angle in the direction (θ,ϕ). Then equation (14) can be further simplified if it is assumed that the transmission loss in the ocean depends only on the distance traversed by the ray entering or leaving the transducer, and not at all on the direction of the ray. Then h and $dS/d\Omega$ are independent of (θ,ϕ), and equation (14) can be written as

$$\bar{G}(t) = \frac{c_0\tau}{2}\frac{F \cdot F'h^2}{4\pi}\frac{dS}{d\Omega}\int mb(\theta,\phi)b'(\theta,\phi)d\Omega. \quad (16)$$

The term $dS/d\Omega$ is placed in front of the integral sign in equation (16) because it is a measure of the transmission loss due to refraction; $dS/d\Omega$ is, in fact, just the reciprocal of the intensity diminution due to normal inverse square divergence plus refraction, according to Chapter 3.

Finally, if it is assumed that scattering in the ocean is independent of the initial ray direction (θ,ϕ), the backward scattering coefficient m can also be removed from under the integral sign. This yields as our end result for the average reverberation intensity

$$\bar{G}(t) = \frac{c_0\tau}{2}\frac{F \cdot F' h^2 m}{4\pi}\frac{dS}{d\Omega}\int b(\theta,\phi)b'(\theta,\phi)d\Omega. \quad (17)$$

These latter assumptions are not always realistic. The assumption that transmission loss in the ocean depends only on the range and oceanographic conditions, and not at all on the initial ray direction, is probably a poor one for volume reverberation in a nondirectional transducer, because transmission loss in a vertical direction may differ appreciably from transmission loss in a horizontal direction. Even in a highly directional transducer of the sort used by the Navy for echo ranging at 24 kc, this assumption may be in error, since at long range rays leaving the projector only a few degrees apart may travel along widely separated paths; such a divergence occurs, for example, when the refraction theory predicts a splitting of the beam. Moreover, when split beams occur, m will not be independent of (θ,ϕ) if the scattering strength of the ocean is not independent of depth. For, if the overall scattering strength of the sea is not the same at all depths, then a pair of rays which become widely separated by the prevailing refraction may reach portions of the ocean with different scattering strengths; in such a case m evidently will depend on (θ,ϕ). Of course m may always be regarded as an average over the entire volume of the ping, and thereby may be removed from under the integral sign in equation (16). But if the scattering strength and transmission loss in the ocean really vary with angle within the main transducer beam, this type of averaging procedure makes the value of m depend on the directivity pattern of the gear; in this event removing m from under the integral sign has little significance.

In equation (17) the dependence of reverberation on the directivity pattern of the gear is contained wholly in the integral, which can be evaluated from the known directivity patterns of the transducer as a projector and a receiver. If the transmission loss obeys the inverse square law, that is, if the losses due to refraction, absorption, and scattering are neglected, then

$$h = \frac{1}{r^2}; \quad \frac{dS}{d\Omega} = r^2$$

and equation (17) becomes

$$\bar{G}(t) = \frac{c_0\tau}{2}\frac{F \cdot F'}{4\pi}\frac{m}{r^2}\int b(\theta,\phi)b'(\theta,\phi)d\Omega, \quad (18)$$

where r, the range of the reverberation, is equal to $c_0 t/2$. In this ideal case, then, the average intensity

of the received volume reverberation varies inversely as the square of the time following midsignal.

In general, however, the ocean is far from ideal and this simple law would not apply. To compare the observed time variation of the received reverberation in the general case with that predicted by the ideal formula (18), the general formula (17) is written as

$$\bar{G}(t) = \frac{c_0\tau}{2}\frac{F \cdot F' m}{r^2}(r^2 h)^2\left(\frac{1}{r^2}\frac{dS}{d\Omega}\right)\frac{1}{4\pi}\int bb'd\Omega \quad (19)$$

or, in decibels

$$10\log \bar{G}(t) = 10\log\left(\frac{c_0\tau}{2}\right) + 10\log (F \cdot F')$$

$$+ 10\log m - 20\log r + J_v + 20\log (r^2 h)$$

$$- 10\log r^2\frac{d\Omega}{dS}, \quad (20)$$

where

$$J_v = 10\log\frac{1}{4\pi}\int b(\theta,\phi)b'(\theta,\phi)d\Omega. \quad (21)$$

The transmission anomaly A is defined (see Section 3.4.1) by

$$A = H - 20\log r.$$

By comparing with equation (8), it is evident that

$$A = -10\log (r^2 h).$$

By substituting this expression for A into equation (20)

$$10\log \bar{G}(t) = 10\log\left(\frac{c_0\tau}{2}\right) + 10\log (F \cdot F')$$

$$+ 10\log m - 20\log r + J_v - 2A + A_1 \quad (22)$$

where $-10\log r^2(d\Omega/dS)$, the transmission anomaly which would result if the normal inverse square divergence were disturbed only by the effect of refraction, has been replaced by A_1. It is apparent from the preceding discussion that the quantities A, A_1, and m in equation (22) must be interpreted as averages over that portion of the effective scattering volume which lies within the main transducer beam.

The quantity A_1, which depends on refraction alone, cannot be measured directly. In principle, A_1 could be computed from the known temperature structure of the sea, according to the methods outlined in Section 3.4. However, this computation is difficult, frequently inaccurate, and often totally impractical because the observed *bathythermograph* [BT] pattern may not extend to a sufficiently great depth. Alternatively, A_1 could be inferred from the observed transmission loss, if the losses due to attenuation and scattering were known. However, it is clear from the discussion in Chapter 5 that these losses are also uncertain.

Equation (22) is the theoretical expression for the average intensity of received volume reverberation and is the one usually used in the comparison of theory with observation. Also, the computation of the scattering coefficient m from observed reverberation intensities is usually done by the use of this equation. It will be remembered that equation (22) was derived on the assumption that the transducer was infinitely far away from the ocean surface or ocean bottom, and therefore that sound traveled from O to X on only one path. In actual measurements, the transducer is always near the ocean surface and may be near the bottom as well. The presence of these surfaces increases the number of paths by which scattered energy from any point in the ocean can reach the transducer. Therefore, equation (22) will give erroneously low values for the reverberation intensity if alternative paths from O to X, of very nearly equal travel time, exist in the ocean. In the following paragraphs, we shall consider the error in equation (22) caused by the existence of such alternative paths. It should be stressed that we are considering here only the increase of *volume* reverberation due to these additional paths. Surface or bottom reverberation will result from scattering when the sound impinges on one of the bounding surfaces, but we are interested here only in the reverberation resulting from the scattering of sound by the volume elements in the interior of the ocean.

Possible combinations of alternative paths are pictured in Figure 3. If the ocean surface is calm, the case of Figure 3A, energy will reach the point X from O not only along the direct path OBX, but also along the path OAX as a result of specular (mirror-like) reflection from the ocean surface at A. If the ocean surface is rough, however, energy may reach X from O along a large, perhaps infinite, number of paths, as indicated in Figure 3B. Because of the principle of reciprocity, the energy returning from X to O also travels along these additional paths.

The existence of these extra paths tends to increase the reverberation intensity received at O at time t. To estimate the amount of increase, we note that for every possible path from O to X and back, there will exist an effective scattering volume of the type of $S_1'S_2'$ in Figure 2, bounded by two closed surfaces from

which scattered energy traveling along that path returns to O at time t. In Figure 3A, illustrating specular reflection, there are four such volumes. One is

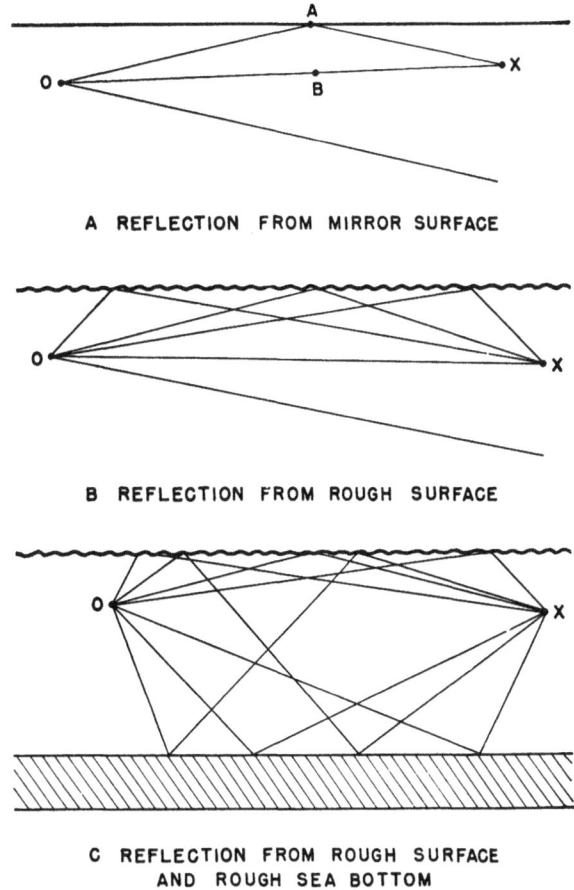

A REFLECTION FROM MIRROR SURFACE

B REFLECTION FROM ROUGH SURFACE

C REFLECTION FROM ROUGH SURFACE AND ROUGH SEA BOTTOM

FIGURE 3.   Alternative paths from transducer to scatterer.

$S_1'S_2'$ defined in preceding paragraphs, corresponding to the path $OBXBO$. The others correspond respectively to the paths $OAXBO$, $OBXAO$, and $OAXAO$. The volumes corresponding to the paths $OAXBO$ and $OBXAO$ are identical, because the travel time does not depend on the direction of travel along the ray; but the volume corresponding to $OAXBO$, the volume corresponding to $OAXAO$, and the volume $S_1'S_2'$ corresponding to $OBXBO$ are in general all different. For each of these volumes there will be an integral similar to that of equation (13), expressing the contribution of the volume to $\bar{G}(t)$. Each such integral can be simplified to a surface integral multiplied by $c_0\tau/2$, as in equation (14). It follows that the average intensity of the volume reverberation should be proportional to the ping length, regardless of whether energy travels between the transducer and a scatterer along one path, or along many paths.

Evaluation of the various integrals of the form (14) corresponding to each possible route from O to X and back, is very difficult. These integrals depend on the reflecting power of the surface, on the depth and orientation of the transducer, and on how the backward scattering coefficient varies with the direction of the incident sound. Also, the possibility must be considered that for small values of t some of the integrals should not be included, since there may not be time for energy to reach the ocean surface along any path and return to O by the time instant t. High seas, with the possibility of a great increase in the number of alternative paths, further complicate the problem.

To avoid these difficulties, the customary procedure has been to assume that equation (22) fully describes the volume reverberation intensity, despite the complications introduced by the ocean surface. The quantity $10 \log m$ then becomes an adjustable parameter which measures not only the actual backward scattering power of the ocean for incident plane waves, but also the effective increase in the volume of scatterers caused by the existence of a number of alternative paths.

If the water is deep and the echo-ranging gear is directional, the ocean surface can complicate the problem only if the main transducer beam strikes the surface. If the transducer beam is directed downward at a sufficiently large angle, the predictions of equation (22) should not be put in error by the presence of the surface. Using a depressed beam has proved to be one of the most convenient ways of studying volume reverberation.

Most reverberation studies, however, have been made with the transducer near the surface, and the beam horizontal. Under those circumstances it is shown in Section 12.5.6 that the value of $10 \log m$ computed from measured volume reverberation intensities and transmission anomalies by means of equation (22) will usually be about 3 db greater than the true value of 10 times the logarithm of the backward scattering coefficient. If the water is shallow enough for rays reflected from the bottom to be important, no simple relation exists between the inferred value of $10 \log m$ from comparison of equation (22) with experiment, and the actual value of the backward scattering coefficient. However, when the bottom is close enough to affect the validity of equation (22), the volume reverberation will almost always be masked by bottom reverberation so that the failure of equation (22) is of only academic interest.

We shall next define the concepts of "reverberation level" and "standard reverberation level," which facilitate the comparison of reverberation measurements performed with different gear and different ping lengths. From equation (13), the average value of the volume reverberation intensity is proportional to the product $F \cdot F'$ where F is the power output of the projector and F' is the receiver sensitivity. It is convenient to eliminate these variables in comparing the reverberation received on different gear. To this end we define the *reverberation level* $R'(t)$ as

$$R'(t) = 10 \log \overline{G}(t) - 10 \log (F \cdot F'). \quad (23)$$

For volume reverberation, we have specifically, from equation (22),

$$R'(t) = 10 \log \frac{c_0 \tau}{2} + 10 \log m - 20 \log r + J_v$$
$$- 2A + A_1. \quad (24)$$

In words, $R'(t)$ is the level of the received reverberation in decibels relative to the power output which would be produced at the terminals of the receiver by an incident plane wave, parallel to the acoustical axis, of intensity equal to the projected intensity on the axis at 1 yd.

It is often convenient to go one step further. Since the intensity of reverberation is in principle proportional to the ping length, it is both desirable and practical to convert all reverberation levels to the same ping length. We define the *standard reverberation level* for the reverberation at the ping length τ as that which would have been received if the ping length had been some standard value τ_0. Let the standard reverberation level be denoted by $R(t)$. Then we have

$$R(t) = 10 \log \overline{G}(t) - 10 \log (F \cdot F') + 10 \log \left(\frac{\tau_0}{\tau}\right). \quad (25)$$

The predicted standard level of volume reverberation is therefore given by

$$R(t) = 10 \log \frac{c_0 \tau_0}{2} + 10 \log m - 20 \log r$$
$$+ J_v - 2A + A_1. \quad (26)$$

The standard ping length τ_0 is usually chosen as 100 milliseconds. It is also frequently useful to convert reverberation levels to reverberation strengths. This is done by adding $40 \log r$ to the computed reverberation levels in equations (24) and (25), thereby obtaining respectively the *reverberation strength* or *standard reverberation strength*.

The quantity J_v, which specifies the relevant directivity characteristics of the transducer, is known as the *volume reverberation index*. For standard Navy gear at 24 kc, J_v is very nearly -25 db. It is shown[3] that for transducers which are circular pistons J_v can be very closely approximated by

$$J_v = 20 \log y - 42.6, \qquad (27)$$

where y is defined as in Figure 4. Numerically, y is half the angle in degrees in the plane $\theta = 0$ between those two rays of the composite directivity pattern

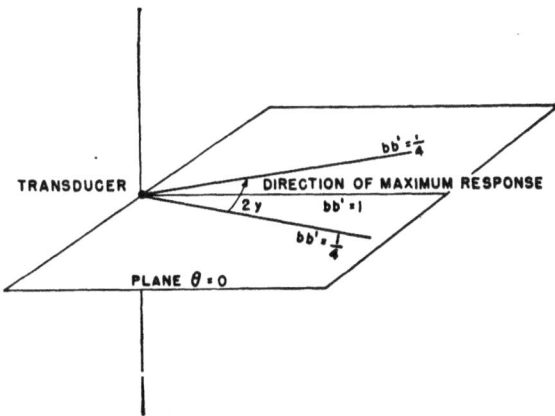

FIGURE 4. Half-width y for circular piston transducers.

for which the product bb' is 0.25. Thus for a transducer in which $b = b'$, y is half the angle between the two rays for which the response as a projector or receiver is 3 db less than the response on the transducer axis. The angle y is known as the "half width" of the composite directivity pattern bb'. Reference 3 also gives methods for calculating J_v for transducers which are not circular pistons.

12.3 SURFACE REVERBERATION

Surface reverberation is defined as the totality of sound scattered back to the transducer by scattering centers in or near the ocean surface. This simple definition is not completely adequate, since it would make surface reverberation a particular part of volume reverberation. We differentiate between these two types of reverberation by assuming that the surface reverberation arises from a thin surface layer of scatterers. The scatterers in this surface layer are assumed to owe their existence to the proximity of the surface and therefore differ in character from the

volume scatterers which supposedly may be found anywhere in the volume of the ocean.

The strength of these surface scatterers would be expected to be a function of the state of the sea surface, increasing with increasing agitation of the sea surface. In practice, surface and volume reverberation are frequently distinguished from each other in just this way; surface reverberation is regarded as that part of the received reverberation which seems to depend on the sea state.

We now derive an expression for the intensity of surface reverberation as a function of range and gear parameters, with the aid of Figure 5. We may pro-

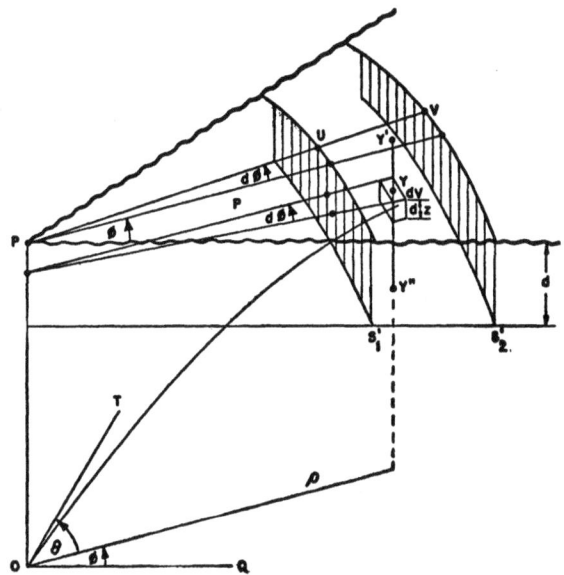

FIGURE 5. Coordinate system used in derivation of surface reverberation formula.

ceed exactly as in the development for volume reverberation, and arrive finally at an equation similar to equation (13). This equation for the surface reverberation intensity $\bar{G}(t)$ is

$$\bar{G}(t) = \frac{F \cdot F'}{4\pi} \int m h^2 b(\theta,\phi) b'(\theta,\phi) dV, \qquad (28)$$

where the integral is taken over that section of the volume $S_1'S_2'$ which contains the surface scatterers. This section need not be of uniform depth, although it is drawn so in Figure 5. The factor m in equation (28) is the backward scattering coefficient in the surface scattering layer and is very probably a function of the depth below the surface. In equation (28), reflection from the surface is explicitly neglected; that is, the ray paths are assumed to go directly from

O to the volume elements at Y without hitting the surface.

To evaluate the expression (28), we must first specify the volume element dV. It is not convenient to define the element dV in the same way as was done for volume reverberation. Instead, we set up a cylindrical coordinate system (ρ,ϕ,z) whose axis OP is the vertical line through the transducer, as in Figure 5; and we define dV as the infinitesimal volume lying between ρ and $\rho + d\rho$, ϕ and $\phi + d\phi$, z and $z + dz$. Then the value of dV is

$$dV = \rho \, d\phi dz d\rho.$$

We integrate over the intersection of the volume $S_1'S_2'$ with the surface scattering layer, which is assumed to have depth d. This volume is an annulus (ring-shaped figure) determined by z varying from zero to d, ϕ varying from 0 to 2π, and ρ varying from S_1' to S_2'. Then equation (28) becomes

$$\overline{G}(t) = \frac{F \cdot F'}{4\pi} \int_0^d dz \int_0^{2\pi} d\phi \int_{S_1'}^{S_2'} \rho m h^2 bb' d\rho, \quad (29)$$

where ρ is integrated from S_1' to S_2'. In the ocean, it can be assumed that on the average sound rays are bent only in a vertical direction. Then, the distance in the ρ direction from S_1' to S_2' is independent of the polar angle ϕ, but may depend on the depth z.

In order to put equation (29) in a form suitable for calculations, we shall have to make several additional simplifying assumptions. First, we assume that the only factor in the expression (29) which depends on the depth coordinate z is the scattering coefficient m, and that m depends only on z. Then equation (29) becomes

$$\overline{G}(t) = \frac{F \cdot F'}{4\pi} \left(\int_0^d m dz \right) \int_0^{2\pi} d\phi \int_{S_1'}^{S_2'} \rho h^2 bb' d\rho. \quad (30)$$

This assumption is readily defended. Since the depth of the surface scattering layer is usually small compared to the horizontal range from the transducer, there is little difference in initial ray direction between the ray which reaches a point Y' in the ocean surface and the ray which reaches the point Y'' a depth d below Y' (Figure 5). Therefore, the product bb' is practically independent of z in our volume of integration. Again, since the depth of the surface scattering layer is usually small, the horizontal distance traversed by a ray in its passage from S_1' to S_2' changes but little from the top to the bottom of the layer. For the same reason it can be assumed that there is usually little difference in transmission loss among the various paths from the transducer to

points in the volume of integration.[a] There is little reason for the scattering coefficient to vary with anything but depth, as long as the grazing angles of the rays on the surface do not vary appreciably over the volume of integration; this will be the case if the ping length is sufficiently short compared to the range of the reverberation.

We may rewrite equation (30) as

$$\overline{G}(t) = \frac{F \cdot F'}{4\pi} m' \int_0^{2\pi} d\phi \int_{S_1'}^{S_2'} \rho h^2 bb' d\rho, \quad (31)$$

where

$$m' = \int_0^d m dz.$$

It should be remarked that the disappearance of d in equation (31) is of little consequence. There is ordinarily no way to accurately estimate d in any particular case; it is just the depth down to which scatterers which depend on sea state appear in significant quantity. Without committing ourselves as to the exact size of d, we may give the factor m' real physical meaning by redefining it as

$$m' = \int_0^{\infty} m dz \quad (32)$$

where m is the backward scattering coefficient of the scatterers causing surface reverberation, is dependent on sea state, and is negligible below some unspecified depth. It seems likely that the depth at which m becomes negligible is usually small enough so that the lack of dependence on z of $h^2 bb'$ can be assumed. If not, or if for any other reason the assumptions used to derive equation (31) from equation (28) are not satisfied, the first integral in equation (30) cannot be regarded as a separate factor, and the concept of an overall surface scattering coefficient m' has no meaning. One situation in which equation (31) does not apply, while equation (28) does, is for surface reverberation in the presence of sharp negative temperature gradients near the range where the limiting ray leaves the surface. This situation is pictured in Figure 6. Strictly speaking, in this case the surface S_2' does not intersect the ocean surface at all; but S_2' may be drawn to intersect the surface, as in Figure 6, with the understanding that the transmission loss is infinite to the shaded volume. Under these circumstances, the transmission loss varies rapidly with depth in some portion of the volume of integration, and the assumptions used to derive equation (31) from equation (28) are not satisfied.

We next make assumptions which enable us to integrate out the variable ρ in equation (31). The integral in equation (31) is taken over an annulus whose horizontal cross section is the ring-shaped area between the circles of radii PU and PV in Figure 5.

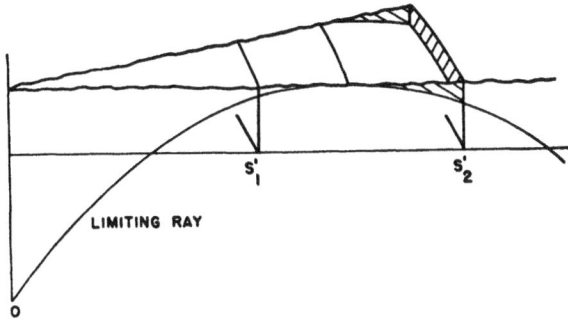

FIGURE 6. Region of surface scattering in presence of sharp downward refraction.

If the ping length is assumed to be sufficiently short compared to the range, then h^2bb' varies but little in the distance from U to V, and equation (31) becomes

$$\bar{G}(t) = \frac{F \cdot F'}{4\pi} m' \int_0^{2\pi} h^2bb'd\phi \int_{S_1'}^{S_2} \rho d\rho \qquad (33)$$

$$= \frac{F \cdot F'}{4\pi} m'h^2 \int_0^{2\pi} b(\theta,\phi)b'(\theta,\phi)d\phi \int_{S_1'}^{S_2'} \rho d\rho. \qquad (34)$$

The step from equation (33) to equation (34) is justified only if the transmission loss h is independent of the polar angle ϕ. With rays bending only in the vertical direction, there is ordinarily no reason why the average transmission loss should depend on this variable. In equation (34), θ is some average angle of elevation of rays which strike the surface between the two circles of radii PU and PV in Figure 5. If the ping length is sufficiently short compared to the range, this average value of θ may be assumed to be the angle of elevation of the ray which leaves the projector at midsignal and hits the surface after a travel time $t/2$. In other words, this average value of θ is the angle of elevation of that ray which passes through the curve of intersection of the ocean surface and the surface S_3 defined in Figure 2.

Now, by simple calculus,

$$\int_{S_1'}^{S_2'} \rho d\rho = \left(\frac{\rho^2}{2}\right)_{S_2'} - \left(\frac{\rho^2}{2}\right)_{S_1'}$$

$$= \frac{(PU + UV)^2}{2} - \frac{(PU)^2}{2} \quad \text{in Figure 5}$$

$$= UV\left(PU + \frac{UV}{2}\right) = \bar{\rho}(UV) \qquad (35)$$

where, if the ping length is sufficiently short compared to the range, $\bar{\rho}$, the mean value of ρ in the annulus, may be assumed equal to the value of ρ where S_3 intersects the ocean surface. UV is the distance on the surface from S_1' to S_2'.

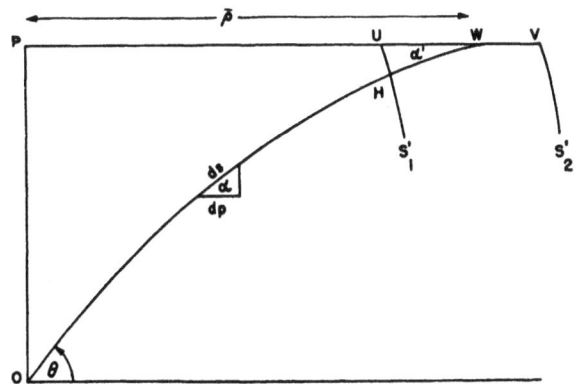

FIGURE 7. Expanded drawing of ray between projector and ocean surface.

Next, we evaluate $\bar{\rho}(UV)$. Referring to Figure 7, if $PW = \bar{\rho}$, if ds is the increment of arc along the ray from O to W in the time interval dt, and if $d\rho$ is the corresponding increment of horizontal range, then

$$d\rho = ds \cos \alpha$$

and

$$\bar{\rho} = PW = \int_0^{t/2} ds(t) \cos \alpha(t) = \int_0^{t/2} c \cos \alpha dt,$$

since cdt is always equal to ds. Also, since the bending is in the vertical direction only, we have by Snell's law

$$\frac{\cos \alpha}{c} = \frac{\cos \alpha'}{c'},$$

where c' is the velocity of sound at the surface, and α' is the angle of elevation of the ray OW at W. It follows that

$$\bar{\rho} = \int_0^{t/2} \frac{c^2}{c'} \cos \alpha' dt$$

$$= \frac{t}{2} \frac{\cos \alpha'}{c'} \bar{c}^2 \qquad (36)$$

where $\bar{c}$ is some average sound velocity. To calculate

UV, the second factor in the expression (35), we notice in Figure 7 that if UH lies in S_1' and if the ping length is sufficiently short, then UWH is very nearly a right triangle with the right angle at H. Thus,

$$UW = \frac{HW}{\cos \alpha'}$$

$$= \frac{c'\tau}{4 \cos \alpha'}$$

since the perpendicular distance from S_1' to S_3 is $c'\tau/4$. With our assumptions,

$$UW = \frac{UV}{2}.$$

Thus

$$UV = \frac{c'\tau}{2 \cos \alpha'}.$$

Substitution of this expression for UV and of expression (36) for $\bar{\rho}$ into equation (35) gives

$$\int_{S_1'}^{S_2'} \rho d\rho = \frac{\bar{c}^2 t \cos \alpha'}{2c'} \frac{c'\tau}{2 \cos \alpha'} \tag{37}$$

$$= \frac{\bar{c}^2 t \tau}{4}.$$

In equation (37), $\bar{c}$, the average sound velocity, can be replaced with little error in any actual situation by c_0, the sound velocity at the transducer. Substituting expression (37), modified by this replacement, in equation (34) gives

$$\bar{G}(t) = \frac{F \cdot F'}{4\pi} m' h^2 \frac{c_0^2 t \tau}{4} \int_0^{2\pi} b(\theta,\phi) b'(\theta,\phi) d\phi. \tag{38}$$

The subsequent procedure is similar to that adopted following equation (17); it is convenient to rewrite the theoretical expression (38) for reverberation in terms of decibels as a function of range. As before we define the range r of the reverberation as $c_0 t/2$; this differs negligibly from the distance along the ray path OW of Figure 7. Proceeding as in Section 12.2, we find

$$10 \log \bar{G}(t) = 10 \log \left(\frac{c_0 \tau}{2}\right) + 10 \log (F \cdot F')$$

$$+ 10 \log \left(\frac{m'}{2}\right) - 30 \log r + J_s(\theta) - 2A, \tag{39}$$

where

$$J_s(\theta) = 10 \log \frac{1}{2\pi} \int_0^{2\pi} b(\theta,\phi) b'(\theta,\phi) d\phi \tag{40}$$

and $2A$ is the two-way transmission anomaly along the ray path.

Equation (38) indicates that the intensity of sur-

face reverberation, like that of volume reverberation, should be proportional to the ping length if the assumptions used to derive equation (38) are satisfied. If these assumptions are not valid in a particular situation, however, the surface reverberation intensity may not be proportional to the ping length. Frequently, these assumptions are not satisfied; thus the proportional dependence on ping length predicted in equation (38) is not as general as the same dependence for volume reverberation predicted by equation (14). For example, if refraction is sharply downward, surface reverberation from ranges near where the limiting ray leaves the surface will not obey the theoretical law (38). Qualitatively, it can be seen from Figure 6 that at a range $c_0 t/2$ somewhat greater than the limiting range, a halving of the ping length may lead to more than a halving of the surface reverberation intensity, since most or all of the shorter ping may be too far from the surface to be effective in scattering. It will be recalled that this situation in which the proportional dependence on ping length predicted by equation (38) is invalid is just the type of situation which had to be ruled out in order to obtain equation (31), which in turn led to the result (38).

In deriving equation (38), surface reflections were explicitly neglected. As explained in Section 12.2, the surface reflections make for alternative ray paths from the transducer to the scatterers. It is shown in Section 12.5.6 that these alternative paths usually cause the value of $10 \log m'$, computed from measured surface reverberation intensities and transmission anomalies by means of equation (39), to be about 6 db greater than the actual value of the backward scattering coefficient of the surface scatterers.

The quantity $J_s(\theta)$ in equation (39) is called the *surface reverberation index*. In general, this index depends on the orientation of the projector relative to the vertical and on the range of the reverberation, and is difficult to calculate for arbitrary transducer orientations. When the transducer beam is nearly horizontal, however, the expression (40) can be evaluated approximately. It is shown [4] that if the transducer is a large rectangular piston in an infinite baffle then, approximately,

$$\int_0^{2\pi} b(\theta,\phi) b'(\theta,\phi) d\phi = \frac{b(\theta - \xi, 0) b'(\theta - \xi, 0)}{\cos \theta} Q(0), \tag{41}$$

where

$$Q(0) = \int_0^{2\pi} b(0,\phi) b'(0,\phi) d\phi, \tag{41a}$$

and ξ is the angle of tilt of the transducer axis relative to the horizontal plane. The relation (41) is probably not valid for angles θ greater than 30 degrees, at which angle the directivity pattern of any actual transducer is likely to differ appreciably from the ideal. Equation (41) was proved in reference 4 only for rectangular pistons. However, even for the circular pistons used in most Navy gear, the use of equation (41) is probably legitimate as long as the correction factor $b(\theta - \xi,0)b'(\theta - \xi,0)(\cos \theta)^{-1}$ does not differ too much from unity. For a horizontal transducer ($\xi = 0$), typical values of the correction

$$10 \log \frac{b(\theta,0)b'(\theta,0)}{\cos \theta}$$

are given in Table 1. This table was computed by the

TABLE 1

θ in degrees	$10 \log \dfrac{b(\theta,0)b'(\theta,0)}{\cos \theta}$ in decibels
0	0
2	-1.0
4	-2.0
6	-6.0
8	-12.0

use of values of b and b' measured for the EB1-1, which is similar to standard 24-kc Navy gear.[5] The use of corrections greater than -12 db is probably not justified. Some further discussion of the use of this correction is given in Chapter 15.

Formulas for $J_s(0) = 10 \log Q(0)/2\pi$ are given in reference 3. For transducers which are circular or rectangular pistons

$$J_s(0) = 10 \log y - 23.8. \tag{42}$$

As was done with the expressions for volume reverberation, we may define the surface reverberation level $R'(t)$ as

$$R'(t) = 10 \log \bar{G}(t) - 10 \log (F \cdot F')$$
$$= 10 \log \left(\frac{c_0\tau}{.2}\right) + 10 \log \left(\frac{m'}{2}\right) - 30 \log r$$
$$+ J_s(\theta) - 2A. \tag{43}$$

Also, we define standard surface reverberation level $R(t)$ as the level of the average reverberation which would have been received at the time t if the ping length had had some standard value τ_0 instead of τ. Then,

$$R(t) = 10 \log \bar{G}(t) - 10 \log (F \cdot F') + 10 \log \left(\frac{\tau_0}{\tau}\right), \tag{44}$$

so that

$$R(t) = 10 \log \left(\frac{c_0\tau_0}{2}\right) + 10 \log \left(\frac{m'}{2}\right) - 30 \log r$$
$$+ J_s(\theta) - 2A. \tag{45}$$

Reverberation strengths can also be defined in a manner similar to that in Section 12.2.

When using equations (39), (43), and (45) it is necessary to remember that A has been defined as the transmission anomaly along the actual ray path to the surface. This transmission anomaly may differ from A', the value of the transmission anomaly measured in the usual experimental determination of transmission loss. Consider specifically that the projector axis is horizontal and that a ray leaving the projector with the angle of elevation θ and an azimuth angle ϕ of zero reaches the surface after covering the slant range r. Then from the definition of A,

$$A = 10 \log F + 10 \log b(\theta,0) - 10 \log I - 20 \log r, \tag{46}$$

where I is the measured intensity in db at the point where the ray strikes the surface. The measured anomaly A' is usually determined from the equation

$$A' = 10 \log F - 10 \log I - 20 \log r' \tag{47}$$

where r' is the horizontal range from the projector to the point where the ray strikes the surface. Neglecting the difference between r and r', we have from equations (46) and (47)

$$A = A' + 10 \log b(\theta,0). \tag{48}$$

Further, from equations (40) and (41) we have for a horizontal transducer

$$J_s(\theta) = J_s(0) + 10 \log b(\theta,0) + 10 \log b'(\theta,0)$$
$$- 10 \log \cos \theta. \tag{49}$$

Substituting equations (48) and (49) in equation (45) gives

$$R(t) = 10 \log \left(\frac{c_0\tau_0}{2}\right) + 10 \log \left(\frac{m'}{2}\right) - 30 \log r$$
$$+ J_s(0) - 2A' - 10 \log b(\theta,0)$$
$$+ 10 \log b'(\theta,0) - 10 \log \cos \theta. \tag{50}$$

Under most circumstances $\cos \theta$ is sufficiently near unity and the projecting and receiving patterns are sufficiently symmetrical, with the result that the last three terms in equation (50) can be neglected. Thus, if the measured transmission anomaly A' defined by equation (47) is used in the analysis of surface reverberation, the correct expression for the standard reverberation level under most circumstances is

$$R(t) = 10 \log \left(\frac{c_0\tau_0}{2}\right) + 10 \log \left(\frac{m'}{2}\right) - 30 \log r$$
$$+ J_s(0) - 2A'. \quad (51)$$

In other words, if A' is used instead of A in equation (45), then the proper surface reverberation index to use is $J_s(0)$ rather than $J_s(\theta)$. The same rule applies, of course, in the evaluation of equation (39) or equation (43)

12.4   BOTTOM REVERBERATION

Bottom reverberation is defined as reverberation arising from sound scattered back to the tranducer by scattering centers in the ocean bottom. These scattering centers are thought to usually lie in a very thin layer on the bottom. Thus the formulas for the dependence on range of bottom reverberation will be similar to those formulas for surface reverberation, and can be derived from them by simple changes of notation. We have then, from formulas (39) through (45),

$$10 \log \overline{G}(t) = 10 \log \left(\frac{c_0\tau}{2}\right) + 10 \log (F \cdot F')$$
$$+ 10 \log \left(\frac{m''}{2}\right) - 30 \log r \quad (52)$$
$$+ J_B(\theta) - 2A,$$

$$\text{with} \quad J_B(\theta) = 10 \log \frac{1}{2\pi} \int_0^{2\pi} b(\theta,\phi)b'(\theta,\phi)d\phi \quad (53)$$

$$R'(t) = 10 \log G(t) - 10 \log (F \cdot F')$$
$$= 10 \log \left(\frac{c_0\tau}{2}\right) + 10 \log \left(\frac{m''}{2}\right) - 30 \log r$$
$$+ J_B(\theta) - 2A. \quad (54)$$

$$R(t) = 10 \log \overline{G}(t) - 10 \log (F \cdot F') + 10 \log \frac{\tau_0}{\tau}$$
$$= 10 \log \left(\frac{c_0\tau_0}{2}\right) + 10 \log \left(\frac{m''}{2}\right) - 30 \log r$$
$$+ J_B(\theta) - 2A. \quad (55)$$

In these formulas m'' is the backward scattering coefficient of the bottom per unit area of the bottom. The coordinate system in equation (53) is similar to that in Figure 5; however, the transducer is usually directed downward instead of upward as in Figure 5. Equation (41) may be used to evaluate $J_B(\theta)$ in equation (53); and as before $J_B(\theta)$ should be replaced by $J_B(0)$ if the transmission anomaly as usually measured is used instead of the actual trans-

mission anomaly along the ray path. The quantity m'' is in general a function of the range since it probably depends on the angle of incidence of the ray on the bottom.

It should be noted that the similarity of the formulas for bottom and surface reverberation does not imply that bottom and surface reverberation arise from similar mechanisms. The bottom scattering originates in irregularities in bottom contour; these irregularities may vary from the fine separations between grains of sand to such macroscopic irregularities as large rocks and underwater cliffs and valleys.

12.5   EXPLICIT AND TACIT ASSUMPTIONS

The preceding derivations of the theoretical formulas for volume, surface, and bottom reverberation were based on many assumptions, not all of which were stated explicitly. In this section we shall discuss briefly the significance and probable validity of these assumptions. Because of present uncertainties regarding scattering sources and the infinite complexity of the ocean, this discussion is partly qualitative and not clear-cut.

It was pointed out in Section 12.1 that though equation (1) governs the propagation of reverberation in the ocean, a complete solution of equation (1) for the reverberation received under given conditions is not obtainable. However, certain general properties of solutions of equation (1) are known, and will be of use here.

The strength of an acoustic disturbance can be expressed either in terms of pressure amplitude or sound intensity. In practical applications, the sound intensity is a more convenient quantity; while in theoretical discussions based on the wave equation (1) the acoustic pressure is more convenient. In equation (1), the sound intensity does not appear explicitly; in fact, it is impossible to derive a simple differential equation, whose dependent variable is the sound intensity, which like equation (1) expresses the fact that the disturbance is a wave traveling through the ocean with the velocity c. In discussing the implications of equation (1), we shall, therefore, be directly concerned with the sound pressure p. To tie in the discussion with the preceding sections of the chapter, we must relate the sound intensity to the sound pressure and also to the voltage generated across the terminals of the receiving circuit.

To simplify the discussion. we shall assume, to begin with, that the instantaneous voltage across the terminals of the receiving circuit is exactly proportional to the instantaneous pressure in any plane wave which is incident on the transducer. This assumption is equivalent to assuming that the receiver introduces no phase shifts, or in other words, that it behaves like a pure resistance or like an ideal infinitely wide band-pass filter. Of course, actual receivers never behave in this way, but it is convenient to postpone temporarily consideration of the effects caused by departure from ideal response in the receiver circuit. Now suppose a plane wave of pressure p is incident on the transducer from a direction defined by the angles (θ, ϕ) of Figure 2. Then the voltage E across the receiver terminal is

$$E = f'\beta'(\theta, \phi)p, \tag{56}$$

where $\beta'(\theta, \phi)$ is the "pressure pattern function" of the receiver, and f' is the voltage across the receiver terminals when a plane wave of unit pressure is incident on the receiver in the direction of its maximum response. Since there is no phase distortion, all the quantities in equation (56) may be assumed real. The rms power output resulting from E, in watts across the receiver terminals, will be

$$\frac{\overline{E^2}}{Z} = \frac{f'^2\beta'^2}{Z}\overline{p^2}, \tag{57}$$

where the receiver is assumed terminated in the pure resistance Z, and the bar indicates a time average over many cycles of the wave.

Now, in a plane wave, the relation between the sound pressure p and the average sound intensity $\overline{I}$, from Section 2.4.3, is just

$$\overline{I} = \frac{\overline{p^2}}{\rho_0 c}, \tag{58}$$

where ρ_0 is the density of water, c is the velocity of sound, and the bar again indicates a time average over many cycles of the wave. Equation (58) remains valid even if the plane wave is being refracted by velocity gradients in the ocean. From equations (57) and (58), we see that the rms power output across the terminals of our ideal receiver, caused by a plane wave incident on the transducer, is proportional to the average sound intensity in the water. Furthermore, by comparing equations (57) and (58) with the definition of F' and $b'(\theta, \phi)$ in Section 12.2, it is evident that

$$F' = \frac{f'^2 \rho_0 c}{Z}, \ b'(\theta, \phi) = \beta'^2(\theta, \phi). \tag{59}$$

However, the scattered sound which produces reverberation reaches the transducer from all directions, and therefore cannot be regarded as a plane wave. For this scattered sound the pressure in the water at any instant is

$$p = \sum_i p_i, \tag{60}$$

where p_i is the pressure in the ith plane wave which arrives at O. The voltage generated across the receiver terminals, by equation (60), is

$$E = f'\sum_i \beta'(\theta_i, \phi_i)p_i \tag{61}$$

where the angles θ_i, ϕ_i define the direction from which the ith plane wave reaches the transducer. The rms intensity resulting from equation (61) is given by

$$\begin{aligned}\frac{\overline{E^2}}{Z} &= \frac{f'^2}{Z}\overline{\left[\sum_i \beta'(\theta_i, \phi_i)p_i\right]\left[\sum_j \beta'(\theta_j, \phi_j)p_j\right]}\\ &= \frac{f'^2}{Z}\left(\sum_i \overline{\beta'^2 p_i^2} + \sum_i\sum_j \overline{\beta_i' \beta_j' p_i p_j}\right) \quad i \neq j\end{aligned} \tag{62}$$

where the double sum includes terms for all values of i and j except i equal to j.

12.5.1 Average Reverberation Intensity

One of the basic assumptions made in Section 12.1 was that the average reverberation intensity is the sum of the average intensities of the individual scattered waves reaching the transducer. Because equation (62) represents the average reverberation intensity, while equation (57) represents the intensity of the individual scattered wave, it is clear that this assumption will be strictly valid only if the double sum on the right-hand side of equation (62) vanishes.

The average in equation (62) is the average over a large number of cycles. All the waves reaching the transducer have very nearly the same frequency when ordinary single-frequency (CW) pings are used. Thus, the value of the double sum on the right of equation (62) depends on the relative phases of the various waves arriving at the transducer. On any one ping the value of this double sum may be positive or negative, and its absolute value may be appreciable or near zero, depending on the phases. Thus, if the expression (62) is averaged over a number of pings, and if the phases vary in a random way from ping to ping, the double sum in equation (62) can be neglected, and we can conclude that the rms reverberation intensity, averaged over a number of pings, will equal the sum of the average intensities received from the individual scatterers in the ocean.

We may expect the phases to vary in a random way because of the properties of equation (1). This equation implies directly that sound propagates through the ocean as a wave with a definite velocity, and that the phase of the returning wave depends on the travel time of the wave from the transducer to the scatterer and back. At 24 kc a phase shift of 2π, amounting to a shift of a complete cycle, results from a relative displacement between two scatterers of about an inch, or a difference in travel time of about 40 μsec. Such phase changes or changes in ray path from ping to ping could result from thermal fluctuations, from the rise and fall of the transducer in the ocean, from wave motion, and from drift of the scatterers and the projecting ship. A relative displacement of one inch in five seconds (the approximate time between pings) corresponds to a relative drift of only 60 ft per hr.

It is worth stressing that the phase shifts discussed in the preceding paragraph are relative phase shifts between waves from different scattering points in the ocean. At any instant sound is being received from many different points on any one scatterer; but if the scatterer is a rigid sphere, for example, the phases of the waves arising at different points on the spherical surface always bear a definite relation to each other. These waves from the different points on the spherical scatterer will always combine to give the same result in equation (62), irrespective of relative displacement between the scatterer and the transducer. Thus, the likelihood that the double sum in equation (62) will average to zero over a number of pings is connected with what might be called the "correlation" between conditions at various points in the ocean. If the ocean were rigid, so that the relative positions and orientations of the scatterers never changed, knowledge of the phase of a returning wave from one point in the ocean would completely determine the phases of returning waves from all other points. In this event the assumption that the double sum in equation (62) averages to zero would be more difficult to maintain. However, since the ocean is not rigid and the positions and orientations of the scatterers change with time, knowledge of the phase of a wave returning from one point determines the phases only of those waves from the immediate neighborhood of the particular point.

The averaging to zero of the double sum in equation (62) is made even more probable by our averaging procedure, which focuses attention on a definite instant relative to the midtime of the emitted signal. As the transducer drifts or otherwise changes its position in the ocean, the reverberation received at a definite instant comes from different points of space on different pings. In many cases, the phases of the scattered waves returning from these two portions of space will be almost completely uncorrelated; in such cases, random phase relations on successive pings are even more likely.

In the absence of definite knowledge about the scatterers responsible for reverberation, it is not possible to make this argument about equation (62) more precise. However, it is apparent from this discussion that in all types of reverberation there are a number of mechanisms that can cause random variations in the phases of the individual returning scattered waves. When these phases are truly random the assumption involved in averaging the individual scattered intensities, to get the average reverberation intensity, is justified.

12.5.2 Definition of Backward Scattering Coefficient

The backward scattering coefficient was defined for a volume small enough so that its relevant properties do not change too sharply with changes of position inside the volume, and large enough that it contains a reasonable number of scatterers. After an explicit assumption that the scattering by such a volume is proportional to the volume, the scattering coefficient of V was defined by formula (4).

It is easy to see that this assumption should be valid if the double sum in equation (62) vanishes. If this term vanishes, the total reverberation intensity will be just the sum of the average intensities of the waves from the individual scatterers; therefore, it should be proportional, on the average, to the size of the scattering volume.

12.5.3 Duration of Scattering by a Scatterer

In Section 12.1, it was assumed that scattering from an individual scatterer begins the instant sound energy begins to arrive at the scatterer and ceases the instant sound energy ceases to arrive.

In considering this assumption, we must recognize

that the discussion to this point has glossed over the fact that neither the outgoing ping nor the scattered sound which reaches the transducer is really single-frequency sound. No sound of finite duration can be a pure single-frequency sound since the latter theoretically lasts an infinite time; it can be shown that the relation between the time duration δt of a ping and the width δf of the frequency band making up the pulse is approximately

$$\delta f \delta t = 1. \qquad (63)$$

In general, the relation between any time-dependent signal $F(t)$ and the frequency spectrum of the signal is given by Fourier's integral theorem.[6] That is, the signal can be written in the form

$$F(t) = \frac{1}{\sqrt{2\pi}} \int_{-\infty}^{\infty} A(\omega) e^{i\omega t} d\omega, \qquad (64)$$

where $A(\omega)$ is determined by the equation

$$A(\omega) = \frac{1}{\sqrt{2\pi}} \int_{-\infty}^{\infty} F(t) e^{-i\omega t} dt. \qquad (65)$$

Equations (64) and (65) are just generalizations of corresponding equations applicable to Fourier series of periodic functions. In these equations $F(t)$ and $A(\omega)$ are generally complex, and ω can be interpreted as equal to $2\pi f$ where f is the frequency of the spectral component corresponding to ω.

It is possible, therefore, to make a frequency analysis of any given ping, using equation (65). This frequency analysis can then be used to obtain a formal solution of equation (1). For, because of the linearity of equation (1), if the scattered sound reaching the receiver as a result of emission of the continuous sound $e^{i\omega t}$ is $B(\omega) e^{i\omega t}$, then the pressure of the scattered sound reaching the transducer as a result of any given ping is

$$p(t) = \frac{1}{\sqrt{2\pi}} \int_{-\infty}^{\infty} A(\omega) B(\omega) e^{i\omega t} d\omega, \qquad (66)$$

where $A(\omega)$ is given by equation (65) in terms of the pressure variation $F(t)$ of the outgoing ping. It is necessary to qualify equation (66). Because the boundary conditions and the velocity of sound at any point are changing with time, the scattered sound which reaches the transducer is not a pure sound, even though a pure sound $e^{i\omega t}$ is emitted. Thus although the pressure of the scattered wave can always be presented as a Fourier integral of the form (64), equation (66) is not rigorously true, if $B(\omega)$ and $A(\omega)$ are defined as above. However, for the purpose of investigating the validity of the assumption under

consideration, these effects of time variation can be neglected, and equation (66) accepted as valid. In addition, for simplicity, the velocity of sound c in equation (1) can be assumed constant.

By using equation (66), the dependence on time of the pressure $p(t)$ of the scattered sound reaching the transducer can be calculated for pings of any length and for various types of scatterers. If $p(t)$ is plotted as a function of the time following the emission of the ping, it turns out that $p(t)$ is always zero until the sound has had time to travel to and from the nearest point on the scatterer. In other words, scattering from an individual scatterer actually does begin at the instant that sound energy begins to arrive at the scatterer. It is not possible to show in general that scattering ceases at the instant the sound energy ceases to arrive at the scatterer. However, for the special case of an infinitely rigid spherical scatterer, it is possible to show that the duration of the scattered sound received from the sphere is the same as the duration τ of the outgoing ping, as long as the relation

$$\frac{D}{c} \ll \tau \qquad (67)$$

is satisfied, where D is the diameter of the sphere. The significance of equation (67) is simple; it means that the scattered sound will have the same duration as the initial ping, provided that the travel time of the sound across the sphere is negligibly short compared to the duration of the initial ping. This result is not unexpected.

The reason why no general proof can be given for the validity of the second part of the assumption under discussion, namely, that scattering ceases at the instant sound energy ceases to arrive at the scatterer, is easily understood. Any real not infinitely rigid scatterer, such as a bubble, will have definite resonant frequencies which will be excited by the incident sound, and the scatterer may continue to radiate sound at its resonant frequencies long after the ping has passed by. Also some sound may enter the scatterer and be reflected back and forth inside the scatterer a number of times before it is scattered back toward the transducer. If the scatterer is large and many such reflections are possible, the duration of the scattered sound will be longer than τ. Despite these difficulties it can be argued that the assumption can be regarded as valid. There is good reason to doubt that resonant bubbles or other resonant scatterers play a large part in reverberation; in any case, the reradiated sound from such scatterers would die out in a very short time compared to the duration of even a 1-msec ping. For most scatterers equation (67) will be satisfied for pings of ordinary length although it may sometimes not be satisfied with scatterers such as rocks on the bottom, for 1-msec pings.

In the light of the discussion in Section 12.5.1, the distance D in equation (67) must be interpreted as the diameter of the volume within which there is appreciable correlation between the phases of waves reflected from different points in the ocean. Scattering volumes separated by so large a distance that there is little correlation may be considered unrelated scatterers. In the absence of definite knowledge about the scatterers, it is difficult to make the argument precise, but it seems unlikely that there will be appreciable correlation over a distance as great as a yard, which is about the length of 1-msec ping.

Along with assumption 2, Section 12.1, it has been tacitly assumed, in the derivation of the theoretical reverberation formulas that the outgoing ping is square-topped (that is, that the intensity rises abruptly to its steady-state value at the beginning of the ping, remains constant until the end of the ping, and then drops suddenly to zero), and that the wave received from any scatterer reproduces the shape of the outgoing ping. It is possible to make the outgoing ping very nearly square-topped, but it is apparent from equation (66) that the shape of the waves returning from each scatterer is not necessarily the same as the shape of the outgoing ping. However, if the ping does not include too wide a frequency band (that is, is not too short) and if the scattering coefficients of the various types of scatterers do not vary too rapidly with frequency, then in equation (66), if sound is being received from only one scatterer, $B(\omega)$ is nearly independent of frequency, and the returning scattered wave does very nearly reproduce the wave form of the outgoing ping. It will be seen in Chapters 4 and 5 that even in a 1,000-c band, scattering coefficients in the ocean apparently change very little, so that, from equation (63), square-topped 1-msec pings should result in square-topped scattered waves.

We can now see the significance of the assumption, made at the beginning of Section 12.5, that the instantaneous voltage induced in the receiving circuit is exactly proportional to the instantaneous pressure of the sound arriving at the transducer. If this is not the case, that is, if there is phase distortion or amplitude distortion in the receiver, then the reverbera-

tion resulting from any one scatterer will not have the square-topped shape of the outgoing ping, and the formulas which have been derived will be in error. Thus, if measured reverberation intensities are to be comparable to the theoretical formulas of Sections 12.2 to 12.4, it is necessary to use flat wide-band systems which have little transient response to the sudden changes of reverberation intensity. The use of narrow-band systems with high transients will usually cause the reverberation received from any scatterer to last longer than the outgoing ping and have a shape different from that of the ping. Since this distortion will decrease with increasing ping length τ, deviation will result from the predicted proportionality of $R'(t)$ on ping length τ in equations (24), (43), and (54). In order to derive appropriate theoretical formulas for such systems it would be necessary to write, using equation (66),

$$E(t) = \frac{1}{\sqrt{2\pi}} \int_{-\infty}^{\infty} A(\omega)B(\omega)C(\omega)e^{i\omega t}d\omega$$

for the voltage induced across the receiver terminals by each scattered wave where $C(\omega)$ describes the frequency response of the gear. The expression for the average reverberation intensity would then involve an integration over the frequency band included in the ping and passed by the equipment. It may be remarked that there is an inherent dependence of response on frequency in any directional transducer, because the pattern functions $b(\theta,\phi)$ and $b'(\theta,\phi)$ are always functions of frequency. Thus, it may be necessary to consider further the effect of directivity on the theoretical reverberation formulas for situations involving very short pings and highly directional transducers.

12.5.4 Neglect of Multiple Scattering

We have assumed that the sound reaching the transducer as reverberation has been scattered only once. It is easy to see that the validity of this assumption depends on the range of the received reverberation. For, as the ping proceeds out from the transducer, it loses more and more energy by scattering; the scattered waves are of course no different from any other sound waves and are themselves scattered. Eventually, therefore, a range is reached at which the ratio between the singly scattered sound returning to the transducer and the multiply scattered sound returning is no longer large. The value of this range depends on the amount of scattering which takes place in the ocean. Thus, the validity of this assumption, like that of all the other assumptions we have made, depends on the properties of the scatterers in the ocean.

In considering the validity of this assumption, we may confine our attention to multiple scattering in the body of the ocean, since there is little likelihood of direct multiple scattering from one surface scatterer or bottom scatterer to another. Experiments on volume reverberation [7] have shown that at short ranges (up to a few hundred yards) multiple scattering in the body of the ocean probably can be neglected. If scattering in the body of the ocean is isotropic, that is, if the backward volume-scattering coefficient m is really the average amount of energy scattered in all directions per unit intensity per unit volume, then it can be concluded from the magnitude of m that multiple scattering is certainly negligible at all ranges of interest in echo ranging.

However, it is possible that forward scattering in the ocean is appreciably greater than backward scattering. It has been suggested that the high attenuation of sound in the ocean at supersonic frequencies results from forward scattering of sound by the temperature microstructure in the ocean. On present evidence, it seems unlikely that forward scattering alone can account for attenuation in the ocean,[b] but if appreciable wide-angle scattering does occur, then at long ranges the neglect of multiple scattering in the theoretical reverberation formulas is not justified. The predicted dependence of volume reverberation on range [equation (24)] would be changed if volume reverberation contained much multiply scattered sound. The evidence discussed in Chapter 14 suggests that multiple scattering in the ocean probably can be neglected at ranges of operational interest in echo ranging. However, more evidence is needed before any definite conclusions can be reached.

12.5.5 Fermat's Principle and the Principle of Reciprocity

It was important to show that multiple scattering can be neglected in the computation of reverberation intensity since that assumption enabled us to delineate the volume $S_1 S_2$ in Figure 1 within which appreciable scattering is taking place. The determination of volume $S_1' S_2'$ (Figure 2) was then based on an application of Fermat's principle. Fermat's principle

is a theorem about the properties of equation (1); it states that when a sound travels between two given points, it always follows a path such that its travel time is a maximum or a minimum.[2] This maximum or minimum value is the same no matter which of the two points is the starting point and which the finishing point. Thus, provided the refraction conditions and boundary conditions are not changing with time, the ray paths and travel times from the transducer out to a scatterer, and from the scatterer back to the transducer, are exactly the same. However, refraction and boundary conditions in the ocean are not constant with time. The existence of thermal fluctuations, and the fact that BT patterns vary from one hour to the next, show that refraction conditions change with time; and surface waves are an example of changing boundary conditions.

Complete elucidation of the effect of these changing refraction and boundary conditions would be highly complicated, and, as usual, lack of information about the scatterers would make it difficult to be precise. However, it seems justifiable to assume that short-term fluctuations in thermal microstructure, or such variations in boundary conditions as waves or random movements of the scatterers, do not modify the average equality of the travel times and ray paths. Moreover, the long-term variations evident on the BT trace are too slow to affect the average equality in a series of pings lasting about a minute. Thus, it appears that the use of Fermat's principle was justifiable in Section 12.2, in the delineation of the effective scattering volume $S_1' S_2'$.

Another theorem about the properties of equation (1) is the principle of reciprocity. In the ocean, transmission loss is thought to be a combination of ordinary inverse square spreading, refraction, absorption, and scattering. According to the principle of reciprocity,[3] if the refraction and boundary conditions are not changing with time, then that part of the transmission loss which is due to inverse square spreading and refraction will be the same for transmission from a nondirectional projector at O to the point X as for transmission from a nondirectional projector at X to O. Of course, the source at O (the echo-ranging projector) is not nondirectional; and the source at X (the scatterer) may not be nondirectional, since scattering is not necessarily the same in all directions. For directional sound sources, the reciprocity theorem requires modification, because of the possibility of reflections and multiple paths between O and X. However, along any definite ray path

---

[b] This point is discussed in Section 5.4.1.

the principle still holds that the transmission losses due to inverse square spreading and refraction from O to X and X to O are the same along that ray, irrespective of the directivities of the sources. In addition, the principle of reciprocity applies also to absorption losses [9] if absorption in the ocean arises from so-called linear processes. Actually, studies of transmission loss show that the processes involved in the transmission of sound in the ocean are only imperfectly understood; but there appears to be no justification at this time for ascribing the absorption losses of sound waves of ordinary amplitudes to nonlinear processes.

Because the scattering coefficients are so small, transmission losses due to scattering may be neglected at all ranges of interest in echo ranging. It follows therefore from the preceding paragraph, and from the definitions of the quantities h and h' in Section 12.2 as transmission losses along the ray, that the principle of reciprocity may be applied to transmission between the points O and X of Figure 2 if refraction and boundary conditions are constant. It is still necessary to consider the effects of the variation of these conditions with time; but by an argument similar to that made in the discussion of Fermat's principle, it seems valid to assume that these time variations will not affect the relation between the transmission losses of the outgoing and incoming sound on a series of pings lasting about a minute. In other words it appears justifiable at this time, in the light of the principle of reciprocity and our present understanding of absorption losses, to assume that on the average the one-way transmission losses h and h' are equal.

12.5.6   **Effect of Surface Reflections**

The complications induced by such variations in boundary conditions as rough seas are exemplified by the difficulties encountered in extending equation (20), derived for an infinite unbounded ocean, to the more nearly realistic semi-infinite ocean. In rough seas (Figure 3B), with the transducer horizontal, it is very difficult to determine exactly the effective volumes from which reverberation is being received at any instant. In calm seas (Figure 3A), however, the volumes corresponding to the various alternative paths should be very nearly identical at ranges greater than a few hundred yards, since at these rather long ranges the path differences between OAX and OBX are usually very small. These volumes will all be about half of the original volume $S_1'S_2'$ because

of the presence of the ocean surface; at long ranges and shallow transducer depths the initial angle of ray elevation θ is restricted almost completely to values between $-\pi/2$ and 0 in the integral (14), instead of varying from $-\pi/2$ to $\pi/2$, as it does when the ocean surface is far away. Since all four of the scattering volumes described in Section 12.2 are very nearly equal, all of the integrals of the form (13) corresponding to these volumes should also be nearly equal, because at these ranges the rays OAX and OBX (Figure 3A) leave the transducer in almost the same direction, and because in a calm sea the reflection coefficient of the surface is very nearly unity. Thus, in a calm sea, with the transducer near the surface and the beam horizontal, the total intensity of the received volume reverberation is obtained by adding up four integrals of the form (13), with the region of integration for each just half the volume $S_1'S_2'$.

Therefore, the presence of the surface increases the received volume reverberation under these conditions to about double the value predicted by equation (17), or to a value about 3 db greater than predicted by equation (22), with the important proviso that the quantities A and A_1 used in that equation are the transmission anomalies that would have been measured if there had been no reflected rays. The transmission anomaly is usually obtained by measuring the transmission loss from a point about 100 yd from the transducer. If so, it is easily seen that with shallow transducers the inferred transmission anomaly is about the same as the transmission anomaly that would have been measured if the surface was far away. It follows, therefore, that in calm seas, with shallow transducers and horizontal beams, the value of $10 \log m$ computed from measured volume reverberation intensities and transmission anomalies by means of equation (22) will be about 3 db greater than the true value of 10 times the logarithm of the backward scattering coefficient.

For rough seas it is not possible to make so precise an analysis. However, it can be argued that the difference between the computed and actual value of $10 \log m$ will be about 3 db in rough seas also since on the average the existence of many paths (Figure 3B) will be compensated for by the loss of reflecting power of the surface. For surface reverberation the existence of surface-reflected paths causes the value of $10 \log m'$ computed from measured surface reverberation intensities by means of equation (39) to be about 6 db greater than the actual value of the backward scattering coefficient of the surface scatterers.

The reason for the 6-db value for the case of surface reverberation is that in equation (28) the volume of integration explicitly includes only the semi-infinite region below the ocean surface. Thus in calm seas the total intensity of the received surface reverberation is obtained by adding up four integrals of the form (28) without any reduction in the volume of integration. However, this conclusion depends on the properties of the surface scattering layer. If the surface scattering layer absorbs sound very strongly, sound may never be able to penetrate the layer to reach the actual air-water interface. In this event, the inferred value of m' from equation (39) will equal the true value of the backward scattering coefficient of the surface scatterers. A situation of this sort in which the surface layer is assumed to consist of a dense layer of resonant bubbles is discussed in Section 14.2.5.

If the water is shallow enough for rays reflected at the bottom to be important, the situation becomes more obscure; as shown in Figure 3C, some of the possible paths between O and X involve both reflection at the sea surface and reflection at the sea bottom. It has already been pointed out that in this situation no simple relation exists between the inferred value of 10 log m and the true value of the backward scattering coefficient. In fact, it may be remarked generally that equations (20), (39), and (52), for reverberation from the volume, surface, and bottom, respectively, are invalid when ray paths involving several reflections between the projector and the scatterer become important.

12.5.7   Overall Evaluation

This section has been concerned with the physical ideas behind the assumptions which have been used to derive the theoretical formulas for reverberation. It has been seen that most of the assumptions used are probably justified, but that no definite proof of their validity is possible at this time. If the assumptions are not satisfied, reverberation may not depend on range and ping length in the manner predicted by the theoretical formulas (24), (43), and (54). None of the considerations of this section affect the possibility of using these formulas as an empirical means of investigating reverberation and computing in each case a value of the backward scattering coefficient from comparison of the theoretical formulas with experiment. However, if the assumptions which have been made are not justified, the magnitude of the backward scattering coefficient deduced in this way will not have the simple physical significance implied in assumption 4 of Section 12.1.

Chapter 13

EXPERIMENTAL PROCEDURES

Tʜɪs ᴄʜᴀᴘᴛᴇʀ describes the principal methods which have been used in the gathering and analyzing of reverberation intensities. It will be seen that the techniques for the study of reverberation have been greatly improved since the first studies. A great many systems have been conceived for such studies; but only those which have actually been put into operation and used extensively in the gathering and processing of data will be discussed here. References to sources which give more detailed information are included in the body of the chapter.

The experimental determination of the frequency characteristics of reverberation will be discussed in Chapter 16.

13.1 EQUIPMENT AND FIELD PROCEDURES

Reverberation measurements have been made under a wide variety of oceanographic conditions, over many different types of bottoms, and at water depths between 10 and 2,500 fathoms. The most common projector depth has been 16 ft, but occasionally various other projector depths have been used. Most of these reverberation studies have been made by UCDWR; quite recently, however, WHOI has undertaken a reverberation program of its own. Although differing in details, the field procedures have in all cases been similar in broad outline.

In the early measurements off San Diego, made aboard the USS *Jasper* (PYc13), this procedure was followed. Upon arrival at the chosen location, the main engines of the vessel were shut down, and the *Jasper* was permitted to drift freely. The rate of drift during the working day varied between 0.5 knot and 2 knots, depending on the wind velocity and ocean currents. The transducer units with their supporting frames were hoisted by means of an electrically driven winch and boom, given the desired orientations, swung over the rail, and lowered into the water to the working depth. With the sound

gear overside, the projector and hydrophone cables were led through a doorway into the wardroom, and attached to the respective pieces of equipment.

In later studies, the depth to which the transducer was to be lowered and the angle the transducer was to make with the vertical were "built in" to the equipment at the time of installation. Hoist train mechanisms were provided, which lowered the transducer to the working depth from sea chests recessed in the keel. The bearing of the transducer in the horizontal plane was adjusted by means of a remote control training system. In these later modifications, the transducers were permanently wired to terminal boards, from which they could be connected to the regular electronic equipment or "sound stack," or to specially constructed research stacks.

After all connections are made, a sound pulse of controllable duration is sent into the water by means of a keying arrangement. As a result of this pulse, scattered sound returns to the transducer and generates a voltage in the receiving circuit. This voltage, after amplification, is recorded as reverberation. Somewhat different methods are used by UCDWR and WHOI for recording the reverberation. Systems for measuring reverberation intensities are discussed in detail below.

13.1.1 Transducers and Electronic Equipment

Most reverberation measurements have been taken at a frequency of 24 kc, which is the prescribed frequency for most Navy echo-ranging gear. Several types of transducer units have been employed at San Diego. Most of the early data [1] were obtained with a pair of similar magnetostrictive units (QCH–3), one used as a projector and the other as a receiver; some of the data reported there were obtained with a crystal transducer, the QB. The QB crystal unit proved to be superior to the QCH–3 units for reverberation studies because of its higher power output

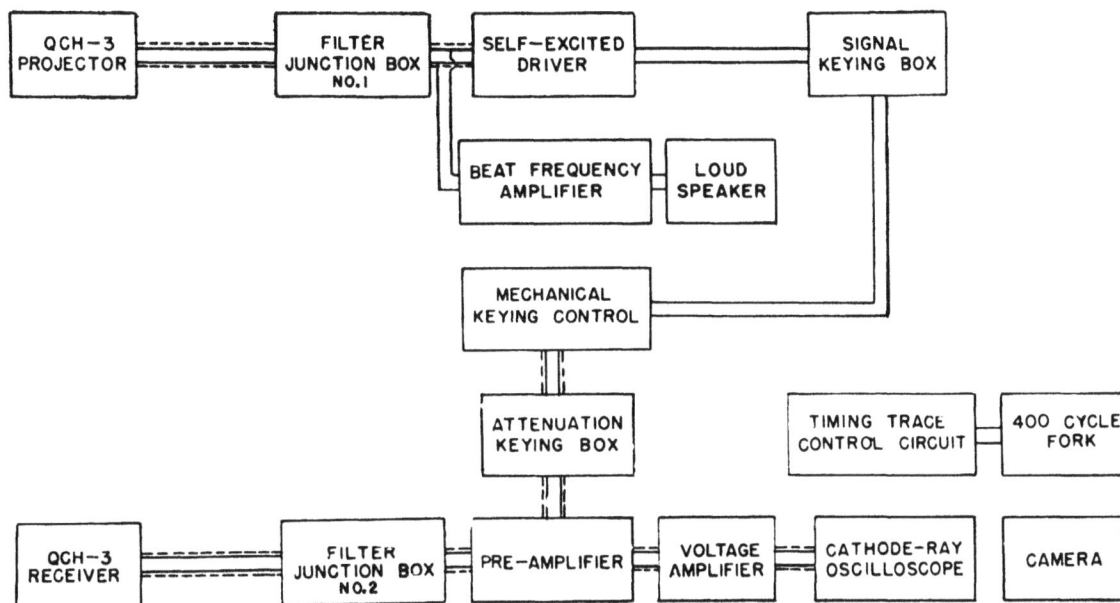

FIGURE 1. Schematic arrangement of apparatus employing QCH-3 units (equipment *A* of text)

when projecting sound and its better response when receiving. Most of the more recent reverberation studies off San Diego have been performed with crystal transducers.

The transducer alone is not capable of sending out pulses and detecting incoming sounds. There must also be equipment which delivers electrical energy to the projector and amplifies and modifies the small electrical impulses at the terminals of the receiver, thereby converting them into a detectable form. In the projector circuit of this auxiliary electric equipment there is an oscillator which generates an electrical signal of the desired frequency, and a power amplifier. In the receiver circuit, there is usually a preamplifier which takes the output at the terminals of the hydrophone and amplifies it somewhat, and then another amplifier, whose output is connected to the recording mechanism. Somewhat different recording techniques have been used by UCDWR and WHOI. At UCDWR, the voltage developed by the returning reverberation is usually fed into a cathode-ray oscillograph so that the instantaneous deflection on the cathode-ray screen is proportional to the instantaneous voltage developed in the receiver. The cathode-ray deflection as a function of time is recorded in permanent form by the use of a camera with continuously moving film. In the technique used until very recently at WHOI, the current generated in the gear by the reverberation activated a galvanometer, which in turn threw a light beam on

a moving roll of sensitized paper. The newest WHOI equipment uses a cathode-ray oscillograph and a camera, but is different from UCDWR equipment in a number of other features. Usually inserted somewhere in the receiving circuit is heterodyning equipment, which converts the incoming high-frequency energy into energy within the range of audible frequencies and thus permits listening to the returning reverberation by ear. This heterodyned signal may be recorded, if desired.

In the following paragraphs, we shall discuss in more detail the principal electronic setups which have been used in making reverberation measurements. For convenience, these setups will be identified by the letters *A*, *B*, *C*, *D*, *E*. Setups *A* and *B* were used at UCDWR prior to January 1943; in later UCDWR studies, setup *C* was used aboard the *Jasper* and setup *D* abroad the *Scripps*. Setup *E* has been used at WHOI.

EQUIPMENT *A*

This equipment [2] employed a pair of QCH-3 transducers, one used as a projector and the other as a receiver. Driven at 23.45 kc, the QCH-3 projector generated a sound pressure on the axis of 88.5 db above 1 dyne per sq cm at 1 yd with a total acoustic power output of 1.4 watts. A block diagram for this system is given in Figure 1. The ping was started and completed by closing and opening the ground circuit in the oscillator-driver stage by use of an electronic keying relay. The ultimate keying control was a synchronous motor-driven pair of shafts. Disks affixed to these shafts operated microswitches which, in turn, controlled the circuit containing the keying relay. By adjusting these disks, it was possible to choose any ping length between zero and several hundred milliseconds, and to control the interval between pings.

Because reverberation invariably decreases rapidly with time, the receiving system must be specially designed to handle a wide voltage range. For this purpose, a variable resistor was built into the receiving preamplifier, the amount of gain being controlled by relays. By a keying arrangement similar to that for controlling the ping length, the equipment could be adjusted so that a predetermined amount of resistance could be removed from the receiving circuit at any desired time after midsignal. Thus, as the reverberation intensity decreased, the gain of the preamplifier was increased in steps.

The output voltage from the receiving amplifiers was fed directly to the plates of the horizontal deflection circuit in a Du Mont Model 175A oscilloscope using a short-persistence screen; the vertical deflection circuit was not connected. A continuous record of the oscilloscope deflections was obtained by the use of a fixed optical system and a camera with moving film. Since the spot on the screen moved horizontally, the film moved vertically downward, taking some time to come up to the desired speed of 12.5 in. per sec. Because the film speed at a given instant was thus not known accurately, an accurate timing record of some sort had to be photographed along with the oscilloscope reflection. The chosen type consisted of the successive images of a slit which were illuminated by the short-duration flashes of a strobotron tube driven by an electrically controlled fork.

When operating properly, this system makes a faithful record of all intensity changes in the received reverberation, since the cathode-ray oscilloscope suffers from no mechanical inertia effects. However, since it is impractical to run the camera at speeds rapid enough to resolve individual cycles, only the time variation of peak reverberation intensity is discernible on the record. Thus this equipment cannot be used to determine the frequency characteristics of the reverberation.

In practice, certain difficulties were experienced with this system. For example, when the projector and receiver were close to each other, difficulty was experienced because of blocking of the receiver amplifier by the received ping, during the period when the projector is radiating its sound pulse. If this blocking is not eliminated, it leads to what may be described as a period of paralysis which lasts for a time after the end of the ping. During this period of paralysis there is serious distortion of the amplitude of the received reverberation. Another problem, also involving blocking effects, was the elimination of transients originating during the keying-in of gain changes. These transients could not be entirely eliminated in the final versions of this equipment.

In the derivation of the theoretical formulas for reverberation in Chapter 12, it was assumed that the projected signal was "square-topped," or, in other words, maintained a constant amplitude for a definite interval. No actual ping has this ideal rectangular shape, since some time is always required for the ping to build up and die away. Figure 2 illustrates the

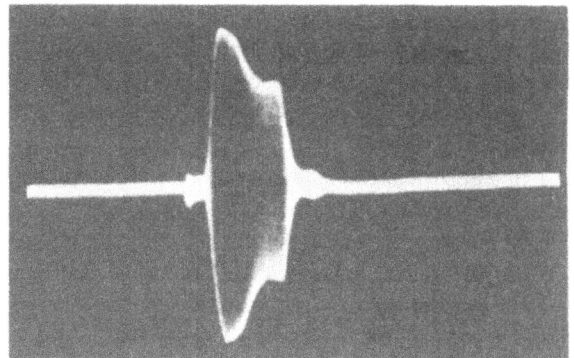

FIGURE 2. Shape of 13 MS ping from QCH-3 projector.

shape of a 13-msec ping sent out by a QCH-3 projector with electronic setup A and recorded by a system with flat frequency response. It will be noted that a definite time, about 1 or 2 msec, elapses before the signal reaches its maximum value. This maximum value is the same as the steady state level for a signal of indefinite duration, but is not held for long; 7 msec from the start of signal emission, the signal level in Figure 2 has diminished below its maximum value by about 4 db. After 13 msec, the signal dies away; however, the rate of decay is measurable.

Other photographs were taken of longer signals, more than 100 msec in length. In all these, the signal attained its maximum in 1 or 2 msec, fell to 3 or 4 db below maximum at 7 msec, and held a fairly steady level 3 to 4 db below maximum between 7 and 50

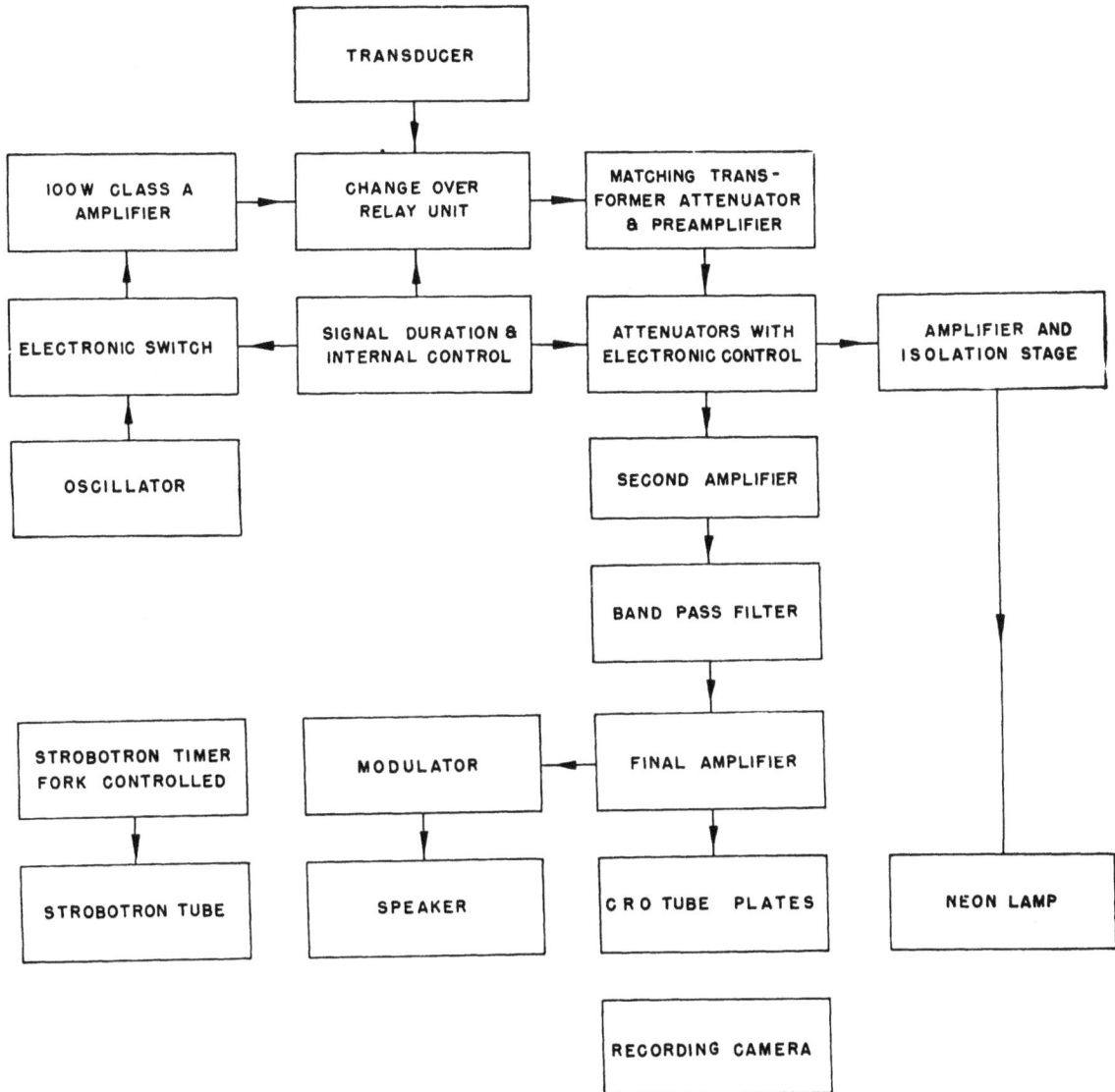

FIGURE 3. Schematic arrangement of apparatus in system used recently at San Diego (equipment C of text).

msec. During the interval 50 to 70 msec, the signal level gradually rose to its fully steady state value, which was maintained for times greater than 70 msec.

With the QCH–3 equipment A, the smallest recordable reverberation level was usually limited by the level of amplifier noise. On occasion, in noisy areas, the ambient water-noise level exceeded the amplifier-noise level.

EQUIPMENT B

This equipment was devised for use with the QB crystal transducer. Since the QB was used both as a projector and as a receiver, the electronic setup had to be somewhat different from the equipment de-

scribed under equipment A. A changeover relay had to be provided to switch the transducer from the projector circuit to the receiver circuit. An improved power amplifier was built for this system, with the result that the projected signal was nearly square-topped in form. Since the receiver circuit is not connected while the projector circuit is in operation, no blocking during the interval of projection was encountered in this system. Even though the receiver circuit is not connected during the ping, a record of the outgoing ping is obtained on the film which records the reverberation; this "ping record" is due to the electrical cross talk generated in the receiver circuit by the high voltages in the projector circuit during the interval of the pulse. Thus, an accurate record of the ping length appears on the film.

With this system, the minimum recordable reverberation signal was limited by amplifier noise during calm water conditions, and by water noise when the sea was choppy.

EQUIPMENT C

In the early part of 1943, new equipment was put into operation by the UCDWR[3] Reverberation Group. This equipment can be used with a wide variety of transducers and was originally provided with four distinct frequency channels—10, 20, 40, and 80 kc. However, any four frequencies between 10 and 80 kc could be used in the projector circuit by proper adjustment of the oscillator resonant circuit.[4] Receiver circuit changes to accommodate different frequencies, such as provision of properly tuned input transformers and band-pass filters, could also be made easily. It appears from later UCDWR memoranda [5,6] that this system was altered to include a 24-kc channel and that this channel has actually been used in the majority of the reverberation runs made with this system.

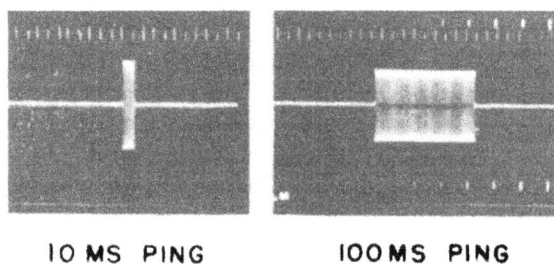

IO MS PING   IOOMS PING

FIGURE 4. Shape of signals sent out by equipment C of text.

The power output with this setup varies with the transducer employed. With the JK transducer at 24 kc, the power output averages about 100 db above 1 dyne per sq cm.[6] A block diagram for this system, assuming a single transducer unit, is given in Figure 3.

This system differs from systems A and B in a number of respects. A major innovation was the use of electronic timing circuits to control the ping length and keying interval, instead of the complicated mechanical motor-driven schemes described previously. The changeover relay circuit, used with a single projector-receiver, is also electronically timed, as are the step attenuators which vary the gain in the receiver

circuit. The pulse projected by this system is practically square-topped. Figure 4 gives photographs of the signal shape for signals of 10 and 100 msec.

The receiver circuit was specially designed for stability in operation. Positions of the gain changes were automatically marked on the film by means of a flashing lamp. It is clear from Figure 5 that the transients during gain changes are not marked enough to be troublesome. In this illustration the timing trace can be seen at the top of the film; the positions of the gain changes are indicated by spots on the film below the oscillograph trace, which precede the actual gain changes by a fixed distance on the record.

Because of its greater convenience, its suitability for a large number of transducers, its elimination of transients, and the square-topped shape of its emitted signal, this system is in many respects a considerable improvement over systems A and B.

EQUIPMENT D

This system is described in references 3 and 7. The projector used with this equipment, the EBI-1, generated a pressure at 1 yd, on the axis, of 104 db above 1 dyne per sq cm. The receiver was a preliminary model, and had a number of faults.[8] Many of the basic features of this equipment are similar to those in systems A, B, and C. However, the detecting mechanism used was not a cathode-ray oscillograph, but a Miller galvanometer mounted in a modified oscillograph camera. Because the galvanometer could not follow 24-kc vibrations, it was operated at 1,000 c by heterodyning the received reverberation to this frequency. One difficulty with this system is that a small percentage change in the i-f oscillator frequency (rated at 251 kc) caused a large deviation in the output frequency from 1,000 c, thereby introducing an error since the response of the recording galvanometer is not wholly independent of frequency. In this system, another galvanometer element was used to record the current fed to the transducer during each ping, while a third galvanometer element was used to make the timing marks.

The Miller galvanometer is naturally resonant at 2,500 c because of its mechanical inertia, and therefore cannot follow the variations in reverberation intensity with the detail possible with the cathode-ray oscilloscope used in systems A, B, and C. However, the Miller galvanometer is convenient to use and is certainly capable of following the variations in reverberation intensity with sufficient detail for the

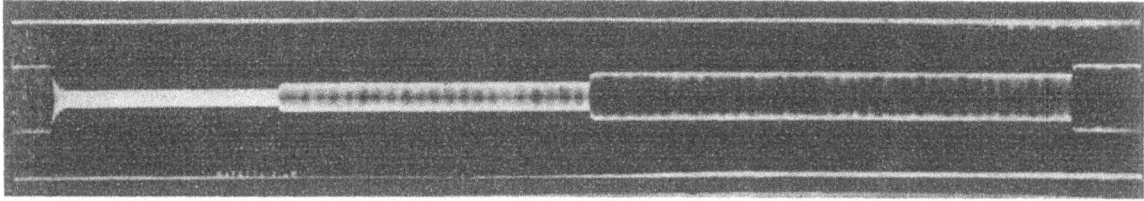

FIGURE 5. Oscillograph record showing negligible transients produced by receiver amplifier gain changes in equipment C (input signal maintained at steady value).

accurate determination of average reverberation levels as a function of time.

EQUIPMENT E

This equipment used by WHOI is described in reference 9. This reference gives only the details of the recording system; presumably in most details the system does not differ essentially from UCDWR systems A to D. One major difference is the introduction of a logarithmic amplifier which makes it possible to record directly in decibels. In the original version of this equipment [9] a seismographic galvanometer with a linear response up to 70 c was the recording device. The deflections of this galvanometer were recorded on photosensitive paper. More recently, the galvanometer has been replaced by a recording arrangement consisting of a cathode-ray oscillograph and a camera. This system has a linear response up to about 1,000 c; at higher frequencies the overall response, through the logarithmic amplifier, falls off rapidly.

With this WHOI equipment, the averaging procedure is simplified by superposing on the same film the records from a number of successive pulses. When the reverberation from a number of pulses is recorded on one film in this manner, the average reverberation curve can be drawn by eye through the densest portions of the trace. Since the film records the reverberation in decibels, the resulting plot is the desired curve of average reverberation level versus time. A sample record, showing the superposition of reverberation from 12 successive pulses recorded with the seismographic galvanometer, is shown in Figure 6. In this illustration, time increases from right to left; the particular features of the galvanometer used by WHOI made it more convenient to present the data in this way.

Table 1 summarizes some of the important information concerning the equipment used in various reverberation studies at UCDWR. Most of the items in the table are self-explanatory. The letters A to D

refer to the electronic setups described in equipment designations; and the figures in parentheses next to the transducer designations in the column labeled "Transducer references" tell where detailed descriptions of these transducers can be found in the bibliography. The column labeled "Reference" tells where the results obtained with the indicated equipment are discussed.

TABLE 1. Equipment used in reverberation measurements.

Reference	Transducer references	Frequency in kc	Electronic setup
1	QCH-3 [1]	24	A
1	QB [1]	24	B
4, 10	GB [10,11]	10	C
4, 10	GA,GB	20	C
4, 10	GA [10,11]	40, 80	C
7	EBI-1 [11]	24	D
5	JK [12]	24	C

13.1.2 Calibration of Projector and Receiver

In order to properly interpret the recorded values of reverberation, and to convert these recorded amplitudes to reverberation levels, it is necessary to know the values of the projector output F and the receiver sensitivity F'. These quantities, which occur in the theoretical formulas of Chapter 2, depend not only on the type of transducer, but also on the electronic equipment used. The procedures used in the determination of F and F' in the field are called "calibration procedures."

The projector is calibrated by measuring the projector output with a standard hydrophone whose response is stable. If an auxiliary projector with stable power output is available, F' can be determined by measuring the output of the receiver when exposed to a pulse from the standard projector. If the output of the auxiliary projector is not accurately

FIGURE 6.   Sample record from Woods Hole reverberation camera (reverberation from 12 successive pulses superposed).

known, the auxiliary projector itself may be calibrated with the aid of the standard receiver, and then used to calibrate the receiver. The echo received from a sphere of known target strength at a known distance has been used by WHOI to determine the product FF'. Present practice at both laboratories is to make a calibration at least once every working day, whenever possible.[6]

Although calibration is simple in principle, experience has shown that there is likely to be considerable inaccuracy in all projector and receiver calibrations. At UCDWR, it was found that the values of F and F' determined by calibration procedures may change unaccountably with time, sometimes changing by nearly 10 db from day to day, and by somewhat lesser amounts during a single day.[6] Some method for detecting calibration errors in the field is desirable since these errors are reflected as errors in the reverberation levels inferred from the measured intensities.

13.1.3   Typical Reverberation Records

Most of the reverberation data obtained by UCDWR are in the form of oscillograms on 35-mm motion picture film. Figure 7 shows a sample record of the reverberation from three successive pings sent out at 8-sec intervals. For convenience in display, each reverberation record was cut into three sections, as shown in the illustration; the three A's make up the first record; the three B's the second; etc. The marks on the upper edge of each record give the time scale; these marks are 2.5 msec apart. The point a represents the emission of the signal; b the onset of reverberation with transients caused by the operation of the changeover relay; and c, d, e, places where the gain was automatically increased by the attenuators. At f, the reverberation has decreased below

background noise. The film speed (12.5 in. per sec) is high enough to show considerable fine structure in the reverberation. However, it is not high enough to resolve individual cycles; thus the trace shown in Figure 7 represents the envelope of the received reverberation.

The records shown in Figure 7 are quite typical and illustrate some of the statements which have been made in this volume about the behavior of reverberation. The recorded amplitude at a given time past midsignal is not constant from record to record, even though these pings were sent out and the reverberation was recorded under the same adjustments of the experimental apparatus. In general, however, whenever reverberation measurements are made, there are major features which persist from record to record. One such characteristic is the point of onset of bottom or surface reverberation. Another is the invariable tendency of reverberation of a given sort (volume, surface, or bottom) to decrease with increasing time, as is predicted by the theoretical formulas of Chapter 12. This decrease makes necessary the provision in the system of gain changes at points such as c, d, and e; without these gain changes it would be impossible to record all the reverberation at measurable amplitudes. Occasionally, successive reverberation records show a systematic increase at certain points. These increases can usually be correlated with the calculated increase due to the onset of surface or bottom reverberation; sometimes they are ascribed to the existence of local regions of high scattering strength.

13.2   ANALYTICAL PROCEDURES

After the field work is done, the films containing the received reverberation records are taken to the

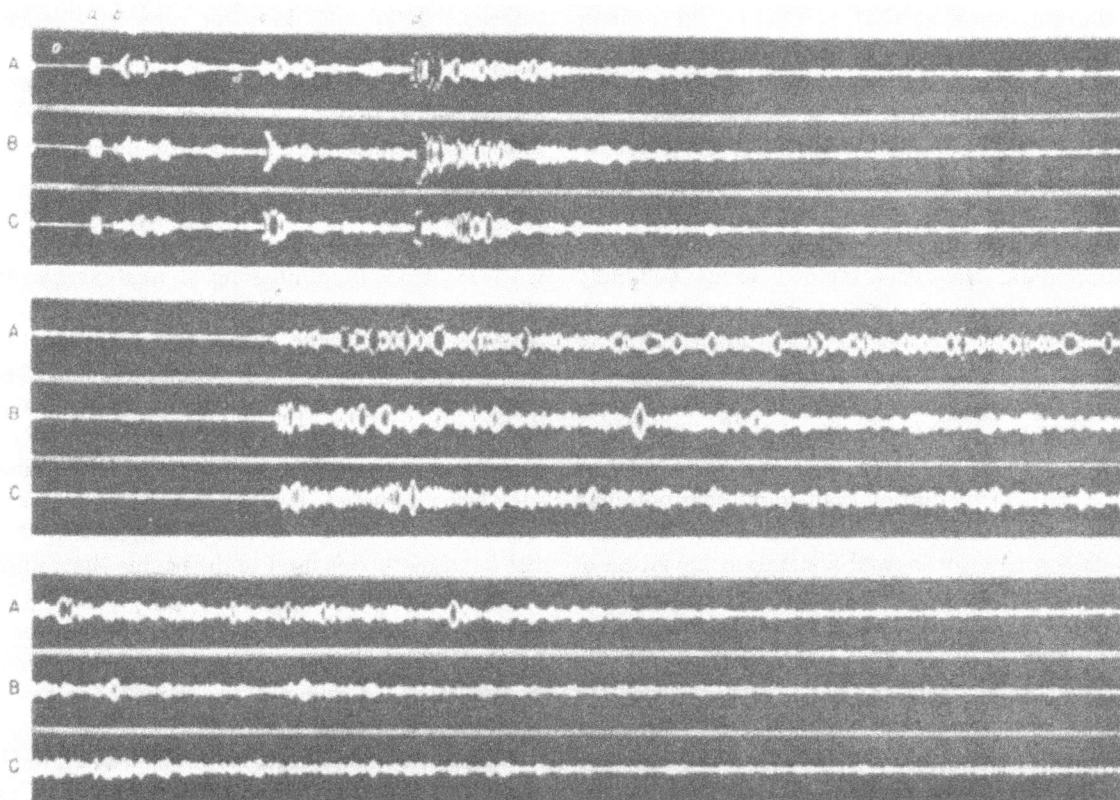

FIGURE 7.   Oscillograph records of reverberation from three successive pings.

laboratory to be analyzed and averaged. These records are divided into sets, each consisting of records taken within a short space of time under similar conditions. The reverberation measurements making up a set are then averaged, and the resultant averages are supposed to represent the expected reverberation under the known external conditions for the set. Obviously the averages cannot be computed for every time instant after midsignal. Times are chosen which are spaced closely enough so that the major systematic changes in reverberation level will be evident.[a] At UCDWR, two methods of averaging have been used: "point method" and "band method."

The point method of averaging is to select a set of points, such as 1, 2, and 3 in Figure 7; measure the amplitude at these points; repeat for all records in the set (usually from 5 to 10); make proper allowance for gain changes and projector-receiver calibration; con-

vert the amplitudes to decibels above the chosen reference level; and finally, plot the resulting average reverberation levels as a function of time or range. Outstanding features, such as reverberation from the bottom or from a suspected deep scattering layer, can be emphasized by choosing many points in their vicinity on the records; and uneventful portions of the record can be passed over with but one or two points to set the general level. Usually the points chosen were spaced so as to give equal intervals on logarithmic coordinate paper.

The alternative method, the band method, was introduced because of the considerable difficulty involved in computing an accurate average with the point method. On some records, the amplitudes are changing very rapidly close to the predetermined point where the amplitude is to be read; and to measure these amplitudes accurately it is necessary to look at the records very closely with appropriate viewing devices. This procedure is both time-consuming and hard on the eyes, especially when the amplitude at the predetermined time is small.

In the band method, a set of points is chosen as in

---

[a] In selecting and manipulating the data, places on the records where obviously extraneous noise showed up have customarily been rejected. Examples of extraneous noises are pings from destroyers, echoes from porpoises, and bursts of ship noises.

the point method. But the amplitude measured is not the amplitude at that point, but the *greatest* amplitude in a band three ping lengths long and centered at the designated point. Corresponding amplitudes are measured for all similar records; allowance is made for gain changes and projector-receiver calibration; finally, after converting to decibels, the average reverberation levels are plotted as a function of time. As a procedure for plotting reverberation data, the band method seems definitely superior to the point method. The amplitude corresponding to a particular point is much easier to obtain with the band method, since it is simpler to pick out the maximum amplitude in an interval than to measure the amplitude at a predetermined point. Also, amplitudes obtained with the band method show much less fluctuation than amplitudes obtained with the point method; an analysis of reverberation records consistently showed a standard deviation of amplitude for the band method of less than 50 per cent of the standard deviation for the point method.[1]

It is difficult to see the exact significance of the averages obtained with the band method. Certainly the band method does not closely resemble the averaging method which was the basis for the theoretical formulas of Chapter 12; the point method, on the other hand, does resemble it. Thus, in order to compare the observational results obtained with the band method with theoretical expectations, the simplest procedure is to correct the band method results to what would have been obtained had the point method been used. The amount of this correction was determined experimentally by comparing the results for many records processed by both the point and band methods. Except at very short ranges, it was found that the band method gives results which average quite consistently 7 db greater than results obtained with the point method. Subtraction of 7 db from the band method results thus gives average reverberation levels which are comparable with the theoretical expectations of Chapter 12. At very short ranges, on the other hand, the reverberation is chang-

ing so rapidly that the band method does not give sufficient detail and does not show any consistent relationship to the point method.

Some more details of the present UCDWR analytical procedure may be of interest to the reader. The individual records are analyzed by placing the films in a viewer against a graph paper background. Vertical and horizontal distances on the film can be measured by counting squares on the graph paper, which is usually ruled in millimeters. In analyzing a record, the analyst first measures the ping length in terms of squares on the graph paper and converts this to milliseconds by comparing millimeters and the distance between points on the timing trace. The number of timing marks in a fixed film length gives the film speed from which a scale of range from midsignal may be constructed. This range scale is set up next to the film in the viewer. At ranges greater than 250 yd, the band method is used to determine the amplitude at the designated range. At ranges of 100 yd and 250 yd, however, the point method is used because at these short ranges, as explained previously, the reverberation is changing too rapidly for the band method to give accurate results.

Despite the simplifications introduced by the band method of averaging, the analysis of a set of UCDWR reverberation records is an arduous and time-consuming process. In the WHOI system E, the final plot of average reverberation level against range can be obtained immediately from the photographic paper, by drawing a curve through the densest area on the superposed reverberation traces. This system is highly convenient for recording and plotting average reverberation levels; but it does not permit any detailed measurement of reverberation fluctuation. Probably the best system for recording reverberation would combine the advantages of both the UCDWR and the WHOI types. This equipment would make a permanent record of all fluctuation on one recording element, while on the other recording element a smoothed trace would be made from which the final reverberation levels could be readily obtained.

Chapter 14

DEEP-WATER REVERBERATION

IT IS CONVENIENT to begin the study of observed reverberation levels by describing the experimental observations in deep water. In deep water it is usually possible to ignore bottom reflections, thereby facilitating comparison of the experimental results with the theoretical formulas of Chapter 12. Also, in deep water, it is frequently possible to eliminate surface scattering and reflections, by directing the beam downward at some angle. When reverberation from the surface and bottom is effectively eliminated, the received reverberation can assuredly be called volume reverberation. The first section of this chapter describes the experimental facts about volume reverberation, as determined by such unambiguous experiments.

In ordinary echo ranging, with the main transducer beam horizontal, part of the received reverberation is surface reverberation and part volume reverberation. It is not easy to make a clear distinction between these two components on observed reverberation records. The distinction between the two is usually made by comparing the measured levels obtained with horizontal beams with the levels observed in unambiguous volume reverberation experiments, and also by observing the dependence of the reverberation levels on sea state. The observed levels with horizontal transducer beams are described in the second section of this chapter.

14.1   TRANSDUCER DIRECTED DOWNWARD

The experimental method for eliminating surface reverberation in deep water has usually been to point a highly directional transducer downward, away from the surface. In this way the main transducer beam does not strike the surface, and the observed reverberation levels are then assumed to be due to volume reverberation. Of course this assumption requires verification, since in the absence of any information about the relative values of the surface and volume backward-scattering coefficients, it is not possible to know in advance how much directivity is necessary to definitely eliminate surface reverberation. However, it may be accepted as a working hypothesis that pointing the main transducer beam down 30 degrees, or more, does eliminate surface reverberation, for standard 24-kc echo-ranging gear. It will be seen later that surface reverberation levels are not usually high enough to contribute to the received reverberation under these circumstances. The following subsections describe the various experimental studies of volume reverberation which have been carried out in this manner.

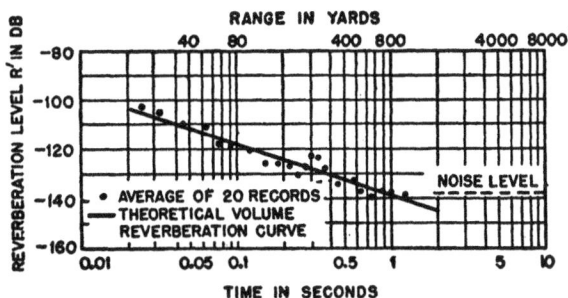

FIGURE 1.   Volume reverberation levels showing inverse square range dependence.

14.1.1   Dependence on Range

According to Chapter 12, equation (22), if the volume scatterers are uniformly distributed, and if the transmission anomaly terms $-2A + A_1$ in equation (22) can be neglected, then the volume reverberation intensity should be inversely proportional to the square of the range, or in other words, the reverberation level should decrease 20 db with a tenfold increase in time. As an example of this dependence, we may refer to Figure 1, which is a plot of data obtained on June 3, 1942.[1] The QCH–3 transducers, projector and hydrophone, were lowered to a depth of 60 ft and tilted downward 60 degrees. The water depth was 600 fathoms and the surface was moder-

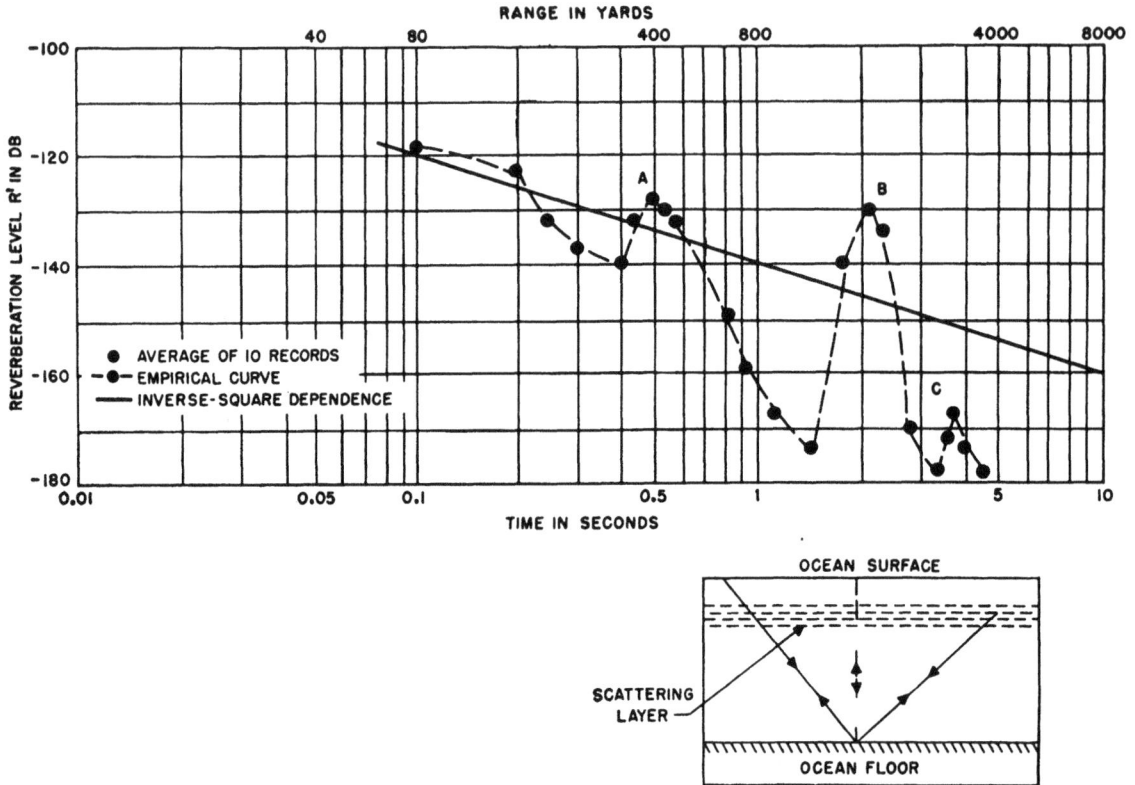

FIGURE 2. Volume reverberation levels with deep scattering layer.

ately calm with the wind velocity averaging 10 mph, and long low ground swells but no whitecaps. A signal length of 10 msec was used. Twenty records were filmed, measured, and averaged, to give the points shown in Figure 1. It is seen that the experimental data fit fairly well the straight-line dependence of R' on log r which is predicted, if all quantities except R' and r are constant, by equation (24) of Chapter 12.

In practice, volume reverberation runs usually show even worse agreement with this simple linear range dependence than do the points shown in Figure 1. In the first place, it is known from transmission measurements that the transmission anomaly terms can rarely be neglected at ranges greater than 1,000 yd (see Chapter 5). Thus the inverse square dependence can be expected only at relatively short ranges. In addition, there is no real reason to expect the volume scatterers to be uniformly distributed in the ocean. However, the fact that an approximately inverse square dependence has been observed in at least a few cases is evidence that our fundamental assumptions about volume reverberation are not altogether wrong. In general, volume reverberation tends to decrease rapidly with increasing range, in at least qualitative agreement with equation (24) of

Chapter 12. However, the detailed dependence on range is frequently observed to be very different from the simple form of that equation; often the dependence of R' on log r is not linear, and when it is linear, a slope of exactly -20 is quite unusual. Such observations are described in the following subsection.

14.1.2 **Dependence on Depth**

Measurements off San Diego with the transducer pointed downward have frequently shown sudden increases in reverberation level which seemingly could only be explained by assuming that in certain deep layers of the ocean the backward volume-scattering coefficient was much larger than at other depths. Figure 2 was drawn from data obtained on July 28, 1942. The QB transducer was pointed downward at an angle of 49 degrees relative to the horizontal, in 660 fathoms of water. Ten records were averaged to give the points in this figure. At the reverberation range indicated by A in the illustration there is a sharp rise of more than 10 db in reverberation level. A comparison with equation (24) of Chapter 12 makes it seem necessary to ascribe this rise to an increase in the backward scattering coefficient m;

certainly it does not seem possible that any change in transmission anomaly could be sufficiently sudden to account for the rise. The geometry of the experiment is shown in the small box of Figure 2, on the assumption that the ray paths are approximately straight lines. The peak at A occurs at a time 0.5 sec after the ping. This corresponds to a reverberation range of 400 yd; thus, the layer of high scattering power must have been centered at a depth of 400 yd $\times$ cos 41°, or about 900 ft. The thickness of the layer, as estimated from the thickness of the bulge at A in Figure 2, was not less than 500 ft. The large increase in reverberation level at B corresponds to the point at which the beam strikes the bottom. The rise at C in Figure 2 could have resulted from scattering of bottom-reflected sound by the deep scattering layer. It could also have resulted from sound which was scattered from the bottom toward the surface, reflected from the surface back to the bottom, then scattered from the bottom back to the transducer. These various possible paths are shown in the small box in Figure 2.

Another record of the many which show the presence of a deep scattering layer is one made August 5,

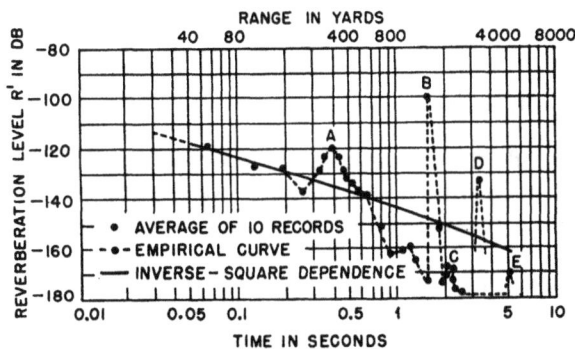

FIGURE 3. Volume reverberation levels with deep scattering layer.

1942. The data, plotted in Figure 3, were obtained with the QB transducer pointed vertically downward in 650 fathoms of water. Figure 3 is an average of 10 pings each 12 msec long. This experimental curve has several important features. The first portion of the curve decreases as 20 log t, indicating uniform distribution of scatterers to a depth of about 500 ft. A deep layer of high scattering power is evident in the vicinity of A in Figure 3; this layer appears to have a mean depth of 1,000 ft and a thickness of about 750 ft. At the position of highest scattering power within the layer, the volume-scattering coefficient is very much greater than its value in the body of the

ocean above the layer. If we use equation (24) of Chapter 12 to estimate 10 log m, assuming that the transmission anomaly terms are small, then 10 log m at A is 16 db greater than 10 log m at points on the line denoting inverse square decay; in other words, m at A is 40 times as great as m at points in the first 500 ft of the ocean. Once the beam is out of the layer, the reverberation level falls off abruptly. At a depth of 2,250 ft, the calculated value of 10 log m, neglecting the transmission anomaly terms in equation (24) of Chapter 12, is 20 db down from the value of 10 log m in the first 500 ft. This difference could not be accounted for by ordinary values of the transmission anomaly terms $-2A + A_1$. It is possible (though not likely) that the sound suffers an abnormally high transmission loss in its two-way passage through the high scattering layer, or there may actually be a layer of low scattering power at the 2,250-ft depth. Echoes from the bottom are noted at B, D, and E in Figure 3. The distance the sound which produces a reverberation peak has traveled can be estimated by noting the time at which the peak appears; it is easily seen that the sound producing the peak at D has gone from the transducer to the bottom, back to the surface, then to the bottom again, and finally back to the transducer. By a similar computation the peak at C is seen to be sound which traversed one of the following two paths: (1) scattered by the layer up to the surface, reflected from the surface to the bottom, and then returned to the transducer, (2) reflected from the bottom up to the surface, reflected back toward the bottom, and then scattered back to the transducer from the deep layer.

Not all scattering layers are at great depths. For example, a scattering layer at a depth of about 200 ft is evident in the reverberation curve of Figure 4. These records were taken with the sound beam directed vertically downward, and with a ping length of 10 msec. Occasionally both shallow and deep scattering layers are present in the ocean simultaneously. An example is given in Figure 5, which is made up of reverberation from 8-msec pings projected at a depression of 60 degrees below the horizontal in 620-fathom water. In that figure, three scattering layers are noted, at A, B, and C. The layer A is at a depth of about 100 ft, B at about 600 ft, and C at about 1,000 ft.

Some of these deep scattering layers appeared to persist for relatively long periods of time. In the same area of operation as that for Figure 2, deep scattering layers were observed at about the same depth over a

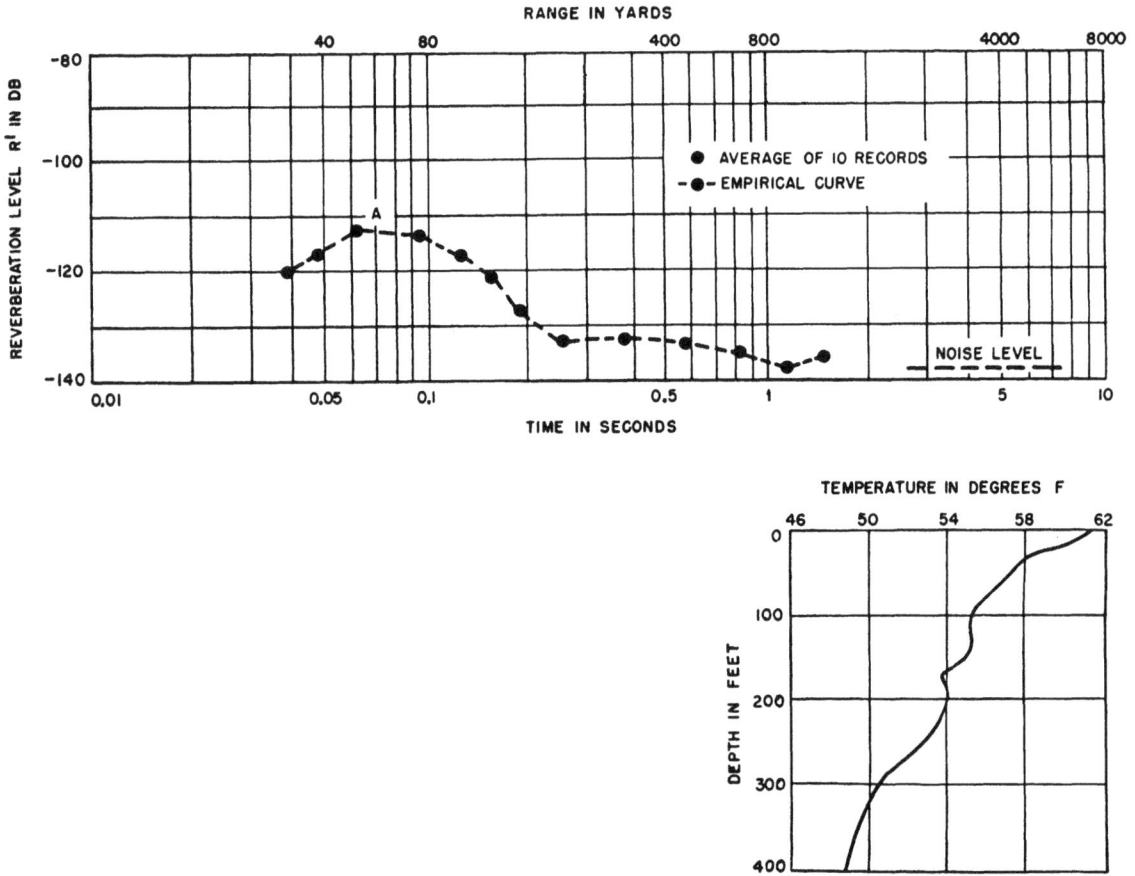

FIGURE 4.   Volume reverberation levels with shallow scattering layer.

period from July 9 to August 5, 1942. On the other hand, on June 16 and 17, a layer was observed at 1,200 ft; but a week later no such layer was detected. Thus the observations indicate that deep scattering layers, in a given area, may sometimes appear and disappear, and at other times persist for periods as long as a month or even longer. Just what these deep scattering layers consist of is not known; they may, for example, be concentrations of fish, bubbles or plankton.[a] The layer of Figure 4 occurs at the same depth as does a temperature inversion on the bathythermograph trace shown in the insert of Figure 4. On the other hand, no inversion is noticed at the depth corresponding to A in Figure 5.

14.1.3 Dependence on Frequency

An extensive series of measurements of volume reverberation in deep water,[2] at frequencies of 10, 20,

40, and 80 kc have been made by UCDWR. These measurements, described later, were made in water depths ranging from 660 to 1,950 fathoms, in the months of January and February 1943. The area of observations extended southwest of San Diego to Guadalupe Island, which is about 250 miles from San Diego and 200 miles off-shore. The various positions at which observations were taken are marked by roman numerals in Figure 6.

Deep scattering layers of the type discussed previously were observed on this cruise. Figures 7 and 8 are plots of typical reverberation records obtained at three positions shown in Figure 6. These data were obtained with the transducers directed vertically downward, sending out 10-msec pings at the four frequencies 10, 20, 40, and 80 kc. Each point on the curves for positions III and VIII is an average of 5 pings, while points on the curves for position IX are an average of 25 pings. It is evident from Figures 7 and 8 that the effective depth of the deep scattering layer does not seem to depend on frequency. This fact is shown somewhat better in Figure 9, which is a

---

[a] The observations reported in this chapter were made during daylight hours. More recent studies show evidence of diurnal migration of the deep scatterers and lend support to the theory of biological origin.

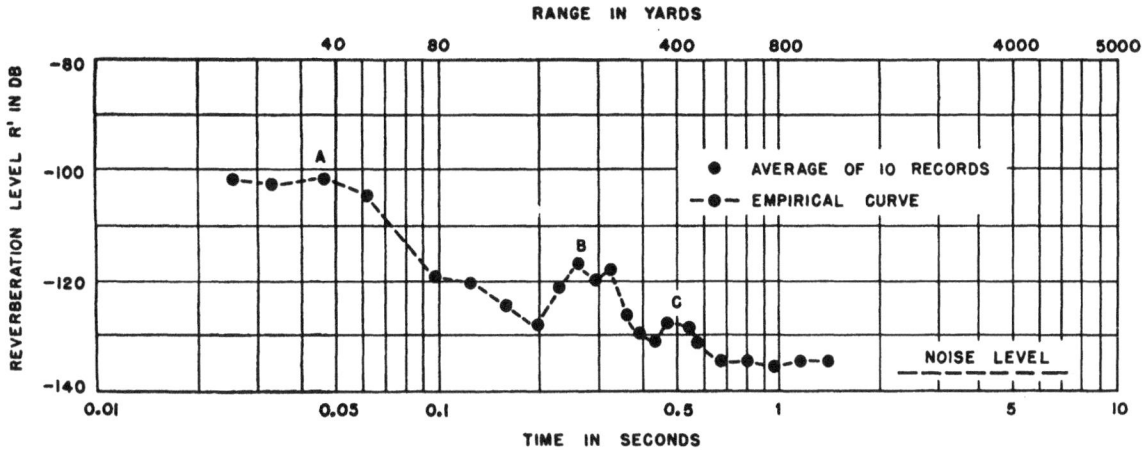

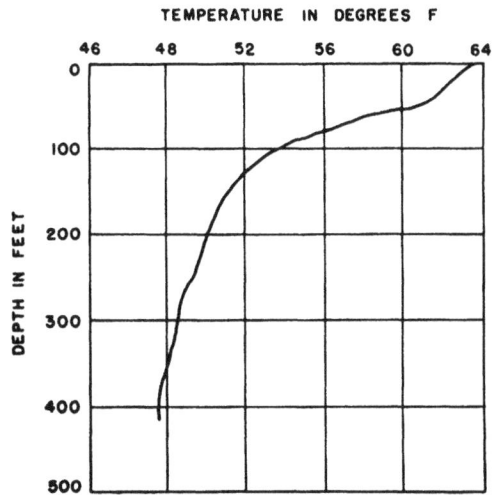

FIGURE 5.   Volume reverberation levels with scattering layer at several depths.

plot of the estimated depth of the deep layer observed for each position and frequency along the line connecting positions III and VIII. Figure 9 illustrates the persistence of the layer throughout the area of observations.

According to equation (24) of Chapter 12, it should be possible to determine log m from the experimentally observed reverberation levels, provided the values of the transmission anomaly term $-2A + A_1$ are known. Since horizontal velocity gradients in the ocean are usually negligible, refraction can be neglected in measurements with a directional transducer pointed vertically downward, and A_1 can thus be set equal to zero. Furthermore, if the acoustic properties of the ocean do not change much with increasing depth, the transmission anomaly resulting from absorption and scattering should be a linear function of range (see Section 5.2.2 of Part I). In other words, if

the ocean is approximately homogeneous, the term $-2A + A_1$ in equation (24) of Chapter 12 should equal $-2ar/1,000$ where r, the range of the reverberation in yards, is equal to the depth of the scatterers giving rise to the reverberation. It follows that if the "uncorrected" scattering coefficient M is determined from the equation

$$R'(t) = 10 \log \frac{c_0 \tau}{2} + 10 \log M - 20 \log r + J_v,$$

then

$$10 \log M = 10 \log m - \frac{2ar}{1,000}. \qquad (1)$$

In equation (1) m, the true value of the scattering coefficient, is constant if the properties of the ocean do not change with depth. Thus, for a homogeneous ocean, with m and a constant with depth, a plot of $10 \log M$ against depth on a linear scale should be a

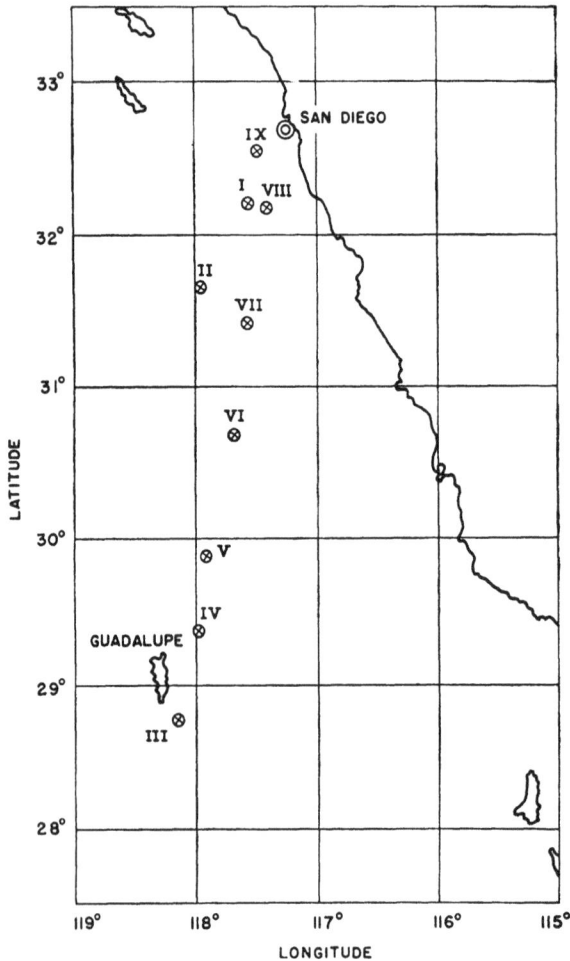

FIGURE 6. Locations where reverberation was measured in 1943 cruise.

straight line. The slope of this line will determine a, the attenuation coefficient in decibels per kiloyard; and the intercept of the line at zero range will determine the value of the true scattering coefficient m.

However, the very existence of the systematic increase in reverberation levels at about 1,000 ft observed in Figures 7 and 8 means that the ocean is probably not homogeneous with depth; thus a straight-line dependence of 10 log M on depth could hardly be expected in this experiment. Figure 10 is a plot of the mean values of 10 log M for the nine sets of records observed in the period January 17 to 20, 1943, at the positions shown in Figure 6. It is obvious from Figure 10 that even if the points in the deep layer between 1,000 and 1,500 yd are ignored, no good fit to the data could be obtained with a straight line.

The failure to obtain a straight-line dependence in

Figure 10 means that either m or a, or both, change with depth. It is possible to obtain further information from Figure 10 by comparing the dependence on depth at different frequencies. From equation (1), for any two frequencies f_1 and f_2,

$$10 \log \frac{M(f_1)}{M(f_2)} = 10 \log \frac{m(f_1)}{m(f_2)}$$
$$- 2[a(f_1) - a(f_2)] \frac{r}{1,000}. \quad (2)$$

If the variations in m are caused only by changes in the *number* of scatterers per unit volume, then $m(f_1)/m(f_2)$ should be independent of depth. Thus, if the attenuation is independent of depth, 10 log $M(f_1)/M(f_2)$ in equation (2) should be a linear function of depth in these runs with the transducer directed vertically downward. Figure 11 is a plot of this ratio against depth, from the data of Figure 10, for the six pairs of frequencies involved. Only three of the ratios are independent; the other three can be calculated from the first three. All the ratios are shown in Figure 11 for comparison. Although most of the graphs show general tendencies to slope in the direction of increasing attenuation at higher frequencies, systematic deviations from the straight line predicted by equation (2) are noted. It appears then that either the *kind* of scatterer changes with depth or the attenuation coefficient varies with depth.

At distances less than 250 ft, attenuation is small, even at 80 kc. Thus the scattering coefficients in the upper 250 ft of the ocean can be computed from vertical reverberation runs without knowledge of the attenuation coefficient. Mean values of 10 log $M \approx$ 10 log m, averaged over seven depths between the surface and 250 ft, are plotted in Figure 12 as a function of frequency, for each of the nine positions of the sending-receiving ship during January 1943 (Figure 6). The solid lines are empirical curves and the dashed lines represent a theoretical relationship discussed later. The shapes of the empirical graphs for the different positions bear little resemblance to each other. However, the two curves for position III, which represent data taken 20 hours apart, reproduce each other almost to within sampling error. The curves for positions I and VIII, which were close together in space but separated by 72 hours in time, are also nearly identical. If these resemblances are not accidental, they suggest that position is a more important factor than time in determining the value

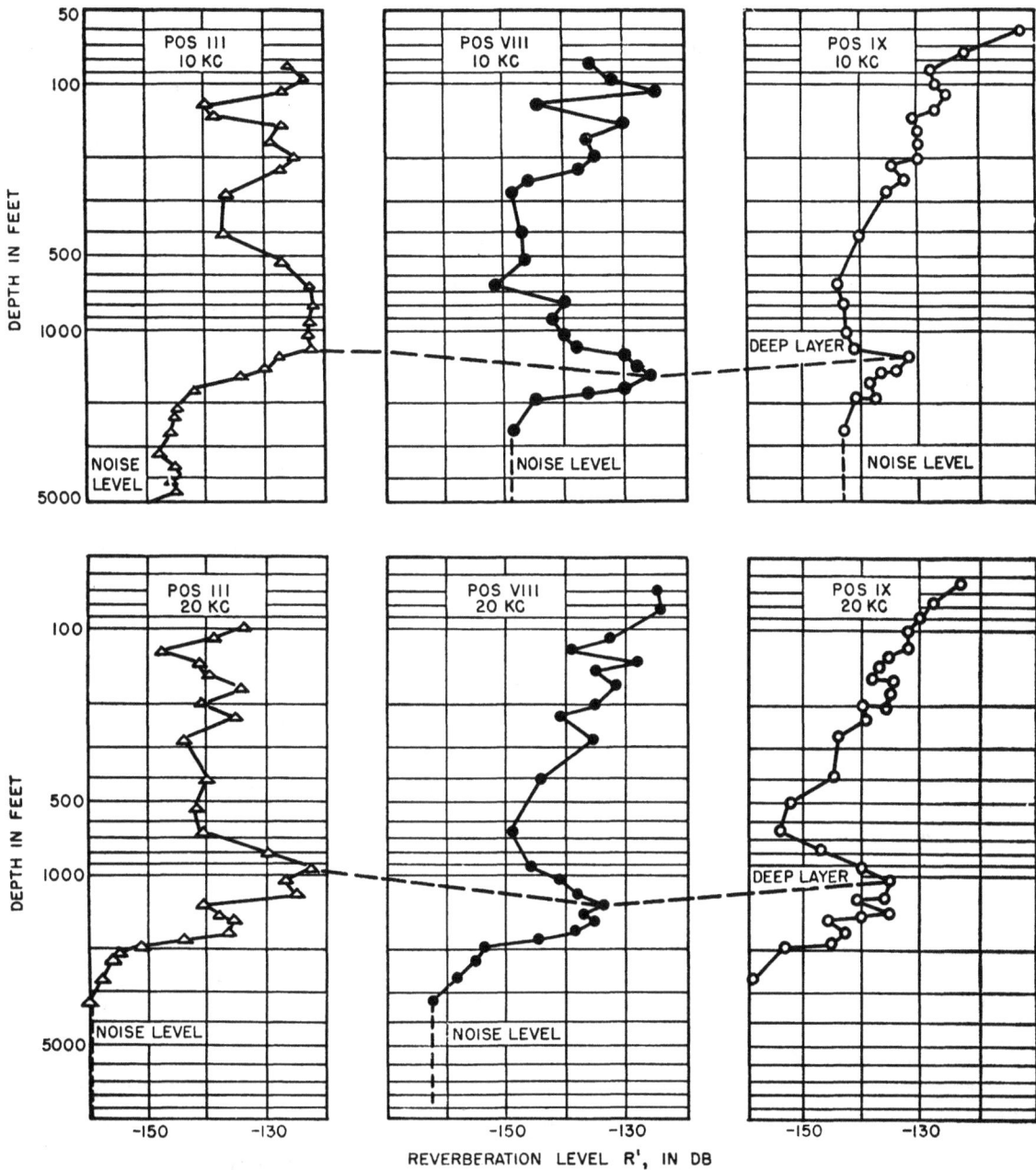

FIGURE 7. Observed volume reverberation levels versus scattering depth; GB units; sound beam vertical.

of the volume-scattering coefficient. It also appears from Figure 12 that the scattering coefficient is not affected in the same way at all frequencies by changes in position. These results, if verifiable, also substantiate the hypothesis that volume reverberation is not an intrinsic property of water as such, but results from scatterers in the ocean whose number and type are affected by oceanographic and climatic conditions. Certainly, if reverberation were a property of

water as such, it is difficult to see how small changes in position could result in the different shapes observable in the curves of Figure 12.

Figure 13 shows the mean values of 10 log M averaged at each frequency over all the positions of Figure 12 and plotted as a function of frequency. The vertical lines in Figure 13 represent mean deviations from this average of the values plotted in Figure 12. Figure 13 shows that, on the average, there

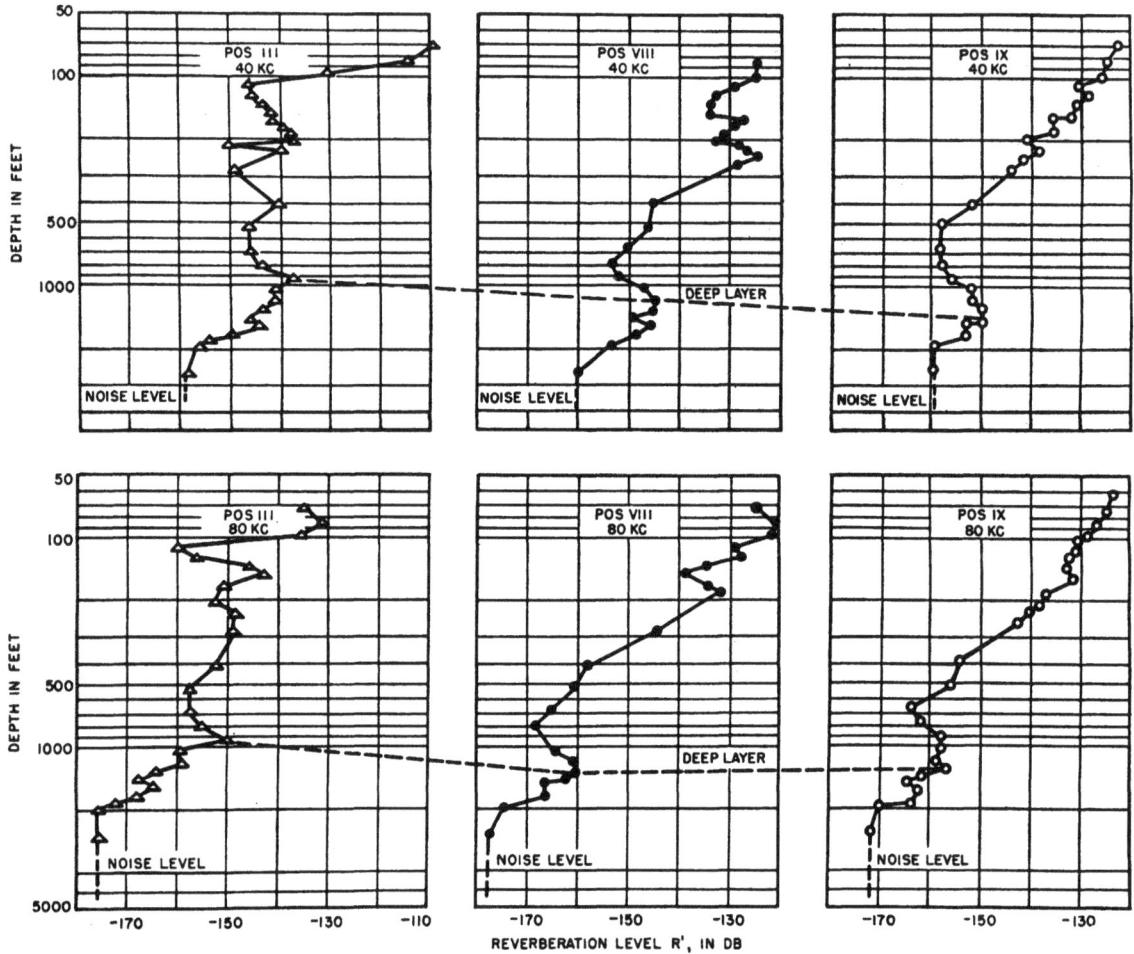

FIGURE 8.   Observed volume reverberation levels versus scattering depth; GB units; sound beam vertical.

is a slight increase in 10 log M with frequency. This systematic increase is small compared to the irregular variation from position to position, but according to reference 2, the observed trend is considerably larger than the sampling error of the measurements, and also somewhat larger than the errors which could be introduced by calibration. The straight line shown in Figure 13 was fitted by least squares; its slope indicates that $M \approx m$ increases as the 0.9 power of the frequency. It seems safe to say that the results of reference 2 do not exclude the possibility that on the average the scattering coefficient is independent of frequency. They admit the possibility also that m may vary as the second power of the frequency but not that it varies as the fourth power of the frequency. The lines

$$m = kf^4 \qquad\qquad (3)$$

are drawn in Figures 12 and 13 for comparison. This fourth-power dependence of the scattering coefficient

on frequency is known as Rayleigh's scattering law and is true for scattering from particles whose dimensions are small compared to the wavelength of the scattered sound.[3]

14.2   TRANSDUCER HORIZONTAL

That short-range reverberation with horizontal pings is often due primarily to scattering from the surface of the sea has been amply demonstrated by experiment. Reverberation intensity has been measured first with the sound beam directed horizontally, and then with it directed vertically downward. In the first transducer position, surface scatterers are irradiated by much of the central portion of the beam; in the second position, they are strongly discriminated against by the directivity of the transducer. When the experiment is performed in a choppy sea with whitecaps, the horizontal reverberation is many decibels higher than the vertical reverberation at

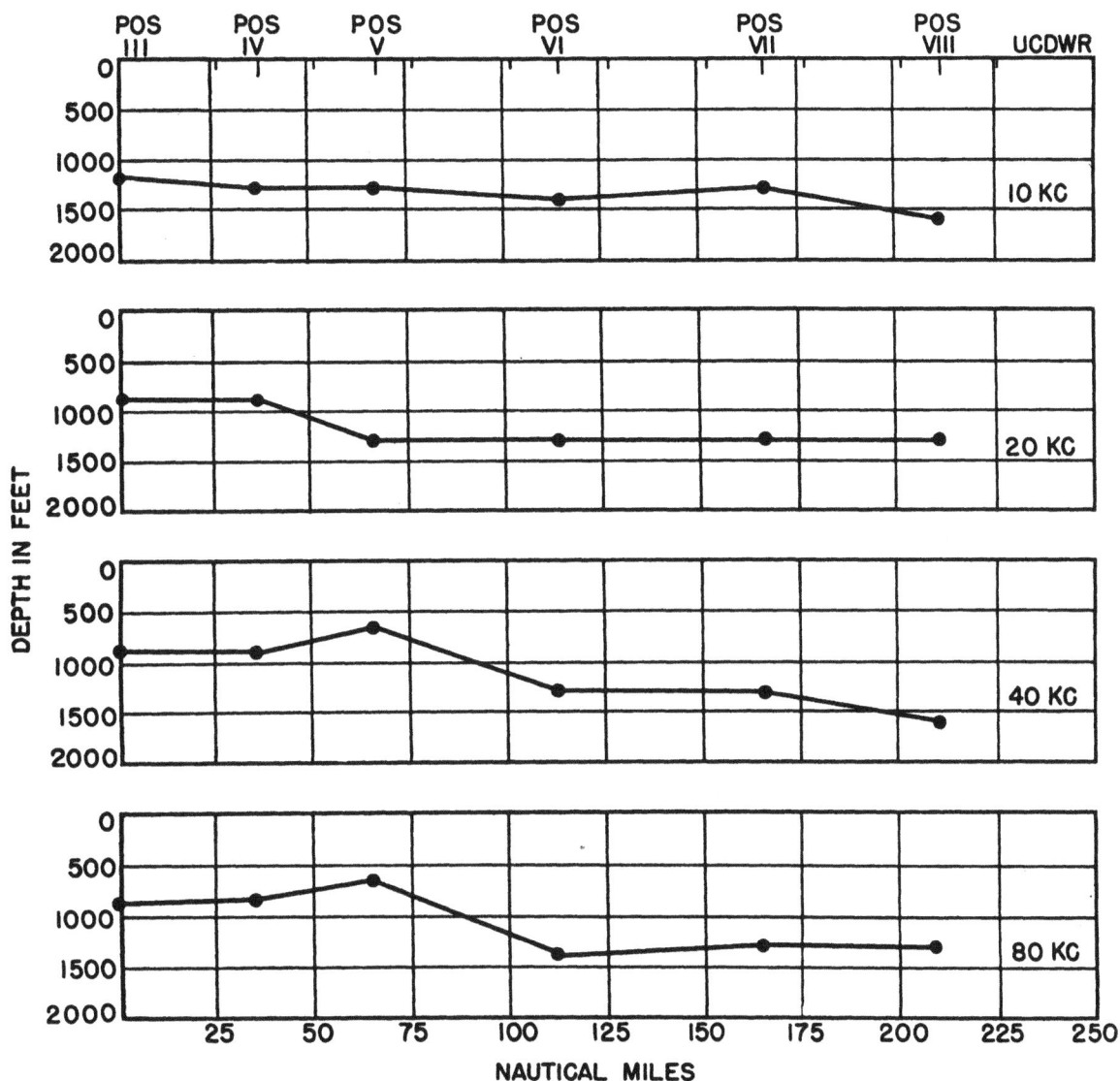

FIGURE 9.   Depth of deep scattering layer at various positions.

short ranges. Furthermore, the difference is usually much greater in rough seas than in calm seas. Of course, it is possible that the volume reverberation obtained with the beam directed downward is less than the volume reverberation with horizontal beams. However, it is pointed out below that observed values of 10 log m with horizontal beams are at most only about 6 db greater than values of 10 log m measured with vertical beams. Thus, the conclusion is inescapable that in rough seas the short-range reverberation from horizontal pings is surface reverberation, that is, reverberation caused by scatterers near the surface of the sea whose number and strength are a function of sea state.

14.2.1 **Dependence on Range and Oceanographic Conditions**

Analysis of reverberation records clearly shows that the range dependence of surface reverberation is itself a marked function of such oceanographic parameters as sea state and temperature gradients. For this reason, it is convenient to treat the effects of all these variables together. The following section summarizes the observational results of studies of surface reverberation, which indicate that surface reverberation tends to fall off with increasing range much faster than predicted by the simple theory of Chapter 12, and that the rate of decay increases

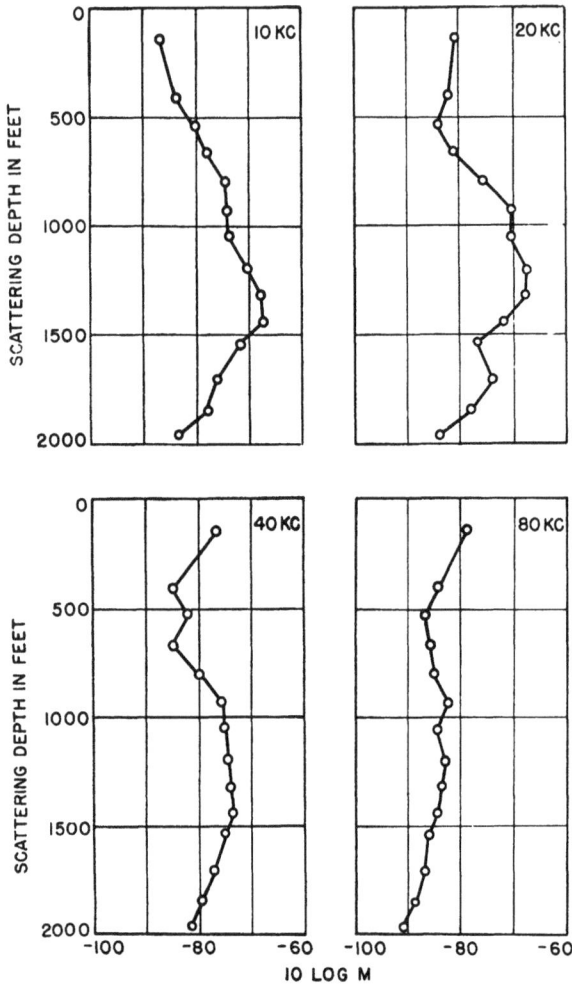

FIGURE 10.   Mean value of 10 log M for all positions in 1943 cruise.

rapidly with increasing sea state. A later section entitled "Possible Explanations" advances some fairly plausible qualifications of the simple theory which may explain away, in part, the seeming discrepancies between theory and observation.

EXPERIMENTAL RESULTS

Figure 14 illustrates the results of one experiment comparing the reverberation with horizontally and vertically directed beams. At the time of the experiment, the sea surface was confused, with whitecaps present and with a wind velocity of 17 mph. A comparison of the two reverberation level curves shows that for times up to 0.1 sec the horizontal reverberation is more than 20 db above the vertical reverberation; for times between 0.1 and about 0.4 sec, it is more than 10 db above; and for times greater than 0.4 sec it is less than 4 db above. Two conclusions are obvious from the figure. One is that on the day of the experiment there was a surface layer of scatterers which was very different in nature from the scatterers in the ocean body. The other is that the reverberation due to surface scatterers decays more rapidly than the reverberation due to volume scatterers. The presence of a deep scattering layer can also be noted at B in Figure 14, and also a shallower scattering layer at A.

According to equation (39) of Chapter 12, the surface-reverberation intensity at short range, where $2A$ in equation (39) can be neglected, should be proportional to the inverse cube of the range, provided

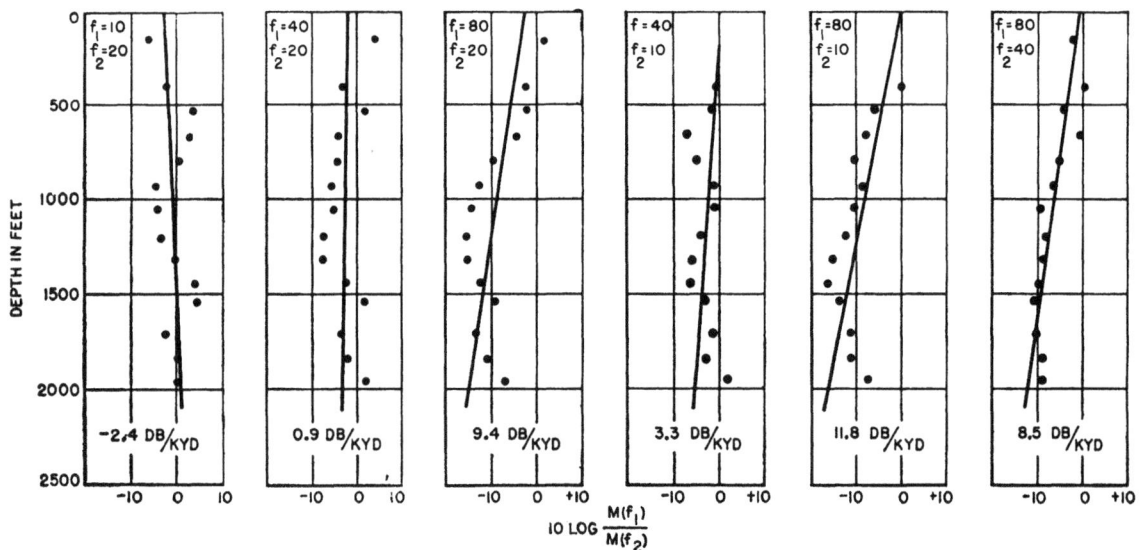

FIGURE 11.   Variation with depth of ratio of scattering coefficients.

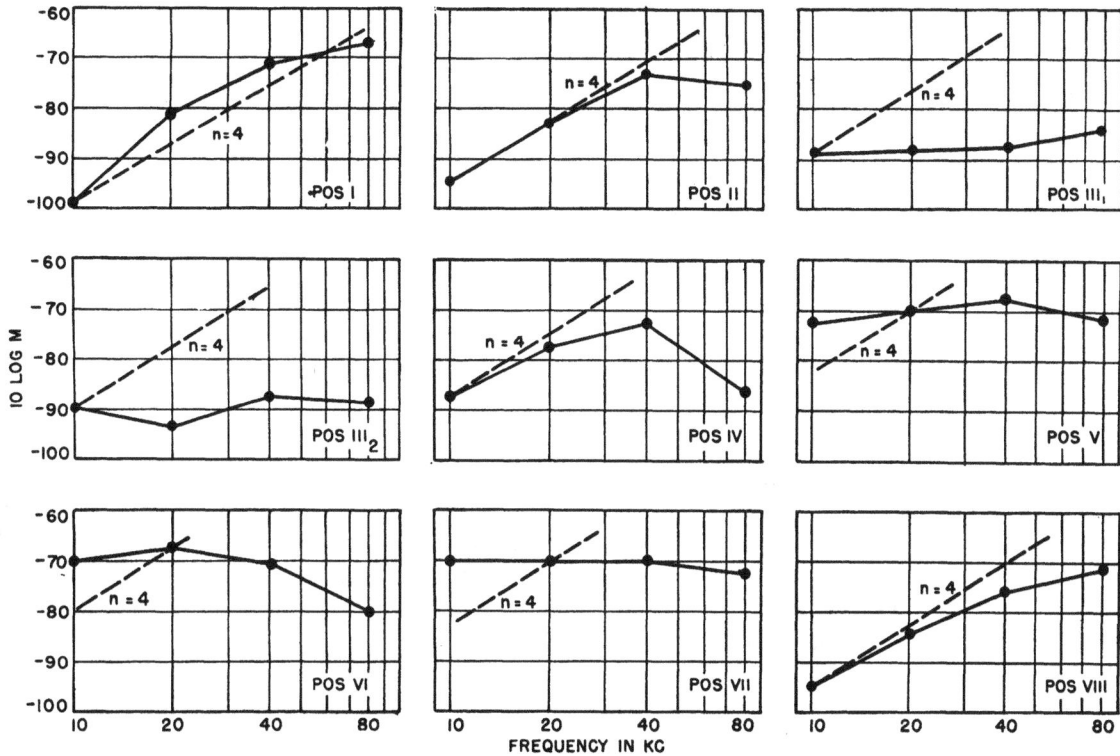

FIGURE 12. Variation of 10 log M with frequency (mean values of 10 log M for depths less than 250 feet). Positions III and III$_2$ refer to measurements made at Position III on two separate days.

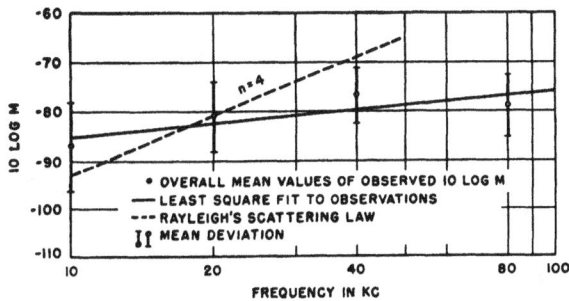

FIGURE 13. Variation of scattering coefficient with frequency (mean values of 10 log M at all nine positions for depths less than 250 feet).

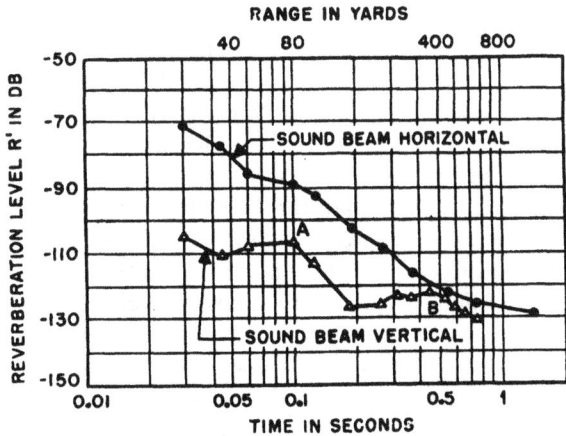

FIGURE 14. Comparison of reverberation from horizontally and vertically directed beams.

10 log $m'/2$ and $J_s(\theta)$ in equation (39) are also independent of range. This simple inverse cube dependence is observed only rarely. Figure 15 shows the reverberation intensities observed on May 8, 1942, with the QCH-3 transducers. On this date ground swells were long and low, and a few whitecaps were forming. The QCH-3 transducers were at a depth of 20 ft with their long dimensions horizontal and with the transducer axes parallel to the sea surface.

In this position the QCH-3 transducers are practically nondirectional in the vertical plane, so that in equation (41), Chapter 12, the correction factor $b(\theta - \xi,0)b'(\theta - \xi,0)/\cos\theta$ is very nearly unity at all angles of importance. Since this correction factor is the only part of $J_s(\theta)$ which can depend on range, it is clear that in these experiments $J_s(\theta)$ was independent of range. With negligible A and constant m' and $J_s(\theta)$, the theoretical equation for surface reverberation, equation (43), Chapter 12, leads to a straight line with a slope corresponding to inverse third-power decay. This simple reverberation decay is indicated by the solid line in Figure 15. The points in Figure 15 are the averages of 36 pings each 8 msec long and agree fairly well with the theoretical straight line.

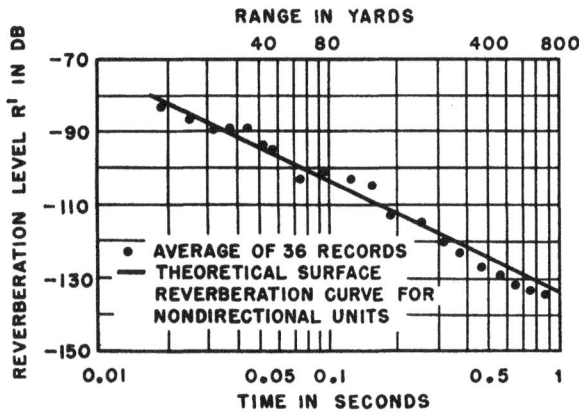

FIGURE 15. Surface reverberation levels showing simple inverse cube dependence. Projector and receiver effectively nondirectional in the vertical plane.

On the same day (May 8, 1942) surface-reverberation measurements were also carried out with the QCH–3 transducers oriented differently. In these experiments the transducers were placed at a depth of 20 ft with the transducer axes parallel to the ocean surface, as before; but the long dimensions of the transducers were vertical instead of horizontal. With this transducer orientation, the correction factor $b(\theta - \xi,0)b'(\theta - \xi,0)/\cos\theta$ cannot be neglected. The values of the correction factor as a function of range were calculated from the known directivity pattern of the QCH–3. A theoretical reverberation curve was then obtained, using equation (43) of Chapter 12, assuming $10 \log m'/2$ independent of range, and neglecting the term $2A$ in that equation. This curve is plotted as the solid line in Figure 16. It can be compared with the points which show the actual reverberation levels observed in this experiment; each point represents the average reverberation from 30 pings each of length 8 msec. Evidently the agreement between theory and experiment is quite good. It may be remarked that at ranges between 80 and

800 yd the observed points in Figures 15 and 16 are in close agreement. This agreement was not to be expected if the scattering coefficient of the surface did not change with time; it is easily verified that the values of $J_s(\theta)$ are quite different for the horizontal and vertical orientations of the QCH–3.[b] Thus the agreement at ranges greater than 80 yd between the observed points in Figures 15 and 16 means that the value of the surface scattering coefficient must have changed during the interval between the two experiments. The value of $10 \log m'/2$ estimated from Figure 15 is about 5 db greater than the value estimated from Figure 16.

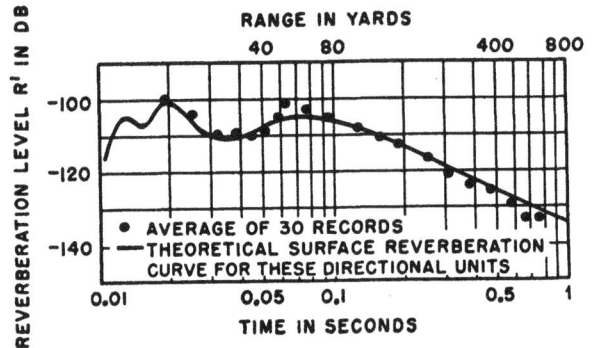

FIGURE 16. Surface reverberation levels showing agreement with theoretical curve. Projector and receiver directional in vertical plane.

Evidently if surface reverberation arises from scattering in a thin layer near the ocean surface, a drop in reverberation is to be expected when the sound beam is bent away from the surface, provided, of course, that the surface reverberation is not masked by volume reverberation at the range where the beam leaves the surface. Under conditions of sharp downward refraction such sudden drops have been observed. For example, Figures 17 and 18 show reverberation levels obtained with the QB transducer on July 24, 1942. On this day ground swells were almost absent, but the ocean surface was scuffed up with a few whitecaps forming. The wind speed was 12 mph. The QB transducer was placed with its axis parallel to the surface at depths of 20 and 60 ft. Twelve consecutive pings, each 11 msec long, were averaged at each depth to give the points shown in Figures 17

[b] The values of $J_s(\theta)$ for both horizontal and vertical orientations of the QCH–3 can be obtained by using the QCH–3 directivity patterns given in Section II of reference 1, in equation (42) of Chapter 12.

RANGE IN YARDS

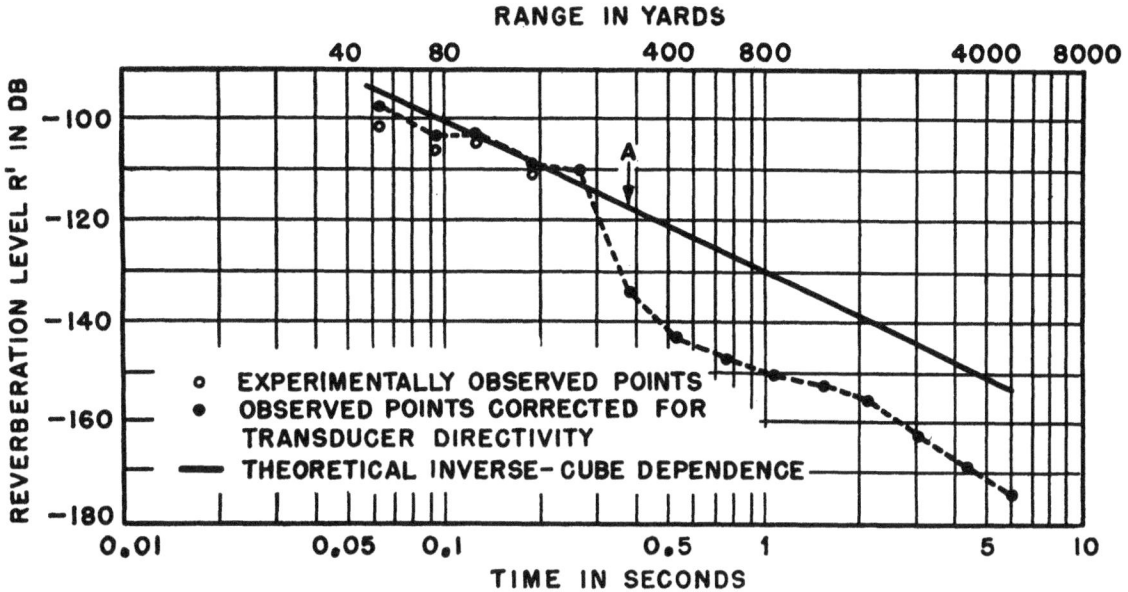

FIGURE 17. Surface reverberation levels with strong downward refraction. Transducer depth 20 feet.

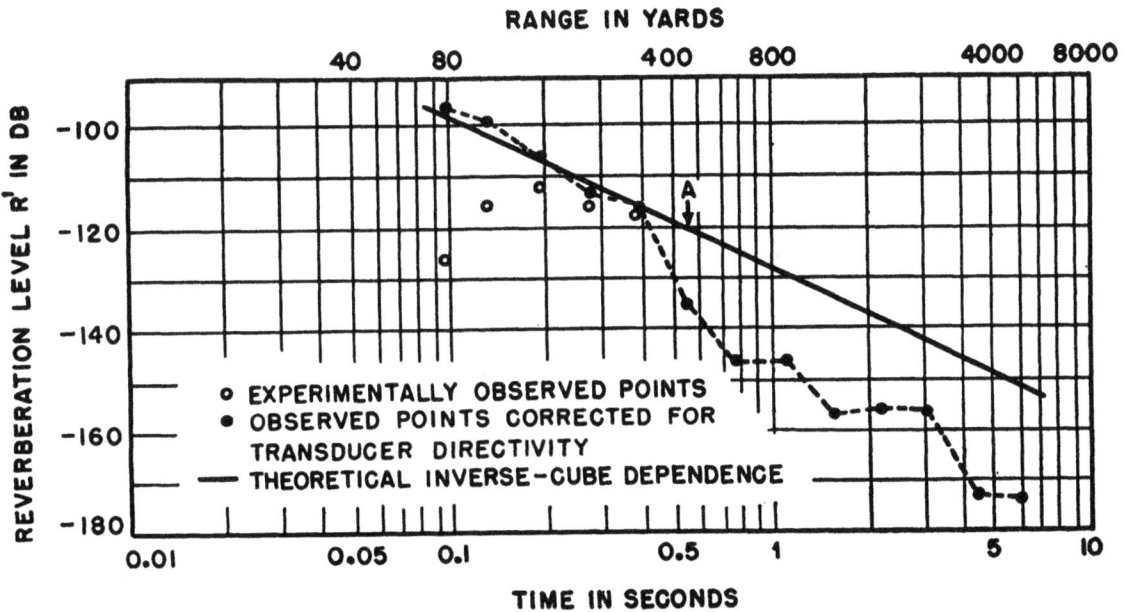

RANGE IN YARDS

FIGURE 18. Surface reverberation levels with strong downward refraction. Transducer depth 60 feet.

and 18. Figure 19 shows the bathythermograph record and the calculated limiting rays for the two depths. The points marked A in Figures 17 and 18 show the range where the limiting ray leaves the surface. Just as in Figure 16, the observed reverberation levels at short range can be expected to differ from a straight line with $-30 \log r$ slope because of the correction factor $b(\theta - \xi)b'(\theta - \xi)/\cos \theta$. To facilitate comparison with this line, the observed levels

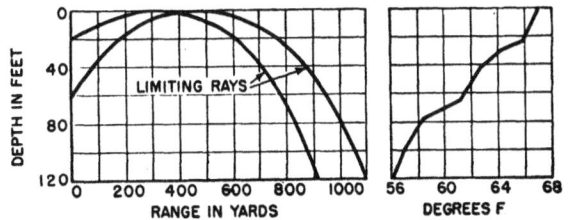

FIGURE 19. Refraction conditions for data in Figures 17 and 18.

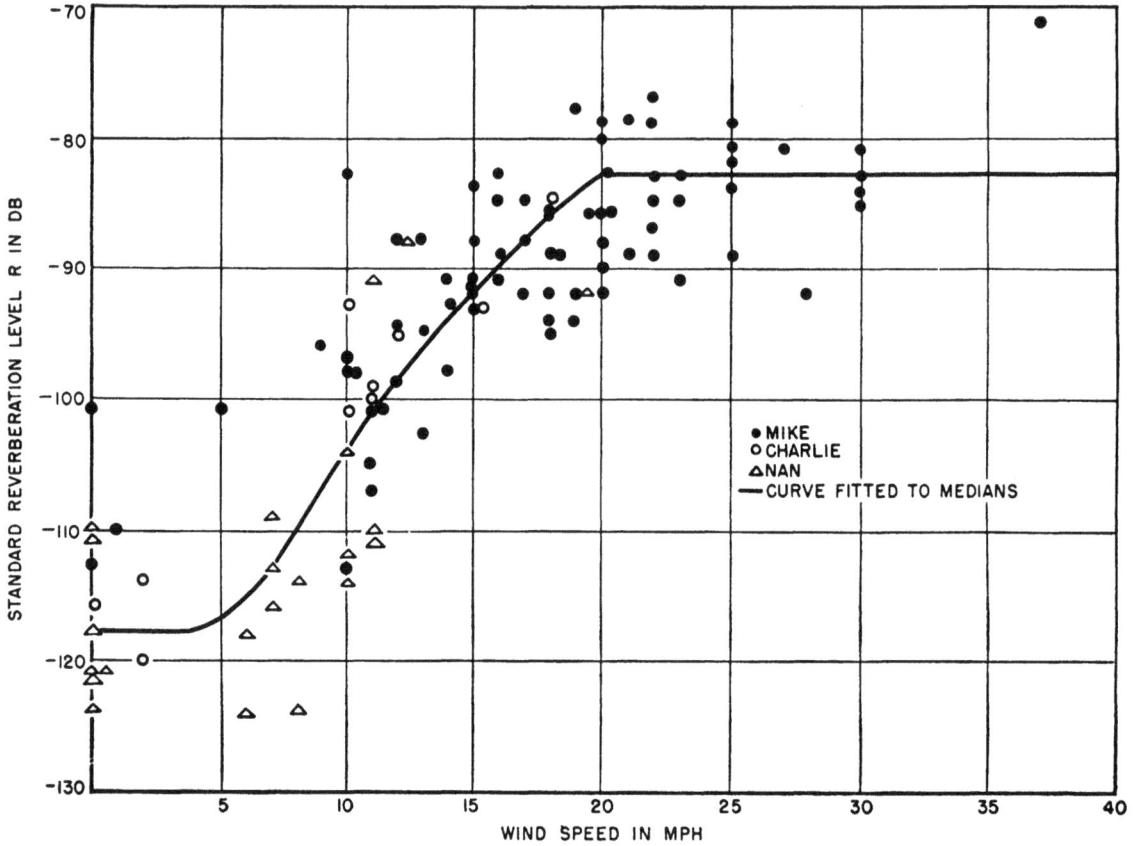

FIGURE 20.   Standard reverberation level at 100 yards as a function of wind speed.

are corrected by increasing the experimental points by the value of the correction factor. That is, in Figures 17 and 18 the solid points are values of

$$R'(t) - 10 \log \frac{b(\theta - \xi,0)b'(\theta - \xi,0)}{\cos \theta}. \qquad (4)$$

By using equations (41) and (43) of Chapter 12, the expression (4) obviously equals

$$10 \log \frac{c_0 \tau}{2} + 10 \log \left(\frac{m'}{2}\right) - 30 \log r + 10 \log \frac{Q(0)}{2\pi}$$
$$- 2A. \quad (5)$$

In equation (5), if A can be neglected and if m' is independent of range, the only dependence on range is contained in the term $-30 \log r$. Thus, with these assumptions the solid points in Figures 17 and 18 should lie on a straight line of $-30 \log r$ slope if refraction has no effect. It is seen that the first few solid points in Figures 17 and 18 do lie on a straight line of $-30 \log r$ slope, but that at a range close to that when the limiting ray leaves the surface, there is a sharp

drop in reverberation level. This drop does not contradict the theory in Chapter 12. It will be recalled from Section 12.3 that equation (43) is not valid and therefore cannot be expected to agree with measured levels at ranges past that at which the limiting ray leaves the surface. In equation (4) the correction factor approaches unity and its logarithm approaches zero as the range is increased, and in both Figures 17 and 18 the correction is practically negligible by the time the sudden drop in intensity occurs. Consequently it is not possible to ascribe the position of the experimental points at ranges past the sudden drop in intensity to uncertainty in the value of the correction factor. Thus the conclusion seems inescapable that the sudden drop is due to the sound rays leaving the surface.

From the evidence in Figures 15 to 19, it appears that our basic assumption, namely that surface reverberation arises from scattering in a thin layer near the ocean surface, is probably correct. The observed $-30 \log r$ slope in Figure 15 and in the long-range portion of Figure 16 also permit the conclusion that

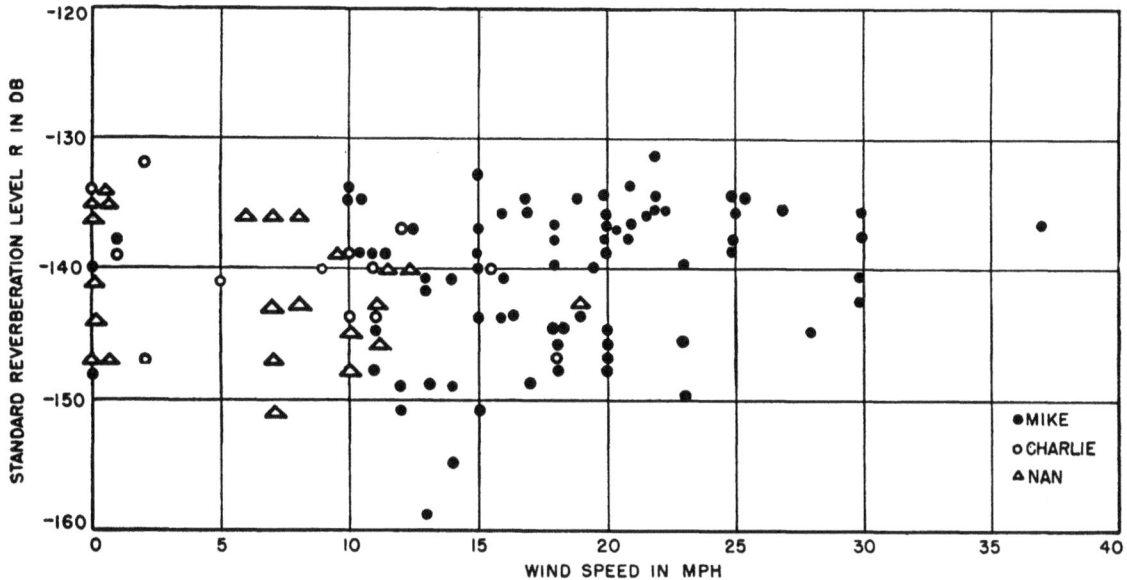

FIGURE 21. Standard reverberation level at 1,500 yards as a function of wind speed.

there are times when the surface scattering coefficient m' is independent of the angle of incidence of the rays on the surface. However, it is not usually possible to fit the observed reverberation intensities with equation (39) of Chapter 12, if m' is assumed independent of range. Thus, while the basic assumptions leading to that equation are probably correct, it cannot be said that the factors involved in surface reverberation are completely understood.

More illustrations of the dependence of surface reverberation on range may be obtained from a memorandum issued by UCDWR,[4] where extensive measurements of observed deep-water reverberation levels at 24 kc are summarized. This summary is based on data obtained on 6 cruises in the period from November 26, 1943 to September 1, 1944. About 110 reverberation curves were obtained, each an average of five successive pings. The ping lengths used varied from 16 to 80 yd, but in the following curves all data have been corrected to the standard 80-yd length; in other words, the following graphs are all plotted in terms of the standard reverberation level. All the data were obtained with the JK projector at a depth of 16 ft on the USS *Jasper* with the transducer axis horizontal.

Figure 20 is a plot against wind speed of all the reverberation levels measured at a range of 100 yd. All seasons of the year are represented. Thermal patterns were of MIKE, CHARLIE, and NAN types (see Chapter 5), represented respectively by dots, circles, and triangles. In Figure 20 a systematic in-

crease in reverberation level of about 35 db is observed as the wind increases in velocity from zero to 20 mph. At wind speeds of 8 mph or less, there is little systematic dependence on wind speed. At greater speeds the level rises sharply, up to speeds of 20 mph or more. Increase of wind speed beyond 20 mph has little systematic effect. This dependence on wind speed is correlated with the roughness of the sea. At 8 mph the wind is strong enough to roughen the surface appreciably; occasionally wavelets may slough over, but no well-developed whitecaps are observed. At about 10 mph small whitecaps begin to appear. When the wind has reached 20 mph the sea is liberally covered with whitecaps. The detailed dependence of the appearances of the sea on wind force is described in a Navy manual.[5] Apparently, as the wind speed increases beyond 20 mph, the resulting increase of whitecaps causes little, if any, additional increase in reverberation.

The median values of the standard reverberation level at 100 yd, as a function of wind speed up to 20 mph, are roughly described by the equation

$$R = -118 + 10 \log (1 + 2.5 \times 10^{-6} u^7) \quad (6)$$

where u is the wind speed in miles per hour. This equation is represented by the solid line in Figure 20. Beyond 20 mph, the function is assumed to be constant at $R = -83$ db.

The reverberation level at long range has a markedly different wind-speed dependence from that at short range. The data in Figure 21 are taken from

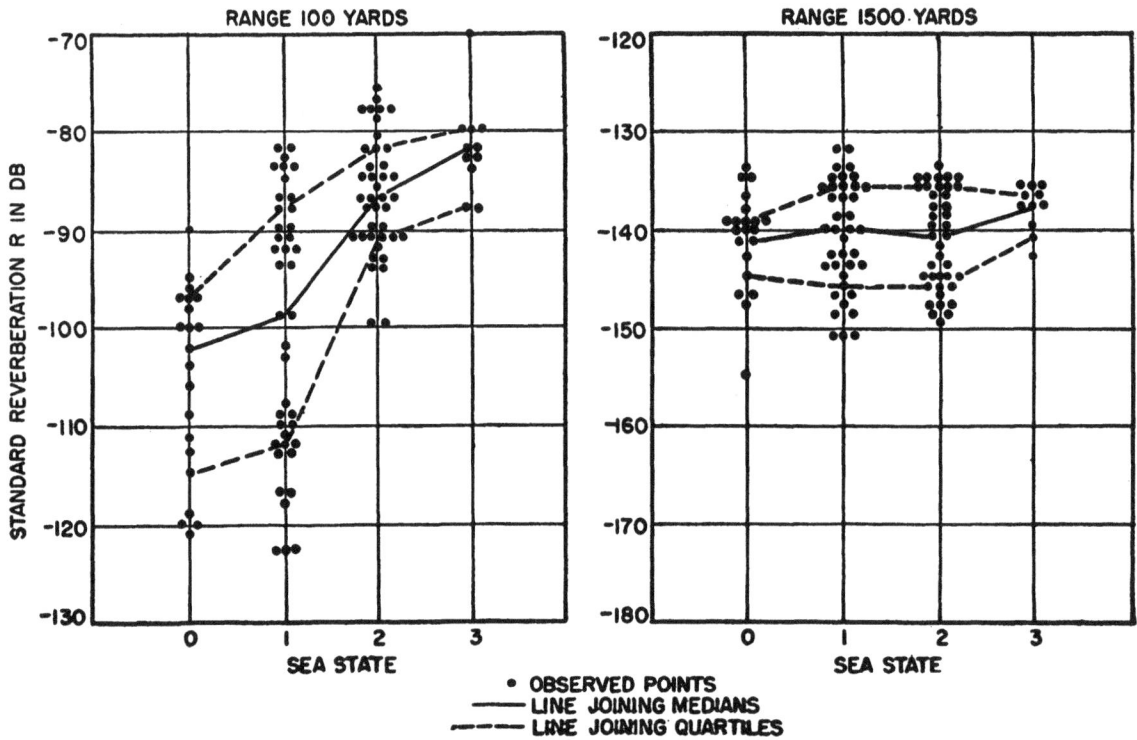

FIGURE 22.   Dependence of standard reverberation level on sea state.

the same reverberation runs as the data in Figure 20, with the exception that the levels shown were measured at 1,500 yd rather than at 100 yd. No significant wind-strength dependence is observed. It seems justifiable to conclude from these data that surface reverberation, which is frequently dominant at 100 yd, has little effect at 1,500 yd; in other words, at 1,500 yd the observed reverberation usually arises from volume scattering. Figure 22 shows the dependence of reverberation on sea state at ranges of 100 and 1,500 yd. The 100-yd levels depend on sea state while the 1,500-yd levels apparently do not; thus the qualitative dependence in Figure 22 is the same as that in Figures 20 and 21. The relation between wind force and sea state is given in the NDRC survey report on ambient noise.[6]

Figure 23 shows the reverberation levels as a function of range, for high and low wind speeds. In this illustration are given the median reverberation levels and the upper and lower quartiles at each range for wind speeds less than 8 mph and greater than 20 mph. It appears from Figure 23 that the reverberation is entirely independent of wind speed at ranges greater than 1,500 yd. Actually, the quartiles at ranges greater than 1,500 yd are not precisely the

same for wind speeds less than 8 mph and wind speeds greater than 20 mph (see Figure 5 of reference 4), but these differences are not thought to be significant. Consequently, in Figure 23, data for all wind speeds are used to determine the range dependence of the reverberation at ranges of 1,500 yd or more.

From Figure 23, for wind speeds greater than 20 mph, the median reverberation level drops 46 db between 100 and 1,000 yd. Thus, for high wind speeds, the reverberation intensities usually drop off nearly as the fifth power of the range, rather than as the third power predicted in Chapter 12 on the assumption that 10 log m' is independent of range. (At 100 yd or more, with the JK at a depth of only 16 ft, the variation of $J_s(\theta)$ with range can be neglected.) Even faster rates of decay than the fifth power are frequently observed. For example, Figure 24 shows a plot of the reverberation levels obtained on the *Point Conception* cruise, on March 2, 1944. The rate of decay in this figure is approximately as the sixth power of the range; that is, R decreases approximately as $-60 \log r$. The wind speed on this run was 37 mph.

Figure 25 shows the reverberation level at 1,500 yd plotted against date and area. The data fall into five

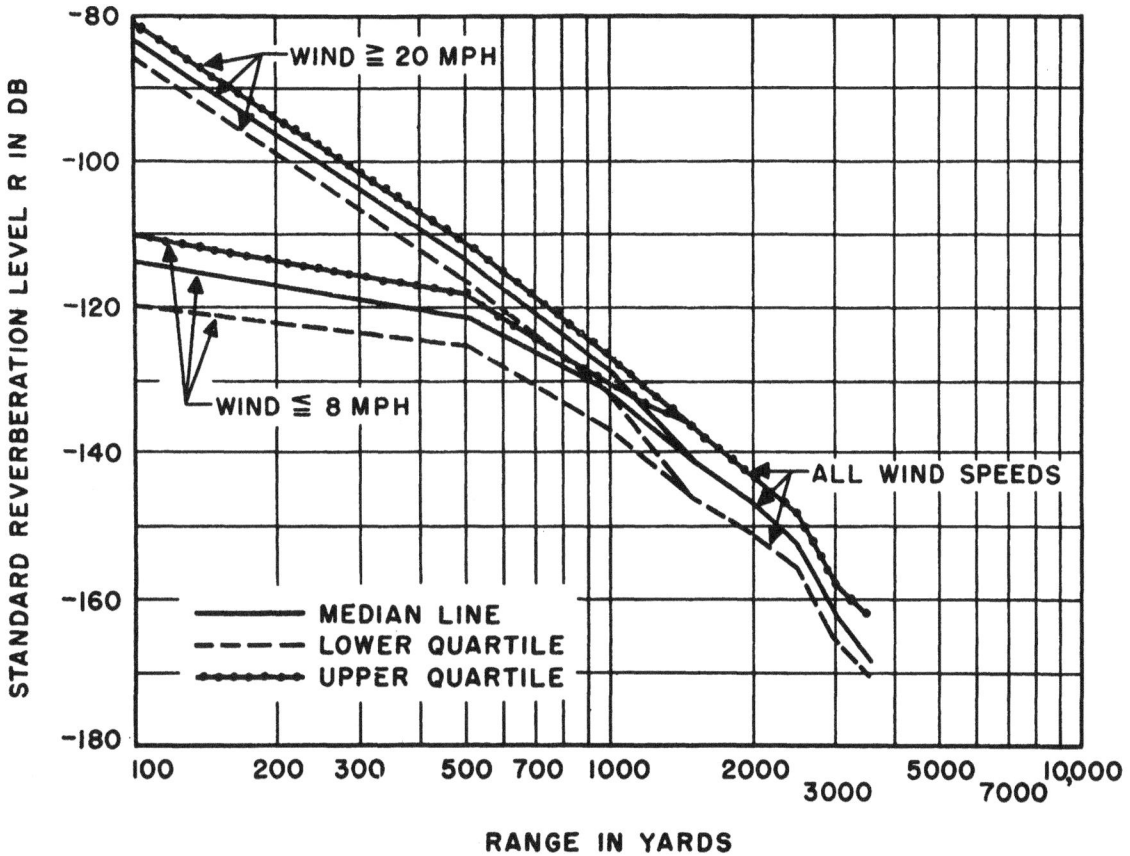

FIGURE 23.   Dependence of standard reverberation level on range and wind speed.

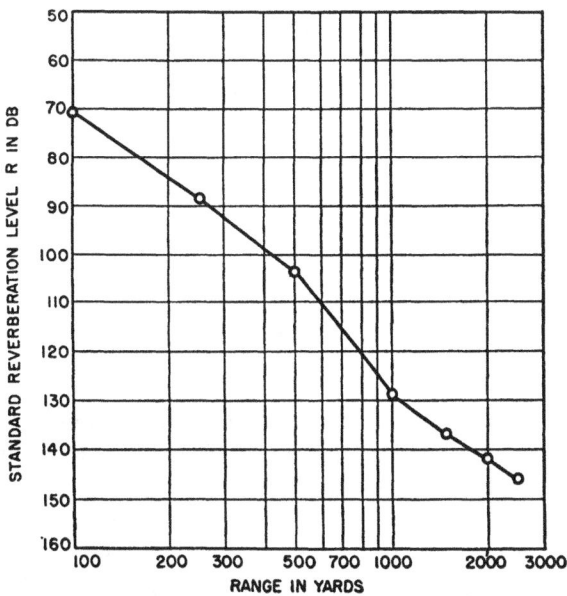

FIGURE 24.   Sample plot of standard reverberation level at high wind speed.

groups, two pairs of which were taken in the same area at different seasons. The data grouped in this way are summarized in Table 1. On the whole, Table 1 shows that the mean reverberation levels at 1,500 yd are independent of season and area, although the *Cedros II* and Point Conception data may indicate some systematic variation.

Figure 26 is an analysis of the dependence of the observed reverberation levels on a parameter which has been found to correlate significantly with transmission studies at 24 kc. This parameter is the depth D_2 in the ocean at which the temperature is 0.3 F less than at the surface.[7] The levels in Figure 26 are referred, for convenience, to an arbitrary zero level which was, however, the same for all curves. Only data obtained after March 22, 1944 were used for this comparison since most of the earlier runs ended at about 2,000 yd. The median curves are seen to be practically the same for D_2 between 5 and 40 ft, but fall off less rapidly for D_2 between 40 and 160 ft. Since, according to reference 7, increasing D_2 means a decreasing transmission anomaly A in equation

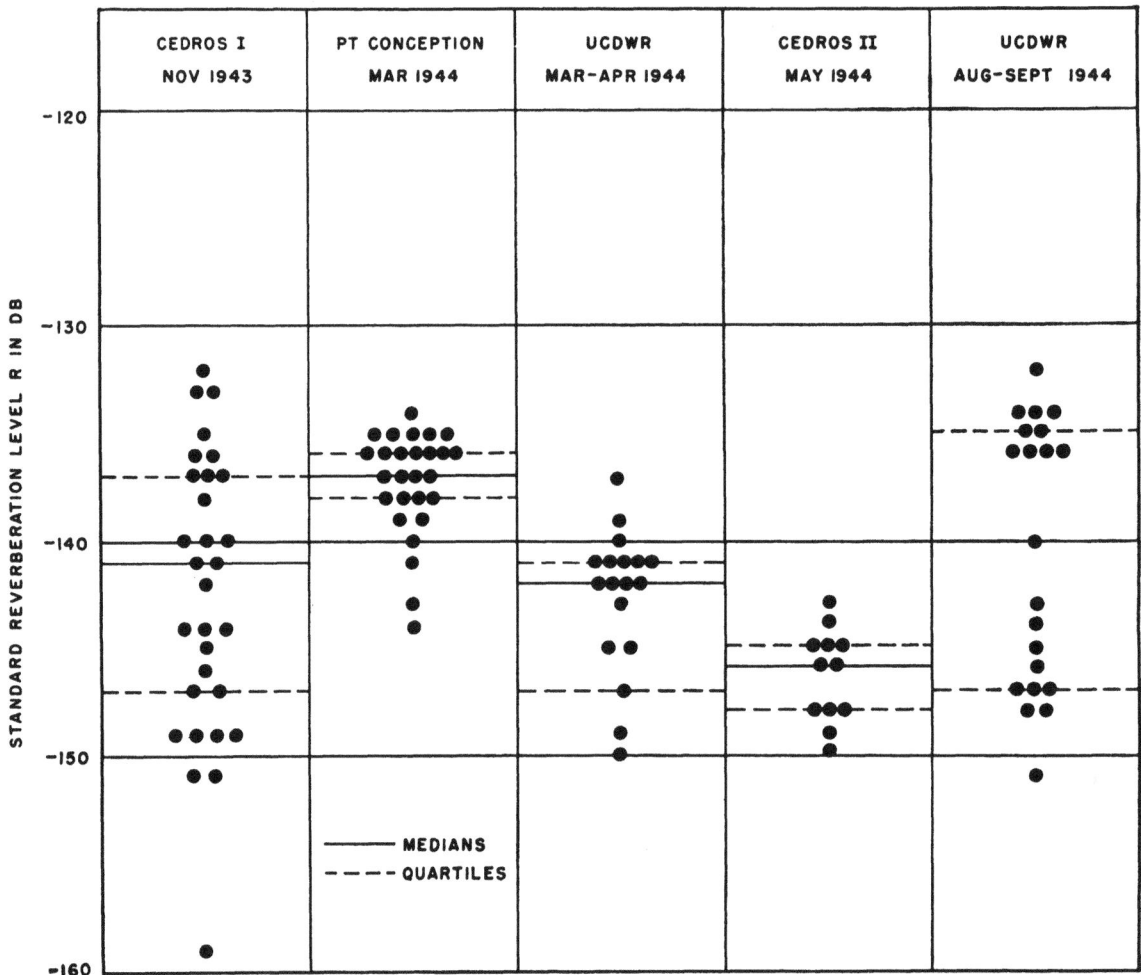

FIGURE 25. Standard reverberation level at 1,500 yards for various dates and areas.

TABLE 1

Area	Month	Median reverberation level at 1,500 yd	Inter-quartile difference	Number of cases
All data	Dec.–Sept.	−140	9	113
Cedros I	November	−141	10	31
Cedros II	May	−146	3	12
Point Conception	March	−137	2	27
San Diego	March–April	−140	6	22
San Diego	August–September	−140	12	21

(26) of Chapter 12, the differences between these curves might be explained as a result of improved transmission to long ranges with larger values of D_2. However, on the whole there is little dependence of the curves on the parameter D_2; certainly no dependence is apparent at 1,500 yards.[c]

[c] In this connection, more recent data obtained by UCDWR are of interest. These data, as yet unpublished, show that with NAN patterns (usually falling in the class

The results of this subsection can be summarized as follows. At ranges less than 1,500 yd, in standard 24-kc echo-ranging gear oriented so that the acoustic

$5 \leq D_2 < 10$) there is frequently a hump in the reverberation curve at a range which corresponds to the deep scattering layer discussed previously. Presumably these humps are most common with NAN patterns because with strong downward refraction the main sound beam usually strikes the deep layer at a well-defined range.

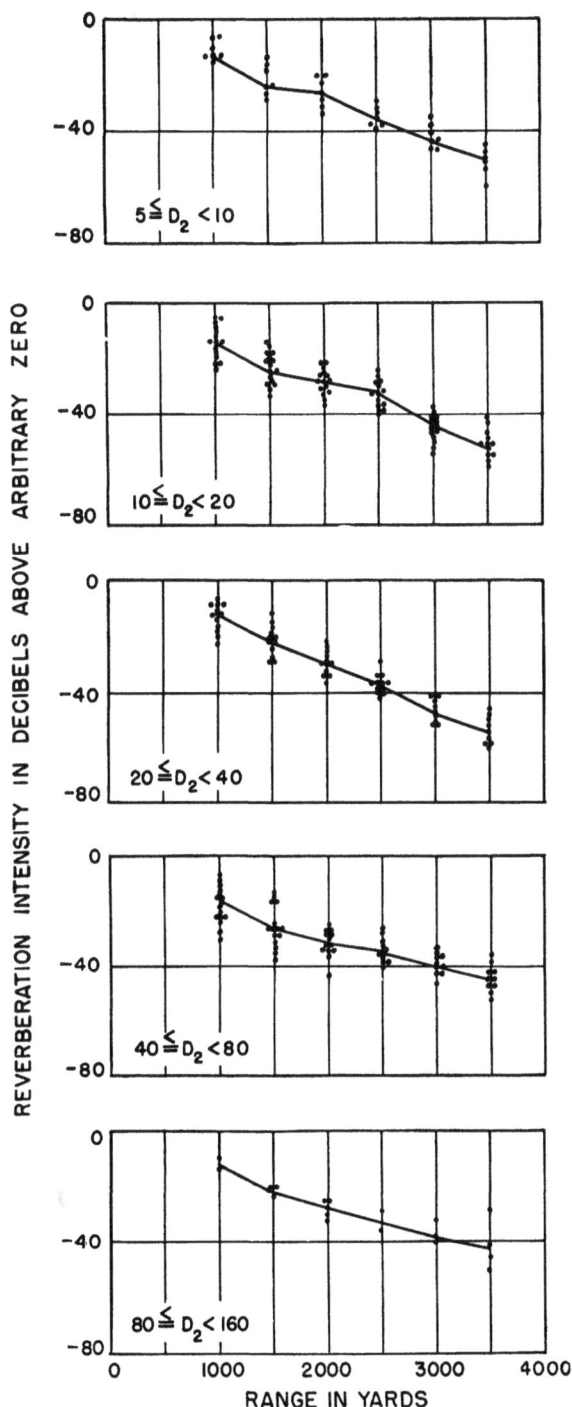

FIGURE 26. Range dependence of reverberation as a function of refraction conditions.

axis is parallel to the surface, the reverberation observed at wind speeds greater than 8 mph is predominantly surface reverberation. At ranges greater than 1,500 yd, the reverberation does not depend significantly on wind speed, location, season, or thermal structure of the ocean. It seems justifiable, therefore, to regard the reverberation at ranges greater than 1,500 yd as the characteristic volume reverberation of the ocean.

POSSIBLE EXPLANATIONS

The lack of dependence of reverberation on wind speed at ranges greater than 1,500 yd is well established, but is nevertheless surprising. The masking of the surface reverberation by volume reverberation at this range is in large part due to the rapid decrease of surface reverberation with range. The reason for this decrease is obscure. The following factors have been suggested, in Section VI of reference 1, as possible causes of this rapid decrease of surface reverberation with increasing range: attenuation, variation of the scattering coefficient with angle of incidence on the surface, shadowing effects of waves, and interference effects in a thin surface layer (Lloyd mirror effect). These possible causes will now be considered briefly.

In Figure 23, the drop between 100 and 1,000 yd in median reverberation level at wind speeds greater than 20 mph is 16 db more than would have been predicted from the $-30 \log r$ dependence of equation (43), in Chapter 12. If this change is due to transmission loss, the term A in equation (43) must have a median value of 8 db per kyd. While this value of the attenuation is not impossible, it is significantly greater than the mean attenuation coefficient observed in transmission studies off San Diego (see Section 5.2.2), especially since with high wind forces the surface layer tends to become isothermal. If the steep slope in the curve of Figure 24 is due to attenuation, the attenuation coefficient would have to be as high as 15 db per kyd. The possibility that some of the increased loss is due to attenuation cannot be ruled out; but on the whole the evidence from transmission studies does not justify regarding attenuation as the primary cause of the rapid decrease of surface reverberation with range.

If the surface scattering layer is very thin, it can be argued that the scattering coefficient m' should decrease at least as rapidly as $\sin \theta$, where θ is the grazing angle of the ray incident on the surface. For the total volume of surface scatterers irradiated by the ping at any instant is proportional to $c_0\tau$, according to Section 12.3. All the energy reaching this volume must pass through the surface whose cross section in the plane of Figure 27 is AB. Thus, if I is intensity at AB, the total energy reaching the surface scatterers per unit time is proportional to $I(AB) = I(c_0\tau) \sin\theta$. If the simple assumption is made that the energy scattered in all directions is the same as the energy scattered in the backward direction, it follows from the definitions of m and m' [equations (4) and (32) of Chapter 12] that the total energy

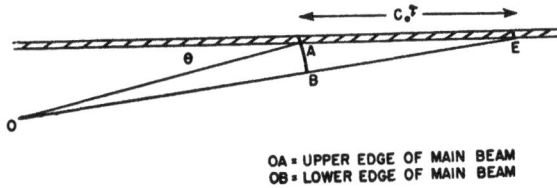

OA = UPPER EDGE OF MAIN BEAM
OB = LOWER EDGE OF MAIN BEAM

FIGURE 27.   Energy reaching surface scatterers.

scattered per unit time is proportional to m'. Thus, the ratio of the total energy scattered per unit time to the total energy reaching the scatterers per unit time is proportional to $m'/\sin\theta$. The former ratio cannot exceed unity; if it is to remain finite at small grazing angles, m' must decrease at least as rapidly as $\sin\theta$. If the scattering from the surface obeys Lambert's law (the law of scattering of light by rough surfaces[8]), then the backward scattering coefficient will be proportional to $\sin^2\theta$. At ranges of 100 yd or more, with transducers at 16 ft, $\sin\theta = \theta$ and is inversely proportional to the range. Thus, comparing with equation (43) of Chapter 12, if the scattering arises in a thin surface layer, and if we can assume that equal amounts of energy are scattered in all directions, the surface reverberation would be expected to fall off as the fourth power of the range, or faster. Supplemented by the added loss due to attenuation, such a variation of scattering coefficient with grazing angle could explain the observed dependence on range in Figures 23 and 24.

However, before we can accept the variation of m' with grazing angle as an explanation of the dependence of surface reverberation on range, we must determine how thin a scattering layer is required for the argument of the previous paragraph to be valid. Figure 28 is a more exact drawing of the situation pictured in Figure 27, drawn so that the layer has appreciable thickness. In Figure 28 the projector O is at depth d, and the scattering layer has thickness h. The scattering volume has a cross section $CADE$ in the plane of the paper, with CD at long range very nearly equal to $c_0\tau$. Energy enters the scattering volume through AE (as in Figure 27) or through AC. From Figure 28 it is easy to obtain a simple criterion

for the validity of the argument of the previous paragraph, if attenuation in the surface layer can be neglected. For, with this approximation, the energies entering through AC and AE are proportional re-

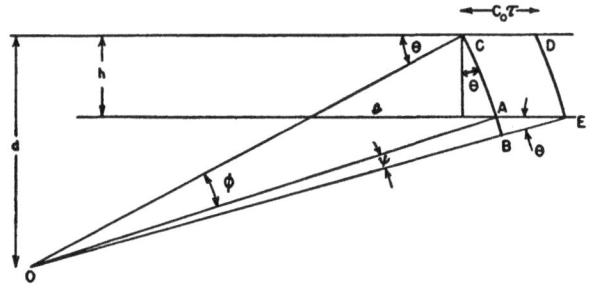

FIGURE 28.   Expanded view of surface layer.

spectively to the solid angles formed by rotating ϕ and ψ in Figure 28 about a vertical axis (in the plane of the paper) through O. At long ranges, θ small, these solid angles are proportional respectively to angles ϕ and ψ. Thus, the calculation in the previous paragraph of the energy entering the scattering volume per unit time is incorrect unless the angle ϕ is very small compared to ψ. At long range we have approximately, with $OC = r$,

$$h = (AC)\cos\theta = r\phi\cos\theta$$
$$AB = r\psi = (AE)\tan\theta = c_0\tau\tan\theta.$$

Thus the condition that ϕ be very small compared to ψ becomes

$$\frac{h}{r\cos\theta} \ll \frac{(c_0\tau)\tan\theta}{r}. \tag{7}$$

For small θ, $\cos\theta = 1$, $\sin\theta = \theta = d/r$. Thus equation (7) becomes

$$h \ll (c_0\tau)\frac{d}{r}. \tag{8}$$

At 1,000 yd, with 100-msec pings and $d = 5$ yd, equation (8) gives $h \ll 30$ in. There are scarcely any data, but it seems likely that the surface layer might frequently be thin enough to satisfy the relation (8). On the other hand, in rough seas it would not be surprising to find that the relation (8) is violated.

Equation (8) was derived neglecting attenuation in the scattering layer; if attenuation is taken into account then it can be shown that the expression (8) must be replaced by

$$1 - e^{-arh/d} \ll (c_0\tau)a, \tag{9}$$

where the attenuation in the layer, in decibels per yard, is $4.34a$. The value of a probably depends primarily on the population of bubbles in the surface layer and is difficult to estimate. If the attenuation is as large as 1 db per yd, a value which is observed in wake measurements (see Chapter 35), equation (9) would be satisfied even with very large values of h. In fact, it is obvious that the relation (9) will always be satisfied if

$$(c_0\tau)a \gg 1. \tag{10}$$

If (8) is not satisfied, it is easy to see that (9) will not be satisfied if (10) is violated.

We may summarize this discussion of the validity of the argument that m' should decrease at least as rapidly as $\sin\theta$ by stating that the argument may be correct, but requires further quantitative information on the thickness of the scattering layer and the attenuation to be expected in the layer. Until this information is forthcoming, the hypothesis that m' decreases with decreasing angle of incidence on the surface is not a wholly acceptable explanation of the variation of surface reverberation with range.

At small grazing angles, the peaks of the water waves are sometimes hidden from the sound source by the troughs. This shadowing effect of waves also causes a reduction of the irradiation of the surface. If the scatterers are largely concentrated in a layer whose depth is small compared with the wave height, a reduction in reverberation might be expected at small grazing angles. Since the grazing angle decreases with increasing range, this effect could account for the rapid decrease of surface reverberation. This hypothesis of the shadowing effects of waves has the added virtue that it explains the increasing rate of decay with increasing sea state, since the larger the waves the more important this effect would be. However, this hypothesis is much too qualitative to be accepted without further study. It can be seen that phenomena in high sea states may actually tend to make the reverberation increase with increasing range rather than decrease. For example, in high sea states, at long range, the sound rays may make large angles of incidence with the wave troughs, thereby increasing considerably the sound returned back to the transducer. A quantitative evaluation of the shadowing effect of waves is difficult and requires a detailed examination of surface roughness. Several papers issued by UCDWR [9-13] are initial attacks on the theory of surface scattering. That the nature of the surface irregularities will affect surface reverberation seems almost intuitively obvious. Another report [14] describes measurements in which definite structure was found in surface reverberation. On this

day there were strong swells with a wind speed of 16 to 19 mph. Distinct blobs were observed in the reverberation, and these blobs altered their range at a rate equal to the rate at which the surface swells were moving. These blobs could be identified much more readily on the chemical recorder than on the oscilloscope record, where the wealth of detail confused the general picture. In Section VII of reference 1, no difference was observed in reverberation measured with the projector beam parallel to and perpendicular to the wave fronts.

A wave in water reflected at the water-air boundary suffers a change in phase of the sound pressure (see Chapter 2). This change of phase results in interference between the direct and surface-reflected rays; the transmission loss between the projector and points near the surface may be increased to a value much greater than the inverse square loss used in deriving equation (43) of Chapter 12. Furthermore, the increase in transmission loss will be a function of the range and of the distance of the scatterer from the surface, and will increase with decreasing depth and increasing range. Thus if the scatterers are located in a thin layer near the surface, this interference between direct and surface-reflected waves may explain the observed rapid decrease of surface reverberation. As with the previous hypothesis of wave shadowing, it is necessary to make a quantitative investigation of this image interference effect before accepting it as an explanation of the range dependence of surface reverberation. For a plane surface, it is shown in Section VI of reference 1 that image interference can lead to a decrease of surface reverberation proportional to the seventh power of the range; consequently, this effect could account for the observed slopes of Figures 23 and 24. However, the surface is not plane. An approximate treatment of the effect of surface roughness in reference 1 shows that the image interference effect becomes less important as the surface roughness is increased. Thus if image interference is causing the range dependence, the slope of surface reverberation should decrease with increasing sea state, which is contrary to what is observed. Another inference from the theory of the image interference effect is that surface reverberation should increase rapidly with frequency at ranges where the interference is important. Unfortunately, there are no experiments on the variation of surface reverberation with frequency.

It may be concluded from this discussion of the range dependence of surface reverberation that the reason for the rapid decrease is not understood, but that there are a number of factors which may play a part. Very likely, all the physical factors which have been discussed previously are included to some degree. In addition, there may well be other causes which have not been considered. It should be noted that at ranges less than 500 yd the measured reverberation for wind speeds less than 8 mph decreases only as the inverse first power of the range in Figure 23. This rate of decay is even slower than the predicted inverse square decay of volume reverberation. No definite explanation has been offered for this feature of Figure 23, but it might be caused by a gradual increase in the value of the volume-scattering coefficient as the deep layer is approached. This effect would, of course, be noticeable only in sea states so low that volume reverberation can be measured at short ranges.

14.2.2 Dependence on Ping Length

The theoretical formulas (22), (39), and (52) for volume, surface, and bottom reverberation in Chapter 12 all have the reverberation intensity proportional to the ping length. The theoretical assumptions required to obtain this result have been discussed in Chapter 12. In this subsection we shall discuss whether or not this strict proportionality may be expected in practice. The only data bearing on this question are reported in reference 1, Section IV, and are summarized later. Unfortunately reference 1 does not state whether the reverberation studied was volume, surface, or bottom reverberation, but reverberation received from ranges as low as 100 yd and as great as 5,000 yd was included in the analysis. However, ranges less than five times the ping length were not included.

Figure 29 shows, qualitatively, that the reverberation intensity increases with increase in the signal length. In that illustration, a record A of reverberation following a 70-msec ping is compared with a record B of reverberation following a 10-msec ping. The attenuator settings were the same for both cases; however, because of the higher level of the 70-msec reverberation, each attenuator step was removed a little later for the 70-msec ping. For this reason, the records are directly comparable only in the intervals 1-1, 2-2, and 3-end in which the amplification is the same for both records. In these intervals, the 70-msec reverberation is clearly much higher than the 10-msec reverberation.

To test quantitatively the predicted relation between reverberation intensity and signal length, sets of data were taken on two successive days of the reverberation following pings of lengths very nearly 10, 20, 40, and 70 msec. Ten pings were measured on each day for each signal length. For each ping length, the average reverberation amplitude was measured at a set of logarithmically equispaced positions, by the band method. The squares of these average amplitudes were assumed proportional to the average reverberation intensities, in accordance with the usual procedure described in Chapter 13.

The agreement between theory and experiment is shown in Figure 30. In that illustration, ten times the logarithm of the ratio of any two ping lengths is taken as the abscissa, and the decibel difference between the corresponding measured reverberation levels is taken as the ordinate.[d] If reverberation intensity were in fact exactly proportional to the ping length, all the observed points should lie on the 45-degree straight line drawn in the figure. In Figure 30 the points for which the longer ping length of a pair was 20, 40, and 70 msec are designated differently so that any systematic departure depending on ping length can be discerned. On the whole, the agreement in Figure 30 between theory and experiment is satisfactory. For some reason, the agreement is better for the ratios involving the shorter signals (10, 20, 40 msec) than for the ratios including the longest signal (70 msec).

The deviations from the straight line in Figure 30 are not too great to be ascribable to experimental error. Thus these data give no reason for doubting the prediction of equations (22), (39), and (52) of Chapter 12, that, under the conditions specified in that chapter, reverberation intensity should be proportional to ping length. However, in view of the importance of knowing the dependence on ping length for comparison of reverberation measurements made with different ping lengths, and for the determination of scattering coefficients, further investigation of this dependence is desirable. The measurements should be repeated for a wider range of ping lengths and for all types of reverberation.

---

[d] The reason that the abscissas of some of the pairs of points in Figure 30 are not the same is that the signal lengths as measured from the film records were not exactly the same on both days. However, the equipment was set on each day for nominal ping lengths of 10, 20, 40, and 70 msec.

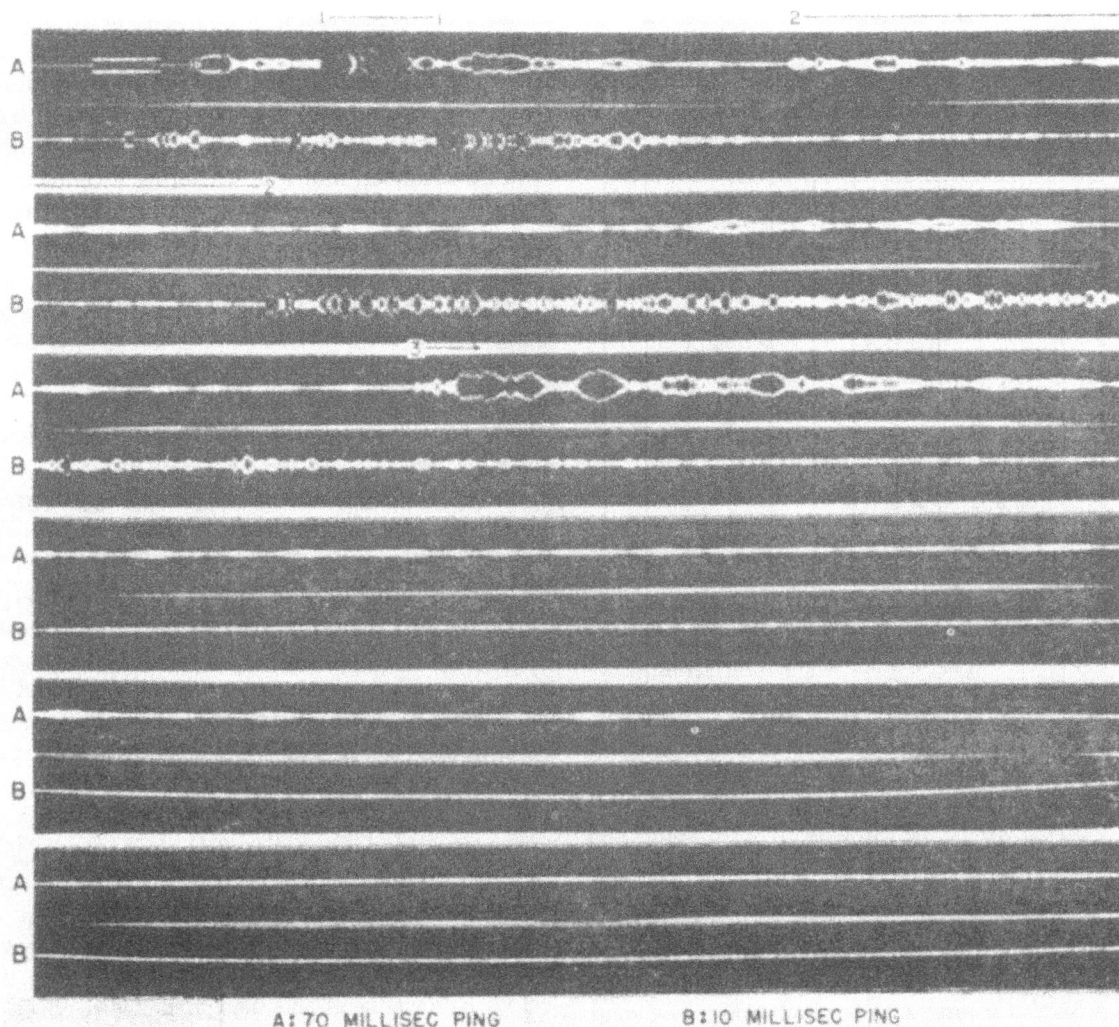

A: 70 MILLISEC PING   B: 10 MILLISEC PING

FIGURE 29. Comparison of reverberation from a 70 MS ping with that from a 10 MS ping.

14.2.3 Unimportance of Multiple Scattering

The theoretical formulas of Chapter 12 are based on the assumption that multiple scattering can be neglected. Experiments designed to measure the amount of multiple scattering are described in a memorandum by UCDWR [15] and summarized below.

The 24-kc, QCH–3 units were mounted 6 ft apart with the long dimension horizontal in such a way that the unit used as a hydrophone could be rotated about a vertical axis. The unit used as a projector was kept in a fixed position. Observations were made with the receiving hydrophone at bearings of 0, 30, 60, and 90 degrees, relative to the bearing of the pro-

jector axis. That is, the receiving hydrophone was rotated so that it faced away from the projector, and the sound received in it was measured. Ping lengths of 15 msec were used, and 5 pings were averaged at each bearing. If the two QCH–3 units had been highly directive, interpretation of the observations would have been straightforward. However, they were not highly directive, so that the portion of the signal projected in the direction in which the receiver was pointing could not be neglected. In order to evaluate the data, therefore, the following procedure was adopted. The expected signal in the receiving hydrophone was calculated from the known directivity patterns of the hydrophone and projector, assuming single scattering was taking place in the ocean. It is easy to see from the derivation of Chapter 12 that this expected signal depends on the integral $\int b(\theta,\phi)$ $b'(\theta,\phi + \alpha)d\Omega$, where b and b' are defined as in Chapter 12, and α is the angle between the projector and hydrophone.

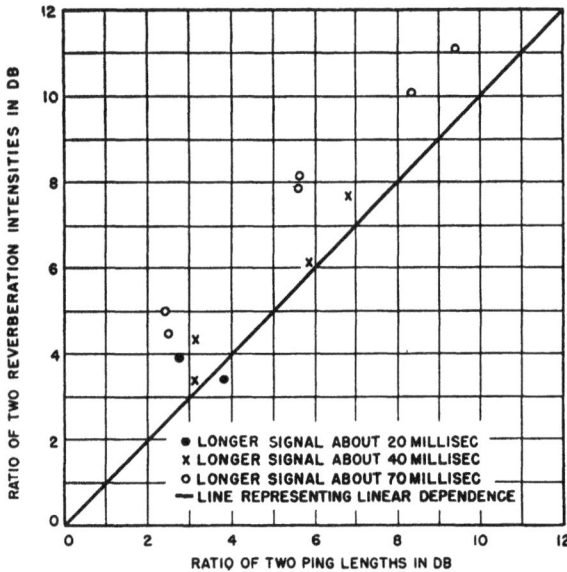

FIGURE 30. Observed dependence of reverberation intensity on ping length.

The values of the above integral were computed for α equal to 0, 30, 60, and 90 degrees, and the predicted reverberation intensities were then compared with the observed intensities. It was found that the calculated levels were within 1 to 2 db of the observed average levels for ranges up to about 200 yd, beyond which no measurements were made. Since the experimental error of the measurements was not less than 1 to 2 db, these results show that at short ranges multiple scattering makes a negligible contribution to the received reverberation. However, these results give no information about the effect of multiple scattering at longer ranges.

It is easy to show that multiple scattering can certainly be neglected if the volume scattering in all directions is the same as in the backward direction. For, in this event, the total energy scattered per second per unit intensity at dV is just mdV (see Section 12.1). The loss in intensity dI in a distance dx of a plane wave of intensity I traveling in the x direction is then

$$dI = mIdx$$

which gives

$$I = I_0 e^{-mx}.$$

Thus under these circumstances the attenuation of a

sound wave by scattering is $4.34 \times 10^3 m$ db per kyd. It is shown later that m is rarely greater than 10^{-5}. By using this value of m, the attenuation due to scattering is 4.34×10^{-2} db per kyd. Now, with any kind of a transducer, but especially with a directional transducer, multiple scattering will not be important in reverberation until the amount of singly scattered energy in the ocean becomes appreciable compared to the amount of energy remaining in the direct sound beam. Obviously, with a scattered energy loss of only 4.34×10^{-2} db per kyd, scattered energy in the ocean is negligible compared to the energy in the direct beam for ranges less than 20,000 yd where the total scattered energy loss is not yet 1 db.

Despite the arguments of the preceding paragraph, more experimental evidence bearing on multiple scattering would be desirable especially since the oblique scattering may be appreciably different from the backward scattering. One way to check the importance of multiple scattering would be to compare with experiment at long ranges the predicted dependence of received reverberation intensity on transducer directivity. If multiple scattering is important, the difference between reverberation levels measured with directional and nondirectional transducers will not be given by J_v [equation (21) of Chapter 12]. No such measurements have been reported; in fact, the whole question of comparing with experiment the dependence of reverberation on the theoretical reverberation indices J_v and J_s seems to have been neglected. Knowledge of this dependence is required for comparison of measurements made with different gear, and also for prediction of the effect on reverberation in echo-ranging gear of changes in gear directivity. There is no reason for doubting the validity of the formulas of Chapter 12 for ordinary gear, but with highly directive gear, multiple scattering and other effects may produce deviations from the theoretical formulas. A UCDWR internal report [16] describes experiments in which the measured vertical directivity patterns in the ocean were very different from the patterns obtained at a calibrating station. Pitch and roll of the echo-ranging vessel will also cause deviations from the predicted reverberation intensities, especially for surface reverberation.[17]

14.2.4   Average Levels

Figure 31 is a summary of the measured 24-kc reverberation levels reported in reference 4. The levels shown are standard reverberation levels, as defined

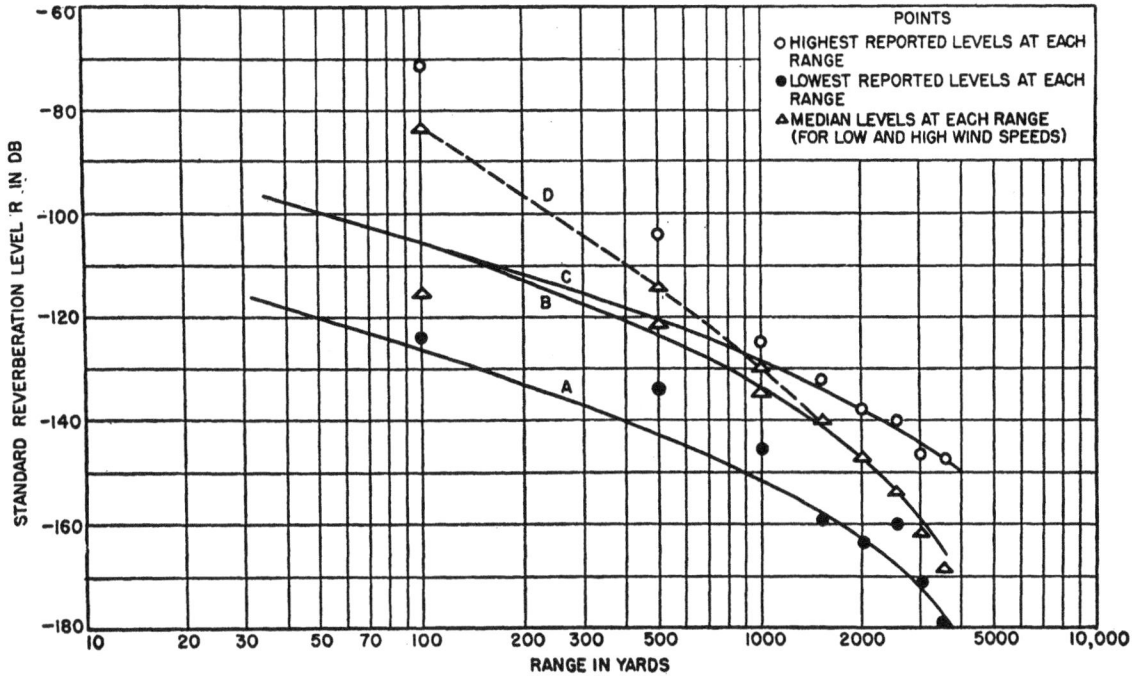

FIGURE 31. Surface and volume reverberation levels.

by equation (25) of Chapter 12. The dots show the lowest reported reverberation levels at each range, and the circles the highest levels. Median values of the data are shown by triangles. At ranges less than 1,500 yd, the lower triangles are the median values for wind speeds less than or equal to 8 mph, and the upper triangles are the median values for wind speeds greater than or equal to 20 mph.

These data are neither an adequate nor a random sample, and detailed analysis of them is not justifiable. However, some interesting inferences, which should be reasonably reliable, can be drawn from the data. The lower solid line in Figure 31 is a plot of equation (26) of Chapter 12 against range with $10 \log m$ set equal to -80 db, A set equal to 3 db per kyd, and A_1 set equal to zero. The upper solid line is drawn with $10 \log m$ set equal to -60 db, with A set equal to 1.5 db per kyd, and A_1 set equal to zero. Both lines were plotted with J_v set equal to -25 db.[18] If all reverberation at ranges greater than or equal to 1,500 yd is volume reverberation, and if the lower levels at shorter ranges are volume reverberation, then the upper and lower solid curves would represent estimated upper and lower limits to 24-kc volume-reverberation levels.

The middle solid curve is drawn with $10 \log m$ set equal to -60 db, and with A set equal to 4 db per kyd. This curve fits the median values of reverbera-

tion for wind speeds less than or equal to 8 mph surprisingly well at ranges from 500 to 3,500 yd; it can probably be assumed that these median values for wind speeds less than or equal to 8 mph represent volume reverberation. The 4 db per kyd value of A is gratifyingly close to the average value measured in transmission studies under good conditions (see Section 5.2.2). These results apparently indicate that a scattering coefficient $10 \log m$ equal to -60 db is a typical value for the volume reverberation from horizontally projected 24-kc sound beams. Occasionally, however, the scattering coefficient becomes very small, $10 \log m$ becoming as low as -80 db.

It does not seem legitimate to attempt any conclusions based on these differences in the values of A required to fit the curves of Figure 31. Both A and m are highly variable quantities, and are known to be a function of depth in the ocean. However, comparison of the upper and median curves at ranges past 1,500 yd suggests that differences in the long-range reverberation levels may be frequently due merely to variations in A.

It has already been pointed out that the deep scattering layers discussed in the first portion of this chapter will tend to increase the reverberation at long range above the levels otherwise expected.[4] Studies of bottom reverberation (see Chapter 15) have shown that the main beam from standard 24-kc gear will usually reach a depth of 1,000 ft at a range of about 2,000 yd. The median curve for volume reverberation in Figure 31 does not show any evidence of an increase in the volume-scattering coefficient at long ranges; it will be recalled that the value of 10 log m in these deep layers frequently exceeded the mean value of 10 log m in the ocean by 15 db or more. However, it appears that this failure to observe the deep layer must have been due to sampling. More recent studies by UCDWR, still unpublished, show that the deep layer is frequently discernible as a very definite bulge on the reverberation curve. These new data also show that the maximum value of 10 log m is −50 db or perhaps even slightly higher rather than the −60 db value indicated by Figure 31.

From Figure 31 and from equation (26) of Chapter 12, we may conclude that the backward volume-scattering coefficient m for horizontally projected 24-kc beams varies between 10^{-6} and 10^{-8} per yd with 10^{-6} the typical value. It will be noted that the dimensions of m in equation (26) of Chapter 12 are per yard; values of m expressed per foot will be one-third or 5 db less. Also, from Chapter 12, we recall that this value of m is about 3 db greater than the "true" value of the backward-scattering coefficient of the ocean. Since equation (45) of Chapter 12 does not describe the range dependence of surface reverberation very well, determination of the surface scattering coefficient m' from comparison of that equation with Figure 31 is not very meaningful. However, if we make the comparison, with A set equal to 4 db per kyd and J_s set equal to −16 db, [18] the median values of 10 log m' at ranges of 100 and 1,000 yd for wind speeds greater than 20 mph are respectively −22 db and −31 db. (Note that m' is a dimensionless quantity.)

It will be recalled that at the beginning of this chapter we assumed that surface reverberation could be eliminated by pointing the transducer downward. Off the main lobe, the response $b(\theta,\phi)$ of standard 24-kc transducers is usually assumed to average about 30 db less than the peak response on the main lobe,[19] but this is only an approximate average value. Using this 30-db estimate, we see from Figure 31 that at ranges of 100 yd, in high sea states, surface reverberation may exceed volume reverberation even with the transducer directed downward, but only if the volume reverberation level is close to the minimum values observed. At ranges greater than about 500 yd, or with wind speeds less than about 15 mph,

pointing the transducer downward should usually eliminate surface reverberation (see Figure 20). This estimate of the wind speeds and ranges at which surface reverberation can be eliminated by pointing the transducer downward assumes, however, that the volume reverberation with the transducer pointed downward is the same as when the transducer is horizontal. Measurements reported in reference 2 suggest that the volume scatterers are anisotropic and, specifically, have a smaller backward-scattering coefficient when the sound arrives from a vertical direction than when the sound arrives from a horizontal direction. However, the observed difference in 10 log m was only about 6 db and thus hardly affects the conclusion stated previously that surface reverberation can almost always be eliminated by pointing the transducers downward.

14.2.5 Scattering Coefficient of a Layer of Bubbles

It is of interest to compare the median values of 10 log m' obtained from Figure 31 with the values expected if the surface consisted of a dense layer of resonant bubbles. The theory of air bubbles in water is given in Chapter 28; the geometry of scattering by

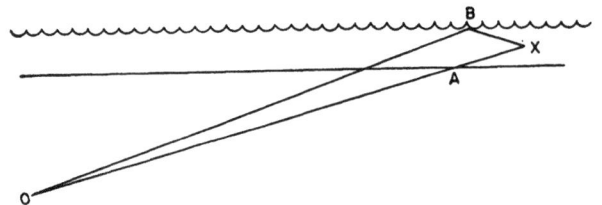

FIGURE 32.   Scattering from surface layer of bubbles.

such a layer is illustrated in Figure 32. By definition, a densely populated layer of bubbles is one in which the attenuation is so high that there is essentially infinite transmission loss through the layer. For this reason, in Figure 32, energy reaches the scatterer at X and returns to the scatterer at O along the direct path OAX only; scattering along a path reflected from the air-water interface, such as OBX, can be neglected since almost no energy reaches B. It follows from bubble theory that multiple scattering can be neglected as well. Neglecting refraction, the expected reverberation intensity can now be calculated directly from equation (29) of Chapter 12, using

$$h = \frac{e^{-ar}}{r^2}e^{-N\sigma_e D} \qquad (11)$$

$$m = N\sigma_s \qquad (12)$$

where N is the number of bubbles per cubic centimeter, σ_e and σ_s are respectively the absorption and scattering cross sections of a resonant bubble (defined in Chapter 28), D is the distance AX, a the usual attenuation coefficient equal to about 4 db per kyd, and r the range.

Evaluating the integral in equation (29) of Chapter 12 with the aid of equations (11) and (12), and, comparing the result with equation (39) of Chapter 12, it is readily found that

$$m' = \frac{\sigma_s}{2\sigma_e} \sin \theta = \frac{\sigma_s d}{2\sigma_e r} \qquad (13)$$

where θ is as usual the angle of elevation of the ray from the projector to the scatterer at range r (Figure 7, Chapter 12), and d the projector depth. For a transducer at 16 ft, equation (15) gives m' equal to -28 db at 100 yd and -38 db at 1,000 yd (using Figure 1 in Chapter 34). These values are about 6 db lower than the median measured values of -22 db and -31 db and are still lower than the highest reported levels at 100 and 1,000 yd (see Figure 31). Since equation (13), derived on the basis of a densely populated surface layer, gives the highest possible value of m' for scattering by bubbles, it seems that measured surface reverberation levels cannot be explained on the hypothesis of scattering by a surface layer of bubbles. It will be noted that, because of the assumed neglect of scattering along OBX (Figure 32), in this situation of scattering by a densely populated surface layer of bubbles the value of m' indicated by equation (13) is the true surface-scattering coefficient. Thus, although the argument presented in Section 12.5.6, that measured values of m' are 6 db greater than the true value of the surface-scattering coefficient, is probably valid, it is evident that the validity of this 6-db relation depends on the physical process which gives rise to the surface reverberation.

Chapter 15

SHALLOW-WATER REVERBERATION

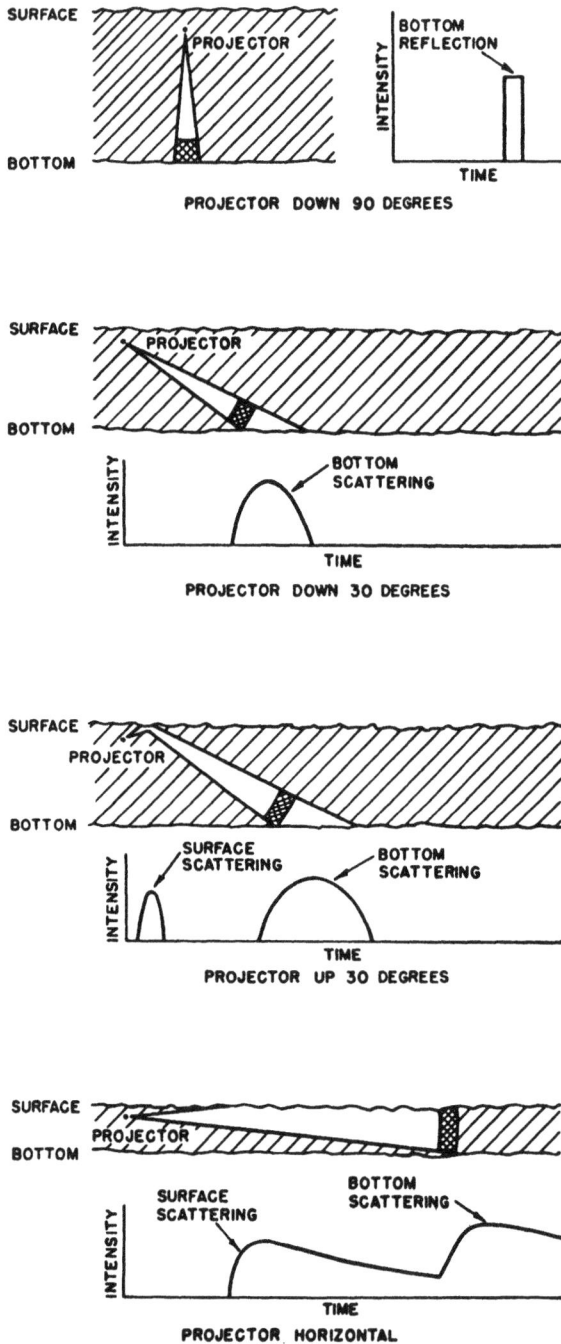

FIGURE 1. Expected behavior of bottom reverberation for idealized projector.

COMPARISON OF THEORY and experiment in bottom reverberation is complicated by the directivity pattern of the transducer and by uncertainty regarding the dependence of the scattering coefficient on the angle of incidence. An additional complication is refraction, which is of considerable importance in determining bottom-reverberation levels. One way in which bottom reverberation differs from the other types of reverberation we have considered is that bottom reverberation is not heard immediately after the initial ping, but appears some time later, usually coming in as a distinct crash. This delay results from the fact that the bottom, unlike the scatterers responsible for surface and volume reverberation, is usually a significant distance from the projector.

15.1 QUALITATIVE DESCRIPTION OF BOTTOM REVERBERATION

Figure 1 illustrates the expected behavior of the bottom reverberation for an idealized projector having constant sound output within 5 degrees of the axis and zero output outside 5 degrees. For illustrative purposes, we may assume there is no refraction. Then for this simple type of sound beam, scattered sound is received, at the time instant t, only from those scatterers included within a sector of a spherical shell centered at the projector and bounded by the limits of the sound beam. The mean radius of this shell is $ct/2$ and its thickness $c\tau/2$, as pointed out in Section 12.2. Bottom reverberation is received whenever this shell cuts off some portion of the bottom.

Bottom reverberation will set in at the time corresponding to the shortest range at which the beam strikes the bottom. Since bottom scattering coefficients are usually relatively large, the total received reverberation will increase sharply at the time of onset of bottom reverberation. Scattering at the bottom will cease, except for sound which is reflected or scattered toward the bottom, at the time the last portion of the beam leaves the bottom.

The case of vertical incidence on the bottom is illustrated in the first box of Figure 1. In this case, as shown in the box, all portions of the beam strike the bottom nearly simultaneously. Thus the reverberation begins and ends very abruptly. The time of onset of the reverberation equals $2d/c$, where d is the depth of the bottom below the projector, and c is the

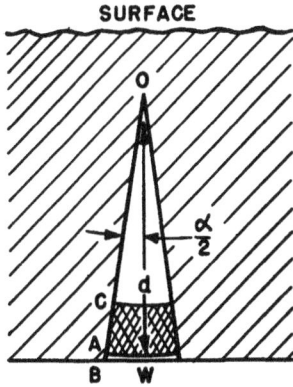

FIGURE 2.   Vertical incidence of ping on bottom.

sound velocity. Evidently, as shown in Figure 2, all portions of the beam do not strike the bottom exactly at the same time so that the duration of the reverberation does depend somewhat on the beam width. For narrow beams, it is easily shown that the reverberation duration is given by

$$\text{Reverberation duration} = \tau + \frac{d\alpha^2}{4c}, \qquad (1)$$

where α is the beam width, shown in Figure 2, and τ the ping duration. Thus, with very short pings incident vertically on the bottom, the duration of the reverberation may appreciably exceed the ping duration.

When the beam is incident on the bottom at some slant angle, all parts of the beam do not strike the bottom at the same time. Consequently the reverberation does not begin or end quite as sharply as in the previous case, and the reverberation duration is greater than the ping duration. This case is illustrated in the second box of Figure 1. In this situation it is easily shown from Figure 3 that the time of onset and duration of the reverberation are given by

$$\text{Time of onset} = \frac{2d}{c}\csc\left(\theta + \frac{\alpha}{2}\right). \qquad (2)$$

$$\text{Reverberation duration} =$$

$$\tau + \frac{2\alpha d}{c}\csc\left(\theta + \frac{\alpha}{2}\right)\cot\left(\theta - \frac{\alpha}{2}\right). \qquad (3)$$

If the beam is pointed up toward the surface, or is horizontal, surface reverberation will be heard before

the bottom reverberation begins to come in. With a horizontal beam, different parts of the beam strike the bottom at widely spaced intervals, so that the reverberation lasts a long time. These statements are illustrated in the third and fourth boxes of Figure 1.

Of course, the sound beam is not actually confined within a cone; some sound is sent in all directions. However, most sound projectors commonly in use confine all but a small fraction of the emitted and received sound within small angles from the beam axis, so that the simple description of Figure 1 should be a good approximation to the observed phenomena. Figure 4 is an experimental illustration of the qualitative predictions of Figure 1. The data making up Figure 4 were obtained in an area where the water depth was 72 ft. It is noticed that the reverberation recorded with the projector directed vertically downward consisted of a single pulse of about the duration of the ping. The reverberation levels recorded with the projector directed 30 degrees down fall on a curve approximating the simple case shown in the second box of Figure 1. There is a rapid rise of reverberation, due to bottom scattering, at the time corresponding approximately to the range at which the main beam first strikes the bottom (easily calculated as 35 yd for a half beam width of 6 degrees) and reaching a peak at about the range where the axis of the beam arrives

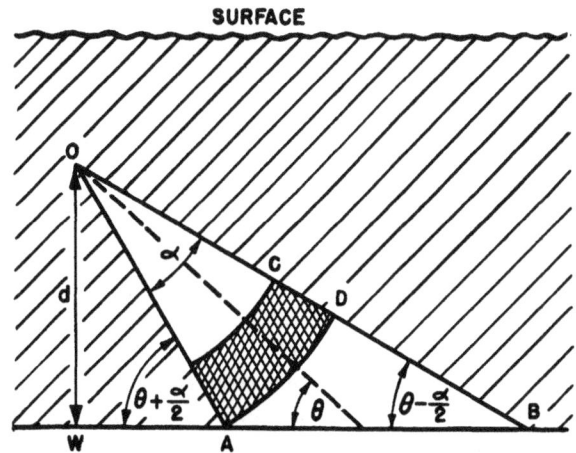

FIGURE 3.   Slanting incidence of ping on bottom.

at the bottom (40 yd). The peak is followed by a rapid decay. The case of the horizontal beam is illustrated in the bottom curve of Figure 4, and is seen to correspond roughly to the bottom curve in Figure 1. The initial reverberation recorded at 40 to 50 yd is received from the bottom on the projector side lobe; this type of reverberation decays rapidly for about

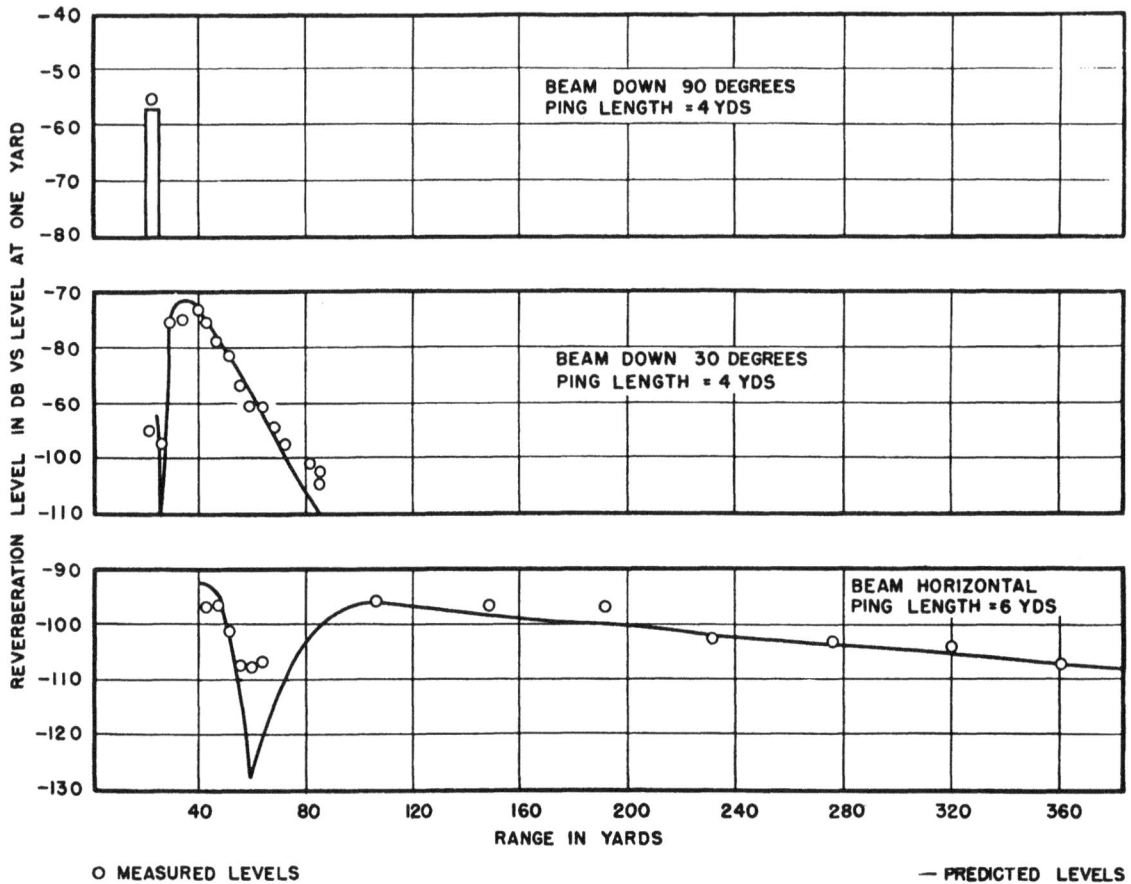

FIGURE 4.   Observed and predicted bottom reverberation levels.

0.01 sec; then there is a rather rapid growth in the reverberation intensity when the main beam reaches the bottom. Further illustrations of the reverberation measured with the beam down 30 degrees are shown in Figure 5. The ocean depth was 48 ft and the projector depth 10 ft so that the peak of the reverberation is observed to come in at a time corresponding to a range of about 25 yd. The ensuing lesser maxima are the result of successive multiple reflections from surface and bottom. These observations were taken over a bottom thickly covered with boulders of the order of one foot in diameter. An interesting feature of these curves is the change in the duration of the main reverberation pulse as the frequency changes. The duration of the pulse decreases progressively with increasing frequency; this effect is due to decrease in beam width. In the case of the projector directed upward at an angle of 30 degrees, it is seen from Figure 5 that the surface reverberation peaks predicted in Figure 1 are lacking. This absence is due to the fact that the reverberation was

measured under conditions which combined very shallow water with a smooth sea surface and a rough sea bottom, so that bottom reverberation masked surface reverberation at practically all ranges. Under other circumstances, with a smooth bottom and a rough sea, or deeper water, the peak of surface reverberation can usually be observed before the crash of bottom reverberation comes in.

These remarks indicate that bottom reverberation, under at least some circumstances, behaves qualitatively as would be expected from a simple geometrical analysis of the time required for the sound to reach the bottom. The next step in the analysis is to attempt to make a quantitative prediction of the expected reverberation levels, and then to compare these theoretical predictions with experiments.

From formula (54) of Chapter 12, it is clear that the received reverberation depends on the transmission loss to and from the bottom, on the transducer directivity, and on the scattering strength of the bottom. It can be assumed that the transducer

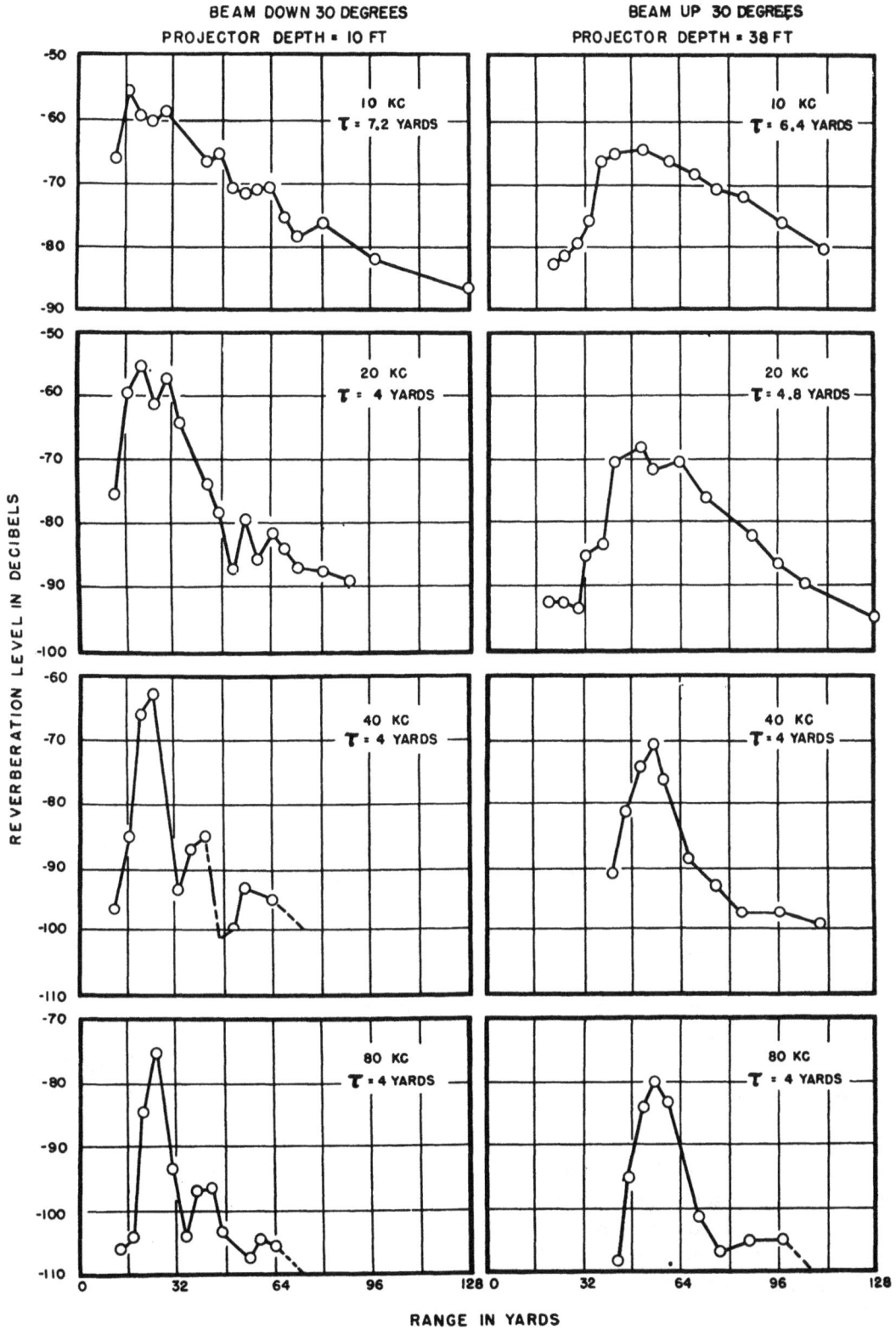

FIGURE 5. Observed reverberation level with beam inclined 30 degrees.

directivity is known in sufficient detail, although in practice it may prove difficult to obtain even this information. The scattering strength of the bottom is in general a function of the angle of incidence of the rays striking the bottom. Temperature gradients in the ocean are frequently sufficiently large that the sound rays in the main transducer beam have suffered appreciable bending by the time they reach the bottom. For this reason bottom reverberation is likely to depend much more strongly on transmission conditions than surface reverberation. To accurately predict bottom reverberation levels, therefore, it is necessary to have detailed knowledge of the ray paths and transmission loss to the bottom.

Unfortunately, there are practically no reported bottom reverberation measurements for which the transmission to the bottom can be regarded as known in detail (including knowledge of the ray paths and the transmission loss along these paths). Consequently, any comparison of predicted reverberation levels with experimental observations must be based on assumptions about the transmission; and detailed agreement in any single experiment between predicted bottom-reverberation levels and observed levels should not be expected. Rather, because of this uncertainty concerning the transmission from the transducer to the bottom, the comparison of theory and experiment becomes even more a purely statistical process than was the case for surface and volume reverberation.

This statistical approach, which is described later, is in some respects justifiable. Transmission studies made to date reveal little likelihood that detailed knowledge of the transmission can be obtained aboard an ordinary echo-ranging warship in any practicable way. What is required is a statement of the average reverberation levels to be expected for various broad classifications of echo-ranging gear, transmission conditions, and bottom types.

15.2 EFFECTS OF REFRACTION

Some of the results of a statistical analysis of bottom reverberation are described in an internal report by UCDWR.[1] In this report, many bottom reverberation curves were plotted against range. The data comprising these curves were all taken at 24 kc with standard Navy gear directed horizontally, but were obtained in a variety of regions, over many different types of bottoms, at differing depths, and with varying refraction conditions. Examination of these data

showed a number of similar features on almost all the curves. In general, the curves showed the following characteristics.

1. A peak which comes in shortly after the outgoing signal and results from surface reverberation.
2. A rapid decay of reverberation with the level reaching a minimum at a range of two times the depth of water.
3. A broad rise in level as the range increases, developing a second peak at a range of about six times the depth of water. This rise is due to bottom reverberation.
4. Beyond the second peak a rapid decrease of intensity, approximately proportional to the inverse fourth power of the range. However, very large variations from this type of decay were observed.

The range of the bottom reverberation peak depends of course on refraction; for this reason, the dependence of the range of this peak on depth becomes a statistical problem. If there were no refraction, in other words, if the sound rays always traveled in straight lines, then the ratio of the range of the peak to the depth (over plane and smooth bottoms) would obviously always be a fixed quantity depending only on the directivity pattern of the transducer. In fact, for standard 24-kc echo-ranging gear, with a beam width of 5 to 6 degrees, the range of the peak would be 10 to 12 times the water depth, if refraction were absent.

In order to judge the usefulness of the statistical study in reference 1, it is necessary to know what kinds of temperature gradients were included, and the extent to which these refraction patterns obtained near San Diego are typical of refraction conditions in other localities. There are reasons for believing that the results of reference 1 may be valid for a wide variety of temperature gradients. Bathythermograph patterns are not completely arbitrary in shape; positive gradients are relatively rare, with the result that most patterns other than isothermal ones show a continuous decrease of temperature between surface and bottom. Furthermore, the effect of refraction is greatest for horizontal or nearly horizontal rays. Once the rays have been bent through an appreciable angle, the amount of additional bending, even by quite sharp negative gradients, is relatively small. For these reasons, the ratio of the range of the peak to the depth may be expected to be relatively constant for a wide variety of gradients excluding isothermal or "nearly isothermal" types, where, for the purpose of this discussion, "nearly isothermal" water

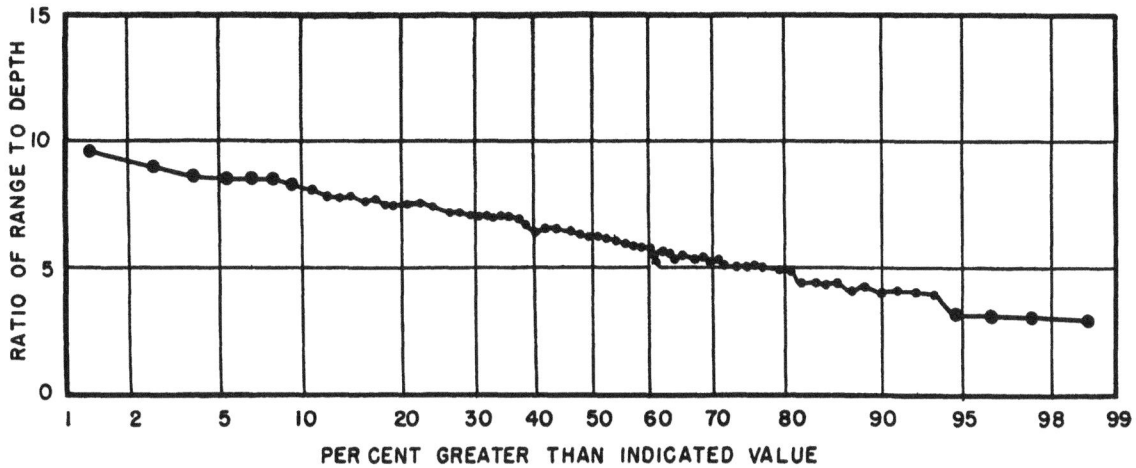

FIGURE 6. Cumulative distribution of observed ratio:

$$\frac{\text{Range to reverberation peak}}{\text{Water depth}}.$$

may be defined as water in which the temperature at the bottom differs from the temperature at the surface by less than five degrees.

Actually, examination of the data in reference 1 shows that relatively few of the reverberation curves analyzed were obtained in isothermal or nearly isothermal water. Thus, the results of that study do not apply to water in which the top-to-bottom temperature change is less than five degrees. This fact helps to account for the disparity between the observed range of the reverberation peak, characteristically about six times the depth, and the predicted value of 10 to 12 times the depth for isothermal water.

After these preliminary remarks, we may examine the San Diego results in more detail. Figure 6 is a cumulative plot, taken from reference 1, of the ratio of the range of the bottom reverberation peak to the water depth. The median point on this curve corresponds to a range-depth ratio of 6.2. Fifty per cent of all peak ranges were found to lie between 5.1 and 7.2 times the depth, and 80 per cent to lie between 4 and 8 times the depth. These results agree well enough with the results of another study by UCDWR.[2] In reference 2, which, however, was based on a smaller number of reverberation curves, the average range to the peak was about 5 times the depth. The difference between these two estimates of the ratio of the range of the peak to the water depth is probably due to sampling and to the fact that the reverberation curves plot only a few isolated points of the measured film. The data discussed in reference 2 are described in somewhat more detail in Section 15.3.1;

they were obtained by using a transducer whose beam pattern was similar to that of standard Navy echo-ranging gear. Bathythermograph data and ray diagrams were available, and it was found that the range to the reverberation peak corresponded to about the range where the 6-degree ray reached the bottom. In another internal report by UCDWR,[3] it was found that the range of the peak usually corresponded to the range at which the 5-degree ray reached the bottom.

For standard Navy gear at 24 kc, the half beam width y defined in Figure 4 of Chapter 12 is close to 6 degrees. Thus, the results described in the preceding paragraph suggest that with standard gear at 24 kc the range where the beam's edge (5- or 6-degree ray) strikes the bottom is the range of the reverberation peak. These results suggest furthermore that in water which is not "nearly isothermal" the range of the reverberation peak is between four and eight times the depth. For simple temperature gradients, it is easy to estimate the range at which various rays will strike the bottom, as a function of the depth to the bottom.[4] Table 1 gives the results of such calculations, for various initial ray angles, water depths, and surface-to-bottom temperature differences; in computing this table it was assumed that the temperature decreased linearly from the surface to the bottom. It is clear from the table that with linear gradients the 6-degree ray always does strike the bottom at a range between 4 and 8 times the water depth, when the temperature difference between the projector and bottom is greater than 5 degrees.

TABLE 1.   Ranges at which rays leaving the projector at different angles strike the bottom.

Temperature difference between projector and bottom in degrees	Depth between projector and bottom in feet	Range at which ray strikes bottom in yards				Ratio of range at which 6-degree ray strikes bottom to water depth
		Angle of ray leaving projector in degrees				
		4	6	8	12	
5	50	171	132	106	74	7.9
	100	345	264	211	148	7.9
	200	696	534	425	299	8.0
	300	1070	809	640	449	8.1
10	50	141	115	96	70	6.9
	100	282	230	194	141	6.9
	200	571	464	385	282	7.0
	300	863	700	580	425	7.0
15	50	123	103	88	67	6.2
	100	244	207	176	133	6.2
	200	491	415	356	268	6.2
	300	743	628	534	402	6.3
20	50	109	94	82	63	5.6
	100	219	188	162	127	5.6
	200	440	373	329	253	5.7
	300	664	568	496	383	5.7

15.3 BOTTOM SCATTERING COEFFICIENTS

Having established the probable limits of the range to the reverberation peak, it is desirable to estimate the height of the reverberation peak. This height has been found to depend markedly on the type of bottom. In general, reverberation is highest over ROCK, less over SAND-AND-MUD or MUD, and least over SAND, although in some cases reverberation over SAND has been reported to be quite high, especially after a storm when rippling of the bottom may be the cause.[1] The relative values of the reverberation over these bottoms may be expressed in terms of the bottom-scattering coefficient m'', by using equation (54) of Chapter 12 if the transducer directivity and the transmission to the bottom are known in detail. Detailed information concerning directivity and transmission has not usually been available, but it has proved possible to determine average values for the bottom scattering coefficients by making reasonable assumptions about the transmission.

In this section, we shall summarize present information concerning the variation of the bottom scattering coefficient m''. The three factors which are expected to be most important in determining the value of m'' are (1) the grazing angle of the sound incident on the bottom, (2) the frequency of the incident sound, and (3) the nature of the bottom. These factors will be considered in the same order.

15.3.1 Dependence on Grazing Angle

Knowledge of the nature of the dependence of m'' on grazing angle is necessary for the detailed prediction of bottom reverberation as a function of range; in addition, the nature of this dependence is of theoretical interest. It has been shown in Chapter 14 that, with simple assumptions about the law of scattering, for a very thin scattering layer the value of the backward scattering coefficient should decrease with decreasing grazing angle at least as rapidly as $\sin \theta$; and for scattering obeying Lambert's law the backward scattering coefficient should be proportional to $\sin^2 \theta$. Thus, determination of the dependence of m'' on angle can furnish information about the law of scattering at the bottom and can also be used to check the validity of our ideas about bottom reverberation.

However, determination of this dependence is not easy. It would appear at first thought that the dependence would be an easy by-product of the analysis of ordinary reverberation runs with horizontal transducers if temperature-depth data were also available.

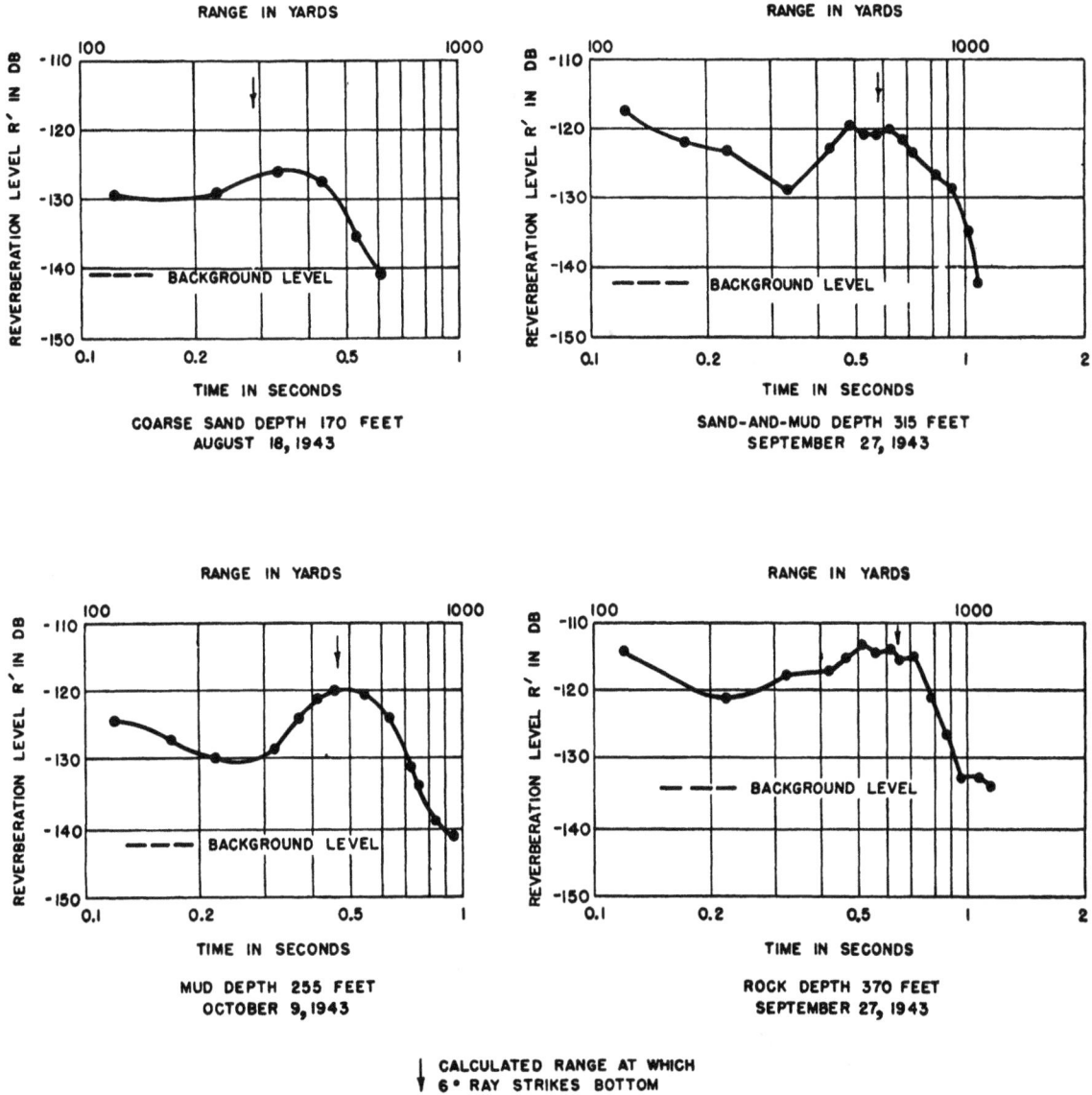

FIGURE 7. Typical bottom reverberation levels with horizontal transducer.

The reverberation level at any range could be translated into the bottom scattering coefficient by use of equation (54) of Chapter 12, and the grazing angle of the sound at this range could be computed from the temperature-depth information. However, it will be seen in the following subsection that this procedure is not workable, briefly because the value of m'' computed in this way is accurate only for the portion of the bottom struck by the central portion of the projected beam, and within this limited area there is not much variation in the grazing angle. Thus, in order to determine the form of the function $m''(\theta)$, specifically designed experiments are necessary.

An Experiment Designed to Measure $m''(\theta)$

Data casting light on the angular dependence of m'' were obtained in a series of experiments, made in August, September, and October of 1943, and described in an internal report by UCDWR.[2] On each day that measurements were taken, the transducer was set either at 0 degrees (main transducer beam horizontal) or at 30 degrees (main transducer beam pointed 30 degrees down from the horizontal), and lowered to a depth of 9 ft. Bottom reverberation for a number of 10-msec 24-kc pings was then recorded as a function of time, by using the equipment described as D in Section 13.1.1. By comparing measurements made with the two different transducer orientations in the same or similar areas, it should be possible to obtain some information about the angular dependence of m''. The following paragraphs describe briefly the analysis of these data made in reference 2; it is convenient to begin by considering the 0-degree data.

For the purpose of analyzing the 0-degree data, the reverberation records obtained were segregated into nineteen groups, each group comprising at least nine records of bottom reverberation taken at nearly the same time on a single day over one of six bottom areas. The records were then measured and averaged over the group to give the mean reverberation amplitude, and the average reverberation level was plotted against range for each group. Typical curves obtained are shown in Figure 7. These curves extend only to the range at which the reverberation becomes comparable to the recording background. On each curve in Figure 7 is shown the range at which the 6-degree ray struck the bottom, as computed from the measured BT pattern. The high levels of the first plotted points at 100-yd range in the curves of Figure 7 are due to surface reverberation.

These reverberation curves for horizontal transducers were then compared with equation (54) of Chapter 12, by using 4 db per kyd for the absorption; in addition, the anomaly due to refraction was computed from the ray diagram drawn from the BT pattern, according to the methods described in Chapter 3. The total correction for the anomaly was found to be small for ranges corresponding to the reverberation peak. The average magnitude of twice the anomaly correction for those ranges was only 2.5 db; but the uncertainty in the transmission anomaly led to uncertain values of m'' corresponding to the reverberation at longer ranges. Nevertheless, by using the data and comparing with equation (54) of Chapter 12, it was possible in this way to obtain for each group of records a curve for m'', the bottom scattering coefficient, as a function of the range of the reverberation. At each range the incident grazing angle of the ray reaching the bottom was computed from the refraction diagram. However, from these curves of m'' against range, as explained in more detail later, the value of m'' was accurately determinable only for grazing angles on the bottom very nearly equal to the grazing angle of the central ray of the main beam. This angle, for all the horizontal projector curves studied in reference 2, lay between 9 and 13 degrees. It is clear therefore that determina-

tion of the angular dependence of m'' was not possible from the 0-degree data alone.

By using the 30-degree data, however, the value of m'' when the grazing angle on the bottom is equal to 30 degrees may readily be determined. With this transducer orientation, the rays in the main beam are only slightly bent by the temperature gradients. Thus they strike the bottom at angles that can be calculated directly from the geometry. Furthermore, since the rays are only slightly bent, the transmission anomaly due to refraction can be ignored. By using equation (54) of Chapter 12, then, and by assuming A equal to 4 db per kyd, the values of m'' at a grazing angle of 30 degrees were determined from comparison with the observed data. This comparison was made on the assumption that the maximum amplitude of bottom reverberation on the 30-degree records corresponded to scattered sound returning along the central ray of the main beam; this assumption is justified from the qualitative discussion in Section 15.1.

The average scattering coefficients determined from these analyses of the data in reference 2 are shown in Table 2. The 10 log m'' values in Table 2

TABLE 2.   Average values of m'' for various bottom areas.

Bottom type	10 log m'' (beam horizontal)	10 log m'' (beam down 30 degrees)	k
COARSE SAND	−33	−24	2.0
FINE SAND	−32	−26	1.5
SAND-MUD	−25	−21	1.0
MUD	−29	−20	2.0
ROCK	−18	−9	2.0
FORAMINIFERAL SAND	−26	. . .	. . .

are of course averages of the values obtained from the nineteen groups into which the original data were subdivided. That is, each group gave a value of 10 log m'' for a definite bottom type, and the entries in Table 2 are each averages over all the groups pertaining to one particular bottom type.

In interpreting these average scattering coefficients, it should be remembered that the 0-degree data were obtained with a horizontal transducer near the surface so that an important part of the received bottom reverberation reached the transducer along paths reflected from the ocean surface. As a result, the values of 10 log m'' inferred from comparison of the 0-degree data with equation (54) of Chapter 12 are 6 db greater than the true value of the bottom scattering coefficient. The values of 10 log m'' shown in the first column of Table 2 are true values, that is, they are 6 db smaller than the values found from comparison of equation (54) of Chapter 12 with the observed reverberation levels. On the other hand, no such 6-db correction for surface reflection was necessary for 30-degree data, since at the latter transducer orientation almost none of the projected sound strikes the surface before reaching the bottom.[a]

Having described the procedure for calculating the 0- and 30-degree columns in Table 2, we now proceed to use these entries for an estimate of the angular dependence of m''. It will be recalled that for all the 0-degree entries the grazing angle of the sound on the bottom lay between 9 and 13 degrees; for present purposes the grazing angle for all the 0-degree entries may be taken as 10 degrees. The grazing angle for all the 30-degree entries may be taken as 30 degrees, since at such a great angle of depression the effect of refraction is negligible. Thus for each of the bottom types considered, we have in Table 2 an m'' for a 10-degree grazing angle and an m'' for a 30-degree grazing angle. By assuming a relationship of the form

$$m'' \sim \sin^k (\theta),$$

it is possible to calculate k for each bottom type. The resulting values of k, rounded off to the nearest half-digit, are displayed in the last column of Table 2.

These values of k are not too reliable, since in order to calculate the individual scattering coefficients in Table 2 a number of assumptions were required about such questions as the proper method for averaging data obtained on different days, and the proper comparison between the point and band methods of averaging when the reverberation levels are changing rapidly. These assumptions, described in detail in reference 2, mean that the results of

Table 2 may be somewhat in error. Nevertheless, Table 2 does indicate that the value of m'' increases at least as rapidly as the first power of sin θ for grazing angles between 10 and 30 degrees.

The data of reference 2 give no information on the nature of $m''(\theta)$ for grazing angles θ less than 10 degrees. It was assumed in reference 1, from which Figure 4 was taken, that m'' is proportional to $\sin^2 \theta$ for angles θ greater than 9 degrees, and was constant independent of θ for angles θ less than 9 degrees. The solid lines in Figure 4 are the reverberation levels predicted on this basis and fit the observed points very well, even at the extreme range of 360 yd on the lower curve, where the grazing angle is only 4 degrees. The very good fit evidenced in Figure 4 seems to indicate that m'' is constant independent of grazing angle at angles less than 10 degrees. However, there is almost no other information on the dependence at angles less than 10 degrees; and a constancy of scattering coefficient as the grazing angle decreased below 10 degrees would make the law of scattering a very complicated function of angle at these small angles. For these reasons it is probably best to regard the dependence of m'' on grazing angle for angles less than 10 degrees as still uncertain. More measurements of this dependence are needed; to obtain values of m'' at small grazing angles, it will be necessary to make measurements in isothermal water.

IMPOSSIBILITY OF DETERMINING $m''(\theta)$ WITH HORIZONTAL BEAMS

We shall now discuss why it was not possible, from the 0-degree data alone, to determine the dependence of m'' on grazing angle on the bottom. Two factors are involved: (1) uncertainty in the beam pattern correction, and (2) the lack of any large variation in the grazing angle, owing to the effect of refraction.

For horizontal transducers, the beam pattern correction as determined from equation (41) of Chapter 12 is small for values of θ less than 6 degrees (see Table 1, Chapter 12). At larger angles the correction increases rapidly because of the sharp decline in the measured beam pattern at the edge of the main lobe. At an angle of 10 degrees, for example, the correction is about 20 db. The application of this large correction appeared to seriously overcorrect the data analyzed in reference 2, giving very large values of m'' at close ranges. It is not difficult to find reasons for this inability to calculate m'' correctly at angles well off the main lobe. In the first place, the very use of equation (41), Chapter 12, for the beam pattern

---

[a] It will be recalled that a 6-db correction was argued in Section 12.5 for surface scattering coefficients. It may be thought that a similar correction should be applied to bottom scattering coefficients, to take account of possible reflections in the layer of scattering material at the bottom. This correction arises, in the case of surface scattering, because the scatterers are thought to extend an appreciable distance into the water side of the air-water boundary; sound can penetrate the scattering layer and strike the air-water boundary at which most of the actual reflection takes place. For bottom scattering, on the other hand, although the bottom scattering layer is not infinitely thin, most of it does lie on the solid side of the twilight region separating the sea volume from the earth's crust. Thus, there is no need to introduce a correction to bottom scattering coefficients due to reflection at the bottom; in fact such a correction, if introduced, would have no physical significance.

correction is questionable at large angles, as has been pointed out previously. In addition, the ship pitches and rolls; at large angles even a small change in orientation of the projector may make a large difference in the received reverberation. There is the further complication that measured beam patterns in the vertical plane have not always agreed with the measured patterns in calibrating stations.[5] These arguments, taken with the overcorrections noted in reference 2, suggest that a correction of 10 db or more is about the maximum which can be safely applied to measured bottom-reverberation levels, if accurate values of the bottom-scattering coefficients are to be expected. We can conclude that the use of equation (41) of Chapter 12 to obtain m'' for rays leaving the projector at large angles (greater than 6 degrees) is quite questionable.

Also, little information about the angular dependence of m'' could be obtained from rays leaving the projector at angles within the main beam, that is, with initial angles less than 6 degrees. For, in these experiments the incident angle at the bottom of the rays within the 6-degree cone was essentially constant; at best, this grazing angle varied only slowly with range. This fact alone would make fruitless any attempt to determine, from the data of reference 2, a detailed degree-by-degree dependence of m'' on grazing angle. Furthermore, the value of the scattering coefficient itself, at ranges beyond the region where the main beam strikes the bottom, becomes more and more doubtful as the range increases, because of uncertainty in the value of the transmission anomaly. These remarks explain why the data of reference 2 were capable of giving m'' accurately only for the angle of incidence on the bottom corresponding to the peak of the reverberation.

It is worth noting that the virtual constancy of the angle of incidence on the bottom, for rays within the 6-degree cone, should be a rather general result with all types of refraction patterns. This conclusion is deduced from Snell's law of refraction, as follows.

Snell's law, which was proved in Chapter 2, tells us that

$$\cos \theta = \frac{c}{c_0} \cos \theta_0 , \qquad (4)$$

where θ is the bottom grazing angle, θ_0 is the angle of the ray at the projector, c is the velocity of sound at the bottom, and c_0 is the velocity of sound at the projector. It is clear from equation (4) that the bottom grazing angle will be smallest for the ray which

leaves the projector at 0 degrees. Thus, the derivative $d\theta/d\theta_0$ equals zero at $\theta_0 = 0$; and θ necessarily varies but little for all the rays leaving the projector within a few degrees of the projector axis.

15.3.2 Dependence on Frequency

A report by UCDWR[6] presents measurements designed to determine the dependence of the bottom-scattering coefficient on frequency. These measurements were made at 10, 20, 40, and 80 kc, with the transducers directed downward at an angle of 30 degrees with the horizontal. The measurements were made in two shallow water areas near San Diego, both with rocky bottoms. The ping lengths used were between 4 and 8 msec. Further details concerning the bottom character and the experimental procedures are given in reference 6. From comparison of the measured reverberation levels with equation (54) of Chapter 12, values of $10 \log m''$ were determined at each frequency and at each of the two positions where measurements were made. These values of $10 \log m''$ were obtained assuming the transmission anomaly A in equation (54) as zero; because measurements were performed in very shallow water, this assumption should introduce very little error. The results obtained in reference 6 are tabulated in Table 3.

An irregular variation of $10 \log m''$ with frequency is noted in Table 3, but according to reference 6 this

TABLE 3. Backward scattering coefficients ($10 \log m''$) as a function of frequency at 30-degree grazing angle.

Frequency in kc	10	20	40	80
Area I	−11	−6	−8	−14
Area II	−22	−17	−21	−15

variation is less than the estimated error of calibration. Also, according to reference 6, change in transducer patterns due to changes in frequency and swinging of the ship at anchor could have introduced errors compared with which the observed variation is not significant. Thus there is no evidence that the bottom scattering coefficient for rocky bottoms at a grazing angle of 30 degrees has any systematic frequency dependence for the frequency range 10 to 80 kc. The mean value of $10 \log m''$, averaged for 10, 20, 40, and 80 kc is -10 ± 3 db for position I and -19 ± 3 db for position II.

The mean values of $10 \log m''$ at a grazing angle of 30 degrees, quoted in the preceding paragraph, should be directly comparable with the 30-degree value of m'' for ROCK in Table 2. It is seen that there is very good agreement with Table 2 for Area I, but there is a difference of 10 db between the value of m'' (30 degrees) obtained at Area II and the value of m'' (30 degrees) in Table 2. This difference could be due to sampling error; it is estimated later that the quartile deviation of m'' for areas of similar bottom classification is $\pm$ 5 db. In this regard it is significant that reference 6 states that the bottom of Area II had patches of SAND–AND–MUD. Thus, it is not too surprising that the mean bottom scattering coefficient in Area II should be lower than in Area I, where, according to reference 6, the bottom was covered with boulders.

The results of reference 6 give no information on the dependence of m'' on frequency for other types of bottom than ROCK. There is no reason to expect that any marked dependence on frequency would be discovered. However, it is necessary to definitely know the frequency dependence, if any, in order to predict the effect on bottom reverberation of varying the frequency of echo-ranging gear. Also, knowledge of this frequency dependence would enable us to assess accurately the present information on the dependence of m'' on bottom type, much of which is based on the assumption that m'' does not depend on frequency. For these reasons it would be desirable to obtain additional measurements over all types of bottom of the dependence of m'' on frequency.

15.3.3 Dependence on Bottom

An analysis of bottom-scattering data obtained with horizontal beams is given in reference 3. These data include many more records than are analyzed in reference 2, among which are data at 10, 20, and 24 kc. A portion of the data was analyzed in a manner similar to that used in reference 2, except that absorption was not included in the transmission anomaly; the transmission anomaly A in equation (54) was computed from the refraction pattern alone. This analysis gave the results listed in Table 4 for a grazing angle at the bottom of 10 degrees. In Table 4, as in Table 2, the values of m'' have been corrected to the true values. Table 4 should, of course, be comparable with the 0-degree column of Table 2, since the grazing angle on the bottom (10 degrees) is the same for both tables.

In reference 3, in addition to this analysis, a more complicated analysis was also attempted to determine the variation of the bottom-scattering coeffi-

cient with angle of incidence. Some evidence that m'' increased with increasing grazing angle was found, but it was not possible to decide which of the three laws, m'' constant, m'' proportional to $\sin \theta$, or m'' proportional to $\sin^2 \theta$, was most nearly representative of the bottom scattering. In view of the inconclusive nature of the results, and also because this analysis rested on some questionable assumptions, these results for the angular dependence of m'' were not included in Section 15.3.1.

TABLE 4. Average values of $10 \log m''$ for various bottom areas for grazing angle at bottom of 10 degrees.

Bottom type	10 kc	20 kc	24 kc
COBBLES	−16	−16	...
ROCK	−23	−23	...
MUD	−32	−38	...
MUD	−31	−34	...
MUD	−34	−36	...
MUD	...	...	−36
MUD	...	...	−37

The value of m'' for the ray leaving the projector at an angle of 5 degrees was relatively independent of the assumptions made. The values of m'' for this ray are given in Table 5. Table 5 includes, for some of the records studied, the grazing angle of the 5-degree ray as calculated from the measured BT pattern. It is seen that in general the 5-degree ray strikes the bottom at a grazing angle of about 10 degrees; thus Table 5 should be comparable with Tables 2 and 4.

From Tables 2, 4, and 5 we can now determine, for each bottom type, the average value of m'' for a grazing angle on the bottom of 10 degrees. To do this, we recall that Tables 4 and 5 were obtained on the assumption that the transmission anomaly was due to refraction alone, that is, that the absorption loss was negligible. However, the values of m'' in Tables 4 and 5 can be corrected for absorption in the following way. It can be assumed, from previous discussions in this chapter, that on the average the ranges at which the data of Tables 4 and 5 were evaluated were six times the depth of the projector above the bottom. These depths are given for the measurements listed in Table 5; for the items in Table 4 they can be obtained from Table 2 of reference 6. Thus, the average absorption loss can be calculated at each frequency for each entry in Tables 4 and 5, by assuming median values of the attenuation coefficient at each frequency (see Figure 17 of Chapter 5). If we increase m'' by the average two-way

TABLE 5. Average values of m'' for various bottom areas for ray leaving projector at an angle of 5 degrees.

Bottom type	Depth of bottom below projector in yards	Calculated grazing angle for 5-degree ray in degrees	10 log m'' for 5-degree ray		
			Frequency in kc		
			10	20	24
COBBLES	38	5.0	−23	−19	...
BOULDERS	10	*	−24	−21	...
ROCK	20	*	−29	−27	...
ROCK	110	8.5	−27	−28	...
MUD	240	*	−36	−35	...
MUD AND SAND	52	10.2	−32	−35	...
FINE SAND	10	*	−33	−36	...
MUD	275	10.5	−31	−37	...
MUD AND SAND	100	*	−38	−38	...
MUD	195	10.0	−35	−39	...
ROCK	15	*	...	...	−26
MUD AND SAND	53	*	...	...	−31
MUD AND SAND	75	*	...	...	−25
MUD	240	11.8	...	...	−25
MEDIUM SAND	215	*	...	...	−31
MUD	230	11.8	...	...	−25

* Angle not calculated.

absorption loss in decibels, the entries in Tables 4 and 5 will be more or less corrected for absorption losses, and the resultant values of m'' will be the best estimates which can be made from the data of reference 6.

Table 6 shows the mean values of m'' determined in this way from the data of Tables 4 and 5. The assumed attenuation coefficients at 10 kc, 20 kc, and 24 kc in db per kyd were 1.3, 3.2, and 4.0 respectively. In Table 6 the mean values for each bottom type were

TABLE 6. Mean values of backward scattering coefficient at 10-degree grazing angle.

Bottom type	10 log m'' from Table 4	10 log m'' from Table 5	10 log m'' from Table 2
ROCK	−17	−24	−18
MUD	−27	−25	−29
SAND–AND–MUD	...	−31	−25
SAND	...	−34	−30

determined by averaging the corrected values of m'' for all three frequencies 10, 20, and 24 kc, giving each entry in Tables 4 and 5 equal weight. The justification for averaging m'' for different frequencies has been discussed in Section 15.3.2.

If data for 10 and 20 kc are not averaged with 24-kc values, a large part of the data of reference 3 has to

be omitted. Another reason for including the 10- and 20-kc data is the following. Examination of the corrected values of Tables 4 and 5 shows that the assumptions which were made concerning the range to the reverberation peak and the value of the attenuation coefficient seem to overcorrect m'' at 24 kc; that is, the corrected values of m'' at 24 kc frequently tend to be abnormally large, especially in deep water where the range to the peak is long. For example, in the last two MUD entries in Table 5, the range of the peak was estimated to be about 1,400 yd, so that the two-way transmission anomaly correction was 11 db. This made the corrected values of 10 log m'' for those entries equal to −14 db, which is much too high for a MUD bottom. The reason that this overcorrection occurred is not clear. It might indicate that the attenuation coefficient for the sound returned as reverberation is less than the attenuation coefficient determined in transmission runs. However, the data of reference 2, which appear quite reasonable, are based on an assumed attenuation coefficient of 4 db per kyd; in the analysis of volume and surface reverberation in Chapter 14, median reverberation levels were fitted quite well by assuming 4 db per kyd as the value of the 24-kc attenuation coefficient. Whatever the reason, the existence of this apparent overcorrection makes it desirable to include values of m'' for 10 and 20 kc in the averages based on the data of reference 3, since at the lower frequencies the corrections for attenuation are not as large.

In Table 6, cobbles and boulders have been grouped under rock; and fine sand, foraminiferal sand, and medium sand have all been grouped under sand. The last column in Table 6 gives the results of averaging a similar grouping of the 0-degree values of Table 2, including coarse sand under sand; the designations sand-mud and mud-and-sand of references 2 and 3 have been replaced by the customary SAND–AND–MUD. If all the entries of Table 6 are averaged with equal weight, we obtain the overall averages in Table 7.

TABLE 7. Overall mean values of backward scattering coefficient at 10-degree grazing angle.

Bottom type	$10 \log m''$
ROCK	-20 ± 5
MUD	-27 ± 5
SAND–AND–MUD	-28 ± 5
SAND	-32 ± 5

The values of Table 7 do not differ significantly from other estimates of the mean bottom scattering coefficients, also based on the data of references 2 and 3. In reference 7 it is estimated that the quartile deviations of the mean bottom scattering coefficients are about 5 db for each bottom type. In view of the crudeness of a classification system which includes all bottom types in only four categories, a quartile deviation of this magnitude is not surprising. Thus this estimate of the deviation from the mean has been included in Table 7. It must be remembered that the values of $10 \log m''$ in Table 7 are true values; that is, they were determined by subtracting 6 db from the values of m'' inferred from comparison of equation (54) of Chapter 12 with the measured reverberation levels. The expected reverberation levels with horizontal beams will therefore be 6 db greater than the levels that would be predicted by the use of equation (54) and the values of $10 \log m''$ in Table 7.

15.4 AVERAGE BOTTOM REVERBERATION INTENSITIES WITH HORIZONTAL TRANSDUCERS

Bottom reverberation levels are a function of range and water depth and in addition depend on refraction conditions, transducer orientation, and bottom type.

For most practical echo-ranging purposes, however, the transducer is oriented so that the transducer axis is horizontal, parallel to the ocean surface. Under these circumstances, over level bottoms, the data which have been presented in this chapter can be used to make some prediction of average bottom reverberation levels. The results of reference 2, discussed in Section 15.3.1, show that under most conditions the transmission anomaly due to refraction is negligible at ranges up to and including the range of the reverberation peak. The results of references 1, 2, and 3, described in Section 15.2, all show that in water other than isothermal or nearly isothermal the range of the reverberation peak tends to be about 6 times the depth between the projector and the bottom, and that this peak corresponds approximately to the range at which the 5- to 6-degree ray from the projector strikes the bottom. The data of references 2 and 3, which were discussed in Sections 15.3.1 and 15.3.3, show that the angle at which this ray strikes the bottom usually is about 10 degrees in nonisothermal water. Thus, knowledge of the average value of m'' at a grazing angle of 10 degrees, coupled with equation (54) of Chapter 12, enables prediction of the average height and range of the reverberation peak over different bottom types in any water depth. It is of course necessary to know the value of A in equation (54). However a value of A equal to 4 db per kyd is probably a good approximation, and for the short ranges at which the reverberation peaks are usually observed, deviations in practice from 4 db per kyd should not be very significant, except possibly when the water is quite deep.

Once the height and range of the bottom reverberation peak are determined, the most significant quantity for echo ranging is the rate at which the reverberation decays as a function of range past the peak. At ranges less than the peak the reverberation usually decreases with decreasing range; such examples as Figure 7 and other similar figures in references 2 and 3 show that the bottom reverberation can hardly increase with decreasing range as rapidly as the expected echo level.[8] At some ranges less than the principal reverberation peak (where the main beam strikes the bottom) reverberation from side lobes can be very high. However, it seems on the whole that bottom reverberation is likely to be most troublesome at ranges past reverberation peak.

At ranges past the reverberation peak in nonisothermal water, the results of reference 1 indicate that on the average the reverberation falls off at about the inverse fourth power of the range, but that large variations from this type of decay are observed. Since the expected echo level also falls off at about the inverse fourth power of the range,[8] the results of reference 1 mean that the possibility of obtaining an echo usually depends on the level of the reverberation peak relative to the expected echo level at the range of the reverberation peak. If the reverberation

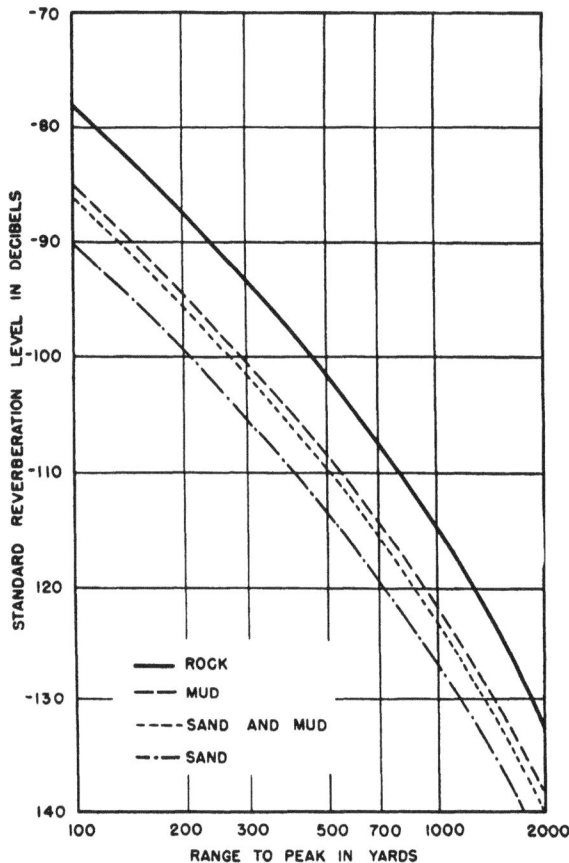

FIGURE 8. Expected level of peak of bottom reverberation as a function of range to peak and bottom type.

peak is high enough to mask the echo at that range, then the echo is not likely to be detected at any range past the reverberation peak. Conversely, if the echo level is well above the reverberation at the range of the reverberation peak, bottom reverberation will probably not limit echo ranging at any range.

The above discussion is not by any means a complete treatment of the problem of echo ranging in shallow water. Many factors are involved in determining the echo-to-reverberation ratio at any range. Also, no account has been taken of the fact that attenuation at long range makes the echo level drop off

much more rapidly than the inverse fourth power of the range. However, attenuation also decreases the received bottom reverberation levels, so that even at long ranges the echo and reverberation levels should decrease at roughly the same rate. In general, it can be said that the problems involved in determining the echo-to-reverberation ratio are so complicated that no satisfactory quantitative treatment has ever been given, although qualitative discussions have been presented in a number of places.[9] It appears then from the foregoing discussion that with present information the best way to characterize bottom reverberation levels is in terms of the level at the principal reverberation peak; this level is determined using equation (54) of Chapter 12 and the known value of the bottom scattering coefficient for 10 degrees grazing incidence.

Figure 8 shows the expected average standard reverberation level at the reverberation peak, as a function of the range to the peak and the bottom type, for ordinary 24-kc echo-ranging gear sending out a horizontal beam. In preparing this diagram it was assumed that the range to the peak is six times the depth of the bottom below the projector. The absorption was taken to be 4 db per kyd, J_b was set equal to -19 db for the 5- to 6-degree ray, $10 \log m''$ was taken from Table 7, and finally the reverberation level for a 100-msec ping was calculated by the use of equation (55) of Chapter 12. Although Figure 8 is the best average curve which can be drawn with present information, the likelihood of deviations in practice from Figure 8 cannot be overstressed. In particular, Figure 8 is not valid in isothermal or nearly isothermal water, when the range to the reverberation peak will usually be very different from six times the distance between the projector and the bottom. It is also not advisable to extend the results in Figure 8 to ranges less than 100 yd because at such short ranges it is again unlikely that the average assumed relationship between range and depth will be valid. It should be noted that the curves in Figure 8 incorporate the 6-db correction for surface reflections discussed in Section 15.3.3. Thus, when the echo-ranging transducer is deep, 6 db should be subtracted from the values in Figure 8 to obtain the expected reverberation levels; in such situations surface reflections will not be important in determining the bottom reverberation level.

In conclusion, we repeat that on the average the reverberation in nonisothermal water seems to fall off at about the inverse fourth power of the range, at ranges past the reverberation peak, but that large variations from this type of decay may be observed. Careful examination of the results of reference 1 shows that the shape of the decay curve is probably not the same over all types of bottoms. In other words, the probability of distinguishing an echo at long range is different over different types of bottoms, even with the same echo-to-reverberation ratio at the range of the reverberation peak. A preliminary study of the shapes of these decay curves over different types of bottoms is being made off San Diego, but unfortunately, the results of that study are not available at this time.

Chapter 16

VARIABILITY AND FREQUENCY CHARACTERISTICS

I<small>N PRECEDING CHAPTERS</small>, we have derived theoretical formulas for the average reverberation intensity, and have compared these formulas with average reverberation intensities observed in practice. By means of this comparison, we have obtained average values of the backward scattering coefficient for volume, surface, and bottom reverberation. In this chapter we shall attempt to analyze the differences between reverberations from successive pings sent out under apparently the same circumstances. These differences may be deviations in amplitude, or deviations in the frequency spectrum of the received reverberation.

There are several reasons for such an analysis. First, it is desirable to know just how well the average curve may be expected to represent individual reverberation curves. Secondly, the deviations from the average depend on the type of mechanism giving rise to the reverberation; thus, analysis of the deviations can give valuable information on the sources of reverberation. Finally, such an analysis may easily reveal significant differences between the behavior of echo fluctuation and reverberation fluctuation since the mechanisms producing these two types of fluctuation are undoubtedly somewhat different. These differences in behavior, if well understood, may be utilized in methods for improving the recognition differential for the echo against a reverberation background.

16.1   FLUCTUATION

In analyzing the amplitude deviations of individual reverberation curves from the average, it is convenient to distinguish between fluctuation and coherence. Fluctuation refers to the deviation from the average of the intensity received at a definite time following the initial ping. This fluctuation is usually measured by the variance

$$\overline{\Delta I^2} \equiv \overline{(I - \bar{I})^2} = \overline{I^2} - (\bar{I})^2 \qquad (1)$$

where $\bar{I}$ is the average intensity at a time t seconds

after midsignal, and $\overline{I^2}$ is the average of the square of this intensity. Average values will be designated by a bar throughout the remainder of this chapter. As discussed in Chapter 12, the average intensity at the time t is to be determined by the following process. A large number of reverberation records are taken under circumstances as nearly identical as possible, and the intensity of the reverberation at a time t seconds after midsignal is read off each record. The average of these intensities is the value referred to by the bar.

Evidently, if all the pings were sent out under exactly the same circumstances, the received intensity at time t should be constant, and $\overline{\Delta I^2}$ in equation (1) would be zero. However, no two pings occur under precisely the same circumstances. There are variations in the power output of the projector; variations in the orientation of the receiver because of ship roll; variations in such oceanographic factors as wave height, wind force, temperature-depth distribution, water depth, and type of bottom material; and overall variations in transmission anomaly. Some of these sources of reverberation fluctuation can be minimized. The power output of the projector can be stabilized to a fraction of a decibel; the ship will not roll on a calm day; and the effects of changing wind force and bottom character can be eliminated by studying only volume reverberation. However, large fluctuations remain no matter how much control is exercised. These remaining fluctuations in volume reverberation are regarded as an inherent property of reverberation.

In the derivation of equation (13) of Chapter 12, expressing the time variation of volume reverberation from a single ping, it was assumed that the reverberation is due to the scattering of sound by a large number of scatterers in the ocean. Fluctuation in the received reverberation is caused by the fact that the total reverberation amplitude is the sum of the amplitudes received from all the individual sources. These individual amplitudes have random phases with respect to each other, and the total amplitude is large or small depending on the degree of reinforcement or interference between these incoming individual amplitudes.

An expression for the fluctuation of intensity caused by the combination of a large number of amplitudes of equal magnitude but random phase was derived by Rayleigh.[1] The probability that the resultant intensity will exceed the value I is given by

$$P = e^{-(I/\bar{I})} \qquad (2)$$

where $\bar{I}$ is the average intensity. A derivation of equation (2) is given in Chapter 7. The actual received reverberation is of course a combination of amplitudes of many different magnitudes, because the individual scatterers are not all of equal strength, because the projector is directional, and because the transmission loss to different portions of the ocean may not be the same. However, it can be shown that equation (2) remains valid, even if all the amplitudes are not of equal magnitude, provided only that there are a large number of amplitudes of each magnitude, and that the number of amplitudes of each magnitude remains essentially constant. Thus, formula (2) is implied by the assumptions used in Chapter 12 to derive the expression for the time variation of the volume reverberation from one ping, with the proviso that the transmission does not change from ping to ping.

The applicability of the Rayleigh distribution function (2) has been tested by the University of California.[2] They first chose sets of ten or more typical reverberation records in deep water; all the records in a given set were taken under similar conditions. The ratio of the observed amplitude to the average amplitude was computed for various times on each set. All told, 420 values of this ratio were obtained for the QB transducer, and 500 values of the ratio for the QCH-3. The results are plotted in Figure 1. There is apparently no major deviation of either experimental curve from the theoretical expression (2). It appears from Figure 1 that the shape of the fluctuation curve does not depend significantly on such factors as the directivity pattern of the transducer, nor on the shape of the pulse sent out; both of these factors are different for the two transducers. It will be noted that equation (2) predicts this independence. Since the times chosen on the records were well distributed, and since no effort was made to distinguish between surface and volume reverberation, it appears that the expression (2) applies fairly

well to all portions of the deep-water reverberation versus range curve.

It is not surprising that formula (2) fits the observed fluctuation of surface reverberation levels about as well as it fits the fluctuation of volume reverberation. If an assumption that the scattering power of the surface remains essentially constant from ping to ping is added to the other assumptions used in the

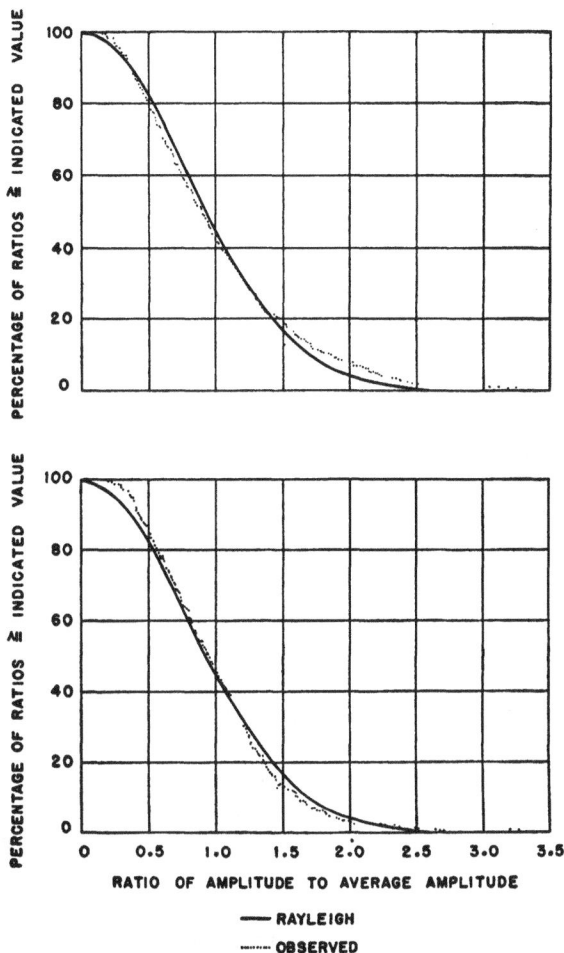

FIGURE 1. Agreement between Rayleigh distribution and cumulative distribution of observed reverberation amplitudes.

derivation of equation (38) of Chapter 12, for the decay of surface reverberation, the fluctuation formula (2) would be predicted for surface reverberation also. Separate tests of the applicability of equation (2) to surface and bottom reverberation have been reported,[3] which show that this equation is a reasonably good fit to the fluctuation of both surface and bottom reverberation levels. These results therefore support the assumption that the surface scattering power can be regarded as a function of certain relatively slowly varying physical parameters such as the wind force and the wave height.

The derivation of equation (2) is based on the assumption that the number of independent interfering amplitudes is large, or, in other words, that the number of scatterers which combine to return sound to the receiver at a particular time is large. However, since this number of scatterers is proportional to the volume of space illuminated by the pulse, it is apparent that the effective number of scatterers depends on the ping length, and cannot be large for very short pulses. In fact, with pings sufficiently short and directional, the received reverberation at any instant will arise from at most one scatterer. With such pings the average number of scatterers per unit volume could be determined from the character of the received reverberation, provided the dimensions of the scatterers are small compared to the average distance between them. For, under these circumstances, the received reverberation would be a series of widely separated echoes from individual scatterers; from the spacing of these echoes the average number of scatterers per unit volume could easily be calculated.

However, it has not yet been possible, and may never be feasible, to reduce the ping dimensions sufficiently to resolve all the individual scatterers responsible for volume reverberation. With standard 24-kc gear, some of the larger individual scatterers are occasionally distinguishable even when pings as long as 5 msec are used.[4, 5] However, as a general rule individual scatterers cannot be identified even with pings as short as 0.1 msec. It has been pointed out [6] that the Rayleigh distribution does not apply when the effective number of scatterers is small, and that it may be possible to determine the average number of scatterers per unit volume from the disparity between the observed distribution function and the Rayleigh form (2). The theoretical distribution depends on the assumptions which are made concerning the total number of scatterers present. The most reasonable assumption is that the number of scatterers obeys a Poisson [7] distribution. If so, the probability $P_n(N)$ that N scatterers are present when the average number present is n is given by

$$P_n(N) = \frac{n^N e^{-n}}{N!}.$$

$$(3)$$

With this assumption, the expected distribution function of the reverberation intensity is calculated in

reference 6, as a function of the average number of scatterers in the portion of the ocean illuminated by the ping at a definite instant. If this average number of scatterers is 10, the deviation of the predicted distribution function from the Rayleigh distribution is of about the same order of magnitude as the deviations of the experimental points from the Rayleigh distribution in Figure 1. However, the points plotted in Figure 1 are not sufficiently numerous for a determination of n. It is estimated in reference 6 that 4,000 points, all taken with the same ping length, are required to say definitely whether an observed distribution more nearly approximates the theoretical curve for 10 scatterers or the Rayleigh distribution.

The average number of scatterers n may also be predicted, in principle, from the percentage fluctuation in intensity. If the actual number of scatterers obeys the Poisson distribution (3), reference 6 gives the following relationship

$$\frac{\overline{\Delta I^2}}{(\bar{I})^2} = 1 + \frac{1}{n}.$$

$$(4)$$

If the intensity is the resultant of a fixed number n of amplitudes whose phases vary at random, it is easy to show that

$$\frac{\overline{\Delta I^2}}{(\bar{I})^2} = 1 - \frac{1}{n}.$$

$$(5)$$

It is readily verified by direct integration that for the Rayleigh distribution (2),

$$\frac{\overline{\Delta I^2}}{(\bar{I})^2} = 1$$

$$(6)$$

in agreement with equations (4) and (5) when n is infinite. A third alternative is to assume that the number of scatterers obeys a Gaussian distribution; this would be the case, for example, if the variability in the number of scatterers were due primarily to accidental variations in the length of the ping emitted by the projector. With a Gaussian law for the number of scatterers, still another formula would be obtained for $\overline{\Delta I^2}/(\bar{I})^2$.

This discussion of reverberation fluctuation has completely neglected the fluctuation due to variability in such factors as transmission loss, projector output, and transducer orientation. With well-functioning equipment such as setup C of Section 13.1.1, variations in projector output occur so slowly that their effect on the short-term fluctuation of reverberation is negligible. Variability in transmission loss is known to be large and rapid; therefore, the fluctuation in reverberation levels must be partly a result of

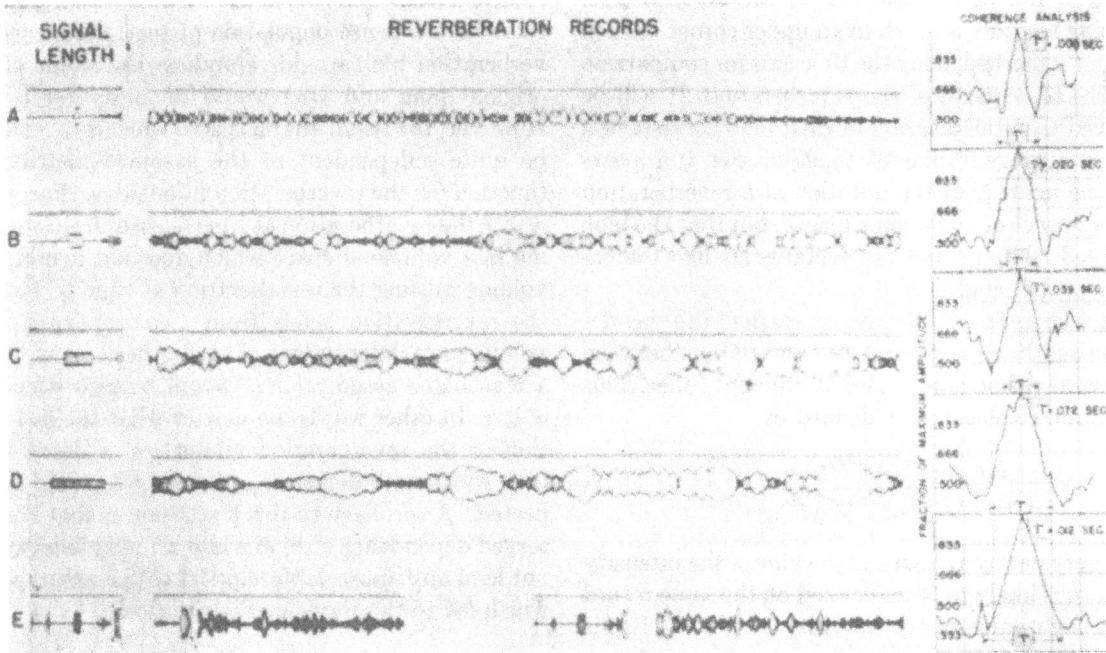

FIGURE 2. The coherence of reverberation.

fluctuation of the transmission loss between the projector and the scattering centers. Transmission fluctuations apparently do not obey a Rayleigh distribution; the standard deviation of the transmitted amplitude is about 42 per cent of the mean transmitted amplitude, in practice, as compared to the 52 per cent predicted from the Rayleigh distribution.[8] Ship roll is another factor, ignored in this sketchy treatment, which is believed to produce significant fluctuations in reverberation.[9] These neglected factors will have to be taken into account before the theory of reverberation fluctuation can be considered at all adequate. Also, the presence of these other sources of fluctuation means that a determination of the average number of scatterers giving rise to reverberation, by a method such as that proposed in reference 6 would not be too reliable even if the distribution function for the number of scatterers were known.

At the present time, average reverberation intensities are determined by averaging between 5 and 12 pings. The validity of this procedure can be estimated from the magnitude of the variance defined by equation (1). It is easy to show that the standard deviation of the average intensity of n pings is just $\sqrt{\overline{\Delta I^2}/n}$. If the distribution function is Rayleigh, then from equation (6) we have

$$\overline{\Delta I^2} = (\bar{I})^2.$$

Thus the average intensity $\bar{I}$ of 5 pings has a standard deviation of $\pm 0.45\bar{I}$, or roughly 1.5 db.

16.2 COHERENCE

The term *coherence* applied to reverberation refers to a tendency of the received reverberation to occur in the form of pulses of the approximate duration of the ping length. The possession of coherence means that if at any instant the reverberation level is high, it is likely that the level will remain high for a little while, and that if the reverberation level is low, it is likely not to become large in a short time.

An experimental study of reverberation coherence has been made by UCDWR, and reported in Section X of reference 2. Ten blobs of about equal amplitude were chosen at random from two QB records of equal ping length, in the time interval between 1.5 and 2.5 sec after midsignal. The amplitude was measured at intervals of about 0.1 ping length from the middle of each blob out to 2 ping lengths on either side. The average amplitude at each of the measuring positions on the ten blobs was then divided by the average amplitude at the middle of the blob and was plotted. The resulting graphs, one for each of several ping lengths, are given in Figure 2, under the heading Coherence Analysis. Time is plotted on the horizontal axis; amplitude relative to the amplitude at the middle of the blob is plotted on the vertical axis. The ping length τ is given in an upper corner of each plot and is marked along the time axis for comparison with the blob width of the reverberation. It will be observed that the diagram of each blob consists of a peak with wings trailing off to either side. It appears from the figure that the duration of a reverberation blob is very close to the ping length, but that the blob is peaked in shape and not rectangular like the envelope of the ping.

The coherence of the reverberation can be described mathematically by the correlation between the reverberation amplitudes at different times. This correlation coefficient ρ is defined by

$$\rho = \frac{\overline{[I(t_1) - \bar{I}(t_1)][I(t_2) - \bar{I}(t_2)]}}{\sqrt{\overline{\Delta I^2(t_1)}}\sqrt{\overline{\Delta I^2(t_2)}}}. \quad (7)$$

If ρ is very nearly 1, then a high value of the intensity at time t_1 is likely to be associated on the same record with a high value of the intensity at time t_2. If ρ is near zero, a given value of the intensity at time t_1 gives no information about the intensity on that record at time t_2. If ρ is very nearly -1, a high value at t_1 implies a likely low value of the intensity at t_2.

It can be shown theoretically that ρ depends, in general, only on the difference in time $|t_1 - t_2|$, provided the average intensities at t_1 and t_2 are not too different.[10] If the outgoing ping is square-topped, and if the average intensities at the times t_1 and t_2 are equal, and if, furthermore, the individual intensities follow the Rayleigh distribution (2), then it can be shown that the correlation coefficient (7) reduces to

$$\rho = \begin{cases} \left(1 - \dfrac{\alpha}{\tau}\right)^2 & \alpha \leqq \tau, \\ 0 & \alpha \geqq \tau \end{cases} \quad (8)$$

where $\alpha = |t_1 - t_2|$ and τ is the ping length.[11] In other words, reverberation levels at times close to the center of a blob will be high, but levels at times more than a ping length away from the center will bear no relation to the intensity at the center of the blob. The value of ρ from the relation (8) is plotted as a function of α/τ in Figure 3. Evidently the result (8) for the correlation coefficient explains the dependence of the blob width on ping length noted in Figure 2.

The precise functional dependence of ρ on α in relation (8) depends on the assumption of a rectangular ping and also depends in part on the assumption of a Rayleigh distribution for the individual intensities. If, for example, reverberation resulted from echoes

returned from widely spaced single scatterers, rather than from a dense population of scatterers, each reverberation blob would reproduce the shape of the original ping and $\rho(\alpha)$ would be unity for $\alpha < \tau$. However, the result that ρ is zero when $\alpha \geqq \tau$ should be quite independent of the assumed distribution function for the reverberation intensities. For, when $\alpha \geqq \tau$ the reverberation at time t_1 arises from scattering in a volume of space which does not overlap the volume causing the reverberation at time t_2. Thus, if the reverberation levels from two nonoverlapping volumes are independent of each other — and this is a reasonable assumption — ρ will be zero whenever $\alpha \geqq \tau$. In other words, no matter what the distribution of the reverberation intensities, a decrease of blob width with decreasing ping length would be expected. A corollary to this discussion is that the observed dependence of blob width on ping length does not lend any appreciable support to the assumptions which led to the Rayleigh distribution.

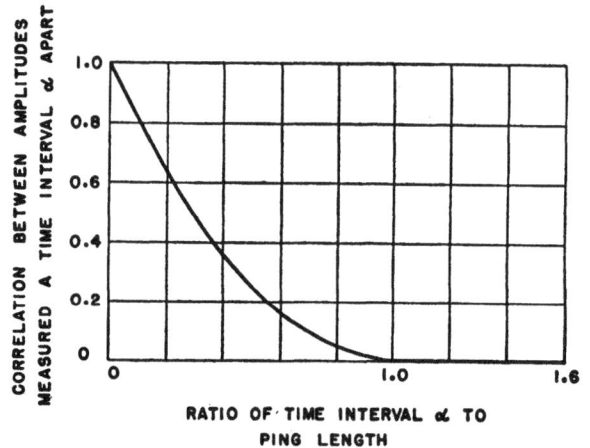

FIGURE 3. Theoretical curve for self-correlation of reverberation intensity.

It is possible to determine all sorts of probability coefficients from the observed reverberation records; these coefficients can then be compared with computations based on various assumptions regarding the sources of reverberation. However, the labor required to analyze the observed records is so great that this method of investigating reverberation has not been considered practical. To illustrate, only one very crude attempt has been made to quantitatively compare the functional dependence predicted by relation (8) with experiment; the agreement cannot be said to have been better than qualitative. Another difficulty in this approach is that the theoretical values of many probability coefficients cannot be computed mathematically. However, a few such coefficients have been computed on the assumption of a Rayleigh distribution for the individual intensities. One of these is the joint probability $P(I_1, I_2, \alpha)$ of obtaining reverberation intensities I_1 and I_2 at the same range on two different pings a time interval α apart.[12] This coefficient is a function of the average velocity of the scatterers relative to the echo-ranging vessel. Motion of the scatterers relative to the transducer changes the ranges and relative positions of the scatterers from one ping to the next, thereby giving rise to fluctuation of the measured reverberation levels. Similarly, the joint probability of obtaining intensities I_1 and I_2 at two different ranges on the same ping has been computed.[13] A general discussion of the significance of these various probability coefficients is given in reference 10. Other references [14-16] may be useful in the analysis of fluctuation and coherence.

It is worth noting that the measured 6- to 7-db difference between the average intensity and the average peak intensity in a band three ping lengths long, referred to in Section 13.2, is another measure of the coherence of reverberation. For example, if the coherence were very poor the average peak height in any finite band would be very large. For in that case the reverberation intensity at any instant would be very nearly independent of the intensity at any other instant; crudely, the band could be divided into a large number of intervals in each of which the probability for a given intensity would follow the simple Rayleigh distribution (2) or some similar distribution. Since the number of intervals would be large, intensities much higher than average would be expected to occur at least once in the band three ping lengths long. So far, it has not proved possible to calculate theoretically, with accuracy, the average peak height in a band.

16.3 FREQUENCY ANALYSIS OF REVERBERATION FROM NARROW-BAND PINGS

The received reverberation is often used as a reference frequency for the estimation of doppler shift in the echo. In order to use reverberation in this way, it is necessary to know the average frequency of the reverberation and the average frequency band width characteristic of reverberation. Such a frequency analysis can also give valuable information about the processes giving rise to reverberation.

The theory of Fourier series tells us that any signal of finite duration can be regarded as made up of single-frequency components of definite amplitudes and phases. A so-called single-frequency ping (CW ping) of the sort emitted by ordinary echo-ranging gear contains not only the nominal frequency of the ping, but also an infinite number of other frequencies (see Section 12.5). However, the band width within which frequency components of significant amplitude lie is usually very narrow for ordinary ping lengths.[a] The returning reverberation is also composed of a band of frequencies. With the assumptions that led to the Rayleigh distribution (2), it can be shown that the band width of the reverberation equals the band width of the outgoing ping. For with these assumptions, the shape of the returning signal from any scatterer is the same as the shape of the ping. The factors which may cause the shape of the returned signal from any scatterer to differ from the ping have been discussed in Section 12.5; in general these factors can be neglected for pings 10 msec or longer. Thus the received reverberation, which is simply the sum of many such signals, must apparently have the same band width as the ping. Other sources of reverberation fluctuation, in addition to the fundamental randomness of phase leading to the Rayleigh distribution, can increase the band width of reverberation. However, it can be shown that these sources can be neglected for pings 100 msec or shorter.

Because of the narrowness of the frequency band, the experimental determination of the frequency spectrum of reverberation is difficult. Besides, the quantity which is measured as a function of frequency and called the frequency spectrum is merely the energy or intensity contained in a narrow band of frequencies; the phases of the individual frequency components are never measured. For many purposes, therefore, the measured spectrum is not the most useful way of describing reverberation. The measured spectrum cannot, for example, give clues as to those time variations of reverberation which cause it to sound like a signal of wavering pitch.

The UCDWR has, however, developed a special device known as the periodmeter for the analysis of

---

[a] The band width may be defined as the frequency band containing half the energy in the ping, or as the frequency band within which intensities of spectral components are no more than 3 db below the intensity of the midfrequency component. For a rectangular ping the band width, defined in either manner, is approximately the reciprocal of the ping duration in seconds; thus, it would be about 10 cycles for a 100-msec ping [see equation (63) of Chapter 12].

the rapid frequency shifts characteristic of reverberation. This device [17] measures the time interval between successive zeros of the alternating signal fed into it. The "instantaneous frequency" of the signal is interpreted as inversely proportional to the measured time interval between successive zeros. This instantaneous frequency is recorded against time on a cathode-ray oscilloscope (which may be photographed); the frequency appears as a spot whose deflection from a base line is a measure of the frequency.

The periodmeter is designed to function in the neighborhood of 800 c; thus reverberation must be heterodyned to about this frequency before frequency

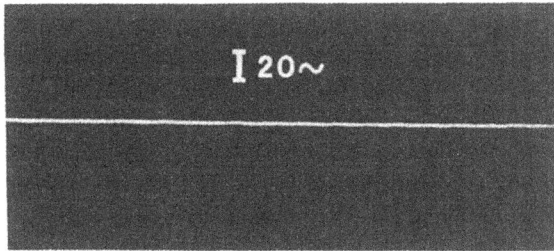

FIGURE 4. Periodmeter record of 800-cycle oscillator tone.

analysis by the periodmeter is possible. Figure 4 is the photograph of a trace obtained by feeding an 800-c oscillator tone into the periodmeter. Figure 5 shows the traces of a ping and its echo from an S-class submarine; the way the echo mirrors the frequency fluctuations in the ping is interesting. Figures 6A and 6B are observed traces for volume reverberation and bottom reverberation, both showing no indication of a definite law for the variation of reverberation frequency with time after midsignal. Such traces are typical of most reverberation. In all these traces, increase in frequency is indicated by a smaller (lower) amplitude on the trace.

The periodmeter has been used to analyze reverberation signals. The results of this analysis will be first summarized and then discussed in more detail. Since all these conclusions are based on more than 1,000 observed points, they have a high degree of statistical probability.

1. There are no obvious systematic changes in spectral character during the decay of a single reverberation. If the average pitch does change with time after midsignal in any systematic way, such a trend is masked by the much larger irregular variations in pitch.

2. The averages of reverberation frequency over time intervals corresponding to the pitch response time of the ear (0.1 sec) show variations large enough to render target doppler discrimination of 1 knot or less highly unreliable. For target speeds of 2 knots or more, target doppler discrimination appears quite reliable.

3. If the outgoing signals are unintentionally frequency-modulated, because of poor design or maladjustment of the transmitting equipment, the rms frequency spread in the reverberation is increased.

4. The frequency spread of the heterodyned reverberation depends on the audio output frequency and the pulse length. The rms frequency spread $\overline{\Delta f}$ was well-fitted by the formula

$$\overline{\Delta f} = K f^{3/4} \tau^{-1/4} \tag{9}$$

where f is the audio output frequency, τ is the pulse length, and K is a constant. Pulses of length 92 msec and frequency 24 kc produced reverberation, which, after being heterodyned to 800 c, had a mean rms frequency spread of 21.5 c.

5. The frequency spread of reverberation does not seem to depend on the frequency of the outgoing signal.

6. The mean observed reverberation frequency agreed with own-doppler values calculated for the axis of the projector. At moderate ship speeds, the finite width of the projector beam did not, through own-doppler effects, cause marked broadening of the beam.

Most of the above results are quite reasonable. The first result indicates that the reverberation arises from a process which is essentially random. The last result indicates that the mean reverberation frequency can be used as a reference for the measurement of target doppler. That is, the mean reverberation frequency f_1 agrees with the following formula for the frequency of an echo received on a moving ship from a stationary target.

$$f_1 = f\left(1 + \frac{2v}{c}\cos\theta\right), \tag{10}$$

where v is the velocity of the echo-ranging ship, θ is the angle between the projector axis and the line of motion of the ship, f is the frequency of the emitted ping, and c is the velocity of sound. This agreement is theoretically expected.[18, 19]

The second result indicates that estimates of target doppler are not reliable unless the target is moving at speeds of 2 knots or more. The third result, that the

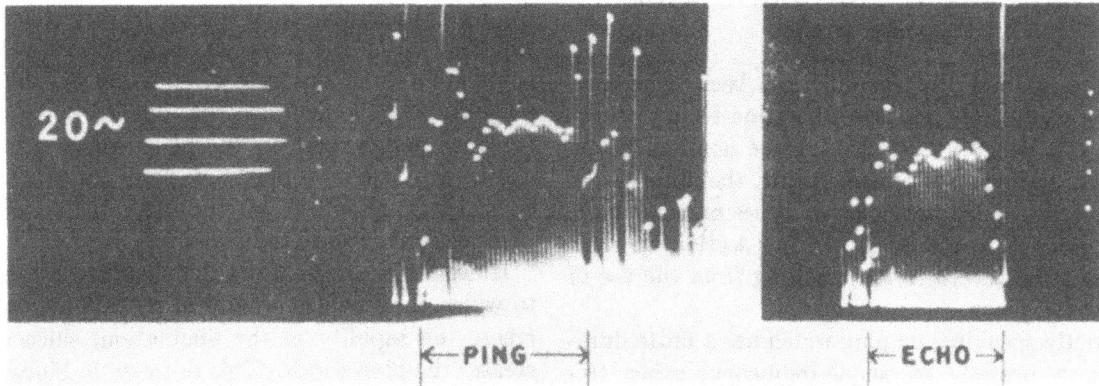

FIGURE 5. Periodmeter record of a ping and its echo.

observed frequency spread in reverberation should be increased by frequency fluctuation in the outgoing ping, is also not surprising.

The dependence of the frequency spread on the audio output frequency, described by equation (9), seems at first sight rather difficult to understand. In principle, heterodyning simply subtracts a constant frequency from the frequencies of all components of the incoming reverberation signal; thus, if there is no distortion, the average frequency spread should not depend on the audio output frequency.[20,21] However, with a little thought, it can be seen that the difficulty arises because the time intervals between successive zeros of a complex signal are not related in any simple way to the frequency spectrum of that signal. Thus, the assumption that the rms deviation of the instantaneous frequency read from the periodmeter should be exactly equal to the true rms frequency spread of the reverberation spectrum is certainly not warranted. The complete elucidation of the relation between periodmeter readings and the reverberation spectrum requires a satisfactory theory of the periodmeter, which has not yet been developed because of mathematical difficulties. A partial analysis of the theory of the periodmeter is given in a report by UCDWR.[17] There it is predicted that the rms spread of the instantaneous frequency should obey the formula

$$\overline{\Delta f} = K f^{1/2} \tau^{-1/2} \qquad (11)$$

which differs somewhat from the observational equation (9). However, the mode of derivation of equation (11) has been subjected to some criticism.[20]

This difficulty in relating the response of the periodmeter to the spectrum of the reverberation

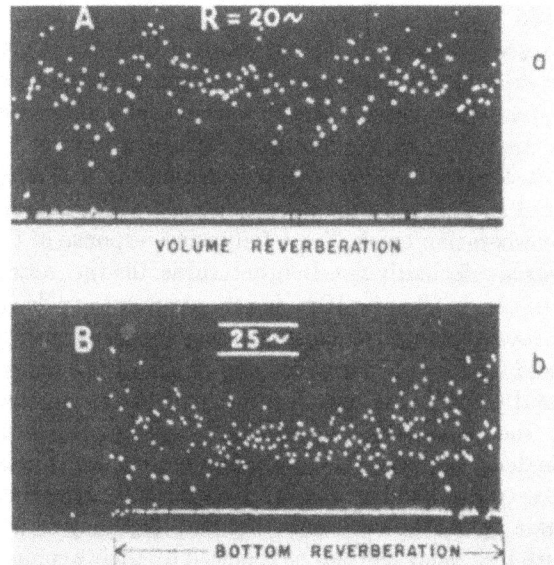

FIGURE 6. Periodmeter records of volume and bottom reverberation.

illustrates the difficulty which is always experienced in predicting the response to reverberation of any complex circuit, such as the ear or other doppler discriminator. It can be argued that, for the ear, the instantaneous frequencies measured by the periodmeter are more significant than the spectrum which gives the intensities in very narrow frequency bands. However, because of the complexity of the ear, this assertion requires proof; for this reason conclusion 2 above, and other similar conclusions, cannot be relied on if based on evidence from the periodmeter alone.

16.4 REVERBERATION FROM WIDE-BAND PINGS

Up until now, this volume has been concerned almost completely with reverberation from narrow-band CW pings (in a CW ping the nominal transmitting frequency is fixed during the interval of transmission). However, other types of pings have been used, as in FM sonar. In this section we shall examine the reverberation resulting from the use of wide-band pings.

Strictly speaking, no ping which has a finite duration can possibly be single-frequency; other frequencies than the nominal one must be present in order that the signal can build up and die off. However, 100-msec CW pings at 24 kc have band widths of the order of 10 c, while wide-band pings often have band widths of 1,000 c or more (for example, a 1-msec CW ping has a band width of 1,000 c). It may be expected that this very real difference in the order of frequency spread will be reflected in a difference in the nature of the returning reverberation.

According to Section 12.5, widening the frequency band in the ping should not seriously affect average reverberation levels, if the frequency response of the gear is sufficiently flat. In other words, the theoretical formulas in Chapter 12 giving the average time decay of reverberation from a single ping should be just as valid (or invalid) for wide-band pings as for narrow-band pings. It is true that many of the parameters in the formulas of Chapter 12 are frequency-dependent, such as the transmission loss, transducer directivity, and the scattering coefficients. However, these quantities vary smoothly and relatively slowly with frequency and can be replaced by their averages over even a 10-kc wide band without introducing much error. Thus the resultant theoretical average reverberation levels for wide-band pings are simply an average, over the frequency band included in the ping, of the levels predicted for the narrow-band pings. A more quantitative discussion of the qualitative ideas in this paragraph can be found in a report by CUDWR.[22]

The average fluctuation, as defined by equation (1), cannot be written simply as the average of the fluctuations of the individual frequency components. Thus, there is reason to expect that the fluctuation of wide-band pings may be different from that of CW pings. The expected magnitude of the fluctuation of wide-band pings will depend on the mechanism hypothesized as responsible for the fluctuation. Con-

fining our attention for the moment to those wide-band pings resulting from the use of very short ping lengths, then, if the Rayleigh distribution is valid, it is apparent from the form of equation (2) that the magnitude of the fluctuation as defined by equation (1) does not depend on the ping length. In other words, with square-topped CW pings the magnitude of the fluctuation will be the same for wide-band pings as for narrow-band pings.

However, the decrease of the ping length required to widen the frequency band of a CW ping will increase the rapidity of the fluctuation, since it decreases the blob width. This decrease in blob width does not necessarily improve the recognizability of a returning echo since the echo length is decreased correspondingly. Studies of echoes resulting from CW pings (see Chapters 21 and 23) indicate that echoes look very much like reverberation blobs, of width about equal to the ping length. This similarity between echoes and reverberation blobs makes it very difficult to devise means for improving the detectability of echoes from CW pings against a reverberation background, other than the obvious but not always feasible procedure of reducing the average reverberation intensity. It is true that a time average over a relatively short interval will include a large number of reverberation blobs and this time average will fluctuate much less rapidly than the unaveraged reverberation. However, because of the similarity of the echo to the reverberation blobs, any averaging procedure applied to the sound returning from a short CW ping is likely to eliminate the echo.

Nevertheless, by adjusting the time interval over which the average is taken, some beneficial effect may be hoped for, especially when the echo intensity is much larger than the average reverberation intensity. Some studies made with very short (0.3 msec) unmodulated pings support this hope. It was found that the use of these very short pings significantly reduced the number of false contacts reported and that the maximum range at which a target could be identified was not affected.

When frequency-modulated pings are used, the signal sent into the water has a continuously varying frequency. With such pings, the received reverberation at any instant may include a wide band of frequencies, depending on the band included in the original ping and on the receiving pass band of the equipment. Theoretical analysis of the expected fluctuation of the reverberation from such pings is difficult, but it has been shown[23] that the envelope of the reverberation from frequency-modulated pings should fluctuate more rapidly than does the echo envelope. Observational evidence that the rapidity of reverberation fluctuation is increased by the use of frequency-modulated pings is quoted in reference 24, and similar observations have been reported in other sources. The echo length resulting from a frequency-modulated ping is equal to the effective ping length, just as with unmodulated pings; the effective ping length is itself determined by the pass band of the equipment.

The only accurate quantitative data on the fluctuation of reverberation from frequency-modulated pings are those given in reference 25. In that report it was found that with frequency-modulated pings the mean standard deviation of the reverberation amplitude is 33 per cent of the mean amplitude, significantly smaller than the 52 per cent value predicted for a Rayleigh distribution. These measurements were made with pings of length 2, 4, and 8 sec, frequency-modulated from 48 to 36 kc. Some results on the coherence of FM reverberation are also reported in reference 25.

The mechanism by which such a decrease in the magnitude of the fluctuation could be accomplished is difficult to visualize; but the possibility of such an effect must be admitted, because our theoretical understanding of the problems involved is not at all complete. Qualitative reports that frequency modulation is effective in reducing the magnitude of the fluctuation cannot be trusted, since they may be based on the results of time averages performed somewhere in the complicated recording equipment.

Recapitulating, in echo ranging the objectionable features of reverberation are twofold. Reverberation masks the echo; also, reverberation simulates the echo, so that false contacts are often obtained. It is apparent from this chapter that a great deal of information still is lacking about such characteristics of reverberation as blob shape, cause and rapidity of fluctuation, and frequency spread.

Schemes which have been suggested to suppress reverberation may be combined with proposals for decreasing the average level of the reverberation by decreasing the ping length or by increasing the hydrophone directivity. Of course, the possibilities of decreasing the ping length or increasing the hydrophone directivity are always limited by other practical considerations. Once the ping length has been decreased as much as practical, and the transducer directivity has been increased to its highest practical value, one must resort to devices which do not reduce the average energy received in the hydrophone, but instead reduce the instrumental effects of reverberation in the detecting mechanism.

Chapter 17

SUMMARY

DEFINITIONS

17.1.1 Reverberation

REVERBERATION IS A COMPONENT of background heard in echo-ranging gear, and is distinguished from the general noise background by the fact that it is directly due to the pulse put into the water by the gear. The traveling ping meets not only the wanted target, but also myriad small scattering centers or other inhomogeneities, each of which returns a tiny echo to the transducer. These tiny unwanted echoes combine to make up reverberation. Thus reverberation, like the echo, is a sound whose pitch is definite and is determined by the frequency of the projected pulse. Often echoes which would be audible over the remainder of the noise background are masked by reverberation. To the operator of echo-ranging gear, reverberation is evident as a quavering ring which sets in as soon as the period of sound emission is finished.

The scatterers producing reverberation may be located near the sea surface, in the main ocean volume, or in the sea bottom. The reverberations produced by these three types of scatterers are called, respectively, volume reverberation, surface reverberation, and bottom reverberation. This distinction is physically meaningful, since these three types of reverberation apparently have different properties and can be experimentally differentiated from each other.

17.1.2 Reverberation Intensity

The strength of the sound heard or recorded as reverberation depends not only on the intensity of the backward scattered sound in the water near the receiver, but also on the nature of the receiving gear. The intensity of the reverberation actually heard or recorded, after the sound in the water has been converted to electrical energy by the receiver, amplified, and passed to the ear or recording scheme, is called the "reverberation intensity" and is given the symbol G. As so defined, G equals the watts output across the terminals of the receiving gear. In general, the reverberation intensity G is a function of time and is related to the sound intensity in the water by such parameters of the receiver system as receiver directivity and receiver gain. Since the quantity G depends on the gear parameters, the absolute magnitude of G is usually not of great significance in studies of the intrinsic character of reverberation or of the mechanisms producing reverberation. In these studies the reverberation level, defined under the next heading, is ordinarily employed.

17.1.3 Reverberation Level

The reverberation level is the decibel equivalent of the reverberation intensity defined in Section 17.1.2, expressed relative to a standard which makes reverberations measured with different gear exactly comparable. Specifically, the reverberation level R' is defined as

$$R' = 10 \log G - 10 \log (F \cdot F') \qquad (1)$$

where F is the projector output at 1 yd in decibels above 1 dyne per sq cm, and F' is the receiver sensitivity in watts of output for a received rms sound pressure of 1 dyne per sq cm. Reverberation levels are much more useful than reverberation intensities for comparing measurements made with different systems, since under identical external conditions two different systems sending out pings of the same length should in principle give the same reverberation level if correction is made for transducer directivity. Reverberation levels usually refer to the average reverberation intensity G found in a succession of pings.

Reverberation intensities are proportional to the ping duration. Often it is desirable to convert reverberation levels to the levels which would be received using a standard ping length. For this purpose we define the standard reverberation level R

$$R = R' + 10 \log \left(\frac{\tau_0}{\tau} \right) \qquad (2)$$

where R' is the observed reverberation level with ping duration τ, and τ_0 is a standard ping duration usually chosen as 100 msec.

17.1.4 Backward Scattering Coefficients

By "backward scattering" is meant scattering back along the incident ray path. If there is only one ray path from the projector to the scatterers, only sound which is scattered directly backward can give rise to reverberation. Thus the more efficient a portion of the ocean is in backward-scattering, the higher will be the level of the received reverberation. The efficiency of a small volume V of the ocean in scattering sound backward is specified in terms of the backward-scattering coefficient m, which is defined by the relation

$$\frac{mV}{4\pi} = \bar{b} \qquad (3)$$

where $\bar{b}$ is the average energy scattered by the volume V per second per unit incident intensity per unit solid angle in the backward direction. The factor 4π is introduced so that in cases where the scattering is the same in all directions, the average energy scattered in all directions per second per unit incident intensity will be just mV.

17.1.5 Fluctuation

The reverberations from two successive pings never reproduce each other exactly. This short-term variability is called "fluctuation." A numerical measure of fluctuation is provided by the variance of the reverberation intensity at a time t seconds after midsignal. More specifically, suppose that a large number n of successive pings are sent out, that records are taken of the n resulting reverberations, and that the n intensities at a time t seconds after midsignal are read off the records. Then if $\bar{I}$ is the average of these n intensities, and $I_1, I_2, \cdots, I_n$ are the n individual intensities, then the fluctuation corresponding to the time t seconds after midsignal is measured by the variance

$$\frac{1}{n} \sum_{i=1}^{n} (I_i - \bar{I})^2. \qquad (4)$$

17.1.6 Coherence

The term "coherence" applied to reverberation refers to a tendency of the received reverberation to occur in the form of pulses or "blobs." The possession of coherence means that if at any instant the reverberation level is high, it is likely that the level will remain high for a little while, and that if the reverberation level is low, it is not likely to become large in a short time. The degree of coherence can be described mathematically in terms of the correlation coefficient ρ between the reverberation intensities at two different times on the same record.

$$\rho = \frac{\overline{[I(t_1) - \overline{I(t_1)}][I(t_2) - \overline{I(t_2)}]}}{\sqrt{\overline{[I(t_1) - \overline{I(t_1)}]^2}} \sqrt{\overline{[I(t_2) - \overline{I(t_2)}]^2}}}. \qquad (5)$$

The bar signifies an average over many successive records.

17.2 DEEP-WATER REVERBERATION LEVELS

In deep water the reverberation heard at ranges past 1,500 yd is almost always volume reverberation. At shorter ranges surface reverberation may exceed volume reverberation, if the sea state is sufficiently high and the transducer beam is horizontal. Pointing a directional transducer downward will usually result in the surface reverberation being less than volume reverberation at all ranges past 100 yd.

17.2.1 Volume Reverberation

The following subsections summarize the known information concerning reverberation from the volume of the ocean. The statements apply only to that portion of the received reverberation resulting from scattering in the ocean volume; the salient facts about reverberation from the sea surface are summarized in section 17.2.2.

THEORETICAL FORMULA FOR VOLUME REVERBERATION LEVEL

The expected volume reverberation level $R'(t)$ at a time t sec after midsignal is given by the formula

$$R'(t) = 10 \log \frac{c_0 \tau}{2} + 10 \log m + J_v$$
$$- 20 \log r - 2A + A_1, \qquad (6)$$

where c_0 is the sound velocity in yards per sec; τ is the ping duration in sec; m is the volume scattering coefficient; J_v is the volume reverberation index; r is the range in yards of the reverberation, $r = \frac{1}{2}c_0 t$; A is the total one-way transmission anomaly to the range r; A_1 is the one-way transmission anomaly to the range r due to the effect of refraction.

Because of surface reflections, not taken into account in equation (6), observed volume reverberation levels with horizontal beams will average about 3 db higher than the levels predicted by that equation.

Dependence on Range

According to equation (6), if the transmission anomaly terms $(-2A + A_1)$ can be neglected, and if the scattering coefficient m is constant throughout the relevant portion of the ocean, then the intensity of volume reverberation should decay with the square of the range; in other words, its level should drop 20 db for each tenfold increase in range. In practice, this simple inverse-square dependence is only seldom observed, because (1) the transmission anomaly terms can rarely be neglected at ranges greater than 1,000 yd; and (2) the value of m often depends on position in the ocean. The well-established deep scattering layers off San Diego, which apparently scatter much more strongly than surrounding regions of the ocean, are examples of the dependence of the scattering coefficient on position.

Though detailed agreement with equation (6) is almost never observed, volume reverberation does tend to decrease rapidly with increasing range, as predicted qualitatively by that equation.

Dependence on Ping Length

According to equation (6), as the ping length is increased the intensity of volume reverberation should increase proportionally. Although more data are needed, measurements to date indicate that equation (6) does describe the dependence of reverberation on ping length. This proportional dependence is also predicted theoretically for surface and bottom reverberation intensities; for these types of reverberation also, more data are needed, but apparently the theoretical relationship is fulfilled.

Dependence on Frequency

The frequency-dependent terms in equation (6) are the volume reverberation index J_v, the transmission anomaly terms $-2A + A_1$, and the scattering coefficient m. The value of the reverberation index can be determined from the pattern function of the transducer by means of equation (27) of Chapter 12.

The transmission anomaly can be estimated by the methods described in Chapter 5. Available data in the frequency range 10 to 80 kc indicate that on the average the scattering coefficient m increases about as the first power of the frequency. However, the data also do not deny the possibility that m is independent of frequency, or that it increases as the square of the frequency.

Magnitude of the Volume-Scattering Coefficient at 24 kc

Observed values of $10 \log m$, inferred from observed reverberation levels, vary between -50 and -80 db, with -60 db as a typical value. The variations of $10 \log m$ have not been correlated, in the studies off San Diego, with variations in any factor other than depth in the ocean. The variation of m with depth off San Diego has not yet been fully explained, but its systematic character seems well established. Using a projector pointed straight down, the measured reverberation off San Diego was found to decrease down to a range of 600 or 700 ft, but then is frequently observed to rise fairly abruptly, and maintain a high value for a depth interval of about 700 ft. The inferred values of $10 \log m$ for depths within the deep scattering layer are often 15 db or more greater than the values of $10 \log m$ at other depths. Some of these deep layers of high scattering power persist in a given area for periods as long as a month or even longer. Although the scatterers in these deep layers have not been definitely identified, it seems probable that they are of biological origin.

17.2.2 Surface Reverberation

The following subsections summarize the known information concerning reverberation from the surface of the ocean. This information applies primarily to reverberation measured with horizontal beams at those ranges where surface reverberation exceeds volume reverberation.

Theoretical Formula for Surface Reverberation Level

The expected surface reverberation level $R'(t)$ at a time t seconds after midsignal is given by the formula

$$R'(t) = 10 \log \frac{c_0 \tau}{2} + 10 \log \frac{m'}{2} + J_s(\theta)$$
$$- 30 \log r - 2A, \quad (7)$$

where m' is the backward scattering coefficient of the surface scattering layer; θ is the angle at the transducer between the sound returning at time t and the horizontal plane; $J_s(\theta)$ is the surface-reverberation index corresponding to the angle of elevation θ; and the other quantities have the meanings given in the section entitled "Theoretical Formula for Volume Reverberation Level" with the further specification that A is the transmission anomaly along the actual ray path to the surface.

Because of reflections from the air-water interface, not taken into account in equation (7), measured surface reverberation levels with horizontal beams will usually be about 6 db higher than the levels predicted by that equation.

DEPENDENCE ON RANGE

According to equation (7), the surface reverberation intensity at short ranges, where the transmission anomaly $2A$ can be neglected, should be proportional to the inverse cube of the range, provided m' and $J_s(\theta)$ also change negligibly with increasing range. This simple inverse cube dependence is observed only rarely. When refraction near the surface is sharply downward, surface reverberation drops abruptly below volume reverberation at the range where the limiting ray dips beneath the surface scattering layer. Moreover, the decay of surface reverberation intensity is usually faster than inverse cube even when downward refraction is weak or absent. For high wind speeds (and therefore high sea states) the decay is especially rapid; for wind speeds greater than 20 mph, the surface reverberation levels usually drop off nearly as the fifth power of the range, and rates of decay as high as the seventh power have sometimes been observed. Factors which may contribute to this unexpectedly high decay rate are: (1) a decrease in the surface scattering coefficient m' as the incident sound ray becomes more nearly horizontal; (2) attenuation; (3) the sound-shadowing effect of surface water waves; and (4) image interference, that is, the interference between direct and surface-reflected waves.

DEPENDENCE ON WIND FORCE

The wind-speed dependence of surface reverberation is most marked at short ranges. At ranges of 1,500 yd or more, with horizontal beams, the received reverberation does not depend on wind speed, and for this reason is ascribed to scattering from the volume of the sea. At a range of 100 yd, as the wind speed increases from 8 to 20 mph, the median reverberation level rises steeply in a manner roughly described by

equation (6) of Chapter 14. With horizontal beams, little increase in level has been observed as the wind speed increases from zero to 8 mph, or as it increases beyond 20 mph.

DEPENDENCE ON FREQUENCY

The frequency-dependent terms in equation (7) are the surface-reverberation index $J_s(\theta)$, the transmission anomaly term $2A$, and possibly the surface-scattering coefficient m'. The value of $J_s(\theta)$ can be determined from the pattern function of the transducer, by equations (40), (41), and (42) of Chapter 12. The transmission anomaly can be estimated by the methods described in Chapter 5 of Part I. Unfortunately, there are no experimental data on the variation of surface scattering coefficients with frequency.

MAGNITUDE OF THE SURFACE-SCATTERING COEFFICIENT

The magnitude of $10 \log m'$ can be obtained from comparison of equation (7) with the measured reverberation at any range. Although this process is open to criticism, since equation (7) does not describe the range dependence of surface reverberation very well, it furnishes us with the only information we now have on the magnitude of the surface scattering coefficient.

By using equation (7), it appears that the increase in surface reverberation at 100 yd as the wind speed increases, noted in the preceding subsection, is due to increases in the surface scattering coefficient m' as the sea becomes rougher. Thus, at a range of 100 yd the median values of $10 \log m'$ obtained by comparing equation (7) with measured levels are -57 db at wind speeds less than or equal to 8 mph, and -22 db at wind speeds greater than 20 mph. At 1,000 yd, for wind speeds greater than 20 mph, $10 \log m'$ averages -31 db. It does not seem possible according to Section 14.2.5 to interpret the reverberation measured at high wind speeds as a result of scattering from a dense layer of bubbles.

17.2.3 Deep-Water Levels with Horizontal 24-kc Beams

For prediction of deep-water, 24-kc reverberation levels with horizontal beams, Figure 31 of Chapter 14 can be used. This figure shows the highest reported reverberation levels, the lowest reported levels, and the median levels, for various ranges and wind speeds.

The main import of this figure is that it indicates both the expected reverberation level and the possible spread in values at any range and wind speed. From the upper and lower limits in the figure were inferred the values of the surface and volume scattering coefficients given in Section 17.2.2.

17.3 BOTTOM REVERBERATION LEVELS

The following subsections summarize the known information concerning reverberation from the ocean bottom. This information is mainly concerned with reverberation from horizontally projected beams in shallow water. Under these circumstances, after a sufficient time has elapsed for the beam to reach the bottom, the received reverberation is preponderantly bottom reverberation.

17.3.1 Theoretical Formula

The expected bottom reverberation level $R'(t)$ at a time t seconds after midsignal is given by the formula

$$R'(t) = 10 \log \frac{c_0 \tau}{2} + 10 \log \frac{m''}{2} + J_b(\theta)$$
$$- 30 \log r - 2A, \quad (8)$$

where m'' is the bottom scattering coefficient, θ is the angle at the transducer between the sound returning at the time t and the horizontal plane, $J_b(\theta)$ is the bottom reverberation index, corresponding to the angle of depression θ, and the other quantities have the meanings given in Section 17.2.1.

With horizontal beams and transducers near the surface, observed bottom reverberation levels will average about 6 db higher than the levels predicted by equation (8), on account of surface reflections.

17.3.2 Dependence on Range

According to equation (8), the bottom reverberation intensity at short ranges, where the transmission anomaly term $2A$ can be neglected, should be proportional to the inverse cube of the range, provided m'' and $J_b(\theta)$ also change negligibly with increasing range. This simple inverse-cube relationship is almost never observed. In the first place, because of the distance between the transducer and the bottom, reverberation from the bottom does not set in until a significant time has elapsed after the emission of the ping. Usually the reverberation then quickly builds up to a peak, corresponding approximately to the time when the edge of the main beam strikes the bottom. After

the peak, the reverberation intensity falls off rapidly, usually about as the fourth power of the range; however, very large deviations from the inverse-fourth power decay have been observed.

The range to the bottom reverberation peak depends on refraction conditions and water depth. In isothermal water, the peak is expected at a range about 12 times the water depth. When the temperature decrease from projector to bottom is greater than 5 degrees, the peak occurs at a range between 4 and 8 times the water depth, depending on the severity of the downward refraction, with a median value of 6 times the depth.

In general, the quantities m'' and $J_b(\theta)$ in equation (8) are dependent on range. However, at ranges past the reverberation peak, both of these quantities usually depend only slightly on range.

17.3.3 Dependence on Frequency

The frequency-dependent terms in equation (8) are the surface reverberation index $J_b(\theta)$, the transmission anomaly $2A$, and the bottom scattering coefficient m''. The value of $J_b(\theta)$ can be determined from the pattern function of the transducer, by equations (53), (41), and (42) of Chapter 12, and, as before, the transmission anomaly can be estimated by the methods of Chapter 5. Measurements on rock bottoms indicate no dependence of the bottom-scattering coefficient m'' on frequency, in the frequency range 10 to 80 kc. It is probable that other bottoms as well would show no dependence of m'' on the frequency of the incident sound, although more data are needed to confirm this point.

17.3.4 Dependence on Bottom

Bottom reverberation levels are not the same over all types of bottoms. Although wide variations are observed, in general the highest reverberation levels are observed over ROCK, lower values over MUD and SAND–AND–MUD, and the smallest values over SAND bottoms. These classifications of bottom type depend on the particle size in the material composing the bottom and are more fully described in Chapter 6.

17.3.5 Bottom Scattering Coefficients at 24 kc

The backward scattering coefficient depends on the angle at which the sound is incident on the bottom.

For a grazing angle of about 10 degrees, a typical value for the angle at which sound in the main beam strikes the bottom, average values for 10 log m'' are -20 db for ROCK, -27 db for MUD, -28 db for SAND–AND–MUD, and -32 db for SAND. Over individual bottoms of a given type, deviations of ± 5 db from these average values may be expected.

There is not much information concerning the dependence of m'' on grazing angle. It appears that for angles between 10 and 30 degrees m'' is roughly proportional to the square of the grazing angle.

17.3.6 Bottom Reverberation Levels with Horizontal 24-kc Beams

Figure 8 of Chapter 15 shows the expected reverberation level at the bottom reverberation peak, as a function of bottom type and of bottom depth below the projector. The height of the peak is a significant quantity in assessing the importance of bottom reverberation in any given situation. For detailed prediction of the levels at ranges past the peak, accurate knowledge is needed of the transmission of sound along the various ray paths to the bottom.

17.4 FLUCTUATION AND FREQUENCY CHARACTERISTICS

17.4.1 Fluctuation

The measured reverberation is probably the resultant of a combination of a large number of small amplitudes of random phase. If so, the probability P that the reverberation intensity will exceed the value I is given by the formula

$$P = e^{-(I/\bar{I})} \qquad (9)$$

where $\bar{I}$ is the average intensity. For the distribution defined by equation (9), the variance defined by equation (4) is $\bar{I}.^2$ Measurements indicate that equation (9) is a fairly good description of the distribution of reverberation intensities. However, the observed fluctuation of reverberation intensity must, in some part, be due to variability in such factors as transmission loss and transducer orientation.

17.4.2 Coherence

Analysis of reverberation records shows that the reverberation tends to occur in the form of pulses or "blobs" of about the length of the ping. For square-topped pings and the intensity distribution defined by equation (9), the correlation coefficient in equation (5) has the value given by

$$\rho = \begin{cases} \left(1 - \dfrac{\alpha}{\tau}\right)^2 & \text{for } \alpha \leqq \tau \\ 0 & \text{for } \alpha \geqq \tau \end{cases} \qquad (10)$$

where τ is the ping length, and $\alpha = |t_1 - t_2|$.

17.4.3 Frequency Spread

For many purposes it is desirable to know the frequency spectrum of reverberation, which gives, as a function of frequency, the energy in each 1-c band. If the reverberation is simply the combination of a large number of individual echoes, each with the same frequency spectrum as the emitted ping, then the resultant reverberation should also have the same spectrum as the ping. This conclusion is probably not far wrong, although precise measurements of the frequency spectrum of reverberation have not often been attempted.

The distribution of the instantaneous frequencies of the reverberation (defined in Section 16.3) is also useful information. This distribution can be measured by an instrument known as the "periodmeter." Periodmeter measurements indicate, among other things, that the spread of instantaneous frequencies in the heterodyned reverberation depends on the audio output frequency and the pulse length, but does not depend on the frequency of the outgoing ping.

17.4.4 Wide-Band Pings

The fluctuation of the reverberation with wideband pings is probably not very much different in magnitude from the fluctuation with narrow-band pings. However, the rapidity of the fluctuation is increased as the frequency band is widened. In general, the average reverberation levels are not affected by widening the frequency band of the outgoing ping.

17.5 FUTURE RESEARCH

Reverberation studies are a powerful tool in the investigation of properties of the ocean. Information from such studies is necessary to determine definitely the nature of the scatterers, is useful in evaluating theories of transmission loss, and can cast light on the temperature microstructure of the ocean. Also, these studies are needed to fill in the gaps in our knowledge of the reverberation levels to be expected under various conditions. In general, measurements of volume reverberation will cast the most light on the fundamental properties of the ocean. Experiments of the following sort are indicated:

1. Measurements of reverberation over a very wide range of frequencies, from sonic frequencies up to several hundred kilocycles.

2. Measurements of the dependence of volume reverberation on transducer directivity, which would help evaluate the importance of multiple scattering.

3. Careful correlation of measured volume reverberation levels with simultaneous measurements of temperature microstructure.

4. Careful correlation of measured reverberation levels with observed transmitted sound levels, especially with such features as sound penetration into predicted shadow zones.

5. Reverberation measurements with deep projectors to demonstrate any fundamental differences between the upper and lower layers of the ocean.

6. A thorough investigation of the deep scattering layers, including the use of underwater photography.

7. Measurements of reverberation in large freshwater lakes.

8. More complete studies of the dependence of reverberation on ping length, especially with very short pings.

9. Investigation of various probability and correlation coefficients of the sort discussed in Chapter 16.

10. Measurements of the dependence of surface and bottom scattering coefficients on the grazing angle of the incident sound.

11. Correlation of measured surface reverberation levels with simultaneous measurements of optical transparency and of entrapped air or other material.

Theoretical investigations of various questions are also required, so that the results of these experiments may be correctly interpreted. Most of these theoretical investigations will be of importance in the subject of transmission as well as reverberation. Typical subjects for theoretical research would be the reflection of sound from a rough surface, and scattering of sound by thermal microstructure.

A final subject of great importance, which requires both theoretical and experimental research, is the development of instrumental means for recording and computing various time averages which are of interest in reverberation studies. Such instrumental procedures would greatly reduce both the time and expense involved in the suggested experiments listed above.

PART III

REFLECTION OF SOUND FROM SUBMARINES AND
SURFACE VESSELS

Chapter 18

INTRODUCTION

OBJECTS MAY BE DETECTED by the echoes they return. In water, sound waves are absorbed and scattered very much less than radio or light waves. Consequently, sound waves are particularly useful in detecting distant objects under water by means of echo-ranging, that is, sending out a sound signal and listening for a returning echo.

The loudness of an echo depends on how much sound is absorbed and how much sound is reflected. As a signal is sent out, the energy spreads; some of it is immediately absorbed by the water and is dissipated as heat energy. The transmission and absorption of underwater sound have been studied extensively in subsurface warfare, and are described in Chapters 1 to 10 of this volume. Some of the energy is scattered at random back to the echo-ranging projector, either by particles or other inhomogeneities in the water, or by the ocean surface or bottom. This scattering gives rise to a phenomenon known as reverberation, which has also been investigated in detail and is treated in Chapters 11 to 17 of this volume. The sound distinctly reflected from an obstacle or *target* in the path of the sound beam — such as a submarine or whale — gives rise to an echo. Chapters 18 to 25 discuss the reflection of sound from various underwater targets.

Many types of targets are encountered in practice. In particular, recognizable echoes have been received from schools of fish, whales, patches of kelp and seaweed, and from sunken wrecks or prominent irregularities on the ocean bottom in shallow water. Certain water conditions give rise to echoes; at very short ranges, echoes from ocean swells have been observed. Wakes, "pillenwerfer," and other types of bubble screens are effective targets. Their acoustic properties have been studied both theoretically and experimentally, and are described in Chapters 26 to 35. In addition, icebergs have been detected by echo-ranging, although no such echoes have been measured.

18.1 TARGET STRENGTH

In subsurface warfare, submarines, surface vessels, and underwater mines are the most important tar-

gets. The reflection of sound from submarines and surface vessels has been investigated in terms of *target strengths*, a quantitative measure of their reflecting characteristics. Submarine target strengths have been studied as a function of the size and shape of the submarine, its orientation with respect to the echo-ranging projector, the distance from the submarine to the projector, and the frequency of the echo-ranging sound. Chapters 18 to 25 summarize all available information along these lines.

Echo-ranging measurements on submerged submarines under more or less controlled conditions have resulted in a large collection of target strength data. In addition, submarine target strengths have been computed theoretically and measured in experiments with scale models. Unfortunately, very little is known about the reflection of sound from surface vessels, as no exhaustive series of tests has been made with this sole object in mind. The data describing surface vessel target strengths are few and scattered; conclusions are tentative and uncertain. No measurements have been made of the reflecting characteristics of mines. However, spheres of various sizes have been frequent experimental targets, and the results of echo-ranging measurements on spheres are probably applicable to small-object location and the detection of mines.

18.2 USES

Tactically, knowledge of how submarines and surface vessels reflect sound is very important. A quantitative evaluation of the contribution which the reflecting characteristics of the target make to the received echo intensity is necessary in order to predict maximum echo ranges accurately. In submarine operations, the reflecting properties of the submarine should be known so that effective evasive maneuvers may be taken to reduce, as far as possible, the chance of sonar contact by enemy antisubmarine vessels. For example, it is known that the strongest echo is obtained when the submarine presents its beam to the echo-ranging vessel. Therefore, keeping the attacking vessel off the beam of the submarine is important in reducing the maximum range at which the enemy vessel can detect an echo from it. It should also be useful for submariners to know under what conditions a submarine is most vulnerable to contact by echo ranging, that is, in what position, at what aspect, or depth, or speed. In addition, any countermeasures designed to reduce the probability of contact, such as by making the submarine a less effective acoustic reflector, require a quantitative knowledge of the reflecting characteristics of the submarine.

Similarly, such knowledge would be useful to antisubmarine vessels in suggesting searching or attacking operations. It is also required for the efficient design and operation of many underwater echo-ranging devices, such as certain decoys or mines. Information on how much sound mines will reflect is important both in the design of mine detection gear and in the evaluation of echo-ranging equipment tests carried out with a particular type of mine.

This report emphasizes how submarines and surface vessels reflect sound under field conditions. Chapter 19 introduces the concept of target strength, defines it in terms of quantities directly measurable, and derives an expression for the target strength of a perfectly reflecting sphere on the basis of ray acoustics. Chapter 20 presents the theoretical background of reflection and scattering of sound from bodies of various shapes on the basis of wave acoustics, and reviews the theoretical calculations of the target strength of a submarine. The technique of the direct, field measurements of submarine target strengths are described in Chapter 21, the indirect measurements in Chapter 22, and all the results of submarine target strength measurements are summarized and discussed in Chapter 23. Finally, Chapter 24 comprises all available information on surface vessel target strengths and Chapter 25 summarizes briefly the reflection of sound from both submarines and surface vessels.

Chapter 19

PRINCIPLES

WHEN A TARGET is in the path of a sound beam, the intensity of the reflected sound measured some distance away will, in general, depend on many factors, such as the intensity of the sound striking the target, the distance from the target to the point where the echo is measured, and the size, shape, and orientation of the target. Often it is desirable to separate these different factors so that the effects of the size, shape, and orientation of the target may be discussed independently of all other factors.

Such a separation is possible only when the radii of curvature of the sound waves striking the target and returned to the receiver are both much larger than the dimensions of the target, in other words, when the waves incident on the target and the waves reflected back to the receiver are essentially plane. In terms of ray acoustics, the incident sound rays must be substantially parallel over the area of the target which they strike, and the reflected sound rays must be parallel over the area of the face of the receiver.

In this chapter target strength is defined quantitatively in terms of the echo level, the source level, and the transmission loss. Then the target strength of a sphere is derived as a function of its radius. Finally, the effect of pulse length on target strength is examined for a simple case. Ray acoustics is employed throughout the chapter, and the arguments are necessarily idealized. No account is taken of the wave character of sound; in other words, all effects attributable to the wave nature of sound such as interference, diffraction, and phase differences are explicitly ignored. The conditions under which this approximation is valid are discussed in Section 19.4. A more detailed theory of target strength in terms of wave acoustics is presented in Chapter 20.

19.1 DEFINITION OF TARGET STRENGTH

Let I_0 be the intensity of the incident sound striking a stationary target, and I_r the intensity of the reflected sound measured at some particular point. If I_0 is doubled, I_r will also be doubled, other factors remaining unchanged; that is, the intensity of the reflected sound will be directly proportional to the intensity of the incident sound.

For a given value of I_0, I_r will depend on the orientation of the target relative to the incident sound and also on where the echo is measured. This dependence of I_r may be quite complicated. In practical echo ranging, however, the problem is simplified because the echo is always measured back at the source — in other words, it is always measured in the same direction as the projected sound, and the target strength depends only on the orientation of the target. Therefore, it will be assumed throughout this chapter that the echo is measured at the source. Although this admittedly is not the most general case, it is the only case of practical importance for echo ranging.

19.1.1 Inverse Square Transmission Loss

The dependence on distance, although complicated near the target, becomes very simple far away from the target, if the sound rays are assumed to travel in straight paths in an ideal medium, with boundaries so far away that their effects on sound propagation can be neglected. It has been shown in Section 2.4.2 that the intensity of sound from a point source, in this ideal case, falls off inversely as the square of the distance from the source. This same inverse square law applies to sound reflected from any target at distances much larger than the dimensions of the target, since at such distances the target behaves as a point source of sound.

Why the inverse square law holds for the intensity of sound reflected from any target, at large but not at small distances, may best be understood by studying Figure 1. Here are shown rays reflected from a target, A to a point near the target, and B to a point far away from the target. Rays reaching a point near the target come from different points on the target and from various directions, if the surface is irregular.

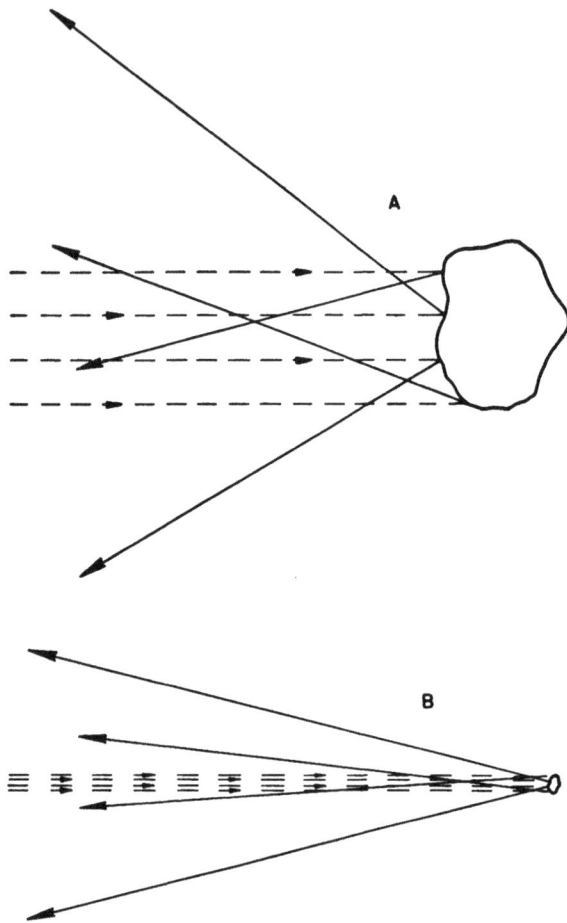

FIGURE 1. Reflected rays at short and long ranges.

Therefore, the way in which the sound intensity near the target varies from point to point is complicated. Rays reaching a point far away from the target all come from essentially the same direction, no matter from what part of the target they are reflected. Thus the target "looks like" a point source, and the inverse square law of intensity will hold. This conclusion, based solely on ray acoustics, is reinforced by considerations of wave acoustics, mentioned in Section 19.4 and described in more detail in Chapter 20.

Sufficiently far away from the target, then, I_r will be not only directly proportional to I_0 but also inversely proportional to the square of the distance r, or

$$I_r = k\frac{I_0}{r^2}. \quad (1)$$

Here k is a constant which in general depends on the size, shape, and orientation of the target. It does not depend on the strength of the sound striking the target, or on the distance from the target, provided

I_r is measured far enough away from the target to make certain that the intensity of the reflected sound will follow the inverse square law. Incidentally, this relation is not valid for explosive sound, which is treated in Chapters 8 and 9.

Now according to equation (89) in Chapter 2, the intensity I_0 of the incident sound striking the target is equal to the intensity F of the projected sound 1 yd away from the source, divided by the square of the distance r from the source to the target, provided that r is much larger than the dimensions of the source. Then

$$I_0 = \frac{F}{r^2}. \quad (2)$$

Substitute equation (2) into equation (1), and

$$I_r = k\frac{F}{r^4}. \quad (3)$$

Equation (3) is particularly interesting because it shows that, for an ideal medium, the intensity of an echo is inversely proportional to the fourth power of the range, as long as the echo is measured at the source and the range is much larger than the dimensions of the target or source. If logarithms are taken and equation (3) expressed in decibels,

$$10 \log I_r = 10 \log k + 10 \log F - 40 \log r. \quad (4)$$

19.1.2 General Transmission Loss

All these equations are derived on the assumption that the medium through which the sound travels is ideal, that all the sound is transmitted freely without refraction, absorption, or scattering, and that the boundaries of the medium are so far away that their effects on the propagation of sound waves may be neglected. In other words, as the sound travels each way, its intensity falls off according to the inverse square law alone. The drop in intensity each way, in decibels, is the *transmission loss H*, which for this ideal case is simply $20 \log r$. The total transmission loss $2H$ to the target and back again is then $40 \log r$.

Generally, however, the intensity of transmitted sound under water does not fall off according to the inverse square law alone. Sound is absorbed and scattered in sea water. It may be bent by temperature gradients and consequently focused or spread out. Often the surface and bottom of the ocean significantly affect both transmitted and reflected sound. Therefore, H will seldom exactly equal $20 \log r$, and

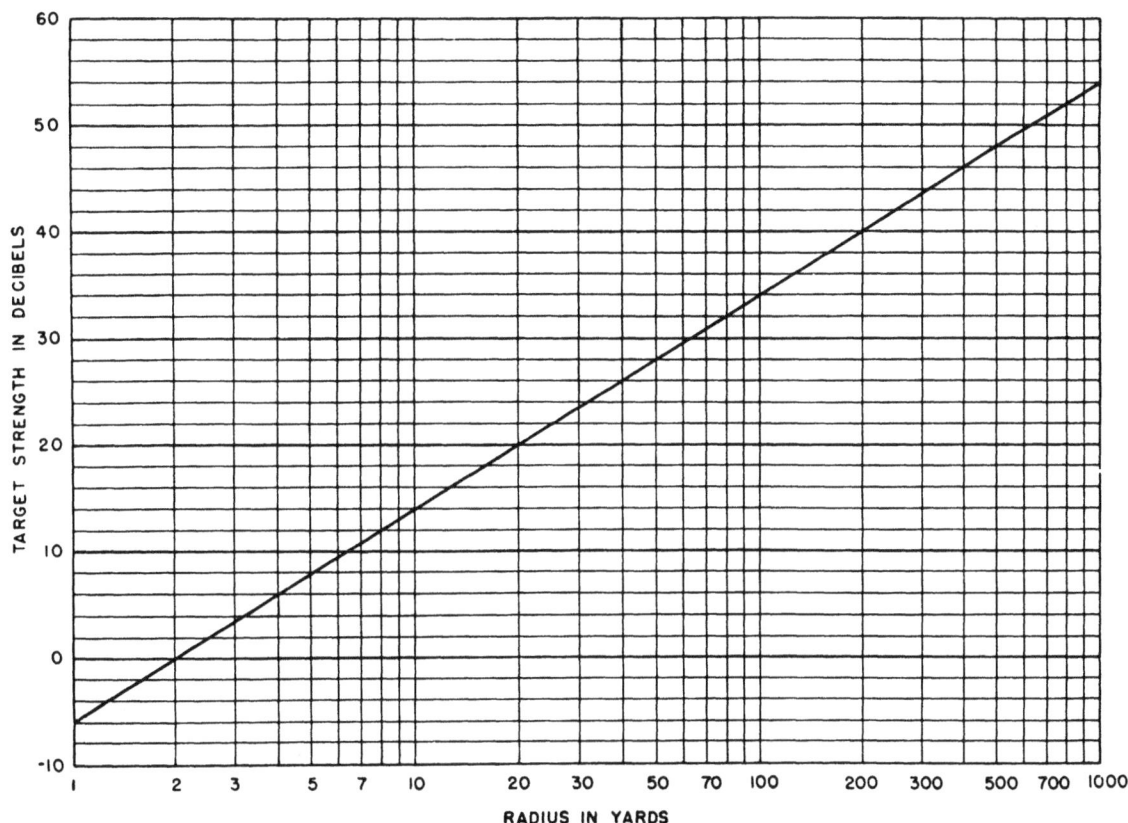

FIGURE 2. Target strength of a sphere.

the two-way transmission loss is more conveniently represented by the more general function $2H$ than by $40 \log r$. Equation (4) then becomes

$$10 \log I_r = 10 \log k + 10 \log F - 2H. \quad (5)$$

The total transmission loss $2H$ cannot be predicted or estimated very reliably because of widely varying oceanographic conditions. Instead, it must actually be measured during the course of the experiment.

19.1.3 Fundamental Definition

By defining the *target strength* T as $10 \log k$, the *echo level* E as $10 \log I_r$, and the *source level* S as $10 \log F$, equation (5) becomes

$$T = E - S + 2H \quad (6)$$

where $2H$ is the total transmission loss in decibels from the source out to the target and back to the source again. Equation (6) is the fundamental definition of target strength. This equation is always used in the computation of target strengths measured at sea since it involves only directly measurable quanti-

ties — that is, echo level, source level, and transmission loss from the source out to the target and back to the source.

Since I_0 and I_r are measured in the same units, it is evident from equation (1) that k has the dimension of an area and the value of T will depend on the units which are used. Since the yard is used in range-prediction work as the unit of length, the source level is defined in terms of the intensity at 1 yd, and the transmission loss, which enters twice into equation (6), is defined in terms of the intensity drop from a range of 1 yd out to the range of the target. Consequently k in equation (3) may be expressed in square yards.

Equation (6) was derived from physical concepts in order to express as a sum of separate terms the effects on the strength of the received echo of (1) the size, shape, and orientation of the target; (2) the intensity of the source; and (3) the range of the target. This separation can be realized only at long ranges, where the sound reflected from the target behaves as if it were emitted from a point source and the target strength becomes independent of the

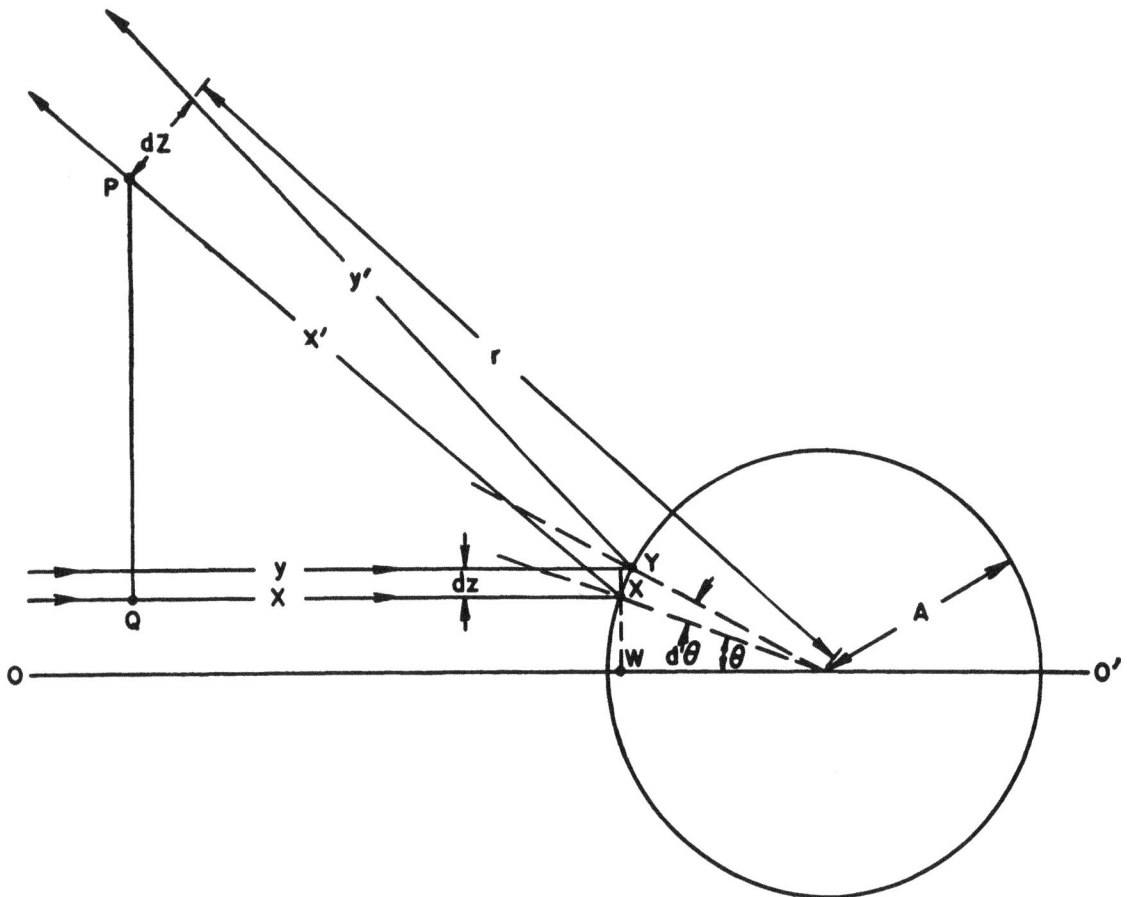

FIGURE 3.   Uniform reflection from a sphere.

range. At long ranges, then, only the transmission loss term depends on the range.

At short ranges, however, the target strength depends on the range as well as on the size, shape, and orientation of the target (see Section 20.4.4). If the source is so close to the target that different parts of the target are struck by sound of different intensities, or if the receiver is so close that the spreading of the sound reflected from the target to it is not the same as the spreading from a point source, the target strength term will depend on range. Therefore, at short ranges equation (6) does not serve primarily to separate the effects of range, transmission conditions, source level, and target characteristics on the echo level, but rather to define target strength under the particular conditions of that measurement.

19.2   TARGET STRENGTH OF A SPHERE

Because a sphere is perfectly symmetrical, the echoes which it returns to a sound source are com-

pletely independent of its own orientation. For this reason, spheres are convenient targets and have frequently served as experimental targets in echo-ranging measurements. In this section, the target strength of a sphere will first be derived simply and intuitively by considering the total intercepted and reflected energy without regard to the angular distribution of energy within the reflected sound beam. Then a more rigorous derivation — within the framework of ray acoustics — will be presented, in which the angular distribution of the reflected energy is considered in detail.

19.2.1   Simple Derivation

Consider a plane wave of sound of intensity I_0 striking a sphere of radius A and cross-sectional area πA^2. Then the total sound energy intercepted by the sphere per unit time will be $\pi A^2 I_0$ and, if reflection is perfect, the total sound reflected from the sphere per unit time will also be $\pi A^2 I_0$.

Now assume that this sound energy is reflected uniformly in all directions. At a distance r from the center of the sphere, it will be spread uniformly over the surface of a sphere of radius r or over the surface area $4\pi r^2$. Since the intensity I_r of the reflected sound equals the total energy $\pi A^2 I_0$ reflected by the target sphere per unit time, divided by the area $4\pi r^2$ over which it is distributed, then at a distance r from the sphere

$$I_r = \frac{\pi A^2}{4\pi r^2} I_0 = \frac{A^2}{4r^2} I_0. \tag{7}$$

But, from equation (1)

$$I_r = k\frac{I_0}{r^2}, \tag{8}$$

where r is the distance from the target to the point where the echo is measured. Therefore by substitution

$$k = \frac{A^2}{4} \tag{9}$$

and

$$T = 10 \log k = 20 \log \left(\frac{A}{2}\right), \tag{10}$$

where T is the target strength and A the radius of the sphere. With the yard chosen as the unit of length, A becomes the radius of the sphere in yards, and from equation (10) it is evident that the target strength is the echo level of the target in decibels above the echo level from a sphere 2 yd in radius. Target strengths for spheres of various radii are shown in Figure 2.

19.2.2 Rigorous Derivation

This derivation explicitly assumed that sound is reflected from a sphere uniformly in all directions. To justify this assumption, consider the same sphere of radius A in Figure 3. Two adjacent rays x and y separated by a distance dz are traveling parallel to OO' and strike the sphere at X and Y respectively, making angles of θ and $\theta + d\theta$ with the sphere radii drawn to the points X and Y. From Figure 4,

$$dz = XY \cos \theta = A \cos \theta d\theta. \tag{11}$$

Now rotate rays x and y about OO'. These rays will describe circular cylinders and dz will generate an area ds between them, where

$$ds = 2\pi XW dz = 2\pi \, (A \sin \theta) \, (A \cos \theta d\theta) \tag{12}$$

or

$$ds = 2\pi A^2 \sin \theta \cos \theta d\theta. \tag{13}$$

The total energy dJ striking the sphere at angles between θ and $\theta + d\theta$ will be the product of the in-

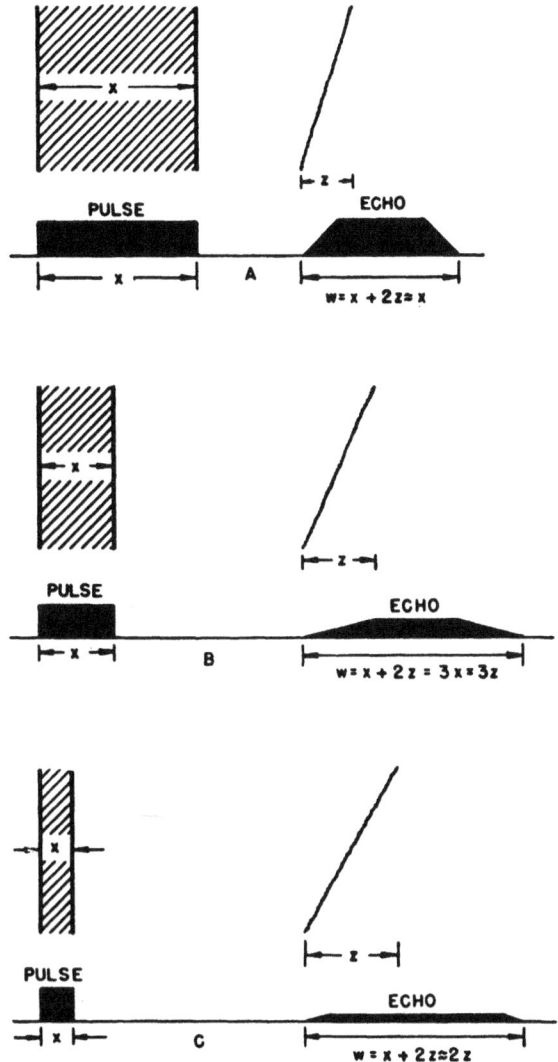

FIGURE 4. Effect of pulse length on target strength, echo length, and echo structure.

tensity I_0 of the incident sound and the cross-sectional area ds, or

$$dJ = I_0 ds = 2\pi A^2 I_0 \sin \theta \cos \theta d\theta. \tag{14}$$

Now consider the reflected rays x' and y' making angles of 2θ and $2\theta + 2d\theta$ with OO' in Figure 3. At a distance r from the center of the sphere, x' and y' will be separated by a distance dZ. At a distance much larger than the radius of the sphere, dZ becomes much greater than XY; and x' and y' may be replaced by r.

$$dZ = r\, 2(d\theta) = 2rd\theta. \tag{15}$$

Again rotate the rays about OO', and dZ will generate the area dS between them, where

$$dS = 2\pi PQ dZ = 2\pi(r \sin 2\theta) \, (2rd\theta) \tag{16}$$

or $\quad dS = 4\pi r^2 \sin 2\theta d\theta = 8\pi r^2 \sin \theta \cos \theta d\theta.$ $\quad$ (17)

Since the intensity of the reflected sound equals the energy reflected per unit time divided by the area over which it is distributed, then

$$I_r = \frac{dJ}{dS} = \frac{2\pi A^2 \sin \theta \cos \theta d\theta}{8\pi r^2 \sin \theta \cos \theta d\theta}I_0 = \frac{A^2}{4r^2}I_0. \quad (18)$$

Thus I_r is independent of θ and therefore is independent of the direction of the reflected sound, and equation (18) is identical with equation (7) derived from a simpler analysis. Rigorously, then

$$T = 10 \log k = 10 \log \left(\frac{A^2}{4}\right) = 20 \log \left(\frac{A}{2}\right), \quad (19)$$

where T is the target strength and A the radius of the sphere.

Equation (19) applies only to target strengths measured far away from the sphere. Close to the sphere, the target strength will also depend on both the direction θ and the range r.

19.3   EFFECT OF PULSE LENGTH

So far it has been tacitly assumed that continuous sound strikes the target and is reflected back to the projector. Usually, however, sound pulses of finite length are sent out, and most target strengths are measured with such sound pulses. In general, target strength will be a function of pulse length, and the dependence of echo intensity on signal length must be investigated.

Consider a curved surface, such as a sphere or an ellipsoid, each part of which reflects sound specularly as a mirror would. This surface is normal to the incident beam at only one point, and only one ray is reflected back to the projector in the direction of the incident ray. Therefore, the echo intensity — and consequently the target strength — will be independent of signal length, and the echo structure will accurately reproduce the signal structure, unless multiple transmission paths which result from surface-reflected or bottom-reflected sound, for example, give rise to multiple echoes. This result, derived on the basis of ray acoustics, is not valid if very short pulses are used, since the wave character of sound must then be considered. However, this result is correct if the pulse is at least several wavelengths long.

On the other hand, consider an extended rough surface, each part of which reflects sound in all directions. A pulse τ seconds long is sent out from a projector a distance r from the target which has an extension z in the direction of the incident beam, as illustrated in Figure 4. Now the first part of the signal will reach the nearest part of the target at a time r/c after it was emitted where c is the velocity of sound and will be returned to the projector at a time $2r/c$. The last part of the signal will leave the projector at a time τ, reaching the nearest part of the target at $\tau + r/c$ and the farthest part of the target at a time $\tau + r/c + z/c$; it will return to the projector at a time $\tau + 2r/c + 2z/c$. The duration of the echo will be the difference between the time when the first part of the signal reaches the nearest part of the target and is returned, and the time when the end of the signal is reflected from the farthest part of the target and is received at the projector. Then if the duration of the echo is σ,

$$\sigma = \tau + 2z/c. \quad (20)$$

19.3.1   Long Pulses

First, let the signal be long compared to the extension of the target (Figure 4A). Then

$$\sigma \approx \tau \quad (21)$$

and the echo length will approximately equal the signal length. Assume that the reflected energy is always directly proportional to the incident energy, and therefore to the product of the signal intensity and the signal length. Then the echo intensity will depend on the signal intensity but not on the signal length.

Now let the signal length equal the depth of the target in the direction of the beam (Figure 4B). Then

$$\sigma = \tau + \frac{2z}{c} = 3\tau, \quad (22)$$

and the echo length will be three times the signal length. The echo will no longer resemble the signal, as the echo intensity grows to a maximum when the target is illuminated by the entire signal.

19.3.2   Short Pulses

Lastly, let the signal be short compared to the depth of the target (Figure 4C). Then

$$\sigma \approx \frac{2z}{c}, \quad (23)$$

and the echo length will approximately equal twice the extension of the target in the direction of the beam. The echo intensity now will depend on the signal length as well as the signal intensity, since the reflected energy will be less for a short signal than a long signal and therefore — as long as the echo length remains constant — the echo intensity will be reduced.

For short pulses, then, the echo intensity and therefore the target strength will depend on the pulse length. In practice, however, fluctuations in the course of each echo, and from echo to echo, tend to obscure this relationship for any individual echo. For long pulses, the echo very closely reproduces the signal envelope, which is usually square-topped. For short pulses, however, fluctuations in echo intensity result in a very irregular hashed structure where a sharp peak or group of peaks stands out clearly against a background which is sometimes 10 db lower.

The peak echo intensity, which is usually used in computing target strengths, is, in general, different from the average echo intensity. Therefore, for short pulses the peak echo intensity may be considerably different from the average echo intensity and may vary in a different way with signal length. Peak and average echo intensities, and how they vary with signal length, are discussed in Sections 21.6.4 and 23.5.1.

19.4 WAVE CHARACTER OF SOUND

Ray acoustics has been used exclusively throughout this chapter in defining target strength, in deriving the target strength of a sphere, and in discussing target strength as a function of experimental variables, just as ray acoustics was employed in Chapter 3 of this volume in treating the transmission of sound through sea water. Experience shows, however, that sound does not always travel in straight lines, and that, for many purposes, ray acoustics is inadequate in explaining and interpreting underwater sound phenomena. An alternative approach in terms of wave acoustics becomes necessary.

Throughout this chapter it has been tacitly assumed that sound is propagated along straight lines as sound rays, and that reflection is wholly specular, in other words, that the angle of reflection always equals the angle of incidence. Many modifications must be introduced if allowance is made for the various wave phenomena affecting echo ranging. Sound is diffracted when it strikes a target or parts of a target whose dimensions approximate its wave length. Thus, the previous discussion applies only to targets considerably larger than the wave length. For the same reason, these results apply only to pulses whose length in the water is at least several wave lengths. In addition, sound reflected from one part of a target may interfere with sound reflected from other parts. Much of the fluctuation commonly encountered in analyzing echoes from underwater targets is attributable to interference. The results in the preceding section on echoes from extended targets are valid only if the interference effects arising from constructive or destructive interference can be eliminated by averaging over several successive echoes. The effect of the wave length of sound on target strength for both specular and nonspecular reflection is discussed in greater detail in Sections 20.4 and 20.6.

Chapter 20

THEORY

I~N CHAPTER~ 19 the concept of target strength was introduced and its meaning defined quantitatively; then the target strength of a perfectly reflecting smooth sphere was derived in terms of ray acoustics. The theoretical background will be presented in this chapter in terms of wave phenomena with a mathematical discussion of the reflection of a sound wave from a target of any shape, and a review of the early theoretical calculations of the target strength of a submarine.

In principle, the reflection of sound from a target can be exactly determined by solving the wave equation derived in Chapter 2 of Part I, as long as the proper boundary conditions at the surface of the target are satisfied. In practice, an exact computation along these lines is mathematically very difficult; the difficulties are most marked for targets large compared to the wavelength of the incident sound. Even for a sphere the rigorous analysis which has been worked out [1, 2] is rather complicated. Numerical applications of these precise formulas have been published [3] for a rigid sphere, whose circumference is from 1 to 10 times the wavelength; the results provide an interesting example of the exact behavior of reflected sound in one simple case. However, even for such relatively small targets the mathematical analysis becomes tedious.

20.1 APPROXIMATIONS

To obtain more general results, various approximations must be made, physical as well as mathematical. The mathematical assumptions made in this chapter are fairly standard and are believed to give essentially correct results. The physical assumptions about the nature of the reflecting surface are more important, however, and require some justification.

In the first place, most of this chapter applies only to targets which are large compared to the wavelength of the sound, and whose surface is smooth; in other words, the radius of curvature of the surface is

also large compared to the wavelength. These restrictions seem legitimate for most targets of practical interest in echo ranging.

In the second place, the material of which the target is composed is assumed to be rigid. In terms of sound reflection, a target is said to be rigid if $\rho_1 c_1 / \rho_2 c_2$ is negligibly small, where ρ_1 and c_1 are the density and sound velocity in the surrounding medium, and ρ_2 and c_2 are the density and sound velocity in the target. When this condition is not fulfilled the problem becomes much more complicated. In most cases of interest to subsurface warfare, the target is bounded by thin metal plates, inside which there may be air or water. The reflection of sound from such plates has been studied,[4, 5] and the results obtained show that even for plates only $\frac{1}{4}$ in. thick, such as generally constitute submarine superstructures, the reflection is practically perfect; transmission and absorption are negligible. Thus, the assumption of perfect reflection from practical targets seems justified.

Some of the additional assumptions which may be made in the discussion of the reflection from targets are discussed in an early British report.[6] This work is particularly interesting because it presents the most complete available application of theory to the target strength of underwater objects.

The essential elements of the theory of target strength, restricted by the physical assumptions which have been made here, are presented in the following sections. First, an approximate but general formula is derived for the pressure of the sound reflected from a target. In Section 20.3, this result is further simplified to give an equation for the target strength of a reflecting surface in terms of the so-called *Fresnel zone* theory. This latter equation is then used to find practical formulas for the target strengths of simple geometrical shapes, which are applicable to the major reflecting properties of submarines; the application to an actual submarine is described in Section 20.5. All this latter analysis applies only to long pulses. The last two sections are devoted to a qualitative discussion of the reflection from targets small compared to the wavelength, and the echoes obtained with very short pulses.

20.2   REFLECTED PRESSURE

Consider first a sound beam striking a surface element dS of a perfectly reflecting, smooth and rigid underwater target. Since this surface is rigid, the primary effect of the target is to prevent the water from moving perpendicularly to dS at the surface of the target. In other words, at the surface of the target, the velocity u of the water, measured along a line perpendicular to the surface, must be zero, or

$$u_z = 0, \tag{1}$$

where the z axis, at the point of incidence, is perpendicular to the target surface. By differentiating equation (1) with respect to the time t,

$$\frac{\partial u_z}{\partial t} = 0. \tag{2}$$

But from equation (17) in Chapter 2 of this volume,

$$-\rho\frac{\partial u_z}{\partial t} = \frac{\partial p}{\partial z} \tag{3}$$

where ρ is the density of the medium, p the pressure of the sound wave, and z the coordinate perpendicular to the surface.

20.2.1   Boundary Condition

Substitution of equation (3) into equation (2) gives

$$\frac{\partial p}{\partial z} = 0, \tag{4}$$

which means that for a rigid target the component of the pressure gradient perpendicular to the surface must vanish at the surface. This is the boundary condition which the solution of the wave equation must satisfy.

In the absence of the target, the sound source will send out a wave whose resulting pressure at any particular point may be denoted by p_1; then p_1 must be a solution of the wave equation [equation (27) in Chapter 2.] In the presence of the target, this pressure p_1 does not satisfy the resulting boundary conditions at the surface of the target. The actual sound pressure p, which must satisfy both the wave equation and the boundary conditions at the target surface, may be written as

$$p = p_1 + p_2, \tag{5}$$

where p_2 constitutes the correction which must be added to the undisturbed sound pressure p_1 in order to satisfy the boundary conditions at the surface of the target.

By differentiating equation (5) with respect to z and by substituting the result into equation (4)

$$\frac{\partial p_1}{\partial z} + \frac{\partial p_2}{\partial z} = 0, \tag{6}$$

which is another way of expressing the boundary condition.

Because the wave equation is a linear homogeneous differential equation, the difference between the two solutions p and p_1 is again a solution, and p_2 by itself must therefore satisfy the wave equation. In other words, the total sound field may be interpreted as the combination of two sound fields. One of these, whose pressure at any specified point is p_1, is called the incident sound; the other, whose pressure at the same point is p_2, is the reflected sound. Each of these quantities satisfies the wave equation, but only their sum satisfies the boundary conditions at the target. In some places, the measured sound pressure may occasionally consist wholly of one or the other of these two sound fields, depending on whether only the sound projected from the source, or only the sound reflected from the target is measured. The problem tackled in this chapter is the evaluation of the reflected sound alone, since it is this quantity which is most important in echo ranging. Therefore, an expression for p_2 must be derived.

20.2.2   Mathematical Formulation

To obtain, rigorously, a general expression for p_2 is usually a very difficult problem. It is comparatively easier to obtain an approximate solution by distributing, over the surface of the target, point sources of sound. Then, if the distribution and strengths of these point sources over the area are correctly chosen, these point sources will emit sound in such a way as to cancel the pressure gradient of the wave incident on the surface, thus satisfying the condition (6).

For a single point source, the solution of the wave equation is

$$p = \frac{B}{r}e^{2\pi i(ft-r/\lambda)} \tag{7}$$

where p is the pressure of the sound field at a distance r from the source, B is a constant which measures the strength of the point source, i is $\sqrt{-1}$, t is the time,

f is the frequency and λ the wavelength of the sound. If the reflecting target itself is considered to be made up of many point sources distributed over its surface, p becomes p_2, the reflected pressure; and the pressure dp_2 produced by all the point sources located in a surface element dS is

$$dp_2 = \frac{G}{r}e^{2\pi i(ft-r/\lambda)}dS, \qquad (8)$$

or the pressure p_2 produced by the entire target becomes

$$p_2 = \int_S \frac{G}{r}e^{2\pi i(ft-r/\lambda)}dS \qquad (9)$$

where G is essentially the average value of B in equation (7) for each individual source multiplied by the number of sources per unit area; the integral is evaluated over the entire area S.

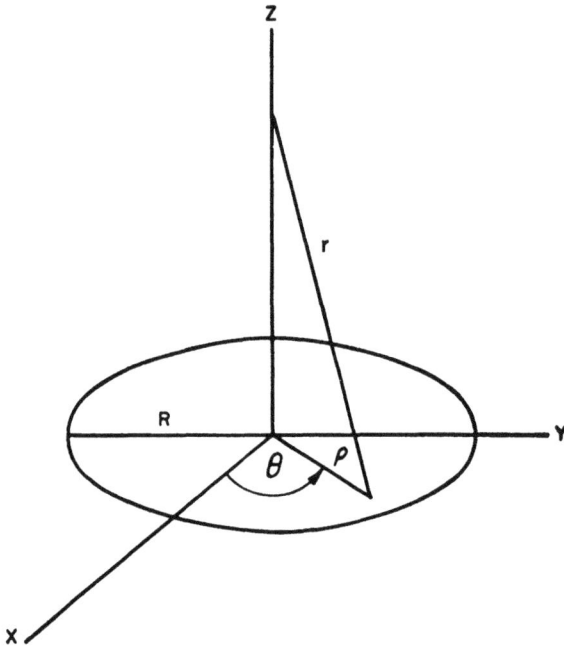

FIGURE 1.   Transformation to polar coordinates.

This quantity G is a measure of the number and strength of the point sources over the area; in general G will vary over the target surface. The function G must be chosen so that the resulting sound pressure p_2 satisfies the boundary condition (6) on the surface of the target.

The value of G at a particular point of the target surface will be assumed to be completely determined by the incident sound pressure at that point. This assumption is not rigorously correct, but it leads to a good approximation if the target has a surface whose radius of curvature is everywhere large compared with the wavelength.

First, a relationship between the value of G at any point and the resulting gradient of p_2 at that point will be derived. Then, the gradient of p_2 may be replaced by minus the gradient of p_1, because of the boundary condition (6). In this manner, a direct relationship will be obtained between the incident sound field p_1 on the target surface, and the value of G required to compensate the gradient of p_1.

Because of the assumption made that the gradient of p_2 at the point on the target surface is determined primarily by the value of G at that point, a particularly simple model may be considered and the result generalized. The pressure gradient at the center of a disk illustrated in Figure 1 will be derived, on the assumption that G is constant over the surface; in other words, the density of point sources on the surface of the disk is assumed to be uniform. If polar coordinates ρ and θ are introduced, the integral (9) for the pressure on the z axis can be transformed as follows:

$$p_2 = 2\pi G \int_{\rho=0}^{R} e^{2\pi i(ft-r/\lambda)}\frac{\rho d\rho}{r},$$

or, since $r^2 = \rho^2 + z^2$,

$$p_2 = 2\pi G \int_{r=z}^{\sqrt{R^2+z^2}} e^{2\pi i(ft-r/\lambda)}dr. \qquad (10)$$

Equation (10) may be integrated directly and gives for the sound pressure on the z axis

$$p_2 = i\lambda G \left[e^{2\pi i\left(ft-\frac{\sqrt{R^2+z^2}}{\lambda}\right)} - e^{2\pi i(ft-z/\lambda)} \right] \qquad (11)$$

and by differentiating p_2 with respect to z, the gradient at p_2 perpendicular to the surface becomes

$$\frac{\partial p_2}{\partial z} = 2\pi G\left[\frac{z}{\sqrt{R^2+z^2}} e^{2\pi i\left(ft-\frac{\sqrt{R^2+z^2}}{\lambda}\right)} - e^{2\pi i(ft-z/\lambda)} \right]. \qquad (12)$$

For the point on the surface where $z = 0$, the gradient reduces to

$$\left(\frac{\partial p_2}{\partial z}\right)_{z=0} = -2\pi Ge^{2\pi ift}, \qquad (13)$$

which is independent of the radius R of the circular surface. This result confirms the assumption that the gradient of p_2 at any point on the surface is determined only by the value of G in the immediate vicinity of that point; thus G is independent of possible variations in G at other points. Actually it is rigorously correct only for a plane surface, but results in a good approximation for other surfaces as long as the curvature is small over the distance of one wavelength.

Consequently, it will be assumed that in general the gradient of p_2 and the value of G are related to each other at each point on the target surface by the equation

$$G = -\frac{1}{2\pi}e^{-2\pi ift}\frac{\partial p_2}{\partial z}. \tag{14}$$

If the boundary condition (6) is to be satisfied, $-\partial p_2/\partial z$ in equation (14) may be replaced by $\partial p_1/\partial z$, and in terms of the incident-sound wave

$$G = \frac{1}{2\pi}e^{-2\pi ift}\frac{\partial p_1}{\partial z} \tag{15}$$

Since the incident sound pressure is usually a harmonic wave, it may be locally described by

$$p_1 = be^{2\pi i(ft-q/\lambda)} \tag{16}$$

where b is the amplitude of the wave, f its frequency and λ its wavelength, and q a coordinate parallel to the direction of propagation. The derivative of p_1 in the direction of propagation is then

$$\frac{\partial p_1}{\partial q} = -\frac{2\pi i}{\lambda}be^{2\pi i(ft-q/\lambda)}. \tag{17}$$

The derivative of the amplitude b has been neglected in this equation since this derivative is usually negligible at distances from the source of many wavelengths.

In any other direction, the derivative will equal expression (17) multiplied by the cosine of the angle between the direction chosen and the direction of propagation q. If the angle between the direction of propagation of the incident sound wave and a line perpendicular to the target surface is θ, then the derivative of p_1 along a line perpendicular to the target surface is

$$\frac{\partial p_1}{\partial z} = -\frac{2\pi i}{\lambda}b\cos\theta\, e^{2\pi i(ft-q/\lambda)}. \tag{18}$$

If this expression is substituted in equation (15), G becomes

$$G = -\frac{i}{\lambda}b\cos\theta\, e^{-2\pi iq/\lambda}. \tag{19}$$

It is particularly interesting to evaluate the wave amplitude b for the case where the incident wave is caused by a point source of sound at a point P, a distance r' from the point of the target surface considered. If at unit distance from P the amplitude of

the incident spherical wave is B, then the local amplitude b equals B/r'; the coordinate q may be replaced by r', and equation (19) assumes the form

$$G = -\frac{i}{\lambda}\frac{B}{r'}\cos\theta\, e^{-2\pi ir'/\lambda}. \tag{20}$$

If this expression for G is substituted into equation (9), the resulting integral for the reflected sound pressure p_2 becomes

$$p_2 = -\frac{i}{\lambda}B\int_S \frac{\cos\theta}{rr'}e^{2\pi i\left(ft-\frac{r+r'}{\lambda}\right)}\, dS \tag{21}$$

where B is the amplitude of the original point source at unit distance, r' is the range from the source to a point on the surface of the target, r the range from that point on the target surface to the point in space where p_2 is to be found, f the frequency and λ the wavelength of the sound. The integration is to be carried out over the whole target surface S of which dS is a surface element.

20.2.3 Physical Interpretation

So far the discussion has been wholly mathematical, without the benefit of a physical argument to support and justify the approximations made. Physically, the analysis is based on the fundamental principle that in the vicinity of a rigid surface the fluid motion in a direction perpendicular to that surface must vanish. If the incident pressure wave made the fluid move so as to violate this condition, the rigid surface would exert a force on the adjacent fluid elements just canceling this motion perpendicular to the surface. This effect may be imagined by replacing each element of area on the target surface by a small piston capable of moving in a direction perpendicular to the surface. In the absence of the boundary condition, each of these pistons would be moved back and forth in rhythm with the motion of the adjacent fluid element. In order to act as parts of a rigid surface, however, these little pistons must each be pushed by a force opposite to that of the motion of the fluid, just sufficient to keep each piston permanently balanced in its original position. This alternating force which each piston exerts on the fluid has the same net effect as the force which a transducer exerts on the surrounding fluid, in other words, each acts as a sound source with spherical wavelets emanating from each individual piston. The appropriate amplitude and phase of these wavelets has been calculated above. The total reflected sound field then represents the superposed effects of all these individual wavelets.

20.3 **FRESNEL ZONES**

In this section, equation (21) will be applied to compute a general formula for the target strength of a smooth and rigid target. Here, smooth means that the radius of curvature of the target surface is large compared to the wavelength of the sound striking it. Moreover, the target is assumed to have a relatively simple shape, always convex, with no marked bumps or protuberances. While this ideal target hardly resembles most actual targets, the consideration of this simple problem gives some insight into the means by which sound waves are actually reflected. Even under these special assumptions, however, it will be shown that the integral in equation (21) can be readily evaluated only by an additional approximation, first suggested by the French physicist, Fresnel.

Consider only the case where the echo is observed back at the sound source; this case corresponds to the situation of chief practical interest, as pointed out in Section 19.1, and in addition simplifies the computations. Then $r = r'$ and equation (21) reduces to

$$p_2 = -\frac{iBe^{2\pi ift}}{\lambda} \int \frac{\cos\theta}{r^2} e^{-4\pi ir/\lambda} dS. \qquad (22)$$

In the integral, however, both θ and r vary over the surface of the target. Therefore, the integral cannot be evaluated by elementary methods, except for certain special cases illustrated in Section 20.4 where an exact integration can be carried out. For most practical purposes, however, an expression for the reflected sound pressure p_2 and, therefore, for the target strength T can be derived by means of an approximate method, which was originally developed in optics and which is known as the method of Fresnel zones.

This method is based on the mathematical analysis developed in the preceding section. Physically, according to equation (21), every point on the target surface which is struck by the incident sound pressure wave becomes in turn a center of outgoing wavelets so that the points on the target surface may be considered "secondary sources" of sound. In optics, this is called Huyghens' principle. Simple addition of the sound pressure in each individual wavelet will give the reflected sound pressure p_2.

Now, every wavelet has a phase depending on the total distance traveled by the sound out to the target and back. In general, these wavelets interfere both constructively and destructively. Destructive interference leads to cancellations due to the phase differences. But a sharp maximum of amplitude — due to constructive interference, where wavelets whose amplitudes are all of the same sign are superimposed — exists in the direction corresponding to *specular reflection*. This is the direction in which the beam is reflected according to ray acoustics. A quantitative calculation of the amplitudes of the different wavelets will show exactly how much energy is reflected in different directions. In this way, wave acoustics can be shown to give the same results as ray acoustics when the wavelength is very short.

20.3.1 **Method**

To compute the amplitudes and phases of the different wavelets, the surface may be divided into successive areas from which all the wavelets emitted are approximately in phase and thus do not interfere destructively. This is the Fresnel method. According to this method, consider a series of wave fronts proceeding outward from a source at the point P, separated by a distance $\lambda/4$ from each other, where λ is the wavelength of the projected sound. When they strike the target, the surface of the target is intersected by these wave fronts in a series of curves which divide the surface into the so-called *Fresnel zones*. The phase of each reflected wavelet, measured back at P, is $2\pi ft - 4\pi r/\lambda$. Since π equals the product of $4\pi/\lambda$ times $\lambda/4$, the distance between two adjacent zones, the wavelets from each zone have an average phase difference of π from the wavelets of the adjacent zones. But a change of phase by the amount π results in multiplication of the amplitude by -1; hence the wavelets from each zone interfere destructively with those from the two adjacent zones. The advantage of the Fresnel-zone approach, as will be shown, is that most of the zones cancel each other, leaving only the effects of the first and last zones to be considered.

While the analysis can be carried out for the Fresnel zones defined by the wave fronts at any one time, it is simplest to take the zones resulting when one of the particular wave fronts considered is just tangent to the closest point on the target. Let R be the value of r at this point, in other words, let R be the distance from the sound source and receiver at P to the nearest point on the target. The first zone is the area on the surface of the target intercepted by the wave front which is a distance $\lambda/4$ from the wave front tangent to the target. In general, the position of the nth zone is then determined by the inequality of equation (23)

$$R + (n - 1)\frac{\lambda}{4} < r < R + n\frac{\lambda}{4} \quad (23)$$

where r is the distance from the source to any point in the zone.

If S_n is the area of the nth zone, equation (22) can be written as a sum of integrals in which each integral extends over only one zone. Then

$$p_2 = -\frac{iB}{\lambda}e^{2\pi i f t}\sum_n \int_{S_n} \frac{\cos \theta}{r^2}e^{-4\pi i r/\lambda}dS; \quad (24)$$

the sum, denoted by the symbol Σ, extends over as many values of n as there are zones. To evaluate the integral in equation (24), define a new variable u_n for the nth zone by

$$u_n = \frac{r - R - (n - 1)\lambda/4}{\lambda/4}. \quad (25)$$

If this equation is substituted in inequality (23), u_n satisfies the relationship

$$0 < u_n < 1. \quad (26)$$

Thus, u_n increases from 0 at the near side of the n zone to 1 at the far side. Equation (24) then becomes

$$p_2 = -\frac{iB}{\lambda}e^{2\pi i(ft - 2R/\lambda)}\sum_n e^{-(n-1)\pi i}\int_{S_n} \frac{\cos \theta}{r^2}e^{-\pi i u_n}dS. \quad (27)$$

Since $e^{-\pi i} = -1$, then $e^{-(n-1)\pi i} = (-1)^{n-1}$, and the reflected sound pressure p_2 may be written

$$p_2 = -\frac{iB}{\lambda}e^{2\pi i(ft - 2R/\lambda)}[P_1 - P_2 + P_3 \cdots + (-1)^{N-1}P_n], \quad (28)$$

where N is the total number of zones and P_n is defined by

$$P_n = \int_{S_n} \frac{\cos \theta}{r^2}e^{-\pi i u_n}dS. \quad (29)$$

In each integral u_n lies between 0 and 1 and obeys inequality (26).

For targets large compared to the wavelength, whose surface is not too sharply curved, there will be a large number of zones and the values of P_n found in successive zones will not change very rapidly as n is changed. The quantity r^2 will scarcely change at all if the distance to the sound source and receiver is much greater than the size of the target. The quantity $\cos \theta$ may decrease from 1 in the first zone to a small value for the higher zones, but if the target is much larger than the wavelength and if its surface is not curving too sharply, the change in $\cos \theta$ from one zone to the next will not be large. Similarly the area S_n of successive zones will not change very rapidly.

The factor $e^{-\pi i u_n}$ varies in the same way in all the zones. Thus, on the average, the partial pressure P_n of the wavelets reflected from the nth zone may be assumed to be equal to the average of the corresponding partial pressure of the wavelets for the preceding and following zones, or

$$P_n = \frac{1}{2}(P_{n-1} + P_{n+1}). \quad (30)$$

Equation (30) forms the basis of the Fresnel approximation. It may be expected to become increasingly accurate for a smooth surface as the wavelength λ decreases indefinitely and the order number n of the particular zone increases indefinitely.

With this approximation, the successive terms in equation (28) cancel out, and the sum of all the P_n's in equation (28) becomes simply

$$P_1 - P_2 + P_3 \cdots + (-1)^N P_N = \frac{1}{2}[P_1 + (-1)^N P_N]. \quad (31)$$

In most practical cases, the value of $\cos \theta$ for the last or Nth zone is zero, since the target surface at this point is tangent to the sound rays. Thus P_N vanishes and the sum of P_n over all the zones is simply one-half the value of P for the first zone. This is a particularly interesting and important result; if only half of the first zone participates in the reflection, and the entire target surface beyond it is neglected, the reflected sound wave is the same as if the entire target were regarded as the reflecting surface. Only a small part of the target surface perpendicular to the sound rays produces the entire reflection; the reflection from this small region is sometimes called a "highlight" as it is in optics. It is this result, derived on the basis of wave acoustics, which corresponds to the specular reflection based on ray acoustics in Chapter 19.

Therefore, set P_N equal to zero in equation (31), substitute the result into equation (28), and p_2 becomes

$$p_2 = -\frac{iB}{2\lambda}e^{2\pi i(ft - 2R/\lambda)}P_1. \quad (32)$$

Now, since the value of r will be almost constant throughout the first zone, unless the source is only a few wavelengths away, r may be replaced by R, the shortest distance from the source to the target. Equation (29) may therefore be written as

$$P_1 = \frac{1}{R^2}\int_{S_1} \cos \theta \, e^{-\pi i u_1}dS. \quad (33)$$

Finally, by combining equations (32) and (33) the pressure p_2 in the reflected wave becomes

$$p_2 = -\frac{iB}{2\lambda R^2}e^{2\pi i(ft - 2R/\lambda)}\int_{S_1} \cos \theta \, e^{-\pi i u_1}dS. \quad (34)$$

20.3.2   Application

To find the target strength corresponding to the pressure of the reflected wave in equation (34), equation (6) in Chapter 19 may be written in the form

$$T = 20 \log |p_2| - 20 \log |B| + 40 \log R \quad (35)$$

where the vertical bars mean that absolute values of the complex quantities involved must be taken. The term $20 \log |p_2|$ is the rms echo level E where p_2 is the actual echo pressure; $20 \log |B|$ is the rms source level S at 1 yd; and $40 \log R$ is twice the transmission loss from 1 yd out to the target at a range R, excluding attenuation losses. Strictly speaking, the rms level is the average value of the square of the real part of the complex quantity rather than the absolute value; however, a more elaborate computation along these lines leads to exactly equation (35). If equation (34) is substituted into equation (35) the target strength T becomes

$$T = 20 \log \left| \frac{1}{2\lambda} \int_{S_1} \cos \theta \, e^{-\pi i u_1} dS \right|, \quad (36)$$

where the bars again denote that an absolute value must be taken. The quantity u_1 in the exponent is defined by

$$u_1 = \frac{4}{\lambda}(r - R). \quad (37)$$

S_1 in equation (36) is the area of the target in which u_1 is less than 1; the integral is evaluated only over those surface elements lying within S_1.

The evaluation of equation (36) provides the solution of the problem presented at the beginning of this section.

20.4   TARGET STRENGTH OF SIMPLE TARGETS

In this section, equation (36) will be used to compute the target strength of relatively simple surfaces, such as spheres, cylinders, and other objects, which have a single highlight. The results obtained may also be applied to more complicated surfaces, as long as the radius of curvature is greater than the wavelength. Whenever several highlights are present, the reflected wave is the sum of the waves reflected from each one separately. In general, they will interfere. However, if an average is taken over a considerable spread of target aspects, and if the highlights are spaced much further apart than the wavelength, the interference will tend to be random; in this situation, the intensity of the echo is simply the sum of the intensities computed for each highlight individually.

20.4.1   Sphere

The target strength of a sphere, on the basis of wave acoustics, may be easily derived from equation (36). The results of this analysis may be used not only for a perfect sphere but also for any target surface whose first Fresnel zone is essentially spherical.

Consider a wave from a source P striking a sphere

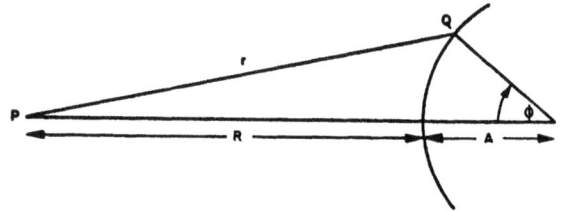

FIGURE 2.   Reflection from a sphere.

of radius A, illustrated in Figure 2, whose nearest point is a distance R away from the source. If ϕ is the angle subtended at the center of the sphere by Q, which bounds an element dS of area, dS is simply

$$dS = 2\pi A^2 \sin \phi \, d\phi. \quad (38)$$

By the law of cosines, the distance r from the source P to the point Q is given by

$$r^2 = (R + A)^2 + A^2 - 2A(R + A)\cos \phi$$

$$= R^2 + 4A(A + R) \sin^2 \frac{\phi}{2}. \quad (39)$$

When R is much greater than A, r is approximately

$$r = R + \frac{2A(A + R)}{R} \sin^2 \frac{\phi}{2}. \quad (40)$$

The quantity u_1 from equation (37) is then

$$u_1 = \frac{8A(A + R)}{\lambda R} \sin^2 \frac{\phi}{2}. \quad (41)$$

For short wavelengths, $\sin \phi/2$ will be very small in the first zone and may be set equal to $\phi/2$; similarly $\cos \theta$ in equation (36) may be replaced by one. Therefore, if equations (38) and (41) are substituted into equation (36), the target strength of a sphere becomes

$$T = 20 \log \frac{1}{2\lambda} \int_0^{\phi_0} e^{-\pi i x \phi^2} 2\pi A^2 \phi \, d\phi, \quad (42)$$

where

$$x = \frac{2A(A + R)}{\lambda R} \quad (43)$$

and

$$x\phi_0^2 = 1. \quad (44)$$

The integration may be carried out and yields

$$\int_0^{\phi_0} e^{-\pi ix\phi^2}\phi d\phi = \left|\frac{e^{-\pi ix\phi^2}}{-2\pi ix}\right|_0^{\phi_0} = \frac{1}{\pi ix}. \quad (45)$$

Thus equation (42) becomes

$$T = 20\log\frac{A^2}{x\lambda}, \quad (46)$$

and if equation (43) is substituted for x, the target strength reduces to

$$T = 20\log\frac{A}{2(1 + A/R)}. \quad (47)$$

This expression is valid only when the distance from the source to the sphere is at least several times greater than the sphere diameter. When R is very much greater than A, equation (47) simply becomes

$$T = 20\log\frac{A}{2}, \quad (48)$$

which is identical to equation (10) in Chapter 19 derived on the basis of ray acoustics. At shorter ranges, equation (36) is still applicable, but must be evaluated more accurately. It may be noted that the value of T in equation (47) is based on the assumption that the transmission loss to the nearest point of the sphere is used in equation (35). If the transmission loss to the center of the sphere is used instead, T must be increased by $40\log(1 + A/R)$ and increases as the range becomes shorter.

As already pointed out, equation (47) may be applied whenever the first Fresnel zone of a reflecting surface is spherical in shape, and has a radius of curvature A much larger than the wavelength λ. The result is independent of the wavelength. Equation (36) could be evaluated more accurately to find a dependence of T on wavelength. This dependence would be appreciable only when the wavelength was no longer much smaller than the sphere radius A, in which case the total number of Fresnel zones would no longer be large. Since the accuracy of the Fresnel method is doubtful under these conditions, the wavelength dependence found in this way would not be very reliable unless confirmed by a much more elaborate investigation.

20.4.2   General Convex Surface

More generally, the curvature of a surface cannot be described by a simple single radius of curvature. In such a case, the boundary of the first Fresnel zone will not be a circle, as was the case for a spherical surface. In a more general case, this boundary will be elliptical in shape, and the surface intersected will have two principal radii of curvature A_1 and A_2, which will usually differ from point to point.

These radii may be defined as follows. Let O be a particular point on the surface and let OC be a line perpendicular to the surface at the point O. Any plane containing OC will intersect the surface in some

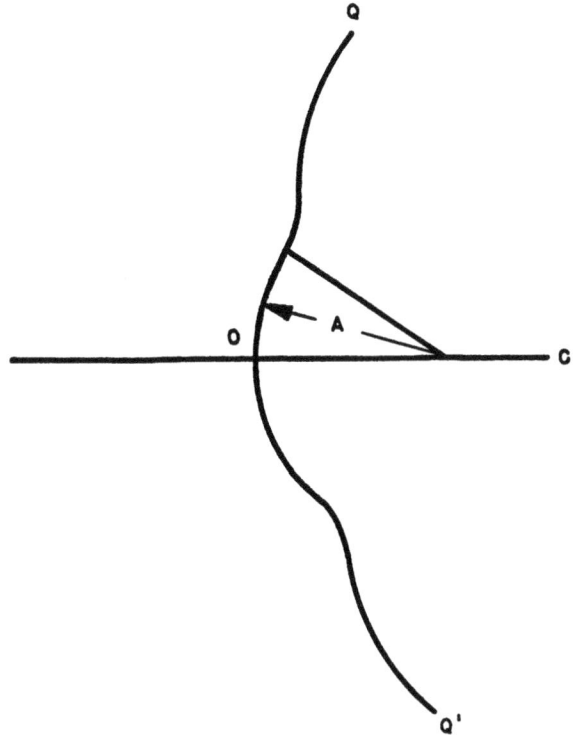

FIGURE 3. Reflection from any convex surface.

line QOQ', as in Figure 3. In the neighborhood of the point O this curve is approximately a circle of radius A. However, as the plane intersecting the target is rotated about the line OC, the radius A of the curve QOQ' will vary. It will have a maximum value A_1 and a minimum value A_2, in general, as the plane rotates through 180 degrees. Furthermore, according to differential geometry, these two radii will be 90 degrees apart. These two quantities A_1 and A_2 are called the *principal radii of curvature* of the surface at the point O. If they do not change rapidly with position on the target surface — more particularly, if they are approximately constant at all points in the first Fresnel zone — the target strength of the surface may be computed.

The derivation is more complicated than that in Section 20.3.1 and will not be given here. The result of the analysis is in the following equation

$$T = 10 \log \frac{1}{2} \frac{A_1 A_2}{(1 + A_1/R)(1 + A_2/R)}, \quad (49)$$

which reduces immediately to equation (47) when A_1 is equal to A_2. While equation (47) was valid only if A_1/R was moderately small, equation (49) is applicable even if A_1/R is very large as long as A_2/R is still small. Equation (49) cannot be used, however, when either A_1 or A_2 approaches the wavelength of sound.

20.4.3 Cylinder

For an infinitely long cylinder, equation (49) may be applied directly by letting one radius of curvature A_1 be infinite. The target strength found for this case reduces to

$$T = 10 \log \frac{1}{2} \left(\frac{A_2 R}{1 + A_2/R} \right), \quad (50)$$

where A_2 is the radius of curvature of the cylinder. This equation is valid only when the wavelength of the sound is much less than the radius of curvature of the cylinder, and when this radius in turn is much less than the range.

For an actual cylinder equation (50) may be used only if the cylinder is perpendicular to the sound beam at some point, and if the cylinder is long enough to include at least the first few Fresnel zones. The expression may therefore be used only at moderate ranges, since with increasing range the length of the first Fresnel zone increases infinitely.

To compute the range beyond which equation (50) cannot be used, let the length of the cylinder be L, and let the sound source lie in a plane which is perpendicular to the axis of the cylinder and bisects the cylinder. Then the path length r to the end of the cylinder is

$$r = R \sqrt{1 + \frac{1}{R^2} \left(\frac{L}{2} \right)^2}. \quad (51)$$

The length of the cylinder will include many Fresnel zones if r given by equation (51) exceeds R by many wavelengths. Therefore equation (50) may be used only as

$$R << \frac{L^2}{\lambda}. \quad (52)$$

For example, for a wavelength of 4 in., corresponding to a frequency of about 15 kc and a cylinder 10 ft long, the range must be much less than 100 yd if equation (41) is to be used.

At long ranges, R is much greater than L^2/λ, and the computed length of the first Fresnel zone exceeds the length of the cylinder. In this case, instead of using the approximation (30) an exact integration of equation (22) over all the zones is possible, provided the variation of $\cos \theta$ is neglected, and the target is far away from the source. Thus in equation (36), instead of one-half the integral over the first zone, we may take the same integral over all the zones. With the same approximation for u made in the previous section, the target strength becomes

$$T = 10 \log \frac{L}{2} \frac{A}{2\lambda(1 + A/R)}. \quad (53)$$

At these longer ranges, the target strength is again independent of the range, in agreement with the comments made in Chapter 19. However, equation (53) presents one case in which the target strength varies appreciably with changing wavelength, even when the wavelength is much smaller than the target.

For intermediate values of the length of the cylinder, both the first and last zones must be considered. A more exact evaluation of equation (22) can be carried through in this special case by use of particular functions called Fresnel integrals, which have been tabulated.

20.4.4 Reflection at Close Ranges

The formulas developed so far in this chapter are applicable to many simple shapes provided the sound source and receiver are not too close to the target. The target strength at close ranges may also be found directly from equation (36). Detailed results have been worked out for cases of this nature, but will not be reproduced here. In general, when R becomes much less than the principal radii A_1 and A_2, the reflection can best be described as reflection from a plane surface. In the limiting case where R/A_2 is negligible,

$$p_2 = \frac{B}{2R} e^{2\pi i(ft - 2R/\lambda)} \quad (54)$$

as long as the sound field this close to the target obeys the inverse square law; for a large directional transducer, this condition is not likely to be satisfied at very close ranges. If equations (35) and (54) are combined, the target strength becomes

$$T = 20 \log \frac{R}{2}. \quad (55)$$

Formulas for the target strength of various types of objects, such as two cones placed base to base, and a circular disk placed at an angle to the sound beam, are given in reference 6.

20.5 REFLECTIONS FROM SUBMARINES

Expressions were developed in the preceding section for the target strengths of various surfaces in terms of the reflected pressures. These formulas were employed in a theoretical study [7] in order to calculate mathematically the target strength of a German submarine.

From an examination of blueprints of the U570,[8] a 517-ton German U-boat captured by the British early in the war and renamed HMS/M *Graph*, the radius of curvature of the surface of the hull and conning tower at different points was obtained. In computing the results, the submarine was approximated by an ellipsoid of revolution, whose semi-axes were 110 and 7 ft. The results were then corrected for the reflections from the conning tower, which was assumed to be a cylinder with a "tear-drop" cross section.

Target strengths were found from equations (49) through (53) in terms of the range and the radii of curvature for different submarine cross sections; ranges of 8, 12, 16, 200, and 1,000 yd were used. At ranges where the conning tower did not include a large number of zones, the Fresnel integrals [obtained when equation (22) is integrated exactly along the length of a cylinder] were used. The calculations were actually carried out in terms of reflection coefficients, which differ somewhat from target strengths derived in this chapter. The results of these computations are presented in Chapter 23 together with the results of the direct and indirect measurements.

20.6 NONSPECULAR REFLECTION

So far only reflections from highlights on a target surface have been discussed. These highlights correspond to specular reflections in optics and give much the same predictions as those found from the ray theory. In particular, the echo is assumed to come only from that region of the target where the surface is nearly perpendicular to the incident sound wave. This section discusses those cases where such reflection cannot occur and where the observations cannot be explained in this way. At the present time, however, different types of nonspecular reflections have not been identified with any observed reflections from actual targets, so that at most this section can only suggest the theoretical expectations.

20.6.1 Rough Surfaces

The most simple type of nonspecular reflection is that from a rough surface, that is, a surface whose irregularities are much larger than the wavelength. Practical formulas applying to this kind of nonspecular reflection from various underwater targets are derived in reference 6. The wavelength of sound is so much greater than that of light, however, that such reflections, which are common in optics, are not to be expected in underwater acoustics. The presence of bubbles on or near the surface of a target can, however, give rise to a diffuse reflection with sound scattered in all directions; the reflection of sound from bubbles is described in detail both theoretically and experimentally in Chapters 26 through 35, which deal with the acoustic properties of wakes.

20.6.2 Diffraction

Another type of nonspecular reflection is that from a surface which has no highlights. Consider, for example, a smooth rigid plane surface in the form of a square, set at an angle relative to the incident rays. This surface will reflect sound specularly, but not back to the sound source. In addition, however, some sound will be reflected in other directions; some of it will be reflected directly backward. This phenomenon corresponds essentially to the diffracted sound observed when a wave passes through a square aperture, and the echo intensity will decrease as $(y/\lambda)^2$, where y is the length of the square and λ is the wavelength. The Fresnel zone theory may again be applied, provided that the effects of both the first and last zones are considered. No results have been worked out along this line, however.

20.6.3 Scattering

A third type of nonspecular reflection is that from objects much smaller than the wavelength. The Fresnel zone theory is not applicable to such small targets, and even the basic equation (21) derived in Section 20.1 is no longer valid, since the derivation assumes that the radius of curvature of the surface is greater than the wavelength. Corresponding analyses have been carried out for targets much smaller than the wavelength; these yield, for a rigid target,

$$T = 20 \log \frac{2\pi V}{\lambda^2}, \tag{56}$$

where V is the volume occupied by the reflecting object. Equation (56) is the so-called Rayleigh scattering law. The echo intensity is directly proportional to the square of the volume of the target and inversely proportional to the fourth power of the wavelength; thus, the echo intensity drops off rapidly as the wavelength increases.

20.7   EFFECT OF PULSE LENGTH

All the previous discussion in this chapter has been concerned with sound waves emitted in an essentially continuous fashion. While Section 2.3 discussed the effect of pulse length in terms of ray acoustics, this section will describe the effect of pulse length on the observed target strength in terms of wave acoustics, developed from the analysis in the preceding sections of this chapter.

It was shown in Section 12.2 that at any instant the scattered sound energy received back at the transducer from the projected pulse comes from a spherical shell of thickness $c\tau/2$, where c is the sound velocity and τ is the duration of the pulse. This result is still true on the more accurate wave theory presented in Section 20.2, as long as the fluid is homo-

geneous and the target is rigid and convex. From a concave target, sound reflected several times may arrive later than singly reflected sound.

This thickness $c\tau/2$ is known as the pulse length. When the pulse length is so long that it includes many Fresnel zones, the echo level will be essentially the same as that observed for continuous sound, provided the echo is measured at a time when the wavelets are arriving from all these zones. At the beginning of the echo, when only the first few zones are contributing, and toward the end, when only the last zones return wavelets to the source, the echo structure is more complicated. However, an application of the Fresnel zone theory would probably give correct results in this case.

When the pulse is only a few Fresnel zones long, the echo structure is presumably more complicated, and the echo duration, for example, may be expected to exceed the duration of the outgoing pulse. The pulse length cannot be less than the thickness of a Fresnel zone, since in that case the outgoing pulse would consist of less than half a cycle, and the wavelength would cease to have much meaning.

Chapter 21

DIRECT MEASUREMENT TECHNIQUES

SUBMARINE TARGET STRENGTHS have been calculated theoretically and measured experimentally. The theoretical calculations described in Section 20.5 are based on assumptions simplifying the geometry of the hull and conning tower, and the way in which the submarine reflects sound. Actual measurements in the field are necessary to verify and amplify these theoretical predictions and to assess their accuracy.

Measurements have been both direct and indirect.[1] Direct measurements consist of echo ranging, with short pulses of supersonic sound, on a submerged submarine at various ranges, depths, and speeds. The intensities of the received echoes are then measured and converted to target strengths. This chapter describes in detail the various experimental procedures and techniques employed by different laboratories in the direct measurements of submarine target strengths.

Indirect measurements, on the other hand, use continuous sound or light reflected from a scale model of a submarine, and interpret these results in terms of supersonic sound reflected from an actual submarine of the same shape; Chapter 22 describes how target strengths are measured indirectly. The results of both the direct and indirect submarine target strength measurements are presented and discussed in Chapter 23 while both the techniques and results of target strength measurements on surface vessels are treated in Chapter 24.

21.1 PRINCIPLES OF DIRECT MEASUREMENT

In order to calculate target strengths, echoes from a submarine may be compared with echoes received at the same time and under the same conditions from a sphere. From the relative intensities of the echoes from the submarine and from the sphere, and from the expression for the target strength of a sphere [equation (10) in Chapter 19], the target strength of the submarine could be readily computed. Since only the relative intensities of two echoes would need to be determined, no absolute measurements or calibrations would be required. But at sea, a sphere large enough to return a strong echo at ranges normally used in echo ranging is too awkward to handle easily and therefore cannot be used in practice to obtain target strengths.

Instead, target strengths are always found by using the fundamental definition [equation (6) in Chapter 19], which defines the target strength of any object in terms of the echo level, the source level, and the two-way transmission loss from the projector to the target and back to the projector again, all expressed in decibels. This expression is simple and easy to use and has the advantage that all the quantities appearing in it may, in principle, be measured directly. Only the difference between the echo level and the source level, and the transmission loss which the signal undergoes as it travels from the projector to the target need to be known in order to find the target strength.

Unfortunately, the difficulties of calibration and other practical problems not yet resolved make the fundamental definition less useful than may be supposed. In particular, the calibration of the transducer, described in Section 21.4 as the measurement of its output as a projector and its sensitivity as a receiver, and the determination of the transmission loss, described in Section 21.5, as well as the large fluctuations and variations normally encountered in underwater sound experiments, introduce numerical uncertainties which cannot be accurately evaluated. Nevertheless, the fundamental definition of target strength introduced in Section 19.1.3 has been used in all direct measurements and has led to reasonably consistent results.

21.2 EXPERIMENTAL PROCEDURES

Four groups have measured submarine target strengths directly. They are: University of California

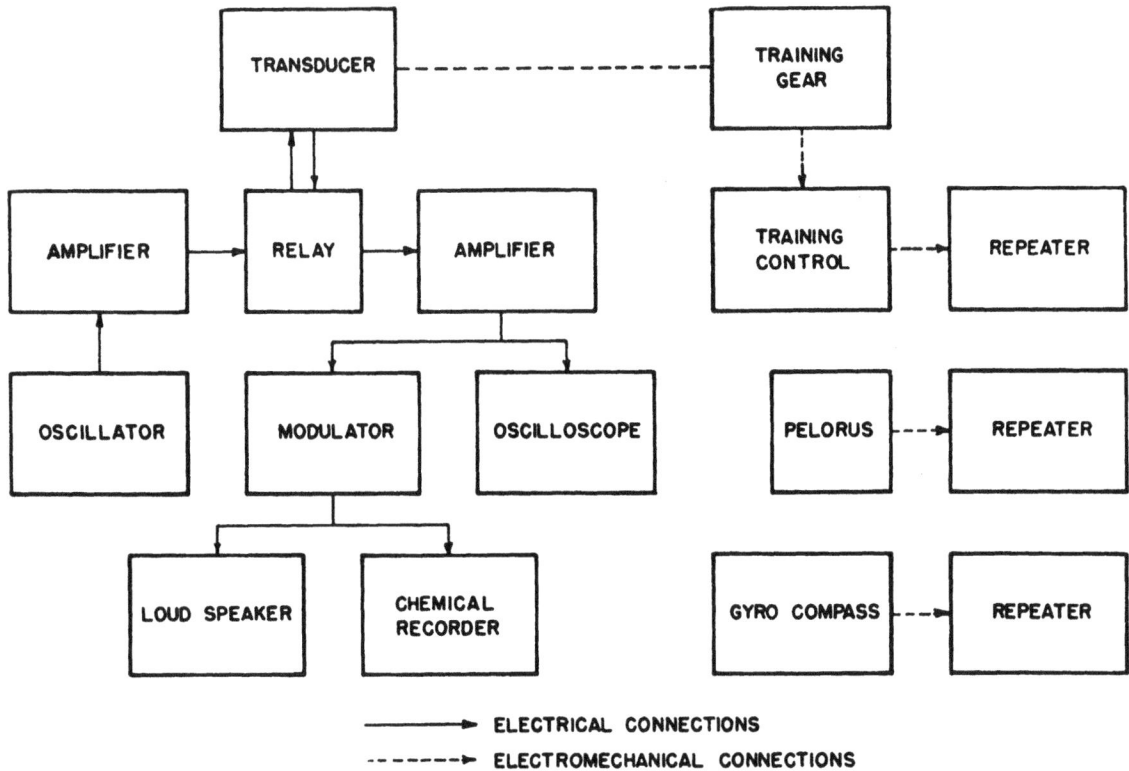

FIGURE 1.   Experimental arrangement.

Division of War Research at the U. S. Navy Electronics Laboratory, formerly the U. S. Navy Radio and Sound Laboratory, San Diego, California [UCDWR]; Columbia University Division of War Research at the U. S. Navy Underwater Sound Laboratory, New London, Connecticut [CUDWR-NLL]; Woods Hole Oceanographic Institution, Woods Hole, Massachusetts [WHOI]; and the Underwater Sound Laboratory, Harvard University, Cambridge, Massachusetts [HUSL]. In addition, various groups at Fort Lauderdale, Florida, have also made measurements of this nature. Widely varying procedures and techniques have been employed by these groups.

21.2.1   San Diego

Most of the direct measurements by UCDWR have been made off the coast of California aboard the USS *Jasper* (PYc13), a converted 135-ft yacht built in 1938, which echo ranged on various S-boats or occasionally on new fleet-type submarines. Square-topped signals from 0.5 to 200 msec long were sent out, usually at a frequency of 24 kc and sometimes at 45 or 60 kc. Early trials used a QCH–3 magneto-

strictive transceiver driven at a frequency of about 24 kc;[2] a few measurements were also made with an experimental model of frequency-modulated sonar gear.[3] Later runs employed standard JK or QC transducers[4] or specially designed equipment.[5]

Most of the echo-ranging equipment was installed in the wardroom of the *Jasper*; a schematic diagram of the installation is shown in Figure 1. A pelorus, an open sighting device attached to a dial and employed in determining bearings, was mounted topside on the flying bridge. An observer visually trained this pelorus on a float towed by the submarine, and the relative bearing of the pelorus was relayed to a repeater dial in the wardroom below. Here, another observer followed the relative bearings of the pelorus and trained the transducer on them; obviously, the bearing accuracy obtainable in this manner was not very high. Then the echoes received by the transducer were amplified and fed into a cathode-ray oscilloscope to be photographed on continuously moving film by a high-speed camera. The echoes were also usually heterodyned and monitored over a loud speaker, and supplementary records were made on the sound-range recorder, where the keying interval was controlled manually as the range changed.

Two types of runs were made. In one, illustrated in Figure 2, the target strength was measured as a function of the aspect of the submarine; the submarine, usually at periscope depth, proceeded at creeping speed while the *Jasper* circled it, trying to maintain a nearly constant range. The other, shown in Figure 3, comprised opening and closing runs, and was used to measure the echo level as a function of

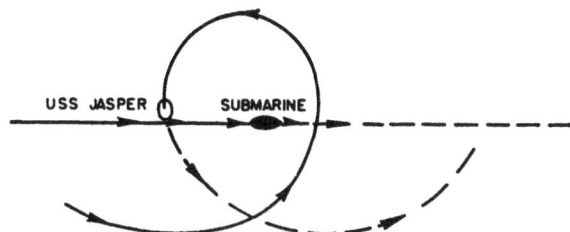

FIGURE 2.   Circling run.

range to determine the transmission loss. Here, the submarine proceeded on a straight course while the *Jasper* followed a divergent course, bearing approximately 60 degrees from the submarine and opened the range until contact was lost; then a closing run was made on a collision course down to a range of several hundred yards. During both opening and closing runs, the speed and course of the *Jasper* and the submarine were held so that the aspect which the submarine presented remained constant.

Since these runs were made, a new type of frequency modulation sonar has been set up at the Sweetwater calibration station of UCDWR for measuring the target strengths of small objects.[6] It is believed that measurements may be made more quickly with this system than with the standard pinging system, but no results are available at the present time.

21.2.2   New London

At New London, tests were made by CUDWR aboard the USS *Sardonyx* (PYcl2) which echo-ranged in Long Island Sound on the USS S–48 (SS159), a 1,000-ton S-boat 267 ft long, first commissioned in 1922.[7] The submarine followed a straight course at a keel depth of 80 ft while the *Sardonyx* circled around it in an arc to maintain an approximately constant range.

A device was used automatically to range on center bearings. The amplified echo intensity was kept constant by manual control of the amplifier gain as the echoes were observed on a cathode-ray oscilloscope. Relative echo intensity was obtained by

recording the amplifier gain settings and by referring to a calibration curve for the system. The bearing, course, and range of both vessels, and the gain settings were recorded about every half-minute. Because complete calibration and transmission data were not available, absolute target strengths could not be computed. Instead, echo intensity was calculated as a function of aspect in decibels relative to the echo level at an arbitrary aspect and plotted for ranges of 600, 1,000, and 1,200 yd.

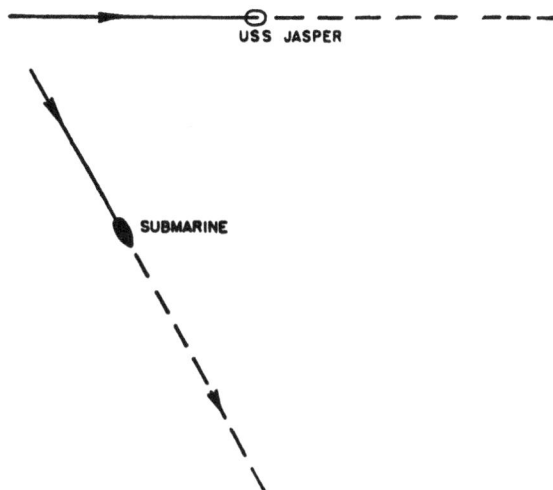

FIGURE 3.   Opening run.

21.2.3   Woods Hole

Target-strength measurements were also made by WHOI observers aboard the USS SC665 just off Fort Lauderdale, Florida.[8, 9] Navy QCU sonar gear was employed, with pulses from 60 to 80 msec long sent out alternately at 12 and 24 kc, at slightly different signal lengths to facilitate separation of the 12-kc data from the 24-kc data. Apparatus was used to range on center bearing.

A hydrophone nondirectional in the horizontal plane was mounted above the conning tower of the 210-ft Italian submarine *Vortice*. Accessory recording equipment was installed aboard the submarine in order to measure the level of the received signals and to determine the transmission loss from the SC665 to the submarine. The submarine proceeded on a straight course, while the SC665 circled the submarine to investigate aspect dependence, and opened and closed the range to investigate range dependence. The submarine also traveled at different speeds and different depths in order to ascertain possible variation of target strength with the speed and depth of the submarine.

21.2.4 **Harvard**

The target strength of the Italian submarine *Vortice* was also measured by HUSL workers using a special sonar first in the area of the Bahama Islands, then off the coast of Florida near Port Everglades. Sonar gear mounted aboard the USS *Cythera* (PY31) echo ranged on the submarine at a frequency of 26 kc.

The first series of tests was made near stern aspect as the *Cythera* and *Vortice* followed parallel courses at speeds from 2 to 6 knots.[10] Cut-ons were obtained by listening to the echoes. Very few data were collected; only 114 echoes were obtained on the *Vortice* during the two days of measurements so that the results cannot be considered conclusive.

During the second and more complete series of tests, the *Cythera* maneuvered around the *Vortice* in order to determine the dependence of target strength on aspect angle, altitude angle, and range.[11] The *Vortice* maintained a speed of 3 knots on a base course at depths of 100, 300, and 400 ft. Echo intensities were obtained for groups of approximately 10 echoes; the source level was measured by training the projector at a monitor transducer, then feeding the voltage across the monitor transducer into a cathode-ray oscilloscope and finding the voltage that had to be applied to the oscilloscope in order to balance it. The speed of the *Cythera* was held close to that of the *Vortice* to prevent bearings from changing too rapidly; training the projector was accomplished by cut-ons. A vertically directional beam from a QHF transducer was used in addition to the original nondirectional beam.

Aspect angles were estimated at intervals of 5 degrees; ranges correct to about 25 yd were read from the sound-range recorder. Altitude angles were not recorded; instead, they were computed from the range, as read from the recorder, and from the depth of the submarine, measured from the ocean surface to the center of the control room about 12 ft above the keel of the submarine.

21.2.5 **Fort Lauderdale**

Three runs were made off the coast of Florida by observers from groups at Fort Lauderdale. In one series of tests, the YP451 remained stationary and echo-ranged on the USS *Pintado* (SS387) and the USS *Pipefish* (SS388), two new fleet-type submarines which ran past the YP451 at prearranged depths, speeds, and ranges. The equipment aboard the YP451 included a crystal transducer, driven at 60 kc, which was suspended on a pendulous pipe so that it was 15 ft below the surface. The platform carrying the transducer was stabilized by an automatic pilot gyro control in one dimension, with its horizontal axis of rotation normal to the axis of the sound beam. In addition, the transducer was automatically trained in elevation. The pendulum and gyro provided a platform which was stabilized in the most critical direction, while the elevation control centered the sound beam on the target vertically; the transducer was trained manually on the target in the horizontal place. The beam width was roughly 25 degrees horizontally and 10 degrees vertically. A 6-string electromagnetic oscillograph recorded the echoes.

Unfortunately, operations with the YP451 were hampered by mechanical difficulties in the alternating current generator and by failure of radio communication with the escort vessel which maintained sound communication with the submarine. Although this lack of communication resulted in unpredictable maneuvers by the submarine, fairly satisfactory data were obtained on echoes from the submarines.

In the second and third runs, signals 30 msec long were sent out at a frequency of 60 kc every 0.6 sec. In the second run, a fleet-type submarine at periscope depth followed a straight course at a speed of 6 knots. The echo-ranging transducer, mounted with accessory equipment in a submerged unit, circled about a fixed point, 230 yd from the course of the submarine, in a radius of 125 ft and at a depth of approximately 35 ft. Aspects were estimated trigonometrically from observed ranges, which had been corrected for the position of the echo-ranging unit in its turning circle.

During the third run, an R-boat was the target, at a keel depth of 100 ft and a speed of 6 knots. The range was decreased continuously; cut-ons were employed in training. Since the echo intensities varied, depending on where the beam struck the submarine, a series of echo maxima was obtained and was used to calculate the target strength. These maxima are illustrated in Figure 4, where the echo level — in decibels below the source level — is plotted against the range; each point represents an individual echo. The target strength was computed from the received-echo voltage, as measured on a film continuously exposed to a cathode-ray oscilloscope, hydrophone sensitivity, total power output into the water, directivity index of the transducer, and the estimated transmission loss.

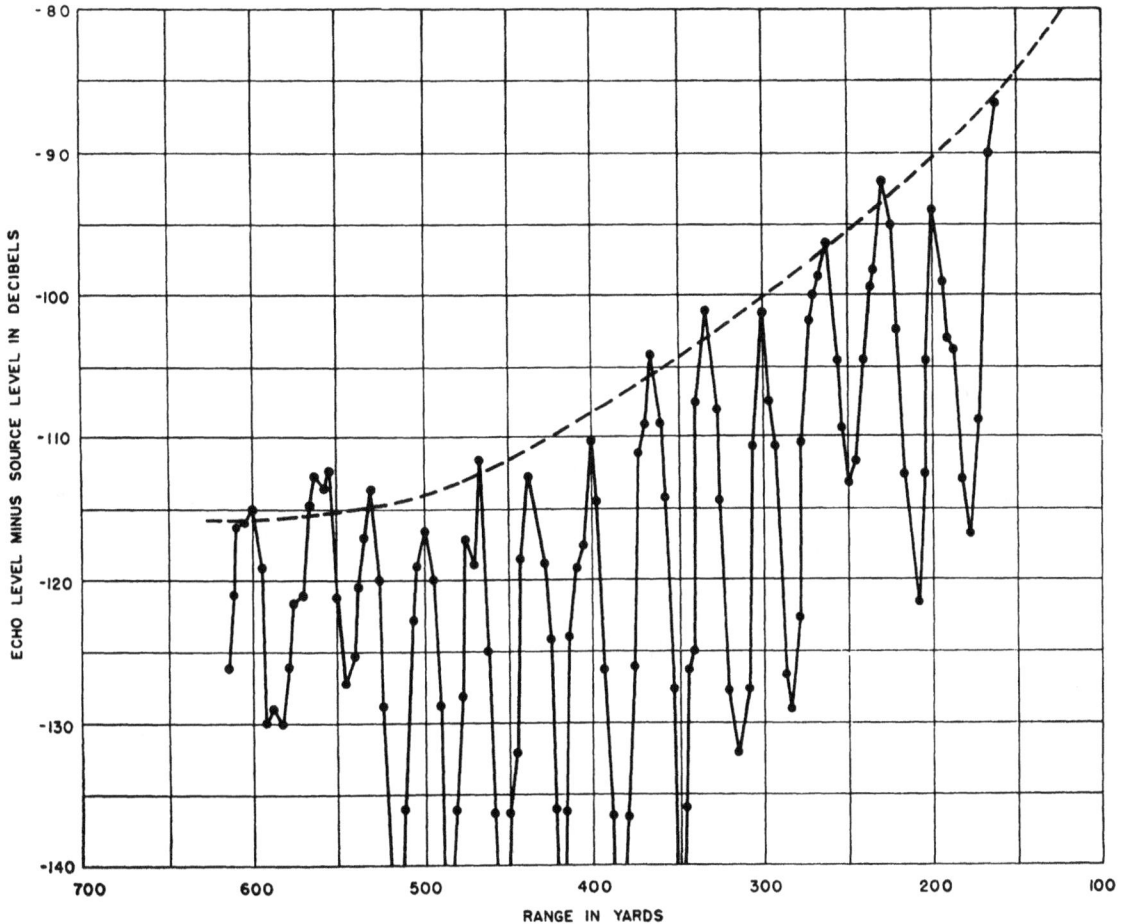

FIGURE 4. Echo maxima at Fort Lauderdale.

21.3   ANALYTICAL PROCEDURES

Target strengths reported here were obtained for the most part from measurements of amplitudes of echoes recorded photographically. Sometimes, however, operating conditions were so poor that upon examination of the photographs it was difficult to identify individual echoes and to distinguish them from noise signals. Therefore echo recognition is important in the study of target strengths and is particularly relevant to a discussion of analytical procedures.

Echo recognition depends not on the intensity of the echo alone, but primarily on the difference between the intensity of the echo and the intensity of the background.[12] At close ranges, echoes are usually strong enough to be easily recognized and are clear and well defined. Distant echoes, however, are often so weak that they cannot be distinguished from the background, and irregular spines and patches may effectively obscure the echo; hence a study of the structure of echoes from submarines and other targets at short ranges may be useful in the recognition and identification of distant echoes from these same targets. Weak echoes may sometimes be attributable to poor training of the transducer or roll and pitch of the echo-ranging vessel, both of which may direct the sound beam away from the target. A high background of reverberation and noise may make an echo hard to recognize. Rough seas and a wide transducer-beam pattern contribute to a high reverberation level, while a surface vessel at moderate speeds or a shallow submarine at high speeds may originate enough self-noise in the transducer dome to mask the echo. Reverberation is treated in detail in Chapters 11 to 17.

Once the echo is recognized and definitely identified as the desired echo, the problem becomes one of measurement and analysis. Various analytical procedures have been employed by different groups in processing the raw material from the oscillogram to the computed target strength.

21.3.1 San Diego

At San Diego, 35-mm film, running at a speed of either 2.5 or 12.5 in. per sec, was exposed to traces on a cathode-ray oscilloscope and then processed and read on an illuminated viewer. Peak echo amplitudes were measured in millimeters, corrected where necessary for the width of the spot of light on the oscilloscope screen, and averaged over a series of echoes. The average was then converted to mean-square pressure level in terms of the calibration constants of the equipments. The transducer and accessory equipment were calibrated before and after each run with an auxiliary calibrated transducer, lowered on a boom from the side of the *Jasper*; Section 21.4 comprises a discussion of calibration errors.

In computing target strengths at San Diego, correction was also made for the deviation of the target from the axis of the sound beam on the basis of beam patterns measured in the laboratory. In addition, the range was found by measuring the distance between the midpoints of the echo and the signal on the film, and referring to index marks recorded every 50 msec at the bottom of the film, corresponding to range intervals of about 40 yd. From calibration data, the source level was calculated, which together with the echo level, and the transmission loss as measured during the opening and closing runs or, less accurately, estimated from prevailing oceanographic conditions but neglecting possible surface reflections, gave the target strength. Simultaneous sound-range recorder records provided a convenient check on the oscillograms.

21.3.2 Fort Lauderdale

A similar procedure was followed by the groups at Fort Lauderdale in analyzing the echoes obtained there. Here, however, the film moved more slowly, at a speed of approximately 1 in. per sec; only 50 ft of film could be accommodated inside the camera. Consequently, the echoes were compressed horizontally and were less detailed, but were still readily measurable.

The fine detail of the oscillograms made at San Diego enabled close determination of echo length as well as a study of echo structure for short pulses; this information was supplemented by an examination of echoes registered on the sound-range recorder. Target strength determinations from the films recorded at Fort Lauderdale, however, may be more accurate than that recorded at San Diego since the motion of the transducer was better controlled, the fluctuations smaller, and the values more consistent.

21.4 CALIBRATION ERRORS

Errors in target strengths measured directly must be due to errors in the echo level, the source level, or the transmission loss, since these target strengths are computed from equation (6) in Chapter 19. Incorrect echo level or source level determinations are usually attributed either to errors in calibration, or to errors in reading the echo level from the trace of the echo recorded oscillographically which UCDWR observers estimate as 2 or 3 db at the most. This section describes errors attributable to calibration of the equipment; uncertainties in the evaluations of the transmission loss are discussed in Section 21.5.

21.4.1 Purpose of Calibration

In target-strength studies, the principal purpose of calibration is not so much the absolute determination of the source level and the absolute determination of the echo level, but rather the measurement of the difference between the two levels. In other words, it is necessary to know only the sum of the transducer output as a projector and response as a receiver if the echo level is measured in terms of the voltage across the terminals of the transducer. Then the difference between the echo level and the source level is simply the difference between (1) the echo level, in decibels above one volt, and (2) the sum of the projector output and receiver response of the transducer.

The latter sum can be obtained by means of auxiliary transducers, without bothering about actual sound pressures. One scheme may employ an auxiliary hydrophone and an auxiliary projector. As a first step, the hydrophone could be lowered from a boom on the echo-ranging vessel, a few yards away from the transducer to be calibrated, and the transducer output measured in terms of the response of the hydrophone. Then the auxiliary hydrophone and the transducer, close together, are both exposed to sound from the auxiliary projector some distance away. Thus, the response of the transducer could be compared with the response of the auxiliary hydrophone; combining the measurements would give the desired calibration of the transducer.

21.4.2 Calibration Techniques at San Diego

The methods of calibration most commonly used in target strength measurements, however, employ calibrated transducers; at San Diego, an auxiliary transducer is lowered over the side of the *Jasper* and used with the standard echo-ranging transducer. First one is used as the projector, then the other, and final calibration is accomplished by referring to the constants of the auxiliary transducer as calibrated at a separate measuring station. Unfortunately, this system is susceptible to errors at every step, so that too much reliance cannot be placed on the accuracy of the calibration.

At San Diego, the greatest error in calibration is believed to be in the measurement of the output of the auxiliary transducer, which is used to calibrate the echo-ranging transducer before and after each run, as mentioned in Section 21.3.1. This auxiliary transducer is calibrated at intervals of roughly four months. Slow drifts of as much as 3 or 4 db have been detected for crystal transducers between calibration checks every three or four months; this drift may be responsible for part of the "variation" observed during target-strength runs, as described in Section 21.6.1. However, since it was not practicable to control or even measure all the factors entering into gear calibration, there is no direct evidence on which to base estimates of the overall calibration error of echo-ranging equipment.

21.4.3 Observed Calibration Errors

Recent indirect evidence suggests, however, that calibration errors as great as 12 db may occur. An example of such large calibration errors is evident in the results of San Diego echo-ranging tests on a sphere.[13-15] The sphere, 1 yd in diameter, was suspended 16 ft below the surface of the ocean at ranges from 24 to 166 yd; echoes from pulses from 0.5 to 7 msec long were received on a JK transducer. Target strengths computed from equation (6) in Chapter 19 varied from −24 to +3 db, approximately 12 db above and below the theoretical value predicted from equation (10) in Chapter 19. Although the very low values are possibly the result of training errors, the very high values seem rather large to be attributed to errors in the estimated transmission loss, especially since the values as high as 3 db were found when the transmission loss was measured directly with a hydrophone placed (1) close to the projector and then

(2) close to the target. However, the possibility that the transmission loss at short ranges fluctuates by 12 db cannot be ruled out at the present time. This large error must result either from large fluctuations in short-range transmission, or from errors inherent in the calibration of the gear, provided that the theoretical formula in Chapter 19 for the target strength of a sphere is applicable to direct measurements.

To provide a check on the validity of this formula, an auxiliary hydrophone was placed a few yards from the sphere during this series of observations and was used to measure both the outgoing pulse and the returning echo. The mean target strength of 350 echoes was found by this method to be −13.3 db, in unusually close agreement with the theoretical value of −12 db. A similar result was obtained at Woods Hole. Thus, the 12-db discrepancy observed when the JK transducer alone was used is undoubtedly the result of errors in the estimated transmission loss, in calibration, or in both. That large systematic errors in these quantities may sometimes be present, even when careful checks are provided, is suggested by the anomalously high values found at San Diego for the target strength of a submarine at 60 kc, and the similar results obtained by Woods Hole at 12 and 24 kc, both reported in Section 23.6.2.

Large errors in calibration may result from (1) large-scale variability of the calibrated auxiliary units employed in methods involving absolute calibration at sea; or (2) gross deviations of the sound field from the theoretical inverse square law in calibration measurements at close ranges, because of interference with reflections from the hull or from other surfaces nearby. Neither of these explanations seems very likely. So far no really satisfactory explanation of the large internal inconsistencies in direct target strength measurements has been advanced. Calibration of ship-mounted gear at sea remains one of the most troublesome of all underwater sound measurements.

21.5 TRANSMISSION LOSS

It has already been pointed out that much of the error in the direct measurements of target strength may be due to errors in the estimated transmission loss; probably a large part of the variability in observed target strengths arises from variability in the transmission loss. This quantity varies widely from hour to hour and from place to place and is seldom known accurately.

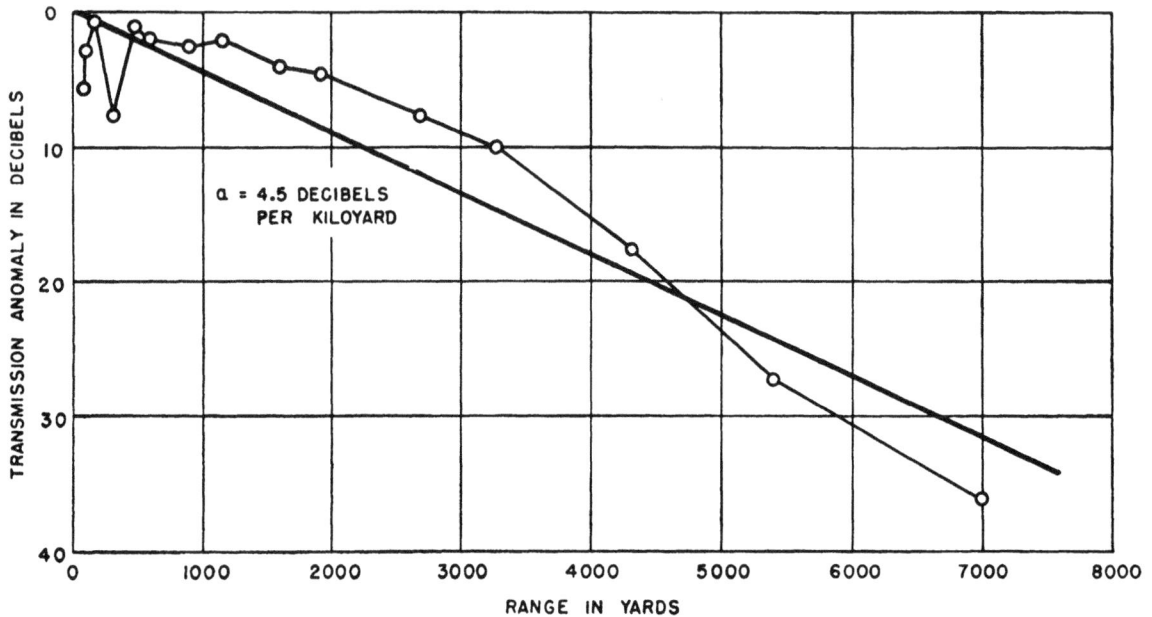

FIGURE 5.   Typical transmission anomaly at 24 kc for an isothermal layer 70 feet deep.

The transmission loss H is defined as the loss in intensity, in decibels, as the sound travels between a point 1 yd on the axis of the sound beam from a small projector, and the target. If the medium through which the sound travels is ideal — if no sound is absorbed, scattered, or refracted, or reflected from the ocean surface or bottom — then the intensity of the sound varies inversely as the square of the distance from the source, as pointed out in Section 19.1.1, and the transmission loss, in decibels, is simply 20 log r, where r is the range in yards. In this case the total transmission loss $2H$ as the sound travels to the target and back to the projector again is simply 40 log r.

This inverse square loss, however, is only a part of the total transmission loss of sound in water. Sound energy is absorbed by the water and dissipated as heat energy. Small particles in the water scatter the sound in all directions. Furthermore, as the beam is refracted by a temperature gradient, it is bent and the cross section of the beam changes in area, changing the intensity of the sound correspondingly.

To account for transmission loss due to absorption, scattering, and divergence arising from refraction, the *transmission anomaly* is defined as the difference between the total measured transmission loss, and the transmission loss due to divergence according to the inverse square law alone. In decibels, then,

$$A = H - 20 \log r, \tag{1}$$

where A is the transmission anomaly, H the total

transmission loss, and r the range. A typical plot of the transmission anomaly against the range is illustrated in Figure 5.

The transmission anomaly has been found to depend rather strongly on the prevailing oceanographic conditions and most particularly on the variation of the temperature of the water with depth. The water in the ocean is usually characterized by a mixed layer of nearly constant temperature down to a certain depth; below that depth, a decrease in temperature with depth, or *thermocline*, will appear. The transmission anomaly depends markedly on the depth to this thermocline as well as on the depth of the hydrophone receiving the echoes.

When the temperature difference in the top 30 ft of water is 0.1 F or less, the transmission anomaly may be considered a linear function of range (see Chapter 5). Hence, it is convenient to define an *attenuation coefficient* as the change in transmission anomaly with range. As a derivative,

$$a = \frac{dA}{dr} \tag{2}$$

where a is the attenuation coefficient, A the transmission anomaly, and r the range. Since in target-strength runs the attenuation coefficient is measured not as a derivative, but as an average over range intervals of 500 or 1,000 yd, a is usually taken as

$$a = \frac{A}{r}. \tag{3}$$

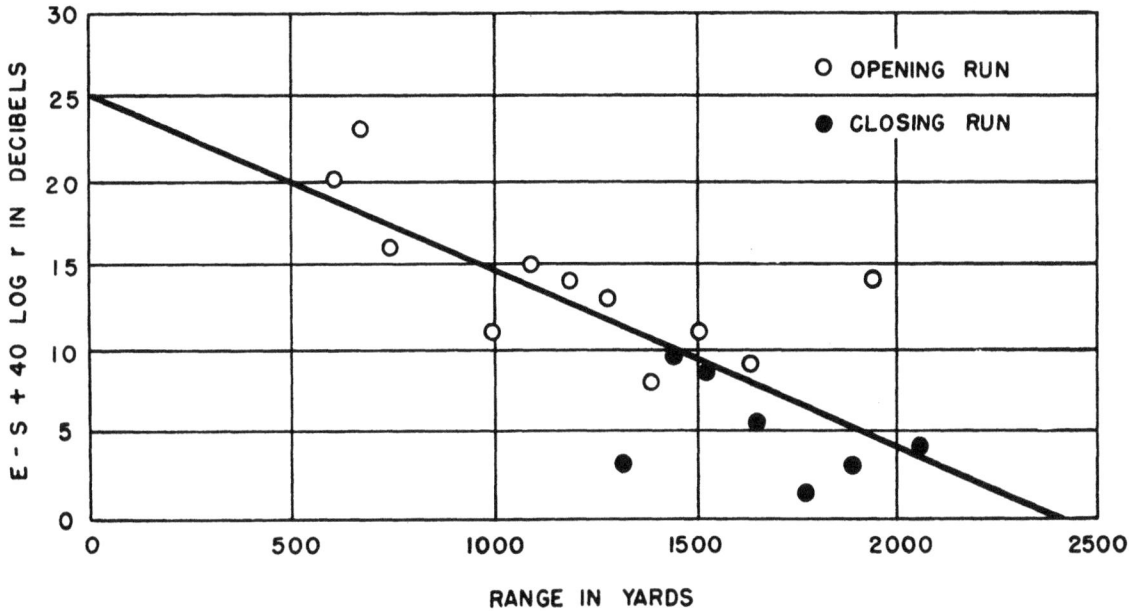

FIGURE 6.   Target strength plot.

In Figure 5, a amounts to about 4.5 db per kyd, at 24 kc. It may be that actually

$$A = ar + b \qquad (4)$$

where b is a constant. Present data indicate, however, that b is probably negligible.[16]

By definition, the transmission anomaly includes the effects of reflection from the ocean surface. So little is known about surface reflection with any degree of certainty, however, that no attempt is made to include in equation (4) an additional term to take it into account. If surface reflection is appreciable, it may cause the constant b in equation (4) to be negative, in effect decreasing the transmission anomaly and therefore the transmission loss itself, as described later in Section 21.5.4.

21.5.1   Methods of Measurement

Transmission loss may be measured in three ways. First, during an opening or closing run, the echo level in decibels above the source level may be corrected for geometrical divergence by adding $40 \log r$; then, the result may be plotted as a function of the range, as long as the submarine maintains a constant aspect. A typical plot of this nature is illustrated in Figure 6. Then the slope of the points represents twice the attenuation coefficient, and the intercept at zero range corresponds to the target strength. Such a de-

termination presupposes that the target strength does not change over the ranges used.

Second, before and after each run on a submarine, a transmission run may be made with an auxiliary surface vessel, in the usual manner, as described in Section 4.3.2.

Third, the signals transmitted by the echo-ranging vessel may be received by a hydrophone mounted on the submarine, amplified, and measured. The transmission anomaly and attenuation coefficient may then be determined readily by correcting the level of the echo above the source for simple geometric divergence, and measuring the slope of the plot of the echo level against the range.

But measuring the transmission loss in any one of these three ways is difficult. Aspects and speeds must be carefully maintained and measured, a procedure particularly difficult for a submerged submarine. For reasons of safety, the echo-ranging vessel is advised not to approach the submarine closer than about 300 or 400 yd, and poor sound conditions often limit echo ranges to 1,000 yd or even less, especially off the coast near San Diego. When a transmission run is made with an auxiliary surface vessel, horizontal temperature gradients may result in a transmission loss between the projector and the hydrophone suspended from the surface vessel which is different from that between the projector and the submarine. Another disadvantage of measuring the

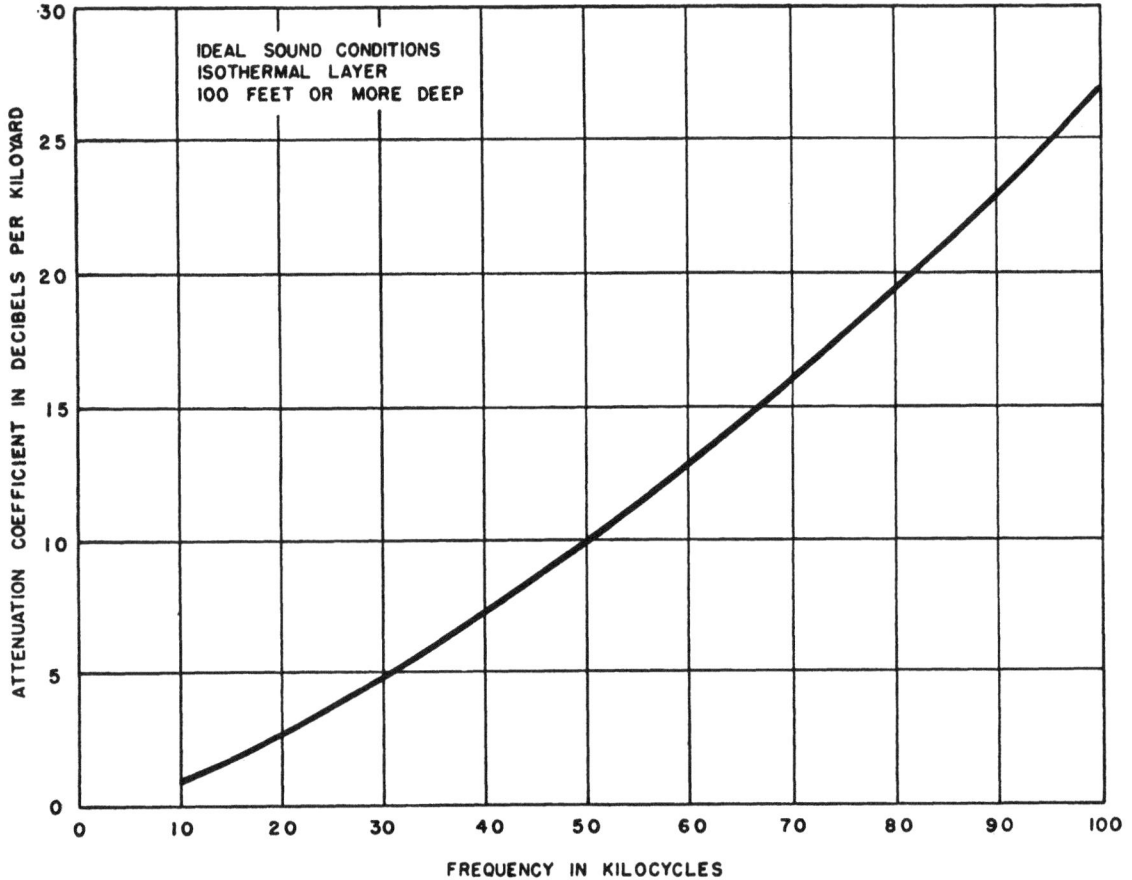

FIGURE 7. Attenuation coefficient as a function of frequency.

transmission loss by the latter method is the presence of four vessels in the operating area, that is, submarine, escort vessel, echo-ranging vessel, and transmission measuring vessel. The use of a hydrophone mounted on a submarine is a definite improvement but introduces new horizontal and vertical directivity problems as well as installation complications.

21.5.2 Inadequacy of Transmission-Loss Measurements

All three methods have been used to measure transmission loss during direct target strength tests. Where ample and consistent data have been taken by any one of these methods, the transmission loss calculated from these data has been used to evaluate the target strength.

Often, however, data have not been consistent. During one run at San Diego, for example, the plot of the echo level, corrected for inverse square law spreading against range, indicated an attenuation coefficient of 19 db per kyd at a frequency of 60 kc while measurements aboard the submarine when analyzed showed a value of only 10 db per kyd. Another identical run the following day gave values for the attenuation coefficient of 11.5 and 16 db per kyd, respectively, as measured by the two methods. Apparently the errors were not systematic. This lack of consistency between two methods was not infrequent. Recent San Diego target strength measurements, however, based on transmission loss measured with a nondirectional hydrophone mounted on the submarine, have been more consistent; this method promises to eliminate much of the uncertainty in the evaluation of the transmission loss. However, the measurements reported [13-15] indicate that even this method does not eliminate systematic error in the determination of target strength, possibly because of peculiarities of transmission at short ranges, possibly because of calibration uncertainties. Certain 60-kc measurements on the USS S-37 (SS142) at San Diego gave a beam target strength of 28.7 db with a standard deviation of 8.5 db when an attenuation coefficient of 20 db per kyd was assumed; when the transmission loss measured aboard the submarine was used in the computations, the beam-target strength rose to 40 db with a much smaller standard deviation of 3.5 db. In most trials reported here, it was necessary to evaluate the transmission loss from an attenuation coefficient, estimated for each run from the echo-ranging frequency employed and sometimes from the prevailing oceanographic conditions.

21.5.3 Estimating the Attenuation Coefficient

The attenuation coefficient in sea water varies widely and depends primarily on the frequency of the echo-ranging sound and on the prevailing oceanographic conditions.[14] For example, in mixed water, or water of constant temperature, at least 50 ft deep, this coefficient is about 5 db per kyd at 24 kc, and in the neighborhood of 15 db per kyd at 60 kc. A plot of the attenuation coefficient against frequency for ideal sound conditions is reproduced in Figure 7 and represents a rough average of observations primarily at 20, 24, 40, and 60 kc; the increase in attenuation coefficient with frequency is quite marked and shows why it is impractical to use very high frequencies for echo ranging.

The attenuation coefficient increases markedly with poor sound conditions. At 24 kc, it may be as high as 15 db per kyd under poor conditions, or even 40 db per kyd under extremely bad conditions.

Very few data are available at 60 kc on the variation of the attenuation coefficient with oceanographic conditions. Empirical formulas have been derived for the attenuation coefficient at 24 kc, however, as a function of the depth of the thermocline. For a hydrophone above the thermocline,

$$a = 3.5 + \frac{170}{D}, \tag{5}$$

and for a hydrophone below the thermocline

$$a = 4.5 + \frac{260}{D}, \tag{6}$$

where a is the attenuation coefficient in decibels per kiloyard and D is the depth in feet to the thermocline. The probable error is about 2 db per kyd.[16] As implied in Chapter 5, these empirical formulas are, in general, less suitable for predicting the attenuation coefficient than other methods based on a more quantitative

classification of the variation of temperature with depth, because transmission anomaly-range graphs significantly depart from straight lines under certain conditions. However, equations (5) and (6) are sufficiently accurate for the present purposes.

Early target strength measurements showed that different values of the transmission loss were obtained by different methods, as described in Section 21.5.2. Therefore, in most calculations representative values of 5 and 20 db per kyd at 24 and 60 kc respectively were taken for the attenuation coefficient. Much of the time no account was taken of the oceanographic conditions which prevailed at the time of the tests, however, with the result that the reported target strengths varied considerably. Examples of this variability are given in Section 21.6 of this chapter.

21.5.4 Surface Reflections

Reflection of sound from the surface of the ocean is neglected in all calculations of target strengths. Such an effect would offset, in part, the loss in intensity caused by spreading and absorption.

Perfect specular reflection from the surface would effectively double the intensity of the sound incident on the target and the intensity of the echo returned to the projector, under ideal conditions. In other words, it would reduce the transmission loss by 3 db each way, or by a total of 6 db from the projector to the target and back again. Thus, in equation (4) the constant b would equal -3 db at ranges of a few hundred yards or more.

Some evidences of surface reflection have been found experimentally. At San Diego a number of oscillograms of echoes from submarines have shown peaks or "spines" at the beginning and end of each echo, separated by a relatively smooth echo of lower intensity; an example is shown in Figure 8 for an S-boat at beam aspect.[17] The first peak is attributed to direct reflection from the hull of the submarine alone when the first part of the pulse strikes the target and is reflected back to the projector along the shortest possible path; the final peak comes from the ray reflected from the submarine to the surface and back to the projector, after the direct echo from the submarine has been received. In other words, the two spines are attributable to reflection along only one path, since there will be a short time at both the beginning and end of the echo when the sound travels only one path back to the transducer. The intensity of the intervening echo is consequently lower because

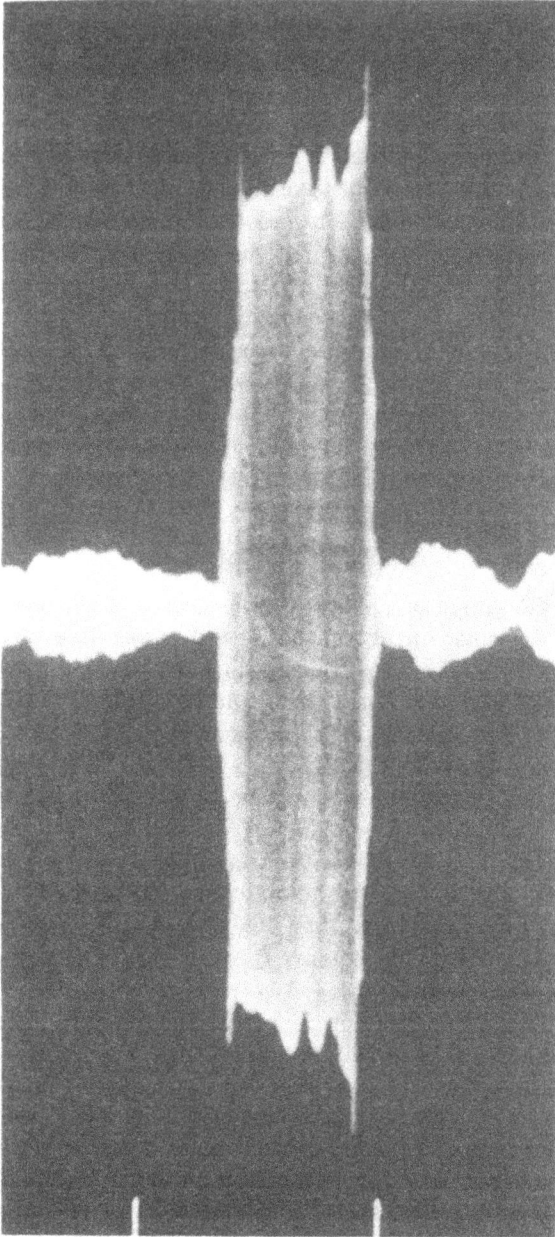

FIGURE 8. Surface-reflected sound.

of a combination of both constructive and destructive interference throughout the duration of the echo, between direct and surface-reflected sound.

A different effect produced by surface-reflected sound is also indicated by more recent information from San Diego.[18] During echo-ranging tests on a submarine from 90 to 200 ft deep, double echoes were observed, under certain conditions, on the chemical recorder and on the oscillograph — a strong primary

echo followed by a faint secondary echo, illustrated in Figure 9. This appearance of double echoes suggests that some of the sound is reflected directly back to the projector to form a primary echo, while some of it is reflected vertically upward to the ocean surface, reflected by the surface back to the submarine and finally back to the projector to form a secondary echo. Quantitative data show that the lapse of time between the primary and secondary echoes is equal to the time necessary for the sound to travel up to the surface and back again, thus confirming this hypothesis.

Although almost all the sound striking the surface is unquestionably reflected back into the water at some angle, the perfect specular reflection expected from a flat surface seems unlikely at sea. The normally rough surface of the ocean and the presence of air bubbles tend to scatter the sound rather than allow perfect specular reflection at the surface. Further evidence minimizing the effects of surface reflection on target strength values is seen in the excellent agreement between the results of the direct measurements computed neglecting surface reflections, and both the indirect measurements and theoretical calculations, where surface-reflected sound either does not appear or may be readily eliminated. Partly for this reason, surface-reflected sound is neglected in all target strength computations in Chapters 18 to 25. However, the results shown in Figure 2 of Chapter 9 and described in Section 9.2.1 suggest that reflection from the ocean surface is frequently very nearly specular. More data are needed to clarify the exact importance of surface-reflected sound in practical echo ranging. At present, the resulting uncertainty of 6 db is about the same as the other uncertainties of observation in target strength measurements.

21.6   VARIABILITY OF ECHOES

Perhaps the largest source of uncertainty in target strength measurements arises from variability of echo intensity. Observed echoes vary widely in two ways (see Section 21.1). Gradual changes in echo intensity over a relatively long period of time from a few minutes to hours are called *variations*. Superimposed on these variations are marked changes which occur from echo to echo and are called *fluctuations*. A large part of the variability of echo intensity is due to variability in the sound-transmitting character-

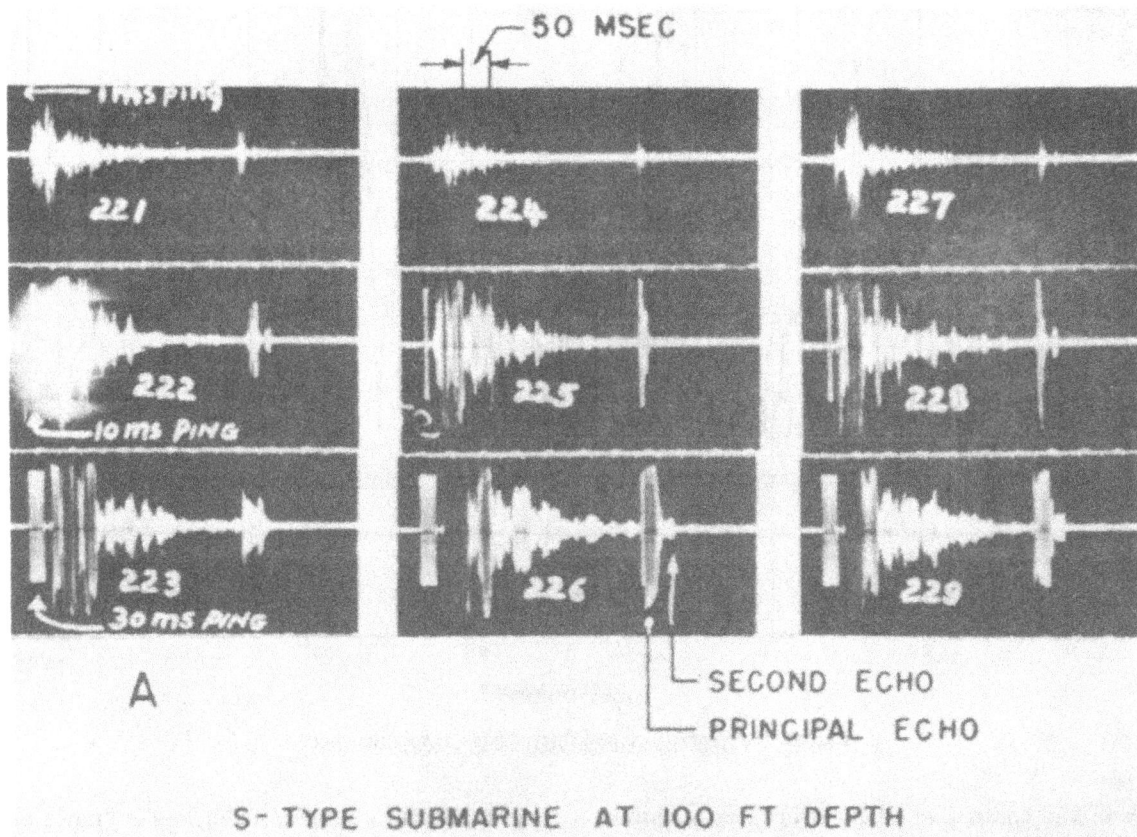

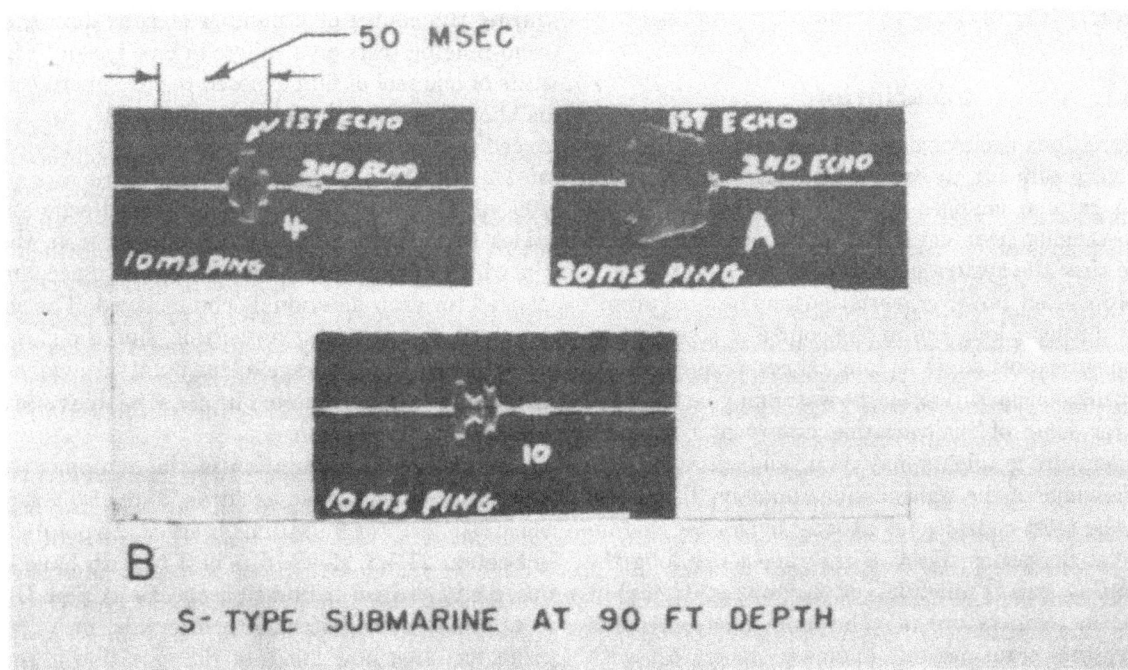

FIGURE 9. Double echoes.

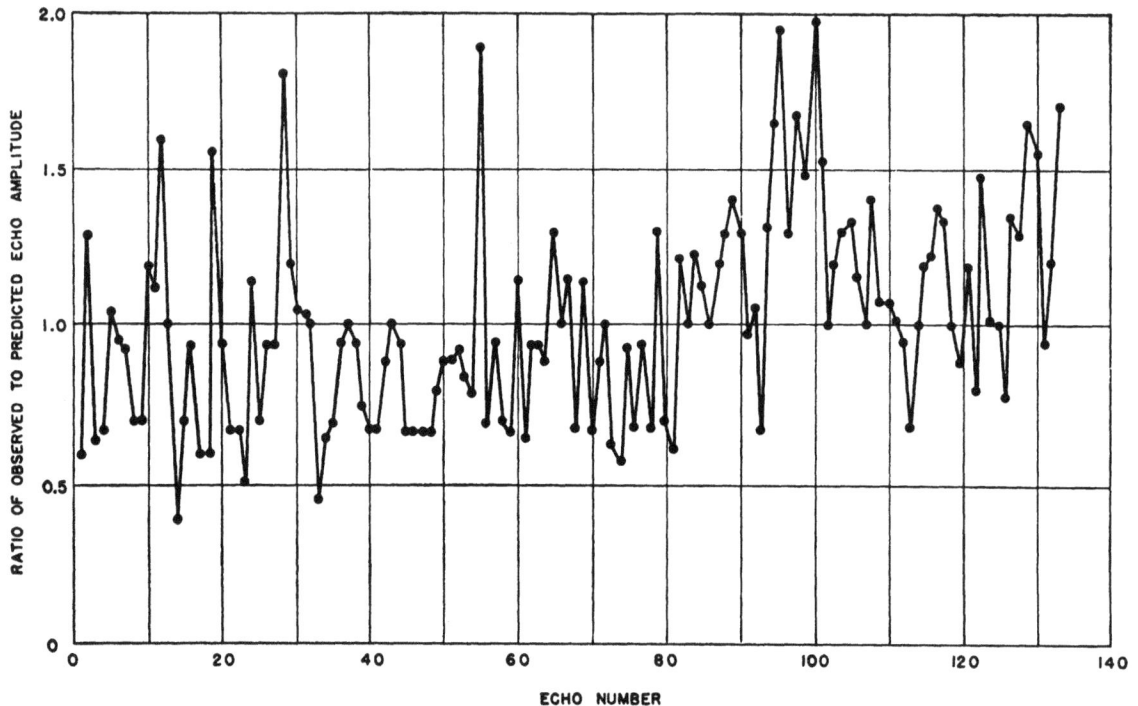

FIGURE 10.   Variations and fluctuations in sphere echoes.

istics of the ocean (see Chapter 7). The remainder may be ascribed to such external causes as changes in the performance of equipment and changes in target aspect.

21.6.1   Variation

Variations occurring over a sufficiently long time are very difficult to detect. Sometimes they result from gradual changes in the characteristics of the echo-ranging gear employed and may be detected each time the system is calibrated.

More often, however, variation may be most prominent during a long run in the course of a single day or on successive days. At long ranges, changes in the transmission conditions in the water may be responsible for some of the variation observed; horizontal temperature gradients may occur and cause changes in the value of the transmission anomaly. This effect may be most conspicuous at long ranges for two interrelated reasons. First, if the ranges are long the operating area is much larger, and horizontal differences in temperature may be more likely. Second, since the transmission anomaly increases with range, variations attributable to slow changes in the transmission anomaly will be greatest at long

ranges. At short ranges, much less is known about variation.

Marked variation in the echo level was observed during the course of a number of runs during early echo-ranging tests on a sphere in San Diego.[2] The results of one reel of film exposed to the sphere echoes, as shown on a cathode-ray oscilloscope, are reproduced in Figure 10; pulses were sent out at intervals of 1.2 sec and the range of the sphere was about 109 yd. Here, the ratio of the observed echo amplitudes to the echo amplitudes predicted from theory (in which transmission loss is taken into account) is plotted for each individual echo received. The short-term changes are most noticeable, but the slow upward slope of the average of the points is evidence of variations as defined here. The cause of this variation, however, is not known.

Changes in the calibration of the equipment over a period of time, known as "drift," are also responsible for some of the variation observed. As pointed out in Section 21.4.2, slow drifts of 3 to 4 db have been observed between calibration checks at San Diego, at approximately four-month intervals, in a crystal projector. Just how much of the variation normally encountered can be attributed to drift, however, cannot be estimated very accurately.

21.6.2 Fluctuation

Many factors contribute to the observed fluctuations of echoes. Much of this rapid change in echo intensity may be ascribed to the roll and pitch of the echo-ranging vessel, which by changing the direction of the sound beam causes the received echoes to vary in intensity. Although gyroscopic stabilization of the transducer was employed at Fort Lauderdale to reduce fluctuations arising from the roll and pitch of the ship, as described in Section 21.2.5, this system has not been used elsewhere for this purpose. Errors in training the echo-ranging transducer toward the submarine have also been responsible for some of the fluctuations encountered; training on the bearing of maximum intensity, by means of cut-ons, is approximate and introduces variability in the received echo intensities by changing the direction of the beam relative to the submarine.

In addition, surface reflection and interference phenomena may be expected to account for part of the fluctuations observed, as the sound beam frequently follows multiple paths to reach the submarine and return back again to the transducer. Chapter 7 of Part I of this volume discusses the evidence showing that transmission fluctuations are very much reduced when surface-reflected sound is minimized. Correlation has been observed between the depth of the transducer below the ocean surface, and the magnitude of the fluctuations observed in echoes from a sphere two ft in diameter;[2] at a range of the order of 65 to 75 yd, elevating the transducer from a depth of 50 to 10 ft below the surface increased the standard deviation of the echo intensity from 18 to 39 per cent. In addition, the overall fluctuation appears to decrease as the signal length increases. At other than beam aspects, interference between echoes from different parts of the submarine is undoubtedly responsible for part of the fluctuation observed, giving rise to an irregular "hashed" echo structure described in Sections 23.8.2 and 23.8.3.

21.6.3 Effects on Echo Level and Echo Structure

Variation affects echo intensity; fluctuation affects both echo intensity and echo structure. Echo envelopes never repeat exactly, and successive echoes at the same range, aspect, frequency, and signal length often appear totally different. This diversity of echo structure not only complicates measurement of the intensity of the echo, but also makes it difficult to resolve the length of the echo and the center of the echo, in effect preventing precise measurement of the range of an individual echo.

Likewise successive echo intensities seldom repeat. As a result, some sort of average must be taken over successive echoes. If target strength is regarded as a measure of the fraction of the incident sound energy reflected by the target, the total reflected energy should be compared to the total transmitted energy. Such an analysis would require squaring the echo amplitudes to give the echo intensities, then integrating the intensities over the duration of the echo to give the total echo energy; this same procedure would be followed with the signal to yield the total signal energy. Such an analysis has not been found practical because of the complex instruments required. In addition, it may be that aural and nonaural detection devices respond more to peak echo intensity rather than to total echo energy, and that therefore peak intensities are more significant.

21.6.4 Peak versus Mean Echo Intensity

Since it has not been feasible to compare the reflected and transmitted energy directly, peak echo amplitudes have been used to compute target strengths. Observations show that these average peak amplitudes do not differ significantly from the rms peak amplitudes, which would correspond to peak intensities. Thus the San Diego results may be regarded as giving average peak intensities. Not only is this method simple and easy to apply, but also it provides values which may be compared directly with recognition differential measurements where peak-echo intensities alone are considered. Peaks, however, fluctuate enormously, especially for off-beam echoes; a sample survey of 100 oscillograms of submarine echoes at San Diego showed a maximum fluctuation of 25 db between peaks, with fluctuations of 10 db not uncommon.

An approximate comparison of reflected and transmitted energy might be made by measuring the mean echo intensity, averaged along the entire length of the echo, and correcting this intensity for the pulse length since the echo length generally is longer than the pulse length. Then the ratio of the signal and echo intensities, based on the same pulse length, would be equal to the ratio of the signal and echo energies.

This procedure was attempted for six echoes

recorded oscillographically at San Diego.[1] The echo amplitude was measured at small intervals along the length of the echo, squared to give the echo intensity, and averaged over the echo length as closely as the echo length could be estimated; then the enclosed area was calculated. Division of this area by the signal length gave a new intensity, the intensity which presumably would have resulted if the echo length had equaled the signal length. This sample analysis, although based on data not sufficient to warrant definitive conclusions, showed an insignificant difference between peak echo intensities and mean echo intensities corrected for signal length.

In general, however, the peak echo intensity differs from the uncorrected mean echo intensity, and this difference is a function of the signal length. It was pointed out in Sections 19.3 and 20.7 that for long pulses, the echo will reproduce the signal envelope while for short pulses fluctuations in intensity will result in an irregular structure, where sharp peaks stand out against a weak background. In the latter case, the peak echo amplitude may be considerably different from the mean echo amplitude, and may vary with signal length quite differently (see Section 23.5.1).

The variability of echoes is responsible for a large part of the uncertainty in the echo level and transmission loss values which are used to compute target strengths. Since echoes are often so irregular that visual estimates of peak intensities are, at best, intelligent guesses, UCDWR observers estimate that systematic errors of as much as 2 or 3 db may result from the difference in personal judgments of different observers.

Because in practice fluctuations and variations behave as very large accidental errors, only a statistical analysis of many echoes may be considered reliable. Hundreds of individual echoes must be carefully averaged, corrected, and analyzed to give target strength results of any significance. At San Diego, a camera has been installed aboard the *Jasper* to record, at the same time as each signal is transmitted, the roll and pitch of the vessel, the true bearing of the ship, the relative bearing of the transducer, and the time and pulse number. Such a record should be useful in analyzing and evaluating each echo, but so far has not been applied to a large number of measurements. So far target strength runs have been analyzed from a reasonably large number of individual observations; first, successive groups of five echoes each have been averaged, then an overall average computed considering changes in transmission loss with range and changes in target aspect. Cumulative distributions and computations of probable errors and quartile deviations have been useful in interpreting the results and assessing their reliability.

Chapter 22

INDIRECT MEASUREMENT TECHNIQUES

REFLECTIONS FROM SUBMARINE models have been studied in order to discover the principal reflecting surfaces on a submarine and to measure submarine target strengths under controlled conditions. Both visible light and supersonic sound have been used in these model tests.

In the investigation of reflection from submarines, models have many advantages over actual submarines. Generally, experimental conditions can be controlled much more easily under laboratory conditions than in the field. Laboratory use of carefully constructed scale models makes possible a reasonably reliable evaluation of target strength as a function of aspect and altitude angles, as well as submarine class, and provides both a theoretical guide and a convenient check on the direct measurements.

22.1 PRINCIPLES OF INDIRECT MEASUREMENT

Three groups have participated in the indirect measurements of reflections from submarine models; University of California Division of War Research at the U. S. Navy Radio and Sound Laboratory, San Diego, California [UCDWR]; Underwater Sound Laboratory, Massachusetts Institute of Technology, Cambridge, Massachusetts [MIT–USL]; and the Underwater Sound Reference Laboratories, Columbia University Division of War Research, Mountain Lakes, New Jersey [USRL]. Only qualitative results were obtained at San Diego while actual target strength values were measured at MIT and at USRL.

22.1.1 San Diego

Early experiments were carried out at San Diego [1] on a 1:60 scale model of the U570 or HMS/M *Graph*, a 517-ton German Type VIIC U-boat which was captured in 1941 off Iceland and served in the British fleet. The model, made of wood and finished with glossy white enamel, was illuminated by a standard projection bulb and photographed in various positions.

The bulb was enclosed in a metal housing with a hole $1\frac{1}{4}$ in. in diameter on one side, and was placed as close as possible to the camera lens so that the angle between the incident and the reflected light at the submarine was only about 3 degrees. Photographs were made at different aspect angles, first with the submarine finished with enamel, then with the submarine covered in part by horizontal and vertical corrugations, and finally with coarse emery cloth covering certain areas on the model. The corrugations and emery cloth were affixed to the submarine in an effort both to reduce prominent reflections and to suggest locations for possible absorption treatment.

These experiments were wholly qualitative, since no measurements were made. The main purpose was to discover the highlights on a submarine which might be largely responsible for strong reflections. Photographs for different aspect angles of the submarine model, without the corrugation or emery cloth, are reproduced in Figure 1.

22.1.2 Massachusetts Institute of Technology

Quantitative experiments using visible light reflected from scale models were conducted at MIT–USL to calculate target strengths of four different submarines.[2-4] A series of measurements were made on models of HMS/M *Graph*, an old S-boat, the USS *Perch* (SS313) and the USS *Sand Lance* (SS381); these models were from 60 to 120 times smaller than the original submarines and were finished with a glossy black enamel.

In order to compute the target strength of one of the submarines, light reflected from the submarine model was compared with light reflected from a sphere also enameled in glossy black. The target strength of the submarine was calculated from the

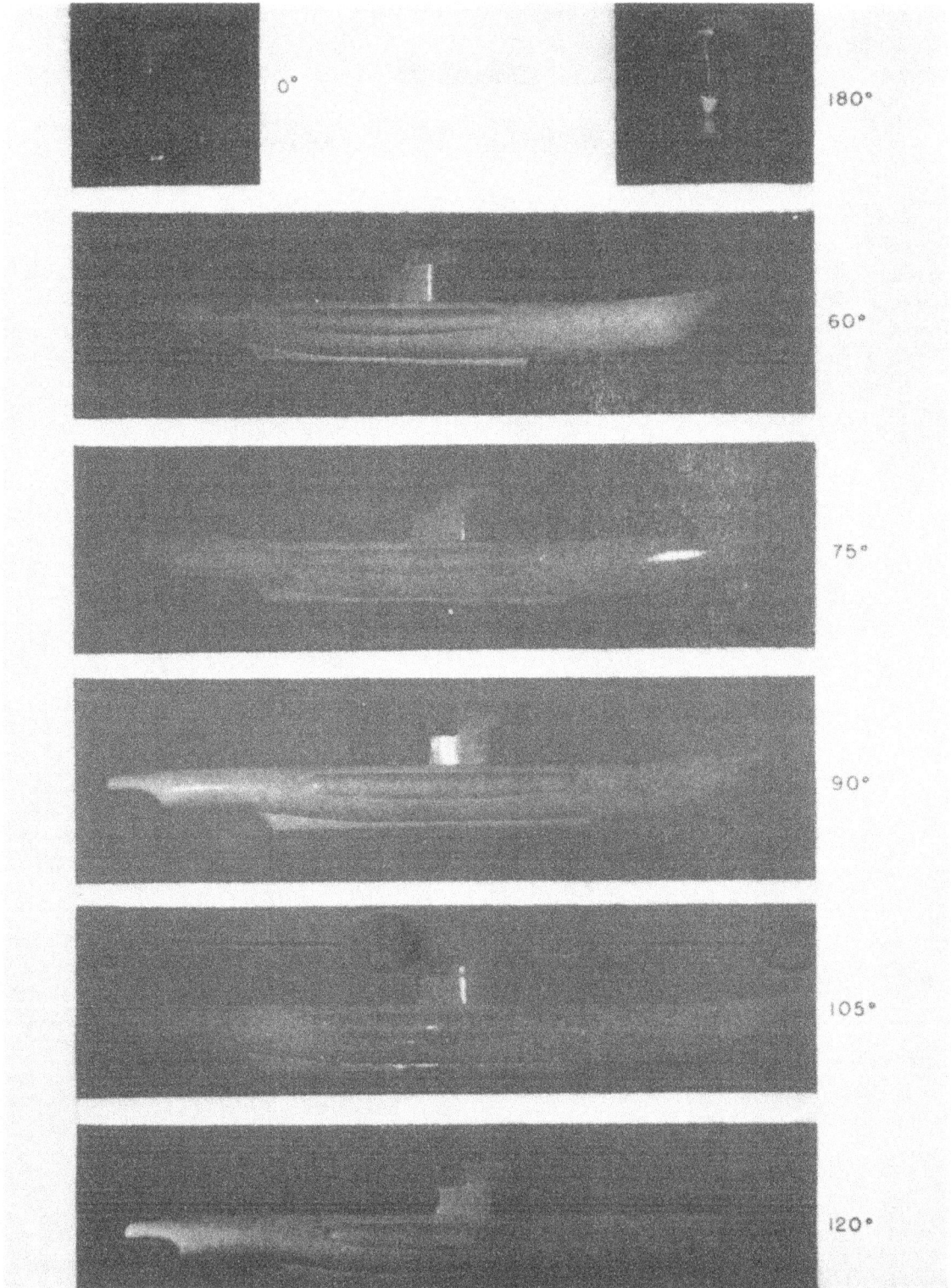

FIGURE 1. Reflection of light from HMS/M *Graph*.

relative intensities of the light reflected from the submarine model and from the sphere, the scale factor of the submarine model, and the expression for the target strength of a sphere [equation (10) in Chapter 19].

The technique of these optical measurements was not simple. Light from a motion picture projector bulb passed through a polarizing element rotated by a synchronous motor and was focused on the submarine model. As a result, the plane of polarization of the incident light rotated at a high speed. Upon reflection from the model, the light passed through a second polarizing element and fell on a photoelectric cell; this second polarizing element was stationary, but adjustable. In effect, the two polarizing elements modulated the intensity of the light incident on the cell and made possible the use of a-c instead of d-c amplifying and measuring equipment. Moreover, the use of modulated polarized light greatly reduced the error caused by light scattered from the walls and other objects in the room in addition to the desired reflected light.

At the same time, the photoelectric cell was also exposed to light from a neon lamp which was supplied with current from both a battery and a step-down transformer. As a result, the neon light contained a small a-c component whose intensity was directly proportional to the alternating current through the lamp, which was measured on a vacuum tube voltmeter. Since the light reflected from the model was adjustable in phase, by use of the second polarizing element, and since the light from the neon lamp was adjustable in magnitude, one was balanced against the other, thus canceling out the a-c component of the light reaching the photocell. When the a-c output from the photocell vanished, this condition of balance was obtained, and the voltmeter reading of the a-c lamp current was then proportional to the intensity of the model-reflected light. The use of this null method made it unnecessary to rely on a calibration of the photoelectric cell.

To compute the target strengths, spheres from 1 to $12\frac{1}{2}$ in. in radius were substituted for the submarine models, and a similar procedure was followed. Photographs were also made at different aspect angles and are illustrated in Figures 2 through 5.

22.1.3 Mountain Lakes

At Mountain Lakes, New Jersey, a model of HMS/M *Graph*, similar to the model used at UCDWR and at MIT, was suspended in water in the path of sound from a supersonic transmitter.[5] The model, built to a 1:60 scale, was constructed of copper 0.5 mm thick, plated with nickel 0.025 mm thick as a protection against corrosion. The model was suspended approximately $2\frac{1}{2}$ ft below the surface of the lake by wires at distances between 1 and 17 ft from the transducers, corresponding to full-scale target ranges between 20 and 340 yd.

Pulses were not used in the indirect measurements at any of the laboratories. At Mountain Lakes, continuous sound was transmitted by a quartz crystal projector, and the echo was received by a separate similar unit which served as a hydrophone. The model scale was 1:60. Since the importance of nonspecular reflection depends on the ratio of the wavelength to the dimensions of the target, it was necessary to scale the wavelength similarly. Consequently, an actual echo-ranging frequency of 24 kc, which is standard for most Navy gear, would require a frequency of 1,440 kc in tests with a 1:60 scale model. However, since the response of the transducers was somewhat higher at higher frequencies, a frequency of 1,565 kc was used most of the time; the corresponding actual echo-ranging frequency was 26 kc.

A beat-frequency oscillator, with a fixed frequency of 15 mc, provided signals between 50 and 3,600 kc, which were amplified and sent through coaxial transmission lines to the projector. The received echo was amplified by a preamplifier in the hydrophone housing, demodulated by the detector circuit and recorded on a continuous strip of paper as the submarine was slowly rotated about a vertical axis. The known calibrations of the transducer and receiver were used, together with an assumed inverse square transmission loss to determine the target strength by using equation (6) of Chapter 19. Under the controlled conditions possible at a reference station on a lake, the calibration is less difficult than it is for gear mounted on a ship at sea; thus the calibration error in these tests was probably small. Also, at such close ranges, temperature gradients and surface reflections are negligible. At a frequency of 1,565 kc, the attenuation coefficient predicted from Figure 7 in Chapter 21 is about 0.6 db per yd. At ranges of only a few feet, this attenuation is negligible and the transmission loss may safely be assumed to obey the inverse square law. At ranges as great as 17 ft, however, this assumption may lead to target strengths which are about 6 db too low.

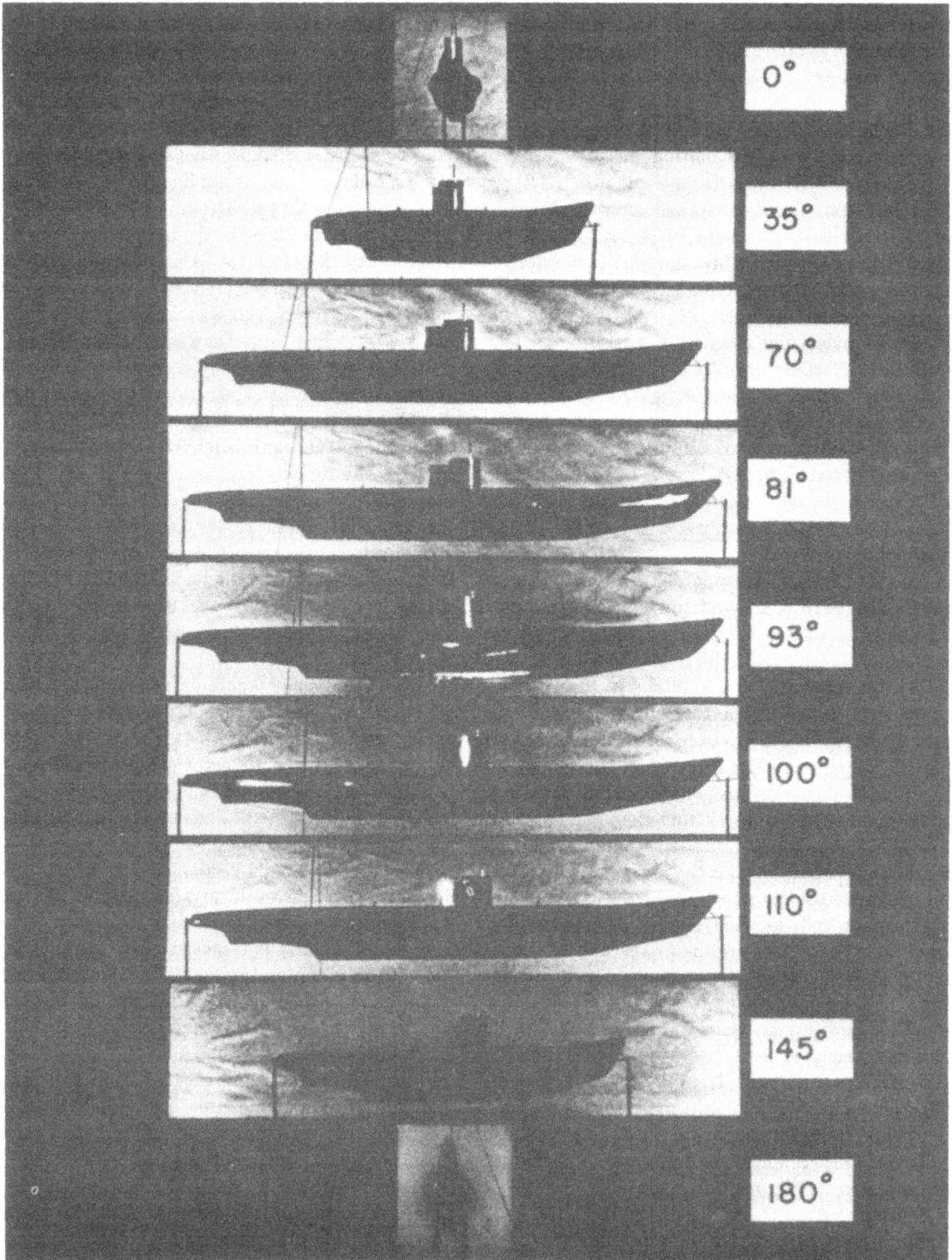

FIGURE 2. Reflection of light from HMS/M *Graph*.

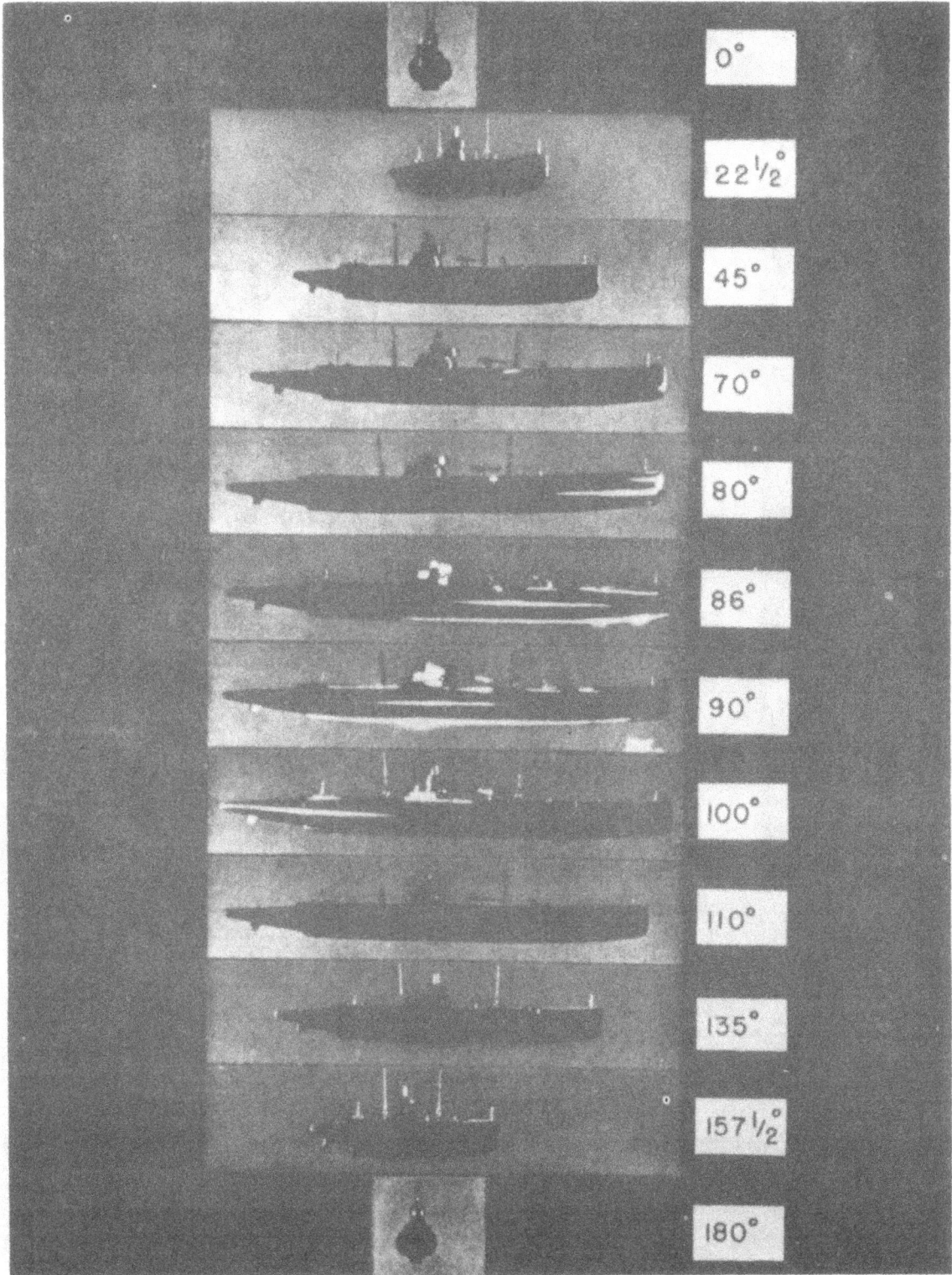

FIGURE 3. Reflection of light from S-type submarine.

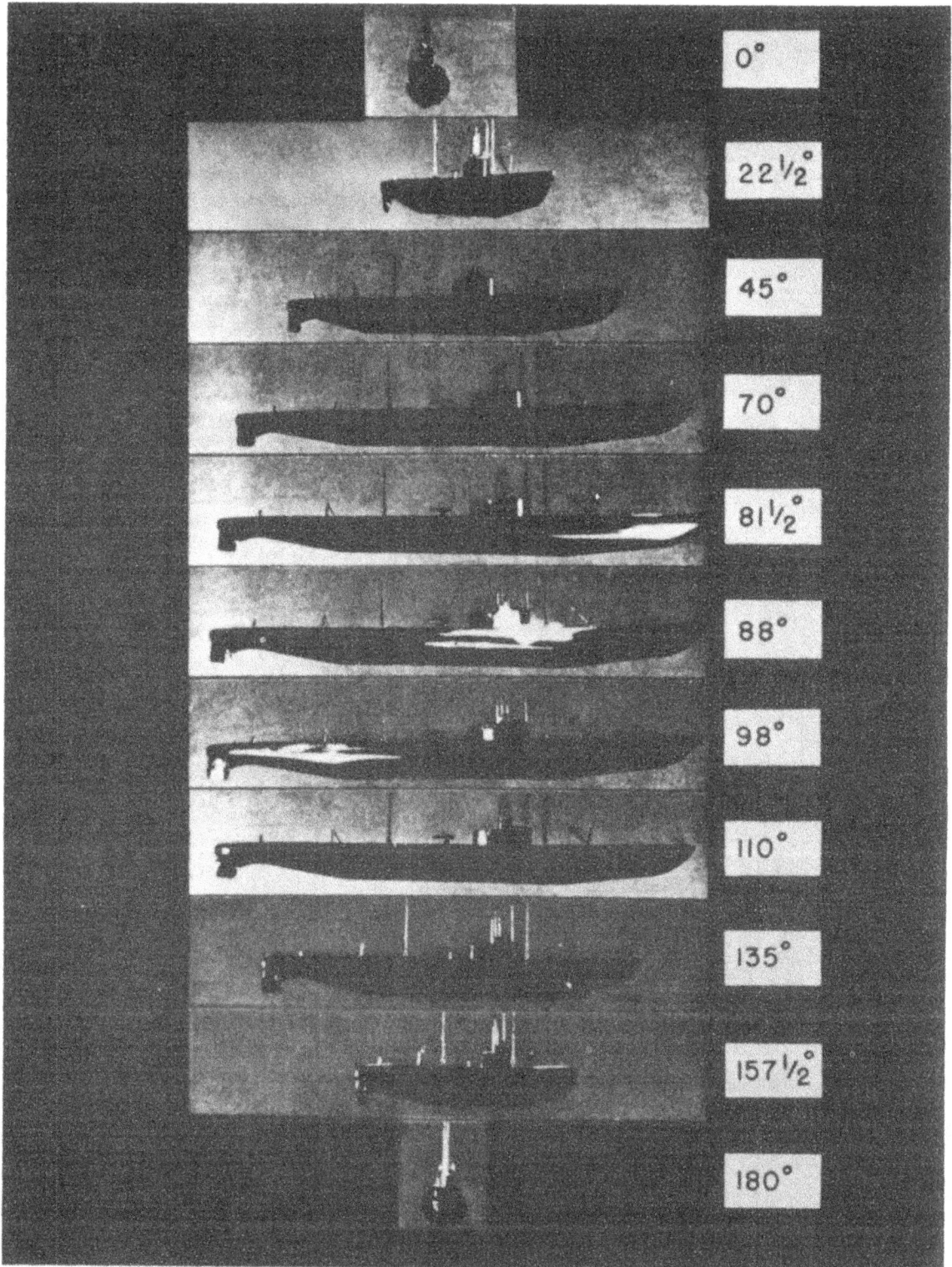

FIGURE 4.   Reflection of light from USS *Perch* (SS313).

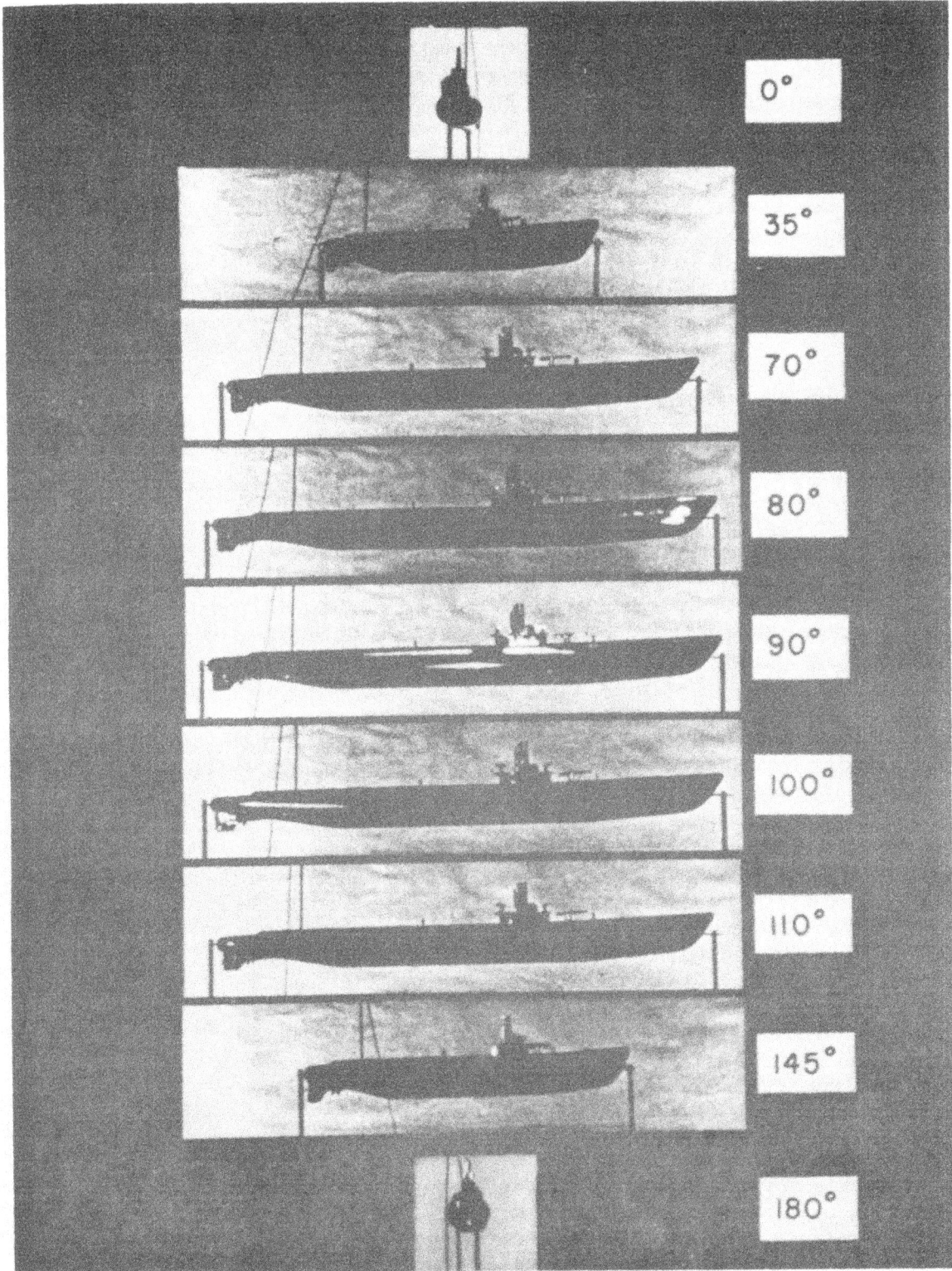

FIGURE 5. Reflection of light from USS *Sand Lance* (SS381).

22.2 SUBMARINE REFLECTIVITY

Since indirect target strength measurements are measurements not on actual submarines but on their scale models, certain possible corrections must be considered before the results can legitimately be compared with the results of the direct measurements. One possible source of error is in the reflectivity of the models used as compared with the reflectivity of actual submarines.

Since the experiments at UCDWR were qualitative in nature and designed only to determine the principal reflecting surfaces on a submarine, the question of absolute reflectivity is unimportant for those tests. The optical experiments at MIT, however, reported specific target strengths. These results were computed from the expression for the target strength of a perfectly reflecting sphere, in other words, a sphere which reflects all the sound striking it without transmission or absorption. Since both the submarine models and the spheres were finished in exactly the same way, these target strength results will apply only to perfectly reflecting submarines.

At USRL, the reflectivity of the hull itself was found to be perfect, within experimental error, over the range of frequencies used. The hollow model was first tested filled with air, then filled with water. No difference was observed in the intensity of the reflected sound for all frequencies between 50 and 2,000 kc. Since reflection from an air-filled hull would be almost perfect, regardless of the transparency of the hull to sound, and since reflection from a water-filled hull submerged in water would come solely from the hull, with no air-water interface to reflect the sound, the experimental results did not justify assuming anything less than perfect reflectivity. Thus both the optical and acoustical indirect measurements are based on perfect reflection of the sound striking the submarine; transmission through the hull and absorption in the steel are neglected.

The steel hull of an actual submarine is also almost perfectly reflecting. Therefore, it appears that the results of the indirect measurements may be interpreted in terms of sound reflected from actual submarines. However, the presence of barnacles, moss, and other marine growth on the hull may appreciably affect the reflectivity. Such an effect would be important for surface vessels or surfaced submarines, where the fouled hulls are exposed to the direct sound beam, but might not be significant for a submerged submarine, since the sound beam might not often strike the lower part of the hull where such growths attach themselves. No measurements have been made to ascertain the effect of barnacles and moss on the reflection of sound, but it is not believed to be significant. Therefore it appears that reflectivity considerations should not greatly affect any comparison between direct and indirect measurements.

22.3 WAVELENGTH EFFECTS

If the indirect measurements of target strengths with submarine models are to be trusted, the experiments must be properly scaled, that is, the dimensions of the models, the ranges and depths at which the tests are made and all the wavelengths must be reduced by the same factor. This factor was 60 for the acoustical measurements at USRL, and all the quantities relevant to the measurements were changed by this factor.

At MIT-USL, however, visible light was used. The models used in the optical experiments were from 60 to 120 times smaller than the submarines they represented. Assume an echo-ranging frequency of 24 kc, and the corresponding scaled wavelengths would be reduced to 0.1 cm for a 1:60 scale or 0.05 cm for a 1:120 scale. Since the actual wavelengths employed were much shorter, errors might be expected in the results.

Two errors in particular might be introduced. At certain aspects where the surface of the submarine subtends only a few Fresnel zones at 24 kc, as described in Sections 20.3 and 20.5, the model subtends many such zones, since the wavelength is much shorter compared with the dimensions of the submarine. As a result, the Fresnel integrals approach their asymptotic values, especially for surfaces of large radius of curvature, such as planes or cylinders, which subtend many Fresnel zones. Since the conning tower on a submarine is relatively flat, the optical measurements with very short wavelengths may overemphasize the effect of the conning tower.

Secondly, nonspecular reflection is less than if the wavelength was properly scaled, by a factor equal to the square root of the ratio of the properly scaled wavelength to the improperly scaled wavelength actually used. This may account for the extremely low target strengths obtained optically at aspects giving very little specular reflection, such as the bow and stern.

Diffuse reflection or scattering may be excessively large optically since the wavelength may be considerably smaller than the surface irregularities. However, this source of error has been minimized by the use of glossy black surfaces.

22.4 DIFFERENCES IN METHODS

Certain errors may arise from the differences inherent between the direct and indirect techniques. As a submarine travels through the water, each surface may be assumed to be screened by a wake of some sort, or at least a turbulent condition in the water, and possibly also by air bubbles surrounding the hull and conning tower.

Although this phenomenon may be present in the direct measurements of target strengths, it is absent in the indirect tests. In the optical methods, no reflecting layer surrounded the submarine model; every effort was made to reduce reflection from dust particles and from other objects in the room. In the acoustical tests, the submarine model was stationary throughout the measurements except for a very slow rotation in the horizontal plane, which could not give rise to wakes or air bubbles. The importance of this effect, of course, depends on the extent to which sound is reflected by turbulence in the water, which is negligible (see Section 34.3.2), or by air bubbles in the vicinity of the submarine (see Section 28.3.5).

Extraneous reflections also may occur during the indirect measurements. Removal of the models, however, has shown that the background level during both the optical and acoustical experiments is negligible compared with the levels of the echoes from the models.

Since continuous signals were employed during both the optical and acoustical measurements, separate transmitters and receivers had to be employed — a moving picture projector bulb and a photoelectric cell in the optical tests, and two similar transducers in the acoustical tests. The distance between them, however, was minimized, so that the angle of incidence and the angle of reflection at the model were as small as possible. At MIT, the bulb and photoelectric cell were approximately 14 in. apart, whereas the model was from 6 to 20 ft distant. At USRL, the two transducers were separated by less than 5 in., while the closest distance of the submarine model was 11 in.; most of the measurements, however, were made with the source and receiver about 17 ft away from the model.

Another difference between the direct and indirect measurements of target strength lies in the method of measurement. In the direct measurements, peak-echo amplitudes were used in all cases, since the echoes were short; in the indirect measurements, however, the echoes were continuous and the results were obtained by using rms intensities. The difference between mean intensities and peak intensities, and the dependence of this difference on pulse length are discussed in Chapter 21.

Other errors may result from discrepancies in the construction of the models. Considerable difference was observed between the two models of HMS/M Graph, one used at MIT and the other at USRL, so that comparison of the two series of measurements is not completely justifiable. At some aspects, a difference of 6 db in the target strengths of the two models was observed when optical measurements were later made on both models of HMS/M Graph; these differences are described in Section 23.2.2. In addition, rudders and propellers were missing from some of the models used at MIT and the model tested at USRL; at certain aspects, they may give rise to strong echoes. The models of the S-boat, the USS Perch and the USS Sand Lance, however, were supplied by the Bureau of Ships and are believed to be accurate.

Chapter 23

SUBMARINE TARGET STRENGTHS

Target strengths of submarines have been computed mathematically from the size and shape of a particular submarine, by the Fresnel zone method outlined in Chapter 20. They have also been measured, both directly and indirectly, by use of the procedures and techniques described in Chapters 21 and

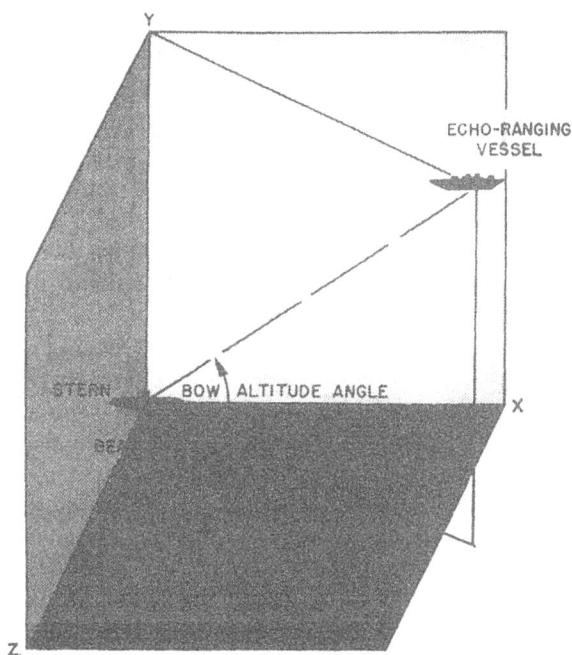

FIGURE 1. Definition of angles.

22, and have been studied in general as a function of orientation, submarine class, speed, range, pulse length, and frequency of the echo-ranging sound. This chapter presents the results of the different methods of determining target strengths of submarines and discusses their applicability to practical echo ranging.

23.1 DEPENDENCE ON ORIENTATION

Since a submarine is irregular in shape, the echoes which it returns depend markedly on its orientation with respect to the echo-ranging beam. The orientation of such an irregular target is conveniently described in terms of aspect and altitude angles, defined in Figure 1.

Consider a system of rectangular coordinates with the origin O at the center of the submarine. The *aspect angle* is defined as the angle between the x axis and the projection of the echo-ranging beam on the horizontal (xz) plane. It is measured in degrees from the bow of the submarine, in a clockwise direction as viewed from above; bow aspect is 0 degree, stern aspect 180 degrees, while beam aspect will be 90 and 270 degrees for the starboard and port beams respectively.

The angle between the echo-ranging beam and its projection on the horizontal (xz) plane is the *altitude angle*. It is measured in degrees, positive when the sound source is above the submarine, negative when it is below the submarine. If the projector is at the same depth as a level submarine, the altitude angle is 0 degree; similarly, if it is directly above a level submarine, the altitude is 90 degrees. The vertex of both aspect and altitude angles is placed at the origin O of the coordinate system, which is taken at the geometric center of the submarine.

23.1.1 Aspect Angle

The strongest echo from a submarine is usually found within 20 degrees of beam aspect — between 70 and 110 degrees, and between 250 and 290 degrees, from the bow of the submarine.[1] These beam and near-beam echoes average about as strong as the echo from a sphere 35 yd in radius and correspond to a target strength of 25 db. Actually, target strengths as low as 7 db and as high as 40 db have been observed at beam aspect, directly and indirectly; most values, however, lie between 20 and 30 db.

At other aspects, the target strength is much smaller and averages between 5 and 15 db, depending on the submarine and the altitude angle. At stern aspect, for example, target strengths measured directly with standard gear vary from 4 to 19 db, de-

388

pending on the submarine [2-4] (see Section 23.2.1). Negative target strengths have been observed in the optical studies at certain aspects and altitudes; for example, at bow and stern aspects the target strength of the German U570 (HMS/M *Graph*) varies from −4 to −6 db when the echo-ranging beam is below the submarine, at altitude angles between −5 and −15 degrees.[5] Since such negative altitude angles are not encountered in practice when echo ranging from a surface vessel on a submerged submarine — because

furthermore, the uncertainty in the aspect angle in some of the measurements was rather large. Consequently, the beam target strengths do not apply to an aspect angle of exactly 90 or 270 degrees. The altitude angle in all cases was small. In the direct measurements reported in this table, the submarine was seldom submerged to a keel depth greater than 100 ft, which at a range of 500 yd corresponds to an altitude angle of 4 degrees, while for the indirect measurements quoted the altitude angle was 0 degree.

TABLE 1.   Submarine target strengths.

	Submarine	Frequency in kc	Beam target strengths and standard deviations in decibels	Bow target strengths and standard deviations in decibels	Stern target strengths and standard deviations in decibels
Theory	U570 (HMS/M *Graph*) [6]	25	25.5	11.5	11.5
Direct * measurements	*Tambor* class [7]	...	18.5	...	8.5† 1.5‡
	USS S-28 (SS133) [8]	24	18	...	...
	USS S-40 (SS145) [2]	24	25 ± 4	12.5 ± 4	12.5 ± 4
	Fleet type [3]	24	24 ± 5§	13 ± 6	19 ± 5
	Fleet type [3]	24	29 ± 3.5‖		
	USS S-34 (SS139) [9]	45	25	...	...
	USS *Tilefish* (SS307) [9]	45	26	...	...
	Fleet type	60	25	...	...
	R class	60	...	...	4
	R class	60	...	...	8.1
	British [4]	18	29	...	...
	Vortice [10]	26	42	40	40
	Vortice [11]	26	...	...	10.5
Indirect measurements	S class [5]	...	30	6	5
	USS *Perch* (SS313) [5]	...	30	6	5
	USS *Sand Lance* (SS381) [5]	...	26	5	9
	U570 (HMS/M *Graph*) [6]	...	27	4	1
	U570 (HMS/M *Graph*) [12]	26	25	6	6

* San Diego measurements at 60 kc are not included here.
† Beam focused on conning tower.
‡ Beam focused on screws.

§ Average of all values in a 30-degree sector centered at beam aspect
‖ Average of maximum values in the sector for each run.

the projector is always above the target — these low target strengths are not very significant. Moreover, they may result from errors in the construction of the model (see Section 23.6.2) or from possible systematic errors inherent in the optical method (see Section 22.4).

Table 1 summarizes submarine target strengths at beam, bow, and stern aspects for the theoretical calculations [6] and for the direct [7-11] and indirect measurements.[2, 12] Certain controversial values discussed later in this chapter are omitted, as, for example, certain San Diego measurements at 60 kc. Most values were averaged in sectors of roughly 10 or 20 degrees;

Ranges varied from 200 to 1,000 yd, and the frequency from 12 to 60 kc for all tests except the optical studies at MIT, where the full-scale frequency was much higher. Although the results of the mathematical studies are only approximate and the results of the direct measurements highly variable, all the data are generally consistent.

The early San Diego values for a fleet-type submarine of the *Tambor* class [7] and an S-boat [8] are not reliable. An early experimental frequency-modulated gear was used to echo range on the *Tambor*-class submarine; since these results are difficult to interpret in terms of standard echo-ranging gear, they cannot be

TARGET STRENGTH IN DECIBELS

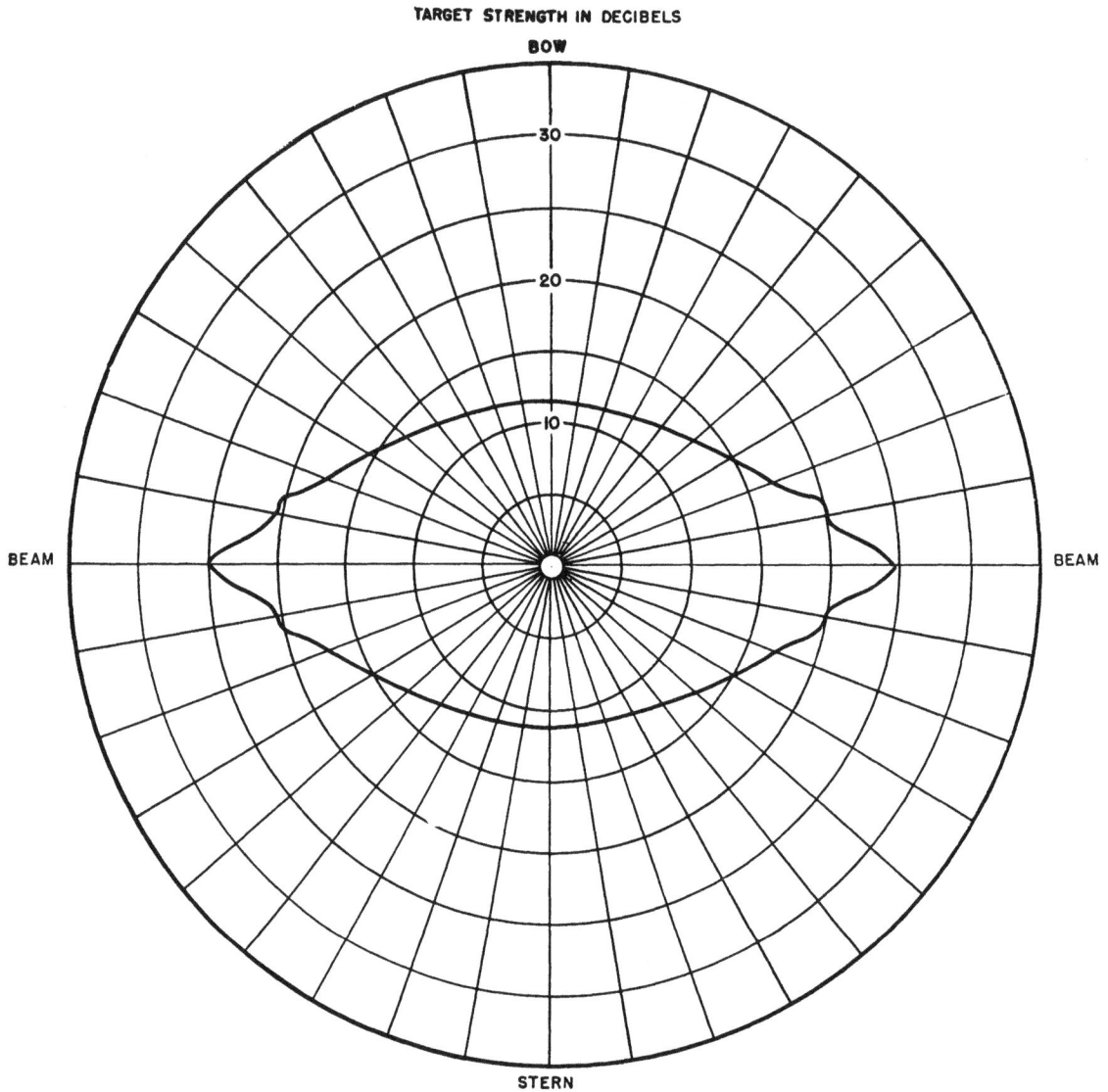

FIGURE 2.　Aspect dependence (theoretical).

weighted as heavily as other measurements. An appreciable variation of target strength with aspect angle was observed, in agreement with other measurements. Also, an observed difference in these experiments of between 10 and 16 db in target strengths at beam and stern aspects, depending on where the beam was focused, seems confirmed by more reliable results. The actual values, however, must be considered doubtful.

The target strengths of the S-class submarine taken from reference 8 were measured by comparing submarine echoes with the echoes from a submerged sphere at a much shorter range than the submarine; fluctuations were large (see Figure 10 in Chapter 21),

and the transmission loss was not known accurately. Furthermore, no significant variation in target strength with aspect angle was observed for the S-boat; this result alone makes the reliability of these measurements very dubious.

The remaining values in Table 1 are in moderately good agreement with each other, especially at beam aspect; values for which no reference is given were found at Fort Lauderdale. The only results out of line are those from the Woods Hole measurements on the Italian submarine *Vortice*.[10]

These measurements on the submarine *Vortice* were made at a range of 1,000 yd and frequencies of 12 and 24 kc; the submarine was proceeding at 6 knots at a

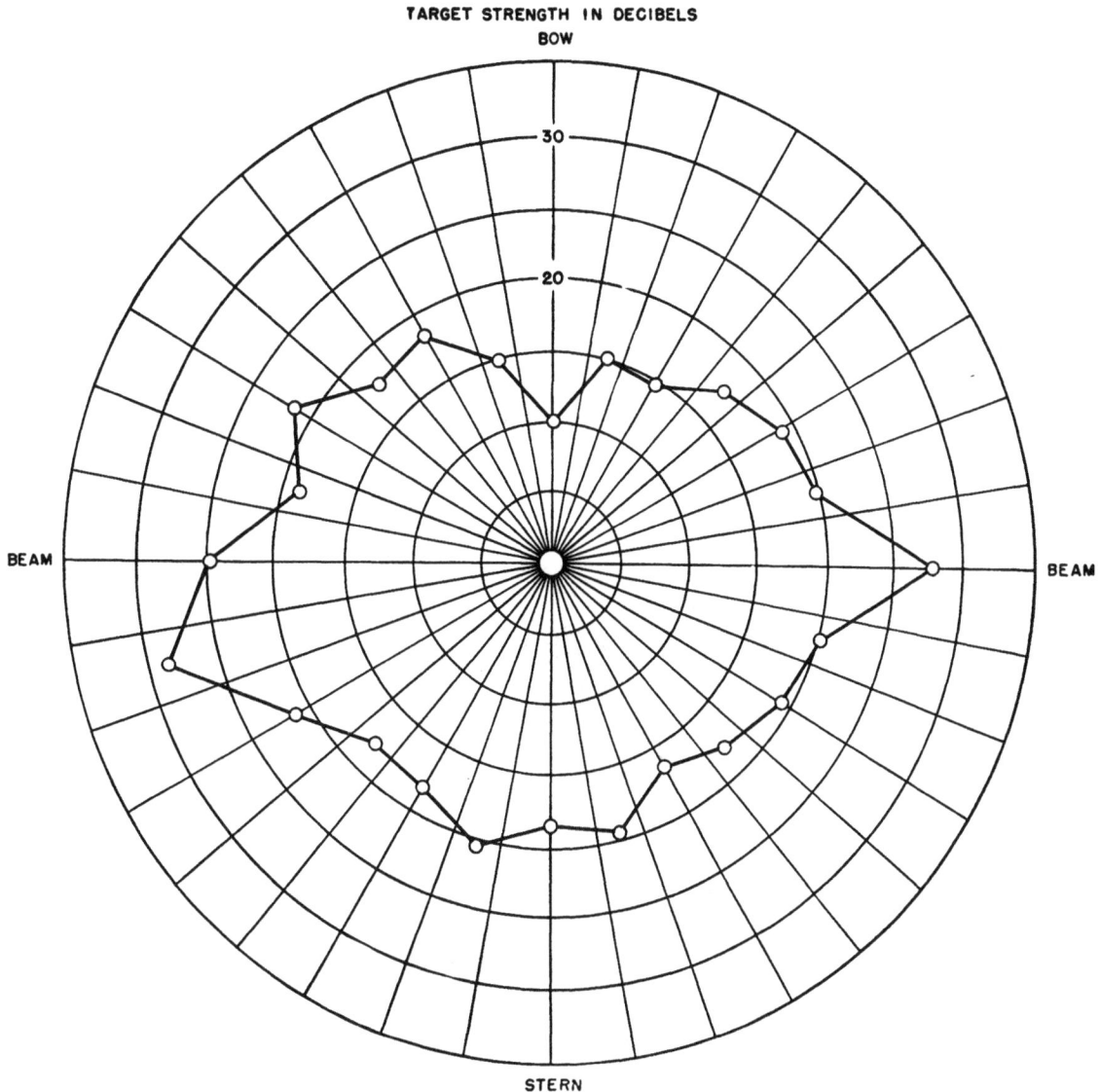

FIGURE 3.   Aspect dependence (San Diego).

depth of 150 ft. The values reported are of the order of 40 db, so much larger than any reported previously by any method that they appear to be the result of faulty calibration; the systematic error of about 20 db at all aspects is difficult to explain in any other way.

It might be pointed out, however, that the transmission loss as measured aboard the submarine was about 15 db greater than that expected from the prevailing oceanographic conditions. Therefore an attenuation coefficient of 4 db per kyd was assumed, in addition to inverse square divergence. In spite of such a transmission loss, however, the target strengths are more than 10 db greater than the highest values previously observed on a similar submarine.

Figures 2 to 6 show typical variations of target strengths with aspect angle. Figure 2 is a plot of the theoretical calculations of the target strength of the U570 (*Graph*) for an echo-ranging frequency of 25 kc and a range of 1,000 yd.[6] They are based on approximating the submarine by an ellipsoid of appropriate dimensions, with a conning tower of "tear drop" cross section; details of this method are described in Section 20.5.

Figure 3 shows the result of a typical target strength run at 24 kc on a fleet-type submarine at

RELATIVE ECHO LEVEL IN DECIBELS

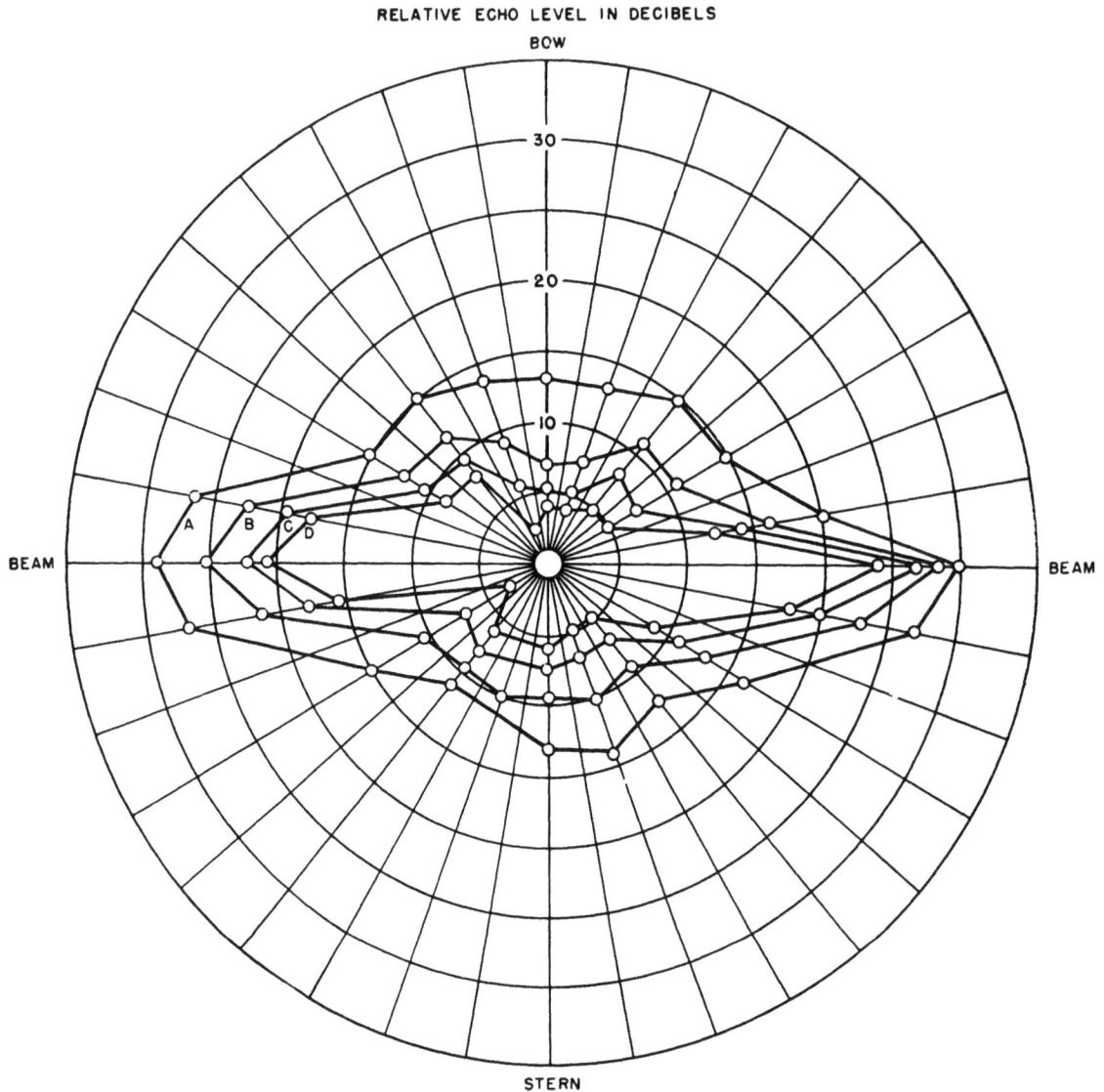

FIGURE 4.   Aspect dependence (New London).   Curves (range in yards): (A) 600; (B) 800; (C) 1,000; (D) 1,200.

San Diego. The submarine followed a straight course at about 2½ knots at periscope depth, while the *Jasper* circled it at a range of about 500 yd (see Figure 2 in Chapter 21). Each point on the curve is the average of all echoes obtained within the 15-degree sector centered at the indicated aspect angle, and represents, on the average, about 40 individual echoes.

Figure 4 is a plot of the echo level against aspect angle for a series of tests made on the USS S–48 (SS159) at New London and described in Section 21.2.2.[13] Each contour represents measurements made at a different range. Since no correction was made either for the transmission loss or for the calibration of the equipment, no absolute target strengths are plotted. These echo level values cannot be compared directly with other target strength values, since they are relative to an arbitrary level. However, the differences in echo levels at different aspects correspond to the differences in target strengths at different aspects, as long as the range remains constant.

Indirect measurements of target strength are illustrated in Figures 5 and 6. Figure 5 shows target strength as a function of aspect angle for a model of the USS *Perch* (SS313), as measured optically at MIT.[2] The altitude angle was 0 degrees and the full-scale range 600 yd. The results of the acoustical tests

TARGET STRENGTH IN DECIBELS

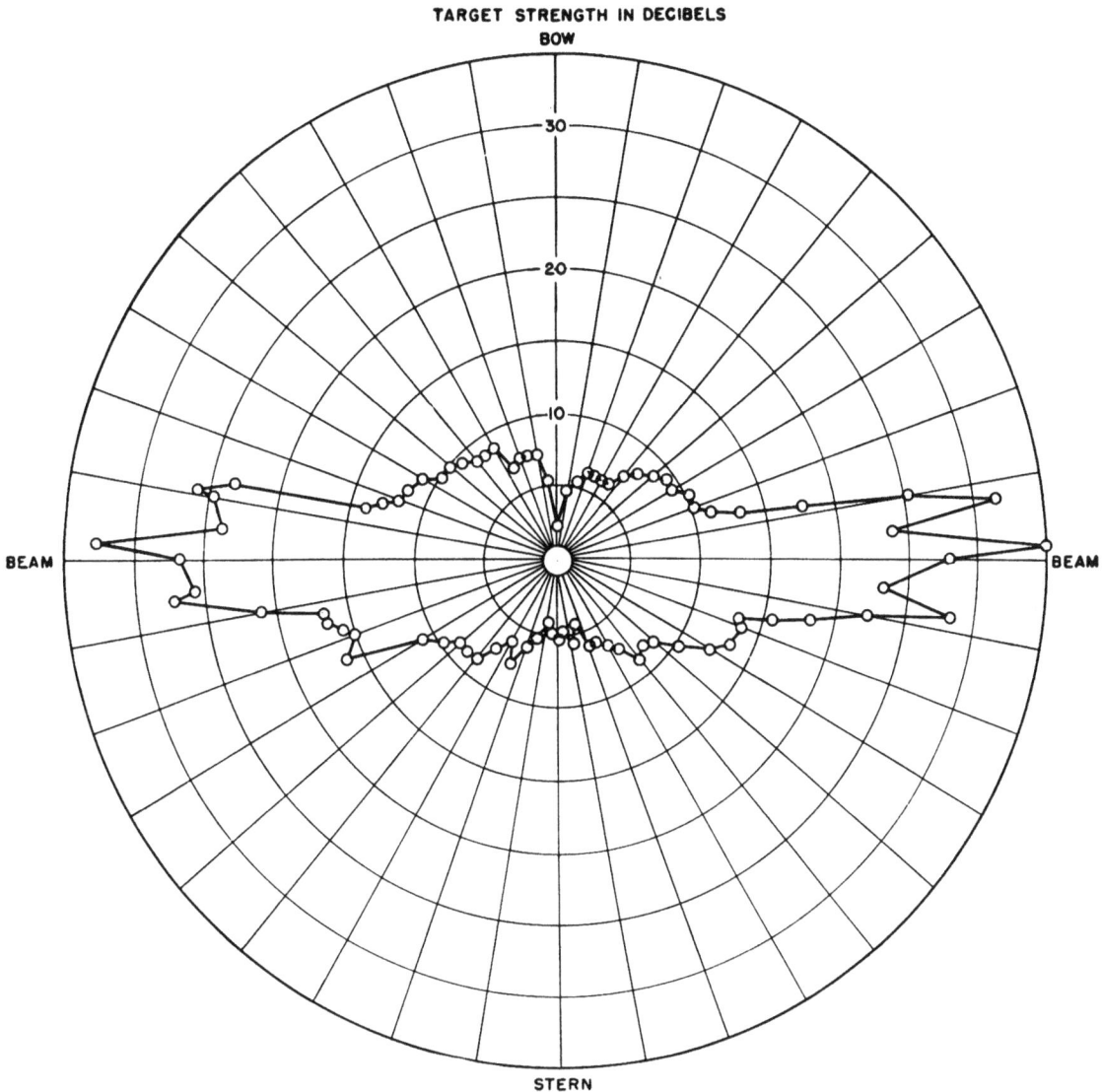

FIGURE 5.   Aspect dependence (MIT).

made at Mountain Lakes on a model of the U570 are reproduced in Figure 6, for an altitude angle of 0 degree, a full-scale range of 340 yd, and a full-scale frequency of 26 kc.[12] Bow and stern target strengths in Figures 5 and 6 are considerably lower than in Figure 2, probably because the ellipsoid used in the theoretical calculations was rounded at either end while the optical and acoustical models were pointed. Other discrepancies at bow and stern are discussed in Sections 23.8.2 and 23.8.3.

23.1.2   Altitude Angle

Target strength varies with altitude angle, but in most cases this variation does not appear to be im-portant practically. Figure 7 illustrates the theoretical predictions of the target strength of the U570 at beam aspect as a function of altitude angle for a range of about 16 yd; [5] since target strengths were found only for certain intervals of the altitude angle, they are represented as sectors in the polar plot of Figure 7. The sharp sectors at particular altitudes are attrib-uted to the sum of two separate reflections — from the blister tank and from the hull itself — neglecting interference phenomena. Figure 8 shows the same plot for the optical measurements made on a model of the USS *Perch* (SS313) for echo-ranging distances; the peak at 90 degrees, when the projector is directly above the submarine, arises from a strong reflection

TARGET STRENGTH IN DECIBELS

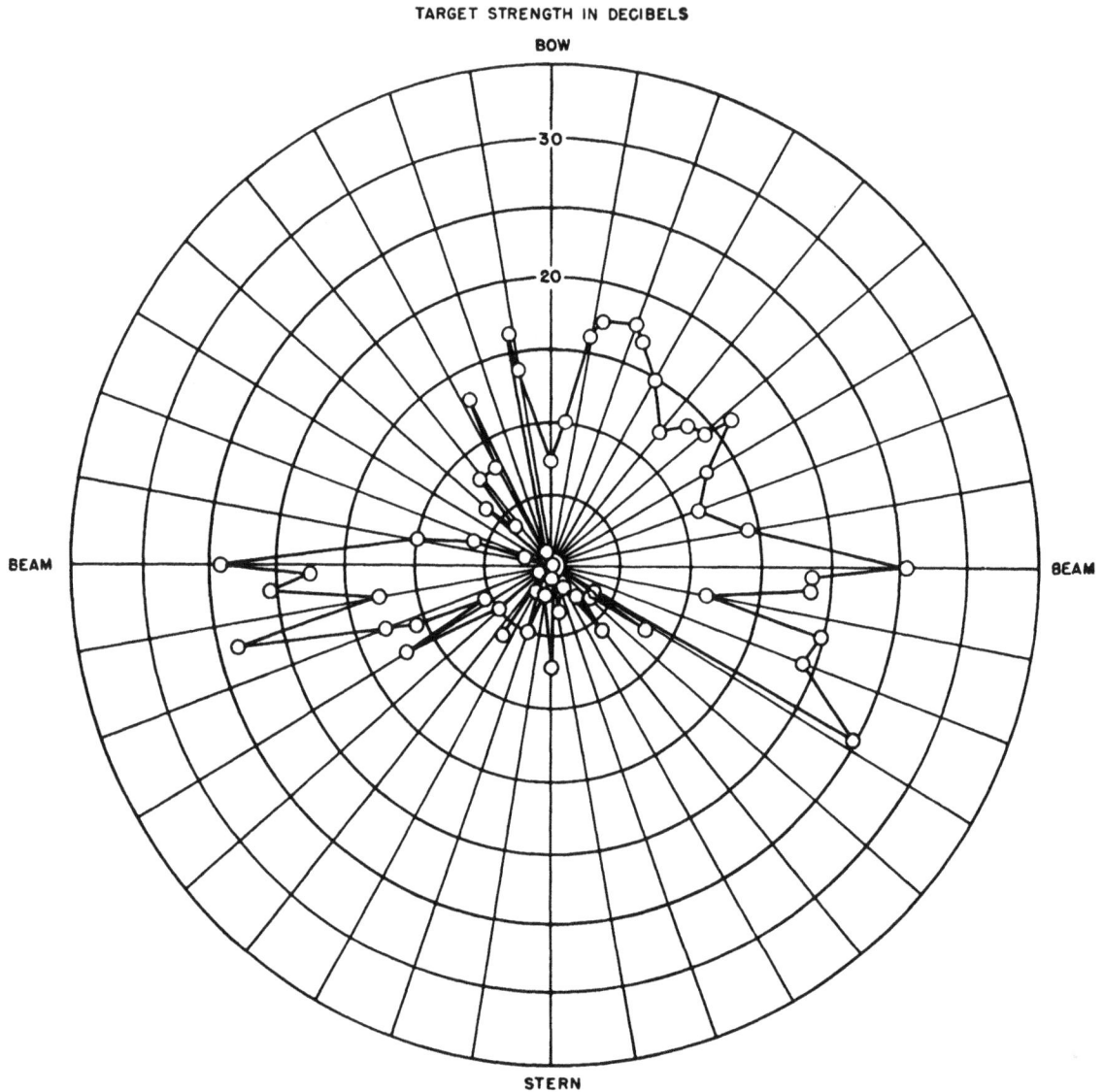

FIGURE 6. Aspect dependence (Mountain Lakes).

from the deck of the submarine. The absolute values in Figures 7 and 8, however, are not comparable because the theoretical calculations were carried out for a projector very close to the submarine, while the optical measurements applied to the ranges of several hundred yards usually encountered in practical echo ranging.

Figure 9 is a smoothed curve showing the relative target strength of the Italian submarine *Vortice* plotted against aspect angle for altitude angles of 0 to 10, 10 to 20, 20 to 45, and 45 to 90 degrees, as measured by Harvard observers; [14] for each curve, the relative target strength at beam aspect was arbitrarily set at 25 db. These data were obtained at a frequency of 26 kc, for submarine depths of 100 to 400 ft and ranges up to 1,000 yd. The aspect dependence apparently becomes less marked and the curve smoother as the altitude angle increases. It might be pointed out, however, that for altitude angles greater than 20 degrees, and a submarine depth of about 400 ft, the sound beam does not completely cover the submarine at near-beam aspects. Therefore the target strength would be expected to show less aspect dependence.

Figures 10 and 11 show target strength aspect curves for different altitude angles, as measured indirectly. Optical measurements on a submarine of the S class are given in Figure 10 for altitude angles of 0,

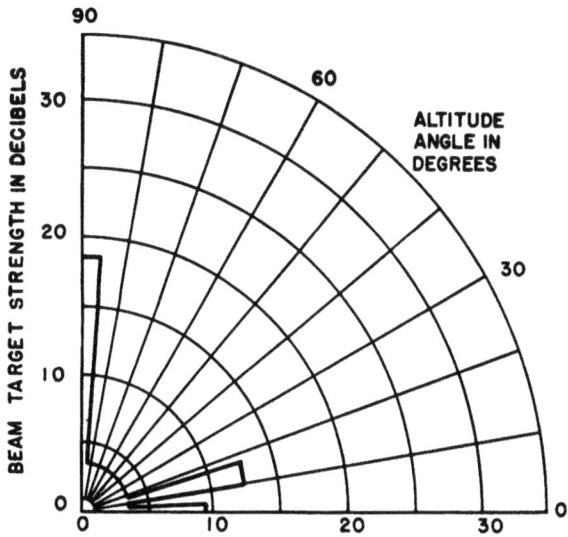

FIGURE 7. Altitude dependence (theoretical), 16-yd range. U570 (HMS/M *Graph*).

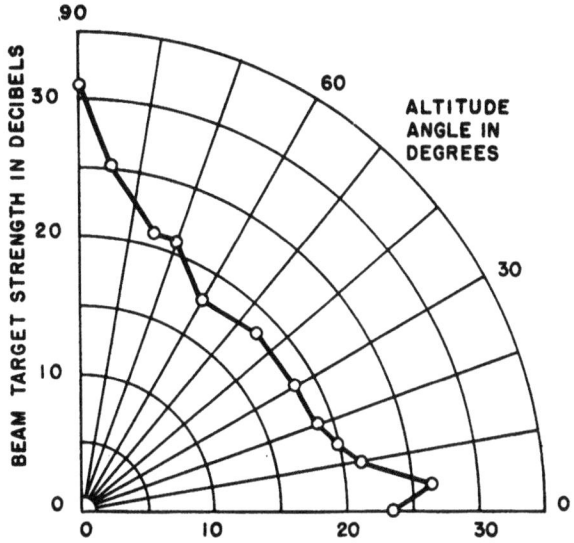

FIGURE 8. Altitude dependence (optical), 193-yd range. USS *Perch* (SS313).

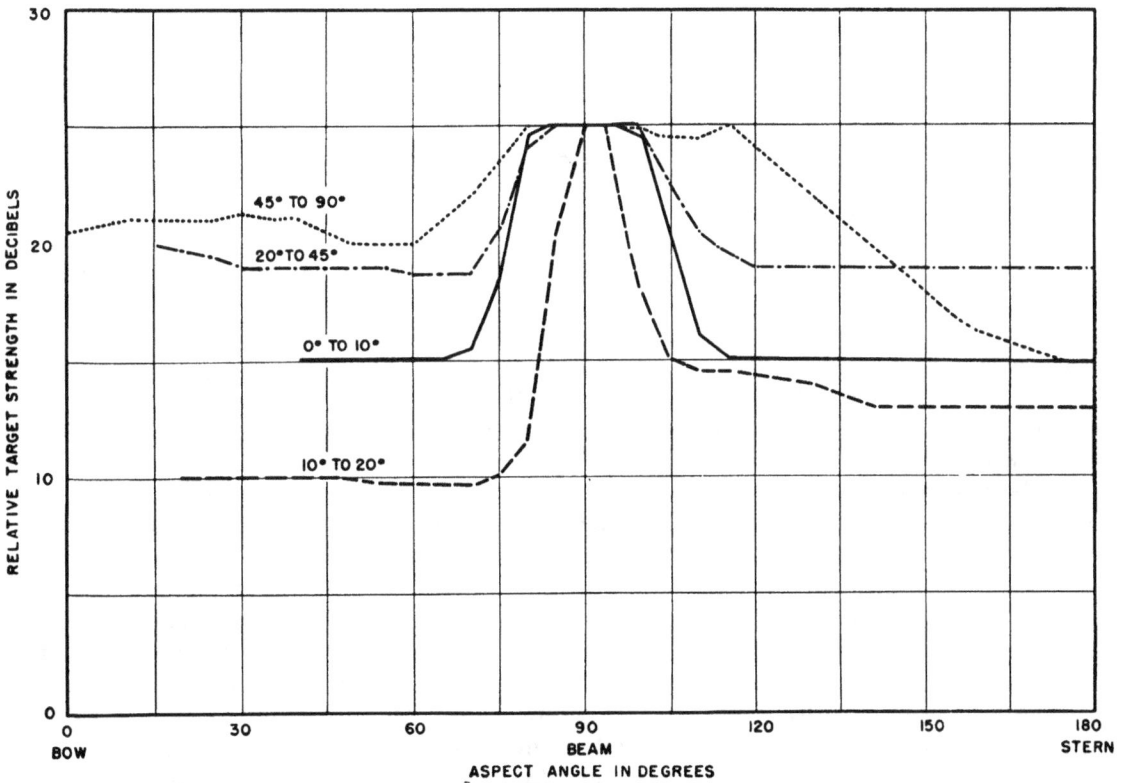

FIGURE 9. Target strength-aspect curves at different altitudes, for the Italian submarine *Vortice* (direct measurements).

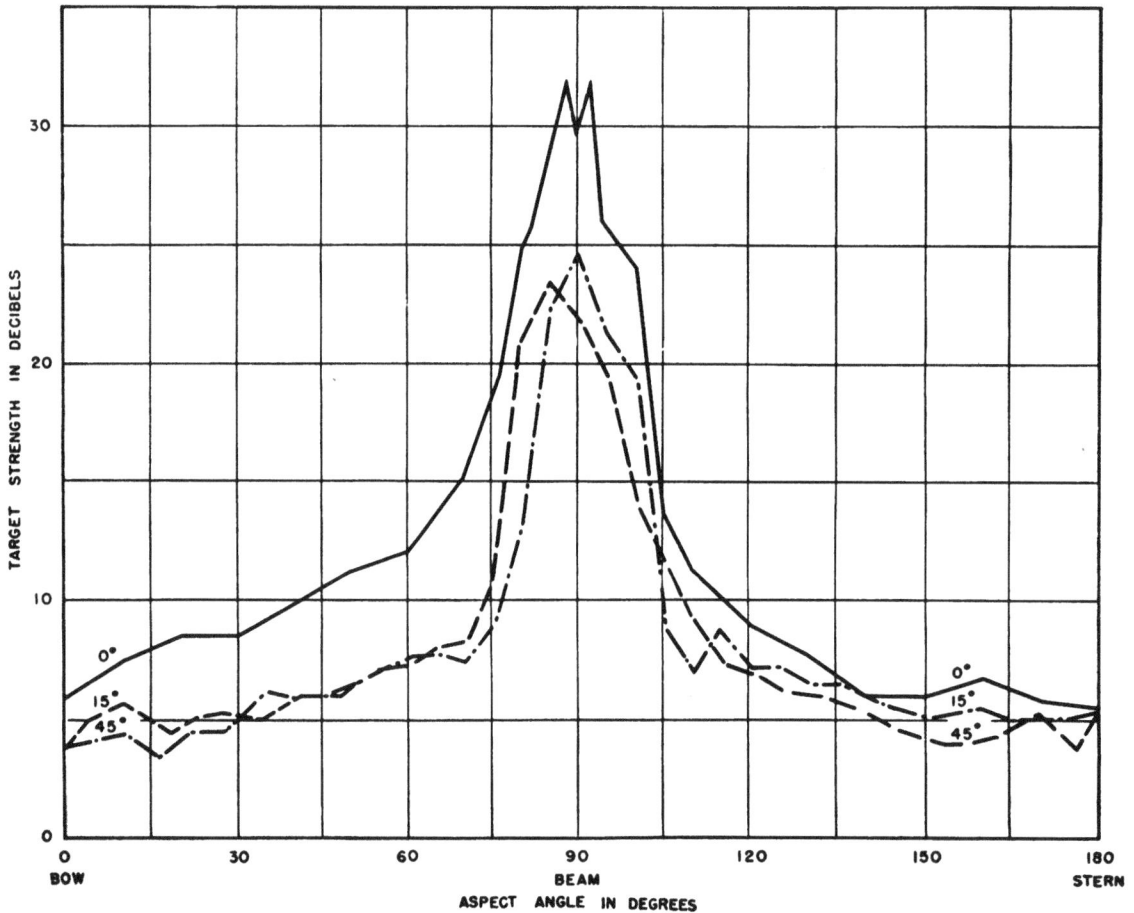

FIGURE 10. Target strength-aspect curves at different altitudes (optical).

15, and 45 degrees. Figure 11 is a similar plot for the acoustical measurements on the model of the U570 at much smaller altitude angles of 0, 1.0, and 1.8 degrees. Occasionally a pronounced maximum of the target strength has been observed at a very small and critical altitude angle, where the conning tower gives a prominent specular reflection or highlight. An example may be seen in Figure 11, at an aspect angle of about 105 degrees. Another example, at only one aspect angle, is shown in Figure 12, where the target strength of the U570 at bow aspect is plotted for five different altitude angles used at Mountain Lakes; here a strong reflection from the conning tower at an altitude of 1.0 degree results in a target strength of 19 db, while the target strengths at altitude angles only a fraction of a degree different are considerably lower. Such maxima, however, are not common, and are generally confined to such a small sector of altitude angles that most of the time they are not likely to be observed in actual echo ranging. Their possible

effect on the direct measurements has not been verified because of the wide fluctuations in echo level tending to obscure such fine detail.

Negative altitude angles are not encountered for surface vessels echo ranging on submarine targets. The low target strengths at these negative angles, measured in the optical studies, have already been mentioned in the preceding section.

At very large positive altitude angles, the differences between beam, bow, and stern target strengths are less marked and the resulting target strength-aspect curve is considerably smoother, as illustrated in Figure 9 for altitude angles between 0 and 90 degrees, and in Figure 10 for altitude angles of 15 and 45 degrees. Direct measurements on a deep submarine were also made at San Diego on the USS *Tilefish* (SS307) at a depth of 400 ft,[9] and gave results not significantly different from measurements at shallower depths except at quarter aspects. Values between 25 and 27 db were obtained at beam aspects in

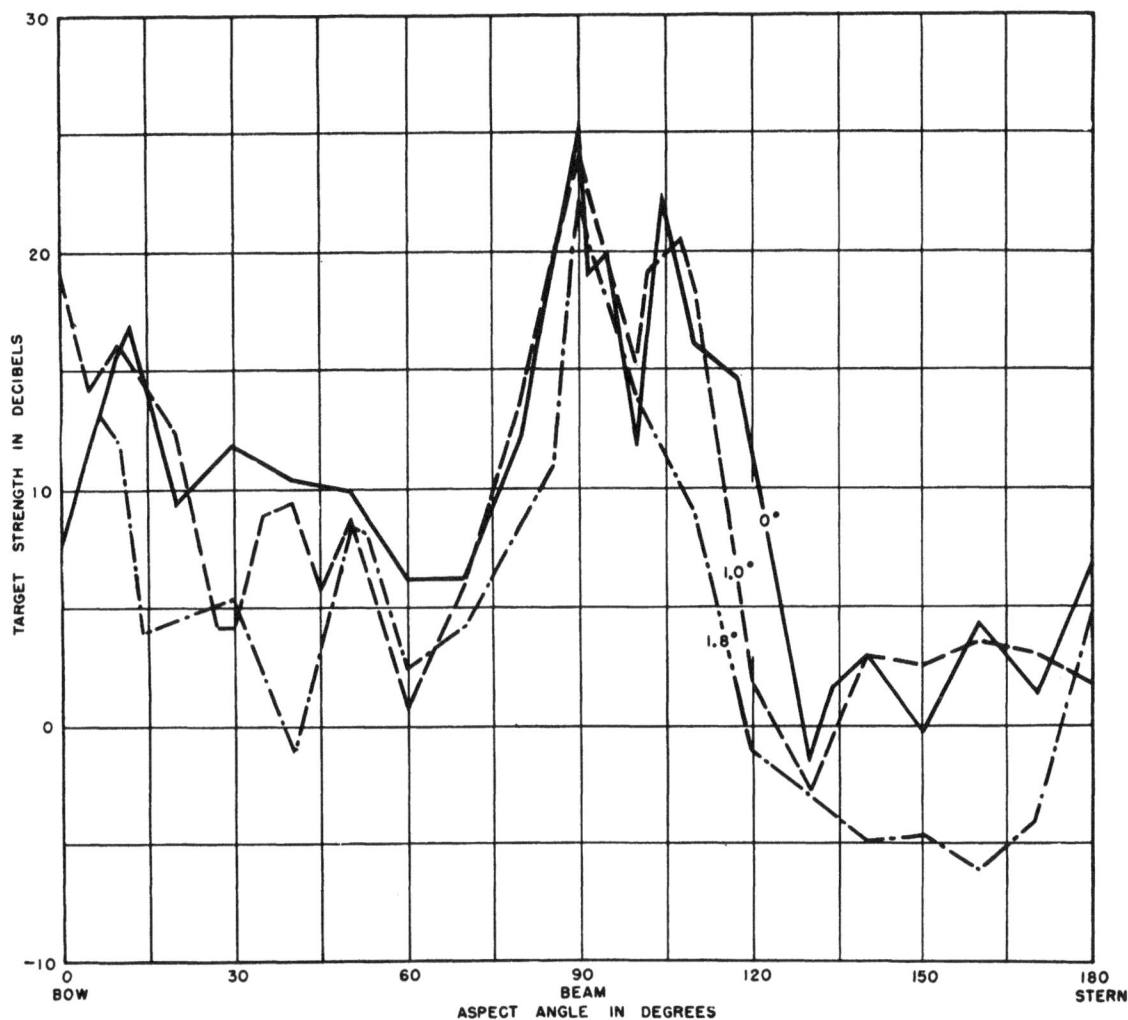

FIGURE 11.   Target strength-aspect curves at different altitudes (acoustical).

good agreement with values at periscope depth. At an aspect angle of 150 degrees, however, a target strength of 27 db was obtained, much higher than any other values reported at that aspect, directly or indirectly. An S-boat, for example, measured in the same series of runs, gave a target strength of 14 db at the same aspect.

This high value of 27 db may result from overcorrecting for the transmission loss by assuming an excessively large attenuation coefficient, since the range of the *Tilefish* was 760 yd compared to 150 yd for the corresponding measurements on the S-boat. The attenuation coefficient assumed was 10 db per kyd of sound travel at a frequency of 45 kc. However, high target strengths may actually be characteristic of deep submarines at certain aspects. It may be men-

tioned that at Fort Lauderdale, one of the strongest series of echoes was obtained when echo ranging at stern aspect on a fleet-type submarine submerged to a depth of 250 ft, at ranges between 220 and 700 yd.

23.2   DEPENDENCE ON CLASS

Submarines of different sizes and shapes may be expected to reflect sound differently; the echoes from an R-boat 186 ft long and those from a new fleet-type submarine more than 300 ft long are not likely to be the same. In particular, specular reflection of sound from a submarine would be expected to depend rather critically on the shape of the hull and especially on the different radii of curvature.

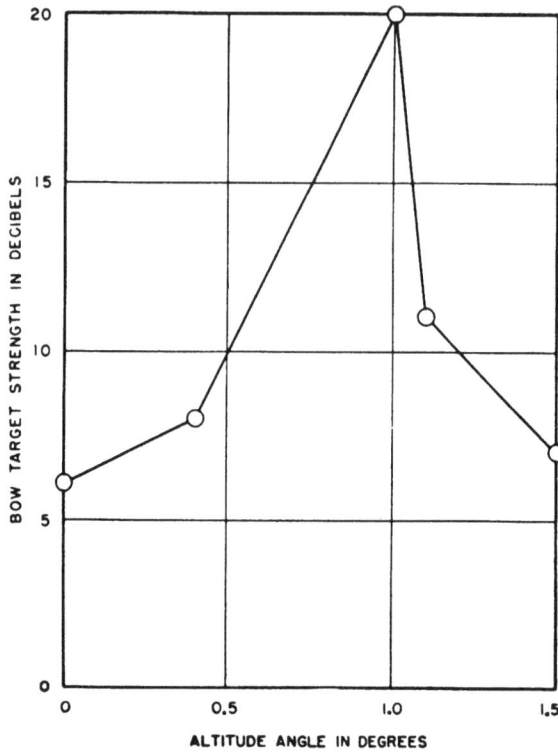

FIGURE 12.   Altitude dependence (Mountain Lakes).

12 db; and for a large fleet-type submarine about 19 db. At beam aspect, the target strengths are more nearly the same.

Unfortunately, no comparative measurements have been made on two different vessels by a single group during a single operation. In view of the systematic errors of 5 to 10 db that may be present in target strength determinations, suggested in Sections 21.4, 21.5, and 21.6, the differences shown in Table 2 are not too conclusive. More accurate data are required to allow any conclusion to be drawn about the variation of target strength between different submarines.

23.2.2   Indirect Measurements

Table 3 lists target strengths for different submarines measured indirectly at an altitude angle of 0 degrees. These values were obtained under controlled conditions and are more self-consistent than the values measured directly; as a result, fluctuations are smaller and the differences between the values are probably more significant than in the direct measurements. Beam target strengths may vary between 25 and 30 db; at stern aspect, the limits are 1 and 9 db.

TABLE 2.   Dependence of target strength on submarine class (direct measurements).

Submarine	Reference	Frequency in kc	Length in feet	Beam target strength and standard deviation in decibels	Bow target strength and standard deviation in decibels	Stern target strength and standard deviation in decibels
R class		60	186	...	...	4
R class		60	186	...	...	8.1
S class	2	24	219	...	12.5 ± 4	12.5 ± 4
S class	Average from Table 4	24	219	19.7 ± 2.5	...	...
Fleet type	3	24	304	24 ± 5	13 ± 6	19 ± 5
British	4	18	300	29	...	...

23.2.1   Direct Measurements

Representative target strengths of different submarines measured directly are listed in Table 2. S-boats, R-boats, and fleet-type submarines have all been targets in direct measurements, but the experimental error is so great that it generally obscures any possible correlation of target strength with submarine class. However, Table 2 shows an apparent dependence of target strength on submarine class at stern aspect. The target strength of R-boats at stern aspect varies from 4 to 8 db; for an S-boat it averages about

The possibility that a fleet-type submarine may return a strong echo at stern aspect, which is suggested in Table 2, is confirmed by the optical measurements on the USS *Sand Lance* (SS381) quoted in Table 3. These measurements give a stern target strength of 9 db, which is less than most similar values measured directly but still considerably larger than for any other submarine tested optically at that aspect. Since the indirect measurements at off-beam aspects are much lower than the direct measurements at those aspects, however, these optical values are probably not too significant.

TABLE 3. Dependence of target strength of submarine class (indirect measurements).

Submarine	Length in feet	Beam target strength in decibels	Bow target strength in decibels	Stern target strength in decibels
S class [5]	227	30	6	5
USS *Perch* (SS313) [5]	308	30	6	5
USS *Sand Lance* (SS381) [5]	310	26	5	9
U570 (HMS/M *Graph*) [5]	220	27	4	1
U570 (HMS/M *Graph*) [12]	220	25	6	6

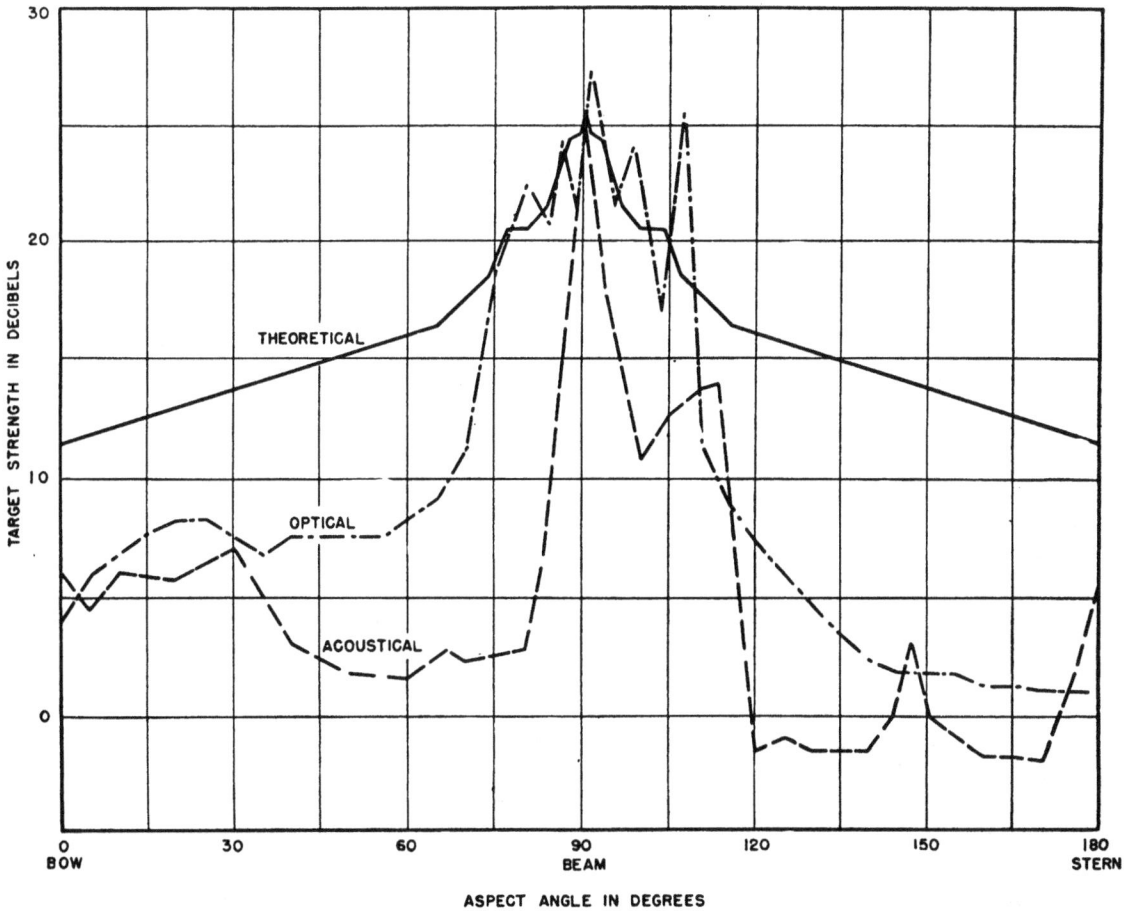

FIGURE 13. Comparison of theoretical, optical, and acoustical target strength of the U570 (HMS/M *Graph*).

Each submarine class will often evidence its own peculiar reflecting characteristics at certain aspects and altitudes. A striking example of this is the value of 20.2 db for the target strength of the U570 at an aspect of 120 degrees, reported in the Mountain Lakes measurements. Contrast this value with a maximum target strength of 26.3 db for beam aspects, 9.4 db astern, and 17.4 db bow-on. This result is attributed to a strong specular echo from the backplate of the conning tower, but is not confirmed by optical tests, perhaps because of differences in the two models.

To test the reliability of these model results for each class of vessel, a comparison may be made between results obtained by different indirect methods on the same type of submarine. Such a comparison is made in Figure 13, which shows target strength plotted against angle for the U570, as calculated theoretically and measured optically and acoustically.

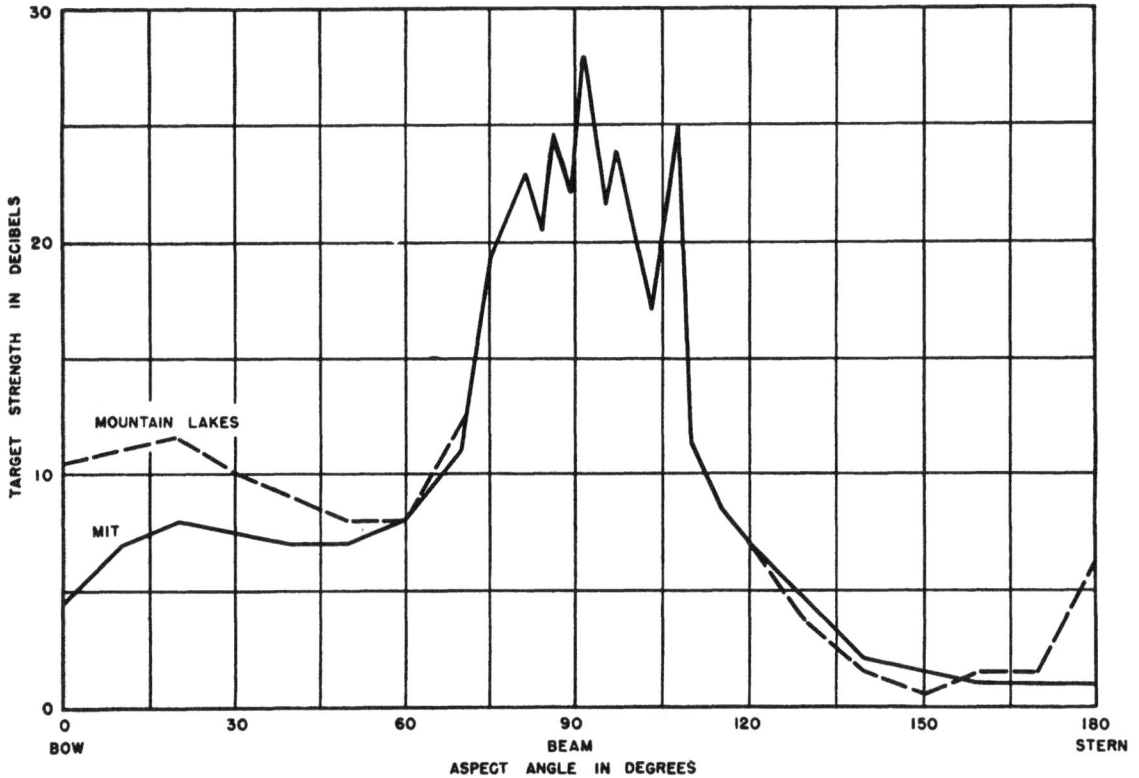

FIGURE 14.   Optical comparison of two models of U570 (HMS/M *Graph*).

The theoretical results at off-beam aspects are not very realistic, since they assume a perfectly smooth and curved reflecting surface. In Figure 13, port and starboard sides were averaged in the optical and acoustical measurements to simplify the illustration. A considerable difference is evident between these three curves, which may be attributed in part to the errors in the models as well as to possible differences arising from the different methods used.

Since different models of the U570 were used at MIT and at Mountain Lakes, the target strength of each model was measured optically, after the model used at Mountain Lakes had been finished in the same way as the MIT model. The results are reproduced in Figure 14 for the model originally used at Mountain Lakes and for the model first tested at MIT. Unfortunately, the Mountain Lakes model was not measured at aspects within 30 degrees of the beam, so that the comparison is not complete. Since the beam target strengths measured both optically and acoustically agree quite well, however, the target strengths of both models are probably the same at beam aspect.

The differences between the two curves in Figure

14 cannot be ascribed to differences in the methods, since both models were tested in the same way — optically. Instead, these differences, which near the bow are as great as 5 db, must be due to differences in the models themselves. Generally higher values were obtained for the Mountain Lakes model. In particular, diving planes at the bow, and horizontal and vertical rudders at the stern increased bow and stern reflections from the Mountain Lakes model. The MIT model had a knife-edge finish at bow and stern. Furthermore, the difference in the profile of the conning towers and the supporting structures for the two models would be expected to reduce somewhat the reflections of the MIT model at stern aspect.

In view of these discrepancies between different models of the same submarine, and between the different indirect methods of determining target strength, too much reliance cannot be placed on the general differences in submarine classes shown in Tables 2 and 3.

23.2.3   Asymmetry

To a first approximation, the shape of a submarine may be represented by an ellipsoid which is sym-

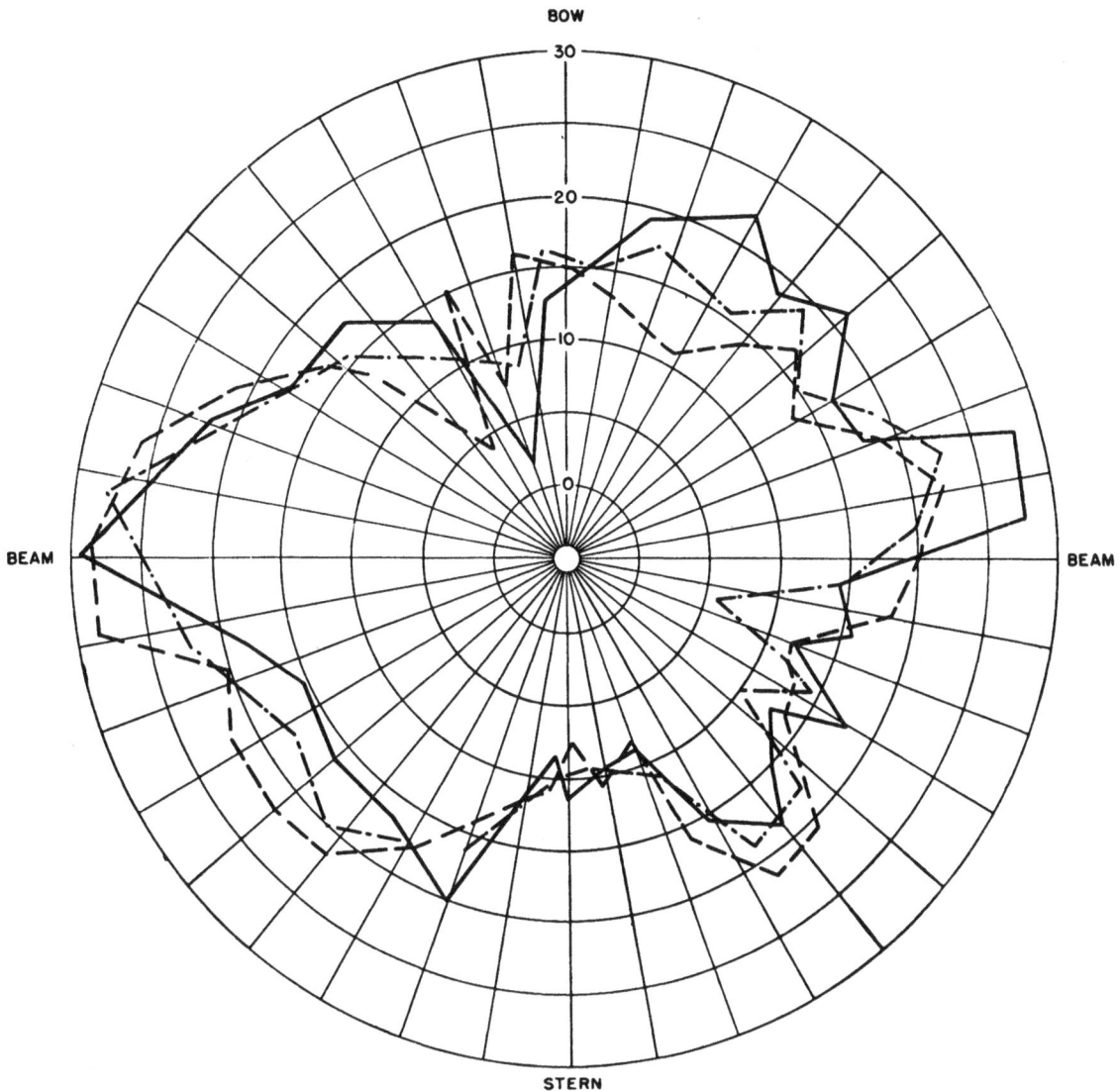

FIGURE 15.   Target strength asymmetry.   Target strength in decibels.

metrical with respect to all three planes in Figure 1. This approximation was the basis of the theoretical predictions described in Section 20.5. If now the conning tower is added to the submarine immediately above its center, the submarine loses its symmetry about the horizontal (xz) plane. This asymmetry about the xz plane is largely responsible for the asymmetry in target strength-altitude angle plots for positive and negative altitude angles. In addition, the shape of fuel and ballast tanks are frequently not symmetrical about the horizontal plane.

Since the shape of the bow of a submarine generally differs from the shape of its stern, perfect symmetry about the yz plane is also absent. However, the port

and starboard side of a submarine are, usually, nearly identical, with only very minor differences, to a first approximation, and therefore symmetry with respect to the xy plane may be expected. In other words, plots of target strengths as a function of aspect should be symmetrical about the longitudinal axis of the submarine.

In the theoretical calculations of the target strength of the U570 as a function of aspect angle, symmetry about the xy plane was assumed, since the submarine was approximated by an ellipsoid of revolution.

A lack of symmetry about the xy plane has been observed in both direct and indirect target strength measurements, and is apparent in Figures 3 to 6. Much of it is due to experimental error, and may not be real. Repeatable asymmetry in target strength-aspect curves has occasionally been found at San Diego and is illustrated in Figure 15 for three runs on an S-boat, where a sharp dip is evident just off the port bow at an aspect angle of about 340 degrees. This dip may be attributable to particular features of the construction of the S-boat, such as the lack of any surfaces normal to the sound beam so that specular reflection cannot occur at that aspect. It is also possible that this decrease in target strength is a characteristic of bow echoes and that the aspect angles were in error by as much as 20 degrees; low bow target strengths are conspicuous in the results of the optical measurements. On the other hand, the variability of echo intensities at other aspects is so large that this dip may not be real, even though it appeared during all three runs.

Horizontal asymmetry in the indirect measurements is apparent in Figures 5 and 6. This asymmetry may be attributed largely to the asymmetrical models used, in other words, the port and starboard sides were not the same. Not only were the models asymmetrical, but, as shown in Section 23.2.2, models of the same submarine appeared to differ rather markedly. Because of these model differences, the observed data cannot be used to confirm the existence of asymmetry in the target strength in the horizontal plane.

23.3   DEPENDENCE ON SPEED

Almost no data are available on the variation in target strength with the speed of the submarine. If echoes come only from the hull and conning tower of the submarine, it can be argued theoretically that the target strength of the submarine itself should not change as the speed is changed. But if a layer of air bubbles immediately surrounding the submarine contributes appreciably to the echoes received, then the target strength would be expected to depend on the speed and depth of the submarine; the scattering of sound from bubbles is discussed in Chapters 26 to 35. In addition, turbulence in the water adjacent to the submarine, which would depend on the speed of the submarine, may be responsible for part of the reflection of sound.

So far, little evidence has been uncovered to identify a layer of air or turbulent water surrounding the submarine as an effective reflector of sound (see Section 33.3), although bubbles have been observed on a submerged submarine traveling through the water[15] (see Section 27.1.1). Most direct measurements have been made on submarines at a creeping speed between 1 and 3 knots; none have been made on a stationary, balanced submarine. Some data describe tests at 6 knots, but no significant difference has been observed between these results and results at lower speeds. At 6 knots, however, reasonably strong echoes were received from a wake behind a fleet-type submarine at periscope depth. As the sound beam crossed the submarine from bow to stern, echoes grew stronger, faded and died out completely for a short time. Then strong echoes were received for several hundred yards behind the submarine, which were attributed to reflection from the wake.

23.4   DEPENDENCE ON RANGE

At long ranges, target strength is practically independent of range. Close to the submarine, however, the target strength will decrease with range. This phenomenon has two causes. First, at very short distances from the submarine, the submarine reflects more like a plane or a cylinder than a sphere, and the inverse fourth power law does not apply. In other words, the target is not equivalent to a point source, and the target strength decreases. This effect depends on the aspect of the submarine and diminishes as the range exceeds the maximum radius of curvature of the submarine.

Secondly, at short ranges the effective portion of the sound beam may cover only part of the area exposed by the submarine. Such an effect would be expected primarily at beam aspect, since geometric foreshortening reduces the area exposed by the submarine at other aspects. For example, a sound beam 12 degrees wide will not cover the entire length of a 300-ft submarine, at beam aspect, at ranges less than 475 yd. However, only those areas on the submarine giving rise to nonspecular reflection would be affected, so that if most of the reflection were specular, arising in a small area amidships, the target strength as measured very close to the submarine from perfectly aimed pulses would not depend on the beam width. From Figures 1 to 5 in Chapter 22, it appears that most of the reflection is concentrated amidships. Thus the effect of beam width, though present, is probably not important except possibly at very short ranges, as long as nonspecular reflection is neglected.

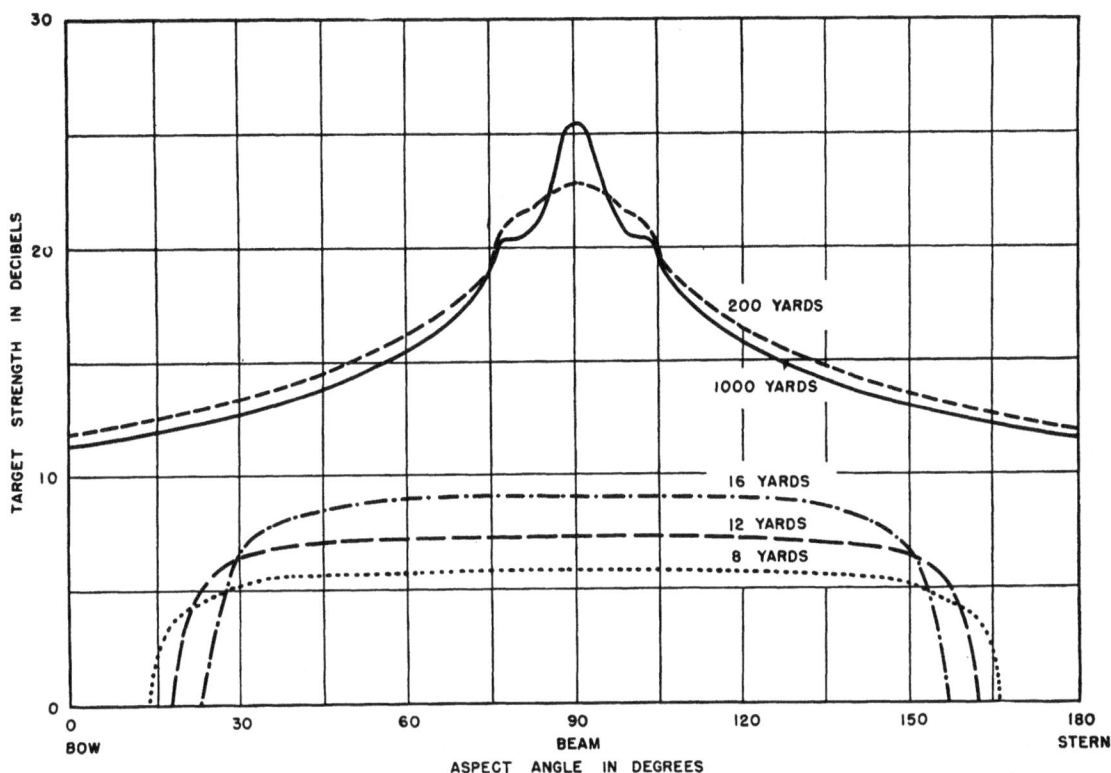

FIGURE 16. Theoretical dependence of target strength on range for the U570 (HMS/M *Graph*).

23.4.1 Theory

In the theoretical target strength studies described in Section 20.5,[6] plans of the *Graph* were employed in calculating target strengths.[16] This submarine has a maximum radius of curvature of about 560 yd. Since target strengths vary with range primarily for ranges less than the maximum radius of curvature of the submarine, the target strength of the *Graph* would be expected to approach a limiting value as the range is increased to 500 or 600 yd.

Actually, the target strength of the *Graph* is very near this limiting value at ranges beyond only a few hundred yards, especially at beam aspect. Target strengths have been computed for ranges of 8, 12, 16, 200, and 1,000 yd on the assumption of a nondirectional source of sound and are plotted against aspect angle in Figure 16. The shorter ranges are the distances from the projector to the nearest part of the submarine. It is apparent from Figure 16 that the variation of target strength with aspect changes markedly as the range is reduced from 200 to 16 yd; no intermediate ranges were used.

These calculations were based on a nondirectional source; therefore the results neglect the effect of

limited coverage of the submarine by a directional beam at close ranges. Such an effect is not important, however, if the reflection is primarily specular and comes from a small area amidships, as discussed in the preceding section.

23.4.2 Observations

Verification of a dependence of target strength on range has not been possible in the direct measurements because the transmission loss has not been known accurately. Instead, it has been assumed, for the ranges used during the target strength measurements — from 200 to 1,000 yd — that the target strength at near-beam aspects remains constant. Such an assumption is necessary in order to calculate the transmission loss; in some of the measurements at San Diego, a constant target strength was assumed at constant aspect, and the transmission loss was computed from a plot of the echo level E plus 40 log r against the range r as the range is opened or closed (see Section 21.5.1). The most recent measurements at San Diego have employed a nondirectional hydrophone mounted on the submarine to measure the transmission loss; this method, if practical, might reduce the usual fluctuations enough to enable an evaluation of target strength as a function of range.

The indirect measurements at MIT, at selected aspect angles, and at full-scale ranges of about 250 and 630 yd, gave about the same values as those at a full-scale range of 190 yd.[17] This result is consistent with the theoretical computations shown in Figure 16, where the maximum change in target strength is only about 3 db at beam aspect, as the range changes from 200 to 1,000 yd. Since for these ranges the target strength was found to be independent of range, it was apparent that the intensity of the light reflected from the models and measured at the receiver varied with range at the same rate as that from a sphere, rather than that from a cylinder or plane, in other words, inversely as the fourth power of the distance. The shortest range was approximately twice the length of the submarine. It is likely that this relation would not hold at much closer ranges where target strengths would be expected to depend strongly on the range.

Target strength was found to depend on the range in the acoustical model experiments at Mountain Lakes, for full-scale ranges of 85, 170, and 250 yd. The submarine model behaved as a cylinder, not as a sphere or plane. For reflection from a cylinder, the echo level should decrease 9 db when the range is doubled as long as the range is not much greater than the length of the cylinder (see Section 20.4.3); for a sphere the same increase in range causes a drop of 12 db. It was found that the echo level at beam aspect actually dropped 8.8 db as the full-scale range was doubled, from 85 to 170 yd. Thus the hull at beam aspect and at short ranges behaves as a cylinder, as might be expected from its large radius of curvature in the horizontal plane.

In general, the indirect measurements agree with the theoretical predictions of the dependence of target strength on range. This predicted dependence should be most marked at ranges less than 200 yd; experimentally, it was observed and verified only at ranges less than 200 yd. However, too much importance cannot be attached to these results, since the measurements were made only at three particular ranges in each of the two indirect measurements, and since experimental errors were so large.

23.5 DEPENDENCE ON PULSE LENGTH

When short pulses are used instead of continuous sound, target strength may depend on the lengths of these pulses. Sections 19.3 and 20.7 discussed in an elementary way the effect of pulse length on measured target strengths. For short pulses — signals whose length in the water is less than the length of the target in the direction of the sound beam — the echo level and therefore the target strength will depend on the signal length. The exact variation of target strength with signal length, however, depends on whether peak echo intensities or average echo intensities are used in computing target strengths.

23.5.1 Theory

In most direct measurements of target strength, peak amplitudes are measured from the oscillograms rather than average amplitudes, because echo profiles are so irregular that the average amplitudes over the length of the echo would be difficult to measure. A simple analysis given in Section 19.3.1 shows that for long pulses the average echo intensity is independent of signal length, while the echo length varies with the signal length. For short pulses on the other hand (see Section 19.3.2), the average echo intensity is approximately directly proportional to the signal length, while the echo length remains constant. This analysis applies to square-topped pulses striking an extended single target.

It is arbitrary whether peak amplitudes or average amplitudes are used to compute target strengths. Peak amplitudes are easier to measure. In addition, most other underwater sound measurements, such as those undertaken in the investigation of recognition differentials, are based on peak amplitudes. It may be, however, that the ear, or the sound range recorder, or other detection devices may respond to the average echo intensity instead of the peak echo intensity, or to the total energy in the echo. Therefore a comparison of average and peak echo amplitudes and their variation with signal length might be a profitable study.

The change of average and peak echo intensities with pulse length has been investigated theoretically,[18] as described in Section 21.6.4. First it is assumed that the length of each individual peak in an echo is approximately equal to the signal length. Then it is assumed that these peaks are statistically independent, or distributed at random throughout the length of the echo. By assuming that the echo is essentially a group of rectangular peaks, each of which follows the Rayleigh distribution for successive echoes and each of which is independent of the

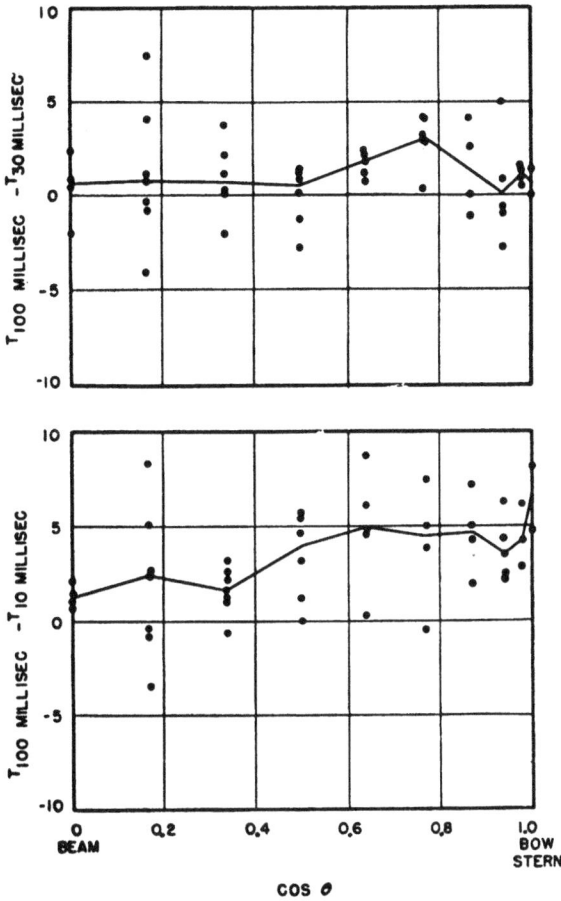

FIGURE 17. Dependence of target strength on signal length for an S-boat at 24 kc.

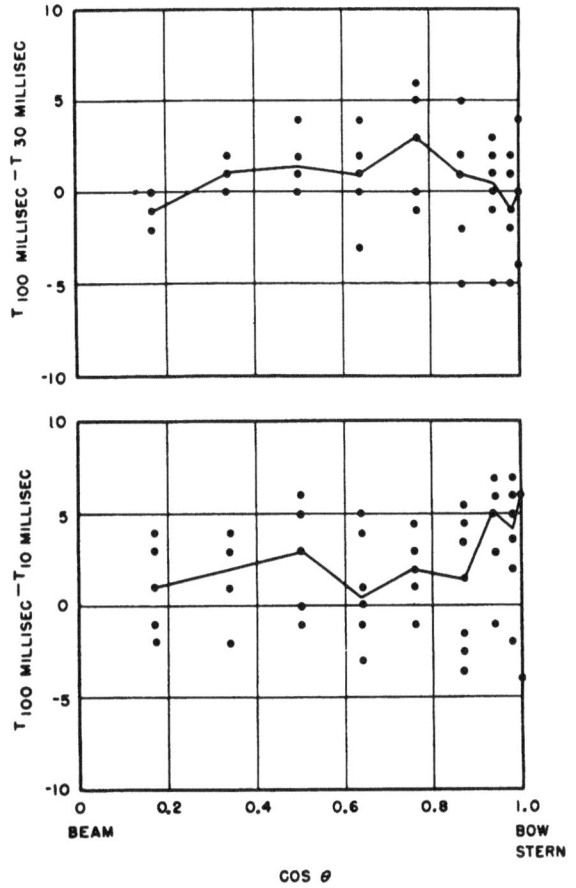

FIGURE 18. Dependence of target strength on signal length for a fleet-type submarine at 24 kc.

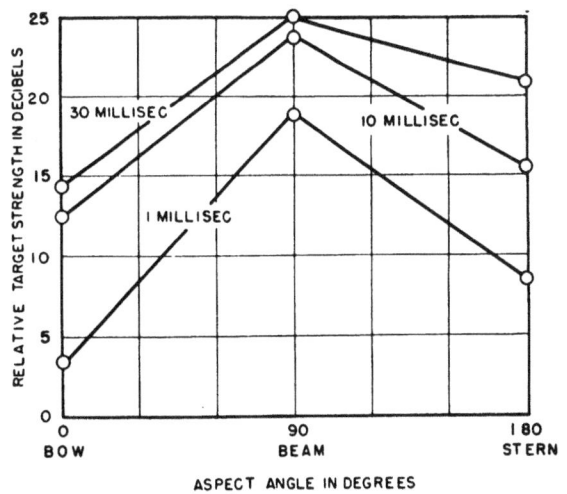

FIGURE 19. Dependence of relative target strength on signal length at different aspects for an S-class submarine at 60 kc.

others, the change in the peak echo intensity, averaged over many echoes, turns out to be considerably less than the change in mean echo intensity, averaged over the length of the echo and then over many echoes, for the same change in pulse length.

For example, it is shown that, for a uniformly reflecting target whose length is 300 ft in the direction of the sound beam, corresponding roughly to a fleet-type submarine at bow or stern aspects, the average peak echo intensity will decrease only 3 db and the average mean echo intensity will decrease 5 db, when the signal length is reduced from 30 to 10 msec. The assumptions on which this analysis is based are so idealized that the exact numerical results are probably not very significant; nevertheless, the general conclusion that the peak amplitude changes less rapidly with the signal length than the average amplitude seems fairly well established theoretically.

TARGET STRENGTH IN DECIBELS

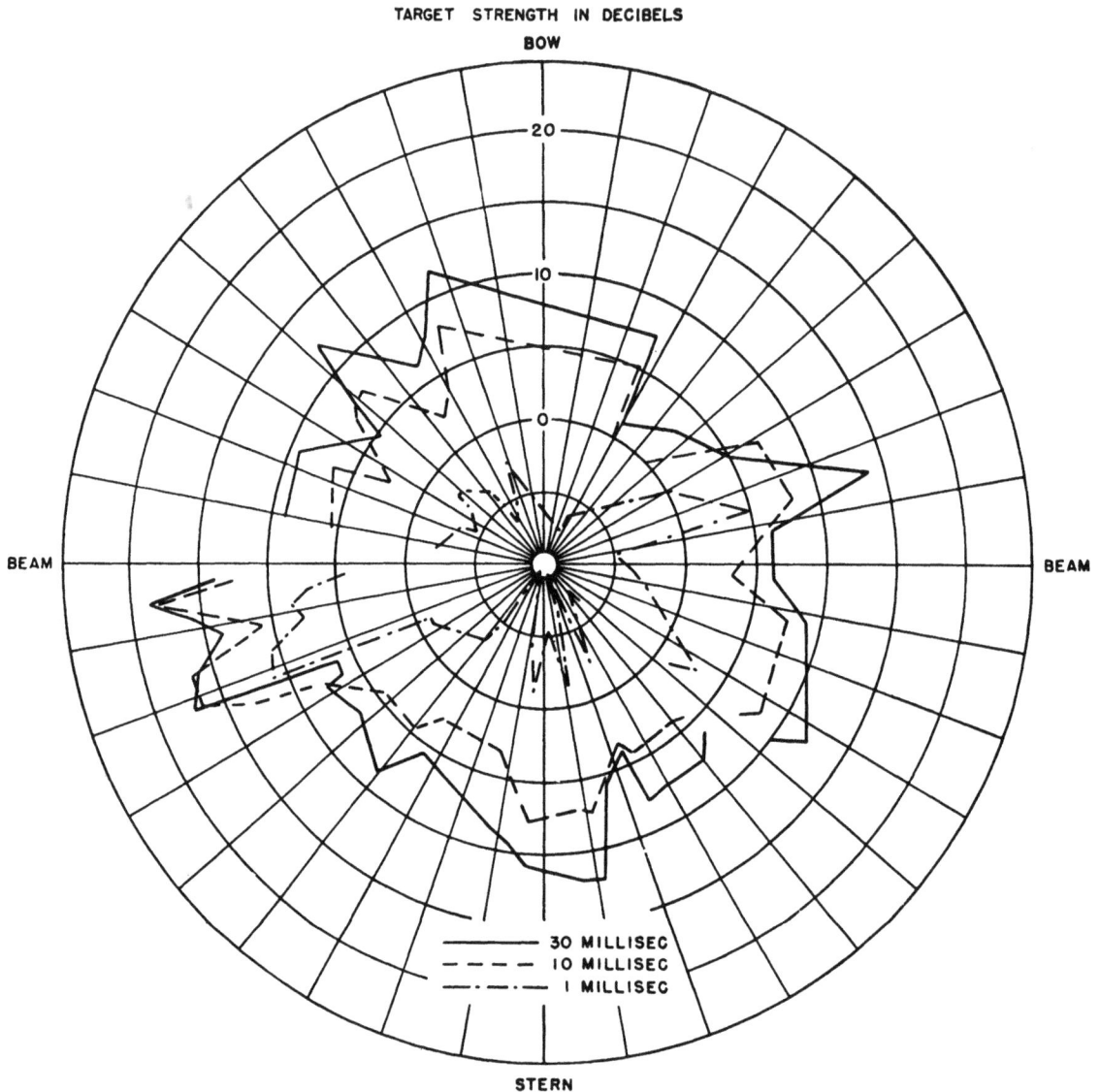

FIGURE 20. Dependence of target strength on signal length for an S-boat at 24 kc.

23.5.2   **Observations**

Since evidence showing a dependence of target
strength on pulse length could be found only in the
direct measurements, these measurements have been
examined and analyzed from this point of view.
Different signal lengths have been employed in
direct target strength measurements at San Diego,
where pulses from 0.5 to 200 msec long have been
used.

Peak echoes from 5-msec signals were found to
average about 4 db lower than echoes from 33-msec
signals, according to early runs at San Diego on an

S-class submarine, using JK gear at a frequency of
24 kc.[19] Further studies showed a minimum dependence of target strength on pulse length at beam
aspects, and a maximum, nearly linear, dependence
at bow and quarter aspects.[20] Later measurements,
however, reported no very significant dependence of
target strength on signal length for signal lengths of
10, 30, and 100 msec.[3] These measurements are illustrated in Figures 17 and 18 for an S-boat and a fleet-
type submarine respectively; the difference in target
strengths between 100- and 30-msec signals and
100- and 10-msec signals were greatest at aspects
near the bow and stern and are plotted against the

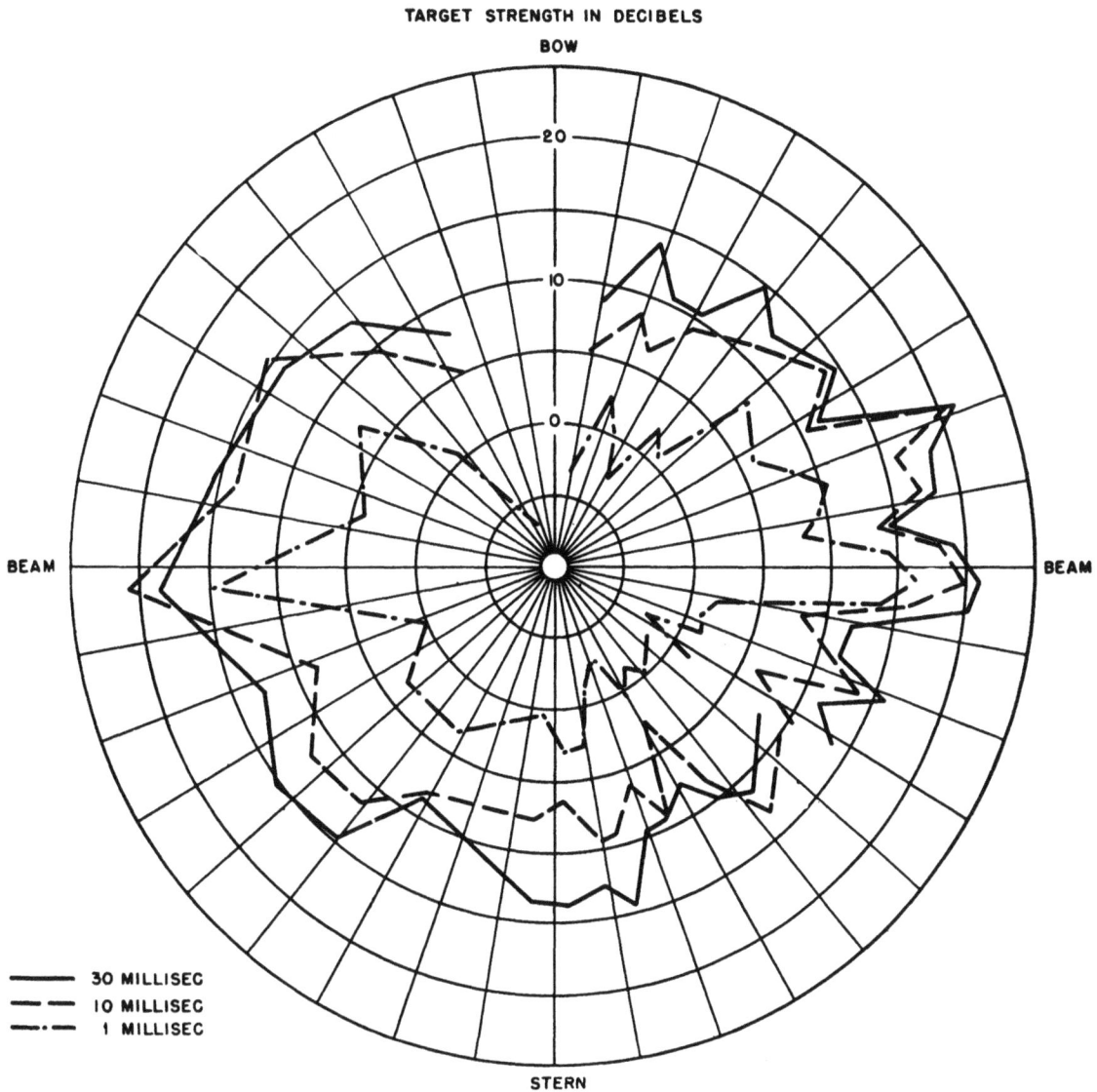

FIGURE 21. Dependence of target strength on signal length for an S-boat at 60 kc.

cosine of the aspect angle. The mean curve connecting the mean points is also indicated. However, this dependence on pulse length is relatively small and not very reliable in view of the large scatter.

Measurements were made at a frequency of 60 kc on another S-boat at creeping speed and a keel depth of 100 ft, with signal lengths of 1, 10, and 30 msec.[21] A significant variation in relative target strength with signal length was observed and plotted in Figure 19 for bow, beam, and stern aspects where each point at beam aspect represents about 40 echoes, at stern aspect about 20 echoes, and at bow aspect only a few. Surprisingly, a large variation with pulse

length is found at beam aspect; here theory would predict a minimum dependence, since the echoes for the most part accurately reproduce the pulses. An attenuation coefficient of 20 db per kyd was assumed in calculating the transmission loss; ranges averaged about 300 yd.

Figures 20 and 21 show target strength as a function of aspect angle for three different signal lengths, 1, 10, and 30 msec, for a submarine of the S class at 24 and 60 kc. A definite dependence on signal length is evident from an examination of the three curves in each figure, although the actual target strength values are rather low. The dependence on aspect angle

TARGET STRENGTH IN DECIBELS

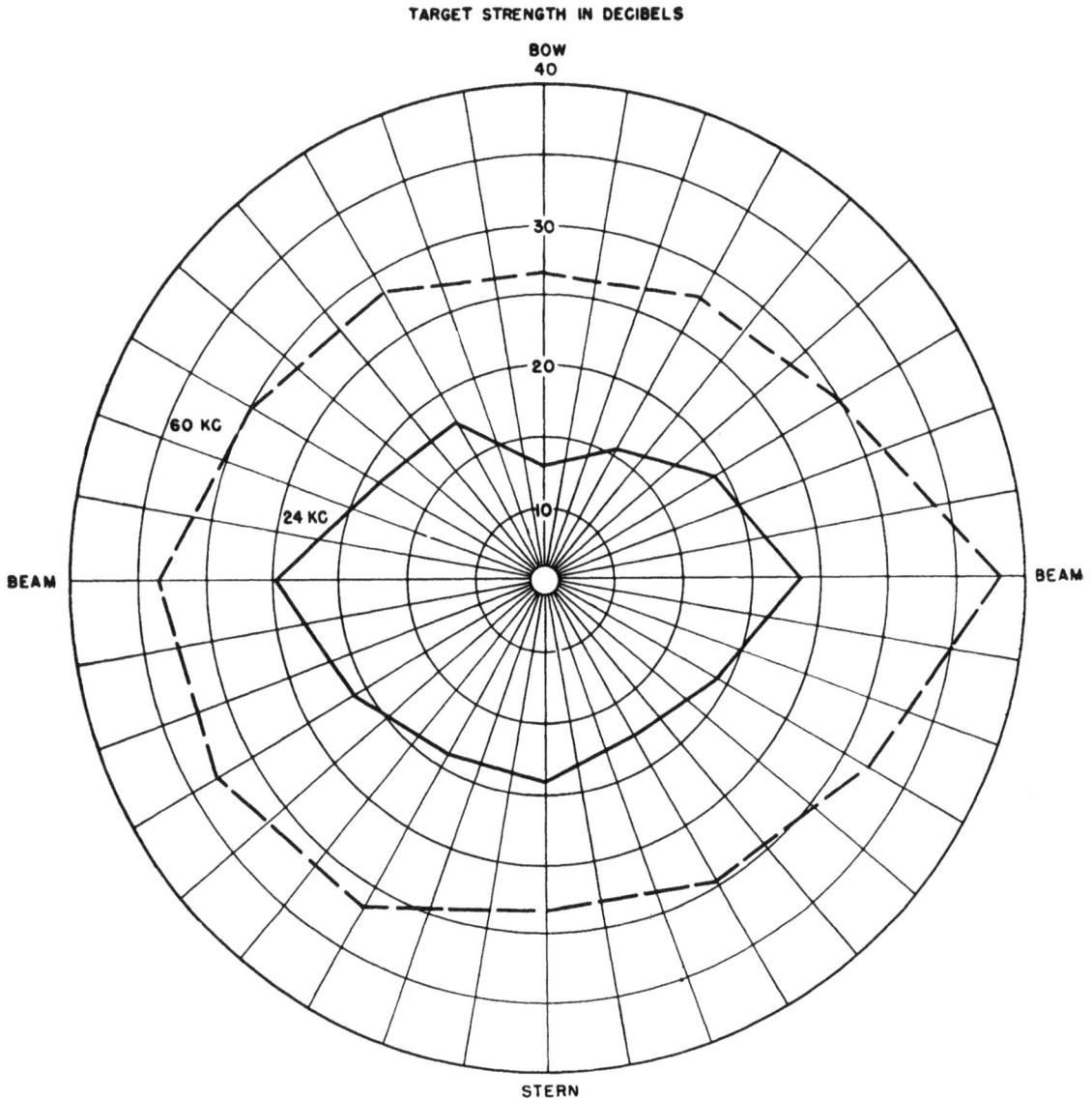

FIGURE 22.   Dependence of target strength on frequency (San Diego).

is somewhat obscured by the fluctuations encountered during these particular runs.

In general, target strength depends on the signal length for short signals although it varies less rapidly than the signal length, or rather, less rapidly than 10 log τ, where τ is the signal length; a decrease in target strength is most marked at signal lengths less than 10 msec and at aspects away from the beam.

23.6   DEPENDENCE ON FREQUENCY

Both sonic and supersonic frequencies have been employed in echo ranging. Lower frequencies are de-

sirable from the point of view of transmission, since the transmission loss increases with frequency. On the other hand, since high-frequency sound is more directive, the bearings of targets may be located more accurately at high frequencies than at low frequencies. As a result, choosing a frequency for echo ranging is always a compromise between these two characteristics.

Target strengths have been predicted and measured directly at frequencies between 12 and 60 kc. Theoretical calculations have been based on a frequency of 25 kc. Most of the direct measurements have been made at 24 kc, since this frequency is standard for most Navy sonar gear, although some measurements have been made at frequencies of 12, 18, 45, and 60 kc. The indirect measurements used scale models; at Mountain Lakes, with a 1:60 scale model of the *Graph*, a frequency of 1,565 kc was employed to simulate an echo-ranging frequency of 26 kc. At MIT visible light was used, and the corresponding full-scale frequency was very much higher than for any of the other measurements.

23.6.1 Theory

The target strength of a submarine depends on frequency according to equation (36) in Chapter 20, especially if nonspecular reflection contributes appreciably to the target strength. Specular reflection depends only slightly on frequency, as described in Section 20.4. Beam echoes result largely from specular reflection, as pointed out in Section 23.8.1. Thus the variation of target strength with frequency at beam aspects may be expected to be slight.

At off-beam aspects, however, specular reflection is much less important and nonspecular reflection may become appreciable; this effect is discussed in Section 23.8.2. At these aspects, the whole submarine appears to scatter sound. If reflections from the superstructure or exterior protuberances on the submarine, such as rails, guns, and periscopes, contribute appreciably to the target strength, the target strength may depend on frequency as long as the dimensions of these scatterers are of the same order of magnitude as the wavelength. Section 20.5 describes the origins of nonspecular reflection.

23.6.2 Direct Measurements

No reliable direct measurements substantiate this expected dependence of target strength on frequency. Figure 22 shows target strength as a function of aspect angle plotted for frequencies of 24 and 60 kc for a fleet-type submarine at San Diego at signal lengths of 10, 30, and 100 msec. Each point represents the average of all observations in a 30-degree sector centered at the point indicated.

A clear-cut dependence of target strength on frequency is apparent in this illustration, as well as in Figures 20 and 21. Figure 21 gives quite reasonable values for the target strength at 60 kc, but Figure 20 shows values about 10 db lower, for a frequency of 24 kc; thus the dependence on frequency is still evident. It is unlikely, however, that the increase in target strength with frequency would be not only so

great but also so nearly uniform at all aspect angles, as Figure 22 shows. Furthermore, the measurements cannot be relied on for two reasons: the transmission loss was not known but estimated, and the calibration of the 60-kc gear was less reliable than the 24-kc gear calibration. The estimated attenuation coefficient of 20 db per kyd used for these computations is larger than that measured elsewhere and is perhaps excessive, since attenuation coefficients of only 10 db per kyd at 60 kc were measured at Fort Lauderdale. However, correcting the high San Diego target strengths at 60 kc by reducing the attenuation coefficient from 20 to 10 db still results in values greater than those obtained elsewhere. The remaining discrepancy may perhaps be attributed to faulty calibration of the gear, described in Section 21.4, since this discrepancy is systematic and apparently independent of aspect angle.

In addition, the high target strengths measured at 60 kc at San Diego do not seem substantiated by target strength measurements at that frequency at Fort Lauderdale. These measurements gave a target strength of 25 db for a fleet-type submarine at beam aspect at 60 kc, assuming an attenuation coefficient of 12 db per kyd. An assumption of an attenuation coefficient of 20 db per kyd would raise this to only 33 db, compared with the maximum value of 44 db recorded at San Diego for the target strength of a fleet-type submarine. Furthermore, wake echoes measured with the same equipment at San Diego and described in Section 33.4.2 were found to be much higher at 60 kc than at 24 kc, contrary to theoretical expectations. These results support the suggestion that calibration errors are responsible for the high values obtained at San Diego. However, in view of the many uncertainties in this subject, the possibility that submarine target strengths are systematically some 10 db higher at 60 kc than at 24 kc cannot be entirely ruled out, even though there is little if any theoretical expectation of such a variation.

No difference was apparent between the measurements at 12 kc and 24 kc made by Woods Hole observers;[10] both target strength-aspect curves were very similar. However, the target strengths reported are so much larger than all other measurements elsewhere that calibration errors were probably present. Therefore, since the calibration at 12 kc was quite different from that at 24 kc, and since all the values seem very uncertain, the lack of any frequency dependence cannot be considered significant.

23.6.3 Indirect Measurements

Visible light from a motion picture projection bulb was used in the optical measurements at MIT. Since the frequency, or the band of frequencies, was not properly scaled to correspond with usual echo-ranging frequencies, it was not practical to investigate the dependence of target strength on frequency. However, since improperly scaled light was used to give results for comparison with direct and other indirect measurements, four frequency effects should be remembered in interpreting the results of the optical measurements.[12]

First, at certain aspects when the insonified surface of the actual submarine subtends only a few Fresnel zones at 24 kc, the surface of the model illuminated by visible light subtends many zones, since the wavelength of the light was much shorter compared with the dimensions of the model than was the wavelength of the sound used in echo ranging compared to the dimensions of an actual submarine.

As a result, for the optical measurements, the expressions for the target strength due to the effects of a group of Fresnel zones approached their asymptotic values, especially for surfaces with at least one large radius of curvature, such as cylinders and planes. On submarines this effect might apply to the conning tower, keel, and top deck, so that in the optical measurements the effect of the conning tower, and the effect of the deck at an altitude angle of 90 degrees, might be overemphasized.

Secondly, nonspecular reflection is too small by a factor of the square root of the ratio of the actual wavelength used to the properly scaled wavelength. In the optical measurements, this factor is about 45, or about 17 db. In other words, nonspecular reflection measured optically is about 17 db too low. This factor may account for the very low target strengths obtained optically at off-beam aspects where nonspecular reflection may be more important.

Thirdly, diffuse reflection or scattering may be too great optically because the wavelength may be much smaller than the surface irregularities. An attempt was made to minimize this error by using glossy black surfaces.

Finally, where two or more specular reflections occur, the light beams do not interfere, as sound beams do, because they are incoherent. Since the incoherent sum is an average of the interference pattern, this sum may actually be more interesting than the detailed interference pattern itself, as the average is more significant, in most applications to practical echo ranging at sea, than the exact pattern.

Thus the results of the optical measurements, though suggestive of what might be encountered in practical echo ranging, cannot be compared directly with the other measurements unless these effects of the wavelength are considered and accounted for.

In the acoustical measurements at Mountain Lakes, no long-term systematic variation with frequency was observed for full-scale frequencies from about 1 to 35 kc. Figure 23 shows relative echo level plotted against frequency for the Mountain Lakes measurements on the *Graph* at beam aspect, at a full-scale range of about 15 yd. The peaks and dips evident in this illustration are largely the result of interference phenomena arising from two specular reflections from the hull and conning tower, as the frequency is changed, and of the response characteristics of the system. Interference phenomena resulting from multiple reflections from several surfaces on the submarine are clearly shown in Figure 24, where the relative beam target strength is plotted against altitude angle for a full-scale frequency of 26 kc. The nearly smooth curves at altitudes of 0 and 90 degrees result from direct specular reflection from the deck and hull respectively. The intricate interference pattern at altitude angles between 10 and 50 degrees and 270 and 350 degrees results from path differences in the sound doubly reflected from the hull and from the blister tank; similar patterns are evident for sound reflected from the bottom of the submarine model.

23.7 OCEANOGRAPHIC CONDITIONS

Target strength measures the reflecting characteristics of a target and is computed from the echo level, source level, and transmission loss from equation (6) in Chapter 19. Since it depends only on the target itself, it is theoretically independent of the medium and its characteristics, independent of the transmission characteristics of sound in water, and therefore independent of oceanographic conditions insofar as they affect transmission.

In practice, however, reported target strengths have been found to depend markedly on the prevailing oceanographic conditions in cases where the transmission loss has not been accurately known. Since target strength is computed from the echo level, source level, and transmission loss, improper appraisal of the transmission loss will appear as an error in the target strength values reported.

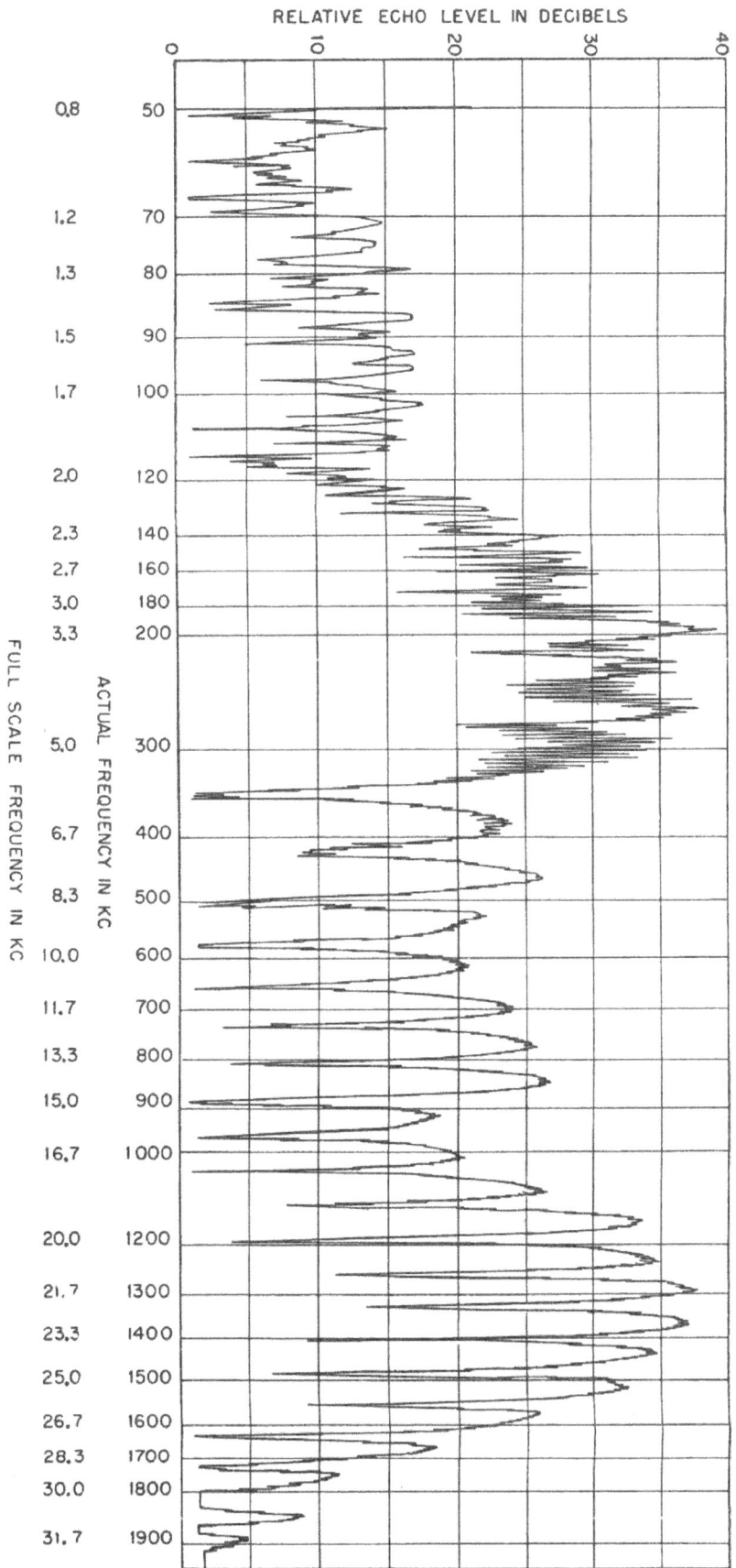

FIGURE 23. Dependence of target strength on frequency (Mountain Lakes).

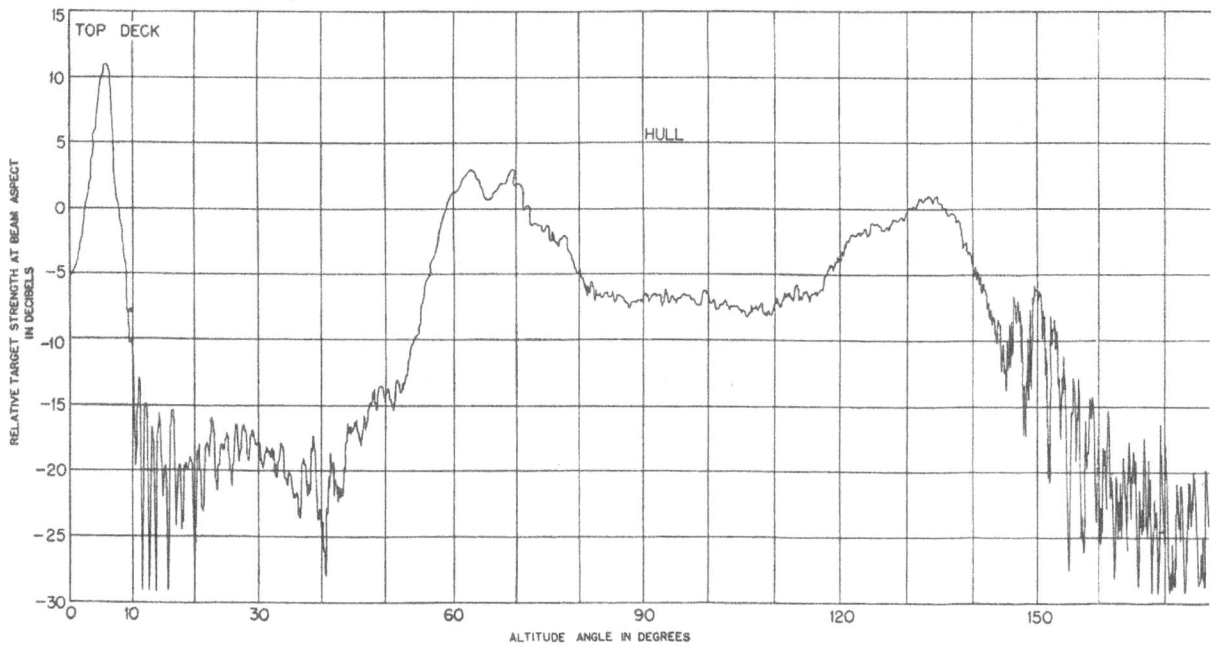

FIGURE 24: Double reflection and interference (Mountain Lakes).

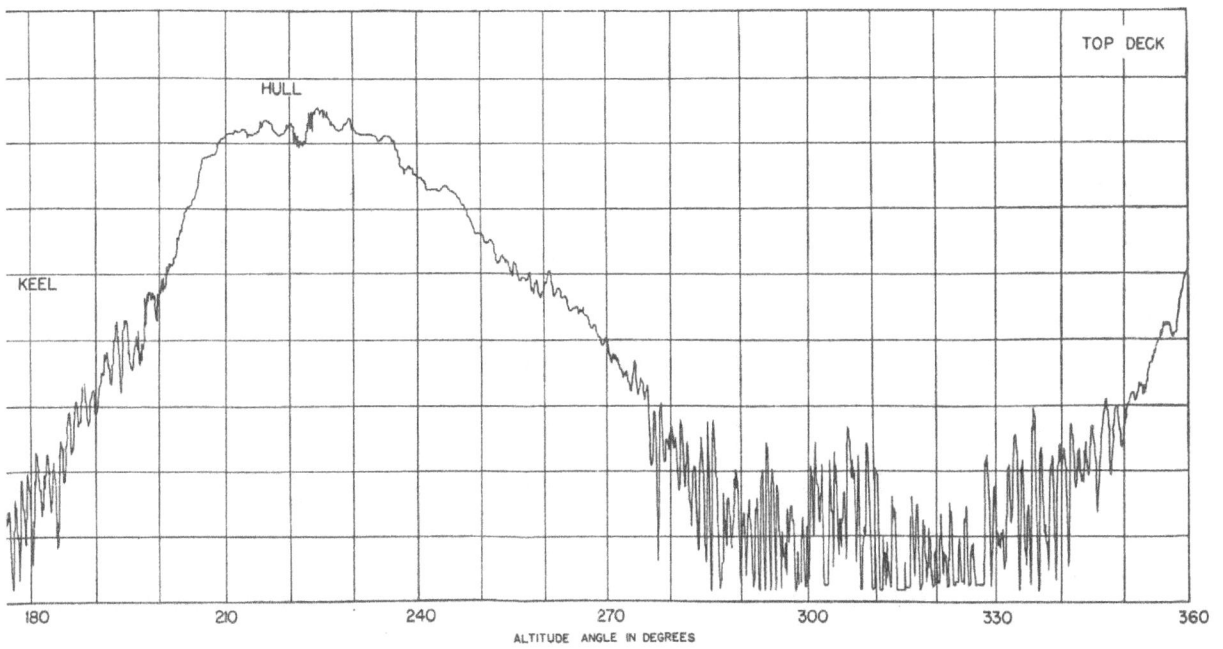

Figure 24 is spread out on left and right hand page.

Section 21.5 introduced the concept of an attenuation coefficient to describe more accurately the transmission loss. This attenuation coefficient varies both with the oceanographic conditions and the frequency of the echo-ranging sound. From a quantity of transmission data at 24 kc two empirical formulas were suggested for estimating the attenuation coefficient,[22] equations (5) and (6) in Chapter 21. They are

$$a = 3.5 + \frac{170}{D} \qquad (1)$$

for the hydrophone above the thermocline, and

$$a = 4.5 + \frac{260}{D} \qquad (2)$$

for the hydrophone below the thermocline, where a is the attenuation coefficient in decibels per kiloyard and D the depth of the thermocline in feet. The probable error of this estimate is about 2 db per kyd.

measurements at 24 kc on different S-boats, calculated by assuming an attenuation coefficient of 5 db per kyd, showed such deviations that a marked dependence on the particular submarine used was suggested.[23] The particular submarines used are designated in the first column of Table 4, while the target strengths, varying from 7 to 25 db, are reproduced in the second column. Investigation of the oceanographic conditions prevailing during the different runs showed an unmistakable correlation between the target strengths and the thermocline depths; when the thermocline depth was only 18 ft, the computed target strength was only 7.3 db, while it rose to 25 db for a thermocline 160 ft deep.

Accordingly, equations (1) and (2) were used to calculate new and presumably more reliable attenuation coefficients, from the thermocline depths, listed in the third column, and the depths of the submarines during the runs, in the fourth column of Table 4.

TABLE 4. Dependence of reported target strengths on attenuation coefficient.

S-boat	Reported beam target strength in decibels	Depth to thermocline in feet	Depth of submarine in feet	Computed attenuation coefficient in decibels per kiloyard	Range in yards	Correction in decibels	Computed beam target strength in decibels
USS S-28 (SS133)	18*	85	45	5.5	350–520	+0.5	18.5
USS S-40 (SS145)	25*	160	90	4.6	300–400	−0.3	24.7
USS S-23 (SS128)	13.7*	50	90	9.7	350–500	+4.7	18.4
USS S-23 (SS128)	15.4*	75	100	5.8	330	+0.3	15.7
USS S-33 (SS138)	7.3*	18	90	19.0	415	+11.6	18.9
USS S-31 (SS136)	22.3†	100	95	5.2	440	0.0	22.3
Mean beam target strength and standard deviation in decibels	16.9 ± 4.8						19.7 ± 2.5

* Assumed attenuation coefficient, 5 db per kyd.
† Measured attenuation coefficient, 5.3 db per kyd.

23.7.1 Effect on Measurements

Direct measurements of the transmission loss during target strength runs on submarines have been difficult and generally unsuccessful, as described in Section 21.5.2. As a result, it has been customary at San Diego to compute the transmission loss from an estimated attenuation coefficient at the particular frequency employed, usually 5 db per kyd at 24 kc and 20 db per kyd at 60 kc.

Target strengths computed in this way show enormous differences. For example, various series of

These new values appear in the fifth column, the appropriate ranges in the sixth column, in the seventh column the corrections resulting from the assumed value of 5 db per kyd of sound travel, not range — assuming the maximum range from the sixth column — and the new target strengths in the last column. Two results are noteworthy: the standard deviation is reduced by a factor of almost two from 4.8 db to 2.5 db, and the mean beam target strength for an S-boat is raised from 16.9 db to 19.7 db. The new value agrees more closely with other measurements.

At 60 kc fewer data are available. An assumption

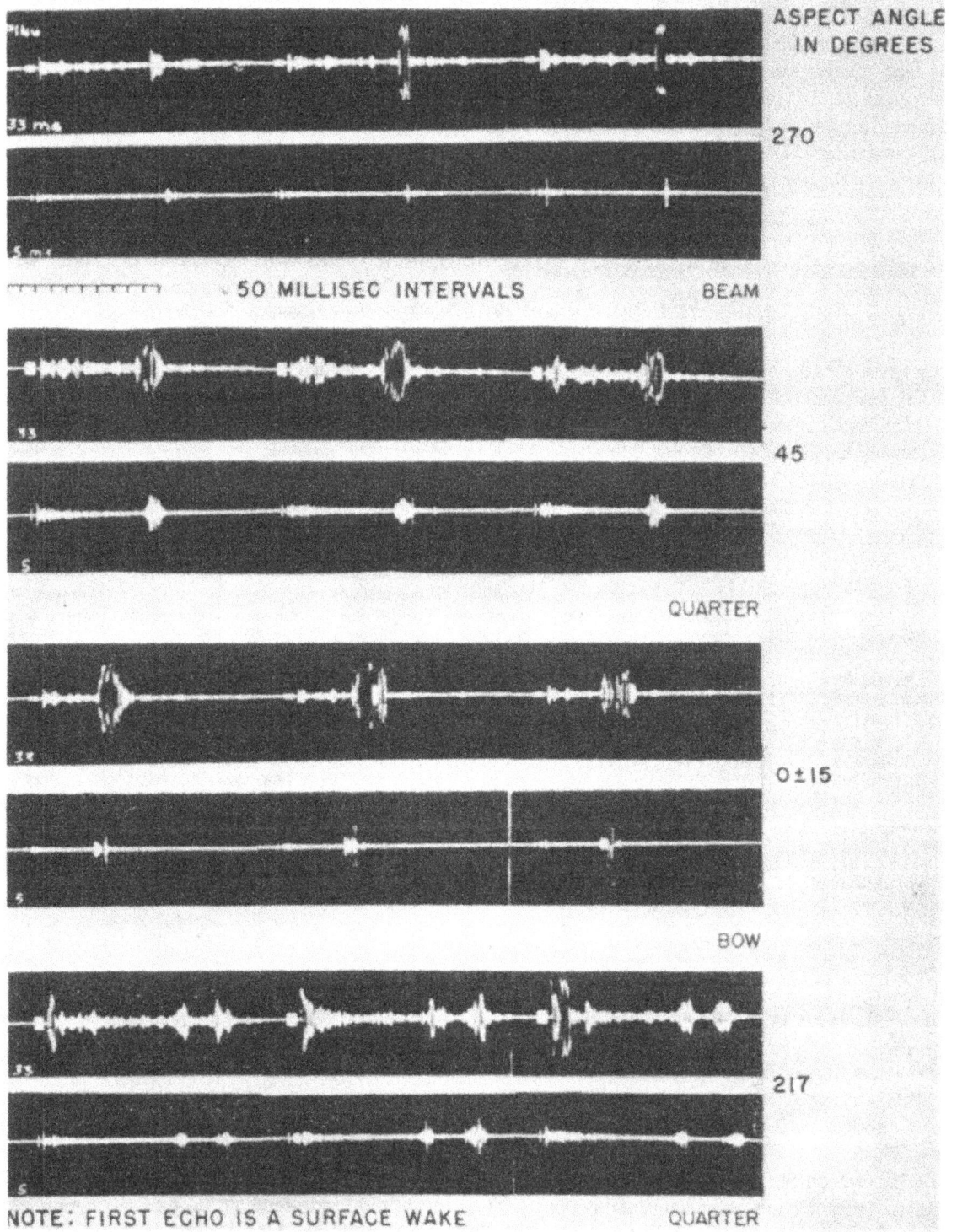

FIGURE 25.   Oscillograms of submarine echoes (S-class submarine).

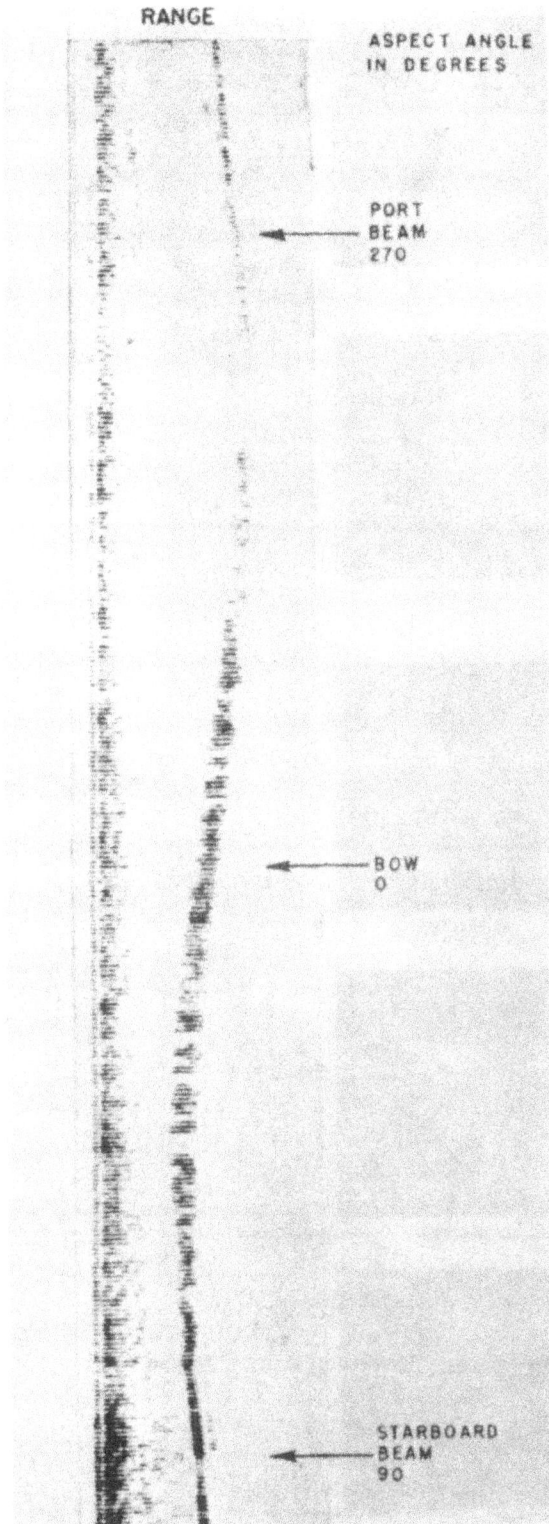

FIGURE 26. Sound range recorder records of submarine echoes from 30-millisecond pulses.

of 20 db per kyd for the attenuation coefficient, although assumed in the San Diego target strength computations, is not substantiated by tests made at Fort Lauderdale which gave results of 9 to 10 db per kyd. These latter measurements were made by plotting the echo level, corrected for inverse square loss, against the range, as the range continuously decreased from about 600 to 150 yd; the submarine was at stern aspect throughout the run. This discrepancy was mentioned in Section 23.6.2. The high value assumed for the attenuation coefficient at San Diego was suggested by early transmission measurements; in these tests, the presence of shallow thermoclines off the coast of California was partly responsible for the high values measured. For deep, mixed water, lower values are more common [24] (see Chapter 5).

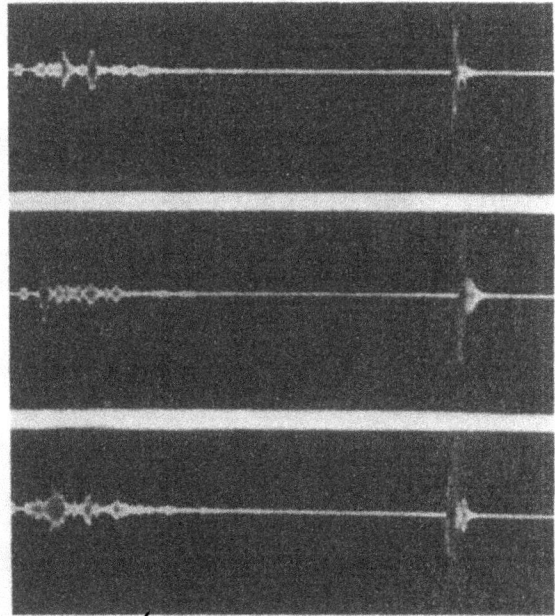

FIGURE 27. Oscillograms of submarine echoes at beam aspect for 10-millisecond pulses.

23.8 STRUCTURE AND ORIGIN OF ECHOES

Most target strengths have been measured from echoes recorded oscillographically on 35-mm motion picture film. At San Diego, this film was run past the oscilloscope at a speed sufficiently high to record the detail of each echo; depending on the signal length and the exact film speed, these echoes may be from 0.2 to 3 or 4 cm long. Accordingly, these echoes have been carefully studied in an attempt to formulate

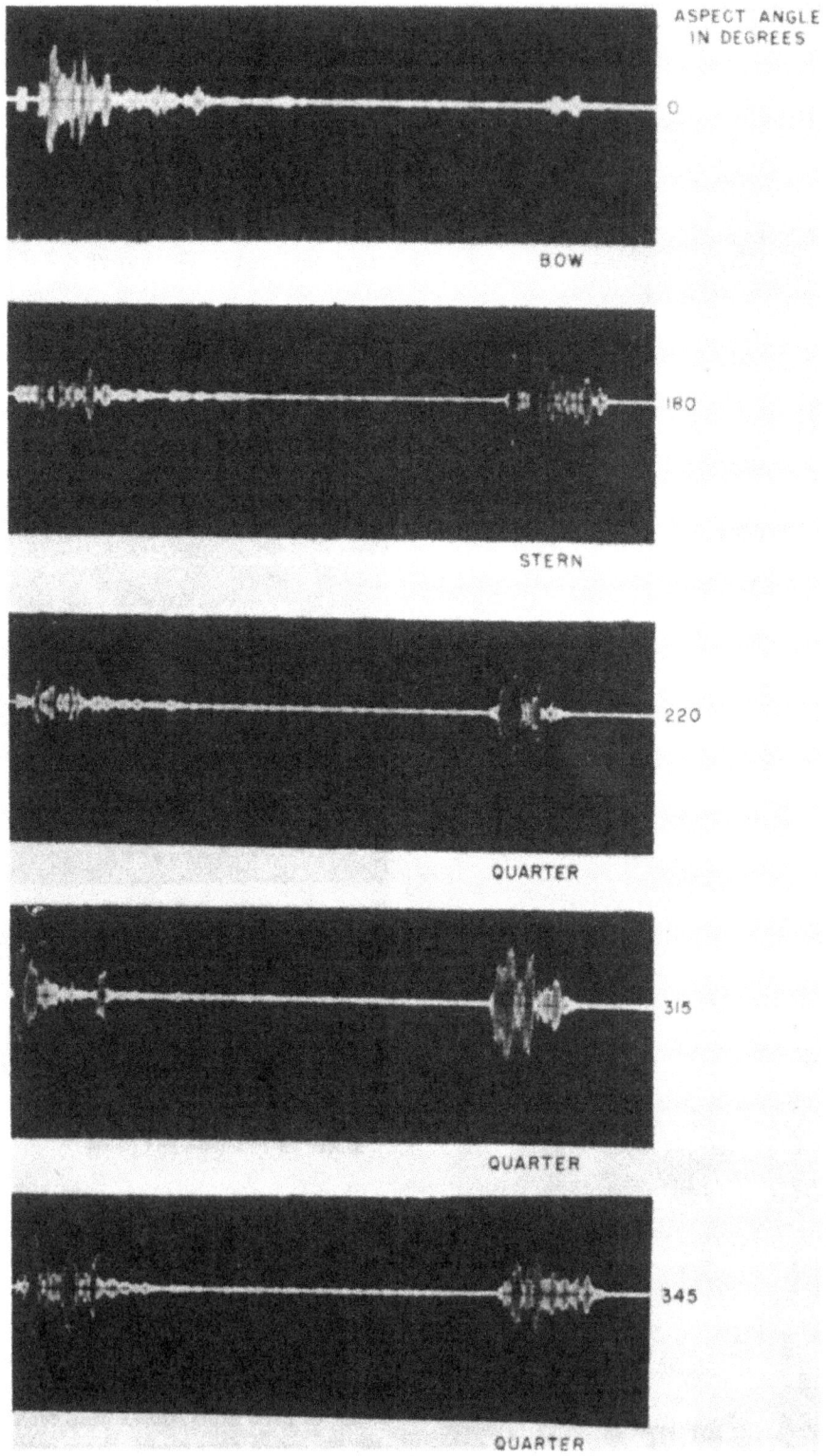

FIGURE 28. Oscillograms of submarine echoes at off-beam aspects for 10-millisecond pulses.

more precise conclusions regarding the process of reflection of sound from submarines.

Examination of hundreds of oscillograms of echoes at San Diego has made possible a separation of echoes into two classes, beam and off-beam, as illustrated in Figure 25. Each class shows its own characteristics and peculiarities, on the basis of which tentative explanations of reflection phenomena have been made. These two types of echoes produce such different traces on the sound range recorder that the appearance of these traces is used tactically to estimate the aspect of the target. Typical sound range recorder traces are illustrated in Figure 26.

23.8.1 Beam Echoes

Beam echoes are always stronger, on the average, than echoes at any other aspect, both according to the theoretical calculations and the direct and indirect measurements. There are four lines of evidence which indicate that reflection is specular and arises primarily from the hull of the submarine.

First, theory predicts strong specular reflection at beam aspect.[6] The theoretical values derived assuming only specular reflection are in excellent agreement with other values measured directly and indirectly. The effect of the conning tower appears to be negligible for the U570 at beam aspect, since it contributes only 0.2 db to the target strength at long ranges.

Secondly, the oscillograms of beam echoes are clear cut and closely resemble the square-topped pulses sent out; further examples are illustrated in Figure 27, which shows oscillograms of three successive echoes from a submarine at beam aspect. These beam echoes contrast sharply with off-beam echoes, which are illustrated in Figure 28. Occasionally, beam echoes show very sharp and narrow peaks at either end, or a short "tail" of lower intensity, which are attributed to two different types of surface reflections described in Section 21.5.4. A typical sound range recorder record of the double echoes at beam aspect is illustrated in Figure 29. Since beam echoes usually equal the signal in length, for signals 10 or more milliseconds long, the effective reflecting surface does not appear to be much extended in the direction of the sound beam; typical oscillograms of beam echoes are illustrated in Figure 30. In other words, the relatively flat area on the hull of the submarine is responsible for almost all the energy in the echoes received at beam aspect. This same specular

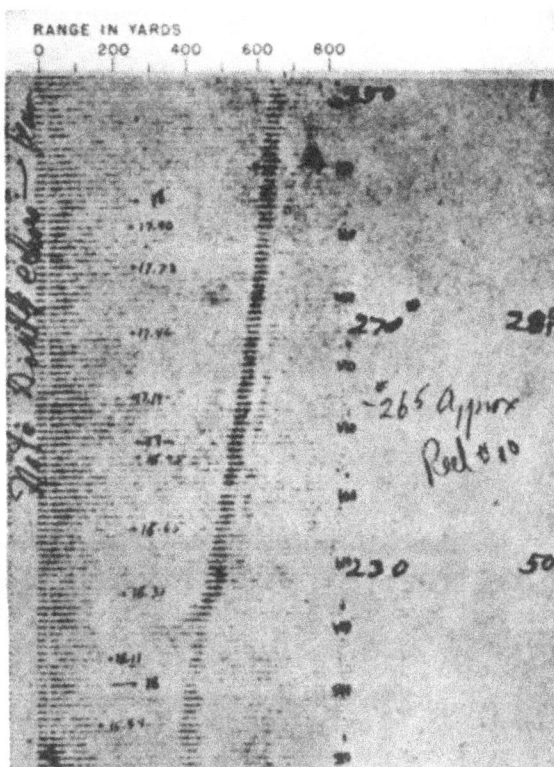

FIGURE 29. Double echoes recorded on the sound range recorder.

reflection may be inferred from Figure 24 where a fine interference pattern is conspicuously absent at beam aspect for altitude angles in the neighborhood of 0 degree.

Thirdly, optical measurements on a model of the *Sand Lance* as well as both optical and acoustical measurements on models of the U570 gave identical target strength at beam aspect with and without the conning tower, over a sector of about 20 degrees, as Figures 31, 32, and 33 show. The conning tower, although important at other aspects, contributes little to reflection at beam aspects.

Fourthly, the importance of hull reflections is evident in Figures 1 through 5 of Chapter 22. Although these photographs refer only to optical illumination of the model and may not apply perfectly to the reflection of supersonic sound from submerged submarines, they may be representative of what happens acoustically.

However, measurements made with short pulses at beam aspect show a detailed echo structure which suggests that not all the reflected sound comes from the submarine hull or ballast tanks. Measurements on

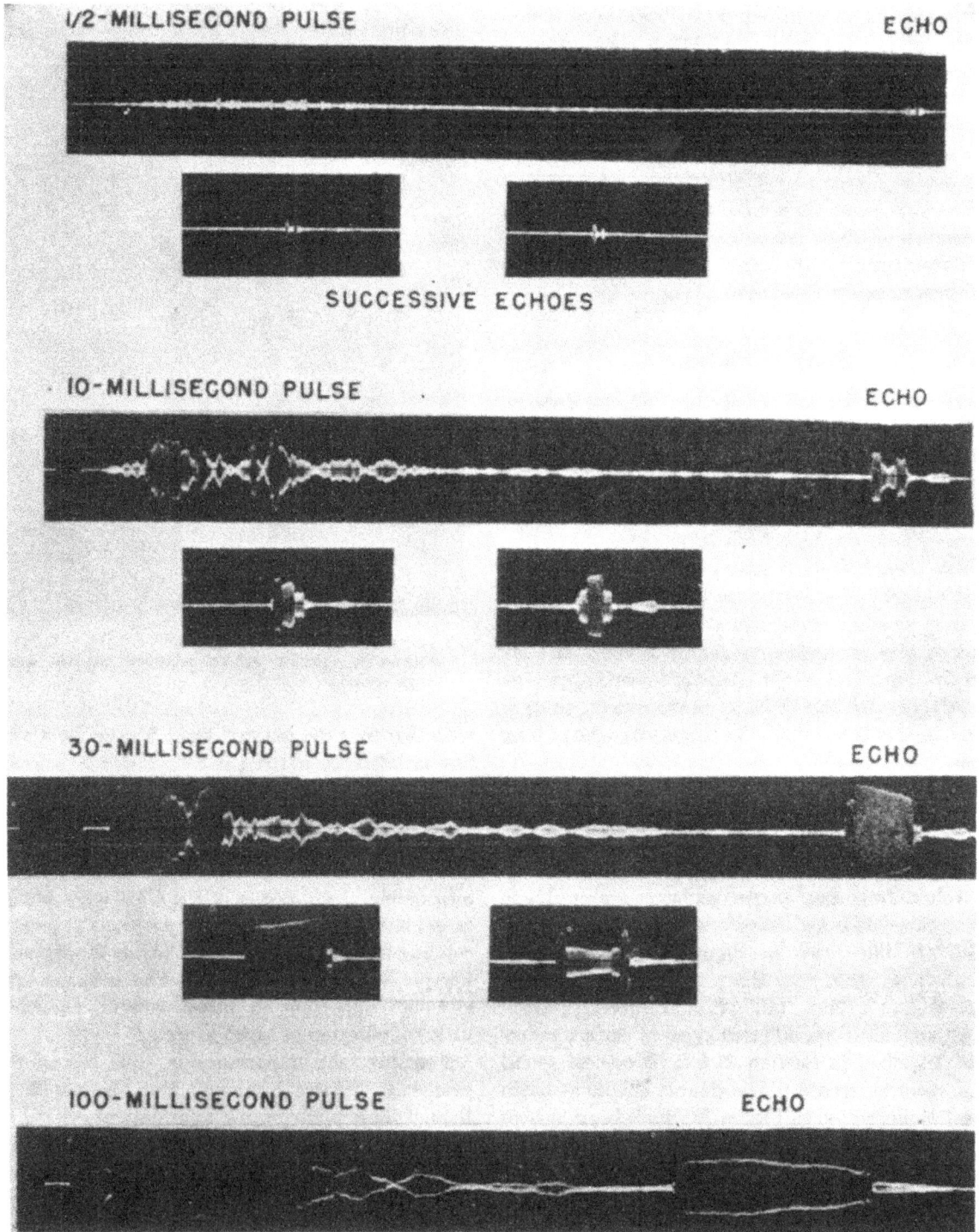

FIGURE 30.   Detailed oscillograms of submarine echoes at beam aspect.

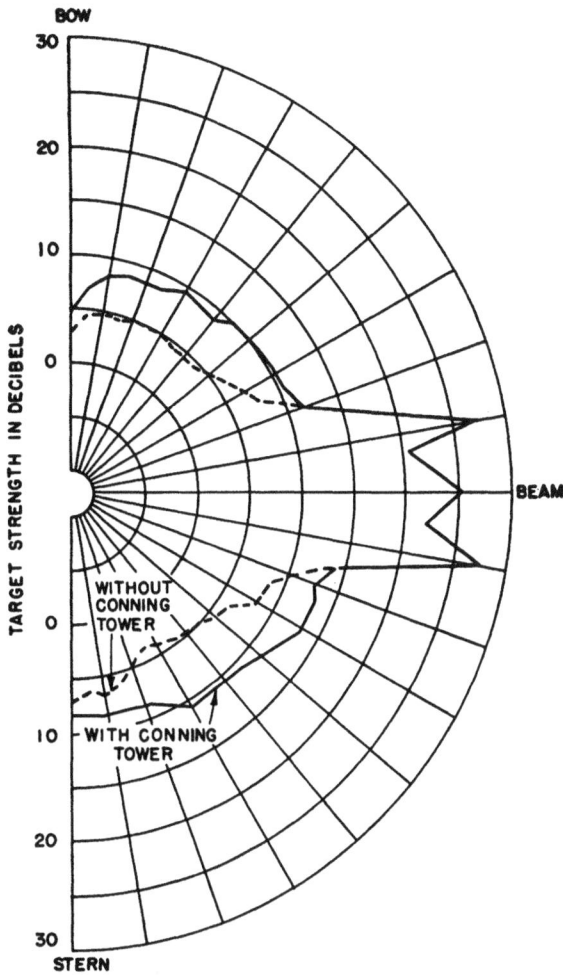

FIGURE 31. Effect of conning tower on optical measurements on USS *Sand Lance* (SS381).

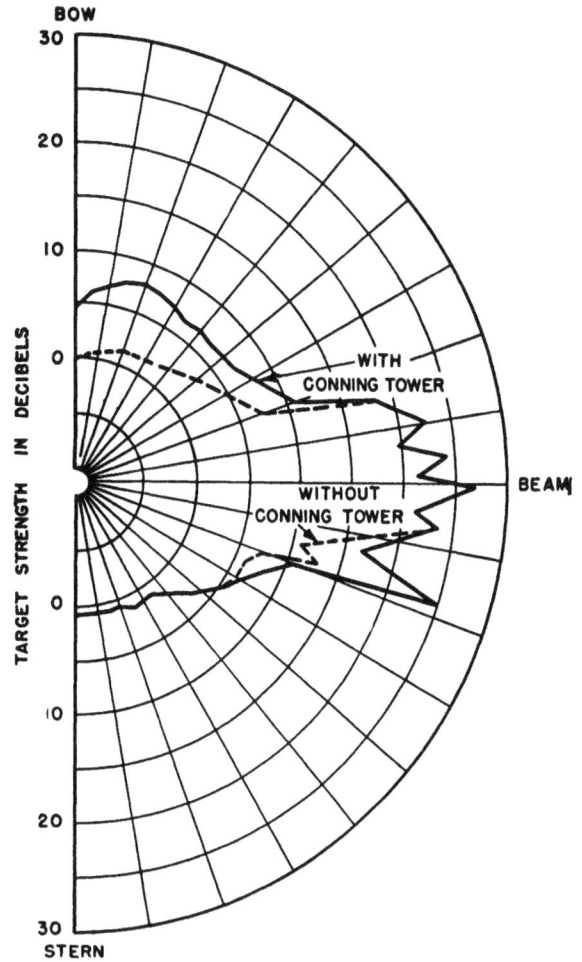

FIGURE 32. Effect of conning tower on optical measurements on U570 (HMS/M *Graph*).

an S-boat at beam aspect for signals from 0.5 msec long resulted in echo oscillograms showing two discrete "blobs" of about equal mean amplitudes with a range separation of about 4 yd.[25,26] Although the finer details of the echo envelopes and the relative values of the peak amplitudes did not repeat from echo to echo, the main features consistently suggested the presence of two distinct reflecting surfaces on the submarine at beam aspect. For signals longer than 4 yd (5 msec), the echo envelopes were almost always resolved into three distinct segments, with the central portion corresponding to the overlap or addition of the two primary echoes found for shorter signals. The amplitude of this central portion presumably varied according to the initial phase difference and amplitude of the two component signals; the difference between the phases changed little during the period of reflection. Typical oscillograms are illustrated in Figure 30; the weak echo following the main one is sound reflected from the submarine up to the surface, back to the submarine and then back to the projector, as discussed in Section 21.5.4. Because the individual echo components in Figure 30 appear to be coherent, at beam aspects the echo is probably specular. Nonspecular or diffuse reflection is apparently unimportant at these aspects, although other surfaces besides the hull and ballast tanks may contribute to the echo.

23.8.2 Off-Beam Echoes

Echoes from aspects other than beam are generally very different from beam echoes. Not only is the echo weaker, but it is also less well defined; examples are shown in Figures 25 and 28. If a system of high resolving power is used, such as the oscilloscope and

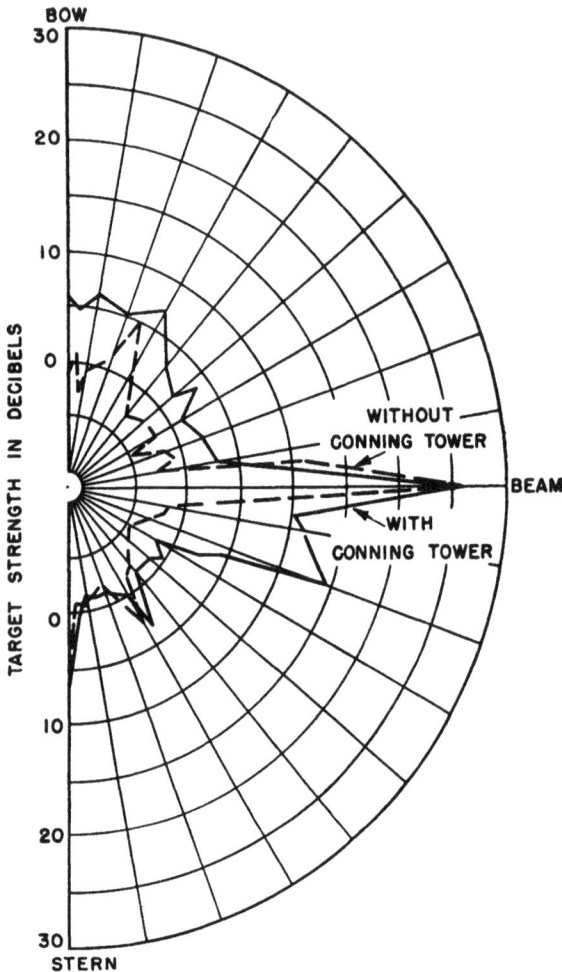

FIGURE 33. Effect of conning tower on acoustical measurements on U570 (HMS/M *Graph*).

high-speed camera at San Diego, the echoes appear to be a group of fine spikes or peaks, although some evidence points to peaks which are found in the same places in successive echoes. A more complete discussion of the detailed structure of off-beam echoes and their origin is postponed to the next section. Here the more general features of off-beam echoes are discussed.

The beginning and end of an echo at off-beam aspect are not clearly defined; usually the amplitude builds up and dies away gradually, blending into the background at either end, so that precise measurement of the echo length is impossible. However, an examination of off-beam oscillograms shows that these echoes are longer than the signals and therefore suggests an extended target.[21]

If the entire length of the submarine is effective in

reflecting sound, as seems indicated for off-beam echoes, the lengths of these off-beam echoes should vary with the aspect angle, depending on the length of the submarine in the direction of the sound beam. In other words, if a submarine is an extended target and scatters sound throughout its length, the length of the echoes which it returns should depend on the aspect which it presents to the sound beam.

Accordingly, the lengths of these off-beam echoes, diminished by the length of the signal used, have been measured or estimated as accurately as possible, then plotted against the cosine of the aspect angle which accounts for the foreshortening of the submarine.[27] Figure 34 illustrates the results of this analysis, where the broken curve connects the measured points, and the solid curve is a polar plot of the cosine of the aspect angle, modified at bow aspect to account for the "shadow" which the forward section casts on the stern section. A similar dip is included at stern aspect, where the after part of the submarine shadows the forward part.

The maximum elongation — echo length minus signal length — was found to occur at quarter aspects, roughly 15 degrees from the bow and stern on either side, and amounted to about 85 yd.[26] The actual length in the direction of the sound beam of a fleet-type submarine 300 ft long, at aspects 15 degrees from bow and stern, is about 96 yd, which confirms the suggestion that the entire surface of the submarine scatters sound. At an aspect angle of 135 degrees, when the target had an extension of 49 yd in the direction of the sound beam, the elongation amounted to about 38 yd.

At bow aspect, this elongation was reduced to 50 yd. This reduction is attributed to the shadow cast by the forward section on the after section. A similar drop, however, was not observed at stern aspect.

These elongation phenomena are apparently independent of signal length and echo-ranging frequency Apparently they are the result of scattering from the entire length of the submarine, instead of reflection from only one or two major surfaces such as the conning tower or screws. They suggest that in addition to specular reflection from the hull, nonspecular reflection or diffuse scattering also occurs, especially at aspects away from the beam. The exact mechanism by which sound is reflected from the entire length of the submarine is unknown.

Similar elongation phenomena were analyzed by British observers in an effort to determine the origin

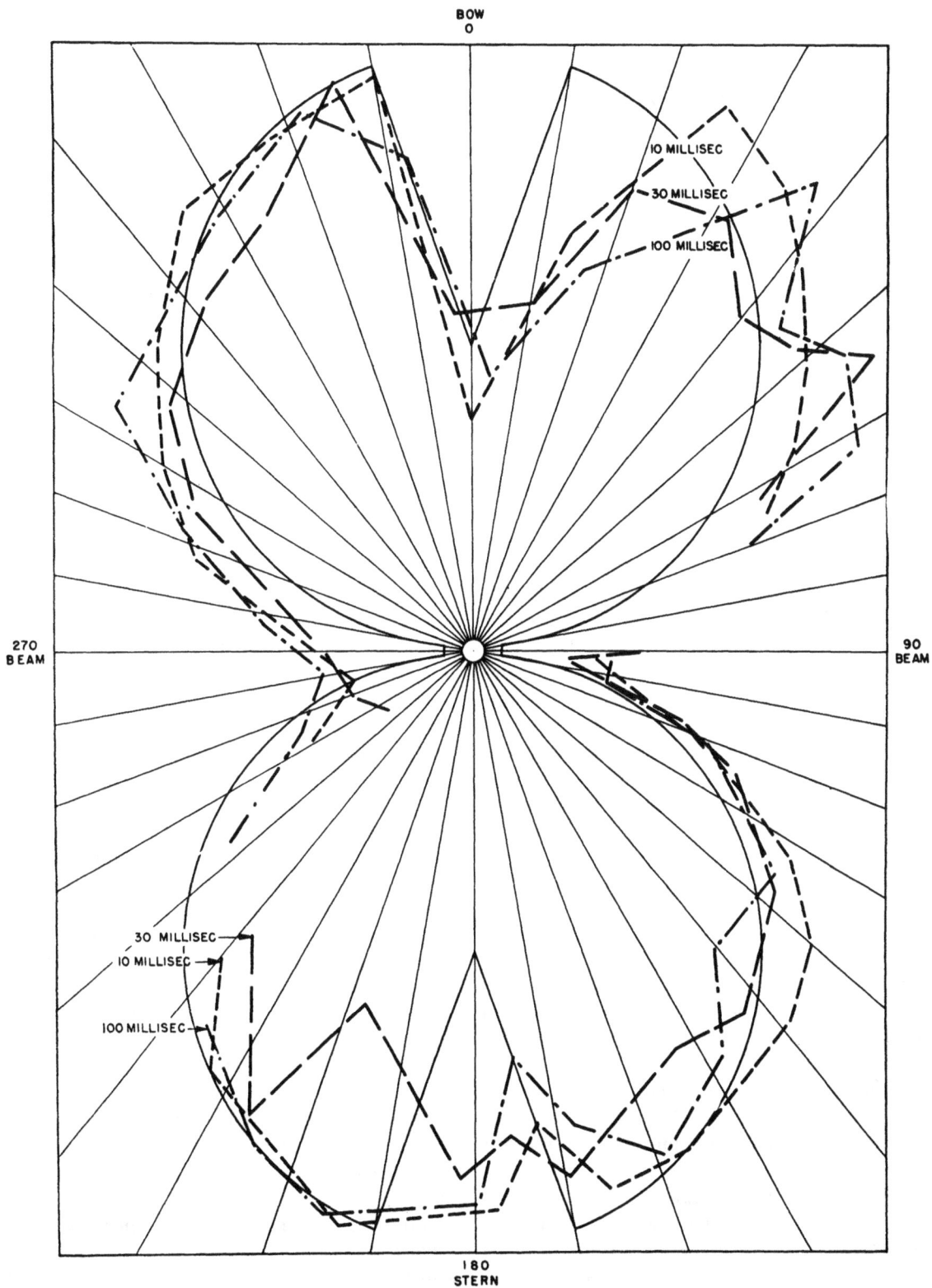

FIGURE 34. Dependence of echo elongation on aspect angle (San Diego).

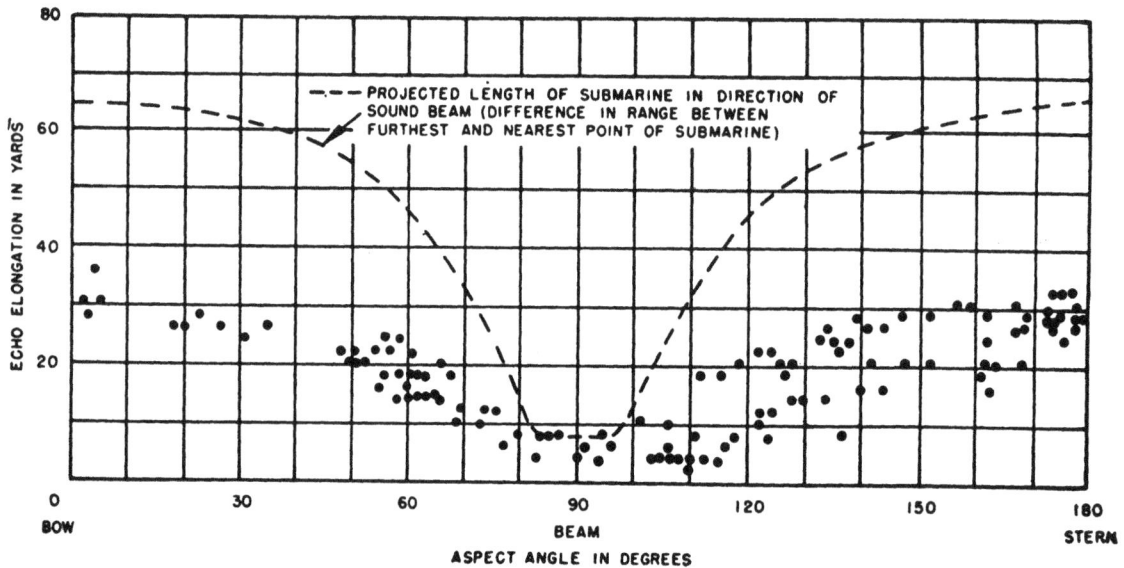

FIGURE 35. Echo elongation and projected length of submarine as a function of aspect angle.

of the nearest echo, but the hypothesis that the entire submarine reflected sound was not confirmed.[28] The elongation, plotted in Figure 35, amounted to only about half the calculated exposed lengths of the submarine, after the pulse lengths had been subtracted from the echo lengths. Similar results have also been obtained in this country using the same technique. It may be pointed out, however, that a sound range recorder, not an oscilloscope, was used in these experiments, and that records from a sound range recorder might be expected not to show the weaker tail part of the echo.

23.8.3   Source of Echoes

Beam echoes originate in large part at the hull of the submarine, as described in Section 23.8.1, with some additional contribution possible from the hull and bilge keel. The echoes are nearly square-topped and result from simple specular reflection from only one or two surfaces on the submarine.

Off-beam echoes, however, apparently come from all parts of the submarine. Oscillograms of these echoes are detailed and show a fine microstructure of peaks and valleys, somewhat similar to reverberation, especially for short pulses. Since study of the elongation phenomena suggests that echoes are returned from most of the submarine, various peaks in the detailed structure of an echo might be correlated with discrete reflecting surfaces on the outside of the submarine. Only short signals could be used, how-

ever; otherwise the signals from individual reflectors on the submarine might overlap.

Accordingly a series of echo oscillograms from signals approximately 0.5 msec (0.4 yd) long were studied at San Diego.[27] The target was a submarine of the S class at quarter aspect, 135 degrees from the bow. The echoes were recorded oscillographically, as usual, but the film was run at a speed of about 13 in. per sec, five times faster than normally. This high speed lengthened each echo and permitted better resolution of the echo structure.

Each echo analyzed consisted of a number of sharp spines, usually between twenty and fifty, which rose clearly above a fuzzy background. The envelope of these spines was roughly cigar-shaped while the envelopes of the less intense parts of the echo peaks were similarly shaped but only about half the amplitude of the spine structure. The distribution of these spines appeared to be random, and no peaks or groups of peaks could be definitely correlated with individual reflecting surfaces, such as the conning tower or ballast tanks. Thus the peaks may be more the result of constructive interference of sound scattered at random from the entire submarine, than of strong reflections from discrete surfaces on the submarine.

Studies of other short-pulse echoes obtained at other aspects from various submarines usually yield somewhat similar results. The echo almost always consists of a succession of peaks rising above the background. These peaks, however, do not always

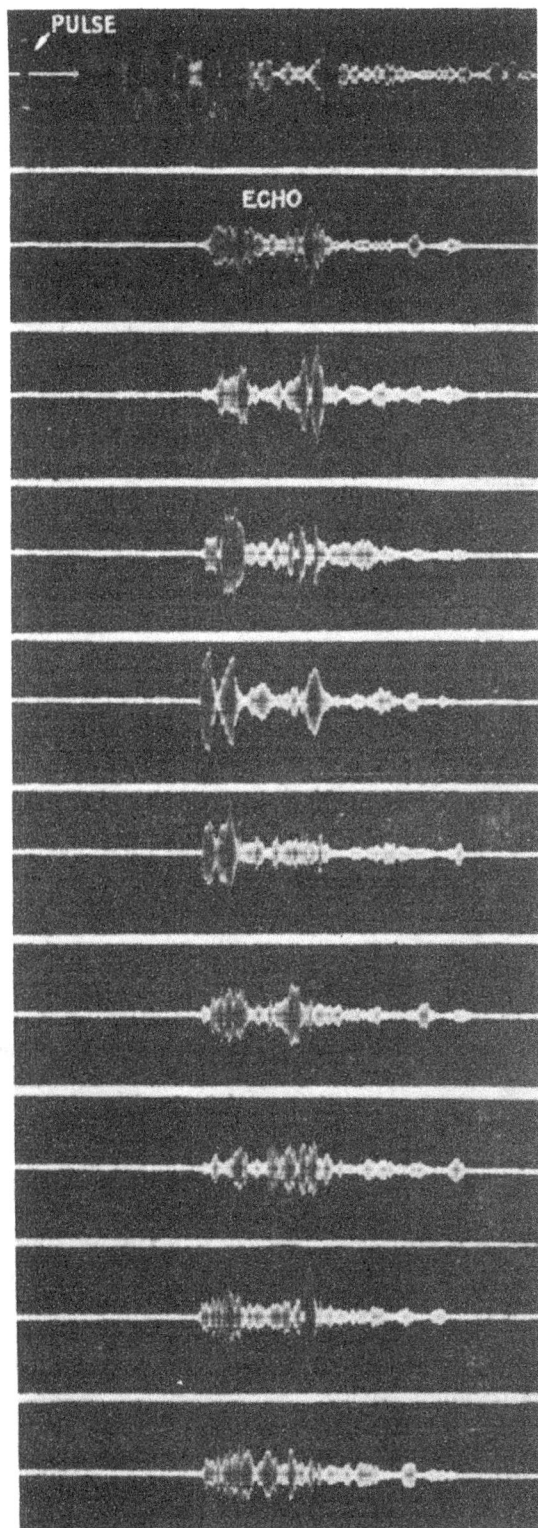

FIGURE 36. Repeatable peaks in submarine echoes.

occur at the same place in successive echoes; rather, they usually appear to be distributed unsystematically though generally nearer the center of the echo than either end.

In some cases, however, repeatable peaks seem to be present in submarine echoes. A series of nine consecutive echoes from 5-msec signals at 60 kc are reproduced in Figure 36 and show two separate peaks or groups of peaks at the same places in each echo. In these measurements the submarine aspect was held nearly constant at about 330 degrees. Thus no definite conclusions can be drawn at the present time as to how often an echo peak will reproduce itself. It is therefore uncertain whether these peaks represent highlights on the submarine or random interference between several reflected sound waves.

In general, the process of reflection of sound from a submerged submarine at off-beam aspects is still imperfectly understood. The entire submarine appears to contribute to the reflected sound, yet specific, repeatable highlights have not been observed in most examinations of echo oscillograms. It is difficult to understand how nonspecular reflection from the submarine hull or from protuberances and fixtures on the outside of the submarine can account for these echoes. Until the origin of these off-beam echoes from actual submarines is satisfactorily explained, the applicability to actual echo ranging of the results obtained with the indirect optical and acoustical tests is open to question.

Chapter 24

SURFACE VESSEL TARGET STRENGTHS

Much less is known about surface vessel target strengths than about submarine target strengths. Few measurements have been made of the sound-reflecting characteristics of ships, and much of the available information has been extracted from experiments where the investigation of the target strength was only incidental to other studies. No mathematical analyses or measurements on scale models have been attempted, so that all target strengths reported here are the results of direct measurements.

Experimental conditions have been far from controlled during these measurements. Ship speed, course, range, and especially aspect angle have been difficult either to estimate accurately or to maintain closely. Many of the tests were made completely at random on vessels happening to pass in the vicinity. Various types of ships served as targets — destroyers, freighters, tankers, coal colliers, transports, and Liberty ships — with the result that although many measurements were made, the data on each ship are too scanty to afford a comparison between different ships. Many variables might have significantly affected the measured echo levels — ship speed, length, width, draft, hull curvature, course, range, aspect angle, sea state, wind force, temperature gradients — so many that a clear-cut separation of variables is out of the question. Furthermore, the results are so few in number, compared with other underwater sound measurements, and the scatter of values is so wide, that only the most tentative and general conclusions may be suggested at the present time.

One of the most important generalizations that may be suggested is the difference between the reflecting properties of moving vessels and still vessels. Ships under way are known to entrain air along their sides as they move through the water (see Section 27.3) With still vessels, on the other hand, entrained air seems less likely. Since small air bubbles are extremely efficient scatterers of sound, it is reasonable to expect that sound striking a moving vessel

might be scattered diffusely by the air bubbles along the sides, just as sound is scattered by the wake laid by the ship. A still vessel, however, might be expected to reflect sound specularly. Such an hypothesis seems to bring some coherence into the observed data. Therefore, it is largely from this point of view that surface vessel target strengths are examined, in this chapter, as a function of aspect angle, range, ship speed, ship type, pulse length, and frequency.

24.1 TECHNIQUES OF SAN DIEGO MEASUREMENTS

Surface vessel target strengths have been measured by only two groups, the University of California Division of War Research at the U. S. Navy Radio and Sound Laboratory, San Diego, California [UCDWR], and Bell Telephone Laboratories, New York, New York [BTL].[1] The measurements off San Diego were made from November 12 to 17, 1943, during a program investigating the acoustical properties of wakes laid by ships at various speeds.[2]

During these tests the USS *Jasper* (PYc13) echo ranged on two flush-decked World War I destroyers, the USS *Crane* (DD109) and the USS *Lamberton* (DMS2, ex-DD119), which followed straight courses at speeds of 10, 15, and 20 knots in deep water. A standard Navy JK transducer was used, sending out pulses 10 msec long at a frequency of 24 kc.

The *Jasper* ranged on the destroyer as it approached; then, just as the beam of the destroyer passed, the *Jasper* began to range on its wake. Consequently no target strengths were measured at aspect angles beyond about 110 degrees from the bow. Errors in the estimated aspect angles were quite large because of unknown deviations of the destroyer from its normal course but could not be evaluated. Ranges varied from 112 to 660 yd.

Echoes from the destroyer and its wake were received and recorded oscillographically on moving picture film, with the equipment employed in the reverberation studies. The average maximum amplitude of five successive echoes, together with the calibration constants of the equipment, was used to compute the echo level at each aspect angle; an auxiliary transducer measured the source level before and after each run. However, the transmission loss was not measured directly. Transmission conditions were fair, since the water was isothermal to a depth of about 50 ft. Accordingly, inverse square divergence and an attenuation coefficient of 5 db per kyd were assumed in estimating the transmission loss. Each target strength was computed from the average echo level of five echoes; each range was the average range over the five echoes, as measured on the oscillograms. Aspect angles were estimated trigonometrically.

24.2 TECHNIQUES OF NEW YORK MEASUREMENTS

Two series of tests were made by BTL on ships in Long Island Sound early in 1944, as part of a specific development project to study the effects of short pulse lengths and receiver bandwidth on echo ranging,[3] and to measure echoes from surface vessels.[4] Because little time was allocated to this part of the program, the work was discontinued as soon as enough data were obtained to establish the range of echo intensities to be expected.

In the earlier measurements, made in Long Island Sound near City Island, Hart's Island, and Execution Light, pulse lengths from 0.05 to 150 msec were used at frequencies between 20 and 30 kc.[3] Echo-ranging gear including a transmitter and a receiving system of adjustable characteristics was mounted aboard a laboratory boat, the *Elcobel*, which was already equipped with a standard Navy projector dome. Targets of these tests were various freighters in the vicinity. No absolute echo levels or target strengths were measured, since the experiment was conducted largely to investigate the effects of pulse and receiver characteristics on reverberation, noise, and echo character. Relative echo amplitudes were found for different pulse lengths, however, and are reported in Section 24.7.

Later studies reported in more detail the reflecting characteristics of a total of twenty surface vessels.[4] In these measurements, a crystal transducer was mounted on the *Elcobel*, a 65-ft boat, in such a way that it could be carried just below the keel while under way, or lowered to a depth of 10 ft for echo

ranging. In the lower position, the transducer could be trained by means of a hand wheel on top of the shaft; however, the speed of the *Elcobel* could not exceed a few knots without interfering with the satisfactory operation of the transducer.

An oscillator aboard the *Elcobel* delivered pulses approximately 3 msec long to the transducer, at a frequency of 27 kc. The echoes received by the transducer were amplified, observed and photographed on the screen of a cathode-ray oscilloscope, whose horizontal sweep was proportional to the time — and therefore to the range of the echo — and whose vertical sweep was proportional to the amplitude of the echo.

In order to obtain target strengths, the average range and the average peak amplitude of between 10 and 60 echoes were measured on the oscillograms. The transmission loss was estimated on the assumption of inverse square divergence and an attenuation coefficient of 7 db per kyd, although such an assumption probably was unrealistic since the water was shallow during these measurements and the ocean bottom was an effective reflector of sound. From the average range, the average peak amplitude, and the transmission loss, the diameter of the equivalent sphere was computed — the sphere which would theoretically return the same echo under the same conditions. Then the target strength was readily determined from the diameter of the equivalent sphere by use of equation (10) in Chapter 19.

24.2.1 Tests on Anchored Vessels

In the first part of the second series of tests, the targets were ships at anchor in the tideway of Long Island Sound, near City Island, New York, where the water was less than 100 ft deep. Echoes from five freighters, a tanker, a Liberty ship, and a small British carrier were measured. During these tests the *Elcobel* was kept under way at a very slow speed, so that both the range and the aspect angle of the target varied in almost all the tests.

24.2.2 Tests on Moving Vessels

The *Elcobel* also ranged on moving ships farther out in Long Island Sound, in the vicinity of Lloyd's Neck, Long Island. These tests were made, without any advance arrangements, on passing ships whose courses brought them close enough to the *Elcobel* to make them satisfactory targets. Unfortunately, the speed of the *Elcobel* had to be held down to a few knots when the transducer was in operation, while the target ships were traveling at least several times faster. Consequently a special procedure was developed to fit these conditions.

As the ship approached, the *Elcobel* maneuvered so that it neared the ship at an aspect angle just off the bow of the target ship. When the range was closed to about 600 yd, the test began, and the *Elcobel* followed a course that kept the transducer constantly aimed at the stern of the ship; the sound beam was wide enough to cover the entire ship even at close ranges. It is possible that echoes were also obtained from the wakes of the ships, although such echoes were probably distinguishable from ship echoes for pulse lengths of 3 msec. Observations were made as frequently as possible, and the range and aspect angle were estimated at the time of the observations. Actually both ships deviated from their nominal courses because of the effects of the wind and sea state as well as inaccuracies in steering, so that the ranges and aspect angles changed rather irregularly.

expected to depend on aspect angle in much the same way as the target strength of a submarine depends on its aspect. In fact, since most surface vessels are more nearly flat at beam aspects than submarines, a sharper dependence might be predicted as long as reflections come exclusively from the hull, in other words, as long as the ship is anchored, or moving through the water very slowly, and gives rise only to specular reflection. If a moving ship reflects sound diffusely, some change of target strength with aspect angle might be expected, but not so marked a change close to beam aspect as results from specula reflection.

Table 1 lists beam and off-beam target strengths for still and moving ships, together with the ranges at which they were measured and the number of individual observations — each comprising at least five echoes — which were averaged to obtain the tabulated results. Here the values given for beam target strength include all measurements at estimated aspect angles between 70 and 110, and between 250 and 290 degrees from the bow, whereas off-beam target strengths include measurements at

TABLE 1   Aspect dependence.

Test	Range in yards	Beam target strengths and standard deviations in decibels	Number of obser- vations	Number of ships	Range in yards	Off-beam target strengths and standard deviations in decibels	Number of obser- vations	Number of ships
Anchored ships (New York)[1]	250–587	37.3 ± 16.3	8	5	168–508	13.3 ± 7.6	23	9
Moving ships (San Diego)[1]	112–640	20.8 ± 5.7	42	2	140–660	17.2 ± 4.9	35	2
Moving ships (New York)[1]	250–490	16.2 ± 6.1	62	12	300–562	13.7 ± 5.1	66	12

24.3   **ASPECT DEPENDENCE**

Surface vessel target strengths have been measured at San Diego and New York for different aspect angles, ranges, speeds, types of ships, pulse lengths, and frequencies. A dependence on aspect angle and on range is suggested by the reported data; in addition, the target strength of ships under way appears to be considerably different from the target strengths of still vessels. Sufficient information, however, is not available to permit evaluation of the effects of the class of ship, pulse length, or frequency on the target strengths measured.

The target strength of a surface vessel might be

all other aspect angles. Ranges, speeds, ship types, pulse lengths, and frequencies are not separated. Table 1 is illustrated graphically in Figure 1.

24.3.1   **Still Vessels**

Only for the anchored ships is the difference between beam and off-beam target strengths roughly the same as the scatter of the observations, as represented by the standard deviation. The dependence of target strength on aspect angle for the individual measurements on these still vessels is shown in Figure 2. Two target strengths at an aspect angle of approximately 100 degrees are conspicuously higher

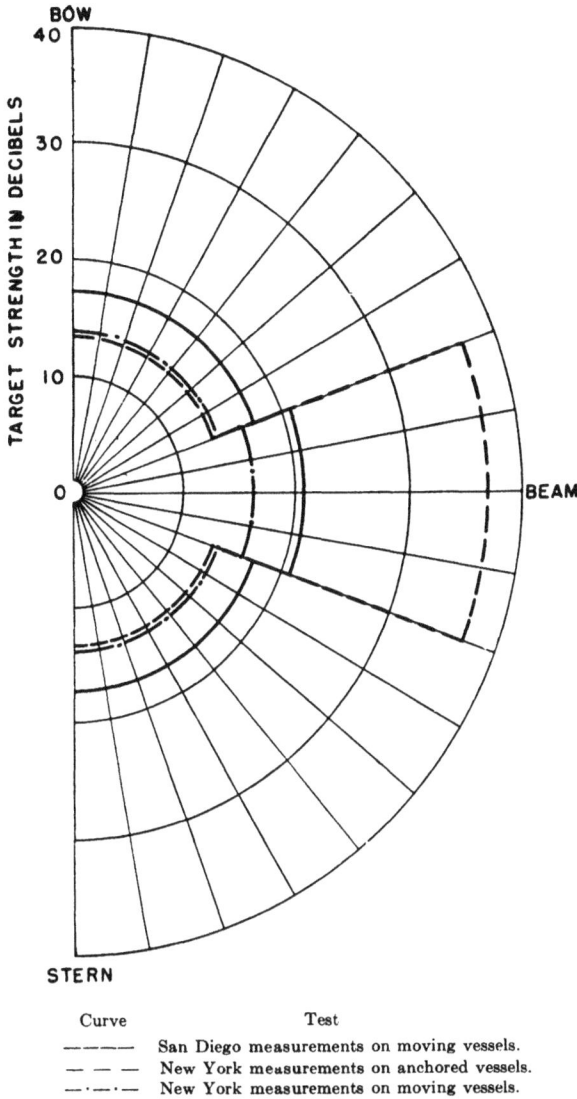

FIGURE 1. Aspect dependence for all tests.

Curve	Test
——————	San Diego measurements on moving vessels.
— — —	New York measurements on anchored vessels.
—·—·—	New York measurements on moving vessels.

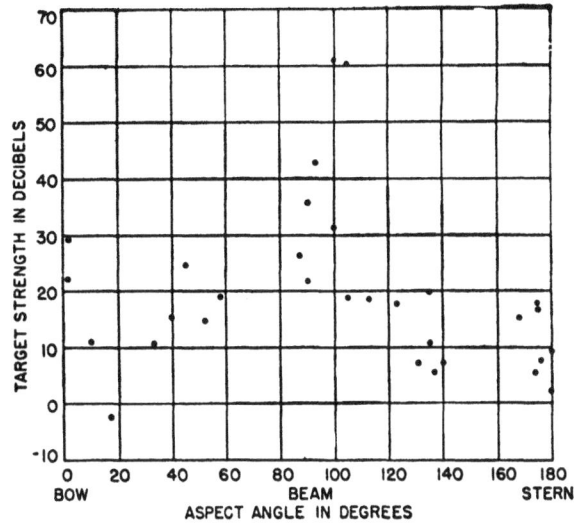

FIGURE 2. Aspect dependence for still vessels (New York).

face vessel target strengths were also found to depend on the range, increasing as the range increased, especially at beam aspect, these high values may be more the result of the effect of the range on the target strength, than the effect of the aspect angle of the target. The data were too few to permit separation of these two factors and independent evaluation of the effect of each on the measured target strengths.

Secondly, so few observations are plotted in Figure 2 that two very marked peaks may not be reliable. Only 31 values were obtained in the New York tests on anchored vessels, and only 8 of these were at aspects within 20 degrees of the beam. In view of the large scatter of values, the observations cannot be considered conclusive, but are at least generally consistent with the theoretical expectation that still vessels reflect sound specularly.

24.3.2 Moving Vessels

The dependence of target strength on aspect angle for moving vessels is much less than the dependence found for still vessels. The small variation in Table 1 is too small to be very significant. For comparison with Figure 2, the individual target strengths for moving ships measured at New York are plotted in Figure 3 and for the destroyers measured at San Diego in Figure 4. The scatter is so large that any possible systematic variation of target strength with aspect angle is largely obscured.

Analysis of the San Diego data on moving vessels

than values at any other aspect, almost 20 db higher than the next highest value, and more than 40 db higher than the average target strength at off-beam aspects. Such a peak would be expected theoretically from specular reflection from the broadside of the ship at beam aspect; at a few degrees away from beam aspect, however, the target strength should be markedly reduced.

This peak in Figure 2 may be exaggerated for two reasons, so that the actual aspect dependence may not be so sharp as it appears. First, the two observations constituting the peak were made at ranges between 500 and 600 yd, but most of the other observations were made at much closer ranges. Since the sur-

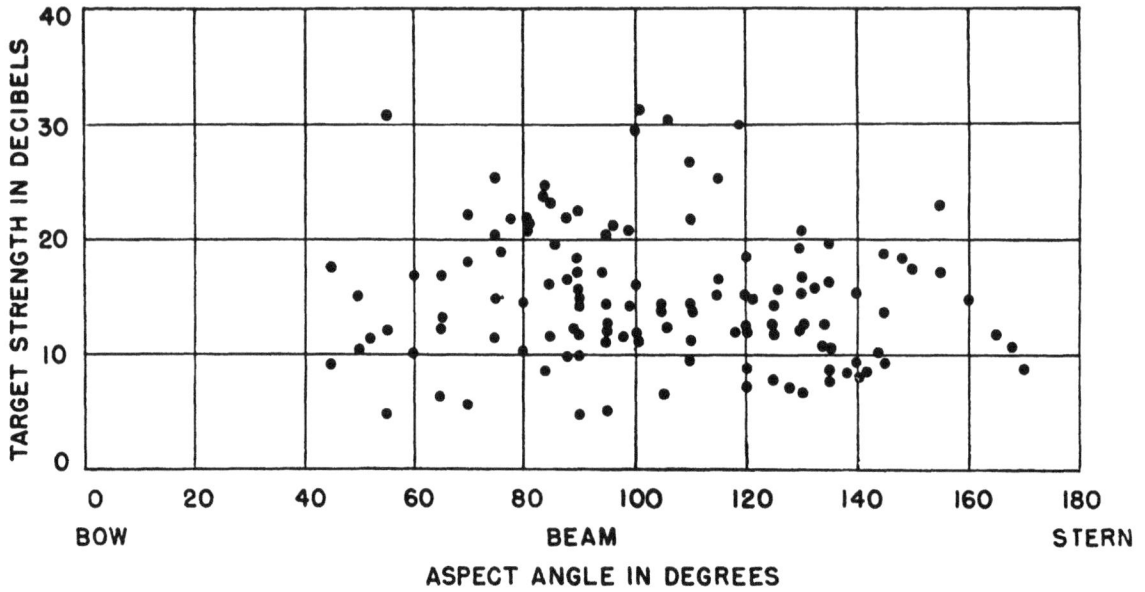

FIGURE 3.   Aspect dependence for moving vessels (New York).

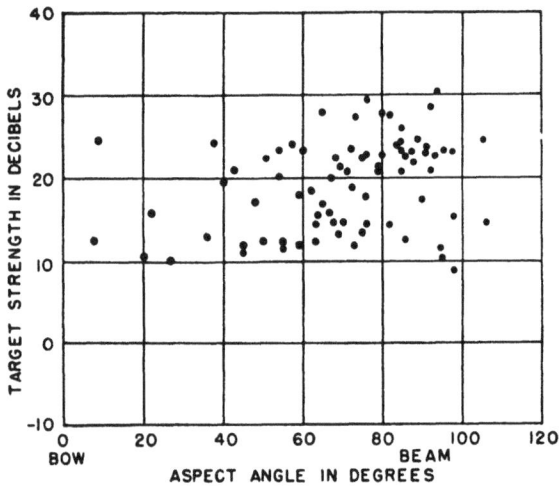

FIGURE 4.   Aspect dependence for moving vessels (San Diego).

TABLE 2.   Aspect dependence at different ranges for moving destroyers (San Diego).

Range in yards	Beam target strength and standard deviation in decibels	Off-beam target strength and standard deviation in decibels
100 to 300	12.3 ± 1.8	13.0 ± 1.7
300 to 500	22.5 ± 3.6	16.9 ± 4.8
500 to 700	23.9 ± 3.6	20.9 ± 3.7

The scatter of the individual values from the averages in Table 2 is less than the scatter from the overall averages in Table 1, and a slight change of target strength with aspect seems significantly shown. Some dependence of this nature might be expected from a diffusely reflecting surface, if all the target were in the path of the sound beam. However, any change with aspect angle as great as that found for submerged submarines seems ruled out by Table 2. These results are generally consistent with the hypothesis that bubbles along the side of the ship are responsible for the echoes observed from moving ships. More accurate data would be required, however, for verification of this theory.

24.4   **RANGE DEPENDENCE**

In Sections 20.4.4 and 23.4.1 it was pointed out that the target strength of submarine depends on the range at ranges less than the maximum radius

leads to much the same results, which are illustrated in Figure 4. Since the target strengths measured at San Diego were found to depend markedly on range (see Section 24.4), they were separated according to range in order to examine the dependence on aspect angle. The data were broken down into three groups, for ranges of 100 to 300 yd, 300 to 500 yd, and 500 to 700 yd, and each group was analyzed for a possible dependence on aspect angle. Average target strengths for aspects within 20 degrees of the beam, and for all other aspects are shown in Table 2 for each range group.

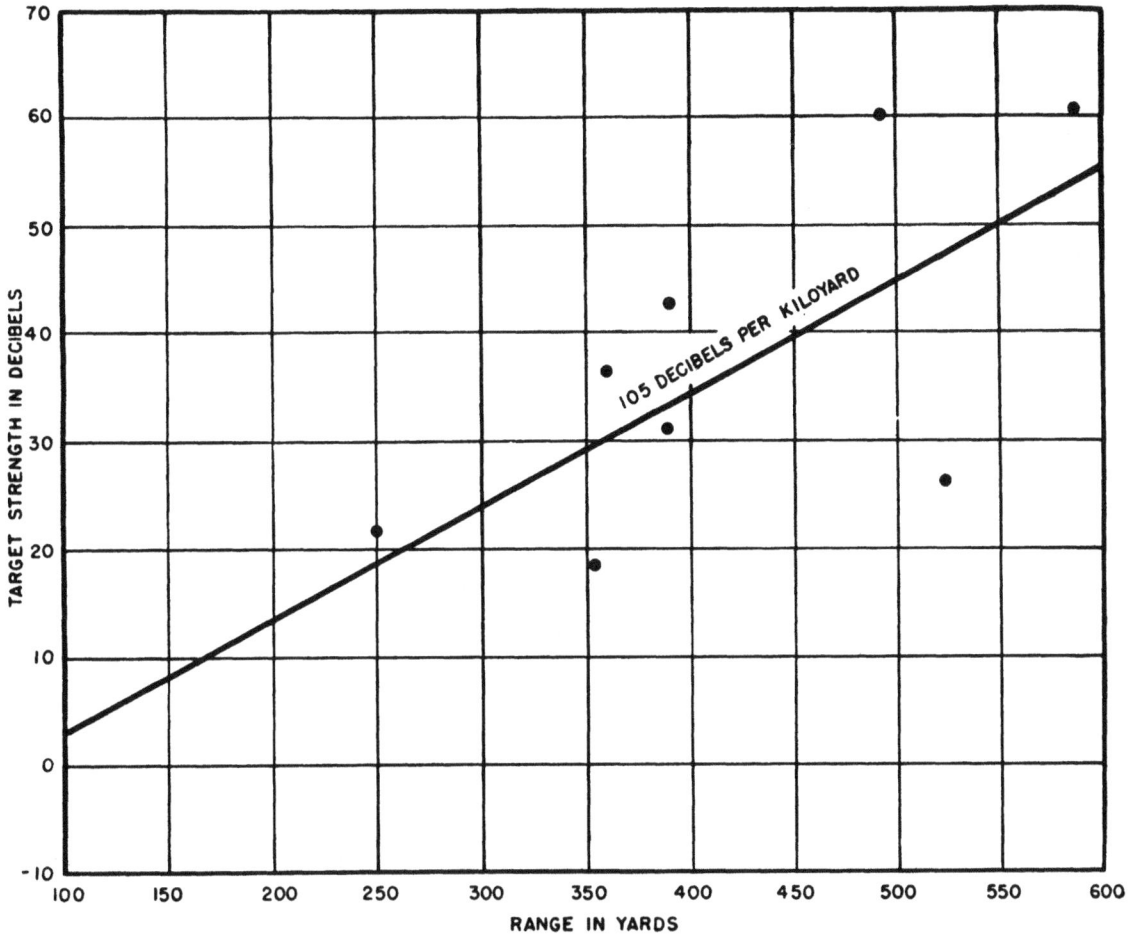

FIGURE 5.   Range dependence at beam aspects for anchored vessels (New York).

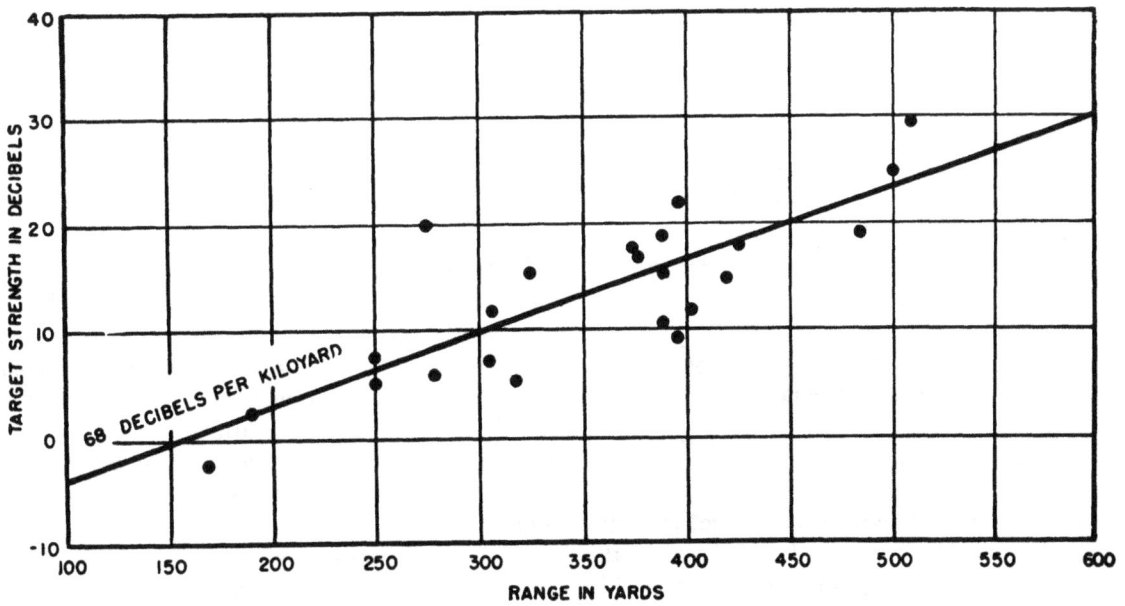

FIGURE 6.   Range dependence at off-beam aspects for anchored vessels (New York).

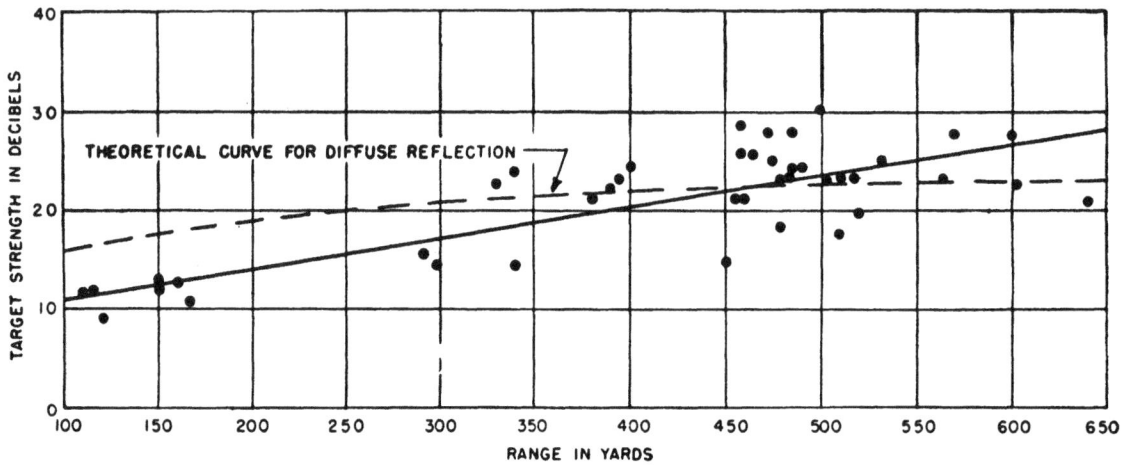

FIGURE 7. Range dependence at beam aspects for moving vessels (San Diego).

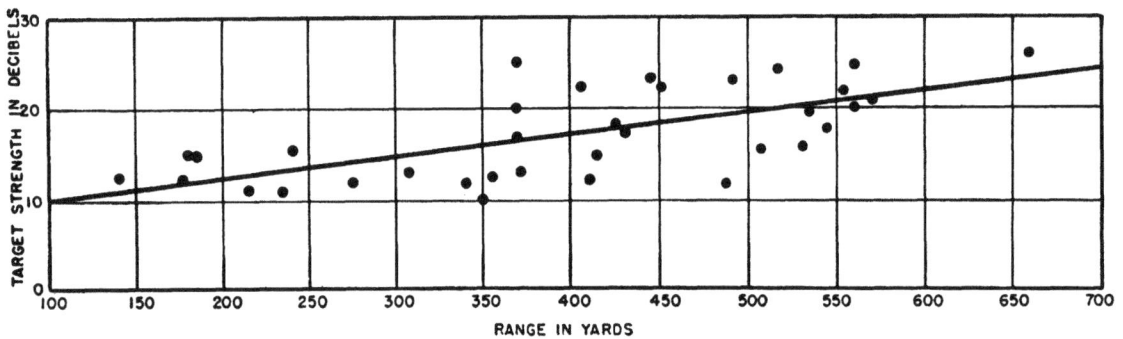

FIGURE 8. Range dependence at off-beam aspects for moving vessels (San Diego).

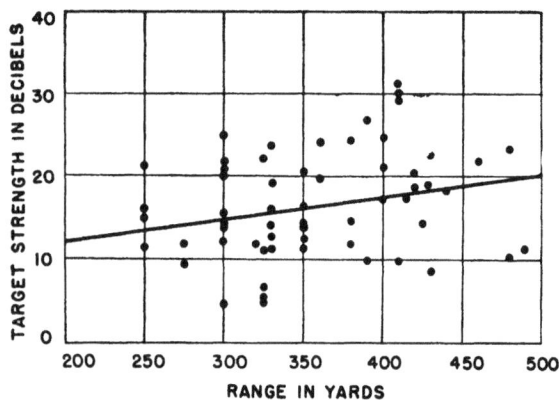

FIGURE 9. Range dependence at beam aspects for moving vessels (New York).

of curvature of the submarine. The target strength of a still ship would be expected to behave in the same way under similar conditions. Because ship hulls may be flatter and may have a larger radius of

curvature than submarines, this dependence on range might extend to much longer ranges than for submarines. On the other hand, the target strength of a moving ship might be expected to increase as the range increases, as more and more of the scattering surface lies in the path of the direct sound beam.

Accordingly, target strength was examined as a function of range, for beam and off-beam echoes, for all three sets of data. The results of this analysis are illustrated in Figures 5 to 10, where in each graph the solid line represents the least squares solution based on an assumed linear relation between the target strength and the range. The slopes of these lines are listed in Table 3. It is apparent that in all cases the dependence of target strength on range is most pronounced (1) for still vessels and (2) at beam aspect.

Three explanations may be suggested to account for the increase in target strength with range: (1) failure of the sound beam to cover the target at short

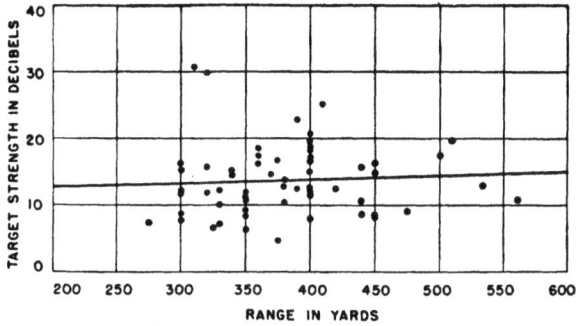

FIGURE 10. Range dependence at off-beam aspects for moving vessels (New York).

ranges; (2) reduced reflection as the range approaches the dimensions of the target; and (3) incorrect evaluation of the transmission loss. The first effect applies only to the measurements on moving vessels at San Diego, since the sound beam used during the New York tests was wide enough to cover the target at all ranges. The second applies primarily to measurements on anchored vessels where specular reflection seems most likely to occur, and the third applies to measurements on both moving and still ships.

TABLE 3. Range dependence.

Test	Slope of target strength-range curve in decibels per kiloyard	
	Beam aspect	Off-beam aspects
Anchored ships (New York)	105.0	68.0
Moving ships (San Diego)	30.5	24.4
Moving ships (New York)	26.3	4.5

At short ranges, how much of the target the sound beam covers depends on the dimensions and aspect angle of the target and on the directivity pattern of the transducer. At San Diego, a standard JK transducer was employed, which had a total beam width of 20 degrees between points on either side of the axis where the response was 10 db lower. If it is assumed that the sound beam was 20 degrees wide and that the destroyer was 300 ft long, then the sound beam did not cover the ship, at beam aspect, at ranges less than about 300 yd. Since many of the beam target strengths were measured at shorter ranges, this failure of the sound beam to cover the ship may account for the decrease in target strength with decreasing range.

24.4.1 Transducer Directivity

To evaluate the effect of the transducer directivity on the target strength-range dependence, the difference between the echo level from a destroyer at beam aspect and from a small target always within the sound beam was calculated, as a function of range, from the directivity pattern of the transducer. This difference is expressed as

$$10 \log \int_0^{x_1} \frac{b^2(\phi)dx}{(x^2 + r^2)^2} - 10 \log \frac{x_1}{r^4}, \qquad (1)$$

where x_1 is the length of the target in a direction perpendicular to the sound beam; $b^2(\phi)$ is the composite directivity pattern of the transducer; and r the range to the center of the target. This difference in decibels between the echo level from the destroyer and the echo level from a small target of the same target strength, as the range is decreased from 650 to 100 yd, is superimposed on Figure 7 as a broken line. The zero level, where the sound beam effectively covers the entire target and the two echo levels are the same, is placed at a target strength of 23.5 db, which is the average beam target strength measured at San Diego at ranges of 450 yd and greater. The difference calculated from equation (1) amounts to about 7 db at a range of 100 yd, and drops to less than 1 db at ranges greater than 500 yd.

This analysis does not take into account the extension of the target by the wake. However, even if the target were assumed to extend infinitely in one direction, the target strength for long pulses would increase only as $10 \log r$ and would not be significantly different from the broken curve in Figure 7. The increase of target strength with range in such a case would be analogous to the similar increase for the target strength of wakes discussed in Section 33.1.1. This failure of the sound beam to cover the entire target, especially at short ranges, is responsible for much of the dependence of target strength on range observed at San Diego at beam aspect. Apparently, however, it is not responsible for all the dependence observed.

Significantly, these echoes from destroyers at beam aspect are approximately as strong as echoes observed from the wakes directly behind the destroyers. In Section 24.1 it was noted that echo-ranging experiments were made on the wakes after the destroyers passed; in these measurements, the sound beam was perpendicular to the axis of the wake. The wake echoes showed the same variation with range as the destroyer echoes, and, like them, showed no significant dependence on speed. Differences as great as 10 db were observed between the wake echoes and the destroyer echoes immediately preceding them, but these differences appeared to be quite random and unsystematic. This equivalency between wake echoes and destroyer echoes is consistent with the theory that both arise from scattering by small bubbles.

The dependence of target strength on range at off-beam aspects from the San Diego results shown in Figure 8 cannot be analyzed very simply. In the first place, the spread of aspect angles covered is very wide; the projected length of the destroyer, measured in a direction perpendicular to the sound beam, varied from 30 ft at bow and stern aspects to 290 ft at an aspect angle 20 degrees from the beam. In the second place, with a pulse length of 10 msec, the entire destroyer was not in the sound beam at the same time, especially at bow and stern aspects; for aspects close to the beam, this effect of pulse length may be neglected, but it becomes important at other aspects. When more of the target comes into the sound beam, the observed echo from a very short pulse will not be stronger but instead will last longer, as pointed out in Section 19.3. Thus the change of target strength with range at off-beam aspects cannot be explained even in part by this simple mechanism.

Transducer directivity is also relatively unimportant in the New York measurements on still and moving vessels since a very wide beam was employed. The total horizontal beam width of the combined projector-hydrophone directivity pattern, between points where the response was 10 db lower than on its axis, was about 40 degrees, which even at a range of 168 yd, the shortest range at which measurements during either test were made, still covers the longest ship at beam aspect. Therefore the decrease in target strength with decreasing range in the New York measurements cannot be explained as a result of the failure of the sound beam to cover the target.

24.4.2 Predicted Dependence

The second explanation which might be suggested for the observed dependence of target strength on range is the predicted decrease of specular reflection with decreasing range for ranges less than the maximum radius of curvature of the target, providing the reflection is specular (see Section 20.4.4). This effect would apply only to echoes from vessels sta-

tionary in the water, which presumably arise primarily from the hull and not from a uniformly scattering layer. However, in the most extreme case, reflection from an infinite plane surface, the target strength will not vary more rapidly than as the square of the range. Such a variation is quite insufficient to account for the large effect observed during echo-ranging trials on still vessels illustrated in Figures 5 and 6. Qualitatively, however, it partly explains the difference in range dependence at beam and off-beam aspects, since the radius of curvature of the ship is greater when it presents its broadside to the incident sound than when it is at bow or stern aspect.

24.4.3 Transmission Loss

A third possible explanation of the observed range dependence is a possible incorrect evaluation of the transmission loss. In none of the measurements was the transmission loss measured directly. Instead, an attempt was made to estimate it from the prevailing conditions on the basis of inverse-square divergence and an additional attenuation proportional to the range.

At San Diego, it was assumed that the intensity of the echo was inversely proportional to the fourth power of the range, weakened by an additional loss of 5 db per kyd of sound travel. The water was isothermal to a depth of 50 ft, so that an assumption of 5 db per kyd for the attenuation coefficient seems somewhat low. Use of equation (1) in Chapter 23 gives an attenuation coefficient of about 7 db per kyd. An attenuation coefficient of about 10 db per kyd would be required to explain the departure of the plotted points from the theoretical curve in Figure 7. Such a high coefficient does not seem very likely when the surface layer is isothermal down to a depth of 50 ft, but it is not impossible.

At New York, the transmission loss was assumed to follow the same inverse square loss with an attenuation coefficient at 27 kc of 7 db per kyd. The temperature conditions of the water were not known; the wind velocity varied from 1 to 23 mph. Whether or not the assumed attenuation coefficient is reliable it is difficult to say. In addition, bottom-reflected sound may have had a marked effect on the transmission loss.

Conditions were very favorable to bottom reflection during these New York tests. The bottom was composed of sand and mud, a mixture which reflects sound very effectively. In addition, the water was very shallow, from 60 to 110 ft deep. The sound beam was not highly directional; the total vertical beam width between points where the response was 10 db down was 20 degrees. Thus if the transducer were level, the sound beam would strike the bottom at a range of only about 126 yd, for water 60 ft deep. Consequently the bottom undoubtedly reflected part of the incident sound in much the same way as the surface, and contributed to the intensity of the echoes received at the transducer.

Assume, for example, that both the surface and bottom reflected sound perfectly, so that at the particular ranges used the sound beam could spread in only one direction — horizontally. In this extreme case, the intensity of the echo would be inversely proportional, not to the fourth power of the range, but to the square of the range. This assumption, of course, is not realistic, but the result suggests that for the New York measurements the actual drop is somewhere between inverse fourth and inverse square; perhaps the echo intensity actually varies more nearly inversely as the cube of the range over a shallow reflecting bottom. This relatively slow increase of transmission loss with increasing range may account for much of the range dependence for moving vessels in the New York data. Even an inverse square dependence of echo level on range fails to account, however, for the observed variation on still vessels shown in Figures 5 and 6, where the echo level actually increases rapidly with increasing range.

Another possibility might account for the dependence of target strength on range for stationary vessels. Sound incident on the hull of the ship will be reflected downward where the hull is curved slightly downward, then reflected upward from the bottom. It is possible that the curvature of the hulls of the surface vessels measured is such that the rays reflected to the bottom will strike the bottom and be reflected back to the transducer only at longer ranges, so that target strengths measured at long ranges will be greater than target strengths measured at short ranges. This explanation may account for the stronger range dependence for stationary vessels than for moving vessels, although it must be regarded as highly tentative in the absence of further substantiating evidence.

In all, it has been well established that the target strength of surface vessels on which measurements have been made apparently increases with range. This increase is much greater at beam aspects than

at off-beam aspects. The exact rate of increase is uncertain because many causes are responsible; measured rates vary from 4.5 to 105 db per kyd at ranges between 200 and 500 yd. The dependence of target strength on range arises from (1) smaller coverage of the target by directive transducers at close ranges; (2) incorrect evaluation of the transmission loss neglecting surface and bottom reflections; and (3) the dimensions and curvature of the target, in so far as they reduce specular reflection at close ranges. Probably none of these effects, however, can explain the enormous observed range dependence for anchored vessels. Further measurements would be required to show the extent to which this observed effect is generally found.

24.5 DEPENDENCE ON SPEED

Very little information is available on the variation of target strength with the speed of the ship, for moving vessels. At San Diego, speeds of 10, 15, and 20 knots were employed; Table 1 lists target strengths without separating the speeds at which they were

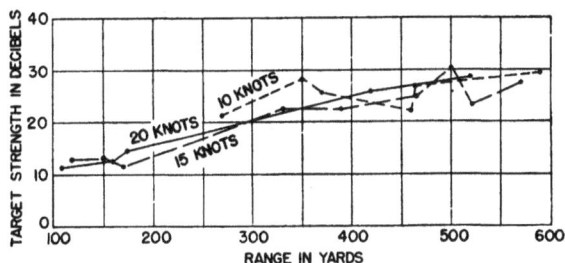

FIGURE 11. Range dependence at different speeds for beam aspect (San Diego).

measured. Figure 11 shows beam target strengths plotted as a function of range for three different speeds, 10, 15, and 20 knots, from the San Diego measurements. The dependence on range is evident, even in only twenty observations, but no significant dependence on ship speed is apparent. Ship speeds were not estimated or measured in the New York tests. As already mentioned, this same lack of dependence on ship speed is characteristic of wake echoes.

24.6 DEPENDENCE ON SHIP TYPE

No clear dependence of target strength on ship type is indicated by the evidence now available. While a large number of different vessels have been

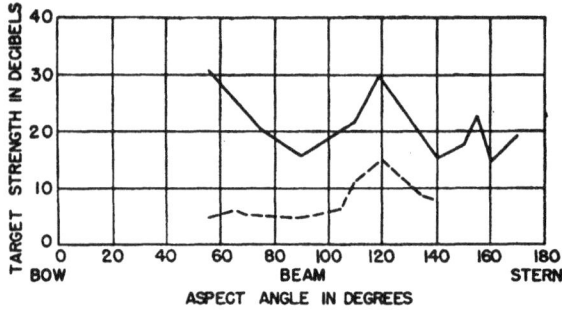

Curve	Ship	Tonnage	Length in feet	Draft in feet	Water depth in feet	Range in yards
———	Navy Transport	12,000	325	25	100	300–510
— — —	Liberty Ship	10,000	300	20	60	300–450

FIGURE 12. Variation in target strength between similar ships (New York).

their estimated tonnages, lengths, and drafts; furthermore, both were measured at roughly the same ranges. It is possible that the fluctuation and variation normally encountered in underwater sound transmission may be responsible for the 10 db difference between the two curves, or that bottom reflection may be the cause, since the transport was under way in water 100 ft deep while the Liberty ship was under way in water almost half as deep. Even perfect bottom reflection, however, cannot account for the observed difference between the two curves, which suggests faulty calibration, widely variable transmission, or large unsuspected systematic differences between the two ships.

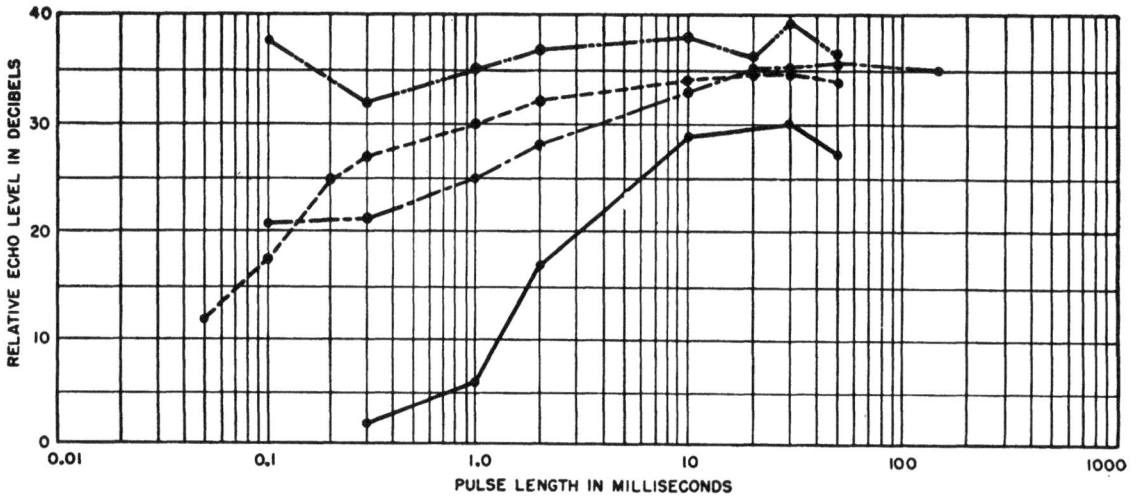

Curve	Range in yards	Approximate aspect angle in degrees
———————	——	180
— — — — —	700	45
— — — — — —	1600	——
— ·· — ·· —	2200	180

FIGURE 13. Effect of pulse length on measured echo levels (New York).

made,[4] the scatter is so great that any correlation between target strength and ship draft and tonnage is obscured.

As an example of the variation in target strength between one ship and another, as measured at New York, Figure 12 illustrates target strength plotted against aspect angle for two large ships of nearly equal dimensions. The difference in their target strengths cannot be attributed to the difference in

24.7 DEPENDENCE ON PULSE LENGTH AND FREQUENCY

Although surface vessel target strengths have not been systematically investigated as a function of pulse length, early studies at New York reported a dependence of echo amplitude on pulse length for pulse lengths of 0.05 to 110 msec.[3] The results of these measurements are reproduced in Figure 13, where the relative echo level in decibels is plotted against the pulse length for four freighters. Little dependence on pulse length is evident for pulses more than 10 msec long, in qualitative agreement with the results described in Section 23.5.2 applying to submerged submarines. However, for pulse lengths of less than 10 msec, the echo level drops rather sharply. More data are required, however, to show how great this dependence will be for any actual vessel.

No information is available on how surface vessel target strengths vary with the frequency of the echo-ranging beam employed. The only tests were made at San Diego at 24 kc and at New York at 27 kc; any difference in the target strengths at these two frequencies would probably be very small, from theoretical predictions, and the actual measured difference is too small to verify any such dependence. For still vessels, if the echo comes from the hull, very little variation of target strength with frequency would be expected (see Sections 20.2 and 20.3). For moving vessels, however, with sound scattered from a layer of bubbles, the target strength would be expected to vary with frequency in accordance with the acoustic properties of small bubbles.

Chapter 25

SUMMARY

25.1 DEFINITION OF TARGET STRENGTH

FOR THE PURPOSES of discussing the reflecting characteristics of different vessels, the *target strength* T of a target is defined by

$$T = E - S + 2H, \tag{1}$$

where E is the echo level, S the source level, and H the one-way transmission loss from the source to the target, all in decibels (see Section 19.1.3). For most targets, T is independent of range at ranges much greater than the dimensions of the target (see Sections 20.4 and 23.4), but may change with the changing orientation of the target relative to the sound beam (see Section 23.1).

25.1.1 Echo Level

The echo level E is defined by

$$E = 20 \log p_e, \tag{2}$$

where p_e is the rms pressure of the echo, in dynes per square centimeter averaged over a few cycles (see Section 19.1.3). If the rms pressure is not constant during the echo, E is defined as the peak rms pressure.

25.1.2 Source Level

For directional supersonic projectors, the rms pressure p of the sound on the axis of a projector is inversely proportional to the square of the range r, as long as the range is much greater than the dimensions of the target and as long as the range is small enough so that attenuation and surface reflection may be neglected. Under these conditions, the source level S is defined by

$$S = 20 \log p + 20 \log r, \tag{3}$$

where p is the pressure of the sound on the axis of the projector, in dynes per square centimeter, at a distance r, in yards, from the projector (see Section 19.1.3).

25.1.3 Transmission Loss

The difference between the pressure level of the transmitted sound at some point, and the source level is called the transmission loss H from the projector to that point (see Section 19.1.2).

25.1.4 Average Values

Since both E and H often fluctuate by as much as 10 db from pulse to pulse, it is customary to use the average echo amplitude in determining E, and the average pressure amplitude at the range r in determining H, in equation (1), where the average amplitude is the average of a number of peak rms amplitudes, if the rms amplitude is not constant over each echo (see Section 21.6.4).

25.1.5 Target Strength of Sphere

A sphere reflects a plane wave equally in all directions (see Section 19.2.2). The target strength of a sphere is

$$T = 20 \log \frac{A}{2} \tag{4}$$

where A is the radius of the sphere in yards (see Sections 19.2.1, 19.2.2 and 20.4.1). This formula is accurate to 0.5 db if the range to the sphere is greater than ten times its radius, and if the wavelength of the sound is less than the radius of the sphere.

25.1.6 Target Strength of a General Convex Surface

The target strength for specular reflection from any convex surface is

$$T = 10 \log \frac{A_1 A_2}{4\left(1 + \dfrac{A_1}{r}\right)\left(1 + \dfrac{A_2}{r}\right)} \tag{5}$$

where A_1 and A_2 are the principal radii of curvature of the target surface at the point where the surface is

434

perpendicular to the sound beam, and r is the range (see Section 20.4.2). This formula is valid only if both A_1 and A_2 are greater than the wavelength of the sound, and if either A_1 or A_2 is much less than r.

25.1.7 Target Strength of a Cylinder

For a cylinder, A_2 is infinite in equation (5) and the target strength becomes

$$T = 10 \log \frac{A_1 r}{4\left(1 + \dfrac{A_1}{r}\right)} . \qquad (6)$$

This equation is valid only when the cylinder radius A_1 is less than r and the wavelength is less than A_1.

25.2 BEAM ECHOES FROM SUBMERGED SUBMARINES

At aspect angles within about 20 degrees of the beam, echoes from submarines are produced primarily by specular reflection from the pressure hull, the fuel and ballast tanks, and the conning tower (see Section 23.8.1) and are much stronger than echoes at other aspects. Oscillograms show that the echo generally reproduces the outgoing pulse (see Figures 25 and 27 in Chapter 23).

25.2.1 Beam Target Strengths

Observed submarine target strengths at beam aspects and at long ranges lie mostly between 20 and 30 db. About 25 db is the average value (see Section 23.1.1); typical values of A_1 or A_2 in equation (4) which would correspond to this target strength would be 500 and 2.5 yd respectively. The observed spread of values may result entirely from experimental errors.

Off-beam target strengths, found at aspects 20 degrees or more away from the beam, are reported in Section 25.3.

VARIATION WITH SUBMARINE CLASS

Observed differences in the target strengths of different submarines measured both directly and indirectly are less than the estimated experimental error in the direct measurements (see Section 25.2.1). Consequently no reliable overall evaluation of the dependence of the target strength on the class of submarine can be made.

VARIATION WITH SUBMARINE SPEED

No significant variation of target strength with submarine speed is expected, since the wake of a sub-

merged submarine is a poor reflector of sound (see Section 33.3). No pronounced variation has been observed in practice for submerged speeds from 1 to 6 knots at keel depths of about 100 ft (see Section 23.3).

VARIATION WITH RANGE

Theoretically, beam target strengths depend on the range at ranges less than the principal radii of curvature of the submarine at beam aspect (see Sections 20.4.4 and 23.4). For a 517-ton German U-boat, approximated by an ellipsoid with principal radii of curvature of 576 and 2.3 yd, the variation of target with range predicted from equation (5) is shown in Table 1. Although no observations are

TABLE 1. Theoretical range variation.

Range in yards	Submarine target strength beam aspect (without conning tower)	Submarine target strength beam aspect (with conning tower)
8	5.5	5.8
12	7.5	7.4
16	8.9	9.1
200	19.2	22.9
1,000	23.2	25.5
∞	25.2	...

available to confirm this variation with range, the result is believed to be reliable.

VARIATION WITH PULSE LENGTH

No marked dependence of target strength on pulse length is expected at beam aspect, since the echo approximately reproduces the pulse (see Section 23.5.1). The available evidence is neither very consistent nor conclusive, but does not demonstrate any sharp variation in the target strength with the pulse length (see Section 23.5.2).

VARIATION WITH FREQUENCY

No variation of target strength with frequency is expected theoretically at beam aspects for specular reflection (see Sections 20.4 and 23.6.1). Observations confirm this prediction (see Section 23.6.2), except for a few measurements at 60 kc; these 60-kc target strengths, however, are so large that calibration errors are believed responsible.

25.2.2 Echo Structure

Generally, beam echoes are square-topped and resemble the outgoing pulses (see Section 23.8.1) For very short pulses, beam echoes from submarines reveal a definite structure. For observations on one S-boat, the main echo consists of two components separated by a distance of about 4 yd; the first component may come from the broadside of the submarine, while the second component may be an echo from the bilge keel or conning tower. After this main echo comes a much weaker secondary echo, presumably resulting from sound reflected from the submarine straight up to the surface, back down to the submarine, and then back to the projector (see Figure 9 in Chapter 21, and Figure 29 in Chapter 23). The presence of this echo structure will be expected to modify slightly the conclusions in the preceding section, since for long pulses the different components will combine. Such a combination will increase the average target strength 3 db at most above its value for very short pulses.

25.2.3   Fluctuation

The fluctuation of beam echoes may be primarily attributed to the fluctuation in the transmission of the outgoing and incoming sound (see Section 21.6). Much of this fluctuation is apparently due to the presence of surface-reflected sound (see Section 21.5.4). Estimates of the fluctuation of transmitted sound are given in Chapters 7 and 10. In addition, for pulses more than a few milliseconds long, interference between the different components of the echo will somewhat increase the fluctuation.

25.3   OFF–BEAM ECHOES FROM SUB-MERGED SUBMARINES

At aspect angles more than about 20 degrees off the beam, echoes from submarines originate along the entire length of the vessel and probably result from both specular and nonspecular reflection (see Section 23.8.2); they are 10 to 15 db weaker than echoes at beam aspect. The echo does not reproduce the outgoing pulse (see Figures 25 and 28 in Chapter 23).

25.3.1   Off-Beam Target Strengths

Observed submarine target strengths at off-beam aspects and at long ranges lie mostly between 5 and 20 db for pulses 100 or more msec long, and usually between 10 and 15 db (see Section 23.1.1). The spread of values is apparently real to some extent, since at different aspect angles echo characteristics are markedly different. At certain off-beam aspects

and altitudes, strong specular reflections from nearly flat surfaces, such as the conning tower, may give target strengths greater than 20 db (see Section 23.2.2); these reflections depend critically on the particular submarine measured.

VARIATION WITH SUBMARINE CLASS

No variation in the off-beam target strengths of different submarines has been observed in either the direct or indirect measurements to be greater than the estimated experimental error in the direct measurements (see Section 25.2.1).

VARIATION WITH SUBMARINE SPEED

No important variation of target strength with submarine speed has been observed at off-beam aspects (see Section 25.2.1).

VARIATION WITH RANGE

At off-beam aspects, submarine target strengths decrease with decreasing range (see Section 25.2.1). At ranges less than the length of the submarine, off-beam target strengths are roughly equal to beam target strengths. Under such conditions, a submarine may be approximated by a cylinder at off-beam aspects except bow and stern, and equation (6) may be used.

VARIATION WITH PULSE LENGTH

Since at off-beam aspects the echo does not usually reproduce the pulse and the echo length considerably exceeds the pulse length, for pulses 100 or more msec long, some variation of target strength with pulse length may be expected (see Section 23.5.1). Observed target strengths decrease with pulse length for signals shorter than 100 msec (see Section 23.5.2). The decrease is most marked for pulses shorter than 10 msec, but even for such short pulses the target strength does not decrease as rapidly as the pulse length, or rather, as rapidly as $10 \log \tau$, where τ is the pulse length in milliseconds.

VARIATION WITH FREQUENCY

No variation of target strength with frequency is expected at off-beam aspects (see Section 23.6.1). This conclusion is contradicted by some target strength measurements at a frequency of 60 kc, which give much higher results than similar measurements at 24 kc (see Section 25.2.1). However, the differences between beam and off-beam target strengths are about the same at 60 kc as at 24 kc, so that if the observed frequency effect is real, it is the same at all aspects.

25.3.2   Echo Structure

At off-beam aspects, echoes from submarines do not reproduce the outgoing pulses because the entire length of the submarine reflects sounds (see Section 23.8.2). The duration of the echo, measured on an oscillogram, may be given by

$$T = \frac{2L}{c} \cos \theta + \tau, \qquad (7)$$

where T is the duration of the echo, L the length of the submarine, c the velocity of sound, θ the aspect angle measured from the bow of the submarine, and τ the pulse length. On a sound range recorder, however, the echo length is about half that given in equation (7), perhaps because only the stronger part of the echo would be expected to show on a recorder using chemically treated paper (see Figure 25 in Chapter 23).

25.3.3   Fluctuation

The fluctuation of echoes at off-beam aspects is due not only to fluctuations in the transmission of the sound each way (see Section 25.2.3), but also to fluctuations resulting from interference phenomena. The echo obtained from a long pulse will be the result of constructive and destructive interference between echoes from individual reflecting surfaces distributed over the length of the submarine. Changes in this interference pattern as the aspect or altitude of the submarine changes slightly will increase the observed fluctuation of echoes.

25.4   ECHOES FROM SURFACE VESSELS

Information on reflection from surface vessels is even more fragmentary than on reflection from submarines. The following conclusions are suggested by the data but cannot all be regarded as confirmed.

25.4.1   Still Vessels

Vessels at anchor seem to behave as targets in the same way as submerged submarines. At aspects close to the beam, target strengths may be very high, as much as 40 db, but at other aspects, for pulses 3 msec long at a frequency of 27 kc, it is usually between 5 and 20 db (see Section 24.3.1). The strong echoes at beam aspect are presumably the result of specular reflection from the hull of the ship.

25.4.2   Moving Vessels

When a vessel is under way, beam echoes are about the same as off-beam echoes (see Section 24.3.2). Observed target strengths of moving destroyers and merchant vessels lie between 10 and 25 db, for pulses 3 and 10 msec long at frequencies of 24 and 27 kc; a systematic difference in the target strengths of different ships is not evident (see Section 24.6). An increase in speed from 10 to 20 knots apparently does not affect the target strength appreciably (see Section 24.5). A decrease in pulse length decreases the resultant target strength, especially for pulse lengths less than 10 msec, but the target strength does not drop as rapidly as $10 \log \tau$, where τ is the pulse length (see Section 24.7).

Echoes from moving vessels may arise from scattering by bubbles of entrained air along the side of the ship. This implies that the echo from a moving ship may be treated as an echo from a short stretch of wake (see Section 33.1.1).

25.4.3   Dependence on Range

Most of the data on target strengths of moving vessels show a marked increase in target strength as the range increases from 200 to 600 yd, in one case amounting to more than 30 db (see Section 24.4). Although some increase is expected from the geometry of the ship (see Sections 24.4.2 and 25.2.1) and from the failure of the sound beam to cover the entire ship at short ranges (see Section 24.4.1), so marked a change seems greater than can be explained on any simple basis; it is quite possibly a statistical accident. Beyond about 600 yd, it is reasonable to assume that the target strength does not depend on the range, and that its value lies within the spread specified for surface vessel target strengths at off-beam aspects in the preceding section.

PART IV

ACOUSTIC PROPERTIES OF WAKES

Chapter 26

INTRODUCTION

26.1   WHAT ARE WAKES?

THE APPEARANCE of a streak of foamy, churned water behind a ship under way, known as the ship's wake, is familiar to every mariner. Because the wake extends along the path of the ship over a length many times the ship's length, it is hard to get a good view of the wake as a whole, even if a somewhat elevated vantage point, such as the bridge or masthead, is chosen. Figure 1 shows the wake of an antisubmarine patrol vessel (PC488) in a quiet sea, as viewed aft from the crow's nest.

The observer in an airplane enjoys ideal conditions for the visual study of wakes. Figures 2 to 6 describe better than verbal descriptions what a wake looks like from a great height. The first four were taken from altitudes of 2,500 to 3,000 ft, the plane in level flight overtaking a destroyer, the USS *Moale* (DD693), which was proceeding on a straight course at constant speeds of 16, 20, 25, and 33 knots. By way of a scale, the ship had an overall length of 376 ft, a beam of 41 ft, and a draft of 13 ft. Figure 6 illustrates what happens to the wake as the ship turns; the foam on the curved section of the path is seen to be visually more dense, especially along the outer edge of the wake. As in turning, acceleration of the ship on a straight course increases the visual density of a wake. Incidentally, the irregular white streaks appearing in Figures 4 to 6 are foam patterns on the water. All the photographs show the delicate bow-wave pattern, fanning out astern with a much greater angle of divergence than the actual wake. Figure 7 is a close-up taken from an altitude of 300 ft, of the bow wave and the wake of another destroyer, the USS *Ringgold* (DD500). The visible structure of the wakes laid by large ships does not differ markedly from that of the wakes of vessels of destroyer size, as may be seen in Figure 8, which gives a view from an altitude of 4,000 ft of a large aircraft carrier, the USS *Saratoga* (CV3).

Beyond the obviously foamy and turbulent nature of wakes, visual inspection does not reveal any of their physical properties. The discovery that "wakes," using this term in a loose sense, are capable of affecting the propagation of sound energy through the water has suggested a new distinction: an *acoustic wake* is defined as a volume of the ocean which has acquired, because of the passage of a ship through it, a greater though transitory capacity for absorbing and scattering sound. The expression "volume of the ocean" is used advisedly, because acoustic wakes have a definite vertical extension, often rather sharply bounded. Acoustic wakes under the surface, originating from submerged submarines, are of particular interest. During a level run at periscope depth, the upper boundary of the wake, spreading out bodily from the screws, does not reach the ocean surface until several hundred yards behind the submarine. Several aerial views of submarine wakes, both during surface runs and after a crash dive, are reproduced in Figures 1 to 6 of Chapter 31.

The temperature distribution in the top layer of the ocean may be disturbed by the passage of a ship in such a manner as to leave a *thermal wake*, detectable by sensitive thermocouples. Evidently, experiments must decide to what extent thermal and acoustic wakes coincide with the body of water called a wake by a visual observer. This problem is discussed in Chapter 31, dealing with the geometry of wakes. However, one interesting feature will be mentioned here: the acoustic and thermal manifestations of a wake may persist over periods of half an hour or more, often long after visible traces of the ship's passage have disappeared.

26.2 NAVAL IMPORTANCE OF WAKES

Wakes can be important in naval warfare in two general ways. In the first place, they may interfere with the successful operation of acoustic devices, by scattering or absorbing sound. In the second place, they may provide a method for detecting, tracking,

FIGURE 1.   Wake of submarine chaser (PC 488), seen from crow's nest.

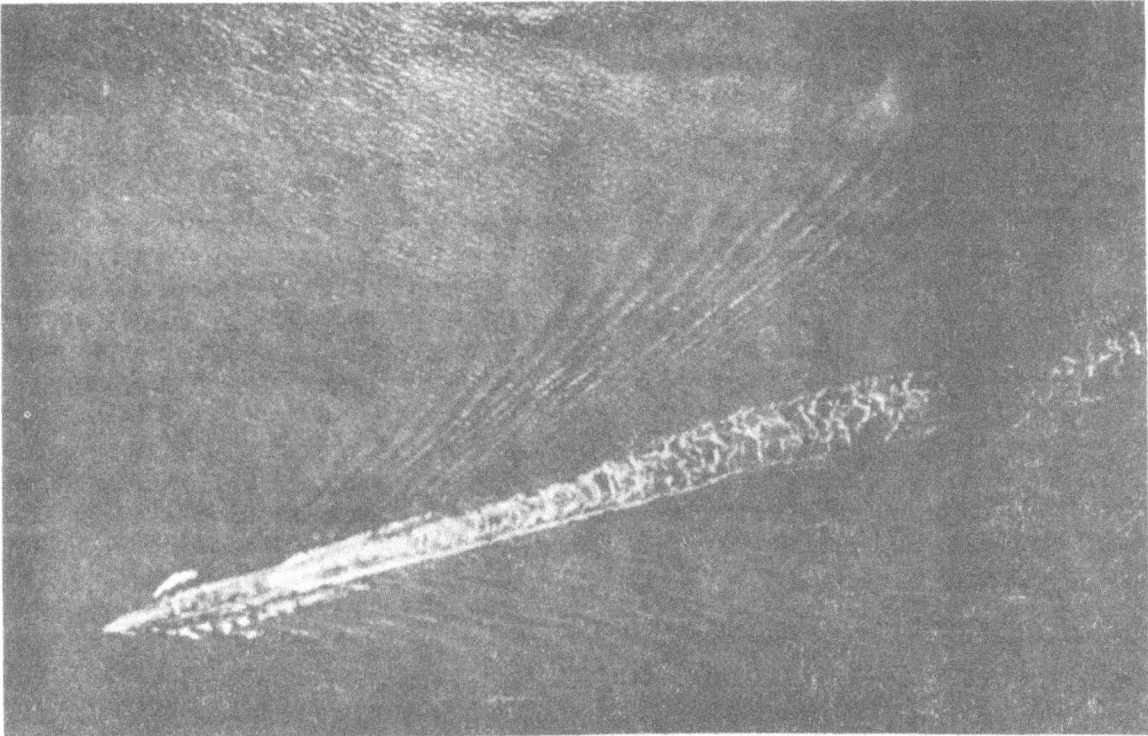

FIGURE 2. Wake of USS *Moale* (DD 693) at 16 knots from 2,500 feet.

or identifying the ship which has produced the wake. Such utilization of wakes in offensive operations comprises visual detection from the air and thermal detection from surface ships or submarines, as well as acoustic detection. However, the present discussion is concerned only with the acoustic properties of wakes and the importance of these acoustic properties in naval warfare.

Acoustic interference produced by wakes is frequently encountered in the operation of sonar gear. False echoes from submarine wakes may confuse the sonar operator on an antisubmarine vessel and may even lead to an attack on a wake knuckle, a disturbance in the water when a submarine suddenly speeds up and turns sharply, while the submarine escapes. During thirty unsuccessful attacks on submarines by United States antisubmarine vessels in 1944, where the presence of a submarine was ascertained but no damage inflicted, 12 per cent of the failures were attributed to attacking wakes, a larger percentage than assigned to any other single cause.

Wakes laid by surface vessels can also be disturbing in antisubmarine warfare. After one or more attacks in an area, echoes from old wakes from surface ships can be confusing. Moreover, a moderately fresh wake is highly absorbent and may shield a shallow target on one side of the wake from detection by a surface vessel on the other side. In fact a projector surrounded by a fresh wake is almost completely useless, since very little sound can escape through the wake. Thus a surface ship will commonly find that its echo-ranging equipment "goes dead" when the ship passes through a fresh wake.

Harbor detection equipment can also be seriously hampered by the presence of wakes. When a destroyer at moderate speed passes in the neighborhood of bottom-mounted supersonic listening gear, ships passing by subsequently cannot be heard for some time. Similarly, sneak craft in the wake of a large surface vessel are very difficult to detect by echo ranging. To reduce the seriousness of these effects in combating submarines, or to use them most effectively in submarine warfare, accurate information is required on the reflection and absorption of sound by wakes under different conditions.

The use of wakes in offensive operations against the wake-laying vessel is a relatively new field. As an example of this utilization of wakes, it was at one

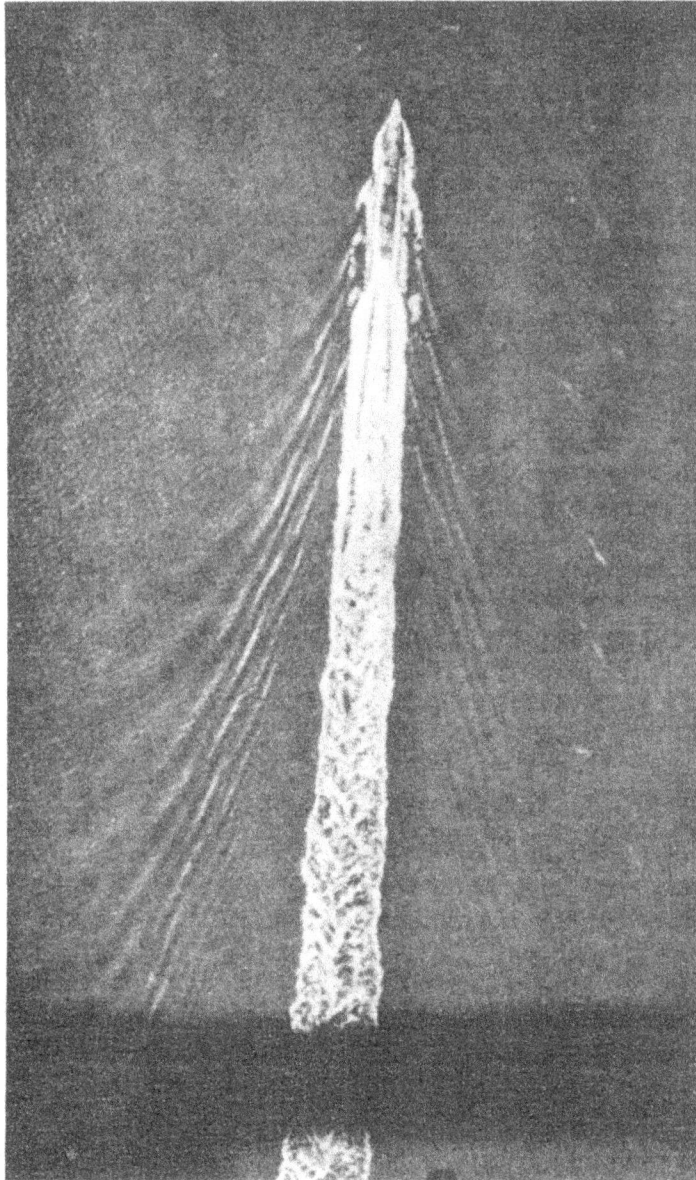

FIGURE 3. Wake of USS *Moale* (DD 693) at 20 knots from 2,500 feet.

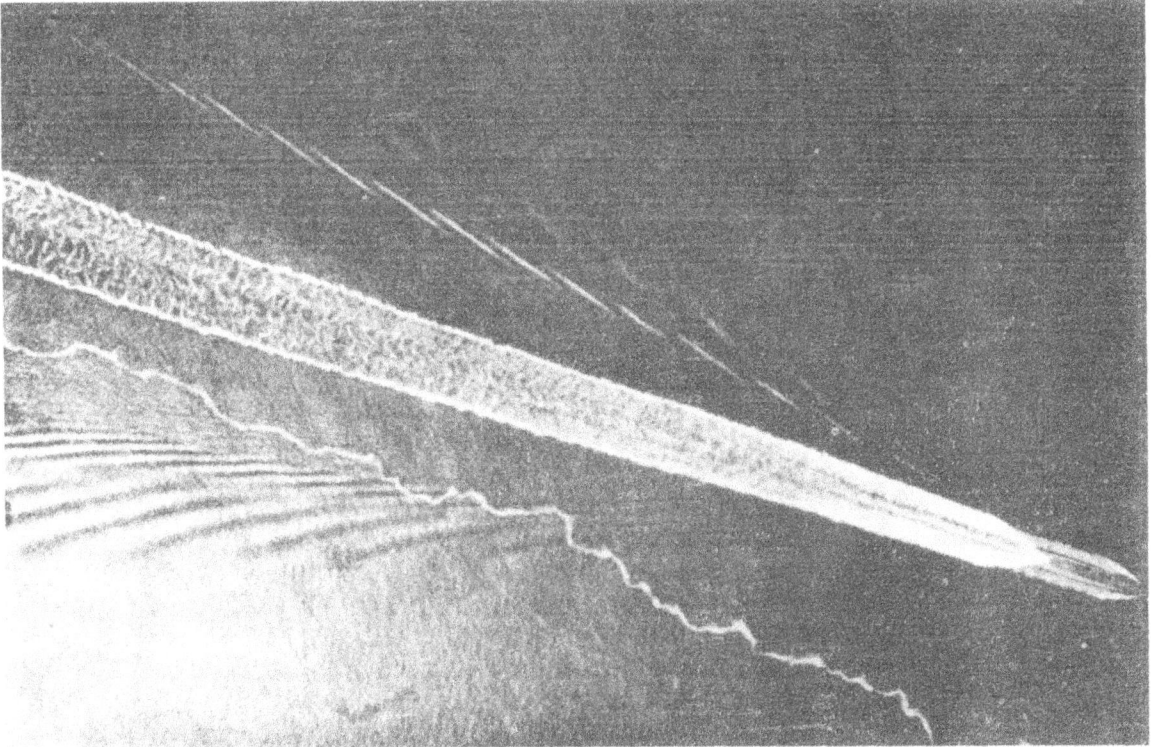

FIGURE 4.   Wake of USS *Moale* (DD 693) at 25 knots from 3,000 feet.

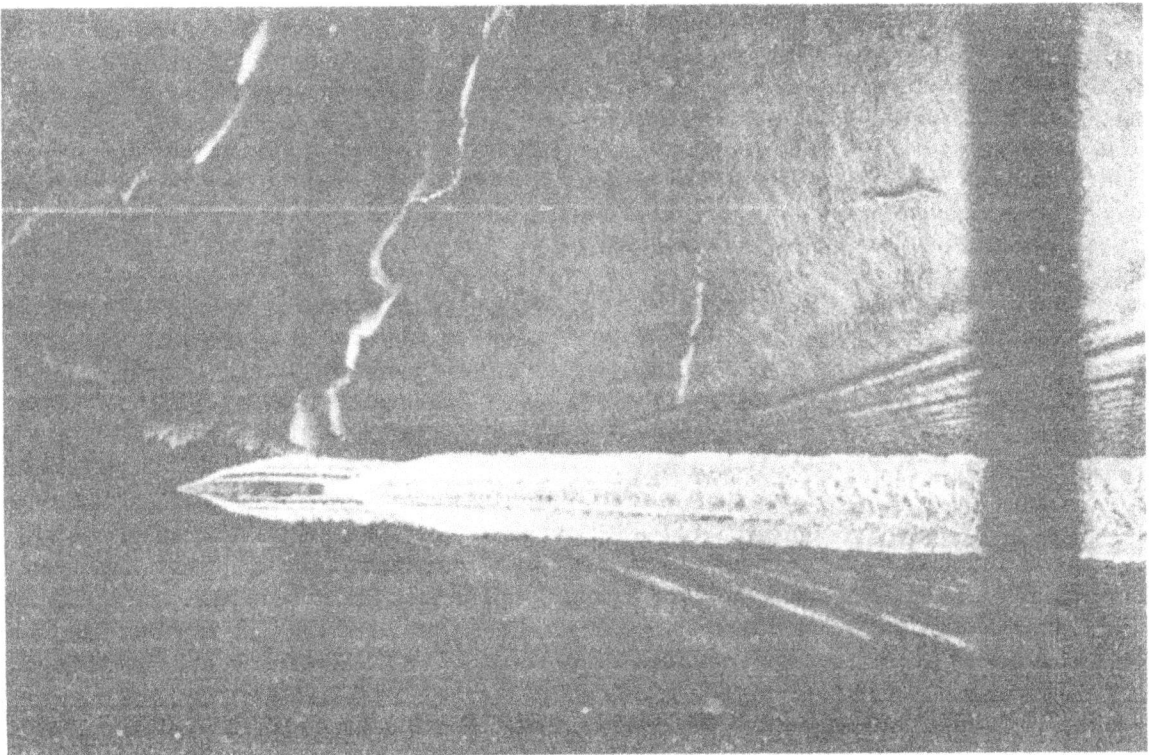

FIGURE 5.   Wake of USS *Moale* (DD 693) at 33 knots from 2,500 feet.

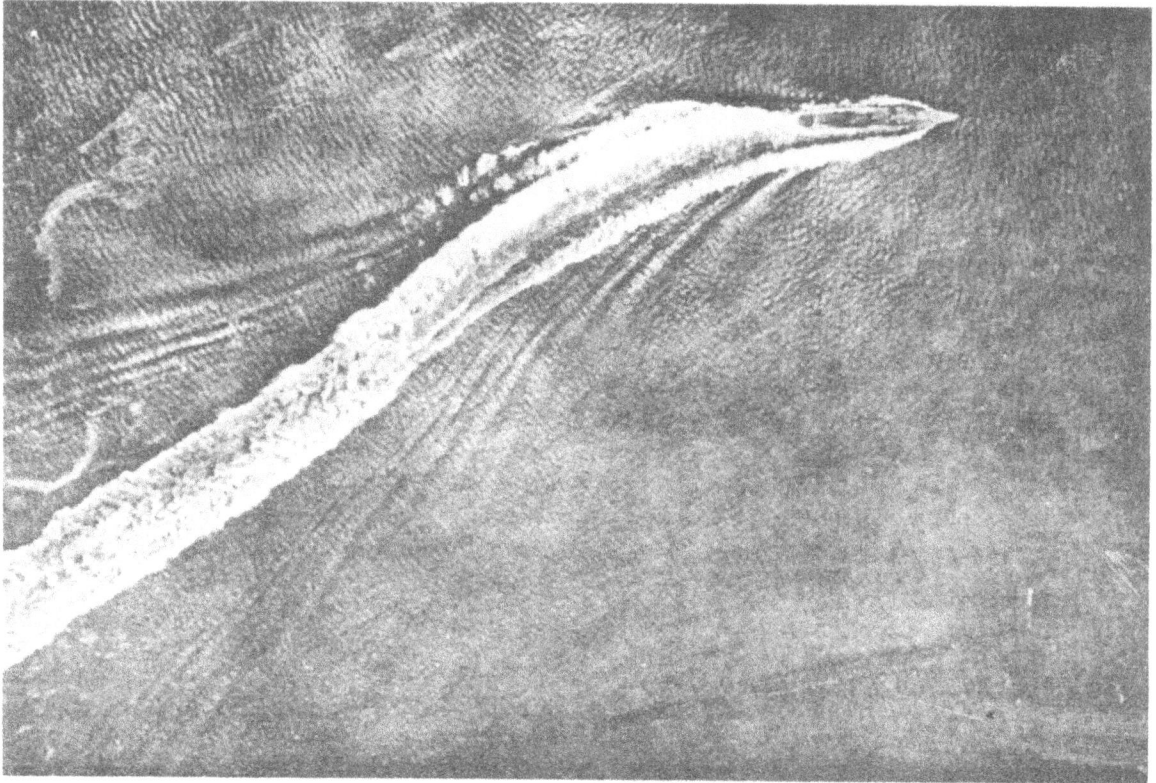

FIGURE 6.   Wake of USS *Moale* (DD 693) as ship turns at 30 knots.

time suggested that attacks on submarines could be made by detecting the wake and then following it until the submarine was reached. This suggested procedure turned out to be impractical, owing to the very low scattering power of the wakes behind slow, deep submarines. The wake laid by a surface vessel reflects sound so strongly and so persistently that acoustic methods might possibly be useful for attacks on such enemy vessels. Obviously a knowledge of the scattering and absorbing power of wakes at different ranges behind a vessel, and at different depths below the surface, would be very useful in the design of equipment' for such methods of attack.

26.3   ACOUSTIC WAKE RESEARCH

The aim of current wake studies is twofold: (1) to explore the overall acoustic properties of wakes with a view to possible tactical applications, and (2) to advance fundamental research on the structure and physical constitution of wakes. The second problem may seem rather academic to those who are primarily interested in the first one. But many questions about wakes presented by naval tactics cannot be answered satisfactorily, at present, for lack of a thorough understanding of the physical constitution of wakes. Thus in the long run, fundamental research is indispensable for developing a comprehensive doctrine of the use of wakes in naval warfare.

The solution of that fundamental problem in itself largely depends on acoustic measurements. Since wake research is still in an early stage, and since only incomplete observations are at hand, it would be impractical to insist upon strict separation of these two aims. Experimental data frequently are relevant from the point of view either of tactical applications or of fundamental research. Accordingly, a certain shift back and forth between practical and theoretical emphasis is unavoidable.

In order to plan, execute, and interpret acoustic measurements on wakes, some working hypothesis concerning the nature of acoustic wakes must be used as a starting point. Three physical explanations of the causes of scattering and absorption of sound in the sea have been suggested. The scattering and absorbing centers have tentatively been identified

FIGURE 7. Close-up of USS *Ringgold* (DD 500), from 300 feet.

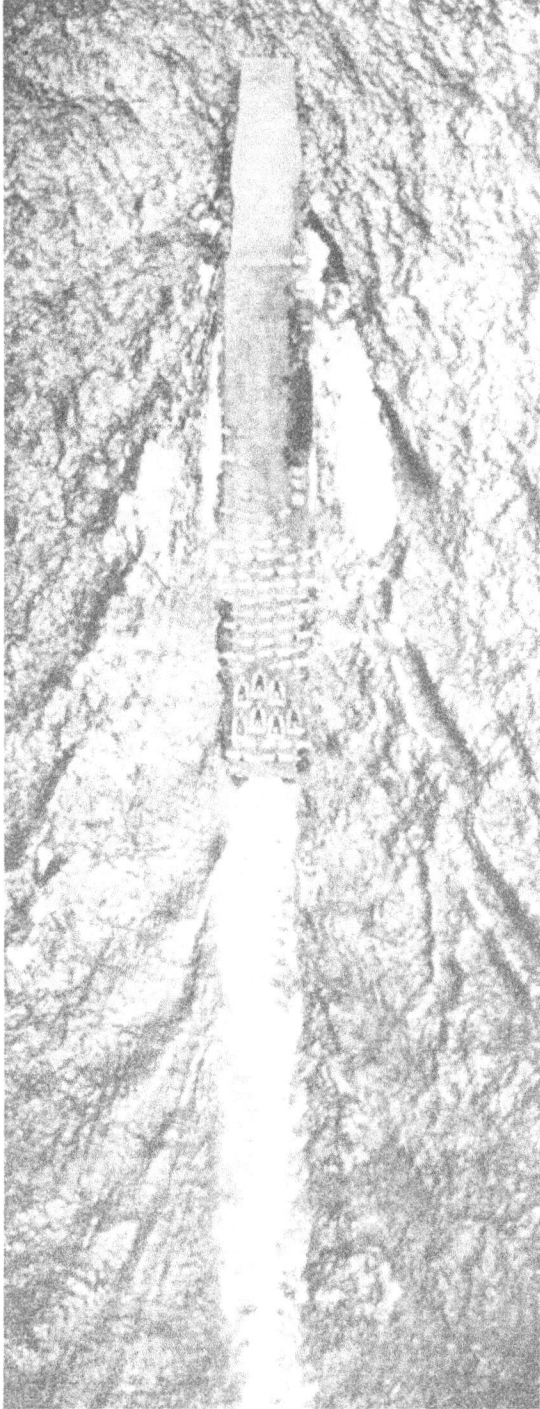

FIGURE 8.   Wake of USS *Saratoga* (CV3) from 4,000 feet.

with (1) air bubbles of widely varying size; (2) turbulent motion in the sea, on a scale small compared with the dimensions of ships; and (3) thermal inhomogeneities or irregularities in the sea, also on a small scale.

Although the bubble theory of acoustic wakes now enjoys general acceptance, it is difficult to put it to a conclusive test; it has been adopted rather by default of the other two explanations. It would seem logical, therefore, to begin by presenting the evidence which shows that the turbulent and thermal microstructure of the sea does not provide an adequate explanation of the acoustic properties of wakes. However, in order to simplify the exposition, it is preferable to discuss first the physical mechanism of the formation and dissolution of bubbles in Chapter 27 and their acoustic properties in Chapter 28, and to defer the necessarily rather cursory treatment of the temperature and velocity structure of the sea until Chapter 29. The theoretical Chapters 27 to 29 comprise the delineation of the working hypothesis which guides current wake research. Then the bulk of this volume (Chapters 30 to 33) describes the technique and the results of acoustic measurements made on wakes. In Chapter 34, the experimental data are interpreted in terms of the bubble theory; in other words, a test of the working hypothesis is undertaken. In the final Chapter 35, some conclusions which should be relevant in practice are drawn from the previous observations. Incomplete as the experimental foundations of some of these conclusions are, it appears useful to formulate some tentative generalizations as to the geometry and acoustic properties of wakes. Pending future research that may fill the conspicuous gaps in our knowledge of wakes, such generalizations should answer at least some of the questions about wakes raised by the demands of naval tactics.

Chapter 27

FORMATION AND DISSOLUTION OF AIR BUBBLES

AIR MAY BE ENTRAPPED mechanically at the ocean surface and dispersed in the form of bubbles; a familiar example is the appearance of white caps on a rough sea. A great deal of air is also trapped along the waterline of any vessel under way. Proof that such entrained air is capable of producing acoustic wakes comes from experiments on the wakes of sailing vessels. Probably the most copious source of bubbles in wakes, however, is propeller cavitation at high speeds.

27.1 FORMATION OF BUBBLES BY CAVITATION

When a cavity is created in water containing dissolved air, gas enters the cavity by diffusion, and when the cavity collapses, this gas remains behind as a bubble. The process of underwater formation of bubbles, therefore, involves two quite different phenomena: (1) the mechanics of cavitation, and (2) the thermodynamics of diffusion and solution of gases in liquids.

27.1.1 Mechanics of Propeller Cavitation

The phenomenon of propeller cavitation has long been known to engineers. According to hydrodynamical theory, cavities in liquids originate when certain patterns of flow produce regions of negative pressure near propellers. Such regions are set up in the vortices formed near the propeller tips, provided that the tip speed exceeds a certain critical limit, and also on the back side of the propeller blade. Hence, it is customary to speak of tip vortex cavitation and blade cavitation. These theoretical deductions have been verified experimentally by taking high-speed photographs of propellers running under water, shown in Figures 1, 2, and 3.

By driving a propeller in an experimental chamber and observing it through a window, the process of cavitation can be followed visually under stroboscopic illumination. When the speed of the propeller is gradually increased, bubbles are seen first to form at the propeller tips, from which they spiral backward in a long stream. Then bubbles begin to cover the part of the blade closest to the tips, forming a sheet on the blade. This phenomenon is sometimes described as *laminar cavitation*, in order to distinguish it from the formation of larger bubbles on the blade face nearer to the hub, called *burbling cavitation*, which starts at still higher speeds. Physically, there is no sharp distinction between laminar and burbling cavitation, and it would be more appropriate to classify them together as blade cavitation.

While persistent cavities are particularly likely to be formed in the tip vortices and on the propeller blades, cavitation also may be produced around sharp projections on the ship's hull, especially during periods of sharp acceleration of the ship. For instance, white foamy spots have been observed visually from a launch on the superstructure of a submerged submarine that passed at shallow depth. The appearance of the white spots did not suggest the release of a stream of entrapped air; hence, the spots were tentatively attributed to cavitation occurring on the superstructure.[1] This result cannot be regarded as general, since the submarine had not been submerged for a long enough time to justify assuming that all surface air entrained during the dive had been dislodged by the time of the observation.

27.1.2 Growing and Shrinking of Bubbles

After a cavity has been formed in sea water which is saturated with air at an external pressure of 1 atmosphere, gas begins to diffuse into the vacuum from the surrounding liquid. Since the diffusion constants for oxygen and nitrogen are nearly equal, the gas collecting in the cavity must have the same composition as that dissolved in the sea water. This com-

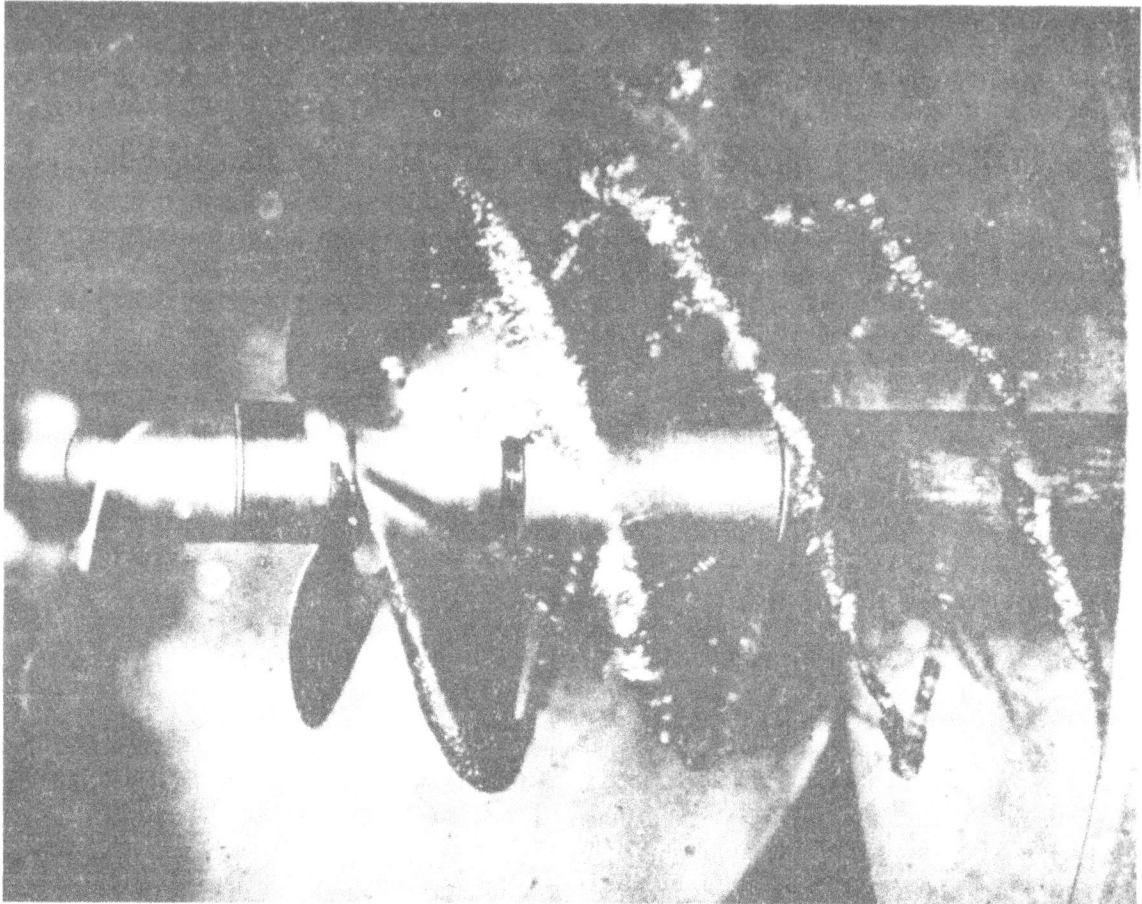

FIGURE 1. Cavitating model propeller. The picture was made with a 1/30,000-sec flash. Note the heavy tip vortices, considerable laminar cavitation near the blade tips, and the start of burbling cavitation of the blade face near the hub. This is a right-hand propeller and the water is flowing from left to right.

position differs markedly from that of atmospheric air because the solubility of nitrogen is twice that of oxygen. Accordingly, the cavitation gas consists of $\frac{1}{3}$ oxygen and $\frac{2}{3}$ nitrogen. The quantity of gas which collects each second in a cavity in moving water is proportional to the surface area of the cavity and to the partial pressure of air dissolved in the surrounding water but is essentially independent of temperature and hydrostatic pressure. The constant of proportionality is roughly 4×10^{-9} mole per sq cm per second per atmosphere.[2]

When the cavity collapses, the gas which has diffused into it will be compressed, and a bubble will be formed with a radius such that the gas pressure inside equals the hydrostatic pressure outside. The cavities formed by blade cavitation collapse so quickly that any air bubbles formed must be very small indeed. However, the cavities originating in the

tip vortices last much longer, since the centrifugal force in the whirling vortex remains high for some time. Thus, presumably it is the tip vortex cavitation that is primarily responsible for most of the air appearing as bubbles in propeller wakes. It has been observed that sea water at all depths contains dissolved oxygen and nitrogen in amounts roughly corresponding to saturation at the surface. For this reason it is undersaturated with respect to a bubble of air or cavitation gas anywhere below the surface, and a bubble of either gas will gradually disappear as the gas reenters the water. The rate of solution agrees with the same simple theory of diffusion as the rate of accumulation of gas in a cavity; indeed, the facts regarding the latter process are largely inferred from a study of the former. The number of moles of air which escape each second from a bubble is approximately proportional to the surface area of the

FIGURE 2. Cavitating model propeller. Picture made with a 1/30,000-sec flash. Shows heavy tip vortices extending down over leading edge and fairly wide area of blade covered by burbling cavitation. This is a right-hand propeller and the water is flowing from left to right.

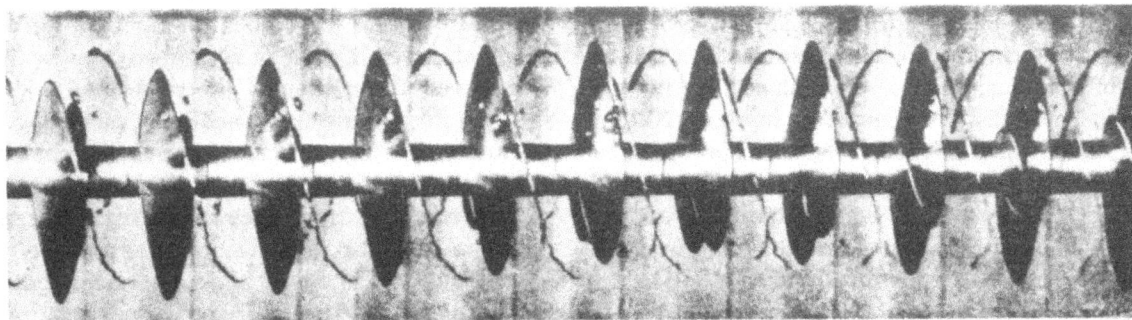

FIGURE 3. Cavitating model propeller. Short sections of the tip vortices are quite clear and the development, growth, progress, and disappearance of individual bubbles in the cavitation on the back of the upper blade can easily be followed. This is a right-hand propeller and the water is flowing from left to right.

bubble and to the difference between the pressure in the bubble and the partial pressure of air dissolved in the water. The constant of proportionality is again 4×10^{-9} mole per sq cm per second per atmosphere. An alternative formulation, assuming a spherical bubble, is in terms of the rate of decrease of the bubble diameter per second. In water saturated with air at the surface, this rate increases from 8×10^{-5} cm per sec at a depth of 5 meters to 18×10^{-5} cm per sec at a depth of 100 to 200 meters. Beyond these depths there is no further significant increase.

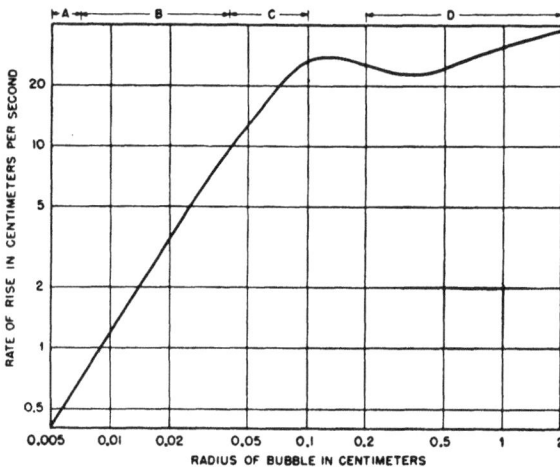

FIGURE 4.   Rate of rise of air bubbles in still water. A. Rectilinear motion, spherical shape.   B. Helical and twisting motion, flattened shape.   C. Irregular. D. Rectilinear motion, distorted mushroom shape.

These theoretical ideas concerning the formation and dissolution of bubbles have been tested in a series of simple experiments;[2] their agreement with the theory appears to be satisfactory. However, it remains uncertain to what extent these conclusions reached are applicable to the conditions prevailing in wakes. According to the experiments, a bubble 0.1 cm in radius, which is the resonant size for 3 kc sound, should dissolve completely in about 20 minutes. If the wake originally contains bubbles of all sizes up to 10^{-2} cm radius, then as the smaller bubbles contract, the larger bubbles also decrease in size; and some bubbles of the smallest size should be found 20 minutes after the formation of the wake. In rough agreement with theoretical expectations, acoustic effects of wakes at supersonic frequencies are observed to persist over periods from 15 to 45 minutes. In a wake, bubbles travel in a field of turbulent motion, rising gradually to the ocean surface where

they may disintegrate; this process constitutes another important factor limiting the lifetime of wakes. The next point to be considered, therefore, is the buoyancy and the rate of ascent of air bubbles in sea water.

27.2 BUOYANCY AND RATE OF ASCENT

The unimpeded rise of bubbles through still water has been analyzed in great detail.[3] From this analysis of all available experimental data and from certain theoretical considerations, a curve was constructed which gives the rate of rise of air bubbles in water as a function of the radius of the bubble and is reproduced in Figure 4. It will be noted that the velocity reaches a maximum at a radius of about 0.1 cm and varies only slightly with the radius thereafter. Several distinct types of motion and shapes of bubbles have been found to be characteristic in various ranges of bubble radii and are shown in Figure 4. No exact delineation of these radius intervals can, however, be made. All observers agree that for very small bubbles the motion is linear. For large bubbles the motion is also approximately linear, although some irregularities have been reported. A noteworthy feature of the velocity curve for radii up to 0.04 cm is that it coincides with the empirical curve for the rate of fall through water of spheres of specific gravity 2. In connection with the laboratory experiments on bubble screens,[4] which will be described in the next chapter, this relation between bubble radius and rate of rise has been tested empirically, and excellent agreement was found over a range of bubble radii from 0.01 to 0.1 cm. These rates of rise of bubbles in still water, as predicted from purely gravitational theory, would lead to the conclusion that all bubbles of acoustically effective size would reach the ocean surface in a time much shorter than the commonly observed lifetime of an acoustic wake.

However, the motion of the ship's hull and the action of its propellers continually set up throughout the wake a strongly turbulent internal motion, which interferes with the streaming of bubbles toward the surface resulting from their buoyancy. This phenomenon is analogous to the transportation of suspended material in rivers. Most suspended material is heavier than water and, therefore, would settle out in nonturbulent flow. But through turbulence this material is maintained in a state of suspension. Similarly, in a wake the bubbles rise toward the surface, while turbulence counteracts this tendency.

FIGURE 5. Bow wave, hull wake, and stern wake of USS *Idaho* (BB42).

FIGURE 6.   Underwater photograph of cavitation spot near bow of a PT boat traveling at 9.5 knots.

The analogy with transport in a river is not complete, since the turbulence at any fixed position in a wake dies out gradually and the bubbles, once they have reached the surface, are likely to disintegrate.

A semi-theoretical analysis of the lifetime of wakes has been presented which aims at finding precisely how much turbulence is needed in order to account for the observed ages of acoustic wakes.[5] In this work, the intensity of turbulence is measured by a certain empirical parameter, and it is shown that the theoretical lifetime of the wake passes through a broad but well-defined maximum if the turbulence parameter is increased steadily.

This theoretical maximum has a simple qualitative physical explanation. While weak or moderately strong turbulence tends to lengthen the lifetime of a wake, as pointed out before, a very large degree of turbulence will speed the decay of a wake by increasing the probability of the bubbles reaching the ocean surface and breaking up, namely, when the average value of the upward components of the turbulent motion exceeds the speed of the rise of bubbles with gravitational force alone. The existence of these opposing effects for very small and very large turbulence accounts for the maximum lifetime reached at some intermediate value of the turbulence parameter. The predicted maximum happens to agree with the average observed lifetime of acoustic wakes, which is from 15 to 45 minutes. Gratifying as this result is, there are not available any measurements of the intensity of turbulence in wakes, and hence the actual value of the turbulence parameter is unknown.

Moveover, should the observations necessary to specify the value of the turbulence parameter be made, the analysis [5] would require some modification before an exact comparison with the observed lifetime of wakes could be made. In particular, the concentration of bubbles at the ocean surface was assumed to vanish, according to the premise that the bubbles reaching the surface are immediately de-

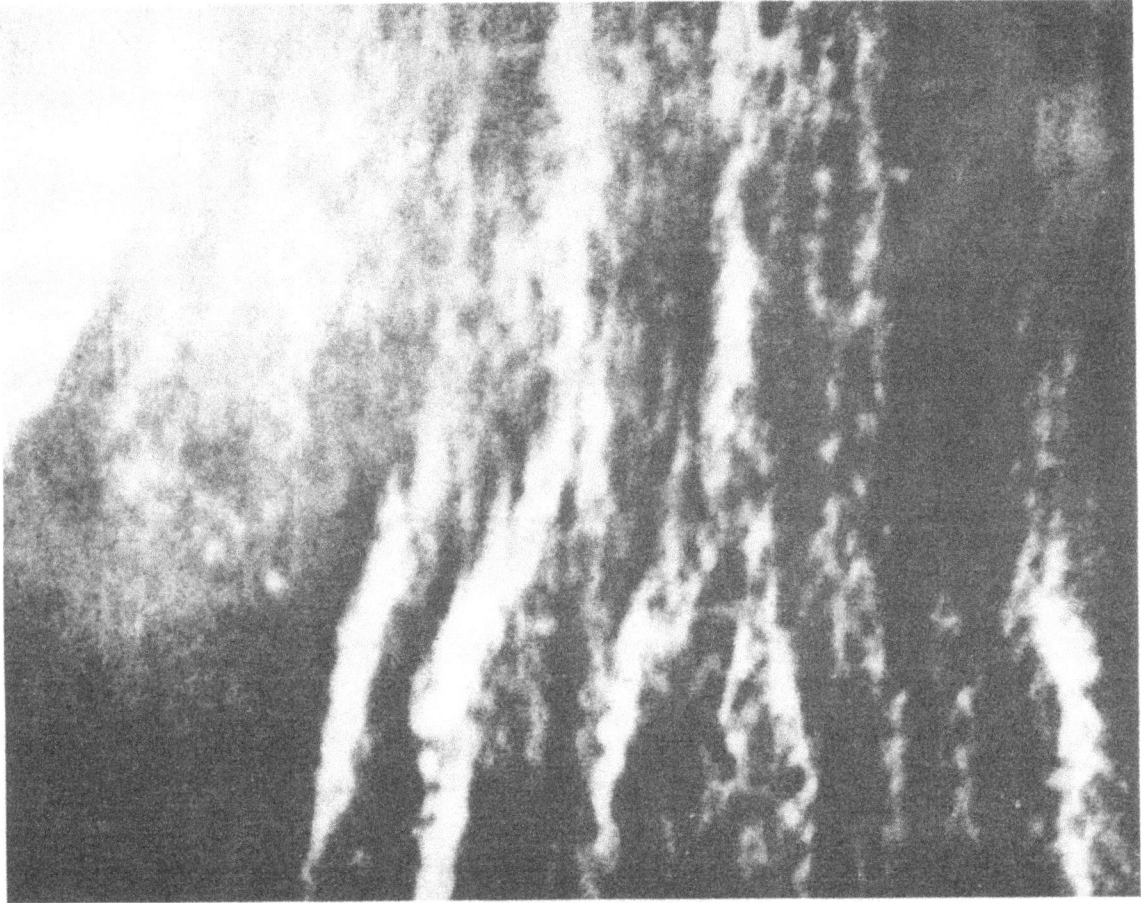

FIGURE 7. Underwater photograph of white water under hull of a PT boat traveling at 9.5 knots.

stroyed and thus removed from the ocean. Even granting the validity of this physical assumption, the removal of bubbles cannot be expressed mathematically by a vanishing bubble density. Inasmuch as the number of bubbles reaching the surface per unit time and per unit area equals the product of the bubble density and their average velocity upward, a vanishing bubble density implies a vanishing number of bubbles reaching the surface and thus does not correspond to the physical situation envisaged. In addition, the decay of turbulence as the wake ages may also have to be considered.

27.3   ENTRAINED AIR

The fact that sailing ships have a conspicuous wake suggests that a good deal of air is trapped along the waterline of any vessel under way. Such air might materially contribute to the mass of bubbles ap-

pearing in the wake of vessels propelled by engines. For instance, if Figure 5 could be relied on, the hull wake on the starboard side of the USS *Idaho* (BB42) would be even stronger than the stern wake. Of course, nothing is known about the extension in depth of the respective foam masses.

Qualitative tests [6] showed that echoes from the wake of a barge towed by a tug alongside could be detected with an NK-1 type shallow depth recorder ranging downward from a launch carried across the wake. However, it was found that this wake was more acoustically transparent than the wakes of vessels propelled by screws and therefore probably had a shorter lifetime.

These conclusions were confirmed by experiments in which sailing furnished the motive power. The ship used was a 104-ft yacht; measurements were made as described for other ships in Section 31.3. The wake when using sail was never found to be

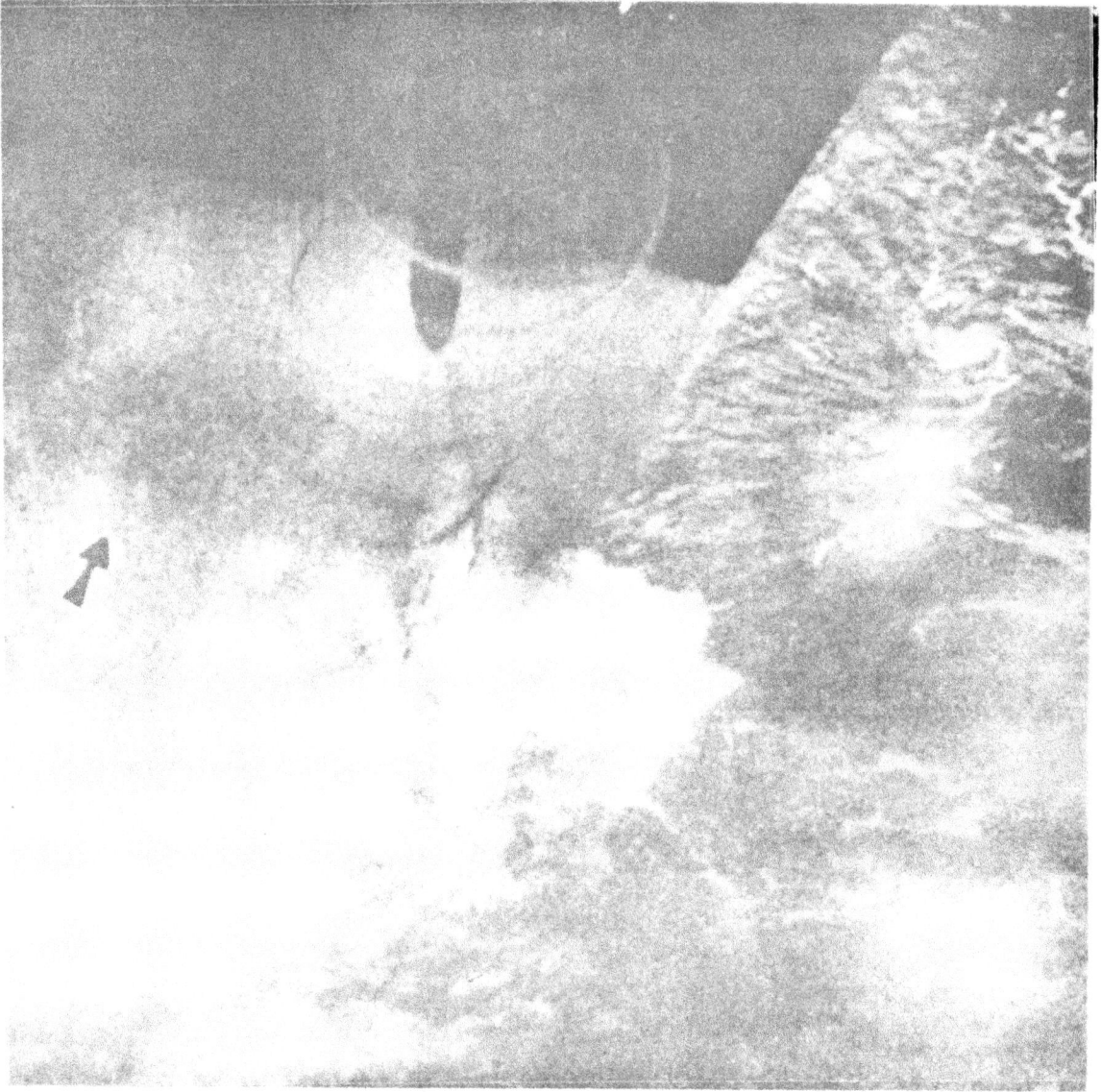

FIGURE 8.   Underwater view from port quarter of a PT boat traveling at 6 knots showing propeller cavitation.

acoustically opaque enough to blank out the bottom of San Diego Bay, where all the experiments were made. The wake thickness did not differ significantly from that observed in runs made with engines only, with the same vessel under comparable weather conditions; the average thickness was 12.4 ft with engines, and 13.0 ft under sail. As far as this scanty evidence goes, the geometric form of the wake seems to be determined primarily by the shape of the hull of the vessel and its speed, and it seems to be immaterial whether the bubbles are produced by entering surface air or by propeller cavitation.

A novel direct approach to the visual study of the subsurface structure of wakes has been made possible by the recent development of underwater motion pictures at the David Taylor Model Basin. This technique should also prove most useful for revealing the distribution of entrained air around the hull. For instance, when a small power boat passed with a speed of 2 to 3 knots over the underwater camera, mounted on the bottom in shallow water, the film shows a strongly foaming, shallow stern wake extending backward from the hull wake over a considerable distance. This wake did not reach down to

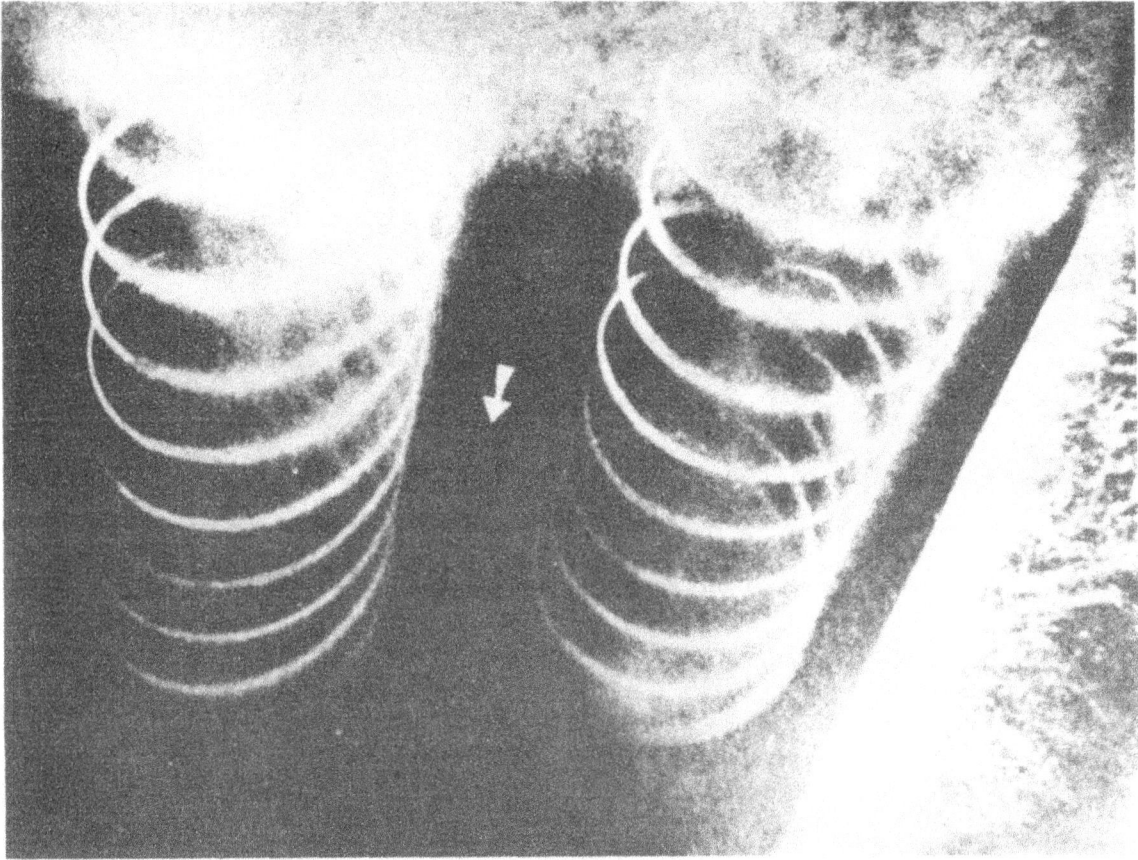

FIGURE 9. Underwater view from starboard quarter of a PT boat traveling at 19 knots showing propeller cavitation.

the depth of the screw of the launch. In fact, no stream of bubbles could be detected as emanating from the screw, which was clearly visible since the launch approached the camera as closely as 12 ft; presumably a speed of only 2 or 3 knots was insufficient to reach the cavitation limit.

Figures 6 to 11 are selected frames from an underwater motion picture showing a PT boat, outfitted with three screws, and its wake. These pictures were taken in water about 40 ft deep, near the Dry Tortugas; the choice of this location was dictated by the need for considerable optical transparency in the ocean. The motion picture camera was mounted, slanting upward, on a steel tower firmly anchored on the ocean bottom, and was operated by a diver. The distance from the camera to the ocean surface was about 15 ft.

At the left in Figure 6, a small amount of entrained air is visible along the water line. Moreover, a sharply outlined cavitation spot is conspicuous; unfortunately no attempt was made to ascertain what sort of unevenness on the hull caused this cavitation. In Figure 7 a large amount of entrained air is seen covering the hull. Both Figures 6 and 7 were made as the vessel traveled at a speed of 9.5 knots; there is no explanation of why the amount of entrained air differs so greatly in these two illustrations.

Figures 8 to 11 illustrate the progressive development of propeller cavitation as the speed increases. They furnish an instructive corollary to Figures 1, 2, and 3, and show that tip-vortex cavitation caused by the screws of a vessel under way at high speeds has the same appearance as that behind a laboratory propeller driven by a stream of moving water.

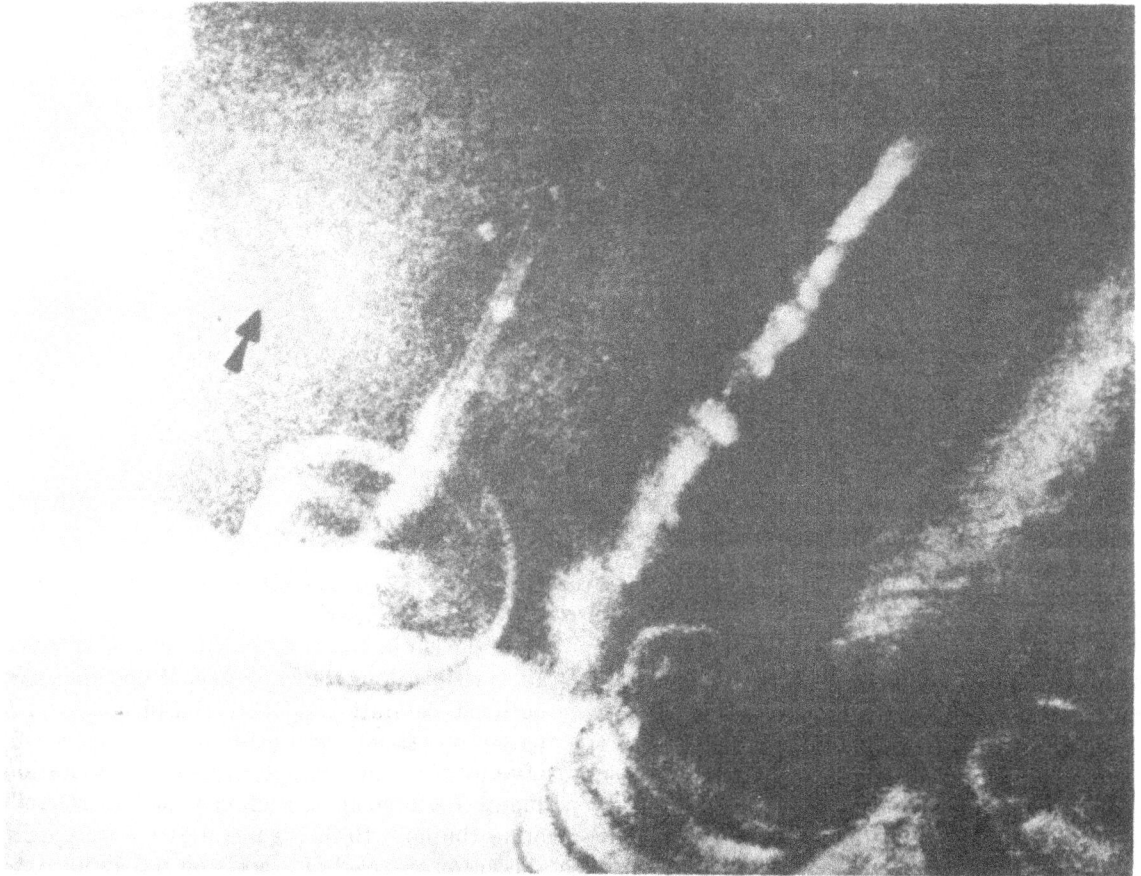

FIGURE 10. Underwater view from starboard quarter of a PT boat traveling at 27 knots showing propeller cavitation.

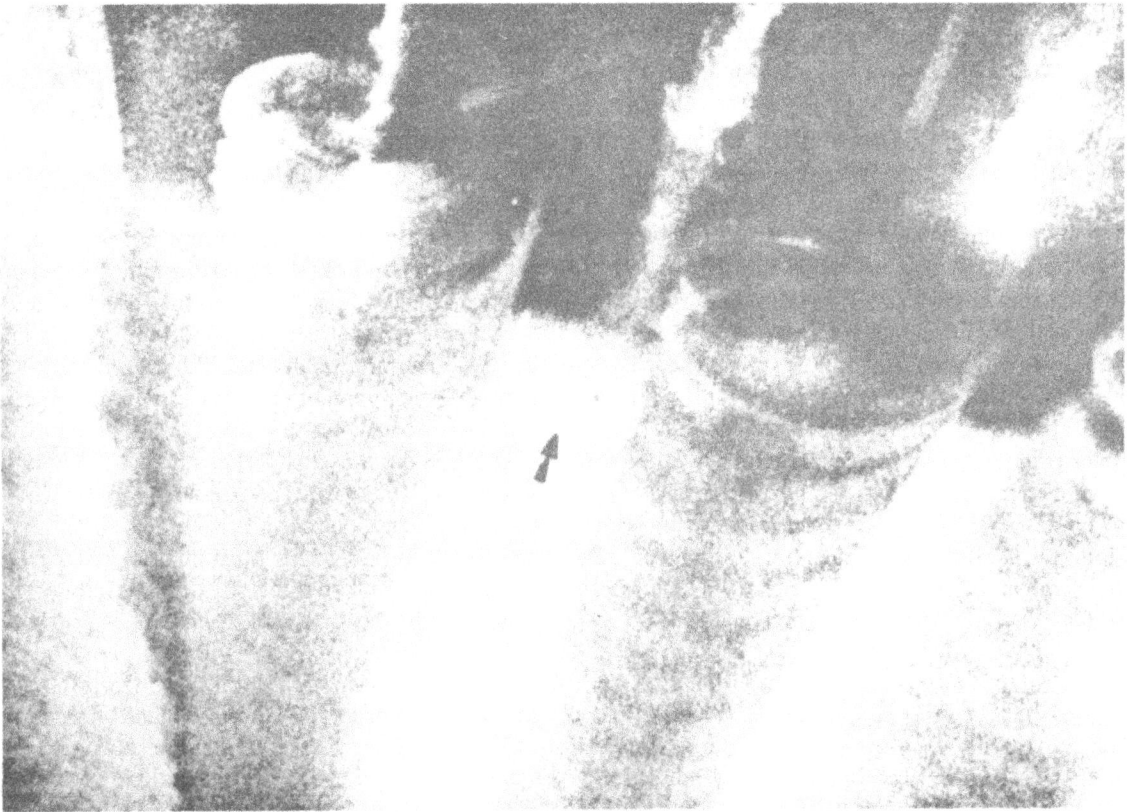

FIGURE 11.   Underwater view from astern of a PT boat traveling at 36 knots showing propeller cavitation.

Chapter 28

ACOUSTIC THEORY OF BUBBLES

THE RIGOROUS TREATMENT of the acoustic characteristics of bubbles, especially of the cumulative effects of a multitude of bubbles, requires a great deal of rather advanced mathematics. For a comprehensive exposition of these theories, reference must be made to several monographs on the subject.[1-4] In this chapter only the principal features of the problem will be sketched, primarily with a view to the later elementary interpretation of the acoustic properties of wakes in Chapter 34. Actual wakes have such a complicated structure that many physical and mathematical refinements incorporated in the rigorous treatment of certain ideal cases have, at present, only academic interest.

The first two sections of this chapter deal with the acoustic properties of individual bubbles. In the third section, the combined acoustic effects of many bubbles are discussed, and the results are applied to the evaluation, from laboratory experiments, of certain physical constants — acoustic cross sections, damping constants, which cannot as yet be predicted from pure theory.

28.1 SCATTERING BY A SINGLE IDEAL AIR BUBBLE

For application to wakes, only those air bubbles need be treated whose radius R is very small compared with the wavelength λ of sound in water, or

$$R \ll \frac{\lambda}{2\pi} \quad \text{or} \quad \eta = \frac{2\pi R}{\lambda} \ll 1, \qquad (1)$$

where η is the ratio of the bubble circumference to the wavelength. Then the pressure amplitude of the incident sound wave can be regarded as constant inside the bubble and in its immediate vicinity. Denoting this amplitude by A, the pressure P_0 of the incident wave can be described by

$$P_0 = A e^{2\pi i f t}. \qquad (2)$$

Although the effect of the sound wave impinging upon the air bubble is a rather complex one, the re-

sulting phenomena may be classified under two principal headings.

First, the periodically variable pressure of this incident sound wave produces a forced vibration of the air inside the bubble, which reacts on the surrounding water and produces in turn an emission of sound waves from the bubble. This secondary sound wave is spherically symmetrical for all practical purposes. Such a process of transforming an incident wave into waves of different pressure distributions is commonly known as *scattering*. The concept of scattering refers solely to the process of redistribution of sound energy — in other words, it is understood that none of the sound energy is converted into other forms of energy. Actually, scattering by a bubble is always accompanied by conversion of part of the impinging sound into heat by any one of a number of processes. These effects are described together under the name of *absorption* of sound. In the present section, scattering by a single bubble will be treated as though it were possible to produce this phenomenon apart from absorption; therefore, the term "ideal air bubble" has been used in the title of Section 28.1.

If the incident sound wave is a plane wave, the rms sound intensity I_0, or the average rate at which sound energy crosses a unit area placed perpendicular to the sound beam, is according to equation (56) in Chapter 2

$$I_0 = \frac{|A|^2}{2\rho c}, \qquad (3)$$

where A is the complex pressure amplitude, c is the velocity of sound in sea water, and ρ is the density of sea water. These waves excite vibrations of the air in the bubble and indirectly excite pressure waves in the surrounding water. In order to compute rigorously the possible types of vibration, the method of normal modes of vibration would have to be applied (see Section 27.1). It can be shown that of the various modes of vibration only the spherically symmetrical ones are significant in the present analysis, and that the other modes, corresponding to directional patterns of scattering, can be neglected.[1,2] This principal mode of spherical symmetry causes the *scattered* sound to be a spherical wave centered upon the bubble, which can be described by the formula,

$$p_s = \frac{B}{r} e^{2i(ft - r/\lambda)}, \qquad (4)$$

where p_s is the pressure of the scattered wave, r is the distance from the center of the bubble to the scattered wave, and B is the complex pressure amplitude of the scattered wave at unit distance; then B/r is the pressure amplitude at the distance r. The intensity of the scattered wave I_s at a distance r is then,

$$I_s = \frac{|B|^2}{2c\rho r^2}. \qquad (5)$$

The problem now is to calculate the intensity I_s of the divergent scattered wave from the intensity I_0 of the plane incident wave. At this point it is convenient to introduce the *cross section* σ_s of the bubble for scattering of sound, which is defined by

$$\sigma_s = \frac{4\pi |B|^2}{|A|^2}. \qquad (6)$$

The physical meaning of σ_s is simple. As the intensity of the scattered sound is given by equation (5), the total scattered energy at this distance r from the bubble center is $4\pi r^2 I_s$. Thus by combining equations (3), (5), and (6),

$$4\pi r^2 I_s = \sigma_s I_0. \qquad (7)$$

Hence, the sound energy flowing through an area σ_s perpendicular to the incident sound beam is equal to the total energy scattered by the bubble in all directions. The bubble itself exposes to the incident wave the cross-sectional area πR^2. If all energy intercepted by this area would be converted into scattered sound, the rate of scattering by the bubble, or the energy scattered per unit time, would be $\pi R^2 I_0$.

Evidently, whether the scattered energy is smaller or greater than the energy geometrically intercepted by the bubble will depend on the ratio $\sigma_s/\pi R^2$. The incident wave excites pulsations of the bubble, which are *forced vibrations* of the frequency f of the incident sound. They will interfere with *free vibrations* of the bubble at resonance, the frequency f_r of which will be computed presently. As is well known, the forced vibrations of any mechanical system become very intense if f is near the frequency f_r of the free vibrations characteristic of this particular system. In this case of resonance the scattered energy can become considerably greater than $\pi R^2 I_0$; thus the scattering

cross section σ_s may far exceed the geometric cross section πR^2 of the bubble.

In order to find σ_s, or $4\pi |B|^2/|A|^2$, the radial velocity of the bubble surface v_R will be computed in two different ways, following the treatment in a report by CUDWR.[5] On one hand, there is a hydrodynamical boundary condition which the air in the bubble has to satisfy during its vibrations, namely that the instantaneous gas pressure inside the bubble must be equal to the external acoustic pressure. On the other hand, there is a relation between pressure and volume (or pressure and radius) which the air in the bubble has to satisfy during its vibration, according to the principles of thermodynamics.

If the vibrations are so rapid that there is no heat exchange between the bubble and its surroundings, it may be assumed that the pulsations are *adiabatic* changes of state. For such changes, it is found from the first law of thermodynamics and from the equation of state of an ideal gas that PV^γ remains constant during the pulsations, where P is the air pressure inside the bubble, V the volume, and γ the ratio of the specific heats, which for air is 1.4. Denoting by P_0 the average hydrostatic pressure in the water, by V_0 and R_0 the volume and radius of the bubble in the state of equilibrium, the condition for adiabatic pulsations with small departures dV and dP from the equilibrium volume and pressure can be obtained by differentiating the relation $PV^\gamma = $ constant:

$$\frac{dP}{P_0} = -\gamma \frac{dV}{V_0}, \quad \frac{1}{P_0}\frac{dP}{dt} = -\frac{\gamma}{V_0}\frac{dV}{dt}. \qquad (8)$$

By expressing the volume of the bubble in terms of its radius R, it is found that

$$V_0 = \frac{4}{3}\pi R_0^3, \quad \frac{dV}{dt} = 4\pi R_0^2 \frac{dR}{dt}, \quad \frac{dR}{dt} = v_R. \quad (9)$$

If p_i denotes the acoustic pressure and A_i the pressure amplitude inside the bubble, the forced vibrations of the air in the bubble are described by

$$p_i = A_i e^{2\pi ift}, \quad \frac{dp_i}{dt} = 2\pi if A_i e^{2\pi ift}. \qquad (10)$$

This internal acoustic pressure p_i is to be identified with the excess gas pressure inside the bubble, dP, which appears in equation (8). Hence, by substitution of equations (9) and (10) into equation (8) it follows that

$$v_R = \frac{dR}{dt} = -\frac{2\pi if R_0 A_i}{3\gamma P_0} e^{2\pi ift}. \qquad (11)$$

The next step is to compute the amplitudes A_i

and B from the given pressure amplitude of A of the incident sound wave and from the hydrodynamical boundary conditions. These boundary conditions are formulated in Section 2.6.1. They require that the pressure p and the component v_R of the particle velocity normal to the surface have no discontinuity at the surface; the continuity of v_R is equivalent to the continuity of $(1/\rho)dp/dR$, which was required in Section 2.6.1.

While the pressure inside the bubble is given by equation (10), the outside pressure is the sum of the incident wave, expression (2), and the scattered wave, expression (4). The pressure of the scattered wave p_s at the surface of the bubble is found from equation (4), with r set equal to R_0. Because of assumption (1), namely that the linear size of the bubble is small compared with the wavelength λ, the term r/λ in the exponent of equation (4) is much smaller than unity in the vicinity of the bubble and, therefore, p_s and its derivative can be replaced approximately by

$$p_s = \left(\frac{B}{R_0} - \frac{2\pi i}{\lambda} B\right) e^{2\pi i f t}, \quad \frac{dp_s}{dr} = -\frac{B}{r^2} e^{2\pi i f t}. \quad (12)$$

Then the continuity of the pressure at the surface of the bubble is expressed by

$$p_i = p_0 + p_s.$$

If the expressions (2), (10), and (12) are substituted into this equation, and the common factor $e^{2\pi i f t}$ canceled, the following equation is found to apply at the bubble surface during the small oscillations usually encountered in practice:

$$A + \frac{B}{R_0} - \frac{2\pi i}{\lambda} B - A_i = 0. \quad (13)$$

The normal component of the velocity must also be continuous at the bubble surface. From equation (11) the normal component of the fluid velocity inside the bubble is known. Its value outside the bubble can be derived from equation (12). The relation between the fluid velocity and the pressure gradient $\partial p/\partial r$ in a certain direction is, according to the argument in Section 2.61,

$$\rho \frac{\partial v_R}{\partial t} = -\frac{\partial p}{\partial r}. \quad (14)$$

By substituting for $\partial p/\partial r$ the derivative of the pressure of the scattered wave from equation (12) with r set equal to R_0 and integrating over dt, equation (14) is transformed into

$$v_R = -\frac{Bi}{2\pi f \rho R_0^2} e^{2\pi i f t}. \quad (15)$$

To this equation the amplitude A of the incident wave does not make any contribution, because the wavelength is much larger than the size of the bubble; since the pressure gradient $\partial p_0/\partial x$ is uniform in the vicinity of the bubble, the velocity corresponding to this uniform pressure gradient is a motion of the entire bubble to and fro rather than an expansion and contraction of the bubble. The continuity of v_R can now be formulated, according to equations (11) and (15), by

$$\frac{B}{2\pi f \rho R_0^2} = \frac{2\pi f R_0 A_i}{3\gamma P_0}. \quad (16)$$

From equations (13) and (16) A_i can be eliminated and a relation between A and B obtained; if the subscript 0 is omitted from R_0, the bubble radius in equilibrium, then

$$B = \frac{AR}{\dfrac{3\gamma P_0}{4\pi^2 f^2 \rho R^2} - 1 + \dfrac{2\pi i R}{\lambda}}. \quad (17)$$

By introducing now the abbreviation f_r, defined by

$$2\pi f_r = \frac{1}{R}\sqrt{\frac{3\gamma P_0}{\rho}}, \quad (18)$$

the physical meaning of which will soon become apparent, equations (17), and (18) and (1) may be combined to give the result

$$B = \frac{RA}{\left(\dfrac{f_r}{f}\right)^2 - 1. + \eta i}. \quad (19)$$

In order to obtain the scattering cross section σ_s of the bubble from equation (6), $|A|^2$ and $|B|^2$ have to be computed from

$$|A|^2 = AA^*, |B|^2 = BB^*, \quad (20)$$

where A^* and B^* are the complex conjugates of A and B respectively. According to equation (19),

$$B^* = \frac{RA^*}{\left(\dfrac{f_r}{f}\right)^2 - 1 - \eta i}. \quad (21)$$

Finally, from equations (6), (19), (20), and (21),

$$\sigma_s = \frac{4\pi R^2}{\left(\dfrac{f_r^2}{f^2} - 1\right)^2 + \eta^2}. \quad (22)$$

For a bubble of a definite radius R, the scattering cross section σ_s has its peak value if f equals f_r; it is then said that the incoming wave is in *resonance* with the pulsations of the bubble, and hence f_r is

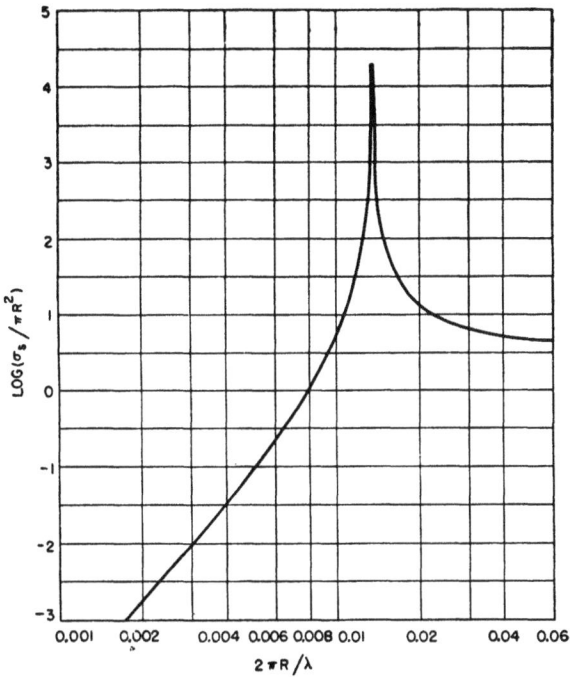

FIGURE 1. Scattering cross section for an ideal bubble.

called the *resonance frequency* for the bubble of radius R.

A plot of $\sigma_s/\pi R^2$ as the function of $\eta = 2\pi R/\lambda = 2\pi R f/c$ is shown in Figure 1, the outstanding feature of which is a sharp peak. This maximum corresponds to the resonance value η_r or according to equation (18)

$$\eta_r = \frac{2\pi R f_r}{c} = \frac{1}{c}\sqrt{\frac{3\gamma P_0}{\rho}} = 1.36 \times 10^{-2}, \quad (23)$$

if P_0 is atmospheric pressure and c is the sound velocity in sea water at 60 F. Thus, at resonance, σ_s is enormously greater than the geometric cross section of the bubble; specifically

$$\frac{\sigma_{sr}}{\pi R^2} = \left(\frac{2}{\eta_r}\right)^2 = 2.16 \times 10^4, \quad (24)$$

where σ_{sr} is the value of σ_s at resonance. Equation (24) can also be expressed in the form

$$\sigma_{sr} = \frac{\lambda^2}{\pi}. \quad (25)$$

While equations (24) and (25) must be considerably modified for an actual bubble, as shown in the next section, the phenomenon of resonance is nevertheless responsible for the great efficiency of bubbles as scattering agents. Moreover, the resonant frequency found from equation (18) is correct for a wide spread

of bubble sizes. This equation has been confirmed by observations at low frequencies, between 1,000 and 6,000 c per sec, [6,7] and also at high supersonic frequencies, between 20 and 35 kc.[7] In each case a single bubble was placed in the sound field, and the sound frequency determined at which the bubble oscillated most violently. The radius of the bubble was then measured either with a microscope, or for the larger bubbles by measurement of the volume of air in the bubble. The values of the resonant frequency f_r found in these measurements for bubbles of air, hydrogen, and oxygen in water at different temperatures agreed with equation (18) within the experimental error of about 5 per cent. Thus within the range from 1 to 50 kc equation (18) may safely be used to predict the resonant radius of bubbles in water. Values computed from this equation are given in Table 1.

TABLE 1. Resonant radius for air bubbles in water.

Frequency in kc		1	5	20	50
Wavelength in centimeters		150	30	7.5	3
Pressure					
Atmospheres	Depth of water in feet				
1	Surface	0.33	0.065	0.016	0.006
2	35	0.47	0.093	0.023	0.009
5	140	0.73	0.15	0.037	0.015
10	300	1.04	0.21	0.052	0.021

For very small bubbles, with radii less than 10^{-3} cm, surface tension becomes important and the compressions and expansions of the gas in the bubble become isothermal instead of adiabatic. No observations for such small bubbles are available, but a theoretical analysis [5] shows that equation (22) is still valid provided that f_r is defined by the equation

$$2\pi f_r = \frac{1}{R}\sqrt{\frac{3gP_0}{\rho}}, \quad (26)$$

where

$$g = 1 + \frac{4T}{3RP_0}; \quad (27)$$

the quantity T is the surface tension of the gas-liquid surface, and other quantities have the same meaning as in equation (18). Equations (26) and (27) should not be used for bubbles of radii greater than 10^{-3} cm.

Equation (22), in addition to predicting the importance of resonance, also gives correctly the scattering coefficient for frequencies considerably greater than the resonant frequency f_r. Since η is less than one, η^2 in equation (22) may be neglected when the ratio f_r/f is much greater than one. Consequently, the scattering cross section for low-frequency sound may be written approximately as

$$\sigma_s = 4\pi R^2 \left(\frac{f}{f_r}\right)^4 = 4\pi R^2 \left(\frac{\lambda_r}{\lambda}\right)^4 . \quad (28)$$

This equation is known as Rayleigh's law of scattering for long-wave radiation. It will be remembered that in optics Rayleigh's law explained the blue color of the sky, as the resonant frequencies characteristic of the atmospheric gases oxygen and nitrogen are far greater than the frequencies of visible light. Hence, the shorter (blue) waves of sunlight are scattered more strongly than the longer (red) waves and reach our eyes with greater intensity. Equation (28) is also applicable to the high-frequency sound commonly used in echo ranging provided that the bubble radius R is very small; if R is less than 10^{-3} cm, however, f_r is given by equation (26) instead of by equation (18).

28.2 SCATTERING AND ABSORPTION BY AN ACTUAL BUBBLE

So far, the attenuation of sound resulting from the absorption of sound energy during the pulsation of the bubble has been neglected. The existence of such an effect is a direct consequence of the second law of thermodynamics, which implies that energy must be extracted from the sound field and dissipated into the surrounding water in the form of heat, in order to maintain the forced pulsation of the bubble against the internal friction of the bubble-water system. In other words, it is thermodynamically inadmissible to treat the pulsation of the bubble as if it were a strictly adiabatic process; therefore it becomes necessary to amend the analysis given in the preceding section for an ideal bubble.

This task is accomplished by adding to equation (13), which expressed the continuity of pressure at the bubble surface, a certain term which takes into account the frictional force modifying the behavior of an actual bubble. Moreover, the exchange of heat between bubble and water by conduction necessitates a modification of·equation (16), which formulated the continuity of velocity at the bubble surface. The treatment of the case of the ideal bubble implicitly assumed that the pulsations are thermodynamically reversible; that is, the work put into the

bubble during compression was supposed to be equal to the work done by the bubble during expansion. Actually, there is heat exchange between bubble and water, but the pulsations are too rapid to permit a complete leveling of temperature at every instant of the cycle. Thus there prevails a continual change of state which is somewhere between the adiabatic and isothermal case.

It is not difficult to see that under such circumstances the pulsation of pressure cannot be in phase with the pulsation of volume. While the bubble is being compressed, the temperature rises steadily; as soon as the rise of temperature becomes appreciable, heat conduction begins to operate and the bubble tends to cool off even before expansion has started. When the minimum volume is reached, the temperature will be decreasing as heat flows from the bubble into the water. Consequently, the temperature maximum will be reached some time before the bubble has been compressed to its minimum volume. Likewise, since the gas pressure is proportional to the temperature, the maximum pressure will not be attained simultaneously with the minimum volume, but some time before. Thus there exists a phase shift between pressure and temperature on one hand, and volume and radial velocity of the bubble on the other hand. For resonant bubbles at frequencies of 100 kc or less, this effect is taken into account[4] by inserting a complex factor $1 - \beta i$ in the right-hand side of equation (11), where β is a positive constant much smaller than one.

The two equations of continuity, (13) and (16), must therefore be replaced, for an actual bubble, by the following ones:

$$p_0 + p_s - p_i = -C_1 \frac{dR}{dt} , \quad (29)$$

or

$$A + B/R_0 - 2\pi i k B - A_i = \frac{C_1 B_i}{2\pi f \rho R_0^2} ,$$

and

$$\frac{B}{2\pi f \rho R_0^2} = \frac{2\pi f R_0 A_i}{3\gamma P_0} (1 - \beta i) . \quad (30)$$

In equation (29) C_1 is a constant measuring the effect of friction, which is assumed to be proportional to the radial velocity dR/dt of the bubble. The term $C_1 dR/dt$ represents the net pressure on the bubble, which is positive when the bubble is contracting $(dR/dt < 0)$; hence, the correction term appearing on the right side of equation (29) must carry a minus sign.

By proceeding exactly as in Section 28.1, the following relation is found instead of equation (19):

$$B = \frac{RA}{\left(\dfrac{f_r^2}{f^2} \cdot \dfrac{1}{1+\beta^2} - 1\right) + i\left(\dfrac{f_r^2}{f^2} \cdot \dfrac{\beta}{1+\beta^2} + \eta + \dfrac{C_1}{c\rho\eta}\right)} \cdot \quad (31)$$

If one neglects β^2 compared to one and defines

$$\delta = \frac{f_r^2}{f^2}\beta + \eta + \frac{C_1}{c\rho\eta}, \quad (32)$$

equation (31) becomes

$$B = \frac{RA}{\left(\dfrac{f_r^2}{f^2} - 1\right) + i\delta(f,R_0)} \cdot \quad (33)$$

Substituting this expression into equation (6), the cross section for scattering by an actual bubble can readily be evaluated:

$$\sigma_s = \frac{4\pi R^2}{\left(\dfrac{f_r^2}{f^2} - 1\right)^2 + \delta^2} \cdot \quad (34)$$

It will be noted that equation (34) is identical with equation (22), which was derived for an ideal bubble, except that δ^2 has replaced η^2 in the denominator. This change affects only the magnitude of the scattering cross section near resonance. Thus the frequency of resonance and the scattering cross section at frequencies far from resonance are correctly given by equation (22), in agreement with the statements made in the previous section.

The knowledge of the scattering cross section does not provide all the information that is wanted in the case of an actual bubble, as the incident flux of energy is reduced both by scattering and absorption of sound. Calling the sum of scattered and absorbed energy the extinguished energy, an *extinction cross section* σ_e can be defined by

$$\sigma_e = \frac{F_e}{I_0}, \quad (35)$$

where F_e is the total energy extinguished by the bubble per unit interval of time and I_0 is the intensity of the incident sound energy. The quantity F_e is equal to the work done, per unit interval of time, on the bubble by the incoming sound beam; this extinguished energy comprises both absorbed and scattered energy. Hence, F_e is equal to

$$F_e = \overline{p_0 \frac{dV}{dt}}, \quad (36)$$

where the bar means the *time average*; p_0 is the pressure of the incident sound wave, and V is the volume of the bubble.

To evaluate equation (36) it is simplest to use real quantities. According to equation (2),

$$p_0 = Ae^{2\pi ift}.$$

Since the initial phase may be chosen arbitrarily, let A be real, and let the sound pressure and sound velocity be represented by the real parts of the expressions developed above. Then

$$p_0 = A\cos 2\pi ft. \quad (37)$$

From equations (9) and (15) it follows that

$$\frac{dV}{dt} = 4\pi R_0^2 v_R = -\frac{2iB}{f\rho}e^{2\pi ift}.$$

Here again only the real part of the entire expression is to be taken. In order to find this real part, split B into its real part B^R and its imaginary part iB^I, and express $e^{2\pi ift}$ in terms of its real and imaginary parts:

$$\frac{dV}{dt} = \frac{2i}{-f\rho}(B^R + iB^I)(\cos 2\pi ft + i\sin 2\pi ft)$$

$$= \frac{2}{-f\rho}[(B^I\cos 2\pi ft + B^R\sin 2\pi ft) \quad (38)$$
$$+ i(B^I\sin 2\pi ft - B^R\cos 2\pi ft)].$$

If equation (37) and the corresponding real part of equation (38) are substituted into equation (36), it is found that

$$F_e = -\frac{2A}{f\rho}\overline{(\cos 2\pi ft)(B^I\cos 2\pi ft + B^R\sin 2\pi ft)}, \quad (39).$$

where the bar denotes an average over many cycles

Since

$$\overline{\cos^2 (2\pi ft)} = \frac{1}{2}$$

and

$$\overline{(\cos 2\pi ft)(\sin 2\pi ft)} = 0,$$

equation (39) becomes finally

$$F_e = -\frac{AB^I}{f\rho} \cdot \quad (40)$$

According to equation (33),

$$B^I = \frac{RA\delta}{\left(\dfrac{f_r^2}{f^2} - 1\right)^2 + \delta^2} \cdot \quad (41)$$

Hence, equation (40) assumes the form

$$F_e = \frac{RA^2\delta}{\left(\dfrac{f_r^2}{f^2} - 1\right)^2 + \delta^2}\frac{1}{f\rho} \cdot \quad (42)$$

By combining this expression with equations (3) and (32), the cross section for extinction is finally obtained:

$$\sigma_e = \frac{4\pi R^2 \left(\dfrac{\delta}{\eta}\right)}{\left(\dfrac{f_r^2}{f^2} - 1\right)^2 + \delta^2} \cdot \qquad (43)$$

The extinguished energy is obviously the sum of the scattered and absorbed energy. Therefore, the *absorption cross section* σ_a of the actual bubble can be defined by the relation

$$\sigma_e = \sigma_s + \sigma_a \qquad (44)$$

and is thus found to be, from equations (34) and (43),

$$\sigma_a = \frac{4\pi R^2 \left(\dfrac{\delta}{\eta} - 1\right)}{\left(\dfrac{f_r^2}{f^2} - 1\right)^2 + \delta^2} \cdot \qquad (45)$$

Note also the simple relation

$$\frac{\sigma_s}{\sigma_e} = \frac{\eta}{\delta} \cdot \qquad (46)$$

A word must be said now about the function δ, defined in equation (32). If β and C_1 are put equal to zero, for the case of an ideal bubble, δ reduces to η, and it is seen that equation (22) is indeed the correct limiting form of equation (34). Numerical values of β and C_1 can be derived by an analysis of the several physical processes known to contribute to the absorption of sound by the bubble — for instance, heat conduction, viscosity, surface tension, and other processes. There are also methods for determining δ empirically from certain observations which will be discussed. Inspection of Figure 2, which shows the damping constant at resonance as a function of frequency, will reveal that the predicted values of δ are much smaller than the observed ones. This discrepancy indicates that some relevant physical processes must have been overlooked in the theoretical analysis of the absorption effects. Hence, theoretical evaluation of β and C_1, although carried out elsewhere,[5] will be omitted from this review, and the empirical values of δ will be used for the interpretation of the acoustic properties of wakes to be given in Chapter 34.

The physical significance of δ can best be visualized by plotting $\sigma_s/4\pi R^2$ against f/f_r. A resonance curve, similar to Figure 1, is thus obtained. The peak value of this graph is, according to equation (34),

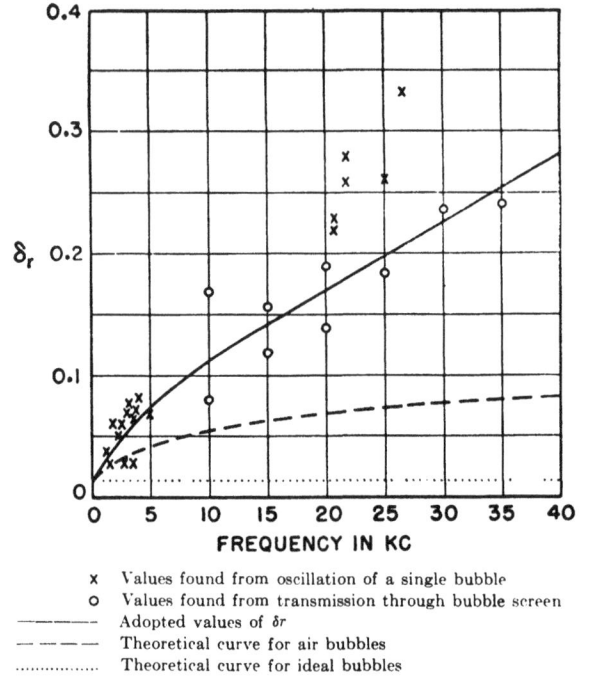

FIGURE 2. Damping constant at resonance.

x   Values found from oscillation of a single bubble
o   Values found from transmission through bubble screen
———   Adopted values of δr
— — —   Theoretical curve for air bubbles
....   Theoretical curve for ideal bubbles

$$\frac{\sigma_{sr}}{4\pi R^2} = \frac{1}{\delta_r^2} \qquad (47)$$

where δ_r is the resonance value of δ shown in Figure 2. If $\sigma_s(f)$ denotes the cross section for any non-resonant frequency, it follows from equations (40) and (47) that

$$\frac{\sigma_s(f)}{\sigma_{sr}} = \frac{\delta_r^2}{\left(\dfrac{f_r^2}{f^2} - 1\right)^2 + \delta^2(f,R)} \cdot \qquad (48)$$

Over a narrow range of frequencies near the peak of the resonance curve, $\delta(f,R)$ in the denominator of equation (48) may be replaced by its resonance value δ_r, and f_r/f is very close to one. Hence, using the abbreviation $q = f_r/f - 1$,

$$\left(\frac{f_r^2}{f^2} - 1\right)^2 = q^2(q+2)^2 = 4q^2 , \qquad (49)$$

and equation (48) becomes approximately

$$\frac{\sigma_s(f)}{\sigma_{sr}} = \frac{1}{1 + \dfrac{4q^2}{\delta_r^2}} ; \qquad (50)$$

in other words, for any given small departure q from the resonance frequency, the decline of σ_s from its peak value is sharper for greater values of $1/\delta_r$ or for smaller values of δ_r itself. The greater the sharpness of the resonance peak is, the smaller is the damping of the pulsation of the bubble. Therefore, δ_r is commonly called the *damping constant*.

28.2.1 Measurement of Damping Constant

The simplest, most direct way to determine the damping constant δ_r is to measure the sharpness of the resonant peak for a bubble in a sound field. Such measurements have been carried out for bubbles in fresh water. In one case,[7] the amplitude of oscillation of a single bubble was observed as the sound frequency was slowly varied. Since σ_s is proportional to the square of the amplitude of oscillation, a plot of these observations yields δ_r directly. Values of δ_r were found by this method for bubbles of hydrogen and bubbles of oxygen, but no systematic difference was found between these two gases.

In another case, the transmission loss through a screen of bubbles all of the same size was observed.[8] To produce this screen, six small *microdispersers* arranged in a line in a laboratory tank were used to produce a stream of bubbles 10 ft below the surface of the water. These bubbles were normally intercepted by a hood, which could, however, be swung to one side for about one second to allow a pulse of bubbles to rise to the surface. Since the larger bubbles arrived at the surface first, and the smaller ones at progressively later times, the bubbles near the surface at any one time were of nearly equal radii. The transmission loss in decibels of sound at a constant frequency crossing this screen was then proportional to σ_e for a single bubble; from a plot of the transmission loss against bubble radius, a value of δ_r could then be determined. A typical set of observed curves obtained with this technique is reproduced in Figure 4. In analyzing these data, account was taken of the variation of δ with bubble radius so that points some distance from resonance could be used as well as those close to resonance.

The values of δ_r found by these two methods are plotted in Figure 2. The dashed line curve shows the theoretical value of δ_r for air bubbles in water, if C_1 is set equal to zero, and values of β are taken from reference 5. It is evident that at the higher frequencies the observed values are much greater than the theoretical values; this discrepancy has already been noted above. The values of δ_r found from a single bubble, which are shown as crosses, are somewhat greater than those determined from the transmission loss of sound through a bubble screen, plotted as circles in Figure 2. In the former set of measurements, the bubble was not free, but was caught on a small wax sphere fastened to a platinum thread, which oscillated to and fro as the bubble expanded and contracted. Since the damping constant may have been increased in this arrangement over its value for a free bubble, these values cannot be relied upon. Thus, the solid line of best fit shown in Figure 2 is based at high frequencies on the values found with the screen of freely rising bubbles. Confirmation of these observed values of δ_r is found in the next section, where the observed data on scattering and absorption of sound by bubble screens are shown to be in moderately good agreement with the theoretical values based on equations (34) and (43) and on the empirical curve of δ_r in Figure 2. For comparison with the observed values, the damping constant δ_r computed for an ideal bubble resonating in water at atmospheric pressure is shown as a dashed line in the figure; the value plotted is taken from equation (23).

28.3 SOUND PROPAGATION IN A LIQUID CONTAINING MANY BUBBLES

The results derived in the preceding sections for a single bubble are only the first step toward the solution of the general problem, the propagation of sound through a medium containing many bubbles. This problem is complicated because the external pressure affecting each bubble is the sum of the pressure in the incident sound wave and the pressures of the sound waves from all the other bubbles. While the mathematics of the problem is complicated, the general results to be anticipated can be presented simply.

28.3.1 General Theory

First, the presence of the bubbles will affect the nature of the medium through which the sound wave is progressing. If the bubbles are spaced much closer to each other than the wavelength, the sound velocity will be appreciably affected by the presence of the bubbles, which alters the compressibility of the medium. In addition, the sound velocity will have a small imaginary part, resulting from the absorption and scattering of sound, and giving rise to an exponential drop of the sound intensity with increasing distance of travel through the aerated water. Thus a sound wave can be reflected, refracted, and attenuated as it passes through water containing bubbles.

On this picture the sound wave behaves as though it were proceeding through a homogeneous medium, in which the sound velocity is a smooth complex function of position.

Secondly, this picture must be supplemented to take scattering into account. The sound waves sent out from the different bubbles produce scattered sound, which goes out in all directions. This scattered radiation may be regarded as resulting from the fact that in a random collection of point scatterers the number of scatterings per unit volume is never constant from one region to another, but shows statistical fluctuations. A theory of the scattering of light in air is given along these lines in a well-known text on statistical mechanics.[9] More simply, the intensity of scattered radiation may be regarded as proportional to the average squared pressure resulting from all the individual bubbles. As the bubbles move around, the relative phases of their scattered wavelets will vary widely, and constructive and destructive interference will be equally likely. With this picture, the average squared pressure may be regarded as simply the sum of the squares of the pressures in each of the scattered wavelets.

In ship wakes the number of bubbles in a small volume is rarely sufficiently great to produce reflection and refraction of sound waves. The gradual attenuation of the incident sound beam and the scattering of sound energy in all directions by each bubble individually are therefore the two effects of greatest interest.

The preceding discussion is, of course, not very rigorous. The results stated here have been proved rather generally, however, in an elegant solution to the general problem.[3,4] This analysis makes certain assumptions, the most important of which are that the bubbles have diameters much smaller than the wavelength of the incident sound, and that the average distance between bubbles is much larger than their dimensions. The solution, as a result of its physical generality, is of considerable mathematical complexity, and therefore will not be reproduced here. But the mode of approach used in this general theory will be briefly sketched.

The chief feature of this theory is its use of *configurational averages*. Different bubbles may be almost anywhere within a certain region. For each distribution of bubbles the sound pressure p at a given time will have some definite value. If now an average value of this pressure is taken for all possible positions of the different bubbles, a configurational average of p,

denoted by $<p>$, results. Thus is usually not equal to the time average of p, since this time average vanishes because of the oscillations of p between positive and negative values. Similarly, $<p^2>$ may be defined as the configurational average of p^2.

The simplified picture presented at the beginning of this section may be given a precise meaning in terms of these configurational averages. The quantity $<p>$ is found to obey the wave equation in a homogeneous medium in which the complex sound velocity is a function of position. This configurational average acts in general as the pressure from a refracted sound wave. Thus $<p>$ gives rise to a transmitted wave; after leaving the scattering region, this transmitted wave bears a definite phase relationship to the incident wave.

For any particular configuration, the value of p may differ from $<p>$. A measure of this difference is provided by the mean square value of $p - <p>$, which is equal to $<p^2> - <p>^2$. The analysis shows that this difference is simply the sum of the squares of the pressures in the sound wave sent out from each of the bubbles. These additional terms therefore represent just the scattered sound, including sound that has been scattered several times. Thus the intensity at any point, which is proportional to p^2, is on the average the sum of two terms; the first term $<p>^2$ represents the coherent wave, propagating through a homogeneous medium in which the sound velocity changes in some way with changing position. The second term, $<p^2> - <p>^2$ represents the sum of the scattered waves from each bubble. At any one time the value of p^2, even when averaged over a few cycles, will usually differ from the sum of these two terms, but as the configuration of bubbles changes, the time average of p^2 should approach the configurational average of p^2. In most practical situations a period of several seconds is usually sufficient to bring the time average of p^2 close to the configurational average. If, then, averages are taken over time intervals of several seconds, the simplified picture presented at the beginning of the section may be taken as correct.

When the average distance between the bubbles becomes very small, or, in other words, as the average number of bubbles per unit volume becomes very large, this simplified picture becomes inadequate. In this case, another term must be included in $<p^2>$, in addition to the two terms representing the refracted (coherent) wave and the scattered (incoherent) waves. This term is difficult to interpret, but contributes to the scattered sound and appears to be due to interference between different scattered wavelets. It is not easy to determine the precise point at which this term becomes important, but it can be shown to be negligible, for resonant bubbles, provided the attenuation per wavelength is less than a few decibels. This is the same condition that must be satisfied if the change which resonant bubbles produce in the sound velocity of the medium is to be relatively small. Since this condition appears to be satisfied in observed wakes, this additional term will therefore be neglected in the following derivation of practical formulas for the attenuation, scattering, and reflection of sound by water containing bubbles.

28.3.2 Transmission

The type of analysis developed in the preceding section will now be applied to find the transmission loss through a region containing bubbles. It will first be assumed that within this region all the bubbles are of the same size. In each cubic centimeter there are assumed to be n bubbles; n may vary from point to point within the region. If I is the intensity in the incident sound beam, the rate at which sound energy is extinguished from the beam by each bubble will be $\sigma_e I$, according to equation (35). Let $I(0)$ be the intensity at the point where the beam enters the region containing bubbles and let $I(r)$ be the intensity after the beam has penetrated a distance r through the region; r is measured along a sound ray. The increment of $I(r)$ after passing an infinitesimal distance dr is, of course, negative and has the value

$$dI = -n(r)\sigma_e I(r)dr. \tag{51}$$

By integration of equation (51) over the path followed by the sound, it is found that at any distance r_1

$$I(r_1) = I(0)e^{-\sigma_e \int_0^{r_1} n(r)dr} = I(0)e^{-\sigma_e N(r_1)}, \tag{52}$$

where $N(r_1)$ is the total number of bubbles in a column of length r_1 and unit cross section. If r_1 is set equal to w, the total thickness of the region, equation (52) gives the total extinction produced by the bubble screen, or the *attenuation* as it is usually called in underwater sound work. Expressing the attenuation on a decibel scale, equation (52) is equivalent to

$$10 \log \frac{I(0)}{I(w)} = 10 \times 0.434 \times \bar{n}\sigma_e w = K_e w, \tag{53}$$

where $\bar{n}$ is the average bubble density in the screen, defined by

$$n = \frac{1}{w} \int_0^w n(r)dr = \frac{1}{w} N(w). \tag{54}$$

The quantity K_e in equation (53) is usually called *coefficient of attenuation*, which is conventionally given in units of decibels per yard. Since $\bar{n}$ and σ_e are usually expressed in units of cm^{-3} and cm^2, respectively, and since there are 91.4 cm to the yard, K_e in decibels per yard becomes

$$K_e = 396.8\bar{n}\sigma_e. \tag{55}$$

The attenuation coefficient K_e is rather easy to determine by acoustic measurements either of a wake (see Chapter 32) or of a bubble screen produced in the laboratory (see Section 28.2). Since σ_e is known for resonant bubbles from the experimental determination of the damping constant δ_r already described in Section 28.2, the bubble density $\bar{n}$ can be computed from K_e and σ_e by equation (55), on the assumption that only bubbles of resonant size are present. However, among copious masses of bubbles, as found in wakes, there will usually be a wide dispersion of bubble sizes. It is important, therefore, to evaluate the attenuation produced by such nonhomogeneous bubble populations.

Let the number of bubbles per cubic centimeter with radii between R and $R + dR$ be denoted by $n(R)dR$, and define S_e as the total extinction cross section per cubic centimeter. From equations (34) and (43), it is then found, by adding up or integrating the cross sections of all bubbles contained in one cubic centimeter,

$$S_e = \int_0^\infty \frac{4\pi R^2 n(R)\left(\dfrac{\delta}{\eta}\right)}{\left(\dfrac{f_r^2}{f^2} - 1\right)^2 + \delta^2} dR. \tag{56}$$

Bubbles of near-resonant radius will make a large contribution to S_e. If $n(R)$ does not change rapidly for radii near resonance, the integral over the resonance peak in equation (56) may readily be evaluated.

This procedure gives the correct value for S_e provided that absorption by bubbles far from resonance can be neglected. Even if the density of bubbles near resonance is comparable with the bubble density at other radii, resonant bubbles will probably make the major contribution to S_e, since σ_e is unquestionably much greater for resonant bubbles than for those of other sizes. However, according to what has been said in Section 28.1.2 about the gradual shrinkage of bubbles, a large number of very small bubbles are likely to be present which may contribute appreciably to the total extinction cross section. Since σ_e for bubbles of sizes far from resonance size is proportional to δ, and since the value of this damping constant is unknown for nonresonant bubbles, it is not possible to state the conditions under which nonresonant absorption may become important. In practical applications it is customary to assume that bubbles near resonance provide the dominant source of attenuation in wakes; as shown in Chapter 34, this assumption appears to lead to agreement with experimental results, and is probably correct at supersonic frequencies for the bubble distributions occurring in wakes.

Then, in order to compute the value of S_e resulting from bubbles near resonant size, $n(R)$, $\eta(R,f)$, and $\delta(R,f)$ in equation (56) may be taken outside the integral and given their values for R equal to the resonant radius R_r. Thus equation (56) is transformed into

$$S_e = \frac{4\pi R_r^2 n(R_r)\delta_r}{\eta_r} \int_0^\infty \frac{dR}{\left(\frac{f_r^2}{f^2} - 1\right)^2 + \delta_r^2} \cdot \quad (57)$$

As the radius of the resonant bubbles struck by a sound beam of the frequency f is R_r, then according to equation (23)

$$(2\pi f_r)^2 = \frac{3\gamma P_0}{\rho R_r^2}, \quad (58)$$

and from equations (23) and (58)

$$\frac{f_r}{f} = \frac{R_r}{R}, \quad (59)$$

and according to equation (49)

$$q = \frac{f_r}{f} - 1 = \frac{R_r}{R} - 1, \quad dq = \frac{-R_r}{R^2} dR. \quad (60)$$

By substituting equation (60) into equation (57), S_e can be expressed by an integration over the variable q. Only the values near the peak (near to q equals 1) make a considerable contribution to the value of the integral. Therefore, the transformations (49) and (50) may be used, and from equations (57) and (60) it follows that

$$S_e = \frac{4\pi R_r^3 n(R_r)\delta_r}{\eta_r} \int_{-\infty}^{+\infty} \frac{dq}{4q^2 + \delta_r^2} \cdot \quad (61)$$

The integral has been extended to infinity. This simplification can be made because on this approximation the contributions which are not very near to the peak can be disregarded. Evaluating the integral in equation (61) gives

$$\int_{-\infty}^{+\infty} \frac{dq}{4q^2 + \delta_r^2} = \frac{\pi}{2\delta_r} \quad (62)$$

and from equations (61) and (62)

$$S_e = \frac{2\pi^2 R_r^3 n(R_r)}{\eta_r} \cdot \quad (63)$$

Let now $u(R)dR$ denote the total volume of air contributed by the bubbles with radii between R and $R + dR$ in 1 cu cm of the air-water mixture. Hence,

$$u(R) = \frac{4\pi}{3} R^3 n(R), \quad (64)$$

and from equations (63) and (64)

$$S_e = \frac{3\pi u(R_r)}{2\eta_r} \cdot \quad (65)$$

The quantity η_r, according to equation (23), has the value 1.36×10^{-2} in sea water at 60 F and at atmospheric pressure. Hence,

$$S_e = 346.5 u(R_r). \quad (66)$$

In computing the attenuation for a region containing bubbles of many sizes, the equations derived at the beginning of this section may be applied directly. It is necessary only to replace the factor $n\sigma_e$ in equation (53) by S_e, taken from equation (66). If this substitution is made, the coefficient of attenuation is

$$K_e = 396.8 \times 346.5 \times u(R_r)$$
$$K_e = 1.4 \times 10^5 u(R_r). \quad (67)$$

This expression is the generalization of equation (55) for bubbles with a wide dispersion in size. It will be used in Chapter 34 to compute the amount of air in wakes from the observed attenuation coefficients.

28.3.3 Scattering

In accordance with the picture for propagation of sound through a region containing bubbles, as presented in Section 28.3.1, the basic equation for scattered sound is very simple. The scattered sound intensity from a region is, on the average, simply the sum of the intensities of the waves scattered by each bubble. For a single bubble, the intensity at a distance r is given by the equation

$$I_s = \frac{\sigma_s}{4\pi r^2} I_0, \quad (68)$$

where I_0 is the intensity of the incident sound at the bubble. This equation may be found from equations

(3), (5), and (6); more simply, it may be written down directly, since by definition $\sigma_s I_0$ is the rate at which sound is scattered by a single bubble, and since the energy is spread out uniformly in all directions, at the distance r it is spread out uniformly over an area $4\pi r^2$. In a small region of volume dV, the number of bubbles is $n\,dV$, where n is the number of bubbles per cubic centimeter.

Equation (68) must be modified to allow for the fact that the scattered sound will be attenuated on its way from the region to a distance r away. Over long distances various sources of attenuation must be considered, such as absorption in the water, scattering by temperature irregularities, and so forth. Over short distances, most of these effects may be neglected, and the transmission loss taken from equation (52). The basic equation for the scattered sound measured a distance r_1 from the region dV then becomes

$$dI_s = \frac{n\sigma_s dV}{4\pi r_1^2} I e^{-\sigma_e \int_0^r n(r)dr} \qquad (69)$$

where I is the intensity at the region dV. If sound from different directions is incident on the region, I must be averaged over all directions for use in equation (69).

Computing the scattered sound intensity from equation (69) is a much more complicated problem than computing the total sound attenuation from equation (51). In the latter case, equation (51) could be integrated along a single sound ray, yielding equation (52) directly. The basic difficulty in solving equation (59) is that the sound intensity I at the volume element dV includes sound scattered in turn from other regions. To consider multiple scattered sound of this type is rather complicated, and leads to integral equations which in general cannot be solved exactly. Methods for treating this problem have been extensively explored in astrophysical literature. [10, 11] The problem of multiple scattering in wakes could probably be studied with success by methods developed for the corresponding optical problem. [12]

Fortunately, bubbles absorb much more sound than they scatter. From equation (46) and Figure 2 it is evident that the ratio σ_s/σ_e for resonant bubbles is less than 1 to 10 for frequencies above 15 kc. For this reason, sound scattered several times from resonant bubbles has usually traveled so far that it is very weak. Multiple scattering will therefore be neglected in all subsequent discussions. In simple cases, the error resulting from this approximation

will be less than half a decibel at frequencies above 15 kc. Even at 5 kc, the error will usually be less than 1 db. For scattering by nonresonant bubbles, multiple scatterings cannot be neglected unless σ_e is much greater than σ_s.

Even with this approximation, the computation of I_s from equation (69) is not simple. The quantity I now becomes the sound intensity incident on the region containing bubbles, and attenuated by its passage through part of the region. However, to compute I_s at any one point the sound arriving from all parts of the screen must be computed; the total scattered sound must be evaluated by summing up the contributions arriving from all different directions. In any practical situation, the directivity of the receiving hydrophone must also be taken into account in order to find the electrical signal received in the measuring equipment. A detailed consideration of these problems in cases of practical importance is given in Chapter 34.

To give insight into fundamental features of the scattering problem, it is desirable to eliminate these geometrical complications as far as possible. Equation (69) is here applied to scattering from a bubble screen, that is, from a layer of aerated water bounded by two parallel planes a distance w apart. Instead of integrating over all directions, we shall compute simply the scattered sound reaching the point P from all directions within a small cone of solid angle $d\Omega$; the quantity $d\Omega$ is simply the area of a cross section of the cone divided by the distance r^2 from P to the cross section. The geometry of this situation is shown in Figure 3.

Let $I(0)$ be the intensity of sound incident on the screen; the incident sound is assumed to be a plane wave, whose rays are inclined at an angle i with a line perpendicular to the boundary of the screen. Within the screen the intensity falls off exponentially; since the path length dr is equal to $\sec i\,dx$, equation (52) gives for the incident sound at a distance x inside the screen

$$I(x) = I(0) \exp\left[-\sigma_e \sec i \cdot \int_0^x n(x)dx \right].$$

As Figure 3 shows, the scattered sound which we are considering makes an angle ϵ with a line perpendicular to the boundary of the screen. Thus in equation (69) the length dr along the path of the scattered sound is $\sec \epsilon\,dx$. Thus we find for the sound scattered from a small element of volume dV, at a distance r_1 from the point P

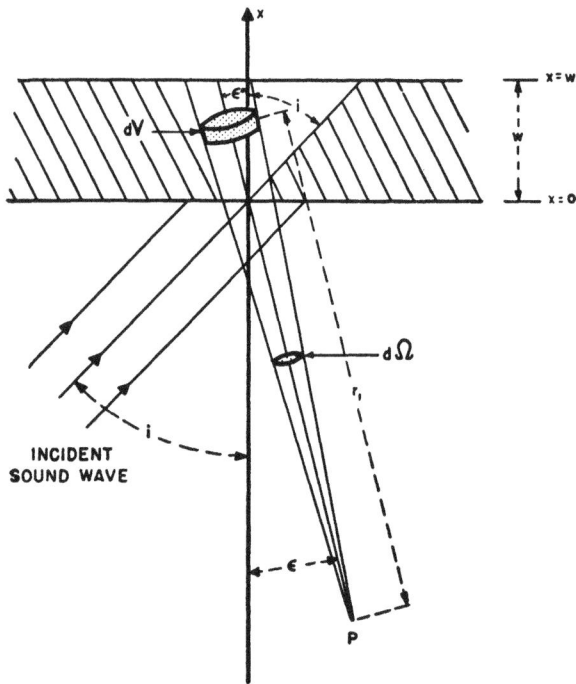

FIGURE 3. Scattering from a bubble screen.

$$dI_s = \frac{n(x)\sigma_s dV}{4\pi r^2} I(0)$$
$$\exp\left[-\sigma_e(\sec i + \sec \epsilon)\int_0^x n(x)dx\right]. \quad (70)$$

For the volume element dV within the cone, we have

$$dV = r_1^2 d\Omega dr;$$

since dr_1 is simply $\sec \epsilon dx$ as before, equation (70) becomes

$$dI_s = \frac{n(x)\sigma_s \sec \epsilon d\Omega dx}{4\pi} I(0)$$
$$\exp\left[-\sigma_c(\sec i + \sec \epsilon)\int_0^x n(x)dx\right]. \quad (71)$$

This equation may be integrated over x from 0 to w, yielding

$$dI_s = \frac{\sigma_s d\Omega}{\sigma_e 4\pi} \cdot \frac{\cos i}{\cos i + \cos \epsilon}$$
$$\left\{1 - \exp\left[-\sigma_e(\sec i + \sec \epsilon)\int_0^w n(x)dx\right]\right\}. \quad (72)$$

It is interesting to note that dI_s in equation (72) is independent of the distance from the screen to the point P where the scattered sound intensity is measured. This apparent contradiction is resolved when it is realized that with increasing distance a larger area of the screen is intercepted within the solid angle $d\Omega$.

Equation (72) has two important limiting cases. When the transmission loss across the screen is large, the second term in the brackets is negligibly small, and

$$dI_s = \frac{\sigma_s d\Omega}{\sigma_e 4\pi} \cdot \frac{\cos i}{\cos i + \cos \epsilon} \ . \quad (73)$$

It may be noted that when ϵ equals i, as is the case for backward scattered sound, $\cos \epsilon$ equals $\cos i$; if also equation (46) is used for σ_s/σ_e, equation (73) yields

$$dI_s = \frac{\eta}{\delta} \cdot \frac{d\Omega}{8\pi} \ . \quad (74)$$

On the other hand, when the transmission loss across the wake is small, it is possible to use the approximate relationship

$$e^{-\alpha} = 1 - \alpha,$$

yielding

$$dI_s = \sigma_s \frac{d\Omega}{4\pi} \sec \epsilon \int_0^w n(x)dx. \quad (75)$$

In terms of the average density $\bar{n}$ introduced in the previous section, equation (75) becomes

$$dI_s = \sigma_s \frac{d\Omega}{4\pi} \sec \epsilon \, w\bar{n}. \quad (76)$$

Thus when the transmission loss across the wake is small, dI_s is proportional to σ_s and $\bar{n}$. But when the transmission loss is great, the scattered sound reaches a constant value, given by equation (73), and is insensitive to changes in $\bar{n}$ or w.

When bubbles of different sizes are present, equation (69) for dI_s may still be used, provided that $n\sigma_e$ is replaced by S_e, the total extinction cross section per cubic centimeter, and $n\sigma_s$ is replaced by S_s, the total scattering cross section per cubic centimeter. The quantity S_e is discussed in the preceding section; equation (65) gives the relationship between S_e and $u(R_r)$, the bubble density at resonance. A similar analysis, considering bubbles only of near-resonant size, leads to the following equation for the total scattering cross section per cubic centimeter:

$$S_s = \frac{3\pi u(R_r)}{2\delta_r}. \quad (77)$$

The consideration of only those bubbles near the resonant size is usually legitimate even if absorption by nonresonant bubbles is appreciable. Since σ_s, the scattering cross section of a single bubble, does not depend on the damping constant δ for nonresonant bubbles, it is possible to evaluate precisely the contribution of bubbles of all sizes. For a single bubble smaller than resonant size, σ_s falls off as the fourth power of the wavelength; hence such small bubbles are not likely to contribute much to S_s unless present in very large numbers. Bubbles larger than resonant size have a scattering cross section about four times their geometrical cross section, but are not likely to be present in greater abundance than smaller bubbles. Thus equation (77) should be valid in a wide range of circumstances.

Since the ratio of S_s/S_e is equal to the ratio of σ_s/σ_e at resonance, equation (74) is still valid when the transmission loss across the screen is large; thus the scattered sound in this case is just the same as if all bubbles were of resonant size. When the transmission loss across the screen is small, however, equation (76) must be used, with S_s substituted in place of $\overline{n\sigma_s}$.

28.3.4 Reflection and Refraction

The presence of bubbles changes the velocity of sound. If the bubble density is sufficiently great, this effect may become practically important, leading to reflection and refraction of the sound beam. Since in ship wakes the number of bubbles present per cubic centimeter is usually not sufficiently great to change the sound velocity very greatly, these effects are not discussed in great detail here. The methods of analysis required to deal with this case are briefly sketched, and the results stated.

The sound velocity is defined by equations (6), (18), and (26) in Chapter 2 as

$$c^2 = \frac{\partial p}{\partial \rho} , \tag{78}$$

where p and ρ are the pressure and density respectively of the bubble mixture. If only a volume V of the mixture is considered, equation (78) may be written in the form

$$c^2 = -\frac{V\frac{\partial p}{\partial t}}{\rho\frac{\partial V}{\partial t}} , \tag{79}$$

using the relation $\rho dV + V d\rho = 0$. The quantities $\partial p/\partial t$ and $\partial V/\partial t$ may be evaluated from the equations in Sections 28.1 and 28.2 yielding the basic equation

$$\frac{c_0^2}{c^2} = 1 \frac{\lambda^2}{\pi} \int \frac{n(R)R dR}{\left(\frac{f_r^2}{f^2} - 1\right) + i\delta} , \tag{80}$$

where c_0 is the sound velocity when no bubbles are present, and $n(R)$ is the number of bubbles per cubic centimeter with radii between R and $R dR$. The integral in equation (80) extends over all bubble sizes. It is assumed that all bubbles present have a radius much smaller than the wavelength, and that the average distance between bubbles is larger than their radius. If these two assumptions are not fulfilled, the theory on the preceding pages breaks down. The details of the derivation of this equation are given in references 3 and 4.

It may be noted that equation (80) is valid only when the density of the liquid-bubble mixture is substantially the same as that of the liquid. Results are given which may be used for any density of bubbles, provided that the bubbles are all much too small to resonate, but much too large for surface tension to become important.[13]

For frequencies far from resonance, the imaginary term in equation (80) may be neglected. For frequencies below resonance, this leads to the equation

$$\frac{c_0^2}{c^2} = 1 + \frac{3u}{\eta_r^2} , \tag{81}$$

where u is the total volume of air present as bubbles in 1 cu cm of the liquid-bubble mixture. Thus u is defined by the equation

$$u = \int \frac{4\pi}{3} R^3 n(R) dR. \tag{82}$$

The quantity η_r in equation (81) is the ratio of the bubble circumference $2\pi R$ to the wavelength λ at resonance, as defined in equation (1). Equation (81) is valid only for bubbles which are sufficiently large that surface tension effects can be neglected; moreover, if the expansion and contraction of the bubble are adiabatic rather than isothermal, the last term in equation (81) must be multiplied by the ratio of the specific heats for the gas in the bubble. It is interesting to note that, subject to these limitations, equation (81) is independent of the bubble radius. Even if u is as low as 10^{-4} parts of air at atmospheric pressure to one part of water, c/c_0 is 0.62.

When the bubbles are all greater than the resonant size, the sound velocity is increased by the presence of the bubbles, and the relation corresponding to equation (81) is

$$\frac{c_0^2}{c^2} = 1 - \frac{3}{\eta_r^2} \int \frac{R_r^2}{R^2} u(R) dR , \tag{83}$$

where $u(R)dR$, defined in equation (64), is the volume of the bubbles present in 1 cu cm of liquid-bubble mixture with radii between R and $R + dR$. Equation (83) has the surprising implication that when $u(R)$ is sufficiently great, c_0^2/c^2 becomes negative, the sound velocity becomes purely imaginary on this approximation, and the attenuation becomes very large. Under these circumstances the imaginary term in equation (80) which was neglected in equation (83), determines the wave velocity and the wavelength. For the case of all bubbles with twice the resonant radius R_r, the critical value of u at which c becomes infinite is 2×10^{-4}, corresponding to a distance between bubbles of roughly thirty times the bubble radius.

When resonant bubbles are present, the imaginary part of the sound velocity becomes important. If an integration is carried out only over bubbles close to resonance, and if $u(R)$ is not changing rapidly with R in this region, the real part of the integral in equation (80) is small and may be neglected, yielding

$$\frac{c_0^2}{c^2} = 1 - \frac{3\pi i R_r u(R_r)}{2\eta_r^2} \cdot \quad (84)$$

This imaginary part of the sound velocity leads to an exponential decay of sound intensity with distance x, since the sound intensity falls off as $e^{-2\pi i f x/c} \cdot$ If the second term on the right-hand side of equation (84) is small, as it is in most practical cases, the resulting attenuation is exactly the same as was found in equations (53) and (67) in Section 28.3.2.

In a region containing bubbles, with any assumed distribution of sizes, and having a sharp boundary, sound incident on this region from bubble-free water will be reflected at the sharp discontinuity. The analysis for this situation is given in Section 2.6.2 where it is shown that the ratio of the amplitude of the reflected and incident waves is given by the equation

$$\frac{A''}{A_0} = \frac{c_1 - c_0 (\cos \epsilon / \cos \iota)}{c_1 + c_0 (\cos \epsilon / \cos \iota)} \cdot \quad (85)$$

This is essentially equation (119) of Chapter 2, with ρ_1 set equal to ρ' and subscripts 0 used for the incident sound wave. The quantity c_0 is the sound velocity in the bubble-free medium, while c_1 is the corresponding quantity across the boundary, where bubbles are present. The energy reflection coefficient γ_e is simply the square of A''/A_0. The angles ι and ϵ are the angles which the incident and refracted sound make with a line perpendicular to the boundary. The ratio of $\cos \epsilon$ to $\cos \iota$ may be found from Snell's law, yielding

$$\frac{\cos^2 \epsilon}{\cos^2 \iota} = 1 + \tan^2 \iota \left(1 - \frac{c_1^2}{c_0^2}\right) \cdot$$

In most cases of practical importance, c_1 is nearly equal to c_0. Thus, $\cos \epsilon$ is essentially equal to $\cos \iota$. By writing equation (81) in the form

$$\frac{c_0^2}{c^2} = 1 + b \, ,$$

the energy reflection coefficient γ_e found by squaring A''/A_0 in equation (85) becomes

$$\gamma_e = \frac{b^2}{16} \, , \quad (86)$$

as long as b is much smaller than 1. This equation may also be used when b is complex but less than 1, provided that the absolute value of b is used. When b is comparable to or larger than 1, the formulas become considerably more complicated.[4]

28.3.5   Observed Acoustic Effects of Bubbles

The effect of a known distribution of bubbles on the propagation of sound through water has been investigated in the laboratory at frequencies from 10 to 35 kc.[8] The method used for producing a screen of bubbles all of the same size has already been described in Section 28.2.1. Special measurements were made to determine the number of bubbles per cubic centimeter at various points in the bubble screen.

The bubble screen was about 17 in. long. Its thickness varied with the bubble radius; for bubbles 0.034 cm in radius, corresponding to a resonant frequency of 10 kc, the thickness was about 3 in., while for bubbles 0.020 cm in radius, corresponding to 17 kc, the thickness was more nearly 5 in. In continuous flow, about 1 cu cm of air per second was fed into the screen, resulting in a total density u of about 10^{-4} parts of air per part of water. When a bubble pulse was formed by turning on the stream of bubbles for 1 sec, however, the bubble densities at the level of the acoustic instruments were much less than this, ranging between 10^{-6} and 10^{-7}.

The acoustic measurements with the bubble pulse consisted in measuring the sound reflected from and transmitted through the screen at a fixed frequency as a function of time since the beginning of the pulse. The transmission loss was measured by reading the sound level in a hydrophone placed on the far side of the bubble screen from the projector. The reflected sound was measured by a hydrophone placed on the same side of the bubble screen as the projector, but separated from the projector by several baffles. The

RADII IN CM OF BUBBLES IN SCREEN
>0.070 0.040 0.030 0.025 0.020 0.017 0.015 0.014 0.013 0.012 0.011

RADII IN CM OF BUBBLES IN SCREEN
>0.070 0.040 0.030 0.025 0.020 0.017 0.015 0.014 0.013 0.012 0.011

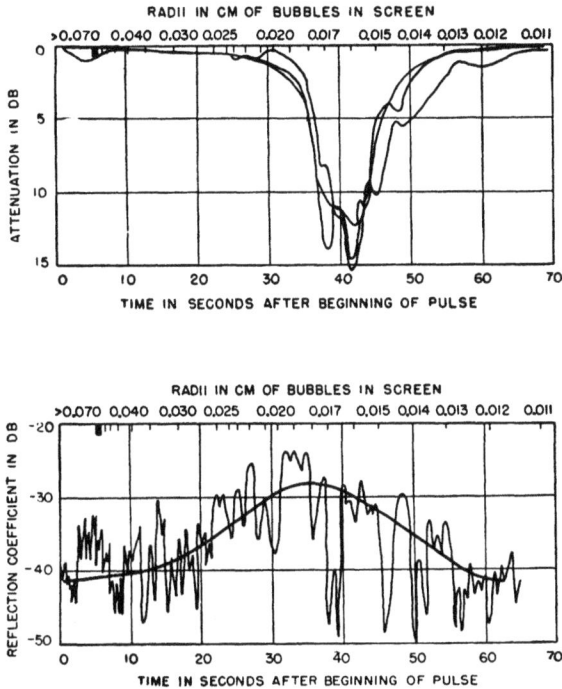

FIGURE 4. Acoustic data taken with bubble pulse screen at 20 kc.

SOUND FREQUENCY IN KC FOR RESONANCE
40 30 20 10

SOUND FREQUENCY IN KC FOR RESONANCE
40 30 20 10

SOUND FREQUENCY IN KC FOR RESONANCE
40 30 20 10

FIGURE 5. Resonant attenuation and reflection with bubble pulse screen.

hydrophone was placed symmetrically with the projector, so that the sound reflected specularly from the screen could reach the hydrophone. However, scattered sound could also reach the reflection hydrophone, and presumably contributed to the so-called reflected sound which was measured.

As the resonant bubbles passed by the level of the acoustic measuring instruments, the transmitted sound intensity showed a sharp dip. The reflected, or scattered, sound showed a very much broader maximum, in agreement with the constant scattered sound intensity predicted by equation (73) whenever the transmission loss through the region is appreciable. However, the reflected sound showed much greater fluctuations than the transmitted sound.

Sample records of transmission through and reflection from bubble screens are reproduced in Figure 4. The curves show the output of the transmission and reflection hydrophones as a function of time elapsed after a 1-sec pulse of bubbles was formed 6 ft below the acoustic equipment. Also shown are the radii of the bubbles arriving at each time. These radii were measured directly by visual means. The upper diagram in Figure 4 shows three transmission runs at 20 kc, superposed on each other. The radius of the resonance peak in this diagram agrees well with the theoretical value of 0.017 cm found from equa-

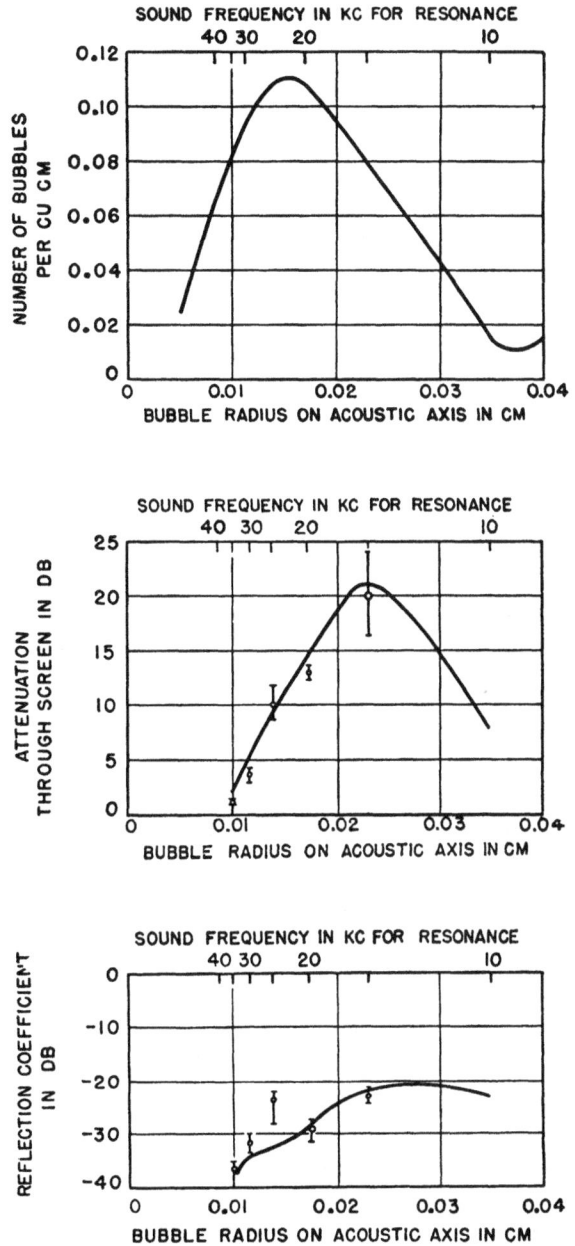

tion (18). The lower diagram shows a reflection run at 20 kc. The measured reflection coefficient is the difference in level between the incident sound at the bubble screen and the sound measured with the reflection hydrophone, placed 2.5 ft from the center of the screen.

Each set of observations was repeated at least three times at each of several frequencies. Before and after each group of acoustic measurements, detailed

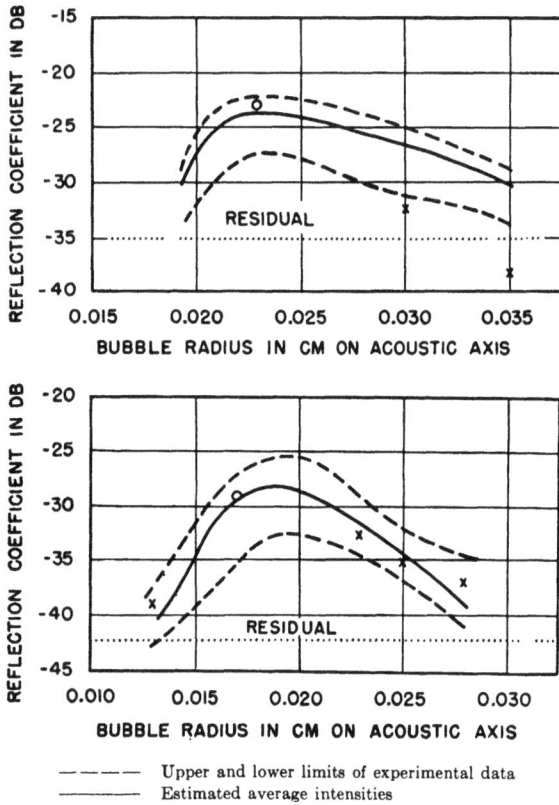

------- Upper and lower limits of experimental data
——— Estimated average intensities

FIGURE 6. Scattering and reflection bubble pulse screen.

observations were made of the number of bubbles of different sizes in the screen, since the operation of the microdispersers producing the bubbles tended to be somewhat erratic. If the physical measurements on the number of bubbles of different sizes in the screen did not give the same results before and after the acoustic measurements, the acoustic data were discarded.

The results of the acoustic measurements on bubble pulses showed moderate agreement with theoretical predictions. Figure 5 illustrates typical results obtained for resonant bubbles. The upper diagram shows the total number of bubbles per cubic centimeter at the level of the transducers at the time when bubbles of each radius reach that level. Since the spread of bubble radii at each time was small compared to the width of the resonance peak for a single bubble, all the bubbles at any one time may be assumed to be of the same size. In the middle diagram, the continuous curve shows the predicted attenuation through the screen, found by substituting in equation (53) the following quantities: the bubble density taken from the upper diagram; the measured thick-

ness of the screen; and the value of σ_e found from equation (43) with f equal to f_r, and with values of δ_r taken from Figure 2. The average observed transmission losses at each frequency are shown by circles, with vertical lines showing the spread of the observations. These experimental points are essentially the maximum difference in sound level produced by the passage of the bubbles; in the middle diagram of Figure 5, for example, the observed resonant transmission loss at 20 kc is about 14 db.

The lower diagram in Figure 5 shows the intensity of the reflected or scattered sound for resonant bubbles. To compute the reflection to be expected from resonant bubbles, the specular reflection was first found from equation (86), with b evaluated for bubbles all of resonant size. To this was then added the scattering to be expected; this scattered sound was found from equation (76), since for resonant bubbles the transmission loss through the screen was always great enough to make this equation applicable. The value of η_r at resonance was taken from equation (23), while values of δ_r at resonance were again found from Figure 2. In the computation of this scattered sound account must be taken of the size of the screen and its distance from the sound projector and hydrophone. The solid curve in Figure 5 shows the theoretical predictions; at 10 kc, specular reflection is most important, while at 30 kc, scattered sound is dominant.

A similar comparison between theory and observation may be made for nonresonant bubbles. The transmission loss measurements yield nothing further of interest, since the width of the observed resonance curve has already been used to find values of δ_r. For bubbles whose size is so far from resonant size that the transmission loss is small, the scattering may be predicted from equation (75), suitably modified to take into account the geometry of the situation. The value of σ_s to be used may be taken from equation (34). Specular reflection from nonresonant bubbles is negligible. Plots of the observed data are shown in Figure 6, where the crosses represent the computed values for nonresonant scattering. The spread of the observations is indicated by the dashed lines, with the solid line showing the estimated average intensities. The circles represent the predicted scattering and reflection from resonant bubbles, already discussed. The dotted line gives the sound level measured at the reflection hydrophone when no bubbles were present.

It is evident that the agreement between theory

and observation shown in Figures 5 and 6 is not bad. Other runs show about the same agreement, with occasional observed transmission losses as low as half or as great as twice the predicted value, and with occasional observed reflection coefficients as much as 6 db outside the spread of the observational data. These discrepancies, which are apparently in random directions, may be the result of irregularities in the bubble-producing devices. It is worth noting that the predicted scattering from nonresonant bubbles should be quite reliable, since the theoretical values are independent of the damping constant. Hence it may be concluded that the agreement of observations with theory is within the observational error, and justifies the practical use of the equations developed in this chapter.

Measurements on continuous-flow bubble screens have also been described; [7] they showed relatively poor agreement with the theoretical predictions. The observed transmission losses rarely exceeded 25 db, while the predicted transmission losses ranged between 50 and 200 db. It is doubtful whether such great transmission losses could be observed, since sound diffracted around the screen would be expected to become important. In addition, in the continuous-flow screen the smaller bubbles extended over a wider region than the larger ones. At the lower supersonic frequencies this halo of small bubbles would not absorb sound, but would reduce the sound velocity, thus tending to bend the sound rays around the screen.

Furthermore, the predicted reflection coefficients for the continuous-flow screen were some 5 to 15 db greater than the observed values. The high specular reflection predicted from theory for these continuous-flow screens would presumably be reduced to a value closer to the observed results if account were taken of the absence of sharp boundaries. In view of the many complexities entering into the explanation of these measurements on the continuous-flow screen, these discrepancies with theory may be disregarded.

An important theoretical question which is not answered by these experiments is the absorption produced by bubbles far from resonance. This nonresonant absorption depends on the variations of δ with bubble radius and sound frequency. Since the values of δ_r are unexplained, the predictions of theory as regards values of δ under other conditions are of little use. The bubble pulse measurements show that the absorption by nonresonant bubbles is usually less than about 5 per cent of the absorption by resonant bubbles. It is not impossible that for some bubble distributions present in wakes nonresonant absorption might be practically important. Further observations under controlled conditions would be required to cast light on this point.

Chapter 29

VELOCITY AND TEMPERATURE STRUCTURE

THE WATER in the wake of a ship is usually in motion relative to the surrounding water. In addition, the temperature of the water at different points in the wake is sometimes characteristically different from the temperatures found outside the wake. The variations of temperature and velocity are important physical properties of wakes, and might be expected to account at least in part for the acoustic effects observed; furthermore, a study of these physical characteristics is of independent military interest. Even if air bubbles are responsible for all the observed acoustic effects of wakes, any theory of the origin and persistence of bubbles must be consistent with known facts about the velocity and temperature structure.

The present chapter summarizes the fragmentary evidence which is available on these two subjects. Sections 29.1 and 29.2 discuss the available data on the velocity and temperature, respectively. In Section 29.3 the resulting acoustic effects are examined. It is shown that scattering from turbulent but wake-free water is negligible; scattering of sound by water with an irregular temperature distribution may sometimes be appreciable, but cannot explain the large acoustic effects observed. Thus, velocity and temperature structure alone cannot account for the observed acoustic properties of wakes.

29.1 VELOCITY STRUCTURE OF WAKES

The simplest wake is that produced by the flow of a fluid past a thin plate parallel to the stream. In this case the plate affects the flow only in a narrow region close to the plate, known as the boundary layer, where the fluid is slowed down. This effect is shown in Figure 1, where the magnitude of the velocity at various points is shown by arrows; for simplicity, only the upper half of the flow pattern is shown. Far behind the plate the velocity distribution still shows the effect of passing by the plate, since the fluid which passed through the boundary layer will be moving less rapidly than the rest of the stream. The arrows in

Figure 1 represent the average velocities of the fluid relative to the plate. Thus these results are applicable directly to the reciprocal situation, when the thin plate (or ship's hull) is moved through still water. In this situation the water in the wake is left moving in the same direction as the plate. It may be noted that in most cases, the flow in the boundary layer becomes turbulent, in which case the flow in the wake will also be turbulent.

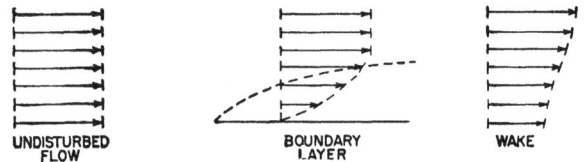

UNDISTURBED FLOW   BOUNDARY LAYER   WAKE

FIGURE 1. Velocity structure.

In addition to the wake produced in this way by passage of a ship through water, there is also the effect produced by the screws. To move the ship forward, the screws exert a forward force on the ship which is somewhat greater than the frictional force produced by the flow of water past the hull; the difference is just equal to the retarding force due to wave action and air resistance. For a submerged submarine, however, the propulsive force is just equal to the frictional force produced by the flow of the water around the hull. To produce this propulsive force on the surface ship or submarine, the screws exert an equal and opposite force on the water, which is forced backward. As a result, the water passing through and around the screws moves in a direction opposite to that of the vessel. The flow of water produced by ship screws has already been discussed in detail in Section 27.1.1 in connection with the formation of air bubbles.

Thus, close to a ship the wake is made of several component parts: one or more *screw wakes*, usually called "slipstreams," moving away from the ship as a result of screw action; and the *hull wake* following the ship as a result of frictional force at the surface of the hull. The backward momentum of the slipstream is nearly canceled out by the forward momentum of the hull wake, except at surface ship speeds so high that wave resistance becomes the most important retarding force on the ship. In the wake of a submerged submarine this cancellation is exact.

At moderately close distances astern, probably much less than a ship length, these different streams become intermingled and confused, giving rise to a turbulent mass of water in which velocities in almost any direction are equally likely. Over a small distance called the *patch size*, the velocity at any one time is reasonably constant, but the velocity at any point fluctuates rapidly. Information on turbulent motion is rather incomplete and no velocity measurements are available in surface ship or submarine wakes. As noted already in Section 27.2, not much is known about the magnitude of the turbulent velocities, the average patch size of the turbulent elements, or the rate at which the turbulence gradually dies away.

29.2 TEMPERATURE STRUCTURE OF WAKES

The water temperature at different points in a wake has been the subject of more study than the water velocity. This is partly because small temperature differences can be measured much more readily at sea than small fluid velocities. By the use of sensitive thermopiles fastened to a surface vessel, temperature fluctuations as small as 0.01 F may be readily recorded. Data obtained with this technique at the U. S. Navy Radio and Sound Laboratory[1] and elsewhere show that the presence or absence of observable temperature structure in wakes depends on the presence of vertical temperature gradients in the sea before the passage of the ship.

29.2.1 Constant Temperature in Surface Layer

When a ship is passing through water all of the same temperature, such as is commonly found in the top 50 ft of the ocean, especially during winter months, no thermal structure in the wake can be observed. Repeated wake crossings under these conditions have failed to show any trace of temperature structure. In such isothermal water, temperature structure could be produced only by the heating action resulting from the passage of the ship. Such heating can readily be shown to be negligible.

To consider an extreme case, suppose a ship at 30 knots is exerting 30,000 hp, and suppose that all this energy goes into heating a wake with a cross section 20 ft square. The increase of temperature resulting in this extreme case is 0.015 F. In most practical cases, the temperature change will be very much smaller. Although small patches of water might be appreciably warmed by water discharged from cooling systems, by dissipation of energy in intense vortices, or by similar processes, most of the wake behind a ship in isothermal water will have a temperature which is practically the same as that of the surrounding ocean.

29.2.2 Temperature Gradient in Surface Layer

When a vertical temperature gradient is observed in the top 20 ft of the ocean, the passage of a ship disturbs the temperature structure and gives rise to a measurable temperature structure in the wake. The thermopiles used in research on this subject have had slow response times, requiring 1 or 2 sec for 80 per cent response; since the surface vessels used in the work were under way at 3 knots or more, changes of temperature over regions less than a few feet in length could not be detected.

The most detailed and quantitative work[1] was carried out with four thermopiles attached to a long pipe mounted vertically on the bow of a small cabin cruiser; the thermopiles were at depths of 4, 6, 8, and 10 ft below the surface. In each thermopile, one set of junctions was thermally exposed to the sea water; the other set was thermally insulated and remained at the average temperature of the surrounding water, averaged over a period of minutes. The output of each thermopile was measured with a self-balancing potentiometer; since these instruments required some 7 sec to reduce an unbalance to zero, these quantitative measurements recorded only the large-scale features of the wake thermal structure.

Results obtained with this technique are shown in Figure 2, obtained in successive crossings of a destroyer wake 8 and 15 minutes old. Accompanying bathythermograph records are also shown. It is evident that the fresh wake consists of warmer water at the two sides, with cooler water in the middle. This distribution probably results from descending currents at the sides, and rising currents in the center; such currents could be produced by the rotation of the slipstreams from the two propellers.

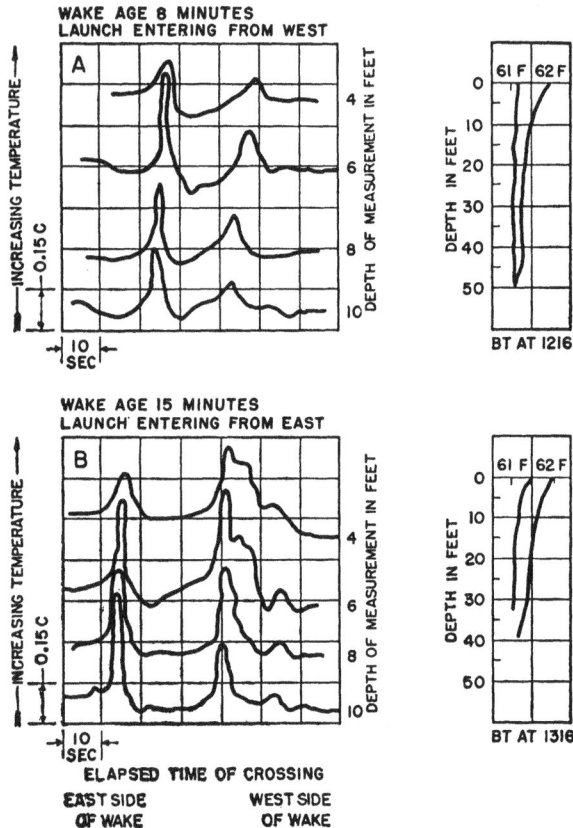

FIGURE 2. Horizontal temperature structure of a destroyer wake.

The thermal structure found for other types of ship wakes is sometimes considerably different from that shown in Figure 2, with single peaks sometimes replacing the double peaks. In general, however, whenever the thermopiles were at the depth of a marked negative gradient — 0.5 degree in 10 ft — as shown on a bathythermograph record outside the wake, the wake near the surface was colder than the surrounding water at the same depth. When the gradient is marked no such general rule may be made. It is interesting to note, however, that thermal wake signals have been readily detected when the gradient outside the wake was almost too weak to be noticed on a bathythermograph record — about 0.2 degree in 20 ft.

Measurements have also been made on the thermal properties of the wake behind a submarine at periscope depth, with a moderate negative gradient present in the surface layer. It was found that effects appeared even at the surface, where the water behind the submarine was found to be a few tenths of a degree cooler than the surrounding water outside the wake. The reason for this rise· of the submarine's thermal wake to the surface is not known.

The persistence of these thermal effects is sometimes quite marked. Identifiable thermal signals have been obtained in crossing wakes an hour or more after these were laid. Not all· wakes exhibit identifiable thermal effects for such a long period, even if the gradient is marked. The limiting factors are the decay of the thermal structure of the wake and the background of thermal irregularities present outside the wake. It is sometimes difficult to distinguish the thermal change found in crossing a wake from those frequently found in sailing through wake-free water. The thermal irregularities in wake-free water also tend to increase with increasing temperature gradients; thus a very strong gradient is not necessarily the best for detecting a wake by its thermal properties.

As shown in the next section, the acoustic effect of thermal structure is greatest for temperature irregularities whose size is about equal to the wavelength of the sound being transmitted through the water. Thus, to compute the scattering of supersonic sound at 24 kc, information on the variation of temperature over regions about 3 in. long would be required. No such information is available, owing to the long time constants of the measuring methods discussed above. Temperature fluctuations over such small regions might be expected in a relatively fresh wake. However, it would be surprising to find such a small-scale temperature structure in a wake more than a few minutes old.

29.3 SCATTERING BY TEMPERATURE AND VELOCITY STRUCTURES

Any region in which the velocity of sound varies with position will affect a sound wave passing through it. For example, theory predicts appreciable reflection from a surface separating two large bodies of water differing considerably in temperature.[2] If variations of the microstructure of the ocean take place over distances not too great compared with the wavelength, an appreciable amount of sound will be scattered in various directions. Although the exact analysis of these effects is complicated, certain results may be derived relatively simply. These results, given below, are sufficient to indicate the general magnitude of the scattering of sound by the temperature and velocity structure of wakes.

Suppose that in some region S the velocity of sound

has some variable value $c + \Delta c$, while in the surrounding water the sound velocity has a constant value c. Suppose also that a plane sound wave, of intensity I_0, and wavelength λ, is progressing through the medium in the x direction. The intensity I_s of the sound scattered from S may be different in different directions, but at long ranges will fall off as the inverse square of the radial distance r from the center of the region S. Since I_s must be directly proportional to I_0,

$$I_s = \frac{k}{r^2} I_0 , \qquad (1)$$

where k is a constant. A more detailed discussion of this equation is given in Section 19.1 of this volume, describing in general the reflection, or scattering, of sound from objects or scattering regions in the sea. The *target strength* T as usually defined is simply $10 \log k$.

The quantity k, which depends on the direction of the scattered sound under consideration, must be related to the values of Δc, the sound velocity fluctuation, at different points in the region S. Only the energy scattered directly backward need be considered here, since this corresponds to the situation of practical interest. It may also be assumed that the scattering is sufficiently small that the sound level at all points in S is practically equal to its value in the incident sound wave in the absence of scattering. This assumption tends to overestimate k; if the scattering is large the sound level will decrease as the wave penetrates the region S, because energy is lost by scattering in the portion of the region S already passed through.

Since the scattering is produced by the relative change in sound velocity, it is reasonable to assume (and, in fact, it can be shown) that the pressure dp_s of the sound scattered from each volume element $dxdydz$ in S is proportional to the value of $\Delta c/c$ for each element. In adding up all the sound from different elements, the differences in phase must be considered. Since sound must travel to the scattering element and then back along the x axis, the difference in phase between two elements separated by a distance x along the x axis will be $4\pi x/\lambda$. Thus to find the pressure of the scattered sound, $\Delta c/c$ must be multiplied by $\cos(4\pi x/\lambda + 2\pi ft)$, where f is the frequency of the sound, and integrated over the entire scattering region S. The scattered sound intensity is then proportional to the square of this integral. In this way it may be shown that the

quantity k in equation (1) is given by the formula

$$k = \left[\frac{2\pi}{\lambda^2} \iiint \frac{\Delta c}{c} \cos\left(4\pi \frac{x}{\lambda} + 2\pi ft \right) dxdydz \right]^2 . \quad (2)$$

By writing

$$\cos\left(\frac{4\pi x}{\lambda} + 2\pi ft \right) = \cos 4\pi \frac{x}{\lambda} \cos 2\pi ft$$
$$- \sin 4\pi \frac{x}{\lambda} \sin 2\pi ft , \quad (3)$$

the integral in equation (2) becomes the sum of two integrals. Now square this sum, and average over the time t, using the relations

$$\overline{\cos^2 2\pi ft} = \overline{\sin^2 2\pi ft} = \tfrac{1}{2} \qquad (4)$$

and

$$\overline{\cos 2\pi ft \sin 2\pi ft} = 0 ,$$

where the bars denote an average over the time t Then the quantity k, which measures the scattered sound intensity, becomes

$$k = \left[\frac{2\pi}{\lambda^2} \iiint \frac{\Delta c}{c} \sin\left(\frac{4\pi x}{\lambda} \right) dxdydz \right]^2$$
$$+ \left[\frac{2\pi}{\lambda^2} \iiint \frac{\Delta c}{c} \cos\left(\frac{4\pi x}{\lambda} \right) dxdydz \right]^2 . \quad (5)$$

As pointed out above, the target strength of the scattering region is $10 \log k$.

When the volume of the scattering region is small compared with the wavelength, the trigonometric functions in equation (5) are constant; since the sum of their squares is unity,

$$k = \frac{4\pi^2}{\lambda^4} \left(\iiint \frac{\Delta c}{c} dxdydz \right)^2 . \quad (6)$$

When $\Delta c/c$ is constant throughout the region, this equation reduces to

$$k = \frac{4\pi^2}{\lambda^4} \left(\frac{\Delta c}{c} \right)^2 V^2,$$

where V is the volume of the region. Equation (6) is the so-called Rayleigh scattering law, which predicts only a small amount of scattered sound. On the other hand, when c is constant over a region large compared with the wavelength, k is again small; as a result of the oscillation of the sine and cosine factors in equation (5) each integral adds up to only a small value.

29.3.1 Effect of Temperature Microstructure

Equation (5) may be used to compute the sound scattered by a mass of water in which the temperature varies rapidly from point to point. For simplicity, suppose that positive and negative values of c are equally likely — that is, that the average temperature of the water is just equal to the temperature outside the scattering medium. Although the distribution of temperature from point to point is a quantity which fluctuates at random, there is a certain patch size over which the temperature does not usually change appreciably. This is represented mathematically by means of the function $\rho(\zeta)$, which is defined by the expression

$$\rho(\zeta) = \frac{\overline{\Delta c(x + \zeta, y, z)\, \Delta c(x, y, z)}}{\overline{\Delta c(x, y, z)^2}}, \qquad (7)$$

where the averaging is to be carried out in space, over all values of x, y, and z in the scattering region. While ζ is a displacement in the x direction in the expression (7) above, the displacement might also be extended in any other direction. The value of the function $\rho(\zeta)$ will depend both on the magnitude and on the direction of ζ. If the displacement is zero, then ρ will equal unity. If the displacement is very large, the values of c at points separated by the distance r show no correlation with each other, and their product is alternately positive and negative, canceling out on the average; thus for large ζ, ρ approaches zero. The patch size is the value of ζ for which ρ becomes small, say less than about $\frac{1}{3}$. The function ρ is called a *self-correlation coefficient*. The temperature microstructure is described as isotropic if $\rho(\zeta)$ is independent of the direction along which ζ is taken.

With some mathematical transformations, equation (5) may be expressed in terms of $\rho(\zeta)$. For an isotropic medium, the resulting equation, which is equivalent to that given in a report by Columbia University Division of War Research [CUDWR],[3] is

$$k = \frac{16\pi^3}{\lambda^4}\overline{\left(\frac{\Delta c}{c}\right)^2} V \int_0^\infty \rho(\zeta)\zeta^2 \frac{\sin(4\pi\zeta/\lambda)}{4\pi\zeta/\lambda}\, d\zeta, \quad (8)$$

where V is the volume of the scattering region. As one fairly general type of possible correlation coefficient, it may be assumed

$$\rho(\zeta) = e^{-\zeta/A}. \qquad (9)$$

By substituting this expression in equation (8), and integrating,

$$\frac{k}{V} = \frac{1}{8\pi A}\overline{\left(\frac{\Delta c}{c}\right)^2}\frac{1}{(1 + \lambda^2/16\pi^2 A^2)^2}. \qquad (10)$$

In actual practice the wavelength λ is usually less

than the patch size A, and the last term in the denominator may be neglected. Correlation coefficients of a form different from equation (10) do not generally give a much greater value of k/V for a given patch size A.

Numerical values may be substituted in equation (10). Fluctuations of 0.5 F with a patch size of 6 in. probably represent a rather extreme assumption. For this situation, k/V is about 3×10^{-7} sq yd per cubic yard of volume. The volume scattering coefficient m discussed in Section 12.1 of this volume is related to k by the equation

$$m = \frac{4\pi k}{V}. \qquad (11)$$

Thus m, in this case, is about 4×10^{-7} per yard. If equal energy were scattered in all directions, m would be the fraction of energy scattered per yard of sound travel through the scattering medium.

Evidently even these extreme assumptions give a very small scattering coefficient. Even if the scattering volume is 10 yd thick, 30 yd across, and 100 yd long, corresponding to the wake in the path of a sound beam, k is about 10^{-3} yd, corresponding to an effective target strength of -30 db. The transmission loss through such a scattering region would be a very small fraction of a decibel. Temperature microstructure cannot explain the strong echoes or the high transmission losses produced by wakes.

29.3.2 ## Effect of Velocity Microstructure

A separate analysis must be carried out for the case where the velocity of the water varies from place to place in the medium. This is a more complicated situation than the one in which the temperature changes, since the fluid velocity has a direction as well as a magnitude. However, it can be shown that equation (5) is still applicable if the component v_x of the fluid velocity in the x direction is used in place of Δc. This seems a reasonable substitution, since it is only the component of the fluid velocity along the direction of the incident sound wave that affects the propagation of this wave.

To compute k, then, integrals of the form

$$\int \sin(4\pi x\lambda)dx \quad \iint v_x dy\, dz \qquad (12)$$

must be evaluated. If the integrals over y and z are computed first, it is easy to see that the entire integral vanishes. The integral of v_x over the yz plane is simply the net rate at which the fluid is flowing across this plane. At any time, the total amount of fluid passing through the yz plane in one direction must be just equal to the amount of fluid passing through in the other direction, and the net flow vanishes. Thus, a random distribution of velocity does not contribute to backward scattering of sound. However, sound may be scattered in other directions, as indicated in reference 3.

Measurements at San Diego [4] and at Orlando [5, 6] are consistent with the result that the sound scat-

tered backward from velocity microstructure is very weak. At San Diego attempts were made to obtain echoes from underwater vortex rings, while at Orlando a mechanical device was used to produce turbulent water in the path of a sound beam and attempts were made to measure the reflected sound. In both cases, no reflected sound could be observed. Although the data do not exclude the possibility that weak echoes may have been present, the combination of measurements and theory point to the conclusion that backward scattering of sound from velocity microstructure may be practically neglected.

Chapter 30

TECHNIQUE OF WAKE MEASUREMENTS

MOST OF THE MEASUREMENTS of submarine and surface vessel wakes discussed in Chapters 26 to 35 have been made by University of California Division of War Research [UCDWR] or by Navy observers at the U. S. Navy Radio and Sound Laboratory [USNRSL] in San Diego. The instruments and physical principles applied to acoustic observations of wakes do not differ essentially from those employed in other underwater sound measurements described in Chapter 4, Chapter 13, and Chapter 21. It is unnecessary, therefore, to introduce here detailed descriptions of instruments and their theory. But before discussing the results, some general features of the experimental work at San Diego on the acoustic properties of wakes will be reviewed.

30.1 LISTENING AND ECHO RANGING

Listening through a wake to a ship under way, or to a mechanical noisemaker, constitutes the simplest type of acoustic observation of a wake. The presence of a wake manifests itself by a reduced sound level at the receiving hydrophone, compared with the same level with no wake interposed. Such observations of the acoustic screening effect are the incidental result of numerous measurements of the sound output of ships. But, in order to obtain quantitative results, it is desirable to use as sound source a transducer or mechanical noisemaker of constant power output, instead of the noise from the screws of a ship. Together with a hydrophone of constant sensitivity, this equipment makes possible determination of the transmission loss which sound undergoes in passing through a wake.

Echoes returned by wakes can be studied by listening or by using objective records of the current generated in the receiving channel of the transducer. While the second method is indispensable for the determination of sound intensities, it does not tell anything about the small changes in frequency that are caused by the relative motion of target and trans-

ducer. The acoustic doppler effect is helpful in distinguishing between the echo from a wake, which is nearly stationary, and the echo from the wake-laying vessel. This distinction is occasionally of practical interest, as in the study of the rather weak wakes produced by submarines in submerged level runs. In such cases it may be useful to preserve an audible record, in the form of a phonograph record, of the wake echo. The supersonic echo obtained aboard the experimental vessel is transmitted by short-wave radio to the laboratory ashore, where the phonographic recording can be done more conveniently than on a rolling and pitching vessel at sea.

30.1.1 Sound Range Recorder Traces

At San Diego it is a standard procedure in all wake work to secure echo records with a sound range recorder of the type in general tactical use. These chemical recorder traces are highly useful for a rapid estimate of the range of the wake and of the decay of its strength. As the chemically treated recording paper is unrolled, with the machine open, the observer makes pencil notes on the margin of the record concerning the work in progress, such as the beginning and ending of the oscillographic recording, changes of the sound frequency used, and other details. Thus the chemical recorder traces also provide a graphical log of the operations.

The general appearance of wake echoes on the sound range recorder paper is illustrated by the photographic reproductions of original records shown in Figures 1 and 2. They are records of wakes laid by the auxiliary yacht *E. W. Scripps* between the echo-ranging vessel, the USS *Jasper* (PYc13) and a target sphere buoyed at a center depth of 6 ft below the surface, in the course of experiments described in detail in Sections 31.2 and 32.3.2. The lower part of Figure 1 shows the sphere echo alone. Immediately after passage of the *Scripps* through the sound beam, there appears a strong wake echo and the strength

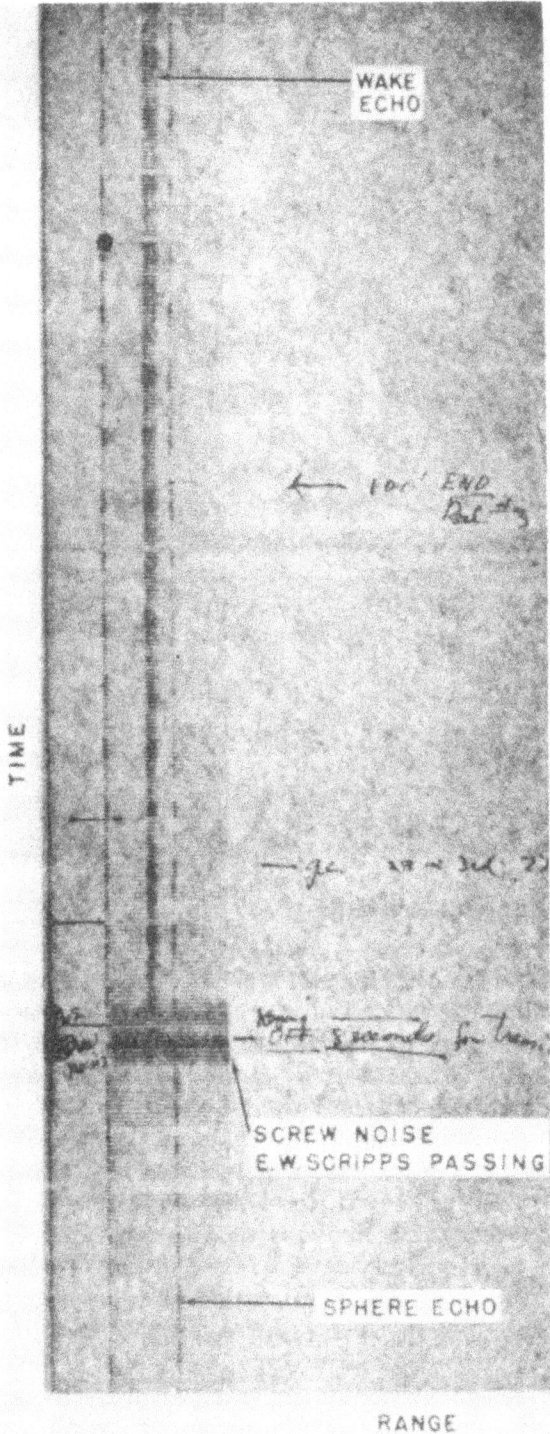

FIGURE 1.   Sound range recorder traces of wake echoes from *E. W. Scripps*.

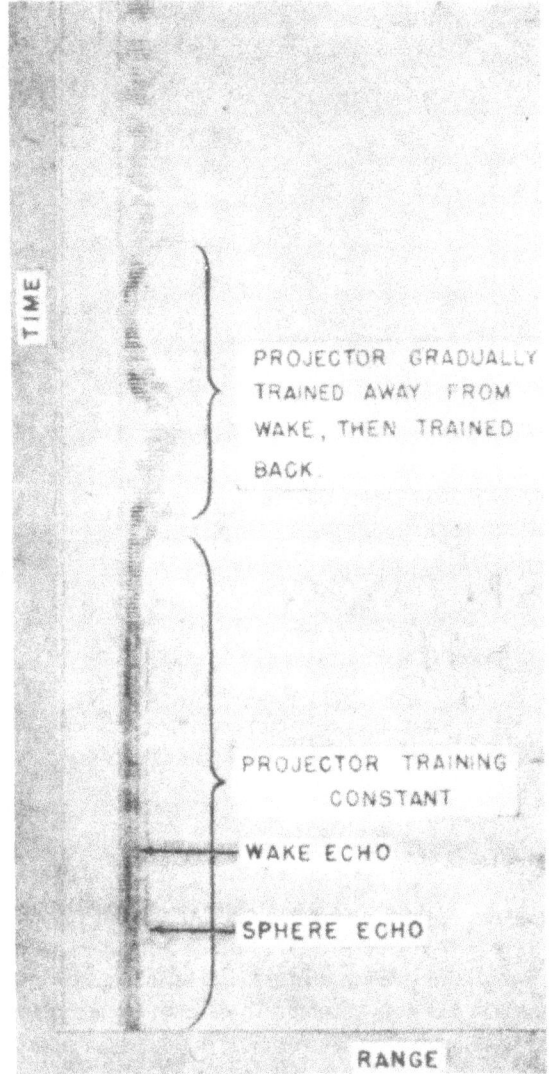

FIGURE 2.   Sound range recorder traces of wake echoes from *E. W. Scripps*.

of the sphere echo is markedly diminished by the two-way transmission loss in the wake. Note the gradual widening of the wake toward the top of the figure, as the wake grows older.

The wakes were laid at right angles to the line connecting the transducer on the *Jasper* with the target sphere. In Figure 1 the projector was kept trained at the sphere in order to study the decay of a fixed part of the wake. Figure 2 shows the effect of gradually changing the training of the projector from its normal training; the range toward the nearest boundary of the wake increases, and since the sound beam now cuts obliquely through the wake, the apparent width of the wake increases proportionally to the secant of the angle included between the sound beam and the normal to the wake. On

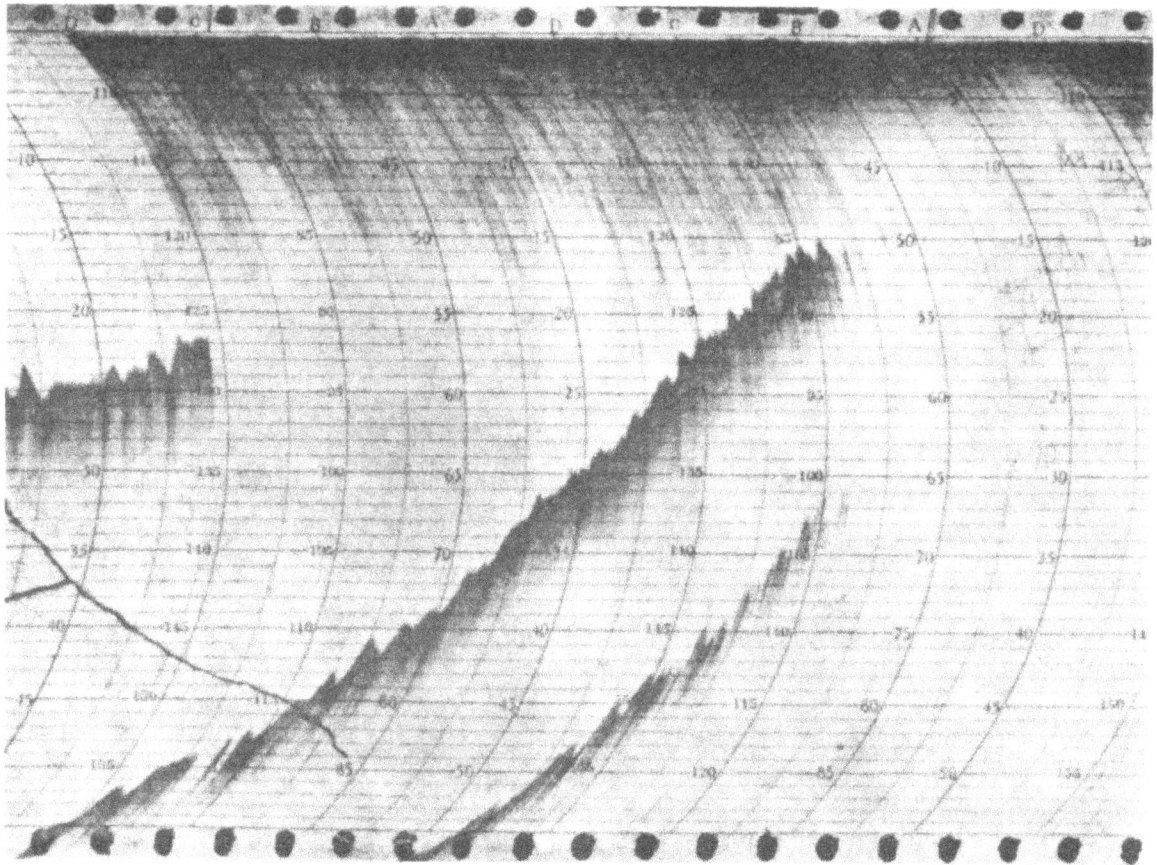

FIGURE 3.   Fathometer record of echoes from ocean surface.

training back the transducer, the effect is reversed, thus causing a symmetrical pattern to appear in Figure 2.

30.1.2   Fathometer Records

Records of a wake, indicating its thickness and transverse structure, are readily obtained with a fathometer carried across the wake by a survey vessel. Such records may be utilized also to compute the vertical transmission loss, as long as a record from a standard target observable through the wake — for instance, the ocean bottom or surface — is also available (see Section 32.3.3).

Early experiments were carried out with the fathometer mounted in the orthodox manner on a surface vessel. Records of the ocean bottom are then blanked out in certain cases when the survey boat enters a surface ship wake. This technique suffers from several disadvantages. It does not give an accurate value for the depth of the wake, since the duration of the wake echo is affected by the beam width, the pulse length,

and other factors as well as by the depth of the wake. Also, the method is not very suitable for the measurement of the transmission loss through the wake, because it requires the echo-ranging vessel to operate in relatively shallow water in order to record the bottom; furthermore, the depth and bottom character may vary considerably while this vessel is moving. If, however, the fathometer is used in the inverted manner, by mounting it on the deck of a submerged submarine, those disadvantages are eliminated; clear strong records are obtained both of the highly reflecting ocean surface and of the surface ship's wake, as illustrated by Figures 3, 4, 5, and 6.

Figure 3 shows a record obtained while the submarine was diving from the surface. The depth scale marked 5 to 50 applies to this dive, with the time axis running from the right to the left. It can readily be verified from the double record in the center of the illustration that the weaker second reflection corresponds to depths that are exactly twice the depth of the stronger first reflection. Thus, the double record

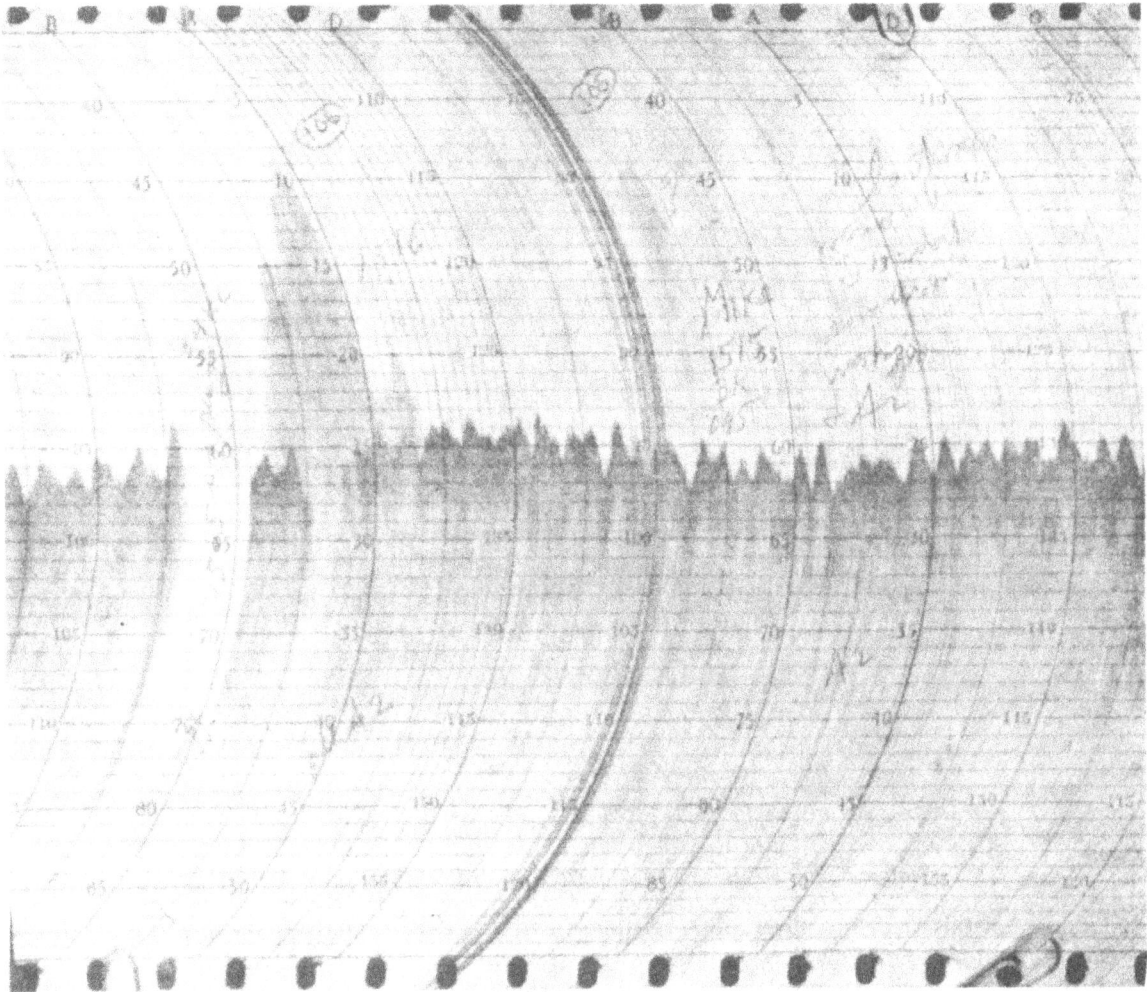

FIGURE 4. Fathometer record of wake echoes from Coast Guard cutter *Ewing*.

is a result of the sound traveling to the ocean surface twice and returning again to the submarine. The dark streaks at the top of this figure result from the acoustically reflecting region formed behind the submarine conning tower, presumably as a result of cavitation originating around the conning tower. The record at the far left is that of the ocean surface after the submarine arrived at a depth corresponding to the scale limit of the recorder and the scale was shifted to bring the record nearer the center of the paper. The small indentations and undulations of the record are produced by the surface swells.

Reflection from a surface ship wake under which the submarine is passing produces in these records a shaded area protruding below the ocean surface, as shown in the next three illustrations. Figure 4 represents the record of a wake laid by the USCGC *Ewing*,

proceeding at 13 knots. The submarine in this case passed under the wake at a point 350 yd behind the *Ewing*. This record was suitable for transmission loss calculations, according to the principles which will be described in Section 32.3.3. The result was a transmission loss of 42 db with 21-kc sound traversing the wake twice. Note that as a result of the large transmission loss, the record of the ocean surface is almost blotted out in the center of the wake, which had a thickness of 15 ft.

The same effect is apparent in Figure 5, showing a wake record originating from the destroyer, USS *Hopewell* (DD681), proceeding at 10 knots. The distance astern is not accurately known, but it is roughly several hundred yards. The transmission loss at 21 kc through the center of this wake was 32 db for the double path. The cause of the extrane-

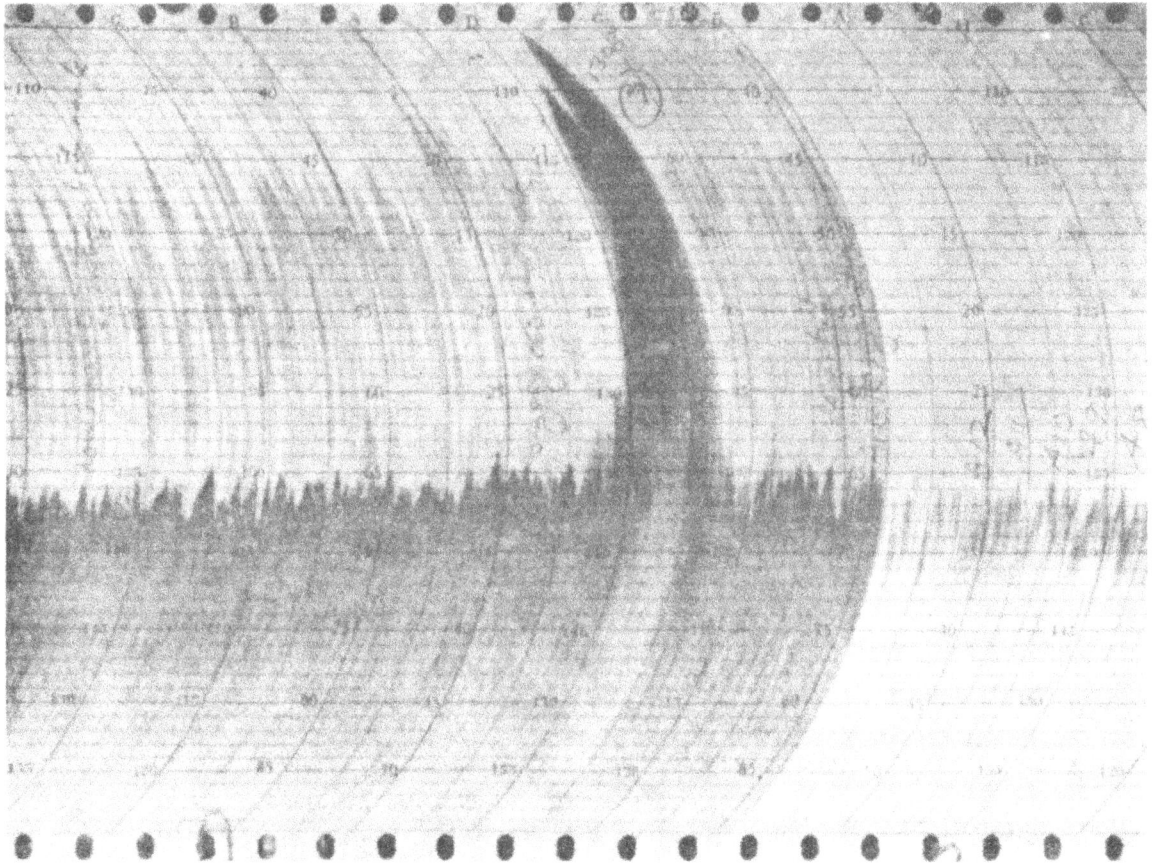

FIGURE 5.   Fathometer record of wake echoes from USS *Hopewell* (DD681).

ous markings on this record is uncertain; probably they are of instrumental origin. The wake is seen to be 30 ft thick at the maximum point.

Figure 6 contains two records of the wake (17 and 11 ft thick, respectively) of the *Ewing*, proceeding at 13 knots; these records were not suitable for transmission loss calculations, since the amplification was increased to record the cross-sectional geometry of the wake. Comparison of Figures 4 and 6 gives an idea of the variations of wake structure occurring in practice; the vessel and speed are the same for both figures. For the proper interpretation of these cross sections, it should be remembered that the sound beam of the customary fathometer is rather broad, including an angle of about 30 degrees, thus causing the fine structure of the cross section to be smoothed out.

30.1.3   Oscillograms

In order to obtain permanent sound intensity records suitable for quantitative measurements, the current generated in the hydrophone is amplified and fed into a cathode-ray oscilloscope, the screen of which is photographed continually by a high-speed camera on standard moving picture film, as described in Section 4.3.3, Section 13.1.1, and Sections 21.2.1 and 21.3.1. The developed negative shows a continuous trace, representing the varying displacement of the luminous spot from its normal position on the oscilloscope screen. Time marks are photographed at suitable intervals as the film moves along steadily. By appropriate design of the electric circuits the displacement of the oscillographic trace is made proportional to the amplitude of the incident sound wave. The square of the amplitude of the oscillographic trace, therefore, is proportional to the intensity of the sound wave, at the face of the hydrophone, multiplied by a factor depending upon the directivity of the hydrophone. If the sensitivity and the directivity pattern of the hydrophone are known, the scale of ordinates on the oscillogram can be calibrated in absolute units to yield the sound pressure in dynes per square centimeter.

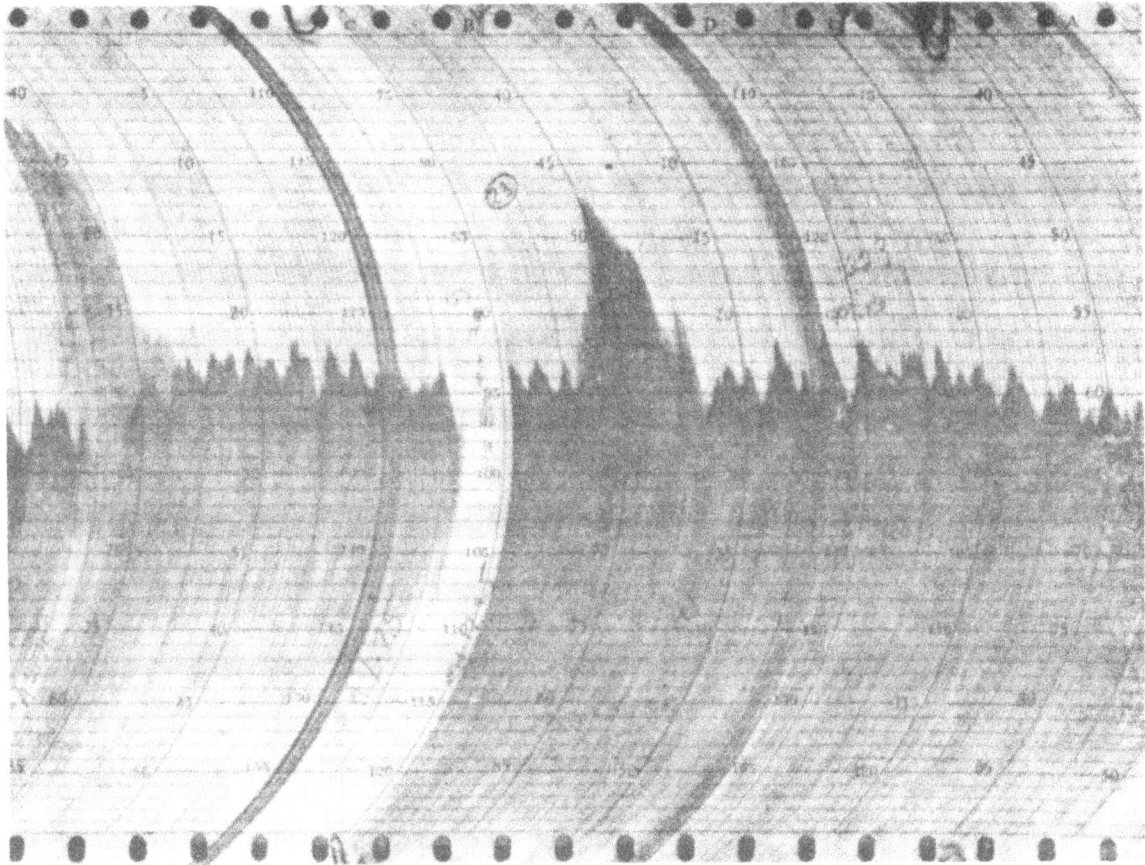

FIGURE 6. Fathometer record of wake echoes from Coast Guard cutter *Ewing*.

This type of recording, which has been used widely in other sound studies, has usually been applied only to the analysis of wake echoes rather than to signals transmitted through wakes. The linear distance on the film from mid-signal to mid-echo provides a convenient record of the range from which the echo was returned, since the distance on the horizontal scale is the product of sound velocity times the time. A number of oscillograms of wake echoes are reproduced below on the scale of the originals. Figure 7 shows three sets of three successive signals, each 3 msec long, and the corresponding echoes both from a wake, laid by the *E. W. Scripps*, and from a target sphere 3 ft in diameter suspended behind the wake at a center depth of 6 ft. The oscillograms were obtained with 24-kc sound during Run 1 of the experiments summarized in Figures 8 and 9 of Chapter 31 and in Table 2 of Chapter 32, which should be consulted for a detailed description of the plan of observations.

The numerical evaluation of wake oscillograms has so far been restricted to the visual measurement of peak amplitudes, described in Section 21.3.1, which generally have been held to be sufficiently representative of the echo as a whole. A more satisfactory though very time-consuming method would be to measure the amplitudes along the entire echo profile, square the amplitudes and integrate them over the time. This integral would be proportional to the total energy contained in the echo. It is possible to design a mechanism which would perform automatically this sequence of procedures. In any event, it would be desirable to supplement and check fundamental wake studies based upon measurement of peak amplitudes by investigating the total energy of echoes.

Current procedure is to place the processed film on an illuminated viewer, read the peak amplitude of the echo with the aid of a transparent scale, and correct the measured amplitude, if necessary, for the finite width of the luminous spot on the oscillograph screen. Averages over five successive echoes are

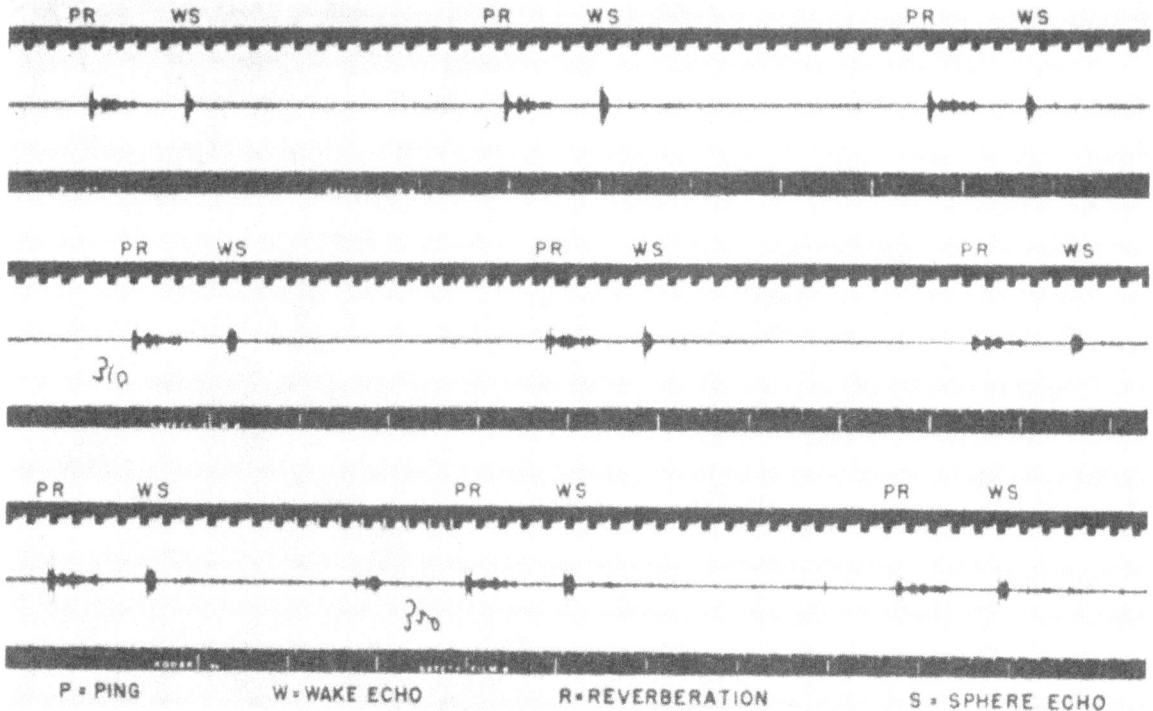

FIGURE 7. Oscillograms of wake echoes from *E. W. Scripps.*

taken, and the averaged peak amplitude is squared to obtain the echo intensity. The resulting average is different both from the average peak echo intensity and the average of the intensity over the entire echo. Since the spread of peak amplitudes may be as much as 10 db, this difference may be appreciable. The difference between average peak amplitudes and average intensities is discussed in Section 34.3.1.

Finally, from the measured peak amplitudes the echo-strength is computed according to the formula:

$$E - S = 20 \log \overline{A_e} - 20 \log k ,$$

where E is the echo level in decibels above 1 dyne per sq cm, S the source level, defined as the sound level 1 yd from the projector on its axis, also in decibels above 1 dyne per sq cm, and $\overline{A_e}$ is the average peak amplitude of the echo as measured on the oscillogram. The constant k on the right side of this equation has to be determined by calibration of the receiving equipment; specifically, k is the amplitude measured on the oscilloscope with an incident wave whose pressure is 1 dyne per sq cm and with the same receiver gain at which $\overline{A_e}$ is recorded. To determine S and k, an auxiliary transducer of known power output and of known sensitivity is used.

30.2 OPERATIONS AND MEASUREMENTS

Besides the acoustic measurements proper, the study of wakes requires the determination of various auxiliary data. In the first place, the geometric coordinates of the part of the wake to which the acoustic data refer have to be known accurately. If the distance from the stern of the ship to the point where the sound beam strikes the wake is known, the age of the wake at the point of measurement may be found by dividing this distance astern by the speed of the wake-laying vessel and computations will be facilitated by use of Figure 3 in Chapter 35. Since the instrumental characteristics of the sound gear employed may undergo slow changes, it may become necessary to calibrate the gear immediately before or after the observation. Furthermore, there are a number of variable oceanographic factors whose instantaneous values have to be taken into account in interpreting the acoustic measurements.

In addition to the wake-laying vessel, acoustic measurements on wakes require one vessel for echo ranging and an additional vessel when a transmission run is made in order to measure the horizontal transmission loss. For measurements of the transmission loss, the use of a second experimental vessel carrying the receiver might be eliminated by echo ranging through the wake at a target sphere and measuring the intensity of the echo returned to the transducer. From echo ranging at wakes, usually the wake-laying vessel proceeds at constant speed on a straight course past the measuring vessel, which either may run a parallel course with different speed or may be hove to. Maintenance of prescribed speeds and course demands accurate seamanship. The relative positions of the two vessels as a function of the time are determined by direct triangulation and dead reckoning. During echo-ranging experiments, an incidental check on those geometric data is obtained by the acoustic ranges.

During transmission runs, the range from the cruising auxiliary vessel, which carries the projector, to the measuring vessel has been accurately determined by the use of airborne sound; simultaneous radio and sound signals are transmitted from the auxiliary vessel, and the difference between the automatically recorded times of arrival of the two, multiplied by the velocity of sound in air, yields the range. Moreover, for transmission runs, the courses of the operating vessels have to be laid out and maintained with great care in order to avoid interference with the acoustic measurements from the auxiliary vessel's own wake.

In working with wakes which are laid and then allowed to age before the measurements or while the measurements are being conducted, the exact location of the wake soon becomes difficult to discern. If the wake-laying vessel lies to, it usually soon drifts enough to be useless as a marker for one end of the wake. The following method has proved helpful when working either with surface craft or submerged submarines, particularly with very long wakes. A small boat lies to at a point near where the initial end of the wake will be laid. As the wake-laying vessel goes by, the small boat moves into the center of the wake and releases a small amount of fluorescein;[a] chrome yellow or any other nonsoluble dye which floats on the surface is not satisfactory for this purpose, since wind drift can move it away from the wake. In the case of a submarine, it submerges as it passes the small boat or if already at periscope depth, the submarine releases the fluorescein. The wake-laying vessel releases fluorescein into the wake at the end of its run. Both this ship and the small boat then keep

their bows touching the dye spot, thus keeping their net drift the same as that of the wake. More fluorescein is dropped off the bow of the marker boats at intervals; one marking will not last when working with wakes older than 20 to 30 minutes. If the observing vessel crosses the wake in the course of its measurements, the wake is located by sighting on the marker boats; the use of a simple optical device for lining up the markers is recommended. One of the marker boats takes a stadimeter range on the vessel which crosses the wake; this procedure aids in computing the wake age at that point. If successive crossings are made, fluorescein is dropped from the observing ship just as it lines up the marker boats. The marker boat closest to this point then moves up to the new dye spot; this insures that the wake between markers remains free of extraneous wakes. Where marker boats are not available, it is helpful to use a mixture of fluorescein and chrome yellow as a marker. The two colors drift apart if any wind is present; the chrome yellow can be seen farther away and is used to locate the fluorescein.

30.2.1 Training Errors

In echo ranging on wakes the trainable transducer is usually operated at a fixed relative bearing. However, in measuring the transmission loss across a wake, it is necessary to keep the trainable projector of the sending vessel aimed at the hydrophone of the receiving vessel; continual changing of the bearing of the transducer is also necessary in echo ranging through the wake at a target sphere. For this purpose, an observer is stationed on the flying bridge to operate a repeating pelorus, to be aimed at the auxiliary vessel or target sphere. A second man stationed at the control stack matches the projector-heading indicating "bug" to the target-bearing repeater. Even so, the projector heading does not hold precisely to the true target bearing. The deviation is partly attributable to the lag in the various linkages of the system and partly to the impossibility of holding the pelorus accurately on the target at all times. Under practical conditions as prevailing on board the USS *Jasper* (PYc13), the estimated errors and their sources are as follows:[1] (1) pelorus aiming error ±2 degrees in fair weather; (2) control stack matching error, projector bearing to target bearing ±2 degrees; (3) lag in training system, gears and projector-heading repeater system ±2 degrees maximum prior to April 1944, when the system was overhauled and the error re-

---

[a] Before using fluorescein in experiments at sea, it should be ascertained whether special authorization by the Area Commander is required.

duced to approximately ±1 degree. The maximum error is, therefore, ±6 degrees and the probable error ±3.5 degrees for data taken prior to April 1944, and ±5 degrees and ±3 degrees, respectively, for data taken subsequently.

At very short ranges there is another correction which may have to be applied because of parallax resulting from horizontal spacing between pelorus position and projector axis. On the *Jasper* this correction amounts to 2.5 degrees for the aft projector and 3.5 degrees for the forward projector, when the target is 100 yd away and bears either 90 or 270 degrees relative to the sending vessel.

Finally there are training errors due to rolling and pitching of the sending vessel. This error can be serious at close range since rolls of 45 degrees have been experienced on the *Jasper* and rolls of over 20 degrees are common in moderate weather. For the same vessel the pitching angle is of considerably smaller magnitude than that of the roll, rarely exceeding 7 degrees. Installation of a device to record angle of roll, pitch, projector heading, target bearing and ship's heading for each sound pulse emitted, has been of great help in recognizing and rejecting acoustic observations that have been impaired by serious training errors.

30.2.2   Field Calibration

The transmitting system's absolute output has to be checked at the beginning and end of each day's operation and also during the operation if excessive variations are encountered. For this purpose an auxiliary transducer, whose performance is known from absolute calibration in the testing laboratory, is lowered into position by means of a special boom which pivots at the rail and swings down to projector depth. To check the actual output of the transducer in use, it is trained on the auxiliary transducer to give a maximum generated voltage; and from the laboratory calibration of the auxiliary transducer, the sound field pressure is determined. For the inverse calibration process, the known power output of the auxiliary transducer is received by the working transducer and the generated current is recorded as in field work. Any appreciable deviation of any of the readings from those normally experienced requires an immediate investigation to determine the source of the difficulty.

According to the experience of the San Diego group with the JK projectors, the standard deviation of the output pressure level was only 1.3 db over a period of 15 months, and 0.4 to 1.0 db for groups of consecutive calibrations within that sequence. However, the standard deviation of the sensitivity of the receiving channel was 3.9 db for the same period and varied from 0.4 to 2.0 db for groups of calibrations within that period. The causes of this variation are unknown. In the course of one day, changes in overall sensitivity, which is the sum of the projector output and the sensitivity of the receiving channel, are negligible; changes in output level did not show any correlation with changes in receiver sensitivity. Also, over the whole period under discussion, changes in output level are not correlated — or at most are weakly correlated — with changes in receiver sensitivity. All these observations refer to 24-kc sound. Incomplete evidence suggests that the performance of 60-kc sound gear is even more variable.

The existence of large calibration errors is also suggested by certain discrepancies among the San Diego data on the target strength of spheres, which are discussed in Section 21.4.3.

30.2.3   Oceanographic Factors

The weather and state of the sea appears to have some influence on the formation and gradual dissolution of wakes. It is advisable, therefore, to keep a careful record of the circumstances prevailing at the time of the observations. The momentary oceanographic conditions have a profound effect also upon the propagation of underwater sound. Hence, it has become a standard practice to secure bathythermograms before and after each set of acoustic observations. The transmission loss in the ocean intervening between sound gear and wake, which must be known in order to correct the measured data, is difficult to determine directly. So far, acoustic observations on wakes have not reached such a high degree of precision as to make it imperative, as in the measurement of target strengths, to determine the transmission loss in the ocean simultaneously with the wake observations. For details on the technique of measuring the transmission loss consult Chapter 4.

Perhaps the most serious disturbances of underwater sound measurements are the rapid and unpredictable changes of the transmission loss, generally referred to as fluctuations and described in Chapter 7, which may amount to many decibels over intervals of only a few seconds. The only way to minimize their influence is to take averages over long series of observations. Even these averages may show·a slow drift with time, sometimes called variation of the transmission loss, but the amplitude of the variations is of a lower order of magnitude than that of the fluctuations. In measuring the transmission loss which sound undergoes while passing across a wake, the fluctuations of the transmission loss in the surrounding ocean mask the effect sought after, or even may entirely obscure it for wakes in an advanced stage of decay. In echo ranging, the sound returned by different parts of the wake undergoes destructive and constructive interference, which together with the gradual change of the internal structure of the wake will invariably cause fluctuations of the wake echoes that are even more rapid than the fluctuations of the transmission loss. Consequently, wake echoes are even more variable in shape than sound signals which have been affected only by fluctuation of the transmission loss in the sea. Figure 7 shows the irregular character of wake echoes resulting from changing interference effects. Fluctuations of the transmission loss in the ocean are also conspicuous in Figure 7; note the change in strength between the last two echoes from the target sphere, in the lower strip of the illustration.

In echo ranging at wakes over short ranges, the reverberation background caused by the scattering of sound in the ocean constitutes an important limiting factor. Under conditions giving very high reverberation, a weak echo may become lost in the background. Strictly speaking, the echo intensities derived from measured echo amplitudes, as described above, include the contribution from reverberation and should be corrected for this superimposed intensity. In practice, this correction may usually be neglected whenever the wake echo is sufficiently strong to be distinguished from the reverberation.

Chapter 31

WAKE GEOMETRY

IN THIS CHAPTER information of rather heterogeneous origin, concerning the dimensions of wakes, is brought together. Some types of acoustic observations are in themselves eminently valuable for determining the geometric characteristics of wakes. However, a good deal has to be known about the geometry of wakes in order to plan and execute their investigation by acoustic methods. Such knowledge has been provided by visual and photographic observations. Brief reference also will be made to thermal wakes, although very little is known about them so far. It is undecided whether or not the visual, acoustic, and thermal manifestations of the same wake agree as to the volume of the sea from which they originate; this problem deserves further study.

31.1 WAKE GEOMETRY FROM AERIAL PHOTOGRAPHS

The serial views of destroyer wakes shown in Figures 2 to 6 of Chapter 26 were selected from a large series of photographs, made available by the Photographic Interpretation Center, U. S. Naval Air Station, Anacostia. They show wakes of the destroyer USS *Moale* (DD693) proceeding on a straight course at constant speeds, ranging from 16 to 34.5 knots. For each speed, photographs of three or more different runs were measured, so that the results represent a fair average. The following conclusions are drawn from measurements made on the original prints.

Immediately behind the screws the wake diverges with an included angle of about 50 degrees. Individual angles measured on different photographs vary between 40 and 60 degrees, but no clear-cut dependence on speed is indicated; these variations may well be spurious. It may be mentioned in passing that the wake of a stationary propeller [1] showed an angle of divergence of about 20 degrees. At a certain distance astern, the wide divergence of the destroyer wake ceases rather abruptly, and thereafter the wake spreads with a total included angle of about 1 degree.

This angle too appears to be independent of the speed with which the wake is laid. However, the distance astern at which the transition from the 50-degree divergence to the 1-degree divergence occurs increases very markedly with the speed of the destroyer. At 16 knots, it is about 65 ft, and at full speed about 280 ft; this variation of distance with speed is not linear, as far as present experience indicates. The numerical values are given in Section 35.1. Observations of several different types, which will be reported in the rest of this chapter, all seem to indicate that the wake spreads out with a large included angle immediately behind the wake-laying vessel, and that at distances astern greater than about 100 yd the wake spreads out with a very small included angle, of the order of 1 degree. However, none of these other observations have the same high intrinsic accuracy as the measurements on aerial photographs. Therefore, the results of these measurements, as incomplete as they are, have been selected for inclusion in Section 35.1. The large initial divergence of a wake is quite conspicuously demonstrated in Figures 1 and 7 of Chapter 26, showing the wakes of a submarine chaser and destroyer, respectively.

Aerial photographs also furnish interesting information on the cross-sectional structures of wakes. For instance, Figures 2 to 5 of Chapter 26 reveal that at short distances astern and at speeds less than 25 knots the destroyer wake has a dense core and edges that stand out conspicuously; with increasing distance, this internal structure gradually fades out. At speeds above 30 knots, the destroyer wake appears to be so strongly turbulent that the core is largely obliterated.

Transverse structure of a different kind is illustrated by the submarine wakes seen in Figures 1 to 6. The wake of a surfaced submarine shows bifurcation, or twin structure, both at 15 and 20 knots in Figures 3 and 4. The same illustrations clearly differentiate a short wake section immediately behind the submarine, which has a large angle of divergence, from

494

FIGURE 1. Wake of surfaced submarine at 6 knots.

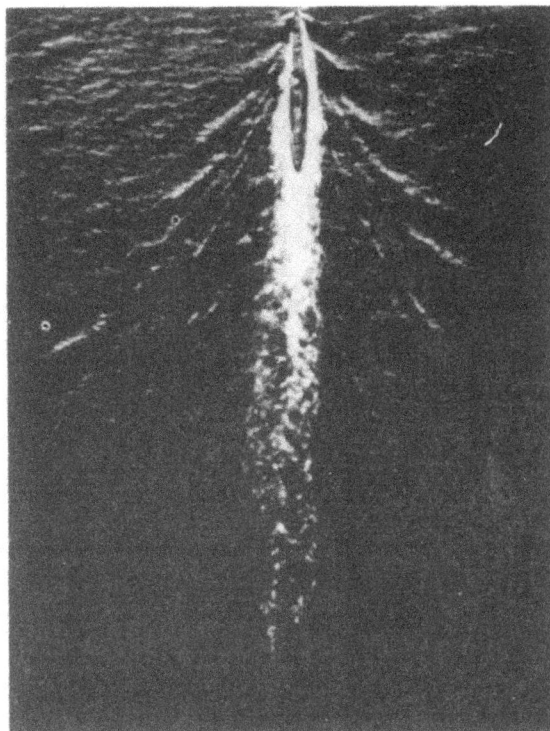

FIGURE 2. Wake of surfaced submarine at 10 knots.

the long wake proper, whose edges show little divergence. Thus, the general wake contour is quite similar for destroyers and surfaced submarines.

Figure 7 gives a remarkable aerial view of a PT boat and its wake. The wake proper is narrow and compact, without visible structure, but the bow wave, for a distance astern of several ship lengths, is visually much longer than the wake.

31.2 RATE OF WIDENING

The rate of widening of an acoustic wake can be determined by measuring the gradual increase of the duration of the echoes obtained with a horizontal sound beam, as long as the signal length is much shorter than the wake width. In practice, subtracting the signal length from the measured length of the echo will correct for the prolongation of the echo length due to the finite signal length and will make possible a direct determination of the wake width.

An analysis along these lines was made for four wakes laid by the E. W. Scripps on November 28, 1944. The Scripps passed between the echo-ranging vessel, the USS Jasper (PYc13), which was hove to, and a target sphere 3 ft in radius buoyed at a center

depth of 6.5 ft. The wakes were laid at right angles to the line connecting the sphere with the Jasper, each run being made in a new location of undisturbed water. All echoes were recorded oscillographically and sound range recorder traces were obtained simultaneously. The signals consisted of pulses of 0.5, 1, and 3 msec long, transmitted in cyclic succession. The duration of the 3-msec echoes was measured both on the oscillograms and on the recorder traces; the results, expressed in yards, are plotted in Figures 8 and 9 as functions of the time elapsed since the Scripps passed. The slope of these curves is the rate of widening. In order to find the width at any time, the plotted values of the echo length should be diminished by 2.4 yd.

The average rate of widening of the Scripps wake, up to the age of 10 minutes, is 5 yd per minute for the chemical recorder traces, and 6 yd per minute for the oscillograms. This difference of 1 yd per minute can hardly be regarded as significant, in view of the differences between the several runs. Note that in both illustrations the graph of Run 2 is located between 5 and 10 yd above the graph of Run 1, though both runs were made with 24-kc sound. The origin of this shift remains obscure, as the sea was unusually calm,

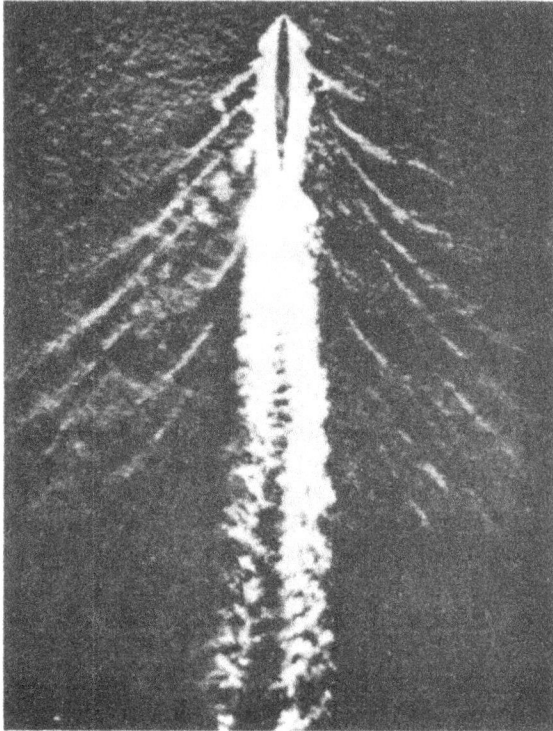

FIGURE 3.   Wake of surfaced submarine at 15 knots.

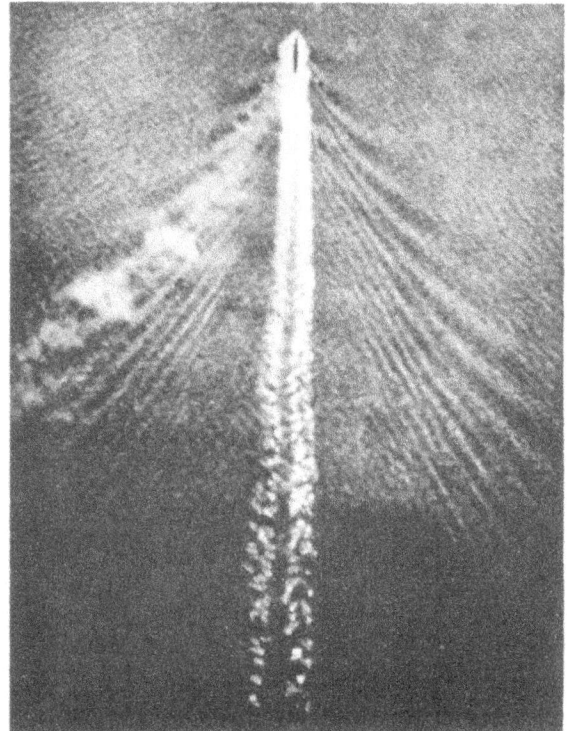

FIGURE 4.   Wake of surfaced submarine at 20 knots.

almost without ripples, during the entire day, and the *Scripps* maintained the same speed of 9.5 knots during all four runs. Evidently, the geometric and physical properties of wakes are difficult to reproduce in repeated experiments, even under ideal weather conditions. The wake laid in Run 2 gave distinct 60-kc echoes at an age of 30 to 40 minutes; during this period the wake width measured on the sound range recorder trace increased from 82 to 100 yd, corresponding to a rate of widening of about 2 yd per minute. No similar tests for the persistence of wakes were made during the other runs.

Comparison of Figures 8 and 9 reveals that there is no systematic difference between the widths found for different sound frequencies. In particular, the alternating use of 45 and 60 kc during Run 4 gave results which are mutually consistent and agree quite well with the 24-kc graphs.

At the stated speed of the *Scripps*, approximately 300 yd per minute, a rate of widening of 5 to 6 yd per minute means that the total angle of divergence of the wake is about 1 degree. This figure is in excellent agreement with the angle of divergence found for destroyer wakes from aerial photographs. If the *Scripps* wake had the same great initial angle of divergence (about 50 degrees) as the destroyer wakes, it could not have been discovered by acoustic width measurement, because this method lacks the necessary "resolving power" along the time axis. At wake ages greater than 10 minutes, the rate of widening appears to decline steadily, and the angle of divergence must decrease correspondingly. However, it should be remembered that these observations were made on a calm sea.

The rate of widening of thermal wakes can be studied by carrying a sensitive thermocouple across the wake at increasing distances astern. These investigations are still in an exploratory stage, but they are mentioned here because preliminary results have been reported for two wakes laid by the *E. W. Scripps*.[2] Thus a comparison of the thermal and acoustic dimensions of the wakes laid by the same vessel became possible. Between the ages of 10 and 60 minutes, the thermal wakes of the *Scripps* showed a linear increase in width from 30 to 50 yd. The rate of widening is about 0.4 yd per minute and the speed of the *Scripps* was 6 knots, or 200 yd per minute. Hence, the angle of divergence of the thermal wake

FIGURE 5. Wake of submarine during crash dive.

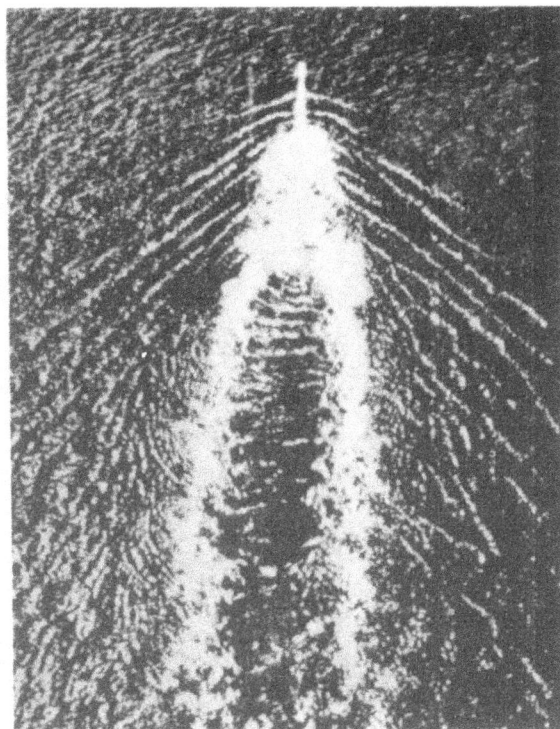

FIGURE 6. Swirl behind submarine after crash dive.

is only about 0.1 degree, or one-tenth of the divergence of the acoustic wake. In addition, the thermal wake appears to be much narrower than the acoustic one.

Since the thermal and acoustic measurements were not made on the same day, it is by no means certain that the thermal and acoustic wakes behave as differently as these observations would seem to suggest. The 1-degree divergence found acoustically applied to wakes less than 10 minutes old, while the available thermal data were apparently all for wakes more than 10 minutes old. Furthermore, the rate of widening may possibly depend on oceanographic factors, such as the temperature gradients in the surface layers of the sea.

31.3   FATHOMETER STUDIES

At the U. S. Navy Radio and Sound Laboratory [USNRSL], numerous measurements with a recording fathometer have been carried out on the wakes of a number of different surface vessels and submarines.[3] Some of these wakes were investigated systematically, and the width and depth of the wake was de-

termined as a function of the distance from the wake-laying vessel and of its speed.

31.3.1   Surface Vessel Wakes

With surface ships, two methods were used for measuring the wake width. When ranging on the wakes of ships which happened to be passing, the survey boat, carrying the fathometer, crossed the wake as nearly perpendicularly as could be judged while speed and distance measurements were made. Some inaccuracy arose in the judgment of the angle of crossing when very far behind the wake vessel. However, the error introduced into the measured width by assuming perpendicular crossing was usually negligible. When the survey boat was still farther astern, a greater error was present in determining the onset and disappearance of the wake record. The duration of recording was measured on chart paper (see Figures 3 to 6 in Chapter 30) by a caliper and rule. From the speed of the chart paper and of the survey boat, the width may be calculated.

In the other method, which was suitable at close range when working with an assigned vessel, the

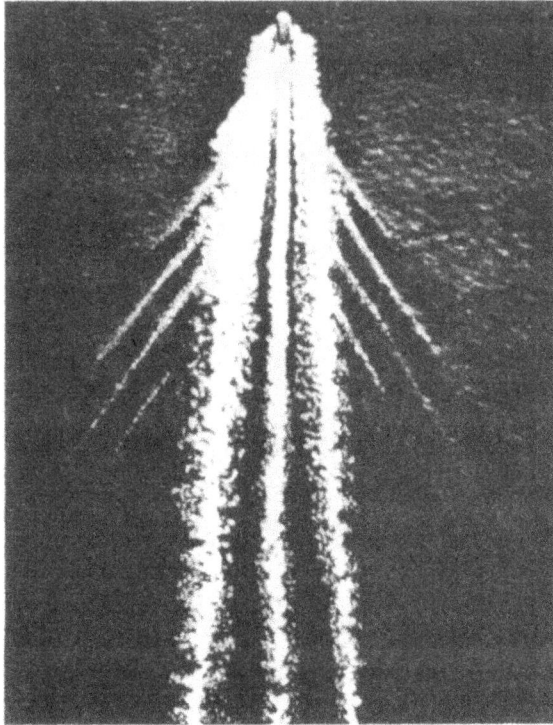

FIGURE 7.   Wake of PT boat at 25 knots.

survey boat was towed by the vessel laying the wake; the latter is called the wake vessel. Measurements can then be made on a wake whose lifetime is effectively constant, in other words, for a constant boat speed, the survey boat is in a wake of the same age at all times. While the wake vessel maintained a steady course, the survey boat under tow was moved in and out of the wake on either side by using the helm. At the moment the survey boat passed the wake edge, as indicated by the fathometer record, the record was marked and a signal was sent to two observers on the wake vessel. One of these observers was on the stern and followed the transverse movement of the survey boat with a pelorus. At the instant of signaling, the angle of the fathometer mounting on the survey boat relative to the axis of the wake ship was noted. The other observer was on the bridge, and at the signal he instantaneously observed the ship's compass course, for use in correcting for the angle of yaw of the wake vessel. The record was marked at the time of signaling the observers on the wake vessel, so that the data could be discarded if it were found that the survey boat was not exactly at the wake edge.

The wake of the *Jasper* (overall length 127 ft, draft 12 ft, beam 23 ft) was studied by the second

method over the range from 50 to 500 ft astern. All the measured widths, expressed in feet, agree with the formula

$$w = w_0 + 0.0625vt, \tag{1}$$

where w_0 is the extrapolated width at the stern of the wake vessel, v the speed of the ship in feet per second, and t the time in seconds since its passage. By differentiating this formula with respect to the time,

$$\frac{1}{v}\frac{dw}{dt} = 0.0625 = \sin\alpha, \tag{2}$$

where α, the total angle of divergence of the wake edges, is 3.5 degrees. For similarly small distances behind destroyers, the wake edges were found to include an angle of about 50 degrees (see Section 31.1). The conspicuous discrepancy between this value and the corresponding one for the *Jasper* is doubtless due to the different type of construction of these ships. The extrapolated value $w_0 = 10$ ft, in formula (1), is very nearly one-half the ship's beam. At distances greater than roughly 5 ship lengths, the divergence of the *Jasper's* wake ceased at a width of perhaps two and a half times the ship beam; only random measurements by the first method were available for this region, however, and thus a small divergence angle of about 1 degree cannot be ruled out as far as large distances behind the *Jasper* are concerned.

The measurement of wake thickness was carried out by proceeding into a wake and either remaining in its center while measuring distances and speeds, or by crisscrossing in order to investigate the thickness at points across the wake. Crisscrossing was necessary in order to locate the wake when operating at distances when the wake was not visible. A given wake will frequently have different acoustic transparencies at different points along its width. In some cases the thickness is the same along the width, and the greater transparency at the edge is caused by its less effective scattering properties. In other cases the wake is thinner at the edges. Some wakes are quite flat at the bottom, others are rounded at top and bottom, or may have one side which sinks below the other at both top and bottom.

A wake cross section asymmetrical in the vertical plane parallel to the beam of the vessel was frequently observed and is apparently correlated with wind direction. Such records were first noted when operating with one engine of a twin-screw vessel and were thought to be the result of this asymmetrical source. The wake of a sailing vessel was investigated next, and was found to have an even more pronounced

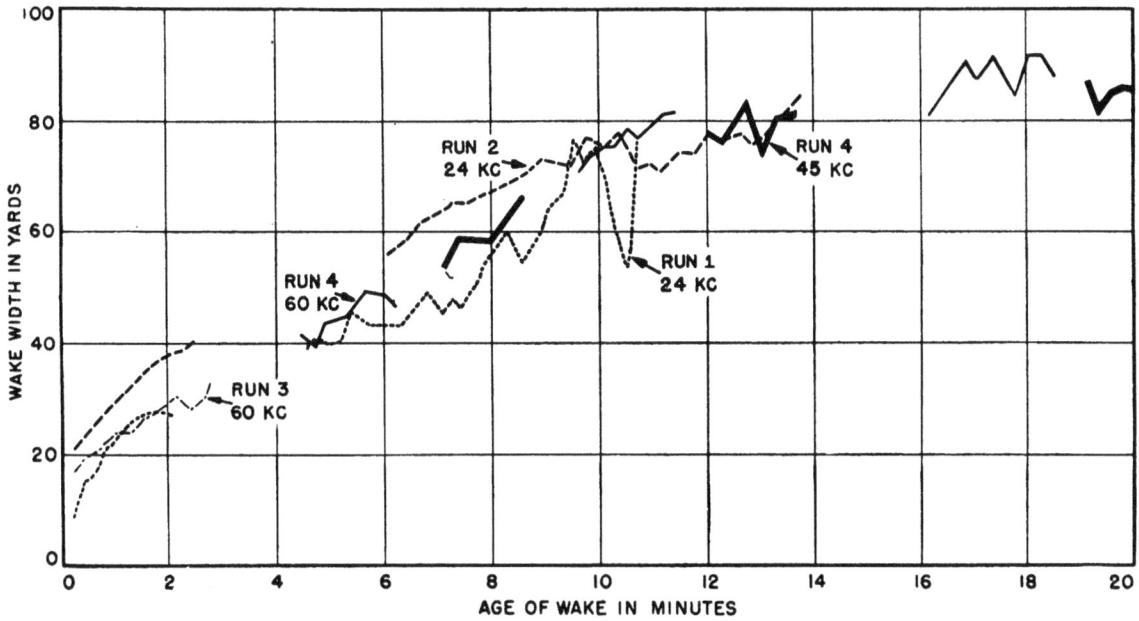

FIGURE 8.   Increase of wake echo duration for *E. W. Scripps* at 9.5 knots.   Measurements of oscillograms.

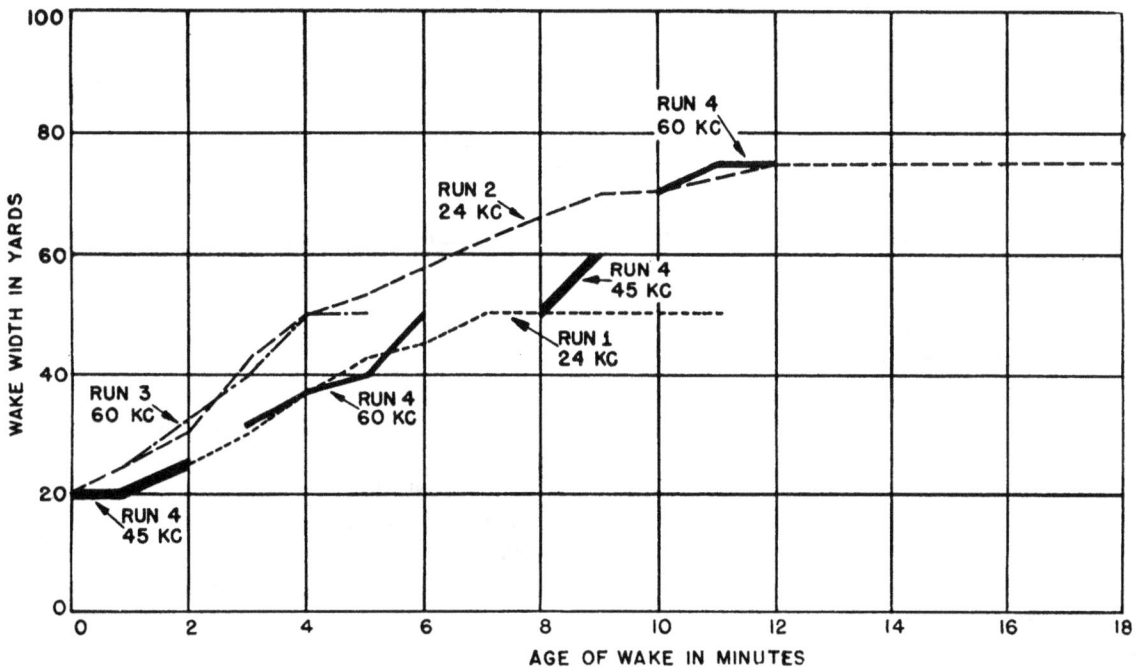

FIGURE 9.   Increase of wake echo duration for *E. W. Scripps* at 9.5 knots.   Measurements on chemical recorder traces.

asymmetry. The sailing vessel heeled over considerably during the runs and the varying area of the hull in contact with the water on either side was considered a cause for asymmetry. The wake of a twin-screw vessel when both screws were turning and when the wind was appreciable was then recorded. Again asymmetrical results were found. The effect is independent of the direction from which the survey boat crosses the wake. A typical value for the slope of the bottom of an asymmetric wake, as found for a 125-ft vessel, is 18 degrees. When the wind shifts from port to starboard, the cross-section geometry of the wake should change to a mirror image of its former geometry, but in the majority of cases this expectation is not entirely confirmed. Perhaps some as yet undiscovered parameter is responsible for this puzzling behavior.

results from actual variations of the wake structure.

The wake thicknesses were plotted as a function of the distance astern and examined for a possible systematic variation. The slope of the bottom of the wake up to 800 yd astern was found to be 5 minutes

TABLE 1. Wake thicknesses.

	USS *Rathburne* (ex-DD113)	USS *Hopewell* (DD681)	USCGC *Ewing*
Speed in knots	10–12	10	13
Thickness of wake in feet for average distance astern of 400 yd	19.5 ± 3.4	23.1 ± 3.6	13.7 ± 3.3
Range of thickness in feet	12–26	10–32	8–20
Ratio * of wake thickness to ship draft	1.63	1.85	1.52

* These ratios are smaller than those previously found[3] for incidental destroyer wakes, and are believed to be more accurate.

Aside from the miscellaneous results just described, an attempt was made to investigate systematically the variation of the thickness h of the wakes of two yachts, the USS *Jasper* (PYc13) and the E. W. *Scripps*, with distance astern up to 3,000 ft and with speed from 3.5 to 11 knots. For either vessel, no systematic changes of h could be noted. A fair average of all measurements of h was 1.70 times the draft, or 2.9 times the screw depth for the *Jasper*, and 1.11 times the draft, or 3.0 times the screw depth for the *Scripps* (overall length 104 ft, draft 12 ft, single screw 8 ft above the keel).

Scattered measurements made on the wakes of numerous large surface vessels of all types gave an average ratio of thickness to draft of 2.02. The wakes of small craft appear to be relatively thicker, with a thickness to draft ratio of the order of four. The only wake depth shallower than the draft was from a carrier wake 4,000 yd from the ship. For a speedboat, h appears to increase considerably with speed.

All these observations were made with a fathometer ranging downward from a measuring boat in the wake being investigated. Later measurements[4] were made with a fathometer mounted on the deck of a submerged submarine, ranging upward at the surface of the ocean. This method, for several reasons mentioned in Section 30.1.2, provided more accurate data than was possible with the former. The accuracy of the individual thickness determination is such that the range of wake thicknesses summarized in Table 1

of arc (or 4 ft per 1,000 yd) upward for the USS *Rathburne* (ex-DD113) and 16 minutes of arc (or 14 ft per 1,000 yd) downward for the *Ewing*. In other words, the differential quotient of the thickness of the wake with respect to the time, which will be required in the later discussion of the decay rate of wake strength, has the following values as upper limits:

$$\frac{1}{h}\frac{dh}{dt} = -0.08 \text{ min}^{-1} \text{ for the } Rathburne \text{ at 10 knots,}$$

$$\frac{1}{h}\frac{dh}{dt} = 0.04 \text{ min}^{-1} \text{ for the } Ewing \text{ at 13 knots.}$$

Additional information on the rate of widening of destroyer wakes is found in a report by UCDWR.[5] Wakes were laid by three different modern destroyers, running past the E. W. *Scripps* at 15 knots. The *Scripps* was hove to and recorded the sound level of a transducer carried repeatedly across the wake by a 50-ft motor launch. The sound level records showed definite breaks whenever the source crossed what may be called the acoustic boundaries of the wake; the time between these breaks was multiplied by the speed of the launch to give the width of the wake, suitable allowance being made for the occasional crossing occurring as much as 30 degrees away from the perpendicular transit. The plot of the entire data collected in this manner (Figure 22 in reference 5) suggested to the experimenters that new wakes widen more rapidly than old ones, with a total included angle of 2 degrees observed as far as 500 yd behind a 15-knot destroyer, and an included angle of 1 degree thereafter. However, a critical examination of the plot reveals such a large quartile deviation that the reality of the differentiation between new and old wakes seems somewhat doubtful. An included angle of $1\frac{1}{3}$ degrees for the entire range of observations, with the maximum distance astern of 2,500 yd, gives a fair representation of the plot. The extrapolated initial width of these destroyer wakes is roughly equal to the beam of the vessel, perhaps somewhat smaller. However, the observations do not cast any light on the very large initial divergence of destroyer wakes, revealed by aerial photographs, because the acoustic measurements did not extend to distances less than 100 yd astern. It should be noted that the angle of spread derived from the acoustic measurements ($1\frac{1}{3}$ degrees) is in fair agreement with that derived from aerial photographs (1 degree), as reported in Section 31.1. It is possible to attribute the difference of $\frac{1}{3}$ degree between the two figures entirely to the inaccuracies inherent in the respective processes of measurement.

31.3.2 Submarine Wakes

Information on the geometry of submarine wakes is less detailed. Among the measurements made with the fathometer ranging downward,[4] an investigation

TABLE 2. Wake of submarine at periscope depth.

Distance from periscope in yards	Wake top in feet	Depth of wake bottom in feet
67	39	70
100	0	27
117	0	40
152	0	36
215	0	31
315	0	25
319	0	28
350	0	30
450	0	23

of the wake of a fleet-type submarine 309 ft long is reported. At periscope depth the keel is submerged to a depth of 60 ft, the screws to a depth of 48 ft, and the deck to a depth of 35 ft below the surface; the speed was 5.5. knots. Table 2 contains the observed depths. The same information for a surfaced submarine of the same class, moving at 7 knots, is given in Table 3.

The maximum distance of 450 yd appearing in Table 2 is not the upper limit of detectability of the wake at a keel depth of 60 ft, as the observers emphasized.

The length of the subsurface wake of an S-class submarine,[6] running at 6 knots, was found to be about 1,000 yd at a depth of 45 ft, 235 yd at a depth of 90 ft, and 100 yd at a depth of 125 ft. These figures give the distances astern of the submarine over which the wake extends before it becomes undetectable by the gear used in these experiments. The bow-mounted 24-kc transducer was trained at a fixed bearing of 30 degrees relative to the *Jasper*,

TABLE 3. Wake of surfaced submarine.

Distance astern in yards	Wake bottom depth in feet
100	32
145	24
180	29
300	26
480	18
660	26
800	21
950	22

which was following the submarine on a parallel course and then gradually fell back. At creeping speed (2 to 3 knots) the length of the acoustically effective wake is less than 30 yd for a fleet-type submarine, according to recent San Diego observations.[7] It would seem, then, that the subsurface wake is not a good scatterer at greater than periscope depth, particularly at slow speeds. Analogous experiments [8] at frequencies of 24 and 45 kc were carried out with a fleet-type submarine, running at speeds up to 9 knots and at depths down to 400 ft. According to the observers, during no run was an echo definitely identified as coming from the wake alone.

From the data summarized in Table 2, it appears that the wake of a fleet-type submarine, running at 5.5 knots at periscope depth, extends to the surface at distances astern greater than 100 yd; the single record at a shorter distance of 67 yd, which suggested a completely submerged wake, unfortunately was uncertain. Later tests, using the same fathometer equipment with the fleet-type submarine USS *Trepang* (SS412), have led to a general confirmation of the previous results.[9] The *Trepang* was running at 8 knots at a keel depth of 60 ft, and the wake appeared at the surface at a distance of 600 ft astern. From this figure, the slope of the top of the wake may be computed, assuming it is constant; the ratio of screw depth to distance of emergence of wake, 48/600 or 0.08, corresponds to a total angle of divergence of 9 degrees at the screws. This value, however, is based on only one record. No clean-cut wake records were obtained at greater depths (200 and 400 ft), but this may be attributed to purely operational difficulties since the submarine found it difficult to pass directly under the stationary launch carrying the fathometer.

Chapter 32

OBSERVED TRANSMISSION THROUGH WAKES

IN CROSSING A WAKE, sound undergoes a transmission loss in addition to that resulting from propagation through the ocean at large. Transmission loss in the ocean is primarily geometric — the sound beam spreads over large distances because of the inverse square law and because of refraction conditions. At frequencies less than 100 kc, the transmission loss from physical causes, such as scattering and absorption, is not very important at the short ranges — a few hundred yards or so — of interest in wake measurements.

The observed transmission loss in wakes, however, is ascribed exclusively to physical causes, scattering and absorption by air bubbles, because the dimensions of wakes are much smaller than the distances over which the geometric effects are particularly important. An exception to this rather sweeping statement may have to be made in the case of sound originating in the wake, as described in Section 32.3. These phenomena, however, are little understood at present, as they have not been sufficiently studied.

32.1 DEFINITIONS

The physics of the transmission of sound through wakes has already been fully discussed in Chapter 28. All that is necessary here is to summarize the conventions concerning the expression and presentation of the measurements of the transmission loss through wakes.

The total transmission loss undergone by a sound beam on traversing a wake, or the *attenuation*, as it is usually called in underwater sound work, is defined by the equation

$$H_w = 10 \log \frac{I(0)}{I(w)}, \qquad (1)$$

where $I(0)$ is the intensity of a parallel beam of underwater sound before entering the wake, and $I(w)$ is its intensity after it has penetrated the entire width w of the wake; the transmission loss in the

wake H_w is distinguished by the subscript w from the transmission loss in the ocean at large, which is commonly denoted by the symbol H. According to equation (53) of Chapter 28, the attenuation by the wake can be represented as a product, namely

$$H_w = K_e w, \qquad (2)$$

where w is the geometric width of the wake, usually measured in yards, and K_e is the so-called *coefficient of attenuation* in decibels per yard. Definitions (1) and (2), as they stand, apply to a sound beam impinging perpendicularly upon the wake; for oblique incidence w obviously has to be replaced by $w \sec \beta$, where β is the angle included between the beam and a line perpendicular to the wake.

Note that equation (1) may be written in the form

$$\frac{I(w)}{I(0)} = 10^{-H_w/10}, \qquad (3)$$

or, by substitution from equation (2),

$$\frac{I(w)}{I(0)} = 10^{-K_e w/10}. \qquad (4)$$

Both H_w and K_e are overall properties of the wake, and it remains to express them as functions of the physical parameters describing the microstructure of the wake, which is known to consist of multitudes of bubbles of all sizes. The acoustic properties of bubbles have been characterized in Chapter 28 by their individual cross sections σ_s, σ_a, σ_e for scattering, absorption, and extinction of sound, respectively. It will be remembered that these quantities vary considerably according to the size of the bubbles, and that, by and large, only bubbles near resonant size make a significant contribution to the average cross section applying to a population of bubbles of all sizes.

Should all the bubbles in the wake happen to have exactly the same size, the coefficient of attenuation would be given by [see equation (55) of Chapter 28]

$$K_e = \frac{H_w}{w} = 4.34 \bar{n} \sigma_e \text{ db per cm} \qquad (5)$$

or
$$K_e = \frac{H_w}{w} = 396.8\bar{n}\sigma_e \text{ db per yd} , \qquad (6)$$

where σ_e in square centimeters is the extinction cross section of this particular size of bubble and $\bar{n}$ in cm^{-3} is the average number of bubbles of this size per cubic centimeter in the wake, as defined by equation (54) of Chapter 28. In the more realistic case of bubbles of many sizes, the attenuation coefficient is given by equation (67) of Chapter 28,

$$K_e = \frac{H_w}{w} = 1.4 \times 10^5 u(R_r) \text{ db per yd} , \qquad (7)$$

where $u(R)dR$ is the total volume of air contributed by bubbles with radii between R and $R + dR$ in 1 cu cm of the air-water mixture, or rather the average of this quantity taken over the entire column in which the sound beam and the wake intersect; R_r in equation (7) is the radius of the resonant bubbles corresponding to the sound frequency used in determining H_w.

The total attenuation corresponding to equation (5) is

$$H_w = 4.34\sigma_e\bar{n}w = 4.34\sigma_e N(w) , \qquad (8)$$

where $N(w) = \bar{n}w$ denotes the total number of bubbles in a column of unit cross section. Thus $N(w)$ is a measure of the total bubble population affecting the sound beam.

Differentiating equation (8) logarithmically with respect to the time, the decay of the transmission loss across the wake is obtained.

$$\frac{1}{H_w}\frac{dH_w}{dt} = \frac{1}{N(w)}\frac{dN(w)}{dt} . \qquad (9)$$

At first, $dN(w)/dt$ perhaps will be positive for sufficiently small bubbles, whose number might be increased rapidly by the gradual dissolution of bubbles of originally larger size. But ultimately, $dN(w)/dt$ must become negative. It is seen then that the decay rate of the transmission loss affords a direct measure of the rate of disintegration of the bubble population.

32.2 EXPERIMENTAL PROCEDURES

In principle, the experimental determination of the transmission loss through the wake requires only relative measurements of sound intensities. If over the period of observations the range from transducer to hydrophone, and the transducer output and hydrophone sensitivity remain constant, then the absolute values of any of these three quantities does not have to be known; the difference of sound levels recorded by the hydrophone before and after the wake has been laid across the sound beam is simply the transmission loss H_w. Whenever, during the course of experiments, the range changes appreciably a correction must be applied, based on the appropriate value of the transmission loss H in the surrounding ocean. Care should be taken to place both transducer and hydrophone at such a depth that they are completely hidden from each other by the wake.

A characteristic feature of transmission measurements of this simplest type is that, for sufficiently short wavelengths, only a very narrow cone of the divergent sound beam emitted from the transducer is utilized, namely the solid angle subtended by the face of the hydrophone at the location of the transducer. Thus, the instantaneously recorded transmission loss is for a sharply bounded layer of the wake. The roll and pitch of the vessel carrying the transducer and hydrophone will raise and lower both instruments and will cause that narrow pencil of sound to traverse the wake at different depths below the ocean surface. Since the acoustic thickness of the wake is likely to vary somewhat vertically, corresponding variations of the measured transmission loss must be expected.

In one respect these variations are even helpful. They afford an automatic smoothing out of the vertical variations of the acoustic thickness and thus produce a better representation of the average state of the wake. The directivity of the sound gear is also important, in so far as rolling and pitching of the vessels carrying the transducer and hydrophone, together with possible training errors, may affect their relative orientation and hence may cause fluctuations in the strength of the signals received. In practice, this effect cannot be separated from other fluctuations of the signals, resulting from changes of the transmission loss in the ocean interposed between transducer and hydrophone. By averaging over long series of signals, these disturbing influences may be minimized, though perhaps not fully eliminated.

32.3 TRANSMISSION LOSS ACROSS WAKES

32.3.1 One-Way Horizontal Transmission Loss

Transmission loss in wakes has been investigated comprehensively only for five vessels of the destroyer

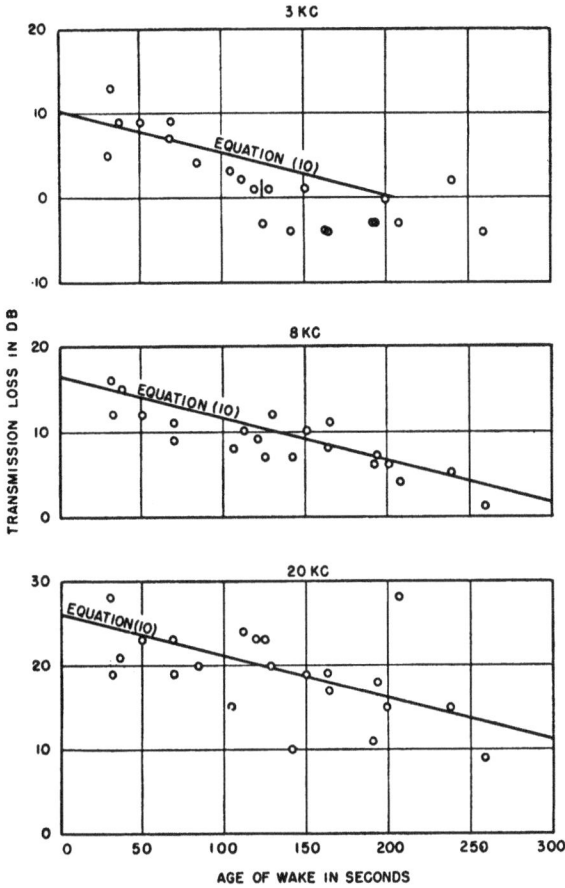

FIGURE 1. Sound transmission loss due to wake versus age of wake. Ship IV, December 30, 1943, 15 knots. Source beyond wake.

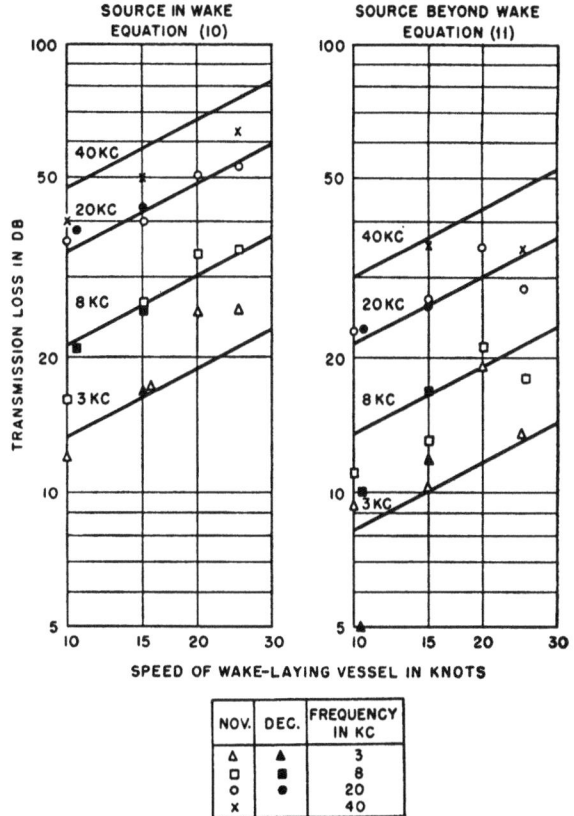

FIGURE 2. Dependence of transmission loss on speed of wake-laying vessel.

type — two old destroyers of the 1916–1917 class, a new destroyer of the *Fletcher* class, and two destroyer escorts.[1] Wakes were laid at speeds of 10, 15, 20, and 25 knots. A 50-ft motor launch repeatedly carried the projectors, mounted at depths of 6 and 7 ft, respectively, across the wake, while the hydrophones were suspended from the bow of the *E. W. Scripps* at a depth of 10 ft, about half the depth of the wake (see Section 31.3). Sound at frequencies of 3, 8, 20, and 40 kc was recorded both with the launch beyond the wake and with the launch inside the wake; while sound recorded when the launch was on the near side of the wake provided reference values.

By applying a correction for the measured average transmission loss in the ocean, all sound levels were reduced to a standard distance of 100 yd, for the three cases of (1) source beyond wake, (2) source in wake, and (3) no wake intervening. The difference between case (3) no wake intervening and case (1) source beyond wake was taken to be the transmission

loss for the source beyond the wake, or H_w as defined in equation (3) of Section 32.1; similarly, the difference between case (3) no wake intervening and case (2) source in wake was taken to be the transmission loss for the source in the wake.

In the original paper, the results are reproduced in separate graphs for each of the several vessels, speeds, frequencies, and locations of the source; one of these is reproduced in Figure 1. However, not all the possible combinations of the different parameters are actually shown. Although not representing the best fit for every single set of observations, the following interpolation formulas are believed to represent adequately most of the data.

Source in wake $H_w' = 2.4(vf)^{\frac{1}{2}} - (4.8 \pm 1.6)t$, (10)

Source beyond wake $H_w = 1.5(vf)^{\frac{1}{2}} - (3.0 \pm 1.4)t$, (11)

where v is the ship's speed in knots, f is the frequency of the sound in kilocycles and t is the time in minutes which has elapsed since the passage of the screws or age of the wake. No standard errors are assigned in the original report to the numerical coefficients of the first terms of equations (10) and (11), but it is stated that the initial values ($t = 0$) of the transmission loss for individual runs show a scatter of the order of 3 db. Figure 2 gives an idea of the accuracy with which equations (10) and (11) represent the initial transmission loss at different speeds and frequencies.

A higher transmission loss for case (2) source in wake, than for case (1) source beyond wake, appears to be well established observationally, but the theoretical explanation for this systematic difference is not at all evident. With the source located inside the aerated water of the wake, air bubbles are likely to be held on the face of the transducer by adsorption. There are theoretical reasons for believing that such a layer of adsorbed gas should reduce, or "quench," the output of the transducer, causing an apparent increase of the transmission loss in case (2). However, it is somewhat surprising that the quenching effect should show a behavior regular enough to follow equation(10).

The difference in the decay rate for case (2) the source in wake and case (1) the source beyond wake is

$$\frac{dH'_w}{dt} - \frac{dH_w}{dt} = 1.8 \text{ db per minute.}$$

This difference may not be significant in view of the standard errors of these quantities. However, if it is accepted at its face value, the relative rates of decay are equal to each other, and given for fresh wakes by the equation

$$\frac{1}{H_w}\frac{dH_w}{dt} = \frac{1}{H'_w}\frac{dH'_w}{dt} = -\frac{2}{(vf)^{\frac{1}{2}}}. \quad (12)$$

This equality between the two rates is evident from equations (10) and (11) in which corresponding coefficients have the same ratio of 5/8. According to equation (9) of Section 32.1, the relative rate of decay is a function solely of the rate of disintegration of the bubble population. The physical significance of the observed decay rate will be discussed in Section 34.4.

Some incidental information on the transmission loss across wakes has been obtained during measurements of the underwater sound output at 5 kc of a destroyer, cruiser, and aircraft carrier [2] observed at varying speeds. Measurements at higher frequencies were also made, but the results are inconclusive as far as the transmission loss across wakes is concerned. All that can be said about the transmission loss at 25 and 60 kc is that it is distinctly higher than at 5 kc;

residual sound intensities, after passage through the wake, in most cases had dropped to the background noise and thus made impossible an evaluation of the transmission loss. Even the 5-kc data, plotted as a function of age of the wake, are rather widely scattered. But for each of these vessels the plot is not inconsistent with tentative predictions made from equation (11) above, a fact that is somewhat surprising in view of the dimensions of two of these three ships listed in Table 1. The initial transmission loss

TABLE 1.   Ship dimensions.

	Destroyer USS *Colhoun* (DD801)	Light Cruiser USS *Trenton* (CL11)	Carrier USS *Hancock* (CV19)
Length in feet	376	555	874
Beam in feet	39	55	93
Draft in feet	13	13	29
Screw depth in feet	11.25	19.5	21.3

($t = 0$) is about 15 db at 5 kc for each vessel, while formula (11) gives 17 db at 25 knots for 5-kc sound. At higher frequencies, the greater absolute value of the transmission loss might facilitate the detection of possible differences between the destroyer and the larger ships.

32.3.2   Two-Way Horizontal Transmission Loss

Another method of measuring the horizontal transmission loss has been tried out in experiments aimed at a simultaneous determination of wake echo strength and transmission loss.[3] The *E. W. Scripps*, running at 9.5 knots on a straight course, laid a wake between the USS *Jasper* (PYc13) and a target sphere 3 ft in diameter buoyed at a center depth of 6.5 ft. The drop in the apparent target strength after the wake was introduced thus was taken to represent the two-way transmission loss across the wake. The wake echo intensity could also be measured on each oscillogram, giving the effectiveness of the wake as a scatterer of sound. A plot of the sphere and wake echo levels for one of the 24-kc runs is shown in Figure 7 of Chapter 33. The results are summarized in Table 2; further reference to the decay rate of the echo strength will be made in Section 33.4.

The 45 and 60-kc data on which the echo strength

recovery and decay rates quoted are based were taken quasi-simultaneously by tuning the sonar equipment alternately to the two frequencies for two-minute intervals. The wake and sphere distances for this run are those quoted in the 45-kc row. The 9-db drop in apparent target strength at 60 kc is based on a separate run, with wake and target distances as stated in the third row of the table.

32.3.3 Two-Way Vertical Transmission Loss

A recording fathometer has been used for the measurement of sound transmission loss in the vertical direction through surface ship wakes.[5] The fathometer was secured on the deck of the submarine USS S–18 (SS123) so as to range upward onto the surface

TABLE 2.   Effect of wake on sphere echoes.

Frequency in kc	Distance to wake center in yards	Distance to sphere in yards	Depth at which sound beam passed through wake in feet	Maximum drop in sphere echo level with wake present in db	Rate of recovery of sphere echoes in db per minute	Rate of decay of wake echoes in db per minute
24	270	350	10	6	1.4	1.5
45	97	162	12	No data with fresh wake		0.7
60	58	98	12	9	0.8	0.7

Earlier measurements of the *Scripps* wake[4] gave a depth of the wake bottom of 13 ft. According to the values quoted in Table 1, the sound beam passed definitely above this bottom depth of 13 ft. However, a short time before the data of Table 2 were obtained, the *Scripps* had been outfitted with a new engine and propeller, so that the wake dimensions may have been altered to some extent. The present propeller is 3.8 ft in diameter and the shaft is 5.5 ft below water line. Therefore, since the *Jasper's* sound projector is 15 ft deep, maximum acoustic shadowing of the sphere by the wake could not be expected immediately after the *Scripps* had passed. These wakes widened laterally, as measured by the wake echo elongation, at about 6 yd per minute. The same rate of spreading may also be applicable in the vertical sense without necessarily implying that a strongly absorbent "core" of the wake ever moves down to an effective screening position in these experiments. This may account for the low magnitude of the observed transmission loss.

Similarly the decay rate of the transmission loss dH_w/dt is one-half the rate of recovery of the sphere echoes; hence dH_w/dt is 0.7 and 0.4 db per minute for 24- and 60-kc sound, respectively. These decay rates are much smaller than that of destroyer wakes, which were found to be independent of frequency — 3.0 db per minute. However, the relative rates of decay are in moderate agreement with those computed from equation (10) for a destroyer speed of 10 knots; numerical values are shown in Table 3.

of the ocean, the echoes being continuously recorded in the control room. With this arrangement, the effect of a surface ship wake is recorded as the submarine passes beneath it. The ocean surface is used as a "standard target." Sample records obtained with this method are shown in Figures 4 to 6 of Chapter 30.

TABLE 3.   Relative rates of decay.

	24 kc	60 kc
$\frac{1}{H_w}\frac{dH_w}{dt}$ for *E. W. Scripps* at 9.5 knots (from Table 2)	0.23 min⁻¹	0.09 min⁻¹
$\frac{1}{H_w}\frac{dH_w}{dt}$ for DD at 10 knots [from equation (10)]	0.13 min⁻¹	0.08 min⁻¹

Quantitative transmission loss results are obtained from the fathometer records by a special procedure of operating the instrument in conjunction with calibration records made in the laboratory. As the submarine passes beneath a surface ship wake, an attenuator in the receiver-amplifier is adjusted so that the effect of the wake plus the effect of the attenuator is such that a light gray "voltage-sensitive" record of the ocean surface echo is produced on the chart paper. Some practice is required, as very little trial-and-error time is available while the submarine is directly below the wake.

The procedure is completed by determining, in effect, the amount of amplifier attenuation required to record the unobscured ocean surface at the same density as that of the record taken below the wake. The difference of attenuator settings in the two cases yields the wake transmission loss directly. The actual procedure was complicated by the lack of a calibrated attenuator; the details of the necessary laboratory calibration of the gain control by matching records for different gain settings and echo levels are described in reference 5. The fathometer record yields an accurate value of wake thickness in each case so that attenuation coefficients can be computed in decibels per foot of wake thickness.

The coefficient of reflection at the ocean surface cancels from the measured transmission loss, because μ affects the sound levels both in and outside the wake in an identical manner, aside from slow variations of the state of the sea. If the ocean surface were a perfect plane, and if the axis of the sound beam impinged upon it perpendicularly, the entire off-axis output of the fathometer would be reflected so as not to return to the transducer. On account of the waves, swells, and other irregularities of the surface, and because of imperfect leveling of the submarine, actually some off-axis sound is reflected back on to the face of the transducer. Hence, it is necessary to keep the submarine at a depth shallow enough to make the central lobe of the sound beam fall entirely inside the wake. This condition was well fulfilled during these experiments, the results of which will now be described.

Sound of 21 kc was found to undergo an average attenuation of 18 ± 3 db during vertical one-way passage through the wakes about 400 yd behind the USS *Rathburne* (APD25, ex-DD113), USS *Hopewell* (DD681) and USCGC *Ewing*, traveling at speeds of 10 to 13 knots. Combining these total attenuations with the wake depths h for the vessels, accurately determined from the same records and already discussed in Section 31.3, average attenuation coefficients in the vertical direction in decibels per yard could be computed and were found to be 3.0 ± 0.6 db per yd for the *Rathburne* and *Hopewell* and 4.8 ± 1.5 db per yd for the *Ewing*. These are grand averages, disregarding differences in the distance astern, which are unknown in many cases, and disregarding deviations of the point of measurements from the center of the wake; moreover, some "knuckles" are included with the straight runs. If only data referring to known distances astern and to the center of straight

wakes are retained, all observations applying to wakes laid by the *Hopewell* are eliminated. Plotting as a function of the distance astern, the attenuation coefficients for the wake of the *Rathburne*, running at a speed of 10 knots (corresponding to a screw-tip speed of 52 ft per sec), the following linear interpolation formula is found for the range from 100 to 800 yd:

$$\frac{H_w}{h} = (3.135 \pm 0.057) + (0.093 \pm 0.018) \times$$

$$\left(\frac{\text{distance astern}}{100 \text{ yd}}\right) \text{ db per yd} . \quad (13)$$

Since there is apparently no correlation between the total transmission loss in the wake H_w and the distance astern, equation (13) implies that the wake becomes thinner in the vertical direction as it ages. A similar plot for the *Ewing*, running at 13 knots, reveals an enormous variation of the attenuation coefficient ranging from 2.4 to 6.6 db per yd without any clear dependence on the distance astern. The distances astern cannot, however, be very accurately determined in these experiments. There is no obvious explanation why the *Ewing* data should show a greater scatter, enough to obliterate any dependence on distance astern. The higher value of the attenuation coefficients for the *Ewing* has been associated tentatively with the greater screw-tip speed (112 ft per sec at 13 knots) of this vessel, compared with the two destroyers. No corroboration for this surmise could be found among the observations of the horizontal transmission loss through wakes laid by different destroyers, already described in Section 32.3.1, although the screw-tip speeds of these vessels ranged from 80 to 137 ft per sec at 15 knots, and from 53 to 95 ft per sec at 10 knots.

It is of interest to compare the attenuation coefficient in the vertical direction with that computed from the total transmission loss measured horizontally. According to equation (11) in Chapter 32, the one-way horizontal transmission loss through the wake 400 yd behind a destroyer traveling at 10 knots is about 20 db for 21-kc sound. The width of this wake is about 75 ft, according to Figure 22 of reference 1. Hence, the horizontal attenuation coefficient is about 0.9 db per yd, or about one-third of the vertical one, as reported above for the destroyers *Rathburne* and *Hopewell*. This discrepancy is probably real, but hardly disturbing. In fact, the average attenuation coefficient would be expected to be smaller horizontally than vertically in case the wake has a strong core and weaker fringes, because the vertical measurements refer to the center of the wakes.

32.4 PROPAGATION ALONG WAKES

On the whole, the methods employed for the study of sound propagation across wakes, described in Section 32.3, have led to apparently consistent results. For sound propagation along wakes, however, the observations do not fit easily into the general picture; they are a few in number and provide insufficient data to permit a complete analysis of all the factors involved.

A mechanical noisemaker [6] was towed both in and below the wake of a destroyer running at 10 and 14 knots, and sound levels were recorded simultaneously by two hydrophones — one towed in the wake by the destroyer and the other suspended at a depth of 10 ft from a boat which was hove to. The destroyer followed a straight course past this boat, while the distance between the noisemaker and the towed hydrophone was steadily increased from 50 ft to 1,200 ft by unreeling the hydrophone cable.

As the cable lengthened the hydrophone gradually descended, ultimately passing below the bottom of the wake, which was assumed to be 20 ft below the surface.[5] The noisemaker was towed 50 ft behind the destroyer, and the hydrophone reached a depth of 20 ft at distances of 400 ft (10 knots) and 1,000 ft (14 knots) behind the noisemaker.

Finding the transmission loss along the wake would require comparing the sound levels recorded by the towed hydrophone with levels recorded by a hydrophone when no wake is present in the direction of the noisemaker. Unfortunately, the levels recorded by the stationary hydrophone, suspended from the boat outside the wake, cannot be used, because the directivity pattern of the noisemaker is unknown. It should be noted that the aspect of the noisemaker, as viewed from the towed hydrophone, is practically constant, while the aspect of the noisemaker relative to the stationary hydrophone changes by about 90 degrees while the destroyer is moving toward, or receding from the point of closest approach.

The sound levels obtained by the towed hydrophone with the noisemaker towed at a depth of 40 ft may serve as an approximate reference level representing the wake-free state, because then most of the path from the noisemaker to the hydrophone runs below the wake. Subtracting these sound levels from the ones applying to the noisemaker towed in a wake, an approximate value for the transmission loss along the wake is found. The numerical values are about 6 db for 3-kc sound and about 13 db for 8-kc sound at a speed of 10 knots; at 14 knots, the values are about 13 db and 30 db for 3-kc and 8-kc sound respectively. These transmission losses are of the same order of magnitude as those found in propagation across wakes. The increase of transmission loss with frequency is also in agreement with what has been learned about sound transmission across wakes.

However, for the entire range covered (100 to 1,000 ft) the transmission loss along the wake does not show the expected increase with distance from hydrophone to noisemaker. The sound levels used as reference values, with the noisemaker 40 ft below the surface, vary inversely as the square of the distance between hydrophone and noisemaker. But the sound levels recorded with the noisemaker in the wake also follow approximately the same inverse square law. In other words, the measured transmission anomalies fail to show any increase with distance behind the noisemaker, which would readily be interpreted as caused by attenuation inside the wake. There is even a slight decrease, perhaps 3 or 4 db, over a range of 1,000 ft; however, this decrease may result from the presence of bottom-reflected sound. Measurements of the destroyer ship sound, with no noisemaker present, gave results similar to those obtained with the noisemaker. These observations are rather puzzling.

The measurements of the sound output of a destroyer, cruiser, and aircraft carrier [2] give additional evidence of a very low transmission loss along wakes. During the so-called Z runs, the vessel to be measured passed the measuring vessel, which was hove to, and then made a turn so that, during the receding run, the axis of the wake coincided with the line connecting the stationary vessel with the receding one. The sound levels recorded were corrected for the transmission loss resulting merely from geometrical divergence according to the inverse square law, and from the corrected levels a transmission anomaly was derived. Attenuation coefficients along the wake were found to be 10 to 80 db per kyd; these attenuation coefficients were not judged sufficiently accurate to warrant a discussion of variation with speed (16 to 30 knots), frequency (5, 25, and 60 kc) and ship type. In order to appreciate fully how small those attenuation coefficients measured along wakes are, it should be remembered that the measured transmission loss across wakes (see Section 32.3) corresponds to attenu-

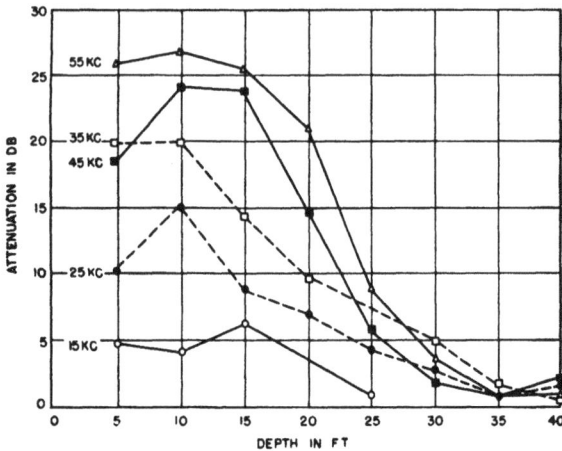

FIGURE 3.   Ten-inch propeller, 1,600 rpm.

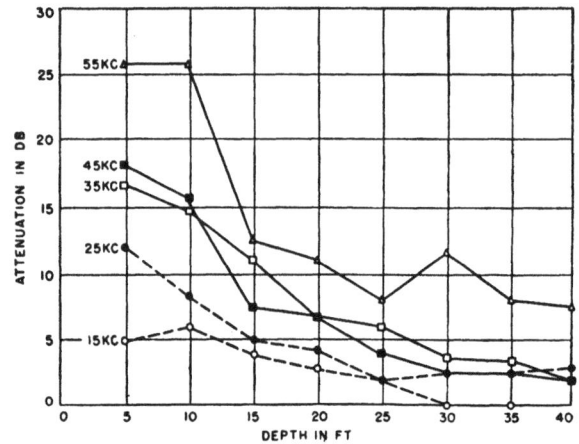

FIGURE 4.   Fourteen-inch propeller, 1,600 rpm.

ation coefficients of 300 to 6,000 db per kyd. This enormous difference is apparently real but has not yet been explained.

32.5 TRANSMISSION LOSS IN MODEL PROPELLER WAKES

Attenuation measurements on wakes of ships under way have been supplemented by experiments with wakes of a stationary model propeller. At the Woods Hole Oceanographic Institution,[2] an electrically operated device was constructed for driving submarine propellers at speeds ranging from 266 to 1,600 revolutions per minute at various depths. This equipment was used in water 70 ft deep.

First the relation between sound output and speed of the propellers at constant depth was studied, and the critical speed marking the onset of cavitation was determined. Four propellers ranging from 10 to 20 in. in diameter were employed. The noise level increased sharply whenever the tip speed of the propeller blades exceeded 33 ft per sec. In earlier experiments with 2-in. propellers, mounted in an experimental chamber, a critical speed of 35 ft per sec had been found at the same hydrostatic pressure. The agreement between these two figures appears quite satisfactory.

Precise measurements of the attenuation were obtained by an arrangement in which the transducer and hydrophone were mounted on opposite sides of the wake on a pipe frame attached to the boom carrying the propeller and held rigid by wire stays. The instruments were 9 ft behind the hub of the propeller. In this way the axis of the wake was made to pass

between the transducer and the hydrophone, which were on opposite sides of it at a fixed distance of 6 ft. This arrangement had the advantage that it was easy to handle and could be used in deep water with complete assurance that the position of the instruments relative to the propeller would not change. However, it did not allow any variation of the distance between the instruments and the propeller. Hence, it was impossible to determine the decay rate along the wake.

With this arrangement measurements of sound attenuation were made systematically at different depths and with different frequencies. Each measurement of attenuation involved the observation of the response of the hydrophone under three conditions: (1) with oscillator on and the propeller at rest; (2) with the oscillator on and the propeller turning; (3) with oscillator off and the propeller turning. By suitable combination of these data it was possible to correct the observations for the noise produced by the propeller. Typical results for the different propellers are illustrated in Figures 3 and 4.

First, it will be noted that the attenuation increases with frequency, being almost absent at 10 kc and rising steadily to 60 kc. This increase with frequency, at any fixed depth, is so steep that it definitely exceeds the increase with frequency of the transmission loss through destroyer wakes, which is approximately proportional to the square root of the frequency [see Section 32.3.1, equations (10) and (11)]. Second, at each frequency, the attenuation diminishes considerably with depth. This effect is more pronounced at the higher frequencies. Since the destroyer wakes have an average depth of 20 ft, and the transmission loss through them is a sort of average over this entire range of depths, the second effect partially cancels the first one. Moreover, the bubble population found in a destroyer wake may be quite different from that in the wake of the stationary propeller (zero slip), and any variation of the relative abundance of bubbles of different sizes is likely to produce frequency-dependent acoustic effects. Hence, the discrepancy noted above is not alarming.

In the case of the 10-in. propeller (Figure 3), the attenuation falls to almost zero at a depth of 40 ft for all frequencies. This phenomenon is consistent with the results obtained in the model chamber mentioned. In the model experiments it was found that fewer nonpersistent cavities were formed at higher pressures. According to the mechanism of bubble formation described in Section 27.1 higher pressure causes a more rapid collapse of the cavities formed, before there is time for a considerable amount of gas to diffuse into them; moreover, cavities containing a given amount of gas are compressed into bubbles of smaller size at higher pressures.

Observations were also made at two frequencies with transducer and hydrophone in the wake, both mounted in the wake axis, but their number is small and no clear-cut conclusions can be drawn from them. There is some indication that for 50-kc sound the output of the transducer may be reduced, or "quenched" by the wake, but for 10 kc the "quenching" effect, if it exists at all, is very much smaller than for 50 kc.

In summary, the observations of wakes produced by a stationary propeller are in reasonable agreement with those of destroyer wakes, as far as the dependence on frequency of the transmission loss is concerned. By dividing the attenuations plotted in Figures 3 and 4 by the distance between transducer and hydrophone (6 ft), attenuation coefficients can be computed. For instance, at a depth of 15 ft attenuation coefficients of 3.6 and 2.3 db per yd are found for 25-kc sound, which is the same order of magnitude as found for destroyer wakes at speeds of 10 to 15 knots. Since the diameter of the wake probably was smaller than the distance from transducer to hydrophone, the values quoted for the attenuation coefficient are actually lower limits; the true attenuation coefficient may have been greater by 50 to 100 per cent.

Chapter 33

OBSERVATIONS OF WAKE ECHOES

Echoes from wakes, like those obtained from other targets, vary considerably with the type of sound gear employed, the prevailing oceanographic conditions, and the physical constitution of the wake. Before the observations are reviewed, it is necessary to outline the theoretical concepts entering into the reduction of the crude data obtained by measurement.

As regards the physical mechanism by which sound is returned from a wake to an echo-ranging transducer, two limiting cases can be imagined. On one hand, the multitude of microscopic scatterers may be spread out so thinly that the phases of the scattered sound waves are distributed at random — that is, so that constructive and destructive interference are equally probable. Then the average power returned to the transducer is obtained by summing up the contributions from the individual scatterers. On the other hand, a wake might reflect sound specularly. This alternative would occur only if the concentration of scatterers near the wake surface increased inwardly very rapidly. It is undecided as yet whether or not specular reflection from wakes does occur; inconclusive evidence on this point will be discussed in Chapter 34. In the present chapter, wake echoes will be treated on the first assumption. Experience has shown that this approach is usually quite satisfactory.

33.1 CONCEPT OF WAKE STRENGTH

33.1.1 Target Strength of a Wake and Wake Strength

Echoes from wakes differ in two important respects from echoes from ships and other small targets. The concept of target strength has been analyzed in Section 19.1 where it was shown that for a target of finite size the target strength becomes independent of range at very long ranges and may be computed from the equation

$$T = E - S + 2H, \qquad (1)$$

where E is the echo level in decibels above 1 dyne per sq cm, S the source level or pressure level 1 yd from the projector, in decibels above 1 dyne per sq cm, and H the one-way transmission loss from the source to the target in decibels. If equation (1) were used to compute the target strength T_w for a wake from the echo level at long ranges, T_w would increase with the range because, for practical purposes, the wake extends infinitely in the horizontal direction; as the range increases, more of the wake becomes exposed to the sound beam, more scattering occurs, and the target strength increases. For the same reasons, a transducer with a broad horizontal beam would yield a higher echo level than a transducer with a narrow pattern beaming sound at the same wake, other things being equal.

It is desirable, therefore, to introduce in place of the target strength of a wake another characteristic, which is essentially the target strength of a 1-yd length of the wake. This quantity is principally a function of the geometric dimensions and of the physical properties of the wake alone and, therefore, will be called *wake strength* and denoted by the symbol W. The wake strength will here be defined in a simple manner for an ideal wake, without regard to the physical structure of actual wakes. In Section 33.1.2, an analysis of this wake strength in terms of the physical constitution of the wake will be given, including the effects originating from the finite length of the sound pulses used in practice.

The wake echo will now be treated as if it were the echo from a plane strip having infinite horizontal extension ($-\infty < y < +\infty$) and a constant vertical height h (depth of the wake) which is supposed to be much smaller than the distance to the transducer. The hypothetical strip is assumed to have a rough surface, so that the reflection of sound by it is nonspecular and perfectly diffuse, with a dimensionless coefficient of reflection s which is the fraction of sound energy returned into a unit solid angle. The fraction of sound energy reflected back, regardless of direction, is then $2\pi s$. By comparison with Section 19.1 it is readily verified that the target strength of one square yard of this wake surface, placed perpendicularly to the sound beam, is 10 log s. Since the depth of the wake is h yards, the target strength of a 1-yd length of wake is 10 log hs. In order to relate this wake strength to the observed echo intensity, considering the directivity of the transducer and the scattering of sound from elements of the wake surface not perpendicular to the sound beam, a more detailed exposition of this simple case is required.

The geometry of this experimental situation is illustrated in Figure 1, from which it is apparent that

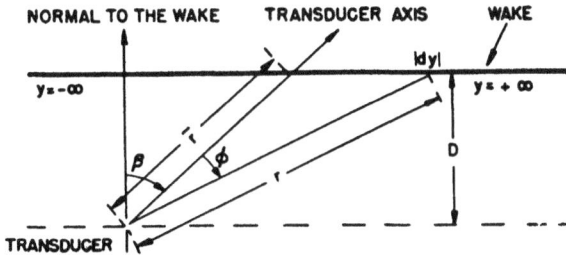

FIGURE 1. Horizontal plane through transducer and wake.

the surface element of the strip having the area hdy receives from the transducer the power

$$I_0 \frac{10^{-0.1ar}}{r^2} b(\phi) \cos(\beta + \phi) hdy, \qquad (2)$$

where I_0 is the output of the transducer on the axis; $b(\phi)$ measures the angular variation of the output around the horizontal plane; r is the distance from the transducer to the surface element hdy; ϕ is the angle between this ray and the axis of the transducer, which subtends the angle β with the normal to the wake so that $\cos(\beta + \phi)$ measures the geometric foreshortening of the insonified area; and a is the coefficient of attenuation in the ocean. If the transmission loss on the return path is taken into account, the equivalent echo intensity I_e is

$$I_e = sI_0 \int_{y=-\infty}^{y=+\infty} \frac{10^{-0.2ar}}{r^4} b(\phi)b'(\phi) \cos(\beta+\phi) hdy, \qquad (3)$$

where $b(\phi)b'(\phi)$ is the composite pattern function of the echo-ranging transducer. The factor $b'(\phi)$ is the ratio between the response of the hydrophone to a signal incident at an angle ϕ to the axis and the response to a signal of equal strength incident on the axis. Thus the equivalent echo intensity I_e is propor-

tional to the output voltage of the transducer acting as hydrophone.

By substitution of the perpendicular range D from transducer to wake,

$$D = r \cos(\beta + \phi)$$
$$y = D \tan(\beta + \phi),$$

equation (3) is transformed into

$$I_e = sI_0 \frac{h}{D^3} \int_{\phi=-\frac{\pi}{2}-\beta}^{\phi=+\frac{\pi}{2}-\beta} b(\phi)b'(\phi) \, 10^{-0.2aD \sec(\beta+\phi)} \cos^3(\beta+\phi) \, d\phi,$$

or, to a very good approximation

$$\frac{I_e}{I_0} 10^{0.2aD \sec \beta} D^3 = hs \int_{\phi=-\frac{\pi}{2}-\beta}^{\phi=+\frac{\pi}{2}-\beta} b(\phi)b'(\phi) \cos^3(\beta+\phi) d\phi \cdot \quad (4)$$

This approximation neglects the variation of the transmission anomaly along the wake, which is insignificant because of the narrow beam pattern of the transducers used in practice. By writing

$$\cos(\beta + \phi) = \cos\beta \, (\cos\phi - \tan\beta \sin\phi)$$

and $$D \cos \beta = \bar{r},$$

where $\bar{r}$ is the range to the wake measured along the transducer axis, equation (4) becomes

$$\frac{I_e}{I_0} 10^{0.2a\bar{r}} \bar{r}^3 = hs \int_{\phi=-\frac{\pi}{2}-\beta}^{\phi=+\frac{\pi}{2}-\beta} b(\phi)b'(\phi)(\cos\phi - \tan\beta \sin\phi)^3 d\phi. \qquad (5)$$

By collecting the terms representing the transmission loss,

$$a\bar{r} + 20 \log \bar{r} = H, \qquad (6)$$

and adopting the abbreviation

$$10 \log \int_{\phi=-\frac{\pi}{2}-\beta}^{\phi=+\frac{\pi}{2}-\beta} b(\phi)b'(\phi)(\cos\phi - \tan\beta \sin\phi)^3 d\phi = \Psi, \qquad (7)$$

equation (5) can be expressed, in decibels, as

$$E - S + 2H - 10 \log \bar{r} - \Psi = 10 \log hs = W. \qquad (8)$$

The quantity Ψ defined in equation (7) will be called the *wake index*, analogous to the *reverberation index* defined in Chapters 11 through 17. The product hs in equation (8) has the dimension of a length. Since the ranges appearing in (8) are customarily measured in yards, the wake strength W is the ratio of hs to one yard, expressed in decibels. Then, by comparing equations (8) and (11), the relation between the wake strength W and the target strength of a wake T_w becomes

$$T_w = W + 10 \log \bar{r} + \Psi, \qquad (9)$$

where $\bar{r}$ is the range to the wake, measured along the transducer axis, and Ψ is the wake index defined by equation (7). The physical meaning of equation (9) has already been noted in the first paragraph of this section.

33.1.2 Dependence of Wake Strength on Physical Parameters

The fundamental definition of wake strength, given in the preceding section, was facilitated by treating the wake as if it were a plane strip with the coefficient of reflection s. In effect, this approach neglected the wake structure along the transverse x axis. Moreover, that analysis tacitly assumed the use of a continuous signal for measuring the wake strength. But if short sound pulses are beamed at a diffusely reflecting plane, the echo profile on the oscillogram, in general, will not reproduce the shape (usually square-topped) of the signal, and the dependence of echo intensity on signal length must be investigated. For a brief theoretical demonstration of this fact see Section 19.3.

On inspection of the sample oscillograms reproduced in Figure 7 of Chapter 30, it will be observed that reflection from a wake also alters the shape of square-topped sound pulses. However, the explanation of this effect is more complicated than that of the variation with pulse length of the echo intensity returned by a plane target. In fact, the echo profile depends on the transverse structure of the wake, which therefore must form an integral part of a comprehensive theory of wake echoes.

According to the working hypothesis adopted in Section 26.3 wake echoes are composed of a multitude of reflections originating throughout the entire wake. Superposition of these scattered waves leads to constructive and destructive interference, because their phases are distributed at random. Consequently the echo intensity measured at any instant will not equal the average value. The difference between the instantaneous and average echo intensity is a rapidly fluctuating quantity, evidently beyond the reach of theoretical analysis, because it depends on the microstructure of the entire wake. Physical significance can be attributed only to the average of many echo profiles recorded in rapid succession. Such averaging is also necessary in order to minimize the effect upon

the echoes of the rapid fluctuations of the transmission loss in the ocean at large, which were discussed in Chapter 7. Accordingly, the theoretical analysis about to be presented refers to "average" echo intensities throughout.

The problem now is to evaluate the relation between the total number, arrangement and physical parameters of the bubbles and the overall reflectivity s of the wake, filling a volume of constant depth h and width w and of infinite length $(-\infty < y < +\infty)$. Let $n(x)$ be the number of bubbles per unit volume at the distance x from the nearest boundary of the wake $(0 < x < w)$; $n(x)$ is supposed not to vary appreciably along the wake axis over a distance of the order of the width of the sound beam, or $\partial n/\partial y \ll \partial n/\partial x$. With σ_s and σ_e representing the cross sections for scattering and extinction defined by equations (34) and (43) of Chapter 28, the echo returned by an individual bubble has the intensity

$$I_e = \frac{\sigma_s}{4\pi} I_0 \frac{10^{-0.2ar}}{r^4} b(\phi)b'(\phi)\, e^{-2\sigma e N(x)\, \sec\,(\beta + \phi)}. \quad (10)$$

The fraction $\sigma_s/4\pi$ of the incoming sound energy is scattered into the unit solid angle, and the term $10^{-0.2ar}/r^4$ represents the two-way transmission loss in the ocean. The sound beam is trained obliquely at the wake, so that the axis of the transducer and the normal to the wake include the angle β. It is the oblique path of the sound beam traversing the wake which accounts for the factor $\sec (\beta + \phi)$ in the exponent expressing the two-way transmission loss inside the wake, which is based on equation (52) of Chapter 28. The geometry of the situation is illustrated in Figure 2. The echo returned by a volume element of the wake is found by multiplying equation (10) by $n(x)hr\,dr\,d\phi$

$$dI_e = \frac{h\sigma_s}{4\pi} I_0 \frac{10^{-0.2ar}}{r^3} b(\phi)b'(\phi)n(x)e^{-2\sigma e N(x)\, \sec\,(\beta + \phi)}dr\,d\phi$$

$$(11)$$

on the assumption that all bubbles have the same size, so that the cross sections σ_s and σ_e are constant throughout the wake. Finally, the echo returned by the entire wake can be evaluated by integration between the appropriate boundaries, which must be chosen carefully; these boundaries are essentially determined by the width of the wake and by the signal length.

When an echo-ranging signal is sent out into the water, the volume from which echoes can be received at any one time fills a spherical shell of thickness $c\tau/2$, where τ is the duration of the square-topped signal and c is the sound velocity. This region, which travels outward at a velocity $c/2$, may for purposes of discussion be referred to as the volume occupied by the echo-ranging signal or pulse. As the pulse travels outward, it will cross the wake, which has roughly the shape of a cylinder of infinite length. The determination of the echo strength requires integration of equation (11) over the volume in which pulse and wake overlap. In the general case, this volume has a rather complicated shape which, moreover, varies with time as the pulse travels across the wake. It should be noted that in most practical situations the depth of the wake is small compared with the range to the transducer, so that the vertical curvature of the pulse boundaries can be neglected; in other words, the pulse will then be treated rather as a cylindrical shell, with the cylinder axis normal to the ocean surface.

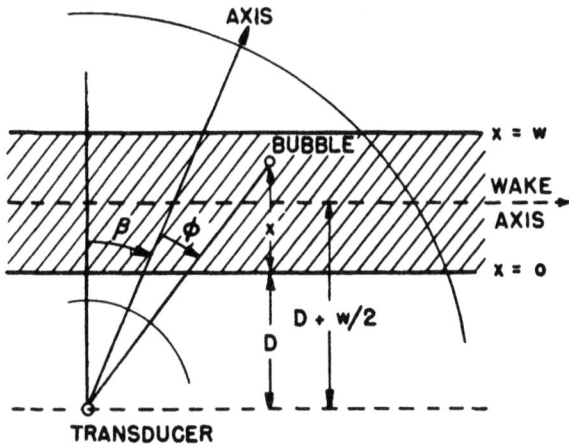

FIGURE 2.   Sound striking wake.

Now there are two simple limiting cases for which the shape of the volume contributing to the echo can readily be visualized, namely when the signal length is either much greater or much smaller than the width of the wake. In the first case, the entire volume of the wake will contribute to the echo for a considerable length of time during which the *average* echo intensity will be constant. In the second case, the echo will come only from a thin spherical shell "cut out" of the wake, so that the *average* echo intensity will vary, while the pulse traverses the wake, without ever attaining a constant value; the mode of this variation is a function of the density distribution $n(x)$ across the wake.

LONG PULSES

The case of very long pulses will be taken up first. If the signal length r_0, which equals $c\tau/2$, is much

greater than what might be called the *slant width* w' of the wake, which equals $w \sec (\beta + \phi)$, then the entire volume of the wake will be intersected by the pulse during a finite interval of time. During this period, the average echo intensity is constant, and can be found by integrating equation (11) over the wake volume. The limits of this integration are most readily given if the variable x is substituted for r, since then x varies between 0 and w, while ϕ varies from $(-\pi/2)-\beta$ to $(+\pi/2)-\beta$ (see Figure 2). By writing

$$r = \left(\bar{r} \cos \beta - \frac{w}{2} + x\right) \sec (\beta + \phi),$$

and

$$\bar{r} = \left(D + \frac{w}{2}\right) \sec \beta, \qquad (12)$$

then

$$dr = \sec (\beta + \phi)dx\,,$$

and the new constant $\bar{r}$ is the range, measured on the axis of the transducer, of what might be called mid-wake — the point of intersection between transducer axis and wake axis. As in the preceding section, the variation along the wake of the transmission anomaly in the ocean will be neglected by setting

$$10^{-0.2ar} = 10^{-0.2a\bar{r}}$$

By substituting equation (12) in equation (11), the integral now reads

$$\frac{I_e}{I_0} \bar{r}^3\, 10^{0.2a\bar{r}} = \frac{h\sigma_s}{4\pi} \int_{\phi=-\frac{\pi}{2}-\beta}^{\phi=+\frac{\pi}{2}-\beta} \int_{x=0}^{x=w} b(\phi)b'(\phi) \cos^2 (\beta + \phi) \sec^3\beta \cdot$$
$$\frac{n(x)e^{-2\sigma e N(x) \sec (\beta + \phi)}}{\left(1 + \dfrac{2x - w}{2F \cos \beta}\right)^3} dx d\phi. \qquad (13)$$

Unfortunately, it is impossible to integrate this expression in closed form. An approximation sufficient for all practical purposes will be given.

Consider first the integral over dx, namely,

$$\int_0^w \left(1 + \frac{2x - w}{2F \cos \beta}\right)^{-3} n(x)e^{-2\sigma e N(x) \sec (\beta + \phi)}dx, \qquad (14)$$

and apply the theorem of the mean value of an integral [a] to the inverse cube term in brackets; in

------

[a] This well-known theorem states

$$\int_{x_1}^{x_2} f(x)g(x)dx = f(\bar{x})\int_{x_1}^{x_2} g(x)dx\,,$$

with

$$x_1 < \bar{x} < x_2\,,$$

as long as $f(x)$ and $g(x)$ are continuous over the range of integration, and $g(x)$ is not negative.

other words, put this term in front of the integral, replacing x by an unspecified, constant, mean value $\bar{x}$. With this procedure the integral over dx in equation (14) can be evaluated without further approximation:

$$\left[1 + \frac{2\bar{x} - w}{2\bar{r}\cos\beta}\right]^{-3} \int_0^w n(x)e^{-2\sigma eN(x)\sec(\beta+\phi)}\,dx$$

$$= \left[1 + \frac{2\bar{x} - w}{2\bar{r}\cos\beta}\right]^{-3} \cdot$$

$$\left[1 - e^{-2\sigma eN(w)\sec(\beta+\phi)}\frac{\cos(\beta+\phi)}{2\sigma_c}\right]. \quad (15)$$

Since, by definition, $\bar{x}$ is confined to the interval between 0 and w, the inverse cube correction factor always lies between the following limits:

$$\left[1 + \frac{w\sec\beta}{2\bar{r}}\right]^{-3} < \left[1 + \frac{2\bar{x} - w}{2\bar{r}\cos\beta}\right]^{-3} < \left[1 - \frac{w\sec\beta}{2\bar{r}}\right]^{-3}.$$

Since $w\sec\beta/2r$ is much less than 1, in most practical echo-ranging situations this correction factor is unimportant, even at short ranges. In view of the limited accuracy obtainable with the current techniques of measuring echo levels, this correction factor will be omitted. The echo integral, equation (13), thus assumes the form

$$\frac{I_e}{I_0}\bar{r}^3 10^{0.2a\bar{r}} = \frac{h\sigma_s}{8\pi\sigma_e}\int_{-\frac{\pi}{2}-\beta}^{+\frac{\pi}{2}-\beta} b(\phi)b'(\phi)[\cos\phi - \tan\beta\sin\phi]^3 \cdot$$

$$[1 - e^{-2\sigma eN(w)\sec(\beta+\phi)}]\,d\phi. \quad (16)$$

Note that

$$10\log e^{2\sigma eN(w)} = 2H_w$$

is the two-way transmission loss in db for perpendicular transit through the wake, according to equation (8) of Chapter 32.

It will be observed that for a high acoustic opacity of the wake $[\sigma_eN(w) \gg 1]$ the exponential in the bracket under the integral is very much less than 1. Hence, in the case of infinite acoustic thickness of the wake, there follows the rigorous formula

$$\frac{I_e}{I_0}\bar{r}^3 10^{0.2a\bar{r}} = \frac{h\sigma_s}{8\pi\sigma_e}\int_{-\frac{\pi}{2}-\beta}^{+\frac{\pi}{2}-\beta} b(\phi)b'(\phi)[\cos\phi - \tan\beta\sin\phi]^3\,d\phi. \quad (17)$$

By comparison with equation (5), the value of the wake index Ψ_∞, where the subscript has been added to emphasize that this index refers to infinite acoustic thickness, can be identified:

$$\Psi_\infty = 10\log\int_{-\frac{\pi}{2}-\beta}^{+\frac{\pi}{2}-\beta} b(\phi)b'(\phi)[\cos\phi - \tan\beta\sin\phi]^3\,d\phi. \quad (18)$$

Conversely, for highly transparent wakes $[\sigma_sN(w) \ll 1]$ the second bracket under the integral in equation (16) may be developed into a series and the quadratic and higher terms neglected, yielding

$$1 - e^{-2\sigma eN(w)\sec(\beta+\phi)} = 2\sigma_sN(w)\sec(\beta+\phi)\cdot$$

The formula for the wake strength of highly transparent wakes then reads

$$\frac{I_e}{I_0}\bar{r}^3 10^{0.2a\bar{r}} = \frac{h\sigma_s}{4\pi}N(w)\sec\beta\int_{-\frac{\pi}{2}-\beta}^{+\frac{\pi}{2}-\beta} b(\phi)b'(\phi)\cdot$$

$$[\cos\phi - \tan\beta\sin\phi]^2\,d\phi,$$

$$\Psi_0 = 10\log\left\{\sec\beta\int_{-\frac{\pi}{2}-\beta}^{+\frac{\pi}{2}-\beta} b(\phi)b'(\phi)[\cos\phi - \tan\beta\sin\phi]^2\,d\phi\right\}. \quad (19)$$

Numerically, the difference between Ψ_∞ in equation (18) and Ψ_0 in equation (19) is negligible for directional transducers as long as β is small. To a very good approximation the general equation (16) can, therefore, be written as

$$\frac{I_e}{I_0}\bar{r}^3 10^{0.2a\bar{r}-\left(\frac{\Psi_\infty}{10}\right)} = \frac{h\sigma_s}{8\pi\sigma_e}[1 - e^{-2\sigma eN(w)\sec\beta}]. \quad (20)$$

Expressed on a decibel scale, equation (20) becomes

$$E - S + 2H + 10\log\bar{r} - \Psi_\infty$$

$$= 10\log\left\{\frac{h\sigma_s}{8\pi\sigma_e}[1 - e^{-2\sigma eN(w)\sec\beta}]\right\}. \quad (21)$$

Hence, the reflectivity of the wake per unit solid angle is

$$s = \frac{1}{8\pi}\cdot\frac{\sigma_s}{\sigma_e}[1 - e^{-2\sigma eN(w)\sec\beta}]. \quad (22)$$

By this equation, the problem proposed at the outset of this section is solved for sufficiently long pulses $(r_0 > w)$. However, it should be remembered that equation (21) does not represent the entire echo profile, but applies merely to its central part which has a constant average intensity because the pulse overlaps the entire wake. The rise and fall of the average echo profile, when only part of the pulse intersects the wake, cannot be represented by a simple formula, because of mathematical difficulties of the same nature as will become apparent presently in the discussion of short-pulse echoes.

SHORT PULSES

For pulse lengths smaller than the wake width $(r_0 < w)$, equation (11) has to be integrated over the volume in which the wake and the cylindrical shell (thickness r_0, inner radius r_1) of the pulse intersect. Hence,

$$I_e = \frac{h\sigma_s}{4\pi} I_0 \int_{\phi_a}^{\phi_b} \int_{r_1}^{r_1+r_0} \frac{10^{-0.2ar}}{r^3} b(\phi) b'(\phi) \cdot n(x) e^{-2\sigma eN(x)\sec(\beta+\phi)} dr d\phi \cdot \quad (23)$$

The limits ϕ_a and ϕ_b of the integral over $d\phi$, unspecified for the time being, are determined by the relative position of pulse and wake and, therefore, vary with time; their explicit form will be evaluated after the discussion of the integral over dr has been finished. Equation (23) is valid only for rectangular pulses — that is, for echo-ranging signals whose intensity is constant for their duration.

The variation of $n(x)$ and $N(x)$ across the wake prevents integration of (23) in closed form. In order to gain any insight at all into the behavior of short-pulse echoes, a drastic simplification becomes necessary. For this reason the discussion is confined to a wake of constant bubble density in the transverse direction. Putting thus

$$n(x) = \text{constant} = \bar{n} = N(w)/w$$

and

$$N(x) = \bar{n}x$$

the integral reads

$$\frac{I_e}{I_0} = \frac{h\sigma_s}{4\pi} \int_{\phi_a}^{\phi_b} \int_{r_1}^{r_1+r_0} \frac{10^{-0.2ar}}{r^3} b(\phi) b'(\phi) \bar{n} e^{-2\sigma e\bar{n}x\sec(\beta+\phi)} dr d\phi \cdot \quad (24)$$

While in the general case ϕ_a and ϕ_b vary as r increases from r_1 to $r_1 + r_0$, for short pulses this change is quite negligible. The integration over dr may then be carried out before the integration over $d\phi$ without difficulty. On account of the geometric relation, from Figure 2,

$$x \sec(\beta+\phi) = r - D\sec(\beta+\phi),$$

the integral over dr in equation (24) becomes

$$\int_{r_1}^{r_1+r_0} \frac{10^{-0.2ar}}{r^3} \bar{n} e^{-2\sigma e\bar{n}[r-D\sec(\beta+\phi)]} dr \cdot \quad (25)$$

After applying the theorem of the mean value of an integral with respect to the factor $10^{-0.2ar}\bar{r}^3$ the integral (25) is transformed into

$$r^{*-3}10^{-0.2ar^*}\frac{1}{2\sigma_e}[1 - e^{-2\sigma e\bar{n}r_0}]e^{-2\sigma e\bar{n}[r_1-D\sec(\beta+\phi)]}.$$

Since r^* differs very little from r_1 (because $r_1 < r^* < r_1 + r_0$, by definition), r^* can be replaced by r_1

without any appreciable loss of accuracy. Thus the echo integral, equation (24), reads

$$\frac{I_e}{I_0}r_1^3 10^{0.2ar_1} = \frac{h\sigma_s}{8\pi\sigma_e}[1 - e^{-2\sigma e\bar{n}r_0}]\int_{\phi_a}^{\phi_b} b(\phi) \cdot b'(\phi) e^{-2\sigma e\bar{n}[r_1-D\sec(\beta+\phi)]} d\phi. \quad (26)$$

The exponential under the integral measures the transmission loss resulting from absorption and scattering inside the wake, since $r_1 - D\sec(\beta+\phi)$ is the distance, along any ray ($\phi = \text{constant}$), from the inner boundary of the pulse to the front of the wake. Now by making the substitution

$$r_1 - D\sec(\beta+\phi) = r_1 - D\sec\beta - D[\sec(\beta+\phi) - \sec\beta],$$

equation (26) assumes the form

$$\frac{I_e}{I_0}r_1^3 10^{0.2ar_1} = \frac{h\sigma_s}{8\pi\sigma_e}[1 - e^{-2\sigma e\bar{n}r_0}]e^{-2\sigma e\bar{n}(r_1-D\sec\beta)}\int_{\phi_a}^{\phi_b} b(\phi) \cdot b'(\phi) e^{2\sigma enD[\sec(\beta+\phi)-\sec\beta]} d\phi. \quad (27)$$

Here the factor $(1 - e^{-2\sigma e\bar{n}r_0})$ comprises the effect of the pulse length on the echo strength. The transmission loss inside the wake has been split into two factors. The first one, namely $e^{-2\sigma e\bar{n}(r_1-D\sec\beta)}$, depends only on the range r_1 of the pulse, which increases with time, but not on the directivity of the transducer; in fact, this first factor is simply the transmission loss, measured along the sound beam axis, from the boundary of the wake to the pulse. The second factor is independent of time and appears as an exponential under the integral over $d\phi$. Using the abbreviation

$$\Psi' = 10\log\left\{\int_{\phi_a}^{\phi_b} b(\phi) b'(\phi) 10^{2\sigma e\bar{n}D[\sec(\beta+\phi)-\sec\beta]} d\phi\right\}, \quad (28)$$

which defines another wake index applying to short-pulse echoes, equation (27) may be written on a decibel scale as follows.

$$E - S + 2H + 10\log r_1 - \Psi'$$
$$= 10\log\left\{\frac{h\sigma_s}{8\pi\sigma_e}[1 - e^{2\sigma e\bar{n}r_0}]e^{-2\sigma e\bar{n}(r_1-D\sec\beta)}\right\}. \quad (29)$$

The quantity $(r_1 - D\sec\beta)$ is the distance, measured along the transducer axis, which the rear boundary of the pulse has penetrated into the wake. Hence, for a directional transducer no appreciable echo intensity will be obtained outside the range of penetration which is given by

$$D\sec\beta < r_1 < (D + w)\sec\beta.$$

During the time interval in which r_1 is confined between the limits stated, equation (27) represents the average profile of short-pulse echoes. Consequently the average echo intensity falls off exponentially from its maximum value, attained immediately after the pulse has fully entered the wake ($r_1 = D \sec \beta$), provided that the value of the integral over $d\phi$ does not vary with time. This condition can indeed be realized in special cases.

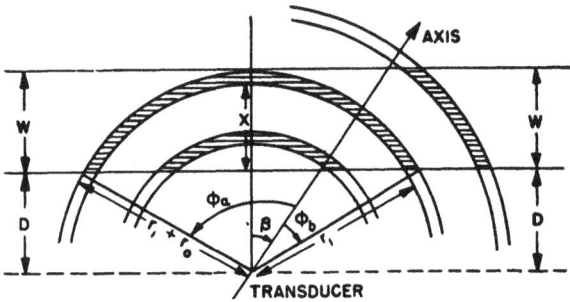

FIGURE 3.   Successive positions of ping inside wake.

Although the integrand in equations (27) and (28) is independent of time, the range of integration is not, in the general case. This fact is illustrated by Figure 3, showing successive positions of a short pulse beamed obliquely at the wake. As the pulse crosses the wake, the limits, ϕ_b and ϕ_a of the integral in equation (28) increase steadily. Thus the wake index is, in principle, a variable quantity, which would seem to impair the usefulness of this concept. However, the directivity of the transducer effectively limits the angular width of the sound beam. If it were possible to place the transducer in such a position that the effective half-width of the sound beam remained smaller at all times than the boundary ϕ_b, the variability of ϕ_b and ϕ_a would become irrelevant for all practical purposes, and the wake index Ψ' would actually be a constant.

In order to formulate this idea quantitatively, note that the boundaries of integration in the wake index are explicitly

$$\phi_b = \cos^{-1}\left[\frac{D}{r_1 + r_0}\right] - \beta$$

$$\phi_a = \cos^{-1}\left[\frac{D}{r_1 + r_0}\right] + \beta \cdot \qquad (30)$$

These formulas can readily be verified by inspecting Figure 3. If the effective angular width of the sound beam be called $2\phi'$, so that the effective half-width is ϕ', the conditions under which Ψ' becomes constant is

$$\phi' \leqslant \phi_b \cdot \qquad (31)$$

By substitution of equation (30) in equation (31), it follows that

$$\cos^{-1}\left[\frac{D}{r_1 + r_0}\right] \geqslant \phi' + \beta,$$

or

$$\frac{D}{r_1 + r_0} \leqslant \cos(\beta + \phi'). \qquad (32)$$

Write

$$r_1 = D + x_1,$$

where, according to Figure 3, x_1 is confined between the following limits

$$0 < x_1 < w, \qquad (33)$$

and substitute in equation (32); then

$$\frac{1}{1 + (x_1 + r_0)/D} \leqslant \cos(\beta + \phi')$$

or approximately, since generally $(x_1 + r_0)/D$ is much smaller than one,

$$1 - \frac{x_1 + r_0}{D} \leqslant \cos(\beta + \phi'). \qquad (34)$$

Hence,

$$D \leqslant \frac{x_1 + r_0}{1 - \cos(\beta + \phi')} \cdot$$

This inequality may be put into a more stringent form by setting x_1 equal to zero, on account of equation (33), so that the desired condition takes the final form

$$D \leqslant \frac{r_0}{1 - \cos(\beta + \phi')}, \qquad (35)$$

which assumes that right from the moment the pulse has entered the wake, or $x_1 = 0$, the effective half-width of the sound beam is smaller than the variable boundary ϕ_b. Any less stringent form of the condition, such as $x_1 \leqslant w$, would distort the representation of the entire echo profile indicated by equation (29).

For numerical evaluation of equation (35), $\phi' = 6$ degrees appears to be a reasonable value for transducers of conventional design used in echo ranging. Results for two typical cases are given in Table 1. Since the shortest signals used in practice correspond to $r_0 = 1$ msec $= 0.8$ yd, condition (35) is easy to maintain in ranging perpendicularly at the wake. Accordingly, the upper limit ϕ_b in equation (28) may be replaced by a practically constant value ϕ' if the range D from the transducer to the nearest point of the wake is chosen so as to comply with the first case in Table 1.

On the other hand, it is evidently impossible to satisfy the second condition in Table 1 — in other words, with markedly oblique incidence of the sound beam the concept of wake loses its usefulness for short signals. The theoretical derivation of the echo profile for short-pulse echoes obtained with an obliquely trained transducer would require extensive numerical integrations of equation (27), taking into account the continual variations of the boundaries ϕ_a and ϕ_b.

TABLE 1. Conditions for constant wake index.

Effective half-width of sound beam $\phi' = 6°$		
I	Sound beam trained perpendicularly at wake $\beta = 0°$	Condition (35) $D \leq 200r_0$
II	Sound beam trained obliquely at wake $\beta = 60°$	Condition (35) $D \leq 1.66r_0$

Summing up the discussion of short-pulse echoes, it should be remembered that the analysis, for mathematical reasons, had to be restricted to wakes having a constant bubble density in the transverse direction. According to the varying transverse structure of actual wakes, their echo profile will deviate somewhat from the exponential shape given by equation (27), even if the sound beam is trained perpendicularly at the wake.

33.1.3 Definition of Wake Strength

The concept of wake strength has been introduced in Section 33.1.1 in the hope of arriving at a wake characteristic that is a function solely of the geometric dimensions and physical parameters of the wake. But the detailed analysis in Section 33.1.2 showed that this aim defies complete realization. The strength of wake echoes depends on the signal length and on the directivity pattern of the transducer in a rather complicated manner; this dependence cannot be formulated mathematically in a simple way without introducing various approximations. These effects are small compared with those resulting from the variability of echo strength with range; this large variation with range can be eliminated from the wake strength by a suitable definition of this quantity.

We now define wake strength by the equation

$$W = E - S + 2H - 10 \log r - \Psi, \quad (36)$$

where E is the echo level, S the source level, H the one-way transmission loss from the transducer to the wake, r the range in yards of the wake, and Ψ is the appropriate wake index. This definition comprises

both equations (21) and (29), referring to long and short signals, respectively, and implicitly disregards the small difference between the ranges defined as $\bar{r}$ and r_1. For all practical purposes the range to be used in the computation of wake strength may be determined from the time interval, purposely recorded on the oscillogram, between midsignal and the instant to which the measured echo amplitude refers. When the difference between the transmission loss H and the inverse-square loss $20 \log r$ increases linearly with range, an alternative way of writing equation (36) is

$$W = E - S + 2ar - 10 \log \mu^2 + 30 \log r - \Psi, \quad (37)$$

where a is the coefficient of attenuation in the ocean expressed in db per kyd, and $(\mu - 1)$ is the reflectivity of the ocean surface, so that the factor μ^2 represents the increase of echo intensity caused by the double path resulting from surface reflection. To be consistent with the standard procedure for computing target strengths, μ should be put equal to 1 (see Section 22.2), so that equation (36) reduces to [b]

$$W = E - S + 2ar + 30 \log r - \Psi. \quad (38)$$

Finally, according to equation (9), the target strength of the wake T_w is given by

$$T_w = W + 10 \log r + \Psi. \quad (39)$$

The wake index Ψ was first defined by equation (7), on the assumption that reflection from a wake can be treated like that from a plane strip. This is, indeed, a good approximation provided that the wake is highly opaque, or $N(w) \gg 1$, and long signals are used, or $r_0 \gg w$ (for example, $r_0 > 2w$), according to equation (18) which turned out to be identical with equation (7):

$$\Psi_\infty = 10 \log \int_{-\frac{\pi}{2} - \beta}^{+\frac{\pi}{2} - \beta} b(\phi)b'(\phi)[\cos \phi - \tan \beta \sin \phi]^3 d\phi. \quad (40)$$

---

[b] All the numerical values of W reported in this chapter have been computed according to the definition given by equation (38). However, when the original publications are consulted, care should be taken in ascertaining what particular definition of wake strength was used by the author. For instance, in one paper,[1] μ is set equal to 2 in correcting echo levels for transmission loss; moreover, a term $10 \log (4\pi)$ is also added to equation (37). The net result is that the values of the wake strength in reference 1 are 5.0 db larger than those computed from equation (38).

Since the reflectivity of a target has been defined, in Section 19.1 of this volume, in terms of the sound reflected into a unit solid angle, it seems desirable to maintain the same convention for the reflectivity s of a wake. Accordingly, the term $-10 \log (4\pi)$ here appears in equations (45) and (46), instead of in equation (37).

However, if the wake is highly transparent, or $N(w) \ll 1$, the echo level E, and also W as computed from Ψ_∞ will be found to increase with the obliqueness β of the impinging sound beam. In this case, the replacement of Ψ_∞ by Ψ_0, according to equation (19), may be expected to give a wake strength independent of β:

$$\Psi_0 = 10 \log \left\{ \sec \beta \int_{-\frac{\pi}{2}-\beta}^{+\frac{\pi}{2}-\beta} b(\phi) b'(\phi) \left[\cos \phi - \tan \beta \sin \phi \right]^2 d\phi \right\}.$$
(41)

For short pulses ($r_0 \ll w$, say $r_0 < w/2$) the appropriate value of Ψ is

$$\Psi' = 10 \log \left\{ \int_{-\phi'}^{+\phi'} e^{2\sigma e n r \, (\sec \phi - 1)} b(\phi) b'(\phi) d\phi \right\}.$$
(42)

This formula is a specialization of equation (28) for $\beta = 0$; hence D in equation (28) is approximately equal to r. Moreover, according to equation (35), the range from transducer to wake must be less than 200 times the pulse length, if equation (42) is to be valid. For a sound beam trained obliquely at the wake — or for $\beta \neq 0$ — equation (28) loses its usefulness, as the wake index becomes a quantity varying with time.

Though their mathematical expressions appear rather different, the numerical discrepancy between Ψ' and Ψ_∞ is quite small, probably never exceeding 1 db for transducers of high directivity. For signals of intermediate length, of the order of the wake width, the mathematical analysis is difficult; it is suggested simply that Ψ_∞ be used. For practical purposes, the differences between the various types of wake indices might be neglected altogether, by writing for perpendicular incidence of the sound beam

$$\Psi = 10 \log \int b(\phi) b'(\phi) d\phi \,.$$
(43)

This approximate formula reveals a close relationship between the wake index and the surface reverberation index, defined in a University of California Division of War Research [UCDWR] report,[2] as

$$J_s = 10 \log \left\{ \frac{1}{2\pi} \int b(\phi) b'(\phi) d\phi \right\}.$$

Hence,

$$J_s + 8db = \Psi.$$
(44)

Furthermore, it is of interest to compare the formulas giving the wake strength as a function of the physical parameters and of the pulse length:

Short pulses, $r_0 \ll w$
$$W = 10 \log hs$$
$$= 10 \log \left\{ \frac{h}{8\pi} \frac{\sigma_s}{\sigma_e} \left[1 - e^{-2\sigma e \bar{n} r_0} \right] e^{-2\sigma e n (r_1 - D \sec \beta)} \right\}.$$
(45)

Long pulses, $r_0 \gg w$
$$W = 10 \log hs = 10 \log \left\{ \frac{h}{8\pi} \frac{\sigma_s}{\sigma_e} \left[1 - e^{-2\sigma e N(w)} \right] \right\}.$$
(46)

It will be noted that the first exponential in equation (45) becomes equal to that in equation (46) for $r_0 = w$, because of equation (54) of Chapter 28, reading

$$\bar{n} w = N(w),$$

which is the definition of $\bar{n}$. Equation (45) is an approximate formula, because in its derivation the assumption $n(x) = \text{constant} = \bar{n}$ had to be made. However, while equation (46) applies to the constant average echo intensity constituting the central part of long-pulse echoes, equation (45) represents the entire average profile of short-pulse echoes.

So far the entire discussion has been restricted to average echo intensities. But, as described in Section 30.1.3, peak echo intensities are customarily measured by the San Diego observers. The measurements reported in Chapters 11 through 17 of this volume on the "band" or "point" method of reading reverberation records imply that about 6 db must be subtracted from the wake strengths computed by equation (38) from peak intensities, in order to express them on the scale of average intensities envisaged in equations (45) and (46). This correction will be applied only in Section 34.3.1, where the interpretation of the observed wake strengths by the acoustic theory of bubbles is discussed. In that context, the wake strength computed from the measured peak amplitudes of short-pulse echoes, and then corrected by subtracting 6 db, will be regarded as corresponding to the maximum of the profile (45), or

$$r_1 - D \sec \beta = 0.$$

33.1.4 Decay Rate of Wake Strength

The decay rate of wake strength, in terms of the physical properties of the wake, is found by differentiating equations (45) and (46) with respect to the time. Before doing so, it is advantageous to make the substitutions

$$\bar{n} = \frac{H_w}{4.34 \sigma_e w}$$

and

$$N(w) = \frac{H_w}{4.34 \sigma_e} \,,$$

where H_w is the one-way transmission loss for horizontal passage of sound through the wake, as defined in equation (8) of Chapter 32. Equations (45) and (46) then read

$$W = 10 \log \left\{ \frac{h}{8\pi} \cdot \frac{\sigma_s}{\sigma_e} [1 - e^{-0.46 H_w r_0/w}] \right\} \quad (r_0 \ll w) \quad (47)$$

$$W = 10 \log \left\{ \frac{h}{8\pi} \cdot \frac{\sigma_s}{\sigma_e} [1 - e^{-0.46 H_w}] \right\} \quad (r_0 \gg w). \quad (48)$$

The result of the differentiation is:

Short pulses, $r_0 \ll w$

$$\frac{dW}{dt} = 4.34 \frac{1}{h} \frac{dh}{dt} + \frac{0.46 e^{-0.46 H_w r_0/w}}{1 - e^{-0.46 H_w r_0/w}} \cdot \frac{r_0}{w} \cdot$$
$$\left[\frac{dH_w}{dt} - \frac{H_w}{w} \cdot \frac{dw}{dt} \right]. \quad (49)$$

Long pulses, $r_0 \gg w$

$$\frac{dW}{dt} = 4.34 \frac{1}{h} \frac{dh}{dt} + \frac{0.46 e^{-0.46 H_w}}{1 - e^{-0.46 H_w}} \cdot \frac{dH_w}{dt}. \quad (50)$$

The first term in these equations is the same for long and short signals. It represents the effect of the change in depth of the wake and is known to be quite small; for two destroyer wakes, according to data reported in Section 31.3.1, $(1/h)(dh/dt)$ was found to be -0.08 and $+0.04$ db per minute, respectively. The second term seems to be the dominant one. While the factor in front of it, containing the exponential, is exceedingly small for fresh wakes, it grows rapidly and approaches infinity for very old wakes; numerical values of this factor can be read from the graph in Figure 4.

The differential quotient dH_w/dt is equal to 3 db per minute, for destroyer wakes, and $(H_w/w)(dw/dt)$ can be estimated from the same data to be of the order of 1 to 2 db per minute. It will be noted that equation (49) would be transformed into equation (50) by setting r_0/w equal to one — except for the term proportional to $(H_w/w)(dw/dt)$ which does not appear in equation (50). The physical meaning of this term is interesting: as the wake ages, it spreads laterally, causing dw/dt to be a positive quantity; consequently, the factor H_w/w is bound to decrease, even if H_w, the total attenuation across the wake, remains constant. According to equations (46) and (48), for long pulses the wake strength is a function of $N(w)$, which is directly proportional to H_w or the *total* attenuation, and which is not affected by a mere spreading laterally of the wake without simultaneous disintegration of the bubble population. But for short pulses

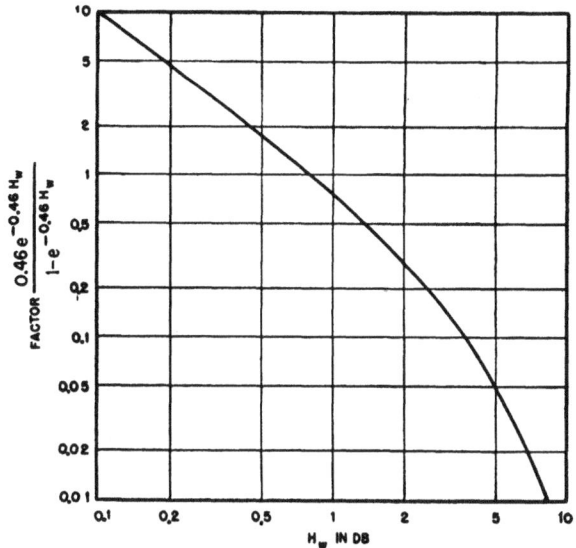

FIGURE 4. Factor appearing in formula (50) for the decay rate of wake strength.

[see equations (45) and (47)] the wake strength is a function of the product of signal length r_0 times the average bubble density $\bar{n}$ which is proportional to the attenuation coefficient. Hence, the decay rate of the wake strength for short pulses is a function of the decay rate $d(H_w/w)/dt$ of the attenuation coefficient which gives origin to a term proportional to (dw/dt) or the lateral spreading, even if dH_w/dt is negligibly small, corresponding to an extremely small physical disintegration of the bubble population.

Summing up, for short pulses, whose volumes do not intersect the entire wake, there exists a progressive decay of wake strength having a purely geometric origin — namely the lateral spreading of the wake. Naturally, for long pulses, which overlap the entire wake, such an effect cannot arise. If the decay of a wake is followed over a very long period of time, and a constant pulse length is employed, it may well happen that the pulse which was chosen, at zero age of the wake, so as to be long will finally become short with respect to the steadily growing width of the wake. At the moment the critical point $w = r_0$ is passed, the term $(H_w/w)(dw/dt)$ suddenly begins to operate, causing an accelerated decay. In order to avoid all unnecessary complications, it may be advisable, therefore, to choose very long signals for the study of the decay rate of wake strength.

The general significance of equations (49) and (50) is that they establish a relation between the decay rates of the transmission loss and the wake strength which can be tested by observation.

33.2 EXPERIMENTAL PARAMETERS

The formulas derived in the preceding section are applied in later sections to the interpretation of echoes from actual wakes. The theoretical results may also be applied to indicate what type of echo-ranging experiment is most suited for fundamental studies of wakes. Certain considerations along this line, especially concerning the choice of transducer directivity, pulse length, and frequency are presented in this section.

33.2.1 Transducer Directivity

The mathematical intricacies of the analysis given in Section 33.1.2 are essentially a consequence of the imperfect directivity of the transducers and the finite range over which the wake is observed. The chief result is a variety of wake indices pertaining to specific experimental situations. Fortunately, the picture is greatly simplified in practice, because of the properties of the transducers customarily employed in echo ranging.

The numerical differences between the different wake indices, for the same directivity pattern, are quite insignificant in proportion to the accuracy attainable in acoustic measurements. As an example, Table 2 gives the wake indices computed from the composite directivity pattern of a particular transducer.

The integral Ψ taken over the composite directivity pattern alone, as defined by equation (43), is given for comparison. It would seem that $\Psi = J_s + 8$ db

TABLE 2. Typical wake indices — UCDWR transducer No. 1917 at 45 kc.

	$\beta = 0°$	$\beta = 60°$
Ψ_∞	−9.75 db	−9.87 db
Ψ_0	−9.75	−6.85
$\Psi' \begin{cases} 2\sigma_e\bar{n}r = 20 \\ 2\sigma_e\bar{n}r = 40 \end{cases}$	−9.58 −9.52	... ...
Ψ	−9.74	−6.74

may be used in place of any of the rigorous values of the wake indices, except for Ψ_0 with obliquely impinging sound beam ($\beta = 60$ degrees). In this particular case the wake index includes the factor sec β, as is physically evident for reflection from a semi-transparent layer of finite thickness. It is concluded, then, that for all practical purposes Ψ may be substituted

for the other wake indices, if the correction factor sec β is applied for oblique incidence of sound on semi-transparent wakes.

Since the wake index Ψ' applying to short-pulse echoes depends on the range and on the attenuation coefficient inside the wake, Table 2 gives a more detailed illustration of the influence of the variable parameters. The effect is seen to be quite small; therefore it does not influence the interpretation of the experiments on the *E. W. Scripps* wake carried out on November 28, 1944, during which the two transducers referred to in Table 3 were employed.

As far as the range is concerned, there is a distinction between short and long pulses. In order to obtain short-pulse echoes of a kind that can be treated by a simple acoustic theory, the sound beam must be trained perpendicularly at the wake and the range must be shorter than 200 times the signal length; equation (35) of Section 33.1.2, which formulates this condition in an exact manner, shows that the exact factor is a function of the directivity pattern of the transducer. Short-pulse echoes produced with a sound beam trained obliquely at the wake defy any simple mathematical analysis and, therefore, are of little use in the study of wakes. As regards long-pulse echoes, however, the range is of minor importance, and the aspect of the wake is of no consequence whatever because the dependence of the wake index on the aspect angle β is fully taken into account by equations (40) and (41). In practice, it should suffice to keep the ratio of wake width w to range r less than about 0.1; this value of w/r makes it possible to neglect the inverse-cube correction factor appearing in equation (15), which was omitted from there on.

33.2.2 Pulse Length

Pulses varying in length from 0.3 to several hundred milliseconds have been employed in echo ranging at wakes. There are some general considerations concerning signal length that apply primarily to the tactical use of wake echoes. In practice, the design of the keying circuits and the build-up time of the transducer set a lower limit to the pulse length. While so far no special study has been made of the optimum conditions for recognition of wake echoes, it may be surmised, from experience with echoes from finite targets,[3] that signals shorter than 10 msec are not suitable for satisfactory recognition by ear. But wake echoes obtained with 1-msec signals, and even with shorter ones, are readily recognized on sound range recorder traces and oscillograms. Reverberation, particularly at long ranges, imposes an upper limit to the practicable pulse length.

Rather different considerations govern the choice of pulse length for fundamental research into the physical constitution of wakes. The aim of such work

which the internal density distribution is undergoing all the time, aside from echo fluctuations due to random interference and variable transmission loss. Only by averaging numerous instantaneous profiles could a truly representative picture of the $n(x)$ distribution be obtained.

TABLE 3.   Wake indices.

Transducer	JK	GD 1143			$2\sigma_e \bar{n} r$
	24 kc	40 kc	50 kc	60 kc	
Ψ_∞ ($\beta = 0°$)	−7.48 db	−6.35 db	−7.36 db	−7.39 db	
Ψ' ($\beta = 0°$)	−7.41	−6.22	−7.28	−7.30	5
	−7.36	−6.14	−7.22	−7.25	10
	−7.31	−6.05	−7.17	−7.19	15
	−7.26	−5.96	−7.12	−7.13	20
	−7.20	−5.87	−7.06	−7.08	25
	−7.15	−5.77	−7.00	−7.02	30
	−7.03	−5.57	−6.88	−6.89	40
	−6.90	−5.35	−6.75	−6.76	50

may be either to establish the overall properties of a wake, or to resolve its microstructure. In the first case the use of long signals, overlapping the entire wake, is indicated, whereas in the second case maximum resolving power is achieved by extremely short pulses. According to equation (40) the wake strength determined with long pulses is a function of (1) the depth of the wake, (2) the average cross section for scattering and extinction by the bubble population, and (3) the acoustic thickness $\sigma_e N(w)$ of the entire wake, which may be determined quite independently by measurement of the horizontal transmission loss. Therefore, a simultaneous observation of the echoes returned by the wake and of the horizontal transmission loss through it offers the greatest promise for testing the adequacy of equation (40) for long signals. The corresponding equation (39) for short pulses has been derived by neglecting the microstructure of the wake, by putting $n(x) = \text{constant} = \bar{n}$. However, on inspection of the rigorous equation (24), it will be seen that the echo profile on the oscillogram is essentially proportional to the function

$$n(x)e^{-2\sigma_e N(x)}$$

ıor the case of extremely short signals, and of an ideal sharp sound beam which could be realized approximately by placing the transducer very close to the wake. Such an analysis of the microstructure of wakes by short-pulse echoes would be of rather limited practical value, because of the rapid changes

As to signals of intermediate length, it may be presumed that equation (39) will represent the variation of W with r_0 reasonably well.

33.2.3   Frequency

Most echo ranging at wakes has been carried out with frequencies between 20 and 60 kc. The available observations suggest a conspicuous variation of wake strength with frequency, but no such dependence can be anticipated theoretically. The dominant factor σ_s/σ_e in the formula for the wake strength does not change much with frequency, for bubbles of resonant size. At present little is known about the relative proportion of resonant bubbles in the total population and how this proportion changes with time; but there is no definite reason to believe that bubbles of nonresonant size predominate, in which case σ_s/σ_e would vary more markedly with frequency. In any event, the influence of σ_s/σ_e on the wake strength would not be expected to account for more than a few decibels. However, some frequency effect may result from the factor $(1 - e^{-2\sigma_e N(w)})$, provided that the wake is highly transparent; otherwise the exponential would be small compared with 1.

33.3   ECHOES FROM SUBMARINE WAKES

Quantitative data on the strength of submarine wakes have recently been computed from the original measurements of echo levels. Some of these have been published before;[4, 5] others were obtained from the files of the San Diego laboratory. During these experiments, the echo-ranging vessel overtook the submarine while proceeding on a parallel course; the observations comprise surface runs, dives, and submerged level runs. In order to reduce the uncertain-

ular ($\beta = 0$ degrees) and oblique ($\beta = 60$ degrees) incidence of the sound beam on the wake. With a few exceptions illustrated in Figures 5 and 6, the observations did not extend over sufficiently long periods of time to reveal the gradual decay of the wakes. Hence only average values of the wake strength W

TABLE 4. Submarine wake strengths.

	Run	β in degrees	Ψ_0 in db	Ping length in msec	Frequency in kc	Wake strength W in db 9.5 knots surfaced	Wake strength W in db 6 knots submerged		Average distance astern in yd
							Depth 45 ft	Depth 90 ft	
USS S-23 (SS128)	1	0	−9.4	30	60	−18		−28	400
	2	0	−9.4	30	60	−16		−25	400
	3	0	−9.4	30	60	−19		−26	400
						Avg −18		Avg −26	
USS S-34 (SS139)	1	0	−8.6	30	45	−12			300
	2	60	−6.5	30	45	−14			300
						Avg −13			
	1	0	−8.6	30	45			−22	200
	2	60	−6.5	30	45			−24	200
								Avg −23	
USS Tilefish (SS307)	1	0	−8.6	30	45	−15		−22	600
	2	0	−8.6	30	45	−13		−20	600
	3	60	−6.5	30	45	−11		−19	600
						Avg −13		Avg −20	
USS S-18 (SS123)	1	60	−6.5	10	45		−34.6		200 to 250
				30	45		−32.6		200 to 250
				100	45		−30.6		200 to 250
							Avg −32.6		
	2*	60	−6.5	10	20		−21.8		10 to 500
				30	20		−18.7		10 to 500
				100	20		−16.7		10 to 500
							Avg −19.0		
	2*	60	−6.5	10	20		−23.6		650 to 850
				30	20		−21.0		650 to 850
				100	20		−19.3		650 to 850
							Avg −21.3		
	2*	60	−6.5	10	20		−1.8		Decay of wake
				30	20		−2.3		Decay of wake
				100	20		−2.6		Decay of wake
							Avg −2.2		

* This run is illustrated in Figure 1. Absolute values of the wake strength W are uncertain because of lack of adequate calibration.

ties of the relative position during the submerged portions of the runs, the submarine towed a marker buoy. Pelorus bearings on this buoy were logged from the echo-ranging vessel. With the aid of the original logs, a diagram was constructed for each run, giving the relative position of submarine and measuring vessel. The distances behind the submarine to which wake echoes belonged were then read from these diagrams. Measurements were made both for perpendic-

are given in Table 4, together with the approximate distances astern to which they refer. The transducers used had a narrow directivity pattern horizontally and a very wide pattern vertically, so that even during the deepest dives — to 400 feet — there was no significant loss of sensitivity.

Although the 0 point of the W scale in Figure 5 is rather uncertain because an adequate calibration of the sound gear is lacking for that particular day, the

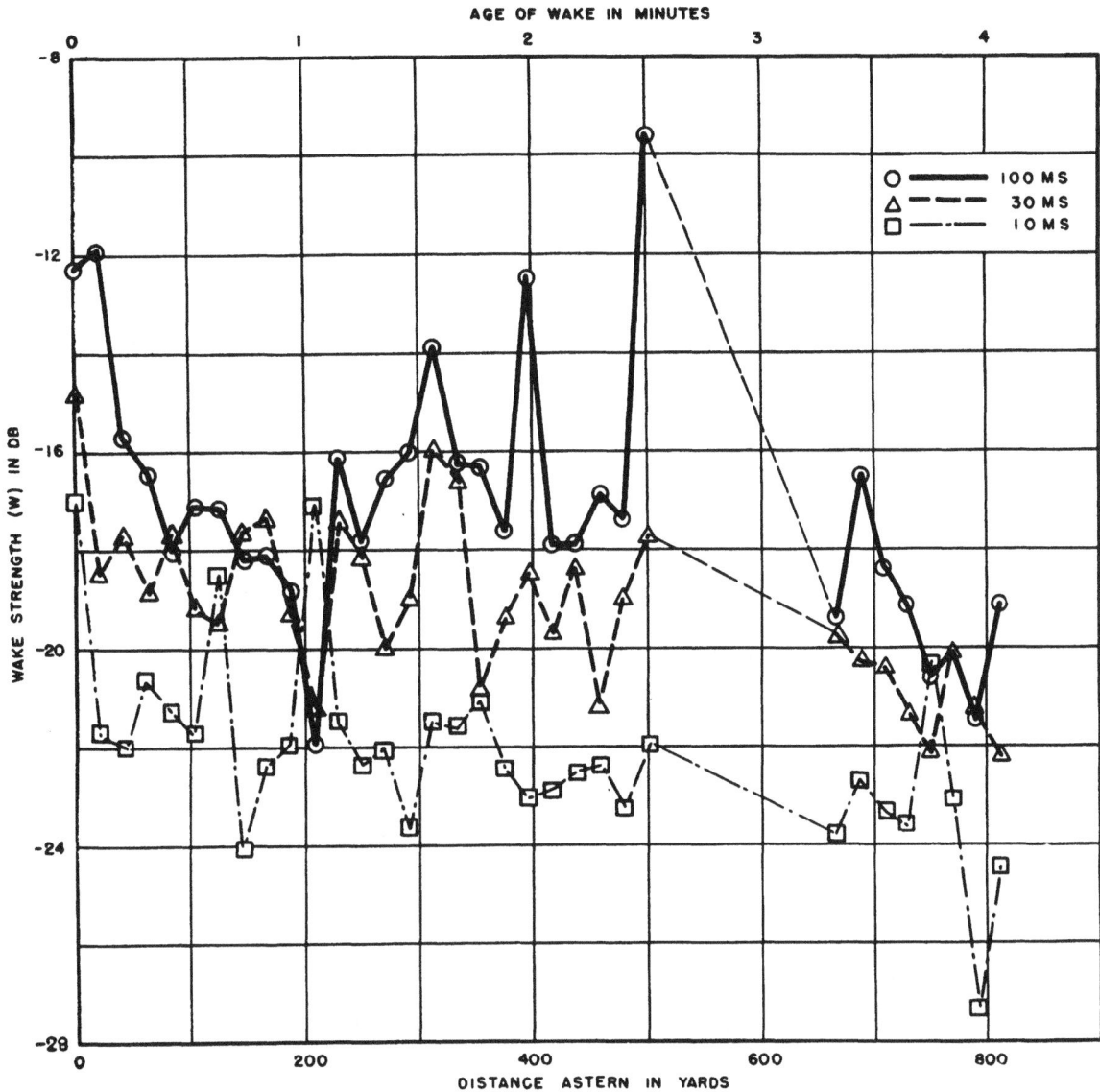

FIGURE 5. Dependence of wake strength on distance astern. Plot for USS S-18, submerged to a depth of 45 ft, for run 2 of Table 1. Echo-ranging vessel and submarine were proceeding on parallel courses at constant speeds of 8 and 6 knots respectively.

plot illustrates some significant features of the observations. The individual points of the diagram are computed from the average of five successive echo levels, and the scattering of these averages gives a good idea of the magnitude of echo fluctuations encountered in practice. Signals 10, 30, and 100 msec long were sent out in cyclic succession, so that the three curves for the different pulse lengths refer to the same wake. Despite the large echo fluctuations, there is good evidence for an increase of W with the signal length r_0. The steep rise of the curves at zero distance astern probably is due to the stern of the submarine.

Up to 500 yd astern — corresponding to a wake age of 2 minutes — the wake strength changes very little, if any. But when the observations were resumed at 670 yd astern, the decay of the wake had definitely set in. The values given in Table 4 suggest that for this wake the decay rate increased with increasing pulse length.

All reliable numerical values of W have been collected in Table 4, together with the values of the wake index used in the individual computations. The latter will permit the computation from W of the corresponding target strength of the wake, if desired. The outstanding feature of the table is the greater strength of the wake laid by surfaced submarines, compared with those from submerged runs. However, during a dive the wake strength does not decline steadily. Instead, repeated peaks occur. Some of the peak values even equal the strength of the surface wakes, as illustrated in Figure 6. These peaks are undoubtedly connected with the diving operations, movement of diving planes, blowing of tanks, and other operations. While the surface values of W are surprisingly consistent — about -15 db — only the order of magnitude of the subsurface strength can be regarded as established, perhaps -25 db to -30 db. The relative acoustic weakness of wakes behind submerged submarines probably results from several causes, such as lack of entrained air and the reduction of cavitation and bubble production at the higher pressure. A small but definite increase of W with pulse length as the pulse length changes from 10 to 100 msec is found for both submerged runs of the USS S-18 (SS123). The increase is small and results largely from the extension of the wake along the axis of the sound beam. Even for a wake whose thickness is less than the signal length, the echo will vary with signal length when the transducer is pointed obliquely at the wake. Only for normal incidence of the sound beam is the change of target strength with pulse length a simple, readily predictable effect.

33.4   ECHOES FROM SURFACE VESSEL WAKES

The San Diego group has studied echoes from wakes laid by numerous surface craft. Early experiments were carried out in San Diego harbor. During 1944, the group carried out a large program of recording echoes from the wakes of a number of surface vessels, including aircraft carriers, destroyers, and some small craft. Wakes for this program were laid on the open sea off San Diego.

33.4.1   Echo Ranging at Wakes in San Diego Harbor, 1943

For these experiments an echo-ranging transducer was mounted on a barge moored to one side of the harbor channel.[1] Most of the measurements were made on wakes laid by a motor launch (length 40 ft, beam 11 ft, draft $2\frac{1}{2}$ ft) traveling at 4 to 6 knots. Incidental results were also obtained by echo ranging at wakes produced by other vessels which happened

to pass; these vessels probably did not travel at full speed in the harbor.

The chief interest of these experiments, which have been reported in detail in reference 1, lies in the fact that short signals — only 9 msec long — were transmitted alternately at 15, 24, and 30 kc. Thus it is possible to analyze the results for a possible dependence on frequency both of the wake strength and its decay rate. The absolute values of the wake strength appear to be less reliable, for two reasons. First, difficulties with the calibration of transducers seem to have been experienced during the early phases of the San Diego wake studies; such would affect the absolute values of W, without impairing the results concerning the dependence of W on frequency. Second, the measurements in the shallow waters of San Diego harbor are likely to have been disturbed by bottom-reflected sound; indeed, an apparent dependence of W on the range was explainable only as caused by some peculiarity in the bottom contour.

The results of these early measurements are summarized in Tables 5 and 6. Values of the wake strength, obtained with 9-msec signals at three different frequencies, are collected in Table 5, which also contains the attenuation coefficients a and the wake indices Ψ_0 used in these computations.

There is no information available on the wake age at which the observations on the larger vessels were made; probably the age did not exceed a few minutes, and the initial wake strength had decayed only slightly. The decay of the wakes laid by the launch was studied systematically over a period of 12 minutes, after which time the echo intensity had dropped to the reverberation level. While the decay of the 15-kc echoes appeared to start immediately after passage of the launch, the echoes at 24 and 30 kc maintained their initial strength for about 2 minutes before they began to decay. The decay of the echo intensity follows a simple exponential law, to a good approximation; thus the strength of the echo expressed on a decibel scale decreases linearly with time. The decay rates found are listed in Table 6. Within the errors of observation, there is no dependence of the decay rate on frequency. But the wake strength W seems to increase with frequency. From the average W for each frequency of the vessels contained in Table 5, excluding the launch and the three fishing boats, the following differences are found:

$$W_{24} - W_{15} = +1.0 \text{ db}$$
$$W_{30} - W_{24} = +4.6 \text{ db}.$$

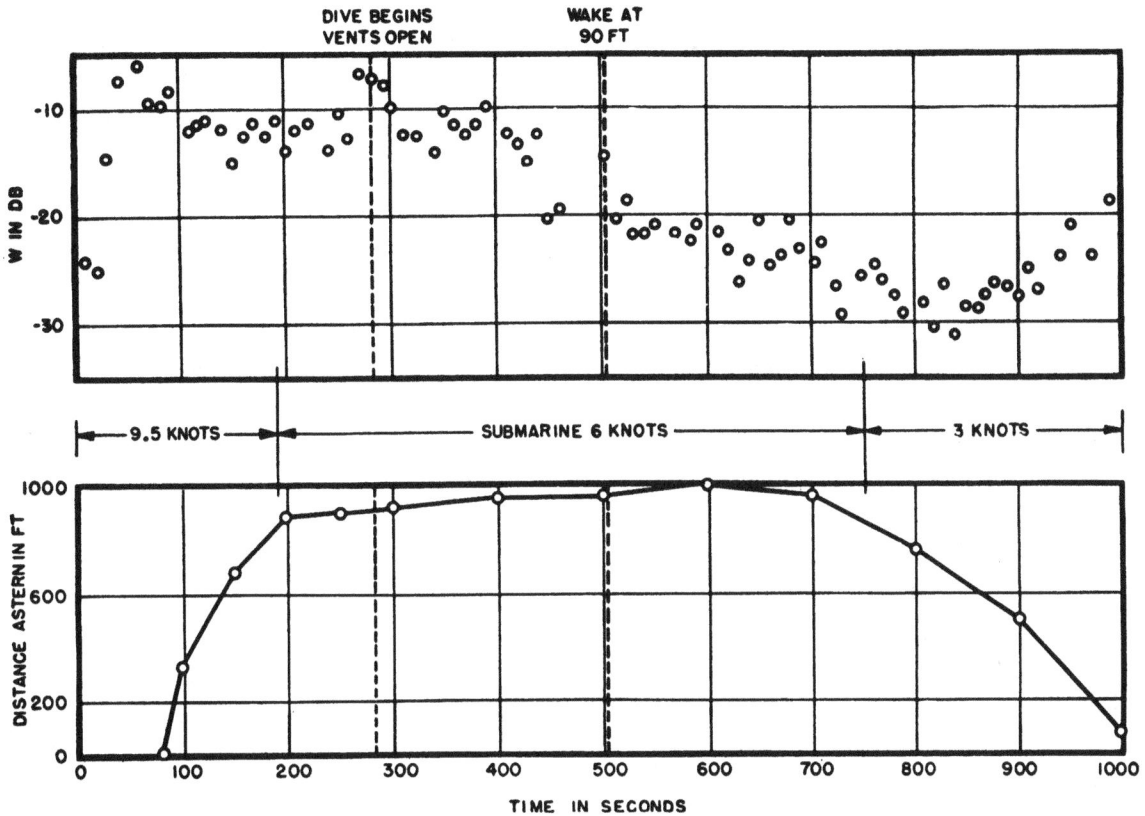

FIGURE 6. Wake strength and distance astern.

TABLE 5. Surface vessel wake strengths.

Vessel	Range in yd	W in db for 9-msec pulses		
		15 kc $a = 3.0$ db per kyd $\Psi_0 = -5.8$ db	24 kc $a = 5.0$ db per kyd $\Psi_0 = -8.0$ db	30 kc $a = 7.0$ db per kyd $\Psi_0 = -8.4$ db
40-ft motor launch	60–150	− 2.9	+ 3.1	+ 8.4
Tanker	330	+ 8.1	+ 7.7	+ 9.0
Fishing boat	298	+ 0.7	+10.7	. . .
Fishing boat	150	0.0	− 0.6	. . .
Fishing boat	152	. . .	− 0.4	. . .
Kelp barge	140	+ 7.3	+ 5.5	+12.2
Kelp barge	190	+ 4.5	+ 9.1	+14.3
50-ft boat	170	+ 2.7	+ 7.9	+12.9
Transport	450	+18.9	+17.5	+22.9
Tank boat	580	+10.9	+10.3	+14.7
Avg (excluding launch and fishing boats)		+ 8.7	+ 9.7	+14.3

The same trend is definitely established by the values of W for the 40-ft launch in Table 5, which are the averages resulting from 16 wakes.

33.4.2 Deep Water

By 1944 considerable progress had been made in standardizing the sound gear at frequencies in the neighborhood of 24 kc. It is believed that the reliability of the absolute calibration of the transducers used for these later measurements is much greater at these frequencies than during the earlier measurements. Moreover, this program was executed in deep water off San Diego, so that interference from bottom-reflected sound was avoided. Pending a comprehen-

sive report on these investigations, a summary of preliminary results has been made available for the
purposes of this volume.

The experiments with craft other than the USS
Jasper (PYc13) itself followed a single pattern. The
recording vessel, the *Jasper*, was lying to in the open
sea, and the wake vessel approached and passed her
while maintaining a straight course. As the wake
vessel approached, the *Jasper* echo-ranged on her,
training the sonar projector with the aid of a pelorus
manned on the flying bridge. When the wake vessel
came abreast of the *Jasper* at the time of closest approach, the training of the sonar projector was halted,
and its true bearing was held fixed and approximately

TABLE 6. Decay rate of wake of 40-ft launch.

Frequency in kc	Mean decay rate and its prob- able error in db per minute	Standard deviation in db per minute	Number of wakes
15	6.8 ± 0.6	4.0	20
24	7.0 ± 0.4	2.5	20
30	7.0 ± 0.5	3.0	20

perpendicular to the wake until the end of the run.
When possible, recording was continued, either continuously or intermittently, until no more echoes
from the wake were detectable above the background
of reverberation.

When the *Jasper* was studying her own wake, a
different technique was necessarily adopted. In this
case, the *Jasper* ran on a straight course at 12 knots
for 10 or 15 minutes; then she turned around and ran
back parallel to her original course with her sonar
projector trained abeam so that the sound beam was
directed normal to the original course. Recording was
continued until a wake echo was no longer discernible
above the reverberation.

A sample graph of the peak echo level received
from a wake against time is shown in Figure 7, which
also contains the echo levels received from a sphere
3 ft in diameter buoyed behind the wake at a depth
of 6 ft. On this run three different signal lengths, repeated in cyclical succession, were used to give wake
echoes. This system of interchanging signal lengths
was used on a number of wakes. On some occasions
the gear used for cycling the pulses was not in order,
and five or six echoes were recorded at one pulse
length before the pulse length was changed manually.
On a few occasions only one pulse length was used
throughout the run.

The initial wake strength is determined by the
initial echo level — the level of the wake echo at the
time (zero time) when the stern of the wake vessel
has just passed out of the sound beam. This time can
only be estimated, and sometimes echoes were not
recorded until some time after the wake was laid.
In such cases the values to be used for initial echo
level are obtained by extrapolating the observations
available back to the zero time. The decay of the
wake is measured as the slope in decibels per minute
of the echo level-time curve. Some thought was given
to the possibility of two decay rates in the wake, the
dividing line between them being rather sharp in time,
but the data were not sufficiently well defined to
allow such a distinction to be made. Therefore, only
one decay rate was obtained for each wake. This is
the rate at which the echoes seemed to decay steadily
for several minutes before they became indistinguishable.

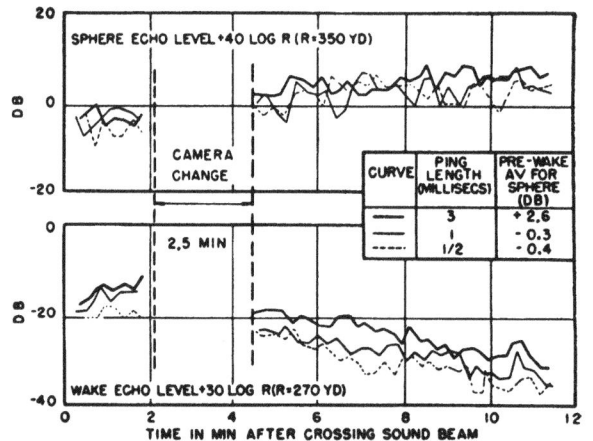

FIGURE 7. Decay of wake from *E. W. Scripps*. Run 1
of Table 8, 24 kc.

It seems fairly certain that there is a systematic
increase in W with the size of the wake vessel, but
Table 7 shows that the magnitude of the effect is not
very large. All that can be said about the speed of
the wake vessels during these experiments is that
they seemed to be running within their normal range
of operating speeds.

These averages include echoes obtained both with
10-msec and 30-msec pulses, as the number of available data was small and the increase of $W_{30\ msec}$ over
$W_{10\ msec}$ is moderate (see Table 8). Presumably, the
standard deviation would not be reduced much by
separating the results according to signal length.

The wake strength appears to increase with fre-

TABLE 7. Dependence of wake strength on wake-laying vessel.

Type of wake vessel	24 kc			60 kc		
	Average W in db	Standard deviation of W in db	Number of wakes	Average W in db	Standard deviation of W in db	Number of wakes
CVE's and AP's	− 7.7	4.1	5	...	...	...
DD's and DE's	− 9.6	6.3	5	+7.9	1.1	2
Laboratory yachts (*Scripps* & *Jasper*)	−13.6	2.6	5	+1.6	3.0	8
Small boats	−18.2	2.0	2	−3.7	2.1	2

TABLE 8. Dependence of wake strength on pulse length.

Type of wake vessel	Frequency in kc	Average difference of wake strength in db		Total number of wakes
		$W_{10 \text{ msec}} - W_{1 \text{ msec}}$	$W_{30 \text{ msec}} - W_{10 \text{ msec}}$	
DD's	24	6.5	3.5	3
CVE's	24	8.5	3.0	3
E. W. Scripps	24	9.0	1.5	1
E. W. Scripps	60	8.0	4.0	2
USS *Jasper* (PYc13)	24	9.5	3.0	3
USS *Jasper* (PYc13)	60	7.5	3.5	3
Avg of all vessels for all frequencies		8.2	3.7	

quency. From Table 7, it can be seen that the average difference between wake strength at 60 kc and wake strength at 24 kc is 16 db. But the reality of this phenomenon is doubtful, because use of this same underwater sound equipment in measurements of target strengths of submarines has yielded results at 60 kc which are also 10 to 20 db above the 24 kc results, thus contradicting theoretical expectations (see Section 23.6.2). It is also important that measurements on submarine wakes made with different equipment and discussed before (see Table 4) show a decrease of W with increasing frequency rather than an increase. In a separate set of careful experiments on surface-vessel wakes, which are summarized in Table 9, a single instance was found where there was a marked difference of opposite sign between wake strength at 60 kc and wake strength at 24 kc. The existence of an isolated but well documented instance like this where an apparent trend is contradicted must be given considerable weight when conclusions are drawn about the frequency dependence of the wake effect.

The decay rate of surface wakes shows very little dependence on frequency between 24 and 60 kc. The average decay rates of the wakes described above are 1.36 db per minute at 24 kc, and 1.18 db per minute

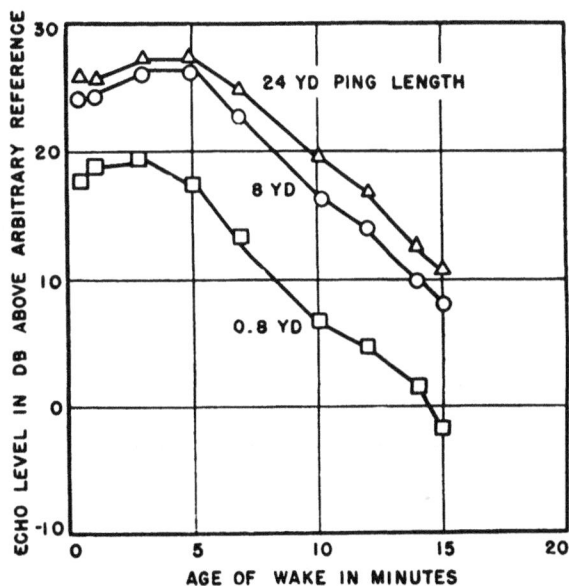

FIGURE 8. Echo level as function of age of wake, for various ping lengths at 24 kc. Wake vessel: Small carrier at about 15 knots.

at 60 kc. The standard deviations of these measurements are 0.59 db per minute at 24 kc and 0.69 db per minute at 60 kc. This difference in averages is so much smaller than the spread of the data that it cannot be said that there is any significant difference between the decay rate at 60 kc and the decay rate at 24 kc.

Figure 8 shows a typical example of the variation of the echo level, or of the wake strength, with signal length. Numerical values of the variation of the echo level, or of the wake strength, with signal length are summarized in Table 8.

This dependence of W on the signal length was predicted from general theoretical considerations (see Section 33.1.2), and the observed magnitude of the effect permits an estimate of the average concentration of bubbles in a wake (see Chapter 34). The number of observed data is too small to warrant any conclusions as to the influence of frequency and size of the wake vessel on the pulse length effect.

TABLE 9.   Wake strength and decay rate, *E. W. Scripps*.

Run	Frequency in kc	Wake index in db	Wake strength W in db at age 0–2 minutes 3-msec pings	Decay rate of wake strength dW/dt in db per minute
1	24	−7.0	− 5	−1.5
2	24	−7.0	−11	
3	60	−7.0	−21	−0.7
4	60	−7.0	−16	
4	45	−6.5	−22	−0.7

Table 9 contains some additional values of W for several wakes laid by *E. W. Scripps* on a day when the sea was unusually calm. Experimental details concerning these observations have already been given in Section 32.3.2, and the transducers used are listed in Table 3; the echo level-time curve of Run 1 of Table 9 is reproduced in Figure 7.

The average W at 24 kc (mean of Runs 1 and 2) is −8 db for 3-msec pulses. In order to make this value comparable with the average value of W at 24 kc for the wakes of the *Scripps* and *Jasper* in Table 7, which is −13.6 db, a correction for the difference in signal lengths used must be made; from Table 8 it may be estimated that $W_{10\,msec} - W_{1\,msec}$ is of the order of +6 db. The corrected $W_{24\,kc}$ of Table 9 is then −2 db, or about 12 db *greater* than $W_{24\,kc}$ in Table 7. After the corresponding correction of the 60-kc data of Run 3 has been made, $W_{60\,kc}$ in Table 9 is still 17 db *smaller* than its counterpart in Table 7; the origin of this serious discrepancy remains unexplained. As for the minor discrepancy at 24 kc it seems worth mentioning that the *E. W. Scripps* had been outfitted with a new propeller and engine in the fall of 1944,

so that the data in Tables 7 and 9 are not strictly comparable. The decay rates in Table 9 do not differ significantly from the averages for all surface vessels quoted before.

33.5   ECHOES FROM MODEL PROPELLER WAKES

At the Woods Hole Oceanographic Institution,[6] a number of experiments were made on the scattering of sound by the wakes of stationary model propellers. Although the published data do not yield absolute values of the wake strength, they give some interesting information on the relative echo intensity as a function of the frequency of sound and of the depth of the propeller.

In order to measure the scattering, the hydrophone and transducer were mounted on the same side of the wake in a horizontal plane including the wake axis. The axis of the hydrophone was vertical and the transducer was directed toward the wake. Both instruments were secured to a pipe frame and were separated by a baffle, in order to reduce the passage of the direct signal from the transducer to the hydrophone. The baffle consisted of a sheet of Celotex 32 in. square and 1/2 in. thick, sheathed with copper; the plane of this sheet was perpendicular to the axis of the wake. This single baffle was found to be preferable to a wedge-shaped baffle composed of two sheets of Celotex making an angle with one another. In order to reduce the direct signal still further, the hydrophone was partially enclosed in a box lined with Celotex and open on the side toward the wake. The perpendicular distance from the instruments to the wake axis was 5 ft; the plane of the baffle, midway between the instruments, was 10 ft from the plane parallel to it through the propeller. With this arrangement, scattering measurements were made in the deep spot 200 ft off the wharf at depths varying from 5 to 60 ft and at frequencies from 30 to 60 kc. At lower frequencies the reflection was too small to measure.

Each determination of the scattering involved the measurement of the signal at the hydrophone under three conditions: (1) with the propeller at rest and the transducer on; (2) with the propeller running and the transducer on; (3) with the propeller running and the transducer off. The results of these three measurements, in decibels, will be referred to by z_1, z_2, and z_3, respectively, with z_1 representing the direct signal from the transducer in the absence of scattering —

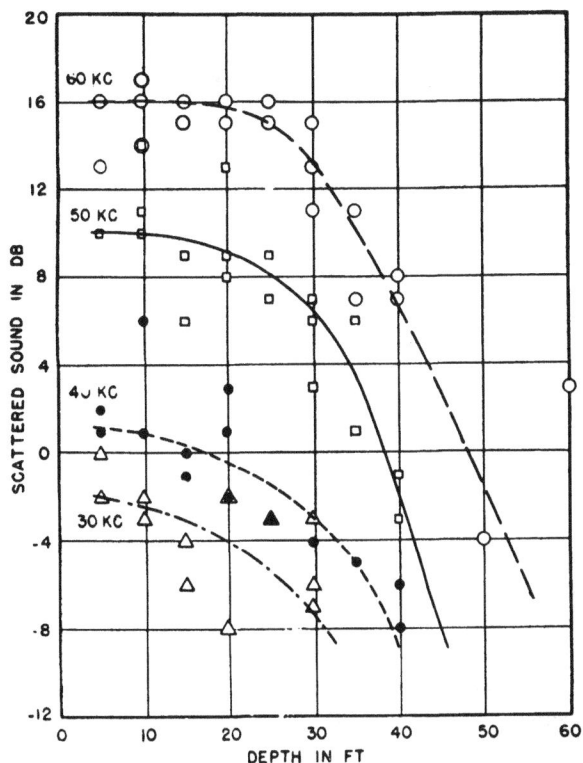

FIGURE 9. Dependence of sound scattered from 10-inch propeller at 1,600 rpm on depth below surface. Direct signal constant for each frequency.

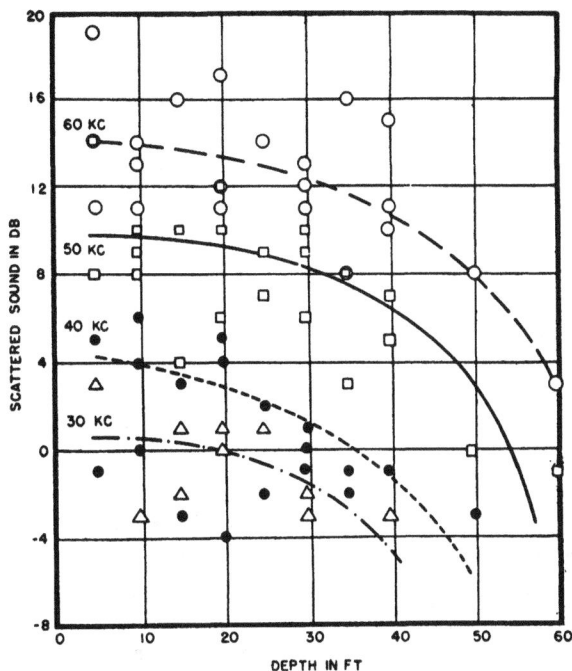

FIGURE 10. Dependence of sound scattered from 14 inch propeller at 1,600 rpm on depth below surface. Direct signal constant for each frequency.

the sound which travels around and through the baffle, and z_3 representing the cavitation and propeller noise. The true value of the scattered sound in decibels, which we call z_r, is in general different from z_2 but may be obtained from it by correction for the direct signal (z_1) and for cavitation and propeller noise (z_3). It is given by the equation

$$10^{z_r/10} = 10^{z_2/10} - 10^{z_1/10} - 10^{z_3/10}.$$

The results presented below were calculated in this way. It should be pointed out that the effect of z_3 was in all cases negligible.

An interval of a minute or a minute and a half was always allowed between successive determinations to make sure that there should be no residual wake from the previous determination to interfere with the following one. Only the 10-in. and 14-in. propellers were used, and only at the highest speed, 1,600 rpm. Under other conditions the scattered sound was too small to measure satisfactorily.

The results of the study are shown in Figures 9 and 10. Although the scatter of the observations is large, particularly with the 14-in. propeller, there can be no question of the general effect. It is evident that there

is a marked decrease in sound scattering with depth. At a frequency of 60 kc the scattered sound is less than $\frac{1}{10}$ as much at 60 ft as at 5 ft. In this respect the situation is similar to that observed in the case of attenuation (see Section 32.5).

The data plotted in Figures 9 and 10 give simply the total reflected sound in decibels at the hydrophone. They take no account of the strength of the direct signal from the transducer. Since the oscillator was always set to give the same output, this signal may be regarded as constant for each frequency. Consequently at each frequency the change in the decibel level of the reflected signal with depth gives the change in the scattering coefficient. Nevertheless, in order to obtain absolute values of the scattering coefficient and to discover its dependence on frequency it is necessary to take into account the strength of the direct signal which would be received by the hydrophone in the absence of a wake at the position of what may be called the "virtual image" of the hydrophone with respect to the wake. This is a point at the same distance from the wake as the hydrophone, but on the opposite side of it. It was estimated to be 6 ft away from the transducer. With this in mind, throughout the study, daily determinations were made of the response of the hydrophone

6 ft in front of the transducer and in the same orientation as in the actual measurements. Such determinations were made for each frequency used in the measurements. The results were found to be independent

TABLE 10. Direct signal as function of frequency.

Frequency in kc	30	40	50	60
Direct signal z_0	22.5	26.2	34.0	36.8

of depth, as would be expected, and were reasonably constant from day to day. Relative minor variations are probably attributable to small differences in the spacing of the two instruments. Values of the direct signal, which will be called z_0, measured in this way are given in Table 10. On the basis of these results, it is a simple task to calculate the relative intensity of the reflected sound. This, in terms of decibels, is simply $z_r - z_0$. Table 11 gives the values of $z_r - z_0$ obtained with each of the two propellers at a depth of 10 ft. In arriving at these results values of z_r were read off the smooth curves of Figures 9 and 10; values of z_0 were taken from Table 10.

The intensities of the scattered sound at other depths are, of course, less than these, in accordance with the way in which the curves of Figures 9 and 10 drop off. It is evident that there is no considerable

TABLE 11. Reflected sound as function of frequency.

	10-in. propeller at 1,600 rpm and depth of 10 ft			
Frequency in kc	30	40	50	60
$z_r - z_0$	−24.5	−25.2	−24.0	−20.8
	14-in. propeller at 1,600 rpm and depth of 10 ft			
Frequency in kc	30	40	50	60
$z_r - z_0$	−22.0	−22.2	−24.0	−22.8

effect of frequency between 30 kc and 60 kc. The decrease of the echo intensity with depth is again a manifestation of the influence of increased pressure on the formation and dissolution of bubbles, as in the decrease of the attenuation with depth described in Section 32.5.

Chapter 34

ROLE OF BUBBLES IN ACOUSTIC WAKES

THE PREVIOUS CHAPTERS have developed a general theoretical background for the study of wakes and have presented the results of acoustic measurements on wakes. In this chapter, a review is first given of the evidence that bubbles are the chief source of the acoustic properties of wakes. Next, the quantitative acoustic measurements are compared with the theoretical formulas derived in Chapter 28. From this comparison, conclusions are drawn as to the amount of air present in wakes. Finally, the rate of decay of acoustic wakes is discussed, and shown to be roughly similar to the rate at which air bubbles disappear in sea water.

34.1 EVIDENCE FOR AIR BUBBLES IN WAKES

At the present time it seems almost certain that small air bubbles are responsible for the observed reflection and absorption of sound by surface ship and submarine wakes. The evidence for this is of two general types, qualitative and quantitative.

From a qualitative standpoint, air bubbles provide the only mechanism yet proposed which can explain the general behavior of wake echoes. In particular, no other explanation seems capable of explaining the very marked dependence of scattering and absorbing power on the depth of the wake. The measurements with the model propeller, described in Sections 32.5 and 33.5, show unmistakably a pronounced weakening of both attenuation and scattering when the propeller is below about 30 ft. Measurements of echoes from submarine wakes show a similar decrease of about 5 to 10 db in wake strength when the submarine dives from the surface to periscope depth. Practical echo-ranging trials confirm the disappearance of wake echoes when the submarine dives below 200 or 300 ft. These observations cannot be explained on the assumption that turbulence or temperature effects are responsible for the acoustic properties of wakes, but they follow naturally from the assumption that bubbles are the important agents.

From a quantitative standpoint, the magnitude of the observed effects is enormously greater than can apparently be explained by any assumed mechanism besides the presence of small bubbles in the wake. It has already been noted, in Chapter 29, that on the basis of present acoustic theory, neither turbulence nor temperature irregularities could account for any appreciable scattering or attenuation by wakes. The absorbing and scattering power of a single resonant bubble, analyzed in Section 28.1, is so great, however, that a relatively small number of bubbles is required to explain the observed acoustic effects.

Any theory of the acoustic properties of wakes cannot be regarded as completely confirmed until reliable quantitative data are shown to be in close numerical agreement with the theoretical predictions. Until independent nonacoustic measurements are made of the bubble density in wakes, or until accurate and reproducible acoustic data can be obtained on wakes under a variety of conditions, it is not possible to verify the "bubble hypothesis" explaining the origin of the acoustic wake. Nevertheless, the general evidence seems sufficiently strong to make this hypothesis highly probable.

34.2 TRANSMISSION THROUGH WAKES

The attenuation of sound by air bubbles in water has been discussed in Section 28.2. The conclusion reached was that probably most of the attenuation is produced by bubbles whose radii are close to the radius R_r of a resonant bubble. Integrating the contributions to the attenuation from all bubbles near resonant size leads to equation (67) of Chapter 28 for K_e, the attenuation coefficient in decibels per yard:

$$K_e = 1.4 \times 10^5 u(R_r), \qquad (1)$$

where $u(R_r)dR$ is the volume of air per cu cm in bubbles whose radii lie between R_r and $R_r + dR$. If K_e is known at all frequencies, equation (1) gives $u(R_r)$ for bubbles of any radius. The total volume u

533

of air in one cu cm of water is then given by the integral

$$u = \int_0^{R\text{max}} u(R_r)dR_r, \qquad (2)$$

where R_{max}, the maximum radius of any bubble present, is assumed to be much less than 1 cm.

Since the attenuation coefficient K_e is directly proportional to the bubble density $u(R_r)$, and since also the damping constant δ discussed in Section 28.2 does not affect K_e, measurements of acoustic attenuation provide a sensitive determination of the amount of air present in wakes. The actual attenuations observed, however, are somewhat complicated by the geometry, since the wake is never sufficiently deep to ensure that no sound reaches the measuring hydro-

To find the absorption in decibels per yard, the resulting transmission losses have been divided by the wake widths for the destroyers given in Section 31.3.1. The values of K_e for a destroyer speed of 15 knots are listed in Table 1, together with the corresponding values of $u(R_r)$. The values of $u(R_r)$ for different ages of the wakes were plotted against R_r for destroyer speeds of 10, 15, 20, and 25 knots, respectively, and the areas under these curves were determined by graphical integration. The resulting values of u, the relative amount of air present, in bubbles of all sizes for different destroyer speeds and wake ages are given in Table 2. Starred values are uncertain, since they are based primarily on the extrapolated parts of the graphs.

Apparently no direct estimates have been made of air present as bubbles in destroyer wakes. The only

TABLE 1. Attenuation coefficient and density of resonant bubbles—destroyer at 15 knots.

| Frequency in kc | Age of wake and distance astern | | | | | | R_r in cm |
| | 1 minute 500 yd astern | | 3 minutes 1,500 yd astern | | 5 minutes 2,500 yd astern | | |
	K_e	$u(R_r)$	K_e	$u(R_r)$	K_e	$u(R_r)$	
3	0.35	2.5×10^{-6}	0.03	2.1×10^{-7}	...	...	0.107
8	0.67	4.8×10^{-6}	0.21	1.5×10^{-6}	0.03	2.1×10^{-7}	0.040
20	1.11	7.9×10^{-6}	0.48	3.4×10^{-6}	0.22	1.6×10^{-6}	0.016
40	1.65	1.18×10^{-5}	0.79	5.6×10^{-6}	0.43	3.1×10^{-6}	0.008

phone below the wake. As a result of this uncertainty, the bubble densities found by use of equations (1) and (2) are somewhat indefinite, though they are probably not in error by a factor of more than two.

Bubble densities may be computed from acoustic measurements for destroyers at different speeds and for different wake ages. They may also be computed for a small high-speed propeller with no forward motion.

34.2.1   Wakes of Destroyers and Destroyer Escorts

The computations for destroyers and similar vessels are based on the extensive transmission measurements across wakes reported in Section 32.3.1. The smoothed curves represented by equations (10) and (11) of Chapter 32 have been used, and an average taken for source outside the wake and source inside the wake, since these represent lower and upper limits to the absorption in the top 10 ft of the wake.

TABLE 2. Fraction of air present as bubbles in destroyer wakes.

| Destroyer speed in knots | u | | |
| | Age of wake | | |
	1 minute	3 minutes	5 minutes
10	5.2×10^{-7}*	1.4×10^{-7}	6.5×10^{-8}
15	7.4×10^{-7}*	2.0×10^{-7}	6.9×10^{-8}
20	7.0×10^{-7}*	2.3×10^{-7}	8.5×10^{-8}
25	9.1×10^{-7}*	2.1×10^{-7}	8.7×10^{-8}

* Uncertain.

attempts to collect bubbles in ship wakes are apparently the attempts made with a 78-ft yacht.[1] About 1 cu cm per minute of air was collected through a ring 8 in. in diameter, 6 ft behind a propeller 38 in. in diameter rotating at tip speeds between 50 and 60 ft per second. Cavitation bubbles could be seen in the water, but the bubble density computed for a slipstream speed of 5 ft per second is only 5×10^{-7} parts of air by volume in 1 part water. This value is in moderately good agreement with the values shown in Table 2.

34.2.2 Wakes of Model Propellers

A similar computation may be carried out for the wakes of small propellers. Measured values of the absorption across a wake are reported in Section 32.5. The cross section of the wake was about 1.5 yd wide at the point where the measurements were made. The values of $u(R_r)$ were computed by use of equation (1) for the 10-in. propeller at 1,600 rpm and for depths of 10 ft, 20 ft, and 30 ft. Somewhat smaller values are found for the 14-in. propeller at the same rpm, possibly as a result of the narrower blades and lower pitch of this propeller. The corresponding values of u — the relative amount of air present in bubbles of all sizes, found directly from these curves — are given in Table 3.

TABLE 3. Fraction of air present as bubbles in wake of 10-in. model propeller.

Depth in feet	u
10	3×10^{-6}
20	2×10^{-6}
30	9×10^{-7}

It is perhaps unexpected that the bubble density in the wake of a 10-in. propeller be from five to ten times as great as the corresponding density in a destroyer wake. Further analysis shows this is not too surprising. The propeller developed 11 hp during operation, with a tip speed of 70 ft per second. When a destroyer is making 15 knots, its two propellers with diameters between 9 and 11 ft, have a comparable tip speed, about 80 ft per second. Moreover the destroyer is moving rapidly, and it is well known that a propeller which is held stationary in the water tends to produce stronger tip vortices than one at the same rpm which pushes itself through the water. Thus the small propeller may be expected to cavitate more vigorously than the propeller of a destroyer at 15 knots. The volumes over which the bubbles produced in one second are spread in these two cases are proportional to the total propeller areas. Thus it would not be surprising to find that the bubble density measured behind the small propeller is greater than the corresponding density in the destroyer wake.

34.3 ECHOES FROM WAKES

The wake strength W is related to the bubble density in a more complicated way than is the attenua-

tion coefficient K_e. In addition, W depends both on the detailed geometrical properties of the wake, and on the physical properties of bubbles of different sizes, and cannot therefore be predicted with any exactness for a known distribution of bubbles. Thus, at most, a rather general agreement can be expected between observed and predicted wake strengths.

The formulas are simplest for long pulses; when bubbles of a single size are present, the wake strength W for long pulses is given by the equation

$$W = 10 \log \left\{ \frac{h\sigma_s}{8\pi\sigma_e} \left[1 - e^{-2\sigma eN(w)} \right] \right\} \qquad (3)$$

taken from equation (48) of Chapter 33. The quantities σ_s and σ_e are the scattering and absorption cross section defined by equations (34) and (43) in Chapter 28, while h is the depth of the wake, measured in yards. $N(w)$ is the total number of bubbles in a column one sq cm in cross section extending through the wake in a direction parallel to the sound beam [see equation (54) of Chapter 28], and the product $\sigma_e N(w)$ is 0.23 times H_w, which is the total transmission loss across the wake measured in decibels. Thus when this transmission loss is high, the exponential term is very small, and W approaches the limiting value

$$W = 10 \log h + 10 \log \left(\frac{\sigma_s}{8\pi\sigma_e} \right). \qquad (4)$$

Equation (46) of Chapter 3 gives the ratio of σ_s to σ_e in terms of δ, the so-called *damping constant*, and η, the ratio of bubble circumference to the wave length of the sound which represents the contribution of radiation damping to the damping constant. Values

TABLE 4. Observed frequency dependence of ratio of scattering to extinction cross sections.

Frequency in kc	1	5	8	13	19	26	36	45
$10 \log \left(\frac{\sigma_s}{8\pi\sigma_e} \right)$	-21	-22	-23	-24	-25	-26	-27	-28

of these two quantities have been taken from Figure 2 and equation (23) of Chapter 28 and the resulting values of $10 \log (\sigma_s/8\pi\sigma_e)$ shown in Figure 1 and Table 4 of this chapter. At 24 kc, this quantity is -26 db, and the maximum value of W is equal to

$$W = 10 \log h - 26. \qquad (5)$$

For a typical wake 10 yd deep this gives a maximum wake strength of -16 db.

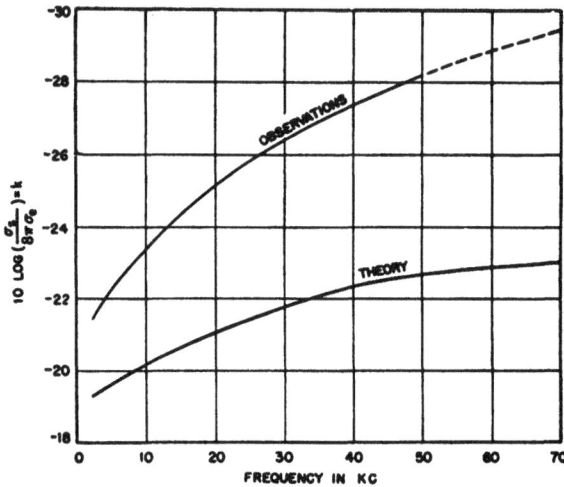

FIGURE 1. Frequency dependence of ratio of scattering to extinction cross sections.

Considering the systematic difference between the observed and theoretical values of δ, as evident in Figure 2 of Chapter 28, it appears highly probable that at 60 kc the damping constant will not be smaller than its theoretically predicted value, since the actual damping by dissipative effects should not be less than

σ_s/σ_e. However, the scattering and absorption cross sections of a resonant bubble are so much greater than those of other sizes that it seems unlikely that bubbles other than those near resonance can contribute appreciably to either the scattering or the absorption. Thus equation (5) may be used for actual wakes, provided that a value of   appropriate to a resonant bubble is taken.

On the other hand, when both the product $\sigma_e N(w)$ and the transmission across the wake are negligible, equation (3) gives for bubbles all of the same size the equation

$$W = 10 \log\left(\frac{hN(w)\sigma_s}{4\pi}\right) = 10 \log\left(\frac{hw\bar{n}\sigma_s}{4\pi}\right) + 19.6, \quad (6)$$

where w is the width of the wake in yards and $\bar{n}$ is the average number of bubbles per cubic centimeter. Since $N(w)$ is the number of bubbles per square centimeter appearing in projection on a plane perpendicular to the sound beam, the equivalent product $\bar{n}w$ in equation (6) must have the same units — that is, square centimeters. It is customary to measure the wake width w in yards, or units of 91.5 cm. Hence, in order to keep equation (6) dimensionally correct, a

TABLE 5. Frequency dependence of damping constant.

Frequency in kc	5	10	15	20	25	30	35	40	45
$10 \log (3/8\delta)$	-6.4	-5.2	-4.2	-3.4	-2.7	-2.1	-1.6	-1.2	-0.8

that resulting from the flow of heat in and out of the oscillating bubble. This predicted value, derived from the theory given in Section 29.2, happens to be about one-third of the observed value of δ at 24 kc. Hence the true damping constant for 60-kc sound very likely is greater than one-third of the observed damping constant at 24 kc. This surmise implies that the theoretically predicted maximum wake strength for 60-kc sound should not exceed the observed value of W at 24 kc by more than 5 db — because η is independent of frequency in this range — unless the effective value of the wake depth h is quite different at the two frequencies.

For the general case of a bubble population comprising all sizes from the largest to the smallest, the analysis is more complicated. If many bubbles of very large radii are present, they will scatter without much absorbing, and σ_s/σ_e will be increased. On the other hand if many bubbles of very small radii are present, these will absorb without much scattering, decreasing

term $10 \log 91.5$, which is equal to $+19.6$, has been added to the right-hand side of equation (6) since w is measured in yards. When bubbles of varying sizes near resonance are considered, equation (6) is modified by the substitution of S_s for $\bar{n}\sigma_s$; S_s is a weighter' mean of σ_s for bubbles near resonance, according to equation (77) of Chapter 28, and is equal to

$$S_s = \frac{3\pi u(R_r)}{2\delta_r}. \quad (7)$$

Equation (6) then may be written

$$W = 10 \log h + 10 \log w + 10 \log u(R_r)$$
$$+ 10 \log\left(\frac{3}{8\delta}\right) + 19.6, \quad (8)$$

where h and w are the depth and width of the wake, respectively, both measured in yards. Values of $10 \log 3/8\delta$ are shown in Table 5 for resonant bubbles at different frequencies. In principle, equation (8) can be used to determine $u(R_r)$ from the observed value of W for any wake across which the transmission loss is less than 1 db. In practice, if the wake strength is less than about -30 db, the echo is difficult to distinguish from the background. Since the theoretical maximum value of W is only -16 db for a wake 10 yd deep, there is a relatively narrow spread of values over which $u(R_r)$ can be varied to give measurable variations in W.

34.3.1 Surface Vessels

For surface ships the transmission loss across the wake is usually large. Thus in theory all surface wakes should exhibit a wake strength W given by equation (5). All wake strengths should be nearly constant and equal to -16 db, except for small variations in 10 log h, presumably not exceeding 3 db at most.

An examination of the surface vessel wake strengths tabulated in Table 5 of Chapter 33 shows that the wake strengths are highly variable. The variability of transmission loss, which could not readily be taken into account in the measurements, probably accounts at least in part for this failure of the wake strengths to remain at a constant level.

Even the average observed values of W, however, cannot be compared directly with the theoretical predictions. In the first place, the measured wake strengths all refer to peak amplitudes. Extensive measurements of reverberation records[2] show that the average peak amplitude is about 7 db higher than the average amplitude; these measurements refer to a segment of reverberation three to six times as long as the signal length. Moreover, since the rms amplitude is about 1 db above the average amplitude, it follows that -6 db should be applied as a net correction. According to the observations, this correction does not change rapidly in magnitude when the length of the reverberation segment analyzed is changed. Since echoes from wakes are structurally similar to reverberation, it is concluded that a correction of -6 db applied to the observed values of W listed in Chapter 33 presumably will suffice to express them on the intensity scale envisaged in equations (3) to (8). In addition, if surface-reflected sound reaching the wake is of the same intensity as the direct sound, the actual transmission anomaly is 3 db less than assumed; another 6 db should then be subtracted from the wake strengths reported in Chapter 33 to give the correct values.

If the correction for surface-reflected sound is neg-

lected, values of the observed wake strengths on an intensity scale may be found by subtracting 6 db from the values of W listed in Table 7 of Chapter 33. The resulting values are shown in Table 6, together with

TABLE 6. Observed and predicted wake strengths.

	Observed wake strengths at 24 kc in db	h in yds	Maximum theoretical wake strength at 24 kc in db
CVE's and AP's	-14	15	-14
DD's and DE's	-16	8	-17
E. W. Scripps	-20	4.4	-20
USS Jasper (PYc13)	-20	6.7	-18
Small boats	-24	2(?)	-23(?)

the wake depths h taken from Chapter 31 and the theoretical limiting values of W found from equation (5). The close agreement between theory and observation for the larger vessels suggests that no large correction is required for the presence of surface-reflected sound. This same conclusion is supported by agreement between direct and indirect determinations of submarine target strength at beam aspect, reported in Sections 21.5.4 and 23.8.1 of this volume.

There are a few cases of anomalously high wake strengths, discussed in Section 33.4. These are difficult to explain on the basis of scattering by bubbles. One possible effect worth considering, that could in principle give rise to very high wake strengths, is the specular reflection of sound from wakes. As pointed out in Section 28.3.4, bubbles not only scatter sound, but also affect the sound velocity. If the boundary of the wake is sufficiently sharp, some sound will be reflected backward. Since the reflected sound rays will go predominantly in the backward direction, rather than out in all directions, the resulting wake echo can be quite high even though the coefficient of reflection is not very great. For the bubble densities found in destroyer wakes, and summarized in Table 2, the reflection coefficient found from equation (85) of Chapter 28 is less than 0.4×10^{-6} and therefore quite negligible. It is possible that higher bubble densities might be present in the highly reflecting wakes of the vessels discussed in Section 33.4, but this seems unlikely. These high values (see Table 5 of Chapter 33) were found in early measurements in shallow harbor waters and have not been reproduced in later, more accurate determinations on wakes of the same vessels. For example, early measurements on a 40-ft motor launch gave a value of 2 db for W; later measurements on the same ship, with more standard equipment, gave a value of -21 db. In view of the failure of the later measurements to reproduce the early high values, these early values can probably be neglected. Until more detailed information is available it may therefore be assumed that on the average the wake strength of large moving surface vessels, measured with long pulses, are all close to the theoretical maximum values found from equation (5); that is, about -16 db for rms amplitudes and -10 db for average peak amplitudes, at 24 kc.

The high values of W found at 60 kc are not easily explained. These values are believed to be less accurate than those at 24 kc, since the equipment had not yet been wholly standardized. It is perhaps significant that in one of the most careful tests — the measurements on the wake of the *Scripps* discussed in Section 33.4 — the value of W at 60 kc was actually less than that at 24 kc (see Table 9 of Chapter 33). Moreover, use of this same underwater sound equipment in measurements of target strengths of submarines has yielded results at 60 kc which are also 10 to 20 db above the 24 kc results, in contradiction to theoretical expectations (see Sections 21.4.3 and 23.6.2). It is also important that measurements on submarine wakes, made with different equipment and discussed below, show a decrease of W with increasing frequency rather than an increase. It is possible that the bubble density at 60 kc is sufficiently high and the wake boundary sufficiently sharp that specular reflection of sound at the wake boundary is sufficient to account for the high wake echoes observed; this possibility has not been investigated theoretically. Until the high wake strengths found at 60 kc can be either explained or shown to be the result of observational error, they will remain a serious discrepancy in the study of wakes.

34.3.2   Submarines

Values of W for submarines both submerged and surfaced are presented in Table 4 of Chapter 33. For surfaced submarines no estimates are available at 20 kc; but at 45 kc, the value of W found for two submarines is -13 db. When 6 db is subtracted to give the wake strength in terms of the average intensity, this value is in close agreement with the maximum wake strength of about -16 db found at 24 kc. While no experimental data are available on the value of the damping constant δ at 45 kc, Figure 1 suggests that

the value of $10 \log (\sigma_s/8\pi\sigma_e)$ at 45 kc does not differ by more than a few decibels from its value at 24 kc. Thus it may be inferred that the wake strength for a surfaced submarine is quite comparable with that for any large moving surface vessel. The decrease of W shown at 60 kc in Table 4 of Chapter 33 is probably not significant.

This same Table 4 shows that the wake of a submerged submarine is a much poorer reflector than the wake of a surfaced submarine. Since the wake strength is less than its maximum value, W should vary with the bubble density, and therefore with submarine depth and speed. While the measurements are not very conclusive, they indicate that for a submarine at 6 knots and a depth of 45 to 90 feet, W is about -25 db at 45 kc; this estimate may well be in error by as much as 5 db. As before, an additional 6 db must be subtracted to convert to an intensity scale, giving -31 db for W. If $10 \log (3/8\delta)$ at 45 kc is taken from Table 5, equation (8) gives

$$10 \log u(R_r) = -31 - 0.8 - 10 \log h$$
$$- 10 \log w - 19.6. \quad (9)$$

If the wake is 10 yd deep and 30 yd across, the bubble density $u(R_r)$ is about 3×10^{-8}, less than a hundredth of the values for destroyer wakes 1 minute old at 15 knots found in Table 1; the assumed wake dimensions are somewhat uncertain, but any reasonable variation of these figures would not change the order of magnitude of $u(R_r)$. If the curve of $u(R_r)$ against R_r were the same as the typical curves for destroyers, the total fraction u of the wake volume occupied by air bubbles would be only about 1×10^{-9}. While no other quantitative measurements are available, practical echo-ranging tests indicate that the bubble density decreases with increasing submarine depth as would be expected at the greater pressure.

In the top 60 ft this decrease is about as rapid as would be expected from the experiments with model propellers. Both the transmission measurements discussed above and the reflection measurements discussed below show that with a 10-in. propeller all acoustic effects are much reduced at depths below 30 ft. However, the reflection measurements indicate that the acoustic effects have largely disappeared at depths below 60 ft, while echoes from submarine wakes have been reported at greater depths. This difference may be due to lack of sensitivity of the acoustical equipment used for the model propeller experiments. Alternatively, the greater size of the full-scale propellers may enable the formation of larger bubbles, which would persist longer at great depths. More complete measurements would be required on submarine wakes of different depths before any detailed conclusions can be drawn as to the variation of wake strength with depth.

34.3.3 **Model Propellers**

The studies of echoes from wakes of model propellers, reported in Section 33.5, are not sufficiently detailed to compare with theoretical predictions. While echo levels were quantitatively determined, neither the geometry of the experiment nor the transducer and hydrophone directivities are well enough known to make possible a prediction of the echo levels from the known properties of air bubbles. These measurements are of theoretical interest, however, because they provide information on the change of wake echoes with depth. This information has already been discussed before.

The data also provide an interesting qualitative confirmation of scattering theory. As is evident from Figures 9 and 10 of Chapter 33, the echo level remains relatively constant in the first 30 ft of increasing depth. In this same depth interval the attenuation in the wake, shown by Figures 3 and 4 of Chapter 32, decreases from a high value near the surface to less than 5 db at 30 ft. This behavior is in accord with equation (3); this equation predicts that as long as the transmission loss is more than a few decibels, the amount of sound scattered from a collection of air bubbles in water will be independent of the density of bubbles in the water.

34.4 **DECAY OF WAKES**

The observations on the decay rate of a wake's acoustic properties should be consistent with the rate of disappearance of bubbles, if bubbles are actually responsible for scattering and attenuation of sound by wakes. Although optical measurements of the bubble density concentration in wakes have been contemplated, at present there are not available any nonacoustic observations of the rate of decay of bubbles. Neither does physical theory permit predicting the rate of wake decay. As set forth in Section 27.2, the turbulent internal motion may be the factor which determines the "life-time" of wakes. However, an adequate theoretical analysis of the effect of turbulence on the rate of rise of bubbles in wakes is still lacking.

Pending the solution of this fundamental problem, the equations (49) and (50) of Chapter 33 suggest a partial test of the theory of decay of acoustic wakes. These equations established a quantitative relation between the decay rate of the wake strength and that of the total transmission loss across the wake; moreover, they do not involve any quantities which are unknown or difficult to determine. With this test in mind, simultaneous observations of the decay rates dH_w/dt and dW/dt were made for a number of wakes laid by the *E. W. Scripps*, on November 28, 1944; these experiments have already been described, and the results were summarized in Table 2 of Chapter 32 and Table 9 of Chapter 33. The results, though insufficient to verify the relationship predicted theoretically, do not seem to be inconsistent with it.

According to the discussion in Section 33.1.4, the following relation should hold for short pulses and fresh wakes:

$$\frac{dW}{dt} = \left[\frac{0.46e^{-0.46H_w r_0/w}}{1 - e^{-0.46H_w r_c/w}}\right]\frac{r_0}{w}\frac{dH_w}{dt} = F\frac{r_0}{w}\frac{dH_w}{dt} \quad . (10)$$

The factor F in brackets can be read from Figure 4 in Chapter 33, using $H_w r_0/w$ as argument; r_0 was equal to 2.4 yd, as 3-msec signals were used. The width of the *Scripps* wake is 45 yd at the age of 5 minutes; hence r_0/w is about 0.05. The results of the numerical test of equation (10) are presented in Table 7. The observed and computed ratios $(dW/dt)/(dH_w/dt)$

TABLE 7. Observed and predicted decay rates.

	Frequency in kc	
	24	60
$\frac{dH_w}{dt}$ in db per minute	0.7	0.4
$\frac{dW}{dt}$ in db per minute	1.5	0.7
$\frac{dW}{dt}/\frac{dH_w}{dt}$ observed	~2	~2
H_w at age of 5 minutes in db	3.0	4.5
$H_w r_0/w$ at age of 5 minutes in db	0.15	0.22
F	7	4
$F r_0/w$	1.05	0.88
$\frac{dW}{dt}/\frac{dH_w}{dt}$ computed	~1	~1

seem to agree as to order of magnitude. Little more can be expected, considering the high sensitivity of the test following from the rapid variation of the function F with H_w.

At any rate, equation (10) and the corresponding one for long pulses, resulting from putting r_0/w equal to 1, seems to account qualitatively for the shape of curves obtained by plotting wake strength against wake age, as illustrated by Figure 8 in Chapter 33. Generally, W remains constant during the first 5 minutes after the wake has been laid, or it may even increase slightly. Thereafter W decreases linearly with time. However, the transmission loss H_w of the wake appears to decrease linearly with time right from the beginning of the wake. The explanation is that for young wakes the factor F in equation (10) is so much smaller than 1 that dW/dt equals 0. After about five minutes H_w seems to have decreased to such an extent that F becomes of the order of one, or dW/dt and dH_w/dt have reached the same order of magnitude.

The observed rates of decay of wake echoes, noted in Chapter 33, are mostly between 1 and 2 db per minute; the much higher values recorded in Table 6 of Chapter 33 may be caused by the rather shallow depth of the wake of the launch, as distinguished from the much deeper wakes of the larger surface vessels. In the interpolation formula for H_w — equation (10) in Chapter 32 — H_w was assumed to decrease linearly with increasing time. An exponential decay would be more consistent with the observations of wake echoes, if equation (10) of this section is fulfilled; the measurements are not sufficiently accurate, however, to indicate which type of decay is actually followed.

Thus it may be concluded that the observed decay rates for scattering and attenuation are mutually consistent, as far as the rather scanty evidence goes. Even if future wake observations would establish beyond doubt that equation (10) is satisfied, these results would by no means suffice to confirm the bubble hypothesis. It should be realized that equation (10) represents a relationship of a quite formal nature and physically does not imply more than the plausible proposition that the acoustic effects of wakes are proportional to the volume density of some unspecified agent.

The total time required for wakes to decay, however, is consistent with the time required for small bubbles to disappear by resolution in sea water. The experiments discussed in Section 27.2.2 indicate that a bubble whose initial radius is 0.10 cm will disappear in about 30 minutes by gradual resolution of air back into the water. Turbulent motion is needed to keep air bubbles from reaching the surface but cannot prolong the life of a wake beyond the time limit set by the resolution process. Thus 30 minutes is an upper limit for the life of an acoustic wake if the greatest air bubbles present are initially 0.10 cm in radius. Since bubbles of this size resonate to sound of 3 kc, the transmission loss observations described in Section 32.3.1 indicate that bubbles of this size are present initially. The observed length of time during which echoes are observed from a surface ship wake averages in the neighborhood of 30 minutes. Thus the observed rate of decay of acoustic wakes is at least generally consistent with the hypothesis that bubbles are responsible for the wake's acoustic properties. Information on turbulence in wakes would be necessary for more detailed comparison. However, this general consistency lends added support to the "bubble hypothesis," especially when added to the data already discussed on (1) the variation with depth of the cross section for scattering and extinction, and (2) the value of the ratio σ_s/σ_e, and its variation with frequency, which affects the absolute values of the wake strength.

Chapter 35

SUMMARY

THE WAKE of a moving ship scatters and attenuates sound. The following sections summarize existing data in the form of rules for predicting the geometry of acoustic wakes — their depths and widths, the attenuation of sound crossing wakes, and the scattering of sound from wakes.

In some cases, these rules are based on few observations. Moreover, the degree of reliability of most of the rules is difficult to assess, and an adequate appraisal of it in most cases can be reached only by study of the detailed expositions given in the preceding chapters.

35.1 WAKE GEOMETRY

For surface ships, the depth h of an acoustic wake is approximately twice the draft of the wake-laying vessel, and is practically constant up to distances of at least 1,000 yd behind the ship (see Section 31.3.1). The depth of the wake laid by a surfaced submarine decreases from about 30 ft at a distance 100 yd behind the screws to about 20 ft at a distance astern of 1,000 yd. The wake of a submerged submarine, running at a periscope depth with a speed of 6 knots, reaches the ocean surface at distances astern greater than 100 yd, corresponding to a half-angle of divergence at the screws of about 5 degrees in the vertical direction (see Section 31.3.2).

The width w of a wake increases with the range r behind the wake-laying vessel. For destroyer and destroyer escort wakes at distances astern greater than 100 yd, the wake fans out laterally in a regular manner, with the wake edges including a total angle of 1 degree (see Section 31.2).

At distances less than 100 yd astern, the wake geometry is less regular and depends upon the speed of the destroyer in a complicated manner. This dependence may be represented by the following equation:

$$w = \frac{w^*}{r^*} r = 0.85r, \qquad (1)$$

which is valid at distances astern r less than r^*. For r^* the values in Table 1, which were deduced from aerial

TABLE 1. Dependence of r^* and w^* on ship speed.

Ship speed in knots	r^* in yards	w^* in yards
16	21	18
20	39	33
25	75	64
33	93	80

photographs (see Section 31.1) of destroyer wakes, should be used. At distances astern greater than r^*, one can compute the wake width by the equation

$$w = w^* + 0.017(r - r^*), \qquad (2)$$

using the same values of r^* and w^* as before.

Acoustic determinations of the width of destroyer wakes (see Section 31.3.1) are much less accurate than the photographic measurements, and seem to be in moderate agreement with the predictions made on the basis of equation (2).

The acoustic properties of the wake apparently vary with position inside the wake, although no definite predictions can yet be made for a particular wake. Outside the boundaries established by the above relationships, the acoustic effects produced by the water are no greater than those typical of the ocean with no wakes present.

35.2 ABSORPTION BY WAKES

When sound from a shallow projector is received on a shallow hydrophone, the transmission loss is increased by an amount H_w if a wake is present between the projector and the hydrophone. This attenuation by the wake H_w may be expressed as

$$H_w = K_e x \qquad (3)$$

where K_e is the attenuation coefficient in decibels per yard, and x is the length in yards of the sound path within the wake [see equation (2) of Chapter 32].

541

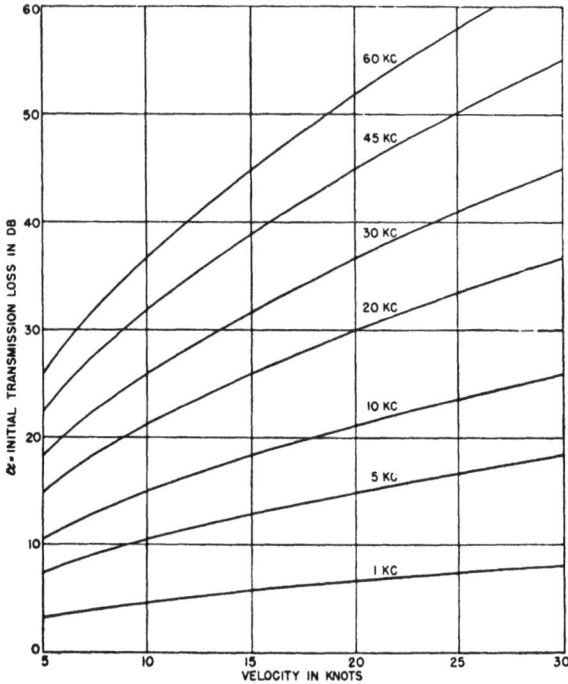

FIGURE 1. Initial transmission loss across destroyer wakes.

the wake, as bubbles with radii between R_r and $R_r + dR$, the attenuation coefficient K_e in decibels per yard may be written [see equation (7) of Chapter 32]

$$K_e = 1.4 \times 10^5 u(R_r) . \qquad (4)$$

In the wake less than 500 yd behind a destroyer or destroyer escort, the attenuation coefficient in the horizontal direction $K_e = H_w/w$ is about 1 db per yd at 20 kc (see Section 32.3). If attenuation at other frequencies by the same wake is taken into account, the total amount of air is about 0.7×10^{-6} cu cm per cu cm of water in the wake of a destroyer at 15 knots, one minute after the passage of the vessel (see Tables 1 and 2 of Chapter 28).

For sound transmitted vertically upward and reflected back by the surface, thus traveling twice through the center of a destroyer wake about 20 ft thick, the attenuation coefficient is found from the equation

$$K_e = \frac{H_w}{2h},$$

where H_w now denotes the two-way attenuation. The

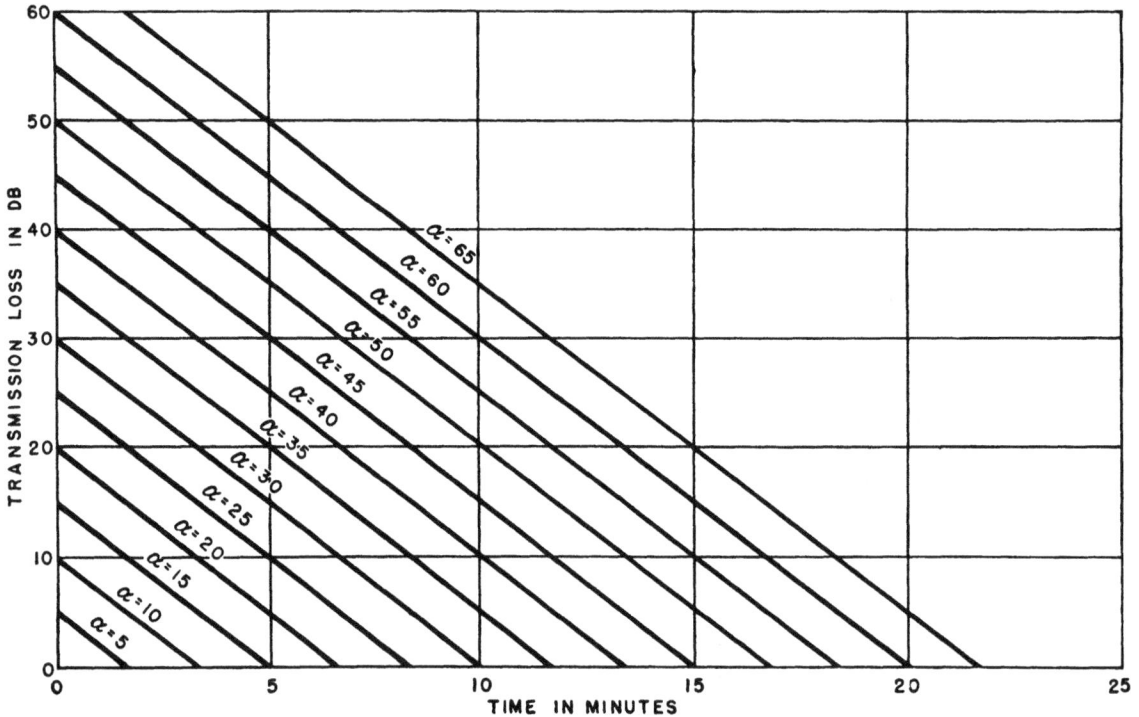

FIGURE 2.   Decay of transmission loss across destroyer wakes. α = initial transmission loss in db.

The attenuation coefficient K_e is determined by the density of air in resonant bubbles of radius R_r. If $u(R_r)dR$ is the fraction of air present, in 1 cu cm of

value of K_e observed in this case is about 3 db per yd at 20 kc [see equation (13) of Chapter 32].

For sound transmitted along a horizontal path per-

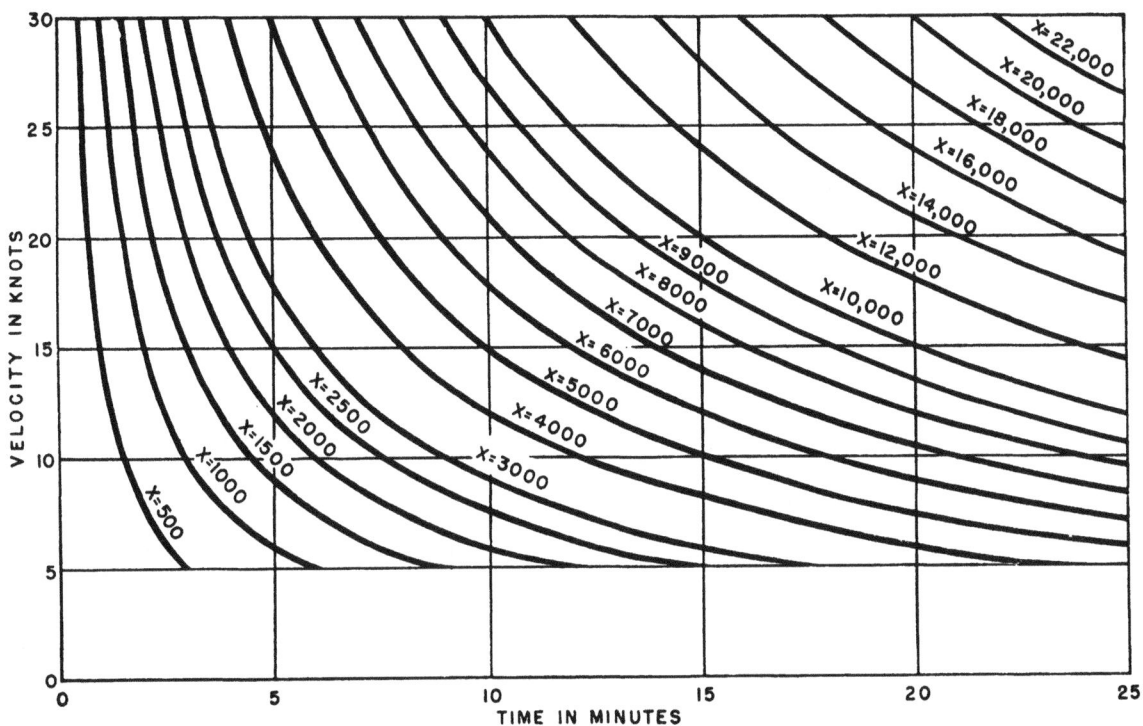

FIGURE 3.   Distance astern in yards as function of wake age and speed of wake-laying vessel.

pendicular to the wake axis, and within 10 feet of the surface, the transmission loss in destroyer wakes is given by the equation (see Section 32.3.1)

$$H_w = 1.5(vf)^{\frac{1}{2}} - 3.0t = \alpha - 3.0t \qquad (5)$$

where f is the frequency of the sound in kc, v the ship's speed in knots, and t the age of the wake in minutes. When the projector is in the wake, the factor 1.5 in equation (5) should be replaced by 2.4; however, the value of H_w in this case may be different for different projectors, since the sound output of the projector may be affected by the presence of the wake. Numerical values of α and H_w resulting from equation (5) can be read from Figures 1 and 2, respectively; Figure 3 may be used to find the distances behind the wake-laying destroyer corresponding to different wake ages and ship speeds.

For the wakes of large surface vessels at speeds between 10 and 25 knots, the value of K_e and H_w are probably much the same as those given by equation (5) applying to destroyers and destroyer escorts.

These values of K_e and H_w are averages over the cross section of the wake and do not take into account possible large changes in these quantities with position in the wake.

For transmission along wakes, equation (3) cannot be used for distances large compared to the depth of the wake, since scattered sound traveling other than straight paths through the wake may become important. In particular, the transmission loss H_w for propeller sounds observed directly behind a ship with a hydrophone at a depth of 10 to 20 ft is of the order of 10 to 100 db per kyd, for frequencies between 5 and 60 kc. This low value may also be due in part to reduction of the absorption coefficient K_e at depths greater than 10 ft in the wake (see Section 32.4).

35.3   ECHOES FROM WAKES

The level E of the echo received from a wake can be determined from the so-called wake strength W using the equation

$$E = S + W - 2H + 10 \log r + \Psi \qquad (6)$$

where S is the level of the rms pressure on the axis of the projector, measured one yard from the projector in decibels above 1 dyne per sq cm; E is the rms pressure level of the echo, again in decibels above 1 dyne per sq cm; r is the range in yards from the projector to the wake, measured along the projector axis, illustrated in Figure 4; H is the transmission loss from the projector to the wake, defined as ten times the

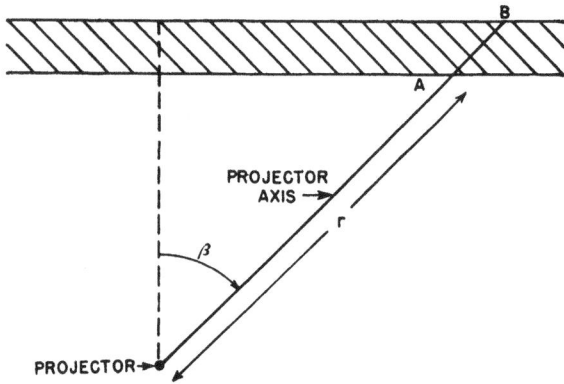

FIGURE 4. Range from transducer to wake.

logarithm of the ratio of rms pressures at a point one yard from the projector and at a point r yards away; and Ψ is a *wake index* based on the transducer pattern which differs for different conditions.

Equation (6) may be written

$$E = S - 2H + T_w \qquad (7)$$

where T_w is the target strength of the wake. Then T_w is related to W, the target strength of a one-yard length of wake, by the equation

$$T_w = W + 10 \log r + \Psi. \qquad (8)$$

While in some ways it is convenient to picture the quantity W, called *wake strength*, as representing the target strength of a one-yard length of wake, Chapter

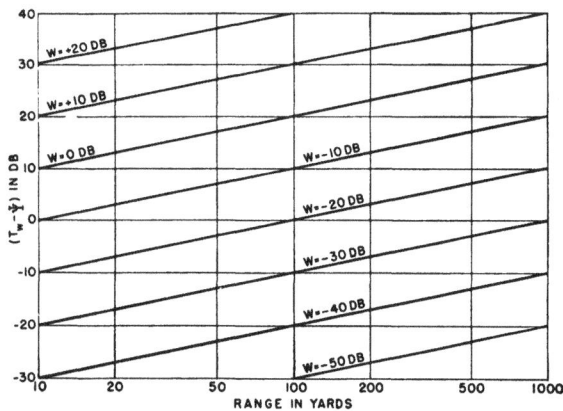

FIGURE 5. Wake target strength as function of wake strength and range.

33 — especially Sections 33.1.1 and 33.1.3 — should be studied for a full understanding of the physical meaning of wake strength. In particular, it should be noted that, according to equation (8), the target strength of the wake T_w depends on the transducer pattern and on the range over which the echoes are

received. Values for T_w for different values of W and r may be found from Figure 5 if Ψ is known. These relationships all assume that both the top and bottom of the wake are in the sound beam.

Since the echo fluctuates, the rms pressure will not be constant within one echo. In this summary, the rms pressure is averaged within each echo and then over several echoes. If in each echo the peak rms pressure recorded is taken and then averaged over several echoes, the wake strength W and the echo level E will be about 6 db higher than the values given here.

35.3.1   Long Pulses

For pulses of duration τ sufficiently long so that $c\tau/2$ exceeds the extension of the wake along the projector axis, reasonably good predictions of the wake strengths of surface vessels and submarines can be made.

If the attenuation H_w across the wake exceeds a few decibels, the wake strength W is given by

$$W = 10 \log s + 10 \log h \qquad (9)$$

where h is the depth of the wake in yards, and s is a function of frequency only, with the values indicated in Table 2. Thus the wake strength of an opaque wake

TABLE 2. Dependence of reflection coefficient s on frequency.

Frequency in kc	$10 \log s$ in db
1	−21
5	−22
8	−23
13	−24
19	−25
26	−26
36	−27
45	−28

10 yd deep is -16 db at 24 kc. As shown in Figure 1 of Chapter 34 (see curve marked OBSERVATIONS), wake strengths at 60 kc are uncertain. Values at frequencies between 10 and 30 kc are probably correct to within about 3 db. A correction of 6 db must be added to the wake strengths computed from equation (9) in order to make them apply to the peak amplitude of the average echo. For a moderately directional transducer, the value of Ψ in the case of long pings is given by

$$\Psi = J_s + 8 , \qquad (10)$$

where J_s is the surface reverberation index, defined by

$$J_s = 10 \log \left[\frac{1}{2\pi} \int b(\phi) b'(\phi) d\phi \right], \qquad (11)$$

where $b(\phi) b'(\phi)$ is the composite pattern function of the echo-ranging transducer. For typical transducers, the surface reverberation index can be computed from the equation

$$J_s = 10 \log y - 23.8, \qquad (12)$$

where $2y$ is the horizontal angular width, measured in degrees, of the sound beam between points down 3 db from the axis.

Thus in this simple case, the target strength of an opaque wake at 24 kc is given by

$$T_w = -26 + 10 \log h + 10 \log r$$
$$+ 10 \log y - 24 + 8 \qquad (13)$$

This equation may be used to predict the initial strength of echoes received from the wakes behind surface vessels.

If the total attenuation H_w across the wake is less than 1 db, the wake strength W may be less than the value found from equation (10). The wake strengths of observed surface wakes are constant for about 2 to 5 minutes, and thereafter decay at a rate of 1 to 2 db per minute; wake echoes can be observed, under good conditions, for 20 to 40 minutes after the passage of a vessel. These times are not inconsistent with what is known of the times required for air bubbles initially 0.1 cm in radius to disappear by diffusion back into sea water. In this situation, Ψ for a directional transducer is

$$\Psi = J_s + 8 + 10 \log \sec \beta \qquad (14)$$

where β is the angle between the projector axis and a line perpendicular to the wake axis. The increase of echo strength with increasing β predicted by equation (14) holds only so long as the ping length is greater than the extension AB of the wake along the projector axis in Figure 4, and so long as the absorption loss along the path AB is less than 1 db.

The wake strengths of submerged submarines at 45 and 90 ft at speeds of 6 knots are about -30 db. Surfaced submarines appear to have about the same wake strength as that predicted for large surface vessels from equation (10) (see Section 33.1.2).

35.3.2 Short Pulses

When the pulse length $c\tau/2$ is less than the extension AB of the wake along the projector axis in Figure 4, the preceding equations are less useful. Although it is possible to predict wake strengths by adding to equation (10) a correction term depending on the signal length, the resulting values of W cannot simply be transformed into echo levels, using equation (7), or into target strengths, using equation (9), unless the echo ranging transducer is beamed perpendicularly at the wake.

For short pulses, the wake strength W decreases with the decreasing ratio of the geometric pulse length r_0 measured in yards to the geometric width w of the wake, and can be predicted from the following equation

$$W = 10 \log s + 10 \log h$$
$$+ 10 \log \left[1 - 10^{-(H_w/5)/(r_0/w)} \right] + 6, \qquad (15)$$

where h is the depth of the wake, measured in yards, and $10 \log s = k$ is the same function of the frequency only as in equation (9), with the values indicated in Table 2. Numerical values of the third term on the right side of equation (15) can be read from Figure 6

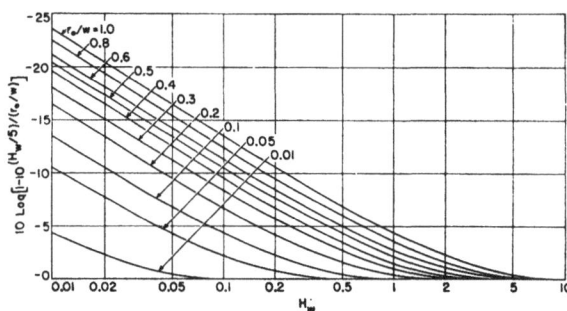

FIGURE 6. Wake strength term as function of attenuation and ratio of ping length to wake width.

as a function of H_w, the total attenuation across the wake, and of r_0/w, the ratio of signal length in yards to wake width. The wake strengths and echo levels computed from equation (15) refer to the peak of the average echo.

Echo levels E and target strengths T_w predicted for values of W computed on the basis of equations (15) and (11) should be quite satisfactory, provided that the sound is beamed at the wake nearly perpendicularly. For lack of anything better, the same predictions may be used in case the sound beam strikes the wake obliquely. The expected discrepancies between observations and predictions, for that case, are believed to be smaller than ± 5 db.

35.3.3 Angular Variation of the Echo Level

When a wake is insonified by a stationary transducer and the echo is recorded by a different hydrophone at several positions, the average echo level thus determined may show moderate variations with position of the hydrophone even after corrections for range to the wake, measured along the hydrophone axis, have been applied. This angular variation of the echo level has not been investigated experimentally; however, a simple theoretical estimate of the order of magnitude of this effect can be made for long pulses and may be useful [see equations (72), (73), and (76) of Chapter 28].

For pulses longer than the width of the wake measured along the sound beam, the echo level should be proportional to

$$\frac{\cos \beta}{\cos \alpha + \cos \beta} \tag{16}$$

if the wake is highly opaque (total attenuation across the wake more than a few decibels); and proportional to

$$\sec \alpha \tag{17}$$

if the wake is acoustically transparent (total attenuation less than 1 db). In equations (16) and (17), β denotes the angle between the transducer axis and a line perpendicular to the wake, as illustrated in Figure 4; α is the corresponding angle between the axis of the hydrophone and a line perpendicular to the wake.

BIBLIOGRAPHY

Numbers such as Div. 6-510-M1 indicate that the document listed has been microfilmed and that its title appears in the microfilm index printed in a separate volume. For access to the index volume and to the microfilm, consult the Army and Navy agency listed on the reverse side of the half-title page.

Chapter 2

1. *Sound Transmission in Sea Water*, Report G1/1184, WHOI, Feb. 1, 1941.   Div. 6-510-M1
2. *An Acoustic Interferometer for the Measurement of Sound Velocity in the Ocean*, Robert J. Urick, Report S-18, USNRSL, Sept. 18, 1944.   Div. 6-510.22-M6
3. *Theory of Sound*, Lord Rayleigh, **2**, The Macmillan Co., 1940.
4. *The Propagation of Underwater Sound at Low Frequencies as a Function of the Acoustic Properties of the Bottom*, J. M. Ide, R. F. Post, and W. J. Fry, Report S-2113, Naval Research Laboratory, Aug. 15, 1943.   Div. 6-510.5-M1

Chapter 3

1. *Higher Mathematics for Engineers and Physicists*, I. S. and E. S. Sokolnikoff, McGraw-Hill Book Co., New York, 1941, p. 146.
2. *Calculation of Sound Ray Paths Using the Refraction Slide Rule*, BuShips-NDRC, NavShips-943 D, May 1943.
3. *Calculation of Sound Ray Paths in Sea Water*, R. H. Fleming, UCDWR, and R. Revelle, USNRSL, Jan. 16, 1942.   Div. 6-510.11-M3
4. *The Sonic Ray Plotter*, L. I. Schiff, UCDWR, NDRC 6.1-sr30-1741, Project NS-140, Report U-246, Aug. 8, 1944.   Div. 6-510.11-M8
5. *Sound Beam Patterns in Sea Water*, NDRC 6.1-sr31-1730, WHOI, Oct. 10, 1944.   Div. 6-510.11-M9
6. *Some Characteristics of the Sound Field in the Sea*, OSRD 546, C4-sr30-083, San Diego Laboratory, UCDWR, Mar. 13, 1942.   Div. 6-510.11-M4
7. *Theory of Diffraction of Sound in the Shadow Zone*, C. L. Pekeris, NDRC 6.1-sr20-846, CUDWR, PAG (A), May 5, 1943.   Div. 6-510.11-M6

Chapter 4

1. *Measurement of Projector and Hydrophone Performance — Definition and Terms*, E. Dietze, NDRC 6.1-sr1130-1833, USRL, Sept. 19, 1944.   Div. 6-551-M12
2. *Apparatus for Recording Reverberation in the Sea*, L. N. Liebermann, OEMsr-31, WHOI, Feb. 23, 1945.   Div. 6-520.2-M3
3. *Operational Procedure and Equipment Used in Sonar Sound Field Studies*, Report U-295, Project NS-140, NDRC 6.1-sr30-2024, UCDWR, Feb. 15, 1945.   Div. 6-510.2-M8
4. *Sonar and Submarine Diving*, Monthly Progress Report for June 1945, WHOI, July 11, 1945, p. 1.   Div. 6-530.22-M21
5. *The Sound Field of Echo-Ranging Gear*, OSRD 2011, NDRC 6.1-sr30-1206, Interim Report U-113, UCDWR, Oct. 1, 1943.   Div. 6-510.22-M3
6. *The Triplane*, D. E. Ross and F. N. D. Kurie, OSRD 1098, NDRC C4-sr30-402, Report U-4, UCDWR, Nov. 23, 1942.   Div. 6-643.21-M2
7. *Supplement to the Triplane*, D. E. Ross and F. N. D. Kurie, OSRD 1689, NDRC 6.1-sr30-961, Report U-4a, UCDWR, June 29, 1943.   Div. 6-643.21-M3
8. *Experimental Surface Model Echo Repeater*, E. M. McMillin and W. A. Myers, June 20, 1942.   Div. 6-643.27-M1
9. *Experimental Underwater Towed Model Echo Repeater*, E. M. McMillin, W. A. Myers, and D. J. Evans, NDRC C4-sr30-515, UCDWR, Sept. 1, 1942.
10. *Asdic Area Trials*, G. E. R. Deacon and H. Wood, British Internal Report 127, HMA/SEE, Fairlie Laboratory, Great Britain, OSRD Liaison Office WA-669-14, May 10, 1943.   Div. 6-570.21-M3
11. *Some Characteristics of the Sound Field in the Sea*, OSRD 546, NDRC C4-sr30-083, Oceanographic Section, UCDWR, Mar. 13, 1942.   Div. 6-510.11-M4
12. *Sound-Ranging Experiments at Key West, July 23–30, 1941*, M. Ewing, OSRD 725, NDRC C4-sr31-130, WHOI, May 23, 1942.   Div. 6-570.21-M1

Chapter 5

1. *Prediction of Sound Ranges from Bathythermograph Observations — Rules for Preparing Sonar Messages*, BuShips, NDRC, NavShips 943-C2, March 1944.
2. *Use of Submarine Bathythermograph Observations*, BuShips, NavShips 900,069, Apr. 25, 1945.
3. *The Oceans*, H. V. Sverdrup, M. W. Johnson, and R. H. Fleming, Prentice-Hall, Inc., 1942.
4. *Sound-Ranging Charts of the Oceans*, U.S. Hydrographic Office.
5. *Submarine Supplements to the Sailing Directions*, U.S. Hydrographic Office.

6. *Measurements of the Horizontal Thermal Structure of the Ocean*, N. J. Holter, Report S-17, USNRSL, Aug. 18, 1944.   Div. 6-540.4-M1

7. *Fluctuation of Transmitted Sound in the Ocean*, NDRC 6.1-sr1131-1883, Technical Memo 6, Sonar Analysis Section, CUDWR-SSG, Jan. 17, 1945.   Div. 6-510.3-M4

8. *Some Experiments on the Transmission of Continuous Sound in 100-fathom to 600-fathom Water*, NDRC 6.1-sr30-1842, Report M-193, Project NS-140, Listening Section UCDWR, Mar. 15, 1944.   Div. 6-510.2-M4

9. *Transmission Measurement in the Vicinity of San Diego, California*, W. B. Snow, H. N. Hoff, J. J. Markham, NDRC 6.1-sr1128-1574, Report D12F/R822, Project NO-163, NLL, Sept. 20, 1944.   Div. 6-510.2-M7

10. *Sound Transmission Measurements Made Off San Diego from August 31 to September 10, 1943*, Projects NS-140, NS-164, and MIT Research Project DIC-6187 [*The Theory of Propagation of Low Frequency Sound in Deep Water*], NDRC 6.1-sr1046-1047, MIT, Feb. 10, 1944.
                                                      Div. 6-510.21-M1

11. *Transmission of Underwater Sound at Lower Frequencies*, Interim Report U-362, Nobs-2074, UCDWR, Nov. 1, 1945.   Div. 6-510.21-M2

12. *Sonic Listening Aboard Submarines*, NDRC 6.1-sr1131-1885, Sonar Analysis Section, Sec. 2.2, CUDWR-SSG, February 1945.   Div. 6-623.1-M8

13. *The Attenuation of Sound in the Sea*, C. F. Eckart, NDRC 6.1-sr30-1532, Report U-236, Project NS-140, UCDWR, July 6, 1944.   Div. 6-510.22-M4

14. *The Influence of Thermal Conditions on Transmission of 24-Kc Sound*, Report U-307, Nobs-2074, Sonar Data Division, CUDWR, Mar. 16, 1945.   Div. 6-510.4-M5

15. *Transmission of Sound in Sea Water. Absorption and Reflection Coefficients and Temperature Gradients*, E. B. Stephenson, Report S-1204, NRL, Oct. 16, 1935.
                                                      Div. 6-510.22-M1

16. *Absorption Coefficients of Sound in Sea Water*, E. B. Stephenson, Report S-1466, NRL, Aug. 12, 1938.
                                                      Div. 6-510.222-M1

17. *Absorption Coefficients of Supersonic Sound in Open Sea Water*, E. B. Stephenson, Report S-1549, NRL, Aug. 2, 1939.   Div. 6-510.222-M2

18. *Attenuation of Underwater Sound*, F. A. Everest and H. T. O'Neil, NDRC C4-sr30-494, UCDWR, Revised July 30, 1942.   Div. 6-510.2-M1

19. *Attenuation of Sound in Sea Water*, G. J. Thiessen, OSRD Liaison Office III-1-830, Report PS-162, CNRC, June 10, 1943.   Div. 6-510.22-M2

20. "Ultrasonic Absorption in Water," F. E. Fox and G. D. Rack, *Journal of the Acoustical Society of America*, 12, 1941, p. 505.

21. "Ultrasonic Interferometry for Liquid Media," F. E. Fox, *Physical Review*, 52, 1937, p. 973.

22. "Ultrasonic Absorption and Velocity Measurements in Numerous Liquids," G. W. Willard, *Journal of the Acoustical Society of America*, 12, 1940, p. 938.

23. *Acoustique-absorption des ondes ultra-sonares par l'eau*, Note (1) De M. B. Biquard, Comptes Rendus, 1931,

pp. 193, 226. (*The Diffusion and Absorption of Ultra Sonics in Liquids*), B. Biquard and R. Lucas.

24. "Absorption of Supersonic Waves in Water and in Aqueous Suspensions," G. K. Hartmann and H. Facke, *Physical Review*, 57, 1940, p. 221.

25. "Absorptions Geschwindigkeits und Entgasungsmessungenism Ultraschallge-beit," C. Sorenson, *Ann. d. Phys.*, 26 [5], 1936, p. 121.

26. "Absorptions of Ultra Sonic Waves in Liquids," J. Claeys, J. Errera, H. Sack, *Faraday Soc. Trans.*, 33, 1936, p. 136.

27. *The Extinction of Sound in Water*, C. F. Eckart, NDRC C4-sr30-621, UCDWR, Aug. 31, 1941.
                                                      Div. 6-510.11-M1

28. *Asdic Area Trials*, G. E. R. Deacon and H. Wood, OSRD Liaison Office WA-669-14, British Internal Report 127, HMA/SEE, Fairlie Laboratory, Great Britain, May 10, 1943.   Div. 6-570.21-M3

29a. Biweekly Report covering period July 25–Aug. 7, 1943, NDRC 6.1-sr31-753, Project NO-140, WHOI, Aug. 11, 1943, pp. 1–2.   Div. 6-510.41-M1

29b. Biweekly Report covering period Sept. 5–18, 1943, NDRC 6.1-sr31-757, Project NO-140, WHOI, Sept. 22, 1943, p. 2.   Div. 6-510.41-M2

29c. Biweekly Report covering period Sept. 19–Oct. 2, 1943, NDRC 6.1-sr31-758, Project NO-140, WHOI, Oct. 6, 1943, pp. 1–2.   Div. 6-510.41-M3

29d. Biweekly Report covering period Oct. 3–16, 1943, NDRC 6.1-sr31-759, Project NO-140, WHOI, Oct. 20, 1943, p. 3.   Div. 6-510.41-M4

29e. Biweekly Report covering period Oct. 17–30, 1943, NDRC 6.1-sr31-1060, Project NO-140, WHOI, Nov. 3, 1943, p. 2.   Div. 6-510.41-M5

29f. Biweekly Report covering period Nov. 14–27, 1943, NDRC 6.1-sr31-1062, Project NO-140, WHOI, Dec. 1, 1943, p. 2.   Div. 6-510.41-M6

30. *A Comparison of Calculated and Observed Intensities for Some Split Beam Sound Field Runs*, R. R. Carhart and L. A. Thacker, Internal Report A-26, Oceanographic Section, UCDWR, Aug. 2, 1944.   Div. 6-510.22-M5

31. *Sound Beam Patterns in Sea Water*, NDRC 6.1-sr31-1730, WHOI, Oct. 10, 1944.   Div. 6-510.11-M9

32. *Layer Effect, Echo-Ranging Section*, R. W. Raitt and M. J. Sheehy, Internal Report A-35, UCDWR, Sept. 9, 1944.   Div. 6-510.41-M7

33. *Layer Effect at 24 Kc and 60 Kc*, M. J. Sheehy, Internal Report A-51, UCDWR, Dec. 27, 1944.
                                                      Div. 6-510.41-M8

34. *The Sound Field of Echo-Ranging Gear*, NDRC 6.1-sr30-1206, Report U-113, UCDWR, Oct. 1, 1943.
                                                      Div. 6-510.22-M3

35. *Sound-Ranging Experiments at Key West*, July 23–30, 1941, M. Ewing, OSRD 725, NDRC C4-sr31-130, WHOI, May 23, 1942.   Div. 6-570.21-M1

36. *Asdic Area Trials*, G. E. R. Deacon and H. Wood, OSRD Liaison Office WA-669-14, British Internal Report 127, HMA/SEE, Fairlie Laboratories, Great Britain, May 10, 1943.   Div. 6-570.21-M3

Chapter 6

1. *Attenuation of Sound in the Sea*, C. F. Eckart, NDRC 6.1-sr30-1532, Report U-236, Project NS-140, UCDWR, July 6, 1944.   Div. 6-510.22-M4

2. *Some Evidence for Specular Bottom Reflections of 24-Kc Sound*, R. R. Carhart, Report A-17, San Diego Laboratory, UCDWR, June 9, 1944.   Div. 6-510.5-M2

3. *Bottom Sediment Charts* [for the guidance of submarines], The Hydrographic Office, July 1944.

4. *Bottom Reverberation. Dependence on Frequency*, NDRC 6.1-sr30-677, Report U-79, UCDWR, June 16, 1943.   Div. 6-520.21-M1

5. *Reverberation Studies at 24 Kc*, OSRD 1098, NDRC 6.1-sr30-401, Report U-7, UCDWR, Nov. 23, 1942.   Div. 6-520-M2

6. *Sonar and Submarine Diving: Monthly Progress Report for June 1945*, Nobs-2083, WHOI, July 11, 1945, pp. 2–4.   Div. 6-530.22-M21

7. *Transmission of 24-Kc and 60-Kc Sound in Very Shallow Water*, M. J. Sheehy, Internal Reports A-31 and A-31a, UCDWR, Aug. 26 and Oct. 23, 1944.   Div. 6-510.221-M1   Div. 6-510.221-M2

8. *Acoustic Properties of Mud Bottoms*, G. P. Woollard, WHOI, Dec. 6, 1944.   Div. 6-510.5-M4

9. *Long Range Sound Transmission*, M. Ewing and J. L. Worzel, Interim Report 1, Nobs 2083, WHOI, Aug. 25, 1945. (See also Chapter 9 of this volume.)   Div. 6-510.1-M4

10. *Some Sound Propagation Measurements in the Fourteenth Naval District*, NDRC 6.1-sr30-1691, Report M-226, Project NS-140, Listening Section, UCDWR, June 19, 1944.   Div. 6-510.2-M6

11. *Some Shallow Water Sound Propagation Measurements in the Thirteenth Naval District*, NDRC 6.1-sr30-1317, Report M-126, Projects NS-140, NS-163, Listening Section, UCDWR, Oct. 26, 1943.   Div. 6-510.2-M2

12. *Transmission of Continuous Sound*, Biweekly Report Covering Period January 23 to February 5, 1944, NDRC 6.1-sr30-1233, Project NS-140, Report U-176, CUDWR, Feb. 11, 1944, pp. 9–12.   Div. 6-510.2-M3

13. *Transmission Survey Block Island Sound*, W. B. Snow, H. B. Hoff, and J. J. Markham, NDRC 6.1-sr1128-1027, Report D12/R616, CUDWR-NLL, Mar. 16, 1944.   Div. 6-510.2-M5

14. *Sonic Listening Aboard Submarines*, NDRC 6.1-sr1131-1885, Sonar Analysis Section, CUDWR-SSG, February 1945.   Div. 6-623.1-M8

15. *Transmission of Underwater Sound over a Sloping Bottom*, R. R. Carhart and K. O. Emery, Internal Report A-39, UCDWR, Oct. 1, 1944.   Div. 6-510.5-M3

16. *Transmission of Continuous Sound*, Biweekly Report Covering Period from January 23 to February 5, 1944 NDRC 6.1-sr30-1233, Project NS-140, Report U-176, UCDWR, Feb. 11, 1944.   Div. 6-510.2-M3

17. *Sonic Listening Aboard Submarines*, NDRC 6.1-sr1131-1885, CUDWR-SSG, February 1945.   Div. 6-623.1-M8

Chapter 7

1. *The Sound Field of Echo-Ranging Gear*, OSRD 2011, NDRC 6.1-sr30-1206, Report U-113, UCDWR, Oct. 1, 1943.   Div. 6-510.22-M3

2. *Amplitude Fluctuations of Transmitted and Reflected Sound Signals in the Ocean*, M. J. Sheehy, Internal Report A-29, UCDWR, Aug. 17, 1944.   Div. 6-510.3-M3

3. *Correlation of Simultaneous Transmission in Deep Water at Different Frequencies*, M. J. Sheehy, Internal Report A-44, UCDWR, Oct. 28, 1944.   Div. 6-510.222-M3

4. *Variation of Signal Amplitude after Transmission in the the Sea*, M. H. Hebb and N. M. Blachman, HUSL, Dec. 19, 1944.   Div. 6-510.11-M10

5. *Detection of an Echo in the Presence of Reverberation*,

C. F. Eckart, OSRD 173, NDRC C4-sr30-175, UCDWR, May 12, 1942.   Div. 6-560.32-M1

6. *Lloyd Mirror Effect in a Variable Velocity Medium*, R. R. Carhart, Memorandum for File 01.92, Report M-140, UCDWR, Oct. 23, 1943.   Div. 6-510.111-M1

7. *Measurements of the Horizontal Thermal Structure of the Ocean*, N. J. Holter, Report S-17, USNRSL, Aug. 18, 1944.   Div. 6-540.4-M1

8. *Fluctuation of Transmitted Sound in the Ocean*, Technical Memorandum 6, NDRC 6.1-sr1131-1883, Sonar Analysis Section, CUDWR, Jan. 17, 1945.   Div. 6-510.3-M4

9. *Theoretical Discussion of Reverberation*, C. L. Pekeris, OSRD 684, NDRC C4-sr20-097, CUDWR, May 29, 1942.   Div. 6-520.1-M7

Chapter 8

1. *Underwater Explosives and Explosions, August 15 to September 15, 1942*, Section B1, NDRC.   Div. 2-130-M1

2. *Relative Pressure Measurements in Shock Wave from Small Underwater Explosions*, M. F. M. Osborne and A. H. Taylor, Report S-2305, NRL, June 10, 1944.   Div. 6-551-M11

3. *Underwater Explosives and Explosions, April 15 to May 15, 1944*, Report UE-21, Division 8, NDRC.   Div. 2-130-M1

4. *Transmission of Explosive Impulses in the Sea*, T. F. Johnston and R. W. Raitt, NDRC C4-sr30-403, Report U-8, UCDWR, Dec. 2, 1942.   Div. 6-510.23-M6

5. *A Study of the Transmission of Explosive Impulses in Sea Water*, T. F. Johnston, OEMsr-30, UCDWR, June 25, 1942.   Div. 6-510.23-M4

6. *Underwater Explosives and Explosions, February 15 to March 15, 1944*, Division 8, NDRC, Report UE-19.   Div. 2-130-M1

7. *Supersonic Flow and Shock Waves*, AMP Report 38.2R, OEMsr-945, AMG-New York University, August 1944.   Div. AMP-101.1-M9

8. *Hydrodynamics*, H. Lamb, Cambridge University Press, Sixth Edition, 1932, pp. 481–489.

Chapter 9

1. *Relative Pressure Measurements in Shock Waves from Small Underwater Explosions*, M. F. M. Osborne and A. H. Taylor, Report S-2305, NRL, June 10, 1944.   Div. 6-551-M11

2. *Development of Single Sweep Equipment for Impulse Work*, T. F. Johnston, OSRD 766, NDRC C4-sr30-189, UCDWR, Apr. 29, 1942.   Div. 6-510.23-M3

3. *The Use of Electrical Cables with Piezoelectric Gauges*, R. H. Cole, Report A-306, OSRD 4561, OEMsr-596, Projects OD-03, NO-144, Division 2, NDRC, WHOI, January 1944.   Div. 2-111.11-M4

4. *Nature of the Pressure Impulse Produced by the Detonation of Explosives Under Water. An Investigation by the Piezo-Electric Cathode-Ray Oscillograph Method*, Report CB-01670-12, OSRD Liaison Office W-201-1E, Admiralty Research Laboratory, Teddington, England, November 1942.   Div. 6-510.23-M1

5. *Propagation of Steep-Fronted Sonic Pulses Through the Sea*, OSRD Liaison Office W-215-5, Internal Report 66, HMA/SEE, Fairlie Laboratory, England, Mar. 17, 1942.   Div. 6-510.23-M2

6. *The Error in the Measurement of Pressure in an Explosion Pressure Wave Due to Finite Gauge Size and to Inadequate Frequency Response of the Recording Amplifier*, Report ADM/219/ARB, OSRD Liaison Office WA-4243-2C, Road Research Laboratory, Great Britain, February 1945.   Div. 6-510.23-M13

7. *Underwater Explosives and Explosions, February 15 to March 15, 1944*, Report UE-19, Division 8, NDRC.   Div. 2-130-M1

8. *A Study of the Transmission of Explosive Impulses in the Sea Water*, T. F. Johnston, OEMsr-30, NDRC UCDWR, June 25, 1942.   Div. 6-510.23-M4

9. *Transmission of Explosive Impulses in the Sea*, T. F. Johnston and R. W. Raitt, NDRC C4-sr30-403, Report U-8, UCDWR, Dec. 2, 1942.   Div. 6-510.23-M6

10. *Solution of Acoustic Boundary Problems*, Parts I to III, L. I. Schiff, University of Pennsylvania, Sept. 4, Oct. 7, and Nov. 2, 1943.   Div. 6-510.1-M3

11. *Explosive Sound Waves in the Sea. Observations with a 2500-cycle Moving-Coil Oscillograph*, T. F. Johnston and R. W. Raitt, Memorandum M-10, OEMsr-30, UCDWR, Sept. 16, 1942.   Div. 6-510.23-M5

12. *Depth Charge Range Meter Tests*, H. B. Hoff, G. R. Perry, et al., Memorandum D50/R1222, Project NS-238, CUDWR-NLL, Nov. 24, 1944.   Div. 6-642.31-M1

13. The experimental points for Figure 5 are taken from reference 7, but the theoretical curves for Figures 5 and 6 are taken from a recent unpublished calculation made by UCDWR.

14. *Theory of Diffraction of Sound in the Shadow Zone*, C. L. Pekeris, NDRC 6.1-sr20-846, CUDWR, May 5, 1943.   Div. 6-510.11-M6

15. *The Sound Field of Echo-Ranging Gear*, OSRD 2011, NDRC 6.1-sr30-1206, Report U-113, UCDWR, Oct. 1, 1943.   Div. 6-510.2-M3

16. *Propagation of Sound in a Medium of Variable Velocity*, C. L. Pekeris, NDRC C4-sr20-001, NLL, Sept. 29, 1941.   Div. 6-510.11-M2

17. *Hydrophone Calibration by Explosion Waves*, J. L. Carter and M. F. M. Osborne, Report S-2179, NRL, Apr. 19, 1944.   Div. 6-510.23-M11

18. *Factors Affecting Long Distance Sound Transmission in Sea Water*, G. P. Woollard, NDRC 6.1-sr31-426, OSRD 1505, WHOI, Mar. 30, 1943.   Div. 6-510.1-M1

19. *Bibliography and Brief Review of Published Material on the Physical Principles of Submarine Detection*, M. F. Manning, NRDC C4, September 1941.

20. *Long Range and Sound Transmission*, Interim Report 1, Mar. 1, 1944–Jan. 20, 1945, M. Ewing and J. L. Worzel, Nobs-2083, WHOI, Aug. 25, 1945.   Div. 6-510.1-M4

21. *Deep Water Sound Transmissions from Shallow Explosions*, J. L. Worzel and M. Ewing, WHOI, (n.d.).   Div. 6-510.23-M14

22. *Explosion Sounds in Shallow Water*, M. Ewing and J. L. Worzel, N111s-38137, NOL and WHOI, Oct. 11, 1944.   Div. 6-510.23-M12

23. *Theory of Propagation of Explosive Sound in Shallow Water*, C. L. Pekeris, OSRD 6545, NDRC 6.1-sr1131-1891, January 1945.   Div. 6-510.12-M5

24. *The Propagation of Underwater Sound at Low Frequencies as a Function of the Acoustic Properties of the Bottom*, J. M. Ide, R. F. Post, and W. J. Fry, Report S-2113, NRL, Aug. 15, 1943.   Div. 6-510.5-M1

25. *Theory of Characteristic Functions in Problems of Anomalous Propagation*, W. H. Furry, Report 680, MIT-RL, Feb. 28, 1945.

Chapter 11

1. *Reverberation in Echo-Ranging: Part I, General Principles*, T. H. Osgood, OSRD 807, NDRC C4-sr20-149, CUDWR, July 28, 1942.   Div. 6-520-M1

2. *Reverberation in Echo-Ranging: Part II, Reverberation Found in Practice*, T. H. Osgood, OSRD 1422, NDRC 6.1-sr20-84C, Project NS-140, CUDWR, Apr. 14, 1943.   Div. 6-520-M3

Chapter 12

1. *Measurements of the Horizontal Thermal Structures of the Ocean*, N. J. Holter, Report S-17, USNRSL, Aug. 18, 1944.   Div. 6-540.4-M1

2. *Theory of Sound*, Lord Rayleigh, The Macmillan Company, New York, 2, 1926, p. 126.

3. *The Discrimination of Transducers Against Reverberation*, OSRD 1761, NDRC 6.1-sr30-968, Report U-75, UCDWR, May 31, 1943.   Div. 6-520.1-M8

4. *Bottom Reverberation. Dependence on Frequency*, NDRC 6.1-sr30-677, Report U-79, UCDWR, June 16, 1943.   Div. 6-520.21-M1

5. *Bottom Reverberation at 24 Kc. E. W. Scripps Data*, R. R. Carhart, Internal Report A-7, UCDWR, May 18, 1944.
Div. 6-520.21-M3

6. *Mathematics of Physics and Chemistry*, H. Margenau and G. Murphy, D. Van Nostrand and Company, New York, 1943, p. 246.

7. *Multiple Scattering*, C. F. Eyring, R. J. Christiensen, and

C. F. Eckart, Memorandum for File No. 01.40, UCDWR, Apr. 18, 1942.
Div. 6-520.11-M2

8. *Relation between Scattering and Absorption of Sound*, Memorandum for File No. 01.40 x 01.72, Report SAS-8, CUDWR-SSG, Dec. 11, 1944.
Div. 6-520.11-M5

9. *Theory of Sound*, Lord Rayleigh, The Macmillan Company, New York, **2**, 1926, p. 145.

Chapter 13

1. *Reverberation Studies at 24 Kc*, OSRD 1098, NDRC 6.1-sr30-401, Report U-7, UCDWR, Nov. 23, 1942.
Div. 6-520-M2

2. *Ibid.*, p. 23.

3. *A System for Recording Reverberation as it Occurs in the Ocean*, NDRC 6.1-sr30-1202, Report M111, UCDWR, Aug. 28, 1943.
Div. 6-520.2-M1

4. *Operational Procedures and Equipment Used in Sonar Sound Field Studies*, NDRC 6.1-sr30-2024, Report U-295, Project NS-140, UCDWR, Feb. 15, 1945, p. 8.
Div. 6-510.2-M8

5. *Limitation of Echo Ranges by Reverberation in Deep Water*, Report M-361, Nobs-2074, Sept. 20, 1945.
Div. 6-520.22-M2

6. *Summary of the Calibration of the Reverberation Equipment, November 24, 1943 to February 23, 1945*, T. H. Schaefer, UCDWR, Apr. 18, 1945.
Div. 6-520.2-M4

7. *Bottom Reverberation at 24 Kc. E. W. Scripps Data*, R. R. Carhart, Internal Report A-7, UCDWR, May 18, 1944.
Div. 6-520.21-M3

8. *Operational Procedures and Equipment Used in Sonar Sound Field Studies*, NDRC 6.1-sr30-2024, Report U-295, Project NS-140, UCDWR, Feb. 15, 1945, p. 18.
Div. 6-510.2-M8

9. *Apparatus for Recording Reverberation in the Sea*, L. N. Liebermann, OEMsr-31, WHOI, Feb. 23, 1945.
Div. 6-520.2-M3

10. *Volume Reverberation. Scattering and Attenuation versus Frequency*, OSRD 1555, NDRC 6.1-sr30-670, Report U-50, UCDWR, Apr. 13, 1943.
Div. 6-520.3-M1

11. *Characteristics of Some Transducers Used by UCDWR*, Report U-23, UCDWR, May 6, 1943.

12. *A Practical Dictionary of Underwater Acoustical Devices* NDRC 6.1-sr20-889, CUDWR-USRL, July 27, 1943.

Chapter 14

1. *Reverberation Studies at 24 Kc*, OSRD 1098, NDRC 6.1-sr30-401, Report U-7, UCDWR, Nov. 23, 1942.
Div. 6-520-M2

2. *Volume Reverberation. Scattering and Attenuation versus Frequency*, OSRD 1555, NDRC 6.1-sr30-670, Report U-50, UCDWR, Apr. 13, 1943.
Div. 6-520.3-M1

3. *Theory of Sound*, Lord Rayleigh, The Macmillan Company, New York, **2**, 1926.

4. *Limitation of Echo Ranges by Reverberation in Deep Water*, Report M-361, Nobs-2074, Sept. 20, 1945.
Div. 6-520.22-M2

5. *Workbook for Prediction of Maximum Echo Ranges*, Bureau of Ships, Navy Department, NavShips 900,055 (labeled NavShips 900,050), December 1944.

6. *Survey of Underwater Sound. Ambient Noise*, V. O. Knudsen, R. S. Alford, and J. W. Emling, OSRD 4333, NDRC 6.1-1848, Report No. 3, Sept. 26, 1944.
Div. 6-580.33-M2

7. *The Influence of Thermal Conditions on Transmission of 24-Kc Sound*, Report U-307, Nobs-2074, UCDWR, Mar. 16, 1945.
Div. 6-510.4-M5

8. *Theoretical Physics*, G. Joos, G. E. Stechert and Company, New York, 1932, p. 581.

9. *The Sea Surface and its Effect on the Reflection of Sound and Light*, C. F. Eckart, Report M-407, Nobs-2074 UCDWR, Mar. 20, 1946.
Div. 6-520.11-M6

10. *Scattering of Sound by the Surface of the Sea*, L. I. Schiff, Project NS-140. Memorandum for file M-217, UCDWR, May 15, 1944.
Div. 6-520.11-M4

11. *Solution of Acoustic Boundary Problems: Part I*, L. I. Schiff, University of Pennsylvania, Sept. 4, 1943.
Div. 6-510.1-M3

12. *Solution of Acoustic Boundary Problems: Part II*, L. I. Schiff, University of Pennsylvania, Oct. 7, 1943.
Div. 6-510.1-M3

13. *Solution of Acoustic Boundary Problems: Part III*, L. I. Schiff, University of Pennsylvania, Nov. 2, 1943.
Div. 6-510.1-M3

14. *Echoes from Swells*, G. E. Duvall, Report A-43, UCDWR, Oct. 27, 1944.
Div. 6-540-M1

15. *Multiple Scattering*, C. F. Eyring, R. J. Christiensen, and C. F. Eckart, Memorandum for File 01.40, UCDWR, Apr. 18, 1942.
Div. 6-520.11-M2

16. *The Short Range Spatial Pattern Measurements on the JK-SK4926 Transducer at 24 Kc*, N. Most, Internal Report No. A-52, UCDWR, Jan. 5, 1945.
Div. 6-510.221-M3

17. *The Effect of the Ship's Roll on Echo Ranging*, J. S. McNown and C. F. Eckart, NDRC 6.1-sr30-1205, Report M-114, UCDWR, Oct. 8, 1943.
Div. 6-510.3-M2

18. *The Discrimination of Transducers Against Reverberation*, OSRD 1761, NDRC 6.1-sr30-968, Report U-75, UCDWR, May 31, 1943.
Div. 6-520.1-M8

19. *Computed Maximum Echo and Detection Ranges for Submarine Echo-Ranging Gear*, W. B. Snow and E. Gerjuoy, NDRC 6.1-sr1131, 1128-1688, CUDWR, July 1944.
Div. 6-570-M2

Chapter 15

1. *Range Limitation in Shallow Water as Controlled by Bottom Character, State of Sea, and Thermal Structure,* F. P. Shepard, Report A-10, UCDWR, May 22, 1944.
   Div. 6-520.21-M4
2. *Bottom Reverberation at 24 Kc. E. W. Scripps Data,* R. R. Carhart, Report A-7, UCDWR, May 18, 1944.
   Div. 6-520.21-M3
3. *Bottom Reverberation,* R. J. Christiensen, Internal Report A-5, UCDWR, May 16, 1944.   Div. 6-520.21-M2
4. *Calculation of Sound Ray Paths Using the Refraction Slide Rule,* NavShips 943, BuShips-NDRC, May 1943.
5. *The Short Range Spatial Pattern Measurements on the JK-SK4926 Transducer at 24 Kc,* N. Most, Internal Report A-52, UCDWR, Jan. 5, 1945.   Div. 6-501.221-M3

6. *Bottom Reverberation. Dependence on Frequency,* NDRC 6.1-sr30-677, Report U-79, UCDWR, June 16, 1943.
   Div. 6-520.21-M1
7. *Maximum Echo Ranges in Shallow Water,* Technical Memorandum 5, CUDWR, Oct. 21, 1944.
   Div. 6-570.1-M5
8. *Computed Maximum Echo and Detection Ranges for Submarine Echo-Ranging Gear,* W. B. Snow and E. Gerjuoy, NDRC 6.1-sr1131, 1128–1688, CUDWR, July 1943.
   Div. 6-570-M2
9. *Bottom Reverberation in Very Shallow Water,* NDRC 6.1-sr30-1845, Report SM-249, Projects NS-140, NS-297, Aug. 18, 1944.   Div. 6-520.21-M5

Chapter 16

1. *Theory of Sound,* Lord Rayleigh, The Macmillan Company, New York, **1**, 1926.
2. *Reverberation Studies at 24 Kc,* OSRD 1098, NDRC 6.1-sr30-401, Report U-7, UCDWR, Nov. 23, 1942.
   Div. 6-520-M2
3. *The Detection of an Echo in the Presence of Reverberation,* C. F. Eckart, OSRD 173, NDRC C4-sr30-175, UCDWR, May 12, 1942.   Div. 6-560.32-M1
4. *Reverberation and Scattering, Series I, Sonar Data,* Report MR-345-I, Nobs-2074, UCDWR, July 1945, pp. 4–6.
   Div. 6-520.22-M1
5. *Reverberation and Scattering, Series I, Sonar Data,* Report MR-365-I, Nobs-2074, UCDWR, September 1945, pp. 4–6.   Div. 6-510.22-M7
6. *Fluctuations in Reverberation Due to Scattering Centers in Water, Part II,* L. I. Schiff, University of Pennsylvania, June 5, 1943.   Div. 6-520.11-M3
7. *Probability and its Engineering Uses,* Fry.
8. *Fluctuation of Transmitted Sound in the Ocean,* Technical Memorandum 6, NDRC 6.1-sr1131-1883, CUDWR, Jan. 17, 1945.   Div. 6-510.3-M4
9. *The Effect of the Ship's Roll on Echo Ranging,* J. S. McNown and C. F. Eckart, Report M-114, NDRC 6.1-sr30-1205, UCDWR, Oct. 8, 1943.   Div. 6-510.3-M2
10. *Theory of Random Processes,* H. Uhlenbeck, Report 454, MIT-RL, Oct. 15, 1943.   Div. 14-125-M7
11. *Coherence of CW Reverberation,* Memorandum for File No. 01.40, Report SAS-11, CUDWR, Dec. 20, 1944.
    Div. 6-520.1-M9
12. *The Fluctuations in Signals Returned by Many Independently Moving Scatterers,* A. J. F. Siegert, MIT-RL, Report No. 465, Nov. 12, 1943.   Div. 14-122.113-M7
13. *The Appearance of the A Scope When the Pulse Travels Through a Homogeneous Distribution of Scatterers,* A. J. F. Siegert, Report 466, MIT-RL, Nov. 9, 1943.
    Div. 14-124.2-M2
14. "Stochastic Problems in Physics and Astronomy," S. Chandrasekhar, *Rev. of Mod. Phys.,* **15**, January 1943.

15. "Mathematical Analysis of Random Noise," S. O. Rice, *Bell System Technical Journal,* **23**, July 1944, p. 289.
15a. "Mathematical Analysis of Random Noise," S. O. Rice, *Bell System Technical Journal,* January 1945.
16. *The Extrapolatory Interpolation and Smoothing of Stationary Time Series,* N. Wiener, NDRC Progress Report No. 19 to the Services, MIT, Feb. 1, 1942.
17. *Frequency Characteristics of Echoes and Reverberation,* W. M. Rayton and R. C. Fisher, OSRD 4159, Project NS-140, NDRC 6.1-sr30-1740, Report U-244, UCDWR, Aug. 9, 1944.   Div. 6-520.3-M2
18. *The Theory of Reverberation and Echo,* C. F. Eckart, NDRC C4-sr30-005, UCDWR, July 7, 1941.
    Div. 6-520.1-M1
19. *Theoretical Discussion of Reverberation,* C. L. Pekeris, OSRD 684, NDRC C4-sr20-097, CUDWR, May 29, 1942.
    Div. 6-520.1-M7
20. *Frequency Spread of Reverberation as Measured with the Periodmeter,* Memorandum for File No. 01.40, Report SAS-15, Sonar Analysis Section, CUDWR-SSG, Jan. 17, 1945.   Div. 6-520.3-M5
21. *Frequency Characteristics of Reverberation,* Memorandum for File No. 01.40, Report SAS-16, Sonar Analysis Section, CUDWR-SSG, Nov. 23, 1944.   Div. 6-520.3-M4
22. *The Dependence of the Operational Efficacy of Echo-Ranging Gear on its Physical Characteristics,* H. Primakoff and M. J. Klein, NDRC 6.1-sr1130-2141, Project NS-182, CUDWR-USRL, March 15, 1945.   Div. 6-551-M14
23. *Frequency Modulation in Echo Ranging,* C. F. Eckart, NDRC C4-sr30-236, UCDWR, July 21, 1942.
    Div. 6-635.1-M3
24. *Observations of Echo Signals Obtained Using Variable Frequency Transmission,* E. M. Macmillan, NDRC C4-sr30-208, UCDWR, July 4, 1942.   Div. 6-510.3-M1
25. *Coherence and Fluctuation of FM Reverberation,* M. J. Sheehy, Report A-37, UCDWR, Sept. 19, 1944.
    Div. 6-520.3-M3

Chapter 20

1. *The Theory of Sound*, Lord Rayleigh, London, 1896.
2. "On the Absorption of Sound Waves in Suspensions and Emulsions," P. S. Epstein, *Theodor von Kármán Anniversary Volume*, California Institute of Technology, May 11, 1941, p. 162.
3. H. Stenzel, *Ann. d. Physik*, Series 5, 41, 1942, p. 245.
4. H. Reissner, *Helvetia Physica Acta*, 11, 1935, p. 140.
5. *The Acoustic Properties of Domes: Part II*, H. Primakoff, NDRC 6.1-sr1130-1366, USRL, Feb. 18, 1944.
   Div. 6-555-M17

6. *Reflection and Scattering of Sound*, H. F. Willis, OSRD Liaison Office WA-92 10f, NDRC C4-brts-501, British Internal Report 50, HMA/SEE, Fairlie Laboratory, Great Britain, Dec. 20, 1941. Div. 6-530.1-M1
7. *Reflections from Submarines*, M. J. Klein and J. B. Kellar, NDRC 6.1-sr1130-1376, Project No. 222, USRL, Apr. 15, 1945. Div. 6-530.1-M3
8. *General Information and Sketch Book for the Engine Room Personnel of German Submarines, Type VII C*, U.S. Navy, DTMB, May 1942. Div. 6-530.22-M1

Chapter 21

1. *An Analysis of Reflections from Submarines*, NDRC 6.1-sr1131-1846, File 01.80, Technical Memorandum 4, Sonar Analysis Section, CUDWR-SSG, Sept. 9, 1944.
   Div. 6-530.22-M9
2. *Reverberation Studies at 24 Kc*, OSRD 1098, NDRC C4-sr30-401, File 01.40, Report U-7, Reverberation Group, UCDWR, Nov. 23, 1942. Div. 6-520-M2
3. *Listening Techniques*, Biweekly Report Covering Period October 4 to October 17, 1942, NDRC C4-sr30-396, UCDWR, Nov. 7, 1942, p. 5. Div. 6-530.22-M2
4. *Target Strength of a Submarine at 24 Kc*, G. E. Duvall, File 01.80, Internal Report A-4, Echo-Ranging Section, UCDWR, May 10, 1944. Div. 6-530.22-M6
5. *Data at 45 Kc on Echoes from a Diving Submarine and its Wake*, [W. M. Rayton], Report M-172a, Project NS-141, NDRC 6.1-sr30-1475, Sonar Section, UCDWR, Mar. 3, 1944. Div. 6-530.22-M4
6. *Internal Waves*, Biweekly Report Covering Period January 21 to February 3, 1945, NDRC 6.1-sr30-2025, Report U-297, UCDWR, Feb. 10, 1945, p. 6.
   Div. 6-501.4-M3
7. *Relative Echo Intensity versus Aspect*, F. E. Gilbert, Jr., and J. K. Nunan, Report P29/R789, CUDWR-NLL, Mar. 10, 1944. Div. 6-530.22-M5
8. *Sonar and Submarine Diving: Monthly Progress Report for May 1945*, Report 3, Nobs-2083, WHOI, May 10, 1945, p. 3. Div. 6-530.22-M19
9. *Sonar and Submarine Diving: Monthly Progress Report for June 1945*, Report 4, Nobs-2083, WHOI, July 11, 1945, pp. 2-4. Div. 6-530.22-M21

10. *Measurements made with 26-Kc DSS on USS Cythera* (Memorandum), C. A. Ewaskio, HUSL, Feb. 21, 1945.
    Div. 6-632.422-M3
11. *Submarine Runs with Directional and Nondirectional Transmitting Beams, 26-Kc DSS on USS Cythera* (Memorandum), C. M. Clay, HUSL, June 18, 1945.
    Div. 6-632.422-M13
12. *Sound Ranges Under the Sea — 1944*, OSRD 4400, NDRC 6.1-sr1131-1880, Sonar Analysis Section, CUDWR-SSG, November 1944. Div. 6-500-M2
13. *Small Object Detection, Sonar Data: Monthly Progress Report, Series I*, Report MR-323-I, Project NS-140, Nobs-2074, UCDWR, May 1945, pp. 9-10. Div. 6-530.22-M18
14. *Reflection of Sound from Targets, Sonar Data: Monthly Progress Report, Series I*, Report MR-334-I, Nobs-2074, UCDWR, June 1945, pp. 10-12. Div. 6-530.22-M20
15. *Reverberation and Scattering, Sonar Data: Monthly Progress Report, Series I*, Report MR-345-I, Nobs-2074, UCDWR, July 1945, pp. 4-6. Div. 6-520.22-M1
16. *The Attenuation of Sound in the Sea*, C. F. Eckart, NDRC 6.1-sr30-1532, Report U-236, File 01.70, Project NS-140, UCDWR, July 6, 1944. Div. 6-510.22-M4
17. *Echoes of Very Short Pings from Submarines*, W. M. Rayton, Report M-301, Project NS-140, Nobs-2074, UCDWR, Mar. 1, 1945. Div. 6-530.22-M16
18. *Surface Reflected Submarine Echoes*, Report M-306, File 01.80, Project NS-140, Nobs-2074, Echo-Ranging Section, UCDWR, Mar. 15, 1945. Div. 6-530.22-M17

Chapter 22

1. *Reflection of Light from a Submarine Model*, R. B. Tibby, Memorandum for File 02.30, Report M-61, UCDWR, May 12, 1943. Div. 6-530.23-M1
2. *Reflections from Submarines at Close Ranges. Model Experiments Using Optical Method*, Project NO-222 and MIT Research Project DIC-6187, MIT-USL, Apr. 8, 1944. Div. 6-530.23-M2
3. *Studies of Optical Reflections from Submarine Models: Part II*, OSRD 3706, NDRC 6.1-sr1046-1053, Project

   NS-222 and MIT Research Project DIC-6187, MIT-USL, Apr. 12, 1944. Div. 6-530.23-M3
4. *Studies of Optical Reflections from Submarine Models: Part II*, OSRD 3706, NDRC 6.1-sr1046-1668, File 07.10, Navy Project NS-222 and MIT Research Project DIC-6187, MIT-USL, Aug. 15, 1944. Div. 6-530.23-M4
5. *Measurement of Reflections from Submarines Using Models and High-Frequency Sound*, J. B. Kellar, OSRD 4439, NDRC 6.1-sr1130-1834, Navy Project NS-140, USRL, Sept. 27, 1944. Div. 6-530.23-M5

Chapter 23

1. *An Analysis of Reflections from Submarines*, NDRC 6.1-sr1131-1846, File 01.80, Technical Memorandum 4, Sonar Analysis Section, CUDWR-SSG, Sept. 9, 1944.
Div. 6-530.22-M9

2. *Target Strength of a Submarine at 24 Kc*, G. E. Duvall, File 01.80, Internal Report A-4, Echo-Ranging Section, UCDWR, May 10, 1944.   Div. 6-530.22-M6

3. *Sonar Sound Field*, Biweekly Report Covering Period September 17 to September 30, 1944, NDRC 6.1-sr30-1862, Report U-262, UCDWR, Oct. 5, 1944, pp. 4–5.
Div. 6-530.22-M10

4. *Pillenwerfer Design*, OSRD Liaison Office WA-328-16, British Internal Report 100, HMA/SEE, Fairlie Laboratory, Great Britain, Sept. 15, 1942.   Div. 6-651-M1

5. *Studies of Optical Reflections from Submarine Models: Part II*, NDRC 6.1-sr1046-1668, Navy Project NS-222 and MIT Project DIC-6187, File 07.10, MIT-USL, Aug. 15, 1944.   Div. 6-530.23-M4

6. *Reflections from Submarines*, M. J. Klein and J. B. Kellar, NDRC 6.1-sr1130-1376, Navy Project NO-222, USRL, Apr. 15, 1944.   Div. 6-530.1-M3

7. *Listening Techniques*, Biweekly Report Covering Period October 4 to October 17, 1942, NDRC C4-sr30-396, UCDWR, Nov. 7, 1942.   Div. 6-530.22-M2

8. *Reverberation Studies at 24 Kc*, OSRD 1098, NDRC C4-sr30-401, File 01.40, Report U-7, Reverberation Group, UCDWR, Nov. 23, 1942.   Div. 6-520-M2

9. *Data at 45 Kc on Echoes from a Diving Submarine and its Wake*, [W. M. Rayton], NDRC 6.1-sr30-1475, Memorandum for File 01.50, Service Project NS-141, Report M-172-A, Sonar Section, UCDWR, Mar. 3, 1944.
Div. 6-530.22-M4

10. *Sonar and Submarine Diving. Monthly Progress Report for June 1945*, Report 4, Nobs-2083, WHOI, July 11, 1945, pp. 2–4.   Div. 6-530.22-M21

11. *Measurements Made with 26-Kc DSS on USS Cythera* (Memorandum), C. A. Ewaskio, HUSL, Feb. 21, 1945.
Div. 6-632.422-M3

12. *Measurement of Reflections from Submarines Using Models and High Frequency Sound*, J. B. Kellar, OSRD 4439, NDRC 6.1-sr1130-1834, Navy Project NS-140, USRL, Sept. 27, 1944.   Div. 6-530.23-M5

13. *Relative Echo Intensity versus Aspect*, F. E. Gilbert, Jr., and J. K. Nunan, Report P29/R789, CUDWR-NLL, Mar. 10, 1944.   Div. 6-530.22-M5

14. *Submarine Runs with Directional and Nondirectional Transmitting Beams, 26-Kc DSS on USS Cythera* (Memorandum) C. M. Clay, HUSL, June 18, 1945.
Div. 6-632.422-M13

15. *Sonar Submerged Submarine Wakes*, [P. H. Hammond], BuShips Problem U2-9CD, Serial S-RS-96, Report ND11/NP22/S68, USNRSL, Aug. 9, 1944.
Div. 6-540.31-M3

16. *General Information and Sketch Book for the Engine Room Personnel of German Submarines, Type VII C*, U.S. Navy, DTMB, May 1942.   Div. 6-530.22-M1

17. *Studies of Optical Reflections from Submarine Models, Part I*, OSRD 3706, NDRC 6.1-sr1046-1053, Navy Project NS-222 and MIT Project DIC-6187, File 07.10, MIT-USL, Apr. 12, 1944.   Div. 6-530.23-M3

18. *Change of Average Peak Echo Intensity with Changing Ping Length*, Lyman Spitzer, Jr., Memorandum for File 01.80, Report SAS-30, Sonar Analysis Section, CUDWR-SSG, Mar. 22, 1945.   Div. 6-530.1-M4

19. *Preparation of Charts of Average Echo-Ranging Conditions*, Biweekly Report Covering Period July 23 to August 5, 1944, NDRC 6.1-sr30-1745, Report U-248, Project NO-140, UCDWR, Aug. 10, 1944, p. 5.
Div. 6-530.22-M7

20. *Preparation of Charts of Average Echo-Ranging Conditions*, Biweekly Report Covering Period August 6 to August 19, 1944, NDRC 6.1-sr30-1750, Report U-253, Project NO-140, UCDWR, Aug. 23, 1944, p. 4.
Div. 6-530.22-M8

21. *Reflectivity of Targets*, Biweekly Report Covering Period October 1 to October 14, 1944, NDRC 6.1-sr30-1865, Report U-264, Project NS-140, UCDWR, Oct. 31, 1944, pp. 3–5.   Div. 6-530.22-M11

22. *The Attenuation of Sound in the Sea*, C. F. Eckart, NDRC 6.1-sr30-1532, Project NS-140, Report U-236, File 01.70, UCDWR, July 6, 1944.   Div. 6-510.22-M4

23. *Reflectivity of Targets*, Biweekly Report Covering Period October 29 to November 11, 1944, NDRC 6.1-sr30-1874, Report U-274, Project NS-140, UCDWR, Nov. 16, 1944, pp. 5–6.   Div. 6-530.22-M13

24. *The Influence of Thermal Conditions on Transmission of 24-Kc Sound*, Sonar Data Division, Problem 2A, Report U-307, Nobs-2074, UCDWR, Mar. 16, 1945.
Div. 6-510.4-M5

25. *Internal Waves*, Biweekly Report Covering Period January 21 to February 3, 1945, NDRC 6.1-sr30-2025, Report U-297, UCDWR, Feb. 10, 1945, p. 6.
Div. 6-501.4-M3

26. *Echoes of Very Short Pings from Submarines*, W. M. Rayton, Problem 2C, Report M-301, File 01.80, Project NS-140, Nobs-2074, CUDWR, Mar. 1, 1945.
Div. 6-530.22-M16

27. *Reflectivity of Targets*, Biweekly Report Covering Period January 7 to January 20, 1945, NDRC 6.1-sr30-2021, Report U-292, Project NS-140, UCDWR, Jan. 26, 1945.
Div. 6-530.22-M14

28. *Origin of Nearest Echo*, W. E. Benton, G. M. Johnson, W. A. Jones, and R. J. W. Morrison, British Internal Report 209, OSRD Liaison Office WA-4297-1 HMA/SEE, Fairlie Laboratory, Great Britain, Feb. 15, 1945.
Div. 6-530.22-M15

Chapter 24

1. *Surface Vessel Target Strengths*, Memorandum for File 01.80, SAG-38, Sonar Analysis Group, CUDWR-SSG, July 5, 1945.   Div. 6-530.21-M3

2. *Oscillograms of 23-Kc Echoes from a Destroyer and its Wake*, [C. F. Eckart], Memorandum for File 01.50, Report M-141, UCDWR, Jan. 3, 1944.   Div. 6-530.21-M1

3. *Status Report on Task No. 5, Effect of Short Pulse Lengths and Receiver Bandwidth on Echo Ranging*, R. W. Kirkland, Report 3510-RWK-HP, BTL, July 15, 1944.   Div. 6-632.03-M5

4. *Underwater Sound Reflecting Characteristics of Surface Ships*, C. Shafer, Jr., Report 2320-CS-PD, BTL, Oct. 6, 1944.   Div. 6-530.21-M2

Chapter 27

1. *Sonar Submerged Submarine Wakes*, P. H. Hammond, BuShips, Problem U2-9CD, Serial S-RS-96, Report ND11/NP22/S68, USNRSL, Aug. 9, 1944, modified as of Nov. 1, 1944.   Div. 6-540.31-M3

2. *Laboratory Studies of the Acoustic Properties of Wakes*, J. Wyman, W. Lehmann, and D. Barnes, NDRC 6.1-sr31-1069, Project NS-141, WHOI, March 1944.   Div. 6-540.3-M3

3. *The Rate of Rise and Diffusion of Air Bubbles in Water*, C. L. Pekeris, OSRD 976, NDRC C4-sr20-326, CUDWR-PAG, Oct. 22, 1942.   Div. 6-540.21-M2

4. *Propagation of Sound through a Liquid Containing Bubbles,*

Part II, Experimental Results and Theoretical Interpretation, E. L. Carstensen and L. L. Foldy, OSRD 3872, NDRC 6.1-sr1130-1629, Project NS-141, USRL, June 23, 1944.   Div. 6-540.3-M4

5. *The Effect of Turbulent Motion on the Rate of Rise of Bubbles in a Wake*, J. S. McNown, NDRC 6.1-sr30-731, File 01.50, Report U-25, UCDWR, Feb. 19, 1943.   Div. 6-540.21-M3

6. *Geometry on Surface Wakes and Experiments on Artificial Wakes*, N. J. Holter, BuShips Problem U2-9CD, Report S-10, USNRSL, May 22, 1943.   Div. 6-540.1-M1

Chapter 28

1. "On the Absorption of Sound Waves in Suspensions and Emulsions," Paul S. Epstein, *Theodor von Kármán Anniversary Volume*, CIT, May 11, 1941, pp. 162–168.

2. *The Stability of Air Bubbles in the Sea and the Effect of Bubbles and Particles on the Extinction of Sound and Light in Sea Water*, P. S. Epstein, NDRC C4-sr30-027, UCDWR, Sept. 1, 1941.   Div. 6-540.21-M1

3. *Propagation of Sound through a Liquid Containing Bubbles: Part I, General Theory*, L. L. Foldy, OSRD 3601, NDRC 6.1-sr1130-1378, Project NS-141, USRL, Apr. 25, 1944.   Div. 6-540.22-M2

4. Leslie L. Foldy, *The Physical Review*, 67, 1945, p. 107.

5. *Acoustic Properties of Gas Bubbles in a Liquid*, Lyman Spitzer, Jr., OSRD 1705, NDRC 6.1-sr20-918, CUDWR, July 15, 1943.   Div. 6-540.22-M1

6. M. Minnaert, *The London, Edinburgh and Dublin Philosophical Magazine and Journal of Science*, 16, 1933, p. 235.

7. E. Meyer and K. Tamm, *Akustische Zeitschrift*, 4, 1939, p. 145.

8. *Propagation of Sound through a Liquid Containing Bubbles, Part II, Experimental Results and Theoretical Interpretation*, E. L. Carstensen and L. L. Foldy, OSRD 3872, NDRC 6.1-sr1130-1629, Project NS-141, USRL, June 23, 1944.   Div. 6-540.3-M4

9. *Statistical Mechanics*, R. H. Fowler, Cambridge University Press, 1929, p. 154.

10. *The Internal Constitution of the Stars*, A. S. Eddington, Cambridge University Press, 1929.

11. *Handbuch der Astrophysik*, Julius Springer, Berlin, 1930.

12. "On the Illumination of a Planet Covered with a Thick Atmosphere," B. P. Gerasimovič, *Bulletin de l'Observatoire Central à Poulkovo* (Russia), 15, No. 127, 1937, p. 4.

13. *A Textbook of Sound*, A. B. Wood, The Macmillan Company, 1941, p. 362.

Chapter 29

1. *Thermal Wake Detection*, D. H. Garber, R. J. Urick, and J. Cryden, Report S-20, USNRSL, Jan. 12, 1945.   Div. 6-540.4-M2

2. *Reflection of Sound in the Ocean from Temperature Changes*, R. R. Carhart, NDRC 6.1-sr30-960, Project NS-140, Report U-74, UCDWR, May 17, 1943.   Div. 6-510.4-M3

3. *Theoretical Discussion of Reverberation*, C. L. Pekeris, OSRD 684, NDRC C4-sr20-097, CUDWR-PAG, May 29, 1942.   Div. 6-520.1-M7

4. *The Geometry of Surface Wakes and Experiments on Artificial Wakes*, N. J. Holter, Report S-10, USNRSL, May 22, 1943.   Div. 6-540.1-M1

5. *Preliminary Measurements on the Acoustic Properties of Disturbed Water*, E. Dietze, NDRC C4-sr20-205, USRL, Sept. 7, 1942.   Div. 6-540.3-M1

6. *Propagation of Sound Through a Liquid Containing Bubbles, Part II, Experimental Results and Theoretical Interpretation*, E. L. Carstensen and L. L. Foldy, OSRD 3872, NDRC 6.1-sr1130-1629, Service Project NS-141, USRL, June 23, 1944.   Div. 6-540.3-M4

Chapter 30

1. *Operational Procedure and Equipment Used in Sonar Sound Field Studies*, NDRC 6.1-sr30-2024, Service Project NS-140, Report U-295, UCDWR, Feb. 15, 1945.
Div. 6-510.2-M8

Chapter 31

1. *Laboratory Studies of the Acoustic Properties of Wakes* (Parts I and II), J. Wyman, W. Lehmann, and D. Barnes, NDRC 6.1-sr31-1069, Service Project NS-141, WHOI, March 1944.
Div. 6-540.3-M3
2. *Thermal Wake Detection*, D. H. Garber, R. J. Urick, and Joseph Cryden, Report S-20, USNRSL, Jan. 12, 1945.
Div. 6-540.4-M2
3. *The Geometry of Surface Wakes and Experiments on Artificial Wakes*, N. J. Holter, Report S-10, USNRSL, May 22, 1943.
Div. 6-540.1-M1
4. *Sound Transmission Loss Through and Thickness of the Wakes of Antisubmarine Vessels*, N. J. Holter, Report S-13, USNRSL, Nov. 22, 1943.
Div. 6-540.32-M2
5. *Sound Transmission Through Destroyer Wakes*, OEMsr-30, Project NS-141, Report M-189, UCDWR, Mar. 8, 1944.
Div. 6-540.32-M3

6. *Chemical Recorder Traces of Submarine Wakes at 24 Kc*, Internal Report A-23, G. E. Duvall, UCDWR, July 18, 1944.
Div. 6-540.31-M2
7. *Reflectivity of Targets*, Biweekly Report Covering Period October 15 to October 28, 1944, NDRC 6.1-sr30-1871, Project NS-140, Report U-271, UCDWR, Oct. 31, 1944, pp. 5-6.
Div. 6-530.22-M12
8. *Wake of a Fleet-Type Submarine*, W. M. Rayton and G. E. Duvall, Internal Report A-34, Echo-Ranging Section, UCDWR, Sept. 5, 1944.
Div. 6-540.31-M4
9. *Sonar Submerged Submarine Wakes*, P. H. Hammond, BuShips Problem U2-9CD, Code 940, Serial S-RS-96, Report ND11/NP22/S68, USNRSL, Aug. 9, 1944.
Div. 6-540.31-M3

Chapter 32

1. *Sound Transmission through Destroyer Wakes*, OEMsr-30, Project NS-141, Report M-189, Listening Section, UCDWR, Mar. 8, 1944.
Div. 6-540.32-M3
2. *Underwater Sound Output of Cruiser, Destroyer, and Aircraft Carrier*, Report SM-268, UCDWR and MIT-USL, Oct. 28, 1944.
Div. 6-580.2-M4
3. *Reflectivity of Targets*, Biweekly Report Covering Period January 7 to January 20, 1945, NDRC 6.1-sr30-2021, Project NS-140, Report U-292, UCDWR, Jan. 26, 1945, pp. 5-6.
Div. 6-530.22-M14
4. *The Geometry of Surface Wakes and Experiments on Artificial Wakes*, N. J. Holter, Report S-10, USNRSL, May 22, 1943.
Div. 6-540.1-M1
5. *Sound Transmission Loss Through and Thickness of the Wakes of Antisubmarine Vessels*, N. J. Holter, Report S-13, USNRSL, Nov. 22, 1943.
Div. 6-540.32-M2
6. *Transmission of Sound Along Wakes*, NDRC 6.1-sr1046-1054, Project NS-141 and MIT Research Project DIC-6187, MIT-USL, July 26, 1944.
Div. 6-540.32-M4
7. *Laboratory Studies of the Acoustic Properties of Wakes*, (Parts I and II), J. Wyman, W. Lehmann, and D. Barnes, NDRC 6.1-sr31-1069, Project NS-141, WHOI, March 1944.
Div. 6-540.3-M3

Chapter 33

1. *Acoustic Measurements on Surface Wakes in San Diego Harbor*, R. R. Carhart and G. E. Duvall, OSRD 1628, NDRC 6.1-sr30-961, Report U-62, UCDWR, May 8, 1943.
Div. 6-540.32-M1
2. *The Discrimination of Transducers Against Reverberation*, OSRD 1761, NDRC 6.1-sr30-968, Report U-75, UCDWR, May 31, 1943.
Div. 6-520.1-M8
3. *Status Report on Task No. 5. Effect of Short Pulse Lengths and Receiver Bandwidth on Echo Ranging*, Robert W. Kirkland, Report 3510-RWK-HP, BTL, July 15, 1944.
Div. 6-632.03-M5

4. *Preliminary Report on Echoes from a Diving Submarine and Its Wake*, Project M-172, Report M-172, Sonar Section, UCDWR, Jan. 22, 1944.
Div. 6-530.22-M3
5. *Data at 45 Kc on Echoes from a Diving Submarine and its Wake*, W. M. Rayton, NDRC 6.1-sr30-1475, Project NS-141, Report M-172a, UCDWR, Mar. 3, 1944.
Div. 6-530.22-M4
6. *Laboratory Studies of the Acoustic Properties of Wakes*, (Parts I and II), J. Wyman, W. Lehmann, and D. Barnes, NDRC 6.1-sr31-1069, Project NS-141, WHOI, March 1944.
Div. 6-540.3-M3

Chapter 34

1. *Laboratory Studies of the Acoustic Properties of Wakes*, J. Wyman, W. Lehmann, and David Barnes, NDRC 6.1-sr31-1069, Project NS-141, WHOI, March 1944.
Div. 6-540.3-M3

2. *Reverberation Studies at 24 Kc*, OSRD 1098, NDRC 6.1-sr30-401, Report U-7, UCDWR, Nov. 23, 1942.
Div. 6-520-M2

CONTRACT NUMBERS, CONTRACTORS, AND SUBJECT OF CONTRACTS

Contract No.	Name and Address of Contractor	Subject
NDCrc-40	Woods Hole Oceanographic Institution Woods Hole, Massachusetts	Studies and experimental investigations in connection with the structure of the superficial layer of the ocean and its effect on the transmission of sonic and supersonic vibrations. Studies and investigations in connection with the oceanographic factors influencing the transmission of sound in sea water.
OEMsr-20	The Trustees of Columbia University in the City of New York New York, New York	Studies and experimental investigations in connection with and for the development of equipment and methods pertaining to submarine warfare.
OEMsr-30	The Regents of the University of California Berkeley, California	Maintain and operate certain laboratories and conduct studies and experimental investigations in connection with submarine and subsurface warfare.
OEMsr-31	Woods Hole Oceanographic Institution Woods Hole, Massachusetts	Studies and experimental investigations in connection with the structure of the superficial layer of the ocean and its effects on the transmission of sonic and supersonic vibrations.
OEMsr-287	President and Fellows of Harvard College Cambridge, Massachusetts	Studies and experimental investigations in connection with the development of equipment and devices relating to subsurface warfare.
OEMsr-346	Western Electric Company, Inc. 120 Broadway New York, New York	Studies and experimental investigations in connection with submarine and subsurface warfare.
OEMsr-1046	Massachusetts Institute of Technology Cambridge, Massachusetts	Studies and experimental investigations in connection with (1) underwater sound transmission and boundary impedance measurements; (2) ship sound surveys at high frequencies; (3) development of devices for the control of underwater sounds; and (4) development of intense underwater sound sources for special purposes.
OEMsr-1128	The Trustees of Columbia University in the City of New York New York, New York	Conduct studies and experimental investigations in connection with and for the development of equipment and methods involved in submarine and subsurface warfare.
OEMsr-1130	The Trustees of Columbia University in the City of New York New York, New York	Conduct studies and experimental investigations in connection with the testing and calibrating of acoustic devices.
OEMsr-1131	The Trustees of Columbia University in the City of New York New York, New York	Conduct studies and investigations in connection with the evaluation of the applicability of data, methods, devices, and systems pertaining to submarine and subsurface warfare.

SERVICE PROJECT NUMBERS

The projects listed below were transmitted to the Executive Secretary, National Defense Research Committee [NDRC], from the Navy Department through the Office of Research and Inventions (formerly the Coordinator of Research and Development), Navy Department. These are the principal Navy projects relating to the physics of sound in the sea.

Service Project Number	Subject
NO-163	Cooperation with the Navy in harbor surveys and surveys of ambient underwater noise conditions in various areas.
NO-222	Acoustic reflection fields of submarines.
NS-140	Acoustic properties of the sea bottom.
NS-140 (Ext.)	Range as a function of oceanographic factors.
NS-141	Acoustic properties of wakes.

INDEX

The subject indexes of all STR volumes are combined in a master index printed in a separate volume. For access to the index volume consult the Army or Navy Agency listed on the reverse of the half-title page.

"Absorption cross section" of bubble, 466
Absorption effect in underwater sound transmission
  absorption coefficient, 97–100
  attenuation measurements, 102–105
  bubble formation, 465–467
  coefficient of attenuation, 100
  frequency ranges, 105–107
  thermal structure, 102
  transmission anomaly, 100–101
  wakes, 541–543
Acoustic interference
  echoes, 377
  intensity, 168–170
  target strength measurements, 410
Acoustic interferometer for sound velocity measurements, 17
Acoustic measurements in underwater transmission, 243–244, 474–477
Acoustic wakes
  see Bubbles in acoustic wakes; Wakes, acoustic
Acoustical axis of sound projector, 26–27
Adiabatic pressure changes during bubble formation, 461
Aerial photographs in acoustic wake geometry, 494–495
Air bubbles in acoustic wakes
  see Bubbles in acoustic wakes
Airey phase of water waves, 232
Anchored ships, target strength measurements, 424–425, 437
Angular variation of echo level, 546
Antinodes of stationary sound waves, 33
Aspect angle, target strength measurements, 388–393, 424
Asymmetry effects on target strength measurements, 400–402
Attenuation coefficient in sonic transmission
  bottom scattering, 320–321
  bubble formation, 469–470
  isothermal water, 100, 104–107
  shadow boundary, 124–125
  target strength measurements, 370, 373, 411–413
  transmission anomaly, 129–131
  wake thickness, 503–504, 508–509
Attenuation of sound
  bubble theory, 533–534
  explosions, 193–197

frequency effects, 209–211
long range transmission, 216–219
propeller wakes, 510–511
scattering layer, 299–301
shadow zone, 67–68
transmission anomaly, 100, 105–107
wake theory, 503–504
wave theory, 27–28
Average layer effect in underwater sound transmission, 112
Averaging methods for reverberation data, 278–280

B-19 H magnetostrictive hydrophone, 74
Backward scattering coefficient of sound, 252, 266, 306, 335
Backward scattering of sound, 254, 483
Band method of averaging reverberation data, 279–280
Bathythermograph
  classification, 92–95
  description, 76
  ray tracing, 60–63
  velocity-depth variations, 197–200
Beam target strengths in echo ranging, 415–417, 435–436
Bell Telephone Laboratories (BTL), surface vessel target strengths, 423–424
Blade cavitation in acoustic wakes, 449
"Blobs" in reverberation of sound, 335
Bottom reverberation of sound, 264
  average intensities, 321–323
  data analysis, 319–321
  deep-water transmission, 86–87
  definition, 264
  description, 308–312
  frequency, 318–319
  grazing angle, 314–318
  refraction, 312–313
  scattering coefficients, 314, 319–321, 338
  summary, 338–339
Bottom scattering coefficients of sound, 314, 319–321, 338
Bottom-reflected sound
  attenuation coefficient, 103–104
  dispersion phenomena, 228–229
  normal modes theory, 222–224
  predictions of ray theory, 224
  ray intensity, 55–56
  reflection coefficient, 219–221

shallow-water transmission, 137–138
simple harmonic propagation, 224–227
summary, 243
supersonic frequencies, 140–141
times of arrival, 221–222
wave equation, 33–34
Boundary conditions in sound propagation
  point source far from surface, 33–34
  point source near surface, 31–33
  reflection and refraction of plane waves, 30–31
  reflection from sea bottom, 33–34
  target strengths, 353
  transition conditions, 28–31
  wake theory, 478
  wave equation, 13–14
BTL (Bell Telephone Laboratories), surface vessel target strengths, 423–424
Bubbles in acoustic wakes
  absorption during bubble pulsation, 464–467
  acoustic effects, 474–477
  attenuation, 469–470
  "bubble hypothesis", 533
  buoyancy, 452–455
  damping constant, 467
  decay of wakes, 539–540
  echo intensities, 514–515
  entrained air, 455–457
  long pulses, 515–516
  multiple scattering, 470–473
  oscillograms, 186–190
  propeller cavitation, 449–452, 539
  reflection, 473–474
  scattering by an ideal bubble, 460–464
  scattering coefficient, 306–307
  short pulses, 516–519
  submarine wake strengths, 538–539
  surface vessel wake strengths, 537–538
  theory, 448, 467–469
  transmission loss, 503–504, 533–535
  wake echoes, 535–537
Bulk modulus of a disturbed fluid, 12
Buoyancy of bubbles in underwater sound, 452–455
Burbling cavitation for bubble formation, 449
"Burning" process in underwater explosions, 173–174

Cable hydrophones, 74–75
Calibration techniques for sound measurements, 492
  reverberation intensities, 277
  target strengths, 368–369
  transmission loss, 76–78
Canadian National Research Council, attenuation measurements, 105
Cathode-ray oscilloscope for acoustic wake measurements, 488–490
Cavitation in bubble formation, 191, 449–450
CHARLIE bathythermograms, 93
Chemical recorder traces in acoustic wake measurements, 484
"Chirp" signal in echo-ranging gear, 23
CN–8 crystal hydrophone, 74
Coherence in sound reverberation, 335
  amplitude, 327–329
  intensity, 339
  transmission, 71
Compression viscosity in attenuation of sound, 28
Configurational averages for acoustic theory of bubbles, 468
Conservation of energy law for secondary sound pressure waves, 186–188
Continuity law in sound wave propagation, 8–10
Continuous-flow bubble screens for acoustic measurements, 477
Convex surface, target strength measurements, 359, 434
"Cross-section" of bubble, 461
CW pings, frequency analysis of reverberation, 329–331
Cylinder surfaces, target strength measurements, 360, 435

Damped vibration, 28
Damping constant, 467, 535–536
Decay rate in sound transmission
  acoustic wakes, 520–521, 539–540
  bottom reverberation, 322
  echo intensity, 526
  shock waves, 184–186
  surface reverberation, 337
Deep-water reverberation of sound
  average reverberation levels, 304–306
  deep scattering layers, 282–284
  definition, 86–89
  echo ranging, 527–530
  frequency effects, 284–288
  multiple scattering effects, 303–304
  oceanographic conditions, 289
  ping length, 302
  range dependence, 289–302
  scattering coefficient, 306–307
  transducer directed downward, 281–288

transducer horizontal, 288–299
  volume reverberation, 281–282, 284–288, 335–337
Deep-water transmission of sound
  see Transmission of sound, deep-water
Density-pressure properties of a disturbed fluid, 11–12
Depth effects in sound transmission
  bathythermograms, 92–95
  bottom reverberation level, 338
  corrections, 49–51
  ray diagrams, 89–90
  temperature gradients, vertical, 90–92
  thermocline transmission, 115–117
  volume reverberation, 282–284
Destroyer wakes, air bubble hypothesis, 534
Detonation process in underwater explosions, 173–175
Diffraction of sound waves
  hypothesis, 201
  nonspecular reflection, 361
  pressure-time records, 204–206
  ray theory, 41
  shadow zones, 65–66, 200–201
  wave equation, 66–68
Direction of sound propagation
  definition, 5–6
  directivity index, 72
  double source, 24–26
  pattern functions, 26–27
  point source, 24
  transducer patterns, 429–430, 522
Doppler effect
  reverberation, 329–331
  wake measurements, 484
Double layer effect in sound propagation, 200
Double sources of sound, 24–26
Drift effect in echo variability, 376

EBI-1 crystal transducer, 276
Echo intensity measurements
  angular variation, 546
  definition of echo level, 434
  long pulses, 515–516
  short pulses, 516–519
  target strength, 347–348, 351, 377
  variability, 374–378
Echo ranging
  equipment, 85
  frequency, 523
  projectors, 241
  pulse length, 522–523
  shallow-water, 321–323
  submarine wakes, 523–526
  surface vessel wakes, 526–530
  target strengths, 343–344, 376
  temperature gradients, 3–4

thermocline, 109–110
  transducer directivity, 522
  wake measurements, 484, 490–493
Echoes, wake
  beam echoes, 415–417, 435–436
  decay, 539–540
  off-beam echoes, 417–420, 436–437
  propellers, 539
  repeater, target training, 85
  source, 420–421
  submerged submarines, 437
  surface vessels, 437
  target strengths, 377, 435
  wake theory, 535–537, 543–546
Eckart, self-correlation coefficient for sound intensity fluctuations, 166
Eikonal wave equation in ray acoustics, 44–45, 64–65
Electromagnetic sources for sonic frequencies, 72
Elongation phenomena of off-beam echoes, 418–420
Entrained air in acoustic wakes, 455–457
Equations for target strengths
  definition, 347
  derivations, 348–350
  reflected pressure, 355
Equations of wave propagation, 8–14, 43–45
  boundary conditions, 13–14
  continuity, 8–10
  differential equations of rays, 45–46
  differential equations of wave fronts, 43–45
  forces in a perfect fluid, 10–11
  initial conditions, 13–14
  motion, 10
  ray paths, 46–47
  state of fluid, 11–12
  wave equation, 12–13
Equipment for reverberation measurements, 272–277
Explosions, underwater, 173–235
  attenuation, 193–197
  bottom reflection, 219
  cavitation, 191
  deep sound channels, 213–216
  diffraction, 200–206
  Fourier analysis, 206–211
  long-range propagation, 211–213, 216–219
  normal mode theory, 224–229
  predictions of ray theory, 222–224
  pressure waves, secondary, 186–190
  reflection coefficients, 220–222
  refraction, 197–200
  shallow-water experiments, 229–235
  shock fronts, 175–177, 182–184
  shock waves, 184–186
  summary, 173–175
  surface reflection, 190–191

transmission, 192–193
variations, 211
wave theory, 178–182
"Extinction cross section" of bubbles, 465–466

Fathometer records for acoustic wakes
submarines, 501–502
surface vessels, 497–501
thickness and structure, 486–488
two-way vertical transmission loss, 507–509
Fermat's theorem of reverberation intensity, 253, 269
Fluctuations in sound transmission
beam echoes, 436
echo intensity, 377
interference, 167–170
magnitude, 158–160
microstructure, lens action, 170–171
off-beam echoes, 437
probability distributions, 160–164
reverberation, 324–327, 335, 339
roll and pitch effects, 167–168
sound pulses, 211
space patterns, 167
supersonic frequencies, 241
time patterns, 164–167
Fluid velocity of sound waves
see Velocity of sound in water
Fluorescein for acoustic measurements of wake-laying vessel, 491
FM sonar, reverberation from wide-band pings, 75, 332–333
Forced vibrations of bubbles, 461
Forces in a perfect fluid, sound wave equation, 10–11
Fort Lauderdale, Florida, target-strength measurements, 366, 368
Forward reverberation of sound, 80
Fourier theory in sound propagation, 23, 30, 206–211, 329
"Free vibrations" of bubbles, 461
Frequency of sound
attenuation, 138
bottom reverberation, 338
characteristics, 23–24
deep-water reverberation, 284–288
echo ranging, 408–410, 523
narrow-band pings, 329–331
periodmeter, 330
shallow-water reverberation, 240, 318–319
sonic, 238–239
supersonic, 238–239
surface reverberation, 337
target strengths, 433
volume reverberation, 336
wide-band pings, 332–333

Fresnel zone theory of target strengths, 356–360
applications, 358
convex surface, 359
cylinder, 360
method, 356–357
sphere, 358–359

Gaussian distribution of sound intensity fluctuations, 161–162, 326
Geometry of acoustic wakes
see Wake geometry in sound transmission
Grazing angle variation in reverberation of sound
bottom scattering coefficients, 314–318
transducer horizontal, 299–301
Ground wave in sound transmission, 230–232
"Group velocity" of a wave train, 227

Harbor detection equipment for submarine wakes, 443
Harmonic waves in sound propagation, 17–18, 22–23
Heterodyned reverberation of sound, 339
"Hidden periodicities" of sound intensity fluctuations, 166–167
"Highlight" in Fresnel zone theory of target strengths, 357, 358
Horizontal transmission of sound
beam echoes, 317–318
bottom reverberation, 321–323, 339
deep-water reverberation, 337–338
transmission loss, 504–507
transmission run, 79
Hugoniot equation for shock fronts, 180, 184
Hull reflections of underwater sound, 415
Hull wake in sound transmission, 478
Huyghen's principle for reflected sound pressure, 356
Hydrodynamic theory of bubble formation, 449
Hydrographic conditions for sound transmission anomalies, 119
Hydrophone depth in sound transmission, 72–74, 148–150

Image effect in sound transmission, 95–97, 190
Image interference in sound field intensity, 32–33, 163, 301
Index of refraction, wave front equations, 44
"Instantaneous frequency" of reverberation, 329–330, 339

Intensity of echoes
see Echo intensity measurements
Intensity of sound, 6
see also Fluctuations in sound transmission
contours, 62–63
experiments, 114–117
formulas, 51–53
interference effects, 168–170
linear gradients, 57–58
phase distribution, 37–38
plane waves, 21
rays, 65–66
reverberation, 265–266, 334
scattering, 532
shadow zone, 65–68
spherical waves, 21–22
thermocline, 112–114
transmission anomaly, 53–54, 58–59
velocity-depth variation, 54–57
wake measurements, 488–490, 504
wave equation, 22
Interference effects in sonic transmission
echoes, 377
intensity, 168–170
target strength measurements, 410
Interferometer for sound velocity measurements, 17
Inverse square law for underwater sound, 6–7, 237, 345–347
Isothermal water, sound transmission
absorption, 97–104
attenuation coefficient, 104–107
bottom reverberation, 313
deep-water transmission, 238–239
echo ranging, 109–110
image effect, 95–97
layer effect at 24 kc, 112–114
layer effect at 60 kc, 117
ray theory, 61
short range transmission, 108–109
temperature-depth pattern, 93
thermocline depth, 114–117
transmission loss, 107–108
transmission runs, 110–111
Isovelocity layer effect, ray acoustics, 56–57

Kennard's theory of propagation of cavitation fronts, 191
Khintchine's theorem, self-correlation coefficient for sound intensity fluctuations, 167

Lambert's law for surface reverberation of sound, 300, 314
Laminar cavitation in bubble formation, 449

Law of conservation of energy for secondary sound pressure waves, 186–188

Law of motion for sound wave equation, 10

Law of similarity for shock waves, 182

Layer effect at 24 kc, underwater sound transmission
ray acoustics, 56–57
theory, 112–114
thermocline depth, 115–117, 238
University of California studies, 114–115

Layer effect at 60 kc, underwater sound transmission, 117

Lens action of microstructure, sound intensity fluctuations, 170–171

Listening equipment for wake measurements, 484

Lloyd Mirror effect in wave acoustics, 32–33, 299, 301

Long Island area survey in sonic transmission, 154–156

Long-range sound channel propagation, 211–219
deep channels, 213–216
experimental results, 216–219
introduction, 211–213

Loops of stationary sound waves, 33

Magnetostrictive effect in sound transmission, 5, 72

Mean echo intensity for target strength measurements, 377–378

Microdispersers for measuring damping constant in sound field, 467

MIKE bathythermograms, 93–95

Motion law for sound wave equation, 10

Motion pictures of subsurface structure of wakes, 456

Moving vessels, target strengths, 425–426, 437

Multiple scattering of sound, 268–269, .303–304

NAN bathythermograms, 93–95

Narrow-band pings, frequency analysis of reverberation, 329–331

Naval warfare, acoustic wakes, 443–448

Navy echo ranging
see Echo ranging

Newton's second law of motion for sound wave equation, 10

NK-1 type shallow depth recorder for acoustic wakes, 455

Nodes of stationary sound waves, 33

Noises, sinusoidal sound vibrations, 23

Nonisothermal water, bottom reverberation of sound, 321–323

Nonresonant bubbles, acoustic measurements, 476, 477

Nonspecular reflections of sound, 361–362, 410

Normal mode theory of sound, 34–38, 222–229
bottom reflection, 222–224
characteristic functions, 35–36
dispersion phenomena, 225–228
general waves, 36–37
intensity of sound, 37–38
plane waves, 34–36
prediction of rays, 224–225
pressure-time records, 228

OAX transducers, 78

Ocean bottoms, acoustics properties, 139–141

Oceanographic conditions for sound transmission
bathythermographs, 76
measurements, 243
target strengths, 411–413
wakes, 492–493

Off-beam target strengths, 417–420, 436–437

One-way horizontal transmission loss, acoustic wakes, 504–506

Optical experiments for target strength measurements, 379–381, 386, 410

Oscillograms for underwater sound data
beam echoes, 415
dispersion phenomena, 233–235
echo intensities, 377
explosive sound, 229–231
ground wave phase of disturbance, 230–233
hydrophone output, 74–76
pressure-time records, 204–206
reverberation data, 278
wake measurements, 488–490

"Overtaking effect" in shock wave theory, 177, 183

"Patch size" of acoustic wake, 479

Pattern function for intensity of backward scattered sound, 254

Peak echo intensity in target strength measurements, 373–374, 377–378

Perfect fluid, law of forces, 10–11

Periodmeter for frequency analysis of reverberation, 330–331, 339

PETER bathythermograms, 93

Phase constant in ray acoustics, 41

Phase distribution in wave acoustics, 37–38, 266

Photographic Interpretation Center, Anacostia, wake acoustics, 494

Physical parameters of acoustic wake strength, 514–519
echo intensity, 514–515
long pulses, 515–516
short pulses, 516–519

Piezoelectric effect in sound transmission, 5, 72

"piling-up" effect in long-range sound transmission, 218

Pings in reverberation theory
coherence, 327–329
duration, 334
narrow-band, 329–331
short pulses, 326
surface reverberation, 302
volume reverberation, 336
wide-band, 332–333

Plane waves, sound propagation
intensity of sound, 21
normal mode theory, 34–36
pressure versus fluid velocity, 19–20
reflection and refraction, 30–31
velocity of sound, 15–17
wave equation, 14–15

Point method of averaging reverberation data, 279–280

Point source of sound
boundary conditions, 33–34
equation, 22, 31–32
image interference effect, 32–33
surface reflection, 32
target strength, 353

Poisson distribution of fluctuations of sound intensity, 326

Power level recorders for underwater sound transmission measurements, 75

Pressure of reflected sound wave, 352–355
boundary conditions, 353
mathematical analysis, 353–355
physical analsyis, 355

"Pressure pattern function" of sound receiver, 265

Pressure versus fluid velocity of sound waves, 19–20

Pressure waves (secondary) in sound propagation, 186–190
oscillatory motion, 186–188
spherical symmetry, 188–190

Pressure waves (nonlinear) Riemann's theory, 178–179

Pressure-density properties of a disturbed fluid, 11–12

Pressure-time curves of shock waves, 184–186, 204–206, 228

Probability coefficients for reverberation levels, 328

Probability distributions of intensity fluctuations of sound field, 160–164

distribution functions, 160–162
Gaussian, 161
image interference, 163
Rayleigh, 161–163
Propagation of progressive waves, 14–15, 17–18
Propeller wakes, sound transmission
bubble density, 535
bubble formation, 449
scattering measurements, 530–532, 539
transmission loss, 510–511
underwater explosions, 173–175
Pulse length, sound measurements
Fresnel zone theory, 362
long pulses, 515–516
short pulses, 516–519
target strengths, 350–351, 404–408, 432
wakes, 522–523, 544–546

QB crystal transducer, 275–276
QCH-3 crystal transducers, 273–275, 290–292

Rankine-Hugoniot theory of shock fronts, 179–181
Rarefractional shock waves in underwater explosions, 180–181
Ray acoustics, 41–68
curvature of ray, 46–47
depth correction, 49–51
diagrams, 59–60, 89–90
eikonal wave fronts, 64–65
general waves, 42–43
intensity along a ray, 51–54
long-range transmission, 216–219
plotter for ray-tracing, 59
ray patterns, equations, 45–46
refraction, 197–200
shadow zones, 65–68
"sound channel" propagation, 211–216
spherical waves, 41–42
temperature-depth patterns, 60–63
transmission anomalies, 58–59
velocity-depth variation, 54–58
vertical velocity gradients, 46–49
wave front equations, 43–45
Ray acoustics, theory of normal modes, 222–229
computations, 224–228
dispersion phenomena, 225, 228–229
predictability, 222–224
Rayleigh's sound scattering law
deep-water reverberation, 288
equation, 325–327, 481
intensity fluctuations, 161–163, 169
nonspecular reflection, 362
radiation, long-wave, 464

Receivers for underwater sounds, 73–76
Reciprocity principle in sound propagation, 38–39, 269–270
Recommendations for sonar research, 241–244, 339–340
acoustic measurements, 244
bottom reflection, 243
oceanographic measurements, 243
reverberation, 339–340
surface reflection, 243
velocity of sound, 242
volume scattering, 242–243
Reflected beam in linear gradient, ray acoustics, 55–56
Reflected wave, 29
Reflection and refraction of plane waves, 30–31
Reflection coefficients for underwater sound
long range transmission, 218–219
ocean bottoms, 220–222
sonic, 137–138
supersonic, 140–141
Reflection of sound
bubble pulses, 473–474
close ranges, 360
submarines, 361, 386
surface of water, 190–191, 373–374
surface vessels, 437
underwater targets, 352–355
Refraction of sound
bottom reflection, 138
bottom reverberation, 312–313, 338
bubbles, 473–474
explosions, 197–200
fluctuations, 170–171
"Resolving time" for short-range sound propagation, 193
Resonant frequency of an air bubble, 462–464, 536–537
Reverberation of sound
see also Bottom reverberation of sound; Deep-water reverberation of sound; Surface reverberation of sound; Volume reverberation of sound
analytical procedures, 278–280
backward scattering coefficients, 266, 335
bottom levels, 310, 338–339
coherence, 327–329, 335, 339
deep-water levels, 335–338
definition, 247, 334
duration, 309
equipment for measuring intensity, 272–278
Fermat's principle, 269
fluctuation, 158–160, 324–327, 335, 339
forward, 80
frequency, 258–259, 329–333, 339

intensity, 252–258, 265–266, 304–306, 334
level, 258–259, 334
peak, 321–323
ping length, 258–259
properties, 247–249
reciprocity theorem, 269–270
recommendations for future research, 338–339
scattering, 250–252, 266–269
strength, 259
surface reflection, 270–271
wakes, 492–493
Riemann's theory for sound waves of finite amplitude, 178–179
Rigorous intensity in ray acoustics, 65–66
Roll and pitch effects on sound intensity fluctuations, 167–168, 377
Rough surface effects on reflection of sound, 361

Salinity effect on sound velocity, 17
Scattered sound
see also Bubbles in acoustic wakes
absorption, 242–243
average levels, 304–306
backward, 266, 335, 483
bubble theory, 306–307, 470–473
deep-water reverberation, 286–288
duration, 266–268
multiple, 268–269, 303–304
nonspecular reflection, 361
propeller wakes, 530–532
shadow zone, 125–129
shallow-water reverberation, 316–317
surface reverberation, 299–302
temperature and velocity of wakes, 480–483
theory, 250–252
Screw wakes, sound transmission, 478
Sea bottoms, acoustic properties, 139–141
"Secondary sources" of sound, 356
Self-correlation coefficient for sound intensity fluctuations, 164–166, 482
Shadow boundary of sound
attenuation coefficient, 124–125
ray theory, 65–68, 89–90
scattered sound, 125–129
zones, 120–122, 200–206
Shadowing effect in surface reverberation, 301
Shallow-water reverberation
see Bottom reverberation of sound
Shallow-water transmission of sound
see Transmission of sound, shallow-water

Ship draft and tonnage, effect on target strengths, 432
Shock wave fronts, sound transmission, 174–186
  law of similarity, 182
  Rankine-Hugoniot theory, 177, 179–181
  Riemann's theory of waves of finite amplitude, 176–179
  structure and decay, 184–186
  thickness of pressure region, 182–184
Short-range sound propagation, 108–109, 193–211
  diffraction hypothesis, 201–206
  Fourier analysis, 206–211
  pulse measurements, 193–197
  refraction effects, 197–200
  shadow zones, 200–201
  transmission variations, 108–109, 211
Similarity law for shock waves, 182
Sinusoidal sound experiments, 192
Slide rule for sound ray tracing, 59
"Slipstreams" in sound transmission, 478
Snell's law of refraction for bottom reverberation of sound, 318
Sonic transmission
  analysis of records, 83–84
  deep-water, 238–239
  frequency effects, 138–139
  listening gear, 87
  Long Island area survey, 155–156
  Pacific Ocean measurements, 156
  shallow-water, 240
  summary, 156–157
"Sound channel" propagation
  deep sound channels, 213–216, 240
  experiments, 216–219
  long range transmission, 211–213
  surface sound channels, 239–240
  temperature gradients, 133–135
Sound field measurements
  see Transmission loss measurements
Sound propagation in liquid containing many bubbles, 467–477
  acoustical observations, 474–477
  reflection, 473–474
  scattering, 470–473
  theory, 467–469
  transmission, 469–470
Sound range recorders for wake measurements, 484
Sound transmission, underwater
  see Fluctuations in sound transmission; Transmission of sound, deep-water; Transmission of sound, shallow-water
Sources of sound
  see also Explosions, underwater
  directivity, 24–27

echoes, 420–421
frequency, 23–24
levels, 347–348, 434
transmission runs, 72–74
Space pattern of fluctuation of sound intensities, 167
Spectrum level in Fourier analysis of explosive sound, 208–209
Specular reflection of sound
  beam echoes, 415–417
  convex surface, 434
  frequency factors, 410
  Fresnel zones, 356
  surface vessel, 430
  target strengths, 373–374
Speed of ship, effects on target strengths, 402, 431
Sphere target strengths
  definition, 434
  derivation, 348–350
  Fresnel zone theory, 358–359
Spherical sound waves, 21–22, 41–42
  intensity, 21–22
  pressure versus fluid velocity, 20
  ray acoustics, 41–42
  wave equation, 18–19
"Spines" of echoes in surface-reflected sound, 373–374
Split-beam patterns in ray acoustics, 61–62
Standard reverberation level of sound, 259
Stationary waves in underwater sound
  see Normal mode theory of sound
Still vessels, target strengths, 424–425, 437
Stoke's hypothesis for attenuation of sound, 28
Submarine reflectivity, 379–381, 386
Submarine tactics in sound transmission, 4
Submarine target strengths, 388–421
  altitude angle, 393–397
  aspect angle, 388–393
  asymmetry, 400–402
  beam echoes, 413–417
  frequency, 408–410
  measurements, 397–400
  oceanography, 410–413
  off-beam echoes, 417–420
  orientation, 388
  pulse length, 404–408
  range, 402–404
  source, 420–421
  speed, 402
Submarine wakes, acoustic measurements
  echoes, 523–526
  experiments, 501–502, 538–539
Supersonic transmission
  data-analysis, 80–84

deep-water, 238–239
frequency effects, 138
listening gear, 87
sea bottoms, 139–141
summary, 153
transmission runs, 141–153
velocity gradients, 142–150
wind force, 152
Surface reverberation of sound
  average levels, 304–306
  definition, 259
  elimination, 281
  grazing angle, 300–302
  index, 262
  intensity, 259–263
  "level" concept, 263–264
  multiple scattering, 303–304
  ping length, 302
  range, 289–299
  reflection effects, 270–271
  scattering coefficient, 306–307, 337
  summary, 336–338
  wind speed, 298–299
Surface vessel target strengths
  aspect angle, 424–428
  deep-water transmission, 527–530
  frequency, 433
  introduction, 422
  measurements, 422–424
  pulse length, 432
  range, 426–431
  reflection, 437
  ship type, 432
  speed, 431
  wake echoes, 497–501
Surface-reflected sound
  fluctuations, 377
  reverberation, 301
  short-range propagation, 196
  summary, 243
  transmission loss, 373–374

Target strength measurement
  approximations, 352–353
  calibration errors, 368–369
  comparison of methods, 387
  computation, 358–360
  convex surface, 434
  cylinder, 435
  definition, 347–348
  echo variability, 374–378
  experiments, 363–366
  Fresnel zones, 356–358
  introduction, 343
  Massachusetts Institute of Technology, 379–381
  mathematical theory, 353–355
  Mountain Lakes, N. J., 381
  nonspecular reflection, 361
  principles, 363–364
  pulse length, 350–351, 362

reflectivity, 353–355, 361, 386
San Diego, 379
scattering, 481–482
spherical, 348–350, 434
summary, 434–435
surface vessels, 422–424
transmission loss, 345–347, 369–374
uses, 343–344
wakes, 512–513
wavelength effects, 386–387
Targets, echo-ranging, 84
Taylor Model Basin, sonic transmission experiments, 188, 456
Temperature gradients in the ocean
introduction, 3
microstructure, 90–92, 482–483
ray diagrams, 89–90
refraction, 312–313
60 kc transmission, 135
surface effects, 239, 296–297
velocity, 15–17
wake structure, 441, 479–480
Temperature gradients (negative) in the ocean
attenuation coefficient, 124–125
ray theory, 61–62
shadow zones, 120–122, 125–129
sharp gradients, 120–129
60 kc transmission, 135
sound channels, 131–133
transmission anomalies, 118–120, 122–123
weak gradients, 129–133
Temperature-depth patterns, ray diagrams, 60–63
Thermal microstructure for sound intensity fluctuations, 169–171
Thermal wakes, sound transmission, 441, 479–480, 496
Thermocline, sound transmission
see also Isothermal water, sound transmission
below isothermal layer, 238
ray theory, 61
submarine target strengths, 411–413
temperature gradients, 89
Thermocouple recorder for sound velocity measurements, 17
Thermodynamic law for absorption of sound, 464
Thickness of acoustic wakes, 498–500
Thickness of shock wave fronts, 177, 182–184
dissipation of energy, 183–184
Hugoniot equation, 184
Riemann overtaking effect, 183
summary, 177
"Time of arrival" of bottom-reflected sound pulses, 221–222
"Time of rise" data for short-range sound propagation, 193

Time patterns of sound intensity fluctuations, 164–167
"hidden periodicities", 166–167
self-correlation coefficient, 164–166
Training errors in echo-ranging on wakes, 491–492
Transducers for acoustic measurements
calibration, 78
directivity, 522
EBI-1; 276
JK, 276
QB-crystal, 275–276
QCH-3; 273–275, 290–292
reverberation intensities, 272–273, 277
target strengths, 429–430
wakes, 492
Transmission anomaly in underwater sound
see also Attenuation of sound
average, 122–123, 131–133
bottom scattering, 319–321
definition, 70–71, 237
image effect, 95–97
isothermal water, 100–104
ray theory, 53–54, 58–59, 67–68
supersonic, 147–150
target strength, 369–371
temperature gradients, 118–120
Transmission loss measurements
attenuation, 373, 503–504
background, 3–4, 69–71
bubble theory, 469–470
echo runs, 84–85
equipment, 76–78
inadequacy, 372
methods, 78–80
observed echo ranges, 85
oceanographic factors, 492–493
one-way horizontal transmission, 504–506
propagation along wakes, 509–510
propeller wakes, 510–511
receivers, 73–76
sources, 72–74
summary, 71–72
supersonic frequencies, 80–84
surface reflections, 373–374
target strengths, 369–372, 411–413, 430–431
two-way horizontal transmission, 506–507
two-way vertical transmission, 507–509
variation, 107–108
wakes, 345–347, 504
Transmission of sound, deep-water
absorption, 97–104
attenuation coefficient, 103–107
bathythermograms, 92–95
characteristics, 86–89
echo-ranging trials, 109–110

image effect, 95–97
introduction, 86
isothermal water, 95, 238–239
layer effect, 112–117
long-range experiments, 216–219
negative temperature gradients, 118–120
scattered sound in shadow zone, 125–129
sharp temperature gradients, 120–125, 239
short-range, 108–109
60 kc transmission, 135
sound channels, 133–135, 239–240
thermocline, 110–111, 238
transmission loss, 107–108
variability of vertical temperature gradients, 90–92
vertical temperature structure, 89–90
weak temperature gradients, 129–133, 239
Transmission of sound, shallow-water
dispersion phenomena, 228–229
experiments, 229–235
reflection coefficient, 140–141
sea bottoms, 137–140
sonic, 154–157
summary, 240–241
supersonic, 139–140
24 kc transmission, 141–143
velocity gradients, 142–150
wind force, 152–154
Transmitted wave, 30
Triangulation in long-range sound transmission, 219
Triplane in echo ranging, 84
Turbulence parameter for acoustic wakes, 452–455
Two-way horizontal transmission loss in acoustic wakes, 506–507

Underwater sound transmission, 236–244
see also Fluctuations in sound transmission; Transmission of sound, deep-water; Transmission of sound, shallow-water
recommendations for future research, 241–244
summary of definitions, 236–238

Variability of echo intensity, 374–378
"Variance of amplitudes" for sound intensity measurements, 237
Variations in sound transmission
short-range propagation, 211
summary, 241
transmission loss, 71, 107–108
Velocity of sound in water
bubble theory, 473–474
microstructure, 482

pressure effects, 19–20
ray equations, 46–49
refraction effects, 197–200
shallow-water transmission, 138
summary, 242
supersonic transmission, 142–150
target depth correction, 49–51
wake theory, 478–479, 480–483
wave equations, 15–17
Velocity-depth variation in ray acoustics, 54–59
beams in linear gradients, 54–58
layer effect, 56–57
transmission anomalies, 58–59
Vertical temperature gradients in the ocean
see Temperature gradients in the ocean
Vertical transmission of underwater sound, 79, 507–509
Viscosity (fluid) effects on sound intensity, 27–28
Volume reverberation of sound
average intensity, 255–256
definition, 253–254
depth, 282–284
frequency, 284–288
index, 259
intensity, 256–258
level, 258, 335–337
range, 281
scattering coefficient, 243, 286, 336–337

Wake geometry in sound transmission
aerial photographs, 494–495
submarines, 501–502

summary, 541–542
surface vessels, 497–501
target strength, 513–514
widening measurements, 495–497
Wake-laying vessel, acoustic measurements, 491
Wakes, acoustic
absorption, 541–543
decay rate, 520–521, 539–540
definition, 441
echo ranging, 484, 543–546
evaluation of research, 443–448
fathometer records, 486–488, 497–501
frequency, 523
geometry, 494–495, 513–514, 541
index, 513, 519–520, 543
listening gear, 484
long pulses, 515–516
measurements, 490–492
oceanographic effects, 492–493
oscillograms, 488–490
physical properties, 514–515
propellers, 530–532
pulse length, 522–523
scattered sound, 480–483
short pulses, 516–519
sound range recorder, 484
submarine, 501–502, 523–526, 538–539
surface vessel, 526–530, 537–538
target strength, 512–513
temperature structure, 479–480
theory, 541–546
thickness, 498–501
training errors, 491–492
transducer directivity, 522
velocity structure, 478–479
widening rate, 495–497

Water wave, sound transmission, 233–235
Wave acoustics
boundary conditions, 13–14, 28–34
equation, 10–13, 43–45, 242
equation of continuity, 8–10
equations of motions, 10
fluid viscosity effects, 27–28
general waves, 36–37
harmonic waves, 17–18, 22–23
intensity, 20–22, 37–38
mathematics, 39–40
normal mode theory, 34–38
plane waves, 14–17, 34–36
pressure versus velocity, 19–20
reciprocity principle, 38–39
sources, 23–27
spherical waves, 18–19
Wave equation for shadow boundary, ray acoustics, 66–67
Wave fronts, ray acoustics
eikonal equation versus general equation, 64–65
equations, differential, 43–46
general waves, 42–43
spherical waves, 41–42
Wave length effects on target strength measurements, 386–387
Wavelets for reflected sound pressure, 356
Wide-band pings, frequency analysis of reverberation, 332–333
Widening rate of acoustic wakes, 495–497
Wind effects in sound transmission
force, 337
speed, 295–299
supersonic, 150

www.ingramcontent.com/pod-product-compliance
Lightning Source LLC
Chambersburg PA
CBHW060945210326
41598CB00031B/4728